This boo...
earlier w... *Chemica*...
with the ...at problem of the relatio...
phology ...chemistry. Of possible a...
this pr... greatest attention is here giv...
bioche... investigation of the morphoger...
mone... study of these chemical subst...
fund... importance for all animal a...
dev... ...ment, including its hereditary aspect...
the greater part of the present book.

The establishment of the existence of...
stances was the result of work done ...
twenty years of the present century; s...
have seen many attempts to iso...
understand their action. What ha...
this and what new problems ha...
by it will be found recorded a...
present work.

In Part I the morphogene...
sidered; that is, the chemi...
ment and all the man...
embryos have so...
phogenetic stimu...
...morphogenetic me...
... is now more th...
..., Francis Balfo...
...vas appointed ...
...hair of Animal...
...been many ...
...hard Assheto...
...liam Bateso...
...bject of m...

BIOCHEMISTRY AND MORPHOGENESIS

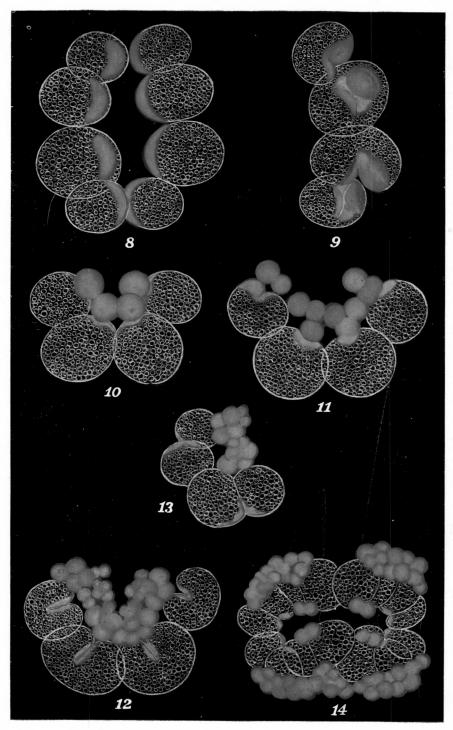

Differences between cytoplasmic cortex and interior in a ctenophore egg seen under dark-ground illumination (*Beroe*), see Fig. 49.

BIOCHEMISTRY
AND
MORPHOGENESIS

by

JOSEPH NEEDHAM
F.R.S.

*Sir William Dunn Reader in Biochemistry
and Fellow of Gonville and Caius College
Cambridge*

CAMBRIDGE
AT THE UNIVERSITY PRESS
1950

PUBLISHED BY
THE SYNDICS OF THE CAMBRIDGE UNIVERSITY PRESS

London Office: Bentley House, N.W. 1
American Branch: New York

Agents for Canada, India, and Pakistan: Macmillan

First Edition 1942
Reprinted 1950

Printed in Great Britain at the University Press, Cambridge
(Brooke Crutchley, University Printer)

孟子曰
博學而
詳說之
將以反
說約也

Mencius said: *extensively learn and in all detail state it, so that later, summarise its essence.*

4th century B.C.

此書紀念

友人 李大桑
鲁桂珍 女士

李約瑟

NOTE ON ILLUSTRATIONS

Of the 328 illustrations in this book a number
(including 4 in colour) are printed as plates
and inserted as follows:

CONTENTS

PART 2. THE MORPHOGENETIC STIMULI

PART 3. THE MORPHOGENETIC MECHANISMS

Bibliographical conventions

Where only one paper or group of papers by an author or group of authors is listed the text contains nothing but the name or names. Where more than one paper or group of papers is referred to, superscript figures are added to the names in the text to allow of identification.

The use of the shortened and (&) indicates collaboration between two or more authors.

Conventions adopted in the chemical formulae

(1) No carbon atoms or their associated hydrogens are shown. Bonds of full length have C at each end, except where other elements occur heterocyclically, or by analogy heterocyclically; such atoms are shown as part of the ring, as if they were C.

(2) Terminal CH_2 is shown simply by the ending of the chain of bonds. Terminal COOH is shown as OOH.

(3) In all chains of bonds a C atom occurs at every angle, together with sufficient H to satisfy its valencies.

(4) It will be seen that many non-cyclic compounds are written as if they were composed of incomplete rings open at one point, rather than elongated chains; in some cases (e.g. vitamin D_2) there is definite evidence for this formulation, but in most cases (e.g. squalene) it is theoretical.

(5) All unsaturated linkages are shown as such.

GENERAL INTRODUCTION

There are three ways in which the great problem of the relation between morphology and biochemistry—the problem to which this book is devoted—can be approached. In the first place we may make a direct attack upon that difficult region lying between the largest chemical particles and the smallest morphological structures which we know. This realm includes the study of paracrystalline aggregates, colloidal micelles, fibrous macro-molecules, protein structure, etc. In recent times the physico-chemical study of the viruses has played a part of cardinal importance in exploration along these lines. The second way in which we may attempt to bridge the gulf between morphology and biochemistry is by studying the chemical changes which go on during embryonic development, a time during which the morphological change is the most obvious variable.

Up to 1931, in which year the author's former book *Chemical Embryology* was published, this was perhaps the only contact between biochemistry and embryology, and the chemist could have relatively little to say about real morphogenesis itself. But it was in that very year that successful embryonic inductions were first obtained after the destruction of the living integrity of the primary organisation-centre of amphibia. Hence there opened the possibility of a third method of approach; the biochemical investigation of what may truly be called the morphogenetic hormones. The study of these chemical substances, of fundamental importance for all animal and plant development, including its hereditary aspect, occupies the greater part of the present book. The establishment of the existence of such substances, the essential elements in the processes of dependent differentiation, known since the time of Wilhelm Roux, took approximately the first twenty years of the present century and will always be associated with the names of Ross G. Harrison, Warren Lewis, Hans Spemann and Otto and Hilde Mangold. The second twenty years of the century, which may come to be spoken of as the period between the two great European wars, were characterised by numerous attempts to isolate the inductor substances and to understand their action. Although much has been achieved, as will be evident from the contents of the present book, we have come to see that the problems are often far more complex than had at first been thought, and that after all the larger part of the mystery remains in that we can as yet form little idea of what constitutes reactivity—the competence to react to the morphogenetic inductor.

In accordance with this distribution of interest, therefore, the present book is divided into three parts. In Part 1 the morphogenetic substratum is considered, i.e. the chemical raw material of development and all the manifold nutritional problems which embryos have solved. Part 2 is devoted to the morphogenetic stimuli themselves, and Part 3 to the morphogenetic mechanisms. Under this head, besides the special metabolism of embryonic life, such general questions as dissociability, heterauxesis, the basis of polarity, etc. are discussed.

Before entering upon the details, however, there are two theoretical questions to which a reference must be made. The relation of biochemistry to morphology is not a banal matter. It has been implicit in countless discussions of theoretical biology and involves themes which have run right through the history of philosophy from the earliest speculations of Greek and Chinese thinkers through the sixteenth, seventeenth and eighteenth centuries, down to our own time. Here I wish only to state my position with regard to two of these, the "irreducibility" of the biological, and the question of form and matter.

Morphologists, turning over the pages of the present book, will probably be inclined to remark that the author finds only "a mass of substances" in the living and developing organism. Yet that is precisely what the organism is, containing in itself innumerable molecules of small size some of which are indeed the morphogenetic inductors, as well as the giant protein molecules which, in responding by new orientations, form the basis of the cellular and organic architecture. Nevertheless the biochemist approaching morphological problems does not forget that all these molecules are ordered and organised in a totally different way to anything which occurs outside living organisms. His difficulties arise rather from the vagueness and lack of quantitative precision common to so much morphological work and perhaps inseparable from it. But everything that can be done to remove this vagueness is being done, and after all, we have already a great many extremely well-defined histological and morphological factors which can be correlated with biochemical ones. Some may feel that the terminology of inductors, competences, etc., is too "cut and dried", but the history of science abundantly shows that in every age, whether of the "word-rectification" school in ancient China, the Arabs, Albertus Magnus, the early Royal Society, or the time of Haller and Boerhaave, the invention and clarification of terms is one of the greatest limiting factors in scientific advance. A clear formulation, even though partial, is better than a vague one; and we must not, in avoiding rigidity, fall into obscurantism. Any who might be led to suppose that words such as "inductor", "evocator", etc., carry animistic undertones, would be under a complete misapprehension. They stand for definite chemical substances the nature of which is as yet not fully elucidated. They conceal no *archaeus* or demon. In the use of the phrase "oxidising agent" the chemist risks no confusion with "passenger agent", and "hydrogen-donators" head no subscription-lists. In this book, the word "organiser" is always used to imply the presence of tissue, whether living or dead; "evocator" refers only to a specific chemical substance; and "inductor" is a looser word available for both these senses. The glossary of definitions on p. 681 will perhaps assist the reader's mental labour. It makes no legislative claims, but only seeks to show the exact senses in which terms are used in the present book, and it ends with a small group of words which now deserve their *requiescat*.

In the previous discussion the words "organised" and "ordered" are the keynotes. We cannot but consider the universe a series of levels of organisation and complexity, ranging from the sub-atomic level, through the atom, the molecule, the colloidal particle, the living nucleus and cell, to the organ and the organism, the psychological and sociological entity. It follows that the laws or regularities which we find at one level cannot be expected to appear at lower levels. The conditions for their appearance do not exist there. It is no use fishing in homogeneous solutions for the laws which hold good for the behaviour of liquid crystals, still less for those which hold for that of living organisms.*

Philosophers have often talked about the reducibility or irreducibility of biological facts to physico-chemical facts. These old controversies are unnecessary if we realise that we are dealing with a series of levels of organisation. We must seek to elucidate the regularities which occur at each of these levels without attempting either to force the higher or coarser processes into the framework of the lower or finer processes, or conversely to explain the lower by the higher. From this point of view, the regularities discovered by experimental morphology will always have their validity and will, in a sense, be unaffected by anything which either biochemistry on the one hand, or psychology on the other, may discover. The behaviour, for example, of an embryonic eye-cup isolated into saline solution, its capacities for self-differentiation, fusion with another eye-cup, lens-induction, regulation, etc., etc. will always remain the same however much our knowledge of biochemistry or biophysics may advance. This is the reason why prediction is possible at a level of organisation which, strictly speaking, we do not yet understand at all, for instance, genetics. But the important point is that although the regularities established at the level of experimental morphology are irrefragable, they will, in the absence of biochemical experimentation, remain for ever meaningless. Meaning can only be introduced into our knowledge of the external universe by the simultaneous prosecution of research at all the levels of complexity and organisation, for only in this way can we hope to understand how one is connected with the others.

This brings us to the ancient distinction between *forma* and *materia*, mainly due to the analytical mind of Aristotle—μορφή and εἶδος against ὕλη. This distinction has had an influence incalculably great on the historical development of biology.† Morphologists for many centuries past have devoted themselves to the study of living form without much consideration of the matter with which it is indissolubly connected. It is not surprising that the numerous devils of vitalism found a congenial abode in the mansions of empty *forma* thus suitably swept and garnished. Moreover, the morphological tradition‡ was to think of matter much too simply, ignoring the vast complexity of chemical structures and the unbroken line of sizes reaching from the hydrogen atom at one end (to say nothing of sub-atomic levels) to the virus molecule at the other. Only in the light of the conception of integrative levels can the saecular gulf between morphology and chemistry be bridged.

* For a fuller discussion of integrative levels, see the present author's Herbert Spencer lecture at Oxford, *Integrative levels; a revaluation of the idea of Progress*. The general picture there drawn has relations with evolutionary naturalism, emergent evolutionism, and dialectical materialism.

† For a brilliant summary of Aristotle's views on these subjects, with convenient references to the texts themselves, see W. D. Ross' "Aristotle".

‡ Originating from the conception of becoming as the privation of one form and the donation of another.

It is true that Aristotle held that there could be form without matter, though no matter without form. But according to him, the only entities which possessed form without matter, were the divine prime mover, the intelligent demiurges that moved the spheres, and perhaps the ψυχῆς διανοητική (the rational soul) of man. Some of these are factors in which experimental science has never been very much interested. On the other hand, he maintained that there could be no matter without form, for however pure the matter was (even the chaotic primal menstrual matter which was the raw material of the embryo), it was always composed of the elements, that is to say, it was always hot, cold, dry or wet, and hence had a minimum of form. In its primitive way this mirrors the standpoint of modern science. Form is not the perquisite of the morphologist. It exists as the essential characteristic of the whole realm of organic chemistry, and cannot be excluded either from "inorganic" chemistry or nuclear physics. But at that level it blends without distinction into order as such, and hence we should do well to give up all the old arguments about form and matter, replacing them by two factors more in accordance with what we now know of the universe, that is to say, Organisation and Energy. From this point of view there can be no sharp distinction between morphology and biochemistry, and we may have every hope that in the future we shall be able to see not only what laws the form of living organisms exhibits at its own level, but also how these laws are related to those which appear at lower and higher levels of organisation.

I terminate this introduction, as in private duty bound, by an expression of my deep indebtedness to all those who have helped me, whether in the train of thought that led to this book, or with specific problems arising out of it. First, to my colleagues and all my collaborators, of many countries and peoples. Secondly, to all my friends in the New World, Americans either by birth or adoption, especially those who gather in the Symposia on Differentiation and Growth; inheritors now of both the two greatest national traditions in causal morphology. From all these have I learnt, and without their discussions many a problem would for me still lie in obscurity. Lastly, to that ancient and unique foundation, the Cambridge University Press, with which I am proud to be connected, and where my friends in every department spare no effort to make a book such as this all that it should typographically be.

Part One

THE MORPHOGENETIC SUBSTRATUM

THE MORPHOGENETIC SUBSTRATUM

1·05. Introduction

In *Chemical Embryology** an attempt was made to summarise all existing knowledge regarding the chemistry of the developing embryo and the raw materials from which it is formed.[†] Ten years ago this summary was inevitably restricted to the actual chemical constitution of the embryo and the raw materials, and to what was known of the metabolism during embryonic development. To-day the interest has shifted to the analysis of the fundamental morphogenetic stimuli which operate in embryonic life. These questions form the subject-matter of Part 2. But it is indispensable to precede them by an account of the main advances during the past ten years in our knowledge of the raw materials from which the embryo is built, and to follow them by a similar account of the new knowledge of embryonic growth and metabolism. These are the subjects of Parts 1 and 3 respectively.

1·1. Problems of maturation

A satisfactory knowledge of the biochemistry of the egg would include not only an exact understanding of the part which each chemical constituent of the egg has to play in the development of the embryo, but also an understanding of the way and order in which these constituent substances are laid down in the preparation of the egg in the ovary. It is likely that without this understanding we shall never be able to comprehend the real nature of those properties of polarity and symmetry for which the egg before development is so remarkable (see Section 3·5). But at present we are far from such a complete knowledge, and we have to content ourselves with fragments of information which we may hope to fit together better later.[‡]

Most of the work on the maturation[§] of the egg has been of a histochemical nature, and of that which has been done with direct chemical methods the greater part concerns the eggs of birds, especially the hen. We shall therefore begin this section with the hen's egg. Like all the others in Parts 1 and 3, it should be read in conjunction with the earlier discussions in *CE*.

The changes occurring in the blood stream of the hen are very considerable. They have been summarised by Riddle[2]. Blood-sugar rises 25 %, blood-calcium 100 %, blood-fat 35 %, and ether-soluble phosphorus 50 %. The weight of the adrenals increases by about 40 % and the oviducts hypertrophy 800 %. Huxley[10] has shown that it is possible to correlate these large physiological variations which follow the rather sudden change from slow to rapid growth of the oocyte (confirmed by Chomković and Orrù[9]) with studies on bird behaviour. Jukes & Kay[3], in their review on yolk proteins, speculated on the mechanism of the rapid deposition of vitellin, pointing out that 99 % of the 3 gm. of protein

---

* Hereinafter referred to as *CE*.

† Since the publication of *CE*, reviews covering the whole or parts of the field have been published as follows: (1934) Paechtner; Ermakov; (1935) Grossfeld; (1936) Ogorodny. Reviews dealing with specialised problems will be referred to in what follows.

‡ I exclude from consideration here the nuclear phenomena of maturation, the reduction divisions, etc., as they are fully discussed in genetical books and monographs.

§ Cf. the brief review of van Dyke.

in an egg-yolk is secreted during the last 5 to 8 days before laying. This protein may be produced from the amino-acids in the plasma or possibly from the plasma proteins by some much less drastic change. Rimington noted that in their distribution of nitrogen (as determined by Crowther & Raistrick) casein and lactoglobulin (serum globulin) are not remotely dissimilar, and himself showed that artificially phosphorised serum globulin has several similarities to casein. Laskowski, in a series of papers, subsequently described the isolation of vitellin from the serum, not only of laying birds, but also of the tortoise and the trout. It would seem, therefore, to be formed in some organ as well as in the ovary. According to Schenck's[2] van Slyke analyses the composition of vitellin from young oocytes differs from that of fully mature eggs, as it contains a higher proportion of hexone bases. Serum vitellin is serologically identical with ovovitellin (Roepke & Hughes; Roepke & Bushnell). After a single dose of pituitary gonadotrophic hormone, serum vitellin appears in the blood (Laskowski); after further administration serum calcium is increased (Riddle & Dotti; Hughes, Titus & Smits), and the ovary greatly hypertrophied (Bates, Lahr & Riddle). The calcium increase is in the non-ultra-filtrable fraction (Correll & Hughes).

By a new method, the administration of labelled (radioactive) sodium phosphate to hens by subcutaneous injection, it has been found that all the lecithin of the yolk is taken up from the plasma, the plasma lecithin being replaced by molecules formed at least mainly in the liver (Hahn & Hevesy; Hevesy & Hahn). It is clear that the variations in the blood of the laying oviparous vertebrate are so large that serologically specific qualities of the blood should be expected during egg-laying; these have been observed by Sasaki on fowls.

During the early, slow, growth phase, the oocyte has a much higher water-content than later on (Romanov[4]), and during the late maturation phase fat and calcium are rapidly deposited, but approximately proportionally. The excess phosphate excretion shown by Common to occur during laying is ascribed by McGowan to the large deposition of calcium as carbonate in the shell during the passage of the egg down the oviduct. What happens to the egg after experimental surgical interference with the various parts of the oviduct has been described in the interesting paper of Asmundson[1], to which the reader must be referred for further details. The view previously held (see CE, p. 235), that an internal secretion of the thymus ("thymovidin") is responsible for the proper formation of the shell, cannot yet be taken as proved, since effective thymus extirpation, owing to the widespread presence of masses of lymphoid tissue which may have the same function, cannot be carried out (Greenwood & Blyth; Riddle & Křížinecký). There are valuable reviews on the factors affecting egg number (Fauré-Fremiet & Kaufman) and egg weight (Graham), and the whole subject of the formation of the hen's egg has been reviewed by Conrad & Scott.

The layers of white yolk which alternate with those of yellow yolk in the hen's egg-yolk* (see CE, p. 286) are said to be due to a deficiency of nutrient materials in the blood stream between 1 and 5 a.m., but they depend to some extent on the nature of the diet of the laying hen, as shown by Conrad & Warren (cit. in Conrad & Scott). Warren & Scott have found that a yolk enters the oviduct about 15 min. after ovulation, passes through the *infundibulum* or funnel in 18 min., spends 3 hr. in traversing the *magnum* or albumen-secreting portion of the oviduct, 1 hr. in the isthmus, and the remainder of the interval, usually 20 to

---

* As is well known they are demonstrable by the ring-like deposition of dyes, which when added to the maternal diet enter the eggs. This has been done on the silkworm (Jucci & Ponseveroni) as well as the hen (Riddle[1]).

24 hr., in the uterus. Peristalsis as well as ciliary action aids the descent of the egg. In the magnum the egg-white is secreted round the yolk as a dense homogeneous gel; this is a network of mucin fibres (R. K. Cole) enclosing a solution of the other egg-white proteins (ovoalbumen, conalbumen, and ovomucoid). Much of our knowledge of the formation of the white and shell has been found out by operations on laying fowls, the details of which may be had in Asmundson & Burmester; Burmester, and Asmundson & Jervis.

The nature of the chalazae has always been mysterious (see *CE*, pp. 117, 135). While Richardson attributes their formation to a special region of the oviduct, Conrad & Phillips explain it as follows. The innermost layer of the magnum gel becomes liquefied as the egg passes down the oviduct, so that the yolk is free to rotate in it. "While it is in the uterus the egg is rotated about its long axis, and since the yolk always tends to float with the blastoderm up, the white is rotated round the yolk. This action wraps the mucin fibres around the yolk and twists them at the ends to form the chalazae. Removal of the mucin fibres from the inner layer of gel liquefies it, and it becomes the inner thin white" (see below, p. 7). Unfortunately, as Burmester & Card point out, yolkless eggs may have chalazae. It has long been known that while in the uterus, the amount of the egg-white is doubled (Pearl & Curtis). The incoming fluid contains mainly inorganic ions, especially K' and HCO_3' (Scott, Hughes & Warren; Beadle, Conrad & Scott), but also (according to McNally[2] but not according to Hughes & Scott) some of the protein fraction known as ovoglobulin, which will be discussed further below (pp. 8, 671). After dissolving out a considerable amount of protein from the thick gel of the magnum, it becomes the outer thin egg-white. Its entry into the egg is slowed and eventually stopped by the deposition of the shell material. The composition of the proteins in the oviduct itself, as related to the secreted egg-white proteins, has been studied by Schenck, who obtained various more or less comparable protein fractions and believes that the percentage of amino-acids changes during the preparation of the proteins for secretion.

While the egg is in the uterus about 5 gm. of pure $CaCO_3$ are deposited on the shell membranes, first as knob-like calcite crystals, the *mammillae*, and then as the spongy layer, also composed of calcite but with much smaller crystals (Richardson; Stewart[1]). The calcium carbonate cannot be secreted in solution, but must appear as an extremely minute suspension (Beadle, Conrad & Scott). During laying, a pullet may be in negative calcium balance (Morgan & Mitchell), and even though this is later repaired, 10 % of the bone Ca" of the adult hen is available for egg-shell formation (Deobald, Lease, Hart & Halpin).

It is interesting that hypocalcaemia and tetany are exceedingly rare in birds. An idiopathic case responded at once to parathyroid hormone given by Hutt & Boyd, who suggest that the rarity of the condition may be due to a special adaptive type of calcium metabolism in birds.

We come now to the contributions of histochemical and cytological type on oogenesis, many of which have come from the schools of Gatenby and of Marza. Unfortunately histochemistry has a long way to go before it can attain the certainty of direct chemical methods, and with the exception of certain special techniques, such as the Feulgen reaction, it is hard to know what to make of its results, especially when, as in some instances, they are expressed in a manner which seems to claim to be quantitative. It is hard, also, to find any common point between the cytological data and the results of chemical methods. The most hopeful line of advance here is undoubtedly the ultra-micro-chemical methods of the Copenhagen school. In the meantime all we can do is to refer to the best studies of histochemists and cytologists on oogenesis. A typical publication of the Irish school is the interesting account by Brambell of the maturation of the hen's egg, but the largest contributions on this subject are those of Marza (Marza & Marza; Marza & Golaescu; Marza & Chiosa; Marza, Marza & Chiosa; de Robertis; Grodziński). The deposition of fat in the fowl oocyte has been studied by Konopacka[2]. All this work and a very large further body of literature is efficiently summarised in Marza's monograph[4]. The exact study of the various types of vitelline globules has great importance, even from the point of view of experimental morphology (cf. p. 217), but they are so poorly characterised, and the methods available so relatively ineffective, that the subject is still in an early phase.* Other histochemical studies are available on the eggs of fishes (Marza[2], Marza & Guthrie; Konopacka[4]), cephalopods (Konopacki[3]), ascidians (Konopacki[4]), various worms and molluscs (Konopacki[1]), and crinoids (Chubb).

Recent additions to our knowledge of the maturation of the amphibian egg (reviewed in CE, p. 1681) have been those of Mestscherskaia, who estimated the respiratory rate of immature oocytes (Q_{O_2} −1 to −2), and Boyd[3], who studied rather completely the deposition of fats and sterols in the maturing oocytes. From her data a series of ratios could be calculated showing the progressive changes in the relative amounts of neutral fat, cholesterol (free or esterified) and phosphatides. Esterification of the cholesterol begins only at a relatively late stage.

One of the most classical observations on egg formation was Miescher's description of the exhaustion of the maternal body of the salmon during the spawning migration owing to the deposition of so much fat and protein in the eggs. This has lately been studied anew by modern methods by Davidson & Shostrom. Following this, Konopacka[3] has studied the appearance of the oil-globules which are so marked a character of fish eggs, and investigations have been made of the chemical changes in the blood during the spawning season in the salmon (Okamura[2]; Miyachi). During the maturation of teleostean eggs there are considerable changes in the fat-fraction (degree of saturation of the acids, length of chain, etc.), but investigators are not yet in full agreement concerning their nature (Channon & Saby; Lovern).

For insects, egg-number has been studied by Alpatov; Fauré-Fremiet & Garrault and Shapiro[1], in whose papers may be found a guide to this special literature. According to de Boissezon, mosquito eggs (*Culex pipiens*) come well to maturity even if the maternal organism has been prevented from biting, provided iron has been supplied to it in its larval stage.

I know of no chemical work on the eggs of mammals (see the book of Pincus), but analyses of the *liquor folliculi* have been made (Tafuri & Testa; Tafuri; Garufi). Marza[4] gives in his monograph an account of the action of hormones, especially the gonadotrophic hormone of the pituitary, in stimulating the growth of the oocyte. The papers of Pincus & Enzmann give access to the literature on these subjects.

1·2. Constitution of the eggs of birds

The fundamental data on which depend our knowledge of the constitution of the hen's egg and those of other species were given in CE (Pt. III, Sections 1·2–1·9). To CE, Table 2, should be added new analyses of the whole egg-contents (duck, Danilova & Nefedjova; turkey, Hepburn & Miraglia).

* Lepeschinskaia makes the (inherently very unlikely) claim that in the eggs of birds some cells arise from certain of the vitelline globules without previous cell-division; her work has been severely criticised by Grodziński[5]. In other organisms, the role of wandering nuclei in the yolk may be indeed considerable (cf. p. 456).

1·21. The shell and shell-membranes

An extremely thin layer of protein, known as the cuticle, exists on the exterior of the shell; it has been described by Romankevitch and studied by Furreg and Marshall & Cruickshank. The last-named authors think that it assists the evaporation of water (see *CE*, p. 874) rather than the reverse, as it is continuous with protein material which fills up the pores in the shell (Almquist[1]). An average hen's egg-shell contains some 7000 of these pores, of mean diameter 0·04–0·05 mm., but the porosity varies with individual hens and is probably an inherited quality (Almquist & Holst). The pores are not more numerous over the air-space than elsewhere. The gas transmission through them may be measured by Romanov's method[6]. There are numerous studies of shell porosity (e.g. Rizzo; Dumansky & Strukova; Swenson & Mottern); it may vary considerably, through *mottled* (Holst, Almquist & Lorenz), where the water-content of the shell is uneven and causes uneven translucency, to *glassy* (Almquist & Burmester), where the number of pores is only half that of the normal egg-shell and the percentage protein in the shell twice the normal value. In such eggs daily water-loss is much reduced. Water-content of the shell is the most important factor in affecting light-transmission through it (Givens, Almquist & Stokstad). According to Almquist[2], the protein actually in the shell resembles collagen rather than keratin. Shell thickness is much the most important factor governing its breaking strength (Stewart[2]; Lund, Heiman & Wilhelm).

For the pigments in the shell, see below, p. 644.

The shell-membranes are generally thought to be composed of keratin fibres matted together. The latest analysis (Calvery[4]) indicates an exceptionally pure and typical keratin. Their structure has recently been re-examined by Szuman, and there is information to be had about their permeability properties (ions, water, membrane potentials, etc.) in the work of Yasumaru; Yasumaru & Sugiyama; Mizutani; Osborne, and Salvatori. Mucin fibres have been detected in them as well as keratin by Moran & Hale.

1·22. The white

The egg-white of the hen's egg (as stated in *CE*, pp. 265 ff.) consists of three distinct layers, an inner thin layer, a thick layer, and an outer thin layer. The first two of these accumulate in the oviduct, the last in the uterus. The layers differ considerably in the proteins which they contain (Hughes & Scott; McNally[1]), but former statements that the water-content is different cannot be substantiated (Holst & Almquist[1]). The great difference in viscosity must therefore be a colloidal phenomenon.

Table 1 (Hughes & Scott). *Egg-white proteins*

	Inner thin	Thick	Outer thin
Ovomucin %	1·10	5·11	1·91
"Ovoglobulin" %	9·59	5·59	3·66
Ovoalbumen %	89·29	89·19	94·43

It is to be noted that the fibrous ovomucin is present mostly in the thick layer, but according to Heringa & Valk keratin fibres are to be found there also. There is no difference in osmotic pressure between thick and thin egg-white, but the diffusion rate of water is quicker in the latter (Smith & Shepherd). The proportion of thick white is definitely an inherited character (Lorenz, Taylor & Almquist; S. S. Munro; Knox & Godfrey), but exactly what part it has to play during embryonic development is not yet known (see below, p. 620). On storage, or spontaneously in some eggs, the condition *watery white* is found (St John[2]; Holst & Almquist[2]); this has nothing to do with water-content of the layers, nor is the thick/thin ratio different, so that the change must be of a subtle colloidal nature. It does not affect the specific rotations of the proteins (Holst & Almquist[3]). It is probably one of the factors accounting for the fall in hatchability with storage time before incubation (see below, Fig. 1). Contrary to earlier statements (St John[1]) there is, according to the reliable measurements of Moran[2], little bound water in either kind of egg-white. Ovoalbumen binds no more than 0·26 gm./gm. protein.

In the egg-white there are four proteins the existence of which is well established, two albumens, ovoalbumen and conalbumen, and two glycoproteins, ovomucoid and ovomucin.* The fifth, ovoglobulin, still remains a matter of dispute, and may continue to do so for some time, since it is extremely difficult to prepare the fractions containing it free from ovoalbumen and ovomucin, and if it is present, it must be so in minute quantities. Conalbumen was shown by Hektoen & Cole[1] to be identical with serum albumen (later confirmed by Bruynoghe), and in the same paper they concluded that ovoglobulin also was an immunologically identifiable antigen. But they admitted that all their preparations of it contained ovomucin and ovoalbumen as well, and the sera which reacted against it reacted against these also. The "ovoglobulin" was not identifiable with serum globulin.†
Later M. Sørensen, applying the orcinol method of Sørensen & Haugaard for sugar estimation to the egg-white proteins, accepted the existence of ovoglobulin, and regarded it as constituting 7 % of the total protein and as containing 4 %

Table 2. *Composition of egg-white proteins*

Moisture and ash-free percentages

	Ovomucin	"Mucin extract"	"Ovo-globulin"	Mother liquor	Ovo-albumen	Ovo-mucoid
Total N	12·6	14·3	14·3	14·3	14·4	12·6
Total S	1·8	0·8	1·1	0·9	1·5	1·4
Cystine	5·6	2·4	3·7	3·8	1·4	3·9
Glucosamine	10·4	3·1	6·4	4·5	1·6	9·7

N.B. "Mucin extract" is probably a mixture of occluded ovoalbumen and dispersed ovomucin; "mother liquor" is probably ovoalbumen contaminated with ovomucin.

In the terminology of K. Meyer's recent review, ovomucoid becomes α-ovomucoid and ovomucin β-ovomucoid.

Subsequent work on the immunology of egg-white proteins, especially with reference to species-differences, has been done by Hektoen & Cole[2] and Hooker & Boyd.

of mannose. Ovoglobulin preparations were found by Böhm & Signer to show flow birefringence (ovoalbumen does not; see below, p. 671), and by Wöhlisch & Belonoschkin to give a higher value in the Gans effect (depolarisation of the Tyndall beam) than any other protein tried. These results, though important, only show that mixtures of ovomucin with other proteins exhibit, as might be expected, strong evidence of anisometric particles.

In his careful separation of the egg-white proteins, E. G. Young[1] decided against the existence of ovoglobulin (see the figures given in Table 2). This is the most recent attack on the problem, but it is fair to say that other workers with special experience in protein chemistry, such as E. C. Bate-Smith, believe in the existence of ovoglobulin.

In E. G. Young's work, ovomucin accounted for 4·0 % of the total protein obtained, ovomucoid 1·4 % and ovoglobulin only 0·1 %. E. C. Bate-Smith[2] gives percentages as follows:

	% total protein
Ovoalbumen	77
Conalbumen	3
Ovomucoid	13
Ovomucin	7
Ovoglobulin	Traces

Young found that the chalazae consist mainly if not entirely of ovomucoid and ovomucin (cf. the work of Conrad & Phillips mentioned above).

Ovoalbumen has naturally been the subject of a large amount of work. Svedberg assesses its molecular weight at 34,500 when purified, but it may be much less in the intact egg, as sedimentation experiments on the undenatured egg-white indicate. Very accurate data are now available for its titration curve, either when undenatured (Loughlin) or denatured (Wu, Liu & Chou); its pH dissociation curve (Kekwick & Cannan); its particle size by ultra-filtration (Elford & Ferry); its amino-acid constitution (Vickery & Shore on the basic amino-acids; complete analysis by Calvery[2]) and its ultra-violet absorption spectrum (Marchlewski & Wierzuchowska). Its fragmentation under the action of enzymes such as papain, pepsin, polypeptidases, etc. has been intensively studied (Annetts; Calvery[5, 6], Calvery & Schock; Calvery, Block & Schock). Its carbohydrate group was said by Sørensen, on the basis of the orcinol method, to be mannose (1·7 % of the protein), but Neuberger, by a new procedure involving tryptic hydrolysis followed by acetylation and deacetylation under the mildest conditions, has obtained a polysaccharide which must be regarded as the prosthetic group. It consists of 2 molecules of glucosamine, 4 molecules of mannose, and 1 molecule of an unidentified nitrogenous constituent. It seems, however, that this carbohydrate group plays but an insignificant role in the immunological behaviour of ovoalbumen (Ferry & Levy).

In ovomucoid there is, of course, a much larger amount of carbohydrate than in ovoalbumen (e.g. 9·2 %, Sørensen; 10·4 %, Young). Zuckerkandl & Messiner-Klebermass regarded this prosthetic polysaccharide as an equimolecular mixture of glucosamine and mannose; Sørensen as 4 glucosamine + 3 mannose + 1 galactose. Masamune & Hoshino now claim it to be an equimolecular mixture of mannose and acetylglucosamine, and this is supported by the work of Iseki and of Fürth, Herrman & Scholl, who when hydrolysing ovomucoid with toluol-sulphonic acid, obtained from this deacetylation $1\frac{1}{2}$–2 mols. reducing sugar for every mol. of acetic acid formed. Mazza has contributed a complete study of the physical and chemical properties of ovomucoid.

In his review on the mucoprotein polysaccharides K. Meyer shows how close their relation is with the uronic polysaccharides of cartilage and umbilical cord. The mucopolysaccharide isolated by Levene & Lopez-Suarez from umbilical cord mucin has turned out to be a hyaluronic acid containing glycuronic acid and acetylglucosamine but no sulphur (Meyer, Palmer & Smith[1]; see also M. Suzuki), in which respect it differs from the chondroitin sulphuric acid of cartilage (cf. p. 443) which contains glycuronic acid and acetylchondrosamine (2-amino-galactose) as well as sulphur. Mucoitin sulphuric acid, like the latter except that acetylglucosamine occurs instead of acetylchondrosamine, is also known. None of these occur in egg-white, but their relation to the ovomucoids may be quite significant. Such polysaccharides form salts with the basic amino-acids of various proteins (Meyer, Palmer & Smith[2]) but may also occur in the dissociated form. Further research on the metabolism and morphogenetic significance of these compounds will be awaited with interest.

How far the various separable proteins of egg-white exist as such in the intact egg is a question not easy to answer. That they are to a large extent artefacts, formed during the application of the isolation methods, is believed by some (Block; Block, Darrow & Cary; Perov), and though their evidence, depending as it does rather largely on relatively small differences in amino-acid estimations, is not altogether convincing, the idea has attracted notable minds (e.g. W. B. Hardy, A. Kossel, etc.) and certainly cannot as yet be said to be disproved. Besides the discussions of Block, there is a review by Alcock on this subject, to which the reader is referred. On the other hand, ovoalbumen, formerly accepted as an entirely homogeneous protein, has been separated into two fractions by cataphoresis (Tiselius & Eriksson-Quensel; Longsworth, Lannau & McInnes). The number of proteins identifiable in egg-white by the acetone methods of Piettre was three, and Orrù[6,8], in studying the changes in electrical conductivity of egg-white with temperature, found evidence for only two; but Young's[2] more recent work with the ultracentrifuge and the Tiselius apparatus suggests five or even six.

On the physical properties of egg-white there are numerous studies, e.g. electrical conductivity (Perov; Dolinov; Romanov & Grover); light transmission (Almquist, Givens & Klose), which depends primarily on the amount of ovomucin present; refractive index and its changes with storage (Romanov & Sullivan), etc. From Brooks & Pace we have an elaborate study of the CO_2-solubility in egg-white. The buffer value per gm. protein is $4·8 \times 10^{-5}$ between pH 6·6 and 7·8; there is little or no carbamino-CO_2 present, and the solubility coefficient is a little higher than that calculated, probably owing to the presence of traces of lipoids. The egg-white contains 4·5 milliequivalents of organic acids (Hosoi); on development this falls but later rises.

Sulphur compounds in egg-white exist only in the proteins, the sulphur-content of which can be entirely accounted for as cystine and methionine (Marlow & King; Toennies).

In later sections we shall consider the part played during development by the constituents of the white, but it may be mentioned here that according to G. Schmidt[2] the development of the embryo on the yolk *in vitro* will only go on if both the proteins and the dialysable substances of the white are present around it. Isotonic salt solutions will not do.

1·23. The vitelline membrane and its physiology

The vitelline membrane of the hen's egg is composed of a network of keratin fibres about $25\,\mu$ thick. It is of particular interest because it separates two systems of widely different properties, and upon their efficient separation the possibility of later normal embryonic development to some extent depends. On one side of the vitelline membrane there is a thick and viscous material containing approximately 50 % water, and on the other a much less viscous phase of which as much as 85 % may be water. The yolk solid is about two-thirds fat and one-third protein; the solid of the white is almost entirely protein. The yolk has double the percentage of ash of the white. Corresponding to these many other differences there is a great difference in osmotic pressure (about 1·5 atmospheres). Not merely is the total osmotic concentration different on the two sides of the membrane, but also there is practically no agreement between the constituent items in that concentration. Thus there is much more potassium, sodium, and chlorine in the yolk than in the white, but the difference is not the same in each case. The osmotic pressure difference is equivalent to about 2 kg./sq. cm., but it is obvious that the membrane is not submitted to any such mechanical pressure. The membrane must be permeable to water, since the yolk swells and osmotic equilibrium is attained in the intact egg if it is stored for some weeks; moreover, yolks readily take up water from dilute solutions (Smith & Shepherd; Orrù[1]). Permeability to some salts, though less pronounced, must exist, for Orrù[4] found that water in which yolks were immersed showed a gain in electrical conductivity, and an increased depression of freezing-point, and reached a pH of 5·7. Orrù[5] also brought forward evidence indicating the passage of glucose, sucrose and raffinose through the membrane into intact yolks. Hevesy, Levi & Rebbe have shown that although heavy water quickly distributes itself between white and yolk when injected, radioactive sodium phosphate does not, but remains for long in the white.

Straub & Hoogerduyn rightly considered that the Schreinemaker equations were inapplicable, and the distribution of ions in yolk and white is so different from what it would be on the Donnan equilibrium theory that the latter must be regarded as not alone sufficient to account for the case of the vitelline membrane of birds' eggs. Indeed, the potential difference between yolk and white, measured by Capraro & Fornaroli, is only 2–3 mV. rather than the 130 mV. required by such equilibria.

As pointed out by Needham & Smith, there are other remarkable facts about the separation of yolk and white. Yolk contains both vitamins B_2 and B_1; white only B_2 (Chick, Copping & Roscoe). Like all the other differences this difference will persist for many weeks in a stored egg, but in time equilibration occurs. Again, at first all the zinc is in the white and all the copper in the yolk (L. K. Wolff), but after a long period of storage this unequal distribution disappears. The pH differs greatly, the yolk is acid, the white alkaline.* The principal facts are summarised in the following table:

Table 3. *Properties of avian yolk and white*

	Yolk	Egg-white	Blood
$\Delta°$, freezing-point depression	−0·55	−0·44	−0·64
Osmotic pressure, atmospheres	6·6	5·3	7·7
% water	50	85	—
Fat % dry weight	66	0	—
Protein % dry weight	33	100	—
Ash % wet weight	1·5	0·7	—
Glucose % wet weight	0·27	0·55	—
pH	4·5–6·0	8·6–9·6	—

Milliequivalents/100 gm. water:

	Yolk	Egg-white
K'	5·81	3·73
Na'	9·30	2·61
Cl'	7·89	5·15

Straub & Hoogerduyn made two important experiments. On putting yolk into a parchment capsule with egg-white outside, osmotic equilibrium was attained in 24 hr. instead of many weeks, as in the intact egg. This was subsequently confirmed by Needham[5]. But Straub & Hoogerduyn also placed the yolk of a fresh egg in diluted egg-white, and after leaving it there for several hours, replaced it in ordinary white, whereupon its osmotic pressure, which had been lowered, rose above, not merely to the same level as, that of the ordinary white. There was thus a recovery of the natural hypertony of the yolk. They concluded that the "living" vitelline membrane tended to encourage the exit of water from the yolk or to impede its entry, or to encourage the entry of salts and impede their exit. This conception fitted in well with the views on the performance of thermodynamic work at living membranes, which were then becoming current (1931; respiratory rate and osmotic work, etc.), and the view of the vitelline membrane as such a membrane was widely adopted, as by Kluyver and A. V. Hill[2]. Hill[1] himself contributed the observation that the osmotic pressure equilibration curves in unincubated eggs (as measured by the vapour-pressure method) were alike whether the eggs had been kept aerobically or anaerobically. If energy were needed, therefore, it must come from some source not involving oxidation in free oxygen.

This led to a series of papers in which the situation was much clarified.† Smith & Shepherd showed that the fundamental experiment of recovery of hypertony by the yolk was to be explained simply on the basis of a temporary

* This is also true, as Imamura[1] and others have shown, for reptilian eggs such as those of marine turtles. During development the yolk becomes more alkaline and the white more acid (Schkljar; Rubinstein; Gaggermeier; Sharp & Powell; Baird & Prentice), confirming *CE*, p. 857.

† Summarised in Needham, Smith, Shepherd, Stephenson & Needham.

heterogeneity of the contents of the vitelline membrane; when the yolk is placed in diluted egg-white or water, water enters, forming blisters of very low osmotic pressure, and not being incorporated in the yolk owing to its great resistance to diffusion.* The recovery of hypertony is therefore an illusion, for the main mass of the yolk has never been diluted at all. Straub later himself adopted this explanation. Stored at 25°, complete osmotic equilibrium between yolk and white is reached after about 70 days; between 0 and 25° the temperature coefficient of this equilibration is between 1·5 and 2. This was later confirmed by Orrù[3]. The shape of the curves and the regular effect of temperature strongly indicate that the primary process is diffusion, hindered by various complex factors. The viability of stored fertile eggs shows a quite different relation to temperature, since there is an optimum at 8–10° and a slow decline of viability on either side of this point (Moran[1]; see Fig. 2). The converse of the Straub-Hoogerduyn experiment was also performed by Smith & Shepherd, namely the concentration of the egg-white by rapid evaporation, but there was no perceptible effort on the part of the yolk to maintain its hypertony.

In contrast to earlier statements in the literature, the infertile unincubated hen's egg gives off no respiratory CO_2 and consumes no O_2 (M. Smith[1]). There is a certain liberation of CO_2 from fresh eggs, but it falls off rapidly to minimal values, and undoubtedly arises from the action of traces of acid substances on the calcium carbonate of the shell. The vitelline membrane itself exhibits no dehydrogenase activity and neither the membrane nor the yolk show any measurable respiration *in vitro* (Needham, Stephenson & Needham). Bacteriologically sterile yolk produces minute traces of lactic acid and of a substance or substances estimatable as ethyl alcohol; the larger amounts described in the older literature were certainly due to bacterial contamination. It is very unlikely that the energy so produced is available for work at the vitelline membrane.

As stated above, yolk and white, separated only by a parchment or collodion membrane, rapidly come to osmotic equilibrium. An apparatus was devised by Needham[5] for investigating the osmotic properties of the isolated vitelline membrane, and it was found that when placed between two salt solutions of different strengths, it opposes no resistance to rapid equilibration. There is thus a paradox which is still not fully resolved. With yolk and white on either side, equilibration is slower than with salt solutions, though not nearly so slow as in the intact egg. It is probable that the physical structure of the yolk and white has some feature which retards the attainment of equilibrium, and which is not wholly destroyed by the mixing of each phase on its removal from the egg. The slow attainment of equilibrium in the intact egg is probably due to the co-operation of the two phases with the membrane. It must be remembered that the yolk is laid down layer upon layer, not poured into the vitelline membrane, and its lipoidal molecules or lipoprotein complexes are almost certainly highly oriented. This was the view of Hardy (in Hale & Hardy), who emphasised the hindrances

* Baldes later believed that the layers of yolk nearest the vitelline membrane have normally a lower osmotic pressure than the central core, so that a continuous gradient would exist from the interior of the yolk to the shell, but this could not be confirmed by Moran & Hale.

to free diffusion in the yolk. Moreover, structures hindering rapid diffusion of water in the white were found by Smith & Shepherd.

The actual existence of a difference in osmotic pressure between yolk and white has been denied by some authors, but mistakenly. Grollman[2], the first of these, used dialysis and the thermo-electric technique for freezing-point determination, but Meyerhof[1], in whose laboratory the work had been done, later showed that his results were vitiated by the absence of a correction for the water initially in the collodion membranes. Bateman[1] and Baldes also vindicated the use of the thermo-electric technique as applied to viscous liquids such as yolk. Howard[2] reinvestigated the question with the Beckmann thermometer and again obtained the same osmotic pressure for yolk and white, but these results were decisively rejected by M. Smith[2], who showed that they were due to plateaus of temperature without the presence of ice in the mixture, caused by a temporary balance between the heat gained by stirring and lost to the cooling bath. Ice forms much more slowly in viscous fatty media than in aqueous solutions, but in a crucial experiment, it formed and persisted at $-0.58°$ but disappeared at $-0.53°$. Similar conclusions were reached by Johlin[1], by Hale & Hardy, and by Weinstein, so that there can be no doubt of the osmotic pressure difference between yolk and white. This can be demonstrated very simply, as Johlin[2] showed, by placing yolk and white in an apparatus such that water can distil over from the more dilute to the more concentrated phase; in such circumstances yolk always gains at the expense of white.

When yolk and white are mixed together, or with solutions of various salts, their physical behaviour is quite anomalous. There occurs on simple mixing a removal of osmotically active substances (Bateman[1]) and some precipitation (Baldes). The changes of viscosity when yolk is diluted with solutions of salts at different pH (Zawadzki), and the changes of electrical conductivity with temperature (Orrù[6]), are very complex.

In the vitelline membrane itself there is said to be some mucin as well as the keratin (Moran & Hale). By the aid of the micro-manipulator three layers can be detected in it, an outer thin layer, the thick middle layer, and an inner thin layer. The thin layers stain as mucin, the main one as keratin. The average bursting strength of the vitelline membrane at laying is about 4500 dynes/sq. cm. (Moran[3]); on storage the membrane weakens but becomes more elastic. As the egg ages water-intake by the yolk when transferred to water becomes more rapid (Orrù[1]). After 40 hr. incubation of a fertile egg the membrane becomes so weak over the germinal area that its bursting strength can no longer be measured. With the growth of the yolk-sac it disappears. Moran[3] recalls in this connection the hatching enzymes of lower vertebrates and invertebrates (see p. 627), but though there may be an evolutionary connection, such an enzymic mechanism can hardly be of any value to the chick embryo. Trypsin rapidly weakens the vitelline membrane, which is interesting in view of the existence of a protease with tryptic properties in the thick egg-white (Balls & Swenson).

1·24. The yolk

The yolk of the hen's egg consists of concentric layers of white and yellow yolk, with a central core of white yolk, which can often be seen in eggs coagulated by boiling. The details of the structure were described in *CE*, p. 235. No new analyses of white and yellow yolk have been made since those previously described (*CE*, p. 286), but further analyses of the white yolk, the "erste Nährung" of the embryo, would be very desirable. Sometimes the peripheral white yolk layer is laid down unevenly, producing a characteristic spotty appearance of the yolk (Schaible, Davidson & Moore).

Egg-yolk proteins have been the subject of an excellent review (up to 1932) by Jukes & Kay[3]. The principal protein of the yolk is the phosphoprotein vitellin. Its amino-acid distribution has been investigated with modern methods by Calvery & White, who obtain higher values for tyrosine, arginine and lysine than previous workers. The most reliable figure for phosphorus-content is 0·92 %. The figure of Calvery & White for tryptophane is somewhat lower than that of Tillmans & Alt. Methionine in vitellin has been estimated by various authors (Baernstein; Scharpenack & Jerjomin); there is about 2 %.

Study of the hydrolysis breakdown products of vitellin has thrown much light on how the phosphorus is combined in the phosphoprotein molecule. Posternak & Posternak were the first to identify serine-phosphoric acid (phosphoserine) as the principal mode of attachment, and this was later fully established by Lipmann & Levene and Levene & Schormuller. Posternak & Posternak obtained an "ovotyrin"* from vitellin containing 10–14 % phosphorus and 28 % *l*-serine.† Such phosphoproteoses seem to occur to a small extent in normal yolk, and it was found that after allowing yolk to autolyse a much increased yield could be obtained. On decomposition the ovotyrin of the Posternaks gave a good deal of pyruvic acid, probably because of the formation of aminoacrylic acid after removal of the phosphoric acid groups from a phosphoserine chain, and subsequent oxidation. The natural phosphoproteoses are rather easily hydrolysed by alkali (heating for 1 min. with *N*/NaOH), an instability which the Posternaks suggested might be of some biological importance in phosphorus transfer from raw materials to embryonic nuclein and bone phosphate (see *CE*, p. 1202).

Measurements of the reaction constant for hydrolysis of vitellin by Lipmann[2] showed it to be identical with that for phosphoserine itself, and Rappoport, using a new method of estimation for serine, concluded that all the P of vitellin is combined as phosphoserine (in contradistinction to casein, where one-third of the P is not in phosphoserine). Agreement has not, however, been finally reached. Blackwood & Wishart noted that pepsin liberates much acid-soluble N and not much acid-soluble P from vitellin, while trypsin does both; they concluded that there must be two kinds of P complex in the molecule, one highly resistant to enzymic attack. Herd obtained phosphopeptones of varying N/P ratios, suggesting that the P is unequally distributed through the molecule. In pepsin there

* Many phosphoproteoses have been isolated from the fragmentation of vitellin, differing slightly in their content of P, Fe, serine, etc., and the names "ovotyrin", "ichthiotyrin", "vitellinic acid", "paranuclein" and "haematogen" are practically synonymous.

† They obtained it also from the vitellin of fish eggs, hence the name "ichthiotyrin". The early workers on vitellin called vitellin "ichthulin" if it came from fish eggs, "batrachiolin" or "ranovin" if it came from amphibian eggs, etc., but these names are quite unnecessary and should be given up. The prefix "ovo-", to vitellin is also obviously superfluous.

is an accumulation of those fragments which contain much P and so are resistant to this enzyme. Herd's results would perhaps not be incompatible with the conception of localised concentrations of phosphoserine chains.

The ultra-violet absorption spectrum of vitellin is given by Marchlewski & Wierzuchowska.

The second protein of egg-yolk is livetin (*CE*, p. 294), a pseudo-globulin containing only 0·067 % P. It occurs to the extent of about 20 % of the total yolk-protein. Its amino-acid distribution has been studied with modern methods by Jukes[1], Jukes & Kay[1], and Kay & Marshall. It contains about the same amount of methionine as vitellin (Baernstein). Jukes & Kay[2] háve shown immunologically that in all probability livetin is identical with serum globulin (cf. the identity of conalbumen with serum albumen). Physical constants of livetin will be found in Kay & Marshall.

Earlier work on protein-bound carbohydrate in the egg had made it very likely that the yolk as well as the white contained a good deal of prosthetic polysaccharide of the mannose-glucosamine type, and Jukes found 4 % in livetin by the Tillmans-Philippi method. It was left, however, for Onoe[1] to isolate from the yolk a third protein, vitellomucoid, analogous with the ovomucoid of egg-white, and containing 10·1% carbohydrate. The work of Orrù[6,8] on the effect of temperature on the electric conductivity of yolk* had already indicated the existence of at least three proteins, and Piettre, by his acetone method, believed he could detect four. Onoe's vitellomucoid proved to have only mannose and glucosamine in its prosthetic group.

Whether these proteins are at any time preferentially absorbed by the embryo we do not know (but see below, p. 620, for some information of this kind on vitellomucoid).

The richness of egg-yolk in neutral fats, phosphatides and sterols has already been pointed out (*CE*, pp. 294, 1218). An egg of average weight (55 gm.) has in its yolk 5·58 gm. neutral fat and 1·28 gm. phosphatides, of which 0·68 gm. is lecithin (Lintzel; Bornmann). The structure of the phosphatides has been closely examined. Rae found that in yolk lecithin the β-form of glycerophosphoric acid predominates (i.e. the fatty acids are arranged symmetrically to the choline-phosphate radicle), in contrast to brain lecithin, in which the α-form predominates, and liver lecithin, which is an equal mixture of the two. There is agreement (Sueyoshi; Sueyoshi & Furukobo; Yokoyama; Riemenschneider, Ellis & Titus; Spadola & Riemenschneider) that the fatty acids of yolk lecithin are mainly isopalmitic, oleic, linoleic, clupanodonic, and 9:10-hexadecenoic, with very little stearic and palmitic.[†] According to Yokoyama, five varieties of the molecule are

---

* Data also in Romanov & Grover.

† The following data are appended:

Acid	No. of C atoms	No. of double bonds
Palmitic	16	0
9 : 10-Hexadecenoic	16	1
Stearic	18	0
Oleic	18	1
Linoleic (Linolic)	18	2
Linolenic	18	3
Clupanodonic	18	8
Arachidonic	20	8

present: dioleic-α-lecithin, di-*iso*palmitic-α-lecithin; oleic-clupanodonic-α-lecithin, oleic-*iso*palmitic-β-lecithin, and dioleic-β-lecithin. In egg-yolk kephalin, on the other hand, Nishimoto found the following varieties: palmitic-stearic-α-kephalin, palmitic-arachidonic-α-kephalin, palmitic-stearic-β-kephalin, palmitic-arachidonic-β-kephalin, and diarachidonic-β-kephalin. The possible significance of these highly unsaturated acids as raw material for embryonic tissues, which may not themselves be capable of desaturating fatty acids, will not escape notice.

Kay believes that the presence of free glycerophosphoric acid in fresh yolk is doubtful; in the fresh egg there is nothing which can act as substrate for phosphatase. But on storage the lipases are active and the acid-soluble phosphorus of the yolk increases (Pine); eventually the inorganic phosphorus of the white also (Janke & Jirak).

The egg-yolk of the hen, and many other birds, contains 1·75 % cholesterol (Miyamori); but that of the duck about twice as much (Gaujoux & Krijanowski; Achard, Levy & Georgiakakis).

Yolk dissolves a good deal more CO_2, according to Brooks & Pace, than white, because of its higher lipoid content. The yolk contains 5·0 milliequivalents of organic acids (Hosoi); on development this rises slightly and later falls.

There is no spermine in eggs (Dudley & Rosenheim). Details on pigments and enzymes will be found in the appropriate sections (see below, pp. 70, 652). Vitamins will also be discussed later (pp. 23, 616), but their presence in yolk and white may be summarised with a select bibliography as follows:

Table 4. *Vitamins in the avian egg*

	Yolk	White	
A	+	−	Russell & Taylor; Cruickshank & Moore
B_1	+	−	Baker & Wright; Plimmer, Raymond & Lowndes
B_2	+	+	Chick, Copping & Roscoe; Aykroyd & Roscoe
P-P	+	−	Goldberger, Wheeler, Lillie & Rogers
C	−	−	Hauge & Carrick; Ray
D	+	−	Guerrant, Kohler, Hunter & Murphy
E	+	−	Card, Mitchell & Hamilton

1·25. Inorganic constituents

New complete analyses of the ash of yolk and white have been made by Straub & Donck (they are in close agreement with those given in *CE*, Table 27) and Yosida has made estimations of total basic ions in yolk and white. These workers were only concerned with elements present in considerable quantity, but there exist studies with modern spectrographic methods which give some information about the rarer elements also.* According to Drea, ash from the whole hen's egg contains Al, Ba, B, Ca, Cu, Fe, Mg, Mn, P, K, Ru, Si, Na, Sr, Ti and V.† Of these, strontium and vanadium appear to be concentrated by the maternal organism in the egg, as the egg contains more than the diet. Boron and manganese occur only in the yolk. Apparently absent were F, Ag, Cr, Pb and

* See below, p. 651, for the results of application of similar methods to foetal tissues.

† It is interesting to compare this list with the trace elements found in newborn mammals and their meconium. Work of Ramage; Sheldon, Ramage & Ramage; and Rusoff & Gaddum shows that these include Al, Ba, Cu, Mn, Sr, Sn, Zn, and sometimes Pb and Ag, but not Sb, Be, Bi, B, Cd, Co, Ln, Th, Ti, V, Yt, or Zr.

Mo. Similar work by Cazzaniga demonstrated Fe, Na, K, Al and Ca in both yolk and white, but Mg only in the white. Love adduces evidence of a heavy metal, probably lead, but this was in Australia, where, as work previously described (*CE*, p. 305) has shown, eggs laid by hens in the vicinity of lead mines may contain considerable amounts of lead (see also p. 556), deposited in the yolk as lead oleate.[*]

The halogen elements in the embryonic raw material have received a good deal of attention. In an elaborate paper on the chloride of yolk and white Mankin[2] concluded that the white Cl′ is usually slightly higher than the yolk Cl′. This is at variance with the results of Straub & Donck, but for this ion the differences between yolk and white are not very great. All investigators are agreed that the iodine content of the egg, though initially very small (about 0·1 mgm./kg.), can be raised many hundreds of times by adding such substances as KI, kelp, etc. to the diet (Almquist & Givens; Wilder, Bethke & Record; Berkesy & Gönczi; Scharrer & Schropp; Jaschik & Kieselbach), though the increase in the eggs laid is not directly proportional to the increase in iodine-content of the diet. Zaitschek found that added iodine raised the hatchability of the eggs 13 % and the number of eggs laid by 12 %, but no such effect was observable by Johnson, Pilkey & Edson. By injecting sodium bromide into laying hens Purjesz, Berkesy & Gönczi were able to increase the bromine of the eggs from 23γ to 7 mgm./egg largely in the yolk. By injecting sodium fluoride into laying hens the fluorine of the eggs could be raised from zero to 40γ/egg. This was confirmed by Phillips, Halpin & Hart, who on adding fluorine-containing mineral supplements to the diet found that the fluorine was deposited in the yolk in ether-soluble combination with lipoid material (like lead). Fluorine in the yolk rose from 0·3 mgm./100 gm. to 1·46. Fluorine toxicosis inhibited laying without reducing the size of the eggs.

By feeding colloidal lead orthophosphate to the laying hen W. B. S. Bishop[4] was able to double the amount of lead in the eggs (0·047 mgm./egg to 0·09 mgm./egg); even in doses subtoxic for the adult the hatchability was much reduced.[†] Selenium will also go into hen's eggs, in quantities sufficient to be dangerous (Franke & Tully; Franke, Moxon, Poley & Tully; Moxon & Poley). Molybdenum has been recognised qualitatively in egg ash (Mankin[1]). In agreement with L. K. Wolff (see above, p. 12), Guérithault found that copper occurs in the yolk to the extent of 1·8 mgm./kg. (wet wt.), but there is none in the white. Manganese also occurs only in the yolk, where there is about 2·5 mgm./kg. (dry wt.), as estimated by Skinner & Peterson, Peterson & Skinner, and McHargue. Finally, silicon has been studied by King, Stantial & Dolan, who find it to be present in both the yolk and white in very small amounts.

By feeding diets of very different ionic composition Schkljar[3] claims to have brought about considerable changes from the normal alkali reserve of both yolk and white.

Doubtless the most important problem which arises with regard to the inorganic constitution of the hen's egg is the state in which iron is found there. In the

---

[*] This work was criticised on technical grounds by H. B. Taylor, but the reply on behalf of the original group of workers by W. B. S. Bishop[5] was convincing.

[†] See p. 556.

earlier literature (cf. *CE*, p. 291) it was often noted that various phosphoproteose preparations, obtained from the breakdown of the vitellin molecule, contained high percentages of iron. This was the reason for the introduction of Bunge's term "Haematogen". The still unsolved riddle of the origin of haemoglobin in the embryo makes a decision as to the state of the iron in the egg of some importance. The modern phase opened when R. Hill[1] in 1930 applied the α-α'-dipyridyl method of iron estimation to egg-yolk. Haematin iron is protected from this reagent, but on adding a reducing agent to egg-yolk in its presence, a deep red colour immediately develops, suggesting that the iron is present as colloidal ferric hydroxide. Since the amount of iron as estimated directly by the dipyridyl method was equal to that estimated after ashing, Hill concluded that the yolk contained no iron in strongly bound organic combination such as haematin. These results were later confirmed by Lenti.

This was widely thought to imply the liquidation of Bunge's haematogen, but McFarlane continued to uphold the view that iron exists in weakly bound organic form for the following reason. After alkaline hydrolysis of lecitho-vitellin (0·25 N/NaOH) a fraction was obtained, not precipitable with trichloracetic acid but precipitable with neutral lead acetate, containing all the iron and 25 % of the copper of the original material. The metals were still present in the material (which was presumably a phosphoproteose, though no phosphorus determinations seem to have been made) after removal of the lead with H_2S (Fe, 1·23 %; Cu, 0·0082 %), and there was no reaction with potassium thiocyanate or sodium diethyldithiocarbamate. McFarlane thought that iron, more loosely bound in vitellin than in haematin, might react with dipyridyl, but the point was left untested. Contributory suggestions that yolk does indeed contain organic as well as inorganic iron arose from two sources, the histochemical work of Marza, Marza & Chiosa, and the fact, established by Elvejhem, Hart & Sherman, that yolk iron is only poorly available for haemoglobin regeneration in rats, its value in this respect, unlike that of most other sources, not corresponding to the dipyridyl assay. Histochemical methods are, however, much too uncertain to decide a question of this nature, and it was subsequently shown that the regeneration discrepancy is due to the sulphur of the egg which, combining in the gut with the copper to form insoluble CuS, withholds from the animal the copper necessary for the catalysis of haemoglobin synthesis (Sherman, Elvejhem & Hart; Rose, Vahlteich & McLeod). The discrepancy disappears if copper is added to the yolk given.

McFarlane's views are not supported by the most recent work of Tompsett using thiolacetic (thioglycollic) acid. In trichloracetic filtrates of yolk there is nothing that interferes with this method of estimation, yet such filtrates show no iron and iron alum added to yolk before the precipitation completely disappears. But if the reaction with thiolacetic acid is carried out before precipitation with trichloracetic acid, then the iron present is readily estimated in the filtrate, and not only does this quantity of iron agree exactly with that found after ashing, but ferric iron added to fresh egg-yolk may be quantitatively recovered. It may therefore be regarded as fairly clearly established that the iron of the yolk is in inorganic form adsorbed in some way on to proteins or lipoprotein complexes. As ordinarily prepared, the "Fe-content" of vitellin is in the neighbourhood of 0·045 % (McFarlane).

McFarlane also believes that copper is organically bound in vitellin (about 0·003 %).

1·3. The effect of pre-developmental factors on development

That the development of the chick embryo is affected by factors operating on and in the egg before development has started admits of no doubt. These factors manifest themselves as differences of hatchability, a subject on which we owe most of our information to American investigations.* The chemical composition of the egg itself can be affected by dietary or genetic influences, and the fate of the egg after laying can be affected by several environmental variables (cf. the reviews by Landauer[23] and by Jukes & Kay[1]).

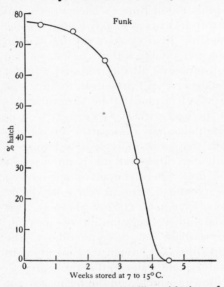

Fig. 1. Decline of hatchability with time of storage before incubation (hen's eggs).

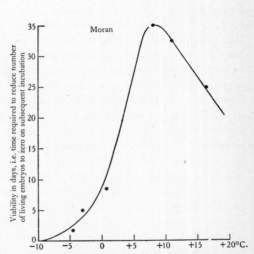

Fig. 2. Optimum temperature for storage of fertile hen's eggs.

The accompanying curve (Fig. 1) taken from Funk's statistical work, confirmed by Kaufman[8] and many others, shows how hatchability falls off with time between laying and the beginning of incubation. As Fig. 2 shows, Moran[1] found an optimum temperature for such storage. The exact interpretation of these findings is not entirely clear, but the unphysiological equilibration of the yolk and white is probably involved.† Hays & Nicolaides believe that eggs laid

* Card[3] has made a useful bibliography of the publications of U.S. Agricultural Experiment Stations on subjects of importance for embryology.

† Loss of water by evaporation is not in itself harmful up to a certain point. Kaufman[8] found that after 5 weeks, when hatchability is reduced to zero, only about 2% water had been lost from the eggs, yet when the water-content of fresh eggs had been reduced by 5% by exposing them to low atmospheric pressures, hatchability is unaffected. Moreover, hatchability could not be restored by soaking the kept eggs before incubating them. The margin of safety for water-loss is therefore, in the hen's egg, appreciable (but see pp. 36, 64).

containing advanced gastrulae have a high probability of successful hatching, while those laid as early pre-gastrulae have a low probability.

Once development has started, the optimum temperature lies between 38·9 and 39·4° C. for natural incubation, and between 37·2 and 37·8° C. for forced-draught incubators with uniform heat distribution. Optimum humidity is also an important variable; it lies between 50 and 60%. There is a considerable literature on the details of these questions because of their industrial importance; it may be approached through the reviews of Landauer[23] and of Romanov[10], and the paper of Byerly[6].

The effect of egg size on hatchability has been subject to some dispute, but it is now generally accepted that hatchability falls off at both extremes of variation from the normal, more especially so at the higher egg-weights (Godfrey; Warren; Funk; Axelsson; Shibata & Murata), cf. Fig. 3. According to Penionskevitch[4] variations of egg-weight are due largely to variations in the total weight of the egg-white. It is interesting, therefore, to find that such characters as

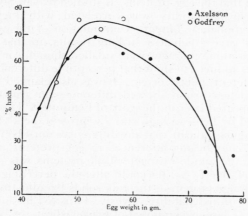

Fig. 3. Relation of egg weight and hatchability.

percentage and physical properties (particularly gel rigidity) of the thick egg-white are inherited (van Wagenen & Hall; van Wagenen, Hall & Wilgus; Hall & van Wagenen; Lorenz, Taylor & Almquist), and that hatchability could be correlated with the properties of the thick egg-white. If it is non-rigid or present in too small amount the embryos die between the 2nd and the 7th day of incubation; if there is too much of it, or of the rest of the egg-white, they die at later stages. Abnormally large and heavy eggs have an abnormally long incubation period (Byerly[5]; McNally & Byerly), and such eggs contain at 48 hr. incubation embryos with more somites than normal. In abnormally small eggs, however, development is not correspondingly retarded. McNally & Byerly found that there is an optimal hatchability in eggs containing 20-somite embryos at 48 hr. incubation.

The characteristics of the egg-shell are also important for hatchability. Excessive porosity will lead to excessive evaporation of water and desiccation of the embryo (Axelsson). Shell porosity is affected by both autosomal and sex-linked genetic factors, but may be artificially produced (as a phenocopy, see p. 392) by withholding calcium from the diet of the laying hen, as in the experiments of Buckner, Martin & Peter. Penionskevitch[5] points out that the shell-membranes are also probably important for hatchability, since they may vary in their capacity for holding water in case of undue desiccation, and hence in their liability to develop a parchment-like condition which makes hatching impossible.

At present there is little evidence that yolk qualities play a role in hatchability. Percentage solids (Bronkhorst & Hall[2]) and the shape index, probably a measure of vitelline membrane elasticity (Hall & van Wagenen), are unrelated to it. There may be some connection between the pigmentation of the yolk and hatchability; this will be referred to below in connection with pigments (pp. 23, 653). Penions-kevitch[4] thinks that "efficiency of yolk utilisation" may be an important factor.

A good deal of work has been done on the effects of diets of different protein composition upon the eggs and the embryos arising from them. In general the results are difficult to interpret. If the composition of the egg-proteins is affected by the protein in the diet, it must be in relatively subtle ways which do not clearly reveal themselves to the still relatively crude technique at our disposal for protein characterisation. Pollard & Carr fed pigeons on different single seed grains and reported that the egg-proteins, as studied by van Slyke analyses, varied considerably. High hatchability was correlated with high melanin (hence probably high tryptophane) content. Continuation of the work later however by Gerber & Carr gave much less striking results, in that the differences in mono-amino and di-amino fractions were smaller, though contributory immuno-logical evidence was adduced to show that perceptible differences existed between egg-proteins of different dietary origin. It was also claimed that embryos from eggs of different dietary origin showed differences in growth rate, though no such differences appeared later in the work of Penquite & Thompson on embryos from eggs produced by hens on high and low protein diets. The experiments of Titus and his colleagues indicate that any differences must be very small. Calvery & Titus found no significant differences in egg-proteins of different dietary origin, except that the amide-nitrogen of all proteins was slightly lower on a soy bean diet. Titus, Byerly & Ellis could alter the percentage dry matter in the yolk and the percentage protein in the yolk solids by adjusting the diet, but this was not correlated with hatchability. Byerly, Titus & Ellis found that on certain diets, especially those largely composed of maize, wheat, soy bean, fish meal, etc., there was a significantly high mortality of the embryos during the 2nd week of incubation. This is not the time at which either of the two normal mortality peaks occur (see p. 64). Whether this was due to a deficiency of essential amino-acids or of vitamins was not elucidated, though all the necessary vitamins were believed to be present in sufficient amount. The largest investigation was that of McFarlane, Fulmer & Jukes, who obtained wide variations in hatchability, and in the stage of development at which the maximal mortality occurred, by employing diets containing buttermilk powder, fish meal, tankage, meat meal, etc. Thus with tankage there was a 50% mortality in the first mortality period (3rd–5th day), while with cod-liver meal the mortality rose to 50% in the second mortality period (18th day onwards). Yet no significant difference was found in the composition of the protein from low hatchability eggs as regards total nitrogen, total amino-nitrogen, tyrosine, tryptophane and cystine content. There were, indeed, variations in total ash and iron-content of the proteins, but they bore no relation to the hatchability. One striking effect of tankage was noticeable, namely an anaemia of the early stages (4th day), haemoglobin failing to form in the blood islands, and an almost colourless fluid circulating through the heart. For this phenomenon no explanation was found. It was certainly not due to iron or copper deficiency. It does not follow that all samples of tankage show these effects, since high hatchabilities were reported with it by Francisco, Chan & Fronda.

Variations may also be induced in the lipoidal fractions of hens' eggs. Cruick-shank has established that unsaturated fatty acids, present in unsaturated oils in the feed, are deposited in the yolk, but that on the contrary an abnormally high proportion of saturated fatty acids in the diet does not affect the normal mixture of saturated and unsaturated fatty acids in the yolk. This is supported by Lorenz, Almquist & Hendry, who obtained positive Halphen tests, indicative of an unsaturated fatty acid, on yolks formed by hens which had eaten malvaceous plants or their seeds (e.g. cotton-seed meal or mallow seed). Almquist, Lorenz & Burmester then found that if, after the depôt fat of a laying hen shows a strongly positive Halphen test, all the eggs larger than $\frac{1}{4}$ in. diameter are removed from the ovary, the yolk subsequently deposited on the remaining eggs will show no positive Halphen test, though the depôt fat will continue to show it. And this is still the case even on a fat-free diet. This ingenious experiment is in agreement with a similarly planned one carried out by Henderson & Wilcke using the dye Sudan III, and it goes far to show that depôt fat is not necessarily used by the hen in laying down the fat of the yolk, a synthesis from carbohydrate and protein being sometimes preferred. On the other hand, as is well known, substances giving unusual flavours and odours do easily become incorporated in eggs (cf. McCammon, Pittman & Wilhelm).

Cruickshank & Moore find that the vitamin A content of egg-yolks (first separated from accompanying carotenoid yolk-pigments by Gillam & Heilbron in 1935) can be raised five times by enriching the diet of the laying hen with it; but as in the case of iodine and other substances, the rate of increase in the egg is far from being directly proportional to the increase in the diet. It is stated by Barancheev and Kirsanov that there is a relation between yolk-colour and hatchability, the darkest yolked eggs developing best, and since the lipochromes are by far the most important coloured substances in yolk (Bisbey, Appleby, Weis & Cover) it may be that lipochromes other than vitamin A have a physiological significance. Deficient vitamin A in eggs, as is generally acknowledged since the work of Holmes, Doolittle & Moore, will produce low hatchability, and up to a certain point the restoration of hatchability will go on in proportion to the vitamin added to the diet of the laying hen (Sherwood & Fraps; Payne & Hughes; R. M. Smith; Bearse & Miller).

It is doubtful whether vitamin B_1 plays any role in hatchability (Ellis, Miller, Titus & Byerly). Vitamin B_2, however, is essential, and, like B_1, can be varied in the egg by feeding different diets (Ellis, Miller, Titus & Byerly; Bethke, Record & Wilder; Lepkovsky, Taylor, Jukes & Almquist; Engel, Phillips & Halpin). As already mentioned (p. 17), egg-white contains only B_2 while yolk contains both vitamins. Norris, Wilgus, Ringrose, Heiman & Heuser have shown that, up to a certain concentration of vitamin B_2 in the diet, there is a direct proportionality between hatchability and concentration, though after the optimal point is reached further additions give no further improvement (see also the work of Ogorodny & Penionskevitch, referred to on p. 389).

The presence of vitamin D in the egg is essential for successful development. It varies greatly in amount with different diets. The question of whether sterol provitamins can be formed by the irradiation of eggs with sunlight or ultra-violet light, discussed in CE, p. 1360, is still not settled, but it seems probable that the penetration of radiant energy through the shell is insufficient to do more than compensate for slight dietary deficiencies (Landauer[9]; Martin, Erikson & Insko; Martin & Insko; Insko & Lyons[2]; Graham, Smith & McFarlane; Payne & Hughes; Murphy, Hunter & Knandel; Carver, Robertson, Brazie, Johnson & St John; Guerrant, Kohler, Hunter & Murphy; Russell & Taylor). By a laborious process

Windaus & Stange separated the provitamin from egg-yolk cholesterol and found it to be ergosterol, present at 0·18% in the crude material. Schoenheimer & Dam and Menschick & Page had already shown that ergosterol, added to the diet of laying hens, is deposited in the eggs produced (for structural formulae, see p. 251). The mechanism of embryonic avitaminosis-D has been more closely examined by Insko & Lyons[2], who found that embryos in such eggs fall more and more behind the normal controls in their calcium and phosphorus content as development progresses. The peak of mortality occurred on the 18th day. As Landauer[15] showed, the presence of the *creeper* gene (see p. 382) causes an earlier onset and a severer expression of D-deficiency symptoms, so that the two mechanisms may be regarded as additive.

Excessive vitamin D in the diet will cause a decline of hatchability even to zero (Branion & Smith), but the effect on the embryos is still very obscure.

It is generally held that vitamin E is essential for successful development. This is difficult to prove, however, for no diets for fowls have so far been devised which do not probably involve other deficiencies also, or cause other disturbances. As we shall see later, interesting specific morphological abnormalities arise in E-deficient embryos (Adamstone[1], cf. below, p. 264), and there are several reports that hatchability is seriously affected by E-deficient diets (Card; Card, Mitchell & Hamilton; Barbas; Barnum; Dols).

Other accessory food factors of hitherto unknown nature may be involved in normal development, as Nestler, Byerly, Ellis & Titus observed low hatchability on diets not deficient in any of the known vitamins.

We come lastly to the more subtle effects on hatchability arising from the condition of the parent birds. From the work of Axelsson and others it may now be considered as established that the hatchability of eggs laid during the first or pullet year is slightly better than that of eggs laid during the second year, and the work of Hyre & Hall strongly suggests that this decrease in hatchability continues during the subsequent years. Somewhat contrary to expectation is the fact that hatchability varies positively with current egg production but not with antecedent egg production; hens able to lay eggs at a high rate on a particular diet will, in general, produce a higher percentage of hatchable eggs than those only able to lay at a low rate on the same diet (Byerly, Titus & Ellis). Under certain conditions (*frizzle* gene, Landauer; sudden falls in external temperature, Funk; individually characteristic low blood-sugar, Bronkhorst & Hall[1]) the laying hen may apparently be unable to deposit all the necessary substances in the eggs, and falls in hatchability may ensue.

1·4. Constitution of the eggs of lower vertebrates and invertebrates

1·41. Reptiles and Amphibia

We have a good deal of information about the composition and metabolic changes during development of the chelonia (especially the turtle *Thalassochelys corticata*; see *CE*, pp. 898 and 1043, etc.), but concerning the composition of saurian eggs extremely little is known. Studies of the egg-white of the loggerhead turtle, *Caretta* (Kondo; Kondo, Yamada & Nagashima), show that it contains two proteins provisionally classified as globulins, and an albumen. This differs from hen's egg-albumen, however, in coagulating 10° higher, containing more tryptophane and tyrosine (spectrophotometrically deduced) and having a different isoelectric point (pH 5·75 instead of 4·89). The egg contains 50% egg-white, 43% yolk and 7 % shell. Analytical data for the percentage composition and proportion of the amino-acids histidine, arginine, lysine, tryptophane, tyrosine and cystine in turtle vitellin are given by Kusui[2]; during development there was no change in the composition of the protein. Recently Fukuda[2,3] has given fairly complete analyses of the eggs of a dozen species of snakes; in general, as would be expected, their composition is similar to those of birds, but they seem to have much less carbohydrate, both free and combined.

In view of the overwhelming importance of the amphibian embryo in experimental embryology, it is very surprising that no studies of its chemical constitution with modern methods have been made. Pierantoni has given an account of the yolk-platelets of amphibian eggs, but there is here an important gap which needs filling. Most of the newer information not contained in *CE* concerns the egg-jelly, composed of fairly pure mucoprotein. Long ago Schulz & Ditthorn and v. Ekenstein & Blanksma isolated galactose from frog egg-jelly mucoprotein, and the recent work of Schulz & Becker has shown, chemically and by fermentation methods, that the prosthetic polysaccharide consists of galactose and glucosamine (cf. p. 9). It is generally known that the jelly is soluble in strong cyanide solutions, which, remarkably enough, do not irreversibly interfere with the subsequent normal development of the embryos, but it is not so well known that exposure to ultra-violet light will also dissolve the egg-jelly (Sypniewski). It has often been supposed (e.g. *CE*, p. 323) that the jelly furnishes nourishment to the emerged tadpoles, which may be seen clinging to it with their suckers, but a special attempt to decide this question (Savage[1]) gives no grounds for such a belief. The mean nitrogen-content of young tadpoles fell in exactly the same way, owing to consumption of yolk, whether they had been allowed to remain on the jellies or not. Disappearance of the jellies is due to bacterial action, not consumption by the tadpoles. It was noted, nevertheless, that the differentiation of the gut seemed to be more advanced in the tadpoles from the jellies than in those from batches where the jellies had been taken away; this contrasts with the statement of Szepsenwol[3] that removal of axolotl tadpoles from the jellies results in an acceleration of neural differentiation owing to the earlier onset of external stimuli. In a later paper, Savage[2] estimated the total carbohydrate in tadpoles from batches with and without jelly, and was able to show that no significant difference exists. The nutritive function of the jelly can therefore probably now be ruled out, but Savage suggests that it may act as a heat-insulator and thermostatic device.

The tension at the surface of a salamander egg (*Triturus virescens*) has been determined by Harvey & Fankhauser.

1·42. Fishes

As regards the eggs of teleostean fishes there is little to add to our former account (*CE*, pp. 331, etc.) except that Mörner has shown that the *dl*-cystine, said to have been isolated from herring eggs by Steudel & Takahashi, was in all probability taurine. Bernheim & Bernheim describe the phosphoprotein of the eggs of the trout *Cynoscion nebulosus*, which, when undenatured, contains free-

–SH groups like myosin or haemoglobin. Konopacka[3] has given an account of the formation of the oil-globule so conspicuous in teleostean eggs, in *Smaris alcedo*; and the surface tension at the interface between such globules and the yolk has been measured by Harvey & Shapiro for *Scomber*, *Phycis* and *Merliccius*.* To account for its remarkably low value it is considered necessary to assume some unknown substance in the yolk of high surface activity, identified in the case of the mackerel as a globulin by Danielli & Harvey.

The protein of the "chorionic" membrane of salmon eggs (*Salmo salar*) was analysed by Young & Inman, who describe it as a pseudo-keratin of mesodermal origin. The permeability properties of the egg-membranes of the trout have been described in an interesting paper by Gray[3], having in mind the work already discussed (p. 12) on the properties of the vitelline membrane of the hen's egg.† The normal trout egg, like those of all organisms developing in fresh water, is impermeable both to water and intracellular electrolytes, the barrier being the (inner) vitelline, not the (outer) chorionic membrane. Later work by Krogh & Ussing, using heavy water, and P. Adler, using dyes, strikingly illustrated this. Injuries to the egg-cell cause exosmosis and a coagulation of the vitellin which spreads rapidly all over the egg from the point of injury. Osmotic stability varies with age, mechanical disturbances, Ca″ concentration of the medium, and the nature and concentration of the exosmotic agent; owing to the impermeability of the membrane no work is required to maintain the normal concentration of salts and water within.‡ Precipitation of the vitellin within the egg can also be brought about by galvanic currents, according to Scheminsky & Scheminsky, anodically if weak, cathodically if strong.

Teleostean eggs contain vitamin A (McWalter & Drummond) and B$_1$ and B$_2$ (Lunde, Kringstad & Olsen).

Analytical work on the eggs of selachian fishes shows that they possess less fatty material than would be expected from their highly lecithic nature. Ray eggs (*Raia oxyrhyncus*) studied by Schmidt-Nielsen & Stene and Schmidt-Nielsen, Aas, Astad & Leonardsen gave the following figures:

	% Wet weight			
	Dry weight	Ash	Protein	Fat
Egg-white	4·0	2·7	1·4	0·0
Yolk	36·0	1·3	28·0	6·7

* And by Harvey & Schoepfle for the oil-globule naturally occurring in daphnid eggs (see also Chambers & Kopač).

† We do not know whether the "chorionic" membrane of the teleost egg is to be regarded as analogous with the shell-membrane or the vitelline membrane of the hen's egg. If it is analogous with the shell-membrane the perivitelline space represents the white. If it is analogous with the vitelline membrane of the hen's egg we might expect to find some protoplasmic layer inside around the hen's egg yolk corresponding with the protoplasmic vitelline membrane of the fish, but for this there is no evidence.

‡ Other fish eggs also have highly impermeable membranes, e.g. *Oryzias* (Ikeda[2,3]), but in this case the embryos within are quite sensitive to external ionic disequilibria. But the marine teleosts have eggs with much more permeable membranes, according to Krogh, Krogh & Wernstedt. The crustacean *Artemia* (the brine-shrimp) is a curious case; the adults live in brines of 25% salinity or more, and show a remarkable constancy of their internal medium (Medvedeva; Warren, Kuenen & Baas-Becking). But the eggs and nauplii need lower osmotic pressures and are sensitive to abnormal ion balances (Boone & Baas-Becking; Jacobi & Baas-Becking; Baas-Becking, Karstens & Kanner; Matthias). Those who wish to pursue these questions further are referred to the monograph of Krogh on osmotic regulation in aquatic animals.

Δ for white was $-2 \cdot 08°$, for yolk $-2 \cdot 90°$. The yolk contained $5 \cdot 89 \%$ urea (see *CE*, p. 343). The eggs of the shark, *Heptranchias deani*, studied by Ono[1], have more fat:

	%
Water	47·10
Fat	23·30
Protein	25·13
Ash	0·81

Ono[2] made a special study of the physico-chemical properties of the fatty fractions in this case. Of the neutral fat 21% was solid acids (isopalmitic, palmitic, stearic) and 79% liquid (oleic, linolenic and iwashic). Unsaponifiable material (12% of the crude fat) included octadecyl and cetyl alcohols as well as the highly unsaturated selachyl alcohol, besides cholesterol (see also below, p. 250). An interesting study of the yolk-proteins of the ray (*Raia batis*) by Fauré-Fremiet[3] confirms the earlier discovery of a livetin in elasmobranch eggs (*CE*, p. 334); in this case it forms about 10% of the total protein. The lecitho-vitellin complex occurs in the form of strongly birefringent rectangular or pyramidal platelets containing 17% phosphatide and $1 \cdot 4 \%$ cholesterol.

Among numerous observational studies on elasmobranch egg-cases that of Aiyar & Nalini on the Indian *Chiloscyllium griseum* may be mentioned, but the only physico-chemical work on them is that of Fauré-Fremiet[7] and his collaborators on the keratin, which they undertook in connection with their researches on protein microfibrils in general (see p. 676). The manner of secretion of this protein in rays is described by Garrault & Filhol. Fauré-Fremiet & Baudouy[2] showed that prokeratin, obtained by Fauré-Fremiet & Garrault from the nidamentary gland of *Raia*, *Scyllium*, *Mustelus*, etc., differs from the protein as found in the completed egg in containing somewhat more sulphur and somewhat less nitrogen. The completed egg-case shows birefringence and a highly oriented micellar constitution (Fauré-Fremiet[7]; Filhol & Garrault). Imbibition properties of selachian egg-case keratin were studied by Baudouy.

We shall have a good deal to say later on, under the heading of the embryo and the environment, about the eggs of selachian fishes.

1·43. Arthropods

Apart from data on the constitution of insect eggs, which will be mentioned incidentally to other questions later in this book (pp. 36, 455), there is little to be said here. At the beginning of development, the egg of the grasshopper, *Melanoplus differentialis* (cf. pp. 578 ff. for a fuller discussion of this embryo), contains $17 - 20 \%$ dry wt. fatty acids, and this amount is reduced by half during development (Slifer[1]). In a later study, Slifer[4] examined the fatty acids from a number of orthopteran eggs. Iodine numbers varied from 128 in the case of *Dissosteira carolina* to 166 in the case of *Arphia sulphurea*, and melting-points from $25 \cdot 5$ to $39 \cdot 5°$ in another series. As the females of all the species were kept at constant temperatures during the egg-laying period, the differences, though possibly due to environmental adaptations, cannot be due to the direct effect of different environmental temperature. In spite of the similarity of iodine values, the fats differ widely in the proportion of solid and liquid fatty acids, and probably in chain length. Sacharov thinks that there is a relation between cold-hardiness of insects and total fat-content of their eggs, but a larger set of comparative figures than those offered by him will be needed to establish this. From the unsaponifiable fraction of silkworm eggs Ongaro has isolated a saturated hydrocarbon $C_{28}H_{58}$ which recalls the hydrocarbons found in fish oils (p. 250).

Remarkable changes in the viscosity of the yolky silkworm egg may be followed, according to Tirelli[1], by normal development. After heating to 50° for $\frac{1}{2}$ hr. the viscosity of the press-juice is 50 % greater than normal, and this change is irreversible, for it persists for two months. Yet there is no interference with normal development.

The inorganic constitution of insect eggs has been little studied, but Melvin has shown that certain insects, such as *Blatta*, accumulate copper, in this case up to 18 mgm./kg. dry wt. in the eggs.

Gause[1] has shown that egg size in *Drosophila* is affected by the maternal diet in larval life.

1·44. Echinoderms

Much the most outstanding contribution in this section is the list of physical and chemical constants for the sea-urchin egg (*Arbacia punctulata*) prepared by E. N. Harvey[4] in 1932. This gives data for all dimensions, densities, viscosity, surface tension, osmotic properties, pH, chemical constitution, rate of development, etc. Unfortunately there is little since that date with which to supplement it.* Costello[1], however, has ascertained the relative volumes occupied by the formed components of twenty-six species of marine eggs (of other groups as well as echinoderms), using a centrifuge method. This volume varies from 40 % in the egg of *Hydroides* to 82 % in that of *Polychoerus*, and in every case it exceeds the volume of the osmotic dead space. This is in agreement with all other authors, such as Lucke[1], who studied the osmotic behaviour of centrifuged *Arbacia* egg fragments and found that it closely paralleled that of the whole egg (though the osmotically inert material was unequally distributed) and Białaszewicz[1]. It is not quite clear whether the "dispersed phase" of Białaszewicz is identical with the osmotic dead space of the American authors. In *Echinometra lucunter*, for example, this osmotically inert volume is 34 % of the initial egg volume (Leitch). A large body of literature exists on the permeability properties of echinoderm eggs, but since most of it is occupied primarily with biophysical problems and so far has thrown little light on morphogenesis, it will not be reviewed here. For further details the reader is referred to the review of Lucke & McCutcheon.

Bank has described the properties of the jelly surrounding the echinoderm egg. He has also made the curious observation that a stratification of the egg constituents, analogous to that produced by centrifuging, can be obtained if the eggs are placed in a 1 % solution of caffein. The phenomena must be similar, for it was noticed that batches of eggs which were unusually resistant to centrifugal stratification were also resistant to stratification by caffein.

Work on the cytology of maturation of the egg of *Echinometra lucunter* (Tennent, Gardiner & Smith; Miller & Smith) was followed by some chemical work by D. E. Smith. Contrary to earlier work on echinoderm eggs (*CE*, p. 1049) a good deal of polysaccharide was found in this one (12·7 % dry wt.). The total solids amounted to 28 % wet wt.; cholesterol was present to the extent of 0·6 % wet wt. and phosphatides 1·8 % wet wt. There was less neutral fat than glycogen (6·3 % dry wt.), so that proteins would account for 80 % dry wt., a striking example of the richness in protein in marine alecithic eggs commented on in *CE*, p. 313. The iodine value of the fat from *Arbacia* eggs was studied by Payne, who found that it varied with the temperature of the water in which the maturation and

---

* Ballentine[2] gives accurate figures for the nitrogen content of *Arbacia* eggs.

spawning had taken place, from 98 at 0° to 150 at 20°. The higher figure is in agreement with one given by Terroine, Hatterer & Roehrig. Navez, who thinks that a fatty acid derivative of echinochrome (see p. 215) may exist, also found evidence of highly unsaturated fatty acids and possibly of squalene (see p. 250).

A short communication by O. Glaser[5], overlooked in former publications, indicates the existence of considerable amounts of copper in echinoderm eggs; this should be further investigated.

1·45. Molluscs and worms

The most peculiar characteristic of molluscan eggs is the presence of relatively large amounts of a polysaccharide formed from d-galactose, not glucose. This galactogen has been studied in a number of papers by May and May & Kordovich. The eggs of the snail, *Helix pomatia*, contain no less than 37·8 % of it (dry wt.) —6·02 % wet wt.—and no glycogen. It has a sinistral rotation $\alpha_D - 22\cdot73°$ and gives no iodine reaction. Both galactogen and glycogen accumulate in the snail's body for hibernation, but only the former is deposited in the eggs. Apart from the gonad, the slime gland is the only other organ which contains galactogen, but as the total egg material has more galactogen than what is present in these organs, it is probable that glycogen from other organs can be converted into it. In starvation the reverse process takes place. Slug eggs also contain galactogen (36·2 % dry wt.), but May and May & Weinbrenner were also able to isolate traces of the polysaccharide from other sources such as the gonads of *Anodonta*, the eggs of the carp (where it makes up 1·2 % of the total polysaccharide, itself only 0·2 % dry wt.), the rabbit ovary (where 7 % of the total polysaccharide is galactogen), and even the human placenta (where 1 % of the total polysaccharide is galactogen). May suggests that galactogen may be, in the higher animals, the mother substance of the milk-lactose galactose and the brain-cerebroside galactose. Progress towards elucidating the exact chemical constitution of galactogen is recorded by Baldwin & Bell. Snail eggs also contain a curious laevorotatory tetrasaccharide (4-d-galactosido-l-glucosan) according to May & Stadelmann.

The X-ray investigations of F. K. Mayer have yielded the fact that in the shells of embryo gastropods, the calcium carbonate is laid down, first in the form of vaterite, then aragonite, and finally calcite. Anomalies in gastropod eggs are described by Löhner and Cardot.

The Japanese gastropod, *Hemifusus tuba*, has been the subject of further researches. Figures for the composition of its vitellin and mucoid are given by Kumon, and the proportion of various amino-acids in the egg-capsule protein has been investigated by Sagara, who takes it to be a keratin. Ankel, on the other hand, finds chitin in the egg-capsules of the European *Nucella (Purpura) lapillus*. These fragmentary data will acquire greater interest when we know more of the chemical constitution of gastropod eggs. During development, according to Kumon, the total nitrogen of the egg interior rises, but he could not find a corresponding fall in the total nitrogen of the egg-capsules (see below, p. 626). Lastly, the eggs of the land-snail, *Achatina fulica*, have recently been analysed by S. Fukuda[3]. These appeared to contain very little fat—only 0·1 % wet wt.—but besides the protein (10·2 %) an equivalent amount of something else which was neither protein nor fat (11·4 %), probably the galactogen of May. There was also, according to Fukuda, an unusually large amount of silicate, but this should perhaps be accepted with caution, as the egg-mass might have been imperfectly freed from earth.

It seems that the eggs of lamellibranch molluscs have fatty acids with the highest iodine values yet recorded for eggs (Terroine, Hatterer & Roehrig). A table given by these authors follows:

Eggs	Iodine value of fatty acids
Hen	103
Frog	142
Sea-urchin	135
Pike	163
Herring	208
Lamellibranch mollusc (*Pecten*)	232

Manganese has been noted as a constant constituent of the eggs of a number of fresh-water lamellibranchs (Bradley; 0·63 % dry wt.).

The first satisfactory analyses of the eggs of cestode worms have been recorded by Smorodintzev, Pavlova & Ivanova.

	% Water	% Organic substance	% Ash	% Protein	% Fat
Diphyllobothrium latum	72·6	26·59	1·15	21·64	4·95
Taeniarhynchus saginatus	77·19	22·30	0·66	15·42	6·88

The low fat-content is worth attention. The adult worms have, of course, a much higher water-content.

1·5. The inheritance of immunity

Among the chemical substances present in eggs antitoxins and antibodies must be included (cf. *CE*, pp. 1444 ff.). While in birds (Ramon; Jukes, Fraser & Orr; Fraser, Jukes, Branion & Halpern) and in reptiles (Grasset & Zoutendyk) both antigens and antibodies are deposited in the eggs, in mammals the placenta is permeable only to antibodies and not to the antigens themselves.* This subject has been well reviewed by Grasset, who gives an account of the literature. It has long been known that at early developmental stages, the power to react to a foreign antigen by the production of active immunity is absent or present only to a slight degree. Thus Freund using rabbits and Kliger & Olitzki using guinea-pigs, in *ad hoc* experiments, found young animals to be poor antibody producers (Jaffé). According to Gebauer-Fuelnegg the formation of antibodies by the chick embryo proceeds only very weakly, and Grasset was quite unable to obtain any antibody formation when diphtheria toxin, either in active or inactive (anatoxin) form, was injected into chick embryos at any time during incubation. The anatoxin did not interfere with normal development, but the toxin quickly killed the embryos. Weinberg & Guelin had exactly the same experience with *B. sporogenes*; the chick embryo can make no agglutinin for it.

In general, therefore, it may be said that embryonic and new-born animals may possess passive immunity, caused either by the deposition of antibodies in the egg or by their passage through the placenta.† During post-natal development this passive immunity disappears and is replaced by the active immunity charac-

---

* This is constantly being confirmed, see e.g. the paper of Tooney on the *B. coli* agglutinin titre of maternal and foetal blood in man.

† Note that this condition resembles that of adult invertebrates, judging from the useful review of Huff. They depend on natural, rather than acquired, immunity.

teristic of adult life, i.e. by the power to form antibodies in response to antigens entering the blood circulation. This general rule may be summarised in the following diagram, from Grasset (Fig. 4).

It is to be noted that embryonic passive immunity falls off before the rise of adult active immunity, so that there should be, as is actually observed, a period of low level of immunity in early post-natal life. The inability of the embryonic and newly born or hatched organism to form antibodies is particularly interesting in view of the lack of species-specificity as between tissues, which embryos of different species or even groups and classes show (see p. 348). Rodolfo has done an admirable piece of work on the permeability of the rabbit placenta to antibodies. This placenta changes in construction during embryonic development.

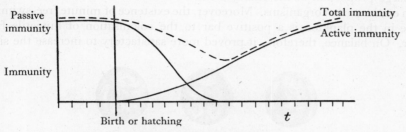

Fig. 4. Embryonic origins of immunity.

On the 10th day the two bloods are separated by three membranes, the foetal endothelium, the foetal mesoderm and the foetal ectoderm; on the 22nd day there are only the first two of these; and on the 29th day only the foetal endothelium. The placenta is thus of the haemo-chorial type at the beginning (see p. 81) but haemo-endothelial at the end. Correspondingly, Rodolfo found that the ratio of titres foetal blood/maternal blood rose from 0·2 at 23 days to 0·8 at 30 days in the case of the agglutinin to *B. abortus* and from 0·05 at 23 days to 1·0 at 30 days in the case of the haemolysin for sheep blood corpuscles.

In this connection the origin of the human blood groups is of interest (Haselhorst; Keeler & Castle; Abe; Kaboth). Oku has shown that they originate in the foetal circulation about the 2nd month of pregnancy, the iso-antibodies coming from the maternal circulation. Keeler & Castle think that small amounts of the iso-agglutinogens also come through the placental barrier, but are immediately destroyed in the foetal blood. A difference between the blood group of maternal organism and foetus is not believed to have any deleterious influence upon the latter, though some authors, such as Garrasi, believe that optimum intra-uterine conditions, as manifested by birth weight and placenta weight, are present when the blood groups are the same.

It may be added that the technical feat of raising chicks from the egg bacteria-free has been accomplished by Balzam.

1·6. The egg and the environment

1·61. The problem of protection; larval compression and ovoviviparity

In what follows we shall have to review a rather complex interplay of factors involved in the relation between the developing embryo and the environment, including various aspects of metabolism and nutrition. But it is all ultimately connected with the devices employed by animal organisms to give their offspring the best chances of survival. The most primitive type of development is probably that in which a large number of eggs with relatively little yolk are laid. Although this has the advantage of facilitating the dispersion of the species, the destruction of a large proportion of the embryos before or after they hatch entails a severe strain on the parent organisms. Moreover, the existence of minute free-swimming larvae in the plankton is a positive bar to the colonisation of fluviatile freshwater. On balance, therefore, it proved more satisfactory to increase the size of

Fig. 5. Developmental stages of the tree-frog *Hylodes martinicensis*.

the eggs and to prolong the developmental period so that the embryo, when it hatched, could face the hostile world with much enhanced probability of survival. This meant compressing larval stages into the embryonic period. The whole range of egg sizes, from that of the sea-urchin to that of the ostrich, is an illustration of this tendency.

A glance at Brinley's table of the egg sizes of fishes is instructive in this connection, for Rass was able to conclude that (among the teleosts at any rate) those species with the largest eggs and the longest developmental periods had the widest geographical distribution. But the larger the eggs became, the more valuable they were to the species, and hence we see arising all kinds of arrangements for the care of the developing embryos and larvae (cf. the reviews of Wunder). High degree of parental care, small number of eggs, large egg size, and long developmental time, all go together, as do the converse phenomena. Thus fishes, marine or fresh-water, which show parental care, lay from 80 to 200 eggs in a season; those which do not, lay from 30,000 to 700,000. For amphibia, parallel figures are 4–270 as against 1000–28,000. Excellent cases of larval compression are found in groups as widely separated as the amphibia—Fig. 5, taken from Wunder, shows the later stages of the development of the tree-frog *Hylodes martinicensis*, in which the tadpole stage is passed entirely within the egg—and the nemertines (Schmidt & Jankovskaia).*

In the course of time it was found that the protection of the eggs could best be assured, not merely by the continued presence of the parent organisms, but by their actually carrying the eggs about with them. This is what happens to-day in the case of the toad *Alytes obstetricans*. But such close contact led inevitably to the complete retention of the eggs within the parent organism, either at some level in the genital tract (e.g. the ovary or the uterus) or in some specially prepared brood-pouch on the ventral or dorsal surface, or even in the mouth, either of the male or female. In some such way ovoviviparity arose. And the process did not stop there, for the maternal organism could carry the offspring for almost unlimited periods if means were taken to reduce the structures interposed between the foetal and maternal blood circulations. So at last true viviparity was attained, as in the mammals.

One of the most elaborate investigations of the origins of ovoviviparity is that of Berrill[2] on ascidians.† The ovoviviparous Molgulids go through a course of development in which the tailed larva is suppressed, and osmotic hatching takes the place of the hatching enzyme (cf. p. 627). For further details the original papers must be consulted.

1·62. The evolution of the cleidoic egg

We now come to the degree of dependence upon the environment shown by the eggs of different organisms. It has for long been recognised (more particularly in *CE*, Sections 6·6, 9·15) that some animals have eggs much more completely closed off from their environment than others, and the word *Cleidoic* (closed box) has come into use to designate the former. A typical non-cleidoic egg would be that of a cephalopod, which is not supplied by the maternal organism with sufficient ash or sufficient water to make one complete embryo, but has to absorb many of its constituents from the sea. The mammalian egg must also be regarded as non-cleidoic, for even its organic constituents are not stored in it at the beginning of development, but have to be provided later on from the maternal organism. A typical cleidoic egg, on the other hand, would be that of a bird, where all that the embryo requires, except the oxygen, is provided within the closed system. Between these extremes many intermediate grades are found (see Table 5).

The most fundamental need of the non-cleidoic egg is for water. This was fully discussed on a previous occasion (*CE*, pp. 889 ff.), but some newer observations justify further comments here. The clearest cases are those of two fresh-water organisms, the trout (*Salmo*) and the axolotl (*Amblystoma punctatum*), for which equations can be drawn up, as in Table 6, showing the amount of

---

* Cf. also the discussion on p. 76. High degree of parental care without large egg size or viviparity can also be attained if the life-span is so adjusted that the earlier generations of offspring can assist in protecting the later ones. In some such way, as Wheeler has taught, arose the social habit among insects.

† See also Child[2].

Table 5. *Evolution of the cleidoic egg*

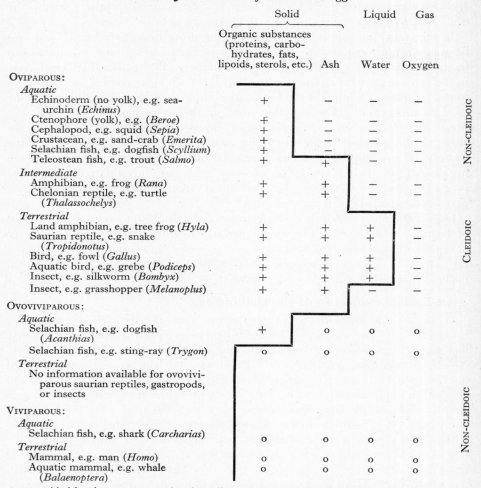

	Solid		Liquid	Gas	
	Organic substances (proteins, carbohydrates, fats, lipoids, sterols, etc.)	Ash	Water	Oxygen	
OVIPAROUS:					
Aquatic					
Echinoderm (no yolk), e.g. sea-urchin (*Echinus*)	+	–	–	–	NON-CLEIDOIC
Ctenophore (yolk), e.g. (*Beroe*)	+	–	–	–	NON-CLEIDOIC
Cephalopod, e.g. squid (*Sepia*)	+	–	–	–	NON-CLEIDOIC
Crustacean, e.g. sand-crab (*Emerita*)	+	–	–	–	NON-CLEIDOIC
Selachian fish, e.g. dogfish (*Scyllium*)	+	–	–	–	NON-CLEIDOIC
Teleostean fish, e.g. trout (*Salmo*)	+	+	–	–	NON-CLEIDOIC
Intermediate					
Amphibian, e.g. frog (*Rana*)	+	+	–	–	
Chelonian reptile, e.g. turtle (*Thalassochelys*)	+	+	–	–	
Terrestrial					
Land amphibian, e.g. tree frog (*Hyla*)	+	+	+	–	CLEIDOIC
Saurian reptile, e.g. snake (*Tropidonotus*)	+	+	+	–	CLEIDOIC
Bird, e.g. fowl (*Gallus*)	+	+	+	–	CLEIDOIC
Aquatic bird, e.g. grebe (*Podiceps*)	+	+	+	–	CLEIDOIC
Insect, e.g. silkworm (*Bombyx*)	+	+	+	–	CLEIDOIC
Insect, e.g. grasshopper (*Melanoplus*)	+	+	–	–	CLEIDOIC
OVOVIVIPAROUS:					
Aquatic					
Selachian fish, e.g. dogfish (*Acanthias*)	+	o	o	o	
Selachian fish, e.g. sting-ray (*Trygon*)	o	o	o	o	
Terrestrial					
No information available for ovoviviparous saurian reptiles, gastropods, or insects					
VIVIPAROUS:					
Aquatic					
Selachian fish, e.g. shark (*Carcharias*)	o	o	o	o	NON-CLEIDOIC
Terrestrial					
Mammal, e.g. man (*Homo*)	o	o	o	o	NON-CLEIDOIC
Aquatic mammal, e.g. whale (*Balaenoptera*)	o	o	o	o	NON-CLEIDOIC

+ = provided by the parent organism in sufficient quantity per egg to make one embryo.
− = not provided by the parent organism in the egg and therefore required directly from the environment.
o = not provided by the parent organism in the egg but transmitted in other ways, i.e. not required directly from the environment.

water which has to be absorbed directly from the environment in order to form the relatively wet tissues of the finished embryo from the relatively dry yolk provided by the maternal organism. There is abundant evidence that all fishes and amphibia stand in like dependence upon the aquatic environment. But starting from the observation of Noble & Brady that the eggs of another species of *Amblystoma* (*opacum*) develop on land in damp leaf-mould, we find that many devices are resorted to by certain groups of amphibia to provide enough water for their embryos while yet not allowing them to develop actually *in* pond or river water. Some of these are described in the reviews of Wunder and Noble. Like the eggs of *A. opacum*, such embryos certainly get the necessary water, but often in remarkable ways. Thus the Japanese frog, *Rhacophorus schlegeli*, lays its eggs in holes in the ground, enclosed in a stable foam which holds much water for long periods. Other species, such as *Natalobatrachus bonebergi*, attach their egg-masses to rocks just above water-level in situations where the humidity of the air must always be considerable. Siedlecki, in his well-known paper on the Javanese tree-frog, *Polypedates reinwardtii*, describes how the mass of eggs is stuck between two leaves, so that when later on the central part of the jelly breaks down, an artificial pond is formed in which the larvae swim. Very many species of tree-frogs, such as *Hyla* and *Phyllomedusa*, show similar adaptations.

Buxton[1], in his book on animal life in deserts, notes it as quite remarkable that amphibia should occur at all in desert regions. It appears nevertheless that they do. Different amphibia inhabiting the Central Australian desert solve the problem of their non-cleidoic eggs in different ways. *Hyla rubella* depends primarily on a much accelerated incubation time, so that the best advantage is taken of waters in temporary pools and streams. So also the Chinese rain-frog, *Kaloula borealis*, according to J. C. Li, gastrulates in 11 hours and hatches in 3 days instead of the usual week or more. Other frogs make use of deep burrows, where humidity will remain high even when the surface has dried out (*Limnodynastes ornatus*), some lay large masses of water-holding jelly in the burrows (*Heleioporus pictus*), and some can store a very large quantity of water in their urinary bladders which may be used for moistening the eggs (*Chiroleptes platycephalus*). In the case of *Heleioporus* the eggs are believed to develop to the tadpole stage and then to "hibernate" till the advent of rain, when the membranes burst and the larvae hatch.

Water, either partly from the parent's bladder, or contained in the sand, is certainly absorbed by the eggs of many chelonian reptiles (see *CE*, p. 898, and lately Imamura). According to new observations (Cunningham & Hurwitz; Cunningham & Huene; Cunningham, Woodward & Pridgeon) the water-intake of the eggs of the turtle, *Caretta caretta*, accounts for the 50 % increase in weight which they show during their development, and the eggs of the diamond-back terrapin (*Malaclemys centrata*), of an initial wet wt. of 10·58 gm., gain 3·07 gm. water while they lose 0·32 gm. in combusted organic substance. These figures are incorporated in Table 6. The content of inorganic substance remains of course unchanged.

It is not quite clear whether we are right in placing all the eggs of saurian reptiles in a definitely cleidoic position as regards water in the diagram. Some species, such as the gecko (Goeldi), are unambiguous, and most lizard eggs are independent of water. But Portmann[2] has made a special examination of the eggs of the grass-snake, *Tropidonotus natrix*, from this point of view and finds that in damp earth there is an uptake of water by the eggs (gain in weight) accompanied by successful incubation, while under the same conditions as hens' eggs there was a rapid loss of water and all the embryos died. The water may enter as vapour; in a saturated atmosphere, an even balance of weight is maintained.

Portmann thinks that many other reptiles occupy such a position, and this would agree with the work of Cunningham & Hurwitz on the American fence-lizard, *Sceloporus undulatus*, and with the old observations of Thilenius on the tuatara (*Sphenodon*; *Hatteria*).*

The water requirements of bird embryos in their cleidoic eggs will be discussed presently in relation to the normal mortality curves which they show (below, p. 64). Here it need only be recalled that (*CE*, p. 876) the water-loss from a bird's egg is proportional to the humidity of the external air. It was formerly thought that the eggs of aquatic birds, such as the grebe (*Podiceps*), which have shells impregnated with wax or oil, were impermeable to water or less permeable than hens' eggs, but this has been contested by Portmann & Jecklin. In deserts the water conservation in birds' eggs must be a problem of great difficulty, and it is interesting to read in Buxton's book that the various species of sandgrouse (*Pterocles*) which inhabit nearly every part of the Great Palaearctic desert (from the Atlantic coast of Africa to the borderland between the Gobi and China) fly regularly many miles to water and return to their nests carrying it in their soaked feathers. The young certainly depend on this for their water, and it is likely that the eggs are watered in the same way. In this connection the "pigeon milk" as a source of water for nidicolous birds should be remembered (see below, p. 75).

Insects present an interesting special case. In structure and metabolic qualities (see below) their eggs are cleidoic, but at any rate in some cases the embryos succeed in getting a certain amount of water from the environment (cf. the previous discussion of this, *CE*, p. 905). For insect embryos, as for those of birds, there is an optimum humidity. In an interesting review (which should be consulted for the special literature), Buxton[2] shows that for many insect eggs, at any rate (e.g. the grasshopper *Camnula pellucida*), the water-loss is directly proportional to the saturation-deficiency.† Percentage hatchability closely follows this water-loss. It may be significant that the eggs of many insects normally occur in places where the humidity is very high and that most of them do not suffer from exposure to saturated air, though a few are known to do so. Excessive dryness, on the other hand, is more commonly fatal; in this case, the embryo may be killed directly, or the shell so hardened that hatching is impossible. Certain eggs, such as those of the South African *Locusta pardalina* (Lounsbury) or the Australian springtail, *Sminthurus viridis* (Davidson), enter a state of dormancy if severely dried, and they may then live for several years and resume their development when wetted. Other eggs, such as those of the yellow-fever mosquito, *Aëdes argenteus*, resist drying, but in a different way, for if drying occurs when they are nearly ready to hatch, the eggs do not shrink, for the membranes have become impermeable to water. In this state they may remain hatchable for 7 months.

---

* But of course such embryos have no opportunity of getting rid of nitrogenous waste-products, see on, p. 50.

† Saturation-deficiency is the amount of water vapour which would have to be added to a sample of air to saturate it, without altering the temperature.

Since the time of Réaumur in 1740 it has been known that the eggs of sawflies, which are inserted into the leaves of plants during their development, perceptibly swell. Réaumur described such eggs as doubling in size as they develop, and queried whether the shell of the egg might act as a "placenta" to draw nourishment from the leaf to which it is fastened. There are numerous modern observations to the same effect (Blunck; Johnson; Roonwal[1]), and dyes have been shown to enter the eggs from the plant. Johnson, in a very detailed study of the development of the egg of the capsid bug *Notostira erratica*, observed that as the eggs swell the yolk extrudes as a plug under the operculum or micropylar cap. He then measured the water-content and found it to rise from 0·052 mgm./egg initially to 0·148 mgm./egg at hatching; correspondingly the dry weight diminished from 0·059 to 0·050 mgm./egg. These figures are included in Table 6.

Bodine[2], however, was the first to show by direct measurement that even in the case of insect eggs lying freely in earth there may be an intake of water. The weight of an egg of the grasshopper, *Melanoplus differentialis*, rises from about 2·5 to 4·7 mgm.; this is due to the imbibition of water. The process remains stationary during diapause. We cannot incorporate these figures in Table 6, as the loss of organic substance was not known. This work was subsequently confirmed on *Locusta migratoria* (Roonwal[1]) and *Popillio japonica* (Ludwig[2]).

The exact mechanism of water-intake in insect eggs remained unclear till recently, especially as Jahn[1] showed that the egg-membranes of the grasshopper are impermeable to salts such as ferricyanide and ferric chloride. Slifer[7], however, has shown that the *Melanoplus* egg has a small, circular, specialised area in the yellow cuticle located at the posterior end; this has been named the "hydropyle" and through it the water exchanges go on. If this hydropyle is covered with any material impermeable to water, the development may be stopped or retarded, although the rest of the surface of the egg is fully permeable to gases. How far this mechanism is present in eggs of other insects and even of other arthropods is discussed in Slifer's paper, to which the reader is referred.

There still remains the further problem of what it is which creates the inward flow of water through the hydropyle. Perhaps the yolk has "deliquescent" properties. It is interesting in connection with this that McLeod states that certain plants (e.g. *Reaumuria hirtella*) secrete salt crystals which deliquesce at night and so provide water which the plant absorbs. It should also be added that though there is evidence that water may enter at the hydropyle, there is no evidence that any nitrogenous waste-products can escape there.

There can be little doubt that insect eggs take advantage of the "microclimates" of small crevices, etc., where the humidity and temperature are different from the surrounding main bodies of air (Leeson & Mellanby; Uvarov).

On the other hand, many insect eggs lose water continuously during development. Thus Tuleschkov found that the water-loss of the eggs of the moths *Lymantria dispar* and *monacha* was of the order of 30 mgm. on 250 mgm. in 2 months.

1·63. The direct transition from the non-cleidoic state to ovoviviparity

Since the first establishment of Table 5 (as *CE*, Table 260) our knowledge of what it implies has been greatly furthered by the excellent work of Ranzi and his collaborators, who have been able to demonstrate, especially in the selachian fishes, all the stages between dependence on the environment as regards water and salts, through approximate equilibrium, to dependence on the maternal organism as regards water, salts and organic substances.

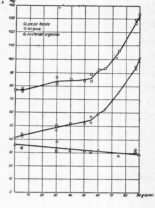

Ranzi's investigations began with the eggs of cephalopods. He was able to show[6] that the weight of the whole egg of the squid *Sepia officinalis* rises during development from about 75 mgm. to about 135 mgm., and that this is accounted for largely by the intake of water. But the total ash rises also from just under 1 mgm. to about $3\frac{1}{2}$ mgm., showing that the cephalopod egg is dependent also on the ash of the sea-water (see Fig. 6). It is not provided either with sufficient water or sufficient inorganic materials for the production of one complete embryo. Following the equation earlier obtained for the trout and the axolotl (shown in Table 6), a similar equation could be drawn up for the squid. This striking intake of water and inorganic material from the environment is doubtless aided by the strong ciliary currents which,

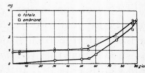

Fig. 6. Dependence of the cephalopod embryo on the environment.

as Ranzi[1] had previously shown, circulate from an early stage of cephalopod development within the perivitelline space. In view of similar findings on the

Table 6. *Dependence of eggs on the aquatic environment*

	Yolk (wet wt.) gm.	+ External water gm.	+ External ash gm.	= Completed embryo (wet wt.) gm.	+ Organic substance combusted gm.	
Trout (*Salmo*)	1·0	0·7	—	1·56	0·14	Gray[1]*
Axolotl (*Amblystoma*)	1·0	0·74	—	1·88	0·065	Dempster
Squid (*Sepia*)	1·0	0·78	0·033	1·727	0·089	Ranzi[6]
Squid (*Loligo vulgaris*)	1·0	0·25	0·008	1·155	0·105	Stolfi[1]
Squid (*Loligo bleekeri*)	1·0	3·38	0·113	4·52	Trace	Kamachi[3]
Dogfish (*Scyllium*)	1·0	1·115	0·029	2·047	0·097	Ranzi[12]
Torpedo (*Torpedo*)	1·0	1·086	0·014	1·972	0·128	Ranzi[12]
Atlantic salmon (*Salmo salar*)	1·0	1·498	—	2·34	0·16	C. R. K. Allen
Turtle (*Malaclemys*)	1·0	0·26	—	1·41	0·03	Cunningham et al.
Capsid bug (*Notostira*)	1·0	0·86	—	1·78	0·081	Johnson

* Trout eggs take in about 18 % of their water immediately on being laid, whether fertilised or not (Manery & Irving). Teleostean embryos in general probably take in far more after hatching than before it (Stroganov; Manery, Warbritton & Irving; Irving & Manery; C. R. K. Allen; S. Smith[1]; Aoki; Bogucki[3]).

eggs of echinoderms, Needham[3] suggested that such marked dependence on the salts of the environment might be a very important barrier to penetration into fresh-water and so help to explain the absence from it of groups such as the echinoderms and cephalopods.*

Ranzi's results on *Sepia* were confirmed for the squid *Loligo* by Stolfi[1] (see Table 6) and independently on a Japanese species of the same genus by Kamachi[3]† (see Table 6). An abundant water-intake was further observed and measured by Jecklin in his study of other cephalopods, especially the octopus, *Alloteuthis*, and the squids, *Sepiola* and *Sepietta*. Kamachi[3], estimating separately a number of ions, found an intake of Ca″, Mg″, SO₄″, Cl′ and, to a lesser extent, Cu and F′. Kumon, working on the eggs of the gastropod *Hemifusus tuba*, obtained substantially the same results. This line of work was carried much further by Ranzi[17] with the help of spectrophotometric methods. He found that the squid embryo can not only concentrate the Na, K, Ca, Mg and P of the sea-water, but also rarer ions such as Cu, Fe, Mn, Mo, V, B, Li, Ni, Si, Sr, Ti and Zn, and even established[23] that a very large proportion of the Cu needed for making the amount of haemocyanin with which the embryo hatches is contained in the sea-water and not given to the egg by the maternal organism. An egg contains $0·8\gamma$ Cu at the beginning of development; it hatches with 12γ. This is not the largest increase (Ca increases eighty times), but in view of the essential nature of the blood-pigment, it is very remarkable. In a later paper Ranzi[24] showed that haemocyanin begins to be formed in considerable quantities about the 60th day of incubation (Naef stage XV–XVI), and correspondingly the absorption of copper from the sea-water occurs mainly from this time onwards.

In none of the cephalopods studied is there any substantial change in osmotic pressure during development (d'Amora; Jecklin).

The work of Ranzi and his collaborators on the elasmobranch fishes has been brought together in several reviews (Ranzi[19, 20, 22]). This group of animals offers special opportunities for the study of the dependence of the embryo upon the environment, for it includes purely oviparous forms (such as the dogfish *Scyllium canicula*), ovoviviparous forms (such as *Mustelus vulgaris*) showing all grades of relationship between the embryo and the maternal uterus, and the viviparous forms (such as *Mustelus laevis*) in which there is a kind of yolk-sac placenta.

---

* A theoretical possibility here is that an egg developing in fresh-water might still be dependent on one particular ion. This seems to be the case as regards calcium with certain molluscs such as *Limnaea* if the circumstantial evidence of S. Bloch is correct, but it is difficult to believe that the fresh-water pulmonates should have any non-cleidoic qualities in view of their probable secondary entry into aquatic life. Yet Burdach long ago, in the dissertation quoted on p. 576, found a little more ash in pond-snail eggs at hatching than when newly deposited. Indications are not wanting that an uptake of calcium may occur even in amphibia. Thus the figures of Iseki & Kumon for the giant salamander *Cryptobranchus japonicus* show a rise of from 5 to 12 mgm. CaO/100 eggs during development, and Takamatsu's for *Hynobius nebulosus* show a rise in total ash. We can hardly dismiss these entirely as due to technical error when there is direct evidence that tadpoles take up calcium salts from the water in which they are developing (Luciani, Filomeni & Severi) and that certain amphibia, such as the toad *Bufo marinus*, will not develop at all in water devoid of calcium (Takano & Iijima). Crayfish can take up Ca″ from fresh-water for calcification (Maluf[4]).

† Kamachi thought that the egg-membranes supplied some of the water, salts and even nitrogen to the embryo, but the figures of Stolfi to the contrary carry more conviction.

Ranzi[12] first showed that the development of oviparous forms such as *Scyllium canicula* exactly resembles that of other marine eggs such as the squid or octopus egg (Table 6) in obtaining some 60% of the inorganic substance which it requires, and some 70% of the water, from the aquatic environment and not from the maternal organism. Now in all the ovoviviparous and viviparous forms this process is repeated[13], as is shown by the figures in Tables 7 and 8, but in

Table 7. *Dependence of eggs on the uterine environment in ovoviviparity and viviparity*

	Yolk (wet wt.) gm.	+	External water gm.	+	External ash gm.	+	External organic substance gm.	=	Completed embryo (wet wt). gm.
Ovoviviparous:									
Dogfishes									
(*Galeus canis*)	1·0		1·05		0·078		0·05		2·18
(*Mustelus antarcticus*)	1·0		6·3		0·013		0·57		8·0
(*Mustelus vulgaris*)	1·0		12·18		0·36		1·77		15·31
(*Trygon violacea*)	1·0		52·1		1·1		8·0		62·2
(*Myliobatis bovina*)	1·0		52·0		1·1		8·2		62·3
Viviparous:									
(*Mustelus laevis*)	1·0		26·9		0·95		5·4		33·25

addition to this, the negative balance of organic substance caused by combustions is gradually changed over into a positive balance as the embryo comes from species to species more closely in contact with the maternal organism. Thus in the spiny dogfish, *Acanthias vulgaris*, there is still a loss of organic substance during development; in *Acanthias blainvillei* the loss by combustions and the gain from the maternal organism just balance, so that there is neither a total loss nor a total gain; and in the smooth dogfish, *Galeus canis*, *Mustelus antarcticus*, *Myliobatis bovina*, etc., we see a gradually increasing deposition of organic substance from the maternal organism in the egg (yolk + embryo) during development. This reaches its maximum in *Pteroplatea micrura*, where the gain in organic substance is some five thousand per cent. At the same time it may be noticed that those elasmobranchs which depend very largely on the maternal organism for organic substance rather than on the yolk, tend also to depend on it for inorganic substance and water, rather than on the sea-water, to a much larger extent than those which carry on their development outside the maternal organism. For example, the water-intake of *Scyllium* is about 200% of the initial store, but in the case of *Myliobatis* and the sting-ray *Trygon* it is several thousands per cent.

Successful correlations were also made by Ranzi[13] between the degree of dependence shown by the egg and (*a*) the nature of the uterine fluids, (*b*) the structure of the uterine wall, and (*c*) the extent to which the stored substances in the liver of the maternal organism are depleted during development. These phenomena may be taken in turn.

As Table 8 shows, there are considerable variations in the composition of the uterine fluid bathing the egg-cases while the embryo develops. In *Torpedo* it is abundant but dilute, in *Scymnus* and *Acanthias* it is less abundant but more concentrated, and at the change-over point between loss or gain of organic substance in the egg, it has a concentration of organic substance of about 3%.

Table 8. *Balance sheet of developing elasmobranch embryos in oviparous, ovoviviparous and viviparous conditions*

Species	t	e	E	Change %	w	W	Change %	a	A	Change %	o	O	Change %	ouf %	uw	R
Oviparous:																
Scyllium canicula	6	1·3	2·7	+108	0·68	2·15	+216	0·013	0·051	+292	0·61	0·48	−21	—	—	—
Ovoviviparous:																
Torpedo ocellata	4	6·8	13·4	+103	2·94	10·31	+300	0·06	0·15	+150	3·78	2·91	−23	1·2	Ib	8·5
Torpedo marmorata	6	144	273	+91	6·56	21·85	+233	0·14	0·40	+186	7·68	5·06	−31	1·5	Ib	9·1
Scymnus lichia	10	130	188	+45	65	131	+100	2	8	+300	63	49	−22	1·6	Ia	6·3
Centrophorus granulosus	10	308	360	+17	142	274	+94	3·4	12	+253	162	74	−54	1·8	Ia	—
Acanthias vulgaris	10	23	41	+78	14·2	35	+147	—	1·8	—	8·6	5·2	−40	2·4	Ia	—
Acanthias blainvillei	9	19·3	29·6	+53	8·4	17·9	+88	0·2	2·7	+1250	10·7	10·8	+1	2·8	Ia	13·6
Galeus canis	10	46·9	102·2	+118	23·5	72·6	+210	0·45	3·6	+700	23	25·5	+11	4·9	II	—
Mustelus antarcticus	10	4·4	32·3	+635	2·1	29·8	+1480	0·05	0·6	+1100	2·3	4·8	+110	7·1	II	—
Mustelus vulgaris	10	3·9	60·6	+1450	1·9	49·8	+2520	0·55	1·5	+173	1·9	8·9	+369	5·1	II	56·1
Pteroplatea micrura	?	—	—	—	—	—	—	—	—	—	0·2	10	+4900	—	III	—
Myliobatis bovina	4	7·2	450	+6150	5·2	381	+7200	0·08	8·0	+9900	1·9	61	+3120	13·3	III	—
Trygon violacea	2	2	118	+5800	0·9	97	+10000	0·2	2·1	+950	0·9	16	+1680	—	III	51·3
Viviparous:																
Carcharias glaucus	10	—	—	—	—	—	—	—	—	—	3·4	32	+840	—	II	—
Mustelus laevis	10	5·5	189	+3340	2·6	152	+5750	0·7	5·3	+660	2·8	32	+1050	9·1	II	37·4*

* In *M. laevis* the weight of the maternal liver may rise again later in development, R is here calculated up to the time when it is at its lowest, i.e. when the embryo weighs 13 gm. This is a further argument for the greater efficiency of viviparity.

Key to Table 8

t   time of incubation or gestation in months.
e   wt. of egg at the beginning of development, gm.
E   wt. of embryo at the end of development, gm.
w   wt. of water in the egg at the beginning of development, gm.
W   wt. of water in the embryo at the end of development, gm.
a   wt. of ash in the egg at the beginning of development, gm.
A   wt. of ash in the embryo at the end of development, gm.
o   wt. of organic substance in the egg at the beginning of development, gm.
O   wt. of organic substance in the embryo at the end of development, gm.
ouf   organic substance in the uterine fluid, %.
uw   type of uterine wall.
R $100\left(1 - \dfrac{FM_0}{MF_0}\right)$, where F_0 is the weight of the maternal liver at the beginning of development and F the weight at the end, M_0 the weight of the maternal body at the beginning of development and M the weight of the maternal body at the end.

Percentages are of the initial value.

When the dependence on the maternal organism is more marked, the uterine fluid becomes turbid and contains formed elements in suspension, as in *Trygon*; it then may reach 13·3 % organic substance. Ranzi[14] found that in some cases, at any rate, leucocytes transport fat from the tissues to the uterine fluid, where later they may disintegrate, leaving the fat in suspension. In *Trygon* uterine fluid contains 8 % fat. The embryo absorbs the uterine fluid through the yolk-sac and also probably in some cases through its enlarged gills; later on injections of Chinese ink show that it passes into the digestive tract through the mouth and spiracles. Proteases have been found in the uterine fluid of *Torpedo* and *Mustelus* (Saviano), but they do not seem to act at the pH of the fluid itself, and perhaps only do so when the fluid has reached some part of the intestinal tract. At an early stage (length 46 mm.) the digestive glands of the *Torpedo* embryo are all functional (Pitotti[1]), and while the uterine fluid is being digested in the stomach, the yolk is being digested in the intestine. Histological details of the selachian yolk-sac will be found in a paper by Macé.

It seems that nourishment of the embryo by means of the elasmobranch "histo-trophe"* is so efficient as to exceed the efficiency of the viviparous types, in which the yolk-sac actually invades the wall of the uterus. As Table 8 shows, the uptake of water, ash and organic substance in *Mustelus laevis*,† the type specimen of selachian viviparity, in percentage of the initial stores, is actually rather less than in some of the most successful ovoviviparous types, such as *Myliobatis* and *Trygon*. There may, however, be many other reasons which have led to the evolution of true viviparity in these fishes, and indeed Ranzi pointed out that the speed of development and absorption is much greater in *M. laevis* than in *M. vulgaris* or *antarcticus*.

This brings us to the question of the nature of the uterine wall. Ranzi's researches show[19] that this differs among the different groups. In *Torpedo* the wall has a great number of villi of one-layered epithelium which contain glands but no mucus-secreting cells (see Fig. 7). In *Scymnus*, *Centrophorus*, *Acanthias*, etc. there are many villi as before (see Fig. 8), but the epithelium is many-layered, and there are a few mucus-secreting cells. In *Mustelus vulgaris* and *Galeus canis* the uterine wall is much less villous, but its smoother surface contains very many cells which secrete a mucous material into the uterine fluid, thus raising its protein-content though still it contains little fat. The protein seems to be mainly mucoprotein, but in addition to this bound carbohydrate, there is about 0·2 % free reducing sugar. As in all other elasmobranch tissue fluids there is abundant urea present. Finally, in *Trygon*, where the uterine fluid is particularly rich in nutrient substances, the villous condition of the uterine wall is carried to an extreme (see Fig. 9). The epithelium is here many-celled, and there are many glands secreting protein and fat. The liquid contains 0·5 % free reducing sugar.

* See on, p. 81, in connection with mammals, and cf. *CE*, p. 1492. Hartman[2,3] sees in these uterine secretions the evolutionary origin of menstruation.

† One can hardly avoid mentioning here the story, told in full by Haberling, that this fish is the γαλεὸς λεῖος of Aristotle, *Hist. Anim.* 565 a & b, who knew it to be viviparous, i.e. placental; but was not believed until the rediscovery of this fact in 1842 by Joh. Müller.

I

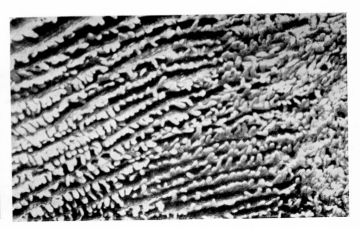

Fig. 7. The uterine wall of *Torpedo*.

Fig. 8. The uterine wall of *Centrophorus*.

Fig. 9. Villi of the uterine wall of *Trygon*.

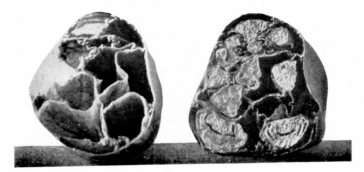

Fig. 10. Egg-cases forming compartments in the uterus in *Mustelus*.

In the true viviparous cases, such as *Mustelus laevis*, the embryos are separated by compartments (Fig. 10) and there is a specially modified condition of the yolk-sac so that it burrows into the maternal tissues (see Figs. 11, opp. p. 46, and 12).*
Correspondingly the content of organic substance in the uterine fluid is diminished with respect to the ovoviviparous forms (Table 8) and fat in particular is missing from it (only 1·25 %). Calzoni[2] has described the placenta of *Carcharias glaucus*.

All these relationships will be found summarised in Fig. 13. The rectangles left of the ordinate indicate the extent to which the egg receives organic substances from the maternal organism; a negative sign means that this does not occur. The rectangles right of the ordinate indicate the organic substance content of the uterine fluid (dotted lines show probabilities). The black rectangles right

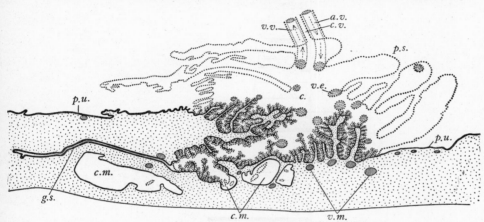

Fig. 12. Epithelio-chorial placenta of *Mustelus* in section. *a.v.*, vitelline artery; *c.*, yolksac cavity; *c.m.*, lumen of mucus-secreting glands; *c.v.*, duct from yolk-sac to intestine; *g.s.*, gland lumen; *p.s.*, yolksac wall; *p.u.*, uterus wall; *v.e.*, foetal blood vessels; *v.m.*, maternal blood vessels; *v.v.*, vitelline vein.

of the ordinate indicate the degree to which the maternal liver is reduced during development (see below), and side by side with each group is placed a diagram of the histological structure of the uterine wall, as described above.

The last of these questions is the reduction of the maternal liver. Ranzi[15] noticed that there was a parallelism between the degree of dependence of the eggs on the mother and the degree to which the liver is reduced in size during their development. The last column in Table 8 shows values for this by a factor, *R*. The rise and fall in liver weight appears to have a close connection with such endocrine glands as the pituitary (Ranzi[21]; Comes; Castigli), the thyroid (Zezza; Guariglia) and the suprarenals (Fancello; Pitotti[2]; Ranzi[22]). It is undoubtedly integrated into the selachian endocrine cycle of reproduction, a discussion of which with relation to the similar cycle in mammals has been given by Ranzi[19]. In mammals the liver changes are much smaller (Biereigel), suggesting that the effects of pregnancy on the adult organism are there less.

* It is noteworthy that the placenta of *Mustelus laevis* resembles the most impermeable placenta type among the mammals; in the Grosser classification, the epithelio-chorial (see p. 81).

The exact nature of the liver changes has also been studied. According to Ranzi & Zezza (confirmed by Falkenheim) there is no destruction of cells during the rise and fall of liver size, simply a swelling followed by a shrinking, so that, as embryonic development progresses, the number of nuclei visible in a single

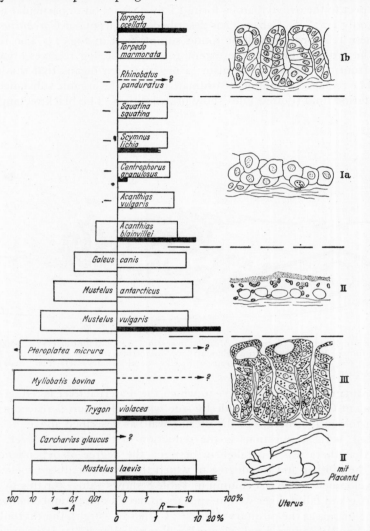

Fig. 13. Diagram summarising the relations between maternal and foetal organism in the elasmobranch fishes.

standard section of the maternal liver constantly rises. The number of visible lipoid droplets correspondingly falls.

We are not without knowledge concerning the nature of these lipoid substances. Leaving out the earlier work on the subject, previously reviewed (*CE*, p. 1180), it may be said that André established the fact that livers of elasmobranch fishes in gestation contain more oil than when non-gestating, e.g.

	% oil	
	Non-gestating	Gestating
Scyllium stellare	20	50
Galeus canis	28	47
Raia batis	22	52

André & Canal then examined the proportion of unsaponifiable matter in the liver oils from adult and young fishes in the pilgrim shark (*Cetorhinus maximus*).

	Fat %	Unsaponifiable %	Cholesterol %	Squalene and pristane %
Liver of adult fish	50	50	2	48
Liver of young fish	64	36	22	18

The young fish has therefore less unsaponifiable matter but relatively more cholesterol. They then compared the liver oil with the egg oil in *Centrophorus granulosus*.

	Fat %	Unsaponifiable %	Cholesterol %	Squalene and pristane %
Liver of adult fish	9	91	1	90
Undeveloped egg	50	50	5	45
Liver of embryo	33	66	4	62

Thus again the embryonic liver has less unsaponifiable matter and relatively more cholesterol. In both animals the embryo and young fish is richer in neutral fat, and this is especially marked in the eggs of *Centrophorus*, where there is as much saponifiable as unsaponifiable matter. André & Canal believed in a conversion of clupanodonic acid to cholesterol and squalene, but of this there is no metabolic proof. Stolfi[2] found that the iodine values of the fat of the maternal liver in *Trygon* are much higher than those of the uterine fluid (especially rich in fat, as we have seen).

	Iodine value
Maternal liver fat	131
Uterine fluid fat	73
Embryo tissue fat	69

It may therefore be concluded that the fatty acids undergo saturation in passing to the embryo, but no information indicating how or where this takes place is yet available.

If the general line of argument so far followed is correct, it might be expected that eggs of related species of different habitat would contain less ash if from sea-water, more if from fresh-water. The only comparative estimations of this kind are contained in a note by Buşniţă & Gavrilescu, who analysed teleost eggs (carp and herring) from different habitats.

	Ash % dry wt.	
	Lake Greaca (fresh)	Lake Tasaul (brackish)
Cyprinus carpio	4·08	3·89
	River Danube (fresh)	Black Sea
Alosa pontica	6·50	5·74

These differences they regarded as significant.

The foregoing investigations are the most complete which have so far been made on the relations between maternal organism and embryo in any non-mammalian group. Viviparity is certainly a very different matter in an aquatic environment to what it is in a terrestrial one, as we shall shortly see (p. 51), and in marine conditions it originates probably as a purely protecting mechanism. However, the same story is found in most groups of animals among the lower vertebrates, and it is worth while to pause for a moment to see how it works out in other fishes and amphibia.

From the excellent review of Wunder[1], which summarises much rather inaccessible literature in the field of natural history, we find that among the teleosts too there has been a move towards ovoviviparity in certain species. While there seems to be nothing corresponding to the yolk-sac placenta of *Mustelus laevis* among teleosts, the ovoviviparous eel, *Zoarces viviparus*, secretes a very fatty and mucous uterine fluid, which places it in a position analogous to *Trygon*. Stuhlman's analysis (1887), the only one ever made, described this fluid as "rich in nutritive substances". In the Cyprinodonts, such as *Girardinus* and *Poecilia*, the eggs develop while still in the ovary, and the embryo hatches from the egg-case later on in the oviduct. Remarkable adaptations to this ovarian life have been described by C. L. Turner; extensive ribbon-shaped proctodaeal processes occur in the embryos of both fresh-water (e.g. *Goodea bilineata*) and marine (e.g. *Parabrotula dentiens*) forms. Some Cyprinodonts seem to be just on the borderline between oviparity and ovoviviparity, for Amemiya & Murayama noted that the eggs of *Oryzias latipes*, which normally develop outside the parent's body, can sometimes do so within it. Unfortunately few biochemical investigations have yet been made from which we can draw any conclusions about the role of maternal nutrition in such cases. C. D. Hsiao, however, working on the deep-sea teleost *Sebastes marinus*, found that, although the incubation is ovarian, there is a pronounced fall in the dry weight of the developing egg, indicating little or no maternal contribution. The other extreme is represented by a cyprinodont teleost *Heterandria formosa*, studied by Scrimshaw, also with ovarian incubation. In this remarkable form, conditions seem to be almost mammalian, for the egg is only 0·39 mm. in diameter at fertilisation and the dry weight increases from 0·017 mgm. to nearly 10 mgm. at birth. A typical teleostean oil-drop is present, occupying 90 % of the space in the minute egg, but it is not utilised until late in development. Absorption of nutritional materials is assisted by a richly vascularised expanded pericardial sac which completely envelops the embryo's head until the later stages. Evidently all grades of maternal dependence occur in teleosts just as they do in selachians. It is interesting and perhaps significant in view of what happens in selachians that in *Gambusia holbrooki* Remotti[2] observed a diminution of liver weight during gestation.

A somewhat less intimate connection may be supposed to exist in those fishes where the eggs are attached in various ways to the external surface of the body of one or other of the parents, as occurs in the fresh-water teleosts *Kurtus gulliveri* (New Guinea) and *Aspredo laevis* (South America). The most striking example of this kind of protection occurs in the Syngnathidae, of which the

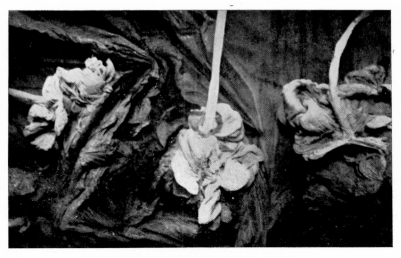

II

Fig. 11. Yolk-sac placentas in *Mustelus*.

Fig. 14. Female of *Nototrema marsupiatum* with the
skin of the back dissected to show the eggs.

sea-horse, *Hippocampus antiquorum*, is a classical case. Here the male fish carries the eggs in a marsupial-like pouch formed of two large folds of skin either on the belly or the tail. On the inside of the pouch in the pipe-fish *Syngnathus typhle* Cohn found secretory glands, as did Kolster in *Hippocampus*, and the fluid contents of the pouch seemed to contain protein and fat, but Comes and Leiner[1], by direct measurements, showed that the embryo of *Hippocampus brevirostris* during its development takes up water and inorganic substances, but not organic substances, from the male organism carrying it.* It compares therefore with *Scymnus* or *Centrophorus* in the selachian series. Some histochemical observations on an embryo of this kind (*Syngnathus acus*), which may be of interest, are contained in a paper by Konopacki & Erecinski.

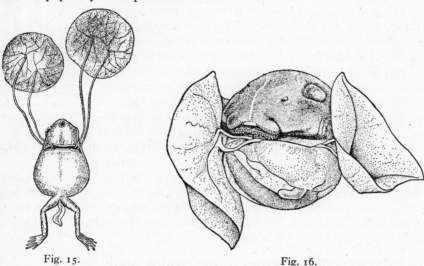

Fig. 15.　　　　　　　　　　　　　Fig. 16.

Fig. 15. Larva of *Nototrema oviferum* with leaf-like gills acting as placentas.
Fig. 16. Larva of *Nototrema marsupiatum* with bell-like gills acting as placentas.

Protection of teleostean embryos may also be carried out by the eggs being held within the mouth of either parent, as in *Arius falcarius* and many other Silurids, as well as among the Cyprinodonts, Trachinids, Serranids and Cichlids, etc. In *Arius* feeding and digestion are in abeyance during the development of the eggs, the fish living on its reserves, but nothing is known about the nutrition of the embryos.

When we turn to the amphibia we find a quite parallel state of affairs. From Wunder's second review[2], which dealt with the amphibia, and from the interesting work of Noble, it can be seen that ovoviviparity, though no true viviparity, occurs. The African frog, *Nectophrynoides tornieri*, retains its embryos to develop in the oviduct, and this is common among salamanders (e.g. *Salamandra atra*). The embryos of the South American blindworm, *Typhlonectes compressicauda*, are

* Leiner's elaborate work[2] may be consulted for an account of the composition and properties of the brood-pouch liquid of *Hippocampus* (see p. 586). It is more acid than sea-water, richer in CO_2, of higher osmotic pressure, and contains a protease which Leiner pessimistically assumes is provided to resorb the dead.

characterised by a special modification of the external gills, which expand into leaf-like plates. It is difficult to believe that these do not have an absorptive function, but no investigations have yet been made into the degree of dependence of any of these forms upon the maternal organism during development. The expanded leaf-like gills may only function in gas-exchange.

Modified gills also occur in embryos of those species which protect their developing eggs by carrying them in pouches of various kinds, similar to the Syngnathid fishes already mentioned. This frequently occurs in tree-frogs such as *Hyla goeldii* or *evansi*, but the best example is perhaps the genus *Nototrema*, where the embryos are carried in a series of pockets on the back of the adult frog. Fig. 14 (opp. p. 46) shows a female of *Nototrema marsupiatum* with the skin of the back dissected to show the eggs, and Fig. 15 a larva of *Nototrema oviferum*, showing its vastly expanded leaf-like gills. Fig. 16 shows a larva of *N. marsupiatum*, showing the bell-like form taken by the gills. Studies of the kind made by Ranzi would be extremely welcome on an amphibian series. The tail may function instead of the gills as an organ of absorption in ovovivi- parous or "marsupial" amphibia; this happens in toads of the type of *Protopipa aspera* or *Pipa americana*. Fig. 17 shows a larva of *Pipa dorsigera*, showing the enlarged and well-vascularised tail.

Fig. 17. Larva of *Pipa dorsigera* with tail expanded to act as placenta.

Finally, buccal incubation occurs also among the amphibia (in *Hylambates brevirostris*), though here it more usually takes the form of allowing the eggs to develop in the vocal sac, as in the Chilian *Rhinoderma darwini*, where the tail is believed to act as an absorptive organ, though metamorphosis is fully accomplished before the larvae leave the sac.

1·64. Metabolic properties of the cleidoic state

We must now return to Table 5. There it can be seen that by far the most self-contained or *cleidoic* eggs are those which develop outside the maternal body in a terrestrial environment. Now far-reaching metabolic changes were associated with the evolution of the cleidoic egg. Closure of the egg system is naturally accompanied by difficulties in the disposal of incombustible nitrogenous waste-products. Thus it has been shown that the eggs of birds, reptiles and insects combust markedly less protein during their embryonic life in percentage of the protein initially present than do the eggs of aquatic animals such as fishes or amphibia. The figures are of the order of 3 % in the former as against 26 % in the latter (see Table 9). Similarly, of the material burned as source of energy during development 6 % is protein in the case of the chick, but 71 % in the case of the frog (see Table 10). On the other hand, the catabolism of fat is correspondingly increased in cleidoic eggs, so that, whereas the aquatic eggs of amphibia and fishes make it some 30 % of their total combustions, the cleidoic ones of birds and insects make it about 80 % (see Table 10). The same relation holds good when the fat combusted is compared with the fat initially present (Table 11).

Table 9 (corresponding to *CE*, Table 162). *Protein combusted as source of energy in % of the initial protein store*

		%	
T.	Chick (*Gallus domesticus*)	4·5	
T.	Silkworm (*Bombyx mori*)	3·9	
T.	Sheep blowfly (*Lucilia sericata*)	0·51	Rainey; Brown
Intermediate:	Turtle (*Thalassochelys corticata*)	16·5	
A.	Trout (*Savelinus fontinalis*)	19·5	
A.	Trout (*Salmo irideus*)	(49·3)	S. Smith[1]
A.	Carp (*Cyprinus carpio*)	41·0	Nowak
A.	Atlantic salmon (*Salmo salar*)	36·3	Hayes[1]
A.	Plaice (*Pleuronectes platessa*)	18·3	
A.	Frog (*Rana temporaria*)	29·6	
A.	Salamander (*Cryptobranchus japonicus*)	27·8	Iseki, Kumon, Takahashi & Yamasaki
A.	Squid (*Loligo vulgaris*)	15·0	Stolfi[1]
A.	Squid (*Sepia officinalis*)	13·7	M. Cori

Averages: **T. 2·9; A. 25·9.**

T. = Terrestrial; A. = Aquatic.

N.B. (1) Detailed references are omitted for those cases already mentioned in *CE*.

(2) The figure of S. Smith may be unduly high owing to the inclusion of a starvation period after the consumption of all the yolk. A value of 32·5 %, calculated from his data, is more correct. If the period before hatching were alone considered, all catabolic losses would be very small, as the teleostean alevin hatches with most of its yolk still unused (Manery, Warbritton & Irving; Irving & Manery).

Table 10 (corresponding to *CE*, Table 161). *Material combusted as source of energy in % of the total material so combusted*

		Carbo-hydrate	Protein	Fat	
T.	Chick (*Gallus domesticus*)	3·02	5·57	91·4	
A.	Frog (*Rana temporaria*)	6·84	70·70	22·4	
T.	Grasshopper (*Melanoplus differentialis*)	—	—	72·5	Boell[1]; Slifer[1,4]
T.	Silkworm (*Bombyx mori*)	26	10	64	
T.	Sheep blowfly (*Lucilia sericata*)	Trace	< 5	> 95	Rainey; Brown
Intermediate:	Turtle (*Thalassochelys corticata*)	—	19	81	
A.	Sea-urchin (*Paracentrotus lividus*)	21	50	29	
A.	Trout (*Savelinus fontinalis*)	—	63	37	
A.	Trout (*Salmo irideus*)	Trace	83·8	16·0	S. Smith[1]
A.	Carp (*Cyprinus carpio*)	—	61·5	23·8	Nowak
A.	Plaice (*Pleuronectes platessa*)	—	90	—	
A.	Squid (*Loligo vulgaris*)	1·0	37·5	61·5	Stolfi[1]
A.	Squid (*Sepia officinalis*)	—	51·5	—	M. Cori
A.	Salamander (*Cryptobranchus japonicus*)	12·3	51·5	36·4	Iseki, Kumon, Takahashi & Yamasaki; Tomita & Fujiwara

| | Averages: **T.** | — | **6·8** | **80·7** |
| | **A.** | — | **62·1** | **32·3** |

T. = Terrestrial; A. = Aquatic.

N.B. (1) The first two sets of figures are those most accurately known.

(2) Detailed references are omitted for those cases already mentioned in *CE*.

(3) Figures for terrestrial molluscs using large amount of stored polysaccharide would be of special interest.

The cleidoic condition may also be connected with other peculiarities, perhaps the most striking of which is the rapidity of avian incubation time as compared with mammalian gestation time, i.e. the fact that it takes considerably longer to make a mammal of given birth weight than a bird of equivalent hatching weight (see *CE*, pp. 480 ff.). For a weight of 1 gm. the mammalian time is 20 % longer; for a weight of 10 or 110 gm. it is twice as long, and for a weight of 1 kg. it is three times as long. It is suggested that this is an adaptive acceleration of the developmental time, permitting rapid passage through the cleidoic state.

Table 11 (corresponding to *CE*, Table 174). *Fat combusted as source of energy in % of the initial fat store*

		%	
T.	Chick (*Gallus domesticus*)	60	
T.	Silkworm (*Bombyx mori*)	48	
T.	Grasshopper (*Melanoplus differentialis*)	54	
T.	Lackey-moth (*Malacosoma americana*)	85	
T.	Sheep blowfly (*Lucilia sericata*)	44·5	Rainey; Brown[3]
T.	Potato beetle (*Leptinotarsa decemlineata*)	36·5	
Intermediate:	Turtle (*Thalassochelys corticata*)	34	
A.	Frog (*Rana temporaria*)	28	
A.	Salamander (*Cryptobranchus japonicus*)	30·8	Tomita & Fujiwara; Kataoka & Takahashi
A.	Primitive salamander (*Hynobius nebulosus*)	25·5	Takamatsu[2]
A.	Torpedo (*Torpedo ocellata*)	24·5	
A.	Plaice (*Pleuronectes platessa*)	10	
A.	Sea-urchin (*Strongylocentrotus lividus*)	20	
A.	Squid (*Loligo vulgaris*)	47·6	Stolfi[1]
A.	Trout (*Salmo irideus*)	51	S. Smith[1]

Averages: **T. 54·6; A. 29·6.**

T. = Terrestrial; A. = Aquatic.

N.B. Detailed references are omitted for those cases already mentioned in *CE*.

Another metabolic characteristic of birds possibly related to the cleidoic nature of their eggs is their great resistance to ketosis; we shall return later to this point (p. 62). But outstanding in birds, insects and saurian reptiles is their possession of uricotelic metabolism.

There are in the animal kingdom three main ways of excreting waste nitrogen from protein catabolism—ammonia, urea and uric acid.* In addition to these substances, many invertebrates excrete amino-acids unchanged, as if the metabolic machinery of the body which separates the amino-acids of the food into quotas for storage and quotas for combustion were inefficient (Delaunay). In certain special groups also there may be special methods or adaptations, such as the excretion of guanine by arachnids (Vajropala; Peschen)† and trimethylamine-oxide by certain teleosts (Grollmann[1]; Grafflin, Gould & Spence), but these must be regarded as minor specialisations.

That the elaborate process of the formation of the purine ring from the catabolic nitrogen had something to do with terrestrial life was early surmised,

---

* The comparative biochemistry of nitrogen excretion has recently been the subject of elaborate tabulation and useful discussion by Heidermanns[1]. I am unfortunately unable, however, to agree with all his conclusions.

Ammonia   Urea   Uric acid   Trimethylamine oxide   Guanine

† The deposition of guanine in the scales of some fishes (Bethe) possibly as an excretion from nuclein catabolism, is well known, but it is not so well known that the guanine crystals from this natural source have for centuries been the basis of the artificial pearl industry, as is described by H. F. Taylor.

but the mammalia, which excrete the whole of their waste nitrogen in the form of urea, were an insuperable obstacle to so simple a generalisation. It was then suggested (*CE*, Section 9·5) that the form of nitrogen excreted by an animal depended primarily on the conditions under which its embryo had to live. Most of the data which have come to light since that time have supported this view. They may be summarised as follows. Between uric acid excretion* and terrestrial oviparity there is a strict correlation, ammonia and urea being associated with aquatic pre-natal life (including development within the mammalian uterus) and uric acid with terrestrial pre-natal life. An inspection of the percentage nitrogen partition in the excreta of representatives of every phylum (*CE*, Table 163) shows the strictness of the correlation.† Birds, saurian reptiles, insects and terrestrial gastropods alike possess the uricotelic habit; mammals, chelonian reptiles, amphibia and fishes excrete their waste nitrogen mainly as ammonia or urea; aquatic invertebrates down to protozoa excrete almost entirely ammonia.‡

The essential point here is that uric acid is the only excretory nitrogenous compound which can be got rid of in the solid form, requiring the loss of little water during its removal. It forms supersaturated colloidal solutions (Young & Musgrave; Young, Musgrave & Graham), from which it readily precipitates in an almost insoluble condition. In insects (Wigglesworth[1]), adult birds (Young & Dreyer) and probably the chick embryo (Needham[4]) there is a cycle of base in its excretion; it passes through the excretory cells in salt form and later the cations are reabsorbed. Urea and ammonia are of course far more soluble and diffusible (solubilities in gm. N/100 gm. water = ammonia 47·0, urea 37·0, uric acid 0·0019).

On these views, as has already been said, the mammalian embryo is an aquatic

---

* It must be understood that we are speaking here only of uricotelic excretion, i.e. the excretion of uric acid as the main end-product of nitrogen metabolism. Uric acid as the end-product of nuclein metabolism is an entirely different phenomenon, quantitatively much less important. Nuclein breakdown may not always go so far as uric acid; thus according to Truszkowski it stops at xanthine in *Hirudo* and *Anodonta*. The mammals have not, of course, lost the power of purine synthesis, in so far as they can construct their own nuclein (cf. p. 635 and Terroine & Mourot; Terroine, Giaja & Bayle).

† H. W. Smith (1932, p. 16; 1936, p. 74) is reluctant to accept the view that uricotelism and terrestrial oviparity are related, but he consistently avoids the difficulty that no other generalisation will explain the lack of uricotelism among the mammals, a plainly terrestrial class. It may be granted that the toxicity of urea has been somewhat overrated in the past, but the impossibility of the alternative of urea excretion need not have been the sole factor leading to uricotelism. A correlation has been suggested between selachian urea retention and cleidoicity (Needham & Needham in 1930; H. W. Smith in 1936). The only class of animals lower than the reptiles, it has been said, which have evolved a structure resembling the cleidoic egg are precisely the elasmobranchs. This may be viewed as an adaptation enabling the embryo to retain its maternal and its own urea in preparation for its adult marine life. Or, alternatively, it may be said that the survival value afforded by the protection of cleidoic egg-envelopes was only a possibility in animals for which the dispersal of nitrogenous end-products during development was, for other reasons, no problem. But this correlation is rather unconvincing, since some invertebrates, such as the cephalopods, have elaborate egg-cases and yet show neither urea retention nor uricotelism. Several authors, e.g. E. Starkenstein, have been keenly aware of the puzzling problems presented by uricotelism in relation to the mammals, but no one has proposed an alternative explanation which fits all the facts.

‡ Confirmation of this is abundant, as in the work of Lvov & Roukhelman; Lawrie; Doyle & Harding, and others on various protozoa; Braconnier-Fayemendy on the leech, *Hirudo medicinalis*; Spitzer on numerous molluscs; and Mollitor on the crab *Eriocheir sinensis*. Teleostean nitrogen excretion is mainly as ammonia, selachian as urea; hence Manderscheid's failure to find the ornithine cycle in the liver of the trout was to be expected.

organism, its metabolism being adjusted to the unlimited water-intake of vivi-parity and to the easy removal of waste-products (by means of the maternal alimentary canal and the placenta and maternal kidney respectively). It is therefore not surprising that mammals are urea-producing organisms. And when we come to consider the phylogenetic relationships involved, we find they can be represented diagrammatically in a scheme somewhat similar to that shown in Fig. 18. The ornithine cycle of Krebs & Henseleit, by means of which urea is made

The ornithine cycle.

The mechanism of the uricase complex.

from catabolic ammonia nitrogen in the mammals, reaches back, we know, into the realm of the fishes and amphibia (Manderscheid);* and it is therefore extremely probable that this was the manner in which urea was formed by the primitive pre-reptiles or stegocephalic amphibia from which modern mammals and modern sauropsida are alike derived. There were then at least four possi-bilities open: (1) to take on a terrestrial life, but by adopting true viviparity to retain the ancient system of urea production; this was the course taken by the Therapsids (Broom) and led to the mammals of the present day; (2) to take on a terrestrial life and to make use of the completely cleidoic egg, correspondingly passing to uricotelic metabolism, as did the sauropsida; (3) to retain a semi-aquatic mode of existence and to continue to lay non-cleidoic eggs outside the

* But not into the invertebrates, where urea is not formed by the ornithine cycle, either in the snail (Baldwin & J. Needham) or the earthworm (Heidermanns[2]). We do not know whether all parts of the cycle are missing in the invertebrates. It may be significant that arginine desimidase (the enzyme producing citrullin, CO_2 and NH_3 from arginine) is absent in the hepatopancreas of *Astacus* (Ackermann).

body; this led to the modern chelonia; (4) to burn no protein at all during embryonic life and hence to avoid the problem of incombustible toxic nitrogenous residues. This course alone would have permitted the existence of the terrestrial

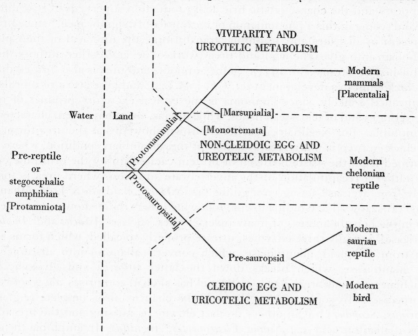

Fig. 18. Scheme to illustrate the origins of viviparity and uricotelic metabolism.

A very similar scheme was arrived at independently by Portmann[3]. He remarks that it contradicts the common statement that the mammalian egg has undergone a secondary diminution of yolk, being "derived from" a large egg of meroblastic sauropsid type.

H. W. Smith (1939, p. 62) advances another scheme. He suggests that by assuming warm-bloodedness the earliest mammals were committed to filtering off a copious glomerular urine from the blood because of their higher blood pressure, and hence to absorbing 90 % of it as man now does. Hence they had less need of uricotelism. But would it not have given them an additional advantage over the reptiles? If urea formation has no connection with oviparity and viviparity, why should not the birds, when they became warm-blooded and had to reabsorb water in their tubules, have reverted to urea formation? Evolution is irreversible only as regards regainments, not losses (cf. Needham[26,31]).

cleidoic egg without uricotelic metabolism, but so far as our knowledge goes to-day, life without protein combustion is impossible. The three courses actually taken are illustrated in Fig. 18.*

The position may be stated in another way by referring to Przyłęcki's rule[1] (1926), which observes that in the animal kingdom uricase (the enzyme which breaks down uric acid) and uricoligase (the enzyme complex which forms it from catabolic nitrogen) are never found together. This is an important support for the view already expressed, that uricotelic metabolism is essentially an adaptation.

* On the information at present available to us, the acquisition of uricotelic metabolism must have occurred three times in evolution, once in the case of terrestrial molluscs, once in the case of terrestrial arthropods, and once in the case of the sauropsida.

That it is an adaptation to terrestrial oviparity follows from other considerations. It is of particular interest that during the first week of the chick embryo's development, an active uricase is present (Przyłęcki & Rogalski), but this soon disappears and the characteristic uric acid production supervenes. It is difficult to avoid seeing in this a phenomenon of recapitulatory significance,* since uricase is present in all fishes and amphibia, transforming the uric acid of their purine catabolism into glyoxylic acid and urea (work of several earlier authors, firmly established by Stransky in 1933). Some ammonia and urea is also formed during the chick embryo's development (cf. *CE*, Fig. 324), and here also a recapitulatory significance is likely, since ammonia is the typical excretory product of many invertebrates and of teleostean fishes, while urea is, as we have seen, characteristic of amphibia. But Needham, Brachet & Brown showed that the urea formed by the chick embryo is not a recapitulation of the amphibian condition, where urea is formed from the waste nitrogen of protein catabolism by the ornithine cycle; it is rather a recapitulation of the invertebrate situation, where urea is formed solely from arginase acting on arginine (as in *Helix*, Baldwin & J. Needham).

As Truszkowski & Goldmanówna, following others, have shown, uricase exists also in the hepatopancreas of many invertebrates, such as *Astacus* and *Anodonta*. "Uricase" is really three enzymes, uricase properly so called, which forms allantoin from uric acid, allantoinase, which converts allantoin into allantoic acid, and allantoicase, which breaks down the latter to urea and glyoxylic acid. Selachian uric acid breakdown, as Brunel has shown, comprises the whole chain of processes. In some teleosts (Brunel) the chain is also complete (e.g. *Esox*, *Cyprinus*, *Scomber*) but in others the last enzyme is missing and the breakdown stops at allantoic acid (e.g. *Salmo*, *Pleuronectes*, *Anguilla*). In amphibia the chain is again complete (Brunel), but in mammals, of course, the breakdown only goes as far as allantoin, with the exception of certain primates and the Dalmatian breed of dog, in which it does not occur at all. Lack of space forbids a treatment of this subject, one of the most interesting direct links between biochemistry and genetics (cf. Onslow; Rheinberger; Starkenstein; Young, Conway & Crandall; Trimble & Keeler).

Returning to Fig. 18, the modern Chelonia stand in an interesting intermediate position. It is known that alone among the reptiles they may show a considerable excretion of urea (cf. *CE*, Table 163, where urea excretion in turtles and tortoises is seen to be variable and high, in contrast to the saurian reptiles, where there is an overwhelming preponderance of uric acid). A recent finding illustrating the latter point is that of Benedict, who found 82 % of the solid urine of *Python* was uric acid, and in addition about 30 % of the nitrogen of the liquid urine. The difference between chelonia and other reptiles has been recently verified by Kuhl, and in conformity with this, Manderscheid found the ornithine cycle in the liver of *Testudo graeca*.[†] She noticed, however, that in the absence of ornithine, added

---

* It would be interesting to investigate other embryos from this point of view. There is a short note by Beck & Truszkowski stating that no uricase is to be found in early human embryos.

[†] Clementi[2], again, has recently shown that on autolysis urea is formed by the livers of mammals, amphibia, fishes *and* chelonia (*Testudo*); not by birds and saurians (*Lacerta* and *Tropidonotus*).

lactate inhibited urea production, as if ammonia and lactate were forming some other compound; a hint that in the land chelonia both urea formation and uric acid formation occur in the same organ. This has since been shown to be the case by Heidermanns[3] and Münzel, who were the first to study chelonia of aquatic and terrestrial habit side by side. Thus the urine of the semi-terrestrial *Testudo* contains relatively much uric acid, while that of the semi-aquatic *Emys* contains relatively little. Now it seems that the course taken by the catabolic nitrogen in the chelonian liver varies with the temperature, for while at 20° much urea is formed by *Testudo* liver-slices, at 37° the uric acid formation is double the urea formation. There may be oecological significance in this, since terrestrial conditions would generally be warmer than aquatic ones, but for the proper interpretation of the situation in the chelonia we need to know a good deal more about the nitrogen metabolism of their embryos. As we have seen (p. 35), their eggs are non-cleidoic as regards water, but they may be, and probably are, cleidoic as regards the exit of nitrogenous waste-products.

It will be remembered however that in one marine turtle Tomita[1] showed (*CE*, p. 1103) that the predominant form of nitrogen excreted during embryonic life is urea. There must therefore here be a way out for diffusible substances as well as a way in for water. Imamura's analyses[2] of the allantoic liquid of this animal are too fragmentary to help us much.

In the liver of *Testudo* the urea production does not fall off as the uric acid formation rises to its maximum, as it would do were urea an intermediate stage in the production of uric acid. On the contrary, it rises to a maximum of its own at a lower level. This adds further support to the conclusion (already well established by the work of Clementi[1] and his school, to which references will be found in Needham, Brachet & Brown) that the uric acid synthesis of the sauropsida does not go by way of urea. Its exact nature is still uncertain, but it was shown independently by Schuler & Reindel and by Benzinger & Krebs that in pigeons the collaboration of two organs is necessary for uric acid formation, the liver synthesising from ammonia and carbon atoms a precursor substance which the kidney converts into uric acid. In the hen and the goose both processes can go on in the liver. Edson, Krebs & Model later showed that the precursor substance passing to the kidney is hypoxanthine. Its conversion to uric acid is accomplished by xanthine oxidase and does not need the cell-integrity of the kidney tissue, but the synthesis of hypoxanthine in the liver does need the normal cell structure and hence will only proceed in tissue-slices. The mechanism of this synthesis is still quite obscure, but Örström, Örström & Krebs have shown that it is specifically accelerated by oxaloacetate, which perhaps provides the carbon chain, and by glutamine, which perhaps provides the ammonia. Glutamine may therefore be an important intermediate substance. Örström, Örström, Krebs & Eggleston find that of ammonia given directly to liver tissue 50 % is converted to glutamine and 20 % directly to hypoxanthine.

The Prototheria (Monotremes), like the chelonia, stand in an intermediate position (see Fig. 18). As already pointed out (*CE*, p. 1145) *Echidna* has a typically mammalian urinary nitrogen-partition (and this has since been confirmed by

Mitchell), but it is marsupial and the pouch must act as a uterus more successfully than it does in *Hippocampus* (see p. 47), where the eggs are very yolky. Here the finding of Hill & Hill is significant; the shell of the marsupial monotreme egg has only the basal layer and no calcareous accumulation, as in *Ornithorhyncus*. What we need to know is the urinary nitrogen-partition of the non-marsupial monotreme, *Ornithorhyncus*, and this the literature does not divulge.

We come now to terrestrial oviparity and the cleidoic state among the invertebrates. It should be associated with uricotelic metabolism, and so it is, as the two important groups of insects* and terrestrial gastropod molluscs testify.

That the nitrogen excretion of insects consists almost wholly of uric acid has been noted again and again.† Uricotelic metabolism begins already at the "link" between insects and annelids, as we know from Manton & Heatley's work on *Peripatopsis*. But just as we have seen that the chick embryo in its early stages excretes a little ammonia and urea, so too the aquatic larvae of insects excrete ammonia (e.g. the may-fly larva, Fox and S. Smith[2]). Ammonia excretion was early reported for the larvae of muscid flies, but this was disregarded owing to the enormous quantities of bacteria in their guts, until Hobson (confirmed by Michelbacher, Hoskins & Herms and by Lennox) demonstrated that even when fly larvae were raised in scrupulously sterile conditions there was a considerable ammonia production.‡ The phenomenon was studied in detail by Brown[1] on *Calliphora erythrocephala* and *Wohlfahrtia vigil*; he found that the uric acid excretion was only $\frac{1}{20}$th that of the ammonia, but that both rose to a peak corresponding to the growth maximum and then fell to zero at the onset of pupation. During pupation uric acid alone is formed (cf. the discussion of Metamorphosis on p. 470, where it is shown that the metabolism of pupae has cleidoic characteristics). Correspondingly, Truszkowski & Chajkinówna found uricase in the housefly and the blowfly (*Musca domestica* and *carnaria*). Somewhat puzzled, they concluded that Przyłęcki's rule does not hold for the insects, but the subject was restored to rationality by Brown[3], who was able to show that in the fleshfly or sheep blowfly, *Lucilia sericata* (sterile), the uricase present in the larvae suddenly disappears at the beginning of pupation, and as suddenly appears again when the adult emerges from the pupal state. In *Lucilia* the following figures hold good:

	Ammonia	Uric acid	Allantoin
Embryonic period, mgm./gm.	0·009	0·178	0·106
Larval period, mgm./100 individuals	41	0·5	2·5
Pupal period, mgm./100 individuals	2	27	3
Adult period, 1st week of life, mgm./100 individuals	1	8	2

* We ought perhaps to add the arachnids with their guanine excretion, for it is practically certain that this is a special type of "uricotelic" metabolism. The nitrogen excretion of all arthropods has been well reviewed by Maluf[3].

† Newly quantitatively confirmed by Heller & Aremówna on the hawk-moth, *Deilephila euphorbiae*; Leifert on the moth, *Antheraea pernyi*; Brown[2] on the grasshopper, *Melanoplus bivittatus*; Truszkowski & Chajkinówna on the black-beetle, *Blatta orientalis*, the cockroach, *Blatella germanica*, and the may-beetle, *Melolontha vulgaris*; Becker[2] on the hornet, *Vespa crabro*, and the yellow butterfly, *Gonepteryx rhamni*; Wigglesworth[1] on the blood-sucking bug, *Rhodnius prolixus*; etc. etc. Becker points out that the pterin pigments of insect wings (see p. 655), though purine derivatives, never occur in insect excreta.

‡ The larvae, according to Tomita & Kumon, contain urease.

Thus during embryonic life about twenty times as much uric acid is produced as ammonia, while during larval life about eighty times more ammonia is produced than urea. The pupa reverts to the embryonic relationship, and the adult to the larval. Brown remarked that the semi-aquatic mode of life of muscid larvae would eliminate the need of an excretory product adapted for terrestrial existence and thereby render available the energy yielded by oxidising uric acid to allantoin.* Similarly, the existence of uricase in the adult may be regarded as a secondary adaptation making the excreta of other animals available to the insect as sources of energy. Rocco's report of uricase in dytiscids and some orthoptera, and of allantoinase in carabids, requires confirmation.

About the mechanism of uric acid synthesis in insects no information is available except from the work of Leifert, who studied it in the moth *Antheraea pernyi*, but as the experiments were made on breis in the doubtful absence of bacteria, her conclusion that urea is an intermediate must be regarded as only provisional.†

A few words must now be said regarding the terrestrial molluscs. Strohl[1], in his valuable review of the excretory mechanisms of molluscs, came to the general conclusion in 1914 that the gastropods are characterised by positive findings of uric acid in their nephridia, but that in the lamellibranchs this substance is always absent, and its place taken by urea. This is obviously in rough agreement with all that has so far been said, seeing that the former embody the large class of terrestrial pulmonates and that the latter are wholly aquatic. But this generalisation rested only on qualitative tests scattered through the literature, unsatisfactory because a positive uric acid test might imply no more than the presence of a nuclein catabolism of the usual type and nothing about the presence of uricotelic metabolism. It remained for later investigators (Wolf; Baldwin & J. Needham; Baldwin[2]; Grah) to demonstrate with varying degrees of success by *in vitro* experiments that the snail does actually form uric acid from added ammonia or amino-acids.‡

---

* The beneficial effects of maggots in the healing of certain wounds are to be ascribed partly to the allantoin and urea which they excrete (Robinson; Brown[2]) and partly to the sterilising effect of the alkaline pH caused by their ammonia excretion. Brown & Farber have studied an interesting anaerobic deaminase in fleshfly larvae which acts only on polypeptides and not on any simpler protein breakdown-products.

† A part of the work of the Bonn school seems to me to be vitiated by the use of enormous amounts of precursor substance relatively to the end-products estimated. Thus a rise from 1 mgm. to 2 mgm. of the end-product after an hour's incubation with 500 mgm. of the precursor in the presence of the tissue is claimed as a 100 % increase. One wonders whether a stricter statistical treatment is not necessary in the case of these heterogeneous breis.

‡ It seems rather unlikely that the course of synthesis is the same in birds and gastropods, because Baldwin & J. Needham found nothing in the snail hepatopancreas which the pigeon kidney could convert into uric acid, and observed that the snail hepatopancreas could not make uric acid from the products of the pigeon liver. For the reasons given in the preceding footnote, Wolf's contention that urea is an intermediate in uric acid synthesis in *Helix* cannot be accepted; the similar result of Baldwin[2] is slightly more convincing because the quantities of precursor added were of the same order as the syntheses obtained. The question, however, is by no means closed. As in birds, the integrity of cell structure is necessary for uric acid synthesis by the gastropod hepatopancreas (Grah). Arginase may be connected with it, for Baldwin[1] has found very large amounts of this enzyme in the uricotelic gastropods and none in the others.

That the terrestrial pulmonate gastropods were really uricotelic, however, had long been probable by reason of the extraordinary accumulations of uric acid in their nephridia. Quantitative estimations give figures such as 660, 720 or 810 mgm. per gm. dry wt. for the nephridium of the edible or vineyard snail, *Helix pomatia*, so that no less than three-quarters of the dry weight of the organ may consist of uric acid. Starting from this point a comparative survey of the uric acid retentions of many gastropods was made by Needham[18] in the hope that the presence of uricotelic metabolism might reveal itself by high uric acid figures.* It must be emphasised that in the absence of information regarding the speed of excretion in each of the species studied, the results cannot be made the basis of any final scheme, but there are reasons for believing that the excretory rate of all molluscs is very slow. The data are worth examination, because like those obtained for the selachian fishes, they illustrate a wide range of oecological, physiological and biochemical differences all within one group of animals (see Fig. 19).

The general results were clear. Beginning with eight species of marine gastropods (operculates) occupying the primitive aquatic environment, it was found that in all cases their uric-acid content was very low† (N. 2–4½, W.B. 0·2–0·5). This was no more than might be expected from nuclein catabolism. On the other hand, we may contrast with these, as the opposite extreme, the fully adapted terrestrial gastropods—pulmonates such as the English *Helix* and *Helicella* or the Paraguayan *Bulimulus*; or the operculate *Cyclostoma*. Here the retentions were found to be extremely high, ranging from N. 150, W.B. 4 in the case of *Helicella* to N. 700, W.B. 10 in the case of *Helix pomatia* and N. 1000, W.B. 80 in the case of *Cyclostoma*. For these figures it seems quite impossible that nuclein catabolism could be responsible, even if no excretion of uric acid ever took place throughout the organism's life cycle; uricotelic metabolism must be present. It is sometimes supposed that these terrestrial gastropods made their way on to the land by direct littoral penetration; if so, the modern series of periwinkles shows the process in action. Comparative estimations of uric acid in the Littorinidae showed that whereas the most marine of them (*L. littorea*) occupies a position quite analogous to the marine operculates (N. 1·5, W.B. 0·1), the most terrestrial of them (*L. neritoides*) approaches the terrestrial level (W.B. 2·4).

The data on the gastropods of fresh-water were rather complicated and offered difficulties of interpretation discussed elsewhere (Needham[31]). Some of the fresh-water pulmonates, such as *Limnaea stagnalis*, always regarded as secondarily inhabiting fresh-water, agreed closely with the land pulmonates (N. 100, W.B. 3), and Baldwin[3] has shown that their embryos, like those of *Helix*, accumulate uric acid as they develop. On the other hand, there are cases, such as *Planorbis* or *Ancylastrum*, where the nephridial uric acid is very low. These it is easy to explain as due to a more vigorous flow of water through their bodies, but such an explanation will not meet the interesting cases of *Paludina* (*Viviparus*) or *Bithynia*,

* The subsequent work of Spitzer went independently over part of the same ground, with substantially the same results.

† Abbreviations: N. =nephridium; W.B. =whole body; figures in mgm. uric acid per gm. dry wt.

fresh-water operculates which, owing to their gills and lack of mantle vascularisation, must surely have reached their present position by colonisation up the rivers. For these operculates contain considerable amounts of uric acid in their nephridia (N. 35, W.B. 5). Possibly they originated not directly from the sea, but by way of a period of semi-terrestrial or terrestrial life, analogous to that of *Cyclostoma*, with its vestigial gill, and so acquired uricotelic metabolism.

It is pertinent to enquire in this connection how far the eggs of terrestrial gastropods are truly cleidoic. Not much is known about them. The observation of Baldwin[3], that *Limnaea* eggs accumulate uric acid, has just been mentioned,

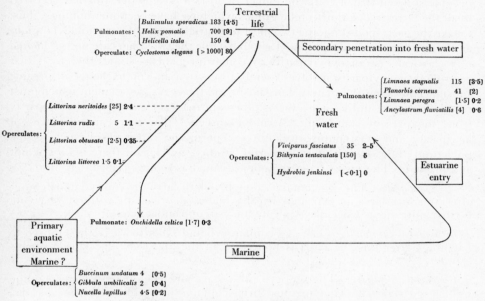

Fig. 19. Uricotelism in the gastropods.

Nephridium figures in ordinary type, whole body figures in black type. Unbracketed figures are experimental, bracketed are calculated; see Needham[31].

and a comparison may be made between the eggs of *Helix* and of the hen. The former accumulate 0·3 mgm. % of the final dry weight in uric acid by the time of hatching; the latter some 0·7 % — values of the same order. The snail embryo loses 19 % of the dry weight of the egg by combustion; the chick embryo 18 %. On the other hand, the snail embryo loses 48 % of the wet weight of the egg by evaporation as against the 15 % similar loss by the chick embryo. But Carmichael & Rivers have shown that the snail embryo will still develop normally after a dehydration of the eggs amounting to 85 %. There is certainly no evidence that the snail egg absorbs water or anything else but oxygen from the environment, and it may therefore be that the snail, instead of providing the relatively impermeable shell of the bird's egg, provides a much greater excess of water in its eggs above the embryo's minimum requirements.

Lastly, it may be mentioned that the presence of uricase (contrary to Przyłęcki's rule) in the gastropod hepatopancreas has been claimed by some workers (Spitzer; Grah; Plum). Of these the only convincing paper is that by Plum, who isolated allantoin from hepatopancreas, but as the evidence even in this case depends upon autolysing breis we have no guarantee that the uricase ever functions as such in the intact gastropod, and every likelihood to the contrary.

All that has so far been said does not exclude the possibility that ovoviviparity and even viviparity might arise late in evolution after the uricotelic habit had become so fixed that it could no longer be lost. It seems that this has occurred in all the terrestrial and uricotelic groups which we have been mentioning. Thus, to take the gastropods first, the two most terrestrial species of periwinkles (*L. rudis* and *neritoides*) are ovoviviparous (Lebour), while the two others are not. *Paludina* is a classical case of ovoviviparity. We may probably assume, therefore, that gastropod ovoviviparity involves merely protection and no exchange of organic substance between the maternal organism and the eggs—a statement which must remain tentative until some exact studies are made of gastropod ovoviviparity from the present standpoint.* Among insects,† too, there are instances of larviparity and even pupiparity, but except in the rarest cases, such as the polyctinid‡ bug, *Hesperocterus fumarius*, studied by Hagan, there is no attempt at a placenta. The insect types may be classified as follows (Hagan; Keilin[1]):

Ovoviviparity. The egg contains enough yolk to carry the embryo to hatching; after this the larva is extruded by the maternal organism without receiving further nutriment (Coccidae, Coleoptera, Sarcophagidae, etc.).

Intussuctio-viviparity. The egg contains enough yolk to carry the embryo to hatching; after this the larva is nourished in the uterus by specialised organs (Diptera).

Exgenito-viviparity. The embryo, in a stage of development corresponding to the egg stage of ovoviviparous forms, obtains nourishment by means of a trophamnion, trophoserosa, or trophochorion. Development occurs in the haemocoele, not in the uterus (Strepsiptera).

Pseudo-placento-viviparity. A placenta-like organ is present. Only two cases known: *Hemimerus* (Dermaptera) and *Hesperocterus* (Hemiptera).

It must be emphasised that all these cases are exceptional, but biochemical studies of their nitrogen excretion would be curious and interesting.

---

* It is true that according to Litwer, the gill epithelium which helps to form the brood-pouch in sphaerid lamellibranchs, has histologically a secretory appearance. For details of their development, see K. Okada.

† It used to be thought that the embryos in the brood-pouch of *Daphnia* were nourished by the maternal organism, but this crustacean ovoviviparity has been disproved by Rammer. Ovoviviparity is common among crustacea, especially cladocera and isopods. Nevertheless Gravier's review of crustacean ovoviviparity shows that structures suspiciously like secretory organs occur in some crustacean brood-pouches, e.g. *Polyphemus*. These questions do not affect the issue of nitrogen metabolism, except as regards land-crabs, which deserve a study they have never received.

‡ For an account of this strange and rare group see Ferris & Usinger.

When we turn to the vertebrates, we find that though there are no ovoviviparous birds, ovoviviparous or viviparous reptiles are by no means unknown (cf. the reviews of Wunder; Graham-Kerr, etc.). Giacomini discovered a kind of viviparity in the Italian lizard *Seps chalcides*, and Flynn found it to be widespread among Australian reptiles. The detailed descriptions of Tencatehoedemaker for *Seps* (see Fig. 20) and of Weekes for the Australian lizards *Egernia cunninghami*, *Tiliqua scincoides* and *Lygosoma* spp., among others, indicate that the contact between maternal and foetal tissues is, in all cases, like that of the placental dogfish, more or less epithelio-chorial. Certain Australian snakes, however, such as *Denisonia superba* (Weekes), seem to have a more intimate connection between maternal and foetal tissues than this. It is lamentable that no information about the nitrogen metabolism of such forms exists, but we may predict that it is uricotelic, and that the viviparity is of late evolutionary origin.*

The course of this discussion may be summarised as follows. The only guiding thread† through the bewildering mass of facts in the comparative physiology and biochemistry of nitrogen excretion seems to be the conditions under which the embryos of the species have to live. We may formulate it in the statement that uricotelic metabolism and terrestrial oviparity are strictly correlated, and this is perhaps the most remarkable instance of the far-reaching metabolic effects of a particular embryonic mode of life. It is not, of course, the whole of the story, for a number of minor specialisations such as the excretion of guanine or trimethylamine exist, and ovoviviparity or even viviparity may in rare instances supervene upon uricotelic metabolism after this has become so fixed that its loss is impossible. Nevertheless, the correlation in general abundantly justifies itself.

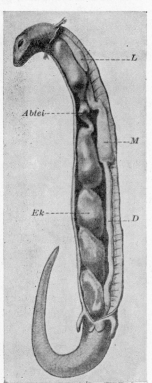

Fig. 20. Pregnant female of the lizard *Seps chalcides*. *Abtei*, abortive embryo; *Ek*, embryo compartment; *D*, intestine; *M*, stomach; *L*, lung.

We will now turn for a moment to the last of the metabolic properties of the cleidoic state, namely the preponderating reliance on fat catabolism (cf. Table 10). As already suggested (*CE*, p. 1186), the provision of metabolic water (the copious hydrogen in the paraffin chain gives about an equivalent weight of water for every fatty acid molecule burned) may be connected with this, where water needs are

* This view has the approval of one of the authorities on this subject, T. T. Flynn, expressed to the author in a private conversation.

† It is important in this subject not to be led astray by false trails. Thus, for his statement that uricotelic metabolism may exist in marine copepods, Maluf ([3], p. 52) relies on a statement made in 1858, and Heidermanns ([1], p. 266) has nothing later than 1872 in favour of the occurrence of uric acid excretion in ascidians.

great, as in terrestrial eggs. Allied with it is the property of unusually high resistance to ketosis (Needham[13]).

From the data given in *CE*, pp. 920 ff., the diet of the chick embryo each day during its development can easily be calculated. This is shown in Fig. 21, in terms of carbohydrate, fat and protein in percentage of the total material absorbed, from which it will be evident that by far the largest proportion of what is absorbed by the embryo is fat and protein. At certain stages, indeed (about the 10th and 19th days), twice as much fat as protein is absorbed. Moreover, as we shall see (p. 592), the material combusted during the last half of development seems to be almost entirely fat.

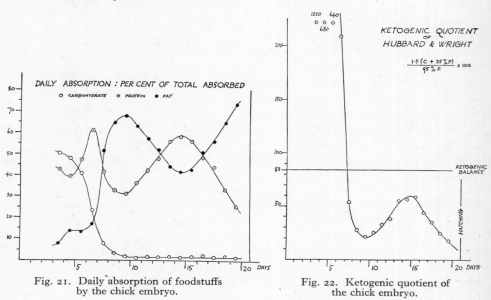

Fig. 21. Daily absorption of foodstuffs by the chick embryo.

Fig. 22. Ketogenic quotient of the chick embryo.

Now the excretion of aceto-acetic acid and β-hydroxybutyric acid is associated, as has long been known, with starvation, severe diabetes, and the restriction of carbohydrate intake. Of these three conditions, the third is eminently observable in the "unbalanced" diet of the chick embryo, and if the ketogenic balance is calculated according to the method of Hubbard & Wright, as in Fig. 22, it can be seen that from the 7th day of incubation onwards the chick embryo is below the danger line. In spite of this fact, however, the chick embryo develops absolutely no ketosis; no keto-acids can be found in the allantoic liquid or elsewhere (Needham[13]), though K. Takahashi[1] detected minute traces of acetaldehyde and acetone. As a search of the literature fails to discover any case of ketosis in an adult bird, even in a diabetic state, it is possible that immunity to dietary ketosis may be another special adaptation to cleidoic development with its intensification of fat absorption and fat catabolism. At the same time, it is true that certain mammals, such as the pig, and even certain types of human being, such as the Eskimo, have remarkable powers of resistance to dietary ketosis. Further work along these lines would be desirable.

In the course of this section, which, in following a single line of argument, has ranged over a wide variety of topics, three points have been raised which require further treatment. These are (1) the normal curves of mortality in avian development, which may throw light on the special qualities of the cleidoic egg, (2) the course of embryonic nutrition in birds, and (3) the biochemistry of the highest form of viviparity, the mammalian placental mechanism. These will be the subjects of the following three sections, and with them we shall complete what has to be said about the nutritional substratum of morphogenesis.

1·7. Standard mortality curves

Payne was the first to make a detailed study of the points in normal embryonic development under optimum conditions at which the maximum mortality occurs, and he has been followed by numerous workers (e.g. Byerly[2]). The fact that under such conditions there is always a certain loss of embryos indicates that the cleidoic egg of birds is not a 100 % efficient vehicle of reproduction. And the legitimate hope has been entertained that if we could find out the reason for these particularly dangerous periods, we should gain a better understanding of the nature of embryonic metabolism in its cleidoic form.

It was early recognised that the two main mortality peaks occur at about 3–4 days and 18–19 days of incubation. Fig. 23 shows the mortality distribution of the fowl embryo (*Gallus*) compared with that of the common pigeon (*Columba*), the ring dove (*Streptopelia*) and various wild Peristeridae (*Turtur, Spilopelia* and *Zenoidura*). The mortality of the turkey, the ring-necked pheasant, and the bob-white quail, are also given. In all cases, notwithstanding the rather different total incubation times (14–28 days), the peaks occur at equivalent moments of the embryonic life history. Riddle[3] has given the most probable explanations of the peaks in saying that the first is connected with respiratory maladjustments and the second with difficulties of water metabolism. "The method of avian reproduction", he wrote, "requires that water be stored in such quantity as will still be adequate at the end of incubation; but the very means—thick shells—which the egg uses to restrict the rate of water-loss may, if carried too far, adversely restrict the particular rate of gaseous exchange which is most favourable for the early developmental stages." Hence mortality might be expected to be most severe at the beginning, before the establishment of the area vasculosa, the allantois, etc., and at the end, when the water-supply, owing to continuous and unavoidable evaporation, is running low.

In favour of the respiratory nature of the first peak there are many facts (mentioned in *CE*, p. 1417 etc.), such as the death of embryos in eggs accidentally retained in the oviduct, the marked sensitivity of embryos before the 4th day of incubation to abnormally high or low oxygen pressures, etc.

That the second peak is connected with water-loss is also fairly well established. For terrestrial embryos the supply of water is, as we have seen, an important problem. The egg-white of reptiles and birds is the equivalent of the jelly surrounding amphibian and dipnoan eggs. The cleidoic egg was an obvious device to prevent undue evaporation of water. It includes, moreover, means whereby the pressure-head of water for the embryo is kept at a relatively constant level, for during development the egg-white is acidified (cf. *CE*, p. 881) and so brought to its isoelectric point, allowing the liberation of water by degrees from the colloidal albumen. The safety margin of water-loss in birds' eggs may be small. On the one hand, it is true that in Weldon's experiments the addition of water to the incubating egg, in such a manner as not to interfere with the processes of respiration while replacing that normally lost by evaporation, led to malformations of the amnion and the death of the embryo. But on the other hand Riddle[3] conclusively showed, using the eggs of tropical birds, that an increase of the normal loss of water by evaporation leads to the death of the embryo. It is impossible to

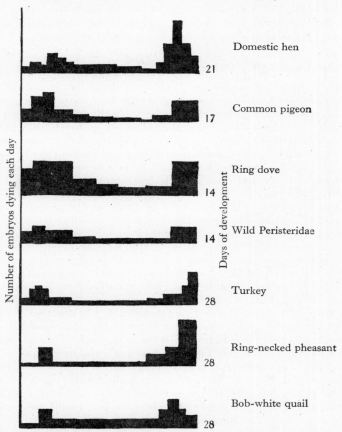

Fig. 23. Distribution polygons showing normal mortalities of seven kinds of birds (the first four from Riddle, the fifth from Martin & Insko and the last two from Romanov).

hatch the eggs of pigeons from the climate of Venezuela (the Orinoco region) in the temperate moderately dry air of North America, but if placed partly in water, they would develop normally and hatch. Death in the former circumstances usually took place during the final two days of incubation and was accompanied by abnormally large air-sacs, indicating excessive evaporation. Indeed, Lippincott & du Puy were able to hatch hen's eggs successfully if they were incubated lying in a layer of distilled water $\frac{1}{2}$ in. deep. Calculation shows that one hen's egg loses by evaporation 9·8 gm. water out of an original 38·4 gm. and gains from the combustion of fat 2 gm. Hence the latter quota may be of importance (cf. above, p. 48 and p. 62).

It is doubtful, however, whether the standard mortality peaks are wholly to be accounted for by the above-mentioned causes. As we have seen (p. 22), they are affected by the diet of the laying hen (Byerly, Titus & Ellis; McFarlane, Fulmer & Jukes), and they may also be affected by toxic substances introduced directly into the egg. When Landauer[4] introduced toxic doses of Mg″ and Li′ salts into developing eggs on the 6th day of incubation there was no effect on the embryonic death rate until the second standard mortality peak, when a much greater mortality than usual took place. A slightly higher mortality of male than female embryos in the fowl is said to exist (Landauer & Landauer), but we do not know at which of the two peaks this selective elimination takes place.

Temperature also has a slight effect upon the peaks. The optimum temperature and optimum humidity for the development of the chick embryo are now very accurately known (see the review of Landauer[23]; the papers of Romanov and his collaborators; Penionskevitch & Retanov, etc.). The optimum temperature is not a constant but varies with the humidity of the air; as the latter increases, the former decreases. There is general agreement that the susceptibility of the chick embryo to periods of chilling increases steadily and continuously through development (Kaufman[7]; Taylor, Gunns & Moses). Romanov, Smith & Sullivan have shown that the second mortality peak is affected by all kinds of variations of temperature away from the optimum.

Contributory to the peaks of mortality, though only to a small degree, are the various "malpositions" which have been much studied in America (Byerly & Olsen; Olsen & Byerly; Waters; Hutt & Pilkey; Cavers & Hutt; Hutt & Cavers; Hutt & Greenwood; Dove). These are classified as follows:

I.	Head between legs.	IV. Head buried away from air-space.
II.	Head in small end of egg.	V. Feet over head.
III.	Beak under left wing.	VI. Beak over right wing.

The first and third of these occur with especial frequency when the egg is incubated large end up; the second and fourth when it is incubated horizontally. The first is invariably lethal, the second probably so (50 %), the fourth is nearly always lethal and is determined during the 2nd week of incubation. Olsen & Byerly believe that the pointed end of the egg should be caudal, and if this relation fails to arise during the preparation of the egg, malpositions are likely to result. Waters thinks that malpositions are often semi-normal stages which most embryos pass through but which in some cases are retained and so prevent hatching. The interesting phenomenon of *heterotaxia*, i.e. the position of the embryo lying on its right side instead of on its left, as normally, occurs to the extent of 0·163 % embryos and has no effect on hatchability (L. W. Taylor). Malpositions have also been studied in the turkey (Martin & Insko; Insko & Martin).

It would be very interesting to have data regarding the standard mortality curves of animals other than birds. Fragmentary data exist for certain fishes (*CE*, p. 1385), and for the silkworm Grandori has classified three kinds of "dead-in-shell" embryos. There are those which have gone no further than the blastoderm stage, there are others which have produced many tissues but in which no metameric segmentation has appeared, and still others which have produced abnormally formed organs. Unfortunately no statistical figures or times are available here. For a good discussion of foetal death in man the reviews of Pfaundler and Sterling may be referred to. Rodents are more fortunate in that aseptic resorption of dead embryos following atrophy is carried out with remarkable facility (Crosman; Corey[2]), cf. *CE*, p. 1329, and this may even happen in cattle (C. W. Turner). As Kosaka has shown, foetal atrophy usually begins in the liver, which becomes progressively less prone to autolysis (as judged by the extent to which it has proceeded in a 72 hr. period) during development. It must be a faculty widespread in the animal world, since Guareschi[3] observed it happening in dead embryos in the modified ovary* of the teleostean fish *Gambusia*.†

1·8. Embryonic nutrition

1·81. The scaffolding of the chick embryo

The temporary structures used by the embryos of birds during their development are three in number, the yolk-sac through which the absorption of nutrient materials from yolk and white is carried on, the amnion which encloses the

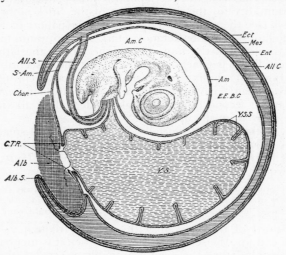

Fig. 24. The membranes of the chick embryo. *Alb.*, albumen; *Alb.S.*, albumen-sac; *All.C.*, allantoic cavity; *All.S.*, allantoic stalk; *Am.*, amnion; *Am.C.*, amniotic cavity; *Chor.*, ecto-mesodermal portion of extra-embryonic blastoderm associated neither with yolk-sac nor with allantois ("chorion"); *C.T.R.*, mesodermal ring surrounding a small area of the yolk which the blastoderm fails to enclose. The remains of the vitelline membrane can be seen; *Ect.*, ectoderm; *E.E.B.C.*, extra-embryonic "body-cavity"; *Ent.*, endoderm; *Mes.*, Mesoderm; *S-Am.*, sero-amniotic connection, later perforated; *Y.S.*, yolk-sac; *Y.S.S.*, villi of the yolk-sac wall (see Fig. 28).

embryo in its "private pond", and the allantois which receives the excreta from the cloaca and allows the exchange of respiratory gases through its blood-vessels closely applied to the inner shell-membrane. We have no need to add to the excellent descriptions of them already available, such as that of Lillie[8], whose diagram is reproduced as Fig. 24; but as to their physiology there is a good deal to be said in amplification of previous accounts (*CE*, pp. 917 ff.).

Byerly[4] was the first to study the membranes of the chick embryo quantitatively, measuring wet and dry weight, and his treatment of the question remains the best. The yolk-sac grows steadily from the 1st day onward to reach a maximum

* See p. 46.

† But see p. 260, where the unshed unfertilised eggs seem to give rise to tumours in worms.

wet weight of about 3·5 gm. on the 15th day of incubation, after which its weight falls to approximately 2·5 gm. at hatching. The allantois does not appear until the 4th day, but it then grows more quickly than the yolk-sac to reach a maximum wet weight of about 1·4 gm. by the 10th day, after which it remains roughly constant in weight till the time of hatching. The third membrane, the amnion, has a much less bulk than either of the other two. Byerly's most striking chart (Fig. 25) demonstrates that not until half-way through development does the

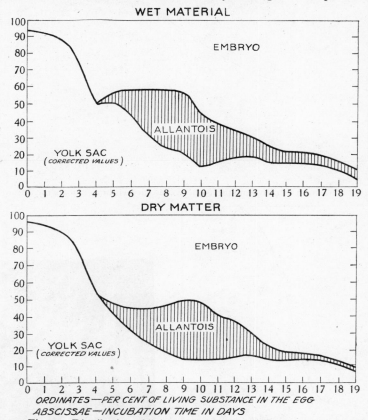

Fig. 25. Distribution of living matter in the developing hen's egg.

embryo itself (whether considered as wet or dry weight) attain a bulk equal to more than half the total living matter in the egg. This is a remarkable illustration of the importance of the scaffolding mechanisms required by highly lecithic, cleidoic, terrestrial eggs.* For determinations of ash and total nitrogen in the membranes we are indebted to Penquite. Fig. 26 shows from a different angle, that of nitrogen storage, the importance of the membranes in the earlier stages.

* Compare with this the closely analogous role of the pulp in avian feather development (F. R. Lillie[10]). And the few quantitative data that we have on the development of galls (see p. 106) indicate that the growth of the gall is at first enormously greater than that of the nymph, while later it almost ceases though that of the insect continues (B. W. Wells). Here, of course, the biology of the process is primarily protection, not nutrition.

When mgm. total N in the membranes are plotted against mgm. total N in the embryo on a double log. grid, it can be seen that not until the 15th day does the

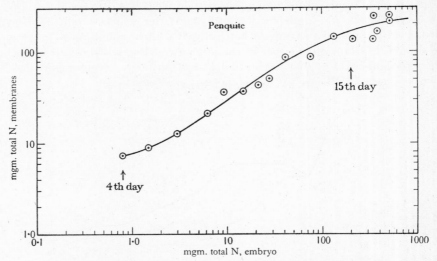

Fig. 26. Relation of nitrogen storage between chick embryo and its membranes.

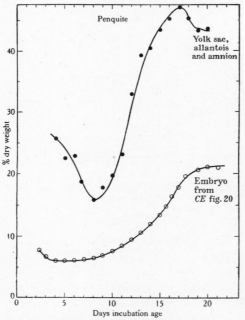

Fig. 27. Hydration and dehydration of the chick embryo's membranes.

Fig. 28. Interior surface of chick yolk-sac showing villous blood-vessels.

embryo overhaul the membranes (embryo 200 mgm., membranes 160 mgm.). The curve then approaches the horizontal, for the embryo continues to grow while the membranes do not. Earlier, at the 4th day, the embryo has ten times less nitrogen than the membranes.

As a result of Penquite's work we know now the percentages of dry weight and ash in the embryonic membranes. Fig. 27, calculated from his data, shows that unlike the embryo, which as is well known shows a progressively rising dry weight %, the yolk-sac's dry weight % passes through a minimum on the 8th day, rising thereafter to a point near hatching at which nearly half its weight is dry substance. This is no doubt connected with the presence of such large amounts of raw materials from the yolk being absorbed by the yolk-sac and transmitted to the embryo, and it is interesting that the maximum hydration of the yolk-sac occurs during the period of its maximum growth rate. As the yolk-sac ceases to grow, and when the yolk has been fully enclosed, a mass of radiating lamellae or villi appear on its interior surface; these play an important part in the intake of food, and are illustrated in Fig. 28, taken from Remotti's study of them[3].

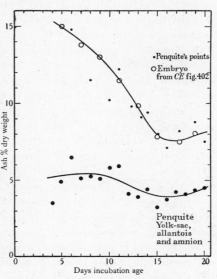

Fig. 29. Ash content of chick embryo and membranes.

During its development the chick embryo (like all other embryonic organisms, cf. p. 546) becomes progressively poorer in ash, related to dry weight. As Fig. 29 shows, however, the content of the membranes in inorganic substances remains roughly constant (Penquite).

A comparison of the absorptive work carried out by the yolk-sac and its metabolism was made by Needham[6]. Defining the Membrane Absorption Rate as

$$\frac{y_i + a_i + e_i + c_i}{y + a} \times 100,$$

and the Membrane Transit Rate as

$$\frac{a_i + e_i + c_i}{y} \times 100,$$

where y_i is the daily increment (wet weight) of the yolk-sac,
　　　a_i is the daily increment (wet weight) of the allantois,
　　　e_i is the daily increment (wet weight) of the embryo,
　　　c_i is the daily increment of material combusted and lost from the egg,
　　　y is the wet weight of the yolk-sac,
　　　a is the wet weight of the allantois,

it was possible to calculate the amount of material retained each day for the construction or maintenance of the yolk-sac and that passed on for the construction and maintenance of the embryo. At the beginning of development the yolk-sac absorbs more than three times its own weight of material each day, but by the 9th day an equilibrium has been attained and after that time it absorbs about its own weight of material each day. The high Absorption Rate of early development is almost exclusively concerned with the construction of the yolk-sac itself. The Transit Rate indicates that after the 9th day almost all the material absorbed is transmitted to the embryo.

Respiratory quotients of all the membranes were measured in glucose medium each day during development[7,14]. As Fig. 316 demonstrates, the R.Q. of the yolk-sac falls from about 0·9 on the 2nd day to 0·6 on the 7th, after which it remains approximately constant a little below 0·6 till the 14th. After the 14th day accurate quotients are difficult to obtain, because the respiratory rate is very low and the bound CO_2, probably because of localised deposits of calcium carbonate, very high. Nevertheless, it is clear that there is a correlation between the growth period of the yolk-sac and its metabolism, for while it is rapidly growing and developing, its R.Q. is high, but when it ceases to do so and begins to pass on all that it absorbs to the embryo, its R.Q. is very low. These low quotients were found not to be due to the conversion of fat to carbohydrate or the desaturation of fatty acids or the formation of keto-bodies. They probably indicate a catabolism of fat with certain modifications.

In the same diagram data regarding the R.Q. of other parts of the egg system will be found. The R.Q. of the allantois falls slowly from about 0·95 on the 5th day to about 0·82 on the 20th. At the time of its origin from the embryo, therefore, its catabolism most closely resembles that of the embryo, and later progressively departs from it. So long as the size of the embryo remains sufficiently small to allow of the manometric measurement of its R.Q. this is found to be unity, but later on the R.Q. must fall, for at the time of hatching it is certainly in the neighbourhood of 0·7. This subject will be referred to again later (p. 592).

A good deal of work has been done on the histology of the yolk-sac. It contains cells of reticulo-endothelial type and cells which phagocytose (Coppini; Zanoni; Steinmüller) and has been studied in tissue culture (J. A. Thomas[3,4]). It may continue to live for some time after the death of the embryo (Remotti[4]). Its vascular cells have peculiar physiological properties (Nakano).

We come now to the enzymic equipment of yolk, white and yolk-sac in relation to the absorption of the raw materials. As we have already seen, the materials which enter the embryo are, after the 5th day, fat and protein in overwhelming preponderance. This may be illustrated in an interesting way by the method of I. Fisher. If a right-angled triangle is drawn, the three vertices representing protein, carbohydrate and fat, the composition of any given diet may be represented by a point. In the case of the point O in the small diagram of Fig. 30, representing a given percentage of fat and protein, the proportion of carbohydrate in the diet is represented by the area of the triangle OPF, and the larger the proportion of carbohydrate in the diet, the nearer O would approach C. Similarly, the proportion of protein in the diet is given by the area of the triangle OCF and that of fat by the area of the triangle OCP. Passing to the large diagram (calculated from *CE*, Section 6·9), it will be seen that the diet of the chick embryo only contains appreciable amounts of carbohydrate in the early stages of development. About the end of the 1st week the line of points approaches the *PF* line and there oscillates during the remainder of incubation in the manner shown in Fig. 30.

As stated in *CE*, pp. 1303 ff., proteases and lipases have long been known to occur in the yolk and white. Yolk-protease is believed to reach a maximum of activity about the 10th day, i.e. at the time of maximal protein catabolism (Remotti[1] on hen's eggs; Calzoni[1] on duck's eggs), and yolk-lipase about the 16th day, i.e. the time of the second maximum of fat absorption. Amylase is also present in yolk (Tateishi[2]), and white (Rotini).

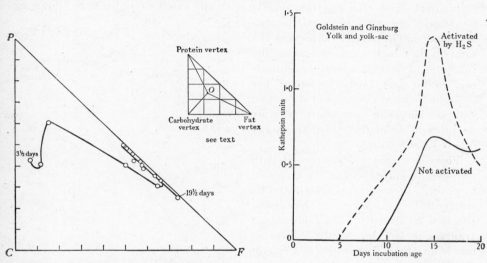

Fig. 30. Diet of the chick embryo (I. Fisher's method)—in gm. not calories.

Fig. 31. Kathepsin activity of chick yolk and yolk-sac.

The kathepsin activity of the yolk and yolk-sac together is shown in Fig. 31. This work (Goldstein[2]; Goldstein & Ginzburg; Goldstein & Millgram) shows that the activity of the enzyme increases up to the 15th day and then remains constant, though if the activity realisable by activation with H_2S is considered, there is a peak of activity at the 15th day. This picture so much resembles that of the growth of the yolk-sac itself that the latter is probably the largest factor in it. All investigators are agreed that during the first half of development the activity of the proteolytic enzymes in the yolk-sac is much greater than that of those in the embryo. Thus Borger & Peters found that the proteinase of the yolk-sac (optimum pH 4·0) on the 8th day was twelve times as active as that of the embryo (in the allantois practically absent), and Krebs found a P.Q.* of 13·2

* P.Q. $= \dfrac{\text{c.mm. amino-N}}{V \dfrac{a}{(a+b)x} \times \text{hr.}}$, where V is the vol. of the active extract, a the vol. of tissue used, b the vol. of glycerol used, and x a factor depending on the percentage water in the tissue.

in the yolk-sac of 8-day embryos as against 1·24 in the embryos themselves. Mystkowski[2] found a P.Q. of about 2·5 (unactivated) and 6·5 (activated by cystein) both at early stages (4 days of incubation) and at late ones (18 days) for the yolk-sac, as against 0·25, unactivated, and 0·35, activated, for the embryo itself. The same is true of dipeptidase. Found by Borger & Peters in the yolk-sac, and to a slight extent in the allantois, it was estimated by Donegan as about three times more active on the 5th day than that in the embryo. There is also amino-polypeptidase in the yolk-sac (Borger & Peters). We may conclude that the heavy burden of digestive work borne by the yolk-sac in the earlier stages involves a powerful enzymic equipment.

The fact that some of the kathepsin in the yolk-sac can be activated by sulphydryl shows that a certain reserve normally exists. Goldstein & Ginzburg found that the ratio between the un-activated and activated values was the same in the chick's yolk-sac as in the placentas of several mammals (guinea-pig, rabbit and rat). Goldstein[2] has brought forward evidence to show that yolk-sac kathepsin acts more strongly on proteins from the egg than on any others save gelatin.

There are enzymes concerned with nitrogenous substances in the yolk-sac, however, which cannot be activated by sulphydryl, e.g. arginase (Needham, Brachet & Brown). The amount of arginase, in any case, in the yolk-sac is much less than that in the embryo (at the 5th day, e.g. $Q_H^{38°}$ 0·32 for the embryo, 0·025 for the yolk-sac).*

The lipases in the egg, particularly esterase (methylbutyrase) and lipase (tributyrinase), have been the subject of an interesting study by Ammon &

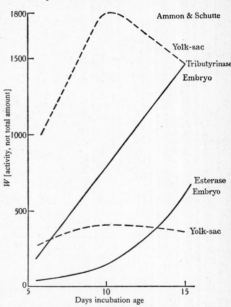

Fig. 32. Lipase activity of chick embryo and yolk-sac.

Schütte. The unincubated yolk has more of these enzymes than the white; yolk esterase 46 W,[†] yolk lipase 88 W; white esterase 5 W, white lipase 15 W. No change occurs in the enzymes of the yolk or white on incubating eggs, whether fertilised or not, but the activity of both enzymes rises regularly in the developing embryo, as shown in Fig. 32. But in the yolk-sac the activity is particularly marked; it is many times higher than the embryo in the early stages, and thirty times that of the yolk, even at the earliest stages (see Fig. 32). In the allantois it is low and remains relatively constant at esterase 50 W and lipase 300 W. That there is a synthesis of the enzymes as a whole appears from the number of units

* $Q_H^{38°} = \dfrac{\text{c.mm. urea } CO_2}{\text{mgm. dry wt.} \times \text{hr.}}$ (manometrically).　　　　　† W = units/c.c. extract.

contained in the entire egg system, which rises from 132 to 1125 in the case of esterase and from 264 to 3102 in the case of lipase.

Choline esterase, on the other hand, rises in activity in the yolk itself (from 96 W to 294 W on the 15th day) as well as in the embryo (392 W to 1120 W). In the yolk-sac it remains constant at 170 W and in the allantois at 100 W (cf. the work of Nachmansohn referred to on p. 643). Physostigmine inhibits it.

Enzymes of the egg-white have not been much studied recently. Apart from the older observations mentioned in *CE*, p. 1304, van Manen & Rimington have recognised two erepsins (*p*H optima at 5·5 and 7·5) and deny that egg-white is autolysable, while Balls & Swenson think that the thick egg-white can autolyse and contains a trypsin-like proteinase. The last-named authors have studied a tryptic inhibitor present in the thin egg-white, which they believe to be a sulphur-containing polypeptide, the action of which can be reversed by adding enterokinase.

This is the place to mention the very remarkable substance lysozyme, a bacteriolytic protein of low molecular weight with the properties of an enzyme, which occurs richly in egg-white, and has lately been crystallised (Abraham & Robinson). Lysis manifest to the naked eye is performed by egg-white diluted 1 in 1 million. Details concerning it will be found in the excellent review of Fleming, but what part any of these substances play in embryonic metabolism is still quite obscure.

Utilisation of the raw materials proceeds most efficiently at the optimum temperature of development (Penionskevitch; Romanov, Smith & Sullivan). At the end of the incubation period the yolk-sac does not degenerate; the vessels simply constrict and the tissue dies "in complete possession of its physiological irritability and anatomical integrity" (*CE*, p. 725). The process of this atrophy and means of preventing it have been studied in detail by Remotti[3]. The removal of the "spare yolk" (*CE*, p. 939) with which the embryo hatches, sometimes called "deutectomy", by no means leads to its death. Barott, Byerly & Pringle have shown that in that case the chick simply lives on the reserves of its own body, and loses more weight than normal chicks at all temperatures observed. Its R.Q. also is unaffected. But whereas the heat production of normal chicks is constant for a given temperature, in chicks without spare yolks it continuously falls off. Of course if no food and water are received from without, after 3 days the chick enters a starvation phase, and its body temperature begins to drop (Card).

Of the amniotic membrane little is known. Remotti[5] assesses the endoamniotic pressure at 5·5 cm. Hg; it does not alter during development and is partly due to a tonus of the muscle fibres in the amnion. Up to the 17th day there is a progressive increase in its mechanical resistance and a progressive decrease in its speed of autolysis, but after that time the process is reversed and there is a rapid change of both properties in the opposite direction.

1·82. Yolk utilisation in the lower vertebrates and invertebrates

We have very little information about the process of yolk utilisation in animals other than birds. Except for what has so far been said about the selachians (p. 40), there is only a note by v. d. Ghinst[3] on the trout, reporting the presence of amylase in the syncitial vitelline cells at the blastula stage. For the amphibia, apart from the rather enigmatic cytological work of Kedrovsky and the Kono-packis, there seem to be only the observations of Dorris, who combined delicate micro-estimations of enzyme activity with histological sections of the developing gut in the axolotl. Amylase is the first enzyme to appear, at Harrison stage 40, when the gut is still histologically undifferentiated. The proteolytic enzymes, pepsin and trypsin, appear at stage 43, just as the differentiation of the stomach glands and the acini of the dorsal pancreatic rudiment are coming to completion. Such enzymes will account for the digestion of what remains of the yolk, but of the processes going on in every cell of the embryo at much earlier periods we know literally nothing.

A quantity of miscellaneous data about the utilisation of yolk in the silkworm egg has been collected in a review by Tirelli[5].

Yolk utilisation probably accounts for a very peculiar phenomenon the meaning of which is at present obscure, namely, the rhythmic contractions of the fish blastoderm from an early stage in development. These were first extensively studied by Ransom in 1854. Formerly (*CE*, p. 897) this was mentioned under the heading of water absorption, it being thought to have some connection with the need of non-cleidoic embryos for water, but we now know that the contraction of micro-fibrils plays a quite fundamental morphogenetic role in insect develop-ment (see p. 463) and may even be not without importance in that of amphibia (see p. 146). However, since only the highly lecithic types of egg among the fishes show it, it is probably a stirring mechanism.

The most thorough study of blastoderm contraction in a teleost is that of Yamamoto[1] on *Oryzias latipes*. Rhythmic waves of contraction are seen first when the blastoderm has enclosed nearly one-third of the egg, and the embryonic shield has just become visible. The blastoderm does not maintain a fixed hemi-spherical shape, but fluctuates continually. Waves of origin dextral to the embryonic axis alternate with waves of origin sinistral to it, though some-times both may occur at the same moment and progress till they meet. Cell boundaries in the blastoderm, straight when at rest, become wavy on con-traction. On studying the effects of temperature, Yamamoto obtained for the frequency of contraction, temperature characteristics (see *CE*, p. 524) none of which were the same as those typical either of rhythmic processes or of oxidation reactions, but which more resembled the figures for rate of embryonic development. When the temperature characteristics for the velocity of the wave and for the pause between waves were obtained, it was found that as temperature rises the time required for each wave to travel over the blastoderm decreases, and the time taken by each pause also decreases; in such a way that the frequency is the intermediate between them. The meaning of the temperature characteristics could, as usual, not be clearly ascertained. The effect of most ions in concentra-tions above the normal was to accelerate the rate of contraction and finally to stop it; this result is reversible if the embryos are returned to normal conditions before too long a time of exposure. Osmotic pressure, on the other hand, had no effect. Intermediate pH levels produce no effect, but at pH 3·0 the movements are accelerated, then stopped; at pH 11·0 the movements are slowed, then accelerated, then stopped. CO_2 first accelerates, then slows and stops. In anaerobic conditions the movements can continue for about $1\frac{1}{2}$ hr. but eventually come to an end. Even after 20 hr. anaerobiosis they may be resumed and development continue normally. Some rH indicators are reduced before the movement stops; others afterwards, thus fixing the point of cessation of movement at rH 6–9. As in muscle the energy required for a certain amount of contraction may be produced anaerobically, but the process cannot continue indefinitely without oxygen.

These data, though interesting, do not answer the one essential question, Can normal development go on in the absence of the contractions? While it may be that they are mainly connected with the absorption of food material from the yolk, we cannot as yet deny them a morphogenetic significance.

Apparently many other fishes exhibit the phenomena. They have been described in the goldfish *Carassius auratus*, where the protoplasmic peripheral layer contracts before the formation of the blastoderm, and the trout, by Yamamoto[2] (with review of the literature); in the stickleback *Gasterosteus aculeatus*, where even the blastomeres contract (Wintrebert & K. C. Yung; Painlevé, Wintrebert & K. C. Yung); in the pike *Esox lucius* (Kasansky); and newly in the eggs of Brazilian fishes such as *Trachycorystes* and *Hoplias* (v. Ihering). They have been filmed by W. Kuhl. They probably account for the rhythmic changes in electric impedance which trout eggs show (Hubbard & Rothschild).

Among the embryos of invertebrates all kinds of movements occur, ranging from the amoeboid motion of the eggs of *Hydra* to the slow pulsations in the eggs of molluscs (*Limnaea*; Comandon & de Fonbrune[2]) and of worms (*Nereis*; Hoadley[6]). We are still far from any understanding of their meaning, but it will be remembered that protoplasmic contractility plays a large part in echinoderm fertilisation, as in the starfish "fertilisation cones" studied by Fol; Chambers[1]; Sugiyama[2] and others.

1·83. Avian "milk" and mammalian colostrum

Since the time of John Hunter's essay "On a Secretion in the Crop of Breeding Pigeons" in 1786, it has been known that in the development of nidicolous birds, such as the pigeon, which do not hatch from their eggs in a fully developed state, the embryos rely not only on the yolk and white but also on a special secretion which they receive from their parents after hatching. This so-called "pigeon milk" is a white slimy substance secreted by a special area in the digestive tract of both parents, the crop gland, and regurgitated to the squabs (Beams & Meyer). It thus represents a "histotrophe" for the embryo prepared at a site other than the genital system (see pp. 42, 81).*

Analyses of pigeon milk by modern methods have frequently been made. All investigators (Dąbrowska; Carr & James; Reed, Mendel, Vickery & Carlisle; Dulzetto & li Volsi) are agreed that it has the following approximate composition:

	%
Water	65–81
Protein	13·3–18·8
Fat	6·9–12·7
Ash	1·5

It is therefore a very much more efficient nutriment than mammalian milk, for its composition resembles that of rabbit or cat milk, yet the pigeon squab doubles its hatching weight in only 2 days, whereas the rabbit doubles its birth weight in 6 and the cat in 9. Pigeon milk contains a large amount of desquamated

---

* Can it be that the gelatinous cement of the nests of certain swifts (Cypselidae), *Collocalia* spp., which has rendered them a famous Chinese delicacy, recommended by traditional dieticians for its nutritional properties when made into soup (*yen wo t'ang*), is really a crop-gland secretion? If so, the gland would thus subserve another phase of reproductive activities.

cell material, which incorporates much of the fat. The hypertrophy of the crop to form the crop gland is remarkable. According to Carr & James the parent birds can be fed on maize while the squabs are being fed from the crop gland, which argues a synthesis of essential amino-acids in the organ. The pigeon milk contains no carbohydrate (Dąbrowska), but has active amylase and invertase (Dulzetto & li Volsi). It also contains adequate supplies of vitamins A, B_1 and B_2 (Reed *et al.*). Its ash (Białaszewicz & Lewis; W. L. Davies) differs greatly from the ash of mammalian milk, having much more sodium and much less calcium. Dulzetto & li Volsi found the low iodine value of 40 for the fatty acids in it, and recognised the larger part of the protein as a phosphoprotein. The presence of serum vitellin in the blood of the *male* pigeon at such times should therefore be worth looking for. The squab gets up to 10 gm. of the milk a day.

A good deal of work has been done on the endocrine control of the crop gland. Like lactation proper, the crop-gland secretion is stimulated by prolactin, the mammotrophic hormone of the anterior pituitary (Riddle & Braucher; Riddle, Bates & Dykshorn; Bates, Riddle & Lahr), which is also responsible for inducing broodiness in birds (Riddle, Bates & Lahr) and which begins its action on the crop by stimulating mitoses in a certain area (Lahr & Riddle), though this action could not be reproduced *in vitro* (Salle & Schechmeister). Prolactin is a protein, very heat-stable and of low molecular weight (Riddle; Bates & Riddle; Bates, Riddle & Lahr). The amounts of it required are exceedingly low (Lyons & Page; Lyons). Hypophysectomy greatly reduces (Schooley, Riddle & Bates) or abolishes (Gomez & Turner) the response of the presumptive crop gland, but adrenalectomy and thyroidectomy have no effect on it. Castration of the male eventually stops the reaction (Patel; Kaufman & Dąbrowska), but only after some time and as the result of the changes in the testis, since the crop gland will still respond to prolactin (Riddle & Dykshorn). Lastly, like lactation proper, the crop-gland secretion is inhibited by oestradiol monobenzoate (Folley & White).

The significance of "pigeon milk" cannot be understood save in the light of the evolution of the sauropsida, well discussed by Portmann[3]. The nidifugous habit, where the embryo is hatched as nearly alike to the adult bird as possible after a relatively long incubation period, is undoubtedly primitive. Portmann divides the birds into the following groups:

Table 12. *"Larval compression" in Birds*

	Time of learning flight	Condition of feathers	Visual condition at hatching	Nutritional dependence on parents	Example
1. Nidifugous	Quick	Feathers	Can see	None	Fowls, etc.
2. Nidifugous	Quick	Down	Can see	None	Geese, etc.
3. Nidifugous	Slow	Down	Can see	Present	Seagulls, etc.
4. Nidifugous	Slow	Down	Can see	Marked	Birds of prey
5. Nidicolous	Very slow	Down	Blind	Marked	Penguins, petrels
6. Nidicolous	Very slow	Reduced down	Blind	"Milk"	Pigeons, etc.
7. Nidicolous	Very slow	Very reduced down	Blind	Chirp reflex, regurgitation, etc.	Singing-birds, sparrows, etc.

He has himself studied the relative growth of organ weights in nidicolous as compared with nidifugous birds (Portmann[5]), from which it is clear that there are considerable differences. It should be noted that in mammals there is a parallel series (cf. Table 13). The epithelio-chorial placenta, with its long gestation time and fully developed foetus at birth, contrasts with the haemo-chorial type (see p. 81) with short gestation time (apart from man, a very exceptional case) and undeveloped foetal condition at birth. There is also a parallel in that some mammals give substances of importance to the young animal through the colostrum after birth while others do not, but the parallel cannot be carried further because colostrum is associated rather with the epithelio-chorial placenta on account of what, owing to its many cell-layers, it has failed to let through, than with the haemo-chorial placenta, transmission through which is very efficient (*CE*, p. 1497). A concrete instance is that of vitamin A, which, as Dann[2,3] has shown, is contained in high concentration in the colostrum of ungulates such as the cow (epithelio-chorial placenta) where 1 day's feed of colostrum is equivalent to 50 days of milk, but only insignificantly in human colostrum (haemo-chorial placenta). Even in human milk, however, there may be substances supplementing the infants' capabilities, for it is said that at first salivary amylase is aided by amylase in the milk (Schlack & Scharfnagel). All kinds of enzymes—catalase, amylase, invertase, monobutyrinase, tributyrinase, etc.—(Katsu) are present in human colostrum (analyses by Widdows, Lowenfeld, Bond, Shiskin & Taylor).

In sum, therefore, among both birds and mammals "larval forms" are to be found, though never so morphologically deviating from the adult form as in the lower vertebrates and invertebrates.

1·84. Maternal diet and embryonic constitution in mammals

Under the general heading of nutrition the question may well be asked whether and to what extent the nutritional conditions prevailing during intra-uterine life in mammals affect the growth and development of the foetus. This raises a problem analogous to that discussed in connection with the chick embryo in Section 1·3. Avoiding the rather large and unsatisfactory semi-medical literature mention will only be made here of a few orienting papers (e.g. for the mouse, Agduhr; for the rat, Wan & Wu; for the rabbit, Rosahn & Greene; Rosahn, Greene & Hu; Kopeć; Kopeć & Latyszewski; for the pig, Mitchell, Carroll, Hamilton & Hunt; for cattle, Eckles; Haigh, Moulton & Trowbridge).

In general it may be said that mammalian development is an all-or-none phenomenon. It may be interrupted by insufficient calorie intake or lack of one or other of the essential dietary materials in the food of the maternal organism; in this case foetal resorption (see p. 65) or abortion will ensue. For example, this will happen in the rat if there is less than 5 % protein in the diet (Guilbert & Goss). But if the process can be carried to its normal termination at all, the composition and viability of the embryos will be approximately normal. In order to study the effects of extreme under-nutrition during pregnancy, Jonen subjected pregnant rabbits to absolute fasting for various periods, with the result that at birth the weights of the foetuses were much more abnormal than their chemical composition. Diminutions of weight of 33 % were not uncommon. But as the following figures show:—

At the 24th day:	Wet wt. gm.	Foetus Dry subst.%	Fat % dry wt.	Maternal organism fat % dry wt.
Normal	11·21	12·48	23·1	48·06
Starvation for 10 days during pregnancy	7·43	12·82	19·0	33·0

there may be a great diminution in total weight while the relative amount of dry substance in the foetus remains unchanged and the fat-content is only diminished to a minor extent. The fat reserves of the maternal organism will be of course much depleted. On the other hand, if the starvation period ceases late in pregnancy, it may be too late for the foetus to be able to take advantage of the nutrient materials now newly circulating in the maternal blood.

Conditions so extreme are never met with in man, though a seasonal variation in birth weight has been reported by Toverud; Abels, and others; infants born during late summer and early autumn being significantly heavier than those born during other months.* Such slight changes will always be difficult to establish and hard to interpret. Sontag, Pyle & Cape could not observe any differences in birth weights according to the dietary level of the mother, and German authors in general regard the birth weight as not being influenced by the maternal diet, even during war famines. Nevertheless, Gerschenson, basing his view on evidence collected in South Russia, where about 1917 the war famines were much worse than in Germany, has shown that very severe hunger conditions do reduce the birth weight in man. In general, mammalian development may be regarded as an all-or-none phenomenon provided the limits of variation of the maternal diet are small; if they exceed a certain limit, then the regulating mechanisms break down and such embryos as are born will be markedly abnormal.†

Very few experiments to raise artificially some foetal constituent have been done. From the work of Dann[1], however, we know that a great excess of dietary vitamin A, though it may enrich the maternal liver of the rat, cannot increase the store in the foetus. Only after birth can it pass through to the young animal in abnormally large quantity via the milk.‡

For human pregnancy there have been a number of studies dealing with the utilisation of dietary constituents by the foetus; it will be sufficient to refer here to those of the Michigan school (Macy & Hunscher on nitrogen and minerals; Macy, Hunscher, Nims & McCosh on calcium; Hunscher, Donelson, Nims, Kenyon & Macy on nitrogen storage; Hummel, Hunscher, Bates, Bonner & Macy on metabolic balance; and Johnston, Hunscher, Hummel, Bates, Bonner & Macy on basal metabolic rate), through which the literature can be reached, as also through the descriptive bibliography of L. Zuntz[4].

* Cf. the mysterious seasonal variation in twin births established by Cole & Rodolfo in cattle, where there is a maximum in August and a minimum in March.

† It is interesting that according to Enzmann, Saphir & Pincus the marked delay in pregnancy caused by lactation does no harm to the blastocysts whose implantation into the uterine wall is delayed. If the delay lasts longer than a certain time, however, the blastocysts degenerate and never become implanted.

‡ The importance of each vitamin for the maintenance of foetal development has been dealt with in a useful review by K. E. Mason[2].

Lastly, it has been conclusively shown that in multiparous animals there is a definite relation between litter size and birth weight (Wishart & Hammond; Hammond[2]; Watt; Crozier & Enzmann; Enzmann & Crozier). There seem to be certain strict limitations on the amount of material which the maternal organism in a given species can devote to embryonic growth and therefore the larger the number of embryos in the litter the smaller each one of them is.

1·9. The mammalian placenta

The tendency to retain the egg and embryo for protection during its development within the maternal body, and to furnish it with extra nourishment, whether of water, salts or organic substance, not provided in the yolk, comes to its climax in the mammalian placenta. In morphological complexity and physiological interest this organ is unique, and it can be no matter for surprise that it has given

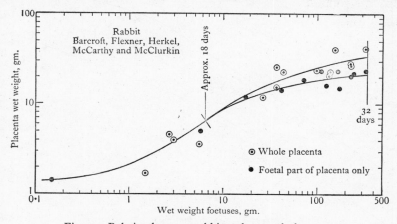

Fig. 33. Relation between rabbit embryo and placenta.

birth to a vast literature, partly, of course, with a medical bearing, but also largely the product of a genuine biological outlook. Since the placenta is of mainly nutritional significance and has relatively little to do with morphogenesis, we cannot allow it much space in the present book, but by referring to reviews and to those papers which themselves contain summaries of previous work, access to the literature may be facilitated for those specially interested in placental problems.

As we have already seen (p. 67), the yolk-sac and other membranes in the eggs of other amniotes, such as the chick, at first outstrip in their growth the embryo itself. The same process takes place in mammals. Fig. 33, constructed from the data of Barcroft, Flexner, Herkel, McCarthy & McClurkin, shows the relation between the embryo and the foetal and maternal parts of the placenta in the rabbit; it is directly comparable with Fig. 26, for it is plotted on a double logarithmic scale. It will be seen that during the first half of development the placenta weighs much more than the foetus, but that about the 18th day, at a weight of 6½ gm. each, the foetus overhauls the placenta and the curve then flattens out as the embryo continues to grow much faster than its scaffolding.

A similar set of data has been published for the rabbit by Hammond[3]. Hammond's[1] figures for the cow are shown in Fig. 34. The relationships are the same—equality in weight between foetus and placenta is reached at a weight of about 300 gm., after which the curve flattens, but equality in weight between the foetus and the foetal liquids (amniotic and allantoic) is not reached until a much later stage in development, at a weight of some 5 kg. The significance of the deviating last point for the placenta is not clear. Similar graphs could be made for all mammals, see e.g. the data on the merino sheep (Cloete; Malan & Curson).

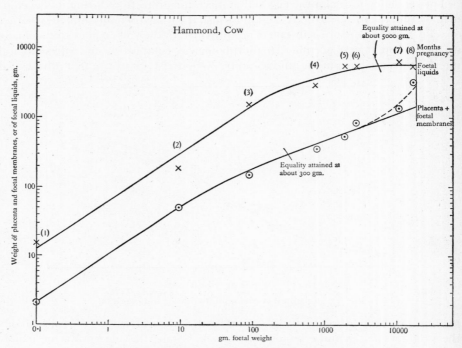

Fig. 34. Relation between cow embryo and placenta.

There is a marked positive correlation between foetal weight and placenta weight at term in the rabbit (Rosahn & Greene; Hammond[3]), in man (de Aberle, Morse, Thompson & Pitney) and probably most mammals.

Although the placenta in many animals gives the impression of being a new organ, it is after all no more than the tissue formed at the points where the two blood circulations of embryo and maternal organism come into contact. Since its significance is almost wholly nutritional, it is logical that the most satisfactory classification of placentas should be a physiological one, and this admirable work was accomplished by Grosser[1,3] in his distinction between the number of membranes or cell-layers which intervene between the two circulations. The following table (simplified from CE, Table 223) illustrates his scheme, somewhat modified by later writers.

Table 13. *Placenta types*

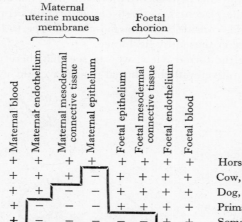

Type of placenta	Maternal blood	Maternal endothelium	Maternal mesodermal connective tissue	Maternal epithelium	Foetal epithelium	Foetal mesodermal connective tissue	Foetal endothelium	Foetal blood	Examples
Epithelio-chorial*	+	+	+	+	+	+	+	+	Horse, pig, donkey
Syndesmo-chorial	+	+	+	−	+	+	+	+	Cow, sheep, goat
Endothelio-chorial	+	+	−	−	+	+	+	+	Dog, cat
Haemo-chorial†	+	−	−	−	+	+	+	+	Primates, some rodents
Haemo-endothelial	+	−	−	−	−	−	+	+	Some rodents

* Recent elaborate description by Brambel.
† Recent elaborate description by Spanner and Kearns; tissue culture studies by Sengupta and Nagayama.

We have to deal, therefore, with a progressive breaking down of the barriers between the foetal and the maternal circulations. It may be considered as established that the condition of the barrier, whether substantial, as in the epithelio-chorial placentas, or very thin, as in the haemo-chorial and haemo-endothelial ones, may be correlated with (1) the transmissibility of substances of various molecular weights from one circulation to another (*CE*, p. 1505), (2) the presence or absence of a large allantois as the receptacle for foetal excretions (*CE*, p. 1495), (3) the relative significance of the uterine "milk"* (*CE*, p. 1492), (4) the relative importance of the colostrum as the vehicle for substances of importance which may not have been supplied sufficiently readily by the placental path (*CE*, p. 1497). In this connection it should be remembered that, as Mossman points out, the placental type may change during the development of the embryo, beginning with the looser connection and going on to a more intimate one. An extreme example is the case of the rabbit, where at first the placenta is haemo-chorial in type, but later becomes "haemo-mesothelial" and later still haemo-endothelial. We have already had occasion to note in the work of Rodolfo the significance which these changes may have in placental function (p. 31).

The question of the evolution of the placenta in mammals is a much disputed one. Benazzi, and Mossman, in his splendid monograph on the comparative

* This secretion from the uterine glands, the "histotrophe", is copious in those animals with epithelio-chorial placentas and scarce in those with haemo-chorial ones. Brambel and Hamlett[2] have well shown how it accounts for the "hippomanes"; packets of uterine secretion surrounded by trophoblast and invaginated into the allantoic cavity. There was a good deal of confusion about these in antiquity as a glance at Aristotle shows (*Hist. Anim.* 572 a 21, 577 a 10, 605 a 2), and like the uterine milk itself they were first properly described by Walter Needham in 1667. Their function, if any, is still obscure.

morphogenesis of the foetal membranes, regard the epithelio-chorial type as the most primitive, and this is in accordance with what has already been said in this book regarding the placentas of dogfishes and reptiles (pp. 43, 61). Portmann[4], on the other hand, considers the haemo-chorial type to be the most primitive; his arguments are criticised by de Lange.

Adopting here the former view, we can see that there were several directions in which the primitive epithelio-chorial placenta could evolve (see Fig. 35). It

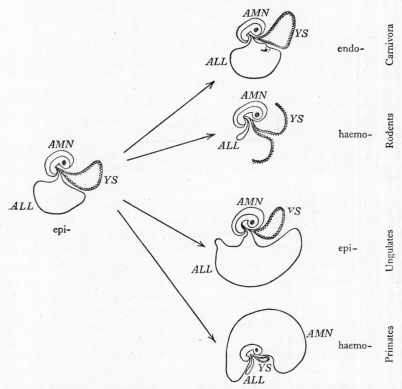

Fig. 35. Sketch of the origin of the several types of placenta. *ALL*, allantois; *AMN*, amnion; *YS*, yolk-sac; after each hyphen read "-chorial".

probably had a large yolk-sac and large allantois, with chorio-allantoic attachment to the uterine wall. Retention of these primitive characters with disappearance of maternal epithelium and connective tissue led to the modern carnivora. But now the allantois could be greatly reduced in size or, on the other hand, greatly enlarged. In association with the former tendency there went an opening up of the yolk-sac so that by the disappearance of its wall most distal to the embryo, the inner endodermal layers (with their ancient yolk-absorbing villi, cf. Fig. 28) could be apposed to the maternal uterine wall.* This process led to the haemo-chorial placentas of the insectivora and rodents. On the other hand, an excessive enlarge-

* This is the famous "inversion of the germ-layers".

ment of the allantois meant contact with the maternal tissues over a greater area and hence the avoidance of the necessity for a very close juxtaposition of the bloods. This led to the modern epithelio-chorial and syndesmo-chorial placentas of the ungulates and all their related groups. But there was also a fourth possibility, namely to reduce drastically the size of both yolk-sac and allantois and to establish over the chorio-amniotic area a close haemo-chorial connection. This course led to the primates and man (see Fig. 36).

The contact area between maternal and foetal circulations in man is very great; Christoffersen has recently assessed it at between 10 and 13 sq. metres, considerably larger than was formerly thought. The volume of foetal blood increases in proportion to the combined weight of foetus and placenta (Elliott, Hall & Huggett) and the relative amounts of it in foetus and placenta have been studied by

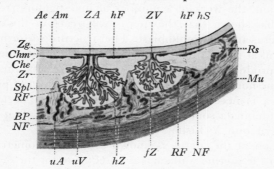

Fig. 36. Structure of the human placenta (from Fischel).

Ae, amniotic epithelium.
Am, amniotic mesoderm.
BP, basal plate.
Che, chorionic epithelium.
Chm, chorionic mesoderm.
fZ, free villus.
hF, hypochorial fibrinoid striae.
hS, hypochorial rim of uterine tissue.
hZ, anchored villi.
Mu, uterine muscle.
NF, Nitabuch's fibrinoid striae.

RF, Rohr's fibrinoid striae.
Rs, peripheral blood sinus.
Spl, septum placentae.
uA, uterine arterioles.
uV, uterine venules.
Zg, loose mesodermal connective tissue between amnion and chorion.
ZA, foetal arteriole.
ZV, foetal venule.
Zr, intervillous blood sinus.

Barcroft & Kennedy on the sheep. At 100 days about half the total blood volume is in the placenta, at 120 days about a third and at 140 days about a fifth. This is due to the increasing total blood volume and foetal size; there is no diminution of the blood in the placenta. Some of the blood in the placenta at the end is saved for the foetus (Barcroft & Gotseff). Grosser[2] has drawn attention to the stagnation of the maternal blood in a villous haemo-chorial placenta such as that of man, and to the contractions which squeeze the blood out of it like a sponge. But as these may occur only once an hour, there is plenty of time for the action of placental enzymes on substances coming to the placenta from the maternal blood. In labyrinthine epithelio-chorial placentas no stagnation of this sort occurs. The mean volume of blood in the uterine vessels rises greatly during development (from 2 to 32 c.c. in rabbits) parallel with the weight of the placenta, not the embryo (Barcroft & Rothschild; Barcroft, Herkel & Hill). During pregnancy the spleen contracts, presumably to supply blood to the uterus (Barcroft & Stevens); this contraction is abolished by denervation (Barcroft).

Contrary to the general belief that the chorio-allantoic and chorio-amniotic regions are the only important seats of exchange in mammals, Everett found that vital dyes accumulate in the highly inverted yolk-sac placenta of the rat from the maternal circulation. He made the interesting incidental observation that on the 12th day of development toluidin blue is instantly reduced on entering the yolk-sac placenta.

Until recent times practically nothing was known of the experimental morphology of the placenta. But starting from the pioneer experiments in which L. Loeb[1] in 1907 was able to produce the maternal part of the placenta in the guinea-pig by applying mechanical stimuli to the uterus in the presence of a favourable endocrine situation, the work of a number of investigators (reviewed by Selye & McKeown[2]) has shown that the uterus will react in this way without the presence of a foetus and its trophoblast. The structure so formed is known as a "deciduoma" and at its maximal size will fill the whole uterine lumen. A deciduoma can be produced during certain periods of lactation, but not during pregnancy. In the normal sexual cycle the active secretion of progesterone from the corpora lutea is necessary for the production of a deciduoma; oestrone inhibits it. Deciduomata sometimes occur spontaneously, e.g. in animals on a diet deficient in vitamin A (Bishop & Morgan).

Another reaction of the uterus is known as the "endometrial mole". Mild traumatisation of the uterine wall such as is requisite for the deciduoma leads to no specific reaction in the oestrone-treated ovariectomised rat, but if the uterine wall is slit, a tumour appears structurally resembling the hydatidiform mole occasionally spontaneously produced by the foetal part of the placenta (Selye, Harlow & McKeown; Selye & Friedman). Oestrone is necessary for this reaction. Hydatidiform moles contain large amounts of oestrone (Philipp), and these uterine tumour-like structures remind us of the connection between the sex-hormones and the carcinogens, to be referred to below (p. 253), especially as regards mammary cancer.

The general upshot of these findings, therefore, is that we must suppose the stimulus of the foetal trophoblast not to be very specific. Given the favourable endocrine situation, the uterus will respond even to mechanical stimuli. Competence (see p. 112) is the variable factor. It is interesting that in the later stages of a deciduoma there is even a breaking down of the endothelium of the maternal blood-vessels, so that a blood sinus similar to that of the inter-villous space of the normal placenta is formed.

If we trace back the placenta to its first origins, we come, of course, to the formation of the foetal trophoblast by the expansion of the blastocyst. The valuable investigations of Pincus and his collaborators on the expansion of the blastocyst* (review by Pincus[2]) have shown that this crucially important stage in mammalian development is under the control of a hormone secreted by the

---

* I am unable to understand why these authors persistently term this phenomenon "ovum growth"; a practice which calls to mind Napoleon's comment on the Holy Roman Empire.

maternal organism, progesterone (see p. 252). The rabbit embryo, for example, expands from 0·12 mm. diameter in the Fallopian tube to nearly 8·0 mm. diameter 6½ days later at the time of implantation. Corner[2] first found that in the uteri of ovariectomised rabbits this expansion will not occur, and Pincus & Werthessen[1] observed that when eggs are cultured *in vitro* it will not occur either, unless progesterone and glutathione are both present. These substances prevent the collapse or herniation of the blastocyst and render its expansion nearly normal. The oestrogens show both stimulatory and inhibitory effects, for which Pincus & Werthessen[2] have advanced explanations.

1·91. Placental constitution

Practically every constituent substance for which search has been made in the placenta has been found there—carbohydrates, proteins, fats, enzymes,* vitamins,† hormones,‡ etc. (cf. *CE*, Section 20). Yet we are still a long way from having a complete account, even for the placenta of one single mammalian species, of the way in which its chemical constitution varies with the development of the foetus.

Perhaps the most classical function of the placenta in this sense is its action as a store of glycogen before the embryonic liver is in a position to take on its glycogenic function. This was known already in the middle of the last century through the work of Claude Bernard (see *CE*, pp. 1019 ff.). It has recently been elaborated by Szendi[3,4], who brings forward evidence to show that there are, during the development of both rabbit and man, a succession of peaks of glycogen-content, first in the maternal part of the placenta, then in the foetal part, then in the foetal lung and lastly in the foetal liver. Szendi & Papp have also found that glycogen is richly contained in the histotrophe (the secretion of the uterine glands). That the placenta behaves much more like liver than muscle was well demonstrated by the experiments of Davy & Huggett, in which human placenta, on autolysis, was found to produce practically no lactic acid but only glucose. In some animals, such as the rat, the peak of glycogen storage must come quite early in development, judging by the results of Corey[3] and Eufinger. In accordance with earlier work, the placental glycogen seems to be insensitive to endocrine control, for Corey[4] found that adrenalin and insulin affected the glycogen-content of both maternal and foetal liver, but not of the placenta.§

As regards the protein constitution of the placenta, there is little to add to former accounts. Drage & Sandstrom, however, have shown that the amino-acid distribution in human placenta globulin is essentially the same as that in human serum globulin, and Graff & Graff have disposed of the old story that placenta proteins were unusually rich in arginine.‖

* E.g. deaminase, human (Abe[2]); kathepsin, human (Abe[2]); catalase, human (Bidone); nuclease, human (Edlbacher & Kutscher); urease, human (Abe[2]); amylase, maltase, invertase, lipase, kathepsin, rabbit (Abe[2]). Other enzymes will be referred to below, p. 89.

† E.g. vitamin A, human (Gaehtgens; Dann[2]; Goldhammer & Kuen); vitamin C, human, histochemically (Tonutti & Plate), chemically (Neuweiler[1,2]).

‡ E.g. thyroxin, human (da Re) and see below.

§ There is also histochemical work on glycogen in the placenta by Silvestri (rabbit) and Melandri and Szendi[1] (man).

‖ See also Laufberger's work on amino-acids of the human placenta, and Tenney's on its collagen (histol.).

Apart from histochemical studies of fats and lipoids in the placenta* there has been some valuable work. It is agreed (Boyd & Wilson; Effkeman) that the conveyance of fatty acids to the foetus is almost wholly in phosphatide form. Towards the end of development in man, the daily utilisation of the foetus from the blood is 35–40 gm. phosphatides, about 7 gm. each of free cholesterol and cholesterol esters, and only 3 gm. of neutral fat. Starting from this point, Boyd[1] argued that if the placenta was acting purely as an organ of transmission its constitution should not change as the constitution of the embryo changed, but that if it were absorbing lipoidal materials and then secreting them in suitable proportions there should be a variation of this kind. The results of a careful study showed that the lipoidal constitution of the rabbit placenta does indeed change considerably, but not always parallel with the constitution of the foetus. We may take the various fractions in order. In percentage phosphatides there is a steady rise in the foetal body, from about 100 to 1000 as embryo wet weight rises from 1 to 51 gm., but in the placenta there is a marked fall, from 1600 to 1100, followed by a rise parallel with that of the foetus, ending at 1500 mgm. %. For free cholesterol the picture was simpler, foetus rising from 50 to 150 mg. % and placenta from 200 to 300 mgm. %,[†] but for cholesterol esters the picture was different. The foetal level remained constant at 25 mgm. %, but that of the placenta rose from 150 to a maximum at 400 and fell again to reach 150 at the end of development. This peak corresponded approximately in time to the fall in phosphatides already referred to, but it is by no means clear why at a certain time there should be a special tendency for fatty acids to be esterified by cholesterol. As for neutral fat, placenta and foetus behaved quite differently. The content of fat in the foetus rose rather suddenly towards the end of development from 200 to 1200 mgm. %, while the placenta showed a vaguely defined fall.[‡] This interesting work is the only attempt so far made to correlate the constitution of placenta and embryo. It suggests that the placenta, though for many substances clearly acting purely as an ultra-filter, may in the case of certain water-insoluble substances have an important secretory part to play.

Arising out of the reference to phosphatides, it is of interest that choline, perhaps derived from them, and probably mainly in the form of acetylcholine. is contained in large amounts in the human placenta (Strack & Loeschke; Bischoff, Grab & Kapfhammer; Strack, Geissendorfer & Neubaur; Strack & Geissendorfer; Reynolds & Foster; Hauptstein). Indeed, according to Chang & Gaddum, human placenta is the richest source of choline in the body (28γ/gm.) with the exception of spleen. From the work of the Chinese school (H. C. Chang & A. Wong; Chang; Wong & Chang; Wen, Chang & Wong; Chang, Wen & Wong) we know that acetylcholine is derived wholly from the villous tissue of the foetal trophoblast and that the uterine muscle is certainly sensitive to it. That it plays a part in the muscular contractions of birth is probable enough, and acetylcholine may indeed be used as a stimulant to such contractions, which, unlike those caused by pituitrin, are rhythmical. In placentas from Caesarean sections the acetylcholine content is higher than normal. There appears to be a bound form of acetylcholine in the placenta.

As there seem to be only very small amounts of acetylcholine in the placentas of rodents, ungulates and carnivores, it seems likely that this mechanism has developed in special association with the villous haemo-chorial placenta, and it

---

* E.g. Piccardo; Yamaguchi: see also fragmentary lipoid analyses by Corda; Jowett; and Klaus.

† This agrees with earlier work of J. Suzuki on the same animal.

‡ This contradicts earlier work of J. Suzuki, who found a rise, but the scatter is very large.

would therefore be of much interest to know whether primates other than man show it. In primitive primates with a labyrinthine placenta it should be absent. The placentas of rodents, carnivores and ungulates, however, contain choline.

Of course, it is generally held that much more responsibility in the birth process is to be ascribed to endocrine factors (such as the oxytocic hormone of the posterior pituitary) and to the increasing sensitivity (competence) of the uterine muscles. Apart from reviews on this subject (Gibbons; Tapfer; Tapfer & Haslhofer), more or less illuminating, there is much interesting work by the Scottish group, which will be found in the papers of Bell & Robson; Robson; Robson & Bell; Robson & Schild, etc. but which cannot be described here—see also the papers of Allan & Wiles, Heller & Holtz, and Witherspoon. One of the most interesting possibilities which arises out of this is that the foetus itself may contribute to the initiation of birth, since in pigs and sheep, at any rate, the posterior pituitary at the end of development contains a considerable amount of the hormone. The quantity in the combined foetal glands at least equals that in the maternal gland. But what sensitises the uterine muscle to the hormone is as yet unknown.

As Snyder[1] and many other workers have shown, birth may be inhibited for a considerable time if new corpora lutea are formed during the gestation period. For example, if a rabbit is made to ovulate by a single injection of pregnancy urine extract on the 25th day, birth will fail to occur on the 32nd day as it should do, and will only happen on the 40th day, i.e. at the time when the induced set of corpora lutea degenerate. Under such conditions the uterus, though at full term, will not respond to pituitrin. Such experiments decisively contradict some old-established views, connecting birth with changes in the foetus, senility of the placenta, mechanical distention of the uterus, and the like. In such experiments it is also found that the life of the foetuses can continue until about the 36th day, but not longer; this fixes the reserve capacity of the placenta for supporting life.

Although the corpus luteum is essential for the formation of the placenta in all mammals, its continuing presence throughout pregnancy is only necessary in some (e.g. rabbit, mouse, opossum; not in guinea-pig, cat, man, horse and donkey). The same is true of the pituitary, but only because of its influence on the ovary. Although the sexual cycle is in general obliterated during pregnancy in all mammals, traces of it may persist, and it appears that the periods of gestation are nearly always simple multiples of the sexual period itself.

The question of what hormones exist in the placenta and whether the placenta itself has endocrine functions has attracted much attention. The extensive literature may be approached by way of several excellent reviews, such as those of Clauberg; Collip, Thomson, Browne, McPhail & Williamson; Selye, Collip & Thomson; Newton[2]; Corner[3]; and Snyder[2]. "It is now more or less taken for granted", writes Newton[2], "that the placenta is an endocrine organ, but its function, except in the most general terms, is not understood.... There is a plethora of data on the profound changes which take place during pregnancy, pointing more and more to the placenta as a focus, but the unifying conception which would make everything fall into place is elusive."

The presence of oestrogenic material in the placenta was adumbrated before the war (1913) by Aschner and others, but not proved (because the vaginal smear test was then not known) until 1924 by Doisy, Ralls, Allen & Johnston. Crystalline oestriol (see p. 254) has been isolated from the placenta, and the water-soluble oestrogenic principle discovered by the Montreal school ("emmenin") has turned out to be oestriol glucuronide. In actual amount the placenta contains at least as much and probably more oestrogenic material per gram than the ovary itself.* that the placenta actually secretes oestrogenic hormone depends on the interpretation of experiments in which the excretion of the hormone continues and the secondary organs do not atrophy after the removal of the ovaries from the pregnant organism. For detailed discussion of these points the reviews of Newton[2] and Corner[3] must be consulted; Newton points out that the possibility of a contribution from the foetal gonads should not be overlooked. Nevertheless, there is strong presumptive evidence that the placenta does secrete oestrogenic hormone.

Progesterone, the corpus luteum hormone (see p. 252), has also been repeatedly demonstrated in the human placenta. That it is formed by the placenta appears quite probable. Progesterone seems to be excreted as pregnandiol glucuronate, and it is significant that this substance increases to its maximum excretion towards the end of pregnancy when the placenta is largest.

Our knowledge of the presence of hormones identical with or similar to those of the anterior pituitary in the placenta also begins with Aschner, who in 1913 obtained ripening of Graafian follicles after injection of placenta extracts into young test animals. Since then the presence of something very like the gonadotrophic hormone of the pituitary in the uterus has been firmly established, and there are many observations, detailed in Newton's review, which indicate that the placenta actually produces it (see also Philipp & Huber). Among the more interesting of these are the experiments of Kido[2], who was able to cultivate human placental tissue in the eye-cavity of the rabbit, and found later that the gonadotrophic hormone was circulating in the blood and excreted in the urine. Nagayama and Gey, Seegar & Hellman also claim to have demonstrated the synthesis of both oestrogen and gonadotrophic hormone by placental tissue growing *in vitro*. On the other hand, there are a number of indications that the gonadotrophic substance of the placenta is not exactly the same as that liberated by the pituitary; and it seems to be unable, even in the largest doses, fully to replace the hormone of an animal's own pituitary. This is often thought to be due to the placenta containing only the luteinising and not both the luteinising and the follicle-stimulating components of the gonadotrophic hormone, but the real state of affairs is probably not so simple. Glebova has found the pituitary "ketogenic hormone" in the human placenta.

Finally, androgenic substances have also been found in the human placenta (Goecke, Wirz & Daners), and extracts of it are said to accelerate amphibian metamorphosis (Ganfini).

In endocrinological complexity, therefore, the placenta seems to outdo even the pituitary, and it will probably be a long time before we have a satisfactory understanding of its many functions. It will in particular be interesting to trace down the production of the various hormones from the evolutionary point of view in order to gain some idea of how the extraordinary complexity of the chorio-amniotic villous haemo-chorial placenta of man arose. A good beginning in this direction has been made by Glebova.

Among the other substances present in the placenta† the pigments are notable, but discussion of them will be postponed till later (see p. 647).

---

* Though Parker & Tenney show that the total oestrogen in both foetus and maternal organism exceeds that in the placenta.

† [Descriptive bibliography by L. Zuntz[1].] Glutathione, human (Macciotta); total ash, rabbit (Furuhashi); iron, human, histochemically (Mayeda; Šesler; Roussel & Deflandre[1]), chemically (Hilgenberg). The massive lime deposits in the placentas of ungulates (Höpping) are noteworthy.

Like the yolk-sac of the chick (p. 70), the mammalian placenta may continue alive for some time after the death of the embryo to which it belongs. In the mouse (Newton[1]), and the monkey (G. van Wagenen & Newton) the placentas of foetuses dead or removed are delivered later on at the end of the normal gestation period. Placenta alone is thus quite able to suppress the oestrous cycle. Brooksby & Newton have also shown that the placenta exercises an influence on water retention by the maternal organism, there being a sharp drop in weight when the placentas are delivered. This is not without interest in view of what has already been said about the water relations of terrestrial animals (p. 33), and the loss of amniotic liquid at birth by most mammals is small.

1·92. Placental metabolism

Much less attention has been paid to the metabolism of the placenta itself than to the hormones in it or to the transmission of substances through it. In continuation of the observations discussed in CE, p. 1461, Loeser has made manometric estimations of the metabolism of the human placenta, with the following average results:

	Q_{O_2}	$Q_L^{N_2}$	$Q_L^{O_2}$
1st month	—	+17·6	—
2nd month	−5·0	+ 7·0	+5·1
At term	−1·4	+ 3·5	+0·57

Evidently, like all other ageing tissues, the placenta suffers a decline of metabolic activity, but it is interesting, in view of the invasive activities of the trophoblast, so often suspected as resembling malignant tissue, that there should be an aerobic glycolysis of appreciable extent at the earlier stages. Substantially the same figures were obtained for rabbit and cat placenta. The effects of glucose, adrenalin, anaemia, etc. on these levels have been studied by Matsuhita, but not with very striking results.

The enzymic equipment of the placenta is not at first sight interesting, since practically every enzyme which has been looked for in the placenta has been found there (review by Mayer & Seitz, as well as CE, pp. 1481 ff.). The recent trend in this work is the study of the dehydrogenases, which have been found for succinate (Thunberg; Tesauro), fumarate (da Cunha & Jacobsohn) and glycerophosphate, lactate, citrate, glycerate, glutamate, hexose-diphosphate, etc. (Thunberg). Weak phosphatases attacking diphenyl-pyrophosphate (da Cunha) and phosphoglycerate (Antoniani & Clerici) are also present. Particularly interesting is the fact established by Busse that the phosphatase of placenta attacking glycerophosphate is more active than that of any adult organ and increases in strength during foetal development. Correspondingly, Busse could never get an esterification of phosphate in the placenta under any conditions. He also found that the zymohexase of placenta (the enzyme converting hexose-diphosphate to triosephosphate) was extraordinarily weak, placenta having only $\frac{1}{1000}$th the activity of muscle. Yet when Kutscher, Veith & Sarreither examined the lactic acid production of placenta and muscle under identical conditions, that of the former was found to have only ⅕th the activity of the latter.* The Pasteur reaction is not very marked in placenta.

An active deaminase has been found in placenta by Botella-Llusia[2]. As placenta produces urea from ammonium carbonate during autolysis (Botella-Llusia[1]), we may surmise that the ornithine cycle is functional there, though this has not as yet been directly proved.

1·93. Placental transmission

From the former discussion (*CE*, Section 21) four main points emerged. (1) There is a relationship between placental transmission and molecular size, smaller molecules penetrating through the placental barrier more readily than larger ones, as through an ultra-filter. (2) In this respect the placentas of different mammals differ, mainly in accordance with the number of cell-layers interposed between the foetal and maternal bloods. Probably there are also considerable differences between the behaviour of the placentas of the same mammalian species at different times in development. (3) Transient increases in the concentration of a diffusible substance on one side of the barrier lead to corresponding increases on the other side. (4) In spite of the fact that the placental barrier behaves like an ultra-filter, there are persistent differences in the levels of the various constituents of maternal and foetal bloods which it is at present impossible to account for either by the Donnan, or any other similar theory, but which in all probability have their explanation in physical mechanisms of the kind postulated by such theories. All the work which has been done since 1931 bears out these statements.

On the whole subject of placental transmission there are valuable reviews by Schlossmann[3]; Runge; and F. J. Koch, besides the descriptive bibliography of L. Zuntz[3]. Taking first the properties of the placental barrier as an ultra-filter, the old conclusion that discrete particles never pass through unless, like pathogenic bacteria, they make lesions in the placenta, has been further substantiated (Nakagawa; Szendi[2]). Non-pathogenic bacteria never penetrate; colloidal particles of intermediate size may or may not. Some phages penetrate (Nakagawa) while others cannot do so (Natan-Larrier, Eliava & Richard). As has already been noted (p. 30), antigenic proteins in general do not go through but antibodies do, hence the passive immunity of the foetus. A good example of this is lactalbumen and its antibody (guinea-pig; Natan-Larrier & Grimard-Richard[1]). There are exceptions to this, however, e.g. ovoalbumen, which, according to Natan-Larrier & Richard[1], can pass through the guinea-pig placenta. There is a curious effect of soaps discovered by these investigators; horse-serum proteins will penetrate the guinea-pig placenta, normally impermeable to them, if a quantity of sodium oleate, which may be very small, is injected first (Natan-Larrier & Richard; Natan-Larrier, Noyer & Richard). The same effect was obtained by Natan-Larrier & Grimard-Richard[2] with bile salts and ovo-albumen, but no direct relation could be found between the fall in surface tension of the blood caused by these substances, and their effect on the placental permeability.

* A non-phosphorylating path for carbohydrate breakdown in placenta is therefore indicated (see p. 610).

Of greater interest than the numerous positive or negative observations on placental permeability to one or another substance* are the experiments which show progressive changes in permeability during foetal development (cf. the work of Rodolfo, already mentioned, p. 31). Thus Lell, Liber & Snyder, using phenolsulphonephthalein, found that the rabbit placenta was much more permeable in the foetal → maternal direction at the 22nd than at the 29th day, but conversely the leakage into the maternal circulation from the amniotic liquid was much greater a t the 30th than at the 21st. Similar changes have been found by a number of investigators, e.g. Snyder & Speert showed that the rate of transmission of neoarsphenamine from maternal to foetal circulation in the rabbit greatly increases during development. The As content of the foetus 24 hr. after the injection rises from less than 1 mgm. AsO_3 at the 19th day to 20 mgm. at term. In the rat, too, according to Corey[1], placental permeability to insulin is greatly increased towards the end of the developmental period. The placentas of rabbit, guinea-pig and dog, according to Birguer & Afanasiev, become progressively less permeable to salts of bromine and iodine as development goes on. There is here a wide field for future work.

Passing now to the concentration differences present in the two blood streams,† it may in general be said that the maternal blood is richer in total solids, proteins, total P, lipoid P, phosphatides, neutral fat, cholesterol, and glucose; while the foetal blood is richer in non-protein N, free amino-N, inorganic P and calcium.

---

* The following data, additional to those in *CE*, Section 21, may be found useful:

(1) *Man:* amino-acids + (Bickenbach & Rupp); dipeptides + (Bickenbach & Rupp); glucose + (Abe[3]); phenoisulphonephthalein + (Albano); citrate + (Lenner); oestrone − (Levy-Solal, Walther & Dalsace); barbiturates + (Berutti); lead + (Baumann); arsenic − (Dejust & Vignes; Farès); arsphenamine + (Eastman & Dippel); various metals and metalloid elements (Farès).

(2) *Dog:* amino-acids + (Schlossmann[2]); insulin + (Schlossmann[4]); barbiturates + (Fabre); quinine + (Regnier); parathyroid hormone − (Hoskins & Snyder); thyreotrophic hormone of pituitary + (Döderlein).

(3) *Rat:* glucose + (Corey[1]); thyroxin and thyrotrophic hormone of pituitary + (Thérèsa); insulin + (Corey[1]); adrenalin + (Corey[4]; Ingle & Fisher); androgens and oestrone + (Hain); phenolsulphonephthalein + (Boucek & Renton); fluorine + (M. M. Murray); colloidal thorium dioxide − (Menville & Ané); radioactive sodium + (Flexner & Pohl; Flexner & Roberts).

(4) *Mouse:* bismuth and antimony − (Lebedeva).

(5) *Rabbit:* glucose, pentoses + (Brandstrup[1]); sucrose, lactose − (Brandstrup[1]); amino-acids + (Brandstrup[2]); oestrone + (Skowron & Skarzyński); adrenalin, posterior pituitary, acetylcholine histamine + (Cattaneo[3]; Falaschino); nicotine + (Sergueev).

(6) *Goat:* glucose + (Passmore & Schlossmann); adrenalin + (Schlossmann[1]).

† The following succinct summary of the most recent quantitative work may be of interest; f ≡ the concentration in the foetal blood, m ≡ that in the maternal blood, c ≡ confirmation of earlier work summarised in *CE*, Table 243, nc ≡ contradiction with it. There are some general figures in the review of de Toni.

(1) *Human* (nearly always at term): total solids, m > f (c) (Levy-Solal, Dalsace & Gutman; Gutman & Levy-Solal; Naeslund[1]); total protein N, m > f (c) (Levy-Solal, Dalsace & Gutman; Gutman & Levy-Solal; Naeslund[1]; Pommerenke; Puccioni, de Niederhausen & Roncallo); fibrinogen, m > f (Naeslund[2]); non-protein N, m < f (c) (Naeslund[1]; Pommerenke); amino-acid N, m = f (nc) (Doneddu), m < f (c) (Naeslund[1]; Pommerenke; Legrand); polypeptides, m < f (Legrand); glucose, m > f (c) (Levy-Solal, Dalsace & Gutman; Gutman & Levy-Solal); citrate, m < f (Lenner); choline, m > f (Späth); vitamin A, m > f (Gaehtgens); vitamin C, m < f (Wahren & Rundquist; Manahan & Eastman); neutral fat, m > f (c) (Levy-Solal, Dalsace & Gutman; Gutman & Levy-Solal); cholesterol, m > f (c) (Rosenbloom); carotin, m > f (Gaehtgens); creatinine, m = f (c) (Naeslund[1]); uric acid, m = f (c) (Naeslund[1]); lactic acid, m = f (nc) (Eastman & McLane); glutathione, m < f (Anselmino & Hoffman[1]; Lemeland & Deletang; Buzzi; Oberst & Woods; Sala); fumaric dehydrogenase, m > f (da Cunha & Jacobsohn); choline esterase, m > f (Navratil); catalase, m < f (Anselmino & Hoffman[1]); ash, m < f (Levy-Solal, Dalsace & Gutman; Gutman & Levy-Solal); total base (Thompson & Pommerenke); calcium, m < f (c) (Garofalo; Mull; Otte); alkali reserve, m > f (Puccioni, de Niederhausen & Roncallo); total P, m < f (Mull); inorganic P.

There are certain exceptions to this; for example, in ungulates (epithelio-chorial placentas) the blood-sugar is higher in the foetal than in the maternal circulation. The concentrations of urea, uric acid, creatine, creatinine, chloride, etc. are usually the same. The differences of concentration are not the only differences between the bloods. Thus the foetal blood in man is richer in haemoglobin (Börner) and in red blood corpuscles (Lucas & Dearing; Seckel), a fact which probably accounts for the higher blood glutathione on the foetal side (Lemeland & Deletang; Oberst & Woods). The surface tension of the foetal blood is considerably higher than the maternal (Kisch & Remertz) and the sedimentation rate of red blood corpuscles differs (Bogaert; Puccioni, de Niederhausen & Roncallo), being ten times faster on the maternal side. This phenomenon is due to the plasma, for Bogaert found that maternal red blood corpuscles in foetal plasma are slowed down, while foetal red blood corpuscles in maternal plasma are accelerated to the maternal rate. The colloidal osmotic pressure of foetal and maternal sera in the sheep differs (McCarthy), for the foetal serum has more albumen and exerts a 50 % higher osmotic pressure per gram protein. The cause of this difference is not understood. The plasma protein continues to rise before and after birth (Clark[1]; Clark & Holling; Swanson & Smith) more or less parallel with the blood pressure. The very low fibrinogen content of foetal blood explains the occasional occurrence of temporary *haemophilia neonatorum* (Naeslund[2]).

There can be little doubt that an explanation of the same type as the Donnan equilibrium, but perhaps more complicated, will eventually be found to hold good for the placental barrier.* Luck & Ritter, in suggestive experiments, found that if a cellophane membrane is set up with a protein solution on one side and a solution of amino-acids on the other, neither osmotic equilibrium nor Donnan equilibrium is ever attained, even at the isoelectric point of the protein.† It is nevertheless striking to see, as in the graphs published by Schlossmann[2] and Passmore & Schlossmann, a large rise in the amino-nitrogen-content of both bloods in the dog, or of the glucose-content of both bloods in the goat or sheep, upon injection of the given substance into the maternal circulation, after which

m < f (c) (Timpe); organic P, m > f (c) (Timpe); oestrone, m < f (Siegert & Neumann); pituitrin, m = f (Siegert & Neumann).

(2) *Cow:* total protein N, m > f (c) (Sánta); non-protein N, m < f (c) (Sánta); fibrinogen, m > f (Sánta).

(3) *Pig:* non-protein N, m > f (Dobó).

(4) *Rabbit:* amino-acid N, m < f (c) (Brandstrup[2]).

(5) *Dog:* amino-acid N, m < f (c) (Schlossmann).

(6) *Rat:* plasma phosphatase, m > f (Weil).

(7) *Mouse:* oestrone, m = f (Soule).

* Bickenbach & Rupp[3] believe, however, that the excess of amino-nitrogen in the foetal blood is only an apparent phenomenon due to the fact that the maternal blood with which the foetal blood has usually been compared is venous. On comparing the foetal level with the maternal arterial level the difference was reduced or abolished. Nevertheless, it is very doubtful whether an explanation of this kind can hold good for all the constituents which have been found to be more concentrated in the foetal blood.

† It is interesting to compare the placental barrier with other barriers such as the haemato-encephalic barrier. The formation of the cerebro-spinal fluid in the pig foetus has been studied by Flexner. At first it is in Donnan equilibrium with the plasma, but rather suddenly its composition changes, whether as a result of secretion or as the result of a new form of equilibrium is not clear. Flexner & Stiehler went on to show that this change-over is accompanied by an increase of dehydrogenase and cytochrome oxidase activity, the appearance of a potential difference, etc. in the barrier (the choroid plexus) itself.

the concentrations on both sides of the barrier settle down at about their original levels, but with the foetal side still some 5 mgm. % *higher* than the maternal.

Some ingenious experiments have been made by Bickenbach & Rupp[1] to study the passage of nutrient substances through the placenta. They found, for instance, that if methyl or amyl oleate or palmitate were injected into the maternal circulation (in man), these unusual "glycerides" were found again on the foetal side, and concluded that if there is a hydrolysis at the barrier followed by a synthesis, it must take place so fast that there was no chance whatever of the placenta being able to form normal triglycerides instead of these esters. Or again[4], on injecting brominated serum proteins into the maternal circulation in rabbits, no trace of bromine was ever found in the foetal circulation or tissues, indicating that the serum proteins of the maternal blood are not in danger of being hydrolysed by the placenta to furnish amino-acids for the foetus. Dipeptides pass through without being hydrolysed.

1·94. The amniotic and allantoic liquids

The amniotic liquid of a mammalian species still awaits an exhaustive examination. The data formerly described (*CE*, Section 22) were all of a somewhat unsatisfactory kind, partly because of the use of estimation methods which would hardly command reliance to-day, and partly because of the extremely large statistical variations naturally occurring in the composition of the liquids. The only recent examination of the osmotic pressure, for example, of the amniotic liquid is that of Howard[3], who believes that it approximates to that of the serum in the pig and man ($\Delta -0·55°$), while that of the allantoic liquid is much lower ($\Delta -0·20°$). But as doubt had to be thrown in another connection on the validity of freezing-points taken with this author's methods (see p. 14), we are not much further advanced than we were in 1931. According to Makepeace, Fremont-Smith, Dailey & Carroll the osmotic pressure of the amniotic liquid in man at term is a little below that of the serum. Good exact measurements of the volume of the amniotic liquid in the rabbit are given by Lell. It declines markedly towards the end of gestation, and there is therefore little water-loss at birth (cf. above, p. 88). This decline may even be found in man (Dieckmann & Davis). Snyder & Rosenfeld and others have believed that the amniotic liquid normally enters the foetal lungs in the early respiratory movements, but it is now fairly certain that this never occurs under normal physiological conditions (Windle, Becker, Barth & Schulz), though swallowing of amniotic liquid does. The amnion itself is furnished with muscle fibres, which like those of the chick embryo's amnion (Pierce; Kuo[1]; Revoutskaia) produce long-continuing contractions (Groebbels[3]). Data for the gas pressures in the amniotic liquid of the rabbit have been given by Campbell, and many substances have been found to be present in it.* In man, for example, urea and uric acid accumulate to a slight extent (Guthmann & May; Cantarov, Stuckert & Davis).†

* E.g. for *Man* [descriptive bibliography by L. Zuntz[2]]: general composition (de Laurentis; Makepeace, Fremont-Smith, Dailey & Carroll; Cantarov, Stuckert & Davies; Shrewsbury); glucose (Mohs); glutathione (Guercia); citrate (Genell); erepsin (Abe[4]); adrenalin (Macchiarulo); myotonic and vasopressor substances (Fomina); thyroxin (Contardo); anterior pituitary hormones (Huddleston & Whitehead; Cozzi); oestrone (Loewe; Morrell, Powers & Varley); for the merino *sheep*: general composition (Malan, Malan & Curson).

† A few observations on the liquids of the chick embryo will be found referred to on pp. 606, 662.

Part Two
THE MORPHOGENETIC STIMULI

"Die beschreibende Chemie der lebenden Teile kann somit als eine Fortsetzung der Morphologie in das Reich der kleinen Dimensionen angesehen werden, und es ist wohl der Gedanke berechtigt, beide Wissenszweige mit denselben geistigen Hilfsmittel in zusammenfassender Weise zu bearbeiten."

A. Kossel, 1921.

"The problem of (hereditary) transmission might be merged in the broader problem of the production of form through chemical processes—the central problem of all development."

E. S. Russell, 1916.

THE MORPHOGENETIC STIMULI

2·05. Introduction.

Although our knowledge of the morphogenetic stimuli which operate from one part of the embryo to another during development is relatively new, it has profoundly modified our picture of the mechanism of the developmental process. So important, indeed, are these modifications that an apology at the outset can hardly be dispensed with on the part of one whose training was not morphological and whose only merit is a genuine effort to pass from the familiar ground of chemistry to the difficult country of the morphologist, bringing what help he can.

A discussion such as the following may be modestly compared to one of those narrow-gauge railway lines joining distant places, in which some European countries abound. Sometimes it runs alongside the well-polished tracks of trunk communications, sometimes it strikes across country, winding through pastoral or wooded valleys with more originality; at one point it touches some obscure but charming village, at another it pauses for a moment to make a junction with a main road. In the present state of our knowledge this is all that the biochemist can contribute, and it must emphatically be stated that the following discussion is not intended to replace in any way the consultation of the works of modern experimental embryologists, classical in some cases already. The "chemin-de-fer vicinal" must not be mistaken for the routes of the great expresses.

To these a reference must first be made. The most important event since 1931 is the publication of Hans Spemann's Silliman Lectures[17] at Yale University, in which he has given us an unparalleled account of the principal processes in development, though confined to the development of amphibia. Covering a wider field is the excellent book of Huxley & de Beer[2], which has the advantage of a continuity with the older theories of the American schools.* A different advantage is possessed by the monograph of Dalcq[4], namely an acquaintance at first hand with the phenomena of development in ascidians and hence an attempt to link together all the chordates in a logical scheme. The little book of Weiss[11], with its stimulating discussion of regeneration phenomena,† has somewhat aged, but he has now supplemented it with his valuable *Principles of Development.* The massive compilation of Schleip remains the most useful account of the comparative aspect of experimental embryology, while Albert Brachet's great textbook has been well revised and brought up to date by Dalcq & Gérard. Finally, the border-line between embryology and genetics has been exhaustively surveyed

---

* It has also the disadvantage, however, that all the phenomena of development are forced rather artificially into the framework of the gradient theory. Examples from invertebrates and vertebrates are placed side by side to illustrate principles not yet themselves fully established, so that the result is sometimes confusing. Though admiring the synthetic genius of these authors, I do not feel that the subject can as yet be so clearly systematised, and I shall not attempt the like here.

† To be received with caution, however; see p. 435.

by Goldschmidt[8] in the third incarnation of his classical essay. Books of lesser importance, though not without value, are those of Dürken[2] and May[2]. The vast mass of data in Korschelt & Heider has been conveniently summarised by Richards[2].

A number of smaller monographs should also be mentioned. The most modern is that of Dalcq[7], which adumbrates a comparative experimental, not merely a comparative, embryology, and contains an account of the double-field theory (see p. 220). The most theoretical is that of Bertalanffy & Woodger, which gives a critical review of modern theories of development, and there are characteristically stimulating books by H. Przibram[6] and N. K. Koltzov[3]. Monographs on the experimental embryology of special groups have hardly yet begun to appear, though it is hard to refuse this designation to the inaugural thesis of Lindahl[7] on the development of echinoderms. Poulson's monograph[2] on insect development is almost entirely descriptive, as are the contributions of Pincus[1] and of Fischel on that of mammals. In the experimental embryology of special organ systems the nervous system has received most attention, in the books of Coghill and Detwiler, but there is also Murray's on the development of bones. The monograph of Hatt on cell streams (morphogenetic movements) in early development was valuable. Finally, there are the interesting contributions of J. Brachet[19] and of Waddington[26] on the role of the nucleus in early development.

We shall have occasion below to mention many of these monographs in specific connections.* In the meantime it must be emphasised that none of the sections of this book dealing with experimental morphology are to be taken as exhaustive. Journals such as the *Archiv. f. Entwicklungsmechanik* and other similar publications contain a vast wealth of interesting facts. We shall only be able to consider such of them as allow of discussion along biochemical lines.

2·1. General concepts of causal morphology

There are two ways of penetrating into such a subject as this. The experimental facts may be described in some more or less arbitrary order, with pauses from time to time to define the various concepts to which the facts have given rise. Or, on the other hand, the concepts themselves may first be defined, and the experimental facts later marshalled in the light of these. Our knowledge of morphogenetic stimuli has now, it seems, advanced to such a point that this second, more logical, way of proceeding may be adopted. A little repetition may, however, be unavoidable, as we shall meet with some facts twice over, first in their theoretical connection, and then again as nodes in the factual network.

It is necessary to begin with the concept of *Determination*.

* Apart from reviews mentioned later on, general discussions from which much can still be learnt (though some, owing to later advances, have now to be read with caution) are to be found in the following: 1923, Spemann[11] (the historical connection with Weismann); 1925, O. Mangold[2]; 1926, Przibram[4] (gives the background to the experiment of Spemann & H. Mangold); 1928, O. Mangold[4,5]; 1929, Gilchrist[2] (caution); 1933, Harrison[12] (caution); 1934, Spemann[15]; Stolte[1] (caution); Ermakov[1] (in Russian, caution); 1935, Weiss[14]; Huxley[11] (caution); Schmidt[2] (in Russian); 1936, May[4] (caution); Caullery[2]; Poležaiev[7] (in Russian); O. Mangold[14,15]; 1937, Harrison[17]; Brøndsted[2] (caution); Daniel[2]; 1938, R. S. Lillie[2]; Poležaiev[9] (in Russian); Borovanský (in Czech); 1939, Caullery[3]; Panagiotou (in Greek); 1940, F. E. Lehmann[16] (caution).

2·11. Determination

At the outset we meet with the contributions of Hans Driesch to the theory of morphogenesis. Although his theory of entelechy is now only a matter of historical interest (see p. 119), the investigations which led to it were quite fundamental. We are so accustomed in modern embryology to the concept of determination, i.e. the fixing of the fates of parts of the embryo at a definite time in development, that it is difficult to think back into the state of mind in the last century when it was supposed that fates were all fixed to begin with.

In 1883 W. Roux[1] enunciated the view that development was brought about by a qualitative division of the germ-plasm contained in the nucleus, and that the complicated process of mitotic division was primarily devoted to that end. It was thought that development proceeded by a mosaic-like distribution of potencies to the cells in segmentation, and that for instance the first cleavage furrow separated the material and the potencies of the right side from those of the left. In 1888 Roux[2] announced that, if one of the two first blastomeres of a frog egg was killed by cautery with a hot needle, a half-embryo developed from the uninjured cell. This "mosaic" development seemed to confirm the view of Weismann that the fates of all the parts were fixed at the outset, and, although Roux's particular experiment afterwards turned out to be deceptive, in that the amphibian blastomeres are not necessarily mosaic, a large number of other eggs were subsequently found in which any injury to the uncleaved or cleaving ovum is reflected in injuries or losses in the finished embryo.*

Considerable astonishment was therefore caused by Driesch's announcement[1] in 1891 that he had obtained complete larvae from single blastomeres of the sea-urchin egg isolated at the two-cell stage.† We may give his account of the discovery in his own words:

"The development of our *Echinus* proceeds rather rapidly, the cleavage being complete in about fifteen hours. I quickly noticed on the evening of the first day of the experiment, when the half-embryo was composed of about two hundred cells, that the margin of the hemispherical embryo bent together a little, as if it were about to form a whole sphere of smaller size, and indeed the next morning there was a whole diminutive blastula swimming about. I was so much convinced that I should get the Roux effect in all its features that, even in spite of this whole blastula, I now expected that the next morning would reveal to me the half-organisation of the subject once more; the intestine, I supposed, would come

---

* For a good account of the early work in experimental embryology, see E. S. Russell's admirable history of nineteenth-century biology, *Form and Function*. Mosaic development was described for ascidian eggs by Chabry in 1887 and Conklin[1] in 1905, for ctenophore eggs by Chun in 1892 and Yatsu in 1912, for molluscan eggs by Crampton in 1896 and Wilson[2] in 1904, etc. etc. It has been usual to express the difference between mosaic eggs and regulation eggs by saying that in the former all the determinative processes have been completed before fertilisation, while in the latter they extend throughout development, centering on gastrulation in the amphibian, for example, but lasting on even into the metamorphosis period. The most complete discussion of this is to be found in E. B. Wilson's[3] contribution to the Spemann Festschrift and in Weiss' *Entwick-lungsphysiologie*, pp. 23 ff. Dalcq, however, in his recent book *Form and Causality*, shows that the distinction cannot be considered absolute, as minor degrees of regulation appear to be possible in mosaic eggs (see on, p. 325).
† v. Baer's account[5] of his study of echinoderm development in Trieste in 1847 is well worth hunting out.

out on one side of it, as a half-tube, and the mesodermal ring might be a half one also. But things turned out as they were bound to do, and not as I had expected; there was a typically *whole* gastrula in my dish the next morning, differing from a normal one only by its smaller size, and this small but whole gastrula was followed by a whole and typical small *pluteus* larva."

In the following year[2] he showed that whole embryos could be produced from one or more blastomeres isolated at the four-cell stage, and later that alteration of cleavage by compression[5], the fusion of two eggs into one[3], or the cutting in certain directions of the original egg-cell[4], all permitted of normal embryos being produced.

Driesch[7] introduced the term "Prospective Significance" (*prospektive Bedeutung*) to indicate the actual fate of any part or monad in the original egg. His great discovery lay in the finding that the significance of such a part was not exhausted by its prospective significance, but that it was widely changeable according to

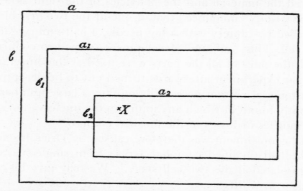

Fig. 37. Driesch's diagram to show the characteristics of a harmonious equipotential system.

various circumstances. And he introduced the term "Prospective Potency" (*prospektive Potenz*) in order to signify the collection of possible fates of such a part. Thus the actual fate (the prospective significance) is chosen from among the possible fates (the prospective potency);* in normal development it will be one constant result, in abnormal development it may be quite a different one. This condition of multiple potency of the parts of the early egg-cell has been termed *Pluripotency*.

These relationships were illustrated by Driesch[7] in the following way (Fig. 37). A plane of the dimensions *a* and *b* represents normal undisturbed development; taking the sides of the plane as fixed localities for orientation, we can say that the actual fate (the prospective significance) of every element of the plane stands in a fixed and definite relation to the length of the two co-ordinates at

---

* The concept of potency has been recently examined by Raven[5], who thinks that the word has frequently been misunderstood since "potency" often means "power" rather than "possibility". But the Aristotelian distinction between *potentia* and *actualitas* was the origin of the embryological use of the term; cf. "Parts of machines while at rest have a sort of potentiality of motion in them, and when any external force puts the first of them in motion, immediately the next is moved in actuality" (*De Gen. An.* 734 b, and cf. Ross' commentary on Aristotle's *Metaphysics*, pp. cxxiv ff.). Raven's other difficulties are, I think, avoided by a careful use of the concepts of competence and induction (see pp. 112 and 104).

right angles. Thus the point X would normally develop into some definite part of the finished embryo, e.g. eye-cup or muscle-cell. But its prospective significance is not as wide as its prospective potency, so that if it formed part of an isolated portion $a_1 b_1$ its relation to the whole would be different and its actual fate would be different. The same applies to its position in another possible isolated portion $a_2 b_2$. These isolated portions, it is to be remembered, regulate themselves in such a way as to reproduce on a smaller scale the relative location of parts seen in the absolutely normal case. They possess, therefore, a non-variable factor of wholeness, represented in Driesch's terminology by the symbol E.* In view of these properties, systems which fulfilled the conditions just laid down for the plane ab were called by Driesch "harmonious equipotential systems".

We realise now, after forty years of experimental research,† that at the beginning of development, parts or monads of the egg, similar to that which has been represented by X in the diagram, are indeed undetermined,‡ and that one of the most fundamental processes in development consists in the closing of doors, i.e. in determination, in the progressive restriction of the possible fates.

By his brilliant extensions of Born's transplantation method,§ Spemann[8] showed in the case of the newt that up to a certain stage of gastrulation the fates of most of the embryonic regions are not irrevocably determined. A piece of presumptive neural tube material removed from one embryo and grafted into another will turn into external gills if it happens to be grafted into the presumptive gill region of the latter. On the other hand, a piece of presumptive skin, if grafted into a suitable region of the presumptive nerve tube of a second embryo, will in due course turn into brain or spinal cord. Up to this stage in gastrulation, therefore, the regions develop always in step with their actual surroundings and without reference to their former surroundings, or their prospective significance if they had not been interfered with. The German word "*ortsgemäss*" was coined to express this. The parts are extremely plastic, like the parts in Driesch's sea-urchin eggs, which would obediently form gut or skeleton according to their position in the whole. Nor does this plasticity or pluripotency stop short at the

---

* Driesch[7], p. 91.   † Summarised in Huxley & de Beer's book.   ‡ See glossary, Determination.

§ Though Gustav Born first discovered the healing powers of amphibian embryos and invented much of the technique of operating upon them, he did not, owing to his early death, follow the possibilities very far. Otto Mangold said to me once "er hatte kein Problem". Spemann's achievements have lain as much in the asking of certain questions as in the answering of them. Grafting and transplantation as such have a long history, but because easier in plants than animals, and in large organisms than in small embryos, did not quickly lead to fundamental discoveries. Skin transplantations in man were known already in the sixteenth century. The following passage strikingly illustrates how nearly the concepts of *ortsgemäss* and *herkunftsgemäss* were approached, though set in a frame of factual error. "A certain inhabitant of Bruxels, in a combat had his nose mowed off, and addressed himself to Tagliacozzus, a famous chirurgeon, living at Bononia, that he might procure a new one; and when he feared the incision of his own arm, he hired a porter to admit it, out of whose arm, having first given the reward agreed upon, at length he digged a new nose. About thirteen moneths after his return to his own countrey, on a sudden the ingrafted nose grew cold, putrefied, and in a few days, dropt off. To those of his friends, that were curious in the exploration of this unexpected misfortune, it was discovered, that the porter expired near about the same punctilio of time, wherein the nose grew frigid and cadaverous" (Walter Charleton, "A Ternary of Paradoxes", London, 1650). Evidently men could speculate on the extent to which a piece of tissue transferred from one body to another, could retain something of its individuality. Autoplastic transplantations appear to go a long way back in the history of surgery beyond Gasparo Tagliacozzi (1546–1599) who brought them to a high level (see Corradi). I owe this interesting reference to my friend Dr Charles Singer.

germ-layers themselves, for mesoderm and ectoderm, for example, are perfectly interchangeable, as O. Mangold[1] demonstrated.

The process of gastrulation, however, is critical, for at this time the original plasticity is lost, and the main fates of the parts are determined. This is irrevocable. The early period of plasticity is followed by one of rigidity, each part being willing now only to undergo a certain special type of development which differs from part to part. If it is grafted into another embryo, it will continue to differentiate in accordance with its inner determination, and not in accordance with the new situation in which it has been artificially placed. In other words, it will develop *"herkunftsgemäss"*. The regions can now only develop towards their presumptive fates. Prospective potency* has been ruthlessly curtailed to prospective significance. The eye region will form an eye whether this points outwards to the external world or inwards to the body-cavity. The invisible

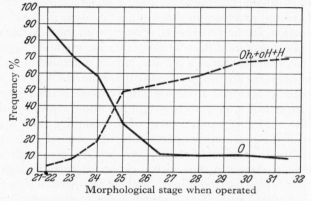

Fig. 38. Determination of the choroidal fissure of the amphibian eye-cup.

process of determination has ushered in the new period of self-differentiation, in which the embryo has become a mosaic of irreplaceable regions, similar to the whole development of certain eggs ("mosaic eggs"), which never manifest a period of plasticity or pluripotency.

As an illustrative case of a statistically well-defined transition from neighbour-wise to selfwise behaviour, we might choose the relatively unimportant choroidal fissure which appears in the lower part of the amphibian eye-cup. T. Sato[2] extirpated the eye-cup at an early stage, turned it upside down and put it back. As Fig. 38 shows, the number of cases showing neighbourwise behaviour, i.e. the appearance of the fissure at the lower part after this treatment, is approximately the reciprocal of those showing selfwise behaviour, i.e. the appearance of the fissure at the "wrong" upper part after this treatment. Determination has thus set in at stage 23 and is complete at stage 26. Similar facts about the ear-vesicle are known from the work of Röhlich and Sidorov.

* There has always been much difficulty in translating the terms *"ortsgemäss"* and *"herkunfts-gemäss"* into English. Grafting, however, has long had English words; there is the stock or plant on which the graft is made, and there is the scion (from the Latin root *seco*, to cut) or piece cut off from another plant and grafted. Hence we might use, as A. L. Peck suggests, "stockwise" for *"ortsgemäss"* and "scionwise" for *"herkunftsgemäss"*. But since there is a confused vulgar use of the word "scion", I prefer "neighbourwise" and "selfwise" and I have adopted them in this book.

One region of the amphibian embryo, however, is much less plastic than the rest during the early part of development. This is the region of the dorsal lip of the blastopore, which has arisen from the grey crescent and will in time form the mesoderm, i.e. notochord, somites, etc. If a piece of this region is grafted into another embryo in the blastula or early gastrula stage, it will "induce" (see p. 104) the neighbouring host tissues to form a secondary embryo, often including nerve tube, brain, eyes, ears, somites, notochord, etc., irrespective of the presumptive fates of those tissues. In other words, it contains within it the influence which determines the fates of regions with which it comes in contact. It is therefore called the *Organiser* (*Organisator* of Spemann). It is easy to demonstrate its fundamental importance in the development of the amphibian egg by separating the first two blastomeres instead of killing one of them, as in Roux's experiments. Their subsequent fate will now depend on whether or not they contain a sufficiency of the region which will afterwards become the organiser region. If both blastomeres contain a portion of it, both will be organised into small but morphologically normal embryos, in accordance with the original experiment of Driesch. If one contains all of it and the other contains none (as will happen if the first cleavage chances to be horizontal), the former will produce a complete embryo and the latter will give up the struggle, as it were, after a few cleavages. These effects were described by Herlitzka as long ago as 1896, but the exact analysis of them awaited the beautiful constriction experiments of Spemann[1] and Ruud & Spemann. A newt's egg constricted into two in the transverse plane will give from the dorsal half a well-formed embryo and from the ventral half a blastula-like ball of yolk-cells developing no further ("*Bauchstück*"). But if it is constricted so as to separate lateral halves, it will give two perfect half-size embryos (see on, p. 228). Schmidt[1] has shown that the same is true for the frog. On the other hand, fusion of two amphibian eggs into one is possible; it was accomplished by O. Mangold & Seidel. The organisation centres may fuse and produce a single neural axis on a giant embryo.

We now understand better the conditions necessary for the fulfilment of Driesch's pluripotency.

2·12. Morphogenetic stimuli

The essentially new thing in Spemann's conception of the organisation centre or Organiser was that it set the process of *Dependent Differentiation* right at the heart of normal development. Only after this process has taken place, and indeed as the response to induction, does self-differentiation proceed. The distinction between dependent differentiation and *Self-differentiation* had originally been made by Roux[2,5], who said that there were some organs, tissues or regions in embryos which possessed within themselves at a given stage all the factors necessary for the realisation of their normal fates, while there were other organs, tissues or regions, which for the realisation of their normal fates required the interposition of factors external to themselves. It is of interest to quote Roux's own words:

"Self-differentiating structures are those which change themselves as the result of causes lying within themselves. This is apart from necessary external factors, which might be called pre-conditions, such as sufficient nourishment, oxygen, and warmth. It must also be supposed that these external factors have no specific effects, controlling the quality, place, time, and magnitude of the change. They do not affect the time, if the change happens no earlier than normally, even though they may be present earlier than normally; nor the place, if they are present in places where the change is not going on. They do not affect the intensity, if in spite of changes in them the magnitude of the embryological change is unaltered; nor the quality, if the change is a morphological one....

Structures undergoing dependent differentiation are those which change in response to causes lying entirely or to a substantial degree outside themselves. The nature of the change, and its time, place, and intensity, may be determined from outside the structure. If all of these depend upon external conditions, the differentiation will resemble the formation of a statue from a marble block, and such extreme dependent differentiation we may call passive differentiation....

A given differentiation may be dependent in respect to one or other of these factors, if not to all; this can be called mixed differentiation. These numerous possibilities make the complete analysis of all the causes of a morphological change exceedingly difficult and complicated."

Thus the distinction was fully made by Roux, but it appears that the kind of factors which he pictured as acting upon embryonic rudiments in dependent differentiation were mainly or entirely the physiological stresses and strains of the later period now known as that of *Functional Differentiation*. Thus the shape of the blood-vessels would be affected by the hydrodynamic forces of the circulating blood, the bones would develop in accordance with certain mechanical stresses, etc. Roux did not clearly envisage the stimuli as chemical.

The concept of morphogenetic stimuli had, however, two other quite different sources. As early as 1858 Virchow developed it in relation to the production of specific types of tumour in man. This subject was further elaborated by Billroth in 1890, who attributed these effects to definite chemical substances elaborated by the stimulating organism, insect, worm or bacillus. We shall shortly return to the interesting and undeservedly neglected subject of gall-formation. But in the meantime the botanists had been deeply engaged in the study of tropism and taxis, and the word *Induction* was used for the first time in this connection by Pfeffer in 1881 in his consideration of the "Induktion spezifischer Gestaltung" by factors external to the plant, such as light or gravity.

The existence of chemical stimuli in embryonic development was first thoroughly brought forward, it seems, in the articles contributed to the *Biologische Centralblatt* in 1894 and 1895 by C. Herbst[2], as the result of his experiments with sea-urchin embryos cultivated in artificial salt solutions from which single ions were missing (see *CE*, Section 13·4). He set out to consider the significance of stimulus-response physiology for ontogenesis and classified all the possible kinds of stimuli, starting with external factors such as light, gravity, temperature, etc. The line of thought followed began with the tropisms of free-swimming organisms and the taxis of plant tendrils, etc., going on to describe the accumulation of cells at the surface of the arthropod embryo to form the blastoderm as an "oxygeno-taxis", and the penetration of vitellophages into the yolk in other forms as a chemotaxis. From this it was not far to think of other chemical substances which might have as their typical response, not movement, but morphological change. A botanical example, the production of swellings by vine tendrils on contact with a solid surface, lay close to hand, though here the stimulus was apparently mechanical, not chemical.

The actual examples which Herbst could give of chemical morphogenetic stimuli were in fact rather few; apart from gall-formation, to which he devoted a good deal of space, there was the origin of the antheridia stalks in *Saprolegnia* (secretion of a diffusible substance from the oogonia leading to the formation of the male antheridium-bearing branches), and the exogastrulation produced by lithium salts in echinoderm embryos, though this last could hardly be regarded as a normal internally produced stimulus. "The exact investigation", he wrote, "of the numberless cases of induction of specific morphological forms by internal factors will form a fundamental aim of future biological researches." But already Herbst was able to give a theoretical outline to problems which have only become acute in our own time. Thus from the study of galls, he drew the conclusion that one stimulus might have quite different responses if it acted on different tissues, or that one stimulus might give rise simultaneously to more than one type of response. On the other hand, some tissues might give a single type of response to a number of chemically different stimuli. Again, the response to stimulus might depend very largely on the stage of development reached by the reacting tissue, which in early stages might have no reactivity, and in still later stages might lose it after gaining it. This was illustrated by the fact that galls are much less commonly produced on fully differentiated tissue than on buds, young leaves, etc. All these suggestions of Herbst will be better appreciated when some of the facts of modern experimental morphology have been described in what follows (cf. p. 261), but they constitute a brilliant intuition of the ideas which studies on hormone specificity have now compelled us to adopt.

Morphogenetic stimuli were at this time definitely "in the air", for in 1893 C. Emery, a student of bacterial tumours, suggested that "some morphogenetic stimulating substances probably exist in the egg, others arise during development at different times and in different organs one after the other". At the turn of the century, however, the stand taken by Herbst was fully justified by the analysis of the development of the amphibian eye undertaken by Spemann[2] and W. H. Lewis[1]. It was found that if the eye-cup were in any way prevented, as by removal or injury, from coming into contact with the overlying head ectoderm, no lens would be formed. Lewis[2] also showed that if the eye-cup was allowed to act on ectoderm from any part of the body, or even on ectoderm from another species of frog, a lens would be perfectly formed. Separated into several parts, the eye-cup pieces still possessed the power of lens induction. In 1901 Herbst's articles, very much rewritten, were published as a book, *Formative Reize in der tierischen Ontogenese*. He had unwittingly postponed the appearance of his book just long enough to enable him to mention these new results of Spemann's.

"Gerade als mein Manuskript so weit fertig war, dass es zur Post gehen könnte, erhielt ich durch die Güte des Herrn Kollegen Spemann den Bericht einer Sitzung der physikalisch-medizinischen Gesellschaft in Würzburg von Anfang Mai 1901, aus dem hervorgeht, dass Spemann die gleiche Ansicht wie ich von der ontogenetischen Entstehung der Linse des Wirbeltierauges hat. Es gelang ihm nämlich, folgende beide, für unsere Frage wichtige, Punkte durch das Experiment sicherzustellen: '(1) Wenn der Augenbecher die Haut nicht erreicht, so bildet sich keine Linse....(2) Sowie der zurückgebliebene Augenbecher die Epidermis berührt, beginnt an der Berührungstelle die Wucherung der Linse.'"

The early history of this subject certainly seems to suggest that the right ideas came some time before the proofs of their rightness.

One passage from W. H. Lewis[1], written in 1903, demands quotation, in the light of modern knowledge:

"It is not unlikely that a new biological chemistry must be developed, its reactions being those between living tissues or between a living tissue and the product of a living tissue. In the example of the lens reaction it seems evident that living ectodermal cells must be acted upon, but it is not so evident that the living optic vesicle cells are necessary, as it may be possible to extract from them substances which will give the lens reaction. By means of such a biological chemistry some of the properties of the various tissues and organs may be discovered and perhaps a deeper insight into the problems of development obtained. The development of a biological chemistry is to be considered as but one of the phases of correlative embryology, and in considering the correlations in development, the mechanical factors must not be neglected."

It would be interesting to know whether these pioneer embryological investigators were at all influenced in their interpretations by the pioneer work which was going on at the same time on the hormones of the adult mammalian body. The first of these studied was the active principle of the adrenal medulla, by Oliver & Schafer in 1894. Its formula was already worked out by Takamine in 1901, and in the following year Bayliss & Starling made the discovery of secretin, after which endocrinology made rapid strides forward to its present position.

2·13. A digression on plant galls

In Herbst's thought we have seen how the earlier investigators were led to the conception of morphogenetic stimuli, and we do indeed find in certain interactions between animals and plants phenomena which must be ascribed to the action of morphogenetic hormones. The whole subject of the plant galls, as, for example, described in the monographs of Küster and of Ross & Hedicke, is the best illustration of this, although up to the present the relation between these forms of "dependent differentiation" and what we now know of organiser phenomena in animals has not attracted much research. A fascinating field lies open here for biochemical exploration and experimental analysis.

As is generally known, galls arise from the abnormal development of meristematic tissues of plants as the result of a stimulus usually caused by an insect parasite. In many cases the exciting chemical substance seems to be contained in the juice injected into the plant by the ovipositing insect at the same time as the egg; in other cases, the substance is apparently produced by the larva as soon as it hatches. But the important point for the present argument is that, although the gall is built up from the tissues of the plant, its morphological form depends upon the nature of the exciting animal. Thus the same plant attacked by different insects will produce very different galls. It is said that in some cases the morphology of the galls may be an easier method of distinguishing between closely related species of insects than the morphological characters of the adults themselves. The usual instance given of this is the American willow, *Salix humilis*, which is attacked by ten different, though closely related, species of gall-gnats (Walsh). Fig. 39 shows one kind of gall produced by such an insect (*Rhabdophaga rosaria*) on the willow. Again, different parts of the same plant may be attacked by the

Fig. 39. Gall produced by the gnat
Rhabdophaga rosaria on the willow.

Fig. 40. Gall produced by the wasp
Rhodites rosae on the rose tree.

same insect with the same result. The currant galls of *Spathegaster* (= *Neuroterus*) *baccarum* occur both on the leaves and the flower stalks of the oak, so that their morphology cannot be a function of the part of the plant that produces them. Finally, the same gall may be produced by the same insect on several species of plant, for example, the sawfly *Micronematus gallicola* produces its brilliant red galls on four separate species of willow, *S. fragilis*, *alba*, *caprea* and *cinerea*. The gall of another hymenopteran, the gall-wasp *Rhodites rosae*, on the rose tree is illustrated in Fig. 40. Few workers have studied the histology of the gall throughout its development; the nicest pictures I have seen are those of B. W. Wells.

All this indicates in the strongest possible manner that in the course of evolution the pursuit of the parasitic habit has involved the preparation of a great number of chemical substances by insects (and some other animals, such as nematodes, also) which have a specific morphogenetic effect upon the tissues of the host plant.* But as yet experiments directed towards proving the truth of this view have not been very many, and their success has been but moderate. The most famous were those of Beijerinck[1], who as far back as 1888 killed the eggs of sawflies by puncturing with a hot needle immediately after their inoculation into the plant by the insect. In spite of this the galls developed normally. In the case of the Cynipid galls, on the other hand, it is certain that the irritating substance is first produced when the larva hatches from the egg, and begins to feed upon the plant. Beijerinck called these substances "growth-enzymes", and in a later paper[2] (1897) applied the idea to the development of form in general, which he thought was determined in the embryo by the diffusion of stimulatory substances.

It appears that no investigator has yet succeeded in preparing a cell-free extract of an insect or its egg by any chemical method which will, when injected into the plant, give rise to the characteristic gall of the species. It is, however, equally true, if we may judge by the review of Küster,† that no investigator has succeeded in causing gall-formation by the application of unspecific stimuli such as the implantation of oils, acids, cantharidin, yeast, proteins, etc. The most probable conclusion is that no one has yet found the correct extraction method, and indeed the literature shows well enough that no very high level of chemical technique has as yet been applied to the problem. Significant, however, is the finding of Laboulbène that the implantation of small pieces of dead larvae successfully produced galls (cf. the persistence of organiser action after the death of the cells of the organiser region), although this worker made the further claim that the injection of water in which the larvae had been washed would also provoke the characteristic gall. The technical difficulties here, it must be remembered, are rather great, for the reactive layer is presumably the cambium, and as this is supposed to be but one cell thick, the problem of directing a sufficiently intense stimulus to the right place is not easy to solve. It may be recalled that the percentage of positive results in chemical experiments with animal evocators is generally not high, for very similar reasons.

Work which has a direct bearing on this was that of K. M. Smith, who, in order to study the nature of the damage produced in apple leaves by the activities of capsid bugs, implanted the salivary glands of these insects beneath the surface of the leaf, and was able to produce in this way all the pathological phenomena usually caused by the presence of the insects themselves. Although these bugs (such as *Plesiocoris rugicollis* and *Lygus pabulinus*) do not form what are, properly speaking, galls, the damage they do is somewhat parallel. The homopteron *Asterolecanium variolosum*, however, does form galls, and Parr was successful in reproducing them by injecting glycerin extracts of the salivary glands. The

---

* The effects of some parasitic worms on animals might be classified as galls, e.g. the hydatid cysts produced in the human liver by the flatworm *Echinococcus granulosus*, or the sarcomas and hepatomas caused by the cysticercus larvae of *Taenia crassicollis*. At this point we reach the borderline between parasitism and true malignant tumours, but so far extractable carcinogens have been sought for in these organisms in vain.

† *Gallen*, pp. 279 ff.

various organs of gall-wasps were investigated by Rössig, who thought he could trace the site of formation of the gall-forming substance to the Malpighian vessels, and this was confirmed later on by Triggerson for the Cynipid which produces the oak hedgehog gall, *Dryophanta erinacei*. Lastly, the structures produced in the larch by the aphids of the *Chermes* group have been proved by Burdon to be due to the injection of a chemical substance, which in this case is relatively unspecific, since all the species produce the same effect.

Fig. 41. Gall produced in five months on a stem of *Phaseolus* by applications of hetero-auxin.

It is now known that the plant-growth hormone auxin, which we shall have occasion to mention again from time to time in the organiser story (pp. 212 and 266), has a good deal to do with galls and plant tumours. In 1912 Molliard produced tuberisation of the roots of pea plants by subjecting them to the action of the products of *Rhizobium radicola* in cell-free extract. Thimann[2] has now proved that the root nodules of leguminous plants, of such fundamental importance to man in the history of agriculture, are almost certainly due to the auxin produced by the invading nitrogen-fixing bacteria. The crown-gall organism, *Pseudomonas tumefaciens*, also produces auxin (Brown & Gardner; Solacolu & Constantinesco). These authors succeeded in obtaining typical crown-galls by applying a hetero-auxin in lanolin paste to a *Phaseolus* stem for five months (see Fig. 41). Similar treatment with hetero-auxin, but at lower concentrations and for a shorter time, allowed la Rue[1] to simulate perfectly the intumescences which arise under conditions of high humidity on the leaves of the poplar, *P. grandidentata*. In a later paper[2] the same author showed that the faeces of larval leafminers contain auxin, and that this produces the proliferations of the host plant.

The whole subject of plant tumours has been reviewed by Levine; Rose, and Gheorghiu. While auxin and related substances doubtless play a great part in the causation of galls and tumours, the high specificity of many insect species in the morphology of their galls suggests that many other specific stimulatory substances must be involved.

2·14. The hierarchy of organisers in animal development

By 1909 the concept of dependent differentiation by the action of chemical morphogenetic stimuli was well known, and Jenkinson, writing in that year, could say:

"Should the results of these extremely interesting investigations be confirmed, the development of some parts, at least, of the vertebrate eye would be processes of 'dependent differentiation', dependent, that is, on causes outside themselves, on stimuli exerted by other parts.... How far such an action of the parts upon one another is of general occurrence as a factor in differentiation, future research alone will show. It seems probable, however, that it will be found to be limited to the first moments in the formation of organs, for we know, from other evidence, that as development proceeds, systems which were equipotential become inequipotential, parts gain independence and the power of self-differentiation, and the value of the correlations between them diminishes."

Thus the morphogenetic stimuli were being linked up with the first process of determination, whereby the fates of the parts of the embryo are fixed, rather than with the further histogenetic elaboration of the parts.

After the discovery in 1924 of the inductor which stimulates the dorsal ectoderm of the amphibian gastrula to form the neural tube, Spemann[12] introduced the term *Primary organiser* for it, since it is the first to work in the developmental process. Seats of processes which occur later on, such as the induction of the lens from the ectoderm by the eye-cup, he termed *Secondary organisers* (or Second-grade organisers), for if no neural tube were formed there would never be any eye-cup. Third-grade organisers also exist (see pp. 301 ff.).

Development, then, consists of a progressive restriction of potencies by determination of parts to pursue fixed fates.* There is a growing conviction that this

---

* Qualms about the determination concept were voiced by Harrison[12] in 1933. His main difficulties were (1) double assurance of eye and operculum, (2) *bedeutungsfremde Selbstdifferenzierung* and (3) the fact that the ascidian, apparently so determined as an egg, is so capable of regeneration as an adult. The first of these is now a very doubtful question (pp. 293, 308); the second has definitely been proved not to happen in neutral medium (p. 157), and the third presents less difficulty if we admit, with Conklin[3], that determination may be reversible. Zhinkin[3] has actually followed the appearance of regenerative and regulative power (*Recuperation*, see p. 442) during ascidian metamorphosis. In amphibian regeneration, as we shall see (p. 435), the generally accepted pluripotency of the blastema rests on insecure foundations. I ally myself in this book with those embryologists who are not prepared to give up the word "Determination", always subject to the conditions of employment given in the glossary.

state of affairs can best be pictured in the manner of a series of equilibrium states. The whole process could then be imagined in the likeness of a series of cones (Fig. 42). At the top of the uppermost cone there is a ball in a position of extremely unstable equilibrium. It will tend to fall along the side of the cone and will reach a point at some one of the 360 degrees of the cone's circumference. Here it will again find itself in a position of unstable equilibrium, only now with respect to a second stage of determination, and will again be pushed in one direction or another, again to occupy a passing equilibrium, and so on until the final stage of absolute stability is reached, i.e. the plan of the adult body. In biological terms a piece of ectoderm will be determined first of all by the primary organiser to form head ectoderm, next it will be determined by the secondary organiser of the eye-cup to form the lens of the eye, and so on. The effect of the primary and secondary organisers in the normal course of events is to give the slight pushes necessary to send the balls down the cones into specific positions of greater stability as time goes on.* A grafted organiser will give a push of this kind to a given ball (plastic region) in a direction contrary to that which it would normally have taken, but for this it

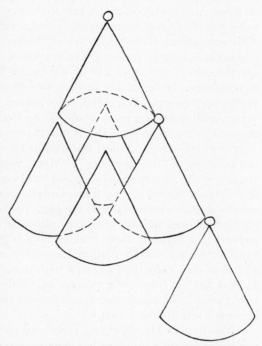

Fig. 42. Action of organisers of successive grades visualised as involving a series of equilibrium states.

is necessary that the ball should be high up the levels of instability, for, if it has already been severely determined, the grafted organiser can have no effect upon it. Conversely, an organiser of the second or third grade may have no effect upon a ball (plastic region) at a higher level of instability than that at which it normally works. The undetermined material is then said to be not yet "competent" to receive its action.

* Roux's idea of a struggle of the parts has been revived in a new form by Raven[5] and others, who picture a "struggle" or competition between the various tendencies in a developing piece of tissue. Raven seems to think that tendencies are more than competences, since they manifest themselves in self-differentiation, but if we adopt the view that there is no self-differentiation without prior induction, competence and induction are the only two concepts required. How exactly an inductor selects and puts into effect its chosen competence is a question which we shall not be able to answer until we know what competence means from a physico-chemical point of view.

2·15. Competence and unstable equilibrium

In all these considerations, we must bear in mind that the state of the reacting tissue is just as important as the nature of the stimulus provided. Determination, and hence, Induction, can only occur if the ectoderm, for example, is in a reactive state; if the ectoderm is too young during the period of application of the stimulus, no effect will be produced; while if the ectoderm has already been determined to be something else, no inducing stimulus can alter or reverse that determination. This state of reactivity* is known as *Competence*.† The roof of the blastocoele of a blastula is not yet competent to react to the stimulus of the primary organiser. On the other hand, the cells of the neural folds were once competent, but are so no longer. In isolated gastrula ectoderm studied by Waddington ([12], p. 81) and Holtfreter[18] (see p. 281), competence disappears after a certain time without any determination having taken place. The ectoderm eventually ceases to be able to respond to an organiser, even if one is implanted.

It is worth while to follow the arguments with which Waddington[1] introduced the concept of "Competence". Much work had been done before that time (1932) by the American school, and has been continued since, in order to study self-differentiation, on the transplantation of pieces of chick blastoderm to the chorio-allantoic membrane of the egg. Such a site was then regarded as a quite indifferent one, which could not influence differentiation. Hoadley's series of papers[1] had made it clear that various regions of the blastoderm possess the capacity for self-differentiation from very early stages of development onward, and that this capacity increases with age. These results were formulated by F. R. Lillie[9] in terms of a process of "differential dichotomy" or "embryonic segregation", in which the capacity of the embryonic cells for self-differentiation became itself differentiated.‡

According to Hoadley[1], an early embryonic cell should only be capable of self-differentiating into a rather generalised cell-type of low organisation. As the process of differential dichotomy goes on, however, the descendants of such a cell acquire more specialised potentialities which have been carved out of the original general potentiality by a process of segregation. Thus, if a piece of tissue was isolated from an early blastoderm the segregation process would cease, and it would differentiate only into epidermis and gut; its weak capacity for self-differentiation would carry it only a short distance along a general path. If a

---

* States of reactivity are important wherever stimulus-response relations exist, as in all neural and humoral activity. Thus in the phenomenon of prepotence, studied by Sherrington, a nervous impulse coming from a nociceptive afferent arc has priority over other reflex impulses for the use of the final common path. The efferent arc could thus perhaps be said to have numerous competences, only one of which can be realised at once. Pharmacology, moreover, abounds in instances where reactivity can be experimentally altered.

† This word supersedes the older "Potence", "Potency" or "Potentiality" discussed elsewhere (p. 100; p. 436 and glossary).

‡ The word segregation was derived from Ray Lankester's discussion of the "gastraea" theory. But he meant by it not a segregation of competences but a segregation of the "physiological molecules" that are going to form "deric and enteric" (i.e. ectodermal and endodermal) material. We can hardly call this a concept of determination, because the possibility of alterations of presumptive fate was not then known (1877), but that is its nearest kin.

similar isolation was made at a later stage, an eye would be obtained; the stronger capacity of the daughter cells carrying them a longer distance along a more special path.

Now the idea of segregation contained two notions; first, that the process gives rise to parts with new potentialities, as Hoadley concluded, and secondly, that it involves a restriction of potentiality from a general to a particular. F. R. Lillie[9] defined it as the "process of origin of parts of the embryo possessing irreversible prospective significance as tested by self-differentiation". But unfortunately self-differentiation techniques do not in fact test irreversible prospective significance. In the newt the ectoderm has the capacity for self-differentiation to epidermis at a stage before it is restricted to this fate. True irreversible prospective significance can only be tested by transplantation experiments in which the tissue is given the chance to prove itself immune to the influences of the host's inductors. But if segregation is limited to being the process of origin of parts possessing irreversible prospective significance as tested by transplantation experiments, it becomes identical with determination, and it loses the first notion mentioned above, namely, that it is responsible for the origin of parts with new potentialities.

New potentialities, however, do in fact arise. In the differentiation of neural plate from ectoderm in the newt, for example, the tissue cannot at first react to the stimulus, it then enters on a period when it can do so, and if the stimulus is applied, the differentiation takes place; if not, the competence may be lost. Whenever competence for specific differentiations arises, new potentialities arise. We shall see later on (p. 155) what can be said about competence and its rise and fall.* But in the meantime, it is clear that since several competences may exist side by side in the same cell region, there is nothing to be gained by referring to their origin as due to dichotomy.†

Lillie's famous paper[9] revealed indeed a very Laocoon striving to dominate the coils of terminology. He introduced the term "specific potency" to mean irreversible prospective significance (i.e. the determined state), thus genuinely confusing the concepts of power and possibility. A piece of an embryo has the *possibility* of a certain fate before determination, and the *power* to pursue it afterwards. The use of "potency" for both these meanings was indefensible. "Segregation" also confused two things, the time at which a certain competence arises and the time at which the determination of a competent tissue is effected. Other expressions introduced by Lillie, such as "open and closed term" for undetermined and determined stages; "isotropic" for pluripotent (though isotropic has a definite meaning in physical chemistry, see p. 662, inapplicable to the egg); and "self-origination" (a word which would have pleased Athanasius)

---

* The idea of competence covered two concepts previously in use by the German school—*reaktionsfähig* and *labil determiniert*, "capable of reaction to a given stimulus" and "possessing labile determination". Labile determination meant that a tissue was capable of differentiating to some extent in a given direction even without further external stimulus. If the external stimulus was given, the normal fate could then be said to be doubly assured. The existence of labile determination and double assurance is, however, very doubtful (see below, pp. 293 and 308).

† There is no reason why sixty different competences should not exist side by side. The oak tree will form some sixty different sorts of galls. We know as yet practically nothing about the origin of competences.

for the alleged appearance of a lens without the stimulus normally coming from the eye-cup (see p. 293); all these are equally unsatisfactory and cannot longer be retained. Lillie was, however, the first to say in so many words that embryonic potencies are a function of the specific proteins in the regions where they occur; a point to which we shall often return (pp. 203 and 670).

Now to be "reaktionsfähig" or competent with respect to several fates is to be in a condition of unstable equilibrium. Whatever provides the stimulus for the realisation of one of the fates which the competences represent will usually have the effect of suppressing the remaining competences or degrees of freedom, and so of inaugurating a condition of more stable equilibrium.*

Fig. 43. Model showing the interpretation of integral curves for the Ross malaria equations as lines of steepest descent on a topographic surface (see text).

The concept of unstable equilibria in early development seems to offer opportunities for advance in the mathematisation of embryology. Lotka, who has devoted much study to the use of equilibrium concepts in biology, gives a list of the various types of equilibria which may be met with. If only two variables are in question, the integral curves may be plotted on rectangular co-ordinates. Families of curves will be obtained passing through the equilibrium points, which will appear as singular points. As an example, Lotka figures the topographic chart of the well-known Ross malaria equations, which describe the course of events in the spread of malaria in a human population by the bites of certain breeds of mosquitos infected with the malaria parasite. In this chart there are two singular points, one at the origin, O, unstable, the other at T, stable. The chart, Lotka points out, obviously suggests stream-lines and a three-dimensional model. Such a model is shown in Fig. 43. The feature of interest is that a singular point like O is represented by a col or saddle-point in the landscape, so that the

* "Competence is a name for the actual state of the tissue at, and before, the time when the instability is resolved and one or other path of development entered upon. Lillie's segregation was a segregation of potencies, which are not in any way descriptions of the state of the tissue at the time of the segregation, but merely formal attributes derived from a consideration of its future behaviour" (Waddington[25], modified).

attainment of stable equilibrium would be represented by the fall of a point from O to the pit T. The model serves, as Lotka says, to bring out an important fact, namely, that there are necessarily certain regularities in the occurrence of the various types of equilibria. Thus two pits of the character of the point T cannot occur without some other type of singular point between them, just as it is not possible for two mountains to rise from a landscape without some kind of valley between them. Now the series of equilibria met with in embryonic determination might be represented by a series of horizontal ledges descending from a point analogous to O, separated perpendicularly from each other by ridges or cols. Such a model, purely qualitative and intuitional, is represented in Fig. 44.

Fig. 44. Qualitative three-dimensional model of embryonic determination, illustrating the transition from unstable to stable equilibria.

The state of "harmonious equipotentiality" would then correspond to the summit point, where the instability is maximal, and a point could descend to the successive levels of instability not in one direction only, but in several, according to its position and other relations to the organiser region.* This is to say, its prospective potency is not limited to its prospective significance. The true mosaic

---

* Stability landscapes are being found useful by various writers on these subjects. Thus Waddington (*Introduction to Modern Genetics*, pp. 182, 186, 190) symbolises the genotype in development as a system of contours enclosing a valley along the bottom of which run the embryonic processes forming the phenotype. Out of the main valleys branch subsidiary valleys; alternatives which, under certain circumstances (changes in the genotype), the embryonic processes would have to pursue, leading of course to a changed phenotype. This analogy is not entirely happy, since side valleys normally branch *into*, not *out of*, the main ones. A river delta would be more appropriate. But the principal distinction between this symbol and the one here proposed is that for a given individual organism not all Waddington's side valleys are live issues, since the genotype only contains a limited number of mutant genes. Some of them are, for we know that by the action of external conditions (temperature, radiant energy, etc.) genic effects can be imitated in a given organism (the "phenocopies", see p. 392). In the mountain model of embryonic determination, however, all the alternatives are live issues, depending on the primary organiser and the chain of inductors. S. Wright[3] has also used the topographic surface or probability landscape for his interesting arguments about genes and evolution; unfortunately he makes the highest hill-tops the points of maximum stability, which reverses the sense of the symbol.

egg, such as that of the polychaete worm *Sabellaria* (Novikov[1]), would never occupy any of the higher ledges.*

Inorganic solutions or vapours may exist in the condition of supersaturation, and they may then be brought to crystallise or to condense by the introduction of a suitable nucleus, the dimensions of which may be minute relative to the system as a whole. This state of affairs is called a metastable state. Thermo-dynamically it is defined by the fact that its potential is a minimum, but not an absolute minimum. The term metastable could justifiably be applied to the plateau-like states which occur in the above model in embryonic determination. The appropriate organiser or inductor would then correspond to the nucleus which ends the state of supersaturation in an inorganic system. It is reminiscent of Bertalanffy's[1] phrase,† that the egg is "charged with morphological form" or has a "morphological charge". Just as inorganic metastable systems are stable in the absence of their nucleus, so, in the absence of the organiser or inductor, normal development and differentiation will not occur, as is beautifully seen in the exo-gastrulation experiments of Holtfreter[8] (see p. 159).

This subject may also be regarded from another angle, i.e. the appearance and accumulation of the chemical substance which is the inductor in any given case. The morphogenetic hormone or inductor will be the product of a chemical reaction, appearing from a precursor substance, or being liberated from inactive combination, and possibly open to destruction by other metabolic reactions so that its effective concentration at any given time may depend on several factors. Fig. 45 shows the situation in generalised form. The horizontal axis is time (during embryonic development); the vertical axis is the concentration of the inductor. The period of competence of a given tissue for a given fate will then be a zone between the times c_1 and c_2. The inductor may be formed or liberated at different rates; if its critical threshold concentration (the horizontal line i, i) is attained before the beginning of competence, as by curve A, then the determination process (d_1, d_2) will begin as soon as competence begins. As an example, the treatment of amphibian larvae with thyroxin from an early stage onwards might be mentioned; metamorphosis will begin as soon as competence to react to the hormone begins (cf. p. 448). In most cases of early embryonic determination, however, the competence is present before the necessary amount of inductor (cf. p. 154), so that curve B represents the state of affairs. Here the determination process will start as soon as the threshold level is attained. Further increases in the amount of the inductor will not add to the morphological effect. If the

---

* It would, of course, do so in pre-fertilisation stages. The equilibria here envisaged, which become ever more stable as development proceeds, approximate to previous uses of the concept in biology. Thus S. J. Holmes and Przibram (*Anorganische Grenzgebiete d. Biol.* pp. 64 ff.) have spoken of the completed organic form of the animal as being an equilibrium which is disturbed by, e.g., the removal of a limb. There ensues a struggle between growth and surface tension or a mass action effect, ending in an equilibrium which *is* the normal form once more. For what is actually known about regeneration in vertebrates, see pp. 430 ff.

† Bertalanffy (*Kritische Theorie d. Formbildung*, p. 223): "So, wie ein Akkumulator mit Elek-trizität geladen ist, so ist gewissermassen das entwicklungsfähige Ei mit 'Form' geladen, die es in Wirklichkeit umzusetzen trachtet." See also his article *Tatsachen u. Theorien d. Formbildung als Weg zum Lebensproblem* in *Erkenntnis*.

necessary amount of inductor fails to reach the tissue before the end of the competent period, as in curve C, no induction will ever take place. Amphibian exo-gastrulation would be a case in point (cf. p. 159), or isolated ageing ectoderm (cf. p. 281). This right-hand region of the graph also represents the state of affairs, as we shall see, in adult tissues, where inductors are present, but no competence (cf. p. 174). We need not regard the period of competence as absolutely fixed; it may be shifted to the right or abolished altogether. Similarly the threshold concentration might be raised, as indicated by the arrow on the right-hand side of the graph. These changes would lead to the suppression of inductions. As we shall see later, many cases of such suppression are known, both in nature (p. 222), experimentally (p. 224) and by the action of mutant genes (p. 368), but we cannot yet say which of these mechanisms is responsible. There is some evidence,

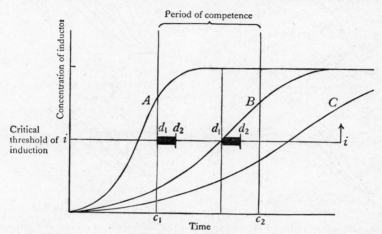

Fig. 45. Generalised scheme of action of an inductor visualised as itself a metabolic product.

derived from studies of avian feather development (see p. 303), that the level of the threshold may be connected with the growth-rate. The faster the part is growing the larger the amount of an inductor is required to affect it. The period of competence might also be shifted to the left; in this case the determination and differentiation of organs would occur before the normal time. Phenomena such as prothetely in insects, where organs characteristic of a later stage may appear in the larva, may perhaps be explained in this way (cf. p. 239).

The above formulation has the advantage that it links up the study of embryonic organisers with the ideas of Goldschmidt on the production of morphogenetic and other hormones in the embryo at rates dependent on the particular genetic equipment of the individual. Animals vary not only as to the presence or absence of a number of organs, but also in the extent to which any one of these organs is formed. An animal might have all its organs complete except one, but that one might be entirely absent. On the other hand, it might have them all, but in incomplete form. Such organisms have been called mosaics* and intermediates respectively.

* Not to be confused with the two other uses of the term. See Glossary.

Shull's work on aphids (*Macrosiphum solanifolii*) led him to use a scheme similar to that just described, which is worth examination because it shows how numerous are the effects which may be produced by one substance if it induces several morphogenetic processes (as for example thyroxin does in amphibian metamorphosis). In Shull's aphids the winged type differs from the wingless in having (*a*) wings, (*b*) three ocelli on the head, (*c*) wing muscles in the thorax, and (*d*) 15–18 sensoria on the third segment of the antennae instead of 4–6. These characters will be referred to as Wg, Oc, Wm, and Sn in what follows. Now the line of demarcation between winged and wingless could be a definite critical

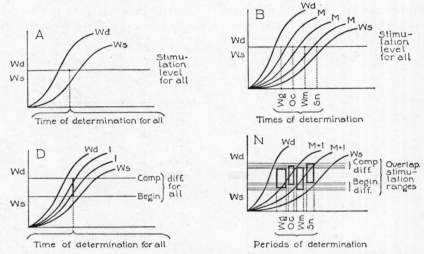

Fig. 46. Schemes of inductor action explaining the occurrence of mosaic and intermediate individuals in aphids.

threshold of inductor concentration, the same for all the structures, and all the structures might have the same moment of determination. In this, the simplest case (Fig. 46 A), a rate of formation of the inductor represented by any curve to the left of the point in question would result in the production of winged individuals (Wd), while a rate of inductor-formation represented by any curve to the right of the point in question would result in the production of wingless individuals (Ws). There would be no intermediates and no mosaics.

Intermediates would occur, however, if subsequent differentiation depended upon the concentration of inductor which had been formed at the moment of determination. This is shown in Fig. 46 D. The effective concentration of inductor would thus be a range, rather than a critical threshold. All curves passing to the left of it would mean winged individuals, all curves passing to the right would mean wingless ones, but intermediate curves would mean the production of true intermediate individuals, which would show greater or less approximation to the winged state depending on the concentration of inductor at the critical time.

Conversely, mosaic individuals would occur if the critical time of determination were replaced by a range or period, as in Fig. 46 B. Here the critical threshold is the same, but moments of determination are different, and the curve immediately to the right of the full "winged" curve would mean the production of individuals which had Oc, Wm and Sn, but no wings. Slower accumulations of the inductor could produce individuals having complete Sn, but nothing else.

These two cases are combined into one in Fig. 46 N, where there are overlapping stimulation ranges, but periods of determination which do not overlap. The second curve from the left would mean the production of individuals which had fully formed Wm and Sn, but only partial Wg and Oc. The third curve from the left would mean the production of individuals which had partially formed Wm and Sn, and no Wg or Oc. The right-hand curve is that for wingless individuals. There would thus be both mosaic and intermediate types. And this is what does actually occur in aphid development. Many other cases are possible, and some are given in the papers of Shull & Stiles.

I shall conclude this section with a few words on the possible biochemical meaning of the condition of competence. Although we have few facts by which to guide our thought, we can at any rate envisage certain reasonable possibilities. Arising out of Wigglesworth's work on the moulting hormone in bugs (see p. 465), came the suggestion that there may be in the cell as many enzyme reaction chains as there are competences; in this case the inductor might be a co-enzyme which would give one such enzyme reaction chain speed priority over the others. Another suggestion, made to me in conversation with L. Rapkine, is that for every competence there might exist a specific protein with its own specific precursor, like trypsin and trypsinogen. The inductor would then be a very minute quantity of the protein itself, which would "seed" the system, and bring about the formation of the characteristic protein by auto-catalysis. A third possibility, in view of the work of Needham, Shen, Needham & Lawrence on myosin and adenylpyrophosphate, is that an inductor may be a substrate, and that during the unavoidable reaction which will follow its entry into the competent cell, the physical configuration of the enzyme protein may itself be changed. In any case, determined differentiation will be due to the direction of the cell metabolism into a specific path.

2·16. The liquidation of the entelechy

The preceding analysis differs very deeply from Driesch's own interpretation of his fundamental experiments[7]. As is well known, he regarded them as affording proof of vitalism, that is to say, of the effective guidance of the embryo's development by a non-material agency. To some it may seem that his views are now of purely historical interest. In my opinion, however, they represent a tendency which has often recurred in the history of scientific thought, a tendency to base a pessimism about the power of scientific analysis on a scholastic and undialectical logic. Only those who have well weighed his arguments are fully entitled to accept the hopes and promises of true physico-chemical embryology.

It will be remembered that reference was made above to the invariable factor of wholeness, E (see p. 101), which is involved in the formation of two whole embryos from two blastomeres which, left to themselves, would only have formed one. "It was not without design", wrote Driesch,* "that I chose the letter E to represent this. Let that factor in life phenomena which we have shown to be a factor of true autonomy be called 'entelechy', though without identifying our doctrine with what Aristotle meant by the word ἐντελέχεια. There is, however, at work a something which bears the end in itself, ὁ ἔχει ἐν ἑαυτῷ τὸ τέλος." The (Drieschian) entelechy was then defined as intensive manifoldness (*intensive Mannigfaltigkeit*) as opposed to the extensive or visible manifoldness which becomes evident as development proceeds.† All order in morphogenesis is exclusively due to the action of entelechy.‡ The entelechy does not exist in space (space-time), but acts into space,§ it is therefore not localisable at any point in space-time.‖ Lacking all the characteristics of quantity, and being incapable of measurement, it cannot be a form of energy,⁋ and it cannot therefore affect the first law of thermodynamics.** Its definition was finally admitted by Driesch[8] himself to be "a system of negations".†† He could, however, describe its action as that of suspending possible reactions,‡‡ one or more of which would otherwise proceed in its absence, and in this sense it could be called an "arranging agent".§§

It seems astonishing to us to-day that Driesch should have had recourse so soon in his analysis to a conception so obscure in all its particulars, so out of harmony with the natural products of the scientific method, so certain to close the door to further experimentation, so resembling the Galenic δύναμις or "faculty".‖‖ We cannot see to-day why it was necessary to place the intensive manifoldness of the egg outside the physical world, in view of the tremendous complexity which the colloidal constitution of protoplasm must involve. Even at the atomic level there would be ample room for it. It must have been evident enough, even in the last century, before the work of Wo. Köhler⁋⁋ and his school, that, when a magnet is divided into two, we are not left with one north pole and one south one, but that an immediate "regeneration" of the whole pattern takes place, and that we have two fields of force, similar morphologically to the original one, but smaller. The particles of the magnet are not, as it were, "determined" to form part of any given pole, but can form part of either, according to their position in the whole. This analogy only needs transposition to the more complicated realm of biological colloids (see below, p. 127).

Again, the *Verschmelzungsversuche*,*** where two eggs are fused together, yet one normal embryo results, have their analogy in the long-known phenomenon of convergent crystallisation (*Sammelkristallisation*). Spatially separated small crystals commonly unite themselves into large ones of the same structure. Still

* Driesch, H., *Gifford Lectures*, p. 106.   † *Ibid.* p. 245.   ‡ *Ibid.* p. 154.
§ *Ibid.* pp. 254, 298.   ‖ *Ibid.* p. 299.   ⁋ *Ibid.* p. 256.
** *Ibid.* p. 257.   †† *Ibid.* p. 300.   ‡‡ *Ibid.* p. 261.

§§ *Ibid.* pp. 292. In this connection it is important that the teleological significance of the entelechy was greatly diminished in the following decade by the discovery of self-differentiation and the realisation that all embryos pass through a stage, introduced by the action of the organisers, where no regulation is possible. The entelechy thus deserts its post when the work is one-third completed. Embryologists were accordingly led to question whether it had ever been there at all.

‖‖ See the valuable edition of Galen, *On the Natural Faculties*, by A. J. Brock.

⁋⁋ Already in 1877 Pflüger and the botanist Vöchting had seen the significance of the magnet analogy for life phenomena.

*** Accomplished for echinoderm eggs by Driesch[3] and for those of the newt by O. Mangold & Seidel. Mosaic eggs, when fused, may give larvae containing two notochords, two nervous systems, etc. (v. Ubisch[4] on ascidia; Ranzi[8] on cephalopods).

more striking is the behaviour of the ammonium oleate crystals of O. Lehmann and Przibram[5], for these soft forms, if side by side, join up together, passing through all the stages corresponding to *duplicitas* greater or lesser in animals.

Driesch's pessimism was often expressed. He did not wish, he said,* to discredit a thorough and detailed study of osmosis or colloid chemistry in their relation to morphogenesis. "But these investigations never give us a solution of the problem." Or again:† "Atoms and molecules by themselves can only account for form arranged, so to speak, according to spatial geometry, they can never account for form such as the skeleton of the hand or foot." But he touched an important point when, in discussing the limits of entelechial action, he granted‡ that "it may depend on some very unimportant peculiarity in the consistency of the protoplasm that the isolated blastomere of the ctenophore egg is not able to restore its simple intimate protoplasmic structure into a small new whole". We know now (as the narrative of Huxley & de Beer[2] well shows) that whether the rearrangement of materials needed for regulation around the main polar axis can take place or not is a matter largely dependent on the viscosity of the protoplasm —an eminently physical factor. The distinction between mosaic and regulation eggs, therefore, can partly be defined in terms of internal friction.

The intensive manifoldness above the atomic level in the uncleaved egg generally receives to-day a formulation in terms of the gradient concept, Child's great contribution to biology.§ The polarities of the egg, brought into being partly by its chance position in the ovary, partly by the point of entry of the spermatozoon, or other similar factors, involve a system of gradients describable in a three-dimensional net of co-ordinates. Quantitative differences in activity along the gradients lead to qualitative differentiation; localisation is due to relative position on a gradient. Even the best cases of harmonious equipotential systems possess limitations unrealised by Driesch; thus the single blastomere isolated from the first four of the sea-urchin regulates because it possesses the full extent of the gradient (cf. pp. 477, 480), for if it is divided transversely, or if the original whole egg is divided transversely (equatorially), no perfect embryo will be formed (Hörstadius[1]; Plough; Schaxel[1]; Conklin[4]). Perfect equipotentiality therefore depends on the possibility of separating a system into parts without segregating regions of different potencies (cf. p. 140).‖

---

* Driesch, H., *Gifford Lectures*, p. 63.   † *Ibid.* p. 101.
‡ *Ibid.* p. 266.

§ So much has been written on biological gradients that it is hard to give a guide to the literature. The subject will be referred to again below in the discussion of fields (p. 129). The value of Child's views on the "metabolic" nature of the gradients will be found discussed in another place (p. 605).

‖ It is of some interest that Driesch's argument, though leading to such unmechanistic conclusions, is fallacious precisely because it adopts such mechanistic premises. To these it *adds* entelechy. The fact that a heap of cells (*summenhafte Gesamtheit*) becomes a whole, in spite of any desired removals and injuries, can only be explained by the entelechy hypothesis, says Driesch. But the premise is not correct; the regulating egg is not a heap of cells (*summenhafte Beieinander*), but an organised system. Regulation can only occur if the system is in a position to reconstitute its original pattern after the removal or shuffling of its cells. The unorganised heap of cells is an abstraction which does not enter into the question, and can be banished from discussion. But if it goes, the entelechy goes with it (Bertalanffy, article in *Erkenntnis, loc. cit.* p. 368). In 1927 Driesch[9] is still writing: "Der Organismus kann zwar in jedem Moment seines Daseins seiner materiellen Seite nach als chemisch-aggregative 'Maschine' gekennzeichnet werden, aber seine andere Seite, der Maschinenbenutzer, darf nicht vergessen werden." This is the purest Cartesianism; it was Descartes who introduced the practice of calling organisms machines, and then postulating transcendent mechanics to drive them.

Driesch's proposition* "that a system, in the course of becoming, is unable to increase its manifoldness of itself" raises, of course, the whole of the age-long controversy between the preformationists and the epigenesists.† But Woodger[1] points out that an organised entity having components standing in internal organising relations to each other *must* have its degree of multiplicity increased if (1) the number of components is increased, (2) the complexity of the relations in which those components stand to one another is increased, (3) the intrinsic patterns of the components become different from one another. Thus, granted the possibility of spatial repetition of a pattern and the possibility of "differentiating division" and "histological elaboration", there is no need to appeal to agents of any kind. If we do not need to invoke the entelechy to account for cell-division, this alone will suffice in principle, since this process must yield an increase in multiplicity in succeeding slices of the "division hierarchy". "Neither the preformationists nor the epigenesists knew anything of spatial repetition of pattern or of histological elaboration, since such processes occur only in living cells, and there was therefore nothing in the experience of the early embryologists to suggest that such things occurred." Here Woodger goes to the bottom of the embryological problem in its biochemical aspect, for our aim must be to discover how it is possible for the formation of duplicate protein (and other) molecules to occur. *The mass-production of replicas is the issue.* Whether it occurs by some form of template method, or by the scission into units of an endless protein chain, we have at present no idea.

In his acute analysis of causation, Woodger[2] has another argument with an important bearing on the theory of the harmonious equipotential system. If two entities are manifestly (i.e. apparently) non-different, we cannot infer that their causal predecessors are intrinsically non-different. This is what makes causal analysis so difficult, because it means that there will always be an element of doubt in the assumption that the two causal predecessors are intrinsically non-different. That two entities are *not* intrinsically different is *always* an assumption. Thus, in the case of pluripotency, it is argued that, since any given monad in the original egg can become any given part of the resulting embryo, therefore the spatial parts or monads of such eggs are intrinsically non-different. In that case it is impossible that differences can subsequently arise in development, since the environment cannot be alone responsible. Hence we are driven to conclude that an entity, the entelechy, which is not a spatial part or monad of the egg, is "at work" producing the differences required for further development. Now it seems clear that such arguments can never be decisive on account of the principle that intrinsic non-difference is always assumed, and consequently it is always open to an unbeliever to say that such experiments do not exclude the possibility of intrinsic differences between the parts concerned *which are not revealed when they are in isolation*. Or, in more biological terminology, the properties of a given monad other than its fate may well be different according to whether it forms part of a two-cell system, or of one of the two cells in isolation.

The meaning of the phenomena of pluripotency has recently been the

---

* Driesch, H., *Gifford Lectures*, p. 319. This is clearly put forward in his *Der Begriff der organischen Form*, pp. 42 ff., where we find: "Grundsatz I: Wie im rein Logischen die Folge nicht mannigfaltiger an (logischem) Inhalt sein kann als der Grund, so kann sich auch im Laufe des Werdens der Grad der Mannigfaltigkeit eines Natursystems nicht 'von selbst' erhöhen. Wer das zulassen würde, würde den Begriff der Kausalität als einer Analogie zur Konsequenz, verletzen." But is this unexplained extension of formal logic to natural phenomena justified? If it is wrong, Driesch's whole theoretical system is wrong.

† See my *History of Embryology*, ch. 4.

subject of an interesting polemic between Mirakel and Wermel. Mirakel purported to bring a proof of the existence of the entelechy logically better than Driesch's, but Wermel had little difficulty in showing that his argument was vitiated by a failure to take account of the dominant position which one part of the egg or egg fragment (the organiser region) has over the remainder. Mirakel considered that better support for the Drieschian hypothesis could be derived from the effects of "*Verlagerungsversuche*", i.e. experiments in which the primary blastomeres of the egg are shuffled at random together and yet give a normal embryo, rather than from the effects of fusion, amputation, or isolation. Some non-spatial agency must be acting upon the shuffled blastomeres in a marshalling manner. But Wermel showed that this conclusion was only true if the material factors which determined the fates of the parts had to act on each one independently and separately; it was not true if one part could determine the fates of the rest. This would then happen no matter what chance arrangement of the blastomeres had taken place. And indeed we know that without the presence of an organiser in the isolated fragment, there can be no "harmonious equipotentiality".*

In conclusion, a word may be said of the remarkable animistic and anthropomorphic tendency noticeable in Driesch's concept of an "arranging agent". For centuries science has struggled to rid itself of the remains of popular demonology.† When Cambridge got its first Professor of Chemistry‡ (in 1703), acids were male and alkalies female; minerals grew, like plants, from seeds; slaked lime protested by giving out heat; and solid bodies cleaved to themselves in cohesion because they preferred the touch of the tangible more than the feeble contact of air. The entelechy belongs shamelessly to pre-Boylian biology. The work of the logistic positivists, Carnap, Schlick and Neurath, clearly illustrates this.§ Their purification of scientific terminology convinced them that the majority of biological concepts, such as cell-division, growth, regeneration, etc., could be reduced to incorporation in the "physical language", i.e. in language based on direct experience. This was not the case, however, with words like "entelechy", which are incapable of formally correct definition, and hence can only occur in meaningless statements.‖ Schlick's prediction that in the future no more books will be written about philosophy, but that all books will be written in a philosophical manner, implies that the concept of entelechy will be searched for in them in vain.¶ And the judgment of another philosopher on the entelechy is

---

* The whole of this argument was foreseen in a somewhat abstract way by Woodger[1]. This is a summary of his reasoning: The possibilities before us in the matter of embryonic development are (1) a Weismannian nuclear preformation; (2) a cytoplasmic preformation; (3) environmental differences during development; (4) appeal to a Drieschian transcendent principle; (5) rejection of the causal postulate. Of these (1) and (3) are excluded as decisively as anything can ever be by experiment. Even Driesch shrinks from (5). No help is forthcoming from (4). Only cytoplasmic preformation remains. "Would it be possible to appeal to just enough cytoplasmic preformation to give us our main axes and the primary germ-layers, and could we thenceforwards appeal on a large scale to relational properties?" (in which organiser phenomena would be included). This is certainly the outcome, in general terms, of modern embryology.

† See the interesting article of J. G. Gregory, *The animate and mechanical models of reality*.

‡ John Francis Vigani, see E. S. Peck.

§ See Carnap, R., *The Unity of Science*, and the trenchant essays of H. Winterstein (*Kausalität u. Vitalismus vom Standpunkt d. Denkökonomie*) and of M. Hartmann (*Die methodologischen Grundlagen d. Biologie*).

‖ Carnap, R., *Überwindung d. Metaphysik durch logische Analyse d. Sprache*.

¶ In Driesch's most recent writings[10] (1935, 1937), he assures us that in his vitalism the existence of formative stimuli was never denied, and retires to the impregnable position that life *is* organisation and that the action of organisers in development must itself be organised.

worth quoting. "There is no magic in mind as such", writes Broad, "which will explain teleology. A mind does not account for anything until it has wit enough to have designs and will enough to carry them out. If you want a mind that will construct its own organism, you may as well postulate God at once, for if he cannot perform such a feat, it is hardly likely that what has been hidden from the wise and prudent will be revealed to entelechies."

2·17. Regional differentiation

We must now return to the Induction process and analyse it a little further.

The classical method of testing whether a given material is able to perform an induction is to place it in the blastocoele cavity of a young gastrula (the *Einsteckung* technique of O. Mangold; see Fig. 62). In such circumstances an induction occurs as a reaction between three factors; the graft, the overlying ectoderm of the host, and the host's organisation centre. During the reaction, the host ectoderm undergoes a process of determination to become part of an embryonic axis, which often includes, besides neural tube, mesodermal organs such as notochord and somites. This determination, and thus the induction as a whole, can only, as we have seen above (p. 112), occur if the ectoderm is in a reactive state; if it is too young during the application of the inducing stimulus, no effect will be produced; while if the ectoderm has already been determined to be something else, such as epidermis, for example, the inducing stimulus cannot alter that determination. In other words, the ectoderm must be competent for the primary organiser. In this classical type of induction experiment, two processes, or rather, two aspects of the same complex process, can be distinguished. Roughly speaking, these two processes can be called (1) the determination that an embryonic axis shall be developed, and (2) the decision as to the regional character of that axis.

An induced embryonic axis often has a "regional" character, i.e. it is a head or a tail or a section of the trunk; some definite slice of the body. Now it is easy to find out, by vital staining or in other ways, what should be the regional character of the implanted organiser. If we compare this presumptive regional character of the graft with the actual regional character of the induced axis, we find first of all that the induced axis nearly always includes a larger region than the presumptive fate of the graft would have led us to expect (Spemann[13] for the newt; Waddington & Schmidt for the chick). Isolated pieces of the organiser region have, in fact, a tendency to regulate themselves to form a whole. We shall see later (p. 283) some striking examples of this. Secondly, we find that the determination of the regional character of the induced axis may be due to the graft, i.e. the regional character of the axis may be exactly the same as the presumptive character of the graft. Thus Spemann[13] showed that organisers with the presumptive fate "head" can induce heads even in the posterior parts of the host body in the newt. Waddington[3] described a similar case in the chick, in which, moreover, the anterior-posterior axes of the host and the induction were opposed. It is therefore clear that sometimes at least the grafted organiser can determine the regional character of the induced axis as well as its presence. But this does not always occur. If we return to our inspection of different cases of induction, we find, thirdly, that the regional character of the induced axis may not be the same as the prospective fate of the graft. Thus Spemann[13] showed that "tail" organisers grafted into the head region of the newt gastrula induce not tails, but heads. He favoured the suggestion that this was due to a particular kind of reactivity in the head ectoderm, but Holtfreter[5] was able to show clearly that no such difference exists between the reactivity of head and tail ectoderm. Holtfreter and Waddington accepted Spemann's alternative suggestion that in such cases the regional character of the induced axis is dictated by the host's organisation centre.

In the classical type of induction we are, then, confronted with two determinations, that of *presence*, and that of *character*. Needham, Waddington & Needham suggested the term *Evocation* for the first type of determination, and *Individuation* for the second.* In the ordinary induction, one of these determinations, evocation, is always performed by the graft; while the other, individuation, is performed by the graft and the host working together, either in a co-operative or an antagonistic manner. We shall see later on (p. 271) that this distinction has proved of considerable value in discussing the new knowledge of organiser phenomena. Evocation is the stimulus to neuralisation, mediated by a chemical substance or substances, acting upon the competent host ectoderm. Its exact histological effects have yet to be elucidated with absolute clearness. Individuation, on the other hand, is the process which builds the neural tissue so formed into a nervous system of which the head end is distinguishable from the tail end. It may well be that chemical substances are involved within the individuation process also, but up to the present time it has not been possible to prove their existence. Primarily individuation is thought of as a "field" phenomenon.† It follows that from the implantation of single chemical substances or fractions one would not expect to get well-formed secondary embryos. As will be seen later (p. 277) the implantation of dead tissues usually leads to the formation of chaotically arranged organs and tissues, while the implantation of single chemical substances or fractions leads only to neural tube formation.

While the exact delimitation of the processes is still not possible, some considerable progress has been made (see p. 278), but the discussion of it must be postponed until the chemical data have been further described. The question is, What is the character of a neural axis which has been evoked but not individuated?

---

* While the word "evocation" has had a long but undistinguished literary history, "individuation" originated in philosophy and passed into biology a long time ago. Commentators on Aristotle, such as Alexander, Simplicius and Themistius, and later the scholastics, discussed the nature of a principle of individuation which differentiates particular objects from others of the same species (see Ross' introduction to Aristotle's *Metaphysics*, p. cxv). Hence the word appears in Jeremy Taylor. Apparently it was S. T. Coleridge, the philosopher-poet, who in his *Theory of Life*, posthumously published in 1848, introduced it into biology. "I define life", he wrote, "as the principle of individuation, or the power which unites a given *all* into a *whole* that is presupposed by its parts" and he spoke of the assimilation of the "lower powers" such as cohesion, elasticity, magnetism, electricity, and chemistry, into the more highly organised living whole. I cannot find that either Herbert Spencer, in his *Principles of Biology*, or Child, in his *Individuality in Organisms*, added much to Coleridge. In any case, the word "individuation" is here used in a much more restricted and well-defined sense (see Glossary).

† Spemann, in his 1921 paper ([10], p. 568), was the first to associate the field concept with the organisation centre.

There were three possibilities: (*a*) that it would consist of a neural tube accompanied by notochord and somites in their proper relative positions, but that the whole structure would be "sausage-like", one end being indistinguishable from the other (see Fig. 47), (*b*) that the axis would be much less definitely arranged, with an abnormal cross-section, or (*c*) that the induced material would consist merely of histologically differentiated neural, chordal, and mesodermal cells, without any definite arrangement at all. The second of these possibilities is now known to be the most nearly true.

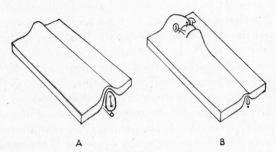

A   B

Fig. 47. Diagram illustrating the concepts of Evocation and Individuation. A, neural axis with no regional differentiation; B, neural axis with regional differentiation.

The notion of the Individuation Field had previously arisen from the experiments of Waddington & Schmidt, in which induction of the primitive streak in birds by heteroplastic grafts had been studied. In the section on Regional Differentiation (2·391), this work will be more fully described; here it is only necessary to outline the concept of the individuation field which was developed in it. It was found that an isolated organiser tends to rearrange its structure so as to include all regions of the embryo. Thus, "an organiser whose presumptive fate was to become the mesodermal part of the middle region of an embryo may rearrange itself so as to become the mesodermal part of a whole embryo....An organiser tends to rearrange the regional structure both of itself and of any tissue lying near it in such a way as to make that tissue part of a complete embryo. This is then a true unit-building activity of the organiser, which works, not by controlling the process of induction, but by modifying its products." The activities of the organisation centre could thus be stated as follows. There are, on the one hand, more or less autonomous processes of induction of neural plate, etc. (Evocation) by which the organiser can prepare from its surroundings all the types of tissue necessary for the formation of a complete embryo. These processes are not under the direct control of a unit-building force, as can be seen by the occurrence of homoiogenetic induction (see p. 154). But in point of fact the amounts of the various sorts of tissue which are normally formed by these processes just suffice for the production of one embryo. Meanwhile a process of a different type, which is expressed in the individuation field, is in control of the regional character of these tissues, and attempts to build them into one complete unit. Waddington & Schmidt described the individuation field as follows: (1) it is strongest near the embryonic axis and falls off in intensity towards the edge of the area pellucida, (2) it is regionally differentiated along the antero-posterior axis of the embryo, (3) more anterior parts dominate over more posterior parts.

The question may be raised whether the action of the individuation field necessarily depends on the presence in it of its active primary organiser. This is now under study by Abercrombie[3].

2·18. Field phenomena

The conviction that the magnetic field has a part to play in guiding our ideas about morphological patterns is now of some seniority.* I pointed out earlier that it fulfills the criterion of divisibility of pattern which was needed for embryology (p. 120). Now if an analogy is a true one, the course of scientific thought will be greatly facilitated; thus Wrinch[1] has referred to the fundamental progress in electrostatics and current electricity due to Clerk-Maxwell's genius in seeing formal resemblances between the conditions to be satisfied in those subjects and those satisfied in already solved problems belonging to other subjects. The masterly apparatus developed by Laplace and Poisson in the treatment of elasticity became available in this way for electrical theory. Can we then obtain knowledge of biological phenomena by means of our knowledge of magnetic phenomena? In other words, is there an identity of formal characters in the two cases? Many, especially in recent times, have thought so, notably Weiss[4]; Gurwitsch; and Rudy. Schultz has caustically remarked that fields without centres, without measurable field strengths, and with no obvious lines of force, cannot invite comparison with the fields of physics. But this criticism is hardly justified, for in William Gilbert's time (1600) certain phenomena of magnetic fields

---

* As a matter of fact it goes back to light-mysticism of Persian and Jewish origin. The Kabala, which was studied by all seventeenth-century biologists, led in one case to the extremely interesting work of Marcus Marci, of Kronland, a Czech. His *Idearum operatricium Idea*, published in 1635, was a mixture of purely scientific contributions to optics, and speculative theories about embryology. Thus he explained the production of manifold complexity from the seed in generation by an analogy with lenses, which will produce complicated beams from simple light sources. The formative force radiates from the geometrical centre of the embryonic body, creating complexity but losing nothing of its own power. Monsters originate from accidental doubling of the radiating centre, or from abnormal reflections or refractions at the periphery (cf. mirror reduplications, accessory organisers, etc.). Marcus Marci, in his forgotten work, thus links together the following trends of thought: (*a*) the old Aristotelian theory of seed and blood, (*b*) the new Cartesian rationalistic mathematical attitude to generation, (*c*) the new experimental approach, in his work on optics, (*d*) the cabbalistic mysticism of light as the fountain and origin of all things; finally, by his brilliant guess of centres of radiant energy, (*e*) he anticipates the field theory of modern embryology. Cf. my *History of Embryology*, ch. 2.

The only parallel to this occurs, it seems, in a quarter far removed from Marci at Prag, but equally devoid of influence upon contemporary thought, namely, the *De Motu Animalium* of Borelli, the founder of the iatro-mathematical school. Chiarugi gives an account of the chapter on generation (Pt. II, ch. xiv). Its interest is that Borelli compares the semen to a magnet arranging iron particles in a field of force. In Harvey, too, a reference to the magnetic field can be found. In the discourse on Conception (1653, p. 539) he says: "The Woman or Female doth seem, after the spermatical contact in Coition, to be affected in the same manner, and to be rendered prolifical, by no sensible corporeal Agent; as the Iron touched by the Loadstone, is presently endowed with the virtue of the Loadstone, and doth draw other iron-bodies unto it."

In the eighteenth century, of course, traces of this are easier to find. Thus from Bourguet, one of the saner ovistic preformationists, may be quoted (1789): "Le mécanisme Organique" which works in generation "n'est autre chose que la Combinaison du Mouvement d'une infinité de Molécules étheriennes, aëriennes, aqueuses, oléagineuses, salines, terrestres, etc. accomodées à des systèmes particuliers, determinés dès le commencement par la Sagesse suprême, et unis chacun à une Activité ou Monade singulière et dominante, à laquelle *celles qui entrent son système sont subordonnées*." (Italics mine.) (*Lettres Philosophiques etc.* Amsterdam, 1729.)

were known although no method was available for measuring field strength. The concept of "biological field" does give a powerful aid to the codification of the *Gestaltungsgesetze*, the rules of morphogenetic order. As complex components (in Roux's phrase) they must first be ordered before they can be analysed.*

The field concept, however, has suffered greatly hitherto by an insufficient accuracy of definition, leading in particular to an uncertainty in the attribution of activity to the fields. For some thinkers the fields have been purely descriptive and symbolic, performing nothing, but permitting us to use a shorthand notation, as it were, for morphological form—for others, the fields have been actively arranging entities, performing morphogenetic work. It is of the first importance, as Waddington[8] has pointed out, that this confusion should be cleared up.

For Gurwitsch in the main the fields are geometrical symbolism, yet from time to time he speaks of their having an action or an influence.[†] For Weiss[4] the fields are more active.[‡] He draws three main conclusions about them. First, if a certain amount of material is removed from the domain of a field, the remainder of the field manifests in due course the same pattern which it would normally have given in larger size (equipotential sea-urchin eggs, regeneration of amphibian limbs). Secondly, if unorganised but organisable material is introduced into the field domain, it is incorporated in it (regeneration of limbs, avian development). Thirdly, two or more fields can fuse to form a larger one (fusion experiments on invertebrate eggs).[§] A similar position is taken by Waddington[26], largely as the results of his experiments on induction in the chick embryo. The "Individuation Field" is a term expressing the tendency of an organiser to rearrange the regional structure both of itself and of any tissue lying near it in such a way as to make that tissue part of a complete embryo. In the chick, for example, a piece of neural plate in the process of being induced in ectoderm lying somewhere within the host's individuation field may follow one of three paths: (1) it may become fused with the host and the two built up into a composite neural plate, (2) it may develop as a separate structure, sharing the regional character of the contiguous parts of the host, (3) its neuralisation may be suppressed, and it may be worked into the host in the form of body ectoderm. The first and second paths are the most usual.

* It should be noted that there is an intimate relation between the field concept and the theory of transformations of d'Arcy Thompson (see p. 531). This would be more generally realised if the Thompsonian method had been applied to the change of form during ontogenesis. That particles do tend to orient themselves on progressively distorted co-ordinates is one of the most essential facts expressed by the term "field".

† His definition is: "Als Feld wird hier ein Raumbezirk verstanden, in welchem durch die Angabe der Coordinaten jedes beliebigen Punktes auch die Gesamtheit der Einwirkungen auf ein in betreffenden Punkte befindliches Objekt in eindeutiger Weise festgesetzt wird." *Loc. cit.*, first paper in *Archiv f. Entw.* p. 392. Rudy repeats this.

‡ Cf. his definitions on p. 24 of his *Morphodynamik*.

§ On p. 293 of his *Principles of Development* the criteria are fuller. A field is (1) bound to a material substratum, (2) an entity, a pattern, not a mosaic, (3) heteroaxial and heteropolar, (4) able to include recognisably separate districts, (5) able to maintain its pattern when its mass is reduced, (6) able to perpetuate its pattern when completely divided into two or more pieces, (7) able to maintain its pattern when its mass is increased, and to fuse with similar pattern entering with the new material if the axial orientations are favourable.

But the amount of thought which has so far been devoted to the concepts of field and gradient is hardly yet sufficient to enable them to bear the load of facts which is often placed upon them.* The gradient concept is of the two the less fundamental, for a system of gradients radiating from a centre would constitute a field, and probably no gradient exists only in two spatial dimensions. In this conceptual structure, then, we find that the term "field" is applied to at least two quite distinct things. In the first place, it is applied to morphological and pre-morphological domains of the character just described in Waddington's words: i.e. wholes actively organising themselves. Here it may not be illegitimate to use such expressions as "the field is a region throughout which some agency is *at work* in a co-ordinated way",† or "the individuation field *sees to it* that"‡ something or other happens. But, in the second place, there is another kind of region also to be borne in mind, i.e. regions "possessing a general determination for the production of certain structures and undergoing progressive regional specification of detail". These, since they only imply *location*, should perhaps be more properly called "districts".§ We shall return to consider them in more detail later (p. 286).

The situation is, of course, complicated. In respect of any given organ, for instance, there are to be considered (1) the district from which it normally develops (the presumptive district), (2) the district from which it will develop if the presumptive district and the surrounding tissue is extirpated, (3) the regions, isolated parts of which will develop into the organ if transplanted into its district, (4) the region from which the organ can be formed by a specific or non-specific stimulus, etc., etc. We need a new vocabulary for these entities just as much as new experiments.

Another, possibly fruitful, way of distinguishing between fields and districts may be derived from a remark of Weiss in his *Morphodynamik*. At a given point on a vector in a field, a field force possesses (*a*) a given quality, (*b*) a given direction, (*c*) a given intensity. In a field proper, such as the individuation field, we could say that every point will possess a different value for each of these three factors, but in a district only the two latter ones will vary; the first will be the same throughout, e.g. limb or heart. Weiss himself, however, did not distinguish between fields and districts.

How can a morphogenetic field be defined? Perhaps it is best thought of in terms of instability. Thus, according to Waddington[8], a field is a system of order such that the position taken up by unstable entities in one portion of the system bears a definite relation to the position taken up by unstable entities in other portions. The field effect is constituted by their several equilibrium positions. This description is in accordance with the conception of development as a series of successive decisions regarding the fates of parts mediated by a succession of organisers or evocators, chemical substances acting as morphogenetic hormones.

---

* See the critical review of Huxley[12].   † Huxley & de Beer[2], p. 276.

‡ Huxley & de Beer[2], p. 319.

§ I reserve the term "area" for the regions of fate-maps. See Glossary.

Instability (high potentiality) is continually giving place to stability (restricted potentiality). A similar thought has been expressed by Whyte: at any moment the components of the factors mathematically describing the equilibrium reached form a continuous spatial function with simple symmetry properties characteristic of the system.

On the other hand, since we know that the primitive layers of cells of which embryos are built are usually indifferent, in the sense that their material may be used for the construction of almost any part, stress may be laid upon the well-known fact that behaviour is a function of position in the whole. Two protein molecules, identical save for their position in the whole, will take part in two quite different structures. Thus, according to Floyd: if to every point of a region of space can be assigned a definite value of a variable whose values are descriptive functions of the organism, so that the properties of the point become a function of position in the region considered, the region can be called a field. In all these questions the principal conceptual difficulty for the biochemist arises from the fact that the cohesive and organisational forces seem to act on what may be called the supra-chemical level. We can as yet hardly form a picture of the way in which the chemical affinities, the colloidal forces, and the large-scale mechanical factors, are integrated in development to produce the morphology of the completed animal.

In any case, it is doubtful whether there is anything in the field concept over and above that of Function, in the mathematical sense. Thus, in an electrostatic field one has simply the idea of the totality of values of the (potential) function associated with a certain set of points. For convenience the equipotential lines or curves on which this function is constant are then often drawn in. The loci along which the function changes with maximum rapidity may also be drawn in. But the equation contains all this information in itself. Would it not be possible, therefore, to arrive at some such equation if we knew the nature and properties of the organiser substances involved, their speeds of diffusion, etc., and also the geometrical distribution of competence or reactivity? In the near future we can hardly hope for such complete knowledge.

2·19. Conclusion

We have now considered the principal concepts necessary for the understanding of morphogenetic stimuli in embryonic development. *Dependent differentiation* implies the *determination* of the fate of a part by *induction* from another part. *Self-differentiation* is the response to the induction stimulus, which may be *primary, secondary*, etc. (of first or second grade, etc.) depending on the time at which it acts in development, but the response can only be given by tissue *competent* to give it. Normal induction involves two factors, *evocation* and *individuation*, the former being the histological change produced by the action of a chemical substance, and the latter the regional organisation which involves, in all probability, field forces, though the action of further chemical substances is likely. When self-differentiation has done all it can do, the stage of *functional* *differentiation** follows, and the activity of hormonal and neural factors effects a *reintegration* (Weiss[16]) of the organism.

2·2.　Mosaic eggs and their biochemistry

As has already been stated, the distinction between mosaic and regulative eggs, which is not an absolute one, really means a difference in the time at which the determination of the main features in the individuation field is accomplished. *Mosaic eggs* are those in which this determination has taken place before fertilisation and cleavage; *regulation eggs* are those in which this determination does not take place until about the time of gastrulation. To the former class belong the eggs of most invertebrate groups, such as worms and molluscs; to the latter belong the eggs of echinoderms and vertebrates. Since it is in regulation eggs that the induction phenomena underlying development can best be studied, and since our most complete knowledge concerns the eggs of amphibia, the succeeding sections will be devoted to an account of the development of this class of animals. Problems of wider significance will there be dealt with as they arise, and only after discussing the chordates shall we return to the invertebrates, especially the echinoderms and insects.

In the meantime something may be said about the mosaic eggs, those which are fully determined already at the time of fertilisation (see the review of Penners[2]). In recent years, two veteran embryologists of the American school have contributed particularly valuable additions to our knowledge in this field. In 1905 Conklin[1] described three different kinds of cytoplasm in the egg of an ascidian, *Styela*, a typical mosaic egg; (1) yellow material, which later forms the muscles of the "tadpole", (2) a slate-grey substance, which later forms the endoderm cells, and (3) the transparent cytoplasm, which later forms the general ectodermal covering of the body. Not until 1931, however, did he publish the results of centrifuging experiments[3]. Here it was found that strong centrifuging would completely dislocate further development. Chaotic differentiation followed; in Fig. 48, for example, the notochord and endoderm are on the outside of the mass, with a stratum of neural tissue between them. All imaginable types of dislocation were produced; in contrast with the echinoderm egg, which, as is so well known, can be centrifuged severely without effect on subsequent development, save an unusual distribution of yolk material among the blastomeres.[†]

The variously coloured regions of the mosaic egg were termed by Conklin[1] ectoplasm, endoplasm, myoplasm and so on, in accordance with their fate, and the cell inclusions (oil-globules, mitochondria, yolk-platelets, etc.) were called "organ-forming substances". This phrase[‡] in later years fell into much disfavour,

---

* Little will be said of functional differentiation in this book as we are quite ignorant of its biochemical aspects. A good monograph on functional differentiation is badly needed; meanwhile reference may be made to the short review of Maschkovzev.

† Disorganisation of mosaic eggs by centrifuging has been not infrequently reported, e.g. Clement recently on the mollusc *Physa*.

‡ The phrase should now perhaps be abandoned since we speak of the chemical substances active in embryonic induction as "morphogenetic substances", i.e. by stimulus given to competent cells. Weiss[16] suggests "organ-specific areas", but we need the term "area" for fate-maps.

since it was evident from centrifuging experiments (Morgan & Lyon; Morgan & Spooner, etc., see Morgan[4]) that not the heavier or lighter cell inclusions themselves, but the cytoplasm normally containing them, was the important factor in the already determined differentiation. Such cytoplasmic regions in mosaic eggs correspond to those parts or regions of regulative embryos which have been determined to form a certain organ or tissue, but have not yet actually done so. But they must be labile before fertilisation, because in 1938 Dalcq[9] succeeded in bisecting unfertilised ascidian eggs meridionally. After fertilisation, both halves developed, giving almost normal larvae of reduced size.

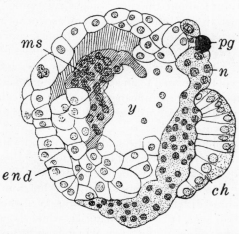

Fig. 48. Eversion of ascidian embryo produced by centrifuging. *ch*, notochord; *n*, neural tissue; *end*, endoderm; *ms*, muscle cells; *pg*, pigmented sensory spot; *y*, yolk.

Another typically mosaic egg is that of the annelid *Chaetopterus*. Though the first work of E. B. Wilson on this egg[1] was done as early as 1882, he published in 1929 some very interesting experiments[3] in which he centrifuged the eggs of this worm at the beginning of the maturation period. They stratified, broke, and gave fragments of different kinds, some hyaline, some yolky. On being fertilised or parthenogenetically activated these fragments showed a close approach to normal development, including even the characteristic rather complicated spiral cleavage, and gave normal trochophores. Yet after fertilisation, this egg, in the normal course of events, would show typical mosaic development. This was a clear demonstration that the primary determination process occurs much earlier in mosaic than in regulative eggs. It should therefore be theoretically possible to find something analogous to the primary organiser in mosaic eggs, acting at the time of the departure of the polar bodies, but this will be an extremely difficult problem, as the removal of parts from a whole without general injury is much harder in the case of a single cell than in the case of a gastrula. This problem has analogies with that raised by the work of Hämmerling (see p. 417).

It was natural that embryologists should ask themselves the question whether anything further could be found out about the different cytoplasms of mosaic

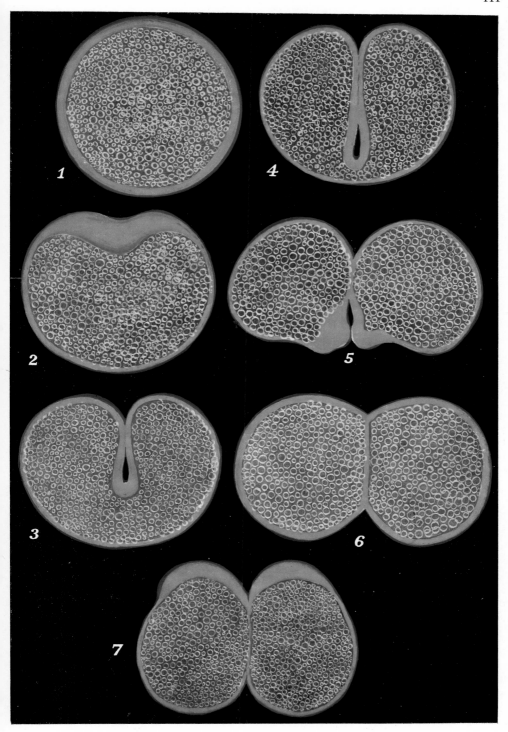

III

Fig. 49. Differences between cytoplasmic cortex and interior in a ctenophore egg (*Beroe*)
seen under dark-ground illumination. For later cleavage stages see frontispiece.

eggs. This has led to a considerable mass of work which is rather difficult to appraise biochemically. Apart from certain specific histochemical work, to be referred to later, the line of least resistance has been to study the appearances which the eggs present after staining by vital dyes. There has been a tendency to interpret the effects as due to differences of pH inside the cells, but as reliable pH indicators rarely penetrate the cell-membrane, and as the dyes that do penetrate it are not usually reliable pH indicators, statements in the literature have to be accepted *cum grano salis*. On the other hand, this literature cannot be entirely neglected, for if a vital stain shows a quite different colour in different regions of a mosaic egg, or in different blastomeres, for example, we can at least say that *some* differences must exist between the cytoplasms of these regions. It must be admitted, however, that we cannot even say that these differences are chemical, for the birefringent and non-birefringent portions of a mammalian muscle fibre react differently to vital dyes, although the evidence goes to show that they differ not chemically but simply in the different arrangement of the myosin fibrils. Moreover, nobody knows the chemical significance of different colours taken up by vital dyes in cells. On the other hand, if it can be shown that polar or asymmetric differences in vital staining appear only in mosaic eggs and not in regulative eggs, this supports the view that determination means a localisation* or *Sonderung* of specific cell elements into different protoplasmic regions. That this is the case is no doubt the most important result of the work here to be described.

It was opened by Spek's examination of the mosaic egg of the ctenophore, *Beroe ovata*, under a microscope with dark-ground illumination. In these conditions, the cortex of the one- and two-cell stages showed a brilliant emerald-green colour (see Fig. 49), and when followed through later development, this green material went almost entirely into the crown of micromeres, which later form the ectoderm (by epiboly). With vital dyes, the difference between the thick cortex and the cytoplasm within it was well marked; thus with neutral red the cortex was yellow and the interior pink, while with nile blue the cortex was pink and the interior blue.

In later work, Spek[3] gave particular study to the eggs of the polychaete worm, *Nereis*. That of the species *dumerilii* possesses a natural indicator, lemon-yellow on the alkaline side, violet on the acid; this showed a bipolar pH difference soon after the extrusion of the polar bodies, the endoderm-forming vegetal pole being acid and the ectoderm-forming animal pole being alkaline. This agreed with the tints taken up by neutral red (yellow at animal pole, red at vegetal pole), and to some extent validates conclusions drawn from neutral red on eggs of other species. The pH "gradient" became intensified during cleavage. The study of the species *limbata*, the eggs of which do not have the natural indicator, confirmed the previous conclusions (Spek[5]), and a similar pH localisation or polarity was found in *Chaetopterus*. On the other hand, when echinoderm eggs were investigated by similar technique, no separation of plasms was to be found (Spek[6]).

* I should like to use the word "segregation" here, but to do so would be to invite confusion with Lillie's theory of differentiation, criticised on p. 113.

A slight difference in vital staining between cortex and cytoplasm was all that could be seen until nearly the time of gastrulation. Thus, at the 64-cell stage, the vegetal pole cells no longer colorise with cresyl violet or nile blue sulphate, while the animal pole cells still do so. Under abnormal conditions, however, such as the use of hypertonic salt solutions, a polar localisation was observable, all the acid material (as judged by methyl red) going to the animal pole. The significance of this is not yet clear, but the general distinction between mosaic eggs and regulation eggs was confirmed in a striking way not by experiment but by direct observation. In Spek's most recent work[8] by vital staining on the amphiblastula of the sponge *Sycandra*, however, there is a discrepancy, for intense localisation was shown, although it has been found that some sponge embryos are quite regulative (Teissier[2] on *Sertularia*).

Spek's line of attack was extended by Ries & Gersch to the egg of a mollusc (*Aplysia limacina*), typically mosaic. In this form, polarity arises at the time of the formation of the polar bodies. Soon a marked stratification has taken place; at the animal pole there is a cytoplasmic region with mitochondria, then a belt of fatty yolk-platelets, then a ring of granules and finally the mass of large yolk-globules. If the egg is placed in reduced methylene blue, the animal pole region turns blue first, but later on all the dye accumulates in the vegetal region, and in the ring of granules. Correspondingly, janus green is reduced to the red derivative throughout the vegetal part of the egg, but not at the animal pole. The first cleavage cuts across the zone of granules, but the micromere contains mostly animal pole material and the macromere mostly vegetal pole material. The position at the two- and six-cell stages is shown in Fig. 50 (opp. p. 146); each blastomere gives a different reaction to the dyes used (vital blue and toluidine blue). Again, Fig. 51 (opp. p. 146) shows how macromeres reduce janus green while micromeres do not; the latter also oxidise reduced methylene blue much more than the former. Trypan blue is taken up only by the micromeres and their descendant tissues. From the data obtained, an approximate value of pH 7·8 was given for the animal pole cytoplasm and about 6·0 for the vegetal pole material, but, although a pH difference may be admitted to exist, the values must be regarded as doubtful. Subsequent work by Gersch & Ries and Cohen & Berrill extended the study of the localisation processes to ascidians such as *Ascidiella*, *Phallusia* and *Ciona*, and to worms such as *Pomatoceros* and *Herpobdella*, with results substantially the same as those already described. Echinoderms, such as *Arbacia* or *Sphaerechinus*, however, gave no sign of localisation of plasms with any of the dyes or reagents employed. The contrast is well seen in the illustration, in which the oxidation of reduced methylene blue is compared between the various species (Fig. 52); in all cases, except the echinoderm, there is to be found a marked localisation of this property to one or two blastomeres.

To what extent are the substances present in the separated plasms identical with those which may be stratified in definite order by centrifuge experiments? Raven[6], and later Ries[3], showed that they are the same by centrifuging mosaic eggs from ovarian stages before localisation and determination had taken place. Vital dyes then gave the same colours to the various stratified layers as had previously

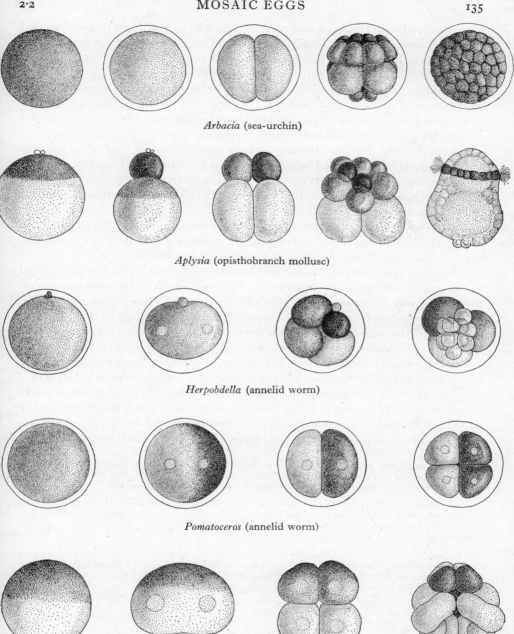

Arbacia (sea-urchin)

Aplysia (opisthobranch mollusc)

Herpobdella (annelid worm)

Pomatoceros (annelid worm)

Ciona (ascidian)

Fig. 52. Methylene blue oxidation during the cleavage stages in a number of species; all the mosaic types show localisation, the regulative echinoderm does not.

been observed with the separated plasms. The heavy centrifugal material tended to be "alkaline", the rest "acid". Raven found little tendency for reorganisation of the egg to take place in accordance with its proper axes after centrifugation (*Chaetopterus, Nereis*, etc.), but it sometimes occurred (*Aplysia*).

The subject was given a more chemical twist by Ries[1,4], who investigated the distribution of ascorbic acid (vitamin C), fixed –SH groups,* benzidine peroxi-dase,† indophenol oxidase (by indophenol blue formation),‡ besides the oxidation of reduced methylene blue and reduction of janus green.§

Except those of *Aplysia*, invertebrate eggs seem to contain little or no ascorbic acid. But in *Aplysia* a marked localisation of this substance was found (cf. Fig. 53). The micromeres get most of it from the outset, save that a small crescentic patch occurs in the macromeres. This they lose altogether, however, to certain descendant cells by the 20-cell stage.

The fixed –SH groups were found to increase in quantity after fertilisation in the echinoderm egg (cf. p. 421) but no localisation was to be seen; in ascidian eggs, on the other hand, an unequal distribution was found.

The peroxidase reaction, too, is very localised in *Aplysia*, being just as confined to the macromeres as ascorbic acid is to the micromeres. This association with the vegetal yolk material persists into the larval stages. In another mosaic egg, that of the ascidian *Ciona*, the localisation is exceedingly sharp (see Fig. 54); it is confined to the vegetal pole blastomeres, and by the time of gastrulation is associated particularly with the muscle-forming mesoderm cells. This is interesting, since we can thus identify the "peroxidase"-containing protoplasm with the yellow "myoplasm" of Conklin, and the long-known "organ-forming substances", thus receive for the first time, however weakly, a biochemical connotation. In later stages, as would be expected, the tail muscles show a strong reaction, but not other parts of the tadpole. Moreover, Konopacki[5] has shown that in the eggs of *Clavellina*, glycogen is also confined to the yellow "myoplasm" and its derivatives.

---

* Ries was under the impression that he was viewing the localisation of glutathione, but as the nitro-prusside test was made after trichloracetic acid fixation, it is likely that the tripeptide was lost by diffusion and that it was the fixed –SH groups which gave his reaction.

† True (thermolabile) peroxidase occurs only in milk. The oxidation of benzidine to a blue derivative by hydrogen peroxide in the presence of a catalyst probably means, therefore, the presence of haematin compounds (thermostable) in minute quantities. They may of course be none the less important for that.

‡ The formation of indophenol blue (see pp. 566, 569) from α-naphthol and p-phenylene-diamine in the presence of a catalyst (indophenol oxidase) is the classical test for the oxygen-activating "Atmungsferment" of Warburg. From the work of Keilin & Hartree, we now know that the reaction proceeds somewhat as follows:

$$\alpha\text{-naphthol} + p\text{-phenylene-diamine} + 4 \text{ cytochrome-C}$$
$$\xrightarrow{\text{spontaneously}} \text{indophenol blue} + 4 \text{ reduced cytochrome-C}$$
$$4 \text{ reduced cytochrome-C} + O_2 \xrightarrow{\text{"Atmungsferment" (cytochrome oxidase)}} 4 \text{ cytochrome-C} + H_2O$$

Absence of colour-formation, therefore, may mean either that the indophenol oxidase is lacking, or that cytochrome-C is lacking, or both.

§ This may well be regarded by biochemists as a "job lot" of processes, but what methods are available depends on technical factors and not upon what we would like to know.

The indophenol blue reaction followed, in *Aplysia*, the distribution of the ascorbic acid rather than that of the peroxidase reaction. It was confined to the micromeres except for a special crescentic area on the macromeres. The echinoderm egg again stood in sharp contrast (*Paracentrotus*), for while the reaction

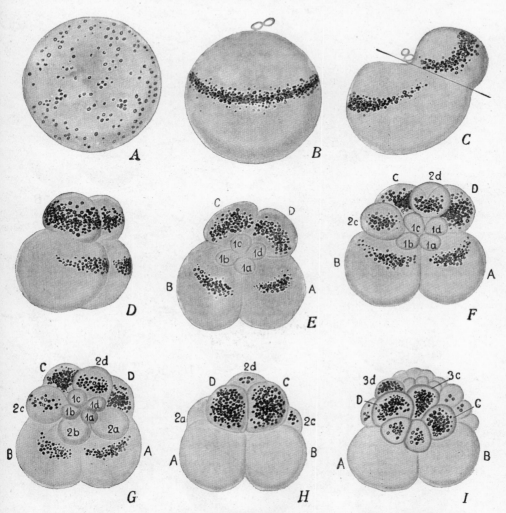

Fig. 53. Localisation of ascorbic acid during the cleavage stages in *Aplysia*. (The numbers indicate the cell-lineage.)

took place especially strongly, every part of the egg or morula reacted to exactly the same extent. In *Ciona* the reaction was again confined to the yellow "myoplasm" and followed this through development. With the oxidation of reduced methylene blue, three reactions were thus found to be specially associated with the yellow plasm.

Now just as Conklin[3] had shown that development of the ascidian egg is interfered with by centrifuging, so Ries[3] showed that this holds true of *Aplysia* eggs,
and the same is reported of *Nereis* eggs by E. N. Harvey[6].

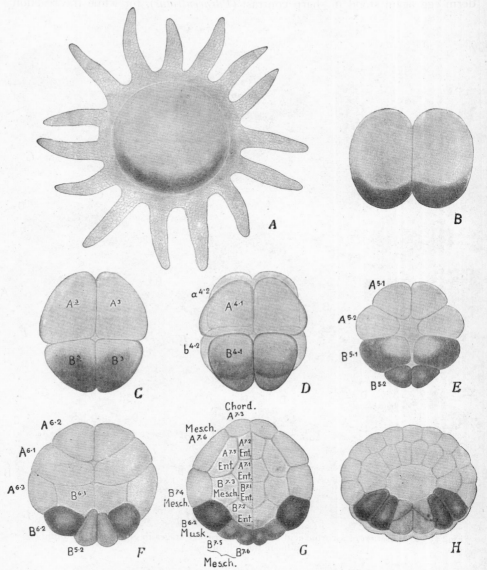

Fig. 54. Localisation of the peroxidase reaction during the development of *Ciona*. *Chord*,
Notochordal material; *Mesch*, mesodermal material; *Musk*, muscle material; *Ent*, endodermal material. Numbers indicate cell-lineage.

The eggs of nematode worms (studied mainly on various species of *Ascaris*)
present a very special case. Of the first two blastomeres, one, the prosomatic

(see below), can produce only ectodermal derivatives; the other, only endo-meso-dermal (stomodaeum and mesoderm) derivatives. But within these two domains far-reaching regulation is possible, including single embryos produced by fusion of eggs, and repair of deficiencies. Dependent differentiation and self-differen-tiation are therefore, as Schleip says (*Determ.* p. 291), combined in nematode development in a peculiar and characteristic way. Correspondingly development remains relatively unaffected even by the severest centrifuging (Beams & King[1]). When Wottge (in whose paper many references to the previous literature will be found) examined *Ascaris* eggs from the point of view of plasmatic localisations, little of this kind could be seen, e.g. glycogen or lipins in histochemical tests on serial sections. The egg, as is well known, is almost completely impermeable once the membranes have established themselves after fertilisation; there is on the outside a protein membrane, then a chitin one and then a membrane which seems to be largely composed of cholesterol.* Vital staining therefore gives no help. *Ascaris* eggs are also very specialised in that they have relatively enormous vacuoles in the cytoplasm, which tend to accumulate at the egg's periphery before the first cleavage. The only visible localisation is a transient accumulation of granules at the vegetal pole before the first cleavage.

Among the mosaic aspects of the *Ascaris* egg, however, is the fact that at the first cleavage it divides into a prosomatic blastomere and a progonadal blastomere; the latter later gives rise to the germ-cells as well as to some somatic material. In the somatic cells deriving from both blastomeres there is a diminution of chromatin at successive divisions from the second to the fifth, but not in those cells from which the germ-cells will arise. This is explained as being due to the local elaboration at the animal pole of a specific chemical substance, a "diminisher", from a precursor localised there; this diminisher, though diffusible, is differentially segregated during cleavage so that the nuclei of some cells have to undergo diminution, but those of others do not. By centrifuging strongly, cleavage, but not nuclear division, is suppressed; the diminisher has therefore free access to all parts of the egg, and the distinction between prosomatic and progonadal cells is abolished, for all the nuclei cast off chromatin into the cytoplasm (King & Beams; cf. the experiments of Moore[7] on the "cleavage-substance" in the echinoderm egg, p. 364).

The general upshot of all the work of Spek, Ries, and their collaborators has been to suggest that in many eggs showing mosaic development† a profound chemical heterogeneity of the parts of the egg sets in very early, parallel with the early determination. In regulative development, on the other hand, a certain homogeneity‡ is preserved until the time of gastrulation. Histochemical and experimental methods confirm each other. The former alone will not carry us much further, but with the micro-chemical technique which should before long be available, it ought to be possible to establish by unimpeachable chemical methods the differences between the determined areas. It should also be possible to do this for areas in regulative eggs between the moment of determination and the onset of histological differentiation.

* This may explain the unique resistance to high pressure shown by nematode eggs, 800 atmo-spheres being needed to stop development (Pease & Marsland).

† Not necessarily in all, for little or no localisation was to be found in the eggs of the polychaete worms *Aricia* and *Chaetopterus* with the methods so far used (Ries[1]).

‡ This is, of course, not to be understood as denying the fundamentally heterogeneous colloidal character of all egg-protoplasm.

As a footnote to the above, it may be mentioned that according to Gersch[1], the dye trypan blue will distinguish between mosaic and regulation eggs, for while the former (e.g. *Aplysia* and *Pomatoceros*) pursue their development unaffected by it, the latter (*Psammechinus* and *Chrysaora*) are completely ruined, giving rise, for instance, to chaotic proliferations of ectoderm.

Among the groups of animals which have been omitted from the foregoing discussion, the cephalopods are perhaps the most remarkable. Alone among invertebrates, their eggs possess so much yolk that cleavage is discoidal, as it is in the fishes. In such eggs the fact that the main mass of nutrient yolk differs in staining reactions or in pH from the blastoderm may not mean that a determinative localisation has taken place. That there is such a difference was shown by Spek[7], who found the yolk (with good indicators) to be strongly acid, and the blastoderm alkaline (5·0 against 7·8), with corresponding staining differences. But whether the cephalopod egg is absolutely mosaic or not is doubtful, as comparatively little work has been devoted to it. The most likely supposition is that it develops mosaically after the first cleavages (after Naef-stage V), i.e. it is much more mosaic than the echinoderm egg but slightly less so than those of annelids or ascidians. This may indeed be true of all molluscan eggs, if we may judge from the discussions of Ranzi[8], who found cases of *duplicitas cruciata* in cephalopods and yet was able to show by explantation experiments that after a relatively early stage full self-differentiation would proceed (Ranzi[9]). Earlier reports in the literature (cited in these just-mentioned papers) of conjoined twin larvae from annelid and molluscan eggs had always been difficult to reconcile with a strict mosaic view, and had been explained away as the result of the fusion of two eggs originally separate. But when Penners[1] in 1922 described the production of twin embryos from the oligochaete *Tubifex rivulorum* after equal cleavage of the egg-cell (instead of the usual unequal cleavage), and when some years later Titlebaum found that *Chaetopterus* twins could be produced fairly regularly by compressing the uncleaved egg, doubt was no longer possible. Later a thorough investigation by Tyler[1] showed that all kinds of agencies (such as compression, abnormally high or low temperatures, centrifuging, and anaerobiosis) will make mosaic eggs give rise to twins. All that these agencies have in common is that they alter the position of the first cleavage spindle. Instead of its ends lying in regions qualitatively different they come to lie in regions qualitatively alike, and hence when the egg divides, two cells are formed each of which has sufficient of each of the various plasms required for complete differentiation. At this stage, therefore, there must still be much plasticity.

The subject has been placed in a clearer light by the admirable experiments of Novikov on the eggs of the annelid worm *Sabellaria*. In exo-gastrulation experiments all parts show perfect self-differentiation (in sharp contrast with what happens in amphibia, see p. 159), and the self-differentiation of parts is unaffected by the transplantation of polar lobes and blastomeres in all possible combinations, even though the contact between the cells may be shown to be close enough to permit of the diffusion of dyes such as nile blue (Novikov[1]). But if by treatment with KCl, cleavage is temporarily inhibited so that both the two first blastomeres receive polar lobe material, then twin embryos, perfect in every particular, result with great regularity (Novikov[2]). It may thus be concluded that the polar lobe does contain an organiser or inductor but that it cannot pass across cell boundaries and can only be transmitted at mitosis. Explanation of this fact would be had if some inductor substances are active in protein combination (a reasonable possibility; cf. the discussion on p. 205). Its significance is that it provides a bridge between the diffusible substances of the type of the amphibian primary evocator and the non-diffusible nuclear inductors described in Section 2·67.

Further work will make more precise the exact degree in which each invertebrate and prochordate egg approaches true mosaic development. In the meantime there is no contradiction between Spek's bipolar pH localisation in the cephalopod egg and the morphological results, since, as we shall see later, a bipolar pH localisation occurs also in the eggs of teleostean fishes (Spek[4]), where the development is eminently regulative. Localisation, in a word, only has a relevance to determination when the cleavage is total.

Before leaving the cephalopods, the possibility of inductive processes in later development ought to be mentioned (cf. p. 325, where it is seen that these occur in the later development of the highly mosaic ascidian egg). The cephalopod eye-cup, so astonishingly similar in organisation to that of vertebrates, develops, as is well known, from an ectodermal invagination, not from the nervous system. Ranzi[4] says that the development of the optic ganglion is a dependent differentiation requiring at first a stimulus from the eye-cup, though later able to proceed unaided. Thus the situation in cephalopods would be exactly the reverse of that in vertebrates.* The cephalopod eye is particularly sensitive to toxic substances, and cyclopia readily occurs (Ranzi[3]).

2·3. Amphibian development and the fundamental principles of morphogenesis

2·31. Description of gastrulation

In order to assist the reader in the discussion which follows, we must first describe very shortly the well-known, but in many ways still rather obscure, process of Gastrulation. Fig. 55 gives a series of longitudinal vertical sections at the successive stages of the process.

After fertilisation, the amphibian egg divides itself into a large number of cells in the cleavage, or segmentation, period. At the vegetal pole, where the mass of yolk has accumulated, the cell-divisions, impeded by this material, occur more slowly, so that at the time when the solid ball (the morula) becomes a hollow one (a blastula, with a blastocoele cavity), the vegetal pole cells are very much bigger than those at the animal pole. There is also a difference of pigmentation. The animal-pole cells are always darker than the vegetal ones, so much so in some cases, such as the common frog (*Rana temporaria*), that the vegetal cells are a pure creamy white and the animal ones a dark coppery brown, almost black. In the newt, *Triton alpestris*, the difference is rather between two browns, and there are some amphibia, such as the newt, *T. cristatus*, whose eggs have no

---

* Cf. the curious effects described by Dragomirov[6], where the amphibian head ectoderm seems to induce a retina in rotated eye-cups.

pigment at all, so that there is little difference between animal and vegetal pole cells. This has been taken advantage of for chemical colour experiments, as will be seen later (p. 203).

The essential happening in gastrulation is that all the marginal zone between the two kinds of cells, and all the vegetal pole cells, become invaginated and enclosed within a ball of ectoderm. The process is seen beginning in Fig. 55.

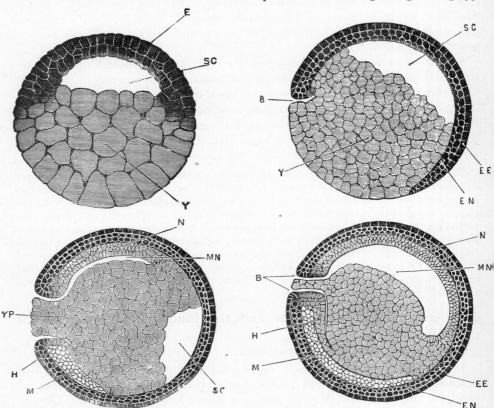

Fig. 55. Longtitudinal vertical sections of the amphibian embryo during gastrulation (from Milnes-Marshall). *B*, blastopore; *E*, roof of the blastocoele; *EE*, *EN*, ectoderm; *H*, ventral lip of blastopore; *M*, mesoderm; *MN*, archenteron; *N*, roof of the archenteron; *SC*, blastocoele cavity; *Y*, yolk-endoderm; *YP*, yolk-plug.

Animal pole cells creep around the vegetal pole yolk-cells on all sides, but with this difference, that whereas at the anterior part of the egg (*EE* and *EN* in Fig. 55) the cells remain continuous with the yolk-cells at the surface; at the posterior part of the egg (*B* in Fig. 55) a groove appears, the dorsal lip of the blastopore. Through this discontinuity the great invagination begins, cell material passing in at the blastopore to form the roof of the primitive gut or archenteron (*N* in Fig. 55). As this happens, the old blastocoele cavity is pushed farther and farther forwards and downwards until it eventually disappears, as can be seen in Fig. 55. At the same time the blastopore progressively closes, passing through a succession of shapes, quarter moon or sickle-shaped, half moon, and finally full moon. The material visible *in* the full moon from outside is the remaining yolk, the yolk-plug (*YP* in Fig. 55). Eventually the full moon narrows to a hardly visible slit. The three germ-layers of classical embryology are now formed; the ectoderm, which encloses the entire embryo; the mesoderm, which forms the archenteron roof and which has been invaginating also from the ventral lip of the blastopore (see Fig. 55); and the endoderm, the mass of white yolk-cells. All the layers are continuous, and by these formative movements the principal organ systems of the future embryo have reached their definitive positions.*

As the blastopore finally closes, the neural folds make their appearance on the dorsal surface of the embryo, defining the edges of the neural plate, which is shaped like a shield, broader at the anterior end than at the posterior.† It is to be noted that the material which passed in first through the dorsal lip of the blastopore will be incorporated into the future head (actually forming part of the pharynx), that which passed in next will form part of the trunk, and the last to enter will form part of the tail. The neural folds now move together towards the mid-line, forming a narrow groove, the neural groove. When the sides of this fuse, the neural tube, the primary axis of the future vertebrate animal, is formed. At the same time, the archenteron roof lying below divides into five longitudinal strips lying side by side. The centre one of these becomes the notochord, and hence the foundation of the eventual axial skeleton, while the strips to the immediate left and right of it cut themselves up into a linear series of blocks, the primitive muscle masses, or somites. Finally, the two outer strips spread out on each side of the body forming the side-plate mesoderm, from which will arise in due course the coelom.‡

2·32. The mapping of the presumptive regions

A knowledge of the exact presumptive regions of the embryo before the accomplishment of the formative movements of gastrulation was a necessary logical preliminary to the study of the results of interference with them. A scheme of the normal fates of the parts (a "fate-map") is required if we are to recognise an abnormal fate when we see it. The earlier workers, such as His and Roux[3], who appreciated the importance of the problem, could only attack it by making small lesions in the embryo, but since healing is very active in young amphibian tissue, they made no progress. It was not until the appearance of the two great papers of Vogt[2], who applied vital stains to the surface of the embryo, that the problem was thoroughly and definitively solved.§

* The foundations of our knowledge of cleavage and germ-layer formation we owe to the work of K. E. v. Baer[1, 2] (1828–1835), whose instructive and witty autobiography should not be neglected by anyone interested in these subjects.

† As the result of cinematographic work, Daniel[1] gives a kind of eyewitness description of this process. With proper lighting the archenteron roof can be seen moving forward like a curtain. Changes then occur in the presumptive neural area in the following order: (1) the cell-layer thins medially, (2) two streaks of pigment accumulate, (3) at the inside edges of these the neural folds arise, extending backward as well as forward. One may be normally several hours behind the other.

‡ The yolk-endoderm has also a part to play at this time. On its upper surface, which forms the floor of the archenteron, a trough appears which later closes, not unlike the neural folds, into another tube, the future intestinal tract.

§ It is interesting that the use of coloured marks in studying space-time changes in living organisms dates back to the eighteenth century when Stephen Hales examined the growth of leaves with marks of red lead and John Hunter used madder for following the deposition of bone.

Minute pieces of agar were soaked in the dye solution (nile blue or neutral red, dil. 1 : 1000) and held against the gastrula for a short time. After the dye has been absorbed by the cells it will be retained by them for a long time without spreading, so that they remain recognisable and can be followed in their movements. The result of marking the outside surface of the early gastrula is seen in Fig. 56 A and B. In the first of these a row of differently coloured marks was placed dorsally to the dorsal lip of the blastopore, running in a series over the animal pole. As gastrulation proceeded, the series passed downwards and in at the blastopore, eventually coming to lie in the mesodermal notochord tissue in the archenteron roof, where it could sometimes be seen in transparency through the outer ectodermal cells. The second experiment showed how marks placed all around the blastopore, but at some distance away from it, moved towards it,

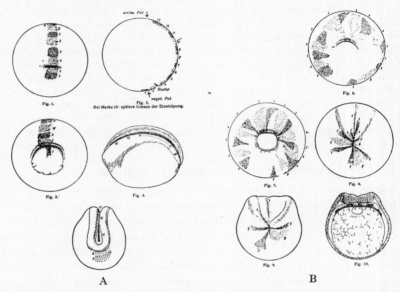

Fig. 56. Movements of stained areas during amphibian gastrulation, showing the cell-streams.

and were invaginated in a way reminiscent of coloured liquids being washed down a drain. It is to be noted that the marks reach the dorsal lip of the blastopore first, since it is there that the invagination is most rapid and earliest. Later on, however, when the yolk-plug has altogether disappeared, and a slit-shaped blastopore alone remains, the marks are seen vanishing into its sides, and even also at its ventral lip. By the time the neural folds have formed, hardly any colour is left on the outside of the gastrula, but as the cross-section shows, it is readily to be found again in archenteron roof and side-plate mesoderm.

The resulting map, which, with minor variations, holds good for all the amphibia, is seen in Fig. 57. The upper map shows the gastrula seen from the left side; the lower one shows it seen from the vegetal pole. Taking the former first, it will be observed that rather more than half the total spherical surface consists of material that will not invaginate, of which presumptive epidermis (lying over the old blastocoele cavity) makes up about one-third, and presumptive neural plate about two-thirds. The endoderm yolk-cells will, of course, also not invaginate, except in the sense that they will be drawn into the interior of the gastrula when the yolk-plug vanishes as the blastopore finally closes. All the rest of the material will invaginate through the blastopore, the beginning of which is seen at J, a point near the old ventral pole but a little farther away from the lowermost point of the egg at its present stage owing to changes in its centre of gravity. At the dorsal side lies the rather large area of future notochord, beneath it and on each side of it the area of the future somites, still farther downwards and ventralwards the future side-plate mesoderm, and filling up the remainder of the space, the future tail mesoderm. The lower map helps in the visualisation, which can, of course, be best attained with the aid of a model.

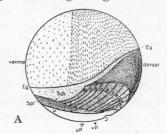

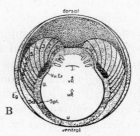

Fig. 57. Vogt's fate-map of the amphibian gastrula. *Eg*, limit of invagination. *Sch*, tail mesoderm. *Spl*, side-plate mesoderm. *u*, endoderm. *Vu.Ex.*, fore-limb. *J*, site of blastopore. *vP*, vegetal pole. *uP*, lowermost point at this stage. A, from the side; B, from below.

It is to be emphasised that we have here a map of the future, a "fate-map", as it were, and that there is nothing to distinguish these various regions, microscopically or to the naked eye, in the early gastrula itself. Whether the regions may be distinguished by any chemical differences is almost entirely a matter for the future, but there are indications that the presumptive mesoderm (the organisation centre) differs from the rest (see p. 203). It is unlikely that any chemical differences will be found among the other, undetermined, areas.

Since the time of completion of this mapping process for the amphibian embryo, it has been confirmed on many forms (e.g. Schechtman[1] in America) and extended to many other vertebrates. From the results general theoretical considerations have been drawn, as by Dalcq, in his monograph; Pasteels, etc. Important though these are for phylogenetic and other problems, we cannot discuss them here, and will refer only to the several maps in the section pertaining to other chordates (Section 2·41).

2·325. The nature of the formative movements

Besides the movement of invagination which has just been described, other movements, equally important, take place during early amphibian development.*

That there must be some inner compulsion to invagination on the part of dorsal lip tissue (*Einrollungstendenz*) is shown by the fact that if a piece of it is transplanted to another place, it rolls in on its own account (Spemann &

* One must remember that movements of migration continue to play an important part till late in development, e.g. the wanderings of neural crest mesectoderm (Harrison[18]) and the migration of placode cells from the acoustico-lateralis rudiment which gives rise to the sense-organs of the lateral line in amphibia (Stone[2]); to say nothing of the development of the peripheral nervous system.

H. Mangold). Moreover, its direction is fixed, and even when the transplanted invaginating cells are oriented in the contrary direction to the host embryo, it forces its way in, as Spemann[8] and F. E. Lehmann[3] showed. Even the amount of invagination is determined in these isolated pieces, for the blastopore lip of a young gastrula will force its way in much farther, according to Spemann, than an old one. This kind of determination has been called "dynamic determination", as opposed to induction.

The mechanism of this invagination tendency is still extremely obscure. If one examines a cross-section of the blastopore lip at its earliest stage, such as that in Fig. 58, one sees that the cells at the base of the furrow have a distinctive bottle-shaped form of their own. These were first described by Ruffini in 1907, who attributed to them secretory functions. Several authors (e.g. Goodale; Vogt[2]; Wintrebert[3,4]; Daniel & Yarwood) have figured these piriform cells and discussed the possibility that they act in a contractile manner. The claim has been made by Wintrebert that by choosing a temperature sufficiently low to inhibit this contraction without inhibiting cell-division ($+3°$ C.), the appearance of the blastopore could be held up until the egg was returned to normal temperature. However this may be, it would be of much interest if contractility could be shown to play a part in gastrulation, in view of the results on insect embryos (discussed in Section 2·81) and the chemical evidence (p. 191) for amphibia. Ruffini also showed that the endodermal cells again become bottle-shaped or piriform when the intestinal tract forms as a furrow on the floor of the archenteron.

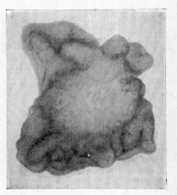

The material which invaginates must come from somewhere. It is provided by the vast expansion of the dorsal and animal pole ectoderm, sometimes called divergence or epiboly. A well-known experiment of Spemann's[17] strikingly illustrates this. When two ventral halves of an early gastrula are grafted together, there does not result a smooth-

Fig. 59. Expansion of ectoderm in two ventral halves of gastrulae grafted together.

surfaced ectodermal ball containing yolk-endoderm, but a most irregular formation, full of lobes and folds on its spreading puckered surface, for the ectoderm of both pieces is expanding. The "organism" progresses no further, of course, owing to the absence of the organiser centre. Such a formation is seen in Fig. 59. Dalcq[6], who has given much attention to the nature of these movements, attributes very plausibly the formation of the amnion in sauropsida and even the proliferation of the trophoblast in mammalian embryos to the same ectodermal tendency to expansion.

What is really difficult, however, to obtain from the morphologists is a satisfactory statement regarding the mechanism of these formative movements. Thus Huxley & de Beer[2]* speak merely of "streaming movements and displacements of the various regions of the embryo" without specifying the process involved,

* *Elements*, p. 25.

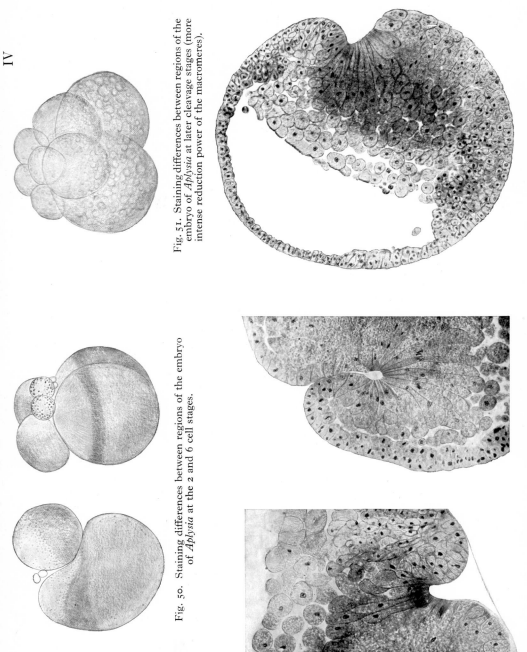

IV

Fig. 51. Staining differences between regions of the embryo of *Aplysia* at later cleavage stages (more intense reduction power of the macromeres).

Fig. 50. Staining differences between regions of the embryo of *Aplysia* at the 2 and 6 cell stages.

Fig. 58. Sections through the dorsal lip of the blastopore showing the piriform cells (from Vogt).

and Pasteels[9], who has worked a great deal on the movements in many different vertebrates, speaks of them just as "déplacements cellulaires" or "migrations". Vogt himself said little on the nature of the process, but believed that cell-division played a certain part ([2], p. 434). Evidently either cell-divisions, perhaps oriented in a specific direction, or changes of shape of the ectoderm cells must be responsible. These processes need not be mutually exclusive.

On the subject of cell-divisions, as estimated by counts of mitoses, the evidence is rather conflicting, and it is difficult to discuss it without reference to embryos other than those of amphibia (see below, p. 329). We must first admit that changes of shape alone can do a great deal. The most cursory dissections of amphibian embryos suffice to show that the roof of the blastula is much thicker than the ectoderm of the later gastrula, whether presumptive neural plate or that lying over the blastocoele cavity. The latter is often extremely thin, more so in the newt than in the frog. A four times diminution in thickness, which is not out of court, would allow for a very substantial expansion of area. It has, however, been maintained by Wintrebert[5] that there does exist in the marginal zone a centre of active cellular proliferation (in *Discoglossus pinctus*), but his evidence, as may be seen from the reply of Pasteels[7], was unsound,* and no mitotic counts were presented. Pasteels denies that mitoses play any part. On the other hand, mitotic counts have recently been made on the gastrula of the toad, *Bufo cognatus*, by Bragg, with the following interesting results.

		Mitotic index (mitoses/100 cells)
Blastula:	Roof of blastocoele	64·8
	Vegetal pole yolk-cells	44·5
Early Gastrula (crescentic blastopore stage):		
	Dorsal lip of blastopore	58·4
	Cap at animal pole	81·1
	Other ectoderm cells	56·0
	Yolk-endoderm cells	12·7
Late Gastrula:	Ectoderm cells	8·8
	Blastopore lips	8·1
	Mesoderm cells	5·9
	Yolk-endoderm	2·6

This table illustrates the higher mitotic rate of animal pole cells than vegetal pole cells before gastrulation, and the great decline in all areas at the end of gastrulation. Particularly interesting was the small area of frequent mitoses central to the expanding ectoderm early in gastrulation. The size of this area, says Bragg, is remarkably small; the ratio of number of cells in it to total cells in the embryo is 1 : 284.

Laborious though work on mitotic counts must be, it would be very desirable to have confirmation of these effects on other amphibia, though in all probability flattening and thinning of the blastula roof plays the greatest part.

The origin of this dynamic determination presents a problem of absorbing interest. In his discussion of it, Spemann[17] points out† that, although it obviously occurs much earlier than the material determination mediated by organisers, it

---

* Wintrebert also spoke of a "mitogenetic" centre, but we may leave this altogether out of account (see p. 422).

† *Embryonic Development*, p. 107.

is capable of wide regulatory changes. Ruud & Spemann, for example, showed that gastrulae from which the entire ventral half has been removed will gastrulate normally and produce neurulae normally shaped but reduced in size. This means, of course, that material which would never normally invaginate will have to do so. More recently, Schechtman[9] has found that the areas of presumptive mesoderm at the sides of the dorsal lip are necessary for its invagination. If they are removed or replaced by presumptive ectoderm, the dorsal lip moves outwards to form one of the "horns" so often described by the earlier workers, instead of invaginating.

Besides the movements already discussed, i.e. invagination of mesoderm and expansion of ectoderm, there are two others of much importance, longitudinal extension (*Streckung*) and dorsal convergence. Longitudinal extension manifests itself in the formation of the tail bud, and occurs in isolation, as Bijtel has shown, when presumptive tail material is explanted. Dorsal convergence appears principally in the approach of the neural folds to each other, and their closure, but it is also seen in the concentration of invaginated mesoderm in greatest thickness in the region of notochord and somites, and at a still deeper level in the closure of the endodermal furrow at the base of the archenteron cavity to form the future intestinal tract.

A fuller description of all these movements will be found in the monograph of Hatt. Bragg's measurements of mitotic rate suggest that in these also cell-divisions play a part, though it may not be a preponderant one.

Finally, Schechtman[3] and Daniel & Schechtman have described a movement of yolk-endoderm cells from the ventral pole upwards in the blastula stage of certain urodeles. This "unipolar ingression" seems to contribute to the convexity of the blastocoele floor, but its significance is otherwise unknown. Were these cells to maintain their connection with the cortex of the egg, they would assume the bottle-shape referred to above, and if the piriform processes contracted, gastrulation might ensue.

2·33. The appearance of the vertebrate axis

2·331. Properties of the primary organiser

As has already been mentioned (p. 101) exchange transplantations (*Austausch-versuche*) performed on the amphibian embryo by Spemann and his school (Spemann[8, 9, 10]; O. Mangold[1, 2, 4]) showed that before gastrulation has begun almost all parts of the embryo are interchangeable with other parts. They used the heteroplastic method, i.e. transplantation between the unpigmented embryos of *Triton cristatus* and the pigmented ones of *T. taeniatus*, taking as their model the classical work of Harrison[1] on the origin of the lateral line organs in 1898 and 1903, about which Spemann[17] was later to write that it taught him "more than almost any other investigation, not only for technique, but also for the methodically advancing analysis". Fig. 60 shows, first, an exchange between presumptive epidermis and presumptive neural plate. Material which would have formed skin formed brain, and *vice versa*.

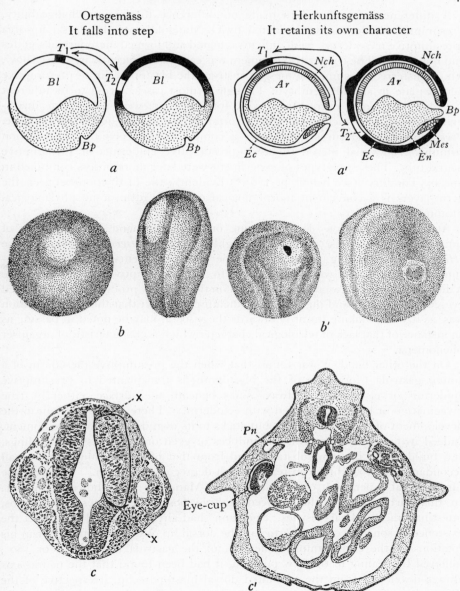

Fig. 60. Exchange transplantations demonstrating the determination process occurring during gastrulation.

a, b, c, exchange of presumptive brain with presumptive epidermis in the early gastrula.
a′, b′, c′, exchange of presumptive eye-cup with presumptive body-wall in the late gastrula.

T_1, T_2, grafted pieces.	Ec, ectoderm.
Bl, blastocoele cavity.	En, endoderm.
Bp, blastopore.	Mes, mesoderm.
Ar, archenteron.	X, X, limits of brain tissue formed by graft.
Nch, notochord.	Pn, pronephros.

A difference of germ-layers made no difference to the interchangeability. Presumptive epidermis, transplanted by O. Mangold[1] into the dorsal lip area (presumptive mesoderm), were duly invaginated and formed mesodermal somites and notochord. The converse experiment was carried out recently by Lopaschov[3], who transferred presumptive mesoderm from the blastopore lip to presumptive neural plate in an older embryo. Though in most cases, owing to its inherent stretching tendency, excrescences were formed, there were some quite unambiguous ones in which it became incorporated into the host's neural tube. Presumptive mesoderm is thus nearly as competent to receive the neuralisation stimulus from axial mesoderm as is presumptive epidermis or presumptive neural tube. These two types of experiment were later more or less combined in one by Töndury[1] (see below, p. 170).* Evidently the distinction between the germ-layers is a convenient morphological one with little or no physiological meaning.†

After the completion of gastrulation, on the other hand, the behaviour of transplanted pieces is no longer neighbourwise (*Ortsgemäss*) but selfwise (*Herkunftsgemäss*). Fig. 60 shows an experiment of Spemann's in which presumptive eye region was transplanted to the presumptive body-wall. Here it did not follow its new environment, but on the contrary produced a well-formed eye-cup with retina and pigmented epithelium. The fact that this new formation was, as it were, looking inwards towards the viscera, may be noted in passing as an instance of that lack of teleological character which is characteristic of organiser phenomena.

On the other hand, it was found that when the presumptive mesoderm of a young gastrula (dorsal lip of the blastopore) is transplanted to presumptive epidermis or elsewhere, it always sinks beneath the surface, either by true invagination or by being covered with ectoderm. "There it does not join in the development of its surroundings, but sticks to its own direction of development. Indeed it does more than this. It induces or even organises its surroundings, and builds from its own substance and from that of its surroundings a small secondary embryo, which can reach a high degree of perfection." This was the fundamental experiment of Spemann & H. Mangold. It is shown in Fig. 61. In this particular case the induced embryo possessed somites, pronephros, intestinal tract, and ear-vesicles. Spemann[8] had already been led to suspect the existence of some special properties in the dorsal lip of the blastopore, for in his rotations of the entire animal pole roof of the gastrula through 90 or 180°, followed by healing in the new position, it had been found that the neural axis always developed from the unrotated dorsal blastopore lip, irrespective of the orientation of the gastrula roof. Or if a young gastrula were cut sagittally (longitudinally), and the two right or the two left halves allowed to heal together,

* And indications that chick germ-layers are also interchangeable are not wanting (Rudnick & Rawles; Waddington & Taylor). There is an admirable review on the whole subject by Oppenheimer[9].

† In this connection it is interesting that, according to Brien[1], the neural tube regularly arises during asexual development in ascidians, not from ectoderm but from endoderm or mesoderm.

double embryos had been formed, each of the neural axes of the twin proceeding from one of the half-lips.

"So ergab sich", wrote Spemann & H. Mangold, "der Begriff des Organisationszentrums als eines Keimbereiches, welcher den übrigen Teilen in der Determination vorangeeilt ist und nun selbst determinierende Wirkungen bestimmten Betrages in bestimmten Richtungen ausgehen lässt. Mit der Analyse dieses Organisationszentrums soll durch die hier darzustellenden Experimente ein erster Anfang gemacht werden."

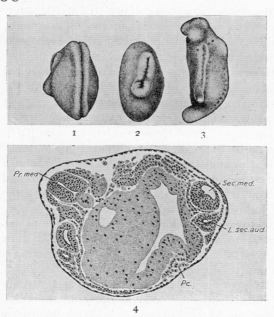

Fig. 61. The fundamental experiment of Spemann & H. Mangold; induction of a secondary embryo by transplantation of the organiser. 1. The host embryo from above. 2. The host embryo from the side, showing the induced neural folds. 3. The same at a later stage. 4. Transverse section showing the two neural tubes. *Pr.med.*, host neural tube; *Sec.med.*, induced neural tube; *L.sec.aud.*, induced ear-vesicle; *Pc*, pericardium.

It is interesting that as far back as 1907 Warren H. Lewis[3] had transplanted the dorsal lip of the blastopore, but only to much older larvae, where it developed in accordance with its presumptive fate into notochord and somites, together with some neural tissue, but without affecting in any way the host tissues, which by that time had become too old to retain any competence for neural inductions.

In the ten years that followed the discovery of Spemann & H. Mangold a considerable number of investigators carried out that large body of work which has made the development of the amphibian embryo more clearly understandable than that of any other organism, although, as we shall see, the sea-urchin egg takes a good second place. Many researches were devoted to the elucidation of the organisation centre as the leader in regulatory processes. Thus Wessel stuck together two dorsal halves of young gastrulae, so that the invaginating mesoderms collided, and not being able to do anything else, diverged one to one side, the other to the other. In this way a cross-shaped twin was produced, and the experimental explanation, in terms of twin organisers, of one of the most puzzling of the classical teratological phenomena, *duplicitas cruciata*, was attained. Spemann & Wessel-Bautzmann investigated the effects of gross excesses or deficits of the material awaiting determination by the organiser. This was done by cutting young gastrulae transversely but not equally and then sticking together the two large pieces and the two small ones. If the operation is made very early in gastrulation perfect regulation results; if later on, all kinds of anomalies are produced. Great deficits in the total amount of presumptive ectoderm were produced by Spemann[1] and Schmidt[1] in their constriction experiments already referred to (p. 103); nevertheless, that part of the embryo which contained the organiser regulated perfectly, so that some material which would not normally have become ectoderm must have done so. Conversely, Waddington[18] showed that considerable regulation will take place when great excesses of presumptive ectoderm are added to a young gastrula.

Perhaps the most thorough of these regulations was that performed by the embryos operated on by S. C. Wang who divided one gastrula sagittally (longitudinally), another frontally (transversely), and stuck the two different pieces together. Since the fused embryo possessed one dorsal half and one left or right half, it had one and a half organisation centres. Nevertheless, in favourable cases these fused and normal neurulae resulted.

Equally important was the exploration of the gastrula to determine the exact limits of the region possessing inductive capacity and its movements (*Abgrenzung*). Spemann & H. Mangold had come near to suggesting that the inductive effect was something to do with the invaginating mesoderm, and soon after their work the *Einsteckung* method of O. Mangold was introduced.* Using this method Geinitz quickly obtained good neural inductions by implantations of dorsal blastopore lip, and Marx[1] followed by showing that the effect could be had equally well with pure archenteron roof. About the same time it was confirmed by F. E. Lehmann[1] that removal of pieces from the invaginating dorsal blastopore lip led to very serious deficiencies in the embryo's nervous system. The most thorough examination of the extent of the inducing area was made by Bautzmann[1], who cut the presumptive mesoderm of the dorsal lip into small pieces with an ingenious "micro-fork" and implanted them, thus surveying the boundaries of the region. His results are shown in Fig. 63, from which it was concluded that most of the presumptive mesoderm possesses inductive power before invagination (most of the dorsal posterior quadrant of the early gastrula). As the early gastrula was used, in which but little invagination can have taken place, it is certain that the inductive power must appear to some extent in the region to be invaginated before invagination has actually begun.

* As we have already seen, this consists in pushing whatever is to be tested into the blastocoele cavity of a young gastrula through a small slit made in its roof (the presumptive epidermis). See Fig. 62. When the cavity disappears, the piece comes to lie between the yolk-endoderm and the ectoderm. Thus it has every opportunity to manifest any inductive power which it may possess. This method has been of inestimable help as a routine test (see below, pp. 172, 176).

The mesoderm of the notochord and somites had been shown to possess inductive power.* Bautzmann[3] next showed that this was concentrated in the central notochordal region rather than in the somites or side-plate mesoderm. This line of thought led to a particularly interesting result when Waddington[12] found that the just-formed side-plate mesoderm in the middle yolk-plug stage can induce the formation of a secondary neural plate when implanted into the blastocoele cavity of a young gastrula. It therefore possesses inductive power. Conversely, the dorsal lip of a young gastrula, substituted for the side-plate mesoderm just mentioned, induces a neural plate on the flank of the embryo. Therefore the flank ectoderm is competent. Yet in normal development no accessory neural plates appear on the sides of the embryo. This remarkable paradox may perhaps be explained by assuming a quantitative diminution in

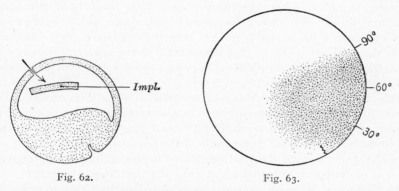

Fig. 62.   Fig. 63.

Fig. 62. The implantation (*Einsteckung*) method of testing for inductive power. *Impl.*, implant.
Fig. 63. Boundaries of the primary organiser region.

inductive power and in competence, so that the side-plate mesoderm induces less strongly than the dorsal lip and the lateral ectoderm reacts less readily than the young gastrula ectoderm. But the suggestion was also made that there might be an active suppression of the inductive power of the side-plate mesoderm by the rest of the mesoderm, in accordance with the individuation field of the whole embryo. Results closely analogous to these were later obtained by Twitty & Bodenstein in a study of the competence of flank ectoderm to form dorsal fin when transplanted heteroplastically to the dorsal region. Graft of species *A* on host of species *A* gave no induction, but on host of species *B* competence showed itself. Conversely a graft from species *B* on host of species *A* showed that some inductive power was present in the latter. These facts teach us that we should always be prepared to state the context in which we are using words such as "competence" and "inductive power".

* The mesodermal archenteron roof lies, of course, *under* the presumptive neural plate. But the practice of calling it the "substrate", as some authors occasionally do, cannot be too strongly condemned, since this word has a meaning in biochemistry which gives it, in this context, the most misleading implications.

The double layer (invaginated mesoderm and presumptive mesoderm), and the invaginated mesoderm alone, was now known to possess inductive power. What of the presumptive neural plate, and the neural plate itself during the earliest stages of its histological transformation? It had been established that the process of determination of the presumptive neural plate went on *pari passu* with the invagination of material towards the future anterior end of the embryo (see (Fig. 64 A). O. Mangold & Spemann now made the discovery that as the determination and histological differentiation of the neural plate proceeds, it acquires the power of inducing another neural plate if implanted into the blastocoele cavity of another embryo (Fig. 64 B). The term *"Homoiogenetic induction"* was introduced to describe this effect. Its immediate importance was to draw attention to the fact that induction cannot be wholly explained as the unit-building activities of an individuation field. If that were so, an implanted piece of neural tube would be expected to call up other structures in its vicinity from the host tissues, not more neural tube.* But these experiments have also a wider significance, the full extent of which has probably not yet been realised (see p. 267). At the same time O. Mangold[7] emphasised that irrespective of the age or provenance of the implant, the induced neural tube always appears and develops synchronously with that of the host (Fig. 64 C).

In this same joint paper, Mangold referred to some experiments, then in progress, directed to answer the question how long the neural tissue would retain its power of induction. These, fully published later (O. Mangold[7]), showed that it could be traced as far as the brain of the free-swimming larva (Fig. 64 D). Such observations formed the starting-point for far-reaching work, reference to which it is more logical to introduce at a slightly later stage (p. 172).

From this work, however, it was clearly apparent that the presence of the stimulus and the presence of the capacity to react to it occupy different time-spans of the embryo's developmental life. Competence and organiser activity do not arise and disappear at the same time. The former appears only for a relatively short time, the latter seems to begin before and to persist for a very long time afterwards. The situation may be summarised in the form of a table (Table 14), constructed partly from the discussion of Spemann ([17], pp. 250–59).

At the blastula stage competence to react to the primary organiser is present nowhere. Organiser is present to a slight extent in the region which will become presumptive mesoderm and will eventually be invaginated.† At the early gastrula stage, when the blastopore is first visible, competence arises in the presumptive neural plate‡ and mesoderm;§ this is followed a little later by the appearance

---

* Later on, Mangold[11] introduced the terms *"autonomous induction"* for cases where the implant stimulates tissue to the forming of new structures, *"complementary induction"* for cases where the implant takes part itself in what is formed. A discussion of this will be found in Spemann ([17], pp. 279 ff.). Homoiogenetic induction by neural tissue is nearly, but not quite always, autonomous. In a famous case of Mangold's, a half forebrain was implanted into the blastocoele cavity; it itself formed brain, one eye-cup, and one olfactory pit, but it induced brain, two olfactory pits, two eyes with lenses, four ear-vesicles, one ganglion and one balancer. Not a very teleological result (see below, p. 231).

† Bautzmann[1], p. 316.   ‡ Spemann & H. Mangold.

§ Lopaschov[3].

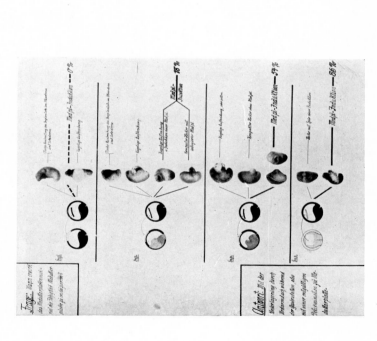

Fig. 64 A. Determination of the neural plate goes on *pari passu* with the invagination of the underlying mesoderm.

Fig. 64 B. Homoiogenetic induction.

Four diagrams of historical interest, prepared by the school of H. Spemann and O. Mangold about 1928, illustrating various aspects of organiser phenomena (continued overleaf).

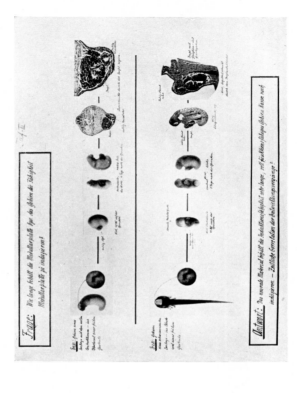

VI

Fig. 64 D. Inductive power is retained in the nervous tissue at least as far as the brain of the free-swimming larva.

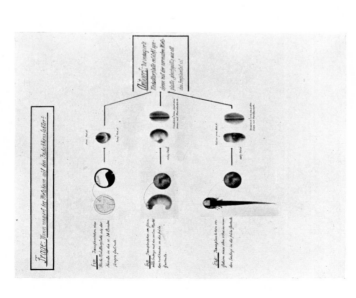

Fig. 64 C. The induced neural tube appears synchronously with that of the host.

Table 14. *Scheme of appearance and disappearance of competence and organiser action in the amphibian gastrula*

	Presumptive mesoderm		Invaginated mesoderm		Presumptive neural plate		Neural plate		Presumptive epidermis		Epidermis		Yolk-endoderm	
	C	O	C	O	C	O	C	O	C	O	C	O	C	O
Blastula ...	−	±	·	·	−	−	·	·	−	−	·	·	−	−
Early gastrula (blastopore first visible)	±	+	−	+	±	+	·	·	−	−	·	·	−	−
Middle gastrula ...	·	+	−	+	+	−	·	·	±	+	·	·	−	−
Late gastrula ...	·	·	−	+	·	·	±	+	·	·	±	+	−	−
Early neurula (neural folds first visible)	·	·	−	+	·	·	+	−	·	·	+	−	−	−
Middle neurula ...	·	·	−	+	·	·	+	−	·	·	+	−	−	−
Late neurula ...	·	·	−	+	·	·	+	−	·	·	−	−	−	−

C = Competence. O = Organiser power.
Presence or absence are indicated by + and − ; a dot indicates that the tissue has disappeared or is not yet formed.

N.B. (1) It will be noted that only in the invaginating mesoderm do competence and organiser power exist side by side. Yet neural tissue does not develop in the archenteron roof. This can only be explained by supposing the competence to be slight, and only manifested when the material is placed in the quite different situation of the presumptive neural area.

(2) Another interesting case is that of the ectoderm which actually covers the just-closed neural tube. It remains ectoderm because its competence is lost just in time. If ectoderm from a younger embryo (gastrula) is substituted for it an additional neural plate is formed (Holtfreter[5]), by homoiogenetic induction (p. 154).

(3) Neural competence exists at much the same intensity all over the presumptive epidermal area and equally throughout the thickness of the cell-layer (Schechtman[6]).

of competence in presumptive epidermis as well.* As the mesoderm is invaginated it loses its competence but retains, indeed intensifies, its organiser action.† As the neural plate is determined and later histologically differentiated, it naturally loses its competence, but at the same time acquires organiser power.‡ Competence in the presumptive epidermis reaches a maximum during gastrulation, but later during neurulation dies away and ceases.§ Presumptive epidermis or epidermis, of course, never possesses organiser power.‖ Finally, yolk-endoderm never possesses either competence or organiser power.

Of equal interest for the future biochemical analysis were the experiments in which Spemann & Geinitz showed that a piece of ectoderm, transplanted into the invaginating mesoderm, and subsequently removed from it and placed in the blastocoele cavity of another embryo, would manifest acquired organiser capacity. The significance of this awakening (*Weckung*) or "infection" will later on be better appreciated ¶ (p. 205).

The question of the species-specificity of the organiser was early taken up. Geinitz, only a year after the basic experiment of Spemann & H. Mangold, found that inductions could readily be obtained by implanting anuran dorsal lips in urodele blastocoele cavities (*Bombinator* → *Triton*). Owing to technical difficulties, the demonstration that events in the anuran gastrula ran quite parallel with those in that of the urodele was for a long time delayed, but in the end successfully accomplished by Schotté[4]; Vintemberger[1]; and Dalcq & Tung. The converse of Geinitz's experiment was then performed by Spemann & Schotté. Again, we shall see later on (p. 172) a wide extension of these experiments. In anurans the organisation centre occupies a smaller area than in urodeles (Schmidt[6]; Schechtman[7]).

2·333. Necessity of the primary organiser

From all the work just reviewed it followed inescapably that the organiser centre was involved in the determination of the main axis of the vertebrate embryo. But how far was it indispensable? We have already seen (p. 103) that a ventral half of an embryo, whether originating by constriction of the egg before cleavage or by equatorial section of the gastrula, will develop no neural axis. Bautzmann[2] found, however, that if a small piece of dorsal blastopore lip was implanted into such a *Bauchstück*, a normal neural axis resulted. Nevertheless, this did not prove that the organiser centre was wholly responsible for neural determination in the normal embryo.

Reasons for thinking that some other factor was necessary arose from two distinct lines of work, but both were afterwards proved to have led to fallacious

---

* O. Mangold[2,4,5].   † Marx[1].   ‡ Mangold & Spemann.
§ W. H. Lewis[3]; O. Mangold[5].   ‖ Bautzmann[1].

¶ *Weckung* was subsequently further explored by Raven[4], who allowed a piece of presumptive ectoderm to remain for only twenty-four hours in the dorsal lip region before being moved again to the ventral region. If it had taken part in the primary invagination it formed notochord and muscle and then induced in itself a little neural plate; if it had not, it formed an ectodermal thickening only. He was thus able to show that *Weckung* depends on invagination, probably because of the special metabolism connected therewith.

conclusions. In the first place, Goerttler[1] believed, as the result of his transplantation experiments, that a successful neural induction in competent ventral or lateral ectoderm could only be obtained if the implanted organiser was so oriented with respect to the host's formative gastrulation movements that its movement tendencies were with them and not against them. Altekrüger thought that successful neural inductions could not be obtained if the isolated organiser was allowed to heal into a ball before implantation. Both these conclusions were decisively proved wrong by Holtfreter[4, 5].

The other line of work had led to results very difficult to understand. Dürken[1] had grafted fragments of the dorsal regions of blastulae or young gastrulae into the eye-cups of free-swimming larvae, where they developed well, causing little or no reaction on the part of their hosts. The pieces were not dissected out from very well-defined regions, but, nevertheless, neural tissue, notochord, cartilage, bone, glands, ganglia, etc. were produced, although the material had never undergone the action of the primary organiser in the normal way. The work was continued by Kusche in such a way as to make certain that no organiser had been taken for implantation into the eye-cavity. Yet presumptive ectoderm, mesoderm and even neural plate gave rise to neural tissue, notochord and muscle. About the same time Holtfreter[1] cultivated gastrula isolates in the peritoneal cavity of older larvae, with similar results. The phenomena were summarised by Bautzmann[5] under the name of *bedeutungsfremde Selbstdifferenzierung*; i.e. self-differentiation in a direction contrary to the prospective significance of the part.*

The suspicion had been growing up, however, that the tissues of older animals might not furnish a perfectly indifferent environment, and so what can be called the search for a neutral medium began. It was found by Holtfreter[2], who noted that healing of isolates would occur well in a salt solution some twenty times weaker than ordinary mammalian Ringer solution, to which a small amount of bicarbonate had been added.† After healing, the pieces could remain in the solution or be transferred to sterile tap-water, and would live and differentiate for many weeks.‡

The presence or absence of the organiser centre was now felt in earnest. Ectoderm, whether presumptive neural plate or presumptive epidermis, isolated from all contact with the presumptive mesoderm, now developed entirely free

---

* Many other transplantation experiments have been made (e.g. rat bone into rat brain (Willis[3]), human placenta into rabbit eye (Kido[1]), thyroid into eye (May[2]), testis into eye (Bayer & Wense), etc.) but, paraphrasing a saying of Claude Bernard's, it is easier to make transplantations than to get any useful information from them. Nevertheless, the general review of Willis[5] is interesting.

† Originally given as NaCl 0·35 %, KCl 0·005 %, $CaCl_2$ 0·01 %, $NaHCO_3$ 0·02 %; $\Delta - 0·22°$; pH 7·8–8·2. This composition is not a perfect approximation to that of the normal perivitelline liquid, as Richards[2] shows; it lacks protein, has too much sodium and chlorine, and is hypertonic still.

‡ Cultures of parts of amphibian embryos have this great advantage over all other forms of tissue culture; the cells contain within themselves their natural stores of nutritive materials, and the latter do not have to be supplied from outside by the more or less hit-and-miss methods of the experimentalist. The contributions of tissue culture in the ordinary sense to the problem of differentiation have been reviewed by Bloom. In studying self-differentiation remarkable results can sometimes be achieved (see e.g. Fell[1, 2, 3, 4]).

from any sign of neural differentiation. Buckled brown masses of cuboidal ecto-dermal cells resulted, such as that shown in Fig. 65. On the other hand, if a piece of blastula or early gastrula was isolated, containing the presumptive mesoderm,

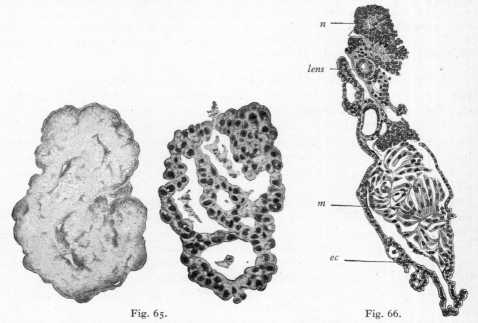

Fig. 65.   Fig. 66.

Fig. 65. Mass of ectodermal cells formed by explantation of ventral ectoderm in neutral medium (external view and cross-section).

Fig. 66. Complex formation of tissues formed by explantation of blastula fragment containing presumptive mesoderm, in neutral medium. n, neural cells; m, muscle cells; ec, ectoderm.

a complicated structure was produced showing all kinds of tissues, muscle, noto-chord, and neural tissue as well as epidermis. Cross-section of such formations[*] are seen in Figs. 66, 67. The situation may thus be summarised as follows:

	Normal	Implanted *in vivo*	Explanted *in vitro*
Presumptive neural plate	Neural tissue	Neural tissue	Skin
Presumptive epidermis	Skin	Neural tissue	Skin

(Skin = epidermis)

Thus, when fully isolated from all organiser action, presumptive neural plate possessed no more tendency to turn into neural tissue than did presumptive epidermis. Other experiments by Schechtman[4] have confirmed these results.[†]

[*] Cf. the discussion on teratomata below, p. 231.

[†] The identification of true neural cells in explants may sometimes be a rather delicate matter. My own experience leads me to concur with the views expressed by Schechtman[4]. "Under the mechanical conditions prevailing in the explant", he said, "some cells become more slender and taper towards one end, thus resembling neural cells. But these have no typical neural plate orientation and disappear as the culture is continued. Typical neural differentiation is charac-terised by a compact mass of elongated cells rather than by occasional ones." (See below, p. 177, regarding "palisade" inductions.)

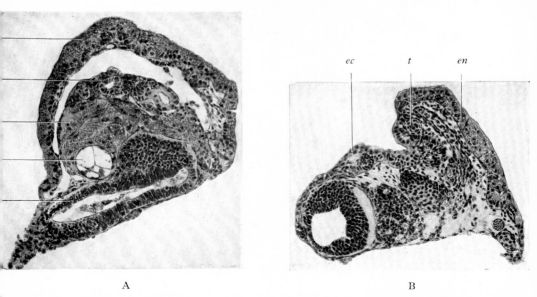

VII

A   B

Fig. 67. Complex formations of tissues arising from piece of early gastrula containing lateral part of organiser. *ec*, ectoderm; *en*, endoderm; *k*, kidney tissue; *m*, muscle; *nch*, notochord; *n*, neural tissue; *t*, tooth papilla.

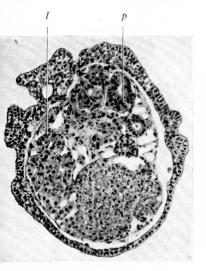

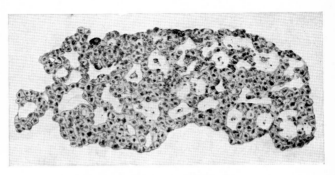

ig. 68. Complex formation of tissues from isolated endodermal region. *l*, liver; *p*, pancreas.

Fig. 70. The ectoderm forsaken by the exoembryo.

The conclusion was in this way reached that the primary organiser was contained in the tissues of the larval host. As we shall see later (p. 173), the investigation of adult tissues has shown by direct experiment that the primary organiser does indeed persist in the tissues into adult life. This fact is of much importance for our understanding of its chemical nature.

Experiments of a quite different type gave the final demonstration that without the organisation centre of the presumptive mesoderm no neural differentiation is possible. For this purpose advantage was taken of a drastic alteration that may occur in the formative movements. Instead of turning in at the dorsal lip of the blastopore, the presumptive mesoderm may turn outwards, evaginating instead

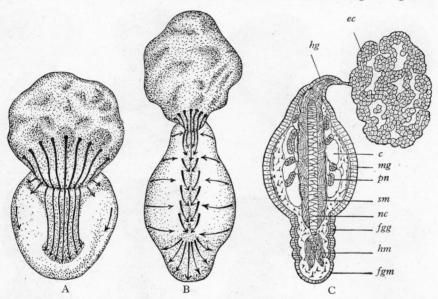

Fig. 69. Diagram of three successive stages in the process of exo-gastrulation. *c*, coelom; *ec*, ectoderm; *fgg*, foregut (gill region); *fgm*, foregut (mouth); *hg*, hindgut; *hm*, head mesoderm; *mg*, midgut; *nc*, notochord; *pn*, pronephros; *sm*, somite muscle.

of invaginating, so that the material which should have filled up the interior of the ectodermal sphere passes away from it and exerts no influence upon it. Such exo-gastrulae, though very undeveloped, were first described by O. Hertwig[1] and C. B. Wilson. Holtfreter[8], who discovered a way of producing them in extreme form at will (cultivation of the blastulae in 0·35 % salt solution*), devoted extended studies to them and described the result in papers already classical.

The process is seen in Fig. 69. The whole of the yolk-endoderm is eventually pulled out by the evaginating mesoderm, forming an endo-mesodermal mass (the "exo-embryo") which, by the time a normal neurula ought to have been formed,

---

* They can also be produced by X-radiation (Curtis, Cameron & Mills). An interesting observation, which links amphibian exo-gastrulation with that of echinoderms (see pp. 477 ff.), is that of Ranzi[5], who produces exo-gastrulation of the lamprey (cyclostome) egg by the action of lithium (see p. 277).

has considerably stretched itself out. Notochord and mesoderm sink within the exo-embryo, which thus becomes entirely surrounded by endodermal cells; these will in time give rise to intestinal glands secreting their products to the exterior, for the whole structure is inside out (Fig. 69 C). Meanwhile, the forsaken ecto-dermal ball proceeds to no differentiation of any kind; presumptive epidermis and presumptive neural plate alike form a buckled mass of brown cuboidal ectodermal cells, as shown in Fig. 70 (opp. p. 158). The external appearance of an embryo in total exo-gastrulation after about a week is seen in Fig. 71. Separation of endo-mesoderm from ectoderm is complete, for the only connection is a thin thread of ectoderm which later may spontaneously break. We have thus a self-isolation of the germinal layers.

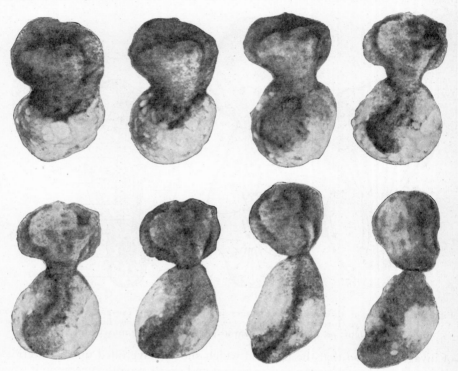

Fig. 71. Successive stages of exo-gastrulation.

In the exo-embryo all kinds of organs develop; pronephros and mesonephros, gonads and smooth muscle, a pulsating heart free of blood-cells (for these develop away from it in its new situation), coelomic cavity and connective tissue (cf. Figs. 72 and 73). Naturally all these structures are entirely without any nerve connections. In the ectodermal part, on the other hand, everything is lacking; there is no neural tube, eye-cup or lens, no mesectoderm, no teeth, no ear-vesicles, etc.

These results brought final confirmation of the view that the underlying of the dorsal ectoderm by the mesoderm is the full, perfect and sufficient cause for its

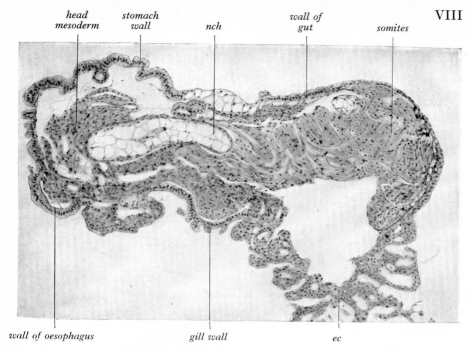

head
mesoderm   stomach
wall   nch   wall of
gut   somites

wall of oesophagus   gill wall   ec

Fig. 72. Sagittal section through the exoembryo.

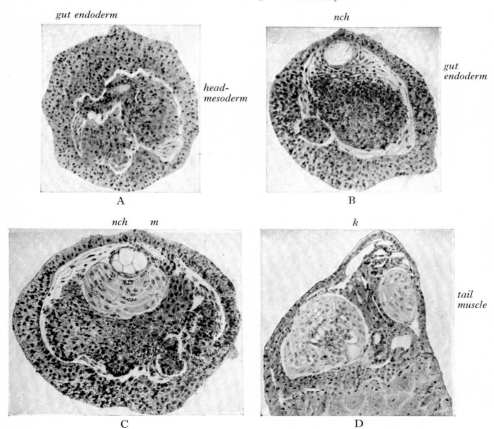

gut endoderm   nch

head-
mesoderm

gut
endoderm

A   B

nch   m   k

tail
muscle

C   D

Fig. 73. Transverse sections through the exoembryo. Conventions as in Fig. 67.

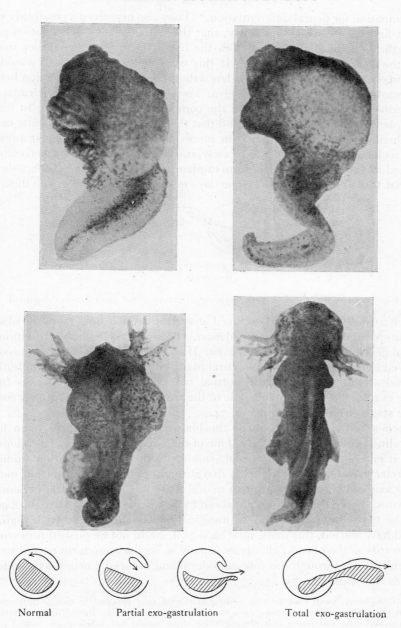

Normal   Partial exo-gastrulation   Total exo-gastrulation

Fig. 74. Stages of partial exo-gastrulation

determination for neural differentiation.* They also negatived a possibility which had at first been canvassed, namely, that the "stream" of determination passed along through the dorsal layer from the lip of the blastopore, rather than up from the roof of the archenteron. If this were the case, inductions would take place in exo-gastrulae, since until a late date there is always a connection between the endo-mesoderm and the ectoderm. Tests were now made of the capacity of the endo-mesoderm to induce and the competence of the ectoderm to react to an implanted organiser. It was found that the endo-mesoderm was quite capable of inducing neural differentiation in pieces of ectoderm laid upon it anywhere along its length (see p. 273). Conversely, it was found that the forsaken ectoderm retained its competence to react to an implanted organiser for a considerable time, but that this property was lost some days or weeks before the isolate died.

| Normal | Exo-gastrulation (Holtfreter) | Ventral induction (Eakin) |

Fig. 75. The three logical possibilities of direction of the gastrulation movement.

Such artificial experiments received abundant confirmation from the observation of natural partial exo-gastrulation. Between total exo-gastrulation and normal development all stages can exist. If a small amount of invagination occurs followed by evagination, a small neural plate will be formed which will build the material around it into a quite normal tail. If invagination proceeds further before evagination occurs, the whole of the trunk may be fairly normally formed. These stages are illustrated in Fig. 74.

When a cell reaches the lip of the blastopore there are three main logical possibilities as to where it may go. One of these happens in normal development, when it enters the archenteron roof, another in the exo-gastrulation studied by Holtfreter[8]. The third possibility is that it might pass ventralwards and induce a neural axis not in the presumptive neural plate but in the ventral presumptive epidermis. This effect was brought about by Eakin[1], who injected a mass of gelatin into the blastocoele cavity before invagination had begun. When invagination should have started, this mass, now hardened, could not be pushed forwards and downwards as the blastocoele cavity usually is, so the mesoderm passed towards the ventral pole through the yolk-endoderm and induced a neural axis under the yolk, as shown in Fig. 75.

* It has, however, been claimed that ventral ectoderm may proceed to neural differentiation without the action of the organiser or a chemical equivalent of it. Barth[3] fused together 2–8 ectoderm explants, and found that although one alone would not differentiate, neural masses, more or less chaotic, appeared in the fused pieces. For this it was necessary to preserve their antero-posterior polarity in performing the fusion. The full publication of these results will be awaited with interest, but it may be noted that the stage from which the pieces are taken is admittedly late in gastrulation and therefore very striking proof will have to be required that the fused mass did not contain small amounts of already determined neural area, which would give homoiogenetic inductions.

2·335. Determination of the endoderm and mesoderm

If we are now prepared to accept the determination of the neural axis by the invaginating endo-mesoderm, it must be admitted that much still remains obscure. The self-differentiation of endo-mesoderm in exogastrulae, or of isolated endoderm (Fig. 68), is remarkable in that no point of origin for this determination can be found. That the dorsal blastopore lip possesses the widest powers of self-differentiation cannot be denied, but we are still impelled to search for the organiser of endodermal and mesodermal organs. Practically nothing can yet be said upon this subject.

As regards the mesoderm, it has usually been assumed that when it sends out its determining influence on the ectoderm it is itself determined. This cannot be so, however, for as Adelmann & McLean established, early somites are very interchangeable. Then Yamada[1] confirmed by transplantations between more widely different mesodermal sites that the mesoderm is still very labile at the end of gastrulation. In the early neurula, somites were transplanted to the pronephros region or to the anterior ventral region. In the former they behaved neighbourwise and formed pronephros; in the latter they behaved neither neighbourwise nor selfwise but formed pronephros tissue there also. This very surprising result (*bedeutungsfremde Selbstdifferenzierung*) could be inhibited if notochord was transplanted at the same time; this, as it were, kept them in order, and maintained them as somitic tissue. If a pronephros-inducing substance could be postulated as existing in the anterior ventral region (though normally having nothing to do there), a suggestion not without plausibility (cf. p. 306), the effect would not be so strange. In the later neurula, somite transplants behave selfwise. But the remarkable thing is that the mesoderm should still be so labile after having neurally determined the ectoderm. As Vogt[5] put it, commenting on Yamada's results: "Das Organisationszentrum führt ohne voranzugehen". Earlier work of Hoadley[5] had shown a similar lability of tail-bud mesoderm in the neurula.

Vogt[5] distinguishes, on the basis of these results, four stages in determination: (1) There first appears the capacity for self-differentiation to a given structure, as tested by isolation into neutral medium; this is present in presumptive notochord and mesoderm before invagination. In the second stage, (2), the capacity to assimilate neighbouring tissues by way of regulation arises; this is tested by defect experiments and appears in the early gastrula. (3) The capacity for harmonious internal regulation (within the germ-layer) appears in the early neurula (cf. Yamada), and (4) as this decays, only disharmonious internal regulation remains possible (late neurula). In the nervous system, as well as the mesoderm, minor degrees of regulation remain possible for a very long time (Detwiler[1, 2, 3, 5]; Coghill[1]; R. M. May[3]).

Yamada[3] later investigated the behaviour of somite mesoderm in isolation. Alone, *in vitro*, it makes no pronephric tissue. If, however, it is implanted into ectodermal balls, it will do so; and here again the presence of notochord will prevent the transition. But if the ectoderm used for the ball is taken from a sufficiently young gastrula, the implanted mesoderm will induce a mass of neural tissue in it, and this will later act like notochord in maintaining the mesoderm as muscle and preventing the transition to pronephros. Now not only will noto-chordal tissue prevent presumptive muscle from turning into pronephros; it will also induce muscle tissue from presumptive pronephros (Yamada[4]). And between presumptive pronephros and presumptive blood there is apparently a relation similar to that between the muscle and the pronephros, for presumptive prone-phros will form blood if isolated into ectodermal balls, while presumptive blood will form pronephros if isolated into ectodermal balls together with notochordal tissue. These relationships may be better visualised with the aid of the following table (Yamada[5]):

	Prospective significance	Result when alone in ectodermal ball	Result when in ectodermal ball in presence of notochord
Presumptive somite mesoderm	Muscle	Pronephros	Muscle
Presumptive pronephros	Pronephros	Blood-cells	Muscle
Presumptive blood	Blood-cells	Blood-cells	Pronephros

This serial system, obviously of great interest, invites the suggestion that the various morphogenetic effects may be due to quantitative variations in the amount or activity of a single inductor rather than to the existence of several. Its weakest concentration would produce blood-cells, a higher concentration would produce pronephros, and the highest would give rise to muscle. Normally, it would have to be pictured as radiating from the notochord, but the presumptive pronephros and blood-forming regions could be regarded as accessory centres of its production. The Dalcq-Pasteels theory of "morphogenetic potential" (see p. 220), to which Yamada[5] himself inclines, is a somewhat less biochemical way of saying the same thing. That something diffusible is at work is strongly indicated by the gradations in the effects obtained, the tissue nearest the notochord frag-ment, for example, being most fully transformed into well-differentiated muscle.

Following a parallel line of thought, Balinsky studied[8,10] the lability of endo-derm. Even when taken from the neurula, yolk-endoderm, if implanted into the blastocoele cavity of a young gastrula, behaves completely neighbourwise, forming whatever organ of the host it happens to be caught up into. Holtfreter's exo-gastrulation experiments[8] were to some extent paralleled by the remarkable operations of Mangold[13], who completely removed all the yolk-endoderm from a young neurula, and cultivated it in salt solution. Such pieces lived on long after all the yolk would have been consumed if they had remained in their normal position, but their self-differentiation was not very striking; they formed only foregut and midgut, no gill clefts, liver or pancreas. Mangold attributed the much greater self-differentiation of the exo-gastrulated material to the presence of mesoderm, which again is significant for the argument of this section.

Mangold also examined the results of replacing endoderm in neurulae after inversion of its dorso-ventral or cephalo-caudal axis, or both. The latter seemed to be much more fixed than the former.

The problem of endo-mesodermal determination is generally passed over in discussions of this subject, and it demands much further work. Is it impossible that there should be appropriate substances present in the invaginated (or evaginated) material, analogous to the primary evocator?

2·34. Biochemistry of the primary organiser

The analysis of the biochemical properties of the primary organisation centre began, as all such analyses must, in the enquiry whether or not the integrity of the living cells responsible, in this case the dorsal lip of the blastopore, was requisite for the inductive effect. At the conclusion of their paper, Spemann & H. Mangold had written "Aus diese Überlegungen ergibt sich das Experiment, eine etwaige Struktur des Organisators zu zerstören und zu prüfen, ob er dann immer noch determinierend wirken kann". Less than ten years later this question was settled affirmatively.

Marx[2] was the first to show that normal respiratory metabolism at least could be excluded, since induction would still occur after the narcotisation of the dorsal lip of the blastopore with trichlorbutyl alcohol and its transplantation to the blastocoele of a host embryo. But Spemann[14] in 1931 went further when he showed that crushing the cells had little or no deleterious effect upon their inducing activity. Some of his material was later worked over by Krämer, and Fig. 76 shows a beautiful secondary embryo produced by the implantation of crushed organiser material.

In the following year, four workers simultaneously, Bautzmann, Holtfreter, Spemann & O. Mangold, showed that the organiser was remarkably resistant to heat, since it could be boiled for some time without loss of activity. This definitely excluded the participation of enzyme effects in the stimulating action. At the same time the stability of the primary organiser of the chick to boiling was demonstrated by Waddington[6], and later on the same proved to be true for the teleostean fish in

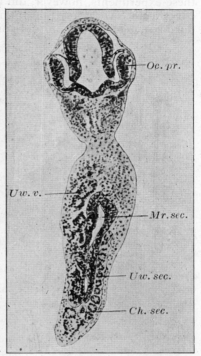

Fig. 76. Induction of secondary embryo by organiser after destruction of its cell-integrity by crushing. *Oc.pr.*, host eye-cup; *Mr.sec.*, induced neural tube; *Uw.sec.*, induced somites; *Ch.sec.*, induced notochord; *Uw.v.*, accessory somite.

Luther's experiments[2]. Holtfreter[6] completed the demonstration in a long and detailed paper which left no doubt that the primary organiser, or, if we confine our attention to the neuralisation stimulus, the evocator (see p. 125, and p. 271), was quite stable at 100° C. Coagulated dead pieces of dorsal lip would induce secondary neural axes just as well as they did when living. Holtfreter's experiments were of particular interest, for besides the classical technique of implanting the material to be tested into the blastocoele cavity of a young gastrula (*Einsteckung*), he also used the methods of surrounding the piece of dead material by two layers or sheets of competent ectoderm (*Umhüllung*), and of laying pieces of competent ectoderm upon a flat piece of dead organiser region (*Auflagerung*). Figs. 77, 78, 79 illustrate the consistency and beauty of the results he obtained.

A digression may be made at this point to emphasise the interest which attaches to this induction of neural structures, not only in the intact egg, but also in spherical masses of ectoderm lying in contact with the source of the evocator. We have here a new "*Gestalt*", a new arrangement of animal form, brought about at will in the plastic embryonic tissue by the experimenter. In none of these experiments, however, did any regional differentiation appear, so that the effect of the evocator acting from the dead inductor could be described as producing so many length units of uniform neural tube in the material at its disposal.

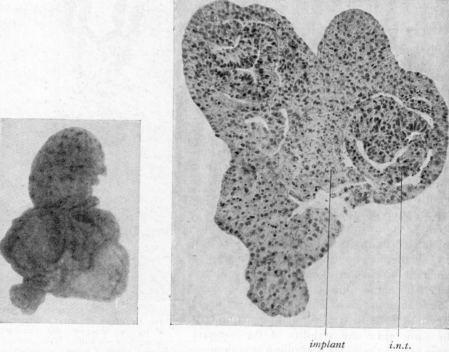

implant   *i.n.t.*

Fig. 78. Induction of two neural tubes by surrounding a piece of boiled organiser with sheets of competent ectoderm.

It thus became certain that the primary morphogenetic stimulus of development is independent of the life of the cell in just the same way as the adrenalin of the suprarenal gland or the hormone of any other endocrine organ is separable from the life of the glands. Not till later was the converse question asked, namely, what degree of integrity must be present in the competent tissue? Here of course there can be no question of destroying the living cells, since the reaction is obviously dependent upon their activity, but normally they are stuck together in a definite sheet, pigmented on the outward surface, and probably covered with a cuticle which makes them more permeable from within than from without.

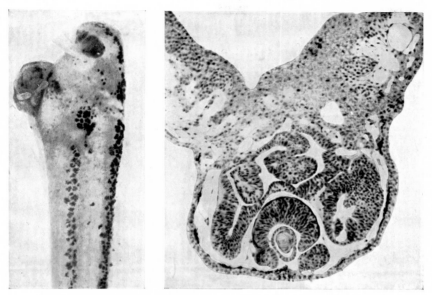

IX

Fig. 77. Induction of secondary brain, eye-cup, etc. by implantation of boiled organiser.

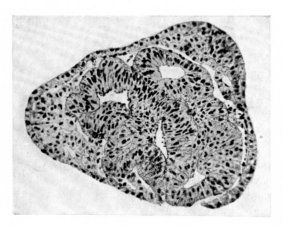

Fig. 80. Chaotic neural formation produced by implantation of boiled organiser into a competent ectodermal ball.

Would induction still occur if these normal relations were interfered with? The question received an affirmative answer in the work of Umanski[2], who succeeded in breaking up competent ectoderm into its constituent cells with a hair-loop without injuring more than a few of them. Into the mass so formed, he implanted a piece of living dorsal lip, with the result that in many cases neural and noto-chordal structures were produced. From his description, the histological changes were more marked and normal than the morphological ones. This is perhaps an indication that the immediate effect of the evocator is upon the cell structure, producing a histological change, and that the morphological changes then follow.

A somewhat similar dissociation between the histological and the morpho-logical or "organogenetic" differentiation was met with by Holtfreter[6] in the experiments where explants, usually healed into more or less spherical form,*

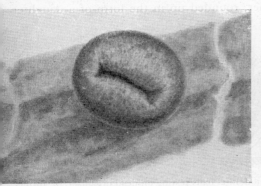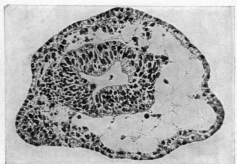

Fig. 79. Induction of neural tube in a ball of competent ectoderm laid on a piece of boiled organiser.

were laid upon a substratum of dead tissue, or received implants of it. "The cytological, tissue-building process", he said, "is more frequently and more typically carried through, than the morphological or organ-building one." Very often brain-like masses of neural cells would be produced, but though they might surround cavities of various shapes, no neural plates or tubes of typical cross-section were to be seen. A typical case of this rather chaotic neural-tube formation is shown in Fig. 80 (opp. p. 166). We shall later return from time to time to this dissociability of histogenesis and morphogenesis, especially in the discussion on evocation and individuation, and the experiments designed to define the difference between them (Section 2·391). In order to disarm the criticism that the difference between the structures produced by dead inductors in explanted isolates of competent ectoderm, and those produced by living inductors in the intact embryo, might be due to the reacting tissue and not to the conditions of

* Isolations of pieces of ventral ectoderm of the gastrula, which then heal themselves into balls and into which implants of all kinds may be placed, have proved of the utmost value (see below, pp. 278 and 280). It is interesting that nature seems sometimes to perform similar experiments, for Bland-Sutton (*Tumours*, etc., p. 618) describes rare ovarian dermoid cysts containing thousands of "epithelial balls" (see below, pp. 232 ff.). The original papers on such cases, however, such as that of Bonney, give rather the impression that the balls are of dead cells, sweat, and hair, and do not contain any true epidermis.

the stimulus, Holtfreter made control experiments in which living dorsal lip pieces were implanted into isolated ectodermal balls. The results were quite definite; perfect neural tubes of considerable length were produced together with tails, sense organs, somites, etc.

Formative movements of the ectodermal isolates were observed by Holtfreter, who summarised them in the way shown in the accompanying diagram (Fig. 81). A short time after the ball was placed on the underlying dead tissue, its outer layer of cells began to buckle on the upper surface, simulating the neural groove. At the same time the surface began to be covered with cells which came out at the lower pole of the isolate and wandered round the outside; these eventually formed the "epidermis", while the others formed the neural tissue. In some cases, these cells issuing from the lower pole spread out on the supporting surface and later, drawing themselves together, rounded up in the form of a daughter

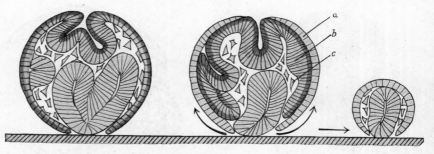

Fig. 81. Formative movements in isolated ectodermal balls resting on dead organiser. *a*, former internal cells; *b*, former outer surface; *c*, mesoderm.

isolate. Holtfreter gave some illustrations of these in section. They contained a small tubular or globular mass of clearly neural tissue, surrounded by an ecto-dermal "skin".

Observations were also made on the time necessary for inductive effects to appear. A twenty-four hour contact between isolate and dead tissue seemed to be the minimum time after which induction would follow. It must be remembered that the whole technique required strictly sterile conditions, so that the boiled inductor was not in the slightest state of decay. Induction was prevented if the ball isolate was allowed to rest on a piece of vitelline membrane, not directly on the boiled tissue.

So far, events had given every promise of a fairly straightforward state of affairs. But Holtfreter[6,11] went on to make the remarkable discovery that parts of the egg which do not possess inductive power in the living condition, acquire it after being boiled. The ventral presumptive epidermis of the gastrula, or the ventral epidermis of the neurula, could thus, once dead, call forth as strong inductions as had previously only been obtainable by living organisers. Even the fat, white, yolky cells of the endoderm would actively induce after being killed by boiling, and so would the fertilised egg and the egg not yet shed from the ovary (see Figs. 84, 85; opp. p. 172).

X

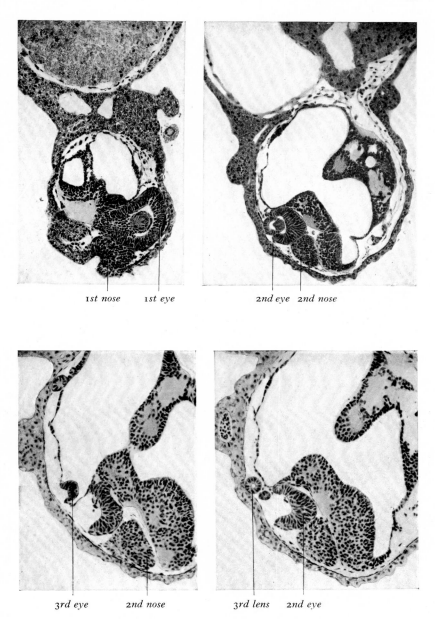

1st nose   1st eye   2nd eye   2nd nose

3rd eye   2nd nose   3rd lens   2nd eye

Fig. 82. Induction of chaotically arranged secondary embryo by implantation of boiled organiser into the blastocoele cavity.

From this extraordinary fact it was at once evident that in normal development the evocator exists all over the egg, but in a *masked* or *inactivated* state.* Only in the dorsal lip of the blastopore is it normally set free, and only in the invaginating roof of the archenteron does it exercise its profound effect upon the competent ectoderm (the presumptive neural plate) waiting for it above. And straightway the means by which it is set free acquires the greatest theoretical importance, for it shows how the most fundamental morphological interactions may be bound up with correctly running metabolic processes. But there is still another point which must be emphasised. Presumptive ventral epidermis contains the evocator in a masked condition, but another name for presumptive ventral epidermis is competent ectoderm, and it is precisely such parts of the egg that we are bound to use as reactants for the biological test of inductive capacity. The seriousness of this fact will become clearer when the progress of the chemical work has been further described.

In the second of Holtfreter's two great papers on this subject[11] a number of new points were raised and questions answered. It was noted that the implantation of dead organisers into the blastocoele cavity of embryos seemed to lead to very disoriented, chaotic, yet well-formed structures. An example is shown in Fig. 82 (opp. p. 168). Here the posterior part of a neural plate was heated to 90° for a few minutes; on implantation a large structure was produced on the host in which were found, besides brain masses, three eyes with lenses and two noses. We shall see other striking examples of this in a later section (p. 233). The differences between inductors and reactors may then be summarised as follows:

Inductor	Reactor	Result
Living organiser	Intact embryo	Perfectly formed secondary embryos, showing lateral symmetry and organs correctly formed and placed
Dead organiser	Intact embryo	Organs well formed, but often in excessive number, and chaotically placed. No marshalling of them into a secondary embryo
Living organiser	Ectodermal isolate	Perfect histogenesis, organs small but usually placed in fairly normal relationships
Dead organiser	Ectodermal isolate	Perfect histogenesis, but little or no organ-formation. Neural cells arranged round irregularly branching cavities

All this can hardly be interpreted save in terms of evocation and individuation (see below, Section 2·391).

Another strange fact was that in some cases a lens alone might be formed from the ectoderm, in the complete absence of any eye-cup. This is another phenomenon which we shall meet with again (p. 299). In the present context, it is noteworthy that it was especially likely to happen if the inductor had been boiled for a long

* An alternative interpretation was advanced shortly after Holtfreter's discovery by Spemann, Fischer & Wehmeier, who suggested the existence of an inhibitory substance (*Hemmungstoff*) in those parts of the embryo which do not induce in the living state. They believed that by treatment with organic solvents, etc. this inhibitory substance could be extracted. There seems no need, however, to have recourse to the hypothesis of an inhibitory substance when the much simpler one of combination of an active principle with a protein, for example, in an inactive complex, is at hand. In a sense, of course, the protein would be the inhibitory substance. But the hypothesis, in the sense in which it was advanced by Spemann *et al.*, regarding the inhibitory substance as one of low molecular weight, has not proved fruitful and is to-day generally abandoned.

time, thus suggesting that the lens-inducing substance might be more stable to heat than the primary evocator (see Fig. 83).

Prolonged boiling, however, does not destroy the inducing action, though Holtfreter[11] found slight indications that after 2 hr. boiling its activity was lessened. More important was the fact that if the tissue were heated to 135° the effects were very greatly diminished, while a temperature of 150° completely abolished them. Ashing naturally gave the same result.* As regards the effect of various chemical agents on the activity of inductors, Bautzmann *et al.* had already reported that a stay in 96 % alcohol for 3 min. had not affected the dorsal lip. This was confirmed by Holtfreter, who found that the activity had not been lost by pieces conserved in strong alcohol for many months, nor even by pieces subsequently treated with xylol and paraffin wax. Activity was unaffected by concentrated hydrochloric acid or by ether.

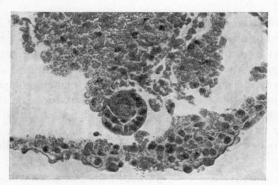

Fig. 83. Induction of a lens in the absence of any eye-cup after implantation of boiled organiser into the blastocoele cavity.

Holtfreter's new contribution[11] was to show not only that the activity of a living inductor was not affected by these treatments, but that activity was conferred by them upon a living non-inductor. Thus presumptive epidermis or yolk-endoderm, after treatment with organic solvents such as alcohol, ether or xylol, would induce just as well as if they had been boiled, and the same result was obtained by freezing followed by drying. This implies that any treatment which denatures the proteins of a non-inductor will cause inducing power to appear, i.e. will liberate the evocator from its inactive combination.

It is often not realised how this fundamental discovery of Holtfreter's clarifies facts of a morphological order which would otherwise be almost inexplicable. Thus Töndury[1] replaced dorsal lip material (presumptive chorda-mesoderm) with presumptive neural plate ectoderm; the new dorsal lip invaginated very well and duly induced a normal neural axis in the overlying tissue. If we did not know that the primary evocator has to be liberated from its inactive complex, and that this is somehow or other associated physiologically with the inrolling at the dorsal blastopore lip, the possible transposition of these two regions would be

* When I reported this fact to Otto Warburg in his laboratory at Berlin-Dahlem in 1933 he remarked, with a characteristic twinkle in his eye, "Dann ist es nicht Eisen".

very hard to understand. Conversely, the uninvaginated presumptive mesoderm could, when transplanted, make neural tube and ectoderm, since its evocator had never been properly set free.

In analysing his results from the morphological point of view Holtfreter[11] noted the significant fact that regional specificity was not conserved by dead inductors. Consideration of this, however, must be postponed till a later section (p. 277). Another striking result was that there was no correlation whatever between the kind of treatment which the killed organisers had received and the morphological structures which they subsequently induced. Whatever the treatment, brains, noses, eyes, ear-vesicles, lenses, balancers, and mesoderm were to be found in each series. When the size of the inductions was taken into consideration, it was found that the activity of "activated" non-inductors was rather less than normal inductors:

	Average size of induction (Holtfreter units)
Dorsal lip region (killed)	1·43
Presumptive epidermis (ventral ectoderm; killed; activated)	1·06
Yolk-endoderm (killed; activated)	0·67

This might authorise the view that there is actually rather more of the evocator present in the region where it is normally liberated or activated than in those regions where it is not.

It was said above that the unfertilised ovarian egg was found by Holtfreter[7] to be active as an inductor if boiled. Since the nucleus is of so relatively large a size in the ovarian egg, this invited a test of relative inducing activity as between nucleus and cytoplasm. Carried out later by Waddington[19], it showed that the nucleus possesses considerably more inductive power than the cytoplasm (see Figs. 84 and 85).

	% inductions	Av. vol. μl. 10^{-4}
Nucleus	57	92
Cytoplasm	28	14

In the meantime, a number of investigators were pushing Geinitz's original discovery of the absence of species-specificity of the primary evocator to much greater lengths. The successful heteroplastic inductions of Waddington & Schmidt on chick and duck were but the counterpart of the older urodele-anuran transplantations, but Waddington[7] was able to induce a secondary embryo in the rabbit blastocyst by a chick organiser, and a secondary chick embryo by a rabbit organiser[11]. In amphibian embryos inductions were produced by chick primitive streak* (Hatt[1]), by teleostean fish organisers† (Oppenheimer[4]), and by lamprey organiser‡ (Bytinski-Salz[4]). The converse implantation of amphibian tissue under competent fish embryo tissue was also successful§ (Luther[2]). Inductions in amphibian ventral ectoderm were brought about by the implantation of a regenerating larval limb bud (Umanski[1 ‖]) and of pieces of rat carcinoma and chicken sarcoma (Woerdeman & Hampe). Krüger[1] alone failed with lens implantations from older embryos, but this was later successful in the hands of Holtfreter[12].

---

* *Gallus domesticus* in *Triton taeniatus*.   † *Danio rerio* in *Triturus torosus*.

‡ *Lampetra planeri* in *Molge vulgaris*.   § *Triton taeniatus* in *Salmo fario*.

‖ He got better results from well-developed buds than from the youngest blastemas.

The ground was therefore in a sense prepared (though this did not lessen the surprise of it) for the discovery of Holtfreter[7, 12] that the evocator was not confined to the developing embryo, but that all adult tissues of all phyla indiscriminately would, if implanted into the blastocoele cavity, induce a secondary embryo. A typical result is shown in Fig. 86, where a secondary brain has been induced in the ectoderm and a secondary gut in the endoderm by the presence of a piece of boiled mouse heart. The following species were tested: two worms, two molluscs, three insects, a crustacean, two fishes, three amphibia, a reptile, two birds, two mammals, and man. From Table 15 emerges the conclusion that although all the tissues tested were active, those of vertebrates were in general more so, both quantitatively and qualitatively, than those of invertebrates. A typical induction by invertebrate tissue is shown in Fig. 87, where a small brain has been induced by the pupal lymph of the hawkmoth *Deilephila euphorbiae*. About the same time the inducing activity of adult newt tissues was established by Needham, Waddington & Needham. Later G. A. Schmidt[12] published further experiments of the same kind, using anuran instead of urodele gastrulae. Holtfreter's series was subsequently completed by Waddington & Wolsky, who tested pieces of the coelenterate *Hydra viridis*, never the possessor of any neural axis itself, as inductors in the amphibian blastocoele, with positive, though rather weak, effects.*

Especially significant was Holtfreter's finding[12] that not all these adult tissues required boiling or other denaturation of their proteins to manifest their activity.† But like the organiser region of the embryo, they did not lose much of their activity, even on prolonged boiling. This point was later more precisely investigated by Chuang, who found that if mouse kidney was kept at 100° C. for 2 hr. practically all its activity was lost. With newt liver especially instructive results were obtained. As the accompanying figures show:

Percentage inductions

Newt liver	Ectodermal					Mesodermal			
	Brain	Nasal placode	Eye	Balancer	Ear-vesicle	Tail	Muscle	Noto-chord	Prone-phros
Fresh	73·2	13·6	9·6	9·0	46·4	47·0	31·3	13·6	13·6
Boiled 2 sec.	34·6	40·1	13·3	28·3	5·5	0	0	0	0
Boiled 5 min.	68·4	68·4	49·3	26·0	16·4	0	0	0	0
Boiled 15 min.	38·1	53·4	23·6	31·2	24·4	0	0	0	0

coagulation of the proteins immediately reduces all mesodermal induction frequencies to zero. The substance or substances responsible must therefore be

---

* This would mean that a new step in evolution may involve the introduction not of a new stimulatory substance but rather of a new reaction to a substance already present. Many instances are now known in which substances having a more or less specific biological effect are found to be remarkably widespread, occurring even in organisms where their effects cannot be manifested (e.g. oestrogenic substances in plants; Skarżynski[1]). D. L. Thomson had already followed this line of thought when he called attention to such facts as these: (1) adrenalin is present in annelids but does nothing there; (2) the posterior pituitary substance causing uterine contractions is found in selachian fishes; (3) there is oestrone in insects though they do not react to it; and conversely, (4) mammalian pituitaries contain the amphibian melanophore-expanding substance, and also the avian prolactin (see p. 76).

† Perhaps it is a failure to realise this which makes Barth & Graff reluctant to admit that the primary evocator is contained in adult tissues and also, in masked form, in those parts of the embryo where its activity is not normally manifested.

h.n.t.

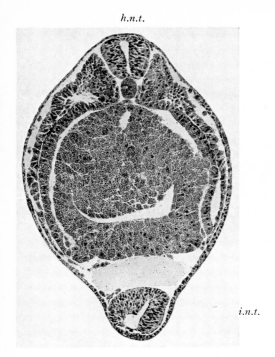

i.n.t.

Fig. 84. Induction of secondary neural tube by implantation of boiled nucleus from an ovarian egg.

h.n.t.

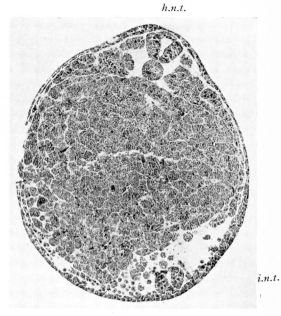

i.n.t.

Fig. 85. Weak induction performed by boiled cytoplasm from an ovarian egg.

h.n.t.

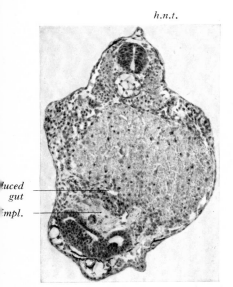

duced
gut

mpl.

Fig. 86. Induction of secondary embryo by implantation of a piece of boiled mouse heart tissue into the blastocoele cavity.

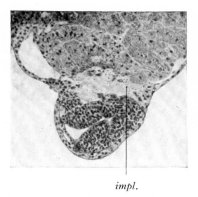

impl.

Fig. 87. Induction of secondary brain by implantation of lepidopteran tissue into the blastocoele cavity.

very heat-labile. For ectodermal inductions there are several alternatives; the frequency may first fall, then rise and finally fall again (as in neural induction itself), or it may first fall and then continuously rise (as in ear-vesicle induction), or it may continuously rise from the beginning (as in balancer induction). The simplest explanation of these differences is that the frequencies are in every case the resultant of two opposing processes, the liberation of the active substances from inactive protein-bound precursor complexes, and the inactivation of the substances themselves by heat. We shall return later to Chuang's important work in connection with individuation and second-grade inductors (p. 279), in which the study of temperature effects was continued.

Table 15. *Distribution of the primary evocator*

	Weak inductions; e.g. neuroid thickenings, palisades, etc.	Strong neural inductions; e.g. small neural tubes or cavities	Very strong neural inductions; e.g. big brains
Mammals			
Homo sapiens	—	—	T, K, L, B, t
Mus musculus	b, F	l, npl	K, L, B, A, H
Bos taurus	—	L	L, Me
Birds			
Gallus domesticus	—	npl	ee, B
Lanius minor	—	T, K, L, S, A, F	—
Reptiles			
Lacerta agilis	—	K, S	L
Amphibia			
Rana esculenta	—	M, r	L, Me
Triton alpestris	r	H, Lb	B, L, C
Salamandra maculosa	L, H	—	B, r
Fishes			
Gasterosteus aculeatus	—	—	H, o, M, L
Insects			
Libellula sp.?	F	N	—
Deilephila euphorbiae	N	pl	—
Vanessa urticae	id	—	—
Molluscs			
Limnaea stagnalis	fM	L	—
Planorbis corneus	fM	—	—
Crustacea			
Daphnia pulex	—	e	—
Worms			
Ligula simplicissima	sob	—	—
Enchytraeus albidus	M	—	—
Coelenterates			
Hydra viridis	sob	—	—

T, thyroid
K, kidney
B, brain
L, liver
t, tongue
H, heart
A, adrenals
r, retina
M, muscle

o, ovarian eggs
e, extract
ee, embryo extract
l, lens
S, testis
F, adipose tissue
N, nerve tissue
b, blood
C, cartilage

Lb, limb-bud
pl, pupal lymph
id, imaginal discs
sob, section of whole body
fM, foot muscle
Me, muscle extract
npl, tumour

What is the meaning of this remarkably wide distribution of the evocator? Taken side by side with the facts already described regarding its presence in masked or inactivated form in all parts of the embryo except the organiser region, the only possible conclusion is that we have to deal with a substance or substances of extremely wide distribution, powerless in the adult owing to the complete loss of the competence for histological and morphological differentiation, and requiring a kind of "protection" in the egg and early embryo except precisely in that region where the neural axis must arise. The situation may be summarised in the form of four logical possibilities, all of which are realised in fact at different parts of the individual and different stages in its history.

	Inducing substance (evocator)	Active inducing substance (evocator)	Competence
Dorsal lip region	+	+	+
Ventral ectoderm	+	−	+
Yolk-endoderm	+	−	−
Adult tissues	+	+	−

Tissues with none of these properties might not be very interesting, but a tissue which possessed competence but no evocator would be exceedingly useful as a test-object. Unfortunately, no such tissue is known.

In Holtfreter's experiments on ovarian and just fertilised eggs, the material was boiled (for technical reasons). This led to the view that the masking complex is not broken down until the appearance of the dorsal lip. But B. Mayer[2] states that blastomeres from the four-cell stage will, if implanted fresh in a competent ectodermal ball, induce a neural tube in it. Masking may therefore not begin until some time during the cleavage stages. The full publication of these results will be awaited with interest.

It is to be noted also how these results of Holtfreter's link up with the old Mangold-Spemann experiment of homoiogenetic induction (p. 154). Neural plate, once formed, will induce more neural plate in competent ectoderm. So will somites or notochord. It is thus clear that at some stage or other, which may be different for each tissue, all the tissues and organs will take on inductive power, i.e. the inducing substance in them will be liberated from its inactive precursor. The homoiogenetic induction itself was the subject of a pretty experiment by Holtfreter[18], in which he showed that mere ageing will not of itself suffice to liberate the evocator in presumptive neural plate. Presumptive neural plate was isolated in salt solution where, of course, it underwent no further differentiation. After two days, when control embryos were well-developed neurulae, it was implanted in the blastocoele of a host. No inductions were ever obtained in this way. Therefore the process of liberation of the active substance must be connected with the processes of determination and histological differentiation. Liver develops from the trunk endoderm, which is inactive to begin with. But at some later stage, the evocator must appear in free form there, since liver tissue is one of the most powerful adult tissues. Muscle, on the other hand, possesses it, like all the mesoderm, from the first.

The structures induced by adult implants are very various. Just as happens with coagulated embryo organisers, there is a tendency for the production of excess formations. Thus in Fig. 88 a piece of boiled salamander brain induced

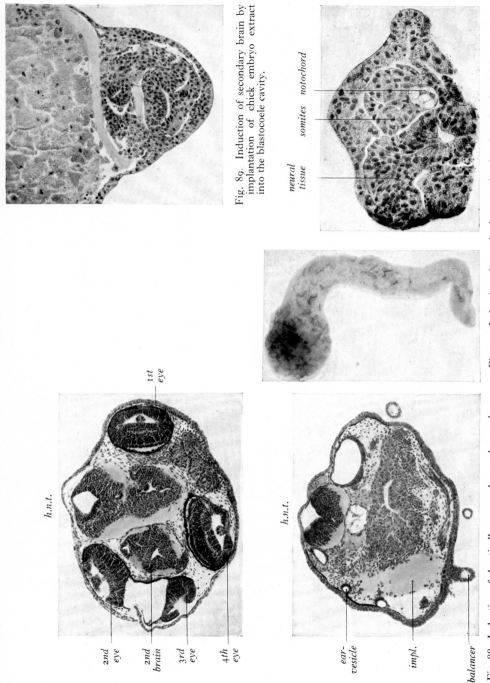

Fig. 89. Induction of secondary brain by implantation of chick embryo extract into the blastocoele cavity.

neural tissue *somites* *notochord*

Fig. 90. Induction of neural tube, notochord and somites by chick embryo extract in a competent ectodermal ball.

1st eye

h.n.t.

2nd eye
2nd brain
3rd eye
4th eye

h.n.t.

ear-vesicle
impl.
balancer

Fig. 88. Induction of chaotically arranged secondary embryo by implantation of boiled adult amphibian brain into the blastocoele cavity.

an accessory brain with four eye-cups, two ear-vesicles and a supernumerary balancer. The curious phenomenon of lens induction without other structures also appears; we shall return to this later on in connection with normal lens induction (p. 299).

One inducing material proved itself particularly fertile in inducing effects (see Fig. 89), namely, 7-day chick embryo extract (in Tyrode solution). In explanted isolates not only neural tissue but also notochord and somites were induced (see Fig. 90). This was the nearest approach to induction by a living amphibian organiser. Among the minor points arising from this work was that no specific differences could be established between the different tissues or the different phyla. Only in general it could be said that in those cases where the largest neural formations appeared there also was it most likely that mesodermal inductions, such as notochord or somite tissue, would occur. Mesodermal inductions were indeed frequent enough. Autolysis, as of adult newt liver, seemed to diminish the inductive power to some extent. Attempts to obtain induction by applying the stimulus from coagulated tissue to the outside of the competent ectoderm rather than to the inside, failed.

Of much interest was the fact that Holtfreter[12] made a large number of implantations of plant materials, e.g. meal, potato, banana, agar, etc., always with negative results. Wax was also without effect. Since that time there have been a number of investigations of plant materials, notably by Ragozina, by Toivonen[1] and by Okada[2], but a study of their published photographs clearly shows that though they themselves interpreted their ectodermal thickenings as positive results, the effects must still be regarded as quite negative.*

The existence of evocators in adult and late embryonic tissue furnished Waddington[10] with a rather serious criticism against the work of an American school (e.g. Hunt[1]; Willier & Rawles[1,2,3], etc.) which studied embryonic determination in the chick by means of transplantations to the chorio-allantoic membrane. He pointed out that just as the presence of active evocators in adult tissues explained the curious results which had been obtained on amphibian transplants before Holtfreter's introduction[2] of the neutral inorganic medium (see p. 157), so it deprived the chorio-allantoic situation of any right to be called neutral. He adduced some inductions in amphibian embryos with chick blood in support of his criticism. Although it was fully justified, there was yet a paradox remaining, for on the chorio-allantoic membrane isolated pieces of blastoderm do not differentiate as fully as they do in the method of explantation on to a plasma clot. From this latter, too, induction might be expected, but does not actually happen. A neutral medium for bird embryos as well as for those of amphibia would be a great desideratum.

Many of the foregoing facts were confirmed by Wehmeier (who had been responsible for the original experiment in which the inductive power was not destroyed by killing a dorsal lip with alcohol, published in Bautzmann et al.) in work simultaneous with Holtfreter's. She obtained quite negative results with agar, gelatin, cholesterol and potato starch; established that treatment of inductors

---

* There is, of course, no reason why plant materials should not produce neural inductions if any of them brought about the right kind of injury to the ventral ectoderm (see on, p. 181).

with organic solvents does not destroy their activity;* and obtained activation of non-inductors with organic solvents. She also investigated a number of adult organs, with results incorporated in Table 15, and confirmed Woerdemann's experiment[4] with tumour tissue. Certain tissues, such as amphibian retina and rat carcinoma, were found by her to be a good deal more active after acetone treatment than before, suggesting that in some cases the evocator is not perfectly liberated in adult life. In her experiments there was a tendency for the formation of enormous neural plates, with abnormal folds and delayed closure. The chaotic disposition of organs, etc. was just as in Holtfreter's work, save that few of her embryos lived long enough to allow it to be well seen.

2·345. Chemistry of the primary organiser

Cell-free extracts showing evocator activity were prepared almost simultaneously by Needham, Waddington & Needham; Holtfreter[11]; and Spemann, Fischer & Wehmeier. The first group of workers ground newt neurulae in a small tube, using the interstitial water as the extraction medium, took up the muddy looking liquid into a capillary, and centrifuged it. When the layers had separated, the tube was cut with a file after heat coagulation at the junctions between them, and the different fractions implanted. Neural inductions of moderate size were obtained from the middle aqueous layer or the aqueous and the fatty layer together. A similar technique was used by Holtfreter, who, using unfertilised eggs, distinguished five layers; centripetally: (1) fatty layer, no inductions, (2) brownish aqueous layer (probably a mixture of (1) and (3)), positive inductions, (3) greyish aqueous layer, positive inductions, (4) black granular layer, not tested, (5) solid substratum of lecitho-vitellin yolk-platelets (CE, Section 1·13), no inductions. These results could hardly give any evidence about the solubility of the evocator, since the aqueous layer would be highly colloidal, but they indicated that it was not bound up either with yolk-platelets or with fat-globules.

Spemann, Fischer & Wehmeier followed another method; they treated unfertilised eggs with acetone and then extracted them with 20 % ammonium acetate, centrifuged out debris and precipitated with alcohol. The evocator was in the precipitate.

From this point of departure different groups of workers were led in different directions. It will be convenient before proceeding further to make quite clear what is, or should be, regarded as a positive result in interpreting the effects produced by the implantation of chemical compounds. Lack of a uniform and reasonable standard has led some workers astray.

In the first place, the competent ectoderm may show no reaction at all. Fig. 91 A shows a case, taken from Needham, Waddington & Needham, where, after implantation of a mixture of tripalmitin and triolein, no reaction of any kind was found on the part of the ectoderm, which simply enclosed the implant in a pouch. On the other hand, there may be ectodermal thickenings of considerable size (Fig. 91 B), made up simply of cuboidal cells very like those of normal ectoderm. These are most frequently lenticular, but they may form large irregular thickenings

---

* This was later confirmed again by Barth & Graff, who make it the basis of their conviction that the primary evocator, and perhaps other inductors also, are of protein nature.

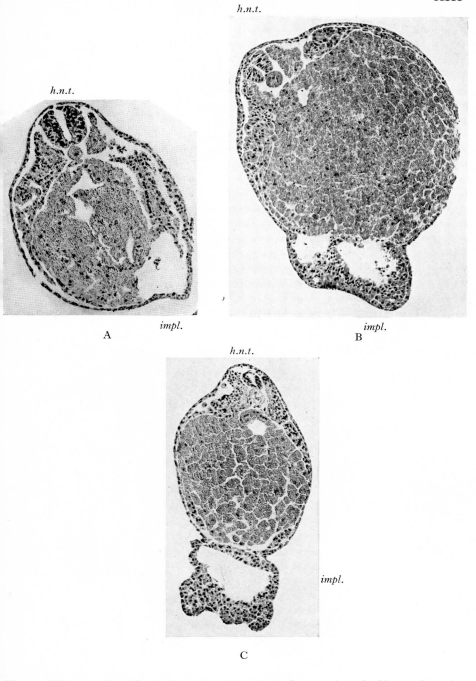

XIII

Fig. 91. Effects produced by implantation of chemical substances into the blastocoele cavity. A. No reaction whatever. B. Ectodermal thickening C. Ectodermal thickening with serrated edge (' sterol bump''). *h.n.t.* or *p.*, host neural tube or plate; *i.n.t.* or *p.*, induced neural tube or plate; *impl.*, implant.

XIV

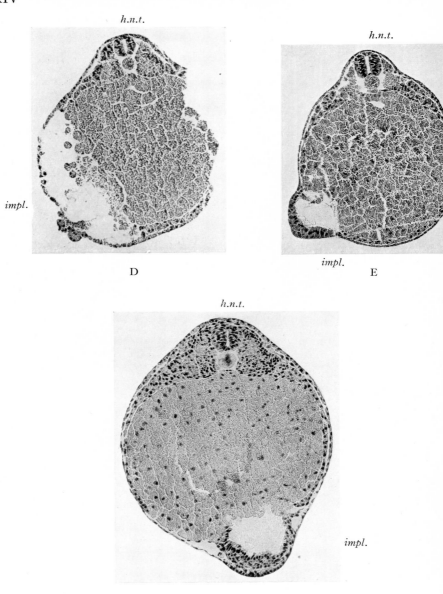

Fig. 91 (*continued*). Effects produced by implantation of chemical substances into the blastocœl cavity. D. Ectodermal reaction occurring just at the point of contact with the implant. E. Weak "palisade" induction. F. Strongly marked "palisade" induction.

presenting a saw-like or serrated edge on the outside; they were associated by Needham *et al.* with the presence of non-effective sterols (such as phytosterol or calciferol) in the implant and were hence called "sterol bumps". An example is given in Fig. 91 C. In later development they form warty structures like those produced by ultra-violet irradiation of the ventral epidermis of the tadpole (Glücksmann[2]). All these types show proliferation without differentiation. Fig. 91 D shows this taking place at a very localised region of the ectoderm touched by an implant of neutral fat and calciferol.

The first stages of neuralisation are seen in the so-called "palisade" inductions. Here the implant is partly or entirely surrounded by histologically differentiated neural cells, columnar in shape, with elongated nuclei, tightly packed together to form a one-layered or two-layered "palisade"-like epithelium. The nuclei are all parallel to each other and to the main axis of the cells, usually in a row at the same level. Such an induction may pass into a true neural tube as successive serial sections are examined, but this is rather rare. On the other hand, there may be enough palisade neural tissue to make two or three neural tubes the same size as that of the host. Examples are shown in Fig. 91 E & F.

A true neural tube is immediately recognisable, as also a neural plate in the act of closing, by their characteristic cell arrangement. Examples have already been seen in Figs. 60 and 61. It is needless to say that care must be taken to distinguish "*Spaltung*" effects, i.e. the mechanical diversion of the formative movements of gastrulation into two streams, by the implant or otherwise, from true inductions. Higher phases of induction, such as the presence of recognisable brain pieces, eyes with lenses, notochord, etc., usually admit of no mistake.

In this connection it is interesting that the mitotic rate in the induced neural tube tissue is the same as that in the normal or host neural tube (Waddington & Shukov; being 4·35 and 4·24 respectively as against 1·46 for ectoderm and 2·62 for ectodermal thickening.

Some of the first investigators of the chemical properties of the primary evocator were inclined to draw conclusions from the fact that dorsal lip pieces retained most of their activity after being allowed to stand for hours or days in organic solvents (e.g. Spemann, Fischer & Wehmeier), but the well-known difficulty of extracting all lipoidal substances from protein material rendered no conclusion possible from such data (cf. the work of Macheboeuf & Sandor on extraction of lipins from blood serum, and the monograph of Macheboeuf). Needham, Waddington & Needham ground newt neurulae to a fine dry powder with anhydrous sodium sulphate and extracted the mass with ether and then with petrol-ether in a micro-Soxhlet apparatus. Extraction was carried on for one day (12 hr.), a period quite insufficient to remove all the evocator, but which it was hoped would remove enough to give positive results, if the evocator was soluble in ether. The ethereal extract was mixed with tripalmitin and triolein and implanted. Positive results were obtained. Ethereal extracts of adult newt viscera also gave positive results.

The subsequent experiments of the Cambridge group of workers were nearly all done using crystalline hen's egg-albumen as the implantation medium, ether-soluble material being dispersed in it by the acetone method of Boyland[1]. They therefore carried out at various times a large number of control implantations from which it appears that very weak neuralisations can occur from crystalline egg-albumen alone. Whereas there is generally no reaction at all (cf. Fig. 91 A) there may occur neuralisations of a weak palisade type (cf. Fig. 91 E).* Control experiments are therefore necessary during each laying season, as the reactivity of competent ectoderm varies from year to year. Without base-lines of this sort interpretation is impossible.

In the next paper of the series, Waddington, Needham, Nowiński & Lemberg obtained small neural inductions from the ethereal extracts of whole newt bodies and mammalian liver, and traced the activity into the digitonin precipitate of the unsaponifiable fraction. This final separation gave rise to a type of induction not seen in any other instance. While the filtrate from the digitonin precipitation caused nothing but ectodermal thickenings of a quite negative character, the precipitate induced rather beautiful palisade formations, which, like full neural tubes, were separated off very sharply from the ventral ectoderm, surrounded the implant closely, and ended abruptly at the junction with the yolk-endoderm (see Fig. 91 F). Judged by the standard of reactivity that year these effects of the unsaponifiable fraction were strikingly positive. Activity in the unsaponifiable material from the ethereal extracts of mammalian liver was observed again by Waddington, Needham, Nowiński, Lemberg & Cohen in a later paper.

Though all these results, owing to the lack of facilities for sufficiently large-scale routine testing, could only be of a preliminary character, the suggestion was nevertheless inescapable that some substance of a steroid nature was indicated. This received considerable support from the finding of Waddington & Needham that fine neural tube inductions could be obtained by the implantation of certain synthetic polycyclic hydrocarbons. The first to give this result were 1 : 9-dimethylphenanthrene and 9 : 10-dihydroxy-9 : 10-di-*n*-butyl-9 : 10-dihydro-1 : 2 : 5 : 6-dibenzanthracene. A typical induction with the latter is shown in Fig. 92 A. These two most active compounds had previously been known only as powerful oestrogens.

The series of polycyclic hydrocarbons and steroids which would act as primary evocators was subsequently extended by Waddington[21], who obtained neural tube inductions with the carcinogens methyl-cholanthrene, styryl blue (Fig. 92 B), and 3 : 4-benzpyrene (Fig. 92 C); and the oestrogens oestrone (Fig. 92 D), 4 : 4'-dihydroxy-diphenyl (Fig. 92 E), 1 : 2-dihydroxy-1 : 2-di-α-naphthyl-ace-naphthene, and 5 : 6-*cyclo*-penteno-1 : 2-benzanthracene.† Particularly good effects were obtained with the water-soluble carcinogenic hydrocarbon sodium-1 : 2 : 5 : 6-dibenzanthracene-endo-α-β-succinate (Fig. 93 A). The hydrocarbons naphthalene (Fig. 93 C) and anthracene (Fig. 93 B), but not chrysene, gave slight positive results, possibly owing to the admixture of some impurity; as also perhaps for the same reason did the sample of cholesterol tested, though cholesterol had given negative results in the earlier work of Needham *et al.* Dihydrocholesterol and equiline were absolutely negative, but inductions of lesser intensity occurred

---

* Small palisades were also obtained from hen's egg-white, though not from egg-yolk, by Wehmeier. If the purified albumen has stood for a long time, tubes may be induced, and the older the specimen the more activity it has; this can be reduced or abolished by extracting it with ether (Shen[2]). There is no change in the titratable –SH of the samples as they age (Shen & Needham).

† The formulae of all these compounds will be given later (pp. 245 ff.).

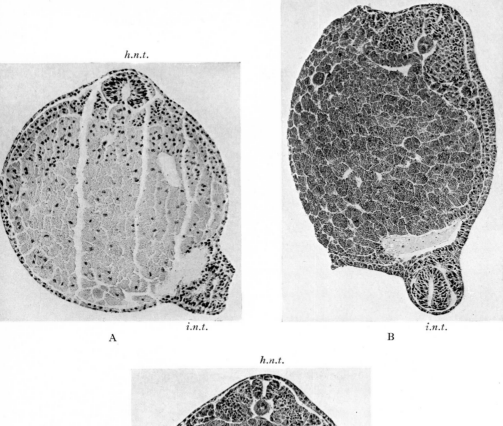

h.n.t.

i.n.t.

A

h.n.t.

i.n.t.

B

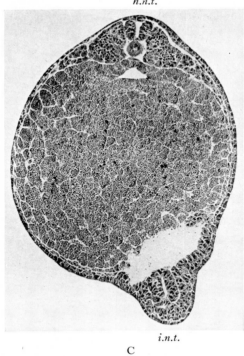

h.n.t.

i.n.t.

C

Fig. 92. Inductions of secondary neural tubes produced by implantations of synthetic chemical substances into the blastocoele cavity. A. 9 : 10-dihydroxy-9 : 10-dihydro-9 : 10-dibutyl-1 : 2 : 5 : 6-dibenzanthracene. B. Styryl blue. C. 3 : 4-benzpyrene. Conventions as in Fig. 91.

h.n.p.

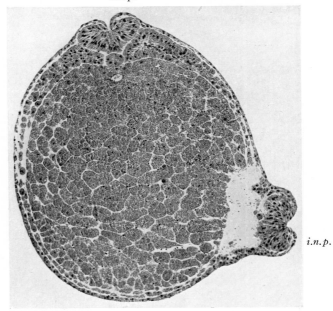

i.n.p.

D

i.n.t.

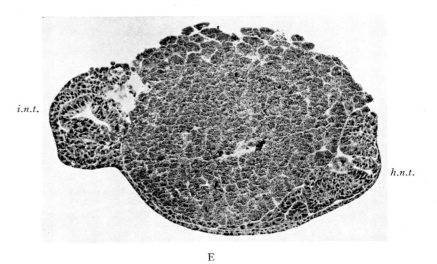

h.n.t.

E

Fig. 92 (*continued*). D. Oestrone. E. 1 : 2-dihydroxy-
1 : 2-di-α-naphthyl-acenaphthene.

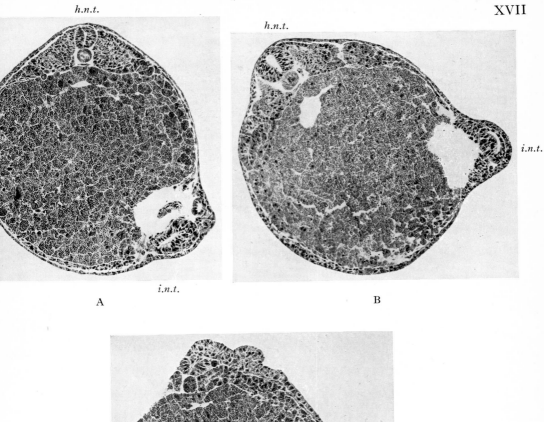

XVII

A   B

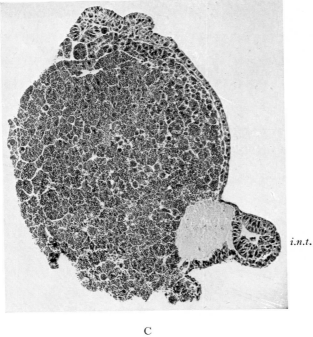

C

Fig. 93. Inductions of secondary neural tubes produced by implantations of synthetic chemical substances into the blastocoele cavity. A. 1 : 2 : 5 : 6-dibenzanthracene-endo-α-β-succinate. B. Anthracene (unpurified). C. Naphthalene (unpurified). Conventions as in Fig. 91.

h.n.t.

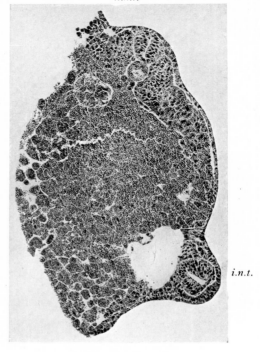

i.n.t.

D

h.n.t.

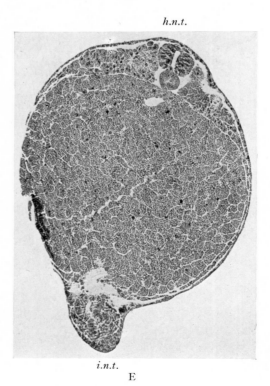

i.n.t.

E

Fig. 93 (*continued*). D. Pregnantriol. E. Squalene.

when sitosterol, pregnandiol and pregnantriol (Fig. 93D) were implanted. The poly-isoprene compound, squalene, also gave inductions of low intensity. Though not a polycyclic derivative, it is one from which such compounds might readily be formed (Fig. 93 E).

A close relation was found in this work between frequency of induction in a series and magnitude of induction produced in each embryo. Fig. 94 shows that the larger the percentage of neural inductions in the case of any substance, the larger the amount of the neural tissue formed in $\mu l. \times 10^{-4}$. The only exception to this was the carcinogen 3 : 4-benzpyrene, which gave rise to a large number of neural tubes, all of which were very small. It is not at present possible to elucidate the meaning of these data, but on the whole they may be taken as indicating the action of a graded dosage of stimu-

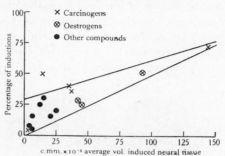

Fig. 94. Relation between frequency and size of neural inductions.

latory effect. We shall shortly see evidence (p. 183) that palisade inductions are indeed the weaker forms of full neural tube induction.

It is likely that we ought to attach considerable importance to the fact that many synthetic polycyclic hydrocarbons and compounds of steroid character are active in neural induction. The general biological significance of this fact will be discussed later (p. 243).

In the meantime other lines of research had been proceeding. Two workers of the Freiburg group, Fischer & Wehmeier, had announced that glycogen, prepared by the usual Pflüger method (i.e. by disintegration of the tissue with boiling potash, followed by alcohol precipitation) would perform neural inductions. It is unfortunate that they never published any photographs of their sections, so that the extent of these inductions is not known, but later on Heatley, Waddington & Needham successfully obtained neural tubes by implantation of glycogen (see p. 278) and the fact itself is not in doubt. Not long afterwards, however, it was shown by Waddington, Needham, Nowiński & Lemberg that if glycogen, prepared in any of the usual ways, is extracted with ether in a Soxhlet apparatus for 50 hr., a very active extract is obtained. The amount of solid material in these extracts must have been very small, since on evaporation of the ether the residue was hardly visible; nevertheless good neural tube inductions were produced, see Fig. 95 (opp. p. 190). However, the fact that the activity of glycogen is due to an ether-soluble substance attached to it in small quantity does not affect the interest of Fischer & Wehmeier's result, which, as we shall see, provided an important clue regarding the nature of the inactive masking complex from which the evocator is normally set free in the dorsal blastopore lip. In the paper of Fischer, Wehmeier & Jühling the suggestion that glycogen itself was the evocating substance was withdrawn, since highly purified specimens were inactive. It may be added that the existence of a very active substance attached to glycogen furnishes additional indirect evidence that the evocator is an unsaponifiable compound, for had it not been so it would have been removed by the strong potash.

About the same time Barth[1] in New York found that beautiful inductions of brain tissue* could be had by the implantation of fairly crude kephalin prepared from mammalian brain. This phenomenon again appeared to be of the nature of contamination of the product with some more active substance, since Waddington, Needham, Nowiński, Lemberg & Cohen noted that the activity of kephalin preparations (though in their hands small) was conserved after saponification and hence destruction of the kephalin. Barth[2] subsequently reported further inductions with purified kephalin, however.

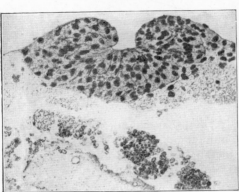 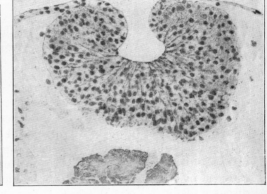

Fig. 96. Neural plate induced by implantation of synthetic oleic acid.

Fig. 97. Neural plate induced by implantation of nucleoprotein preparation (thymo-nuclein).

When Fischer, Wehmeier & Jühling announced that good neural tube inductions could be obtained by implantation of thymonucleic acid and of muscle adenylic acid,† the subject began to enter a more complicated phase. This was followed up by the publication of Fischer, Wehmeier, H. Lehmann, Jühling & Hültzsch, in which neural inductions were produced by many oils, but particularly with oleic acid prepared from crystalline dihydroxystearic acid, and with linolenic acid, repeatedly purified through the bromide. Octadecenic acid, prepared from ricinoleic acid, was also active, and so also oleic acid synthesised from sebacic acid and octyl alcohol. Here again no photographic evidence was published, but Fischer, in a separate communication, adduced two good neural plates, one the result of synthetic oleic acid implantation, the other the result of the implantation of nucleo-protein from calf thymus. These are reproduced in Figs. 96 and 97. Fischer et al. and Woerdeman[6] confirmed the finding of Needham

* Much better than those actually figured by him, as I know from slides which he kindly allowed me to examine.

† If this is true, it might be explained by an intervention of the ribo-purine, related as it is to cozymase, in the course of carbohydrate breakdown, affecting the liberation of the masked evocator. Subsequently, in the hands of Suomalainen & Toivonen, yeast adenylic acid (which, though it differs from muscle adenylic acid only in the position of the phosphoric acid group, does not participate in the enzymic reactions of carbohydrate catabolism in animals) gave no neural inductions whatever.

et al. that active ethereal extracts of embryos or adult tissues could be obtained, but the first-named investigators claimed that this activity was lost on neutralisation of free acids in the extracts. This effect could never be obtained by the workers in Cambridge, in spite of several attempts. Water-soluble fatty acids and solid fatty acids were not found to be active.

In the opinion of Fischer and his collaborators, induction is brought about by the stimulus of a large number of acid substances, owing their evocator power to their acidity. In this concept there is some obscurity, since the buffering power of the tissues would be expected to neutralise acid ions. But it is clear that we have to deal with complex phenomena, and especially with one rather deep-seated methodological difficulty, which has already been referred to (pp. 169, 174).

The difficulty may be stated in this way. We have already seen that the presumptive epidermis which contains the masked evocator, and the competent ectoderm which reacts to the stimulus of an artificially supplied evocator, are one and the same tissue. Thus competence to form histologically neural tissue and a morphologically recognisable neural tube cannot be separated from the existence of a masked evocator. Free evocator can exist without competence, as in adult tissues, but the opposite condition has not yet been discovered. It is therefore logically impossible to maintain, in any case where a given substance is used as the successful stimulus for an induction, that this substance stands in close chemical relation to the substance normally at work in the dorsal lip of the blastopore, for it can always be held that the substance tested simply unmasked or activated the natural evocator already present in the competent tissue.* Thus, in the diagram of Fig. 98, X is the implanted substance; it may act to give the response of induction, R, directly, but it may also act by liberating the masked organiser, (E), which then performs the induction. Analogous difficulties seem to be not unknown in other branches of biology, for pharmacologists, according to Ing, seek to account for the parasympathetic action of pilocarpin by supposing that it stimulates cells, which usually secrete acetylcholine, to do so. Or again, the sympathomimetic effect of tyramine and ephedrine is said to be due to the stimulation of the suprarenal gland by these drugs to produce adrenalin. The application of oleic acid, therefore, to the competent ectoderm, might simply be a more gentle way of setting free the masked evocator than boiling the tissue or killing it in some other manner.

There is one way out of the dilemma, however, namely a close attention to the question of the minimal effective dose. We can probably accept the principle that, other things being equal, the larger the necessary dose of X, the more damage will be done to the tissue, and the more likely it will be that the natural evocator (E) will be unmasked; whereas the smaller the necessary dose, the nearer the substance probably stands in chemical constitution to the evocator itself. If the effective substances are considered in this way, it is found that while the dosage of glycogen, kephalin, oleic acid or thymonucleic acid required to produce an effect has been rather large, that of the ether-soluble sterol-containing fractions

---

* The early investigators of the hormones of adult mammals did not have to contend with a difficulty of this kind.

has been exceedingly small. Thus the active ethereal extract of glycogen already mentioned was implanted at the strength of under o·o1 mgm. per c.c. of implantation material (gelatin or egg-albumen or fat), while glycogen has been used at about 200 mgm. per c.c. and thymonucleic acid much stronger.

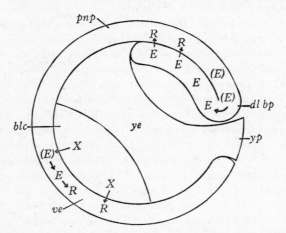

Fig. 98. Diagram to illustrate direct and indirect induction.

X, chemical substance	dl bp, dorsal lip of the blastopore
(E), masked organiser-substance (evocator)	pnp, presumptive neural plate
E, free, active, organiser-substance (evocator)	ve, ventral ectoderm
R, response of neuralisation	ye, yolk endoderm
blc, blastocoele cavity	yp, yolk-plug

The ectoderm is thickened for diagrammatic purposes.

This question was put to a direct test by S. C. Shen[1], who implanted the water-soluble hydrocarbon derivative $1:2:5:6$-dibenzanthracene-endo-α-β-succinate, previously shown to be carcinogenic by Barry et al. and L. D. Parsons, into the blastocoele cavity in carefully graded doses from $1·0\gamma$ to $0·0000125\gamma$ per embryo. The results are shown in Fig. 99.

The first noteworthy point was that the inductive action of this hydrocarbon showed an *optimal* dose at $0·0125\gamma$ per embryo. At this dilution the number of neural tubes induced was a little over 40 % of the number of embryos in which the implant was retained, but on each side of it the percentage of neural tubes fell off. While it is not easy to picture the mechanism of the optimal dose effect in precise chemical terms, the fact that there is an optimal concentration makes it rather unlikely that the substance injures the tissue on which it is acting.* If it did so, and liberated the masked organiser substance, its effect would be expected to increase with concentration.†

Another interesting point is that the palisade inductions rise to a maximum at concentrations lower than the optimum for neural-tube inductions. This supports

* An examination of the slides showed no larger number of degenerating cells and pycnotic nuclei at the higher dosages than at the low.

† There is a parallel here in the work of Koltzov[5] on the production of polyploidy in plants by colchicine (see p. 509), where an optimal dosage is also found.

the impression hitherto generally current, that they are weaker manifestations of the same process (see p. 177).

The parallelism between frequency and size of inductions noted by Waddington[21] was found to hold good in Shen's work also; it will be seen from Fig. 99 that the average volume rises on each side of the optimal concentration to a peak at $85 \, \mu l. \times 10^{-4}$. This level compares with the largest volumes of Waddington's series.

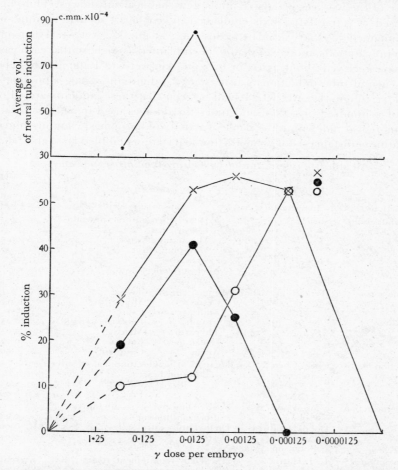

Fig. 99. Relation between dosage, induction frequency and size (with a water-soluble hydro-carbon derivative). × all inductions, ● neural tubes, ○ palisades.

Perhaps the most important point about Shen's investigation, however, is the low level of dose required to produce neural inductions. It is instructive to calculate the doses required in mgm./kg. wet wt., and to compare the resulting figures with others in the literature for various biological responses. Assuming that a newt gastrula weighs approximately 4 mgm., we find from Fig. 99 that the maximum neuralising activity (including both neural tubes and palisade inductions) occurs when the gastrula receives $0·001\gamma$. The volume of the blasto-coele cavity is taken as $\frac{1}{4}$ of the egg volume, namely 1 c.mm., and the implantation mass $\frac{1}{4}$ of the blastocoele cavity. Such a dose corresponds to 0·25 mgm./kg. wet wt. Effects are still obtained, however, at a dilution one hundred times lower, i.e. 0·0025 mgm./kg. wet wt.

This range stands in some contrast with the dosages which have been employed by some of the workers who have implanted chemical substances. Thus Fischer *et al.* used 5 % free fatty acids emulsified in agar-agar, so that 1 c.mm. would have contained 50γ fatty acid, i.e. a dose of 3·15 gm./kg. wet wt. Their nucleo-protein preparations, moreover, had to be implanted as such, so that each gastrula must have received about 250γ, or 62·5 gm./kg. wet wt. Later Barth[2] reported effects (neural-tube inductions?) using 0·05 % digitonin in egg-albumen, i.e. a dose of 31·5 mgm./kg. wet wt. It will be seen that these amounts are all of the order of a hundred or a thousand times the dosages used in Shen's work. The latter, moreover, fall within the range of activity of hormones, vitamins, and other stimulating substances, for which a few data are assembled in Table 16. The

Table 16. *Activities of biological stimulating substances*

Biological response reactions	Investigators	Effects and agents	Effective dose mgm./kg. wet wt.
Neural inductions	Shen[1]	1 : 2 : 5 : 6-Dibenzanthracene-endo-α-β-succinate:	
		Maximum effect	0·25
		Minimum effect	0·0025
	Fischer *et al.*	Fatty acids	3150
		Nucleoprotein	62500
	Barth[2]	Digitonin	31·5
Hormone action	Went & Thimann	Minimal effective doses:	
		Oestriol Mouse	0·1
		Androsterone Capon	0·1
		Insulin Rabbit	0·25
		Thyroxin Man	0·01
		Histamine Cat	0·01
		Auxin Oat plant	0·0001
Drug action	Chen, Chen & Anderson	Digitalis principles, minimal effective doses on heart:	
		Cat	0·2–8·3
		Frog	0·0007–1·5
Oestrogenic action	Cook, Dodds & Lawson	9 : 10 : Dipropyl-9 : 10-dihydroxy-dibenzanthracene	0·25
	Dodds, Golberg, Lawson & Robinson	Diethyl-stilboestrol	0·004
		Oestrone	0·016
Vitamin actions	L. J. Harris[3]	*A* Daily requirement, man	0·016
		B_1 Daily requirement, man	0·011
		C Daily requirement, man	0·39
		D_2 Daily requirement, child	0·0018
Carcinogenic effect	Deelman	1 : 2 : 5 : 6-Dibenzanthracene minimal dose yielding tumours in 34 weeks:	
		Mouse	0·6
	Shear[1]	1 : 2 : 5 : 6-Dibenzanthracene minimal dose yielding tumours in 14 months:	
		Mouse	0·016

N.B. (1) The inductions with digitonin were always of the palisade type, and apparently not much more pronounced than those of controls in which egg-albumen is implanted.

(2) The figures for Fischer's inductions were by a slip printed wrongly in the original table.

comparison between these is, of course, a little difficult, since hormones and drugs are injected in single doses, the vitamin figures are for daily requirements, and the carcinogens, which usually work very slowly, are administered by successive injections. However, the main point, that the dose of water-soluble hydrocarbon producing the maximum inductive effect on the gastrula ectoderm lies in the same range as that of many biological stimulating substances, is clearly established.

Since this is the case, it would seem more likely that the action of this substance (and hence, by implication, of the other hydrocarbons active in neural induction) is *direct*, that is to say, in imitation of the normal organiser, rather than *indirect*, by liberating the organiser substance masked in the ventral ectoderm.

On the other hand, it is equally likely that some results, such as the rather weak inductions reported by Okada[2] on implantation of kaolin, silica, etc., and those reported by Barth[2] on the implantation of very acid and alkaline jellies (pH 3 and 10), are to be interpreted as indirect inductions. In the former case notable numbers of degenerating cells were observed in the vicinity of the implant, and Okada himself explained his effects in this way.*

Unfortunately, the force of arguments about dosage has been somewhat weakened by the remarkable results of the plant physiologists who investigated the hormone, auxin, which causes growth by the elongation of the cellulose cell-walls. Auxin, as isolated from the higher plants, is a cyclo-pentene derivative, the five-membered ring of which has one double bond and three branched side-chains (see below, p. 266). Its most interesting relative is the constituent of the anti-leprosy chaulmoogra oil, chaulmoogric acid, in which, however, the ring is terminal to a long chain. Hetero-auxin, however, as isolated from yeast, is β-indol acetic acid, a substance allied to the common amino-acid tryptophane. Yet, according to Kögl, hetero-auxin is active in some cases in only half the dose of auxin itself, though it produces exactly the same reaction of cell growth.

We have not here the difficulty, however, which would arise if the test-object contained a sufficient amount of *easily liberated* masked auxin to give the reaction (see p. 212). Furthermore, hetero-auxin is not contained in the higher plants.

Another case is that of the stilbene derivatives (see below, p. 256), which are as active oestrogenically as the hormone of the ovary itself. In spite of all, however, the establishment of the minimal effective dose remains a matter of great importance.

* Inductions have been claimed by Finkelstein & Schapiro[2] for carnosine (β-alanyl-histidine). Unlike most workers, they paid attention to dosage, and had 14 % positive results at 0.01γ per gastrula. Only two illustrations are offered, however, in support of this; one shows a mere ecto-dermal thickening, and the other an oblique section cutting both the cephalic and caudal ends of

Carnosine

the neural axis, one of which was perhaps mistaken for an induction. Though it does not seem likely that carnosine is effective, there would be no harm in testing it again. Tests of vitamins and hormones by Suomalainen & Toivonen were all negative. The effects of autolysis or proteases on tissue containing the masked evocator have not yet been investigated.

Until recently the suggestion that the higher fatty acids liberated the evocator in the ectoderm by a generalised injury less severe than the denaturation of its proteins was the only one possible. But it has been shown by both the Amsterdam and Basel schools that testosterone, the active substance of the internal secretion of the testis, closely related to the androsterone which appears in the urine, is dependent for its action on another factor. Its full activity is only attained when it is "activated" by certain substances, among which higher fatty acids, such as elaidic, oleic, crotonic, and especially palmitic acids, are prominent. Optimal action occurs at C_{16}. The activation seems to be unspecific, and that the phenomenon is not one of increased absorption through the intestinal wall can be demonstrated, according to Miescher, Wettstein & Tschopp, by injecting the testosterone on one side of the body and the higher fatty acid on the other. The whole parallel is important, for it supplies a new hint of a relationship between the fatty acids and steroids where stimulatory substances are concerned. It should, moreover, not be forgotten that, as will be seen later (p. 248), oleic acid may possibly itself be a precursor of the sterol ring system.

To summarise, then, it is clear that neural inductions can be brought about by a wide variety of chemical fractions and pure or relatively pure substances. Which, if any, of these, is identical with the primary evocator occurring in the dorsal blastopore lip, and upon which the whole normal process of development depends, is as yet an unanswered question. The question is rendered particularly difficult owing to the presence of the natural substance in masked condition in the very tissue on which alone the activity of a chemical substance can be tested. Studies of dosage, however, suggest that the natural evocator is a steroid substance, for apart from the direct evidence from solubility, etc., the only substance which has so far been shown to act in concentrations of the vitamin or hormone order is a polycyclic hydrocarbon. Many of the other types of substance which have given positive results have probably done so by unmasking the natural evocator. The next step ought to be the isolation of the intact complex itself; this alone will allow of the identification of the natural evocator. Here the ultracentrifuge may be expected to give important help.

There has sometimes existed a certain tendency (e.g. in Spemann[17], pp. 232, 369) to abandon too readily the search for logical consistency in the chemical facts, and to write off the whole induction effect as "unspecific". This is a counsel of despair. Whatever the ultimate value of the facts so far accumulated, there is one thing of which we can be sure, namely that throughout normal development the evocators of the different grades are at work, and the understanding of their mechanisms must be mainly a matter of time. It will not be reached without much more knowledge than we have at present about the characteristics of metabolism in the various regions of the gastrula during the critical time of evocation and individuation. This will be the subject of the succeeding section.

The above interpretation of embryonic inductions by chemical substances differs somewhat from views advanced by other authors such as Woerdeman[6].

Woerdeman's main thesis was that the dead implants or chemical substances activate the cell material in some way, but that the production of the supernumerary embryonic axis is the work of the host itself, which is capable of reacting in this way to any stimulus. There were really two ideas involved in this conclusion: (1) that the host plays an essential part in the "induction" by a dead organiser, and (2) that the stimulus which proceeds from the implant is merely an indifferent activation.

With the first statement Waddington & J. Needham in their commentary agreed; indeed the first mention of this in the literature had been made by Waddington[6] in connection with the action of boiled organiser in the chick embryo. But they pointed out that they had distinguished two aspects of the host's influence; on the one hand, the determination of the histological character of the induced tissues by the particular kind of reactivity present in the ectoderm submitted to the inducing stimulus; thus there is only a certain period of development during which the host ectoderm will react to an inductive stimulus by the formation of neural tissue (*Competence*); and on the other hand, the exertion by a living organiser of influence on the regional character of an embryonic axis induced in its neighbourhood (*Individuation*). It is certain that we must take into account the possible individuating influence of the host's organisation centre on embryonic axes induced by dead implants.

The advantage of analysing the host's influence into these two aspects is a gain in flexibility. Thus we can envisage the possibility of phenomena which involve competence but not individuation. Some of Holtfreter's inductions[16] by dead material in ectoderm removed from the host are an example (see p. 277), but a better one still was attained when Heatley, Waddington & Needham found similar inductions performed by a chemical implant (see p. 278). Woerdeman got into some difficulty in interpreting Holtfreter's results. The basis of his theory was that the dead organiser activates a "field" already existent in the ectoderm, but he also suggested that the fields are somehow bound up with the host's organisation centre and are therefore absent from the ectoderm in question, or at best only weakly developed in it; hence he was left with no explanation for the regular appearance of induced neural tissues in Holtfreter's explants. Here the two concepts of competence and individuation carry us further. The regular formation of induced tissues indicates that the competence of the ectoderm is normal, while the contorted and irregular shape of the inductions points to the weakness of the individuating influences. Further work is required before we can say whether isolated competent tissue retains some power of individuating into organs any tissues which may be induced in it, but the experiments of Holtfreter make it likely that the power, if any, is small.

The reverse of the above case is found when a piece of tissue individuates but does not exhibit competence; a normal induction by a living organiser provides an example, since the living organiser does not itself realise any competence for the formation of neural tissue although it does individuate the neural tissue which it induces. Here we are not concerned merely with the activation of a "field" in Woerdeman's sense, but are bound to consider also the determination of the regional character of the field; and it is exactly this determination which we call individuation. Finally, it may be mentioned that Holtfreter's experiments with transplants of ectoderm on to older larvae (see below, p. 274) have shown that at a stage when the competence to form neural tissue has entirely disappeared from the host ectoderm, a capacity to determine the regional character of induced neural tissue still persists; here again the single process of the activation of a field, spoken of by Woerdeman, has been broken up into two phases.

The second part of Woerdeman's thesis, namely, that the dead organisers provide merely.an indifferent activating stimulus, is still more open to criticism. As we have seen, substances have been described (such as certain sterols) which stimulate the ectoderm to intense proliferation but which have not yet given rise to induced tissues. The hypothesis that in more favourable cases they also will act as inductors remains a hypothesis, and one which does not seem plausible unless some more cogent reasons can be adduced in its favour.

Moreover, the alleged induction of neural tissue by thermal or mechanical stimuli rests on a very slight foundation. There are only two papers in which it is asserted. Gilchrist[1] and Castelnuovo exposed embryos to differential thermal gradients. Gilchrist's pictures show only irregular puckerings of the ectoderm the neural nature of which is uncertain, and one "*Spaltung*". Illustrations of Castelnuovo's results were never published, but she describes her positive cases as "unite caudalmente", which strongly suggests "*Spaltungen*" and not proper inductions. Attempts have recently been renewed to bring about induction by cauterising ventral ectoderm, but the results of Cohen[2] are extremely dubious and Margen & Schechtman report consistent failure to achieve any neural differentiation by mechanical injury. The only convincing evidence of induction by an indifferent stimulus (implantation of celloidin) concerns quite a different process, the formation of limbs (Balinsky[2]), and this, though so often quoted, rests only on one single experiment (see p. 305). While we must not dismiss absolutely the possibility that competent ectoderm can form neural tissue in response to several kinds of stimulus, the more likely hypothesis is that it reacts like other physiological systems which are sensitive to one chemical substance or to a limited class of chemical substances.

The word *Evocation* carries out several useful functions. In the first place, if we separate from "organisation" in general the concept of "regional determination" or "individuation" we require a word for what is left. This was at one time called "induction as such", but is now spoken of as "evocation". Secondly, we give the name "evocators" to all substances which are known to stimulate gastrula ectoderm to neural differentiation, while thirdly, we can speak of the substance which is active in normal development as "the evocator". Some grounds for preferring, instead of the word "activator", a specially invented word, have been given above. But there is an additional reason, namely, that "the evocator" exists in the egg before gastrulation in a masked condition, from which it can be liberated by the action of heat or by less drastic treatments. The term *Activation* is here used to describe this liberation.

In an interesting review Weiss[14] too has critically discussed the organiser problem. Weiss emphasised the significant difference between live and dead organisers and clearly distinguished between the "evocating" activity of the inductor and the "individuating" activity of the host field, "all the organising, i.e. pattern-determining, effects going to the credit of the latter". "It appears", he concludes, "that this is the only stand one can safely take in view of such facts as have come to our knowledge so far" ([14], p. 667). With this we may agree, save that such a formulation of the facts, like that of Woerdeman[6], fails to distinguish clearly between competence and individuation.

2·35. Morphogenesis and metabolism

2·351. Metabolism of the regions of the gastrula

To gain a clear understanding of the mechanism of activation of the primary evocator in normal development (liberation from a protein complex) we need to know as much as possible about the metabolism of the various regions of known embryological properties in the gastrula. Until a few years ago, this was impossible, owing to the delicacy of the necessary technique.

A convenient transition is provided by the fact that it has proved possible to liberate the evocator by methods less drastic than denaturing the tissue-proteins. If a piece of ventral ectoderm is isolated from the gastrula and cultured for a day or two in Holtfreter solution containing a small amount of a dye such as methylene blue, it will on implantation into the blastocoele of another embryo manifest inducing power (Waddington, Needham & Brachet). The graft may also become neuralised itself, since it possessed competence before it acquired inducing power. In one case, after 2 days in $M.10^{-5}$ methylene blue, the graft induced a large brain in a host and at the same time was converted itself into a compact mass of neural tissue. It might thus be expected that if the isolate was cultivated longer in the methylene blue solution it would become neuralised, and this was subsequently found to be the case by Waddington (described in Needham[30]) and by Beatty, de Jong & Zieliński (Figs. 100, 101, opp. p. 190). In accordance with the classical work of Barron and his collaborators on the increase of respiratory rate caused by reversible oxidation-reduction indicators (Barron & Hamburger; Barron & Harrop; Barron & Hoffman), this was at first associated with a metabolic acceleration brought about by the dye. It was later shown, however, by Beatty, de Jong & Zieliński that methylene blue has a relatively small effect upon the oxygen uptake of ventral ectoderm (45 %) and that in all probability other dyes, such as neutral red or janus green, which have a much lower oxidation-reduction potential than methylene blue and have never been shown to affect cell respiration, may also activate the primary evocator. A more probable hypothesis, therefore, may be that these dyes compete with the evocator for its protein carrier, displacing it and so liberating it in active form. Dye-protein complexes have been studied by many workers (e.g. Marston; M. R. Lewis[1] and Lewis & Lewis on viruses; W. W. Smith & H. W. Smith; Commoner on un-fertilised *Chaetopterus* eggs). The action of dyes on the physiological and embryo-logical properties of ventral ectoderm would repay a good deal of further study.

Other substances were also investigated by Waddington *et al.* and Beatty *et al.*, such as dinitro-*o*-cresol, a stimulant of glycolysis (see p. 569), but the few experiments gave mainly negative results. No inductions had been obtained when Finkelstein & Schapiro[1] had implanted this substance into the blastocoele cavity, and Dawson[3], applying it to the intact amphibian embryo during the earlier stages of development, obtained no effects other than a number of persistent yolk-plugs. Its effects on teleostean embryos are slighter still (Waterman[6]).

Initial steps towards the elucidation of special metabolic characteristics of the dorsal lip region were made by histochemical methods. The earlier papers on this subject, e.g. those of Konopacki & Konopacka or Hibbard[1], were entirely cytological and so did not help this particular problem. The first investigation of characteristics of regions seems to have been that of Voss, who found that while the ectoderm of the gastrula, as well as presumptive mesoderm and presumptive neural plate, gave a strong "plasmal" reaction, there was none whatever to be seen in the roof of the archenteron immediately after invagination. As the substance, plasmalogen, responsible for this reaction* now turns out to

* Discovered by Stepp, Feulgen & Voit; cf. the reviews of Feulgen and of Grevenstuk.

be a special type of phosphatide in which aldehydes of higher fatty acids are combined with glycerol, phosphoric acid and amino-ethyl alcohol (Feulgen & Bersin; Feulgen, Bersin & Behrens; Bersin, Willfang & Nafziger), it may be that the dorsal lip of the blastopore is the seat of considerable changes in lipin metabolism. The connection of plasmalogen with cytoplasmic nucleotides (see p. 269 and p. 636) forbids us to dismiss it as unimportant although its function still remains obscure.

A clearer line of work was begun when Woerdeman[2], using histochemical methods, observed a similar disappearance of glycogen from the dorsal lip of *Amblystoma* embryos during invagination. The division between the glycogen-containing cells of the presumptive mesoderm and the just-invaginated glycogen-poor cells was so sharp as to warrant the use of the term *"Glykogen-grenze"*. An example is shown in Fig. 102.* He noted the relation with the organisation centre, and in a second paper[3], continuing work with the same methods, found a diminution of glycogen in the optic cup at the time of induction of the lens. He was thus led to propose some causal connection between glycogen breakdown and inductions, whether of first or second grade. The subject was then taken up by Raven[1], also of Amsterdam, who transplanted a dorsal lip from one embryo to the ventral part of another. In the host it invaginated as usual, taking with it a certain amount of host ectoderm (the process studied by F. E. Lehmann[3]), and, on sectioning, showed a glycogen boundary indistinguishable from that seen in normal invagination. In some cases invagination had proceeded so far that this boundary lay not in the implanted organiser but in the host ectoderm experimentally made to invaginate in this way. The conclusion that some causal connection, and not a mere chronological correspondence, existed between glycogen-loss and invagination was further strengthened[3] by the implantation of presumptive epidermis into the dorsal lip, where, as we have seen, it acquires inducing power and later forms notochord and somites. In such pieces also there was a marked glycogen-loss. On the other hand, pieces of presumptive epidermis implanted into the blastocoele cavity never showed any diminution in the histochemically detectable glycogen; the effect was not due, therefore, to removal of tissue from the surface to the interior of the gastrula. In a further paper, Woerdeman[4] noted that in standard inductions produced by implantation of living dorsal lips, the inductors had always lost a large part of their glycogen. The train of thought that glycolysis was concerned with induction led him to make the experiments with muscle and tumour tissue already referred to (p. 171), though with glycogen he had no success.

Lastly, it is not without interest, in connection with the disappearance both of glycogen and plasmalogen from the dorsal lip, that Voorhoeve had previously noted a similarity in distribution, appearance and disappearance of these two substances in a wide variety of tissues.

The general upshot of all this work was to suggest some kind of causal connection between glycogen-loss and the invagination process. We have already remarked upon the curious flask-shaped cells concerned with invagination (p. 146)

* From a slide kindly given me by Jean Brachet.

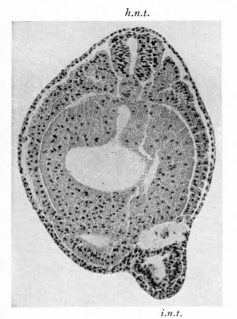

h.n.t.

i.n.t.

Fig. 95. Induction of secondary neural tube by implantation of ether extract of glycogen.

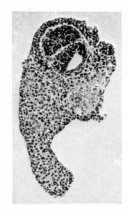 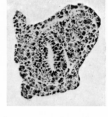

Fig. 101. Effect of methylene blue on explanted competent ectoderm; induction of neural tube; two cases.

Fig. 100. Effect of methylene blue on explanted competent ectoderm; control in Holtfreter solution only.

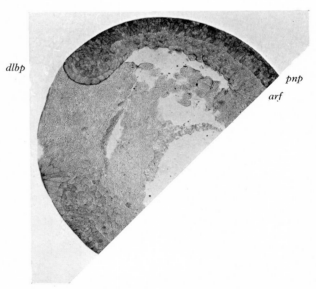

dlbp

pnp

arf

Fig. 102. Sharp disappearance of glycogen from the dorsal lip during invagination (the "*Glykogen-grenze*"). *pnp*, presumptive neural plate; *arf*, roof of the archenteron; *dlbp*, dorsal lip of the blastopore.

and the impression of contractility which they have made upon several observers. Could there be something more than a coincidence in this association of glycogen breakdown and contractility in gastrulation?

Great though the pioneer value of histochemical work may be, it is particularly vulnerable to technical criticism. Such was the case with the glycogen boundary, for Pasteels & Léonard and Pasteels[4] brought serious objections of this kind against the conclusions of Woerdeman and Raven, who in the meantime had been largely confirmed by Tanaka, working in Japan on *Rana nigromaculata*,* and by Brachet[14] on *Discoglossus*. It was held that the loss of glycogen from the archenteron roof was an artefact due to the fixation fluids employed, and that this did not occur if a dioxan method was used. We need not, however, enter into the details of the discussion, for the subject was put on a firm foundation by Heatley[1] and his collaborators, using a direct chemical method. It thus turned out that though the scepticism of Pasteels had been justifiable, Woerdeman was right.

Heatley[1] adopted the micro-chemical methods brought to a high level of perfection by the school of Linderstrøm-Lang at Copenhagen, and worked out a procedure for the determination of glycogen in amounts of tissue of the order of 1 mgm. with a probable error of $\pm 2\gamma$ glycogen, using at the end of the process the thiosulphate micro-titration method of Linderstrøm-Lang & Holter[1] for reducing sugars. By these means Heatley demonstrated a fall of about 35 % in the dorsal lip of the blastopore during invagination.† In a later paper (Heatley & Lindahl) this effect was confirmed and a careful survey of all the

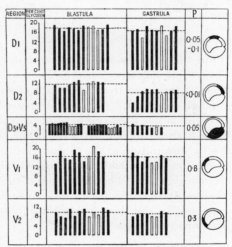

Fig. 103. Microchemical data demonstrating the marked fall in glycogen content of the invaginating dorsal lip of the blastopore.

regions of the embryo showed that the fall of glycogen was inappreciable in regions of the gastrula other than the invaginating material. Fig. 103 shows a summary of their results.

We must therefore accept the existence of a fall in glycogen-content of the invaginating material coincident with gastrulation. Whether this occurs in other types beside the amphibia will be discussed later (p. 334).

That a general decrease occurs in the glycogen-content of the amphibian embryo between the early cleavage stages and the time of hatching had long been known (*CE*, p. 1043), but in work parallel with that of Heatley's it was con-

* Miyajima's paper contributes nothing here.

† This work cannot be passed by without a mention that it was the first in which a direct and reliable chemical method was applied to the various separated regions of a gastrula; hence a landmark in chemical embryology.

firmed by Brachet & Needham on *Rana temporaria* and by Takamatsu[2] on the Japanese newt, *Hynobius nebulosus*. Since the work of Willstätter & Rohdewald[2] it had been known that glycogen exists in cells and tissues in two forms, "lyo-glycogen", readily extractable with boiling water or dilute trichloracetic acid, and "desmo-glycogen", a further fraction only obtainable by breaking down the proteins with concentrated potash in the classical Pflüger method. Brachet & Needham[2] therefore investigated the amphibian embryo from this point of view, thinking it possible that the results would be of interest in relation to the manner in which the evocator is bound in the cells. With the Good-Kramer-Somogyi micro-method the following figures were obtained:

	Mgm./40 embryos total glycogen	Lyo-glycogen % of total
Just fertilised eggs	3·24	91·1
Early gastrulae	3·25	93·2
Early neurulae	3·03	95·1
Advanced neurulae (closing neural folds)	2·69	97·0
Larvae at hatching	2·26	97·4

Thus the glycogen bound to proteins in the desmo-form, i.e. extremely tightly, always formed a very small proportion of the total glycogen. But during development it decreased from about 9 % of the total glycogen to about $2\frac{1}{2}$ %, while the total glycogen (in % dry wt.) decreased from 16·7 to 10·7 %. It was interesting that the fall at the beginning of neurulation (i.e. from the half-moon yolk-plug stage to the visible demarcation of the neural folds) was almost entirely accounted for by diminution of the desmo-glycogen.

Heatley & Lindahl also occupied themselves with determinations of lyo- and desmo-glycogen. They discovered the interesting fact that at gastrulation the desmo-glycogen decreases to a considerable degree in all parts of the embryo, not only in the dorsal lip of the blastopore. Their results are shown in the accompanying table:

Table 17. *Glycogen in the amphibian embryo*

Region	Total glycogen			Desmo-glycogen				
	Blastula	Gastrula	Decrease %	Blastula	% of total	Gastrula	% of total	% Decrease
Presumptive neural plate	17·8	16·5	7	0·90	5	0·48	3	47
Organiser region	12·0	8·3	31	0·50	4	0·32	4	24
Yolk endoderm	4·3	3·9	9	0·32	7·5	0·24	6	26
Presumptive epidermis	16·7	16·5	1	0·76	4·5	0·44	2·5	42
Ventral ectoderm	10·0	9·3	7	0·57	5·5	0·46	5	19

Thus in all the regions of the gastrula there is an absolute, and in most of them a relative, decrease in desmo-glycogen. Moreover, in the presumptive mesoderm which invaginates, the glycogen which disappears is not mainly desmo-glycogen, for the total decrease is nearly 4 mgm. %, while that for desmo-glycogen is less than 0·2 mgm. %. It is evident, therefore, that desmo-glycogen cannot be the only fraction of glycogen to which the evocator is attached in the cells if such an attachment exists.

This conclusion was entirely substantiated by the results of the implantation of desmo-glycogen and lyo-glycogen into the blastocoele cavity of the intact embryo by Heatley, Waddington & Needham. Both these fractions possess the power of induction, i.e. contain the ether-soluble active material in adsorbed condition; and, from a statistical point of view, neither was more active than the other.

2·352. Respiration and glycolysis in the regions of the gastrula

For a long time it had been evident that, side by side with direct estimation of substances in presumptive neural plate, archenteron roof, and ventral ectoderm, a survey of the characteristics of these tissues was necessary along the lines of the Warburg technique, measuring the respiration, fermentation, and respiratory quotient. But the technical difficulties in measuring accurately the gas exchange of a piece of tissue weighing, perhaps, only 50γ dry wt., were very considerable, and in the earlier experiments probably not overcome.

The first investigation was that of Brachet[8]. He compared the respiration of *Rana temporaria* gastrulae, in which the dorsal lip had been destroyed by cautery, with others in which a part of the ventral ectoderm had been similarly destroyed, using Fenn micro-respirometers and a small type of Warburg mano-meter. It resulted that the loss of the dorsal lip region entailed a considerably greater diminution of oxygen consumption than the loss of ventral ectoderm. Comparisons of the carbon-dioxide elimination by the Saunders pH indicator method on isolated pieces of dorsal lip and ventral ectoderm gave similar results, dorsal lip : ventral ectoderm; $1·89 : 1·00$). The experiments could, however, only be regarded as qualitative, since no measurement of the size of the pieces was made.* In a second series of experiments, Brachet[10] modified the Saunders method to include a Linderstrøm-Lang titration, and measured the amount of the isolated pieces of *Discoglossus pinctus* gastrulae by a usual micro-Kjeldahl method giving the total nitrogen. Again the carbon-dioxide elimination was 84% higher in the case of the dorsal lip than in that of the ventral ectoderm. He also made estimations of respiratory quotient by the Meyerhof-Schmitt method, obtaining $1·1$ for the dorsal lip and $0·77$ for the ventral ectoderm. At the same time, Waddington, Needham & Brachet measured the oxygen uptake of *Triton alpestris* isolates from dorsal lip and ventral ectoderm in a modified Gerard-Hartline micro-respirometer, the dry weight of the pieces being obtained on a micro-balance. In this case the oxygen consumption for unit weight proved to be of exactly the same order in the two cases. Brachet & Shapiro devised a very ingenious apparatus in which the intact gastrula was made the centre of two Gerard-Hartline capillaries, so that the respiratory rate of one hemisphere could be compared directly with that of the other (Fig. 104). Using embryos of *Rana sylvatica* a difference of 47% was found in favour of the dorsal lip hemisphere. In a later paper, however (Brachet[22]), where *Rana fusca* was used, the difference was hardly detectable for oxygen though clearly seen for CO_2 (29%), again indicating an R.Q. difference.

* For the same reason we must disregard the later experiments of Stefanelli[3], though he used a refined technique.

Finally, Fischer & Hartwig[2] attacked the problem by the more straightforward but rather prosaic method of using Warburg manometers with cups of relatively small volume, and a large number (sometimes as many as sixty) isolates from dorsal lip and ventral ectoderm. Even then the observed excursions were not very far from the thermo-barometric range, and the experiments had to be continued for as many as 24 hr. Dry weights were obtained directly. Under these conditions dorsal lip region showed an oxygen uptake not more than 20 % higher than ventral ectoderm. The significance of this difference is questionable.

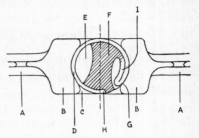

Fig. 104. Brachet-Shapiro apparatus for measuring Q_{O_2} of the intact amphibian gastrula. *A, A,* two Gerard-Hartline capillary manometers; *B, B,* mano-meter chamber; *C,* gelatin to fix the gastrula; *D,* ventral ectoderm; *E,* blastocoele; *F,* yolk endoderm; *G,* dorsal lip of blastopore; *H,* ventral lip of blastopore; *I,* archenteron roof.

The first essential, therefore, if manometric experiments on the regions of the amphibian gastrula were to be carried through with some conviction of reliability, was an ultramicro-manometer 1500 to 2000 times more delicate than the standard Warburg apparatus. This need was met in the summer of 1937 by the suggestion that the Cartesian diver should be employed for the purpose.

The use of the Cartesian diver as an ultramicro-manometer was first proposed by Linderstrøm-Lang[2], and further details of the method as used by the Copenhagen school were given in a later communication (Linderstrøm-Lang & Glick). The Danish workers, however, did not apply the method for use with minute pieces of tissue. It was adapted for the problem of the gastrula by Needham, Boell & Rogers. At the same time an ultramicro-Kjeldahl method capable of accurately determining between 1 and 20γ protein N was worked out by Needham & Boell and employed for measuring the amounts of tissue used in the divers. As will be seen, the Cartesian diver* micromanometer proved of the utmost value for the range of gas exchanges required.

If a bubble of air or gas is enclosed in a small glass vessel so that it floats within a larger vessel, the buoyancy of the diver will vary according to the pressure imposed on the whole system, and it will sink or rise as this pressure rises or falls. If the pressure is approximately constant, the diver will hover in unstable equilibrium. Conversely, if the gas phase in such a diver is increased or diminished in amount by the process of some chemical reaction inside it, the pressure required to maintain it at a given level will correspondingly rise or fall. In this way the diver is equivalent to a constant volume manometer if from time to time it is brought to a zero line and the pressure required to do so noted.

The principle of the method is shown in Fig. 105. For different purposes different arrangements of the divers were used. B shows a diver of the simplest type, used for the measurement of anaerobic glycolysis. The tissue and the physiological medium occupy the bottom of the diver bulb, beneath which there is a glass tail to maintain the diver in a vertical position. The amount of glass in

* It appears that the Cartesian diver has nothing to do with Descartes. It was first described by Raffaele Magiotti, a pupil of Galileo, in a little book on the incompressibility of water (1648), which he dedicated to Lorenzo di Medici. The book contains a diagram of the diver, which was no doubt later called "Cartesian" as a synonym for anything "scientific" or mechanical, and which remained without application, a philosophical toy, until the present time.

the tail is adjusted so that the equilibrium pressure of the diver is fairly near that of the atmosphere. The neck is partly occupied by a drop of oil, so placed that there is a small bubble of air between the oil and the flotation medium. The flotation medium, a strong salt solution giving minimum gas solubility without excessive viscosity (lithium chloride, ρ 1·245°), is seen within the diver vessel in Fig. 105 A, with the diver floating in it. The diver vessel is submerged in an

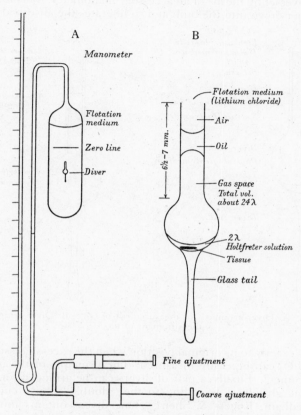

Fig. 105. A. Principle of the Cartesian diver manometer. B. Diver arranged for measurement of $Q_L^{N_2}$.

accurately regulated thermostatic water-bath, and connected to a water mano-meter, the pressure in which can be adjusted (as in a Warburg manometer) by a syringe mechanism. It is convenient to have fine and coarse adjustments. Readings are accurate to 0·5 mm. and it is therefore possible to measure changes of the order of a millionth of a cubic centimetre of gas. To take a reading, it is only necessary to bring the diver to a temporary standstill at the zero line, and to note the pressure.

For measurements of processes other than anaerobic glycolysis, divers variously modified were used. Fig. 106 A shows a diver arranged for measurement of oxygen consumption (Boell & Needham); the neck is waxed and supports beside the oil-seal a drop of alkali to absorb CO_2. B shows a diver arranged for the measurement of respiratory quotient (Boell, Koch & Needham). The neck is somewhat longer than before and waxed throughout its length, it supports a column of acid as well as the oil-seal and the alkali. During the first part of the experiment, CO_2 is absorbed by the alkali and the diver's buoyancy steadily falls, but at the conclusion of the respiratory period an excess pressure of about 12 in. Hg is set up in the diver vessel by means of an accessory manometer. This forces the alkali and acid together down into the bulb and so liberates the absorbed CO_2. As the

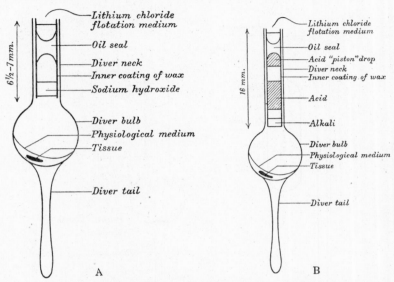

Fig. 106. A. Diver arranged for measurement of Q_{O_2}. B. Diver arranged for measurement of R.Q.

oil-seal returns exactly to its original position when the excess pressure is released, the liberated CO_2 can be easily measured. The principle of the method is that of Dickens & Šimer[1]. Finally, a pair of divers can be arranged for the measurement of aerobic glycolysis according to the "double cubical cup" method of Warburg (Needham, Rogers & Shen). One diver contains much more suspension liquid and less gas phase than the other, hence, owing to the high solubility of CO_2, less of this gas from respiration or aerobic glycolysis enters the gas phase than in the diver with little liquid. From the behaviour of the two divers the aerobic glycolysis may be calculated.

Details of the Cartesian diver methods,* which will probably be of great help in all branches of biological investigation where minute structures have to be examined, will be found in the papers mentioned. The fermentation of $0·1\gamma$ yeast gives an easily measurable long-continuing linear change and the respiratory quotient of a dorsal lip can be determined on an overall gas turnover of $0·1\,\mu$l. The gas change most commonly used in the following measurements of gastrula regions was of the order of 200–300 mμl. in 2 or 3 hr. from pieces of tissue containing from 5 to 15γ protein nitrogen.

* In what follows ml. = cubic centimetre; μl. = cubic millimetre = λ; mμl. = c.mm. × 10^{-3}.

The relation of the Cartesian diver method to other micro-manometric methods can best be appreciated by comparing them in the form of a table (modified from Jahn), as follows:

Table 18. *Comparison of micro-manometers*

Type of manometer	μl. gas change per cm. on the manometer scale	Nearest unit to which meniscus can be read	Approximate sensitivity per such unit, μl.
Standard Warburg (cf. M. Dixon)	20	0·5 mm.	0·5–1·0
Small Warburg, as used for metabolism of tissue cultures (Laser[1]; Meier)	3–7	0·5 mm.	0·15–0·35
Micro-manometers of Fenn (Schmitt; Duryee; Fischer & Hartwig[2])	3–5	0·1 mm.	0·03–0·05
Micro-manometer of Krogh (adapted from Bodine & Orr)	2	0·2 mm.	0·04
Capillary micro-manometer (Howland & Bernstein)	1	10 μ	0·001
Capillary micro-manometer (Gerard & Hartline; Brachet & Shapiro; Waddington, Needham & Brachet) in closed chamber	1·8–2·3	6 μ	0·001
Separate chamber micro-manometer with mica mirrors optically read (Heatley[2]; Heatley, Chain & Berenblum)	0·04–0·08	—	—
Very fine capillary micro-manometer (Stefanelli[1])	0·06–0·12	0·5 mm.	0·003–0·006
Micro-differential of Fenn type (Schmitt)	5	1 μ	0·0005
Cartesian diver manometer (Linderstrøm-Lang[2]; Boell, Needham & Rogers)	0·003–0·022	0·5 mm.	0·00015–0·001

The actual working range of the diver, as of any respirometer, should of course be so chosen that the approximate sensitivity shown in the table forms only a small part of the differences regarded as significant.

With gastrulae of *Rana temporaria* Boell, Needham & Rogers obtained the following figures for anaerobic glycolysis:

	$Q_L^{N_2}$	NH$_3$ production (μl. $10^{-3}/\gamma$ dry wt./5 hr.)
Dorsal lip region	+0·63	2·31
Ventral ectoderm	+0·21	0·97

Another series of experiments on *Triton alpestris* gave a similar, though not quite so large, difference. It was thus clear that the anaerobic glycolytic rate is about three times as high in the dorsal lip of the blastopore as it is in the ventral ectoderm. This substantial difference appeared also in the anaerobic ammonia production. There was no difference between the regions in the ratio between uncompensated and ammonia-compensated carbon-dioxide output. Some ammonia measurements were done on other stages also; they showed what appears to be a rise in ammonia production at the beginning of the action of the primary organiser followed by a fall (see p. 624).

Owing to the replacement of inert yolk by active protoplasm, tadpole tissues showed a $Q_L^{N_2}$ in the neighbourhood of +3·0.

The action of dinitro-*o*-cresol on the anaerobic glycolysis of the regions was studied by Boell & Needham[1] in a separate paper. It had little effect on the ammonia production but raised the uncompensated CO_2 production some 300 %; the $Q_L^{N_2}$, therefore, which depends on both factors, was raised to an intermediate degree. The effect, however, was exerted to a much greater extent upon the ventral ectoderm than upon the dorsal lip region; in the case of the uncompensated CO_2 production, indeed, the ventral ectoderm activity was raised to the level of that of the dorsal lip. This difference was compared by Boell & Needham[1] with that between the Q_{O_2} of diapause and developing grasshopper embryos, and that between Q_{O_2} of unfertilised and fertilised echinoderm eggs.

The former was studied by Bodine & Boell[2] in the course of a series of investigations on the developing orthopteran embryo, details of which will be found in Section 3·32. As is well known, the grasshopper embryo develops actively at first for a few weeks but then enters a state of dormancy, the diapause, in which no morphological development occurs. This lasts several months, after which a period of intense activity, both morphological and metabolic, leads in three weeks to the end of embryonic development. Respiratory rate is considerable during the two active periods, but sinks to a low steady state during the diapause. Since the respiration in the two active periods is powerfully inhibited by KCN and CO, while that of the diapause is not, it is to be supposed that diapause respiration proceeds mainly by "non-ferrous", possibly flavin-catalysed, systems. It was now found that taking normal respiration as 100, the diapause respiration was raised by nitrophenols to 375, while that of the active periods only to 250. The extra respiration was cyanide-sensitive. The fact that the respiratory quotient was raised from the usual diapause level of 0·75 to unity was at first attributed to the extra respiration being at the expense of carbohydrate but it was not fluoride- or iodoacetate-sensitive, and ammonia estimations demonstrated that at least three-quarters of it was due to protein breakdown.

Table 19. *Changes produced in metabolism of amphibian gastrula regions by treatment with dinitro-o-cresol*

		Dorsal lip region	Ventral ectoderm
$Q_L^{N_2}$:			
	Normal	0·63	0·21
	Dinitro-*o*-cresol	0·73	0·54
	Difference	+0·10	+0·33
	% rise	16	157
Uncompensated CO_2 ($\mu l . 10^{-3}/\gamma/5$ hr.):			
	Normal	0·82	0·28
	Dinitro-*o*-cresol	1·11	1·07
	Difference	+0·29	+0·79
	% rise	35	282
Ammonia production ($\mu l . 10^{-3}/\gamma/5$ hr.):			
	Normal	2·31	0·97
	Dinitro-*o*-cresol	2·23	1·62
	Difference	−0·08	+0·65
	% change	−3	+67

In the case of the amphibian gastrula, however, the fact that dinitrocresol did not raise the ammonia production to any extent suggests that here the action is mainly if not entirely upon carbohydrate utilisation. This agrees with the findings of Ronzoni & Ehrenfest on mammalian muscle.

The parallel with the sea-urchin egg is of much the same kind. The respiratory mechanisms here involved will be discussed on pp. 563 ff., but it is sure from the work of Runnström[7] and others, that the respiration of the unfertilised egg is not sensitive to cyanide, while that of the fertilised egg, proceeding at a much higher rate, is so. Later, Runnström[14] found that although pyocyanin, added as an accessory hydrogen transporter, increased the respiratory rate of both un-fertilised and fertilised eggs, its effect was of the order of 200 % on the former and only 80 % on the latter. The effects of nitrophenols on the sea-urchin embryo have been studied by Clowes and his collaborators (see Krahl & Clowes[1,2] for the literature). Most of their attention was devoted to the interesting fact, discovered by them, that concentrations of nitrophenols and halophenols which give the maximal stimulation to respiration also cause a reversible blockage in the mechanism of cell-division. In one place, however (Clowes & Krahl[1]), they state that the effect of nitrophenols on respiration was greater for the unfertilised than for the fertilised eggs—a six times against a four times increase. Thus while the respiration of the fertilised egg is normally four times that of the unfertilised egg, after both had been treated with dinitrophenol, it was only two and a half times as great.

An entirely different type of reagent, dimethyl-*p*-phenylene diamine, was found by Runnström[10] to raise both unfertilised and fertilised egg respiration to the same high level, indicating that the low activity of the unfertilised egg is not due to any deficiency in the indophenol oxidase.

All these phenomena resemble to some extent the state of affairs found for the amphibian gastrula regions. In general we may say that wide variations in metabolic activity exist in correlation with various physiological and morpho-genetic states. The regions or states of high activity may or may not be in a condition of maximum possible activity, but the regions or states of low activity are certainly damped down in some way or other. For respiration, this damped state may be due to a block in the chain of carriers in the Warburg-Keilin system, so that such respiratory activity as there is has to proceed by way of a "non-ferrous" system. For glycolysis we can now begin to visualise the damping mechanism since it has been shown that the glycolytic enzymes are themselves sensitive to oxidation (see the reviews of Rapkine[9] and K. C. Dixon). In the case of the amphibian gastrula, at any rate, the fact that the organiser substance, or evocator, is liberated in a region of high activity, and not in a region physiologically damped, can hardly be without significance.

Oxygen consumption of dorsal lip and ventral ectoderm was investigated by Boell & Needham[2], using the gastrulae of *Discoglossus pictus* and *Amblystoma mexicanum*. In neither case could any appreciable difference between the respiratory rate of dorsal lip and that of ventral ectoderm be observed.

Q'_{O_2}
(related to total N,
not dry weight)

	Discoglossus	*Amblystoma*
Blastula roof	—	− 2·63
Dorsal lip region of gastrula	− 4·80	− 3·21
Ventral ectoderm of gastrula	− 4·93	− 3·18
Closing neural folds	− 3·02	

The measurements in the paper of Waddington, Needham & Brachet were thus confirmed rather than those of Brachet & Shapiro. We may nevertheless accept the view that the difference between the two hemispheres of the gastrula found by Brachet & Shapiro in their double-capillary apparatus is real. But since the gastrula is a relatively complicated morphological structure rather than a homogeneous ball, this only means that there is more of the rapidly respiring tissue in one hemisphere than in the other. It does not mean that the dorsal lip region has a higher intrinsic respiratory rate than the ventral ectoderm. Direct measurements show that the two rates are statistically identical.

Later work by Boell, Nicholas & Sawyer confirmed the above observations and established the existence of zones of respiratory rate, ranging from a maximum in anterior presumptive neural plate and anterior presumptive epidermis, through intermediate values for posterior presumptive epidermis and dorsal lip region, to a minimum in the yolk-plug. This may explain some of the differences between the results of different investigators noted above. The exact region taken to compare with the dorsal lip may have varied a good deal. Intrinsic respiratory rate, however, is probably very similar throughout the gastrula, since direct centrifugation measurements of the amount of yolk in the various regions show an inverse proportionality between yolk-content and respiratory rate.

As a good deal of work has been done on the respiration of intact amphibian gastrulae, it is of interest to compare the figures in the literature with those found by the diver method on isolated regions. The following table summarises them.

Table 20. *Respiration of intact amphibian gastrulae*

	$T°$	μl. O_2 uptake/ 100 embryos/hr.	μl. CO_2 output/ 100 embryos/hr.
Białaszewicz & Blȩdowski:			
Rana temporaria, gastrulae	20	24·5	—
Saunders:			
Rana temporaria, gastrula	14	—	24·4
Parnas & Krasinska:			
Rana temporaria, gastrulae	—	13·7–25·0	—
Wills:			
Amblystoma punctatum, Harrison stage 10, gastrulae	25	34	—
Triturus torosus, Harrison stage 10, gastrulae	25	39	—
Rana pipiens, Harrison stage 10, gastrulae	25	39	—
Stefanelli[1,2]:			
Rana fusca, cleavage stages	?	(9–18) av. 13	—
Brachet & Shapiro:			
Rana sylvatica, gastrulae;	25		
Dorsal lip hemisphere		70·6	—
Ventral ectoderm hemisphere		48·1	—
Brachet[6]:			
Rana temporaria, gastrulae	?	29·9	—

If now we compare these results with those reported by Boell & Needham[2] we find that the respiration of dorsal lip region and ventral ectoderm is a good deal higher than would be expected from the volume of embryo occupied by them. From the data on the axolotl gastrula, the average amount of protein N taken was $11·1\gamma$ or $\frac{1}{32}$nd part of the total protein nitrogen in the embryo. As Q'_{O_2} is of the order of 3, the oxygen consumption for the whole embryo would be

106μl./100 embryos/hr., a figure considerably higher than any given in the table. It should, however, be remarked that in Brachet & Shapiro's measurements, the value of 70μl./100 embryos/hr. was also much above the average for the whole intact embryo. We can certainly conclude that the yolk endoderm must have an extremely low respiratory rate and occupies a large proportion of the total weight and nitrogen of the embryo.

It is interesting to compare the absolute levels of Q_{O_2} reported by Boell & Needham[2] for the amphibian embryo with those for the mammalian egg. For the ripe unfertilised egg of the cow, Dragoiu, Benetato & Opreanu have found values ranging from -21 to -35. The contrast with those of the order of $-0·5$ or under for the amphibian embryo remains great even when we allow for the $17°$ difference in temperature, and must doubtless be due to the fact that the amphibian egg is highly lecithic, and that much yolk is present even in cells deriving from the animal pole (see footnote on p. 157).

The same consideration applies to comparisons between the respiratory rate of gastrula regions and adult amphibian tissues. Thus tadpole epidermis, according to Erdmann & Schmerl and Börnstein & Klee, has Q_{O_2} of from $-1·21$ to $-2·18$ (cf. the $Q_L^{N_2}$ figures reported by Boell, Needham & Rogers for tadpole epidermis of -1 to -2). This contrasts with the $-0·2$ to $-0·3$ for ventral ectoderm (Boell & Needham[2]). Börnstein & Klee found that skin from old frogs had a lowered respiratory rate (though not to the gastrula level) and that explants of skin had a much higher Q_{O_2}, up to $-6·75$. Comparable data for adult nervous tissue will be found in Winterstein & Hirschberg and Gerard.

In connection with the respiratory rate of the presumptive mesoderm and neural tissue it is interesting that Brachet[21] found neural differentiation can go on in $M/1000$ KCN but not in $M/3000$ iodoacetate.

There still remains, however, the observation of Brachet[10, 22] that dorsal lip isolates showed a greater CO_2 elimination than ventral ectoderm pieces. This could only mean, as Brachet himself had found, a difference between the two regions in respiratory quotient. When the respiratory quotient of the two regions was investigated by the new methods in the work of Boell, Koch & Needham, this expectation was confirmed. The dorsal lip of the blastopore shows a much greater trend towards unity than the ventral ectoderm. The figures are given in the accompanying table and in Fig. 107:

	Gastrulae of *Amblystoma mexicanum* R.Q.
Blastula roof	0·75
Half-moon yolk-plug gastrula:	
Dorsal lip region	1·0
Ventral ectoderm	0·83
Three-quarter-moon yolk-plug gastrula:	
Dorsal lip region	1·0
Ventral ectoderm	0·89
Closing neural folds of young neurula	1·0

Thus as soon as gastrulation begins the dorsal lip region attains a respiratory quotient of unity. The respiratory quotient of the ventral ectoderm also rises, but more slowly, and has not exceeded 0·90 by the time when further dissection

of pure ectoderm is practically impossible owing to the expansion of the underlying mesoderm.

The exact significance of these results must be a matter for further research.
We may wish to assume that the respiratory quotient of unity of the dorsal lip
during gastrulation implies a predominantly carbohydrate catabolism and to
associate this both with the breakdown of glycogen there shown by Woerdeman
and Heatley and with the liberation of the primary organiser. But we must
remember that protein breakdown ending in ammonia may also give a respiratory
quotient approximately unity. The quotients of the ventral ectoderm certainly

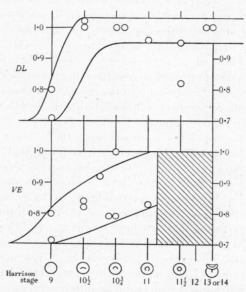

Fig. 107. Respiratory quotient of the regions of the amphibian gastrula. *DL* dorsal lip of the
blastopore; *VE*, ventral ectoderm.

give the impression of a mixed fat-protein catabolism. At all events it is clear
that we must see a change in *quality* of metabolism, rather than in *quantity*,
setting in at the organisation centre at the onset of gastrulation.*

Finally, the aerobic glycolysis was measured by Needham, Rogers & Shen,
and although the double-cup method, whether on the standard or the ultramicro
scale, does not give results of such accuracy as the other manometric methods,
it was established that there is little or no aerobic glycolysis in either of the two
regions of the gastrula. The Pasteur effect is thus equally efficient throughout
the gastrula.

---

* In connection with the differential metabolism of the regions of the gastrula, it will be remembered that according to Bellamy and Bellamy & Child, disintegration begins in the dorsal lip
of the blastopore when embryos are treated with toxic concentrations of mercuric chloride,
potassium cyanide, and ammonia. The claim was contested by Cannon, and the original papers
must be referred to by those who wish to estimate its significance. The dorsal lip of the blastopore
has, of course, long been claimed by Child to be a centre of "high metabolism".

Summing up then, we find as between dorsal lip region and ventral ectoderm a marked difference in anaerobic glycolysis and anaerobic ammonia production, no difference at all in respiratory rate or aerobic glycolysis, and considerable differences in respiratory quotient. The way is now evidently open for much further extension of our knowledge of the metabolism of the gastrula, a knowledge which alone will enable us to understand the mechanism of liberation of the primary evocator from its inactive precursor.

2·353. The role of sulphydryl in gastrular metabolism

The foregoing results may be set beside some interesting work done by Fischer & Hartwig[1] and by Piepho[1], in which it was found that when gastrulae are stained with various vital dyes and then placed under strictly anaerobic conditions, there is a more rapid reduction of the dye in the dorsal lip region, and later in the neural plate, than elsewhere. This difference was seen not only in the intact egg but also between isolates from the regions. Though it is never possible to be very precise regarding the meaning of the intra-cellular reduction of coloured indicators, these results must surely indicate a greater activity of certain oxido-reduction systems in the organisation centre relative to the other parts of the embryo.*

Of considerably more importance is the study of the fixed –SH groups in the gastrula suggested by Rapkine[8] and carried out in interesting work by J. Brachet[18]. If embryos of the unpigmented *Triton cristatus* are used it is possible to ascertain the distribution of fixed sulphydryl by performing the nitroprusside reaction on the whole embryo after denaturation of the proteins with trichloracetic acid. If this is done on the oocyte the large nucleus, but not the cytoplasm, is coloured a deep pink.† At the time of fertilisation the pink colour is seen only in a zone at the animal pole; Brachet[18] supposes that the proteins containing fixed –SH groups have been derived from the nucleo-plasm during maturation. Yolk-platelets never give the reaction. Two hours after fertilisation the grey crescent, not the animal pole, shows the colour when the test is performed, and later on the greater part of the blastula roof gives the reaction. After the appearance of the blastopore, the organisation centre shows it intensely. It is present over the whole presumptive neural plate area and intensified at the cerebral end. In the neurula all the neural area shows it.

Since the total area which colours bright pink increases from fertilisation to the end of neurulation, there would seem to be a synthesis of proteins containing much fixed –SH at the expense of the yolk-proteins.‡ Brachet[18] then noted other differences between the proteins in different gastrula regions. The distribution of granules staining red with the pyronine-methyl-green method was exactly similar to that of the fixed –SH groups. This was not at first understood, but Brachet[24] was later able to show, using a nuclease specific for ribo-nucleotides, that the pyronine-staining, basophilic, material is probably cytoplasmic (phyto-)

---

* Cf. the results by similar methods on the chick (p. 335) and on echinoderms (p. 496).

† High fixed –SH content has been observed also in the nucleus of the echinoderm oocyte by Dulzetto and by Ries[2] (see pp. 136, 567), and in that of the amoeba also by Chalkley[2] (see p. 422).

‡ Cf. the former somewhat similar observations of Chatton, Lvov & Rapkine on ciliates.

nucleoprotein (see Sections 3·434 and 3·44). This linkage of the fixed –SH protein in the organiser region with the cytoplasmic nucleotides may well be of the greatest importance. We are reminded of the high inductor activity of nuclear material (Waddington[19], cf. p. 171) and of the presence of phyto-nucleotides in embryonic and sarcomatous tissue (Claude[2], cf. p. 269). Indeed, it may be no coincidence that the latter are normally combined with plasmalogen, a substance known to disappear from the invaginating material during gastrulation (Voss, p. 189 above).

Centrifugation experiments of Pasteels[11] strengthen the view that the fixed –SH proteins have something to do with the organiser effect. If eggs in the cleavage stages are centrifuged as thoroughly as may be without rendering cleavage abnormal, the fixed –SH proteins are thrown to the animal pole with the fatty layer away from the yolk and pigment. At a later stage, though a blastopore may form, gastrulation will not progress farther; in this case it is found that the arrested embryo will show no fixed –SH proteins at the dorsal lip region. Or the normal induction may be inhibited partly or wholly (see p. 226); in this case no fixed –SH reaction will be obtainable from the inhibited regions.*

It may be added that glutathione and cysteine showed no inducing power in the *Einsteckung* experiments of Finkelstein & Schapiro[1], but the above-mentioned work of Brachet[18, 24] applies to fixed, not free, –SH groups, and the phenomena in normal development are probably very complex. We cannot as yet fit this into its proper place.

In any case, the careful study of inhibitors on gastrulation will be very illuminating, especially where their chemical action is already fairly definitely known. Iodoacetate and fluoride have been used by J. Brachet[21, 22, 23]. Fluoride in doses as high as $M/40$ (sufficient to inhibit glycolysis completely in the chick embryo) has no effect on amphibian development before hatching (we can hardly doubt that it penetrates into the egg) and weak concentrations of iodoacetate ($M/300$) have also no effect, though these would be expected to inhibit glycolysis. On the other hand $M/30$–$M/60$ iodoacetate, while allowing cleavage *and* gastrulation to proceed normally, arrest development at the stage of the full-moon yolk-plug, i.e. just before the appearance of the neural folds and during the induction process. The inactive concentrations of fluoride and iodoacetate have no effect on the respiration of the whole embryo, but the effective concentration of iodoacetate inhibits the respiration of the gastrula 90 %. Respiratory quotient (Meyerhof-Schmidt method on thirty embryos) is lowered in all the inhibitors, whether morphogenetically effective or not, to 0·83; lactic acid production and glycogen breakdown are greatly inhibited. If at this inhibited stage the embryos are replaced in ordinary water, development will continue normally, but if they are allowed to remain in the inhibitor, dissolution will follow.

All this strengthens the suggestion that sulphydryl groups have something to do with neural induction. It may be, of course, that there is a reversible denaturation of the protein in the inactive precursor complex. Or the oxidation-reduction

---

* Criticisms of the technique used in all this work, made by Giolitti, were successfully rebutted by J. Brachet[18].

equilibrium may be affected. The latter possibility is supported by the work of Brachet & Rapkine (only available as yet in preliminary form) who could provoke neural differentiation in explants of ventral ectoderm by immersing them in reducing solutions (–SH; $M/10$ reduced glutathione, $M/10$ thiomalic acid, $M/10$ cystein) and could suppress neural differentiation in dorsal lip explants by immersing them in oxidising solutions (–S-S–; $M/20$ oxidised glutathione; $M/50$ alloxan). Confirmation of this will be of much importance. We may recall that the flavine-adenine compound which acts as the prosthetic group of the amino-acid oxidase will only join up with the protein moiety of the enzyme when in the reduced condition.

In all further development we shall have to bear in mind three problems: (1) the metabolic events which give rise to the active substance, (2) the nature of the active substance itself, and (3) the metabolic events accompanying and associated with the neuralisation response of the competent tissue.

2·355. The nature of the inactive evocator complex

It has been suggested above that the hypothesis which explains most simply the facts known about the distribution and function of the primary evocator is that it is normally inactivated by combination with other substances, probably protein, perhaps polysaccharide also. From this combination it is normally liberated only in the dorsal lip of the blastopore. In other regions of the gastrula it may also be liberated by any treatment which will denature the proteins of the tissue.* Some preliminary evidence exists which suggests that the primary evocator is a substance of steroid nature. Interest therefore attaches to the complexes ("symplexes" or "cénapses") between proteins and lipins which have been more thoroughly studied. In adult tissues the primary evocator is found in the free form, but from the first formation of the oocyte till gastrulation it is held in combination, and thus competent tissues which would otherwise react to it are protected from its influence until their competence is lost.

Treatment less severe than denaturation of the proteins may apparently liberate the evocator. Such an activation carried out by dyes has been described above (p. 189); the dye may act by displacing the evocator from its combination. The inductions obtained by nucleic acid (see above, p. 180) might receive a similar explanation since Stenhagen & Teorell have shown that nucleic acid will displace bilirubin from its association with serum albumen.

The association which we have seen to exist (see p. 190) between the primary evocator and glycogen suggests a further extension of the foregoing hypothesis. Is the inactivating complex one in which glycogen is involved as well as protein? In this case a metabolic attack either on the protein or on the glycogen might break up the complex and liberate the evocator. That the complex cannot

---

* In his recent *Silliman Lectures*, Spemann ([17], p. 244) seems to maintain that the appearance of the evocator in the ventral ectoderm after treatment with heat cannot be the same process as the appearance of the evocator in the dorsal lip of the blastopore during normal metabolic events. For this distinction, however, there are no convincing reasons. The fact that a loose complex may be broken down to liberate an active substance in many different ways renders superfluous a belief in the existence of more than one such loose complex.

be identical with desmo-glycogen was shown by the experiments described on p. 193, but the evocator might be associated with another fraction of intra-cellular glycogen.

That loose complexes exist in the cell has been known since the time of Hoppe-Seyler (1866); perhaps lecitho-vitellin of egg-yolk is the most classical example. Without preliminary treatment with alcohol it is impossible even to approach a complete separation between the protein and the lipin. The monograph of Macheboeuf and the reviews of Przyłęcki[2,3] give an account of these complexes. The phenomenon of "lipo-phanerosis" (see Macheboeuf and Lison), has also long been known to cytologists; under certain treatments there may be a great increase in the amount of lipid histologically detectable. A particularly striking instance is due to Heilbrunn. If echinoderm eggs are treated with

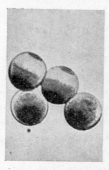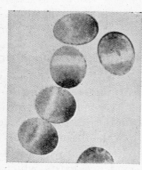

Fig. 108. Lipo-phanerosis in echinoderm eggs demonstrated by centrifuging.

ammonium salts in fairly high concentration and then centrifuged, a large apparent increase in the fatty layer is observed (see Fig. 108), indeed the hyaline zone may be reduced to a mere streak. An analogous effect was obtained on amoebae with ultraviolet irradiation by Heilbrunn & Daugherty.

Further evidence to our purpose may be adduced from the literature as follows.

Complexes of proteins with polysaccharides

Lyo- and desmo-glycogen (Przyłęcki & Białek; Willstätter & Rohdewald[2]; Przyłęcki[3]; Gurfinkel)

Myosin and glycogen (Przyłęcki & Majmin[1]; Mystkowski[1]; Mystkowski, Stiller & Zysman)

Polysaccharides and clupein (Przyłęcki, Gedroyć & Rafałowska)

Starch, dextrin, and various proteins (Przyłęcki & Rafałowska; Rafałowska, Krasnodebski & Mystkowski; Przyłęcki, Andrzejewski & Mystkowski; Przyłęcki & Dobrowolska; Przyłęcki & Grynberg[1]; Bartuszek; Przyłęcki & Targońska)

Studies on the nature of the linkage (Przyłęcki, Kasprzyk & Rafałowska; Przyłęcki, Cichocka & Rafalowska; Przyłęcki, Rafałowska & Cichocka; Gedroyć, Cichocka & Mystkowski)

Complexes of proteins with unsaponifiable lipins

Sterol-protein combinations in blood (Gardner & Gainsborough; Bruger; Mellander; Bills; Theorell)

Sterol-protein combinations in yeast (McCorquodale, Steenbock & Adkins)
Sterol-protein combinations in wool (Schoenheimer & Breusch)
Artificial sterol-protein complexes (Przyłęcki, Hofer & Frajberger-Grynberg) in
 surface films (Rideal & Schulman).
Antigenic steroid-protein substances (Marrack)
Lipochrome-protein complexes (Palmer & Eckles; Kühn & Lederer; Wald)

Complexes of proteins with saponifiable lipins

Lecitho-vitellin
Blood-lipins (Macheboeuf & Sandor; Wu, Liu & Chen; B. Delage; Grigaut)
Artificial protein-fat complexes (Przyłęcki & Hofer). The freer a protein is from
 fatty material the more rapidly is it attacked by proteases (Janicki; Przyłęcki[3]).
 Mixed lipo-protein monolayers have been studied by Schulman; Schulman
 & Hughes; Schulman, Stenhagen & Rideal.

Complexes of saponifiable lipins with polysaccharides

Fatty acids and corn starch (Taylor & Nelson; Taylor & Sherman)
Lecithin and glycogen (Przyłęcki & Majmin[2])
Lipin phosphorus and starch (Posternak)

Complexes of unsaponifiable lipins with polysaccharides

Fatty acids and phytosterol on banana starch (Lehrman & Kabat)

Complexes of unsaponifiable lipins

Sterols and carcinogenic hydrocarbons (Clowes, Davis & Krahl)

Complexes of nitrogenous substances with proteins

Nucleic acid and serum proteins (Przyłęcki & Frajberger; Przyłęcki & Grynberg[2])
Nicotine and proteins (Leontiev & Alexandrovsky)
Uric acid and proteins (Przyłęcki, Grynberg & Szrajber; Przyłęcki & Kisiel;
 Grynberg & Kisiel)

The literature on these complexes is thus a large one, and it is by no means claimed that the above-mentioned papers, or those referred to later in this section, constitute a full list of all the complexes which have been studied. Much uncertainty as to their nature exists. A complex may be a well-defined chemical compound with a stoichiometric relation between its components, it may be a dissociable adsorption complex, or it may merely be the co-existence of two or more substances which the methods hitherto available have failed to separate. Thus cataphoresis has proved to be the only means of separating polysaccharide from flavoprotein. Modern methods of protein chemistry tend more and more to lead away from the vague views of earlier colloid chemistry, and to throw light on the nature of the linkage in the true complexes. As yet, of course, we have hardly any knowledge of the extent to which physiological significance depends upon the nature of the linkage in the complex. At all events no need for hesitation exists in picturing the unmasking or activation of the primary evocator as a liberation of it from inactive combination.

Throughout the field of biochemistry it is becoming more and more clear that the properties of compounds are very greatly dependent on whether they exist free, or in more or less loose combination with other molecules, such as proteins, of which they may be prosthetic groups, or polysaccharides, with which they may form complexes. All the logical possibilities seem to be represented. Thus:

(1) A compound may be "active", or possess a certain property, both in the bound and the free state. In this case,
(1 a) the same activity may be shown, or
(1 b) the nature of the activity may be quite different.
(2) Activity may only be manifested in the bound state.
(3) Activity may only be manifested in the free state.

Examples of (1 a) are the proteolytic enzymes, some of which, such as pepsin and kathepsin, are believed (Willstätter & Rohdewald[1]) to occur in two forms, such as desmo-kathepsin, strongly attached to the protein cell structure, and lyo-kathepsin, much more readily extractable. The enzymic activity is, however, the same in both cases (Bamann & Salzer).* Some drugs, moreover, appear to enter into close combination with proteins, e.g. the trypanocidal Bayer 205 (Boursnell & Wormall).

Examples of (1 b) are very common. Thus lactoflavin, in the free state, is vitamin B_2, but when attached to protein (as ribitylflavin-phosphate), forms the so-called "yellow" enzyme or co-enzyme dehydrogenase of oxidation-reduction processes (cf. Karrer's review[2] and Fig. 307). It seems that the strict line of demarcation between proteins with well-defined prosthetic groups and proteins united with other substances as complexes is breaking down. Warburg[4] has proposed the view that co-enzymes are, in fact, the prosthetic groups of their corresponding enzymes. Thus adenosine triphosphate, which we may represent as adenine-ribose-phosphate-phosphate-phosphate, is the classical phosphorus-transporter in the phosphorylation cycles of glycolysis (see Fig. 322). But co-enzyme I (co-zymase), which may be written adenine-ribose-phosphate-phosphate-ribose-nicotinic acid, is the co-enzyme, and hence the prosthetic group (when the active complex has been formed), of the triose-phosphate dehydrogenase. Co-enzyme II differs only in the possession of a third phosphoric acid group, and would be regarded as the prosthetic group of another series of dehydrogenases. The co-enzyme of the amino-acid oxidase from a structural point of view unites the fields of glycolysis, fermentation and oxidation, for its structure is adenine-ribose-phosphate-phosphate-ribitol-flavin. Another instance is vitamin B_1 (Aneurin or Thiamin). United with two phosphate groups it forms co-carboxylase, presumably the prosthetic group of the carboxylase of yeast. Baumann & Stare, in their review on co-enzymes, make the interesting suggestion that in the body thiamin-pyrophosphate is united with ribose and adenine. This would bring all the co-enzymes into one group. These ideas are extended to pigments in a review by K. G. Stern[2]. Finally, it is known that vitamin B_6 (Pyridoxin or Adermin; the rat anti-dermatitis vitamin) occurs in a protein-bound form, though it has the same activity whether bound or free (Kühn & Wendt).

* We shall see later (p. 610) a case of some importance for embryology in which a whole system of enzymes appears to be of the desmo-type, namely that concerned with the "non-phosphorylating" glucolysis; in contrast to the enzyme systems of the phosphorylation cycles of glycolysis which have long been available for study in tissue extracts.

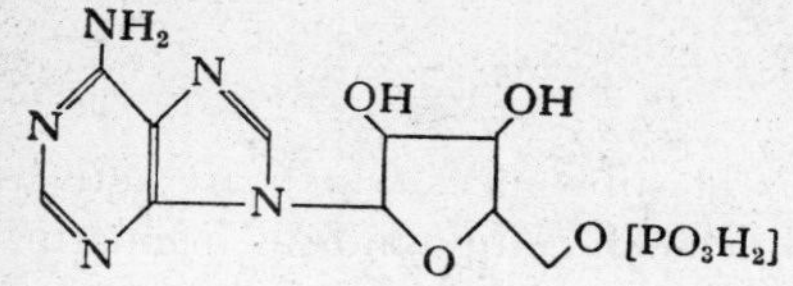

Adenosine monophosphoric acid
(Muscle adenylic acid)

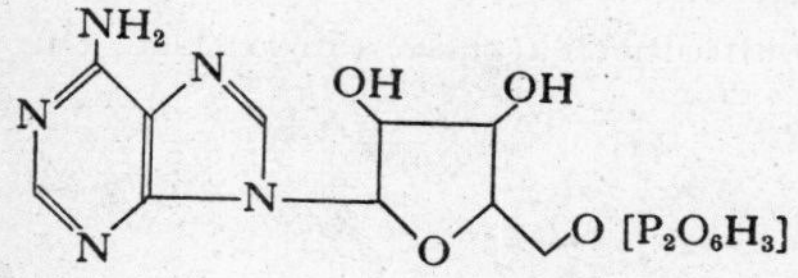

Adenosine diphosphoric acid

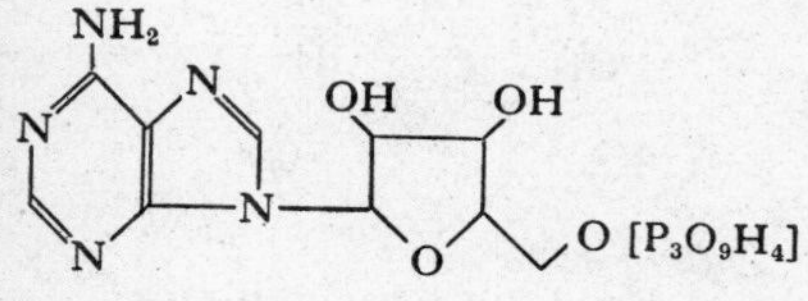

Adenosinetriphosphoric acid

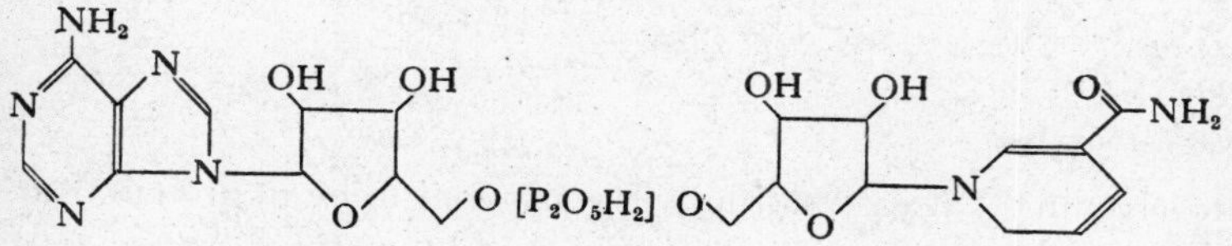

Cozymase
(Coenzyme I)

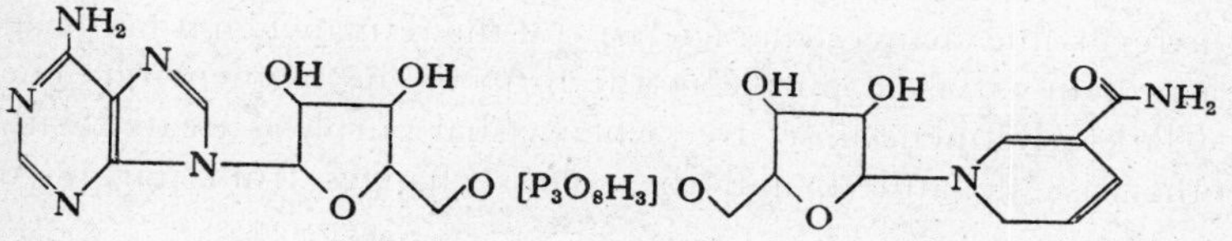

Coenzyme II

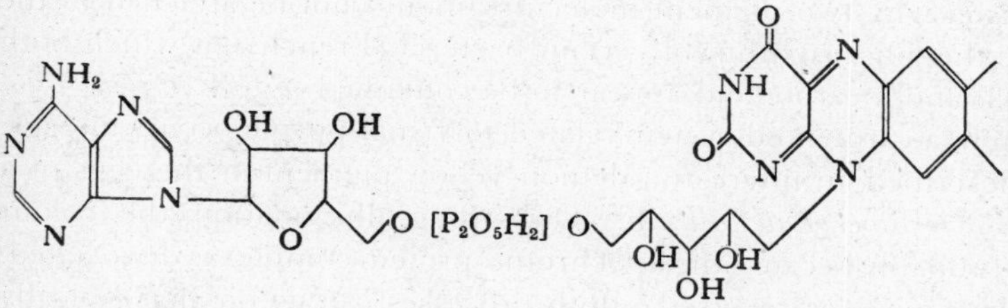

Amino-acid
oxidase coenzyme

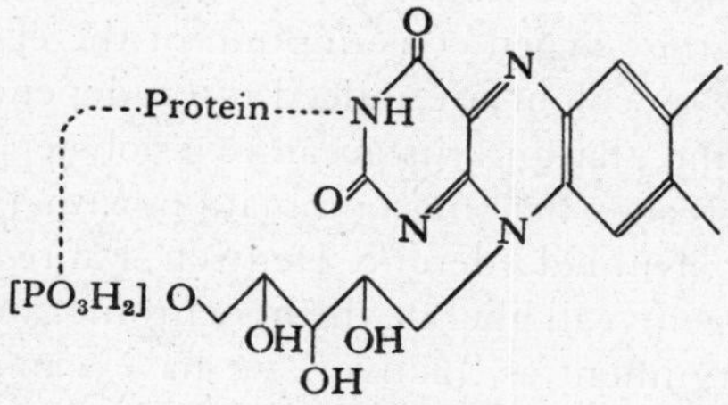

Flavoprotein
("Yellow enzyme")

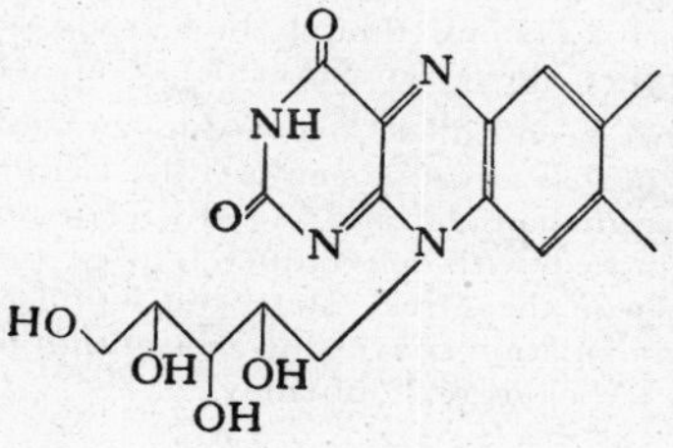

Lactoflavin
(vitamin B_2)

On the other hand, there are many examples of substances which are active *only* in the combined form (case (2) above). While such compounds as haematin have little biological activity, the tetrapyrrol system, once combined with protein, gives such powerfully active substances as cytochrome, catalase, peroxidase, the

Aneurin (vitamin B₁)

Co-carboxylase

Pyridoxin (vitamin B₆)

haemoglobins, etc. (Keilin[3]).* Leaf chlorophyll is also said to be in protein combination (E. L. Smith). In the same way, vitamin A, a lipochrome (see p. 249), only carries on the reversible changes characteristic of the retina when it has been combined with a protein as visual purple in the form of the pigment retinene (Wald; Wald & Clark). Retinene is set free when visual purple is treated with chloroform or otherwise denatured, and the colour changes from purple to greenish-yellow.

There is no reason why two pigment molecules should not be attached to the same protein, e.g. the photosynthin of certain bacteria (French), in which both bacteriochlorophyll and a carotenoid appear to be contained.

Curiously enough, a carotenoid system related to visual purple occurs in eggs. Yamamoto[3] has described an interesting lemon-yellow pigment in the eggs of a polychaete worm, *Ceratocephale osawai*. On strong illumination the colour changes to green, probably because the lipochrome-protein complex is dissociated, and in the dark the process is reversed. Both processes can go on anaerobically and do not affect the oxygen consumption of the eggs.

Properties such as colour are indeed often dependent on complex-formation. Another lipochrome, that of crustacean (e.g. lobster) eggs and carapaces, astacin (see p. 249), is green when in combination with protein, but pink when free (Kühn & Lederer; Kühn, Lederer & Deutsch; Fabre & Lederer; N. A. Sørensen).† Any denaturing agent will liberate the lipochrome, and if we regard the pink form for the sake of argument as "active" we have a model for the liberation of the

* The open tetrapyrrol system, though biologically inactive, also combines with protein, e.g. bilirubin-serum-albumen (Pedersen & Waldenström).

† The bound form has been called "ovoverdin" but this seems a rather unfortunate name, since the term "uteroverdin" has already been used by Lemberg[1] for an open-chain tetrapyrrol derivative, found in the mammalian placenta, closely related to the blue pigment of birds' egg-shells (see p. 647) and identified with biliverdin. It used to be thought that astacin was still united with a fatty acid when in the "free" state, but Kühn & Sørensen have shown that the presumed saponification was an oxidation converting astaxanthin to astacin. Lipochrome-protein complexes also occur in insects (Abeloos & Toumanov).

evocator. Stern & Salomon demonstrated the interesting fact that the liberation of the pigment from its protein shows mild thermal reversibility (at temperatures not above 70°). Still more striking was the observation of Peters & Wakelin that traces of soaps (such as sodium oleate, palmitate, tetradecoate and dodecoate) would act like alcohol or chloroform as liberating agents in concentrations as low as 1×10^{-3} M. The effect was not seen with fatty acids below C_{12}, i.e. below the change from water- to "fat"-solubility. The liberated pigment can be put back on to its protein by the action of 0·03 M $CaCl_2$. It is possible that we have here an explanation of the neural inductions caused by fatty acids in the experiments of Fischer et al. (see p. 180) as it is unlikely that fatty acids would long remain wholly in the free form within, or in close contact with, living tissues.

We thus approach group (3), the substances only active in free form, which most closely resemble the position of the evocator in the amphibian gastrula. The plant growth hormone, auxin (see p. 266), which also in other ways shows interesting parallels with the evocator, is known to exist in inactive, bound, form (Kögl et al.; Allen; Went & Thimann). There is also evidence that vitamin C (ascorbic acid) exists in bound form combined with protein in plant tissues (Guha & Pal; Reedman & McHenry).* Human blood, according to Sinclair, contains a good deal of vitamin B_1 in bound, inactive form, and the same may possibly be true of milk (Houston & Kon). There is also a bound form of vitamin E (Emerson, Lu & Emerson). Acetylcholine, liberated suddenly at cholinergic nerve-endings (cf. the review of G. L. Brown), appears to arise from a pharmacologically inactive non-dialysable and denaturable protein complex precursor, according to Loewi; Kahane & Levy; H. C. Chang, and Mann, Tennenbaum & Quastel.† Sympathin, too, liberated at adrenergic nerve-endings, must exist there in some stored form (cf. the review of Rosenblueth), and adrenalin-protein complexes are being investigated (Utevsky & Levantzeva; Utevsky & Osinskaia). And the water-soluble heat-stable melanophore-controlling crust-acean neuro-humors (Parker) seem also to increase in activity after boiling, as if liberated from protein complexes.

We are thus coming to the general conception of combination with protein as the most important element in biological activity. An "active" substance of low molecular weight will generally require combination with a specific protein in order that its activity may be manifested.‡ And it may be held in inactivated form on another, perhaps equally specific, protein. It may well be that the evocator has to be adsorbed on to a protein in order to carry out its effects, just as it has to be previously liberated from inactive combination. Unfortunately it is as yet

---

* But cf. Fujita & Ebihara.

† There are those, however, e.g. Trethewie, who do not accept the evidence so far advanced for "bound" acetylcholine, but they seem to be fighting a rearguard action.

‡ It is, of course, essential to specify exactly what activity is in question. Thus flavin-adenine-dinucleotide united with one protein will form the amino-acid oxidase; united with another it will form a co-enzyme dehydrogenase. To some extent the tools of the cell seem to involve interchangeable parts; several ends may fit on to one handle or several different sorts of handle on to one end.

extremely hard to picture exactly how the effect of the primary evocator is produced in the competent cells. This problem will be referred to again in Section 3·5.

2·356. The evocator compared with plant hormones

We spoke above of auxin. This well-known plant hormone has been the subject of a large amount of research, ably summarised in the monographs of Boysen-Jensen[1] and of Went & Thimann.* Auxin, formed or liberated at one part of a plant (e.g. the tip of the *Avena* coleoptile), is transported from cell to cell and stimulates longitudinal growth of the cellulose walls at another. Its effects are therefore entirely different from those of the primary evocator in animal development; one stimulatory substance being concerned with growth, the other with differentiation. Nevertheless the parallels between the action of auxin and evocator are so interesting, and the differences so instructive, that it is worth while to give them a brief discussion.[†]

In the first place, it is evident that what we have previously spoken of as competence enters into plant physiology also, as it probably does in any system of stimulus-response reactions. Thus the stem, but not the root, is stimulated by auxin, and there is indeed a reversal of reactivity, for the elongation of the root is inhibited by auxin (W. & T. pp. 14, 213) though the number of roots is increased (W. & T. pp. 183 ff.). Similarly, in the *Avena* coleoptile, the production of auxin and the growth rate at first increase side by side, but later on the growth rate falls off some considerable time before the auxin production shows any signs of decreasing (W. & T. p. 59). Competence to grow therefore fails before the supply of stimulatory substance.

We may next distinguish five points in which auxin and evocator action are analogous, and two points in which they seem to differ.

(1) Bound auxin. By no means all the auxin in a plant is present in the free state (B.J. pp. 58, 63; W. & T. pp. 64, 65, 68, 116, 132). Three states of auxin are defined: the storage form, set free by saponification of, e.g., seed oils, and hence probably an ester; the "bound" form, extractable only by organic solvents such as chloroform; and thirdly, the free form, capable of diffusing out of the plant into an agar block placed in intimate contact with a cut surface. In germination

---

* Referred to in what follows as B.J. and W. & T. respectively. All the literature will be found in these monographs. There are also the valuable reviews of Kögl; Went; Avery; Boysen-Jensen[2]; Thimann[1]; Thimann & Bonner; Haagen-Smit; Erxleben.

† There is also the question of the existence or action of auxin in animal embryos. Agar impregnated with auxin, when implanted into the blastocoele cavity of amphibian embryos, produced only the usual ectodermal thickenings (du Pan & Ramseyer). On the growth of embryonic fibroblasts it has no marked effects (Ramseyer & du Pan). But it is certain that auxin (or heteroauxin) is synthesised during embryonic development. In the amphibian embryo there is none before gastrulation, but it increases rapidly after the closure of the neural folds (Rose & Berrier). It is also definitely synthesised by blowfly and silkworm moth embryos (Berrier). In the chick embryo, as a careful paper by Robinson & Woodside shows, there is a relatively enormous increase of it during development.

the coleoptile grows from a minute bud on the mesocotyl to a length of 4 cm., with few or no cell-divisions. Auxin is continuously produced by the growing tip, which itself does not elongate, and there is to all appearances a close parallel here between the liberation of auxin in free form at the tip and the liberation of the evocator in the dorsal lip of the blastopore. When cut off and placed on agar, the tip continues to produce auxin for many hours, but ceases to do so at last, probably on account of exhaustion of the precursor. When estimation methods as quantitative as those for auxin have been worked out for the evocator, we shall be able to follow similarly the production of free evocator. Normally the tip receives supplies of the precursor from the seed.

(2) Species-specificity. Like the evocator, auxin effects show no differences between species and genera of the plants used as source of stimulus or reactant (B.J. p. 57; W. & T. p. 114).

(3) Just as we have seen reason to believe that a variety of substances may unmask the evocator in the reacting tissue, so certain unspecific influences have been demonstrated to produce their effects by way of the auxin mechanism in plants (B.J. p. 114; W. & T. pp. 130, 191). To this class of effect belong the growth stimuli of acids and of ethylene.

(4) Irradiation with X-rays may inhibit the growth of coleoptiles, but application of auxin later shows that this is not due to interference with the competence (B.J. p. 66). These facts may be compared with those to be discussed shortly regarding the evocator (see p. 222).

(5) The auxin effect involves no change whatever in respiratory rate (B.J. p. 122; W. & T. pp. 128, 164), though the growth process is dependent on respiration in that it will not go on anaerobically and that cyanide inhibits it *pari passu* with respiration. On the other hand, dinitrophenol accelerates respiration but not growth. Here again there is a striking similarity with the differentiation hormone of the gastrula.

The apparent differences between auxin and evocator action are equally interesting. Thus:

(1) Auxin is destroyed by the action of oxidising enzymes at cut surfaces (W. & T. p. 57), or in extracts.

(2) Auxin disappears in the plant metabolically (W. & T. p. 86). In the same way acetylcholine after its action in the adult vertebrate body is destroyed by choline esterase and adrenalin by adrenalin oxidase. On the other hand, there is no information available about the destruction of the evocator. Homoiogenetic induction (p. 154) shows that as the neural tube is determined, so its cells acquire induction power, and in later development the evocator appears free in all tissues. The question of its disappearance will not be solved till adequate estimation methods become available for it.

In the universality of its distribution, the primary evocator resembles another plant-growth hormone, or rather, group of hormones, the Bios complex. In 1901 Wildiers, working on a suggestion of M. Ide, established that yeast cells will not grow and divide normally if the amount in which they are first sown is too small. Subsequent research (well summarised by Kögl & Tönnis) demonstrated the presence of the missing necessary substances in wort,* and Bios was eventually split up into three synergistic, not summative, factors. Bios I is now known to be *i*-inositol (Eastcott).† Bios II (sometimes called biotin) has been crystallised by Kögl & Tönnis and prepared by them from hen's egg yolk and white by a process involving esterification, distillation of the volatile ester, and hydrolysis. This product is active in a concentration of 1 in 4×10^{11}. It has been shown by Kögl & v. Hasselt to be present in all animal tissues, and is perhaps identical with "vitamin H" and "co-enzyme R" (György *et al.*). A similar substance was studied by R. J. Williams and his collaborators and named by them "Pantothenic acid". This substance, now known to be a dipeptide of α-hydroxy-β-β-dimethyl-γ-hydroxybutyric acid and β-alanine (Williams & Major; Williams, Finkelstein, Folkers, Harris, *et al.*), also occurs in all animal, and some plant, tissues and seems to be present in some protein-bound, non-diffusible form in egg-white (Williams, Lyman, Goodyear, Truesdail & Holaday). Purified by brucine, a preparation has been obtained active at 1 part in 2 billion of culture medium (Williams, Truesdail, Weinstock, Rohrmann, Lyman & McBurney). It has an effect on yeast respiration as well as on yeast growth, stimulating the Q_{O_2} of yeast 60 % (Pratt & Williams). It seems to be identical with the chick anti-dermatitis factor (B_3) of Jukes. Bios III is the least well understood of the factors in the group, but Kögl & Tönnis have established that it contains nitrogen though no sulphur or phosphorus.

These facts are mentioned in order to show that cases of highly active substances, widely, if not universally, distributed among living tissues, are already known.

2·357. Induction and fertilisation

Before leaving the discussion of phenomena which may be parallels of embryonic induction, a word must be said about fertilisation. The question whether induction is in any sense related to echinoderm parthenogenesis is a very legitimate one. Unfortunately, the subject of the parthenogenesis of the eggs of invertebrates is at present in a state of some confusion, for apart from the reviews of J. Loeb[3] and Y. Delage (in 1909 and 1910 respectively) nothing is available except that of Just[2] which upholds the "fertilisin" theory of F. R. Lillie[5,7]‡ in a rather partisan way. A detailed account of the literature in this field from a modern point of view is greatly needed.

Space is not available here to embark upon a discussion of even the most important relevant facts (see F. R. Lillie & Just), but it is certain that echinoderm eggs give off a number of active substances into the surrounding water. This "egg-secretion" is not produced by immature eggs, and is no longer produced

* W. Lash Miller, one of the earlier investigators of the Bios factors, met certain criticisms of the work as follows: "All those who say 'no doubt some chemical in wort makes all the difference, but it need not be Bios', have qualified to join the Last Ditch Bacon Club, which holds that Shakespeare's plays were written, not by Shakespeare, but by some other person bearing the same name." This is not without a moral for embryologists.

† Other cycloses are inactive (Kögl & v. Hasselt).

‡ To me, it seems that F. R. Lillie's interpretation of fertilisation along immunological lines has done more harm than good. This mistake would be interesting methodologically as showing that one should not use concepts taken from a level of *unnecessarily* high complexity.

by fertilised or parthenogenetically stimulated eggs. It can be washed out by repeated changes of water, thus rendering the eggs unfertilisable, but their fertilisability can be restored, as Woodward showed, by adding further egg-secretion. During the fertilisable period, a wide variety of agents, including all kinds of chemical substances (Y. Delage), blood or tissue juices (J. Loeb; Bogucki[2]), and changes in osmotic pressure and temperature (R. S. Lillie[1]), will bring about parthenogenesis. Of these the most effective in common use is hypertonic sea-water (double the normal salinity). Loeb's view[3] was that all parthenogenetic agents caused a cytolysis of the superficial layer of the egg, and hence the "release of a catalyser"; and for this view there is still much to be said. In its most modern form, suspicion has fallen upon calcium, the presence of which either outside or inside the egg seems proving to be essential for all types of partheno-genesis (Pasteels[1]; Mazia[2]; Heilbrunn & Wilbur). Other agents would simply affect the threshold to or competence for the action of calcium.

On the whole, it seems that artificial parthenogenesis of echinoderms has more in common with the stimulation of a nerve than with embryonic induction, for the very wide range of successful stimulators always suggested a physical rather than a chemical mechanism of immediate response. The parallel with nerve was in fact used by the earliest student of parthenogenesis, Tichomirov[2], who in 1886 discovered the sulphuric acid parthenogenesis of silkworms. Parthenogenesis with sperm extracts has never been achieved.

The active substances of the egg-secretion have not so far been fully analysed. There is (1) the activator for spawning of the male. This is believed to be gluta-thione or a similar sulphydryl compound in the case of polychaetes (Townsend), but in molluscs its properties are different (Galtsov). There is (2) the activator for sperm motility. In echinoderms this seems to be a volatile substance, since steam distillates of egg-secretion are effective (Clowes & Bachman), as also propyl, amyl, and cinnamyl alcohols. There is then (3) the substance responsible for the chemotactic attraction of the spermatozoa (sometimes called aggregation); this has been recently identified by Hartmann, Schártau, Kühn & Wallenfels as the egg-pigment echinochrome, a polyhydroxy-naphthoquinone; though the claim is contested by Tyler[8]. There is finally (4) the substance or substances called

Echinochrome (1:2:3:4:5:6:8-heptaoxy-7-ethyl-naphthalene)

"fertilisin" by F. R. Lillie[7]. This consists of two factors, the cleavage factor, in the absence of which development cannot proceed, even if a sperm should penetrate into the egg; and the agglutination factor, which causes the spermatozoa to stick together around the egg-jelly. The fact that agglutination power and fertilisability often ran parallel (C. R. Moore; Just[1]) led Lillie to regard the two effects as due to one single substance, but it seems more likely that two are involved, since O. Glaser[4] and Woodward were able to separate them, the former being undialysable and precipitable with ammonium sulphate, the latter being dialysable and precipitable with barium chloride. Yet Carter found .that both effects were obtainable with thyroxin, which restores fertilisability and also agglutinates spermatozoa. The whole subject urgently needs reinvestigation by up-to-date chemical methods, as in the new work of Tyler and Tyler & Fox.*

Parthenogenesis brought about by material obtained from other eggs would be a parallel for induction performed by killed ventral ectoderm. This has been claimed to occur by O. Glaser[1] ("autoparthenogenesis"), but the experiments have never been generally accepted and still require confirmation.

Parthenogenesis, then, resembles nervous irritability rather more than embryonic induction. In the last resort it may be, of course, that all cases of stimulation are ultimately identical, but nerve reaction differs from reaction to a hormone by its much greater unspecificity, probably because rapid physical changes play a greater part in it than the enzymic changes which hormones must occasion.

We must now return to the gastrula and consider the biological aspect of the liberation of the evocator.

2·36. Blastopore formation and the liberation of the evocator

During the preceding discussions the reader must often have asked what it is that governs the place of appearance of the blastopore and hence the liberation of the evocator and the determination of the neural axis. To answer this question we have to go back to a period of development which has received less attention than gastrulation, namely that of the fertilised egg and the embryo during early cleavage.

The dorsal lip of the blastopore and part of the presumptive neural plate area represent the dorsal surface of the future larva. In the Anura the site of the dorsal lip of the blastopore is foreshadowed in the fertilised but uncleaved egg by the so-called "grey crescent", a band of material lighter than the animal pole but darker than the vegetal pole. A similar "dorso-marginal" zone exists in Urodeles but not so obviously. Our problem concerns the origin and significance of this zone, and hence, since it distinguishes dorsal from ventral, the origin of the bilateral symmetry (dorso-ventral asymmetry) of the embryo. Experimental work on the subject has been very confused, but the position has recently been clarified.

Everyone is agreed that in the fertilised egg there are two gradients,[†] the animal-vegetal one, obvious on account of the accumulation of yolk towards the

---

* A word may be added concerning interesting recent work on copulation-stimulating substances in algae (Moewus; Kühn, Moewus & Jerchel). The cells of *Chlamydomonas eugametos* produce, in light of all wave-lengths, a substance which accelerates the movements necessary before copulation. This is crocin, the digentiobiose glycoside of crocetin, a lipochrome, and it is said to be effective in dilutions of 1 in $2 \cdot 5 \times 10^{12}$. Cis-crocetin-dimethyl-ester, in mixtures of various proportions with trans-crocetin-dimethyl-ester, forms the male and female substances. These are effective in dilutions of 1 in 3×10^8.

† I use the word here with no "arrières pensées" or presuppositions about metabolic questions.

vegetal pole and the corresponding reduced division rate there*; and the dorso-
ventral one, with the grey crescent as its apex (see Fig. 109). It is generally held
that the animal-vegetal gradient originates from the position of the egg in the
ovary, receiving its reserve materials from a capillary vessel reaching one point only
on its surface. Bellamy's proof of this and its subsequent withdrawal are of the
less importance because on general grounds such a mechanism may be taken as
extremely probable. But the origin of the other gradient, or, if you will, the
position of the grey crescent, has been difficult to determine. In 1883 Roux[1,4]
believed that he had proved a strict correspondence between the point of entry

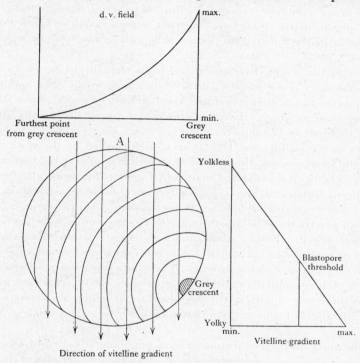

Fig. 109. Diagram to illustrate the double gradient of the amphibian egg.

of the spermatozoon and the bilateral symmetry in the frog. As in the case of
the regulation question (see p. 99) the prince of theoreticians was beggared by
the difficulties of experiment. Investigations with more modern technique, however (T. C. Tung[1]; Pasteels[10]), have shown that there is no causal relation between
the point of entry and the position of the grey crescent. In some cases, on the
contrary, where a moderate statistical agreement may exist, it is likely that the
plan of bilateral symmetry determines to some extent the point of entry of the
spermatozoon. A. Brachet[1], too, had shown that frog eggs parthenogenetically

---

* It is always assumed that this is due to greater accumulation of yolk-platelets at the vegetal pole,
but recent work of Daniel[3] raises the suspicion that the continuous phase may be more responsible
than the platelets.

stimulated by pricking with a needle showed bilateral symmetry entirely independent of the point of pricking. Meanwhile the same independence of sperm entry point had been conclusively demonstrated for the Urodeles by B. G. Smith; Vogt[1], and Bánki. It has even been shown by Ancel & Vintemberger[2] (who think that the direction of the incident light has something to do with the determination of the plane of bilateral symmetry) that the plane may be altered after the spermatozoon has penetrated far into the egg.

There has been much reluctance on the part of embryologists to give up the spermatozoon as a determining factor in symmetry, probably because no other causative factor could be visualised. But there is no reason for denying the maternal organism the power to construct a paracrystalline aggregate having quite complicated symmetry properties. The yolk gradient alone would be too easy a task.*

Another factor which has sometimes been thought to be connected with bilateral symmetry is the position of the first cleavage plane. The authors named above, however, and others, such as Ancel & Vintemberger[2], have shown that there is no relation here; the grey crescent is usually unequally distributed between the first two blastomeres.

The key to the problem, as Dalcq[8] says,† is furnished by the work which has followed the strange result obtained by O. Schultze in 1894, who discovered that a frog's egg, when compressed and maintained upside down just after fertilisation and during the early cleavage stages, undergoes abnormal gastrulation, and may often give rise to a twinned embryo. What happens is simple enough; under the influence of gravity the yolk falls through the egg and comes to lie in the previous animal pole region. Later on the anomalous gastrulation takes the form of Y-shaped or even double blastopores which invaginate to produce various types of monster, or else well-defined twins. When the experiments were repeated by Penners & Schleip and Penners[3] on a larger scale thirty years later, it became clear that there is *no* point on the egg's surface at which a blastopore may not form. But it always forms at the junction between yolky and yolkless-pigmented cells.

The meaning of these phenomena was not understood until Pasteels[11, 12, 13] carried out a thorough reinvestigation of Schultze's experiment. It was then found that after rearrangement of the egg by drastic changes in the yolk gradient, the blastopore always forms at the junction of yolky and pigmented cells but oriented towards the original site of the grey crescent, though this may have become invisible. Subsidiary blastopores may form later at the side of the redistributed yolk-mass.‡ Blastopore formation therefore depends on two factors, (1) the contact between animal micromeres and yolk-cells, (2) the grey crescent. And as the effect of the grey crescent appears, though

* In this connection it may be well to recall that in one group of animals at any rate, the insects, we have known, since the work of Hallez in 1886, that the egg's axial orientation is exactly the same as that of the maternal organism in the ovaries of which it was formed.

† In his *Form and Causality*, etc., pp. 79 ff.

‡ They need not face the grey crescent, though one will, but may face the opposite direction. The field determines their existence, not their orientation.

weaker and slower, at remoter parts of the egg's surface, it is convenient to think of the grey crescent as the centre of a field, rather than as the apex of a gradient.

These facts explained much that had previously been hard to understand. Confusion had primarily been caused by the simple-minded identification of grey crescent or corresponding region with the organisation centre. It was not realised that the organisation centre above the formed blastopore, with its liberated evocator, may be a very different thing from the grey crescent with its masked evocator. There is indeed a pre-determination of the dorsal zone, but not a pre-formation of the organiser. Thus Fankhauser[2] had shown that ligatures of the uncleaved newt egg would produce the same results as those of Ruud & Spemann in their classical constriction experiments on the gastrula (p. 103); and Pasteels himself[2], by surgical destruction of the grey crescent region of the frog egg (following the pioneer work of Moszkowski in 1903), found inhibitions like those produced by similar technique at the blastula stage by S. Suzuki. Hence the too hasty conclusion that the active organisation centre itself, not merely the dorsoventral gradient or field, was present in the earliest cleavage stages.

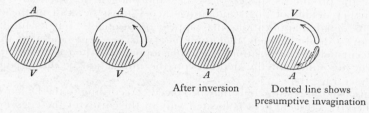

After inversion   Dotted line shows
                     presumptive invagination

Fig. 110. Schultze's experiment in its extremest form (see text).

The importance of the junction between yolky and yolkless areas was likewise demonstrated. Votquenne first showed that the micromeres of the frog's egg (stage VIII) form later all the presumptive mesoderm and hence the organisation centre. Yet when Vintemberger[2] isolated them, no morphogenesis was produced. Only when he combined them[3] with a mass of yolk-cells from another embryo were axial organs and structures later formed. Presence of yolk-cells was therefore shown to be necessary for the formation of the organisation centre and the liberation of the evocator.

Schultze's experiment was also performed on an oriental newt, *Hynobius lichenatus*, by Motomura[2]. In this case the development is very slow, giving time for all the yolk to reach the former animal pole when the egg is inverted, before cell-divisions prevent its redistribution. When the blastopore forms it forms nearly in its normal place, but exactly inverted, and invagination goes on in mirror image of normal invagination (see Fig. 110). This is the extreme case in which normality is returned to after passage through a series of abnormal intermediate types. In most amphibia, the yolk does not fall evenly through the egg, but leaves streaks and irregular masses behind. Some will remain fixed to the cortex at the former vegetal pole, and hence two blastopores will form in the grey crescent field instead of one.

These facts have been made the basis of a general theory of early embryonic fields by Dalcq & Pasteels and by Dalcq[7] in his monograph. These authors apply it with much boldness to the development of all vertebrates and even of some invertebrates. Whether or no their wider conclusions stand the test of time (for certain criticisms, see Hörstadius[10]), it is sure that they are laying the foundations of comparative experimental, as opposed to morphological, embryology. Their view is briefly as follows. The animal-vegetal gradient (yolkless-yolky) they regard as pervading the egg's substance; the dorso-ventral gradient they regard as localised in the egg's cortex. But the dorso-ventral axis is thus a surface field rather than a gradient, and Dalcq & Pasteels ingeniously derive many of the phenomena of later development, including even the dynamically determined Vogtian areas, from the topography of the co-ordinate system so formed. They picture the existence of some chemical reaction between the cortex and the yolk (factors C and V), the product of which would be distributed in a regular way as a series of thresholds to which the morphogenetic processes would be sensitive or competent. Thus they even describe the metamerism of the somites in terms of the numerical values of the co-ordinates of their system. The blastopore will form wherever the product of the reaction reaches a certain threshold. The amount of the product is spoken of as the "morphogenetic potential".

The further details of this stimulating theory must be read in the original. Some of its detailed applications may appear questionable, but its fundamental basis is not in doubt, namely that there is a yolk gradient, and that there is a grey-crescent field. The position of the blastopore, and hence of the liberation of the evocator, undoubtedly depends on the interaction of this gradient and this field (see Fig. 109).

One criticism interesting to us may be voiced. According to the Dalcq-Pasteels theory, the grey-crescent field is located only in the cortex of the egg (*Form and Causality*, pp. 83, 85, 146). At first sight the only evidence for this view seems to be that the field is unaffected by maintained inversion and by centrifuging. And, as is well known (see p. 660), the echinoderm egg has symmetry and polarity properties which are unaffected by the entire redistribution of the movable "ballast" and which have been thought to be referable to a paracrystalline condition of the cytoplasm, not of the cortex only. A liquid crystal re-forms again with its molecules in characteristic orientation after the passage of a foreign body. It may be, therefore, that the grey-crescent field runs also through the substance of the egg as well as in its cortex. Against this, however, it may be argued rather convincingly that in that case phenomena ought sometimes to be met with in which an invagination would begin, not from the surface of the egg, but from some interior point at the junction of yolky and yolkless cells. But this never happens; the formation of the blastopore is essentially a cortex phenomenon. Furthermore, in Dalcq's view[7], once the field of "morphogenetic potentials" is fully established, there is no doubt that it permeates the internal cytoplasm, for different regions of it have different inclinations and different inductor capacities.

Another criticism, due to Holtfreter,* is that if the presumptive organiser cytoplasm had a specific gravity intermediate between that of the yolk and that of animal pole cytoplasm, it would always tend to take up a position between them, and there would be no need to postulate a chemical reaction between the yolk itself and other constituents.

* In conversation. This is reminiscent of the Hippocratic writer's theory of development (*CE*, p. 57).

The question may be asked whether the vitelline or animal-vegetal gradient has any physico-chemical properties analogous to those which, as we saw in Section 2·2, are possessed by the severer localisations seen in mosaic eggs. Owing to the pigmentation of most amphibian eggs, vital staining methods have not been available, but Dorfman[3] has applied electrometric technique. According to him, the vegetal pole is negative to the animal pole in the eggs of *Rana temporaria* and *arvalis*, changing sign at fertilisation, but as the eggs of *esculenta* showed the opposite potential difference, and a correspondingly opposite change of sign, our doubts are aroused. Later papers (Dorfman[4]; Dorfman & Grodsensky) give the following figures for polar material withdrawn into capillary tubes for measurement:

	Vegetal pole	Animal pole
pH	5·64	6·18
E_h in mv. (aerobic)	+262	+216

Animal pole material would thus be more alkaline and have a more negative reduction potential (the E_h figures agree with an average for the whole egg given by J. Brachet[6]). This is perhaps the kind of work which cannot be appraised without a personal acquaintance with the technique used.

Disorganisation of the embryo by centrifuging has been carried out by many experimenters over many years, and even some chemical analysis of the redistributed materials has been done (*CE*, p. 347). We must, however, pass over the earlier work, and only mention results which bear on the present problems. Just as the yolk, but not the grey-crescent field, is displaced by maintained inversion, so centrifuging in the cleavage stages leads to double or even multiple blastopores, and later to all kinds of monsters (Beams, King & Risley on *Rana pipiens*; Ch'ou[2] on Chinese species; Konopacki[2] on *Rana fusca*; Motomura[2] on *Bufo formosus*). A case of double invagination without induction was observed by Beams, King & Risley, there being two notochords but no neural tube. Similar duplications, particularly accessory tails, were seen on centrifuging advanced blastulae of *Hyla regilla* by Schechtman[5], and young gastrulae of *Rana fusca* by Pasquini & Reverberi, but in these later stages, after the organiser region has been formed, an abnormal contact between inducing mesoderm and competent ectoderm will explain matters without the appearance of subsidiary organiser centres.

The best analysed centrifuge experiments are those of Pasteels[11,14]. The strongest centrifugation which the young morula can stand without cleaving abnormally gives three types of result: (1) inertia of the yolk-cells leading to spina bifida or exo-gastrulation, (2) partial or total inhibition of invagination, leading to deficiencies in the head organiser, (3) invagination without induction. Only the second and third type need be mentioned. The invagination paralysis of the second type was shown by Pasteels not to be due to removal of glycogen from the dorsal lip area by the centrifuging, and its causation is obscure. The invagination without induction of the third type is of the greatest interest, and we shall return to it in the next section (cf. Fig. 116).

Dislocation of the normal relations of yolk gradient and grey crescent field may be brought about by other means than the redistribution of the yolk. Witschi[1], in his studies on the development of over-ripe amphibian eggs, to which we shall have occasion to refer again (p. 255), found that in some cases, multiple blastopores were formed, giving rise to twins and various types of monsters. Since the yolk had not been redistributed, this effect would appear to be one directly on the grey crescent field, possibly a decay of the original dorso-ventral organisation of the egg; a reversion to randomness of molecules previously oriented in space.

2·37. Anomalies of organiser function
2·371. Organiser inhibition

In the foregoing section the normal process, where liberation of the evocator accompanies blastopore formation, has alone been considered. But if the former could be inhibited without the latter, invagination would take place but no induction of neural structures in the overlying ectoderm. Such an effect has been several times described.*

The earlier localised irradiations of amphibian embryos with ultra-violet light[†] (W. M. Baldwin; Dürken[3]) produced little of interest,[‡] but in later work Dürken[4] succeeded in obtaining a few clear cases of invagination without any sign of induction after localised treatment of the dorsal lip of the blastopore with ultra-violet light. In his discussion he envisaged only the actual destruction of the evocator substance and thought it strange that a substance so stable to many chemical treatments should be sensitive to ultra-violet rays. In view of the classical formation of calciferol (vitamin D) by irradiation of ergosterol, this is perhaps not so remarkable. But even more probably the radiation may be supposed to inhibit the liberation (activation) processes. The effect of ultra-violet rays on carbohydrate catabolism appears to be complex, for while Stiven found increased lactic acid production in muscle extracts, a destructive effect upon amylase was observed by Kumanomido and by Nadson & Stern. Perhaps this may be resolved by findings such as that of Hutchinson & Ashton that shorter wave-lengths of ultra-violet destroy amylase, while longer wave-lengths accelerate its action. A formal analogy to inhibition of evocator liberation might be derived from the work of v. d. Laan on auxin, the formation of which from its precursor is inhibited by ethylene.

Reith[4] subsequently tested the inductive capacity of dorsal lip pieces which had been irradiated with ultra-violet *in situ*, and found, probably because the range between complete inhibition of the liberating processes and necrosis or

---

* Of course in every *Bauchstück* (see p. 103) mesoderm gets inside but as it comes from the ventral lip it does not induce.

† Other references to the effects of various forms of radiation on embryos will be found later in the present book, e.g. in connection with inherited abnormalities (pp. 368 ff.). The mere establishment of variations in susceptibility during development received some treatment in *CE* (Section 18), and will not be taken up again here because its interpretation remains quite enigmatic.

‡ Except in regard to the origin of *spina bifida*, which, as Baronovsky & Schechtman have now shown, can be produced at will by localised irradiation of the lateral blastopore lips in anurans.

death is very narrow, that in all his cases the inducing power was present.* Behaviour in the host, however, showed certain peculiarities which will be mentioned again later in connection with regional differentiation (p. 277). Similar work is being continued by Brandes.

In the preceding section it was mentioned that Pasteels[11], by carefully dosed centrifuging during the cleavage stages, also obtained invagination without induction, no trace of notochord or neural plate being visible. The only indication of the dorsal surface of these embryos was the excentric position of the endodermal gut. Now in the work of Brachet[18], which was proceeding at the same time, it was found that the distribution of fixed –SH groups was quite different in the centrifuged eggs from the normal. The protein carrying them passed, unlike the yolk-platelets and the pigment, to the animal pole, together with the fat and aqueous cytoplasm. Hence in the dorsal lip of embryos invaginating without induction, none of the protein was to be found. This seems to be almost complete proof that the protein plays a part in evocator liberation, and might indeed itself be the protein of the masking complex.

Fig. 111. Cross-section through an anidian blastoderm (chick).

The primary organiser may also be suppressed in the sauropsid egg (see p. 330). It has long been known that in the sauropsida, there are occasionally found blastoderms where an active proliferation of cells goes on but no trace of axiation appears. Primitive streak and neural groove never form. These blastoderms are called *Anidians*, and have recently been described anew by Grodziński[3]. It is extremely likely that in these cases there has been a failure either of the formation, or more probably the liberation, of the primary evocator. The natural cause of this is unknown, but X-radiation will produce them (Ancel & Vintemberger[1]), and Edwards long ago was able to obtain anidians regularly by incubating hen's eggs at subnormal temperatures. This is the usual way of obtaining them. The work of Tur[2], to be referred to again in connection with cancer (p. 258), on the extraordinary ectodermal thickenings of some early lizard and bird blastoderms, is also interesting in this connection, for these never develop an embryonic axis (Fig. 111). Anidians may develop an area vasculosa.

* In the *Annual Rev. Physiol.* 1938 I suggested that this result implied that Dürken's effect had been on the competence of the presumptive neural region rather than on the evocator. But as his irradiations were localised to the dorsal lip of the young gastrula, i.e. to presumptive mesoderm, this explanation can hardly hold good.

Perhaps the most striking instances of the effect of a definite external influence in suppressing organiser action have come from the line of work pursued by F. E. Lehmann. Lehmann discovered a number of specific inhibitions of morphogenesis produced by artificial agents which leave all the surrounding tissues unaffected. The theoretical basis for such work lay in the conviction that not all the tissues or organs of the body pass through their points of maximum growth

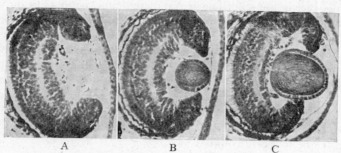

Fig. 112. Suppression of lens-induction by trichlorbutyl alcohol.
A. 1 in 900. B. 1 in 1200. C. Normal.

rate or the most difficult phases of their development at one and the same time. There exist "critical points" for each structure (Stockard; and *CE*, p. 1409). Thus Lehmann[5] found that trichlorbutyl alcohol (the same narcotic which, as we saw (p. 165), has no influence on the primary evocator) can abolish specifically the second-grade evocation of the lens by the eye-cup in amphibia (see pp. 291 ff.). Fig. 112 shows three of the results he obtained; on the left an embryo maintained in a strong concentration of the narcotic, in the middle an embryo maintained in a weaker concentration of it; and on the right a control embryo. It will be seen that the lens is entirely suppressed on the left, abnormally small in the middle, and normal on the right. Between the size of the lens induced and the concentration of the narcotic there was a far-reaching parallelism. Fig. 113 shows the relation between lens size and concentration of narcotic when the period of application of the narcotic corresponded to the neurulation period. At other periods the effects were less marked and less regularly

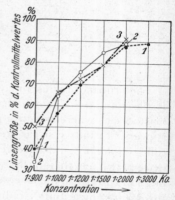

Fig. 113. Relation between concentration of narcotic and extent of lens-suppression.

dependent on concentration. After the appearance of the lens-forming ectodermal thickening the process of lens-formation could no longer be inhibited. Lehmann[9] went on to study the effect of another variable, temperature. He found that the inhibitory effect was more pronounced the lower the temperature (between 10 and 22°).

As regards the mechanism of this inhibition no definite decision is as yet possible, but Lehmann[8] regarded it as a diminution of ectodermal competence rather than an action on the inducing substance or its precursor. He recalled that according to Manuilova & Kislov the frequency of successful lens inductions by eye-cups in belly epidermis is not so high as in the normal lens area, suggesting a lower level of competence in the unusual situation.

These facts raise the question of the naturally-occurring blindness of cave-dwelling animals. There can be little doubt that it involves a series of effects due to interference with the inductor chain brain→eye-cup→lens→cornea (see p. 290) at various different points. Thus Schlampp found that in *Proteus anguineus* a thick wad of connective-tissue mesoderm intervenes between the eye-cup and the epidermis, inhibiting cornea induction. In other forms, such as the blind cave-fish *Anoptichthys jordani*, there is a spontaneous failure of lens-induction (Gresser & Breder). From the work of a considerable number of investigators (e.g. Eigenmann; Eigenmann & Denny, etc.) the blind forms can be seen to present a graded series ranging from (1) normal development followed by secondary degeneration of the retina, through (2) considerable size-reduction of the eye accompanied by overgrowth of the eyelids,* and (3) partial or drastic failure of induction of eye-cup, lens, or cornea, to (4) complete absence of the lens or cornea and failure of eye-cup differentiation. Such a series of fishes would be *Lucifuga-Chologaster-Amblyopsis-Troglichthys-Typhlichthys-Anoptich-thys*. A similar series among blind amphibia would run: *Spelerpes-Typhlotriton-Typhlomolge*, and lizards such as *Amphisbaena* and *Rhineura* lack lens and cornea. Interference with the visual inductor chain occurs also in abnormalities due to lethal or semi-lethal genes, to which we shall return in Section 2·65, and in nuclear abnormalities such as androgenesis (Porter; see Section 2·64). It must be remembered that among the inverte-brates also there are many blind cave-forms, the literature on which may be approached by way of Wolsky's interest-ing paper[5] on the blind gammarid *Niphargus*.

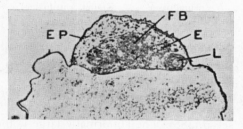

Results of interest in this connection are those reported by Solberg[2], who irradiated eggs of the teleostean fish *Oryzias* with X-rays. A suitable dose will almost completely suppress the development of the eye-cup, but apparently not before it has exerted

Fig. 114. Suppression of eye-cup and brain development by irradiation of teleostean em-bryo with X-rays, without suppression of lens-induction. *EP*, Epidermis; *FB*, fore-brain; *E*, eye debris; *L*, lens.

sufficient lens-inducing stimulus on the ectoderm. Solitary lenses thus appear, reminding us of those found in Holtfreter's experiments (see p. 170): Fig. 114. The suppression of the eye-cup is associated with a spherulation of the cells, and this is also the case when the neural tube is caused to degenerate by X-radiation. Suitable doses would also suppress the appearance of gills and mouth, but not the ear-vesicle.

* In preliminary experiments, Twitty[7] has been able to produce this overgrowth of the eyelids by immobilising the eye-ball or substituting one of a smaller species.

Another remarkable finding of Lehmann's[6] was that by the action of suitable concentrations of lithium* and potassium chloride, embryos could be produced completely lacking notochords. Fig. 115 A (opp. p. 234) shows one of these. Above is the normal cross-section, showing the somites on each side of the notochord and the neural tube above it; below is one of the notochordless embryos. Fig. 115 B illustrates a later stage of development. The surrounding morphological and histological differentiation looks remarkably normal. Here there has evidently been no interference with the primary evocator, but the differentiation of the invaginated mesoderm has been profoundly altered. In the ultra-violet irradiation experiments the formation of the notochord was eliminated as well as the primary induction of the neural axis.

The absence of notochord was found by Lehmann[13], however, to lead in many cases to quite abnormally formed neural tubes (see later, under *Nachbarschaft*, pp. 284). Frequently the spinal ganglia developed normally. More surprisingly, Lehmann & Ris observed a much delayed but fairly normal development of cartilaginous axial skeleton in the regions where the notochord was absent.

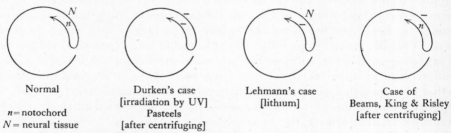

| Normal | Durken's case [irradiation by UV] Pasteels [after centrifuging] | Lehmann's case [lithium] | Case of Beams, King & Risley [after centrifuging] |

n = notochord
N = neural tissue

Fig. 116. Cases of intereference with primary organiser function in amphibia.

Full investigation of the missing notochord region by Lehmann[13] with vital dyes showed that the presumptive notochord material had been converted wholly into somite material.†

The converse of Lehmann's notochord suppression was obtained by Ranzi & Tamini using sodium thiocyanate (an "animalising" agent in the echinoderm, see p. 486). The abnormally large notochords were explained by these authors as due to inhibition of the protein metabolism of regions surrounding the dorsal lip, while leaving carbohydrate metabolism in the lip itself unimpaired.

Lehmann has interpreted the general significance of his findings in several interesting reviews[8,10,11]. The logical relationship of the cases of interference with primary organiser function in amphibia is shown in Fig. 116.

* The strange morphological effects of Li′ will often be referred to in succeeding sections (e.g. pp. 277, 283 and 486). The ion also induces manifold abnormalities of shape in bacteria (Enderlein; Kühn & Sternberg; P. Hadley).

† The legitimacy of speaking of this process as "mesodermisation", as Lehmann[13] does, is not quite clear, since both notochord and somites are generally regarded as mesodermal. He was anxious to approximate the effects of lithium on amphibian to those on echinoderm development (see pp. 477 ff.), but it is perhaps as yet too early to attempt this until we know more about the metabolism of both these groups in the gastrulation period. Cohen, if I understand him aright, thinks such missing notochords may be "endodermised" and lost in the gut roof.

Anatomical and genetical literature is indeed full of cases in which the stimulus for the development of a certain structure has failed. For example, the paper of Dankmeijer describes the congenital complete absence of the tibia in man. Not only may the inductive stimulus of eye-cup on ectoderm fail; there may be a failure of those as yet unknown factors in regional differentiation which bring about the formation of the eye-cups themselves. Such failures were discussed by v. Baer[6] already in 1862, who saw their significance for cave-blindness and inherited anomalies.

Among external causes of this, vitamin A may be mentioned. Vitamin A is a lipochrome allied to carotin, astacin, and the retinene of the eye; and its structure is that of a long carbon chain, containing four unsaturated linkages, and ending in a hexatomic ring with one double bond (see p. 249). Information is now beginning to come through that this vitamin may be connected in some way or other with the formation of organs in embryonic development. Thus it seems that in the absence of sufficient vitamin A from the diet of some mammals during gestation, there may be a complete failure to form the eye-cup, and hence no induction

Fig. 117. Suppression of eye-cup formation in the pig in vitamin A deficiency.

of lens from the adjacent ectoderm. In 1932 at the Texas State Agricultural Experiment Station a Duroc-Jersey gilt receiving a vitamin-A deficient ration farrowed eleven pigs, all of which were completely devoid of eye-balls. Hale, who reported the case, calculated that the chance of such a result being due to inheritance was exceedingly small, and he was subsequently able to repeat the result with other gilts, and sows, by placing them on a ration containing insufficient amounts of the vitamin. Fig. 117 shows one of the pigs born with complete absence of the eye-cup.* Such pigs as had eye-cups suffered from defects of vision, including total blindness. In view of the fact that vitamin A is closely related chemically to the lipochrome constituent of visual purple, the experiments have additional interest.†

Lack of vitamin A has also been shown to be responsible for a disturbance in the mechanism of normal limb-bud formation. In 1921 Zilva, Golding,

---

* "Eyeless" rabbits are sometimes ascribed to Guyer & Smith, but their work only involved certain apparently inherited defects such as opacity of the lens. In *Drosophila*, of course, there is a well-known recessive gene for "eyeless" (see p. 376).

† Wolbach, in an interesting lecture, has emphasised the relation between vitamin effects and embryological problems.

Drummond & Coward found that on an A-deficient diet some of their sows produced little pigs in which the limbs were completely missing. Fig. 118 shows how the hind-limbs are represented by nothing but minute stringy appendages.

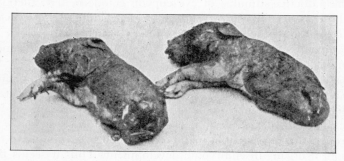

Fig. 118. Suppression of limb-bud development in the pig in vitamin A deficiency.

2·375. Organiser excess; Twinning

If the grey crescent field of an amphibian embryo, or the individuation field of a chick, for example, were to split into two, the primary evocator would be liberated in active form at two places instead of one, and a twin would result. Such phenomena must lie at the basis of many kinds of twinning and poly-embryony. The literature on this subject is a large one; v. Baer[4] worked on it over a century ago, and in recent times we have the monographs of Newman[1] (though written before 1924) and the review of Patterson.

In the ordinary way, one egg gives rise to one embryo. But in polyembryony, one egg may give rise to a number of embryos, even up to two thousand, if we take the case of some parasitic insects. This must ultimately be related to the fact that, as we have already seen, the destinies of the parts of the embryo are not fixed at the beginning, but only become so as development proceeds, owing to the ordered action of a series of determining agents, or organisers. While identical twins or quadruplets are a relatively rare occurrence in most species, polyembryony of a much higher order has established itself as a specific character in a few. This may involve either the retention of pluripotence through many cleavages, followed by separation of blastomeres, or the scission of the individuation field at a later stage followed by the appearance of many organisation centres instead of one. It seems that the exact time of splitting varies very greatly in different organisms. Thus in the earthworms, such as *Lumbricus trapezoides*, there takes place a curious kind of double gastrulation so that the two gastrulae have their blastopores facing, and presently break apart (Kleinenberg[1]). In the parasitic hymenoptera (*Platygaster, Litomastix*, etc.), the processes are very odd and clearly highly specialised, but may roughly be described as the separation of many early blastomeres, followed by isolated differentiation (Leiby). The case of the Texan nine-banded armadillo (*Dasypus novemcinctus*), which normally produces four, and occasionally five, identical-twin embryos from a single egg, is the closest to our present discussion, for here the blastocyst, which is similar in every respect to that of other mammals, regularly develops four embryonic axes instead of one. The deduction that a highly controlled liberation of the primary evocator at four separate points instead of one takes place is too plausible to be doubted.* In Fig. 119 is shown a diagram of the armadillo blastocyst illustrating the description of Newman & Patterson.

In human monozygotic (identical) twins, a common chorion and sometimes even a common amnion is found. This is regarded by Arey as proof that the twinning process occurred subsequently to the differentiation of the ovum into an inner cell-mass and an outer trophectoderm, as it would do if there was a reduplication of the primary organiser. The only flaw in this argument is that, according to Greulich, chorionic walls may fuse over the area where they meet. The presence of only one chorion at birth, therefore, does not absolutely disprove the view that a separation of blastomeres occurred.

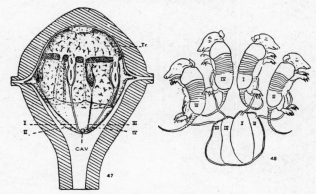

Fig. 119. Normal formation of identical quadruplets in the armadillo.

What happens normally in the armadillo happens occasionally in other mammals and in birds. Assheton[1] figures two entirely separate germinal areas in a blasto-dermic vesicle of the sheep, and the later results of lesser degrees of separation can be seen in the review of Keller & Niedoba. O'Donoghue photographed a chick blastoderm in which two quite separate embryonic areas had developed, lying head to head, while Tannreuther found two distinct embryos only united at the extreme posterior end of the primitive streak (Fig. 120). The parallel between such a picture and that of a blastoderm in which a secondary embryo has been *induced* artificially by the transplantation of an organiser is striking (see Fig. 121, opp. p. 234, from Waddington & Schmidt). The latest experimental study of twinning in bird blastoderms by injury is that of Twiesselmann, who gives the previous literature.†

* This would be called in the old terminology "fission", not "blastotomy" or "budding". Newman[1] spoke of the "isolation of four growing-points". That this comes about through temporary developmental arrest in accordance with theories of earlier American authors is unlikely, as Hamlett[1] has shown that the quiescent period happens at the wrong time. The general views expressed in this section have since been advocated also by Dalcq[13].

† The interesting reviews of Hildebrand and of Strohl[2] give details of natural twinning in turtles and snakes; Tung & Tung[2] describe it in anurans.

It is interesting in connection with evidence to be given later regarding genetical factors, that this division of the individuation field into two or more fields may be under the control of inheritable factors, since according to Davenport[1], the production of human identical twins appears with great frequency in certain families, and may therefore be connected with a specific gene or gene-group. Against this there is the conclusion of Greulich's recent review, that only dizygotic (fraternal) twinning in man is hereditary. But for birds it has been shown by Byerly & Olsen[2] that the incidence of twinning varies according to the breed, a fact which will be difficult to explain on other than genetic lines.

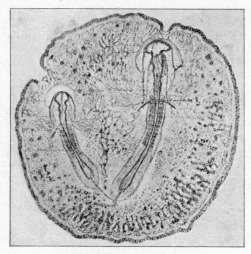

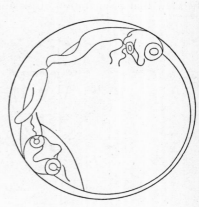

Fig. 120.   Fig. 122.

Fig. 120. Naturally occurring complete twinning in the chick, probably due to separate endo-derm invagination.

Fig. 122. Twinning in the minnow produced by ultra-violet irradiation.

The study of developmental abnormalities also indicates that the extreme type of twinning may be brought about by various influences. Without pursuing the large literature on this subject, we may refer to the paper of Harman on the appearance of double notochords running parallel in chick embryos after incuba-tion at abnormally high temperatures in the early stages. Fish embryos seem to be particularly prone to scission of the individuation field and production of two organiser centres instead of one. Stockard, in his well-known studies on critical points in development (see *CE*, p. 1409) produced all varieties of twins in trout, minnows, etc. by the action of magnesium chloride, insufficient oxygenation and other means.* Ultra-violet irradiation, too, will produce twins completely separated, and hence not to be ascribed to interference with gastrulation move-ments (Hinrichs and Hinrichs & Genther on *Fundulus heteroclitus*); see Fig. 122. Finally, Enzmann & Haskins[3] find that neutron bombardment produces all kinds of reduplications of organs in *Drosophila*, quite unlike the effects of X-rays.

* Cf. the new analysis of fish twins and polyembryos by Lynn.

2·376. Organiser excess and anomalous competence; Teratomata

We now come to the Teratomata. We must first of all make a distinction which is sometimes overlooked in the literature. The word τέρατα in Greek means monsters, in the sense of signs and wonders, like *prodigium* or *portentum* in Latin. It is a commonplace of embryology that terata may be produced by every conceivable kind of deviation from normal development, and their causation is certainly very unspecific. On such deviations the monograph of Dareste, though sixty years old, is still classical. But terata are not teratomata. A teratoma may be defined as a more or less malignant assembly of tissues, well differentiated histologically, but showing varying degrees of morphological differentiation, and embedded in the body of an otherwise normal fully developed organism. The really typical character of such an assembly is its approximation to chaos; as if the tissues and morphological forms had arisen from the interplay of almost entirely unregulated forces—resulting in the state described by Schiller:

> "wo rohe Kräfte sinnlos walten,
>   da kann sich kein Gebild gestalten".
>
> *(Lied von der Glocke)*

The literature on the subject of teratomata is extremely large. The writings of pathologists on teratomata, moreover, seem to have been marked by an unusual degree of unscientific speculation, inaccurate description and logical mistakes, as may be seen from the long and remarkable series of critical reviews by Nicholson[1,2]. A great deal of this, however, may perhaps be put down to the fact that for the last fifty years, at least, pathologists have approached the study of teratomata wearing, if we may say so, the spectacles of an unjustifiable hypothesis about the nature of these tumours. This hypothesis is expressed in the words of Bland-Sutton[2], who, defining teratomata, says: "A teratoma is an irregular conglomerate mass containing the tissues and fragments of viscera belonging to a suppressed foetus attached to an otherwise normal individual." As the result of this basic misconception, chaotic and truly innominate structures have been given the names of nipples, limbs, vulvae, jaws, and all manner of dignities to which they really have no claim.* And this is all the more unfortunate as in most cases the brief anatomical descriptions have not been accompanied with that thorough histological analysis which the study of serial sections alone can give.

Before going further, it would be as well to make the discussion more concrete by giving a few examples of the kind of phenomenon we are considering. First of all, as is generally known, the morphological forms attained by teratomata may be very complete. Fig. 123 shows an ovarian dermoid teratoma. According to Bland-Sutton's description[3], it was removed from a single woman twenty-six years of age. It was the size of a coco-nut and contained a mixture of grease and hair. At the lower pole of the tumour there was a patch of reddish epidermis covered with reddish hair, and out of which appeared two teeth, an incisor and a bicuspid, fully erupted. Underneath this skin was a mass of nerve tissue. The large size of such cysts is supposed to be due to the pressure of the sweat and

---

* The homologies of Wilms, who wrote in 1895, resemble nothing so much as the unfortunate error of William Croone (1672), whose identification of a torn piece of vitelline membrane with the preformed chick embryo has recently been unkindly made public again by F. J. Cole.

grease secreted over a long period of time,* and the growth of the hair. The presence of well-formed teeth in teratomata is not uncommon; Fig. 124 shows an X-ray photograph of an ovarian one, described by Sprawson.

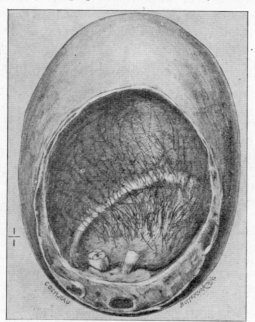

Fig. 123. Ovarian dermoid teratoma (human).

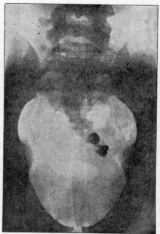

Fig. 124. Well-formed teeth in an ovarian teratoma (X-ray photograph).

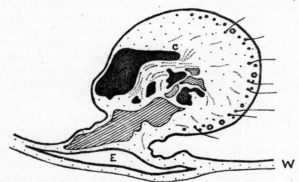

Fig. 125. Tissues in an eminence within an ovarian dermoid teratoma. Black = bone; shaded = neural tissue; stippled = ganglia; dotted lines = connective tissue; C, cap of cartilage; E, cleft in W, cyst-wall.

The same chaotic mixture of tissues occurs in a case described by Willis[2], who gives the instructive cross-section shown in Fig. 125. This is a section of an

* The fatty material in ovarian dermoid cysts may be of a very odd composition; Behmel finds large quantities of cholesterol esters in it but no vitamin A or sex-hormones, while Dimter isolated 20 % of squalene, or some very similar hydrocarbon, from it.

eminence which projected into an ovarian dermoid cyst about the same size as that just described. The tumour originated from a single woman aged thirty-eight. It can be seen that the eminence contained bone, central nervous tissue, ganglia, nerves, and connective tissue. At one point the bone had a cap of cartilage. Another of Willis' cases had masses of cerebellar tissue with beautifully differentiated Purkinje cells. But in these pictures, as in all teratomata, the general character is the same, namely a complete absence of organisation of all these tissues into a whole, or we might say, an absence of individuation.

The following case provides an opportunity for the transition to experimental embryology. Barnard described a multiple teratoma of the peritoneum in a woman of forty, probably originating from an ovarian dermoid cyst. The growths, irregular, nodular, and partly cystic, occupied most of the internal surface of the peritoneum, and when sectioned revealed the usual chaotic formation of tissues. There were undifferentiated cells, spindle cells arranged in groups and bands, islands of cartilage or trabeculae of bone, squamous and columnar mucus-secreting epithelium, and well-developed brain tissue. In one of the masses on the omentum a perfectly formed tooth was found. In Fig. 126 is shown a typical collocation of tissues; in one and the same section can be seen brain tissue, nerve ganglion, cartilage, and squamous epithelium.

Such a picture at once calls to mind the kind of chaotic distribution of tissues which is seen when for some reason the individuation field in a young embryo is thrown out of gear. As we have seen, the normal induction of the primary axis in an embryo by a living piece of organisation centre involves both evocation, the stimulus of a specific hormone-like chemical substance, and individuation, the regional differentiation of the axis so called into being. In all probability a dead piece of organisation centre carries the evocator but not the individuation field (see below, p. 277). It might therefore be expected that the implantation of a dead organiser would call forth the appearance of more chaotically arranged structures than that of a living one, always provided that the individuation field of the host was not sufficiently strong to control and order the newly appearing differentiations. Let us therefore look at Fig. 127, which is taken from Holtfreter's work[12] on the distribution of the evocator in the animal kingdom (see p. 172). Embryo extract was prepared from chick embryos as is usual in tissue culture work, coagulated by heat, and then implanted into the blastocoele cavity of a newt embryo. Here in due course two neural tubes were induced, three separate lots of notochord tissue, several somites, and at least two ear-vesicles. Or consider Fig. 128, in which is shown the disorderly but impressive array of no less than five ear-vesicles induced in the ventral ectoderm of a newt embryo by the implantation of a small piece of heat-coagulated mouse liver. Surrounded by these lies the great irregular induced brain (Holtfreter[7]).

It will surely be admitted that these parallels are too striking to be illusory. They lead us directly to the basic assumption that what lies behind all the strange phenomena which we see in the teratomata is the failure of the individuation field, at some point early in development, to control the action of evocating substances.

It is interesting that there has long been an intuition that teratomata would not be explained until our knowledge of experimental embryology was much further advanced. Thus Benda in 1895 attempted something in this direction, but at that time so many concepts subsequently found to be misleading (such as post-generation and universal mosaic development) were in the field, that little could be done. The notion of "formative stimuli" had, it is true, been advanced, but nothing, or nearly nothing, was known about them (see pp. 103 ff.). It seems that the credit of first realising the application of our knowledge of organiser phenomena to teratomata should rest with Budde, who only two years after the critical experiment of Spemann & H. Mangold wrote a paper along these lines in describing a case of teratoma. What he called an "abgesprengter Organisator" might, he thought, have been at work. Later on Nicholson, Langdon-Brown, and doubtless others, have realised that organiser phenomena provide the only basis for a modern theory of the origin of teratomata. "What I miss", wrote Nicholson, "in the established teratoma is the co-ordinating action of the whole upon its parts; in more scientific language, evidence for the action at any stage of development of the dominant organiser for a body."

If, then, we can only explain the occurrence of teratomata as the result of un-coordinated organiser action, it is none the less necessary to point out that besides the stimulus there is also the reacting tissue. In addition to the evocator, we must also take the competence into account. Obviously we have to consider two fundamentally important factors: firstly, the liberation of evocators of first, second, third, or lower grades at times and places where they are not normally liberated; secondly, the persistence of competence to respond to them beyond the time at which it is normally lost. We know that the primary evocator is present in most adult tissues of animals in the free state. It is therefore likely that the second of these factors may be more important for the formation of teratomata than the first. But it is evident that from them both, by all sorts of combinations and permutations, everything that we know of under the name of the teratomata may be produced.

There remain for consideration the reasons for which in some cases a teratoma manifests malignancy. As Willis[1] and Nicholson[2] point out, the majority of teratomata are benign rather than malignant. Ovarian ones are generally benign, while those of the testis are almost always frankly malignant, "seldom failing to yield metastases and to kill their hosts". It is noted that when a benign ovarian tumour becomes malignant, the change affects only one component, such as the epidermis, so that a squamous-celled carcinoma quite similar to any epidermal carcinoma is produced. In the testis, on the other hand, the malignancy of the teratoma is a property of the whole growth, so that metastases frequently exhibit two or more components of the primary growth. Occasionally, as appears from one of Willis' cases, one component of an otherwise relatively quiescent teratoma of the testis "assumes disproportionate powers of proliferation" and invades the remainder as a malignant neoplasm. There can be no doubt that the terms "benign" and "malignant" are but relative, and that there may be grades of malignancy depending ultimately on the inherent growth energy of the various components; thus in another of Willis' cases, the neuro-epithelial tissue appeared to be growing more rapidly than the glandular portion, while that in turn was outstripping the cartilaginous material.

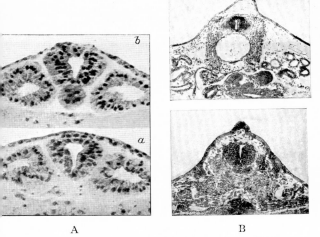

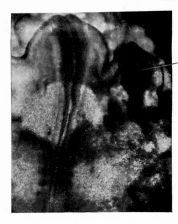

XX

A   B

Fig. 115. Suppression of notochord formation by lithium.
A. Early stage. B. Late stage.

Fig. 121. Twinning in the chick produced
by implantation of organiser.

*induced
embryo*

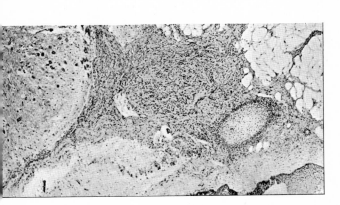

Fig. 126. Tissues in a multiple teratoma of the peritoneum.

impl.   *neural
tissue*   *ear-
vesicle*

somite muscle   *notochord tissue*

Fig. 127. Chaotic assemblage of tissues
induced in an amphibian embryo by
implantation of chick embryo extract
into the blastocoele cavity.

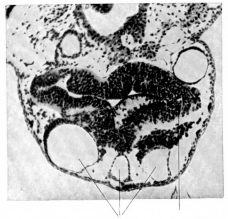

induced ear-vesicles   *induced brain*

Fig. 128. Numerous ear-vesicles around a brain induced by boiled mouse
liver implanted into the blastocoele cavity of an amphibian embryo.

In all this we can dimly see the highly complicated operation of many stimulatory substances. Evidence has been adduced indicating that the primary organiser of axial structures in the embryo may be of steroid character; it is certain that primary induction can be performed by synthetic hydrocarbons allied to the sterols and that similar hydrocarbons are powerfully carcinogenic. When we remember that it is likely that carcinogenic substances are produced in the body by deviations in sterol metabolism, we are prepared to picture that the wild growth of malignant tissues, as well as the highly organised differentiation of structures in early embryonic development, depend on sterol metabolism for their controlling "hormones". In addition, we know that there is a relation between the carcinogenic substances and the sex-hormones, so that we can set side by side with this the relatively high frequency of teratomata in the mammalian ovary and testis. The day may come when from a knowledge of the transformations of the cyclo-penteno-phenanthrene ring in a given instance, together with an understanding of what is really meant by the term "competence", we should be able to predict just what tissues and morphological forms should arise in a teratoma, what parts of it should be malignant, and how malignant they should be.

The denial that the teratoma is a distorted foetus is important. Budde; Nicholson; and Willis have all pointed out that no teratoma has ever been found to show signs of axiation or metameric segmentation.* Chordal tissue has never been proved to exist in a teratoma. In general the picture is, as has already been said, one of chaotic tissue-formation, implying the presence of all sorts of morphogenetic stimuli *except* the individuation field and its governance of *regional* differentiation. This is expressed by Willis when he says that teratomata possess no "true somatic regions", i.e. masses of nervous tissue but no brain, masses of bone but no bones, renal tissue but no kidney. The condition, indeed, is even more chaotic than that produced by implanting a dead organiser into a newt embryo, as in Holtfreter's experiments. "We do not accept foetiformity in a single object", says Nicholson, among those teratomata which have been thought by their discoverers to include limbs or vertebral columns. The multiplicity of certain constituents in teratomata is also significant. Willis enumerates cases in which many separate "nervous systems" were present, and cases in which dozens of separate "tonsils" in dozens of separate little patches of "pharynx". Specimens of teratomata containing thousands of respiratory or alimentary cysts have been described, and the number of teeth may much exceed those normally contained in one adult body. We are reminded of the numerous ear-vesicles of Fig. 128. The only possible conclusion is that either the appropriate evocator has been liberated in a large number of places or that competence has appeared or been retained in a sporadic chaotic way.

One interesting point that arises here is that just as the muscles and glands of an induced embryo in the amphibia, or of an un-coordinated mixture of tissues induced by a dead organiser, become functional if the host lives long enough,

---

* The only exceptions are the peculiar "embryonic emboli" or teratoma metastases found in the blood-vessels adjacent to an affected testis (Peyron; Peyron & Limousin). Said to resemble axiated blastocysts, they require further critical evaluation.

so the tissues of a teratoma may also be functional. Teratomatous sweat-glands, choroid plexus, haemopoietic red marrow, all actively secrete and perform their normal functions. Teratomatous muscle contracts. Particularly significant in connection with the remarks just made on sterol metabolism is the fact that patients with testicular and other malignant teratomata often develop a strongly positive Zondek-Aschheim reaction (Willis[4]; Crew[2]; Freed & Coppock), which suggests that the teratomatous tissue may be producing condensed ring compounds similar to oestrone.

Finally, Willis and Nicholson have studied what they call "tissue correlations" in teratomata. In a succeeding section we shall study these lower-grade organiser effects in normal development. Almost all the smooth muscle in teratomata is clearly associated with glandular cavities. The impression is given that the development of the muscle about the gland elements is determined by the presence of the epithelium. Or again, there is a significant correlation between islands of cartilage with areas of differentiating nervous tissue. We know from the work of Filatov[1]; A. Luther; and Guareschi[1] that in the amphibia the cartilaginous auditory capsule fails to appear if the ear-vesicle has previously been extirpated. Conversely, a grafted ear-vesicle will produce a cartilaginous capsule around it. A similar influence seems to be at work in the teratomata. And stimuli of this kind are undoubtedly involved in all the other processes, tooth formation, origin of meninges, etc.

One class of teratomata can be explained in a particularly straightforward way from what we know of experimental embryology; namely, the sacro-coccygeal tumours known as sacral parasites or sacral tera-tomata. A drawing, taken from Schwalbe, is shown in Fig. 129. Holtfreter[3] has pointed out that like all the varieties of spina bifida, this condition can be almost certainly referred back to an abnormality of gastrulation. Amphibian exo-gastrulation has already been described (p. 159). Instead of the cells moving in at the blastopore as they should, they will move outwards and away from the ecto-derm, so that, if the process is complete, the ectoderm and endo-mesoderm become completely separated. This extreme condition is shown in Fig. 130 *a* diagrammatically. The process can occur in all degrees, depending on how long a time exo-gastrulation takes place while the cell stream is moving. Such a series of stages corresponds well with the varying degrees of *Acardius* or *Acephalus* monsters in mammals, where the posterior parts of the body are well formed, but not the anterior

Fig. 129. Human sacral teratoma.

and cephalic parts. v. Baer[3] described acephaly in a pig in 1828, and human cases are always turning up, e.g. the pretty example of le Vine & Wolf. The amount of neural tissue which is formed in the ectoderm depends exactly upon how much it is underlain by the endomesoderm, and if none of this underlying takes place at all, no neural material ever will be formed. Now if we compare a mammalian sacral teratoma with the partial exo-gastrulation of Fig. 130 *b* we see that there is something of a parallel, for in both cases there is an appendage attached in the neighbourhood of the anus, containing a variety of tissues. The only important difference is that human sacral teratomata often contain epidermis and neural tissue. This may be explained by the differences between the process of gastrulation in mammalian embryos and that in the embryos of amphibia. The contact of endo-mesoderm with ectoderm is all that is necessary to produce these neural differentiations. For instance, if a piece of competent ectoderm be placed upon the evaginated part of a completely exo-gastrulated newt embryo, the evocator will spread to it and induce nerve tissue and mesenchyme in it (see p. 273).

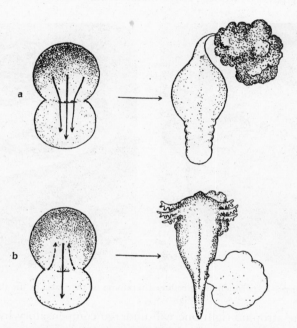

Fig. 130. Diagrams of amphibian exo-gastrulation, to explain the origin of sacral teratomata in man (*a*, complete; *b*, partial).

Until recently all known cases of teratoma were spontaneous ones. But Michalovsky and later Bagg[3] have studied the effects produced by injecting 5 % zinc chloride into the testis of the fowl. In many cases teratomata result, sometimes so large as to occupy most of the visceral cavity. A typical case is shown in Fig. 131. These tumours may contain cartilage, bone, muscle, fat, connective tissue, nervous tissue, glands, epithelium, feather follicles, and lumps of adenocarcinomatous or carcinomatous tissue. The action of the injected metal is intensified by injections of pituitary gonadotrophic hormone, but this alone has no such action. Dare we say that the metallic ion has had the effect of restoring all embryonic competences at one blow, and that the evocating substances, normally present but functionless in the adult tissue, have then done their work?

Falin, at any rate, in a recent description of such tumours, accedes to this view, suggesting that presumptive spermatocytes are the responsible cells. The experimental production of teratomata opens up all kinds of valuable possibilities.

In a later paper, Bagg[4] found that if autolysed testis tissue is injected at the same time as the zinc chloride, a teratoma develops subsequently at the former site of injection, not the latter. The metal itself appears to be eliminated quickly from the testis.* Michalovsky-Bagg teratomata are transplantable to other fowls (Anissimova).

Perhaps these facts are connected with the curious phenomena of "seminoma" (Champy, Lavedan & Marquez; Kirschbaum & Jacobs). If the left testis of a fowl

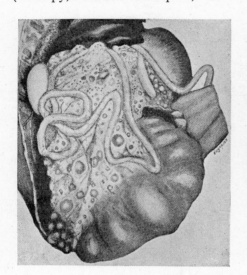 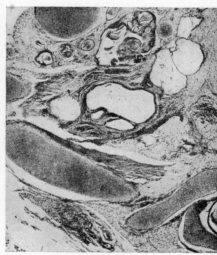

Fig. 131. Experimentally produced teratoma of the testis in the fowl.

is removed, the atrophic right one will undergo compensatory hypertrophy, but often this is overdone and a characteristic tumour of large size results. In such over-regeneration there is thus an escape from the individuation field.

This section shows how important is evocator control in normal development. Appearance of supernumerary evocators may be a normal process, as in some sorts of polyembryony, or it may be a pathological one. If the primary organiser is concerned, the result will be a complete twin or a complex monster, if second- or third-grade organisers are concerned, the result will present itself to the morbid anatomist as a teratoma. In this latter case we have also to consider the persistence or recuperation of competence in tissues whence it should have disappeared, but it is clear that the possible permutations and combinations of irresponsible organisers and perversely competent tissues will provide everything that is contained in the literature on teratomata; the "foetiformity" theory notwith-

* It may be remembered that zinc is used as a protein precipitant. Would other denaturing agents have similar effects? Other zinc salts are active (Falin & Gromzeva).

standing. Pathologists have recently begun to apply this point of view successfully in various specialised directions, e.g. multiple meningeal and perineural tumours (Worster-Drought, Carnegie-Dickson & McMenemey); odontomata (Sprawson); benign ovarian teratomata (Willis[4]), the rare chordomata (Richards & King), etc.

Other phenomena besides teratomata suggest the anomalous appearance of competence where it ought not to be. In the abnormality in insects, sometimes of genetic origin, known as *"prothetely"*, a caterpillar may form pupal antennae (Goldschmidt[5])—see Fig. 132—or pupal wings (Ferwerda). Goldschmidt[8] (*Physiological Genetics*, p. 173) suggests that as the development of the imaginal discs is under hormonic control, the threshold may have been prematurely lowered. This is perhaps another way of saying that competence has appeared where it should not have done. The converse of this is seen in insect intersexes, where competence seems to be unduly damped, with the result that the development of the imaginal discs begins extremely late. As chitinisation of the soft parts is the end-stage in insect development, these organs become chitinised as imperfect stages or even as imaginal discs instead of completed organs. Hence the embryonic structure becomes "fossilised" and appears in the adult (Goldschmidt[4]). This subject will be taken up again in the section on determination in insect development and metamorphosis (p. 466). We have now to pursue further the malignancy problem.

Fig. 132. Prothetely.

2·38. Organisers and cancer
2·381. The escape from the individuation field

The foregoing discussion has brought us to the verge of the problem of malignancy. In normal development, differentiation and growth are highly organised. In the last section, a short review of the problem of teratomata showed what happens when differentiation proceeds in an unorganised way. We are unfortunately only too well acquainted with circumstances in which growth also becomes unorganised, and certain types of cell, proliferating in superabundance, invade the territories of peaceful tissues performing their specified normal functions.* At first sight organiser phenomena have little to do with cancer, for if one thing is more characteristic than any other of evocator substances, it is that they are concerned with the stimulus to a specific differentiation, not a general growth.† But evocation, as has already been pointed out (p. 125),

---

* The growth rate of neoplasms is quite comparable with that of embryos (Schrek; Sugiura & Benedict).

† This is well expressed by Fischer-Wasels: "Das Wachstum der malignen Geschwulst unterscheidet sich prinzipiell und wesentlich von dem Wachstum der embryonalen Zelle. Die Schnelligkeit der Zellbildung ist nur ein äusserliches Merkmal. Das Wesen des malignen Wachstums dagegen, das nämlich irgendeinem höheren Plane, insbesondere dem Organisationsplane des Organismus, nicht unterstellt ist, findet sich bei der Embryonalzelle nicht, ja wir finden hier

is only the half of organisation. The organiser has a certain relation to the plan of the whole body which we may speak of as existence in an individuation field. The main characteristic of an individuation field is that all tissue lying within it tends to be built up into one complete embryo, and in any one part of the field all tissue tends to be built up into the organ corresponding to that part. One of the most important aspects of the cancerous growth is that it is an *escape* from this controlling pattern or field.* This manner of stating the problem may in time be rather fruitful.

It is at once evident that a great deal will depend on the extent to which the individuation field (the assumption of which is indispensable in embryology) *persists into the adult condition*. It may be suggested that power of regeneration is a suitable measure of this. If so, the mammalian individuation field has been wholly lost during development. Nevertheless there are organisms which possess remarkable powers of regeneration in the adult state, such as the newts, on which so much experimental work has been done (cf. Section 2·71). In the adult newt, the leg area is still so potent that it can mould into a limb any mass of competent tissue either grafted into it, or formed on it as a regeneration bud. We are led at once to ask whether there is any difference in susceptibility to cancer between Urodeles which can regenerate lost limbs and tails so well, and Anura and reptiles which have no such capacity. Unfortunately, it seems that this is not known. Cramer[1], in his review on cancer in animals, points out that the relatively short lives of many animals preclude the clear appearance of many tumours which may have made a beginning. It may perhaps be of significance that a search through the literature on amphibian cancer gives far more cases of neoplastic growth reported for reptiles and anurans than for urodeles. As the data have not so far been assembled, a list (as complete as possible) is given in the accompanying table (Table 21). The cases are unfortunately too few in number to allow of any statistically reliable conclusions; for example, it must be remembered that for many years past the frog has been a much more common laboratory animal than the newt. In illustration of the table, I give two pictures, the first, from Eberth's classical paper of 1868, shows the external appearance of a frog suffering from multiple nodular adenomata of the skin (Fig. 133); the second shows a cross-section of Murray's carcinomatous newt[1] (Fig. 134). It may be observed that the ectodermal tissues are alone affected, and that there is no infiltration of the underlying muscle.

Perhaps the best-known amphibian tumour is the carcinoma in the leopard frogs of New England (*Rana pipiens*), described by Lucké[2] and Lucké & Schlumberger. The kidneys of some 2 % of the population show non-encapsulated

sogar das grossartigste Ineinandergreifen kompliziertester Gestaltungsvorgänge, das wir in der Natur kennen. Alles embryonale Wachstum und Geschehen ist beherrscht von dem Organisationsplan des Körpers, von der Einheit des Individuums. Also eine wesentliche Ähnlichkeit der beiden Wachstumsformen ist überhaupt nicht vorhanden" (p. 1388). We may leave out of account the many speculations (such as may be found in the pages of W. R. Williams) on the origin of cancer from "embryonic cell-rests", etc. Fischer-Wasels discusses them at length, and the recent progress along biochemical lines has deprived them of most of their interest.

* A morphological escape. Histologically and functionally, of course, tumours are quite faithful to the tissues from which they take their origin.

Table 21. *Neoplastic growth in Reptiles and Amphibia*

Author	Date	Species	Nature of tumour		No. of cases	Remarks
			Class	Site		
REPTILES						
Bland-Sutton, J.[1]	1885	Indian monitor lizard, *Python sebae*	Chondroma	Limb	1	
Pettit, A. & Vaillant, L.	1902	*Python* sp.?	Carcinoma	Ovary	1	
Pick, L. & Poll, H.	1903	Brazilian tortoise, *Platemys geoffroyana*	Fibroma	Stomach	1	
Koch, M.	1904	Lizard, *Lacerta agilis*	Benign tumour	Thyroid	1	
Plimmer, J.	1912 & 1913	Tortoises	Papilloma	Occipital skin	1	2 in 6000 autopsies
			Carcinomata	?	2	
Schwarz, E.	1923	American lizard, *Tupinambis teguixin*	Carcinoma	Hand	1	No metastases; no infiltration
Scott, H. H.	1927	*Crocodilus porosus*	Sarcoma	Liver	1	Metastases
Anonymous	1930	*Crocodilus* sp.?	"Non-cancerous"	?	1	
Smith, G. M. & Coates, C. W.	1938	*Cheloma mydas*	Fibro-epitheliomata	?	4	
ANURANS						
Eberth, C. J.	1868	Frog	Multiple adenomata	Skin	Many	
Smallwood, W. M.	1905	Frog	Columnar-celled carcinomata	Adrenals	1	
Plehn, M.	1906	Frog (*R. esc.*)	Multiple carcinomata	Ovary	1	No infiltration
Murray, J. A.[1]	1908	Frog (*R. esc.*)	Adeno-carcinomata	Skin of thigh	2	No metastases
Carl, W.	1913	Frog (*R. esc.*)	Hypernephroma	Adrenals and kidney	1	
Pentimalli, F.[1]	1914	Frog	Adenoma	Skin	1	No penetration of deeper layers, no metastases
Secher, K.	1919	Frog	Cystadenoma	Skin	1	Infiltration, no metastases
Lucké, B.[2]	1935	Frog (*R. pip.*)	Adeno-carcinomata	Kidneys	300	
URODELES						
Pettit, A. & Vaillant, L.	1902	Salamander, *Cryptobranchus* sp.?	Fibropapilloma	Hand	1	
Pick, L. & Poll, H.	1903	*Cryptobranchus japonicus*	Carcinomatous cystoma	Testis	1	Metastases
Murray, J. A.[1]	1908	Newt, *Triton cristatus*	Nodular carcinoma	Skin of head and tail	1	No infiltration
Schwarz, E.	1923	Salamander, *Megalobatrachus maximus*	Fibroma	Hand	1	
Champy, C. & Champy, C.	1935	Newt, *Triton*	Epitheliomata		Many	

NOTE. Teutschlander mentions a melanosarcoma of axolotl in a table but there is no reference to it in his text. Whether one of the above cases is meant must remain uncertain.

adeno-carcinomata infiltrating and destroying the surrounding renal tissue, and often attaining a size so large as to cause much displacement of the abdominal viscera. Metastases are rare, but the tumour is transmissible experimentally.

It is obvious that opportunities for experimental attack along these lines are inviting and numerous. Unfortunately amphibian tissue seems to be exceptionally refractory to carcinogenesis by hydrocarbons (Duran-Reynals; van Heyningen[1]), though success has been achieved, both in the adult newt (Koch, Schreiber & Schreiber) and the frog tadpole (Briggs[2]). But if amphibian cancerous tissue transplantable on to newt tissues could be found, it would be possible to see

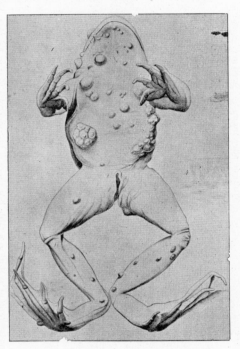

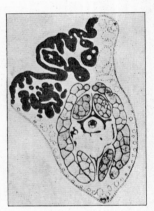

Fig. 133. Multiple nodular adenomata in the frog.

Fig. 134. Carcinoma of the tail in the newt.

whether such wildly growing material could be mastered by the individuation field controlling a limb area. If so, it should be organised into a normal limb. Enquiry could also be made into the degree of competence still possessed by such cancerous tissue. By subjecting it *in vitro* to the action either of a living organiser or of evocator substances, its potentialities might be revealed. The influence of carcinogenic agents on regeneration is also an obvious direction of study. Here experiments have again so far been negative; thus Martella painted tar upon the skin of the newt *Molge* near the tail without any effect, Tokin found no influence of the synthetic carcinogenic hydrocarbons which he tried on the regeneration of *Hydra*, and Slizyński obtained equally negative results when painting tar on *Drosophila*. One might picture that if the powerful growth stimulus of a carcinogenic substance acted within a powerful individuation field, the structures produced would show some sort of harmoniously regulated differentiation.

The nearest approach to a verification of the idea expressed in this section occurs, perhaps, in Pflugfelder's remarkable work on the stick insect, *Dixippus morosus*. If the *corpora allata* (see p. 465) are removed, the regeneration of appendages after amputation or autotomy no longer follows a normal course, giving rise instead to all kinds of wild, and even teratomatous, proliferations. Still more extraordinary, the structure which appears after excision of the faceted eye is an ocellar type of eye which this species never normally possesses, though other phasmids do (cf. the balancer-sucker situation, p. 342). The hormones of the *corpora allata* in this insect may therefore have a profound effect both on the individuation field and on the competences of the blastema cells. The further investigation of the whole question seems to promise much.

In a succeeding section (p. 430) we shall have something to say about regeneration phenomena as a whole. From the foregoing line of thought it follows that regeneration tissue ought to show some of the properties of embryonic tissue, pluripotence, competence, etc. That it does so has been widely believed, but the final proof appears hardly yet to have been given. As we shall see, lens competence, at any rate, has been shown to be present in the regenerating blastema (Schotté & Hummel); and Schotté[5] has stated that he and his collaborators have obtained inductions of ear-vesicles by brain grafts in the regenerating tail tissues of tadpoles. More still, he reports that implantations of pieces of mouse embryos into mouse fibrosarcomata led to teratomatous arrays of organised structures. The full publication of these results will be awaited with interest.

2·383. Evocators, carcinogens and oestrogens

The second way in which embryonic evocators are connected with the problem of cancer is due to the fact that synthetic carcinogenic substances have been found to act strongly as primary evocators in the amphibian embryo (Waddington & Needham; Waddington[21]; Shen[1,2]), cf. pp. 178 ff. This necessitates a discussion of the biological effects of substances of steroid and polycyclic hydrocarbon character. It will be brief, since there exist many valuable reviews of this literature.

We shall begin with the chemical carcinogens.* In 1910, when the researches now to be described began, it had long been known that manual workers who came into contact with tar, especially when originating from gasworks, were particularly liable to suffer from cancer. The credit for demonstrating that cancer could be produced experimentally in animals by painting tar or other chemical substances on the skin is usually ascribed to Yamagiwa & Ichikawa, who in 1915 obtained cancerous growths in the ears of rabbits by long-continued applications of tar. Some years earlier, however, Bayon had mixed a watery extract of gasworks tar with sterilised lanoline, and, on injecting this into the rabbit's ear, obtained enormous epithelial proliferations simulating cancer. These experiments

---

* The term "carcinogen" is here loosely used, but it is common to speak of the chemical carcinogens as producing sarcomata if injected intra-muscularly, or epitheliomata if painted on the skin. Some compounds (e.g. 10-methyl-1:2-benzanthracene) produce only the former, others (e.g. 9:10-dimethyl-1:2-benzanthracene) only the latter (Badger *et al.*).

are now somewhat difficult to evaluate, since although Bayon demonstrated negative controls with lanoline alone, there is reason to believe that lanoline may sometimes contain carcinogenic substances (Peacock[1]), and moreover the carcinogenic substances of tar are not themselves water-soluble. Bayon's work, however, has historical interest.

The two Japanese investigators were soon followed by two Swiss, Bloch & Dreifuss, who showed that the carcinogenic activity was due to neutral nitrogen-free substances possessing a high boiling-point, and the fractionation of tar then began in earnest under the aegis of Kennaway, Hieger, Cook, and their collaborators in London. Isolation of the active fractions was facilitated by Mayneord's discovery that the fluorescence spectra of carcinogenic tars differed from those of non-carcinogenic tars. The first carcinogenic hydrocarbons prepared from tar in a pure state were described by Cook, Hewett & Hieger, and later they were synthesised in the laboratory (Barry, Cook, Haslewood, Hewett, Hieger & Kennaway). They all contained the tricyclic phenanthrene condensed ring system, similar to that of the long-known animal and plant sterols. The most active of these substances is 3 : 4-benzpyrene, but 5 : 6-cyclopenteno-1 : 2-benzanthracene is also very efficient. These produce squamous-celled carcinomata in rodents and even metastases are not uncommon. 1 : 2 : 5 : 6-Dibenzanthracene is less efficient than the two foregoing compounds in producing carcinomata, but when injected subcutaneously into rodents it gives rise to sarcomata. Nitrogen may be introduced into the ring skeleton, as in 1 : 2 : 5 : 6-dibenzacridine or 3 : 4 : 5 : 6-dibenzcarbazole, without abolishing the carcinogenic activity. One of the parent hydrocarbons of all these substances, chrysene, has been claimed to be active, but this is uncertain, and its activity cannot be very marked (Barry & Cook). It often contains 3 : 4-benzpyrene as an impurity (Schürch & Winterstein).

The following description of chemically induced cancer is given by Murray[2]. If a mouse is painted on the back twice weekly the hair is first destroyed and the covering epithelium of the skin thickens, the lower layers showing an infiltration with lymphocytes, mast cells, and polymorph leucocytes. Then discrete warty prominences appear with the histological structure of benign papillomata. If the painting is now stopped (15 weeks) these warts may regress and drop off, but if the substance painted has been highly active they will now grow more and more rapidly, adhere to the deeper tissues, which will become infiltrated, and metastasise into the adjacent lymph nodes, the lungs and sometimes into other organs.

This description recalls the conception of progressive determination used in experimental morphology. Many other causes which produce uncontrolled cancerous growth are of course known; the secretions of some animal parasites, the application of cold and of X-rays, may be mentioned. But in spite of all pleas to the contrary, the simplest working hypothesis seems to be that these less precisely known carcinogenic agents act by liberating within the tissues substances more or less closely chemically related to the carcinogenic hydrocarbons isolated from tar. A number of interesting facts are relevant to this.

Early in the development of the subject it was found that carcinogenic substances were produced by exposing yeast and human skin to high temperatures (e.g. 920° C.), with the formation of tars (Kennaway). This showed that such substances could be formed from chemical compounds normally present in the living tissues. But later on a more elegant demonstration of this was given by Cook & Haslewood and by Wieland & Dane, who found that it was possible, starting from a sterol compound (the bile acid, desoxycholic acid), to synthesise through various intermediary stages (12-keto-cholanic acid and dehydro-nor-cholene) one of the most strongly active carcinogenic hydrocarbons known;

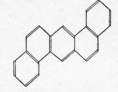

1 : 2-Benzanthracene

1 : 2 : 5 : 6-Dibenzanthracene (C)

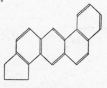

5 : 6-Cyclopenteno-
1 : 2-benzanthracene (C, O)

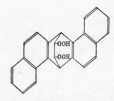

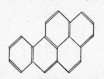

3 : 4-Benzpyrene (C, E, O)

1 : 2 : 5 : 6-Dibenzanthracene,
endo-α-β-succinate (C, E)

10-Methyl-
1 : 2-benzanthracene (C)

5 : 10-Dimethyl-
1 : 2-benzanthracene (C)

5 : 9-Dimethyl-
1 : 2-benzanthracene (C)

as powerful as methylcholanthrene

Note: In the accompanying tables of formulae abbreviations are used as follows: C, carcinogenic; O, oestrogenic; E, active as primary evocator.

Reviews on the Steroids

General Review: Kühn[2]
Sterols and related compounds: Fieser[1]; Lettré & Imhoffen; Rosenheim & King; Sobotka[2]; Windaus[2]; Schoenheimer & Evans; Mozołowski
Bile acids: Sobotka
Vitamin D: Bills; Brockmann; Friedmann[2]; Rominger
Plant heart poisons: Jacobs & Elderfield[2]; Elderfield; Tschesche[2]
Toad heart poisons: Fieser[1]; Tschesche[2]
Saponins: Fieser[1]; Tschesche[2]
Alkaloids: Fieser[1]
Male sex-hormones: Ruzicka[1]; Dodds; F. C. Koch
Adrenal cortex hormone: Wintersteiner & Smith
Oestrogens: Dodds; Dodds & Lawson[3]; Schmidt-Thomé; Pincus & Werthessen[3]
Carcinogens: Cook[3]; Fieser[1, 3]; Cook, Haslewood, Hewett, Hieger, Kennaway & Mayneord; Cook & Kennaway; L. Loeb[5]
Polyterpenes; Ruzicka[2]; Jacobs & Elderfield
Scymnol: Tschesche[1]

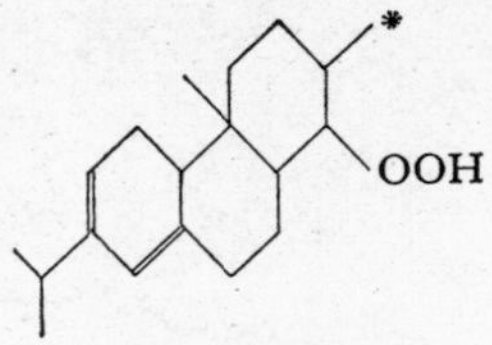

* Position of this methyl group uncertain.

Abietic acid

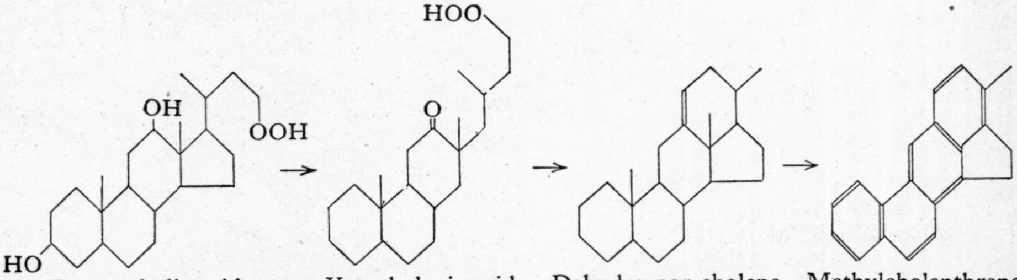

Desoxycholic acid → 12-Ketocholanic acid → Dehydro-nor-cholene → Methylcholanthrene (C, E)

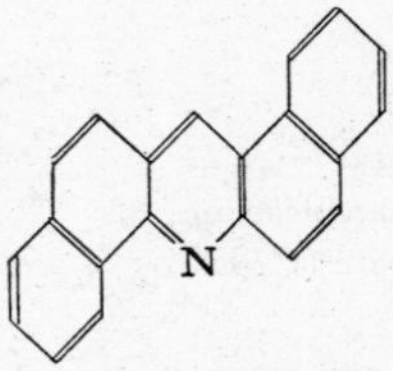

1 : 2 : 5 : 6-Dibenzacridine (C) 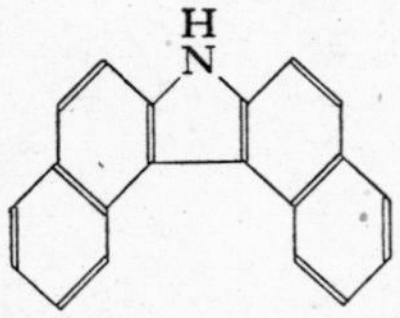   3 : 4 : 5 : 6-Dibenzcarbazole (C) 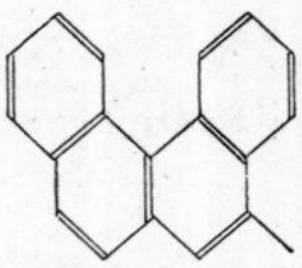   2-Methyl-3 : 4-benzphenanthrene (C)

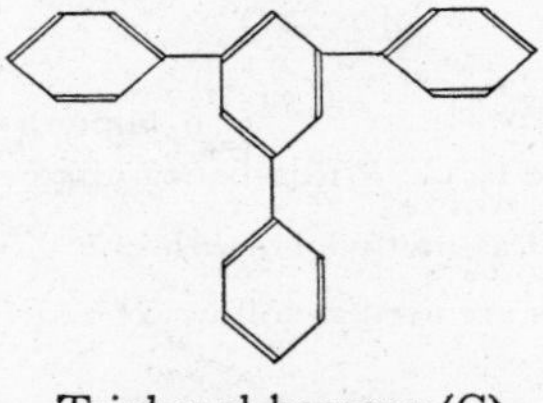

Triphenyl-benzene (C)

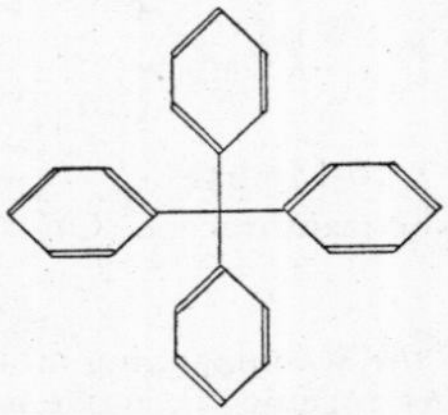

Tetraphenyl-methane (C)

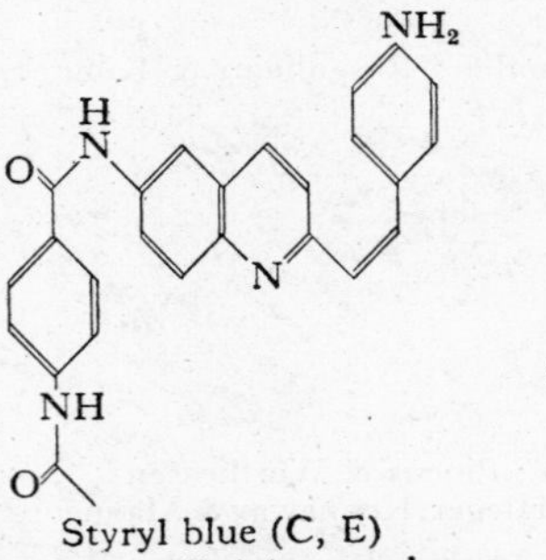

Styryl blue (C, E)
[2-(p-aminostyryl)-6-(p-acetylamino-benzoylamino)-quinoline methoacetate]

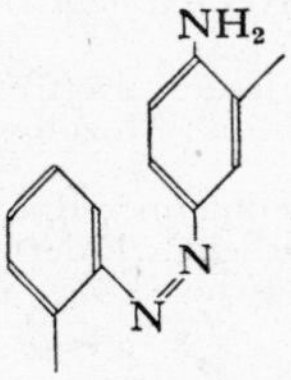

2-Amino-5-azo-toluene (C)

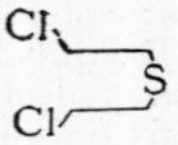

Dichlorethyl sulphide (anti-C)

methylcholanthrene. It was subsequently found by Ghiron and by Cook, Kennaway & Kennaway that desoxycholic acid itself has a slight carcinogenic activity. All this indicates, though of course it does not prove, that from sterols normally occurring in the body, a deranged metabolism may produce specific carcinogenic substances which will stimulate certain cells to uncontrolled proliferation (Cook[1, 2]). In preliminary work Burrows & Mayneord and Polletini have obtained cancer from the X-ray irradiation products of cholesterol and ergosterol. Bergmann[2] suggests that the side-chain of cholesterol might participate directly in a cyclisation.

Furthermore, it has been found that polycyclic carcinogenic hydrocarbons are oxidised photochemically to give water-soluble products (Allsopp). These products have an inhibitory action upon important respiratory enzymes such as lactic dehydrogenase, and roughly speaking the action is greater the more active the hydrocarbon is in cancer production (Boyland[2]). This action resembles that of arsenious acid, a substance which is known to be carcinogenic, so that we may envisage the following classification of substances: (a) the polycyclic hydrocarbons themselves, (b) substances (or influences) which stimulate the tissues to produce their own polycyclic carcinogens, (c) substances of very divergent chemical composition which perhaps stimulate the cells or inhibit processes within them at a point different to that affected by the polycyclic carcinogens. It has also been shown that the oxidation products of the active hydrocarbons inhibit both respiration and glycolysis in every tissue investigated (Boyland & Boyland[1]). In this way we begin to see how work on the chemical stimulating mechanism may link up with work on the metabolic peculiarities of malignant as opposed to normal tissue, already pushed so far by the school of Warburg (cf. CE, p. 758).

The reviews of Cook[3]; Cook, Haslewood et al.; Cook & Kennaway and Loeb[5], and the discussion opened by Murray[2], give full details of the chemistry of the compounds so far found to be carcinogenic. Fieser[1, 2, 3] has reviewed the relation between chemical structure and function. But it is now clear that the condensed ring system is not indispensable in chemical carcinogenesis. Morton, Clapp & Branch found symmetrical triphenyl-benzene and tetraphenyl-methane to be moderately active.* Similarly Browning, Gulbranson & Niven showed that the dye styryl blue, which contains one heterocyclic ring as well as three benzene rings, was strongly carcinogenic. Even a compound with only two rings may be active, such as the 2-amino-5-azo-toluene of Shear and Kinosita, which produces liver carcinomata. The effects produced by styryl blue on respiration and phosphorylating glycolysis have been studied in interesting work by Pourbaix.

The relationships between the many ring compounds of biological activity will become clearer if we glance at their formulae. Those who wish to investigate any point in further detail will find what they want in the excellent monographs of Fieser[1] and of Lettré & Imhoffen.

The parent hydrocarbons of the whole series are anthracene, phenanthrene, and chrysene. Phenanthrene, isomeric with anthracene, was discovered by Fittig & Ostermayer in 1872 in tar. But if these are the structural parents, cholesterol itself is the starting-point of our knowledge of the sterols. A constituent of all animal tissues and cells, it occurs in particularly large amounts in brain, adrenal cortex, and bile, and has been known since the eighteenth century and the days of Chevreul and Thudichum. It has undoubtedly a large part to play in the cell structure (see p. 663). It is synthesised by animals. The sterols of plants, ergosterol and stigmasterol, differ by having a double bond in the side-chain and (in the former) ring II more aromatised; they are readily synthesised in the plant but not much absorbed from the intestinal tract of animals (Schoenheimer[1]). It is inter-

* This is contested by the London school (Bachmann, Cook et al.; Badger, Cook et al.).

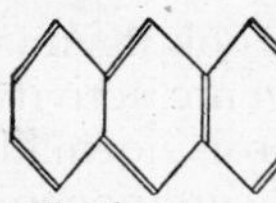

Anthracene

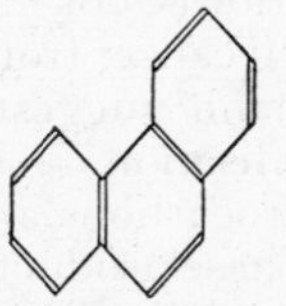

Phenanthrene

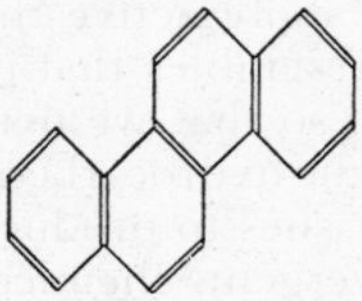

Chrysene

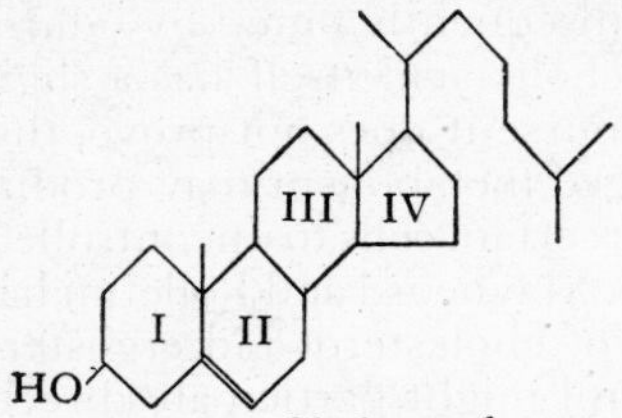

Cholesterol

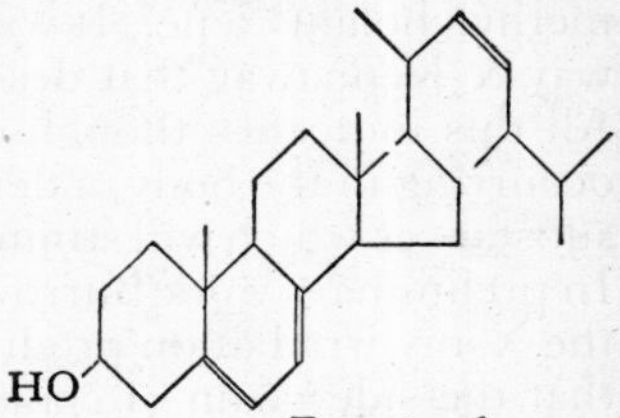

Ergosterol

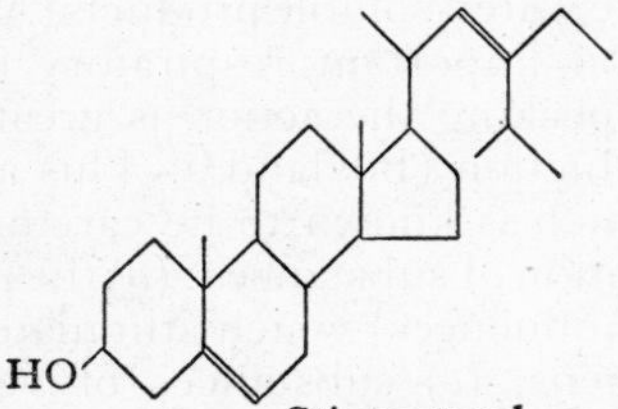

Stigmasterol

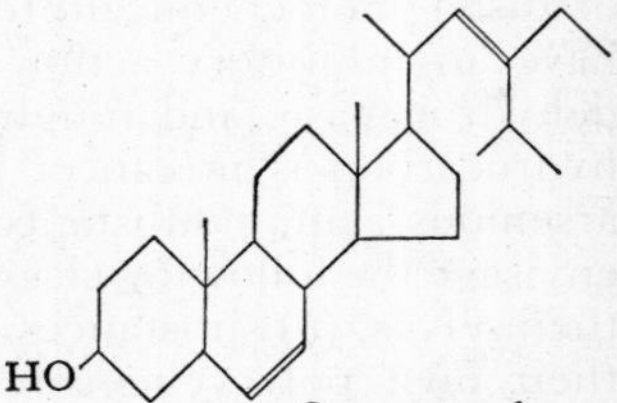

Ostreasterol

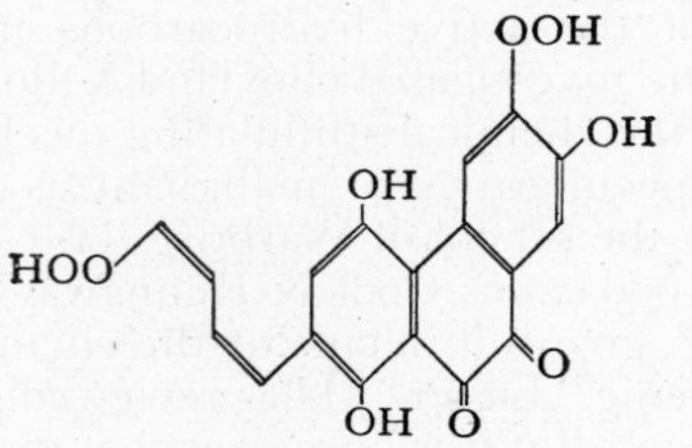

Telephoric acid

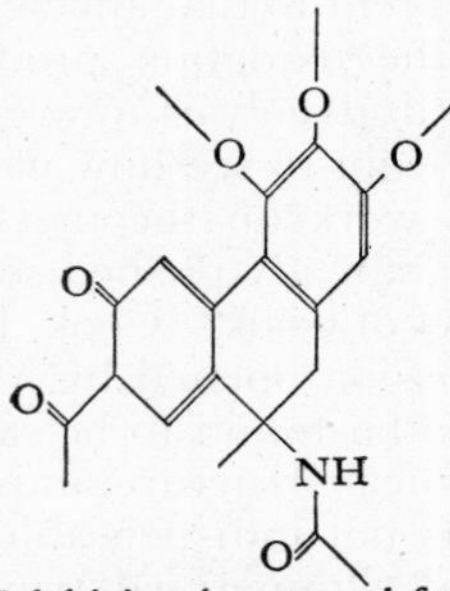

Colchicine (suggested formula)

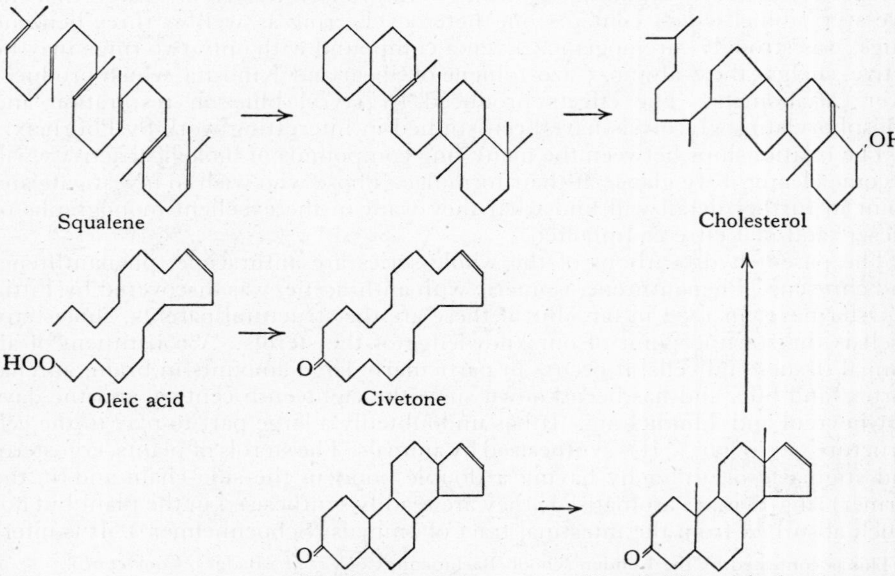

Squalene

Cholesterol

Oleic acid

Civetone

Hexadehydrocivetone

Dimethyl-hexadehydrocivetone

esting that among the invertebrates are found sterols, such as ostreasterol from the oyster (Bergmann[1]), which resemble those of plants in having a double bond in the side-chain, but are no more aromatised in ring II than the animal sterols. The phenanthrene ring occurs more widely than the sterol ring system; thus the curious orange pigment isolated from certain moulds, telephoric acid, has a phenanthrene structure; and so also the drug colchicine, which has a powerful inhibiting action on cell-division, arresting it in metaphase (Brues & Cohen; Boyland & Boyland[3]; see p. 509). Many alkaloids, too, such as morphine, codeine, and thebaine, possess the phenanthrene ring system.

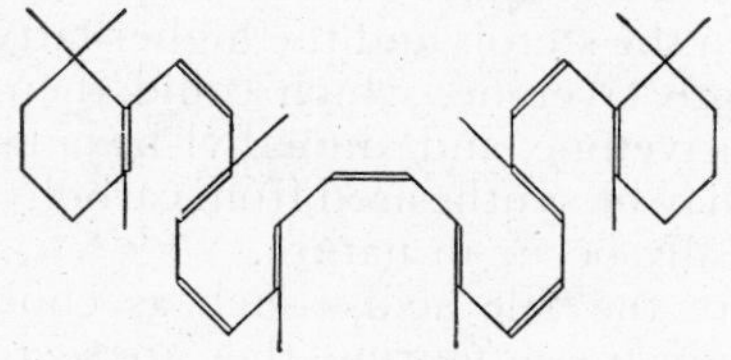

β-Carotene

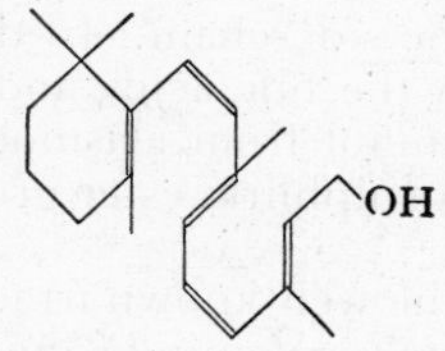

Vitamin A

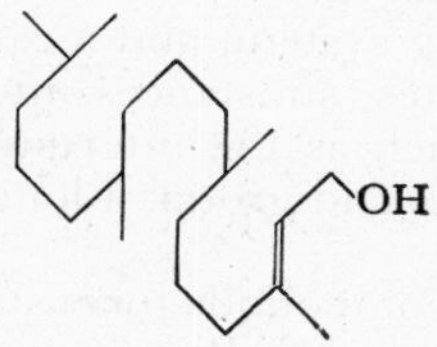

Phytol

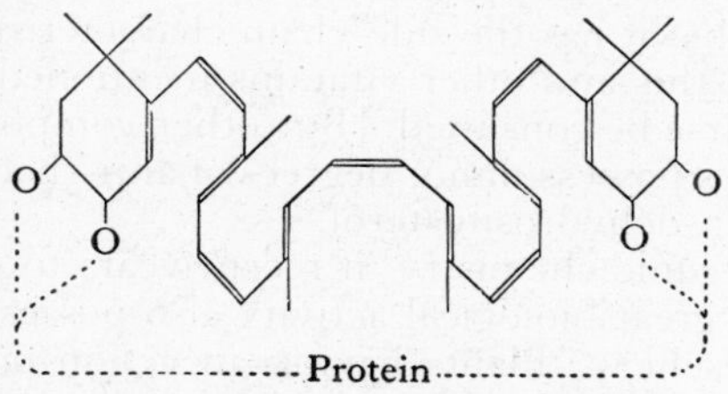

Ovoverdin (green)

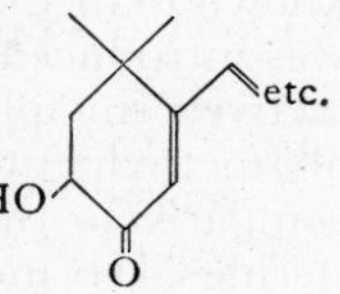

Protein

Astaxanthin (the naturally occurring red carotenoid)

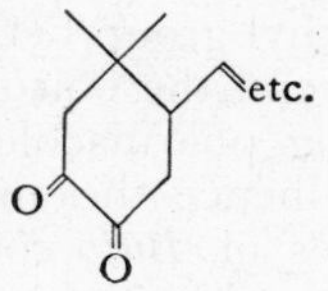

Astacin (formerly thought to be the naturally occurring red one)

Nothing certain is known regarding the mode of synthesis of the sterols, nor the nature of the substances to which they are most closely related. Some have imagined a connection with the polyterpenes. Thus the non-cyclic polyterpene, squalene, an oily substance present in large amounts in the livers of certain elasmobranch fishes, has been transformed by cyclisation into tetrahydrosqualene which bears a certain relation to cholesterol, though rather a distant one (Karrer & Helfenstein), and a tetracyclic triterpene into a pentacyclic triterpene (Beynon, Heilbron & Spring). If this relationship existed, there would be a connection between the steroids and the lipochrome pigments, which are all polyterpenes, such as carotin, astacin (cf. pp. 210, 652), vitamin A, and the alcohol phytol, attached to the tetrapyrrol-magnesium framework in chlorophyll (cf. the review of Karrer[2]). Physiological experiments, such as those of Channon, which have essayed to prove this relationship, have not yet quite succeeded in doing so. Others have imagined a connection between the sterols and the higher fatty acids, and certainly oleic acid can be cyclised into civetone, which could then give a sterol ring system through hexadehydrocivetone and dimethyl-hexadehydro-civetone (Windaus). In plants the sterols may be synthesised from carbohydrates, if the interesting schemes of J. A. Hall really occur in nature.

Another class of steroid compounds are the bile acids, such as cholic acid (reviewed by Sobotka[1]). Combined with glycine or taurine, they are secreted by the liver in all vertebrates. They are characterised by having at least two hydroxyl groups on the ring system, and a carboxyl group on the side-chain. In the bile of lower vertebrates, substances intermediate between the bile acids and other steroids occur, such as the interesting compound scymnol from elasmobranch bile, which has no acid group, a long side-chain, and a sulphuric ester grouping terminal to it.

Anti-rhachitic compounds form still another class. The well-known irradiation of ergosterol (reviewed by Bills and by Friedmann[2]) leads to vitamin D_2, in which the side-chain is unaltered, but the ring system has broken at ring II. The natural vitamin, D_3, is identical with this save that it has the side-chain characteristic of cholesterol. For details of the action of this and other vitamins mentioned, the monographs of L. J. Harris[2] will of course be consulted. But other compounds besides the natural D vitamins themselves possess minor degrees of anti-rhachitic potency, e.g. the irradiation product of 7-dehydrositosterol.

It has been one of the triumphs of organic chemistry in recent years to show that other, more obscure, substances of great biological activity also possess the sterol skeleton (see the review of Tschesche[2]). Plants having an action on the heart have been known for many centuries, and an extract of *Strophanthus hispidus* is used as an arrow poison by primitive tribes. Foxglove was known to mediaeval medicine. Digitalis, the "sheet-anchor of cardiac therapy", was introduced into the modern pharmacopoeia by Wm. Withering in 1785. The active principles of plant heart poisons consist of sterols united with sugars, mainly methylpentoses. Digitoxigenin is typical. The hydroxyl on ring I is placed similarly to that in cholesterol, but there is an additional one at the junction of rings III and IV, and in the side-chain there is an unsaturated lactone ring. Strophanthidin is unique in having an aldehydic group instead of the usual methyl group between rings I and II. Oxygen in the side-chain is also characteristic of the poisons of toad venom, known from remotest antiquity and possessing pharmacological properties similar to those of the plant cardiac glycosides; hence the ancient Chinese drug, *ch'an su*, prepared from toads. Many species of *Bufo* contain bufotoxin in skin-glands, a compound of a fatty acid, suberic acid, arginine, and the steroid bufotalin. Here the heterocyclic ring in the side-chain is six-, not five-membered. The saponins, too, are of steroid or polyterpene nature. Plants which give soapy or foamy extracts have been known for centuries (cf. common names such as soapwort) and are often used for washing purposes, as for instance

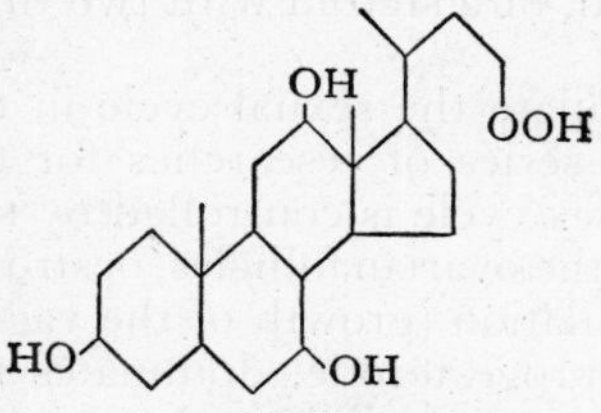

Cholic acid

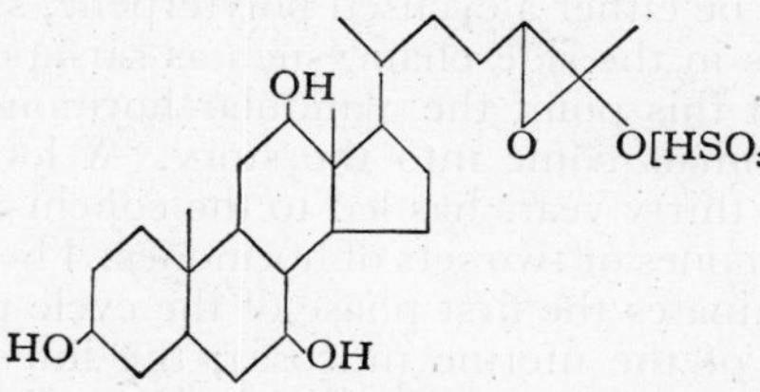

Scymnol

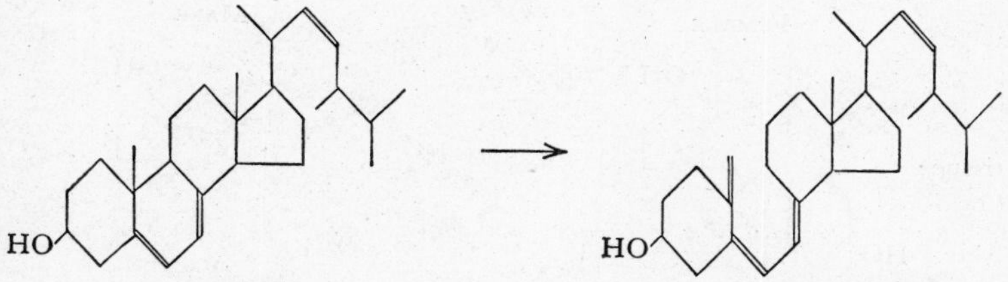

Ergosterol $\longrightarrow$ Vitamin D$_2$

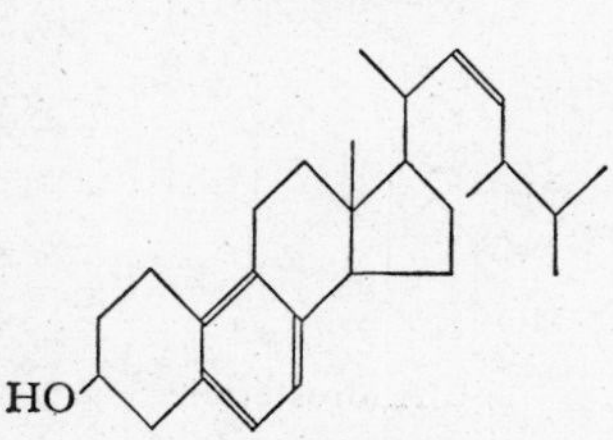

Neo-ergosterol

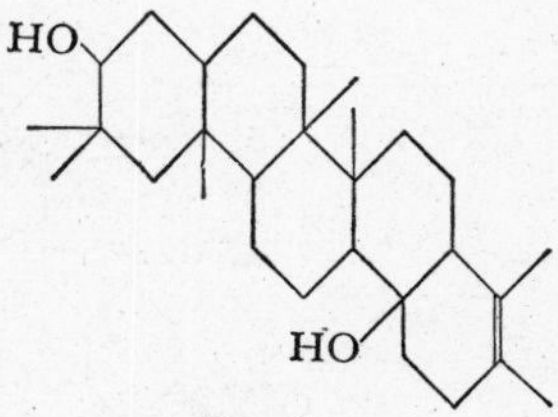

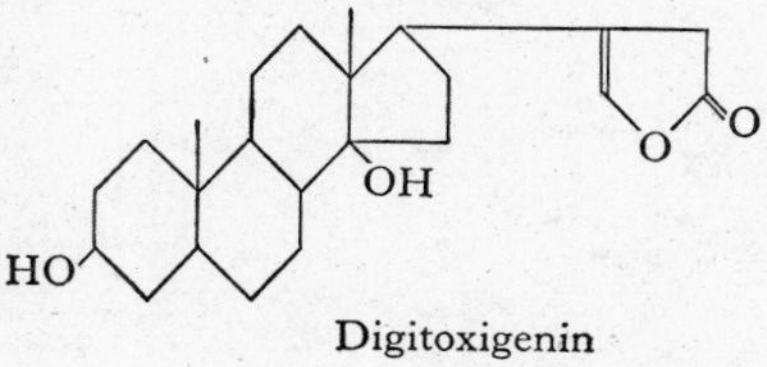

Digitoxigenin

Betulin

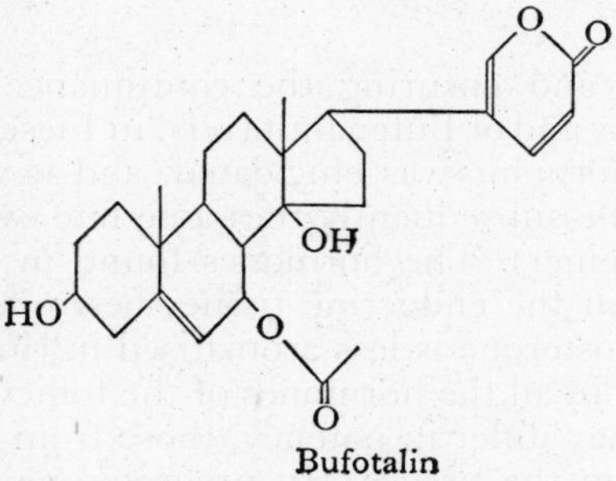

Bufotalin

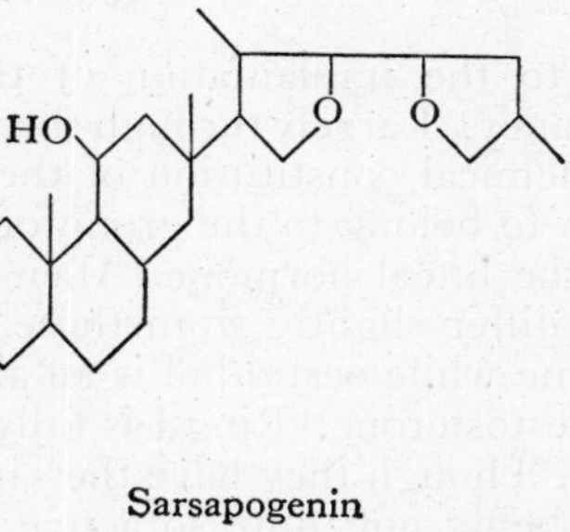

Sarsapogenin

by the Arizona Indians. Saponins are glycosides and the sapogenin or aglycone may be either a cyclised polyterpene, such as betulin, or a steroid with two furyl rings in the side-chain, such as sarsapogenin.

At this point the glandular hormones which regulate the sexual cycle in the mammals come into the story. A long-continued series of researches for the past thirty years has led to the conclusion that the sex cycle is controlled by two hormones or two sets of hormones. The hormone of the ovarian follicles, oestrone, dominates the first phase of the cycle preceding ovulation (growth of the vagina and of the uterine mucosa); the luteal hormone, progesterone, dominates the second phase, permitting the formation of the placenta (including the changes

FEMALE | MALE

In tissue

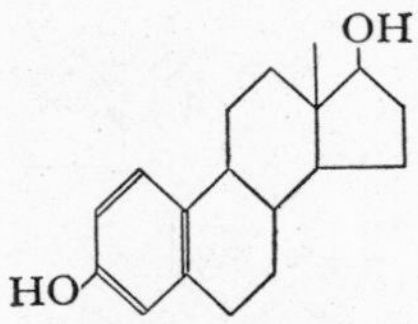

Oestradiol (O)

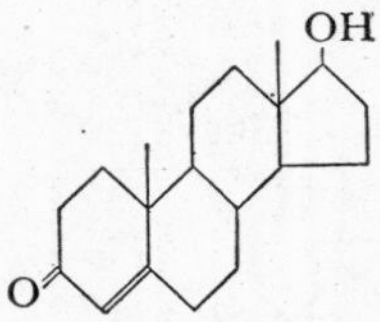

Testosterone

In urine

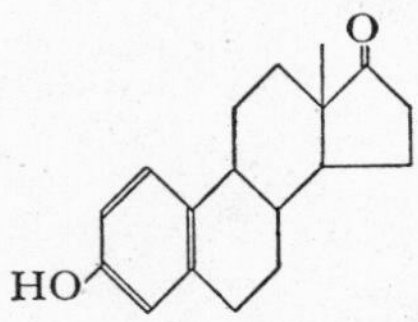

Oestrone (O, E)

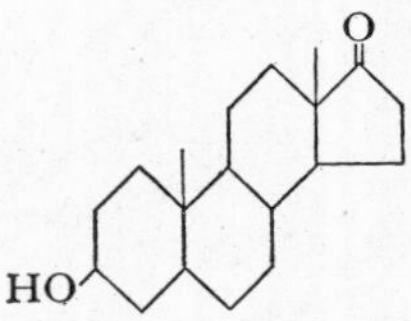

Androsterone

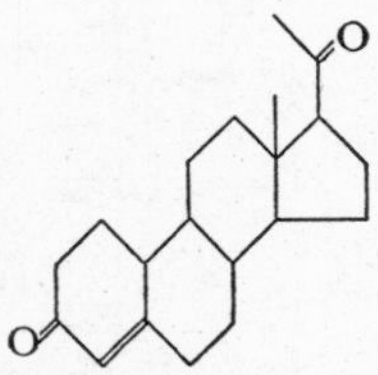

Luteosterone (progesterone)

prior to the implantation of the embryo) and ensuring the continuance of pregnancy. Largely through the work of Doisy and of Butenandt (refs. in Fieser[1]), the chemical constitution of the follicular hormone was elucidated, and it was shown to belong to the group of sterols, while subsequently the same fate over-took the luteal hormone (Allen & Wintersteiner). The hormones found in the urine differ slightly from those isolated from the endocrine tissue; oestrone is a ketone while oestradiol is an alcohol; androsterone is less aromatised in ring I than testosterone. Ring I is fully aromatised in all the hormones of the follicular group. Though they have the same effects they differ in potency, those from the urine being not quite so active as those from the tissues. In progesterone the side-chain is not so short as in the others. If ring IV has two hydroxyl groups instead of a keto or one hydroxyl group, as in oestriol (p. 254), occurring in human pregnancy urine, the activity is reduced 100 times.

The structural relationship between many of these compounds is seen if starting with cholesterol we cut down the side-chain (p. 254). First pregnandiol is produced, an oestrogenically inactive compound isolated from human pregnancy urine, then androsterone. If ring I of pregnandiol is aromatised progesterone (luteosterone) results. If pregnandiol is aromatised in ring I, and its side-chain at the same time cut down, we obtain testosterone, then on further aromatisation oestrone, and when ring II is also aromatised, equilenin, a weakly active substance isolated from horse pregnancy urine. Further aromatisation leads to the carcinogenic substances such as methylcholanthrene.

Corticosterone (p. 254), the hormone of the adrenal cortex, has since been shown to have a structure very similar to that of progesterone (review by Wintersteiner & Smith).

It now appeared that an overlap between the stimulus for cancer production and the stimuli of the sexual cycle existed. Aschheim & Hohlweg made the interesting discovery* that substances are present in extracts of bituminous material such as coal and mineral oil, which have considerable oestrogenic potency. Obviously this might be due to substances with a similar ring structure to the naturally occurring sex hormones. In accordance with this line of thought Cook, Dodds, Hewett & Lawson and Cook, Dodds & Greenwood examined a large number of synthetic hydrocarbons (the carcinogenic activities of which were known) as to their ability to produce oestrus when injected into castrated rodents, or to change the feathers in capons from the characteristic male to the characteristic female colour. It was thus found that high oestrogenic potency was possessed by bodies such as 1-keto-1 : 2 : 3 : 4-tetrahydro-phenanthrene. But some compounds were both carcinogenic and oestrogenic, such as 3 : 4-benzpyrene and 5 : 6-cyclopenteno-1 : 2-benzanthracene. Substances may thus be carcinogenic or oestrogenic or both (cf. the review of Schmidt-Thomé). This overlap is also found on the borderline between the oestrogenic substances and the sterols, for neo-ergosterol is oestrogenically active. Oestrogens were later found by Donahue & Jennings in ovaries of invertebrates, and by Skarżyński[1] and others in plant tissues, which, though apparently unaffected by oestrone in any way, can destroy it and therefore perhaps make it (v. Euler & Zondek). Even a progesterone-like substance has been obtained from plant tissues (de Suto-Nagy).

Pursuing further this chain of relationships, it has been possible to show that the ovarian hormones, interacting with hereditary factors, are responsible for the most frequent tumour of rodents, mammary carcinoma. L. Loeb and his collaborators were able to show that if in mice belonging to strains with a known high incidence of cancer the ovaries were extirpated at the age of three months the cancer incidence fell to zero (Lathrop & L. Loeb; L. Loeb[2]; C. F. Cori; Burrows[1]; L. Loeb, Burns, Suntzer & Moskop). If extirpation was performed at successively later periods, the cancer incidence correspondingly rose. Or if the animals were subsequently given oestrone (Lacassagne[1]) or a synthetic oestrogen such as triphenylethylene (Robson & Bonser) or stilboestrol (Shimkin & Grady), tumours developed. Seeing that the normal oestrogens of the body induce periodical growth of the mammary gland, it is of great interest that given certain hereditary predispositions, pathological growth may be induced there. Natural or synthetic oestrogens, however, will also induce uterine fibromata (Lipschütz & Vargas).

Since the first discovery of oestrogenic substances other than the naturally occurring hormones and their derivatives, many other substances not possessing

* Later often confirmed, as by Arthus & Provoost.

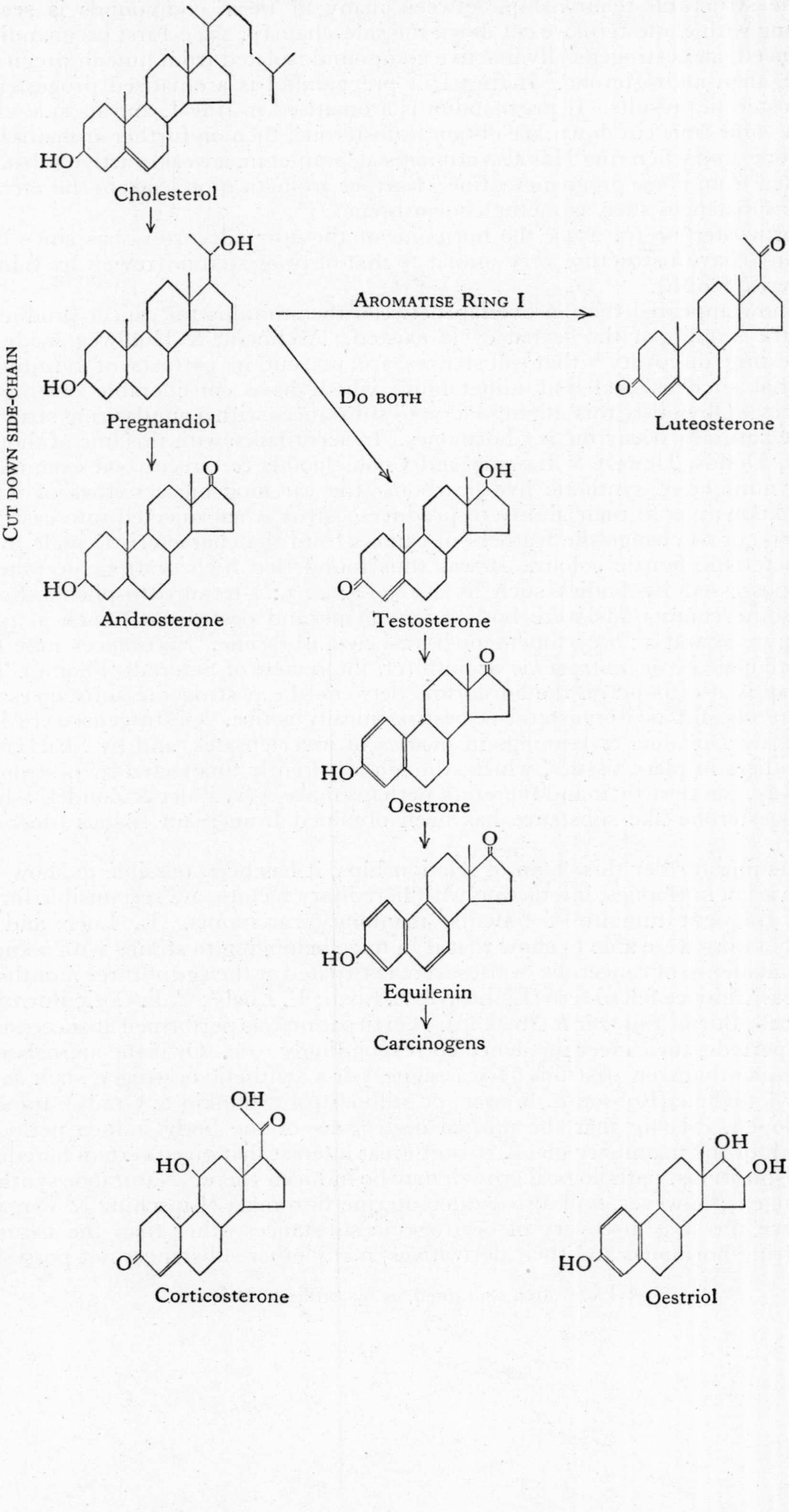
HO
Cholesterol
OH
HO
Pregnandiol
AROMATISE RING I
O
O
Luteosterone
CUT DOWN SIDE-CHAIN
DO BOTH
O
HO
Androsterone
OH
O
Testosterone
O
HO
Oestrone
O
HO
Equilenin
Carcinogens
OH
O
HO
O
Corticosterone
OH
OH
HO
Oestriol

the phenanthrene ring system have been found to be active (see p. 256). Knowledge of oestrogens has thus developed in the same direction as that of the carcinogens. The simplest of the simple polycyclic systems is the double ring of the diphenyls and stilbenes (Dodds, Golberg, Lawson & Robinson; Campbell), such as 4 : 4′-dihydroxy-diphenyl and 4 : 4′-dihydroxydistilbene. Later Dodds, Golberg, Lawson & Robinson reported that the two substances 4 : 4′-dihydroxy-α : β-diethyl-stilbene and 4 : 4′-dihydroxy-γ : δ-diphenyl-β : δ-hexadiene were several times more potent than oestrone and at least as potent as oestradiol. The acenaphthenes are more complicated. Some of them have been found to possess evocator activity and most are polyploidogens (see p. 509).

The search for a monocyclic oestrogen led to the examination of *p*-oxyphenyl-ethyl alcohol, which showed a weak activity. Strong activity on the part of *p*-propenylphenol (anol) was later traced to the presence of polymerisation products (Dodds & Lawson), such as di-anethole (Campbell, Dodds & Lawson).

What has this mass of facts to do with organiser phenomena? The answer (or perhaps the problem) is that polycyclic hydrocarbons act on competent ventral ectoderm like the primary evocator. Waddington & Needham (cf. p. 178), as soon as the first evidence in favour of the steroid nature of the evocator itself was available, tested 1 : 9-dimethylphenanthrene and 9 : 10-dihydroxy-9 : 10-dialkyl-1 : 2 : 5 : 6-dibenzanthracene (using the di-*n*-butyl derivative), substances strongly oestrogenic, with positive results (Fig. 92). Later the series was extended by Waddington [21], and the compounds found to be active as evocators are marked (E) in the accompanying tables of formulae, with the exception of naphthalene, cholesterol, anthracene, squalene, sitosterol and 5 : 6-cyclopen-teno-1 : 2-benzanthracene, where the effects were either very small or probably due to contamination with other substances. Chrysene, dihydrocholesterol and equiline were tested at the same time with quite negative results. Subsequently S. C. Shen obtained many fine inductions with the water-soluble 1 : 2 : 5 : 6-dibenzanthracene-endo-α-β-succinate, as already described (p. 182), and the minute dosage at which this substance was active suggests that the action of the hydrocarbons is direct rather than indirect.*

Whatever relations there may be between evocators, carcinogenic hydrocarbons, and steroid compounds, there are in the biological literature some curious facts which would be simply explained if a close chemical relationship were to exist, and which in any case have a significance of their own.

The first is the suggestive series of researches of Witschi [1]. Briefly the facts are as follows: If frog's eggs are kept for various lengths of time before being fertilised with sperm, those which are kept a short time are abnormal in their sex-determining mechanism and give an abnormal sex-ratio. Under the best conditions nearly 90 % males may be produced, and similar effects are known in fish development from the work of Mřsic; Huxley [2], and others. Those eggs which are kept rather longer produce many double monsters, indicating a disturbance in the normal processes of production or liberation of the evocator. Those which are kept longer still produce teratomatous proliferations whose malignant nature was tested by the formation of metastases on transplantation to older larvae. Now since we know that the principal sex-hormones belong to the group

---

* The steroids may, of course, have other, minor, morphogenetic effects, such as the asymmetries of neural fold formation described by Töndury [3], when testosterone, cholestenone, etc. act upon the young neurula from the outside.

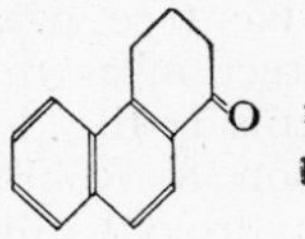

1, Keto-1 : 2 : 3 : 4-tetrahydro-phenanthrene (O, E)

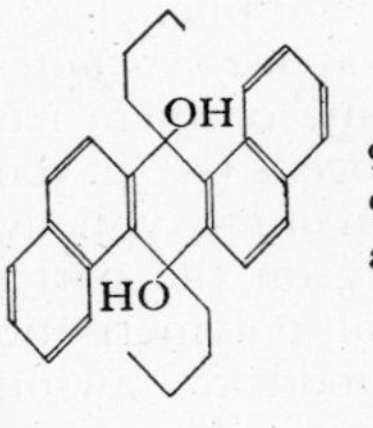

9 : 10-Dihydroxy-9 : 10-dihydro-9 : 10-dialkyl-1 : 2 : 5 : 6-Dibenz-anthracene (O, E)

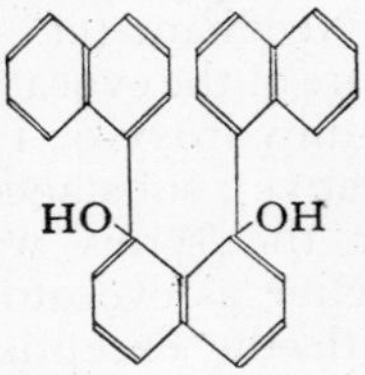

1 : 2-Dihydroxy-1 : 2-di-α-naphthyl-acenaphthene (O, E)

4 : 4'-Dihydroxy-diphenyl (O, E)

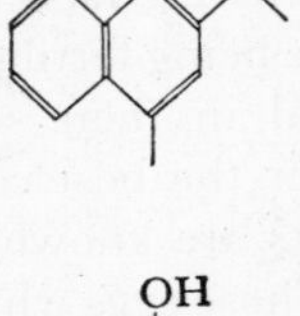

1 : 9-Dimethyl-phenanthrene (O, E)

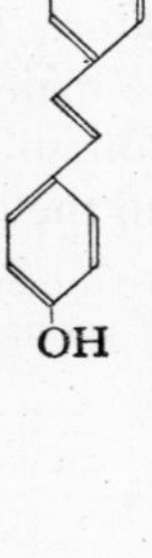

4 : 4'-Dihydroxydistilbene (O)

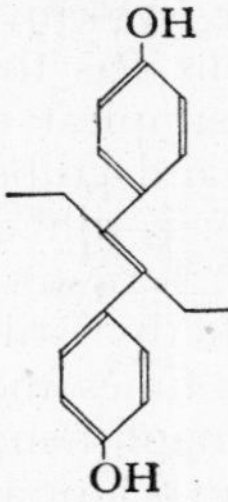

4 : 4'-Dihydroxy-α : β-diethyl stilbene (diethylstilboestrol) (O)

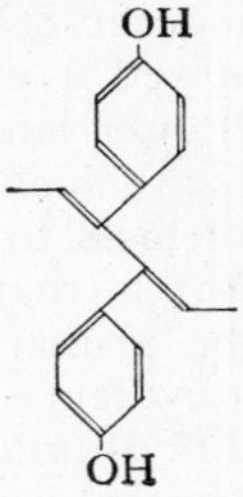

4 : 4'-Dihydroxy-γ : δ-diphenyl-β : δ-hexadiene (O)

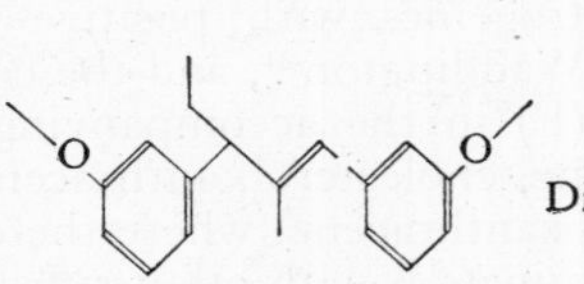

Di-anethole (O)

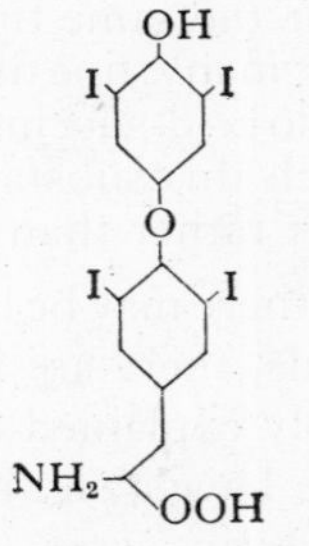

Thyroxin

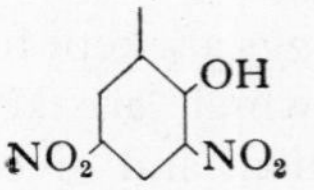

Dinitro-o-cresol

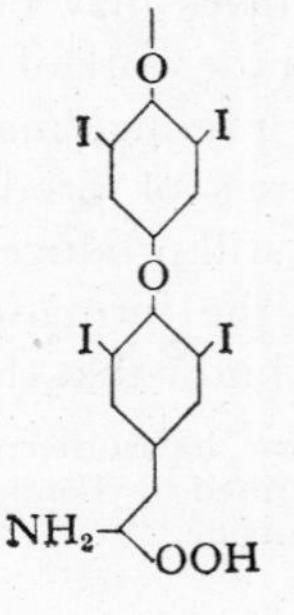

o-Acetyl-thyroxin

of sterols, since we have certain indications that the primary evocator also belongs to this group, and since we can be fairly certain that carcinogenic agents have a related ring-structure, the whole series of effects could be explained on a unitary hypothesis, namely, that over-ripeness involves a progressive disturbance of steroid metabolism.

Let us look a little more closely at the malformations which follow medium over-ripeness. The simplest were duplicated forms, as shown in Fig. 135, and in these no doubt the neural tubes were joined. Such cases may arise from an obstacle in the way of the cell streams during gastrulation, and cannot be certainly regarded as the work of accessory organisers. But as the other drawings in the figure show, many cases were obtained in which secondary embryos grew out of the flank or the belly, giving an exterior appearance very similar to that of induced embryos caused by implantation of organiser material or of active evocator substances. In a later paper, Witschi[1] tells us that in a late-fertilised embryo there may be seen, at gastrulation, the formation of multiple blastopores. A double breakdown of the normal correlating mechanism has therefore occurred, first, the inhibitory functions of the grey crescent individuation field, which normally bring it about that the blastopore appears at one point and one point only of the egg's surface, have failed to act, so that several more or less successful invaginations begin. Secondly, the metabolic processes in the embryo, which normally lead to the liberation of the evocator only in one region, have got out of control and liberate it in improper places, such as the lateral and ventral regions. Over-ripeness acts, therefore, like heat or an organic solvent, in liberating the evocator indiscriminately. In many of Witschi's cases the secondary neural tubes and notochords were quite separated from those of the "host", proving that accessory organisers had been involved.

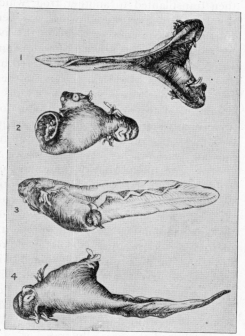

Fig. 135. Malformations caused by fertilisation of over-ripe eggs in the frog.

But suppose that the over-ripeness has proceeded still farther. According to Witschi's description the ectoderm becomes much thickened, consisting of large cuboidal cells, and "often by a wild growth, gives rise to epitheliomata". The neural tube may be completely missing and in its place there may only be a mass of incompletely neuralised cells. If the brain is formed at all, there may be disturbances of second-grade induction, so that a malformed optic cup growing out may cause no lens formation in the skin. Large tumour-like masses appear in the endoderm, and the separation of mesoderm into notochord and somites may not even be attempted. Early death ensues.

Could these very interesting tissues be kept alive by transplantation into normal animals? Witschi[1] found that they could, and that in many cases they grew by infiltrating the reticular system, coming to resemble in appearance the malignant melanosarcoma of man. Then in a just metamorphosed frog which had carried a strongly growing implant for 22 days, a metastatic nodule was found in the connective tissue. The nodule and some other tissues were then transplanted to another frog, which some weeks later developed a large tumour on the peritoneum which was penetrating the bladder and destroying it. Further metastases were found in the liver. In this way, the embryonic tissue, by reason of its clearly acquired malignancy, had lived 92 days instead of the 14 or so which it would have survived had it been left in its original isolation.

Three stages would exist, then, in the disturbance of embryonic sterol metabolism by over-ripeness: (1) deviations in the sex-hormones, (2) uncontrolled liberation of the evocator, (3) production of a carcinogenic agent. Embryonic abnormalities are, indeed, of fairly frequent occurrence after late fertilisation* (see the paper of Blandau & Young on guinea-pig eggs and Lillie's work on annelid development, p. 511 below).

It is unfortunate that no photographs of Witschi's epitheliomata have been published, but it is of interest that during the investigation of evocator action by chemical methods, similar conditions were observed (Needham, Waddington & Needham). In the study of the controls, by implantation of pure neutral fats, proteins, and animal and plant sterols, it was seen that whereas the implantation of triglycerides produces absolutely no reaction in the ectoderm (Fig. 91 A), the implantation of triglycerides mixed with small amounts of sterols, such as cholesterol, gave rise to swellings of a characteristic cuboidal epithelium. There is in such cases a random distribution of cells, and at the edges the cuboidal shape is not lost, so that a saw-like or serrated edge is presented to the exterior. The condition was illustrated in Fig. 91 C. We used to refer to these forms as "sterol bumps", and the case illustrated is by no means the most extensive of such thickenings, which sometimes attained a great size. This is certainly proliferation without differentiation. It would be of considerable interest to see whether tissue of this sort would show any malignancy when implanted into a larva or an adult frog.

Other investigators have had similar experiences. Tur[2], who has made a study of the "anidian" blastoderms of sauropsida (see above, p. 223), speaks of "neoplasmoid" blastoderms in birds and reptiles. In these cases, the blastoderm may form an enormously thick mass of ectodermal cells, incapable of normal differentiation, but situated at the centre of an otherwise normal area pellucida. The normal ectoderm would be from 15 to 90 μ in thickness; but this material had an average thickness of 180 μ. Tur's discussion[3] attributes a cancerous quality to

---

* Zimmerman & Rugh could not reproduce Witschi's results, but they worked with different amphibian species.

these curious formations, which require much further investigation.* One is depicted in Fig. 111.

It is at any rate sure that the chick embryo and especially the allantois has competence to react to the Rous sarcoma agent. As Rous & Murphy showed, cell-free extract of the sarcoma, applied to the allantoic membrane, causes the production of sessile nodules as much as a centimetre or more in diameter in the course of a few days. Later, Rothbard & Herman could not get benzpyrene-induced tumours to grow in the chorio-allantoic position, but another, "virus", tumour, the Stubbs-Furth sarcoma, grew there readily, even when the stimulus was applied as cell-free extract.†

A similar line of investigation, which has not been followed up much in recent times, was that of workers who implanted embryonic tissues into the tissues of adult animals. In some cases cancerous growths, it is claimed, have been produced by these means. There is, for instance, the long memoir of Belogolowy in 1918, illustrated with drawings and micro-photographs remarkable for their obscurity. Belogolowy took morulae and blastulae of anuran amphibia and implanted them into the tissues of adult frogs, or into the body-cavity. In many cases, as would be expected, self-differentiation followed, but quite often the "*Laichball*" came to lie in such a position that when a large number of cuboidal epithelial cells were formed, they penetrated the surrounding tissues by infiltration and presented the appearance of a round-celled sarcoma. Later, work of the same kind was carried out by v. Tiesenhausen and Skubiszewski on the fowl. Hashed chick embryos implanted intramuscularly or into other tissues produced, in a few cases, appearances which this author classified also as "round-celled sarcoma". At this lapse of time, it is perhaps

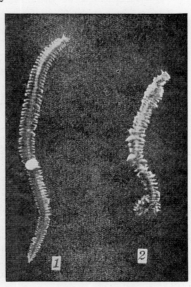

Fig. 136. Spontaneous tumours in the annelid *Nereis*.

difficult to appraise the meaning of such work, but although it was undertaken, no doubt, at the instigation of a theory of "cell-rests" which does not much interest us to-day, it may nevertheless have resulted in the discovery of some interesting facts, and might with advantage be further pursued.

Now cases of natural implantation of embryonic material are known, namely, the retention within the body of eggs which should have been shed. While working at the marine biological laboratory of Roscoff in France some fifteen

---

* Kaestner is sometimes cited as having found neoplastic growth in the early chick embryo, but his illustrations show nothing more remarkable than irregular bulges on the neural tube. Tumours of *late* foetal life are, of course, well known, though rare (Fischer-Wasels; Wells).

† It was Rous & Murphy's work which led to the use of the chorio-allantoic membrane as a site for the culture of viruses; another story altogether. Key references here are the original paper of Woodruff & Goodpasture, and the monograph of Burnet.

years ago, my wife and I were struck by the frequency with which polychaete worms of the genus *Nereis* carried what seemed to be tumour growths, either of a whole segment, or of the parapodia. This phenomenon was subsequently the subject of a couple of studies by Thomas[1], who showed that it was due to the degeneration of the oocytes which had not been shed as fertilisable eggs. The condition is shown in Fig. 136. Later on, the worm often breaks in two at this point, and two normal individuals are formed. The cross-section of Fig. 137 shows the early stages of a tumour. The degenerating eggs have stimulated the growth of a connective tissue which has infiltrated and already quite disorganised the left dorsal muscle. On the ventral surface a proliferation can also be seen, but much less advanced.*

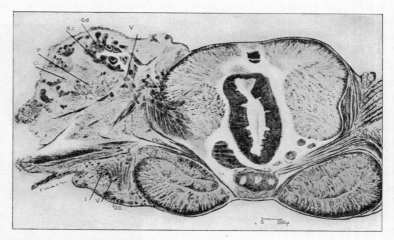

Fig. 137. Cross-section of a tumour in the annelid *Nereis*. *V*, Blood-vessels; *Od*, degenerating eggs; *Rc*, proliferating connective tissue; *P*, parenchyma; *Cl*, free cells; *I*, clump of cells.

An interesting review on cancer in invertebrates was subsequently written by Thomas[2].

Non-embryonic autoplastic transplants have recently been made by A. Fischer[6], who after repeatedly moving a piece of mammary gland around from flank to neck, and so on, allowing it to heal in each time, obtained at last more or less malignant carcinomata. He regards metastases as natural autoplastic grafts. All kinds of metabolic derangements might, of course, be expected, in such regrafted tissues, not only deviations of steroid metabolism. Goerttler[2], moving in the same borderland, has implanted pieces of amnion and uterine wall into adult muscle. He claims to have observed the development of epithelial cysts and even of hair follicles. He is probably right in thinking that such tissues still contain some curious potencies.

* Cf. the suggestions of a connection between milk stagnation and mammary cancer in mammals (Fekete & Green; Bagg[5]).

2·384. The nature of biological specificity

The facts just discussed have during the last ten years compelled a thorough overhaul of our ideas of what is meant by biological specificity. As Dodds has insisted in several reviews, there are, beside the normally occurring active substance—the key itself—a number of "skeleton" or even "pass" keys. The meaning of this fact is still unknown. It may be that the normal active substance does not act as such at its site of action, but has to be broken down in metabolism to something chemically simpler, and that all the other substances which act as well are all capable of being broken down to the identical active grouping. Or it may be that in order to act the active substance must combine into the form of some complex with a specific protein, and that all the active substances are capable of forming this complex and so presenting their active grouping to whatever mechanism it affects. Efforts directed hitherto, however, to determining some common grouping in all the substances which will bring about a given effect, have not so far been attended with success (Friedmann[1]; Cook & Dodds; Cook, Dodds & Warren).

In any case, it is clear that there is a considerable overlap between the activities of various substances. The situation may be represented by a diagram such as that in Fig. 138. A given biological effect can be produced at will, not by one chemical substance, but by any one of a group of more or less related chemical substances. We may therefore represent this by a contour line enclosing a domain. But as some of the substances in this domain will bring about other biological effects also, the domains must be drawn overlapping with one another. Such a representation takes no account of the relative potencies of the various compounds, and it would indeed be preferable to show the domains not as overlapping areas, but on a three-dimensional diagram as interpenetrating cones with potency as the vertical axis. Since the number of compounds seriously studied, however, is still fairly restricted, and since in some cases, such as organiser action itself, no quantitative method of dosage has yet been applicable, the two-dimensional presentation must for the time being suffice.*

The central domain of the diagram represents the sterols and bile acids. Flanking it in various directions lie the domains of the alkaloids, the resin acids, the saponin aglycones, and the adrenal cortex hormone. Overlapping with it are domains of the anti-rhachitic compounds and the cardiac aglycones; the former because substances (2), (40) and (41) have a weak anti-rhachitic action (Sherman & Sherman), and the latter because substances such as (44) are known with a weak digitalis-like action. Lower down and to the left lie the domains of the "male" and "female" stimulating substances. The male domain overlaps with the female domain because some substances exist, such as (10), (12) and (13), which have both male, and even both types of female, stimulating properties (cf. Deanesly & Parkes; Emmens & Parkes; Korenchevsky & Hall). Substance (14) is neo-ergosterol, which has a relatively weak oestrogenic potency (Cook,

---

* A domain which could also have been included on the diagram is that of the substances inducing polyploidy (see p. 509) for both colchicine and the oestrogenic acenaphthene derivatives are strongly active in this respect.

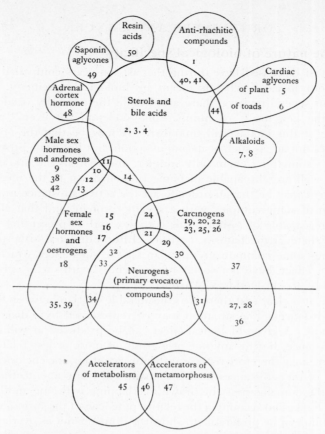

Fig. 138. The overlapping domains of biological specificity. Above the horizontal line, condensed polycyclic compounds; below it, simple polycyclic and monocyclic compounds.

(1) Calciferol (vitamin D)
(2) β-Cholic acid
(3) Cholesterol
(4) Scymnol
(5) Digitoxigenin
(6) Bufotalin
(7) Morphine
(8) Codeine
(9) Androsterone
(10) Testosterone
(11) Anhydro-oxy-progesterone
(12) Δ⁵-Androstenediol
(13) Trans-dehydroandrosterone
(14) Neo-ergosterol
(15) Oestrone
(16) Oestradiol
(17) Progesterone
(18) 1-Keto-1 : 2 : 3 : 4-tetrahydro-
phenanthrene
(19) 10-Methyl-1 : 2-benzanthracene (39)
(20) 5 : 10-Dimethyl-benzanthracene
(21) 3 : 4-Benzpyrene (79)
(22) 5 : 9-Dimethyl-1 : 2-benzanthracene
(23) 3 : 4 : 5 : 6-Dibenzcarbazole (33)
(24) 5 : 6-Cyclopenteno-1 : 2-dibenzan-
thracene (48)
(25) 1 : 2 : 5 : 6-Dibenzanthracene (36)

(26) 1 : 2 : 5 : 6-Dibenzacridine
(27) Triphenylbenzene
(28) Tetraphenylmethane
(29) 1 : 2 : 5 : 6-Dibenzanthracene-endo-
α-β-succinate
(30) Methylcholanthrene (101)
(31) Styryl blue
(32) 1 : 9-Dimethyl phenanthrene
(33) 9 : 10-Dihydroxy etc. (see p. 256)
(34) 4 : 4′-Dihydroxy-diphenyl
(35) 4 : 4′-Dihydroxy-distilbene
(36) 2-Amino-5-azo-toluol
(37) 2-Methyl-3 : 4-benzphenanthrene
(38) Androstane-diol
(39) Di-anethole
(40) Irradiation product of 7-dehydrositosterol
(41) Irradiation product of 22-dehydroergo-
sterol
(42) Δ⁴-Androstene-dione
(43) Colchicine
(44) Desoxycholic acid
(45) 4 : 6-Dinitro-o-cresol
(46) Thyroxin
(47) Acetyl-thyroxin
(48) Corticosterone
(49) Sarsapogenin
(50) Abietic acid

N.B. Figures in heavy type show relative carcinogenic activity in Cook's units[3].

Dodds, Hewett & Lawson). The carcinogenic domain does not overlap with the sterol domain, for up to the present no substance with a typically sterolic ring structure has been shown to possess carcinogenic activity. On the other hand, both the oestrogenic and the carcinogenic domains are prolonged downwards to cross the frontier separating the condensed polycyclic hydrocarbons (above) from the simple polycyclic and monocyclic structures (below). There are simple polycyclic oestrogens, (35), (39), just as there are simple polycyclic carcinogens, (27), (28), (36).

The compounds which have proved active as "neurogens", i.e. which have stimulated the ventral ectoderm of a gastrula to neural differentiation, are shown as forming an intermediate domain between the oestrogens and the carcinogens. All possible overlaps have been found. Thus one compound, (21), possesses all three forms of activity; two, (29) and (30), are carcinogenic and neurogenic; two, (32) and (33), are oestrogenic and neurogenic, and the overlap is repeated in the case of the simple polycyclic compounds just as in the case of the condensed polycyclic ones. The question whether these hydrocarbons act upon the ventral ectoderm directly, imitating the action of the natural primary evocator, or indirectly, liberating the evocator present masked in the competent tissue, still remains open, though we have noted evidence (p. 183) that the action is likely to be a direct one, since compound (29) acts in such minute dosage.

The overlap between the male and female domains is one of great physiological importance, though it cannot be dwelt upon here. It is now clear that in the physiology of sex-hormones a development has taken place analogous to that which has occurred in the study of organisers. As the chemical specificity decreases, so more weight has to be thrown on to the reactivity, i.e. the competence, of the tissues. Zuckerman[2] has pointed out that the testis can produce oestrogens as well as the ovary, and it is appreciated that the richest source of oestrone is the urine of the stallion (Zondek). The observation that after ovariectomy the implantation of a small piece of testis will restore the uterus to normal condition is now an old one. Both male and female urine and tissues contain both oestrogens and androgens (Fee, Marrian & Parkes; Womack & Koch). In an interesting series of papers Burrows[2]; Burrows & Kennaway and G. van Wagenen have studied the effects produced in male mammals by injection of oestrogens, such as separation of the pubic bones, testicular hernia, and hypertrophy of the prostate. Conversely, androgens affect the female mammary gland (G. van Wagenen & Folley). Zuckerman[1,2] has shown that these effects can be explained by different reactivities on the part of anatomical derivatives of the Mullerian (primary female) and Wolffian (primary male) ducts (see p. 311). Thus the hydatids of Morgagni (the vestigial Fallopian tubes of the male) respond well to the injection of oestrone (Zuckerman & Krohn). Specific competence may therefore be retained by vestigial organs in virtue of their embryonic origin. Conversely, some embryonic tissues seem not to have acquired this specific competence. Hain has shown that partial hypospadias of the female rat foetus can be caused not only by oestrone but also by dihydroxyandrosterone.

The diagram has been amplified by the inclusion, in the region of the simple polycyclic compounds, of two domains, one representing accelerators of amphibian metamorphosis and one representing accelerators of metabolism. The most classical case of a specific morphogenetic stimulation produced by a glandular hormone is no doubt the effect of thyroxin upon amphibian metamorphosis, discovered by Gudernatsch[1] and investigated by many workers, such as Huxley[5]. After the establishment of the circulation and the development of the thyroid gland, thyroxin is poured into the blood at the time of metamorphosis. The processes of growth and differentiation which form the hind-limbs are thus activated, the resorption of the tail and gills takes place, and a host of other changes, such as the perforation of the operculum (in anurans) and the development of the outer ear, follow. All these changes are not due to the specific action of thyroxin, though they take place much earlier than they would normally if thyroxin is fed. Many of them involve local specific stimuli of the nature of third- or fourth-grade organisers (see pp. 447 ff.). It is likely, however, that most of them are to some extent under thyroid control.

Now it is known that in mammals thyroxin considerably raises the basal metabolic rate, but has little or no effect upon the isolated tissue-slice (Dodds). Thyroxin also increases the gaseous exchange and probably the heat production of the tadpole (Huxley[5]; Schwartzbach & Uhlenluth). But the effect of such compounds as dinitro-o-cresol, which in man have a hyperthermising tendency like thyroxin, and which greatly increase both the respiration and the glycolysis of tissue-slices, has been tried on metamorphosis by Cutting & Tainter and found to be negative. Thyroxin cannot be replaced by them. Conversely, acetyl-thyroxin, which accelerates amphibian metamorphosis, has no hyperthermising effect in mammals (Swingle, Helff & Zwemer). We see, therefore, that just as the situation in the steroid realm can be represented by a system of overlapping domains, so also can that in the realm of accelerators of metabolism and metamorphosis.

In all these questions, the conception of competence must never be forgotten. Evidently the tissues of the tadpole are competent to react to thyroxin, since if it is fed, metamorphosis will be accelerated, but this is only the case after a certain developmental stage has passed. Under other names, the conception of competence is being widely used at the present time in the experimental pathology of cancer. Thus amino-azo-toluene produces its tumours not at the point of injection, but only in the liver. The following cases illustrate the concept in use.

One of the most peculiar of the effects of avitaminosis is that described by Adamstone[1] in the chick embryo if it develops in an egg produced by a hen suffering from lack of vitamin E.* About the end of the fourth day of incubation pathological structures arise in the extra-embryonic blastoderm, forming what is called a "lethal ring". Intensive cell proliferation in the mesoderm results in the choking of the vitelline blood-vessels, so that the embryo perishes from anoxaemia and lack of nutritive materials (see Fig. 139). There is also serious bleeding into the exo-coelom, usually from the atrium of the heart, at and near the

* It is doubtful whether the deficiency here was a pure one.

rupture of which peculiar histiocytic mesenchyme cells occur. The general picture resembles that described by Evans, Burr & Althausen for the lethal effect of avitaminosis E on the developing rat embryo, where wild proliferation of mesoderm occurs, obliterating the placental vessels, together with weakening of the vessel walls and extravasation of blood.

These relationships might be expressed in the following way. In presence of vitamin E the extra-embryonic blastoderm is not competent to react to the carcinogenic stimulus, but lack of the vitamin makes it more sensitive.* Similarly, adult mammalian epithelium is competent to react to carcinogens, but as the work of Berenblum has shown, this competence can be abolished by the action of mustard gas (β-β-dichlorethyl-sulphide) in subtoxic doses. Such doses, according to Berenblum, Kendal & Orr, inhibit glycolysis both aerobic and

A   B

Fig. 139. "Lethal ring" in the chick embryo in vitamin E deficiency. A. Total mount of 110-hr. embryo showing lethal ring. B. Radial section of blastoderm through the lethal ring (96-hr. embryo). *ect*, Ectoderm; *som*, somatic mesoderm; *spl*, splanchnic mesoderm; *nom*, non-cellular layer of mesoderm; *va*, vitelline artery; *vv*, vitelline vein; *apm*, advanced area of proliferating mesoderm.

anaerobic. Chloracetal and its related compounds, too, inhibit glycolysis and reduce competence to react to carcinogens (Crabtree[4]). As Berenblum & Wormall have shown by immunological tests, mustard gas effects changes in proteins sufficient to convert them into proteins foreign to the organism, probably by combining with the free amino groups. This gives us a glimpse into the meaning of competence, for apart from any action of the substance on enzymes, protein molecules which could otherwise have been oriented into new configurations could no longer be so (cf. p. 119). It would be of the greatest importance for embryology

---

* Adamstone[2] subsequently described lymphoblastoma commonly occurring in young chicks on vitamin E free diets. The formula for vitamin E will be found on p. 266 (and see the review by L. I. Smith). By opening the heterocyclic ring and converting both oxygen atoms into quinonoid form, a substance active as vitamin K (necessary for blood-coagulation) may be obtained. This is the first case in which one vitamin has been transformed into another by a simple chemical process. The natural vitamin K is a naphthoquinone derivative with a phytyl side-chain (see the review of Weidel), and many naphthoquinones possess some of its activity.

to possess agents capable of enhancing or diminishing the competence of tissue regions for specific differentiations.

The most natural interpretation of the action of the carcinogens would be to regard them simply as growth-promoting substances acting on single tissues and forcing them to break the bounds of the animal's individuation field. But Haddow[3] has shown on the contrary that though the carcinogenic hydrocarbons do not affect the later growth of a hydrocarbon-induced tumour, they actually *inhibit* the growth of transplanted tumours of chemical origin (Haddow & Robinson[1,2]) and of spontaneous tumours (Haddow[2]), and inhibit the growth of

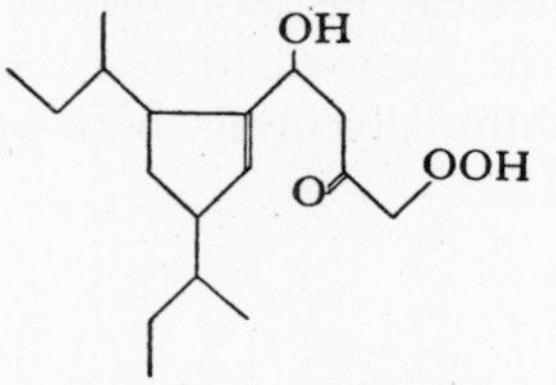

Auxin b (auxenolonic acid)

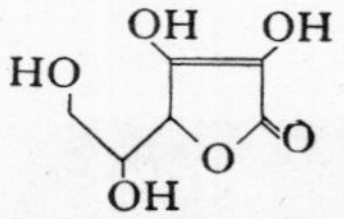

Ascorbic acid (vitamin C)

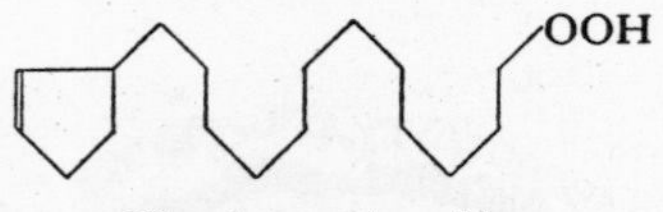

Chaulmoogric acid

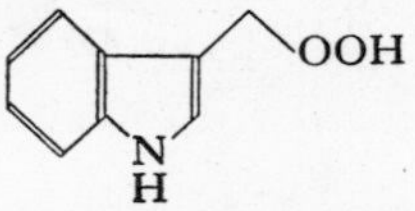

3-Indol-acetic acid (hetero-auxin)

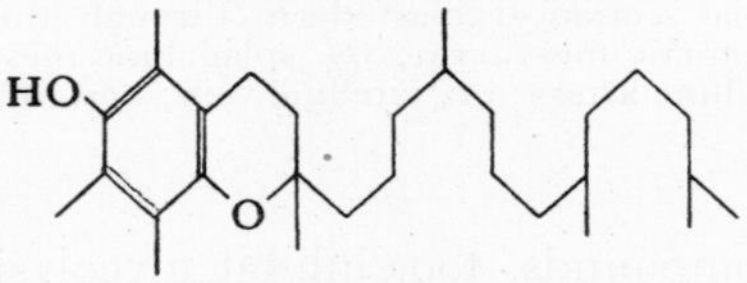

dl-α-Tocopherol (vitamin E)

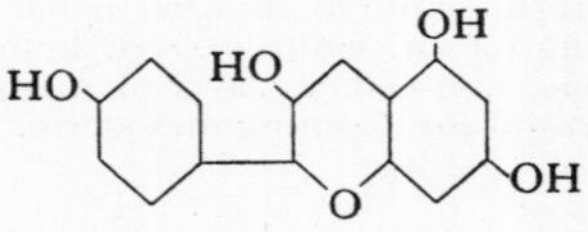

Pelargonidin

the whole animal bearing them (Haddow, Scott & Scott), without interfering with the animal's metabolic rate or food-consumption (Lees). In general, non-carcinogenic hydrocarbons do not show these effects. In order to account for this, Haddow[1] proposed an analogy with the formation of secondary colonies in bacteria, differing in their fermentative properties from the primary colony. According to this view, carcinogens primarily inhibit growth, until after a while some cells, by a kind of mutation, develop resistance to the growth-inhibiting effect and begin to proliferate uncontrollably. It is doubtful whether this theory will prove acceptable.* The long latent period (often of months) required

* A newly discovered fact which shows how cautious we must be in interpreting the Haddow effect is that the tumour-inhibiting power of the carcinogens has been shown to be very dependent on the nature of the solvent which is used as a vehicle for them (Morelli & Dansi).

before the carcinogens take effect is certainly mysterious enough. While the view of Haddow[1] would account for the reparation of any inhibition of local, or the restoration of general, growth to the normal, it hardly seems to account in any precise way for the escape from the organism's individuation field and consequent proliferation *over and above* the normal which the carcinogens initiate.*

The only hint we have as to the mechanism of the Haddow effect arises from the work of White & White who find that the growth-inhibiting action of hydrocarbons, not, in their experiments, entirely confined to those with marked carcinogenic power, can be fully reversed by the administration of the –SH compounds, cystine and methionine. They suggest that the detoxication of the hydrocarbons employs sulphur which would otherwise have been available in the body, e.g. for the sulphydryl stimulation of growth (p. 420). If this were the case, we should expect that small doses would show a growth-accelerating action while large ones would show a growth-inhibiting one. This is exactly what has been found experimentally by Hearne-Creech in careful work on explants of mouse fibroblasts, and there is confirmatory evidence (Morogami; for yeast, Cook, Hart & Joly). That small doses have a different effect from large ones is after all a common phenomenon in biological reactions, and is shown in cancer itself by the effects of X-rays.

2·385. Homoiogenetic induction and cancer

In the three foregoing sections, we have considered the relation between organiser phenomena and cancer, first with regard to the escape from the individuation field, and secondly with regard to the chemical carcinogens. But there is a third point of contact between experimental morphology and cancer research. Since the work of Rous[1] in 1910 it has been known that some tumours can be transmitted from animal to animal, not merely by transplantation but by the injection of rigorously cell-free extracts. This has led to a very large amount of work (well reviewed by Foulds[1]) and to the hypothesis that a filtrable virus is the causative agent of these tumours, and perhaps of all tumours. There has been some difficulty in reconciling this view with the facts which have come to light in the study of the chemical carcinogens. As we have seen, there are fair grounds for picturing the production of carcinogenic substances in the body by deviations in the normal metabolism of the steroid group.

The foundations of the virus theory appear to be two in number, first that cancer phenomena exhibit many immunological properties, and secondly that the "virus" multiplies in the host, since after a certain period of tumour growth much more of the active agent can be obtained from the tumour than what was originally injected. Now it is interesting that a formal analogy exists between these types of cancer transmissible by cell-free extracts, and homoiogenetic

---

* It is to be noted that, like embryonic inductors, the carcinogens do not seem to affect the process once it has been set in motion.

induction.* From the work of Mangold & Spemann (see p. 154) we know that as soon as the neural plate of the amphibian embryo is induced by the organising influence of the roof of the archenteron, this neural tissue itself acquires the power of inducing a secondary neural plate, when implanted into the blastocoele cavity of another embryo. A whole series of neural plates could therefore be "propagated" in this way, and it is sure that the process could be carried on by implantation not of neural plates but of cell-free extracts of them. There seems thus to be an analogy between the induction of a neural tube, followed by the homoiogenetic induction of successive neural tubes by cell-free extracts of each other, and the induction of a neoplastic growth by tar or a carcinogen, followed by the successive propagation of this cancer by cell-free extracts. That chemically induced cancer *can* be transmitted by cell-free extracts is not perhaps as yet fully established, but the original experiments of McIntosh have been confirmed and extended (Parsons; McIntosh & Selbie).

The analogy depends upon the assumption that in the competent ectoderm there exist stores of the active substance masked probably by combination with a specific protein. The same may be true of adult animals with regard to an endogenous carcinogen. Or the synthesis of further supplies may be stimulated autocatalytically like trypsin from trypsinogen or plant viruses in their hosts (Northrop; Bawden & Pirie). The active chemical carcinogenic agents may not be the hydrocarbons themselves, but more or less loose combinations of them with specific protein. This hypothesis has two advantages. In the first place it would probably allow for all the immunological and serological phenomena which have led to the postulation of a cancer "virus". It is not for biochemists or experimental morphologists to criticise the literature of pathologists, but the study of such works as the book of Gye & Purdy or the lectures of Rous[2] or of Andrewes has failed to convince one reader at any rate that it is either necessary or helpful to speak of the causative agent of cancer as if it were a living organism.† This particular controversy is no doubt a barren one, since, as Pirie[1] has held, any combination of properties may be expected at the borderline between the smallest particles which we do not hesitate to call living and the largest ones which we do not hesitate to call non-living. But of the reality of the immunological evidence there is no doubt, so that specific proteins are clearly involved. Studies such as those

---

* This analogy arose in a conversation with C. H. Waddington and was suggested in a letter to the *British Medical Journal* in 1936. We afterwards found that the parallel between organiser phenomena and filterable cancer-producing substances had been drawn by L. Loeb[4] in 1931 and the idea that hydrocarbon-protein complexes may be involved had been rather vaguely put forward by Pentimalli[4] in 1935. It seems to have puzzled some reviewers. Thus F. E. Lehmann[12] (in the *Ber. u. d. ges. Biol.*) thought that the active carcinogen-protein complex was being compared with an *active* evocator-protein complex. A glance at p. 208 will show that combination with protein may either inactivate or activate a compound. Foulds[2] (in *Amer. Journ. Cancer*) thought that neural evocation could not be compared with carcinogenesis because the latter involved individuation. A glance at pp. 234, 235 shows that this is the last thing carcinogenesis does! It will be admitted that stimulus to differentiation is a very different thing from stimulus to uncontrolled growth.

† It is an instructive exercise to read through the writings on the virus theory of cancer, substituting the words "active agent" or "active extract" for "virus", wherever it occurs. The results are illuminating.

of Beard & Wyckoff, along lines which have been so fruitful with the plant viruses, will help more than anything else. Perhaps the hypotheses of active and inactive hydrocarbon-protein complexes will also be found helpful in cancer research.

In this connection we may remember the sterol-protein complexes with antigenic properties, reviewed by Marrack, and the compounds of proteins with carcinogenic hydrocarbons made by Creech & Franks. There are investigators of the Rous sarcoma virus who claim that part of its active agent may pass through organic solvents (Jobling, Sproul & Stevens; Menke), but this is not as yet generally accepted. Nevertheless Claude[1] has isolated from the Rous sarcoma extract an extremely active fraction which consists of two parts, a lipoid part accounting for 35 % and a protein part accounting for 65 %. The former resembles the aldehydic phosphatide plasmalogen (see p. 190) and the latter is a ribose-containing nucleoprotein. Most remarkable, however, is the fact that Claude[2] has isolated a very similar or identical fraction from 8-day chick embryos, possessing no carcinogenic activity. The question is, he writes, "whether the main constituents of the purified chicken tumour fraction represent inert elements existing also in normal cells; or whether the substance found in normal chick embryo tissue represents a precursor of the tumour principle, which could assume, under certain conditions, the self-perpetuating properties of the tumour agent."* Later Claude[3] succeeded in isolating the same fraction from mammalian embryos, and finds that it is associated with particles of a particular size (50–200 $\mu\mu$ diam.) indicating mitochondria or their fragments. On denaturation by heat or radiant energy, the nucleic acid spectrum appears (Claude & Rothen). These discoveries give great significance to the finding of Sturm, Gates & Murphy that the inactivation spectrum of the tumour agent is that of nucleic acid, and suggest that the chick embryo material may be identical with the growth-promoting factor described on p. 628; a possibility now under examination by Tennant, Liebow & Stern.

The formation of complexes with proteins might perhaps explain the curious results which many workers have recently been obtaining with the chemical carcinogens. As detected by ultra-violet absorption spectrography they disappear from fowl muscle a few days after injection (Chalmers) and from the circulating blood (Peacock[2]). The bulk, says Chalmers, is transformed or undergoes chemical change. Chalmers & Peacock found that most of the amount injected is eliminated, and wrote: "It is possible that the action of the chemical carcinogens is far more rapid than has generally been assumed, and that essential changes occur in the cell long before malignant multiplication commences. If the carcinogens act as such and over a prolonged time, then the amounts required for carcinogenesis must be very minute, and they must become fixed in some way at the site of injection so as to be protected from the process of elimination." This agrees well with the view that a specific hydrocarbon-protein complex has to be

* Cf. the observation of Sharp, Taylor, Finkelstein & Beard that a macromolecular protein fraction very similar to that of horse encephalomyelitis may be obtained by centrifugation from normal 11-day chick embryos.

formed.* The minuteness of the amount probably combining with protein may account for the failure of Berenblum & Kendal to find more than traces of the hydrocarbons some time after injection when digesting the whole animal with potash and examining benzene extracts spectrographically. Moreover, attachment to protein might render the substances unstable to caustic potash. Finally, in all such experiments it must be remembered that Sannié has shown that the typical spectrum and fluorescence of the hydrocarbons may be quite extinguished in the presence of other substances, including proteins.

The objection to the virus theory of cancer lies not so much in that it endows with the properties of living matter particles which possess demonstrably only some of these properties. An autocatalytic specific protein with a hydrocarbon as prosthetic group, capable of replicating itself from some precursor substance, may be termed a "virus" if this is thought desirable, although some word with fewer biological undertones ought surely to be preferred. Plant viruses can now be obtained in the paracrystalline and even in the crystalline state. Why should we not call the primary evocator a virus, since it makes more of itself parallel with the histological change which it produces? The objection to the virus theory of cancer lies rather in the fact that it emphasises primarily the exogenous origin of uncontrolled specific proliferation. It has to avoid the difficulty that very few cancers are infectious by postulating the existence, perhaps in symbiosis, of a virus or viruses in most of the cells of the organism.† But on grounds of methodological simplicity ought we not to prefer the metabolism existing also in all the cells of the organism? Rous[2] himself has put the matter well. "The factual obstacles to a parasitic cause for tumours have been whittled down, but one theoretical difficulty bulks as large as ever. It takes this form 'all tumours must have the same cause. Certain benign tumours cannot possibly be due to a parasite, those manifestly referable to developmental anomalies (teratomata) for example. It is impossible to tell of some tumours whether they are benign or malignant. And so...no malignant growth can have a parasitic cause.'" Rous prefers of course the view that all tumours have not the same type of cause. An observer is entitled to doubt whether any proof of this negative has yet been given.

The relationships between chemically induced and transmissible tumours are now under close examination. The infective Shope papilloma of rabbit skin can be transmitted from rabbit to rabbit in cell-free extract, but when malignancy occurs it cannot be so transmitted. But if the active principle of the papilloma is injected into a rabbit the ears of which have been tarred until the wart stage, fulminant carcinogenesis occurs and true malignant tumours quickly develop (Rous & Kidd). Parallel experiments have been made with the active principle of the Shope fibroma. Though dibenzanthracene tumours are generally not transmissible in cell-free filtrate, they contain an antigen which is capable of

---

* In this connection it is interesting that the effects of the Rous sarcoma agent are greatly enhanced by the simultaneous injection of the testis spreading factor, now regarded as a mucinase (or of azoproteins, i.e. serum protein coupled with p-diazobenzenesulphonic acid, which have the same effect), as Claude[4] has shown.

† Is an "indigenous virus" which cannot be cultured by bacteriological methods outside the body anything else than just some unknown substance of protein properties found *in* the body?

giving rise to antibodies neutralising the active principle of the Rous sarcoma (Foulds[3]). If a dibenzanthracene tumour is growing in a fowl which also bears Rous sarcomata, the active principle of the latter can be isolated from the former (Mellanby[1]) so that a transplant of the former will reproduce itself but a cell-free extract of the former will reproduce the latter. However, all the normal tissues of a Rous-sarcoma-bearing fowl also contain the active agent (Mellanby[2])—a possible parallel with the presence of evocators in them but no competence. These results parallel those of others (Shabad; Neufach & Shabad; Kleinenberg, Neufach & Shabad), who have obtained tumours in mice by injecting benzene extracts of the livers or bile of cancerous humans; the tumours were transplantable but not transmissible in cell-free extract. In their most recent reviews Gye and Cramer[2] lay much emphasis on the question of competence, or as they term it, susceptibility. They also regard the difference between transmissible and non-transmissible tumours as very mysterious, but is it any more so than the capacity which some enzyme systems possess, and others do not, of coming out into cell-free extracts of tissues?

We may take leave of the cancer problem with a reference to the view which Lockhart-Mummery has gone about to prove, and to which Haddow[1] inclines, namely that tumours are produced by mutations of genes within normal somatic cells. In a review, Haldane[2] pointed out that "there is a common and very orderly type of change which satisfies Lockhart-Mummery's criterion of mutation, namely the differentiation of cells in the normal course of development. There is no evidence that changes of this kind are due to the alteration of a single gene, and a great deal to the contrary. Until it is shown that differentiation is due to gene mutations, it seems reasonable to regard carcinogenesis as anomalous differentiation rather than mutation." This sentence provides the justification for the discussion of cancer phenomena in a book on chemical embryology.

2·39. Regional differentiation
2·391. Evocation and individuation

In this section we now return to the consideration of normal amphibian development. In the introduction we saw (pp. 124 ff.) that it has been found profitable to distinguish between two components in embryonic induction: (1) *Evocation*, the chemical stimulus to axial neuralisation, and (2) *Individuation*, the field forces which render the ends of this axis asymmetrical by differentiating head end and tail end. What are the facts upon which this distinction is based?

Although the importance of the prechordal plate (part of the "head-organiser") in the morphogenesis of the head had long been appreciated (Kingsbury & Adelmann; Adelmann[1]), the subject was first approached experimentally in Spemann's investigation[13] of regional specificity in the organisation centre. Organiser material which invaginates first will eventually act as head-organiser and itself form part of the wall of the pharynx; organiser material which invaginates last will eventually act as tail organiser. Position on the axis implies therefore a time difference in the earliest stages, and by taking advantage of this, Spemann was able to remove organiser material at varying times from the onset of gastrulation and so have material of varying regional properties available for transplantation. He then implanted "head-organiser" or "tail organiser"* at different antero-posterior levels in host embryos. The four logical possibilities gave him the following results.

Operation	Result
Head-organiser at head level	Secondary head, with eyes and ear-vesicles
Head-organiser at tail level	Complete secondary embryo, including head, from presumptive trunk and tail ectoderm
Tail-organiser at head level	Complete secondary embryo, including head
Tail-organiser at tail level	Trunk and tail only

These relationships are shown also in the accompanying diagram (Fig. 140). Head-organiser can thus form a head at both head and tail levels, but tail-organiser

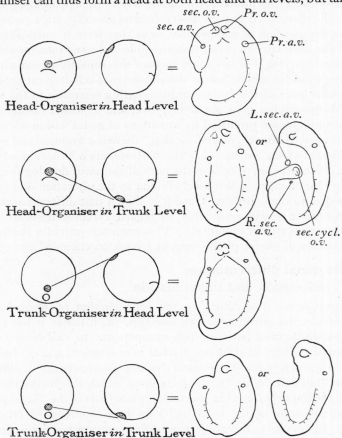

Fig. 140. Spemann's original experiment on the implantation of head- and tail-organiser at different antero-posterior levels. *Pr.o.v.*, Host eye-cup; *Pr.a.v.*, host ear-vesicle; *sec.o.v.*, induced eye-cup; *sec.a.v.*, induced ear-vesicle; *L*, left; *R*, right; *cycl.*, cyclopic. (From Huxley & de Beer[2].)

can only form a head at head level. The reaction of the host tissues to tail-organiser at head level is to form a head and at tail level to form a tail. These experiments

* It is inaccurate to use the word "tail" instead of "trunk" in this discussion, since at the stages in question the tail bud has hardly formed, but we do so for the sake of emphasis.

of Spemann's were made using the presumptive mesoderm as such. In later developmental stages, however, the same regional properties exist, as was shown for the notochord by Bautzmann[4] and for homoiogenetic induction by the neural plate by Mangold[7].

By 1933, therefore, the paradox had been reached that while on the one hand it was clear that a chemical substance or morphogenetic hormone was involved, it was also clear that there was some regional differentiation present in the organisation centre. It was this which led to the distinction between evocation and individuation. We must now try to analyse the two more closely.

When Holtfreter's[8] exo-gastrulae (see p. 159) were under intensive study, the opportunity was taken to find out whether the evaginated endo-mesoderm possessed regional differentiation. Tests were made in the manner shown in

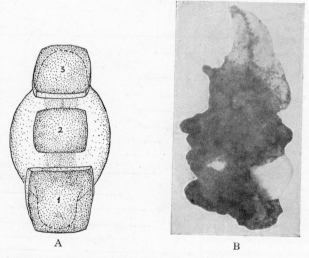

A   B

Fig. 141. Tests demonstrating the regionally specific inductive properties of the exo-embryo. A. Pieces of competent ectoderm laid upon the evaginated endomesoderm. B. Tail induced in such a piece by contact with the posterior end of the exo-embryo.

Fig. 141. Pieces of ectoderm from various ventral regions of the young gastrula were laid upon the evaginated material. After healing on to it they proceeded to develop entirely in accordance with the regional position of the underlying material; thus a piece placed on the end which had first evaginated produced brain, eyes, and ear-vesicles, while a piece placed on the caudal end (that nearest the abandoned ectodermal mantle) produced a tail. Antero-posterior regional specificity in induction was thus demonstrated[9] again.

In all experiments where an implantation takes place, as, for example, into the blastocoele cavity, there are three factors to be taken into account: (1) the influence of the host axis, (2) the specificity of the implanted material, and (3) any inherent specificity in the reacting material, i.e. the ventral ectoderm. In order to clarify the results which have just been described the third of these

factors had, if possible, to be ruled out, and this was done by Holtfreter[5] in his transplantations of pieces of ectoderm from various regions of the gastrula to various regions of the neurula. Here the competent ectoderm, from whatever region originating, showed itself capable of being induced to all kinds of structures. The distribution of these is shown in Fig. 142. Thus an open neural plate would be formed in the gill region, or at the level of the ear-vesicle there would arise a piece of brain with an olfactory pit and an eye with a lens. At other places glands, mesenchyme, cartilage, and muscle developed. Towards the tail end head organs were never formed, but notochord, nephridia, dorsal fin, and tails were induced. Fig. 143 shows an implant in the trunk region which was

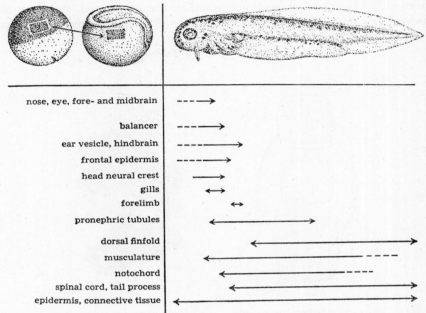

Fig. 142. Diagram showing the results of transplanting presumptive ectoderm or presumptive neural tube tissue from the gastrula to various regions on the flank of the neurula.

induced to form muscle, notochord, and a piece of nervous tissue, all surrounded by epidermis. The statistical relations between percentage induction of a given structure and various given regions of the host are also shown below in Fig. 160, to which reference should be made. As there were no differences between the behaviour of various regions of the gastrula, even between presumptive epidermis and presumptive neural plate, this work definitely proved that there is no regional specificity in the reacting gastrula ectoderm.

This being the case, we have to consider in transplantation experiments only the mutual influence of host individuation field and whatever specificities the graft brings with it. Normally, as the foregoing experiments have shown, the head region exercises some kind of "dominance", in that all other factors notwithstanding, a head tends to be induced in the host's head region, and head-organiser

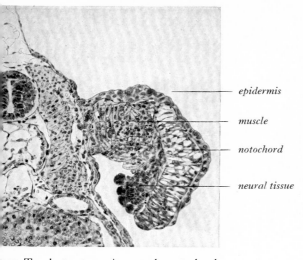

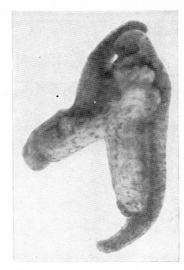

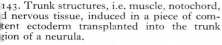

epidermis

muscle

notochord

neural tissue

143. Trunk structures, i.e. muscle, notochord, d nervous tissue, induced in a piece of competent ectoderm transplanted into the trunk zion of a neurula.

Fig. 144. Secondary embryo oriented almost diametrically opposite to the primary one.

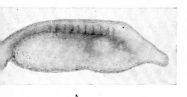

| A | B | C |

:45. Exchange of head- and tail-organisers in the primary axis (Hall's experiment). A. The embryo with louble "tail". B. Reconstruction of the neural tube. C. Cross-section through the anterior region where he head and brain should have been.

neural tissue formed in Triton *ectoderm*

neural tissue formed from Bombinator *implant*

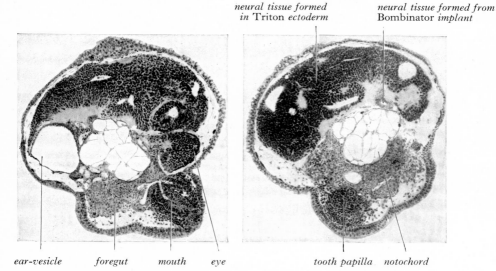

ear-vesicle   foregut   mouth   eye   tooth papilla   notochord

Fig. 148. Head structures induced in a competent ectodermal ball by implantation of living head-organiser.

tends to induce head even in the host's tail region. But cases are known where this dominance breaks down. Thus Fig. 144 shows an induced twin obtained by Holtfreter[9] by the implantation of a piece of living dorsal lip. The induced embryo was complete in almost every particular, but oriented in almost exactly the opposite direction to the antero-posterior axis of the host. The host's individuation field may therefore sometimes be incapable of overcoming the field of the implanted organiser. A similar case had been obtained in the chick on grafting the posterior two-thirds of the primitive streak by Waddington[3], and a number of naturally occurring twin embryos described in the monographs of Newman[1] and Dareste, the papers of Mitrophanov, etc., show more or less opposed embryonic axes.

So far we have discussed only the effects produced when two embryonic fields are involved, that of the host, and that of the secondary embryo induced by the implant. But if, instead of transplanting early-invaginating and late-invaginating organisers to other gastrulae, we were to exchange them with one another, we should have only the primary embryonic axis to deal with. This was done in the interesting experiments of E. K. Hall[2], who thus put tail-organiser where the head-organiser should have been, and head-organiser where the tail-organiser should have been. The result was that embryos were produced entirely lacking heads, and there was no development of brain vesicles or eye-cups; instead, the anterior part of the neural tube was prolonged forwards into an excrescence like an "elephant-trunk" (see Fig. 145). Thus tail-organiser had induced a tail-like neural tube at head-level. Correspondingly, no head structures were induced at tail-level although the head-organiser was present there. If Spemann's experiment[13] indicated a head dominance, therefore, that of Hall[2] almost indicated a tail dominance. But the two forms

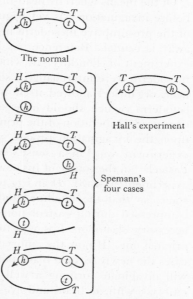

The normal

Hall's experiment

Spemann's four cases

Fig. 146. Comparison of the experiments of Spemann and Hall on regional differentiation. Inductors are indicated by small letters and the result by capitals.

of dominance are not the same. If normal head-organiser is present in its normal place in the primary axis it will, as we have seen, usually overcome tail-organiser at that level in a secondary axis. If there is no secondary axis, tail-organiser cannot take its place in the primary axis. The regional specificity therefore resides, as so many experiments have shown, in the inducing mesoderm. Head-organiser in the tail region of a secondary axis will overcome the tail-organiser of the primary axis and produce a head. But head-organiser in the tail region of the primary axis, when no secondary axis is present, can only produce tail (see Fig. 146).

These facts, though fascinating, remain enigmatic. Hall's only explanation was a loss of head-inducing properties on the part of head-organiser when situated in the tail region of the primary axis. But this is only a statement of the problem. Moreover, there are some further possible permutations which have not yet been tried. For example, what would a head-organiser do at tail level if the primary axis was represented there by a head-organiser, potent as regards neural-tube induction but impotent as regards regional specificity? Or what would a head-organiser do at head level if the primary axis was represented there by a tail-organiser forming an "elephant-trunk" excrescence and suppressing brain development? We must await further experimentation.* But in the meantime it is clear that the cautious will avoid giving assent to any theories of "physiological dominance" purporting to explain the original experiments of Spemann.

Of the means whereby head-organiser comes to be different from tail-organiser before invagination has yet begun we know extremely little, save that the part of the presumptive mesoderm nearest the blastopore *is* cephalic and that further away is caudal. Reference has been made already (p. 170) to the remarkable experiment of Töndury[1] in which presumptive neural tissue and presumptive mesoderm were interchanged, so that the former dorsal lip (which should have been organiser) allowed itself to be formed into neural tube, and the former ectoderm (which should have suffered a neural fate) invaginated and actively organised. Liberation of the primary evocator is thus in life absolutely dependent upon the invagination process with its special metabolic characteristics. The experiment was, in a sense, extended to antero-posterior differentiation by Pasteels[13] in the course of his new analysis of "maintained inversion" (Schultze's experiment; p. 218). If an egg which has been held upside down in the one-cell stage, so that the vitelline gradient has been reversed or altered, is severely compressed during gastrulation, it will fail to invaginate and after some hours accessory blastopore lips will form further away from the yolk-mass than the original one. If now the embryo is released from its compression, in favourable cases the newly formed blastopore will have the advantage of the former one and will invaginate forming a well-developed embryo. But the head of this new embryo, with its brain and eye-cups, will have been formed from what would have been, had the original blastopore persisted, "tail-organiser". Thus, wrote Pasteels, "a tail-organiser, maintained artificially in pre-gastrular conditions, may 'ripen' and become a head-organiser ".[†]

According to Lehmann[14], head-organiser and tail-organiser differ in their susceptibility to lithium chloride. The accompanying graph (Fig. 147) shows that

* Hall[2] used pieces from younger and older gastrulae; Waddington[27] tried to use pieces from gastrulae of the same age. But the dynamic determination was too strong. Head-organiser insisted on invaginating in the trunk position and tail-organiser would not invaginate at all. Hence there was little regulation; in the former case a rod of pure notochord stuck up through the neural tube. Perhaps a similar effect would explain the *pigtail* lethal in mice of Crew & Auerbach[2].

† Dalcq & Pasteels believe that the difference between head- and tail-organisers is primarily a quantitative one, and they express it by varying values of their C/V product. Certainly each point on the main axis during its formation must have a different value in terms of the vitelline gradient and grey crescent field co-ordinate system. But this explanation will remain verbal until some concrete differences, whether chemical or physical, have been demonstrated between different antero-posterior levels.

each section of the antero-posterior axis has its period of maximal susceptibility to a 6 hr. treatment. It might be thought that this would correspond with the time of inrolling through the dorsal lip, but that does not seem to be the case because the anterior head-organiser, for example, is particularly sensitive some time after it has passed through the blastopore lip.

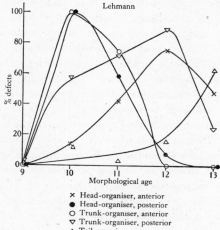

In the analysis of regional differentiation much help has been forthcoming from the use of pieces of ectoderm cultivated in isolation away from the individuation field of the donor. These round up into balls suitable for receiving an implant. Holtfreter[16] implanted pieces of head-organiser and tail-organiser into them, with the result that inductions were produced entirely according to the specificity of the inductor. Examples are shown in Figs. 148 (opp. p. 274) and 149 (opp. p. 278). In the first of these, head-organiser has induced a large piece of brain, an eye-cup, an ear-vesicle, and a mouth-rudiment with tooth-papillae. In the second, tail-organiser has induced muscle, notochord, and tail neural tube.

× Head-organiser, anterior
● Head-organiser, posterior
○ Trunk-organiser, anterior
▽ Trunk-organiser, posterior
△ Tail-organiser

Fig. 147. Specific sensitivities of head- and tail-organisers to lithium.

This regional specificity of the inducing mesoderm was retained for periods up to a week if the inductor was cultivated in isolation during that time.

On the other hand, Holtfreter's experiments[16] with pieces of organiser region after heat-coagulation showed equally clearly that the regional specificity was now lost.* Some of his experiments consisted of implantations of dead organisers into the blastocoele cavity of other gastrulae. In these cases the following figures were obtained:

	Brains	Noses	Eyes	Ear-vesicles	Lenses	Balancers	Meso-derm
Dead organisers:							
Head-organiser	19	5	5	1	2	9	2
Tail-organiser	22	3	2	6	0	8	0
Ventral ectoderm	12	3	0	4	2	7	1
Endoderm	7	1	0	2	0	3	2

There was no relation between structure produced and antero-posterior level in the host. Thus any dead part of the gastrula would induce almost any organ in the living gastrula. It is striking that in these experiments tail-organiser would induce cerebral structures, though, as we have seen from the work of Hall[2], it cannot do this if it is substituted while living for the normal head-organiser.

No relation was discernible between the manner of killing and the kind of structures produced by the dead organisers.

With implantations into isolated ectodermal balls Holtfreter obtained exactly the same results. Here the inductions were generally purely neural. An example

* Reith[4] thinks that his ultra-violet irradiated dorsal lips had lost their regional specificity, but this requires confirmation.

is shown in Fig. 80; it was produced by implantation of a piece of heat-coagulated ventral ectoderm. Such an effect might be regarded as pure evocation, but some kind of specificity might still be carried by the dead inductor. In order to carry the matter further it was desirable to implant into an ectodermal ball isolated from the donor's individuation field, not tissue from an embryo, but a purely chemical evocator, and this was done by Heatley, Waddington & Needham. Glycogen prepared from mammalian liver (see p. 179), with the ether-soluble evocator of unknown constitution adsorbed on it or loosely combined with it, was used as inductor. The results are shown in Fig. 150. They fairly closely resemble that of Holtfreter just described. Massive neural inductions were produced, but these had not the character either of recognisable parts of the nervous system nor of sections of the normal neural tube. On the other hand, the cells were not oriented entirely at random. They tended to arrange themselves radially surrounding hollow tubular spaces, of which there might be as many as six, branching or independent, in one induction. Following on such inductions of neural tissue, the repulsion of overlying ectoderm and the formation of melanophores was observed, doubtless dependent differentiations brought about by the presence of the neural tissue as such.*

Further evidence of the same kind was later obtained by S. C. Shen[2] in a rather more elegant way, namely, by immersing competent ectodermal balls in very dilute solutions of the water-soluble hydrocarbon derivative, $1:2:5:6$-dibenzanthracene-α-β-endo-succinate. Although in some cases fairly typical cross-sections were seen (as in Fig. 151) they were due to balls, not tubes. No further differentiation was found. These experiments, in which a patternless inductor acts on patternless ectoderm, are of much interest, as ruling out all mechanical effects due to the solid implant.

The question asked on p. 125 was thus answered. Pure neural evocation does not involve the formation of a regionally undifferentiated neural tube with normal cross-section; it simply means histological differentiation together with a tendency to radial arrangement around cavities.

2·392. Individuation as the organised action of eidogens

How we are to interpret the remaining factor of induction, individuation, is one of the most difficult problems confronting embryology to-day. While it is convenient for purposes of discussion to speak of the individuation field, we must be careful that this concept is not used in such a way as to sterilise further research. The possibility is by no means excluded that other chemical substances may be at work in the formation of the normal neural axis besides the primary evocator. The existence of second-grade inductor substances is, of course, beyond doubt. Thus, as we shall see in Section 2·395, Lopaschov[5] has reported that if the eye-cups are removed from several neurulae, coagulated by heat and implanted into isolated ectodermal balls, they may induce new well-shaped eye-

---

* Spongy polytubular masses of neural tissue rather reminiscent of these may also be produced if the determined neural tube is removed after its closure and cultivated *in vitro* without the underlying notochord. This type of anomalous differentiation was described by Monroy[1] (see also p. 284).

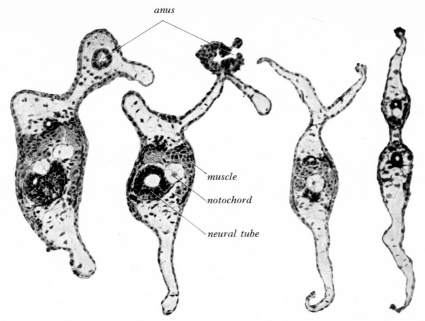

anus

muscle

notochord

neural tube

Fig. 149. Tail structures induced in a competent ectodermal ball by
implantation of living tail-organiser.

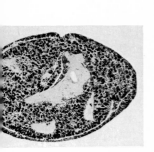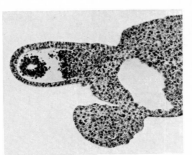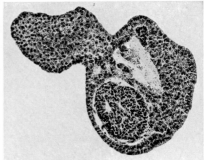

Fig. 150. Neural inductions produced in competent ectodermal balls by
implantation of the glycogen-evocator complex.

Fig. 151. Neural induction produced
in competent ectoderma ball by
immersion in very dilute 1 : 2 : 5 :
6-dibenzanthracene-α-β-endo-suc-
cinate solution.

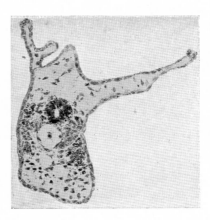

Fig. 152. Induction produced by implantation of
adult tissue into competent ectodermal ball.

cups.* This must be regarded as a homoiogenetic induction by a second-grade inductor. But we may also find it worth while to distinguish between substances which stimulate tissues to form an organ (inductors) and substances which modify in various ways the shape of organs so induced.

If such substances exist, it is at any rate clear that the position in which they are normally liberated is so exactly controlled by the individuation field of the inducing tissues that the regional correspondence of inducer and induced is secured. Waddington[20] has discussed the possible existence of these substances or *Eidogens* in a short review.† The most important thing which head-organiser does as opposed to tail-organiser is to modify the cross-section of the neural tube to form brain rather than spinal cord. Inductors and eidogens‡ would thus be the mechanism by which the individuation field of the inducer controls that of the induced.

On the analogy with evocators of all grades (see p. 302) eidogens might be expected to be present in adult tissues, and the way was therefore open for H. H. Chuang[1] to make the interesting experiment of implanting pieces of adult organs (which as we have seen have inducing power; p. 172) into isolated ecto-dermal balls. Using fresh adult newt liver and mouse kidney, he found that the types of induction produced by these implants were significantly different. Not merely evocated neural tissue appeared, but neural tubes, eye-cups, elongated notochords, nasal grooves, etc. etc. An example is given in Fig. 152.

"Here a completely heterogeneous inductor," said Chuang[1], "not itself taking part in the induction, brings into being typically formed, complex, and highly individuated structures" in the ectodermal ball. The statistics were as follows:

	Brains	Eyes	Noses	Ear-vesicles	Balancers	Tail	Muscle	Noto-chord
Newt liver	85·2	7·4	16·5	74·1	22·2	48·1	59·2	51·8
Mouse kidney	100·0	23·7	44·3	23·7	11·3	0	0	0

% Inductions

Thus contrary to expectation, liver tended to give rise to more caudal organs and kidney to more cephalic ones. When a similar statistical analysis was made of the results of implanting liver and kidney into the blastocoele cavity of intact embryos, the effects of the two organs were almost identical. The individuation field of

---

* Lopaschov found that no definite eyes were induced if less than a certain quantity of dead material was employed, but it does not seem possible to attribute the lack of definite organs in the experiments of Heatley, Waddington & Needham to a quantitative lack of evocator, since the masses of neural material obtained often filled up the greater part of the explant.

† The word "modulator" was suggested by Waddington[20] in ignorance of the fact that Bloom and Weiss[14] had proposed the use of the term "modulation" to describe all reversible physiological changes in histological structure, such as those occurring at the final stages of differentiation. These may be far-reaching enough, as in the spurious de-differentiation shown by some tissue cultures. I shall therefore use the word "eidogens" here to mean substances (possibly allied forms of one evocator) which differentiate further the product of an evocator stimulus in the sense of the induction field in question, or which modify the shape and form of organs already induced (from εἶδος, form).

‡ If such eidogens are present it is strange that in Hall's experiment they fail completely to act when transposed.

the host was capable, therefore, of completely masking the specificity which had been revealed in the previous experiments.

It is also important that the effectiveness of these adult tissues depends in a systematic way on the length of time for which they have been boiled. While most of the inductions fall off statistically, eye and nose inductions actually increase up to 15 min. boiling and then fall off, as if an inductor was first liberated from inactive combination (see Fig. 153). Notochord and pronephros inductors, moreover, are more sensitive to heat than tail and muscle inductors, and these in turn

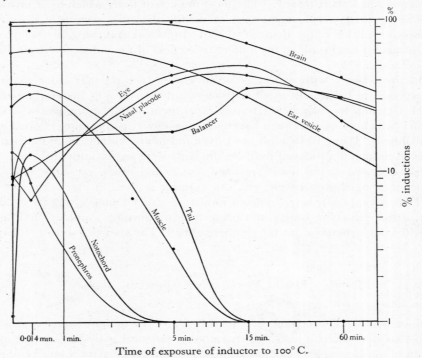

Fig. 153. Relative sensitivity of the various inductors to prolonged exposure to 100° C.

much more so than the inductors of brain and ear. Everything points to the existence of a series of compounds or complexes differing in their reactions to high temperature (cf. pp. 172, 173).

Chuang's work was independently continued by Toivonen[2], who made only implantations of alcohol-coagulated tissues into the blastocoele. It was found nevertheless that liver (whether from fish (*Perca*), snake (*Vipera*), jay (*Perisoreus*) or guinea-pig (*Cavia*)) always had a tendency to induce eyes, noses, forebrains, midbrains, and balancers, while kidney induced hindbrains, ear-vesicles, gills, notochord, somites and tails, and heart muscle induced all organs. Guinea-pig thymus proved to be a powerful and specific inductor of solitary lenses (see p. 299), with little other activity.

There is no doubt that the actual mechanism of the individuation field is a subtle interplay of evocators and competences. At first, gastrula ectoderm is omni-competent, it may even accept a mesodermal or endodermal fate. Later it is competent for neural-tube formation; later still, only for the formation of such structures as lens, balancer, or gills. Waddington[12] has drawn attention to the situation in the flank of the embryo (cf. p. 153). The just-formed side-plate meso-derm from the gastrula in the middle yolk-plug stage can induce the formation of a secondary neural plate when implanted into the blastocoele cavity of a young gastrula. It therefore possesses inductive power. Conversely, the dorsal lip of a young gastrula, substituted for the side-plate mesoderm just mentioned, evocates a neural plate on the flank of the embryo. Therefore the flank ectoderm is competent. Yet in normal development a neural plate does not appear on both sides of the embryo as well as on its dorsal surface. This remarkable paradoxical result may be explained by assuming a quantitative diminution in inductive power and in competence, so that the side-plate mesoderm evocates less strongly than the dorsal lip, and the lateral ectoderm reacts less readily than the young gastrula ectoderm. This would be part of the mechanism of the individuation field, but it is quite possible that there may exist an active suppression of the inductive power of the side-plate mesoderm by the rest of the mesoderm.

The loss of competence in ectoderm does not seem, however, to be dependent upon its continuous existence in the organism's individuation field. This loss proceeds just as well when the ectoderm is isolated and cultivated apart in neutral saline medium. In connection with the experiments on the situation in the lateral body-wall which have just been described, Waddington[12] cultivated young gastrula ectoderm *in vitro* until the controls had reached the open neural plate stage, then implanted archenteron roof into it, and obtained some neural inductions. The ectoderm seemed, therefore, to retain its competence somewhat longer in isola-tion than *in vivo*. But in Holtfreter's exo-gastrulation experiments[8], all neural competence was eventually lost by the abandoned ectodermal mantle. Lopaschov[1], after allowing isolated gastrula ectoderm to age *in vitro*, transplanted it on to various situations of a neurula, obtaining balancers and no neural inductions. His experiments were few, however, and it was left for Holtfreter[18], in a work especially devoted to this question, to give a comprehensive timetable of the loss of various competences in isolated ectoderm. This table was as follows:

Inductions following transplantation to neurula situations

Hours from isolation	Brain	Neuroid vesicle	Neural cells	Ear-vesicle	Mesenchyme	Balancer	Muscle
0	+	+	+	+	+	+	+
15	−	+	+	+	+	+	−
24	−	−	+	−	+	+	−
46	−	−	+	−	+	−	−
54–96	−	−	−	−	−	−	−

Thus after 15 hr. isolation competence to form brain masses was lost, but com-petence to form chaotic groups of neural cells was retained till the 46th hr. Competence to form an ear-vesicle was retained longer than that to form muscle. In general this agrees with what takes place in the intact embryo; it seems therefore that the loss of competence is not due to its active suppression by the organism's individuation field.

But that inhibitory effects do occur in normal development as part of the individuation field seems very likely. For instance, Balinsky[7] has described the appearance of supernumerary balancers, even up to eleven on one embryo, after removal of most of the endoderm. This he interprets as being due to the removal

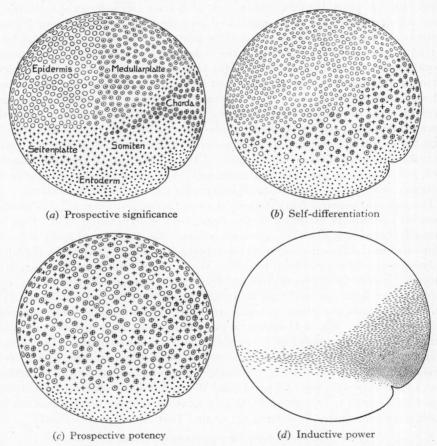

(a) Prospective significance

(b) Self-differentiation

(c) Prospective potency

(d) Inductive power

Fig. 154. Holtfreter's maps of the properties of regions in the early amphibian gastrula.

of inhibitory influences. There is also the well-known fact that any piece of anuran ectoderm when isolated tends to produce one or more suckers. How far these inhibitory effects are due to mechanical relationships which operations disturb, cannot as yet be decided.

One genuine inhibitory effect, at any rate, is the influence exerted by the prechordal plate (the anterior end of the notochord) on the presumptive eye-districts lying above it. After a great deal of confusion about the origin of cyclopia, Adelmann[3] succeeded in showing that normally the two eye-districts overlap and that the development of this intermediate material into eye is inhibited by the prechordal plate. If this is interfered with, whether mechanically or by the action of lithium (see p. 226), cyclopia results. Adelmann regards this inhibitory effect as analogous to the contiguity effects described below.

Another way of examining the action of the individuation field is the study of the regulations which the isolated organiser region is capable of effecting. In the experiments of Holtfreter[19] (p. 158) we have already seen that very many structures can be produced by the isolated organiser region. Holtfreter[20] brought together the facts in the following way (Fig. 154), using the classical terminology. The prospective significance of the parts (*a*) follows naturally the Vogtian fate-map (cf. p. 145). The prospective potency of the parts (*c*) is very loose; apart from the yolk-endoderm, any piece of the early gastrula can turn into any later derivative (cf. Bruns; Töndury[1]). The region where the primary evocator will be liberated (*d*) is, of course, the dorsal blastopore lip. But the self-differentiation map (*b*) shows that while yolk-endoderm will always remain endoderm, and while both presumptive epidermis and presumptive neural plate will always remain ectoderm, the organiser region has the power of producing notochord, neural plate, ectoderm and somites. Lopaschov[2] has now shown that exactly what will be produced by isolated dorsal lip regions depends upon their size.

If one dorsal lip region from one embryo is isolated, only notochord and muscle fragments usually appear. If, however, several pieces of such material, derived from several gastrulae, are placed together, they fuse and later produce a large number of tissues, regulating, indeed, in the direction of a whole embryo. The following table shows that a direct relation exists between the size of the isolated mass at the outset and the degree of individuated complexity afterwards produced:

No. of dorsal lips fused together	Notochord and muscle	Epidermis	Pigment cells	Neural tissue	Brain	Eye	Lens	Ear-vesicle
1–4	+	–	–	–	–	–	–	–
4–5	+	+	+	–	–	–	–	–
5	+	+	+	+	–	–	–	–
6–10	+	+	+	+	+	+	+	+

The accompanying photographs (opp. p. 290) show a fused "organism" made of seven dorsal lips. In the first (Fig. 155 A) two brains, a notochord and an ear-vesicle are seen; in the second, further along the organism, there are three notochords, traces of neural tissue, a piece of gill and an eye with a lens (Fig. 155 B).

The Russian school of experimental embryology has done good service in drawing attention to the importance of the quantitative aspect. Thus Dragomirov[3] found that pieces of pigment-epithelium of the eye, if sufficiently large, would, when isolated, regulate into complete eye-cups, but if below a certain size would merely form bladders of pigment epithelium. Similar facts have been noted in the case of regenerating limbs.

One of the most remarkable properties of the individuation field is its capacity of regenerating bilateral symmetry when this is destroyed. Thus it was often found in the earlier work (and confirmed in special *ad hoc* investigations by Ekman[2] and B. Mayer[1]) that the lateral half of a dorsal lip, placed in the blastocoele cavity, induces a bilaterally symmetrical secondary embryo with fully paired organs.

2·393. Contiguity and affinity effects

Included in the conception of the individuation field must be all those eidogenic influences of one organ or structure upon another which have been described under the head of *Nachbarschaft*; "neighbourhood" or contiguity effects. Holtfreter[10] found, from his numerous explantation experiments in which presumptive tissues were combined in all sorts of permutations, that the normal configuration of the neural tube, for example (*typische Gestalt*), was only reached under certain conditions. Consider Fig. 156. If neural tissue differentiates in isolation the cells round up into a spherical or rod-like structure, with the nuclei at the exterior (*a*). If they differentiate within a mass of mesoderm a hollow ball or tube is formed, with the nuclei usually at the interior (*b*). If the neural induction is performed by muscle, the tube formed will tend to show an excentric lumen (*d*). If there is

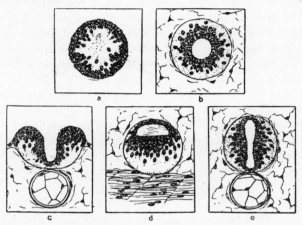

Fig. 156. Contiguity effects governing the attainment of the normal configuration of the neural tube.

insufficient mesoderm the tube will not close (*c*).* Only if the induction is performed by notochord in the presence of sufficient mesoderm will the normal cross-section of the tube be shown (*e*). Some of these effects have already been seen in certain cases.

In later work, Holtfreter[21] considerably extended this line of thought, developing it into a general system[22] of "tissue affinities" by the study of isolates and various combinations of isolates. Isolated endoderm alone shows first a tendency to round up into a ball of cells, and later spreads out on the bottom of the dish, suggesting "intestinal" epithelium (Fig. 157; 1*a*, 1*b*). If a little ventral ectoderm and endoderm are isolated together with a small amount of mesoderm, the spreading tendency is abolished, and the spherical object holds together with endoderm forming part of the periphery and ectoderm the remainder (Fig. 157; 2). But if more ectoderm is provided, sufficient to enclose the whole, then the endoderm will form a tubular structure within the mesoderm, thus showing that

* See below, p. 371, in the section on genetic factors, where failures of neural tube closure occur; and for general contiguity effects, the hybrid chimaerae of Hadorn[3] (p. 359).

its polarity has been reversed (Fig. 157; 3), and simply embedding it in a mass of mesoderm will give the same result. It may be noted that this "gut"-formation has proceeded without any prior invagination.

Now what happens when ectoderm and endoderm are isolated together without mesoderm? At first they cling together, but later they tend to separate (Fig. 157; 4), and this self-isolation always occurs after a certain time. The converse experiment of separate cultivation followed by fusion also succeeds; after a certain time, this cannot be done, showing that an autonomous ageing process of "chemical differentiation" or "increasing physiological incompatibility" has gone on. Thus

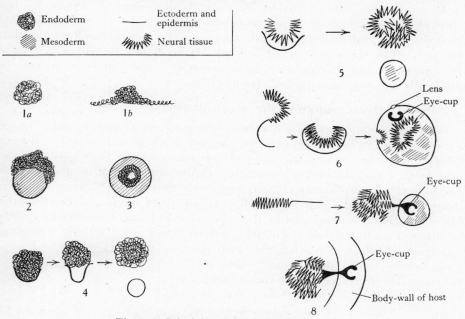

Fig. 157. Principles of tissue affinity (see text).

mesoderm behaves as a "connective tissue" in the truest sense of the words; without it the building of an organism is impossible. Of the meaning and mechanism of these important facts nothing definite is yet known, but the problem seems to be essentially one of cohesion or congruence of surfaces, and urgently invites attack from the point of view of surface chemistry.*

Holtfreter went on to study the relations between neural tissue and ventral ectoderm in isolate combinations. If much neural plate is combined with little ectoderm, the epidermis will eventually separate from it, as if from endoderm, and the neural tissue will form a naked brain (Fig. 157; 5). But if a better supply of ectoderm is made available, the neural tissue will embed itself in an ectodermal ball, in which it may induce an accessory neural plate and lenses (Fig. 157; 6). The lumen of the ball will fill with neural crest mesenchyme. In the former case, should an eye-cup form, the epidermis will not separate itself entirely from the

* Cf. Rashevsky[1].

neural mass, but will enclose the eye-cup, suggesting that the newly formed sense organ, though not the nervous system, has an affinity for epidermis (Fig. 157; 7). Exactly similar results may be obtained by implanting a neural mass into the coelom of a tadpole; should an eye-cup be formed, it will bury itself in the body-wall of the host, as if attracted by the host's epidermis (Fig. 157; 8). Ear-vesicle and nasal placode also "prefer" close contact with epidermis. The repulsion between nervous tissue and ectoderm is well seen both in normal development and in various kinds of induction (Fig. 150), where the ectoderm seems to lift itself off, well away from the nervous tissue below.

All this had been adumbrated by W. Roux[6] in his theories of "cytotaxis", but only Holtfreter's brilliant explantation technique could have demonstrated experimentally that such effects never take place at a distance, as Roux thought, but only when the tissues are in contact. It need only be added that these principles of tissue affinity are quite non-species-specific and may be seen in practically any desired xenoplastic combination.

2·394. Individuation and the organ-districts

What then are the successive manifestations of the individuation field? We see it first in the interaction between the vitelline gradient and the grey crescent cortical field, determining the point of appearance of the blastopore. At the beginning of gastrulation the organisation centre acquires (rather gradually, as Lopaschov[4] believes) its antero-posterior regional differentiation. During induction this regional character manifests itself in the shape of the neural plate. At the neurula stage it is complicated by the appearance of numerous subsidiary fields or districts each associated with the differentiation of a specific structure.

Lehmann's exchange experiments[2] between presumptive regions show these coming into existence. Holtfreter's grafts[5] of young gastrula tissue on to neurulae reveal, wrote Spemann ([17], p. 309), "the existence of persistent embryonic fields. In the same way as a neural field may be induced by a piece of isolated notochord in an embryo out of which the normal neural plate has already been induced, or as the same effect may be produced by a piece of neural tube, nay, even by a piece of brain from a swimming larva; thus also the whole larva, even of later stages, is permeated

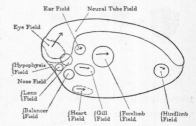

Fig. 158. The organ-districts of an amphibian neurula, viewed from the side. The arrows indicate districts that are polarised from the first moment of their appearance (from Huxley & de Beer[2]).

by embryonic fields which become manifest as soon as a piece of reacting material has been brought under their influence. Even in the adult animal such fields have not ceased to exist; their existence can always be demonstrated by a regenerating blastema which has been exposed to their influence" (Schaxel[2]; Milojević; Guyénot[2]; Weiss[7]).

Fig. 158 gives a diagram of an amphibian neurula showing these fields or districts. They represent the parts of the embryo where the presumptive rudiments of eye, limb, gill, etc. will become determined. Their appearance has been termed by Weiss[16] *Emancipation*. Under no circumstance can a limb be produced from a lens district, but the outer edges of the districts are indistinct, and they may slightly overlap with one another.* Here the full concept of field activity does not appear to be justified, although when the limb, for instance, has begun to grow out, it may constitute a sort of field, a portion of the whole individuation field, since indifferent material added to it will be built into its own form.†

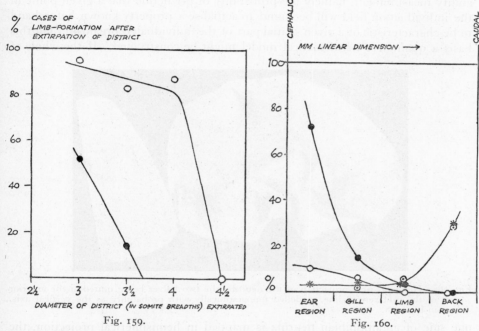

Fig. 159. Relation between size of area extirpated from a district, and frequency of subsequent manifestation of its property (here limb-formation). ● Wound cleaned; ○ wound not cleaned.

Fig. 160. Relation between site of implantation of competent ectoderm on to the neurula, and its subsequent fate. ● Ear; ○ balancer; * muscle; ⊙ pronephros.

The districts represent, in fact, a rather different form of order. Organ-forming potency falls off from their centres in a definite measurable way. Thus Harrison's work[3] on the limb buds of amphibia will permit of graphical formulation (see Fig. 159). The percentage of positive cases of organ formation after the excision of different sized circular pieces can be plotted against the diameter of the pieces (measured in somite breadth), i.e. the distance of their circumferences from the

---

* As the organ districts are forming, certain general determinations are proceeding, such as that on account of which the ectoderm can now no longer be turned inside out and yet develop normal skin (Luther[1]).

† One of the most remarkable results of Weiss' work[11] on the regeneration of limbs is the smallness of the pieces of tissue which can carry the field, e.g. tiny discs of arm attached to legs produce arms, not legs, on regeneration (see Fig. 247).

presumed centre of the limb district.* In such a case we are really measuring degrees of "limb probability", and could erect probability contours which would fix the configuration of the district round its centre. Similarly, Holtfreter[5], as we have seen (p. 274), studied the results of grafting bits of presumptive epidermis or presumptive neural plate from a gastrula into various regions of the flank of a neurula. The grafts undergo differentiation into various structures, and the frequency with which a given result is obtained can be plotted against the linear dimension of the body (see Fig. 160). Here again we have essentially a probability measurement, namely the probability of prediction that a given point in the individuation field will be found to actualise a property known beforehand to be characteristic of a given spatial part of the individuated organism. On this basis a qualitative mathematical model might be constructed. If the whole of

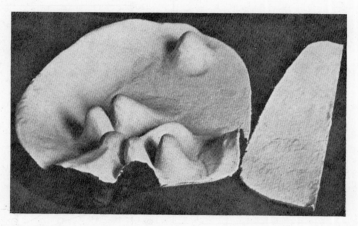

Fig. 161. Hemispherical projection of the neurula, the two halves being united at the anterior part of the mid-ventral line, modelled topographically with probability as the vertical axis. Key in Fig. 158.

one side of an amphibian neurula is mapped in hemispherical projection, the probabilities of organ formation can be represented by a landscape of hills, the contour lines being the "isobars", as it were, of probability (see Fig. 161). The status of any given point depends on the probability of bets that (a) an added piece of indifferent tissue will contribute to, (b) an excised piece will inhibit, the formation of such and such an organ. There must, of course, be some underlying factor, which we could call "limb-forming intensity". Let us try to picture what this could be.

Implicit in the foregoing discussion is the idea that each of the organ districts is associated with, is, in fact, the expression of, an inductor or a group of inductors, possessing hormone-like properties.

* The converse experiment to Harrison's, namely the transplantation of discs of varying sizes from the limb district to the flank, was made by Takaya. Grafts less than 4 somite breadths in diameter were usually absorbed, but those of 5 and 6 somite breadths diameter produced quite normal limbs. Normal development was more likely to happen if a small piece of the dorsal neighbouring ectoderm was included in the transplant, suggesting an asymmetry of the district boundaries. The two sets of experiments agree well.

As Huxley[13] has told us, the classical concept of hormone can now be divided into several logical subdivisions. (1) There are the metabolites with physiological functions, transported by the circulating blood, e.g. CO_2 or lactic acid. These, without possessing any high degree of specificity, certainly co-ordinate the activities of the body. (2) There are the true hormones, organic molecules specially manufactured in *ad hoc* glands, and possessing great specificity at their sites of action, which again they reach transported by the blood. (3) There are the "diffusing hormones", such as the water-soluble heat-stable substances (described by G. H. Parker[2]) which appear to be liberated at the end-organs of nerves and to control the contracting and expanding of chromatophores in fish and crustacea. In this class would come the substances generated at cholinergic and adrenergic nerve-endings. (4) There are next the "contact-hormones", i.e. the numerous evocators of different grades which determine the fates of the various parts of the embryo. The distinction between these last two classes is not very clear-cut, for the necessity of contact between the active and the passive tissues may only mean that the substance involved possesses very feeble powers of diffusion.* Lastly (5) come substances of morphogenetic nature produced directly by genes and perhaps only acting in the cell which makes them (see Section 2·67).

A theoretical possibility is now of interest. It does not seem so far to have been envisaged that a substance might be a hormone without diffusing at all. It is possible to picture a single molecule or a molecular aggregate (perhaps of a paracrystalline nature) exerting a polarising influence around itself in all (or some) directions of space for a considerable distance, even into microscopic dimensions if the argument of Hardy[2] (see p. 675) is remembered. The concentric zones which would then exist would not be zones of decreasing concentration of hormone (as we now generally assume) but zones of increasing randomness of arrangement of its own molecules or of other particles. The gradient in a given direction would not be due to the falling off of the amount of the substance but to the falling off of its oriented arrangement.† Limb probability might thus be ultimately probability of molecular orientation. But a gradient in amount of inductor substance is perhaps no less likely.

Recognition of the existence of organ districts in the neurula brings us to the subject of the second-grade organisers, to which the following sections will be devoted.

Before leaving district differentiation, however, one should note that it shows us some striking examples of field fusion. The heart, for example, in amphibia, arises from the two lateral mesoderm plates, which, as Dwinnell has shown, may begin to contract some hours before they fuse. But if their fusion is prevented, or if each is isolated, two almost perfect hearts will be produced (Ekman[1] on amphibia; Jolly on mammals).

* Huxley's classification actually involved the use of the term Activator; but as we have already seen (p. 188), it may be well to reserve this for whatever it is that liberates an organiser substance from inactive combination. It is unfortunate that Weiss[14, 16] also uses the term as almost synonymous with Evocator.

† In some such way, perhaps, as the gradual tapering off of cortex into cytoplasm in the echinoderm egg.

2·395. Second-grade organisers

The organisers of the second grade are so described because they come into play after the activity of the primary organiser, the laying down of the main axis of the body, is completed. We cannot here, for lack of space, examine them in too great detail. Some treatment of them, however, is indispensable for two reasons, first, because in certain cases the biochemical analysis of their mechanism has already begun, and secondly, because some rather important theoretical ideas, such as that of "double assurance", have originated from the study of second- and third-grade organisers.

Spemann's conception of the chains of inductors which operate in organ differentiation is illustrated in Fig. 162 (adapted from Holtfreter[19]). Neural

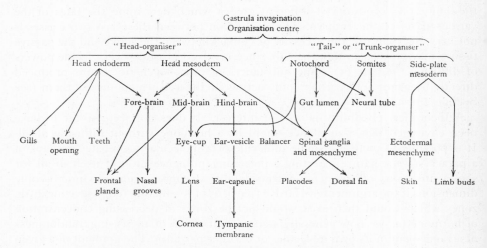

Fig. 162. The succession of inductors of second and lower grades.

induction, as we have hitherto seen it, is shown by the action of the head endoderm (at the junction of the archenteron roof with the yolk-endoderm) and the head mesoderm in inducing the parts of the brain; and by the action of the notochord and somites in inducing the neural tube. These effects are indicated by arrows in the diagram. Second-grade organiser effects are seen in the induction of the lens by the eye-cup, and the ear-capsule by the ear-vesicle. The induction of the tympanic membrane by the ear-capsule (strictly speaking, the tympanic cartilage) is an example of a third-grade induction.

An instance of a very late, e.g. fourth-grade, induction, is afforded by the discovery that peripheral tissues induce nerve-endings of an appropriate kind in the nerves reaching them (Dijkstra; Weiss[15]).

In the present section it will be convenient to begin with the most classical and typical case of second-grade organiser, that which operates in the formation of the lens by the eye-cup. I shall then describe shortly other cases and end by dealing with the most peculiar problem of limb-bud induction.

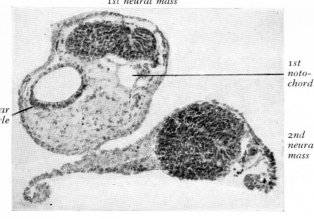

XXIII

1st neural mass

1st noto-chord

2nd neural mass

~~ar~~
~~le~~

A

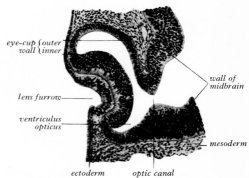

eye-cup {outer / inner} wall

lens furrow

ventriculus opticus

wall of midbrain

mesoderm

ectoderm   *optic canal*

Fig. 163. Induction of the lens from competent ectoderm by the eye-cup.

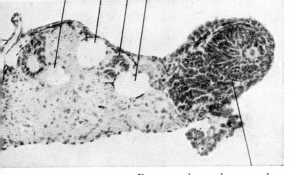

| 1st noto-chord | 3rd noto-chord | 2nd neural mass | 2nd noto-chord |

B   *3rd neural mass and eye*

g. 155. "Organism" formed by the development of seven dorsal blastopore lips fused together.

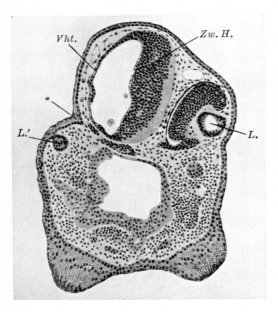

Vht.   *Zw. H.*

L'.   *L.*

Fig. 164. Apparent development of lens without the eye-cup stimulus in *Rana esculenta*. *Vht*, Wound epithelium of operated side; *Zw.H.*, midbrain; *L*, lens of normal side; *L'*, lens of operated side.

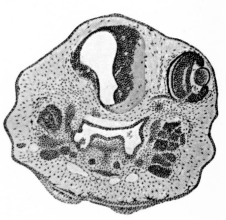

Fig. 165. Absence of lens on the operated side after removal of the eye-cup in *Bombinator pachypus*.

2·3951. Eye-cup and lens induction

The morphogenetic stimulus emanating from the eye-cup which induces the lens from the ectoderm with which it comes in contact has already been referred to (p. 224). Fig. 163 illustrates the process. As we have also seen (p. 105) Spemann[2] showed, at the beginning of the century, that if the presumptive eye-cup rudiment was destroyed no lens subsequently developed, although it was certain that the presumptive lens tissue itself had not been injured (*Rana fusca*). This result has been confirmed hundreds of times on many different amphibia by later investigators (see the review[8] and papers[3,4] of O. Mangold). Defect experiments were followed by transplantation experiments, first carried out (independently) by Warren H. Lewis[1]. The eye-cup was exposed, cut off, and pushed backwards under the ectoderm so that it would act on ectoderm never normally subjected to lens-inducing influences. New lenses were excellently formed (*Rana sylvatica* and *Rana palustris*). Filatov[6] and Woerdeman[10] subsequently showed that a stay of only 24 hours suffices for the eye-cup to exert its lens-inducing power on the ectoderm. The converse experiment, that of transplanting ectoderm from outside the presumptive lens district on to a site over the eye-cup, was also successfully made by Lewis[2]. It was thus proved that the eye-cup acts as a second-grade organiser, inducing a lens from the ectoderm overlying it. We shall return later to the question of the nature of the inductive influence.

But before long it was found that the question was not so simple as it sounds from the above description; the historical development of the subject brought it into difficulties.

"In my first experiments on *Rana fusca* in 1901", wrote Spemann[17] long afterwards (p. 60), "lens-formation and the clearing of the deep black epidermis into cornea did not take place if the eye-rudiment in the neural plate had been removed. Since a direct injury of the primary lens-forming cells could with certainty be excluded, their failure to form a lens seemed to prove that the lens formation is only possible under the influence of the eye-cup. It seemed natural to extend this result to include all vertebrates. But soon cases of lens formation in the complete absence of the eye-cup became known. It was shown by Mencl in 1903 that the lens occurred spontaneously in *Salmo*,* and after experimental removal of the eye-cup by King in *Rana palustris*. Sure of my case with *Rana fusca* and biased by the idea that all animal forms must behave in a similar way, I did not at first allow myself to be shaken in my conviction. The lenses in King's experiment were not very distinct; Mencl's cases seemed to admit of another explanation. Nevertheless, I repeated my experiments on another object, *Rana esculenta*, and now it appeared to my great surprise that after a neat removal of the eye-rudiment from the open neural plate, lenses of great perfection can originate...."

Spemann[6] was therefore forced to conclude that while in some species the lens owes its origin absolutely to the stimulus from the eye-cup, in others the presumptive lens district is determined in some other way at an earlier stage, and will self-differentiate without the help of the eye-cup. The contrast is shown in Figs. 164, 165, where on the one hand a lens is shown without an eye-cup in a

---

* In a case of *duplicitas anterior* one head had two lenses but no eye-cups.

Rana esculenta embryo, and on the other hand the complete absence of a lens in an embryo of *Bombinator*.

Later investigations have shown that amphibian embryos may be divided into several groups, according to the degree of determination of the lens district before the eye-cup comes to act upon it. Spemann classifies them as follows:

(1) A highly differentiated lens develops from the lens district without the assistance of the eye-cup:

> *Rana esculenta*   Spemann[6]

(2) Ectodermal thickenings or vesicles of questionable significance develop from the lens district without the assistance of the eye-cup:

> *Rana palustris*   H. D. King
> *Bombinator pachypus*   Spemann[4]; v. Ubisch[3]
> *Rana fusca*   v. Ubisch[3]
> *Rana catesbiana*   Pasquini[2]

(3) Absolutely nothing takes place in the lens district without the assistance of the eye-cup:

> *Pleurodeles waltlii*   Pasquini[1]
> *Triton taeniatus*   O. Mangold[3]
> *Triton alpestris*   O. Mangold[3]
> *Pelobates fuscus, Bombina*   Popov, Kislov, Nikitenko &
> *bombina, Triton cristatus*   Canturisvili

The second and third of these groups may for the present purpose be taken together, as it was in *Rana fusca*, after all, that the first proof of the eye-cup induction was obtained. The cases of *Rana esculenta* and to a lesser extent of *Amblystoma punctatum* are the real difficulties.*

Two other peculiarities have been noted which differentiate these two species from the others in the list, first the lack of dependence of the size of the lens upon the size of the eye-cup, and secondly, the distribution of lens competence over the neurula ectoderm. With regard to the first of these, Spemann[17] has observed that in *Rana esculenta* a diminution in the size of the eye-cup does not bring about any corresponding diminution in the size of the lens, whereas in a species such as *Bombinator*, where the formation of the lens is very highly dependent upon the eye-cup, we find that the smaller the eye-cup or piece of eye-cup, the smaller the lens. The detailed references on which these facts are based will be found in Spemann's book ([17], p. 67). This property fits in quite well with the degree of inductive dependence shown by the lens district. With regard to the second peculiarity, embryos of different amphibian species differ in the extent to which

---

* Spemann[17] lists *Amblystoma punctatum* also among those forms in which a highly differentiated lens develops from the lens district without the assistance of the eye-cup. But this is not altogether convincing, for the following reasons. Le Cron showed that removal of the eye-cup resulted in failure of lens formation. Then Harrison[4] transplanted presumptive lens-district ectoderm to other regions of the head and found that well-formed lenses resulted. But before removal this ectoderm had been for a short time in the near neighbourhood of the eye-cup. Harrison wrote: "Contact between eye-cup and ectoderm has at this time been established, though the two are not adherent and may be readily separated without cells from one layer sticking to the other." One has only to assume therefore that in this species the action of the lens-inducing substance is particularly rapid. If the lens ectoderm was removed in earlier stages "small and not fully differentiated lenses" sometimes but not always resulted. The form belongs therefore rather to group (2).

lens-forming competence is distributed over the ectoderm. In *Rana fusca** it is everywhere. But in *Bombinator*, though head ectoderm from any site will form lens, trunk ectoderm will not. And in *Rana esculenta* lens-forming competence is restricted completely to the presumptive lens district (Spemann[5]).† This again fits in with the degree of inductive dependence shown by the lens district.‡

On the other hand, the eye-cup of *Rana esculenta* is undoubtedly giving off its morphogenetic stimulus, as was shown by the experiments of Filatov[2] and Pasquini[2]. Ventral ectoderm from a species with widely extended lens competence (*Bufo vulgaris*) was transplanted over the eye-cup of *Rana esculenta*, with the result that an excellent lens was produced. We know that without such a stimulus this ectoderm would never have reacted in this way. The paradox thus arose that the eye-cup of *Rana esculenta* appears to give out a stimulus which its ectoderm stands in no need of.

This fact has been the principal mainstay, though not the first origin, of the conception of "double assurance". This interesting idea, taken from engineering, was first used by Rhumbler[1] in his mechanical descriptions of cell-division about 1897 (*mehrfache Sicherheit*), and was used by Spemann[12] for the present case and other cases in embryology (*doppelte Sicherung*). Just as an engineer plans a bridge to carry loads and to withstand wind pressures much larger than any which it will normally meet with, so the forces of evolution have brought it about that some ontogenetic processes involve more than one chain of causation. This may mean (1) that in the event of one process failing, another one, previously in reserve, will accomplish the same end, or (2) that normally two or more processes collaborate towards the production of the same result.

Spemann (p. 92)[17], believes that the principle of double assurance is well established, and that it is of far-reaching importance, not only in development, but also in the functions of the adult organism. In its first form, as stated above, it may lead to strange conclusions. Some biologists have not failed to find it rather bizarre that nature should have taken such elaborate precautions through long ages, in the case of *Rana esculenta*, against the operative methods of future experimental embryologists. It is true that nature does many strange things; thus we have seen already many inductors with nothing to evocate, and many instances of slumbering competence which is never awakened by the appropriate inductor. In its second form, however, it is with difficulty gainsaid. Who can doubt, as F. E. Lehmann's[4] "synergetical principle of development" (*kombinative Einheitsleistung*) asserts,§ that very many causes must have contributed towards the normal configuration of a perfect neural tube or a perfect eye? But nevertheless it is always our duty, in conformity with the immortal

---

* And in many other forms, e.g. *Triton vittatus* (Sikharulidze).

† Not confirmed by Gosteieva on the Russian variety of *Rana esculenta*; a varietal difference? See also the work of Manuilova[2].

‡ But according to Popov[1] these differences are only quantitative. Body ectoderm of *Bufo viridis*, which, when tested by implantation of eye-cups under it, has no lens competence, will, when implanted into the eye-cup cavity as for Wolffian regeneration experiments (see p. 295), give a lens[5].

§ See also Bytinski-Salz[3].

principle of William of Ockham, to search for the simplest possible causation chains. And I am not convinced that the principle of double assurance has yet, so far as the development of the eye is concerned, been put on a basis quite free from criticism.

There are two ways in which the anomalous behaviour of the lens district in *Rana esculenta* could be explained so as to bring it within the realm of straight-forward inductions. It could be said that the presumptive eye-cup region, even while yet still undifferentiated from the anterior part of the neural plate, possessed in these species lens-inducing power, so that the determination of the lens, though occurring very early, was due to the eye rudiment all the same. This was the suggestion of de Beer ([1], pp. 183 ff.), but Spemann ([17], pp. 94 ff.), perhaps justifiably, regards it as unconvincing, since no parallel for such an "induction-at-a-distance" is known and he himself had shown that even a thin layer of mesoderm lying between eye-cup and competent ectoderm would block the passage of the inducing stimulus.

The other suggestion is a new one. Let us suppose for the sake of argument that the inducing stimulus is mediated by a definite chemical substance (for facts bearing on this, see p. 298). In most species this substance would be sharply restricted to the eye-cup itself, but in some, such as *Rana esculenta*, it might escape from the eye-cup by diffusion or otherwise—at any rate it might be present in the mesoderm surrounding.* It this were so, removal of the eye-cup would not imply removal of all of the effective substance, and a lens would develop later on, giving the impression that it had self-differentiated. The crucial experiment would therefore be the cultivation of *Rana esculenta* presumptive ectoderm in true isolation, as for instance in Holtfreter solution. If the theory of double assurance is right, a lens should be produced even in these conditions; but if the material stimulus is necessary, there should be no lens. When this test was made by Perri[2] in 1934 with numerous explants, no lenses resulted. On *Bombinator* isolation experiments were also carried out by v. Ubisch[3]; in this case two-layered vesicles with no sign of lenses were formed. Harrison's trans-plantations[4] of presumptive lens district into the head ectoderm of other axolotl larvae do not help here, since the new environment was not strictly neutral. But strangely enough, even *Rana esculenta* lens districts under such conditions did not develop lenses.†

Unless, therefore, Perri's crucial experiment of keeping presumptive lens district ectoderm of *Rana esculenta* in isolation from an early stage onwards should be repeated and found to be wrong, the theory of double assurance must

* A parallel to this could be found in the anti-coagulation substance, heparin, allied to chondroitin sulphuric acid (see p. 443), which is found in the tissue around the great blood-vessels (Wilander).

† After I had written this paragraph I found that a similar idea had been mooted by others. Thus Mikami[3] points out that in some cases head mesoderm has been shown to possess lens-inducing power (Okada & Mikami). He himself found that on transplanting presumptive lens district to other sites, lenses were only formed in the head and gill regions, not elsewhere. Holt-freter[13], too, takes the same point of view, since in some of his xenoplastic transplantations done for quite other reasons, lenses arose without eye-cups in the close neighbourhood of other head organs ([13], p. 404). Only by explantations into neutral medium can the problem be finally settled.

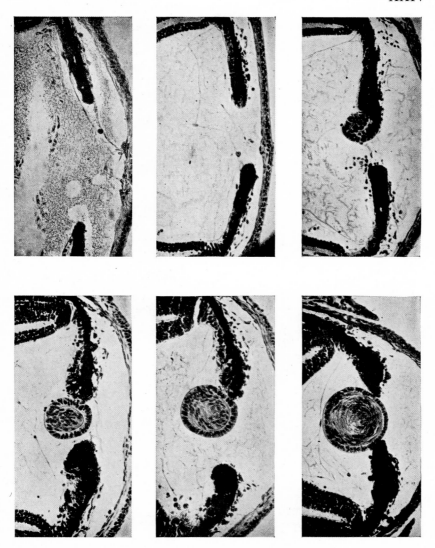

XXIV

Fig. 166. Wolffian regeneration of the amphibian lens from the margin of the iris.

be held to be disproved. Its application to the primary organiser (in the work of Goerttler[1]) has been generally abandoned (see p. 157), and reasons will later be given for doubting its applicability to the opercular perforation (see below, p. 308), and the induction of the ear (p. 302).

In Fig. 158 we have seen the approximate position of the lens district in the neurula, but we know also that lens competence, in most species at any rate, is much more widely distributed over the ectoderm. It appears also to be present in some strange places. It is to be found in the last place, perhaps, where it might have been looked for, namely the eye-cup itself. In the nineties of the last century the discovery was made independently by Colucci and by G. Wolff that if in larval or even in adult amphibia the lens is removed, a new one will form from the upper margin of the iris ("Wolffian regeneration"). The beautiful series of pictures taken from T. Sato[1] show this process in operation (Fig. 166). The new lens usually arises from the upper margin, whether or not the eye has been previously turned upside down (T. Sato[3]), but if this operation is performed later on, then the new lens regenerates from the lower margin (the previous upper one), as Wachs[2] has shown. In other words, competence for lens formation persists into adult life in the iris, but only in a half-segment of it, and the area of this persistence of competence is itself determined when the dorso-ventral axis of the eye is determined. But this competence of the eye-cup may under certain circumstances come into play even in early development, as it did in Spemann's experiment[3] where mesoderm intervened between the eye-cup and the ectoderm, or in that of Beckwith, where the lens competence was absent from the covering ectoderm,* or in that of Perri[2], where eye-cups were explanted *in vitro* without any ectoderm.

Wolffian regeneration led to a method of study of the nature of the inducing influence which has not been without results. At Spemann's suggestion Wachs[1] cut off small pieces of the iris and pushed them into the eye-chamber (the *corpus vitreum*); there, quite without contact with any tissue, they were induced to form well-shaped lenses. Later on, the same experiment was successfully made with pieces of ectoderm by T. Sato[1] and by Nikitenko[1], Wachs[1] having previously controlled the process by finding no change when pieces of iris were implanted in the labyrinth of the ear. According to Törö[1], iris tissue cultures can be converted into lens (chick). If the results of Popov[1] are to be accepted, lens competence is very widely distributed. In some chance cases he noticed[3] lentoidal bodies produced in mesoderm and even in endoderm adjacent to implanted eye-cups of *Rana esculenta*, so he implanted neural plate tissue, eye-cup itself, ectoderm, somite mesoderm, and ear-vesicle into the eye from which the lens had been removed[6]. All of these formed lenses (eye-cup and ectoderm not surprisingly), with the exception of the ear-vesicle. But the relations of the eye-cup with the ear-vesicle seem to be rather peculiar, for Dragomirov[1] found that eye-cup will

---

* The possibility that lens formation over transplanted eye-cups might really be by Wolffian regeneration and not from the belly ectoderm was examined in an *ad hoc* investigation by Adelmann[2] and ruled out, though such Wolffan regeneration is possible if the eye is placed so as to face inwards (see also Mikami[5]).

induce lens from ear-vesicle instead of ectoderm if it should come into contact with it. Ear-vesicle, too, according to Dragomirov[4], may be diverted to form retina (Figs. 167, 168, opp. p. 300). Lens competence in epidermis seems to last into very late stages in some amphibian larvae (Popov, Eudokimova & Krymova), and it has been shown by Schotté & Hummel to exist in both ectoderm and mesoderm of the regenerating limb-bud blastema* (see also p. 442). During metamorphosis there is temporary decline in the capacity for Wolffian regeneration (Monroy[2]).

It is interesting that Wolffian regeneration fails in the presence of a normal lens. Although a chemical stimulus must be involved, no induction takes place if a piece of iris is placed in an eye-cup which retains its lens (Wachs[1]). Kobayashi tried to influence lens regeneration by injection of lens-protein into the circulation, but unsuccessfully. T. Sato[4], however, threw further light on the problem by

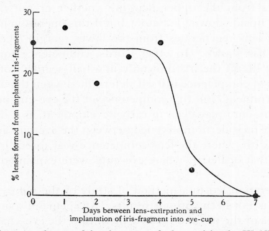

Fig. 169. Relation between degree of development of a lens arising by Wolffian regeneration, and its power of inhibiting lens-formation from an implanted iris-fragment.

finding out exactly how far lens regeneration from the iris had to go before it could inhibit the formation of a lens from a piece of iris or ectoderm implanted into the eye-cup. The results are shown in Fig. 169. During the first four days after lens extirpation no change is seen in the iris and during this time normal regeneration from implanted iris fragments was quite uninhibited. On the fifth day, however, the first sign of Wolffian regeneration is seen in the depigmentation of the iris; this coincides with a great reduction in the number of fragments which give a lens. At later stages, regeneration from fragments entirely fails, and such fragments often differentiate into retina-like tissue.

Can a dead lens exert the inhibitory action? From the work of Kesselyak it appears not, so a protein is indicated as the active factor rather than a small molecule insensitive to denaturation.†

* Contested by Stone & Sapir. Schotté[6] has replied and the question is not yet quite settled.

† About all we know of the embryology of the lens-proteins is that the characteristic β-crystallin does not appear till the first differentiation of the lens-fibres (Sauer[2]).

Wolffian regeneration also fails normally in some amphibia. In the Oriental newt, *Hynobius unnangso*, for example, it does not occur, but Ikeda[1] showed by embedding eye-cups in mesoderm that it is nevertheless possible in very early stages. It is not normally seen because the induction of the lens in the ectoderm is a much faster process. Competence for lens or lentoid formation appears in all parts of the eye-cup simultaneously but at different strengths. Like Dragomirov, Ikeda found that structures other than ectoderm or eye-cup itself, such as the nasal placodes, possess lens competence and perhaps the lens inductor.

The significance of the induction of lenses from small fragments of iris or other tissue implanted into the eye-cup is, of course, that they are lying free there, and that the lens-inducing substance must therefore be diffusing across a considerable distance of vitreous or aqueous humour. Mikami[1] has shown that the exact

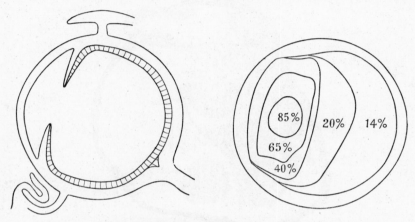

Fig. 170. Topographical distribution of frequency of lens-formation from implanted iris-fragments.

position of the implanted fragment is a matter of some importance. If a sagittal section of the eye be compared with a frequency contour map based on the number of successful lens inductions obtained, it is seen that the most favourable position is about that of the normal lens itself, though either in front of it or some way behind it are favourable positions, while the furthest recesses of the cavity are not very favourable (see Fig. 170). This may be a hint regarding the origin of the inducing stimulus.

We may now turn to the experiments with injured and killed eye-cups. Poležaiev[5] (following the example of Spemann[14] and Krämer) subjected eye-cups to crushing between two glass plates, then implanted them under flank ectoderm of the late gastrula. Sometimes a fairly well-formed eye-cup would regulate itself out of a small piece which had escaped the crushing; in this case, a lens would be induced, but no lenses were induced by the squashed material if regulation failed. On the other hand in two cases ectodermal thickenings, possibly early stages of lens formation, were induced after implantation of a heat-coagulated eye-cup. Implantation of boiled mouse liver gave negative results. Similar experiments were made, in larger series, by Popov & Nikitenko, but they too never obtained true lens induction from killed eye-cups implanted into gastrulae.

Lopaschov's experiments[5] were more systematic. They consisted in the implantation of from one to seven eye-cups, living or heat-coagulated, either into the blastocoele or into isolated ectodermal balls. In the blastocoele even five living eye-cups induced only one lens. Neural inductions were found only in the isolated balls. A single dead eye-cup gave no induction, but two to seven always induced atypically shaped eye-cups; these then induced lenses from the ectoderm.

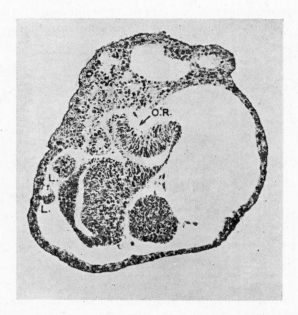

Fig. 171. Imperfect lenses induced by eye-cups themselves the result of neural inductions performed by dead eye-cups on competent ectoderm. O.R. eye-cup; L, lentoids.

In one experiment seven dead eye-cups produced three small eye-cups. The results suggested, therefore, that not only was the lens-inducing substance present in the killed tissue, but also an eye-cup-inducing substance, perhaps present in the neural plate as soon as it arises, and distinguishing (among other things) head-organiser from tail-organiser. This point has already been mentioned (p. 278). The hypothesis of an eye-cup-inducing substance, emanating from head mesoderm or from presumptive brain tissue, already discussed by O. Mangold ([8], p. 232), was elaborated by Lopaschov[7] in a later paper. He took head mesoderm from an early neurula, allowed it to fuse with a piece of hind-brain (the eye-cups derive from the mid-brain), and surrounded the mass with a piece of belly ectoderm. In a considerable number of cases excellent eyes resulted. This is good evidence for the existence of a specific eye-cup-inducing substance in the head mesoderm.

The subject was further pursued by C. D. van Cleave and by Mikami[2]. The sensitivity of van Cleave's material (*Amblystoma punctatum* instead of *Triton taeniatus*) did not permit him to implant more than one boiled eye-cup into one ectodermal ball. In most cases no effect was produced, but in some there was a differentiation of the ectoderm into neural masses with cavities, and in three cases a new eye-cup was formed which induced a lens in the ectoderm near it (Fig. 171). Ethereal and acetone extracts of eye-cups, implanted in egg-albumen into ectodermal balls, gave no neural inductions and no lens formation, but the experiments were not very numerous or varied. Mikami had better success (see below).

The question of the chemical nature of the lens inductor and the eye-cup inductor therefore stands now in about the same position as the question of the nature of the primary evocator stood in 1931. It invites much further research.

It will be noticed that in most of these experiments with dead eyes, lenses were not usually found unless an eye-cup itself had first been induced. This may mean that the lens evocator is less stable to boiling than the eye-cup evocator. One very interesting fact which bears on this is the induction of solitary lenses, i.e. lenses without any neural or eye-cup-like structures, brought about by the implantation of adult tissues into the blastocoele cavity (Holtfreter[12]; Toivonen[2]). Fig. 172 shows two such solitary lenses produced by the implantation respectively of fresh salamander liver and boiled salamander heart. These observations, though incontestable, have given rise to much misgiving in the minds of those who have sought to unify facts and hypotheses in this field. Thus Spemann ([17], pp. 220 & 370) has commented: "One thing is evident from the beginning, that a cooked heart, for instance, which induced an isolated lens, cannot have brought about this effect by some specific lens-forming agency." Now the simplest assumption is, as we have seen, that primary organiser action involves a chemical substance, and that since all adult tissues act, they must contain it, capacity to react to it having long been lost. If they contain the primary evocator, why not the secondary ones also? The disinclination to postulate a considerable number of active substances is, I believe, misguided, for there is no lack of chemical compounds isolated from living tissues for which no function has ever been found or even suggested.* It is therefore not evident that coagulated heart tissue cannot contain the lens-inducing substance.

Holtfreter's results were afterwards confirmed by Okada & Mikami, who found, however, that the lenses produced after implantation of heart and liver were slightly atypical in form and structure. Nose rudiment, head mesoderm, and neural plate or brain also have the power of inducing lenses, and protein-denaturation (boiling or alcohol) does not destroy it (Mikami[4]).

---

* *i*-Inositol is an excellent case in point. A universal cell constituent, synthesised by the chick during its development (*CE*, p. 1229, 1240), it had been known for more than half a century, but no function had ever been attributable to it (cf. the review of Needham[1]) till Eastcott's discovery that it is one of the components of the Bios group of phytohormones (see p. 214). Another excellent case is kynurenin (see p. 410). The general views here adopted are in sharp contrast to those recently expressed by Woerdeman[9].

If the phenomenon of lens induction is due to a chemical substance, it might be possible to exhaust the store of it in the eye-cup by repeated extirpations of the lens followed by successive regenerations. This interesting experiment has been carried out, and it was indeed found that the third lens of such a series was very ill-developed, after which the power of further lens formation fell to zero (Manuilova[1]; Manuilova, Machabeli & Sikharulidze). But after a prolonged rest, the eye could again form a lens from inserted material. Development of the power of restitution (synthetic formation of the substance?) therefore requires a period of time. Eye-cups of different species seem to have different stores, for Nikitenko[2] and Ciaccio obtained regeneration seven times in succession.

Of the metabolism of the eye-cup during lens induction practically nothing is known. The only hint arises from the work of Woerdeman[3], who, stimulated by what he had seen in the dorsal lip region after invagination (p. 190), examined the eye-cup histochemically for glycogen. During lens induction the central part of the eye-cup was found to lose its glycogen almost entirely. J. Mori states that the retinal portion is early enriched with cholesterol esters. Much further work will be needed to elucidate the meaning of this, and to establish the metabolic changes going on in the eye-cup.

As the eye-cup is an outgrowth from the neural axis, and as new neural plates may be formed by the action of the latter in homoiogenetic induction (p. 154), it might be thought that eye-cups implanted into the blastocoele would induce secondary neural axes. This, however, when specially tested by Schmidt & Ragozina and by Holtfreter[7], was found only to occur extremely rarely, if at all.

Before leaving the subject of lens induction, we must refer to a question of much theoretical importance which work on lens induction has settled. It may be asked whether a tissue which has never come under the influence of the primary organiser can react successfully to that of a secondary organiser. Could isolated ectoderm taken from the ventral region of the early gastrula, for instance, react to a lens-inducing stimulus? Although it would have held a certain position in the pre-gastrular individuation field (Dalcq & Pasteels), it would never have suffered the remote effects of primary neural induction, and would never have felt the quite powerful influence of the individuation field of the late gastrula and early neurula stage.

The question was answered by three investigators using three different methods. Waddington[13] isolated presumptive ectoderm from the young gastrula, cultivated it for some time, and then implanted into it presumptive eye-cup material from older embryos. Well-formed eye-cups resulted, together with lenses from the ectoderm (see Fig. 173). Filatov[8] implanted eye-cups into a *Bauchstück*, i.e. an embryo from which the organisation centre had previously been removed and which was nothing, therefore, but a mass of endoderm cells surrounded by a layer of ectoderm. Here again, though no previous neural induction had occurred, lenses were induced. Woerdeman's procedure[8] was more complicated than either of these; he implanted eye-cups into the flank of a late neurula as far away as possible from the head ectoderm, and then grafted over these eye-cups ectoderm of various ages from much younger embryos. It was

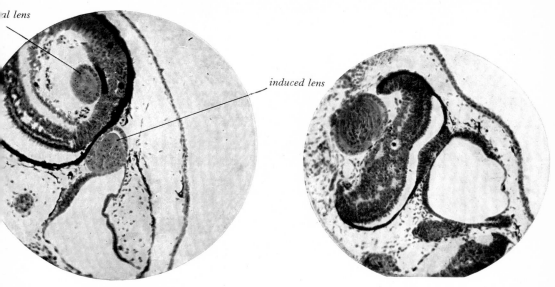

al lens

induced lens

Fig. 167. Induction of lens from the wall of the
ear-vesicle by the eye-cup.

Fig. 168. Diversion of the wall of the ear-vesicle
to form additional retina.

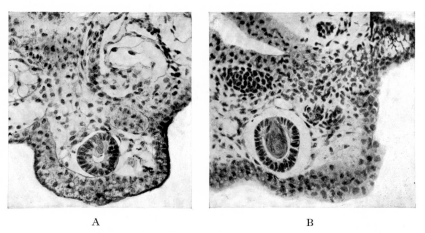

A   B

Fig. 172. Solitary lenses (without eye-cups) induced from ectoderm by implantation of adult tissue
into the blastocoele cavity. A. Salamander liver (fresh). B. Salamander heart (boiled).

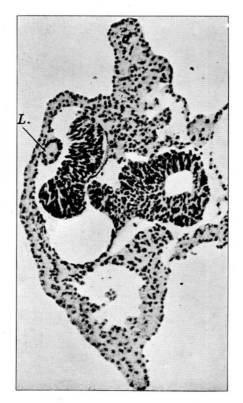

XXVI

Fig. 173. Induction of a lens (*L.*) from ectoderm which had never been subjected to the gastrular individuation field.

found that ectoderm from the youngest gastrula stages would form lenses. There can therefore be little doubt that competence to react to a secondary organiser stimulus arises before the embryo has been stabilised by the effective action of the primary organiser.

2·3952. Other second-grade and third-grade inductors

We shall have little to say about other second-grade and third-grade organisers, for up to the present time little progress has been made with them from the chemical point of view. Much information regarding them will be found in the chapter on self-differentiation (the "mosaic" phase of development) in the book of Huxley & de Beer[2]. A list of them (which makes no claims to be either complete or final) is given in Table 22; it provides a rich field for the biochemistry of the future.

Under the head of second-grade organisers we ought perhaps to include the chemical stimuli which have been believed by some to underlie the phenomena of neurotropism, i.e. the attraction of outgrowing nerve-cell axons in specific directions. Little work has been done in this field recently, though before 1914 it was much studied by Ariens Kappers and Ramon y Cajal; reviews up to 1928 by Tello and O. Mangold[5] should be referred to, together with Harrison's Croonian lecture[14].

A large amount of work has been done on the determination of the cornea; it will be found in the review of Mangold ([8], p. 294). Popov[2] has shown that cornea competence is extremely widespread. All the workers are agreed that the stimulus from the eye-cup is chemical, though mechanical extension of the skin over an implanted ball of glass (W. H. Cole) or wax (Baurmann) has a slight effect. That this is due to the curvature was shown by Cole, who obtained nothing at all when a flat piece of glass was implanted under the skin.

The inductor for the blood (see pp. 164, 378, 648 ff.) is not definitely known, but it forms in a small region just anterior to the heart, and if this is extirpated, completely bloodless larvae can be produced (Frederici; Słonimski). Remarkable powers of blood-vessel construction are shown by the blood cells, to judge from the observations of Parker[3] and Hueper & Russell on the formation of capillaries.

Cartilage of the visceral skeleton is formed from the mesenchyme of the neural crests (Stone), which possesses a power of neural induction never normally employed (Raven[2]), since it meets no competence in its migrations. According to Fischer[3], cartilage can make more of itself from added fibroblasts, and so should be regarded as possessing a specific inductor. This could not be confirmed by Rumjantzev & Berezkina (see below, p. 438).

The formation of the ear-vesicle, it is known, depends on a stimulus from the hind-brain and head mesoderm. From the work of Filatov[1]; Guareschi[1], Lopaschov[7] and others it is known that the formation of the cartilaginous auditory capsule is dependent on a stimulus from the ear-vesicle. Like most secondary inductors, this is quite non-species-specific (Albaum & Nestler; G. A. Schmidt[7]). But other inductions also take place before the final state of the auditory system is reached. Probably stimulated by thyroxin, the metapterygoid region of the quadrate buds

Table 22. *Second and lower grade organiser effects*

Induced structures	Inductor	References
SECOND-GRADE ORGANISER EFFECTS		
Eye-cup	Mid-brain and head mesoderm	Lopaschov[7](1); Adelmann[3]; Alderman
Lens	Eye-cup	Spemann[2]; W. H. Lewis[1]
Nasal placodes	Fore-brain	Raven[2]; Lopaschov[6](1); Kawakami
Ear-vesicles	Hind-brain and head mesoderm (6)	Dalcq[2]; Guareschi[5]; Mangold[16]; Harrison[13](2); Lopaschov[6](1); Ponomareva[1]; Gorbunova; Kogan
Posterior pituitary	Notochord (anterior end)	Bedniakova
Gills	Branchial endoderm	Ichikawa[2]; Severinghaus; T. H. Shen; Mangold[13](3); Lopaschov[6](1)
Spinal ganglia	?	Harrison[18]
Pronephros	Mesoderm	Machemer
Limb buds	Mesoderm	Mangold[16]
Teeth } Mouth-invagination}	Pharyngeal endoderm	Ströer; Riesinger(4); Holtfreter[14]
THIRD-GRADE ORGANISER EFFECTS		
Dorsal fin	Spinal ganglia mesenchyme	Harrison[7]; Holtfreter[19]
Cornea	Eye-cup (all parts) and lens	Mangold[8]
Retina	Eye-cup	Dragomirov[5]
Ear-capsule	Ear-vesicles	Filatov[1]; Guareschi[1,2]; Kaan
Tympanic membrane	Annular cartilage from quadrate	Helff[6]
Heart and blood	Endoderm	Balinsky[10]
Suckers } Balancers}	Fore-brain and fore-gut endoderm	Lopaschov[6](1); Holtfreter[19,20](7); T. Yamada[1]
Anterior pituitary	Posterior pituitary and brain (8)	Gaillard[3]; Burch; Eakin[2](10)
Sense organs of lateral line	Brain or ear-rudiment	Harrison[1](5); Stone[1]
Anus	Tail-bud mesenchyme	Risley; Schechtman[8]
Ciliary polarity	Mesoderm	T. C. Tung & F. Y. Tung[1](9)
Mesonephros(13)	Pronephric (Wolffian) duct	Waddington[17]; Grünwald; O'Connor[1]
Cloacal diverticula (chick)	Pronephric (Wolffian) duct	Boyden[1]; O'Connor[2]

Notes to Table 22

(1) This was assured by Lopaschov's experiments[6] in which the whole of the neural plate of a young neurula was removed and replaced with belly epidermis no longer neurally competent. Completely aneural embryos were produced, in which balancers and gills but *no* nasal placodes, ear-vesicles, or eyes, appeared. The gills and balancers, however, were abnormally formed, probably owing to lack of mesenchyme from the neural crest (a *Nachbarschaft* effect). Zwilling believes that the archenteron roof rather than the fore-brain is the inductor of the nasal placodes.

(2) This case led Harrison[13] to suggest the term "triple assurance". Both hind-brain (*a*) and its underlying mesoderm (*b*) possess inductive power for ear-vesicle; moreover ear-vesicle competence (*c*) seems to be more intense in the presumptive ear-vesicle district than in other ectoderm. But the inductor may well be transmitted by diffusion from the mesoderm to the hind-brain, and variations of competence in different regions are not uncommon. The prime factor is therefore the head mesoderm.

(3) This was assured by O. Mangold's experiments[13], in which the whole of the endoderm was removed from a young neurula, leaving only the meso-ectodermal-neural "mantle".

(4) If the endoderm of the foregut is denied access to the skin by transplantation of a piece of belly endoderm, no mouth or mouth-parts ever form.

(5) R. G. Harrison's work[1] on the development of the lateral-line sense organs was the first in which two embryos of different species were grafted together with a problem of causal morphology in view; its historical importance has already been referred to (p. 148). Stone's papers[1] contain a detailed analysis of the remarkable backward migration which the post-auditory placode carries out in forming these sense-organs.

(6) The role of the mesoderm in inducing ear-vesicles is shown in a beautiful experiment of Pasteels[14] where after centrifuging in the cleavage stages (see p. 221) the neural tube and notochord were almost entirely suppressed; nevertheless two well-formed ear-vesicles were induced at the anterior end of the mesoderm (see also Zwilling[2] and O. I. Schmalhausen). Ponomareva[2] finds that competence for this induction persists until the end of the late tail-bud stage.

(7) But Holtfreter[19,20] believes that anuran ventral ectoderm alone in isolation will produce suckers, and so will presumptive neural plate, though for good shape and typical form the endodermal stimulus is necessary. Here, then, the inhibitory action of the individuation field may be at work (cf. p. 282).

(8) Should this induction be interfered with, as in the centrifuge experiments of Beams & King[2] or the transplantations of Burch, severe endocrine disturbances will attend the absence of the anterior pituitary, including *inter alia* an absolute inhibition of metamorphosis (see p. 454), and the absence of the normal black coloration. This is due to the maintained contraction of the melanophores, not to any interference with melanin synthesis. Spontaneous cases of failure of this induction have been reported (Holt in a pig; Fawcett in a dogfish); cf. p. 454 f.n.

(9) Cf. p. 337.

(10) Atwell & Taft's autoplastic heterotopic grafts of presumptive anterior pituitary apparently differentiated without the stimulus from the posterior (neural) part. But the stimulus might have been effective before their experiments began. Etkin & Lotkin believe that the remainder of the gland exercises an inhibitory effect on the growth of the *pars intermedia*.

(11) It has been suggested that in mammals the mammary ducts are induced by the epithelial ingrowths around the nipple (Deakin); and the presence of accessory nipples on all parts of the body (Hartman[1]; Landauer[26]) suggests a widespread ectodermal competence to a, sometimes errant, inductor.

(12) In birds inductor phenomena are suspected in the development of the feather rudiments, so thoroughly studied by the Chicago school (Lillie[10]; Lillie & Juhn; Lillie & Wang; Fraps & Juhn; Juhn & Gustavson; Juhn, Faulkner & Gustavson). It seems that the rachis can induce barbs and a new ventral locus, for twinning and considerable regulation can occur in feather development.

(13) See p. 319.

off a piece of cartilage which moves backwards and nears the skin. This is the annular tympanic cartilage, and it induces in the skin the tympanic membrane of the outer ear. Fig. 174 shows an experiment of Helff[6], who transplanted the cartilage to parts of the skin other than that where it normally forms, and was always able to obtain by this means a typical tympanic membrane. Without the ring of cartilage no membrane will ever form. Moreover, the property of inducing tympanic membrane appears to be common to many other cartilages in the body, but in normal development the annular tympanic cartilage is the only one not shielded by muscle, and so the power is never exercised (Helff[11]). The yellow fibrous part of the *lamina propria* of the tympanic membrane itself also appears to depend for its origin on the presence of the tip of the columella (Helff[7]). The inductive power of the annular tympanic cartilage lasts into adult life, and more surprisingly, so does the competence of the epidermis (Helff[6]; I. Sato). Once the membrane is formed, it does not regress after transplantation to the back (Helff[12]).

So strong is the inductive power of the annular tympanic cartilage that it can even apparently override a prior determination. Helff[8] showed that the site of the dermal plicae on the head is determined rather early, but they do not appear till the thyroxin period. Transplanted back to non-metamorphosed larvae the plicae regress; then if moved again over the annular tympanic cartilage, they will form tympanic membrane (Helff[10]; Helff & Stark). It seems that this is only possible in heteroplastic transplants; in homoioplastic ones regression occurs but tympanic membrane competence is not retained (Helff[13]).

Lastly, Helff[15] has shown that the membranes induced by dead tympanic cartilages are nearly as typical as those induced by living ones. A diffusible substance is therefore almost certainly involved.

The inductor of the gills is the branchial endoderm. The gill district of the ectoderm moves into place over it before the induction begins (Murtasi[1]). The gradual restriction of gill-forming competence to the gill district has been the subject of particularly careful work by Ichikawa[1], which lends itself to quantitative description. As would be expected from the work of Holtfreter[5] (see p. 274), gill-forming competence is at first present all over the surface of the neurula, except on the neural plate itself. Thus Harrison[11] obtained gills from presumptive limb-bud ectoderm. Later on it gradually disappears from behind forward, concentrating finally in the gill district. This was ascertained by transplanting

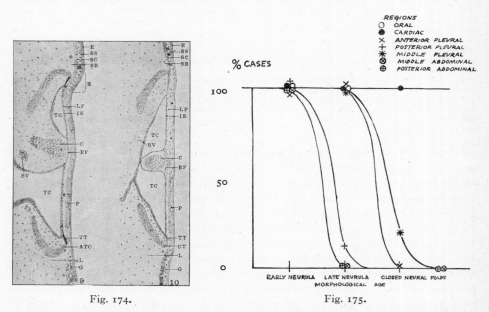

Fig. 174.   Fig. 175.

Fig. 174. Induction of the tympanic membrane by the annular tympanic cartilage. *E*, epidermis; *SS, stratum spongiosum*; *SC, stratum compactum*; *SB*, subcutaneous connective tissue; *LP, lamina propria*; *IE*, inner epithelial layer; *P*, pigment masses; *C*, tip of columella; *EF*, elastic fibre region of *LP*; *TT, tensor tympani*; *TC*, tympanic cavity; *ATC*, normal annular tympanic cartilage; *CT*, annular tympanic cartilage transplanted; *G*, skin glands; *L*, lymphocytes; *BV*, blood vessels; *X*, transition region between membrane and skin.

Fig. 175. Progressive falling off of gill competence outside the gill-district (embryos of *Triturus pyrrhogaster*).

pieces from various areas to the gill area and observing whether or not gills were formed. By plotting the percentage number of positive cases against the time (or rather, the stage of morphological development) we obtain a picture showing the decline of gill-forming competence (Fig. 175).

Like the ear-vesicle inductor, the gill inductor is not species-specific (T. H. Shen; Rotmann[1]).

The only other work of biochemical significance on second- and third-grade inductors is that of Lopaschov[7]. Having found, as we have seen (p. 298), that dead eye-cups would produce further eye-cups from isolated ectodermal balls, and that living eye-cups, placed in the blastocoele, produced only lenses, he tried the effect of placing dead nasal rudiments and ear-vesicles in the blastocoele. No inductions of any kind resulted. He concluded therefore that the substances responsible for inducing those structures were exhausted or not further diffusible.

The interplay of competences and inductors at the beginning of the self-differentiation period must be a very complicated one, but the potentialities revealed by transplantation experiments probably rarely come into action in nature. Thus lens competence is present in the gill district (Spemann[5]), the gill-district ectoderm can act as tooth inductor (Adams), but belly ectoderm, unlike mouth ectoderm, has no tooth competence (Ströer), and head ectoderm cannot respond to the balancer (Carpenter) or the limb-bud (N. K. Arnold) inductor (cf. p. 437).

2·3953. Limb-bud induction

We come now to the question of limb-bud induction. This problem has given rise to a considerable literature of its own, which can be explored through the reviews of Mangold[6] and Balinsky[4,5] and the books of Spemann[17] and of Huxley & de Beer[2]. The determination of the polarity of the limb-buds, in particular, has brought to light facts of the greatest interest, for the different spatial axes of the limb-buds are not determined at the same time. We shall, however, postpone the consideration of this till p. 666, where it will be examined in connection with the properties of liquid crystals. For a general discussion of limb-bud determination the reader must be referred to the reviews already mentioned, since the subject has not as yet reached the stage at which biochemical experiments have been performed. It may suffice to say that in the earlier stages of limb-bud development far-reaching regulation is possible, and as many as three limbs may be produced by dividing one bud (Tornier[1] on *Pelobates*; Lecamp on *Alytes*). At the same time the work done included one famous experiment which has proved rather a stumbling-block, and which must be mentioned here.

As Balinsky[1] has said, limb-bud induction differs from other phenomena dealt with in this section in that no obvious source of the induction is to be seen. If limb-bud corresponds to lens, nothing exists corresponding to eye-cup. The side-plate mesoderm, which itself takes part in the formation of the limb, would at first sight, at any rate, not be suspected.

Modern work on limb-bud induction arose in connection with the induction of the cartilaginous auditory capsule by the ear-vesicle. About 1914 Filatov[1] showed that the ear-vesicle, if transplanted to other parts of the body, would induce an auditory capsule from mesoderm which would never normally have formed it. After the war a younger Russian worker, Balinsky[1], tried to repeat this, using side-plate mesoderm as the source for the auditory capsule, but found instead to his astonishment that a supernumerary limb was produced over the transplanted ear-vesicle. This was confirmed by Filatov[2] and has since then been repeated hundreds of times with the same result.* Balinsky[3] suspected the intervention of inflammation processes and tried therefore to reproduce the result by the implantation of a foreign body. In one case, and in one case only, a super-

* See e.g. Machabeli.

numerary limb was formed after the implantation of a small piece of celloidin. Balinsky himself has never laid undue emphasis on this single result, which, however, has loomed very large in the literature of experimental morphology; Spemann rightly says that it deserves further investigation ([17], p. 217).

The pertinent criticism of Stone[2] that the limb-bud, after implantation of ear-vesicle, develops from skeletogenous mesenchyme inadvertently transplanted with the latter, was at first only rebutted by the care with which Balinsky dissected the ear-vesicle, but later Balinsky[2] obtained successful heteroplastic inductions using an ear-vesicle of *Hyla* in the body-wall of *Triton* and showing that the resulting limb was formed of *Triton* tissue. The criticism lost further force when the nasal grooves were found by B. Glick and Choi to be an even more powerful limb inductor than the ear-vesicle, though they are situated farther away from the limb district than the ear-vesicle. Conversely, the workers of the Russian school could never get limb inductions from pronephros tissue, though this is situated nearer to the limb district than is the ear-vesicle. Balinsky[6] now uses nasal tissue rather than ear-vesicle for preference. Eye-cups, on the other hand, were found by Detwiler & van Dyke to be ineffective in limb-bud induction, and the pituitary rudiment by Filatov[5] to be effective.

Later interesting papers by Balinsky[5] have been devoted to the spatial and time relations of limb-bud induction. Using both ear and nose transplants he studied systematically the frequency of limb inductions at different antero-posterior levels. The results are shown in Fig. 176. The frequency of induction of free limbs falls off fairly continuously from the fourth to the fourteenth segment, but from the eleventh posteriorly the frequency of inductions of pelvic girdles rises steeply.* This is a beautiful illustration of the quantitative falling off of specific organ-building intensity at the "edges" of the various organ districts (cf. p. 286 and Fig. 161). Limb induction by nose-rudiment was further studied in Balinsky's counts of mitotic rate[6]; he found that the nose has a higher rate than the ear-vesicle, and is inclined to attach much importance to this, interpreting the whole effect in a Childian way. He has also presented valuable data on the decline of competence for limb-bud induction with time, and finds that the induced limbs always appear somewhat later than the normal ones, and later, too, the more caudal the segment in which they are being induced.† If nasal placode is present as inductor, the early stages of limb formation will proceed after transplantation of side-plate mesoderm and ectoderm to the head region (Poležaiev[11]), and also after their isolation *in vitro* (Poležaiev[10]).

Putting all these facts together, we may well ask what is their meaning. Their discoverers, experimental morphologists, have up till now been content only with the vague and unsatisfactory hypotheses of "gradients of metabolic rate", "bio-electric potentials", etc. It can certainly do no harm to formulate more

---

* As Balinsky[5] and Mangold[6] have pointed out, there is a phylogenetic interest in this wide area of the limb district, since anciently every metamere had great developmental expression (cf. the segmental appendages of polychaetes to-day). There are still species-differences among amphibia as regards the ease with which limbs may be induced (Balinsky[9]).

† Interesting, though puzzling, experiments on limb induction in late larval and even adult life have been described by Nassonov. We shall return to this subject later on (p. 438) in connection with regeneration.

precise working hypotheses, and the biochemist, undeterred by the famous celloidin experiment, will first of all propose the examination of the embryo for specific substances. Only if this should fail shall we be entitled to fall back on more complex explanations. It is of course obvious that the normal inductor for the limb-bud is neither the ear-vesicle nor the nasal placode, but why should

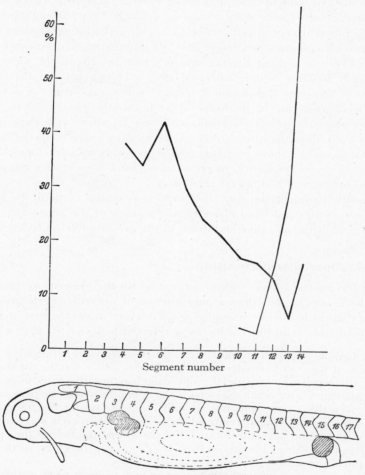

Fig. 176. Frequency of induction of fore- and hind-limbs by implanted ear and nose tissue, according to the fore- and hind-limb districts. Thick line, free limbs; thin line, pelvic girdles.

these structures not contain a substance which, while not identical with that normally present at the centre of the limb-bud area in the mesoderm, would carry out the same morphological stimulus? That such a substance *does* normally appear in the limb district is suggested by G. Hertwig's homoiogenetic inductions[2] in which transplanted limb-buds cause the production of new limb-buds. We have already seen (pp. 262 ff.) many cases of overlapping physiological effects caused by substances the chemical constitutions of which resemble each other in various ways. We have already seen (pp. 172 ff.) that evocators exist in many places where they seem to have nothing whatever to do, such as adult tissue. Had Balinsky's transplantations never been made, the ear substance would never, in nature, have had the opportunity of demonstrating one of its properties, but the same may be said of oestrone, never present, as far as we know, in the amphibian egg, but capable of inducing a secondary embryo if brought into contact with the competent ventral ectoderm. As for the celloidin experiment, we have already seen that injuries to a tissue may liberate the masked evocator which it contains. There is nothing against, and much in favour of, the view that the normal limb evocator is only liberated after a certain time in the side-plate mesoderm. This is also the view of Poležaiev[6], who speaks of an "X-substance". Hence the isolated case of Balinsky's celloidin would correspond to the weak inductions of neural plates by Okada's kaolin. To allow such experiments to seduce us from following up the obvious chemical meaning of all the dependent differentiations which have been unravelled in embryonic development during fifty years of the experimental morphologists' work is to render nugatory that wonderful intellectual and operative effort.

The next experiments on limb-bud induction should certainly be the study of dead ear-vesicles and nasal placodes, and a study of the effects produced by implanting adult tissues, alive and dead, in the presumptive limb districts.

2·3954. Opercular perforation

Before leaving the subject of limb formation for the present, we must devote a brief discussion to the curious phenomenon of opercular perforation, since, although it has little intrinsic morphological interest, it was the first origin of the theory of "double assurance" which was referred to above (p. 293).

In late tadpole stages in anurans the gills are overgrown by a fold of skin, the operculum, at the posterior edge of which there is a hole serving for the passage of water in respiration. Inside the opercular cavities the fore-limbs develop. In order that they may reach the outside the operculum must be perforated; this happens over the elbow, and was at first supposed to be due to mechanical pressure. But in 1906 Braus[1] removed the limb-bud at an early stage from inside the cavity, and found nevertheless (in a few cases) that the hole in the operculum appeared just the same. The perforation would occur, therefore, in the absence of mechanical pressure.

Unfortunately, it is certain that many more factors are at work in this complex situation. The histolytic products of gill degeneration may be decisive in opercular perforation. In addition to this, there are numerous cutaneous glands within the cavity, both in the skin of the limb-bud and in that of adjacent areas; these may also have a part to play. And finally, there is the general internal medium, with its thyroxin effect, to be considered. We have therefore five factors at least to contend with: (1) perforation as a self-differentiation phenomenon of the opercular skin, (2) mechanical pressure from the limb, (3) secretion of the cutaneous glands, (4) products of the autolysing gills, (5) thyroxin in the blood circulation.

Facts have accumulated as follows. Helff[3] at first placed most responsibility on the autolysing gills, for if transplanted to the back, skin perforations occurred over them there. Transplanted histolysing muscle acted similarly. In later work[14], however, he found that opercular skin, transplanted to the back, showed a general, not a local, histolysis, and that skin from side or back transplanted to the opercular site, histolysed and perforated there. Then Weber directed attention to possible products of the skin glands. It is doubtful if they would all be removed by transplantations of the limb-bud, even including the shoulder girdle, to a site under the skin of the back, as was done by Blacher, Liosner & Voronzova, who found that the limb-bud did not perforate on the back. These workers, like Helff, obtained perforation on the back, however, if the degenerating gills were placed under the skin there. But then they found that opercular skin, grafted by itself to a site on the back, also perforates in due course; and this was controlled by transplantations from other places such as ventral surface and flank; these never perforated. Finally, skin from back or side transplanted to the opercular site refused to perforate, and the limb was permanently retained within the cavity (contradiction of Helff). Later Liosner & Voronzova[1] used larvae of different ages. Opercular skin transplanted forwards to older hosts perforated sooner than that of the donor controls, though later than that of the hosts, thus showing the influence of the internal medium. Transplanted backwards to younger hosts, it perforated sooner than theirs; here no influence of internal medium would be expected. Finally, limbless perforation is by no means a constant phenomenon, and its frequency varies from species to species. Injected thyroxin (in the hands of Liosner & Voronzova[2]) accelerated without dissociating the processes, but an interesting shift of the gearing was obtained in this way by Alphonse & Baumann using 1/50,000 thyroxin. The time relations were changed so that sometimes the operculum disappeared before the gills had degenerated at all, and in other cases the gills degenerated without affecting the operculum. Alphonse & Baumann noted that the operculum is supplied with blood from the fifth and sixth branchial arches, and put forward the view that the real cause is the atrophy of the blood-supply brought about by thyroxin.

From this account it will be quite evident that the complexity of the causative factors is in the case of the opercular perforation as yet too great to have been disentangled by experiment. That morphogenetic stimuli are involved we cannot doubt, but we need to know much more about the latent periods of such stimuli and the properties of the competence involved. Only when it has been clearly proved that at some stage in the process two or more leading factors have simultaneous causal responsibility shall we be entitled to have recourse to the notion of "double assurance".

2·396. Second-grade organisers and caste differentiation in social animals

All the inductors so far considered operate when their due time comes in every normal ontogeny. If, however, certain second-, third- or fourth-grade inductors were to operate only in certain individuals, differences of a social order would be set up, and it may be that something very like this happens in caste differentiation among social insects. The ant colony itself may be said to have an ontogenesis of its own (Wheeler[1], p. 438). Moreover, since such substances would seem to enter the developing organisms mainly by the mouth, the line between inductors

and vitamins becomes as thin as that between inductors and hormones. The
"royal jelly" on which the developing queen bee is nourished* (analyses by
Melampy & Jones; Townsend & Lucas) has been thought to be rich in vitamin E
(claimed by Hill & Burdett; denied by Mason & Melampy and by Evans, G. A.
Emerson & Eckert) and in gonadotrophic hormone (claimed by Heyl; denied by
Melampy & Stanley). Whatever the properties of the royal jelly may be, recent
investigations (G. B. Castle; Light, Hartman & O. H. Emerson; reviewed by
A. E. Emerson[2]) have led to the remarkable discovery that alcohol or ether
extracts of termite queens or soldiers, fed to the population of the hive or
colony, are just as effective as the living queens or soldiers themselves in preventing the appearance of further queens or soldiers. Such inhibitory substances
could be called worker-inductors, or they might be regarded as inhibitors of the
queen- and soldier-inductors which in ordinary circumstances would regenerate
these types in the competent community. If there are really such "social
hormones", and the evidence for them is impressive, they would readily spread
through the hive since the colonial insects are constantly licking each other's
secretions, a process which Wheeler ([2], p. 230) called *Trophallaxis*.

2·397. Sex determination

The subject of sex determination perhaps hardly falls under the head of
morphogenesis, since compared with the main construction of the body its effects
are relatively trivial, however important for the individual and the species they
may be. But as the early development of the gonad undoubtedly involves
reactions of inductor type, a brief account of it is necessary. Indeed it may turn
out that the inductor substances here involved will be the first second-grade
organisers to which we can assign a definite chemical constitution.

That the sex of vertebrates is genetically conditioned is now a commonplace.
Reference to any standard account, such as that of Waddington[22], will show that
since 1902, when McClung discovered the sex chromosomes, it has been understood that the two sexes carry different factors whereby maleness or femaleness
is determined. But it is only more recently that the mechanism of action of these
chromosomes has been discovered. As we learn from reviews, such as those of
Willier[1] and Witschi[8], the prospective germ-cells arise (taking the amphibian as
representative) from the dorsal wall of the gut endoderm and become embedded
in the splanchnic mesoderm as it moves in from the two sides to form the dorsal
mesentery.[†] The median plate so formed then splits longitudinally into two
bundles which move outwards. The next step is the separation of chains of cells
from the blastema of the mesonephros, which migrate into the gonad to form

---

* Since it is apparently a salivary secretion, it recalls the "pigeon milk" mentioned on p. 75.
Convergent evolution? The caste development, moreover, strongly resembles a determination
process (Wheeler[2], p. 199) in that if larvae are taken from worker cells after different lengths of
time and fed on royal jelly, the resulting queen differentiation is progressively more and more
imperfect. It is interesting that nature, too, performs experiments which interfere with normal
caste differentiation, as in the well-known parasitism of ants by *Lomechusa* and other staphylinid
beetles, in which forms intermediate between the proper castes occur in large numbers.

† In vertebrates other than amphibia the germ-cells may have more migration to do than this
(see the reviews of Heys and Danchakova[2]).

a central core, or medulla. Thus the endodermal germ-cells are associated with two sorts of mesoderm, a cortex with female potentialities and a medulla with male potentialities. As has been known since the pioneer work of Laulanié, their fate depends upon the genetic constitution of the animal. *In a genetic female the cortex develops* at the expense of the medulla and induces the formation of eggs; *in a genetic male the medulla develops* at the expense of the cortex and induces the formation of spermatozoa. No other mesoderm has competence to form either cortex or medulla. In addition to the gonads there are of course also the sex duct systems.* The Wolffian duct,[†] which has male potentialities, arises as the pronephric or segmental duct on each side of the body; the Mullerian duct,[‡] with female potentialities, originates generally quite independently from the coelomic epithelium. Out of these all the later sex ducts are formed, and although both appear in every organism only one system persists and develops, according to which hormone is poured into the blood stream, the other degenerating into vestigial structures. Such vestigial structures are (in man) the hydatids of Morgagni (cf. p. 263), which are the remains of the Mullerian system in the male, or the Epoöphoron and the canals of Gärtner, which are the remains of the Wolffian in the female.[§] To complete the picture, the secondary sex characters must be mentioned. Pigmentation, spurs, combs, wattles, external genitalia, accessory sex glands, etc. are all, as has been known for many years, under the control of the sex-hormones.[‖]

It may thus be seen that in sex differentiation there is a continuous succession of inductors, first the nuclear inductors (cf. p. 408), which determine whether cortex or medulla shall develop, next the inductors of the cortex and medulla themselves, which settle the fate of the germ-cells and the gonadal tissue, and lastly the sex-hormones produced by the formed testis or ovary, which control the development of the sex duct systems and the secondary sexual characters. About the first of these nothing is known. About the last we know a great deal, some of the chemistry of which has already been discussed (p. 252). What can be said of the inductors of the cortex and medulla?

That the fate of the germ-cells depends on which part of the gonad gets the upper hand, cortex or medulla, was proved by the transplantation experiments of Humphrey. At a stage earlier than that of any gonadal differentiation, the flank mesoderm on one side was exchanged orthotopically between different embryos. After completing development it was found that the sex of the gonad on the trans-

---

* Of these a very convenient introductory account will be found in the little book of Wiesner[2].

† Named after C. F. Wolff (1738–1794), whose *Theoria Generationis* appeared in 1759.

‡ Named after Johannes Muller (1801–1858), whose *Bildungsgeschichte d. Genitalien* appeared in 1830.

§ The suppressed system may sometimes manifest itself abnormally in later life, as when the epoöphoron gives rise to parovarial cysts. The contents of these, which may attain a quantity such as 6 litres, seems to be an ultrafiltrate from the blood (Dierks & Becker).

‖ Chick embryos of both sexes develop combs and spurs, but in females these soon degenerate (Louvier; Danchakova & Kinderis[1]). This would be expected, since in birds the homogametic sex is the male (see below). Precocious development of these characters may be obtained by injecting male hormone (Domm & van Dyke). This is an instance of competence normally preceding the appearance of the inductor.

planted side was often different from that on the normal side, so that on the transplanted side the germ-cells had come to lie in a gonad with a different genetical determination from that on the normal side. They may therefore develop at variance with their genetic constitution, according to the induction to which they are subject.

That the mutual influence of cortex and medulla is carried out by substances has been proved largely by parabiosis experiments, the literature of which has been ably reviewed by Witschi[4]. When two amphibian larvae of opposite sexes are grafted together side by side in early stages, they live a long time and proceed far towards sexual maturity. But the gonads antagonise one another. In toads, it is true, the effects are small, presumably because the substances are not very diffusible.* But in frogs, the male medulla will suppress the cortex of the ovary nearest to it, though the diffusibility is insufficient to affect the further ovary. Evidently the substances travel through the tissues rather than through the blood stream. Moreover, if the distance is too great, as it is when the animals are grafted in chain position rather than side by side, no effects occur. In urodeles, however, the substances are more diffusible still and both gonads of the female partner are affected, their development being suppressed rather than masculinised. This male dominance may, however, be reversed if species of different sizes are used (Burns[1]), and a male of small species is united with a female of large species; in this case the ovaries will suppress the testes, showing that there is a quantitative factor in the balance.

The "struggle" between the cortex and the medulla may be affected in various ways. Thus high temperature favours the medulla at the expense of the cortex (Witschi[2]),† and we have already seen (p. 255) that over-ripeness of the eggs has a profound effect in the same direction.

There has been much discussion regarding the nature of the two factors, which have been called "cortexin" and "medullarin". They are certainly not species-specific in amphibia. Their low diffusibility suggests that they are not identical with the sex-hormones as we know them in later development (Witschi[6]). But in the last few years it has become possible to apply high concentrations of crystalline sex-hormones to embryos still in the indifferent stage of development,‡ and the

* That this is so appears also from the fact that the anterior third of the gonad in toads never develops any medulla and hence always becomes of ovarian character. This is the long-known Bidder's Organ. The Esthonian F. H. Bidder described this curious phenomenon in 1846. If the rest of the gonad is removed in later life, this vestigial ovary will awaken and produce fertilisable eggs (Guyénot & Ponse[1]). Another remarkable occurrence is the presence of "testicular eggs" in toad and frog testes which have undergone some degeneration, due, for example, to transplantation (Welti). This is to be explained by persistence of the cortical inductor in some suppressed form.

† The effects of temperature on these inductors are very remarkable. The involuted accessory sex organs of castrated male mice may be restored to normal by grafting an *ovary* into the ear (R. T. Hill; Hill & Strong). The ovary gives off a little androgen, enough to effect the restoration. But implantation into the (warmer) abdomen will not do. Thus low temperature (as of ear or scrotum) favours androgen production; just contrary to what is found in amphibia.

‡ Earlier investigators (e.g. Minoura; Greenwood; Kemp; Dennis; Willier[4]; and Willier & Yuh) were unable to obtain sex reversal in the chick using chorio-allantoic grafts of gonads. Presumably the amounts of sex-hormone so introduced are insufficient to perform the necessary gonadal inductions.

sex reversals which have been obtained have been very complete. Cortexin and medullarin may perhaps be protein complexes of oestradiol and testosterone (see pp. 206, 252, 316). It is at any rate clear from the work of Witschi[8] and his school that the effects of the hormones (applied by injection or absorption) and the inductors (applied by parabiosis) may differ. Thus salamander tadpoles treated with oestrogens show feminisation, but in parabiosis the male partner suppresses ovarian development. And frog tadpoles treated with androgens show masculinisation, but in parabiosis the partners have no effect on each other.

Among the most successful investigators of these sex reversals have been Danchakova and her Lithuanian collaborators; E. Wolff and the Strasbourg group; and Willier *et al.* in Chicago. It has become clear, however, that the effects of the hormones differ a great deal according to the class of vertebrates studied. This can be seen from Table 23. The full feminisation of genetic male bird embryos was accomplished by Danchakova & Massavicius by introducing oestrone dissolved in oil into the allantoic cavity on the 4th day of incubation.* But when testosterone propionate was similarly introduced by Danchakova[3], the embryos could not survive it. Part of the mechanism of this lethal action seemed to be the precocious transformation of the mesonephros into epididymis, with a consequent fatal oedema following on the suppression of normal excretory function. Conversely, Danchakova[5] was able to produce full masculinisation of the mammalian embryo (guinea-pig) by introducing testosterone propionate in an oil-drop into the amniotic cavity very early in gestation.† This provided a perfect experimental imitation of the classical "freemartin" effect, first described by F. R. Lillie[6] in 1916. When the foetal membranes of male and female calf embryos become united *in utero* so that the circulations are continuous, the female partner becomes greatly modified in the male direction, forming an intersex with testes and male ducts. This can occur also in pigs (W. Hughes), but does not in certain mammals such as the peludo (*Dasypus villosus*) where monochorial twinning with vascular anastomoses is a regular thing (Fernandez). In the freemartin the stimulating substance may, of course, be medullarin and not testosterone. For the mammalian embryo, oestrone, Danchakova[4] claims, in doses sufficient to give sex reversal, is quite lethal.‡

It is impossible not to correlate the fact that mammals and birds behave thus so differently with the fact that the heterogametic sex is different in the two classes of vertebrates. In mammals there is always a tendency to revert to female characters, and in the early stages of sex differentiation the female hormone does

* In later life after hatching, such genetic males (in the absence of further injections) revert in the male direction giving intersexes (Danchakova, Vaškovičuté, Jankovsky, Massavicius & Kinderis).

† This may also be done, but not so surely, by injecting male hormone into the pregnant mammal (see work reviewed by Korenchevsky).

‡ The lethality of the homogametic hormone is, however, regarded by other workers in this field with some scepticism. In view of facts to be mentioned a few paragraphs below it is certainly difficult to understand. Several investigators, of whom Raynaud is the most successful, have obtained feminisation by injecting oestrone into the maternal organism, but it has to be admitted that all such effects are on the sex duct systems and not on the gonads, hence less deep-seated than the reversals obtained by Danchakova.

Table 23. *Summary of sex-reversal effects on embryos of different vertebrate groups*

	Basic type	Hetero-gametic sex	Presence of oestrogen in eggs and embryos	Results of treatment with crystalline sex-hormones in the earliest possible stages			
				♀ hormone		♂ hormone	
				♀ sex	♂ sex	♀ sex	♂ sex
Mammals	♀	♂	+	Semi-lethal or no effect	Semi-lethal (some feminisation)	Maximal masculinisation (the freemartin)	Mild hyper-masculinisation
Birds	♂	♀	−?	No effect or hyper-feminisation	Maximal feminisation	Semi-lethal (some masculinisation)	Semi-lethal or feminisation if androsterone used
Reptiles	♂	♀		Hyper-feminisation	Feminisation	—	—
Amphibia	♀ but variable	Urodeles ♀ Frogs ♂ Toads ♀	+	No effect	Maximal feminisation	Maximal masculinisation	No effect
Teleostean fishes			±	—	Feminisation	Masculinisation	—

REFERENCES

N.B. (1) The basic type indicates that sex away from which reversals occur most easily.
(2) When the heterogametic sex is male, half of the sperm have only one X-chromosome; when it is female, half of the eggs have only one X-chromosome. In amphibia, where both sexes have all the chromosomes, there is an XY pair.

General. Witschi[3,4,8]; Burns[4]; Spratt & Willier
Mammals. Effects of ♀ hormone; Danchakova[5]; Korenchevsky; Greene, Burrill & Ivy. Effects of ♂ hormone; Danchakova[4]; Raynaud.
Birds. Effects of ♀ hormone; Danchakova & Massavicius; Wehefritz & Gierhake; Willier, Gallagher & Koch; Wolff & Ginglinger; Gaarenstroom; van Oordt & Rinkel. Effects of ♂ hormone; Danchakova[3]; Willier, Gallagher & Koch; Kozelka & Gallagher; Gaarenstroom.
Reptiles. Effects of ♀ and ♂ hormones; Danchakova[7]; Danchakova & Kinderis[2]
Amphibia. Effects of ♀ hormone; Burns[2]; Gallien; Foote & Witschi. Effects of ♂ hormone; Burns; Gallien[3]; Foote & Witschi; Uchida.
Fishes. Effects of ♀ and ♂ hormones; Padoa[2]; Witschi & Crown; Berkowitz.

not seem to be needed (hence the prepotence of the male in freemartins). But in birds there is always a tendency to revert to male characters, and in the early stages of sex differentiation the male hormone does not seem to be needed. Were these respective hormones present normally in the early stages, they could hardly be so toxic when artificially introduced. The reptiles in this respect would be expected to behave like the birds, and this was established experimentally by Danchakova[7] and Danchakova & Kinderis[2]. In amphibia, the basic type seems to be variable. It has long been known that in some species and races of amphibia all individuals develop as males and in others all as females until metamorphosis, when equilibrium is attained by 50 % spontaneous sex reversals. This is known as Pflügerian hermaphroditism. Quite often these reversals fail to go to completion, giving adult amphibian intersexes (T. H. Ch'êng). However, in amphibia, complete sex reversal in both directions has been obtained by injection of crystalline sex-hormones into the young tadpole (Burns[2,3]; Gallien; Foote & Witschi).

Prior to the discovery of these facts, Wiesner[1] had already advanced the view that only one of the hormones is an active agent in development, that of the heterogametic sex, male in mammals, female in birds. Differentiation of the organs of the homogametic sex would thus proceed without hormonal stimulation, although they would retain the competence for reacting to the opposite hormone. This would agree well enough with a lethality of the homogametic hormone, but is open to a number of criticisms (Raynaud; Zuckerman & Groome).

In order to clear up these difficulties and confusions the data on the presence of sex-hormones in embryo, placenta, etc. will have to be re-examined, and a great deal more direct biochemical work done. As we have already seen (p. 88), the mammalian placenta, both of male and female foetuses, contain much oestrone; and there is every likelihood that it actually secretes it (E. Allen; Allen, Pratt & Doisy; Brouha & Simonnet; Morrell, McHenry & Powers; Parkes & Bellerby). The amniotic fluid of cows slaughtered during pregnancy has been used as a source of commercial oestrone preparations (cf. Loewe[1]). The permeability of the placenta to oestrone was at first disputed (lit. in Zuckerman & G. van Wagenen), but there is now little doubt of it, since Skowron & Skarżyński showed quantitatively that the oestrogen of the foetal tissues rises when large doses are given to the maternal organism, and Soule finds that oestrone is always at the same level in maternal and foetal bloods. According to Cole, Hart, Lyons & Catchpole and Parker & Tenney, it occurs in all foetal tissues, male or female, after a certain stage, and even largely in the testes of the foetal horse.*

On the other hand, the placenta (of man at any rate) also contains androgenic substances (Goecke; Goecke, Wirz & Daners; Greene, Burrill & Ivy), though only if the embryo is female. This apparently paradoxical result was confirmed on fraternal twins of opposite sexes. It seems, therefore, that we must envisage a balance between oestrogen and androgen in the embryonic adnexa.

---

* There is a curious observation of Meyerhof & Zironi that mammalian embryo extract containing no oestrogens, inhibits male gland activity if injected into the mammal.

About organisms other than mammals still less is known. Fellner's early claim to have found much oestrogen in the hen's egg was decisively disproved (Loewe[2]; Allen, Whitsett, Hardy & Kneibert; Doisy, Ralls, Allen & Johnston), but there may be a trace, for Riboulleau now states that the undeveloped hen's egg contains 2γ/gm. oestrone, and indirect evidence of its presence is offered by Altmann & Hutt. During development oestrogenic hormone is certainly synthesised (Serono, Montezemolo & Balboni), though according to Serono & Montezemolo it declines in amount after reaching a maximum of 10 rat units per egg on the 11th day of incubation. We do not know whether the synthesis occurs only in eggs containing female embryos.

In amphibian eggs oestrogen has been found in well-estimatable quantities (Loewe, Lange & Käer), and its presence in teleostean eggs has been claimed (Sereni, Ashbel & Rabinowitz) and denied (Weisman, Coates & Moses).

It seems certain, then, that as far as mammals are concerned, development of both sexes takes place in a medium comparatively rich in female hormone, though this is probably not true of birds.* How, therefore, it may be asked, is any male development ever possible? Moreover, if the homogametic hormone is toxic to embryos, how is any female development ever possible? These questions have hardly yet received a final answer. It may be that oestrogens and androgens in or around foetal tissues are combined with protein and so inactivated. After making this suggestion, I found that the binding of sex hormones to proteins in normally-occurring inactive complexes has actually been demonstrated (Mühlbock; Dingemanse & Mühlbock). Or it may be that although oestrone is normally present, female differentiation does not need it, while male differentiation has involved the suppression of the duct system which possesses female competence at an early stage, so that there is nothing left for the oestrone to act upon. At the same time, the female systems must retain male competence, as is shown by spontaneous and experimental freemartins. Nevertheless we must still suppose that in the earliest stages of mammalian female development, when the cortex of the gonad is overcoming the medulla, oestrogen is there at work, since in birds oestrone (when injected sufficiently early) can act as cortexin.

Arising out of the foregoing discussion there are a number of interesting points. If cortexin and medullarin are really identical with oestradiol and testosterone, or at any rate very similar in constitution, the induction of a female gonad from endodermal germ-cells and mesodermal medulla will constitute a case of homoiogenetic induction (see p. 154), as Danchakova[4] has pointed out. For an oestrogenic sterol will be stimulating the formation of cells which will later produce more oestrogenic sterol. Obviously this holds conversely for "medullarin".

One of the most acceptable arguments which may be brought forward against the identity of cortexin and medullarin with the sex-hormones is that in the parabiosis experiments there is so much suppression of gonadal differentiation of the opposite sex, and relatively less reversal of direction of development, though this does occur, especially in frogs. Witschi[7] has therefore been led to

---

* A differential embryonic mortality as the result of this, bearing most heavily on male chicks, has been reported by Landauer & Landauer, but denied by Byerly & Jull[2].

postulate two sets of substances, inhibitive as well as stimulative.* The former would be more diffusible than the latter. But it remains to be seen whether all the effects cannot be obtained with sex-hormones of different constitutions.

Among the experiments on the action of hormones, not a few paradoxical effects have been obtained. Of these perhaps the most peculiar is that of Padoa[1], who on administering oestrone to a type of amphibian which normally develops 100 % female up to metamorphosis (*Rana esculenta*) obtained a 100 % transformation of the tadpoles into males. Gallien on *Rana temporaria* and Witschi & Crown on *Rana pipiens* were unable to reproduce this result, but in view of the great differences between amphibian species this cannot be taken as a lack of confirmation. They obtained results more in agreement with theory, the male hormone masculinising the tadpoles and the female one feminising them. E. Wolff[2] then showed that androsterone has a masculinising effect on female chick embryos and a feminising effect on male ones, and though at first doubt was thrown on this finding, it was later confirmed for androsterone and trans-dehydroandrosterone (Willier[3]; Willier, Rawles & Koch) and for methyl-17-androstanol-17-one-3 (Wolff[4]; Wolff & Wolff[1]). Testosterone acetate behaves similarly but not so strongly (Wolff & Wolff[2]). It has, however, already been mentioned that substances are known with both masculinising and feminising properties (see p. 261). Furthermore, in later life, testosterone was found by Danchakova[8] to have feminising effects on the guinea-pig, and it is recognised in clinical practice as possessing truly bisexual properties (Korenchevsky & Hall).

The action of the pituitary has been excluded in these sex-reversal effects. After its complete destruction by localised irradiation (Wolff & Kantor) or its extirpation (Fugo) the gonads show no deviations from normal development (Wolff & Stoll; Fugo & Witschi), and sex reversal is still quite possible (Wolff[3]). On the other hand, it seems that towards the end of incubation the gonads can respond to injected gonadotrophic hormone, if the conclusions of Domm & Dennis are correct. Daineko[2] gives data on the appearance and accumulation of the gonadotrophic hormone in the pituitary.

In sum, it must be admitted that the question of sex differentiation in the higher vertebrates is exceedingly complex, and will not be fully resolved until we know much more than at present about the distribution of the sex-hormones and the competence to react to them.

The invertebrates as well as the vertebrates have been the subject of an immense amount of research on sex determination. Here Goldschmidt's work[4] is classical. In the gypsy-moth *Lymantria* gynandromorphs and intersexes are common. The former do not throw much light on sex differentiation, for they are really sexual chimaerae, in which one part of the body is of one sex, and another of the other. They arise from abnormalities in nuclear divisions whereby the female-determining *X*-chromosomes get lost, and they may consist of a complex mosaic. But intersexes are animals in which all the organs begin their development as if

---

* Witschi ([4], p. 482) analogises this inhibition with the inactivation of the primary evocator in all regions of the gastrula other than the dorsal lip of the blastopore (see p. 174). But it resembles more closely certain inhibitive effects of individuation fields (see p. 282).

the animal was a male, and later switch over to complete it as if the animal were a female, or *vice versa*. This is analogous to the Pflügerian hermaphroditism of some amphibia, already mentioned, except that conditions in the moths are more flexible. The sex factors of the nucleus, in Goldschmidt's view, produce male and female hormones, and development follows whichever of these is produced in excess. Hence, according to the speeds of the metabolic reactions producing the substances, the intersex switch-over point will occur at various times. The same considerations apply to the development of other insects, such as *Drosophila* (Bridges[2]). Unfortunately, nothing whatever is known of the nature of the substances in question. Attempts to influence the sex differentiation of insects by the mammalian sex-hormones (e.g. Danchakova & Vaškovičuté) have not so far been successful, though the presence of oestrogen in adult insects has been established (Loewe, Raudenbusch, Voss & van Heurn).

No discussion on sex determination would be complete without a mention of the remarkable case of the gephyrean worm, *Bonellia*, which has been studied by Baltzer[2]; Herbst[4], and their collaborators over a number of years. Convenient reviews of the subject are those of Baltzer[5] and Nowiński[2].

The female worm has a body about the size of a walnut and a long proboscis projecting about a yard for collecting the mud on which it feeds. The male is extremely small, never more than 3 mm. in length, and lives as a commensal in the female's intestine and uterus. The egg develops into a sexually indifferent larva, which after a short free-swimming life begins to proceed in either the male or the female direction. Which it does depends upon whether or not it becomes attached to the proboscis of an adult female; if attachment takes place, it will become a male, if not it will become a female (Baltzer[2]). At the same time it should be mentioned that a very small number of larvae do spontaneously develop into males without the help of the proboscis, but these genetic males do not exceed 1 % of the larval population. The process of transformation of the larvae into the male line of development can be clearly followed during their stay on the proboscis (Glaus), to which they are stuck by an adhesive secretion which certain larval glands produce (Loosli). Workers in this field thus speak of the number of "proboscis-hours" in an individual's history.

Clearly the sex determination in these worms is exceedingly labile; just the opposite of the strongly genetically conditioned insects. Experiments on the nature of the inductor substance in the proboscis have begun, but owing to the comparative rarity of experimental material, have not been carried very far. Baltzer showed that aqueous extracts of the proboscis are effective, and Nowiński[1] found that none of the active substance in proboscis tissue was lost to acetone, but could be obtained by subsequent extraction with water. The substance is quite stable to boiling. At the same time it seems that the larval system is rather sensitive to a number of external factors; thus Herbst[4] was able to obtain fairly high proportions of males by low pH, by traces of copper, by glycerol, and by variations of ion concentration such as K' and Mg". Heydenreich did the same with high CO_2-concentrations. On the other hand Mutscheller was able to show by direct estimation of the copper content of the proboscis that it contains about one hundred times too little to allow the assumption that copper is the naturally occurring masculinising factor.

It would appear, therefore, that a number of external conditions can set the development going in male or female direction. Perhaps it is not too rash to suggest that what these conditions do is to liberate naturally occurring male or female hormones—which may be identical with the products of the nucleus—from inactive combination. The role of the proboscis might therefore be that of supplying an activator rather than an inductor itself (cf. the discussion on amphibian neural induction, p. 188).

2·399. Organisers and recapitulation

In modern biology we speak of "recapitulation phenomena" rather than of the recapitulation theory, since the facts of persistence of structures in ontogeny characteristic of embryos of species lower and earlier in the phylogenetic scale are accepted by everyone. In the development of one of the higher animals, organs which were important and functional in its ancestors may be found retained in modified, and rudimentary or vestigial, form. The difficult question is why a vestigial organ should be retained at all. Kleinenberg[2] was the first to suggest that vestigial organs are retained because they are in fact indispensable in paving the way during embryonic development for the appearance of later organs which play an essential part in the animal's economy. In an earlier discussion (*CE*, p. 1635) I made the idea more precise in view of the results of modern experimental embryology, and suggested that a vestigial organ is retained because it provides a morphogenetic stimulus which the animal cannot do without.*

The notochord is not a clear case of this. Still functional in the amphibian larva as part of the skeleton, it has become vestigial in birds and mammals, and hence it might be suggested that it has been retained solely for its power of inducing the neural plate. But the induction is believed to be performed by the whole of the axial mesoderm rather than by the notochord alone.

A clearer case has since then been found in the study of the development of the urino-genital system. By the union of the pronephric tubules the Wolffian duct is formed; it then grows backwards. The pronephros[†] is quite vestigial and functionless. If the pronephric rudiment is removed (in amphibia) no duct arises, and the mesonephros either fails to appear (Miura[1]; Waddington[17]) or is almost completely absent (Shimasaki; O'Connor[1]).[‡]   Similarly, Waddington[17] and

---

* Others later gave support to this view (e.g. de Beer[3]; Shumway; Filatov[9]; F. E. Lehmann[15]; Luntz; H. J. Muller). Doubts, which I share, as to whether *all* cases of vestigial organs and especially vestigial behaviour patterns, can be explained in this way, have been voiced by A. E. Emerson[1].

† First described by Joh. Muller in amphibia (1829) and by Balfour & Sedgwick in birds (1878). In amphibia the mesonephros is functional and essential to life (Howland); in birds it is functional but perhaps not physiologically essential. However, the pressure of what it secretes is important for the normal development of the allantois (Boyden[1]).

‡ In this induction there is apparently a graded series of species differences in the degree of dependence which the mesonephros has upon the pronephric duct. The misgivings of O'Connor[1] arising out of this are perhaps unnecessary if we remember the explanation advanced above (p. 292) for the exactly parallel case of lens induction.

Grünwald have shown that in the chick, if the Wolffian duct is cut, mesonephros, metanephros and Mullerian duct never subsequently develop. It is not necessary to speak of double assurance in cases where a trace of the latter tissue is formed until we know that traces of the responsible substances do not diffuse from the pronephros.

It may thus be said that the functionless pronephros induces the mesonephros and metanephros by means of the Wolffian duct as an intermediary. It will be interesting to see whether the whole range of recapitulation phenomena, which have given rise to so much bewilderment and so many pages of speculation, will turn out to be intelligible in terms of morphogenetic stimuli, and so ultimately of biochemistry.

The wider questions which arise at this point, of the relations between organiser phenomena and phylogeny, especially as regards homologies, cannot be dealt with here. De Beer (in a book[2] and essay[3]) brings evidence to show that (1) characters controlled by identical genes need not be homologous, (2) homologous characters need not be controlled by identical genes, (3) structures can owe their origin to different organisers without forfeiting their homology, and (4) homologous structures do not necessarily arise from the same situations in the gastrula, and perhaps not even from the same germ-layers. Bertalanffy[4], on the other hand, seems to maintain the converse of (3).

Another way of expressing the view here put forward about recapitulation phenomena would be to say that the retained, vestigial, or transitory, structures, are examples of "correlated variation". Often discussed by Charles Darwin (*Origin of Species*, p. 8; *Variation of Animals and Plants*, pp. 311 ff.), the occurrence of characters, whose inward relation we do not as yet understand, has long been taken notice of. Some instances of this, wrote Darwin, are quite whimsical, such as the deafness of white cats with blue eyes. And Dollo spoke of the useless characters fixed through long periods of evolutionary change in the marine turtles, because they were "corrélatifs d'une structure indispensable".

2·4. Organisers and fate-maps in Prochordates, Fishes, Sauropsida and Mammals

Until recently it was difficult to distinguish anything in common between the different modes of development among the chordates. The strictly mosaic development of the ascidians seemed to be in sharp contradiction with the regulative development of amphibia, and the early rearrangements of presumptive parts seemed impossible to reconcile in the flat blastodermal system of birds with the spherical system of amphibia and fishes. Largely owing to the synthetic work of Dalcq[7]; Pasteels[15] and Peter[3], however,* we are now able to see that there are gradual relationships between all these different groups. Although relatively little biochemical work has been done on organiser phenomena in groups other than amphibia, we shall give a brief description of the happenings in these other types.

---

* Between the schools of Dalcq and of Peter there are still certain differences of opinion.

2·41. A comparison of fate-maps

First of all, let us look again at the map of the presumptive parts of an amphibian embryo at the beginning of gastrulation. Fig. 177 (*a*) shows the fate-map of an anuran, the toad *Discoglossus pinctus* (Pasteels[7]); it closely resembles the map shown already in Fig. 57 A for the newt, derived from the classical work of Vogt,

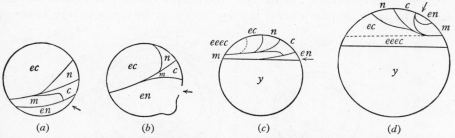

(*a*)   (*b*)   (*c*)   (*d*)

Fig. 177. A comparison of fate-maps (lateral). (*a*) Fate-map of the toad (*Discoglossus*); (*b*) fate-map of the lamprey (*Lampetra*); (*c*) fate-map of a teleostean fish; (*d*) fate-map of a sauropsid. *n*, Presumptive neural area; *c*, presumptive notochord; *m*, presumptive mesoderm; *en*, presumptive endoderm; *ec*, presumptive ectoderm; *eeec*, extra-embryonic ectoderm; *y*, yolk. The arrows indicate the position of the blastopore.

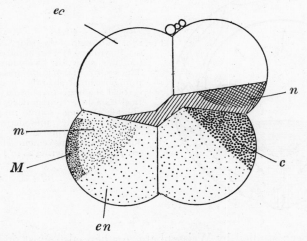

Fig. 178. Fate-map of the ascidian *Ascidiella* (lateral). Lettering as in preceding figure.
M, Myoplasm (see p. 131).

save that the presumptive neural area is smaller. If now we compare this with a map for the ascidian *Ascidiella scabra* (Fig. 178), drawn by Vandebroek[2] in the eight-cell stage (cit. in Dalcq[7], p. 37), we see a considerable resemblance. The presumptive neural area can be seen just above the presumptive notochord area, and the rest of the embryo's surface is divided between presumptive ectoderm and endoderm. "Who would have imagined", wrote Dalcq, "that when Conklin described the famous pigmented crescents of *Styela* and *Ciona*, nature then revealed to him the essential features of the germinal organisation in chordates?" These pigmented plasms have already been discussed (p. 131). No vital staining experiments have as yet been made on *Amphioxus*, but Dalcq believes that its arrangement of presumptive areas would be much the same.

It will be noticed that whereas in the amphibian fate-map there is a "meso-dermal girdle" stretching right round the embryo and dividing presumptive ectoderm from endoderm, the simpler prochordate system has no such arrange-ment. The same simpler scheme is found in the cyclostomes, where *Lampetra*

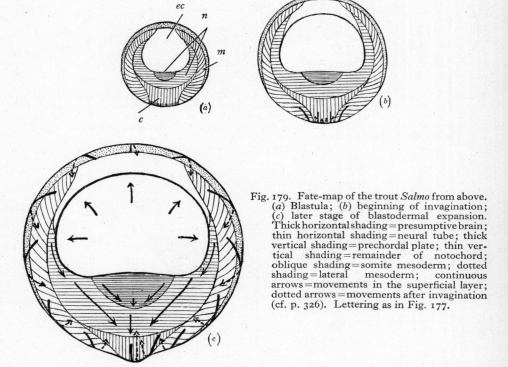

Fig. 179. Fate-map of the trout *Salmo* from above. (*a*) Blastula; (*b*) beginning of invagination; (*c*) later stage of blastodermal expansion. Thick horizontal shading = presumptive brain; thin horizontal shading = neural tube; thick vertical shading = prechordal plate; thin ver-tical shading = remainder of notochord; oblique shading = somite mesoderm; dotted shading = lateral mesoderm; continuous arrows = movements in the superficial layer; dotted arrows = movements after invagination (cf. p. 326). Lettering as in Fig. 177.

fluviatilis has been studied by Weissenberg, for here also most of the embryo's surface is taken up by presumptive ectoderm and endoderm, only small crescentic areas being left for presumptive neural tissue and mesoderm (Fig. 177 *b*).

When we leave the holoblastic eggs and have to deal with eggs in which the cleavages in the yolk are few or absent, as in fishes, we find just the opposite state of affairs, for of the area of the cap of cells which sits on top of the yolk-mass, the greater part is taken up by presumptive neural tissue and presumptive mesoderm. Gastrulation takes place by the expansion of this blastoderm (especially its ectodermal part) so as to envelop the whole of the rest of the egg (the yolk-mass), and, as the expansion proceeds, the mesoderm is tucked in underneath the neural axis, first the prechordal plate, then the notochord and somites, like the unrolling of a carpet, but with lateral contributions. The distribution of the areas on the fate-map of the blastoderm before gastrulation, as established by the work of Pasteels[6], is shown in Figs. 179 and 177c, for the trout *Salmo irideus*.

This was shortly afterwards confirmed by Oppenheimer's work[5] on *Fundulus*. Selachian fishes, though they have much more yolky eggs, seem to fall into line, as the diagram of Fig. 180 for the dogfish, *Scyllium canicula*, taken from Vande-broek[1], shows. These have, however, some of the characteristics of amniote eggs.

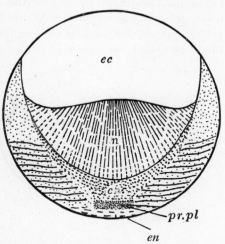

Fig. 180. Fate-map of the dogfish *Scyllium* (from above). *pr.pl.*, Prechordal plate. Lettering as in Fig. 177.

The sauropsida have been, of course, the most difficult of the groups to bring into a consistent series, for the gastrulation of amniotes is no longer marginal. The edges of the blastoderm, sweeping down, round, and under the yolk-mass, no longer represent the edges of the blastopore. The blastoderm has, in fact, faced with the colossal work of absorption of the nutrient materials from the yolk, divided itself into two portions, the "embryonic" and the "extra-embryonic"; and while the latter is engaged in enveloping the yolk, the former has developed a new blastopore, at which invagination of the endo-mesoderm takes place, followed by the appearance of the neural axis just as in amphibia. This is shown in Fig. 181c, the fate-map for the tortoise *Clemmys leprosa*, established by Pasteels[8]. The typical pattern of the amphibian dorsal region can be made out, for the presumptive notochord area is sandwiched between the presumptive neural area anteriorly and the presumptive endodermal area posteriorly, with the presumptive somite areas on each side of it. Anterior to the presumptive neural area is the embryonic ectoderm, and the whole system is surrounded on all sides by the extra-embryonic ectoderm. Invagination itself takes place in a way extremely similar to that of the anamniotes.

In the birds the reduction of the presumptive endoderm area has gone much further than in reptiles. Fig. 181d shows the map of the presumptive areas proposed by Pasteels[5,9], as the result of vital staining experiments, for the chick. The embryonic area is here again surrounded by a vast area of extra-embryonic ecto-derm, and the gastrulation process, which has become very complicated, differs from that in reptiles. The primitive streak, the absence of typical archenteron and archenteric roof, and the movements of Hensen's node, are all new features. For further details the reader must have recourse to the morphological literature.

If the endodermal area of the amphibian fate-map (Fig. 177a) were to be reduced enormously in size, one can see that it would appear completely enclosed by presumptive mesoderm, as is almost the case in the birds. In this case, if the presumptive ectoderm were to expand continuously as the volume of the egg expanded, it would come to surround the other, small, presumptive areas as it does in the sauropsida (Fig. 177 d). But this does not explain the change in gastrulation mechanism. Why should the edges of the blastoderm in meroblastic fishes preserve the significance of the blastopore lips in enveloping the yolk and yet the corresponding edges in the amniotes not do so? Anamniote blastopores might be called "peridiscal"; amniote ones "intradiscal". It is perhaps only a

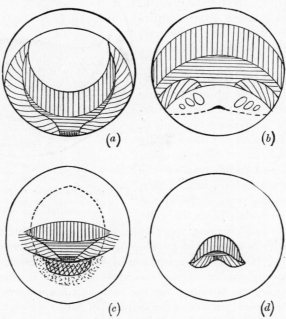

Fig. 181. A comparison of fate-maps. (a) Fate-map of the trout (*Salmo*); (b) fate-map of the toad (*Discoglossus*); (c) fate-map of the tortoise (*Clemmys*); (d) fate-map of the chick (*Gallus*). (a, from above; the others, facing the blastopore, b, from behind, c and d from above.) Vertically shaded = n; horizontally shaded = c; obliquely shaded = m; vertical-horizontal cross-hatched = prechordal endoderm; obliquely cross-hatched = en; dotted = lateral and extra-embryonic mesoderm. Lettering as in Fig. 177.

verbal explanation to say that with the enormous increase in the amount of stored yolk material, there has come about a dissociation between the yolk-enclosing and the embryo-forming functions of the blastopore. This structure has therefore been divided into two: (a) the yolk-enclosing edges of the blastoderm and (b) the points of invagination of the endo-mesoderm in the embryonic area. For an attempted explanation of this in terms of the Dalcq-Pasteels theory (cf. p. 220), the monograph of Dalcq[7] must be consulted. As has been mentioned (p. 146), the formation of the amnion may perhaps be referred to the characteristic ectodermal tendency to expansion (epiboly) seen in amphibia. Enclosure of the embryo within the amnion would necessitate the formation of the allantoic diverticulum of the gut as a further consequence for respiratory needs.

With the mammals we reach a spectacular reversal of the yolk-accumulation process. The minute yolk-mass is vestigial. But as eutherian mammals are probably descended from ancestors having foetal membranes analogous to those of the sauropsida, we should expect to find analogous processes still at work. It is suggested that the expansionist tendencies of the ectoderm, still preserved, account for the formation of the trophoblastic villosities (and hence ultimately the placenta), which precedes the formation of the amnion, just as does the expansion of the extra-embryonic blastoderm all over the yolk-mass in birds. Up to the present time no definitive fate-map has been established for a mammal, but it will no doubt resemble to some extent those of the sauropsida.

2·42. Determination in Prochordates

The ascidians, ever since the classical work of Chabry[1] in 1887, have been regarded as undergoing typical mosaic development.* But the production of half-embryos from one of the two first blastomeres is no longer considered proof that no regulation occurs. Dalcq[9], in numerous merogony and blastomere-shuffling experiments, has shown that a considerable degree of regulation is possible, although the neural rod is certainly not determined by the notochord, as the neural tube is in amphibia. However, in later development, several cases of induction have been found. Whatever the origin of the ectoderm covering it, the anterior endoderm induces the adhesive papillae (T. C. Tung[2]), and later on it is these papillae which, in some cases, give off a stimulus necessary for the resorption of the tail prior to the beginning of sessile life (Holmgren; Zhinkin[2]). Similarly, the "eye" and otocyst of the ascidian tadpole are dependent upon the presence of that part of the endoderm which normally underlies the brain (Rose; Vandebroek[2]) and the otolith upon the presence of the tail (Berrill[2]). Induction phenomena exist, therefore, in ascidian development, but it is mosaic in so far as the action of the primary organiser may be said to be carried out before cleavage.

Of the biochemistry of ascidian induction nothing is known, if we except the isolated observation of Zhinkin[1] that certain dyes greatly accelerate ascidian metamorphosis. Conklin[3] has the impression that during metamorphosis the tadpole tissues become acid, but this needs further investigation.

2·43. Determination in Cyclostomes and Fishes

In cyclostomes and fishes the inductive origin of the neural axis has been placed beyond doubt by the admirable work of Bytinski-Salz[4] on the lamprey, *Lampetra planeri*,† Oppenheimer[1,3] on the perch and minnow, *Perca fluviatilis* and *Fundulus heteroclitus*, and Luther[2] on the trout, *Salmo irideus*. In the case of the lamprey, with its holoblastic cleavage, conditions were not far removed from the amphibia, and Bytinski-Salz was able to reproduce the twin embryos by inductions in almost all particulars similar to those of the classical type. But in the case of the fish eggs, with their very large amount of liquid yolk, their oil-globules, and their tough "chorionic" membranes, the operational technique was much more difficult. Oppenheimer, the first to succeed, was quickly followed by Luther.

* See p. 99. † Confirmed later in independent work by T. Yamada[2]. See also p. 451.

Normal teleostean development may present a beautiful picture because of the relative transparency of the embryonic tissues and the yolk. The classical description of it is usually said to be that of H. V. Wilson[1] on *Serranus*, the sea-bass, but many modern descriptions are valuable, among which may be mentioned those of Price[1] on the whitefish, *Coregonus clupeiformis*, and of Solberg[1] on *Fundulus*. Some of the latter's photographs are here reproduced (Fig. 182). During gastrulation the whole blastoderm expands to cover the yolk with a thin membrane,* but invagination, as has been said, occurs only at a localised region of its edge, along a line (at right angles to the advancing edge) which lays down the future neural axis. The axis (the embryonic shield), and also the advancing edge (the germ ring), are thicker than the rest of the blastoderm. Earlier investigators, restricted to defect experiments, cautery, etc., had noticed that injuries to the embryonic shield or the dorsal lip of the invagination were followed by great deficiencies in

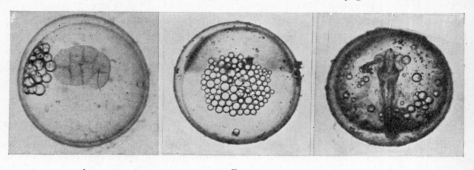

A   B   C

Fig. 182. Stages in the development of a typical teleostean fish (*Fundulus heteroclitus*). A, 2½ hr.; 8-cell cleavage stage. B, 18 hr.; blastoderm growing down over the yolk. C, 42 hr.; neural axis fully formed. *A* and *C* from above, *B* from the side.

the embryo or its complete absence (Kopsch; Sumner; Hoadley[2], etc.). More remarkable was the conclusion of Lereboullet as far back as 1855 (brought to light by Oppenheimer[6]) as the result of his study of fish twinning, that the dorsal lip region had a leading part to play in the formation of the embryo: "Le bourrelet embryogène", he wrote, "doit donc être considéré comme un amas, une sorte de magasin d'éléments organisateurs, et comme le point de départ de toutes formations embryonnaires, regulières ou anormales."

Fig. 183, taken from Luther[2], shows the way the induction experiments are performed. If the just-invaginated mesoderm is transplanted to a point under the ventral edge of the blastoderm diametrically opposite to the natural neural axis, a secondary neural axis will form just as in the amphibia. Fig. 184, taken from Oppenheimer[3], shows experimentally produced twins in later development. In other experiments, Luther[4] rotated the whole of the upper layer, and

* In connection with conditions in mosaic eggs (see Section 2·2) it is interesting that teleostean yolk is considerably more acid than the animal pole plasm and later the blastoderm (pH 5·6 against 7·6, Spek[4]; cf. *CE*, p. 850). A fine picture is obtained when teleostean eggs are stained with cresyl violet; the yolk is blue, the oil-globules yellow and the blastoderm red. Trifonova[1] finds that the blastoderm becomes slightly more acid (neutral red staining experiments) as development goes on, whether normally or parthenogenetically begun.

again obtained a twin, for the originally invaginated endo-mesoderm induced a
new axis in the presumptive ectoderm , and the old neural axis completed itself
by further invagination on the opposite side of the egg.

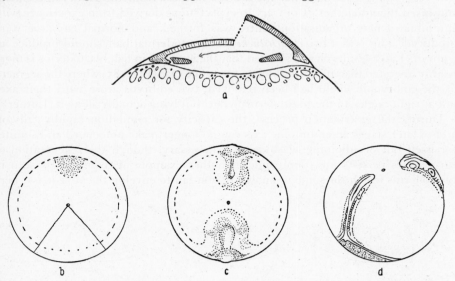

Fig. 183. Method of performing induction experiments in the fish egg. (*a*) Transplantation of
invaginated mesoderm (side view); (*b*) limits of operation (from above); (*c*) resulting twin
embryonic axis (from above); (*d*) the same; later stage (from the side).

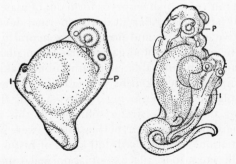

Fig. 184. Induction of secondary embryos in a teleostean fish by implantation of organiser.
P, host embryo; *I*, induced embryo.

In general, events proceed much as in the amphibia. By removing a segment
sufficiently large to include the whole of the presumptive embryonic area, Luther[5]
obtained anidian blastoderms where no trace of axiation was to be found.
Smaller removals permitted various degrees of regulation,* some of which were so
remarkable as to give the impression that evocator can be liberated in regions outside

* Pieces of blastoderm transplanted on to the yolk-sac of older larvae according to a method of
O. Mangold[9] showed some *bedeutungsfremd* differentiation (Luther[3]), but this is probably not
important in view of what has been said above (p. 157) regarding the neutral medium. By this
method, however, the decline in competence of the extra-embryonic areas is well shown[7].

the embryonic shield where this normally does not occur. With this Oppenheimer[7] agrees. In order to test it further, Luther[5] took the extra-embryonic halves of two very young gastrulae, i.e. the sectors farthest from the embryonic shield, and fused them together. Contrary to expectations derived from experience with amphibia, a new organisation centre was formed, and normal embryos were produced. This is as if two *Bauchstücke*, or an ectodermal ball, should produce an embryo. Later results clearly showed that the evocator in fish blastoderms is more readily liberated than it is in amphibia, for in one case (Fig. 185) where the segment of the embryonic shield had been removed, two embryos arose with their axes not at right angles to the blastodermal edge, but lying actually along it (Luther[6]).

During the gastrulation process, the capacity for regulation rapidly falls off and in later stages an embryonic axis can no longer make notochord and somites for itself out of adjoining blastoderm if necessary, though the regeneration of ear-vesicles persists for a long time. Another example of regulation is seen in blastoderms which for some unknown reason fail to envelop the yolk-mass; here

Fig. 185. Formation of twin neural axes in fish embryo along the edges of the embryonic shield.

highly differentiated but very small embryos form out of material the presumptive fates of which would have been different during normal development. As in the amphibia, presumptive epidermis and presumptive neural tissue are quite interchangeable before the action of the organiser has taken place (Luther[2]). Head- and tail-organisers are distinguishable, as in amphibia (Eakin[3]; Dragomirov).

An original experiment undertaken by Oppenheimer[2] was to make teleostean development "holoblastic" by removing embryos from their yolk at various stages and cultivating them in Holtfreter solution. If this was done before the 32-cell stage, a variety of chaotic "hyperblastulae" were produced with little histological differentiation. If afterwards, all kinds of histological differentiation (neural, notochordal, etc.) took place, but abnormal topographical relations, possibly owing to the abnormal water metabolism caused by absence of the yolk (see *CE*, p. 890), occurred. The evocator is thus not liberated when the blastoderm is isolated from the yolk before gastrulation (an interesting comment on the Dalcq-Pasteels theory, cf. p. 220). In the later explants the head structures were commonly better formed than the trunk ones, no doubt because, being made first, they are the less dependent on the migration of the blastoderm over the yolk. In later stages the main morphogenetic function of the yolk is doubtless mechanical.

Oppenheimer's experiments[2] were the logical converse of those made by O. Hertwig[2] in 1899, who rendered amphibian eggs meroblastic by centrifuging down the yolk-platelets into a solid mass; animal pole cleavages occurred but gastrulation was impossible. This has also been done by the use of colchicine (see p. 509; Keppel & Dawson). She subsequently found[7] a curious capacity for morphological without histological differentiation in presumptive tail buds isolated at an early stage from fish blastoderms. After the 4–8 somite stage, however, capacity for self-differentiation set in; this could only have been due to the passage of some influence around the germ-ring from the embryonic axis.

Up to the present time little biochemical work has been done on the organisers of fish eggs, but the subject has been opened by Luther's demonstration[2] that the dorsal lip will carry out inductions (though weakly) after being boiled, and that the brain tissue of a young fish of the same species, as also the liver tissue of a newt, are effective inductors.

With regard to the formative movements of gastrulation in fish, there is general agreement that they cannot possibly be accounted for by differences in mitotic rate. Pasteels[3] has published data for the trout showing that there is no significant difference between the mitotic rate of the various parts of the embryonic shield, though it is higher there than in the extra-embryonic area:

	Mitoses/c.mm.
Embryonic area: average	375
Ranging from	
posterior node (dorsal lip)	317
to	
cephalic part of neural axis	475
Extra-embryonic area: average	260

These conclusions seem to be supported by the results of American workers on *Fundulus*, *Coregonus* and *Gambusia* (Richards[1]; Richards & Porter; Richards & Schumacher; Self). The blastoderm certainly thins very greatly during expansion.

Second-grade organisers in the development of fishes have not as yet received much attention. Under eye-cup and lens induction (see the review of O. Mangold[8]), the remarkable results of Werber should be mentioned. After treating *Fundulus* eggs with various semi-toxic agents, such as dilute solutions of acetone, well-differentiated but completely solitary eyes with lenses appeared on the blastoderm, far away in some cases from the main embryonic axis (see Fig. 186). Whether this is due to some kind of fragmentation of an already determined region of the blastoderm, or to the liberation of eye-cup inductor at an unusual place, is not known. As regards the problem of lens induction, we have already seen that Mencl's case of spontaneous lens without eye-cup played a historical part in the development of our knowledge of lens induction (see p. 291). Such lenses lacking eye-cups have often been reported (review of O. Mangold[8], cf. Fig. 187). In attempting to explain their presence Werber believed that a fragment of eye-cup always accompanied them, though it might be too small to be detected; but a better explanation is at hand in the hypothesis that the restriction of lens inductor to the eye-cup is not a *necessary* condition, and that the lens inductor may be found free in the head mesoderm, as indeed the experiments of Okada & Mikami have proved is the case. Free fish lenses appear not to occur outside the head region, but several small ones may arise side by side. Direct proof that the fish eye-cup induces a lens in presumptive epidermis has not yet been given. Cornea induction may apparently take place from the lens alone, as Werber found cases in which the cornea had formed properly over free lenses. Wolffian regeneration in the fish eye remains doubtful.

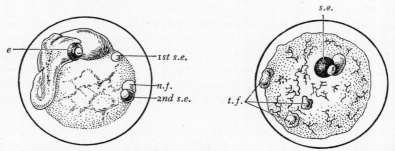

Fig. 186. Development of solitary eyes on fish blastoderms. *e.*, eye of embryonic axis; *s.e.*, solitary eye; *n.f.*, neural fragment; *t. f.*, tissue fragments.

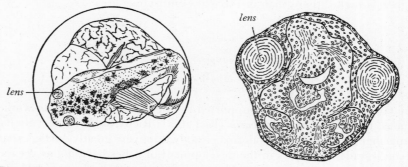

Fig. 187. Formation of lenses in teleostean fish embryos in the absence of eye-cups.

Lastly, according to Filatov[7], the induction of the auditory capsule by the ear-vesicle does not occur in teleostean fishes. The question of the induction of lateral line sense organs by the lateral line nerve in fishes is complicated, and the recent paper of Bailey on this should be consulted.

2·44. Determination in Birds

The localisation of the presumptive areas on the avian blastoderm has already been given (p. 324). But the morphogenetic movements by which these masses of material move into their proper places are still obscure and have given rise to an exceptionally large amount of controversy (Gräper[2]; Wetzel; Kopsch; Marbach; all cited and discussed in Pasteels[15]; and Jacobson). The blastoderm consists of two parts, the transparent *area pellucida*, surrounded by the opaque yolky extra-embryonic blastoderm, which will later form the yolk-sac (*area opaca*). Briefly, what is seen on the surface of the flat area pellucida is first the appearance of a furrow, the primitive streak, followed by its extension to a maximal length. It then begins to shorten,* and its anterior end (Hensen's node) moves backwards followed by the neural folds of the embryonic axis itself, as if it were drawing

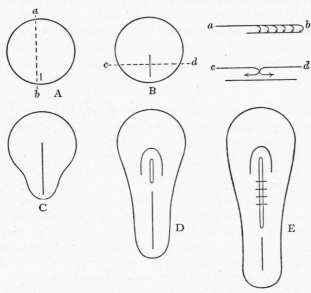

Fig. 188. The morphogenetic movements of the chick blastoderm (see text).

the latter as a pencil would do (see Fig. 188). In due course the somites and the head-fold appear on the embryonic axis. We now realise that the gastrulation process has been chronologically dissociated in the birds as compared with the amphibia. Long before the primitive streak has appeared, the endoderm has formed, not, as used to be thought, by invagination forwards at the posterior edge of the area pellucida, but by a general downward migration of cells (Fig. 188 *ab*), which, according to Jacobson[1], takes place from a rather well-defined area. Nevertheless, the primitive streak does represent an extremely elongated blastopore. Cells converge towards it, partly directly, and partly in the famous "polonaise movement" from the outer edges of the area, and are invaginated, forming the notochord and mesoderm (Fig. 188 *cd*) (Jacobson[2]). Towards the conclusion of this process (and not during it, as in the amphibia), the notochord

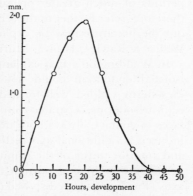

Fig. 189. Elongation and shortening of the chick primitive streak.

* The regularity of the extension and retraction of the primitive streak was appreciated before its function was understood, and the best quantitative data are still those of Edwards, collected in 1902 (Fig. 189). Retraction must not be thought of as active *contraction*, what happens is rather the passage of a wave of morphological change, convergent ectoderm obliterating the streak and then being induced to form neural tissue.

begins to stretch, inducing first the cephalic end of the neural tube, and later the more posterior parts. The withdrawal of Hensen's node is the outward and visible sign of this. It should be noted that in birds nothing corresponding to the yolk-plug is seen, and the blastopore can never be said to be "open".

There seems to be agreement (Pasteels[5]; Derrick[2]) that the morphogenetic movements are not primarily due to mitosis, as the following figures show:

	Pasteels	Derrick
Mitotic rate: primitive streak	2·43	6·1
non-axial ectoderm	2·45	6·2

But it is not clear why the two sets of figures should be so divergent, nor does it seem to have been shown that the area pellucida thins during gastrulation as much as would be expected. This subject merits a more thorough examination, especially as, according to Derrick, subsidiary centres of rapid cell-division exist on both sides of the primitive streak which "feed cells into the groove and the elongating area pellucida".

That inductors play a part in bird embryos parallel to their action in other vertebrates is now well established, largely by the work of Waddington[1] and his collaborators. Two principal methods have been used for studying the experimental morphology of avian development, grafting pieces of blastoderm to another egg at the chorio-allantoic membrane where in time they become vascularised by the host's allantoic circulation; and cultivating operated blastoderms *in vitro*. The former has been much used in America, but it suffers from the criticism already noted (p. 175) that inductors may be present in the blood of the host.* The latter is not so free from criticism as the amphibian salt-solution explantations of Holtfreter, since plasma clot has to be used as a supporting medium, but it permits of much greater operative possibilities than the former. We shall not, therefore, refer further to the self-differentiations obtained by the former method, but the literature may be explored through the following references (Rawles[1]; Willier & Rawles; Hunt[1]; Murray & Selby; Stcherbatov; Dalton[2]; Hoadley[1]).

In Waddington's first explantations[1] of blastoderms of two days' incubation, isolated endoderm could not be successfully cultivated, but isolated ecto-mesoderm yielded well-formed neural groove, notochord, somites, etc. Two ecto-mesoderms, separated from endoderm, and cultivated with their mesodermal faces in contact, always both developed neural grooves, and one axis frequently induced an additional axis in the companion blastoderm. Grafts of the middle and anterior thirds of the primitive streak, placed between endoderm and ectoderm in a peripheral portion of the area pellucida, induced secondary neural axes (Fig. 190). This disposed of the suggestion previously made that Hensen's node alone was the inducing centre. Next, an experiment similar to that which we have described for teleost embryos was done; the endoderm was separated from the ecto-mesoderm, rotated 180° so that the longitudinal axes were diametrically opposed, and replaced for cultivation (Waddington[2]). That part of

---

* The parallel with implantations of amphibian tissue into eye-cup or body-cavity is especially close, for Hunt[2] has obtained *bedeutungsfremd* self-differentiation of blastoderm ectoderm in chorio-allantoic grafts, endodermal organs arising from mesectoderm.

the endoderm which had lain under the normal neural axis frequently induced a secondary embryo in the area pellucida in the same line as the primary, but this process invariably involved the formation of a primitive streak first. Endoderm does not therefore induce directly, but only by calling into being the morphogenetic movements of primitive streak formation. In the heteroplastic transplantations of Waddington & Schmidt (chick/duck and *vice versa*), it was definitely proved that parts of the primitive streak which do not contain Hensen's node induce; other aspects of their work have already been considered under the head of regional differentiation (pp. 124 and 271). Homoiogenetic induction by neural plate has been found (Waddington[3]). Competence for neural axis formation exists not only in the area pellucida, but in the area opaca also (Waddington[5]); in the former it falls off steadily during the decline of the primitive streak, and by the late head-fold stage it has quite disappeared, only ectodermal thickenings

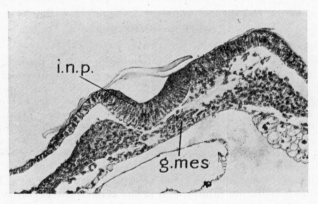

Fig. 190. Neural induction in the chick. *g.mes.*, graft mesoderm; *i.n.p.*, induced neural plate.

being then produced by primitive streak grafts (Woodside[1]). If presumptive epidermis, presumptive notochord mesoderm, or presumptive neural plate pieces are put between endoderm and ecto-mesoderm under the primitive streak, all are induced to form neural tissue (Abercrombie[1]), a fact which recalls the work of Lopaschov and Töndury on the amphibia, already described (p. 170). If primitive streak pieces are placed in this position, a "struggle" between the two axes, taking various forms, occurs (Abercrombie & Waddington). The previous antero-posterior axis of the graft may be reversed, its presumptive regional differentiation may be shifted so as to correspond to that of the host, or its tissues may be incorporated into those of the host; on the other hand, if the axes "side-step", the graft may perform an induction on the host, either by mesodermal evocation of new neural tissue, or by starting a new primitive streak and all that that implies. The converse of Abercrombie's experiment (making presumptive mesoderm into neural tissue), namely, making presumptive epidermis into mesoderm, can also happen (Waddington & Taylor), if a fragment of ectoderm is implanted into the primitive streak itself.

Relatively little work has been done on the biochemistry of the primary organiser in the chick, but simultaneously with the first work of this kind on amphibia it was found by Waddington[6] that primitive streak would still induce after having been killed by boiling (see Fig. 191, opp. p. 342). Later, Abercrombie[2] obtained good inductions with 1 : 2 : 5 : 6-dibenzanthracene-α-β-endo-succinate and other hydrocarbons. He also showed that coagulated presumptive ectoderm was active. These facts have been supplemented only by the experiments of Morita, who, using boiled pieces of frog and mouse viscera, and of 14-day chick embryo tissues, induced many fairly well-formed twins and even triplet embryos. Pieces of tissue killed by organic solvents, such as acetone, alcohol, or xylol, were about equally effective, but inductions were also produced by blocks of paraffin wax, agar-agar, and colophonium resin. The significance of this is not yet clear, but though no micro-photographs are yet available, it appears that the secondary embryos tended to form along torn edges of the blastoderm; they may therefore be regarded rather as regulation phenomena than as inductions. Though we know that any part of the primitive streak may be regenerated after removal (Waddington[1]), Morita's agar inductions could not be confirmed by P. Gray.

One of the most interesting recent approaches is that of Mazia & Davis. If the blastoderm is irradiated with ultra-violet light at the primitive streak stage, the neural folds arising later will never close. Study of the wave-length effectivity revealed a typical steroid absorption spectrum (cf. Warburg's work identifying cytochrome oxidase as a pyrrol-protein and that of Sturm, Gates & Murphy identifying the chicken tumour agent as a nucleoprotein). Since, however, neural differentiation was histologically unaffected, Mazia & Davis were probably dealing with something other than the evocator itself (see p. 370 where genetic failures of the neural plate to close are discussed).

Amphibian histochemical work has had a parallel in the chick, for Jacobson[2] has studied the distribution of glycogen in the cells of the germ-layers. He finds that there is a marked loss of glycogen from the cells of the outer layer, *both* at the time of endoderm formation *and* at the time of mesoderm formation through the primitive streak. There would thus be a biochemical parallel between amphibian and avian gastrulation. Application of the Smith-Dietrich technique, however (said to reveal the presence of sterols, phosphatides or cerebrosides), brought to light a new phenomenon, namely a large concentration of these substances in the primitive streak itself.* There must thus be an increase of them in the cells approaching the streak and a decrease in the cells as they move away laterally in the mesoderm. This decrease does not seem to take place, however, in the head-process mesoderm, moving cranially and about to induce the anterior part of the nervous system. Use of the Ciaccio modification enabled Jacobson[2] to state that these substances are steroidal rather than lipoidal in constitution.

It is also said that a crescentic area anterior to the primitive streak (roughly corresponding to the presumptive neural area of Jacobson[2] and Pasteels[9,15]) also stains darkly with the Smith-Dietrich technique. If this is confirmed, it will be the first indication that biochemical differences may exist between the presumptive areas on fate-maps.†

* The concentration of these substances may affect conclusions as to cell streams drawn from movement of dyes with lipin affinity, such as nile blue, as Jacobson warns.

† At this time also, or somewhat later, there may be differences between the regions in the nature of their ash (Horning & Scott).

As we have seen (p. 203) the organisation centre of the amphibian gastrula has proved to be a region which reduces certain dyes more rapidly than other regions (Piepho[1]). Experiments of this kind have been made on the chick embryo by Rulon, who finds that with janus green, reduction occurs first at the edges of the primitive streak and around Hensen's node; later in the fore-brain and neural groove. This is in agreement with the amphibian results. Preliminary tests by Rulon also indicate that the nitroprusside test for fixed –SH is given most strongly by the chick's organiser region, again in agreement with J. Brachet's findings on amphibia[18]. Barnett & Bourne have studied the distribution of ascorbic acid in early development histochemically. At first only in the extra-embryonic blastoderm, it appears in the nervous system on the fourth day; perhaps because at earlier stages the supply cannot keep pace with the utilisation. We lack knowledge of the metabolism in the various regions of the blastoderm; the methods used up to the present (Needham[7]; Philips[2]) have been too crude to detect any local differences.

We come now to the second-grade organisers in chick development. The classical self-differentiation experiment of Strangeways & Fell on the eye gave of course no information regarding the eye-cup inductor. Hoadley's work[1], in which, after transplantation of presumptive eye district to chorio-allantois, eye-cups were first not formed at all, and then more and more perfectly formed, according to the advancing age of the donor blastoderm, suffers from the general criticism of chorio-allantoic work. But it is probable that the eye-cup is induced by the fore-brain and head mesoderm as in amphibia. Lens induction occurs much as in amphibia* (Reverberi; Waddington & Cohen; van Deth). Lens competence appears to be very widespread (Danchakova[1]; Alexander), and as many as five lenses may be induced by a single transplanted eye-cup. Van Deth finds that the area of competence shrinks progressively, as in the case of the amphibian gill (p. 304). The inductor acts on the competent ectoderm just as well from the outside as from the inside. Wolffian regeneration can occur (Alexander). Other second- and third-grade inductions have been investigated by Waddington & Cohen, who found that fore-brain may induce a nasal placode in non-presumptive ectoderm, and by Szepsenwol[2] and Waddington[15], who studied the induction of the ear-vesicles. The former identified the acoustico-facial ganglion as the ear inductor, but Waddington showed that though it may have an effect it is not the sole inductor since the neural tube at ear level will also act.† Ear competence is present around the presumptive ear district, and even (according to Szepsenwol[1]) in the endoderm. The characteristic head curvature of the chick embryo is directly dependent on heart differentiation (Waddington[14]), and so, apparently, is the formation of the liver (Willier & Rawles[2]).

---

* According to Waddington & Cohen, the chick, as regards self-differentiation of a lentoid ecto-dermal thickening after removal of the optic cup, belongs to Group 2 (p. 292), but these thickenings are not at all like lenses, and may be due to a trace of the inductor remaining behind (see p. 294).

† Here the rare case reported by Goldby is of interest. Not one ear-vesicle only, but a set of ectodermal placodes in series with it, appeared on the side of the head of a chick embryo. If these were traces of the acoustico-lateral system of the anamniota they were more probably due to an atavistic inductor distribution in the neural axis than to an atavistic competence distribution, for the latter is normally so widespread.

2·45. Determination in Mammals

Naturally, work on organiser phenomena in the mammalian embryo has been much delayed by technical difficulties. A remarkable lack of serological specificity as between birds and mammals, first discovered by Japanese investigators (Kiyono & Sueyoshi; Hiraiwa), has helped, however, to overcome this, for Nicholas & Rudnick[2] in 1933 succeeded in cultivating very young rat embryos on the chick chorio-allantois. If a good placental circulation was achieved, good correlative development followed, but if the tissues were vascularised directly by the allantoic circulation, excellent histological differentiation might occur, but the topographical relations would be chaotic. At the same time Waddington & Waterman cultivated rabbit embryos *in vitro* on chick plasma clots, obtaining brains, hearts, neural tubes, and somites, starting with primitive streak stages. In the following year Waddington[7] obtained inductions in rabbit blastoderms by chick primitive streaks, and later *vice versa* (Waddington[11, 16]). But inductions by rabbit primitive streak in rabbit have not yet been obtained, though Törö[3] has produced striking homoiogenetic inductions by neural tube in rat embryos (Fig. 192, opp. p. 342).

Future progress in this important field will no doubt be furthered by the new apparatus of Nicholas[2] for maintaining the development of the mammalian blastocyst *in vitro* by a circulating medium. Waterman[2]; Nicholas[2]; Nicholas & Rudnick[1, 3] and others have cultivated mammalian blastocysts in mammalian plasma, but so far without advancing our knowledge of the inductions proceeding in them.

Work analogous to that of Richards and his collaborators (see p. 329) has been carried out on the rat embryo by Preto, but no correlation between mitotic rate and morphogenetic movement was found.

The mammalian egg (rodents) has been investigated in the spirit of Spek's work with vital stains by Arnold (cf. p. 133). In conformity with its regulative character no differences between the regions of the egg or the morula could be detected, but as soon as the blastocyst was formed, the cells of the embryonic mass gave reactions quite different from those of the blastocyst wall. In so far as could be ascertained from the rather uncertain pH indicators which alone would penetrate into the cells, the embryonic mass was on the alkaline side of neutrality and the blastocyst wall cells on the acid side (see p. 141). The fluid of the blastocyst cavity appeared to be above pH 8·5.

2·5. Temperature and morphogenetic processes

Acceleration of embryonic development with rise of temperature is in itself a banal phenomenon, and its mathematical analysis leaves us only with further enigmas. It will not here be reconsidered (cf. *CE*, pp. 515 ff.).* In so far as the whole pattern of the organism reacts alike, accomplishing development more quickly, little further understanding of the fundamental processes is attained. Only

---

* Among the most careful recent papers here are those of Atlas[1] on amphibia; Worley; Price[2]; J. A. Moore[1]; Merriman on fishes; Pritzker on the chick.

when temperature changes dislocate the developmental processes in an observable way can any progress be made. At the same time it may be useful to show that specially sensitive periods exist with regard to temperature, one process, for example, being much more sensitive to temperature change than another.

For these reasons, the most interesting work on temperature effects in morphogenesis (the determination problem) is probably that of Twitty[1]. The determination process may be shifted experimentally in time, and so thrown out of gear with the growth process. Twitty found that the polarity of the ciliary beat of the ectodermal cells of the amphibian embryo is normally determined during the closure of the neural folds (cf. Austoni), for the cilia of ectodermal grafts rotated

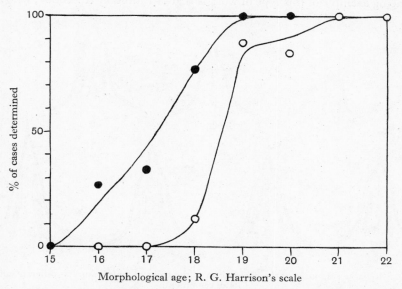

Fig. 193. Influence of temperature on time of determination of ciliary polarity in *Amblystoma*. Embryos kept: at 15° C. ○, at 7° C.; ● before the operation.

through 180° before that stage beat in the same direction as the cilia of the adjacent host ectoderm (neighbourwise), while ectodermal grafts rotated and transplanted later retained their original direction of ciliary waving (selfwise). But in embryos allowed to develop at lower temperatures than normally, Twitty found that the ciliary polarity appeared to be determined at a much earlier stage relatively to general morphological development. As Fig. 193 shows, the percentage of transplants showing determined ciliary polarity rose to its maximum earlier relatively to morphological age in the cooled embryos than in those maintained at normal temperature. If, then, the determination process itself could be thought of as having a temperature characteristic, it would be lower than development, i.e. visible morphological differentiation, as a whole. Conversely, a reaction inhibitory to the determination process would have a temperature characteristic higher than that of development as a whole. The fact that determination can be affected in this way by temperature is a hopeful prognosis for a better physico-chemical understanding of it.

Another interesting fact is that according to Lopaschov[8], neural induction by dead inductors is much more effective at low temperature than at high. The explanation of this is unknown. But on the contrary, regulation in homoio-plastically grafted eyes in axolotls is less effective at low temperatures than at high, or may even fail altogether (Twitty[7]).

Differences of temperature characteristic between different stages of develop-ment have long been reported, e.g. for the development of the sea-urchin embryo (*CE*, p. 507), though they can hardly perhaps as yet be said to be established. A fairly convincing example of this type of work is that of Ljubitzky & Svetlov, who noted

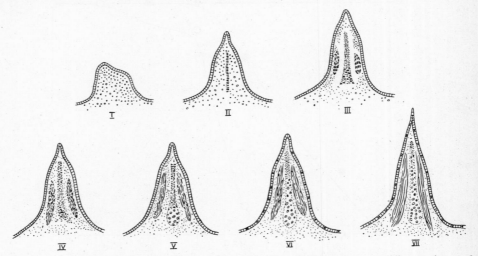

Fig. 194. Relative responses of the pectoral fin of the trout to temperature at different phases of its development.

Stage	I	II	III	IV	V	VI	VII
Q_{10}	2·3	1·0	5·9	3·0	1·0	1·0	1·0

the relative responses of the pectoral fin of the trout to rise of temperature at different periods of its differentiation. Their results are summarised in Fig. 194. In the earliest stage, that of the undifferentiated blastema, the temperature acceleration was considerable, then it dropped, to rise again at the period of rapid histological differentiation of the skeleton and muscle masses. When the histo-logical differentiation was over, the effect of temperature again became negligible. Similar, but less clear-cut results, have been described by Ljubitzky for the period of blastoderm formation, yolk encirclement, and axiation in the trout, and by Latinik-Vetulani[1] for the early development of the amphibian embryo. Her results are difficult to interpret owing to the high supra-maximal temperatures used; Hoadley[7], studying the same kind of material histologically, noted that persistence of the blastopore and microcephaly, due to some interference with the head-organiser, were the most common results.

A good deal of work has been done on the *localised* application of heat to the amphibian embryo, in which the embryo is subjected to a "temperature gradient".* The main effect of this seems to be an acceleration of cleavage, so that the usual differences between the size of animal and vegetal pole cells in the blastula, for instance, are lessened or abolished. Later stages may show fairly marked asymmetry, one neural fold being larger than the other. Eventually great differences between the development attained by head and tail ends may be found (Vogt[3]; Huxley[†]; Dean, Shaw & Tazelaar).[†] But the only investigator who has drawn new conclusions about determination from such experiments is Gilchrist. His first experiments (Gilchrist[1]), did not go beyond the asymmetries, puckerings, etc. of the previous work. In later work, however (Gilchrist[3]), he adopted the plan of heating first on one side, then on the other, so that each side should have been equally subject to the acceleration produced by rise of temperature, though this would have occurred at different times. The gradient was usually applied for about 50 hr., with a reversal at the mid-point of the period, and a succession of these periods on different batches of eggs covered the whole of development from the earliest blastula stage to just before the closure of the neural folds. The argument now ran as follows: if one-sided heating produces an increase in the size of the neural fold on that side, then the period during which the heating was applied must have been that at which the determination of the neural plate was going on. And so similarly for other structures. The results may be tabulated thus:

Process	Period during which determination would be proceeding
Morphogenetic movements of gastrulation	Cleavage stages—early blastula
Determination of neural plate	Early blastula—late blastula
Determination of size of brain and head	Late blastula—half-moon gastrula
Determination of pronephros	Full-moon gastrula—open neural plate

There is thus a certain contradiction between these results and those already described (p. 154), for while, according to the accepted view, the determination of the neural plate proceeds during invagination by the action of the archenteron roof, here it appeared to take place before invagination had begun. Gilchrist[3] proposed, as an explanation of this difference, that we should distinguish between "weak" and "strong" tests of determination. "Strong" tests are those in which the neighbourwise or selfwise behaviour, and hence the determination of an embryonic part, is tested by transplantation to an unusual region of a host. "Weak" tests are those which do not subject the part to such new surroundings; e.g. the method of reversed thermal gradients. If then, the determination process takes an appreciable time from its first onset until its final completion, we should expect that the stronger the test, "that is, the more potent it is in bringing the developing material into new and active relationships, the later will the time of determination appear to be". But this distinction between "strong" and "weak" tests is not acceptable. The definition of determination is the firm capacity of the tissue for self-differentiation, from which it will not be deterred *whatever its environment*, if it lives. "Weak" tests simply do not test this at all.

It is unfortunate that Gilchrist published no clear histological sections of his asymmetric neural plates, for we do not really know what a "larger" neural fold

---

* We have already noted (p. 188) that extreme applications of this have been made in the effort to produce inductions, but so far without success.

† Asymmetries of various kinds, interpreted in gradient terms, have also been obtained by the use of a variety of unrelated chemical substances in the water surrounding amphibian gastrulae (Wolsky, Tazelaar & Huxley; Svetlov[2]).

on one side means. It may be that more cells are derived from the adjoining presumptive epidermis, but it might also merely mean that there is less floor in these plates, i.e. a simple distortion of the normal cross-section.

For the sake of argument, however, let us grant that Gilchrist's larger neural folds contain cells which should have been epidermis. It is true that for cells at the border of the neural area, determination of the size of the plate is the same thing as determination of the existence of the plate. But there is nothing to prove that the effect of localised warming on determination is immediate. It might affect processes preliminary to evocator liberation. Or it might exert a delayed effect on such border-line cells so that the inducing stimulus, when it came later on, spread further than usual in that direction. In other words, competence, not determination, would be affected.

Gilchrist's experiments, therefore, have not proved that determination is a process requiring an appreciable amount of time, and that there is an intermediate stage of labile determination. But to quarrel with this view would be to dispute a truism. All statistical curves of determination processes show an intermediate period when sometimes one result is obtained, sometimes the other. This had been pointed out by Brandt some five years before Gilchrist's work, as the result of a consideration of the determination of limb-buds (discussed on p. 666), where the various spatial axes are determined at different times. Exactly how quickly the determination process proceeds in a given tissue is, of course, a matter for much future research. A new notation is really required, describing for any given piece of tissue at any given developmental stage its subsequent behaviour in all possible new situations, including self-differentiation *in vitro*. This would involve a list of its competences.

Second-grade organisers have been the subject of a few experiments in which temperature was studied. Lehmann[9] found that the action of trichlorbutyl alcohol in suppressing lens formation in amphibia (see p. 224) was maximal at $14°$ C.; less marked at lower and higher temperatures. After some earlier work, Nakamura[1] has established that Wolffian regeneration is especially sensitive to temperature; in the oriental species studied by him (*Triturus pyrrhogaster*) winter temperatures of $16°$ C. inhibit it completely. Light intensity and nutritive conditions were ruled out. Further relevant facts will be referred to later (pp. 450, 516).

2·6. Genes and organiser phenomena

We now enter upon the widely ramifying subject of the relations between genetics and embryology. It may be separated into a number of divisions. In the first place, it is known that organisers will act across genetic boundaries, i.e. material of quite different genetic constitution will react to the same organiser stimulus, though the ways in which it does so will differ according to this constitution. Organisers must indeed be regarded to a large extent as the intermediary mechanisms between the gene equipment and the final form and properties of the developed animal. It is therefore important to know how genes and organisers are related in normal development. But in order to do this advantage must be taken of those instructive cases in which abnormal development occurs. If development is only very slightly abnormal and leads to a changed organism, we may trace the change back to a mutation in the genome but the case may be difficult to analyse embryologically owing to the smallness of the change. All such changes are not small, however; hence the importance of lethal or sub-lethal genes. At the same time advantage must also be taken of the abnormalities which occur when the nuclear substance of either gamete is experimentally interfered with before or about the time of fertilisation. There are also to be considered various interferences with the normal plan of growth rates of the parts of the body, which we may classify as disturbances of the individuation field.

In the following discussions we shall no longer be in a realm where the end-result is purely morphological. It is difficult to consider the actions of the genes which cause morphological effects in abstraction from the actions of those which govern reaction rates, the expression of which is in terms of the production of coloured substances, or the accumulation of chemically defined storage materials.

The first section, then, will deal with the action of organisers across genetic boundaries, that is to say, the effects of an inductor of one genetic species upon competent tissue belonging to another. From pp. 171 and 301 it will by now be evident that the primary organiser of vertebrates, and at least some of the second-grade organisers, have no species-specificity. A piece of chick primitive streak in the blastocoele cavity of an amphibian embryo inducing a secondary embryonic axis can obviously induce only an amphibian one. In such extreme cases it is easy to see that the induction can only be "autonomous", not "complementary". But in cases where the inductor and the competent tissue are more closely related, it might happen that the inductor would overcome the genetic determination of the reactor in favour of its own characteristics.

2·61. The action of organisers across genetic boundaries

The question was first posed by Spemann[10] in 1921; he found that when presumptive neural tissue of *Triton taeniatus* was grafted into the presumptive gill district of *Triton cristatus*, it helped to form the gills while at the same time retaining its own histological character. Later Rotmann[5], a member of Spemann's school, devoted himself entirely to following up this interesting line of work. The experiments on the gill were first confirmed (Rotmann[1]; Schmidt[10]) and extended to the limb buds,* but a more decisive case was found in balancer induction (see p. 302). If an interchange was made between presumptive balancer ectoderm of *T. taeniatus* and *cristatus* the balancers subsequently formed, though quite normally induced (by the host's endoderm), "showed exclusively the characters of the species from which the ectoderm had been taken". Balancers of *cristatus* not only have a thickened ectoderm compared to the others, but they enclose more endoderm and point forward, while the others point backward (Rotmann[2]). Lens-forming ectoderm interchanged between the two species behaves

---

* Ectoderm of *cristatus* put into the limb-bud district of *taeniatus* formed a limb composed of *cristatus* ectoderm and *taeniatus* mesoderm; it had the shape of a *taeniatus* limb. On the other hand *cristatus* presumptive mesoderm was so implanted that it invaginated with the *taeniatus* mesoderm; thus a limb composed of *taeniatus* ectoderm and *cristatus* mesoderm was produced (on a *taeniatus* embryo); it had the shape of a *cristatus* limb. Hence the mesodermal part of the limb is much more important in determining its form than the ectodermal part (Rotmann[1]; Harrison[10]). Such ectodermal grafts persist in showing the characteristics of the donor's skin long after metamorphosis (Rotmann & McDougald).

similarly; on induction it will give a lens, but the size of the lens will be in accordance with the species from which the competent ectoderm is derived, and will not be related to the size of the inducing eye-cup (Rotmann[4]).

It is thus clear that the competent tissue gives the required answer correctly, but "in its own language", to use an anthropomorphic analogy. The most striking case of the kind so far described is that of the mouth parts of amphibia. Urodele larvae have genuine teeth in their mouths, of the same structure and origin as the teeth of all vertebrates; those of anurans, on the other hand, have only horny jaws and stumps quite unlike teeth. If urodele and anuran ectoderm could be so exchanged as to come under the influence of the head endoderm in mouth induction, what would be formed? Correspondingly, urodele larvae possess, somewhat laterally and behind the mouth, a pair of balancers, while anuran larvae have two areas of thickened ectoderm secreting a sticky substance and called suckers, behind the mouth in the median line. Again an exchange of urodele and anuran ectoderm was required.

The second question was answered before the first one. In 1932 Spemann & Schotté transplanted presumptive epidermis from any region of an early gastrula of the frog *Rana esculenta* into the balancer district of *T. taeniatus*, with the result that very well-shaped and functional suckers were formed* (Fig. 195). The converse experiment was made by Rotmann[3], who moved presumptive ectoderm from a young *Triton* gastrula to the sucker district of *Bombinator pachypus*, with the result that well-formed balancers were produced (Fig. 196). In one case, where the graft was one-sided, a sucker was formed in its normal place in the host from host material, and on the other side a balancer was formed from graft material. Many other cases of this kind have since been obtained (e.g. by Schmidt[8]). We therefore have in the endoderm of these districts a common stimulus producing a different result according to the competence of the tissue subjected to it. Spemann has written "one might assume that the stimulus releasing the development of either organ would still be the same as in that organ of the ancestors from which the two recent organs have been derived; whereas the reaction of the ectoderm would have changed" ([17], p. 359).

In this connection a certain experiment of Mangold's[10] is very instructive. The axolotl larva, though a urodele, does not possess a balancer. Hence ectoderm from a suitable area of the axolotl larva would not produce a balancer when grafted into the balancer district of *Triton*. But on the contrary, presumptive balancer ectoderm from *Triton*, removed long before determination could have occurred, formed a balancer after being grafted on to the appropriate region of *Amblystoma*. Therefore it seems certain that in the course of evolution the ectodermal competence, but not the endodermal inductor, has been lost in the axolotl.

The first question, that of the teeth as opposed to horny jaws, was settled by Schotté (cited in Spemann[17]) and by Holtfreter[13,14]. Genuine teeth will be produced in urodele ectoderm after the action of anuran mouth-organiser, the normal effect of which is horny jaws (Fig. 197). Horny jaws will be produced in

* Often confirmed, as by Schotté & Edds.

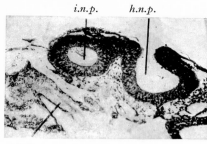

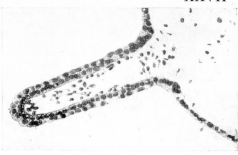

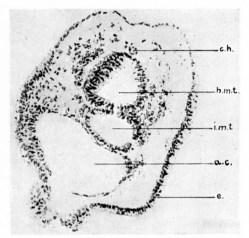

Fig. 191. Induction by dead chick organiser in the chick. *h.n.p.*, host neural plate; *i.n.p.*, induced neural plate; *co.gr.*, coagulated graft.

Fig. 196. Balancer formed in urodele ectoderm on the anuran body, by the action of the anuran sucker inductor.

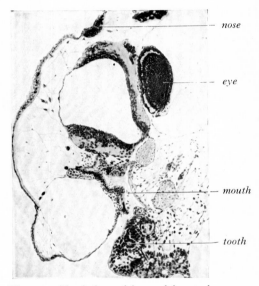

Fig. 192. Homoiogenetic induction in mammals (rat). Implantation of a neural plate fragment of a 9-day embryo into the proamniotic cavity of a 7-day embryo, followed by 70 hr. cultivation *in vitro*. *c.h.*, notochord; *h.m.t.*, host neural tube; *i.m.t.*, induced neural tube; *a.c.*, amniotic cavity; *e.*, endoderm.

Fig. 197. Teeth formed in urodele ectoderm on the anuran body, by the action of the anuran mouth-inductor which normally produces horny jaws.

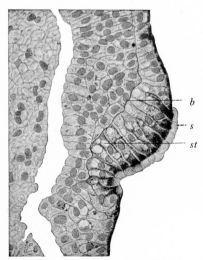

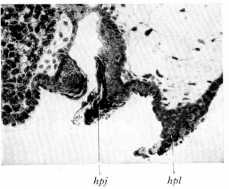

Fig. 195. Functional sucker formed in anuran ectoderm on the urodele body, by the action of the urodele balancer inductor. *b*, boundary of the outer ectodermal layer; *s*, secretion; *st*, droplets of secretion.

Fig. 198. Horny jaws formed in anuran ectoderm on the urodele body by the action of the urodele mouth-inductor which normally produces teeth. *hpj*, horny prominence of jaw; *hpl*, horny prominence of lip.

anuran ectoderm after the action of urodele mouth-organiser, the normal effect of which is teeth (Fig. 198). It is thus certain that competences for organs which the host does not possess at all may react to stimuli which normally act upon other competences. "It is as if", says Spemann[17], "the cue were quite general—'mouth armament', and at this cue the needed organs would be formed, as they are provided in the genotype of the reacting tissue." In terms of modern knowledge of biochemical specificity, this may well mean that the same key (the chemical substance, the inductor) may be able to unlock several different locks.* This may also have a bearing upon the induction of the limb-bud by ear-vesicle or nasal placode, for there is no reason for supposing that the normal inductors in those organs may not be able to unlock the limb-bud system, although naturally such an overlap of specificity never comes into play in the normally developing organism, and does no harm there.

2·62. Genes and the later stages of individuation

The experiments just described have given a clear answer to the question whether the determination of a tissue of one genetic composition can be brought about by the action of a tissue of another genetic composition. Yet with the first histological differentiation of an organ the development of that organ is not finished. Growth processes have to occur, and the onset of physiological function has to be successfully passed through. The further question may therefore be asked in the case of a transplanted organ; would its growth rate and its physiological characters be determined by its host or by the donor? In other words, as we have seen, a piece of tissue of one genetic composition will behave *ortsgemäss* (neighbourwise) to tissue of another genetic composition as regards the organ induced, but *herkunftsgemäss* (selfwise) as regards the "style" of the new organ, which will be in accord with the graft's genetic composition. If now an organ fully determined as regards its histological differentiation, such as an eye or a limb-bud, is transplanted, will it continue to show the growth rate characteristic of its donor, thus behaving selfwise in accordance with its genetic composition, even though this should lead to serious disequilibria in the plan of the host's body; or will it be in some way regulated to correspond with the latter? Correspondingly, when the histological differentiation of an organ is determined, is its later physiological behaviour also determined, or can this still be altered by transplantation to a host the analogous organ of which shows a different physiological behaviour? It will be seen that we have here the link between the determination problem and the problem of heterauxetic growth (Section 3·2).

These questions have been the study of Harrison and his colleagues over a long period of years. They presented certain special difficulties, for example, the difficulty of accurate measurements of organ sizes without killing the animal, and the complications produced by different levels of food-intake, etc. For these details the reviews of Harrison[11] and Twitty[3] must be consulted; here the require-

---

* Acetylcholine is a parallel to this. Though present in many invertebrate groups, the reaction to it on the part of the effector systems is by no means the same. And we know of substances, such as 1 : 3-dimethylbutyl-ethyl-barbiturate (Swanson & Chen) or harmine (Chen & Chen) which cause convulsions in mammals but depression and anaesthesia in amphibia.

ments of space limit us to a short account of the main results. Reference has already been made to the classical experiment of Harrison[1] in 1898 on the hetero-plastic grafting of differently pigmented anuran larvae for the study of the

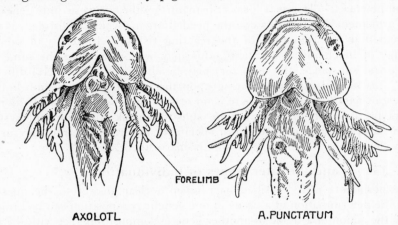

FORELIMB

AXOLOTL   A.PUNCTATUM

Fig. 199. Differences in limb-development at the beginning of the larval stage in the two species of *Amblystoma, tigrinum* and *punctatum.*

development of the lateral line (p. 148); from this the later work was a natural development.

The two species of axolotl, *Amblystoma punctatum* and *tigrinum,* differ in the time of appearance of their fore-limbs. At the beginning of the larval period, when the yolk has been used and the animals are ready to feed, the limbs of *punctatum* are almost fully developed, having arm, forearm and hand with two long and two rudimentary digits. The limbs of *tigrinum,* on the other hand, are at the same age only minute buds (see Fig. 199). When transplanta-tions of the presumptive limb areas were made (Harrison[6]; Twitty & Schwind) it was found that the growth rates were quite determined, *punctatum* limb on *tigrinum* host developing rapidly and *tigrinum* limb on *punctatum* host developing slowly. Owing to the fact that *A. tigrinum*

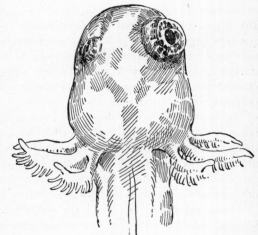

Fig. 200. Larva of *A. punctatum* carrying an eye of the larger *tigrinum* species.

attains in adult life a larger size than *A. punctatum* there is later a secondary reversal of limb sizes.*

* Note that the Q_{O_2} of *tigrinum* embryo tissue is higher than that of *punctatum* (Wills).

Essentially similar results have been obtained for the eye by Harrison[9]; Twitty & Schwind and Stone. The eye-cups of *tigrinum* are normally very much larger than those of *punctatum*; Fig. 200 shows a larva of the latter species carrying an eye of the former on the right side. But on transplantation the eye preserves almost exactly the same dimensions as it would have done in the donor (see Fig. 201, taken from Twitty & Schwind). Towards metamorphosis there is, however,

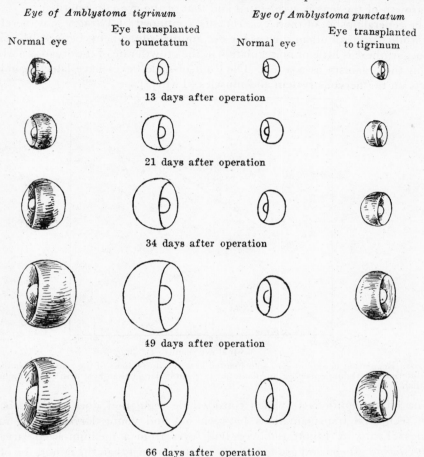

Eye of Amblystoma tigrinum   *Eye of Amblystoma punctatum*

Normal eye   Eye transplanted to punctatum   Normal eye   Eye transplanted to tigrinum

13 days after operation

21 days after operation

34 days after operation

49 days after operation

66 days after operation

Fig. 201. Lack of regulation in axolotl eyes transplanted between species of different eye-size.

a very slight regulation of eye size in *punctatum* eyes transplanted to *tigrinum*. The experiments were further elaborated by Harrison[9] who transplanted eye-cup and lens ectoderm separately and found that *tigrinum* eye-cup on *punctatum* host, with a lens which it has itself induced from *punctatum* ectoderm, still grew more rapidly than the *punctatum* eye though less so than the transplanted *tigrinum* eye-cup with *tigrinum* lens. The small *punctatum* lens could therefore exert a certain inhibiting influence. More remarkable, in the converse experiment, the *punctatum* eye-cup was accelerated in its growth by the *tigrinum* lens which it had induced but which was really too large for it. Lens ectoderm alone grafted from *tigrinum* to *punctatum* suffers induction, and produces a lens which is too large for the eye-cup, hence the growth of the lens is retarded but that of the eye-cup is accelerated. Conversely, lens ectoderm alone grafted from *punctatum* to *tigrinum* suffers induction, and produces a lens too small for the eye-cup, hence the growth of the lens is accelerated and that of the eye-cup retarded.

The general upshot of these experiments, continued by Lazarev, was therefore that while within an organ such as the eye, mutual adjustments between the parts can occur, there is little or no regulation of the growth rate of the organ as a whole to suit the organism as a whole. The "out-of-scale" eyes have later secondary effects on the nervous system and muscles (Twitty[2]).

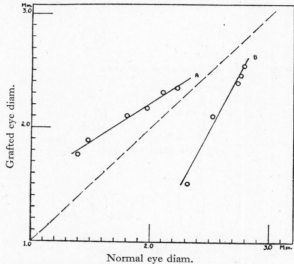

Fig. 202. Regulation in axolotl eyes transplanted between individuals of the same species but of different age (homoioplastic transplantation). *A*, graft larger than host eye; *B*, graft smaller than host eye.

The only conditions in which considerable regulation does seem to occur is in reciprocal transplantations between old and young larvae of the same species. Twitty & Elliott removed both eyes from a medium-sized larva of *Amblystoma tigrinum* and grafted one to a larger animal than the donor, the other to a smaller. There followed in every case an accelerated growth of the graft in the former case ("catching-up") and a retarded growth in the latter ("marking time"), so that after some while the normal proportions were restored. This is shown in Fig. 202. Here, however, genetic size differences are not involved. In his review Twitty[3] explains this homoioplastic regulation effect on the basis of the Robb partition-coefficient theory of heterauxetic growth* (see p. 536). Twitty & Elliott showed that similar age regulations could also be observed in heteroplastic eye grafts though in such cases the position of equilibrium finally reached

* See the explanatory blood analyses of Twitty & van Wagtendonk.

is not, of course, the symmetrical one proper to the unoperated host. We shall later have occasion to note some other homoioplastic transplantations* between younger and older animals with reference to regeneration (p. 439) and metamorphosis (pp. 449, 454, 466) phenomena, and some other manifestations of the "catching-up" effect (p. 589).

The extremest cases of regulatory or neighbourwise development seem to be those studied by Detwiler[2], where the larger part of the spinal cord is transplanted. After implantation of a piece of *tigrinum* spinal cord into a *punctatum* embryo at the corresponding position, the grafted cord is for a considerable time too long, too thick, and too full of cells. But eventually a far-reaching regulation takes place so that it fits precisely into the nervous system of the host. This is explained

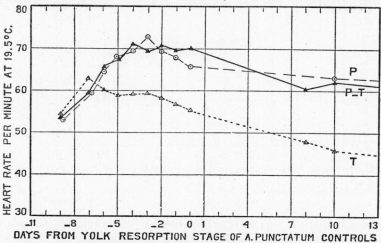

Fig. 203. Physiological selfwise behaviour of the pulse rate in exchanged axolotl hearts (*punctatum* grafted to *tigrinum*).

as "functional differentiation" due to the modifying influence of the stimuli reaching the central nervous system from all parts of the *punctatum* body. With this case we link the determinative period of development with the period of functional differentiation, and reach the utmost confines of the former.

The experiments of Copenhaver on the heart-beat rate of transplants between *Amblystoma* species are of particular interest, as they bring physiological function also within the framework of the concepts of neighbourwise and selfwise development. At the end of yolk absorption the normal rate of beat for the *punctatum* heart is 60/min. at 19·5° C.; that for the *tigrinum* heart is about 40/min. As Fig. 203 shows, the pulse curve of the *punctatum* heart grafted to a *tigrinum* embryo closely follows the rate of the donor species. When portions of hearts were transplanted, it was found that the posterior part sets the pace. As Fig. 204 shows, the pulse curve of the heart in a *punctatum* embryo which contained sinus and atrium from *tigrinum*, followed exactly the pulse curve of the *tigrinum* embryo.

* The gill and the ear-vesicle have also been studied from this angle, but these cases are complex, and the reader must be referred to Harrison's review[11].

No doubt the most marvellous case of physiological selfwise development is that reported by Giersberg, who exchanged at an early stage the brains of different species of toads and frogs (e.g. *Pelobates fuscus* with *Rana arvalis*). An animal, when adult, possessing a *Pelobates* brain in a *Rana* body, showed the strong digging instinct characteristic of *Pelobates*.

Much of what has been described in this section holds good also of insects. Bodenstein[1,13] has shown that regulation of growth-rates only occurs in transplantations between young and old animals.

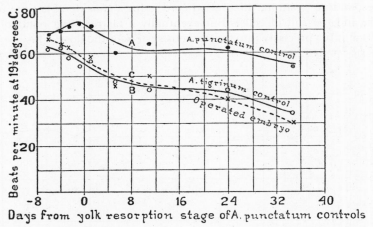

Fig. 204. Physiological neighbourwise behaviour of the pulse rate in exchanged axolotl hearts (*punctatum* under the influence of sinus and atrium of *tigrinum*).

But how it is that the very bold xenoplastic and heteroplastic transplantations so far described have been possible? Is there no such thing as "species incompatibility" and immunological resistance to transplants? This we must examine in the following section.

2·63. The origin of species-specificity

This section is to some extent a digression. It is only intended to call attention to the generalisation that tolerance to transplants decreases both phylogenetically as one ascends the taxonomic scale, and ontogenetically as the earliest stages of development are passed through. A wide literature could be drawn upon to illustrate these statements; here we shall simply adduce a number of experiments which make the relationship clear. The phenomenon is an example of the "law of recapitulation" in its simplest form (as described in *CE*, p. 1638), namely that the simple precedes the complex both in ontogeny and in phylogeny. The capacity of cells to live side by side though of different genetic, specific or even generic constitution, depends on the degree to which mutual protein intoxication occurs, and the older a given animal, or the higher a given animal in the taxonomic scale, the more certain is such intoxication to happen. For both age and phylogenetic status involve increased complexity, and hence increased specificity, of protein molecules as of morphological organisation.

Taking first the phylogenetic series, a large body of work (ably reviewed by L. Loeb[3]) has shown that there is a graded series of reactions to foreign tissues, from complete tolerance through leucocytic infiltration and gradual resorption to rapid disintegration and sloughing of the graft. In mammals there is intolerance between individual organisms of the same species and variety, even, although to a lesser degree, between individual organisms of close blood relationship. In amphibia these individual differences are not seen. But in spite of the far-reaching fusions which, as we have already found (pp. 156 ff.), are possible between urodele and anuran amphibian embryos, it was early found by Weigl that xenoplastic urodele/anuran skin transplantations are impossible in the adult. Heteroplastic skin transplantations between adult or nearly adult anurans succeed only with difficulty (W. Schultz[1]; C. E. Hadley), though it is possible to improve them by interposing a period of *in vitro* culture between removal from the donor and implantation into or on to the host (R. Erdmann). Cotronei and his colleagues have devoted a considerable amount of work to analysing the reasons for the incompatibility between anurans and urodeles, and have established the curious fact that anuran transplants into urodeles take better than urodele transplants into anurans (Perri[1]; Cotronei & Perri; Spirito[1]; Cotronei & Spirito; Cotronei & Guareschi). They are of the opinion that where embryonic transplants offer difficulty, yolk absorption is responsible.

As regards the ontogenetic series, we have the numerous facts due to Idzumi; Kritchewski and others (reviewed in *CE*, p. 1444), which show that the immunological character of organisms varies during their development.[*] Antigens come into being at successive points in time.[†] Here should be quoted the pioneer experiments of Braus[2] in 1906. He found that adult amphibia (*Bombinator pachypus*) contain an antigen which would produce a precipitin in rabbit blood, but that this precipitin could not be formed by frog embryos and tadpoles of the same species. They therefore do not contain the specific protein characteristic of older stages. Recently a complete piece of work of this kind has been carried out on the egg, embryo, larva and adult of the bee by Avrech & Heronimus, with similar conclusions.

Parallels to this may be found in the literature on chorio-allantoic grafting. The work of Murphy and his collaborators (reviewed in *CE*, p. 1454) showed that minced rat tumours may grow well on the chorio-allantoic membrane of the chick, but that towards the 18th day of development some change takes place[‡] in the host which leads to the leucocytic infiltration and death of the piece of tissue on the chorio-allantois. That other mammalian tissues and mammalian embryos may grow and develop in this situation has already been made clear

---

[*] Cf. Section 1·5 above. The classical work of Nuttall in 1904 on immunological differences between animal species is now being extended on precise genetical lines (Irwin; Irwin & Cole; Irwin, Cole & Gordon). Cf. all the work on blood groups in man.

[†] Cf. the recent work of Baumann[2]; Baumann & Witebsky; Witebsky & Szepsenwol.

[‡] Murphy traced this effect in part to the spleen; an organ which has since been confirmed to have a powerful inhibiting effect on the growth of transplanted tumours (Chaletzkaia[1, 2]). There is, indeed, a large literature on the anti-carcinogenic action of the spleen.

(p. 336).* That the embryonic tissues of birds have little specificity has also appeared from many grafting experiments (p. 333). C. J. Sandstrom[1] has investigated this point further. He found that only young undifferentiated duck kidney tissue would grow in chick eggs as a chorio-allantoic graft. Young duck metanephric tissue which would do this, however, would not heal properly when implanted into hatched chicks, forming large tumour-like nodules instead. Sandstrom & J. T. Kauer then found that minced duck kidney tissue from embryos of 13 days' incubation would grow well on the chick chorio-allantois, while that from 21–27 days' incubation would not. Transplantation of minced duck kidney tissue from ducks 1 week after hatching caused the death of the host within 48 hr. These experiments give an unusually clear glimpse into the development of tissue species-specificity. Not all tissues would do so; thus duck embryo cartilage, which did not become vascularised, developed on the chick chorio-allantois without causing any disturbances (Sandstrom & G. L. Kauer).

In a series of papers Waterman[3] has compared the development of rat and rabbit embryos on the chick chorio-allantois and implanted into the omentum or under the kidney capsule of adults of their own or related species. He concludes that "embryonic tissue of an animal of another class is apparently more tolerant than adult tissue of a closely related host". Perhaps the most surprising exemplification of this is the transplantations of Oppenheimer[4] already referred to (p. 171) in which half-blastulae of *Danio rerio* (the zebra fish) were implanted into the blastocoele cavity of amphibian eggs (*Triturus torosus*). The fish tissue developed wherever located, and its mesenchyme cells even mingled with those of the amphibian to form a composite tissue. Neural induction of course occurred, as has been described. Oppenheimer[8] later showed that though fish and amphibian epidermis and cartilage may grow into a continuum, fish grafts are never vascularised.

If embryonic tissues have so little species-specificity it should follow that their raw materials could be exchanged with little injury. How far this could be pushed is at present unknown, but there are statements that bird embryos can use the yolk and white of other bird species (*CE*, p. 1444) and this has recently been confirmed by Ermakov[2], who replaced part of the chick's egg-white with that of the pigeon, and by T. Li, who interchanged the albumens of the gastropods *Limnaea* and *Agriolimax*. Moreover, the growth-promoting factor in embryo extract (see p. 628) is not species-specific, as Kiaer[1] and others have shown.

It is generally accepted that the substances responsible for tissue incompatibility, when it appears, are protein in nature. An indication that this may not always be so is provided by the curious finding of Twitty & Elliott, in the transplantations already referred to between *Amblystoma tigrinum* and *Triturus torosus*, that *Triturus* eyes grafted into *Amblystoma* cause complete paralysis of the hosts, from which they do not recover until a few days before they begin to feed. Development is otherwise entirely normal, both for graft and host.

---

* Moreover, the chorio-allantoic membrane is an excellent site for the cultivation of viruses (according to a mass of work summarised in Burnet's monograph), though many of these will not grow in the adult fowl.

Parabiosis experiments showed that the *Amblystoma* paralysis continues until the complete loss of yolk from the *Triturus* partner (Twitty & Johnson). Correspondingly, it was found that *Triturus* egg extracts cause paralysis when injected into *Amblystoma*, but not extracts of the larvae. Adult female blood, doubtless because of the conveyance of the reserve materials to the ovaries, was also active in this respect. The substance is not active when administered by the mouth; it is destroyed by boiling and not soluble in organic solvents. When compared with the active principles from the skin glands of certain toads (cf. formulae on p. 251), it was found to have a quite distinct physiological action, i.e. not on the heart rate, but, as was ascertained by ingenious transplantation experiments, exclusively upon the nervous system (Twitty[4]). Nevertheless, this case does not resemble the common effects of tissue incompatibility, for the eye and other tissues heal well in these interspecific transplantations, and it may be that a non-protein substance is involved.

What may be another example of this is the case described by Brinley & Jenkins where in transplantations of heart, but not of eye, liver or spleen, between embryos of two different species of *Fundulus*, the host's heart ceases after a time to beat and death ensues. The fact that the heart of the host stops beating first makes one suspect some incompatible influence of a humoral character. Yet another toxin discovered by transplantations is described by Humphrey & Burns.

We have seen, then, that embryonic compatibility and organiser influences transcend genetic boundaries, but that the response is always conditioned by the genetic character of the reacting tissue. What else does the nucleus do in early development, and how are the characters of the phenotype brought into being in the embryo?

2·64. The role of the nucleus in development

To-day we are accustomed to think of the nucleus as the repository of the chromosomes and genes which during embryonic life develop the individual characteristics of the completed organism, giving it blue fur instead of white, or long bristles instead of short ones. But how far the nuclear equipment is responsible for the specific, generic, or even class characters of the organism is hardly as yet settled. The attempt to throw light on this has led to the very large amount of work on cross-fertilisation (hybridisation), but it has of course been seriously hampered by the impossibility of obtaining crosses from widely different organisms, and the tendency to weakness and early death of crosses from even fairly nearly related organisms. Lack of space will prevent any full consideration of this literature here, except in so far as it assists us to envisage the processes of embryonic development. Moreover, the uncertainty as to exactly what the nucleus is doing in the earliest stages of embryonic development has led to experiments in which every kind of combination and permutation of gametic and zygotic material has been investigated. In general it may be said that the trend of recent work has been to show that what was not thought viable before, may actually be so, a fact which gives much hope of further similar advances in the future.

The terminology is here a little complicated. Without going into the history of the various terms, we shall follow in the main Fankhauser's recent review[5]. Table 24 and Fig. 205 summarise the principal possible relationships between nucleus and cytoplasm in early development. True hybridisation involves fertilisation by foreign sperm followed by nuclear fusion (1), but foreign sperm may also be used to fertilise an egg the nucleus of which has previously been put out of action (4a), or a fragment of an egg containing a normal egg nucleus (5a), or a fragment of an egg not containing the egg nucleus (6a), or, finally, foreign sperm may be made to fertilise an egg, and the egg nucleus, together with a certain amount of cytoplasm, may be removed before nuclear fusion has occurred (7a). Parthenogenetic fertilisation may be carried out, either on the normal egg (2), or on an egg fragment containing the egg nucleus (8), or even on an egg fragment containing no nucleus at all (9). An egg may develop while carrying only sperm nucleus and its derivative nuclei, either because the egg nucleus has been put out of action beforehand (4), or because a fragment of the egg not containing the egg nucleus has been fertilised (6). This latter case (with the special modification of the removal of as little cytoplasm as possible with the egg nucleus) has afforded, as we shall see, one of the most fruitful ways of penetrating into the question of the role of the nucleus in early development.

An exhaustive survey of the literature of these many types of development cannot here be attempted, but an approach to it will be facilitated through the papers quoted and reviews such as that of P. Hertwig[2] on hybridisation in general. It will be convenient to take the different cases in the order in which they occur in the table, and in Fig. 205.

The general characteristic of all the haploid types is the abnormality of their development, and their tendency to early death. Those embryos that survive hatching form larvae showing stunted growth, delayed differentiation, reduced motility and reactivity to stimuli, and general oedematous swelling of the body. The detailed exploration of these anomalies should throw a great deal of light on the normal mechanisms which have here been interfered with.

The most obvious way of retaining the haploid condition of the unfertilised egg is by *Parthenogenesis*, thus entirely excluding the sperm chromatin. The action of the spermatozoon may be partly replaced by pricking with a glass needle, the action of chemicals, etc. which "activate" the egg. The reviews of Parmenter, Peacock and Rostand on parthenogenesis should be consulted here. In the invertebrates it occurs naturally, and just as polyembryony has become fixed as a species characteristic in many parasitic hymenoptera (cf. p. 228), so parthenogenesis regularly occurs in many hymenoptera and coccids (see the review of Schrader & Schrader.* In vertebrates, only the faintest traces of spontaneous parthenogenesis have been met with, but experimental partheno-genesis has been achieved even for mammals (Pincus & Shapiro).

After the successes of J. Loeb[3] in 1899 with echinoderm eggs, Bataillon attacked the problem for the amphibia at the turn of the century—references to this older literature will be found in Loeb's review. It is now known that all kinds of agents will bring about parthenogenetic activation in amphibian eggs; traumatism (pricking), hypertonic solutions, heating followed by cooling, or

* For the chemical parthenogenesis of silkworm eggs, see p. 583.

vice versa, exposure to electric induction shocks, organic solvents, etc. But a distinction is made between "activation", which determines a certain number of the preliminary phenomena of development, and "regulation", which permits cleavage and further development. The treatments described above only lead to "activation", i.e. they cause orientation of the egg, the appearance of the second polar body, the appearance of the membrane, the onset of unfertilisability, and the formation of the grey crescent. In order to add the "regulation" factor it is absolutely necessary to inject into the egg an active substance present in the blood and tissue fluids of the adult. It is easy to see how erratic the earlier results were, for a slight contamination of the exterior of the egg would lead to the belief that full parthenogenetic development had been obtained by pricking. This discovery was due to Guyer's experiments of 1907, and though at first Bataillon[1] opposed Guyer's views, he later demonstrated himself[2] that absolutely clean eggs would never develop after pricking alone (1911).*

The blood factor was soon found to be quite non-species-specific among the vertebrates[†] (Bataillon[2]) and to be present in all organ extracts and juices (Bogucki[1]). Cell-free extracts of blastulae and gastrulae are as effective as blood, but not similar extracts of unfertilised eggs or two-cell stages (Bogucki[1])—a remarkable parallel to the liberation of the organiser substance (cf. pp. 174 ff.). The factor is only contained in nucleated cells; thus the erythrocytes of mammalian blood alone are inactive, while the leucocytes are active, but the nucleated erythrocytes of sauropsid or amphibian blood possess the factor (Bataillon[4]). The factor seems to be present in some invertebrate tissues; thus positive results were obtained with the testes of *Lumbricus* and the hermaphrodite gland of *Helix* but not with crustacean blood or sperm or *Dytiscus* sperm (Bataillon[4]), nor with any protozoan, bacterial, or plant material (Rostand). There are variations in the state of activity of the factor. Molluscan testis tissue extracts allow of cleavage but never of gastrulation, while horse leucocytes allow gastrulation but not usually hatching, and with bird leucocytes large numbers of tadpoles may be produced.

The blood factor is not stable to boiling, but is destroyed at about 55° C. (Einsele[1]), although the motility of amphibian spermatozoa ceases already at 37°. It is stable to desiccation over sulphuric acid (Rostand), but it is destroyed by the action of denaturing agents such as organic solvents; hence it is probably a protein and perhaps an enzyme.[‡] The molecule cannot, however, be very large, as it passed through collodion ultra-filters in the experiments of Parat. No successful effects have so far been obtained by injecting into the egg solutions of purified enzymes such as pepsin, rennin, etc. A nuclein prepared from erythrocytes by Bogucki[1] was inactive, and many other substances similarly injected by Rostand gave no results.

---

* The delay in the acceptance of Guyer's facts was perhaps due to his own rather fantastic theory of leucocyte proliferation within the stimulated egg.

† Hence the paradoxical result reached by Bataillon[3], that much better development took place from a *Bufo* egg parthenogenetically fertilised by injection of frog blood, than from one properly fertilised by frog sperm, including nuclear fusion. Inadequate nuclear fusion (hybridisation) could not in this case equal the haploid parthenogenetic state.

‡ But a very short heating to 75°, which coagulated most of the protein in a blood extract, did not, according to Einsele[1], inactivate it.

Table 24. *Terminology of possible relationships between nuclei and cytoplasm in development*

		Term employed	
(1) Egg fertilised by foreign sperm	Diploid nucleus divides	Hybridisation (cross-fertilisation)	D
(2) Egg fertilised by artificial means	Egg nucleus alone divides	Parthenogenesis*	H
(3) Egg fertilised by sperm so treated that nuclear fusion cannot ensue	Egg nucleus alone divides	Gynogenesis	H
(3a) Egg fertilised by foreign sperm incapable of nuclear fusion	Egg nucleus alone divides	Hybrid gynogenesis ("false hybrids")	H
(4) Egg, the nucleus of which has been so treated that nuclear fusion cannot ensue, fertilised by normal sperm	Sperm nucleus alone divides	Androgenesis	H
(4a) The same, with foreign sperm	Sperm nucleus alone divides	Hybrid androgenesis	H
(5) Fragment of an egg, containing the egg nucleus, fertilised by sperm	Diploid nucleus divides	Diploid merogony	D
(5a) The same, with foreign sperm	Diploid nucleus divides	Hybrid diploid merogony	D
(6) Fragment of an egg, not containing the egg nucleus, fertilised by sperm	Sperm nucleus alone divides	Andro-merogony†	H
(6a) The same, with foreign sperm	Sperm nucleus alone divides	Hybrid andro-merogony†	H
(7) Fragment of an egg, containing the egg nucleus, removed from an egg fertilised by sperm, before nuclear fusion	Egg nucleus alone divides	Gyno-merogony	H
(7a) The same, with foreign sperm	Egg nucleus alone divides	Hybrid gyno-merogony	H
(8) The same, fertilised by artificial means	Egg nucleus alone divides	Parthenogenetic gyno-merogony	H
(9) Fragment of an egg, not containing the egg nucleus, fertilised‡ by artificial means	No nucleus present at all	Parthenogenetic merogony	NN

* Parthenogenesis in the invertebrates which is diploid because the reduction division has been omitted is not considered here (see Waddington[22], p. 59).

† If the nucleus is removed with but a minimum of cytoplasm, this approximates to (4) and (4a), save that no remains of the egg nucleus are present.

‡ Many writers use the term "activated" for some of the above cases; I reserve it, as Bataillon[4] did, for those processes which can be made to happen in the absence of the "blood factor". Thus fertilisation = activation + regulation.

N.B. The letters in the right-hand column indicate the resulting situation; D = diploid, H = haploid, NN = non-nuclear.

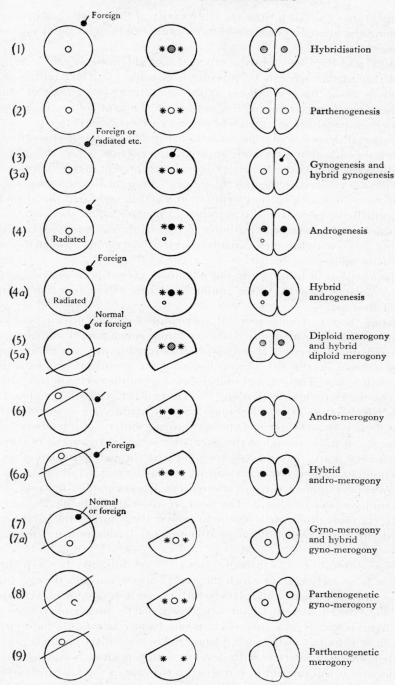

Fig. 205. Possible relationships between nucleus and cytoplasm in development.

In immature eggs, taken from the body-cavity of the female, parthenogenesis by pricking and injection of the blood factor cannot be brought about successfully (Barthélemy).

Bogucki[2] and Rapkine[1] subsequently extended this work to echinoderms and showed that parthenogenesis by pricking is possible for echinoderm eggs also, if blood or tissue juice is present at the same time, or extracts of embryos from the blastula stage onwards but not before. This, of course, is only one of many treatments which will give echinoderm parthenogenesis (see p. 214).

The general upshot of all this work is that in normal fertilisation the sperm nucleus brings in with it something, apparently an enzyme, which, though present in all adult cells which have nuclei, is either absent, present in insufficient amount, or in some way masked, in the unfertilised egg. Since the isolated acrosome alone was effective, producing haploid fertilisation, Parat thought that the blood factor must normally be present in this extreme "cephalic" cap of the spermatozoon. How exactly this substance contributes to normal development, or rather, how it permits the normal cleavages, gastrulation, etc. which cannot go on in its absence, still remains altogether obscure.

Parthenogenesis in fishes, birds and mammals has not been analysed anything like as completely as it has in the amphibia; details will be found in the monograph of Rostand.

We come now to *Gynogenesis*. If spermatozoa, before insemination of the eggs, are treated with moderate doses of radium emanation, X-radiation, ultraviolet irradiation, or certain chemicals such as trypaflavin, their ability to enter the egg will not be affected, but no fusion of the sperm nucleus with the egg nucleus will follow. The cytology and embryology of the effects of this have been studied by the Hertwigs (O. Hertwig[3]; G. Hertwig[1]; P. Hertwig[1]) and by Dalcq[1] and his collaborators (Dalcq & Simon; Simon[1], etc.). The outstanding fact of interest to us is that the gynogenetic eggs usually suffer a characteristic impediment to gastrulation, though if the irradiation or other treatment of the sperm has been very short, they may successfully pass through gastrulation and give free-swimming haploid larvae which may live as long as a fortnight. As Rugh has shown, development is never affected before gastrulation, however severely the sperm have been irradiated. The more we know about the metabolism of the gastrula, the more profitable it will be to analyse the nature of the paralysed state (delayed effect) into which it enters after this pre-fertilisation treatment of the sperm.

Gynogenesis can also be brought about by hybridisation between different species of frogs and toads, in which the foreign sperm activates the egg, but fails to fuse with the egg nucleus, and plays no further part in development (G. Hertwig[1]; S. Ch'ou[1]; Bataillon & S. Ch'ou; Rugh & Exner). Such embryos are called "false hybrids"; a typical case is Hertwig's *Rana esculenta* ♀ × *Bufo viridis* ♂. They can be definitely distinguished from true hybrids if the sperm is previously irradiated; this will interfere with development if nuclear fusion usually takes place because it will prevent it, but if, as in the case of the false hybrid, it does not take place, no additional troubles in development will appear.

In all these cases, regulation from haploidy to diploidy may occur, either by a division of chromosomes not followed by cell-division or by a resumption of the chromatin material of the second polar body.*

Lastly, gynogenesis may be brought about by treatment of the egg with cold. Crosses of *Bufo vulgaris* ♀ × *Rana temporaria* ♂ normally fail to gastrulate, but Rostand was able to prevent nuclear fusion by exposing the eggs to low temperatures; in this way they developed "parthenogenetically" under the influence of the egg nucleus only and gave free-swimming larvae.

In *Androgenesis*, the egg develops solely under the influence of the sperm nucleus. Treatment of unfertilised amphibian eggs with radium seems to have relatively little effect on the cytoplasm, but it puts the egg nucleus out of action, and so prevents nuclear fusion (G. Hertwig[1]; P. Hertwig[2]; Dalcq[4]). Or the egg nucleus may be removed with a minute pipette before fertilisation (Curry; Kaylor). These androgenetic eggs suffer a marked delay at gastrulation, but providing their nuclei have the exact haploid chromosome number, they frequently continue to develop and up to the present time have been cultivated as long as the stage of fore-limb-bud development (Kaylor; Porter).

2·645. Merogons and chimaerae.

So far we have been considering only those cases in which the main mass of the egg cytoplasm participates in the development of the organism. Development of egg fragments is called *Merogony*. The simplest case is that in which the fusion of the egg and sperm nuclei is prevented by ligaturing the egg into two parts. This was first accomplished for the newt, *Triton taeniatus*, by Spemann[7] in 1914; the part containing the sperm nucleus will develop into a haploid larva. It was later investigated by Baltzer[1] and by Fankhauser[2]. In this method not only is the number of chromosomes reduced to one-half but the amount of egg cytoplasm is correspondingly reduced. The fact that such embryos become abnormal and die disposes, therefore, of G. Hertwig's suggestion that the cause of the difficulties of haploid embryos lies in the excessive amount of yolk which they contain. Indeed, an exactly opposite explanation was advanced by Ch'ou. In his gynogenetic embryos already mentioned, it was observed that yolk utilisation proceeded even faster than in the diploid controls; his suggestion therefore was that as the total nuclear surface was greater owing to the smaller volume of each nucleus, an abnormally high interaction rate between nucleus and cytoplasm led to an abnormally high metabolic rate, sufficient to kill the animals. So far none of these theories has been put to experimental test with accurate methods. Another factor in the haploid "syndrome" is the oedema which, as Dalcq & Simon have pointed out, cannot be due to insufficiencies of the earliest excretory organs as such, since it begins already in the gastrula stage. It must be remembered, however, that gastrular and neurular oedema is quite commonly met with in the cultivation of diploid embryos. A genetical interpretation of the haploid "syndrome" has been suggested by Darlington, who thinks that all haploid

---

* Apparently it may go a great deal further. Kawamura describes polyploid frog larvae obtained by parthenogenesis—mostly viable—but as weak as haploid ones if hexaploid or over.

embryos may have at least one lethal factor hidden in their chromosomes. Unmasked by the absence of its normal allele, this will kill the embryo sooner or later. Unfortunately such an explanation will not account for the healthy behaviour of haploid embryos which have regulated to diploid by a nuclear division not followed by a cell-division (Parmenter). These should now be homozygous for the lethal factor, and hence should be less instead of more healthy than the haploid embryos. So King & Slifer have shown that partheno-genetic grasshoppers are not viable unless regulation to diploidy takes place. One fact at any rate is undoubted, namely that the high rate of mortality of merogonic embryos during gastrulation is associated with the frequent occurrence of cells with a sub-haploid or otherwise unbalanced chromosome number (Fankhauser[3]). What exactly this means in terms of physiology and biochemistry remains to be determined.

If an egg is divided into two after fertilisation but before nuclear fusion, the part containing the egg nucleus alone develops much less well than that containing the sperm nucleus alone, generally passing through only a few cleavages (Woerdeman[5], on echinoderms), but Fankhauser[4] has recently obtained very abnormal blastulae from such fragments in amphibia (*Gyno-merogons*). In Spemann's original work, the fragment of the egg other than that containing the sperm nucleus alone contained an egg nucleus fused with a sperm nucleus, since in the species with which he was working polyspermy normally occurs.

The greatest advances in our knowledge of the role of the nucleus in early development have arisen from the study of *Hybrid andro-merogons* by Baltzer, Hadorn and their collaborators.* Andro-merogons were first obtained by Boveri[1] in 1889 on sea-urchin eggs, and many subsequent workers have studied them in the hope of deciding what specific characters, if any, are carried by the egg cytoplasm, but in general they do not live long enough to permit of the exact determination of the species characters. Spemann's experiment above mentioned was extended in 1920 by Baltzer[1] to include cross-fertilisation and the study of the resulting hybrid andro-merogons. Such combinations are written with the sex sign of the egg in brackets to show that the egg nucleus has been removed, thus;

Triton taeniatus (♀) × *Triton palmatus* ♂.

It was soon established that these embryos would pass through the gastrulation stages without difficulty, but died in a characteristic way as neurulae. Although many tissues and organs (such as epidermis, neural tube, and notochord) remained healthy, the head mesoderm lying between notochord and foregut began to degenerate at a certain stage, the appearance of pycnotic nuclei preluding complete histolysis. As a result of the spread of this degeneration to the rest of the body, death of the embryos followed.

The problem arose, therefore, of analysing a little more deeply the causes of this "disease", and the first contribution of Hadorn[1] was devoted to showing that tissues other than presumptive head mesoderm, isolated from a haploid embryo of the cross *T. palmatus* (♀) × *T. cristatus* ♂, would, when transplanted into a

* Reviews by Baltzer[3, 4, 6].

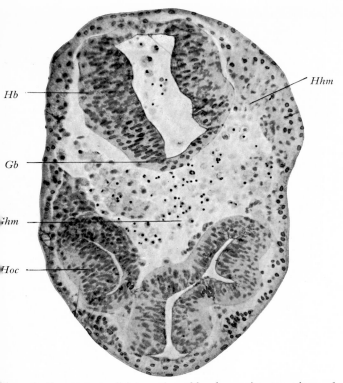

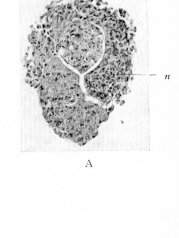

XXVIII

A

Fig. 206. Spontaneous disintegration of head-organiser mesoderm of hybrid andro-merogon even when surrounded by normal tissue. *Hb,* host portion of brain; *Gb,* graft portion of brain; *Hhm,* healthy host mesoderm; *Ghm,* degenerating graft mesoderm; *Hoc,* host eye.

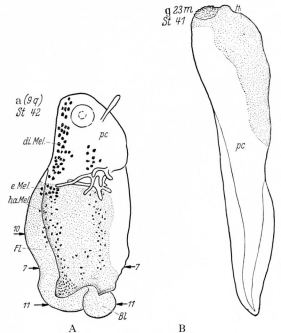

A   B

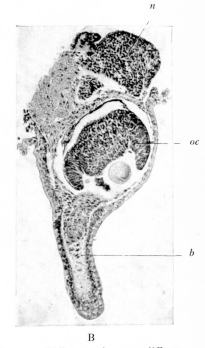

B

Fig. 208. Chimaerae in which hybrid andro-merogon tissue (stippled) combined with normal diploid tissue (white). A. The normal anterior end can produce a fairly normal trunk but not a normal tail in the haploid tissue. B. The normal posterior end forms a normal tail, but the haploid anterior end can do nothing. *Fl,* fin; *ha.Mel.,* haploid melanophores; *.Mel.,* diploid melanophores; *e.Mel.,* diploid melanophores migrated to the haploid tissue; *Bl.,* epidermal blister in tail-region; *H,* functional haploid heart.

Fig. 207. Differences between differentiations in explants. A. Material from a hybrid andro-merogon makes only chaotic neural tissue (*n*). B. Material from a hybrid diploid embryo makes brain (*n*), eye-cup (*oc*), lens, and balancer (*b*).

normal diploid embryo, develop perfectly normally. Thus presumptive epidermis went to form normal skin, ear-vesicles, and blood-cells; presumptive neural tissue went to form normal eye-cups and brain; even presumptive trunk somite material gave normal somites. Only presumptive head mesoderm persisted in degenerating in its new environment. Fig. 206, taken from Hadorn's paper, shows the disintegration of the head-organiser tissue. These facts demonstrated two things, first that there is a specific "haploid disease" of head mesoderm, which transplantation into a normal host cannot cure, and secondly that the remaining tissues of the haploid embryo are perfectly capable of indefinitely long life and differentiation if they are removed from the influence of the disintegrating head mesoderm.*

How far are the neighbouring tissues of the host responsible for this high differentiation capacity of the haploid tissue? This was answered by Hadorn[2], who, adopting the explantation methods of Holtfreter (see p. 157), cultivated various parts of these andro-merogonic haploid gastrulae *in vitro*. In this way it was possible to show that parts other than presumptive head mesoderm would live much longer than if they had remained in the andro-merogon itself. In some cases, such as presumptive notochord, they would differentiate histologically very fully. But in no case was the differentiation seen in these explants as good as that, either of such parts transplanted into a diploid host, or of parallel explants from a non-hybrid andro-merogon, or of parallel explants from a hybrid but not merogonic embryo. Thus Fig. 207 shows that while the hybrid andro-merogonic material made only chaotic neural tissue in explant, the hybrid diploid tissue formed a good eye-cup with lens, and a balancer. Thus not haploidy as such, nor hybridisation as such, could be held responsible for the low self-differentiation capacity of the tissue, but presumably the mixture of both factors. And what was remarkable was that this low differentiation capacity could be so much enhanced by *contiguity* with normal diploid embryonic tissues.

In order to get further light on the hypothetical "histogenetic substances" which it seemed necessary to postulate to explain this effect (*entwicklungs-fördernde Hilfe*), Hadorn[3] next combined halves of hybrid andro-merogonic embryos with halves of normal diploid embryos to form various beautiful types of "chimaerae". The results may be summarised by saying that provided a normal head-organiser is present, the rest of the tissues will develop normally, and that the influence of the normal diploid tissues seems to spread out in a kind of gradient into the others. We may first compare the difference between an embryo of which the anterior half is haploid tissue and one of which the anterior half is diploid tissue. The former, inhibited by its degenerating head mesoderm, has formed a tail and trunk structures but nothing else; the latter has produced a well-formed brain, eye, and balancer, besides fairly normal trunk structures (Fig. 208). The influence favouring differentiation which seems to radiate from the diploid tissue is well seen in Fig. 209. This chimaera had a diploid anterior

---

* We are reminded here of the specific mesoderm-inhibitory substance in malt extract discovered by Heaton and carefully investigated by Medawar[1]. It appears to be of carbohydrate nature and to possess aldehydic groups (see p. 629), and it has no effect at all on ectoderm and epithelia. It, or something very like it, can be extracted from adult liver (Brues, Jackson & Aub). Medawar[2] thinks it inhibits the cell-movements at ana-telophase.

end and a haploid posterior end. The diagram shows that some distance caudally to the established junction between the two types of tissue, the form of the neural tube was entirely normal. Still further posteriorly, however, it became very abnormal and finally disappeared altogether.

Two other types of chimaerae were made, one in which sagittal halves were combined, in which case considerable lengths of normal neural tube tissue were formed from the haploid tissue; and another in which diploid organiser material was enveloped in haploid presumptive ectoderm. These latter developed quite normally, giving larvae which even passed through metamorphosis, although their skin was purely haploid. Needless to say, the reverse combination gave nothing except that an eye-cup was induced, which degenerated before it had had time to form properly.

Subsequent cases proved even more difficult to analyse than the *Triton palmatus* (♀) × *T. cristatus* ♂ andro-merogons of Hadorn. The cross *Triton alpestris* (♀) × *T. palmatus* ♂ was studied by de Roche. Here the andro-merogonic haploid embryos develop to the closed neural tube stage before dying, but no special area of degeneration is to be found in them, such as the head mesoderm previously described. All the tissues fail at the same time. But on the other hand transplantation into normal diploid *palmatus* gastrulae showed that every part of the haploid embryos, including the head mesoderm, possesses full powers of differentiation in the host. A similar lack of specific disease centre was found in the work of Baltzer, Schönmann, Lüthi & Boehringer on crosses of less similar animals, e.g. *Triton palmatus* (♀) × *Salamandra maculosa* ♂. The true hybrid cross here always perished in the gastrula stage, and measurements of the number of abnormal nuclei showed a wave of these in the late blastula preceding the rapid increase in the number of degenerated nuclei which leads to death (Fig. 210; Schönmann). The andro-merogonic cross fared still worse, seldom surpassing the middle blastula stage owing to the general nuclear disintegration (Boehringer). Transplantations of blastula roof from the andro-merogon showed no capacity for further development, but those from the true hybrid cross, though normally doomed to die in the gastrula stage, were found to be capable, in a normal diploid *palmatus* host, of every imaginable kind of normal differentiation (Lüthi). In some way or other the lethal tissue can be "vitalised" by a "histogenetic stimulus" from the normal host.

Effects quite analogous to those just described were seen by Strasburger in crosses of the ladybird beetles *Epilachna chrysomelina* × *E. capensis*. During blasto-derm formation the nuclei degenerate and the embryo dies.

The whole subject has been well summarised by Baltzer & de Roche. Apart from Hadorn's disintegration centre in the head mesoderm, other differentiation-inhibiting factors must exist in hybrid andro-merogons, and in some cases, as we have seen, they arise in the absence of any disintegration centre. Their inhibition, however, is reversible by contact with normal tissue; that of the disintegration centre is not. Such a reversibility (cf. Fig. 209) is reminiscent of homoiogenetic induction (cf. p. 154). Whether the inhibition affects mainly the power of inductors or the competence of tissues to react to them is not yet known.

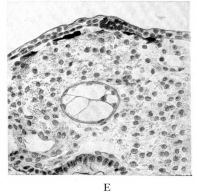

XXIX

Fig. 209. Favourable effects on hybrid andro-merogen tissue resulting from contiguity with normal
  diploid tissue. Anterior end diploid, posterior end haploid (A, B); junction at G–G. The cross-
  section at a–a (shown in C) is still normal; that at b–b (shown in D) is no longer so; and at c–c
  (shown in E), the neural tube has disappeared altogether. *ngl*, nuclear ganglion layer; *f*, fibre
  layers; *n*, neural mass replacing normal neural tube; *AA*, eyes.

But in any case this merogonic lethality has an obvious relation to the many cases known of lethal genes (see below, pp. 365 ff.), where a gene or group of genes with mendelising properties brings about abnormal development. In the first place, both types of lethality act at a definite time in development, as in the cases of the *yellow* mouse (Kirkham[2]), the *short-tailed* mouse (P. Chesley) and the *creeper* chick (Landauer[23]). In the second place, a specific part of the embryo is affected, as in the *short-tailed* mouse (Chesley) or the hereditary tumours of the fruit-fly (Gowen; Stark). Thirdly, the lethality can be overcome by transplantation of some of the organs to a more favourable environment.

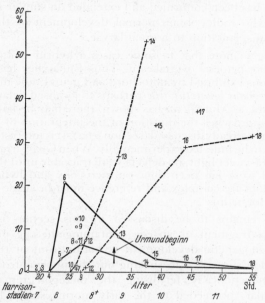

Fig. 210. Nuclear disorganisation in the late blastula stage preceding death of the gastrula in hybrids of *Triton palmatus* and *Salamandra maculosa*. The zone between the continuous lines shows the percentages of abnormal nuclei and mitoses; that between the dotted lines shows the percentages of pycnotic and degenerated nuclei.

This has been shown in many organisms other than amphibia. It was first proved for sphingid moths (Bytinski-Salz[1]; Bytinski-Salz & Gunther) in 1930. Although the reciprocal cross is healthy, *Celerio euphorbiae* ♀ × *C. gallii* ♂ causes the death of all females as pupae. Yet ovaries and wing buds lived on normally after transplantation to normal animals.

Ephrussi[5] carried the matter further by using tissue-culture technique on the homozygous lethal *short-tailed* mouse. Tissues taken from the posterior regions of such an embryo, which normally dies *in utero* on the 10th day of development, could be cultivated *in vitro* for as long as two months and differentiated into cartilage. Whether this is a reversal of intrinsic differentiation-inhibition or a removal from the degenerating influence of a "disintegration centre" is not yet clear. Since then Hadorn[4] has described a parallel case. *Lethal-giant* fruit-fly larvae, at first normal, fail to pupate properly, and survive for a time in a

transparent oedematous state before they die. During this time the imaginal
discs degenerate completely. In order to see the extent of this lethality, Hadorn
transplanted the ovaries (which develop early like the larval organs but which
persist into the adult state) into genetically normal larvae. The ovaries then were
found to develop much further than they do in the lethal larvae, even producing
eggs, though not reaching full adult development. It was subsequently shown
that the mechanism of the delay in pupation in *lethal-giant* larvae is a failure of
the *corpora allata* or homologous structures to secrete the pupation hormone*
(Hadorn[5]; Hadorn & Neel; Scharrer & Hadorn; Burtt).

Hadorn's work has been confirmed and extended on similar *Drosophila* lethals
by Medvedev[3], who could obtain normal development of the imaginal buds
indefinitely by transplantation to normal larvae.

It is interesting to note that in some cases a hybrid embryo may develop
relatively far before nemesis overtakes it. Thus Rubaschev has reported on the
fantastic contortions exhibited by notochord and neural tube in the trout crosses
Salmo fario × *Coregonus lavaretus*, due perhaps to localised delays at gastrulation.
And in toad crosses, as Montalenti has shown, there may be failures of limb-bud
formation. This approaches the question of the integration of organs in foreign
hosts, which we have already discussed (section 2·62). It is interesting, too, that such
delayed lethality can be imitated experimentally. When Geigy irradiated *Drosophila*
embryos with ultra-violet light, no defects were traceable until the imaginal stage.

A good deal of work has been done on merogony and hybridisation in sea-
urchins. The results are complex and reference must be made to the reviews of
v. Ubisch[6] and Hörstadius[6].

All that now remains to be discussed in this section is *Parthenogenetic
merogony*. That a fragment of an egg containing the egg nucleus should be
capable of parthenogenetic development is nothing out of the ordinary. But
in 1907 McClendon succeeded in sucking out the nucleus from an echinoderm
egg and parthenogenetically fertilising the merogon remaining. Cleavage
(somewhat irregular) proceeded to the early morula stage. Thirty years later
E. B. Harvey[6] treated echinoderm eggs in E. N. Harvey's centrifuge-microscope
(descriptions; E. B. Harvey[2,3]; E. N. Harvey[3]) so that they fragmented into
halves or quarters. The non-nucleate fragments could be parthenogenetically
fertilised and would develop as far as free-swimming blastulae.

E. B. Harvey's previous work[5] on centrifuging echinoderm eggs at high speed
had shown that although the distribution of the layers of granules, yolk, etc.
varies from species to species, all the fragments retain their fertilisability, with
the result that diploid merogons and andro-merogons develop in large numbers
when the eggs are fertilised with normal sperm. Cross-fertilisation succeeds to
approximately the same extent in the merogons as in the whole eggs, so that with
foreign sperm hybrid diploid merogons and hybrid andro-merogons are produced. Fig. 211A shows unfertilised *Arbacia* eggs stratifying and pulling apart
as observed in the centrifuge microscope. Below are the red non-nucleate halves;
at the right one of the red halves is fragmenting to quarters, at the extreme right
the separation of quarters has just taken place. Apparently quite normal plutei

* For an account of this see below, p. 465.

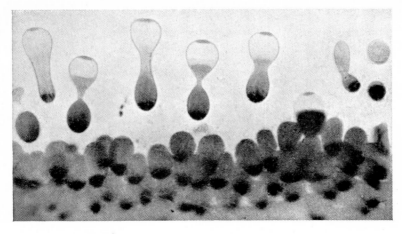

XXX

Fig. 211 A. Unfertilised echinoderm eggs stratifying and pulling apart under the influence of strong centrifugal force.

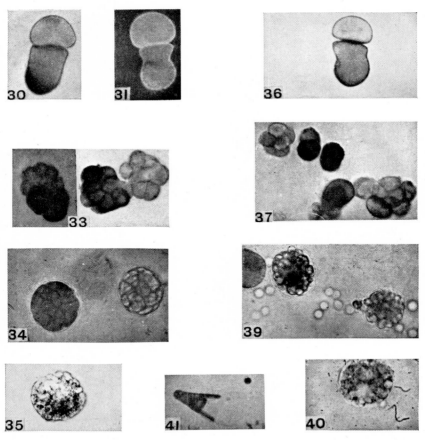

Fig. 212. Development of non-nucleate parthenogenetic merogons as far as the blastula stage in the sea-urchin *Arbacia*.

were frequently raised from the andro-merogons, which in some cases (e.g. *Sphaerechinus granularis*) developed better than the diploid merogons of the same species.

In the work on parthenogenetic merogony *Arbacia* eggs were centrifuged into four quarters, as follows, centripetally: (1) clear quarter, containing oil-cap, nucleus, and hyaline cytoplasm, (2) granular quarter, containing some yolk, (3) yolk quarter, and (4) pigment quarter (Fig. 211 B). The nucleate quarter gave viable parthenogenetic gyno-merogons, but the granular quarter was difficult to obtain, so most of the parthenogenetic merogons were produced from the third and fourth quarters. Hypertonic salt solutions were used as activating agents. The non-nuclear merogons throw off fertilisation membranes, and the successive

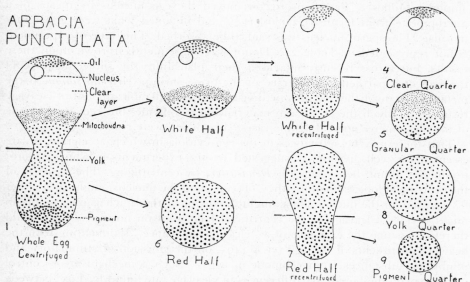

Fig. 211 B. Diagram of the stratification of an echinoderm egg and its separation into merogons.

cleavages are regularly preluded by the appearance of cytasters. Eventually blastulae with as many as 500 cells are reached, but development goes no further. Fig. 212 gives a general survey of the development.

These results, confirmed and extended on European species of echinoderms[7] and on *Chaetopterus* (E. B. Harvey[8]), and by the use of ultra-violet light as a parthenogenetic agent for the non-nucleate fragments (E. B. Harvey & Hollaender), demonstrate that the egg-cytoplasm has within itself the potentialities for at least the early stages of development. Gastrulation and further differentiation, however, seem to require the intervention of the chromatin material.*

* The parthenogenetic merogons are devoid of any Feulgen-staining material, but we cannot conclude from this that the nucleus has not set free any of its contents during the stretching of the egg, since at this time nearly all the nuclear material of the egg is of the ribo-nucleotide type (see p. 633). There must be much of this in the cytoplasm anyway. But it cannot carry the merogon into gastrulation, and the external addition of nucleotides, vitamins, and hormones gives it no further help.

In connection with this, reference may be made to the question whether cleavage rate in echinoderms is a function of the cytoplasm or the nucleus. A. R. Moore[3] reviews the older phases of this, and describes his own experiments with the cross *Dendraster* ♀ × *Strongylocentrotus* ♂. The hybrid embryos showed the cleavage rate characteristic of the egg species. By micro-dissection nucleate and non-nucleate fragments were made, and fertilised either with *Dendraster* or with *Strongylocentrotus* sperm, thus giving homospermic and hybrid diploid and andro-merogons. In every case the cleavage rate was that characteristic of the cytoplasm, even when only a foreign sperm nucleus was present. Substantially similar conclusions, but less certain because based only on cross-fertilisations, were reached by de Francesco, and for amphibia by J. A. Moore[2].

With regard to the question of the cleavage of non-nucleate merogons in amphibia, Jollos & Péterfi in 1923 removed the egg nucleus by means of a micropipette from fertilised axolotl eggs, and obtained what were apparently morulae. The sperm nucleus was said to be absorbed. This type of experiment is not represented in Table 24. Doubt was subsequently thrown upon these results, but Gross recently increased their plausibility by subjecting to cold the eggs of a parthenogenetic race of the brine-shrimp, *Artemia salina*, at the metaphase state of the first meiotic division. A considerable amount of cleavage resulted, in which the chromatin material did not divide, but remained in one of the blastomeres only. In some cases "blastulae" were reached.

That there may be substances in the egg-cytoplasm which accelerate, or are essential for, cell-division, is suggested by the experiments in which Moore[7] produced "dumb-bell" eggs (in *Dendraster*) by centrifuging. The cleavage of the centripetal lobe, containing the "oil-cap" and the nucleus, might then either be retarded, or the nucleus alone might divide, or neither nucleus nor cytoplasm might divide. Cleavage of the centrifugal lobe went on normally. One could therefore assume the existence of a "cleavage substance" sedimenting with the heavier constituents of the egg (cf. p. 493, on the emission of stimulatory and inhibitory substances from echinoderm eggs). Such a mechanism of control might explain the hereditary differences in cleavage rate found by Fox[1] between sea-urchin races from different latitudes.

In reading through these sections, the biochemist will have been particularly impressed with the large number of chemical substances (important in many phases of development), the existence of which has been as good as proved by the methods of experimental morphology. In the case of the "blood factor" of fertilisation a preliminary survey of the properties of the substance has already been made. In the case of the "histogenetic stimulators", which the haploid embryo lacks and which contiguity with diploid tissues can provide, the substance or substances are still to some extent hypothetical. We shall shortly have occasion to describe some more of these substances, under the head of "nuclear inductors" (section 2·67). But before going on to these other substances, clearly produced by the activity of the nucleus, it will be convenient to study the question of genic lethality, since this has been raised in so acute a form by the experiments already described on haploid lethality.

2·65. Lethal Genes and organisers

Early in the history of genetics it was noticed that among the offspring of certain crosses, all those which would have been expected to contain a certain segregation of genes appeared to be missing. Investigation showed that they were missing because they had all died before the time of hatching or birth. In this way arose the conception of genes which are lethal to those animals which carry them. Here we shall not find it necessary to go into the genetics of the question; this may be found in the genetical monographs and reviews (e.g. Sinnott & Dunn; Mohr[1]; Landauer[23]; Wipprecht & Horlacher; Morgan, Bridges & Sturtevant). All that I wish to demonstrate is that in many of those cases where the mechanism of genic lethality has been investigated from the embryological point of view, there is ground for the belief that the effect is due to interferences with organiser phenomena.

It is evident that within the limits of what we call the normal, there are many different characters possible for the individual, some relatively insignificant (such as eye-colour), others less so. We know that all these are controlled by genes. At one boundary of the normal there are advantageous variations; at the other there are disadvantageous ones. Disadvantageous variation may be pushed so far as to achieve lethality, so that the organism bearing such genes will be killed, and this may happen at any stage of embryonic or post-natal development. Such genes are called lethal, or if the effect occurs late, sub-lethal. There can thus be no sharp line of demarcation between all these genes, nor can their effects be sharply distinguished from the teratological anomalies accidentally produced in development, except in so far as the regular inheritance of the condition is demonstrated.

Since development is an epigenetic chain of processes, the earlier a gene exerts its effects in development, the more serious its effects will be. And since all animal development is a highly integrated process, it is not difficult to understand why most variations should be disadvantageous, why, in fact, evolution has proceeded "in the teeth of a storm of adverse mutations". The more highly integrated a process is, the more difficult it is to alter it without injuring it. At the same time, a very drastic mutation affecting early embryonic stages may occasionally be viable and allow of the occupation of a new oecological niche. Goldschmidt[9] has called such organisms "hopeful monsters", and the possibility of their occurrence in relation to the origin of species was the motive long ago in Bateson's classical collection of morphological abnormalities in his *Materials for the Study of Variation*.

The time at which genes act in the life cycle is indeed variable. Some genes do not act in the life cycle of the organism carrying them, but upon the immediate descendants of the first generation, for instance those genes which affect egg characteristics ("maternal inheritance"). The time at which genes act has been the subject of an ingenious essay by Haldane[1], who suggested the following classification, so far as animals are concerned:

MZ Genes which affect the maternal zygote of the organism carrying them. Influences of the foetus upon the mother would be examples of this; none have so far been studied, but such genetically controlled influences must almost certainly exist.

Z_1 Genes which act early in embryonic development. Often lethal. Examples of these will be given later.

Z_2 Genes which act in larval development. Examples later.

Z_3 Genes which act before the attainment of sexual maturity. The majority of genes which have been studied in animals fall into this class.

Z_4 Genes which affect reproductive organs or behaviour, or secondary sexual characters. Examples are genes for milk yield or egg yield, and those which modify the metabolism of the sex-hormones.

Z_5 Genes which affect egg-shell, egg-white or other maternal egg or uterine structures. An example is the egg-shell shape and texture of silkworms (Toyama). "In the evolution of the mammal a critical part was doubtless played by Z_5 genes which interact with MZ genes in the foetus to determine a suitable response of the uterus to its contents."

DG Delayed gametic genes, i.e. those which affect gametes, not necessarily carrying them, borne by zygotes carrying them. Example: yellowness of yolk in silkworm eggs, associated with yellowness of maternal blood (Toyama), and genes affecting egg-shape and fecundity in the fruit-fly (Crew & Auerbach).

DZ Delayed zygotic genes, i.e. those which affect zygotes, not necessarily carrying them, borne by zygotes carrying them. This is true "maternal inheritance". Blueness in silkworm embryos is due to pigment formed in the serosa, a zygotic structure, but is unaffected by the paternal genes (Toyama). Dextrality and sinistrality of coiling in gastropod eggs such as *Limnaea* are DZ characters, probably depending on the molecular structure of the unfertilised egg (Boycott, Diver, Garstang & Turner; Diver & Anderson-Kottö). In *Drosophila* there is a lethal DZ gene affecting female embryos more than males, so that "whether a given female lives or dies depends not on her own genetic constitution but on that of her mother" (H. Redfield). As has been noted already (p. 364) cleavage rate is a property of the egg-cytoplasm in echinoderms, so that if we may leave cytoplasmic inheritance out of account, this also would be controlled by DZ genes.

Lethal genes are always harmless in the heterozygous condition, and so are transmitted and perpetuated by organisms which do not suffer from them. In the homozygous condition they produce their fatal effects. It does not seem to matter how many lethal genes an animal carries in the heterozygous state, as an organism will be quite normal though carrying four or more lethal factors, the presence of one alone of which in double dose would kill it (J. Schultz[1] on the fruit-fly; C. V. Green on the mouse).* Lethal genes are important for the embryologist because, as has been well said, they give us "experiments in embryology in the reversed direction. While in experimental embryology the experiment is known, but the result is doubtful, here we have to study the results first and by going backwards to find out what sort of an experiment nature has performed."

* Some genes *must* be present; an animal some of whose cells lack them entirely ("homozygous for deficiency") will often show local suppressions of these cell groups; these are the "cell lethals" of Demerec. Lack of space prevents further consideration of them here, but see below, p. 408.

The following discussion naturally arranges itself in certain sections. We shall discuss (1) genes causing the death of the embryo before the action of the primary organiser, (2) genes interfering with the action of the primary organiser, (3) genes interfering with the action of second- and third-grade organisers, (4) genes producing anomalies of the individuation field by changes in growth rates of parts, (5) genes producing anomalies of embryonic metabolism, and (6) genes killing or deforming the embryo by unknown mechanisms. The fourth class, genes giving rise to anomalies of the individuation field, may be divided into two. Changes in growth rate may be such as to affect the general plan of the organism and so have serious morphogenetic consequences, or they may occur quite uniformly and smoothly so that they have no such consequences. They may also involve the proliferation of one particular tissue at the expense of all the others, and so give rise to tumours.

2·651. Genes causing the death of the embryo before the action of the primary organiser

It so happens that the classical case of this effect was also the first case of genic lethality to be discovered. Cuénot[2] in 1905 noticed that no homozygous yellow mice were obtainable, and later Castle & Little showed that *yellow* is a gene which invariably leads to the early death of the embryos if present in the homozygous state. After a good deal of purely genetical work, it was established by Ibsen & Steigleder, and particularly by Kirkham[1, 2], that most of the morulae and blastocysts degenerate and die before implantation has occurred. Such blastocysts have been described by Corner[1]. Should implantation take place the blastocyst is devoured by phagocytes. The causes of these events still remain quite obscure, but as can be seen from modern work on the physiology of the mammalian blastocyst (p. 84) many possible mechanisms exist. Another case is that of the lethal mouse studied by Glücksohn-Schoenheimer[2], which survives implantation successfully but goes no further owing to a complete failure to form any mesoderm (cf. p. 359 and p. 629). Since it is associated with *short-tail* strains (p. 374), it may be regarded as a very extreme form of the mesodermal (notochordal) anomaly exhibited by them.

In insects a large number of lethal genes have been described, but few have been submitted to embryological analysis. In *Drosophila* some of these seem to kill the embryo at an extremely early stage, thus N. P. S. Dobzhansky found that eggs homozygous for *star* develop a black spot in the region of the micropyle and die before any appreciable development has taken place. Chromosome deficiencies and aberrations, in particular, such as *nullo-X* and *nullo-IV*, inhibit development at the earliest stages (J. C. Li[1]; Brehme). Similarly in silkworms, some of the lethals described by Nishikawa, by Astaurov, and by Efroimson & Rilova kill the embryos at the very beginning of development. Poulson[3] alone has investigated the way in which these serious chromosomal abnormalities affect the formation of the germ-layers in the embryo. When the *X*-chromosome is totally absent (*nullo-X*), the nuclei fail to divide in the posterior end of the egg, and even in the anterior part they remain densely clumped so that no cell boundaries are ever formed (see the account of insect development later; p. 455). A slightly less severe suppression of development is seen when only half the chromosome is present (*half-X*); here the nuclei migrate to the egg's surface but no cell-divisions appear and no blastoderm. In *notch*-8 the blastoderm forms normally and neural tissue hypertrophies, but little endoderm and no mesoderm ever appears. All such cases will well repay further investigation. It has already been found that the respiration of *nullo-X* embryos is markedly reduced (Boell & Poulson).

Serious genic abnormalities, such as the total absence of chromosomes or parts of chromosomes, absence of genes, unusual translocations and segregations, etc., may also have something to do with the development of tumours (see the discussion below, p. 385, on hereditary tumours, and the previous discussions on cancer, pp. 239 ff.). Jones has drawn attention to the probable genetic basis for fasciations and some galls (see p. 106) in plants. Conversely, Kostov & Kendall have demonstrated irregular chromosome behaviour in galls produced by insects. Certain varieties of apples are characterised by well-marked mole-like projections on the surface of the ripe fruit, due to the egg-laying of the plum curculio (*Conotrachellus nenuphar*). The abnormal tissue, an unregulated mass of cells, must be different from the fruit tissue, because the scab fungus, for instance, will spread up to the margin of the galls but no farther. Some varieties of apple respond to the insect's stimulus (are "competent") but others do not, and this tendency is transmitted to some of the seedlings.

It has long been known, too, that X-rays, radium, ultra-violet light, heat, and other physical agents, produce chromosomal irregularities. But in addition to these, non-disjunction of chromosomes has been obtained (e.g. by Mottram) as an effect of treatment with the dye gentian violet, with tar, or with the dibenzanthracene derivatives, which we have already mentioned as being related both to the sterols, and perhaps to the primary evocator (see p. 243). This raises the question of whether nuclear anomalies are cause or effect in malignant tissues. Boveri[3], in a classical monograph, allotted them an important role, but as may be seen from the review of W. H. Lewis[4], the question is still quite undecided.

2·652. Genes interfering with the action of the primary organiser

Here the cases known fall into two groups, those affecting the head-organiser and those affecting the tail-organiser (see pp. 271 ff.). It must be emphasised that the embryological analyses have in general not been pushed back sufficiently far to establish with certainty that this kind of interference has taken place, but the disturbances are in some cases so deep-seated that it is difficult to doubt it.

The most striking case of far-reaching disorganisation of the head-organiser is that of the hereditary *otocephaly* in guinea-pigs described by Wright & Eaton and later anatomised by Wright & Wagner. Disorganisation is found in many different degrees (see Figs. 213 and 214). First the ears approach each other till a median throat opening is formed. In grade 5 the mouth opening is altogether lost, in grade 6 the nostrils fuse, in grade 8 the eyes fuse (cyclopia), in grade 10 the eyes completely disappear (anopthalmia), and eventually there is no sign of a head at all. In all cases the remainder of the body is well formed. This wonderful series of progressive head defects (the genetics of which was later further studied by Wright[4]) must involve head-organiser in the first instance,

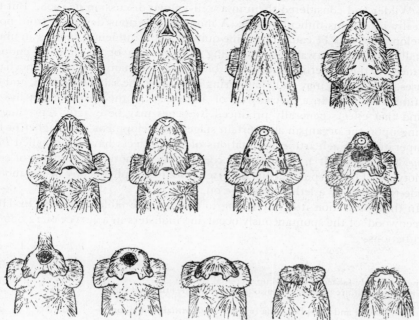

Fig. 213. Otocephaly in guinea-pigs.

Fig. 214. Photograph of otocephaly in guinea-pigs.

but also the foregut as mouth-organiser, and the mesectodermal cells of the neural crest which normally make the head skeleton.*

* Perhaps similar genes exist in all vertebrates: cf. Fasten's headless sheep, the *agnathic* lethal absence of jaw in cattle described by Ely, Hull & Morrison, and the human cases of Redenz. Otocephaly can be brought about experimentally by the action of lithium in amphibia; there it is regarded by F. E. Lehmann[7] as due to disturbances of the head-organiser. In rodents direct irradiation of the foetus *in utero* will produce many profound head defects (Job, Leibold & Fitzmaurice).

Wright[5] seems to have felt that the border-line between monsters and genetic abnormalities raised some difficult problems. The American embryologists Mall and Wilder had considered them in a well-known discussion in 1908. But there is really no serious difficulty here. A monster, a serious deviation from normal development, may be caused by some quite external influence, such as an unusual metallic ion in the water surrounding a frog's egg or an unusual endocrine balance in the maternal and foetal blood, or by many relatively mechanical causes. But a gene may be able to bring about just the same deviation, and to do so time after time in a given stock of animals.* In the same way, it has been found that effects generally produced by genes may be exactly reproduced by acting upon the organism at a certain stage of development with higher or lower temperatures. Such artificial imitations of gene effects have been called *Phenocopies* (Goldschmidt[8]), because they copy in the phenotype the effects of certain genotypes. Thus a spontaneous monster would be a phenocopy of the monsters made regularly by a lethal gene (see on, p. 392).

In this connection it is interesting to examine the table drawn up by Hutt & Greenwood[2] of the spontaneously occurring monsters in a survey of 12,000 chick incubations.

	Percentage of total chicks	Percentage of total anomalies
Hyperencephaly (absence of eye-cups, cranial roof, etc.)	1·52	41
Exencephaly (failure of skull formation; brain extrudes through the meninges)	0·84	23
Microphthalmia and Anophthalmia (eye-cups abnormally small or entirely absent)	0·81	22
Exencephaly and Microphthalmia	0·25	7
Miscellaneous (including ectopic viscera, malformed limbs, otocephaly, complete absence of head, and various duplicities)	0·25	7

Thus in spontaneous anomalies, failure of the eye-cup inductor seems to be the most common, while the otocephalic abnormalities, up to complete failure of head-organiser and so of all cephalic structures, are relatively rare.

Head abnormalities may also be the result of the segregation of an induced translocation (Snell[2]; Snell, Bodemann & Hollander; Snell & Picken). When male mice were X-rayed (with a dose of 600 Röntgen units) and then mated to normal females, about a quarter of their progeny were semi-sterile, producing small litters. These were found to be due to a lethality producing gross malformations of the anterior part of the neural tube in many of the embryos. There is a failure in the closure of the neural folds so that the whole of the fore-brain, mid-brain, and hind-brain may be open (see Fig. 215). The cranial roof is represented only by centres of ossification at the sides of the head, and those parts

---

* In an interesting review Wright[5] has elaborated a conception of genetic otocephaly dependent on rates of development (following the thought of Child or perhaps rather Stockard). He drew a graph showing the time at which developmental inhibitions would have to become effective to produce the various effects observed. This is not inacceptable, but R. G. Harrison, in the discussion of his paper, rightly drew attention to the existence of specific inductor substances in the head the action of which the genes must interfere with. Wright & Wagner admitted that the primary factor may be the prechordal plate, the importance of which for head development had been well appreciated as long ago as 1924 by Kingsbury & Adelmann, although at that time little or nothing was known of the role of the mesoderm in neural induction.

of the brain that are formed sit later upon the animal's head like a beret (Fig. 216).
Such a pseudencephaly appeared about the same time as the American work of
Snell and his colleagues in a strain of mice in Norway as a lethal mutation, and

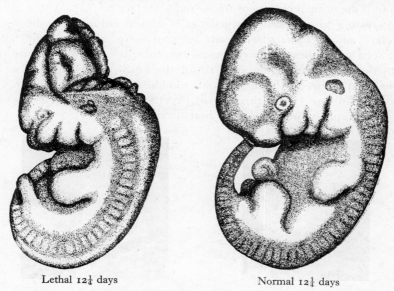

Lethal 12¼ days   Normal 12¼ days

Fig. 215. Failure of mesencephalon and rhombencephalon to close in progeny of X-rayed mice.

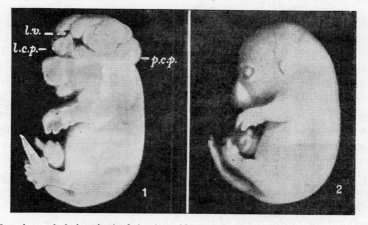

Fig. 216. Pseudencephaly in mice (14½ days). 1. Abnormal strain. 2. Normal. *l.v.*, lateral ventricle;
l.c.p., lateral choroid plexus; *p.c.p.*, posterior choroid plexus.

was the subject of an embryological analysis by Bonnevie[2]. She succeeded in
tracing the anomaly back to the open neural plate stages, in which it appears
that the notochord and mesoderm in the head region remain adherent to the
neural tissue much longer than in normal embryos. If the lethals do not die at
this stage, pseudencephaly results. The neural tissue seems to be unduly abundant

in relation to the head mesoderm. Even so, the exact nature of the failure of the brain folds to close remains obscure,* and an understanding of it will probably have to await a better knowledge than we have at present about the processes going on in the normal closure of the neural folds (see p. 334). At the same time it is interesting to compare this genetically caused failure with the results of Holtfreter[10] on *Nachbarschaft* already referred to (p. 284), and the failure of closure seen sometimes in embryological experiments. It is almost sure that we have to deal here with some anomaly of the head-organiser. Lastly, it may be added that failure of the neural folds to close has been described in a spontaneous human case (1-month embryo) by Orts. There is indeed a possibility that pseudencephaly in man may be hereditary.

Fig. 217. Lower jaw reduction in progeny of X-rayed mice.

Fig. 218. Head anomalies in progeny of X-rayed mice.

Now it is interesting that an extremely minor degree of the same phenomenon (lack of complete closure of the neural canal) has been shown to be largely responsible for one of the best-known genetic lethal multiple anomalies, the Bagg-Little foot and eye "syndrome" in mice, which itself involves far-reaching and serious defects. In 1924 Little & Bagg obtained a strain of mice after X-ray irradiation which showed the presence of lethal genes causing the following effects: (*a*) anomalies of the eye and face, (*b*) anomalies of the limbs, including club-foot and digital reduplications, (*c*) anomalies of the hair in the saddle region, and (*d*) disorganisations of the head, including severe reductions of the jaw. These latter gave some of the embryos a peculiar "elephant-trunk"-like appearance (see Figs. 217, 218). Some years later Bagg[2] showed that the limb anomalies were associated with the presence of blisters and blood clots at the extremities, which interfered with the fore- and hind-limb "fields".

At this point the embryology of the condition was taken up by Bonnevie[1]. She showed that although the development of the nervous system proceeds almost perfectly, there is a remarkable expulsion of cerebro-spinal fluid from a foramen in the roof of the hind-brain. Such a foramen is normal in some mammals, but in the lethal mice, it allows a large quantity of the fluid to escape, forming blisters under the epidermis, first on the dorsal surface of the head and then gradually

* Experimental transplantations are really needed to elucidate the anomaly in its earliest stages, but of course these would be very difficult to do on mammalian material.

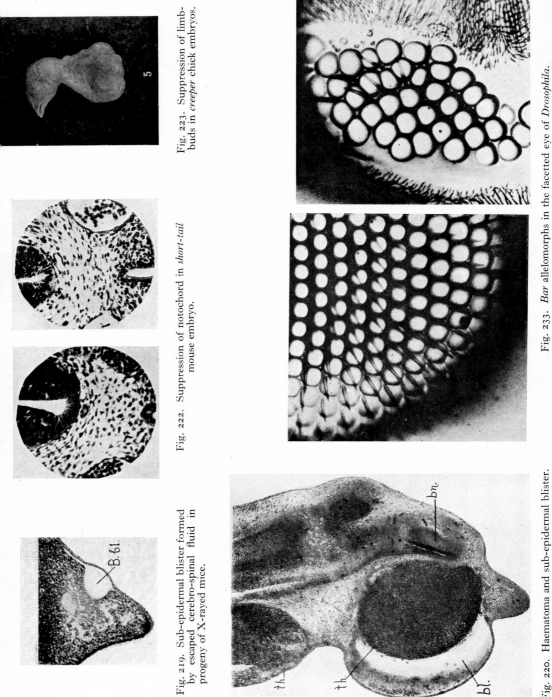

Fig. 219. Sub-epidermal blister formed by escaped cerebro-spinal fluid in progeny of X-rayed mice.

Fig. 222. Suppression of notochord in *short-tail* mouse embryo.

Fig. 223. Suppression of limb-buds in *creeper* chick embryos.

Fig. 233. *Bar* allelomorphs in the facetted eye of *Drosophila*.

Fig. 220. Haematoma and sub-epidermal blister.

spreading all over the surface of the body. Some of the blisters move into the limb-bud districts and so are carried into the limbs, where they give rise to syndactyly, polydactyly, etc. and interfere in various ways with the normal limb pattern (see Fig. 219). Others move in front of the eyes and so prevent the induction of the cornea by the eye-cup and lens. As time goes on the capillaries on the floor of a blister weaken, discharging their blood into it, and forming in many cases haematomata. Fig. 220, taken from Plagens, who attached much importance to the bleeding, shows such a haematoma pressing into a fixated blister. The genetic control of these effects is complicated, for it seems to be possible to breed a strain in which eye defects predominate over limb-bud defects, and another in which the reverse is true (Little[2]; Little & McPheters). Moreover, in the strain of Bagg & Little, the eye defects were much more severe than in the material studied by Bonnevie, including even the absence or atrophy of both eye-cups, which could not possibly be referred back to the effects of blisters. Similarly, in one of Bagg's strains (Bagg[1]), the foot and eye anomalies were accompanied by the complete absence of one or both kidneys, and this could be intensified by selection.

In this interesting case, therefore, we have to deal with a number of effects, some of which can be brought together as the results of one deviation from normal development, namely, the expulsion of cerebro-spinal fluid from a weak spot on the roof of the hind-brain. For this reason we have classified the Bagg-Little strains under the heading of primary organiser anomalies, but it is clear that second-grade organisers seem to be involved too, particularly the eye-cup organiser, the pronephros-metanephros organiser system, and the inductor of the cornea.

It may be noted that cerebral hernia, a mild degree of pseudencephaly, has been found to exist as an inherited sub-lethal factor in pigs (Nordby) and cattle (Shaw). What is to some extent the converse, hereditary hydrocephaly, is inheritable in mice (F. H. Clark; Zimmermann).

Another case which gives us a transition to interferences with second-grade organisers is that of the *shaker-short* mice. "Waltzing" mice occur in various well-known genetical types, such as "circlers" and "shakers", each type having a characteristic movement disturbance and a characteristic deafness. The "short-tailed shaker" type arose as a mutation and according to Dunn[4] is due to one recessive gene. Its embryological analysis, undertaken by Bonnevie[3], showed that, owing to a weakness of the roof of the brain (cf. the pseudencephaly and Bagg-Little types just described), marked hernia containing cerebro-spinal fluid and blood occurs. Hence the normal pressure of cerebro-spinal fluid within the brain cavities does not arise, and there is no outward movement of the brain walls, as in normal embryos (see Fig. 221). Hence the ear-vesicles, instead of being compressed to a more discoidal shape, tend to remain spherical, as they were when they were first induced. From the experimental work of Kaan we know that although the ear-vesicle may after its first induction take the initial steps in the formation of the labyrinth, the formation of a normal organ depends on its relations to neighbouring structures. And so all the later distortions of the internal ear in *shaker-short* mice were found by Bonnevie to follow from this primary failure of expansion on the part of the brain. Thus they form no endolymphatic appendage, which is probably normally induced by the hind-brain.

We come now to the genetic disturbances associated with the caudal end of the body; interferences with tail-organiser. The well-known *short-tail* gene in mice is lethal when in the homozygous condition. An account of its first discovery and a review of its genetics has been provided by Zavadskaia, Koboziev & Veretennikov. What we know of its embryology is mainly due to P. Chesley, who found that homozygous *short-tail* embryos die at 11 days gestation and are resorbed. The first signs of anomaly appear at $8\frac{1}{2}$ days as small blisters on each side of the mid-line, which later disappear. There then arise marked irregularities of the neural tube, which sends out branches some containing a lumen, some not; the somites are not well formed, and posteriorly completely absent; there are no hind-limb buds. In section, it can be seen that the differentiation of the

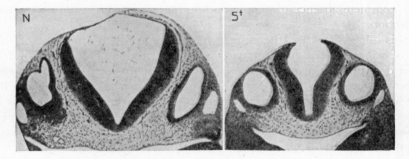

Fig. 221. Cross-sections through the head of a normal mouse embryo
and a *shaker-short* one.

notochord is very deficient (see Fig. 222, opp. p. 372); the picture is thus rather reminiscent of the notochords suppressed by lithium in Lehmann's work[6] (p. 226). Here, however, disorganised neurulation follows. We are bound to recognise an interference with the primary organiser.

Heterozygous *short-tail* embryos have been investigated by Glücksohn-Schoenheimer[1]. Chesley had shown that their tails atrophy as the result of a sharply marked constriction, and Chesley & Dunn had studied their genetics. Glücksohn-Schoenheimer found that this constriction appears at the 11th day, but is histologically visible one day earlier, for sagittal sections of the tail bud reveal that although the neural tube has grown back some way into the tail, the notochord has not. Here again an anomaly of the notochord, the primary inductor, leads to the failure of an axial structure.

Numerous lesser degrees of inherited tail abnormalities have been described. The tail of the *flexed-tail* mouse (Hunt & Permar; Hunt, Mixter & Permar) is fixed in one or more right-angle bends due to the unilateral fusion of adjacent vertebrae (Kamenov), owing to the presumptive inter-vertebral discs becoming bone instead of fibre. Kamenov, however, could not show that the anomaly was caused by any anomaly of the notochord. *Flexed-tail* involves, for reasons at present unknown, considerable anaemia (Mixter & Hunt), due both to haemoglobin deficiency and lack of erythrocytes. A gene with similar effects, *wry-tail*, exists in cattle (Atkeson & Warren).

Trunk-organiser may be affected as well as the primary organisers of head and tail, if we may interpret in this way the *short-spine* anomaly in cattle described by Hutt[2].

The other well-known case which might be classified as tail-organiser failure is the gene *rumpless* in fowls, the genetics of which have been studied by Landauer[12] and by Dunn & Landauer[2]. Known since the seventeenth century, this condition involves the absence of all the synsacro-caudal vertebrae posterior to the third, and the complete suppression of the five free caudal vertebrae, together with the uropygial (oil) gland and other neighbouring structures. From the work of du Toit we know that the segments missing from the adult fowl are missing from the tail bud of the 3rd day embryo. Intermediate grades, also hereditary, between complete rumplessness and the normal, exist (Landauer[2]). It is interesting that according to Dunn[2] rumplessness occurs spontaneously from time to time in large chick populations, and Danforth[2] has succeeded in producing phenocopies of *rumpless* by subjecting developing eggs of normal chicks to abnormally high temperatures during the early incubation stages.

2·653. Genes interfering with the action of second-grade organisers

In the course of the preceding discussion we have already seen a number of cases of this, mostly concerning the organs of the head. The extreme cases of otocephaly of Wright & Eaton must involve failure of eye, ear, and mouth inductors. The eye-cup suppressions in some of the Bagg-Little strains may be due to failure of the eye-cup inductor, and the kidney suppressions may, indeed must, be interpreted in a similar way. Certain hybrid guinea-pigs also may show complete anophthalmia (Eaton; Pictet & Ferrero) and mice too (Chase & Chase). Genetic eye-cup suppression may be compared with that already noted as occurring in vitamin-B deficiency in pigs (p. 227); possibly the gene and the vitamin act upon the same mechanism in different ways.

Other inherited defects in head organs are the useless chick mandibles described by Asmundson[2] (this is a lethal gene, since feeding is not possible), and the deficient teeth in the *grey-lethal* mice of Grüneberg[1], which will be referred to again. Congenital absence of all teeth in man is known (Thadani). The failure of cornea induction in the Bagg-Little strains as analysed by Bonnevie[1] was due to mechanical separation of inductor and competent tissue by the travelling blister,* but cases of more subtle interference with cornea induction are known, such as that reported by Whitlock. In a certain guinea-pig strain, the presumptive cornea, though normally underlain by lens and eye-cup, never achieves the transparency typical of the normal cornea, and differentiates into epidermis

---

* Cf. the identical separation of inductor and reactor in the blindness of cave forms, already referred to (p. 225).

·instead. A particularly interesting genetic anomaly, which involves the persistence of the gill-slits in sheep, giving fistulae instead of wattles, is described by Vassin.

In insects also the development of a head organ, such as the eye, may fail, as in *Drosophila* (Derrick[1]) and *Habrobracon* (Speicher). An *eyeless* fruit-fly has the two inner optic ganglia on each side but the outer one has disappeared; there may be no ommatidia and no facets (Richards & Furrow). Derrick showed that the eye-rudiment is greatly reduced in size from its first appearance. Moreover the moment of first appearance is later than normal.

But the most many-sided genetic deficiencies in a second-grade district are those of the limb-buds. They may be entirely suppressed, as sometimes occurs in *creeper* chick embryos (Landauer[17])—see Fig. 223, opp. p. 372—to which we shall return, or in the cat (Kirkham & Haggard) or man (Ilberg). Or distal parts may be suppressed to varying extents, as in the absence of hands or feet (acheiropodia) or of the digits (brachydactyly), and the digits may fail to separate (syndactyly). On the other hand, reduplications may occur, as in polydactyly.

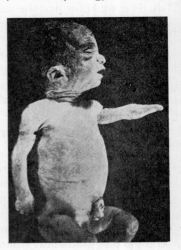

Fig. 224. Suppression of limbs in human development.

Suppression of distal parts of limbs in mammals used to be interpreted as being due to "intra-uterine amputation", i.e. strangulation of the limb by amniotic folds, umbilical cord, etc. But Streeter has conclusively shown that such "amputation" is really a failure of formation, due to defective limb-bud histogenesis followed by de-differentiation and healing of the stump. In man (where the end-result may be as pictured in Fig. 224) defects tend to be present in all four extremities if in one, and there is some likelihood that the condition may be hereditary. In the rabbit genetic brachydactyly and acheiropodia has been studied by Greene & Saxton. Nothing is visibly abnormal before the 18th day of gestation, but then there occurs vascular break-down and extravasation of blood followed by necrosis. In this case, therefore, a limb-bud organiser cannot be incriminated, and we are reminded rather of the vascular degeneration which follows vitamin-E deficiency (see p. 265) in embryos.* Genetic failure of the limb-buds to form beyond the first joint has been described in cattle by Wriedt & Mohr. This was associated with extraordinary deformations of the face, giving the calves the appearance of parrots.

---

* The appearance of limb buds followed by their complete or almost complete resorption may occur not as the result of a distinct lethal gene, but in crosses. Thus Hamburger[1] has reported such phenomena in newt hybrids (*taeniatus* ♀ × *cristatus* ♂ and *palmatus* ♀ × *cristatus* ♂), and Montalenti in toad crosses. Suppression of limb buds, as it imitates what may happen genetically, may be looked on as a "phenocopy" phenomenon (see p. 392). It has been brought about by selenium injections in the case of the chick (Franke, Moxon, Poley & Tully) and by 10 % sucrose in the case of amphibia (Sladden).

That lethal genes may involve organiser excess as well as organiser suppression appears clearly from the work of Wright[2] on polydactylous guinea-pigs. Guinea-pigs usually have four toes on the front foot and three on the hind foot. By selection within a race containing a gene producing supernumerary fingers and toes, Wright obtained a strain in which the homozygotes invariably died during the foetal period, after producing paddle-shaped feet bearing from nine to twelve digits each, while the heterozygotes showed the pentadactyl foot. In the homozygous condition, the embryos also had head defects including micropthalmia. These interesting embryos were studied by J. P. Scott, who was only able to find, however, that the supernumerary centres of digit-formation appeared in the earliest stage possible. He interpreted the effects as due to an arrest of morphogenesis and alteration of relative growth rates, but the evidence for this is not entirely convincing and growth rates alone will hardly account for reduplications of morphological entities. It must be noted that the polydactyly is not the cause of death, which occurs about the 26th day of development. The cause of death is the bursting of blood-vessels in the region of the shoulder-hump, a mass of connective tissue behind the brain which is the site of a large fat body in the normal newborn animal. The haemorrhage quickly spreads to other parts of the body. Genetic polydactyly in mice is described by Danforth[1] and by Droogleever-Fortuyn; it apparently involves slight corresponding abnormalities of the central nervous system (Y. C. Tsang), and is not due locally to travelling blisters but rather to abnormally numerous growth centres (T. K. Chang).

Polydactyly occurs rarely in man, where it may possibly be hereditary (cf. the discussion in Scott's papers). Club-fingers in man are certainly hereditary (Witherspoon[2]). Club-fingers produced by disease might be regarded as phenocopies of this.

In insects genes causing distortions of the appendages are well known. Thus Hoge discovered a gene producing reduplications of the legs on each segment in *Drosophila*, and the gene *crippled*, described by Komai for the same animal, causes all kinds of leg abnormalities: twisting, shortening, forking, etc. and in some cases complete suppression (cf. "amputation"). But the most remarkable phenomena of aberrant leg effects in insects are perhaps those of "hetero-morphosis", especially *aristapedia*. Regenerative hetero-morphosis has long been known. After extirpation of the eye, an antenna may be regenerated instead of the missing eye (Herbst[3] on *Palaemon*; Křiźinecký on *Tenebrio*), and the new organ may be functional (Lissmann & Wolsky). After removal of an antenna a tarsus may be regenerated instead (Przibram[2] on *Sphodromantis*; Cuénot[3]; Brecher; and Borchardt on *Dixippus*). But Balkaschina discovered a recessive gene in *Drosophila* which regularly produces a leg or tarsus from the presumptive antenna rudiment. Normally segmentation of the antenna in its imaginal disc begins on the fourth day of larval life while that of the tarsus begins on the second day. In the *aristapedia* flies, however, segmentation of the antenna begins two days earlier than it should. This time-difference might account for the change, as suggested by Goldschmidt[8] and Braun, by giving one inductor opportunity to act out of its due order.

One of the most interesting features of *aristapedia* is the fact established by Waddington[25] that genes which normally act upon legs act just as well upon the *aristapedia* leg. Conversely, an animal may be produced in which only the proximal part of the antennal appendage is a leg and the distal part is an antenna; it is then found that genes such as *thread*, acting on antennae, exert their effect perfectly in this very circumscribed and isolated locality.*

An analogous anomaly is that of *proboscipedia*, where the mouth organs always assume the character of tarsi (Bridges & Dobzhansky).

The experimental embryology of amphibia has shown that we may speak of a blood inductor (see pp. 164, 301 ff.). Perhaps such a second-grade inductor is affected in the lethal gene *anaemic* of mice first discovered by Little[1]. That it kills embryos bearing it at birth or in the first post-natal week was established by de Aberle and Detlefsen. Analysis of the effect by Gowen & Gay† showed that although the haemoglobin content of the erythrocytes is normal, only one-third of the normal number of these is present (see pp. 648 ff.). This certainly points to a deficiency in the inductor-reactor system of the blood cells. The lethal embryos are not deficient in iron, and feeding iron or liver to the mothers has no beneficial effect. Repeated injection of blood into the peritoneal cavity of the new-born embryos was found, however, to prolong life almost indefinitely (cf. the administration of pituitary to genetic dwarfs, p. 390).

Yet another condition indicating an induction failure is that of "nakedness" in organisms normally covered with a hairy or furry coat or feathers. Genes for this, usually sub-lethal, are particularly common; they have been described, for example, in the rabbit (W. E. Castle), the pig (Roberts & Carroll), the mouse (Crew & Mirskaia), the rat (Roberts[2]), the cow (Mohr & Wriedt), and the chick (Hutt[3]; Hutt & Sturkie). In the last-named case, the gene for *naked* completely inhibits the eruption of the feathers from their follicles, and is 50 % lethal during the last 2 days of incubation. Feeding cysteine and cystine to the laying hen had no effect. Skin from naked chicks transplanted to normals was found to be determined—no feathers could be produced from it; and *vice versa*. In the *hairless* rat, the condition is associated with disturbances of the oestrous cycle, and the organ weights differ considerably from normal animals (F. E. Emery), probably because, as Benedict & Fox have shown is the case in the *hairless* mouse, the basal metabolism is very considerably raised above normal. To the histology of these genic effects an elaborate study was devoted by L. T. David, who found that there could be either imperfect keratinisation breaking off the hairs as they emerged from the follicles; or degeneration of the follicles; or a failure of the formation of the follicles.

More serious skin anomalies than nakedness are also known. For example, in cattle, the gene or gene-complex *epithelial* causes defects in the skin and mucous membranes at sites which soon become infected after birth (Hadley & Warwick). In man the outstanding example is the condition known as *ichthyosis congenita*, where a colossal hyperkeratosis transforms the skin of the new-born child into

---

* Phenocopies of aristapedia have been produced by Rapoport[2] using boron compounds.

† See also Smith & Bogart for biochemical data.

a leathery coat, traversed by bleeding fissures (see Fig. 225 from Lesser). The literature on this and similar anomalies can be traced through the review of Mohr[1].

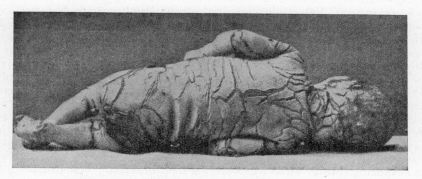

Fig. 225. The skin anomaly *ichthyosis congenita* in man.

2·654. Genes producing anomalies of the individuation field by changes in the growth rates of parts

In the first place it must be realised that changes in growth rate during development may be genetically brought about in such a way as to cause no morphological effects. Pure magnification or reduction of the pattern of the organism need involve no distortion. Thus in *Drosophila, giant** (Gabritschevsky & Bridges), where the larvae continue to feed and grow longer than normal larvae, may be regarded as the opposite of *vestigial* (Alpatov[1]; Dunn & Coyne), where larval growth is stunted. In neither case is morphogenesis interfered with. The same applies to *dwarf* in mice (see p. 390).

Alternatively, the shape of the organism may be changed by a change in shape of some part, such as the muscle masses, without appreciable disorganisation of morphogenesis. Such seems to be the case in *Drosophila* with the gene *chubby* (T. Dobzhansky & Duncan), where the shape of the eggs is the same, the growth rate of the embryo is the same, but the shape at birth is different (shorter and broader), because the muscles are laid down in shorter and broader form. Another gene, *rudimentary*-12 (N. P. S. Dobzhansky), simply shortens the wings, but has no effect on general growth rate.[†]

But if changes in growth rate affect directly one developing part and not the others, or if a general change in growth rate occurs at a time when some processes are proceeding much more rapidly than others, so that the former are preferentially affected, then serious disturbances of morphogenesis will ensue. Of this character appears to be the lethal homozygous *creeper* gene in fowls, perhaps the most thoroughly analysed of any lethal hitherto discovered, owing to the work of Landauer, Dunn, and their collaborators.

* But cf. *lethal-giant* already mentioned (p. 361).

† The most outstanding cases of genetic shape distortion without morphogenetic disturbance are found in plants, as Sinnott's studies[1] of fruit shape show.

Chondrodystrophy (formerly wrongly called "foetal rickets")* is a malformation due to rather subtle disturbances in the formation of cartilage and the rate of perichondral ossification. Long known to occur in all the higher animals, it was exhaustively examined in spontaneously occurring cases in the chick by Landauer[1] in 1927. Externally, chondrodystrophic embryos are characterised by a shortening of the legs, curvature of the long bones, and a shortening of the base of the skull (cf. the radiograms in Fig. 226). The syndrome is always lethal. Investigations of Dunn[3] established by extensive breeding experiments that this malformation is not hereditary, but occurs in all breeds of fowl, and probably has some relation to the vitamin-D content of the egg (cf. the work of Hart, Steenbock and their collaborators on irradiation of eggs and laying hens, with improvement of hatchability; described in CE, pp. 1361 ff.). The presence of certain plant proteins, such as soya-bean meal and cotton-seed meal, in the diet of the laying hen, increases the frequency of chondrodystrophic embryos (Byerly, Titus & Ellis[1]), and it is the finding of many authors (e.g. Munro[1]) that there is a seasonal variation in the incidence of chondrodystrophic embryos, the frequency decreasing with advancing spring. Hens which during their first laying year produce a high percentage of chondrodystrophic embryos do not continue to do so in subsequent laying seasons (Byerly, Titus & Ellis[1]). All in all, there can be little doubt that chondrodystrophy is a malformation produced by environmental factors, especially vitamin-D deficiency.[†]

Concerning its physiology we possess a certain amount of knowledge. Dunn[3] found that chondrodystrophic chicks are seemingly unable to absorb calcium from the shell as normal chicks do (CE, p. 1363). Endocrine glands are all normal with the exception of the pituitary and thyroid which are very small and histologically undifferentiated (Landauer[3]; Sun[2]). The blood picture of the chondrodystrophic chick is markedly different from that of the normal, showing an increase in eosinophile and basophile leucocytes and in megaloblasts and erythroblasts (Landauer & Thigpen), together with chemical changes such as low blood phosphate (Ogorodny[2]). Since the nutrition of the adult laying hen seemed to be implicated in the causation of chondrodystrophy, an amino-acid deficiency presented itself as a possibility in addition to the vitamin question, and since gelatin, the best-known protein in cartilage, contains some 22 % of glycine, Patton & Palmer were led to estimate glycine in normal and chondrodystrophic embryos. Thus corn (maize) meal, fed by Byerly, Titus & Ellis[1] and found to lead to a high incidence of chondrodystrophy, contains no glycine at all. The results were as follows:

	Glycine % of total protein
Eggs from hens on optimum ration	2·21
Eggs from hens on glycine-deficient ration	2·16
Normal chick embryos at hatching	9·3
Chondrodystrophic chick embryos at hatching	6·5

Thus a low glycine ration does not significantly diminish the glycine in the eggs laid (cf. p. 22). But it does seem to inhibit the power of the hen to transmit to the embryo the enzymic mechanisms necessary for carrying out that synthesis of glycine which, as the table shows, normally takes place during development. These facts were later confirmed by Patton, who found that chondrodystrophic embryos frequently had only half the amount of glycine contained in normal embryos.

* True foetal rickets does occur, e.g. in man under famine conditions (Maxwell, Hu & Turnbull).

† But the causation is doubtless complicated, as chondrodystrophy seems to be favoured by diets low in manganese (Lyons & Insko).

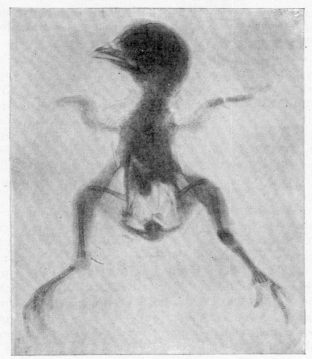

A. Normal

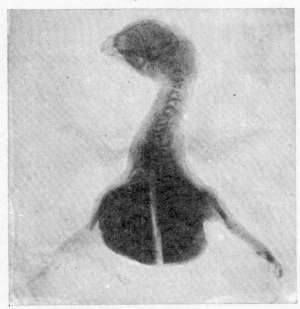

B. Chondrodystrophic

Fig. 226. Comparison of normal and chondrodystrophic chick embryos (X-ray photographs).

Chondrodystrophy may take milder forms, of which the nutritional micromelia described by Byerly, Titus, Ellis & Landauer, and anatomised by Landauer[19], for instance, may be an example.

With these nutritional malformations as a background, we may now consider the genetically produced extreme form of chondrodystrophy known as homozygous *creeper*. The Creeper (Scots Dumpie, Courtes Pattes, or Krüperhuhn) is a long-known breed of fowl characterised by a striking shortness of its extremities. The shaft of the tibia is bent, and a complete fibula (normally rudimentary) is present; the toes are curled. In the shafts of the long bones there is no endochondral ossification. That the *creeper* gene was lethal in the homozygous state was first suggested by I. E. Cutler, and definitely established in later breeding experiments (Dunn & Landauer[1]; Landauer[5, 10]; Landauer & Dunn[2]). The case is thus analogous to the *yellow* mouse, which never breeds true.

Landauer made a detailed anatomical and histological study of the development of the skeletal system in both heterozygous and homozygous *creeper* chicks (Landauer[6, 11]). From the first of these it was suspected that growth rate changes played a very significant part, particularly since the greater was the normal length of a long bone, the greater was its relative reduction in the Creeper. Exact measurements of this kind were carried out by Landauer[14] who found that the long bones of *creeper* chicks were always shorter than normals, and that the more distal they were, i.e. the later they arose in development, the more serious was their relative reduction. Lerner's heterauxetic calculations[1] confirmed this conclusion. Study of the homozygous lethal revealed more far-reaching anomalies; great reduction of wings and limbs, reaching almost to their suppression; fusion of radius-ulna and tibia-fibula; failure of formation of epiphysial cartilages; no trace of ossification. Eye abnormalities accompanied the skeletal ones, especially coloboma, to which Landauer devoted a special paper[8].

When Landauer[7] examined the earliest stages of the homozygous *creeper* embryo, he found that it was always demonstrably smaller than normal, even as early as 36 hr. incubation (cf. Fig. 227). He concluded that the primary action of the *creeper* gene was to produce a general inhibition of growth, and that the limb-buds, which were growing most rapidly at the time of inhibition, consequently suffered most severely. This view was tested by Fell & Landauer, who cultivated limb-buds and mandible rudiments from 3–5-day normal chick embryos in a growth-restricting medium (plasma-saline with very little added embryo extract). Such limb-buds showed after a fortnight's cultivation many of the abnormalities characteristic of *creeper*, e.g. tibia-fibula fusion, retardation of cartilage differentiation, absence of perichondral ossification.*

An interesting attempt to get further light on the chemical mechanisms involved in *creeper* deficiencies was made by Landauer, Upham, Rubin & Robison, who estimated the bone phosphatase in normal and *creeper* chicks. But there was no difference between the two. Nor did the injection of aqueous

---

* Mandible rudiments, however, developed normally, just as they do in *creeper* homozygotes thus showing that conditions which prevent normal ossification of cartilage do not prevent ossification of membrane bone.

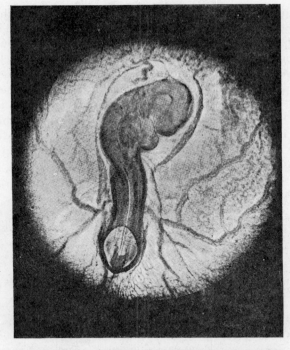

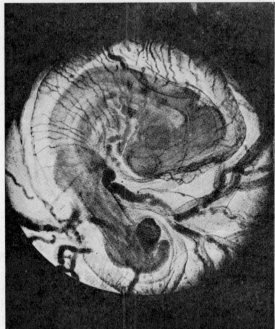

Fig. 227. Retardation of growth in *creeper* chick embryos.

bone extract to heterozygous *creeper* chicks have any effect on the growth of their limb bones. Sex hormones are perhaps unlikely to be involved in the syndrome, as castration has no effect on either normal or *creeper* bone growth (Landauer[21]).

The relation between vitamin-D deficiency and *creeper* chondrodystrophy has also been studied by Landauer[15], who found that *creeper* chicks show symptoms of rickets some time before normal ones, and that their rickets is more pronounced. The chondrodystrophic condition prevents the ameliorating responses to D-deficiency which occur in normal embryos. Similarly, Landauer[9] showed that irradiating eggs with ultra-violet light had no effect on homozygous *creeper* embryos, but improved the hatch of the heterozygous ones to the same extent as that of normal embryos.

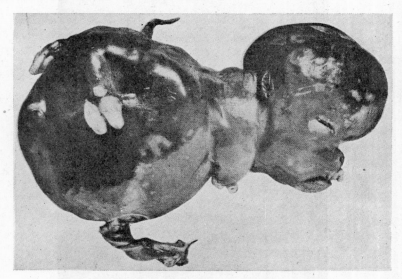

Fig. 228. Achondroplasia in the "bull-dog" calf.

The homozygous *creeper* condition is, as has already been said, lethal. It therefore invited explantation or transplantation experiments, designed, like those of Hadorn and Ephrussi already described (p. 359), to test what powers of normal differentiation existed in the homozygote unable to manifest themselves on account of its lethality. This problem was studied by P. R. David, who found that isolated fragments of the homozygote embryo, representing practically every tissue of the body, would, when cultivated *in vitro* or as chorio-allantoic grafts, survive long beyond the ordinary lethal period of their donor. Heart fragments showed an unusually low growth rate, but the other tissues behaved normally in this respect. David concluded that the death of the homozygotes was due, not to cellular inviability, but to some dislocation of normal correlations caused by the growth-rate changes and the differential reactions to them.

Analogous experiments, due to Hamburger[2], consisted in transplanting homozygous *creeper* limb buds at $2\frac{1}{2}$ days' development to normal embryos. Survival, long after the presumptive death of the donor, was observed, but marked *creeper* effects, such as phocomelia, occurred. In control transplants (normal to normal) growth inhibition took place but no *creeper* effects. All these experiments show that the cause of lethality resides outside the limb area. Whether a general inhibition of growth will really account for the effects is still doubtful.

It seems that there is a general similarity between the Creeper breed, Dark Cornish (Landauer[18]) and Japanese Bantam (Landauer[24]) in homozygotic lethality.

In mammals achondroplasia is of course well known, and the great monograph of Rischbieth & Barrington proved that in man it is inherited. Achondroplasic dwarfs have a complete absence of the cartilage columns in growth zones of the long bones, and a mucoid degeneration of the matrix. In cattle, the Irish Dexter breed has long been known to produce achondroplasic foetuses, which are called "bull-dog" calves. Fig. 228, taken from Crew's anatomical study of them[1], shows the strange appearance which they possess, with the feet close to the body owing to the absence of limb elongation. The heterozygous Dexter animals show the same blood picture as that of chondrodystrophic chicks (Landauer & Thigpen). Bull-dog calves occur in other breeds besides the Dexter (Carmichael) and Lipsett reports achondroplasia in urodeles.

2·6545. Hereditary tumours

The *creeper* condition is demonstrably one in which a general growth retardation bears particularly heavily upon those organs which are growing most rapidly at the time when it occurs, namely, the limb-buds. When we turn to the second class of individuation field anomaly, the hereditary tumour, we find an opposite state of affairs, namely a specific growth acceleration of certain small groups of cells.

It is in *Drosophila* that the classical hereditary tumours occur. In 1916 Bridges[1] noticed that a certain sex-linked gene, *lethal*-7, caused the appearance of intense black spots within the larvae, which died in great numbers. Closer analysis of the condition was undertaken by Stark[1], who found that the black spots were growths resembling melanoma. They originate mainly in the imaginal discs (Fig. 229 shows a tumour developing in the two anterior discs), but also in the proventricular ganglion or the salivary glands. A single larva may carry as many as fifteen tumours, and metastases along the dorsal aorta occur. When minute pieces of the tumours were transplanted into adult flies, all died from extraneous infection, but in two cases considerable growth of the tumour occurred first. *In vitro* the tumour continues to form pigment. Stark was unable to isolate any organism or virus from the tumours, but found on the contrary that they appeared just as usual when the rearing of the larvae was done under aseptic conditions. They appeared to be unaffected by X-radiation.

Several years after the first enquiry of Stark, the *lethal*-7 stock was re-examined (Morgan, Bridges & Sturtevant) and the surprising discovery was made that although the gene was still lethal, the tumours had disappeared. Nevertheless the tumour could always be brought out again by crossing with other stocks. Appearance of the tumours, therefore, is not a *necessary* condition of this lethality. In a recent paper, Stark[3] has given improved microphotographs of the tumours, but proposes a theory of their origin from embryonic "rests" which is not very illuminating. However, when it is remembered that extensive infiltration of surrounding tissues from the tumours seems to occur, it is hard to understand why the statement is so often made that cancer does not occur in invertebrates (cf. also the tumours of annelids, described on p. 259).

In 1919 Stark[2] discovered another kind of tumour in *Drosophila* as a mutant, also a melanoma, but not sex-linked and not lethal. It develops in early or late larval stages, and may replace a wing or a leg, but its persistence into the adult stage in abdomen or thorax does not seem to decrease the length of life. Aseptic transplantation to normal larvae did no harm to either; the donors completed their metamorphosis without tumours, and the hosts continued to live with the melanoma within them. The genetics of this strain were subsequently studied by Stark & Bridges.

Still another hereditary melanoma of the fruit-fly was discovered by Gowen[2].* It is sex-linked and lethal to adult flies, which carry it only in their appendages; it does not appear in the pupa at all and is certainly not contagious. This tumour was of X-ray mutant origin; Rapoport[1] describes several further types, all lethal, and similarly produced. X-ray treatment, in the hands of Enzmann & Haskins[1], has produced a fourth melanoma strain, in which the tumour occurs only between the post-alar bristles and varies from a small black button to a large horn-shaped excrescence as big as the fly itself. This tumour, on the other hand, was found to be sensitive to X-rays.† Enzmann & Haskins[2] also found that a persistent larval spiracle may, under the action of X-rays, express itself either as a melanotic

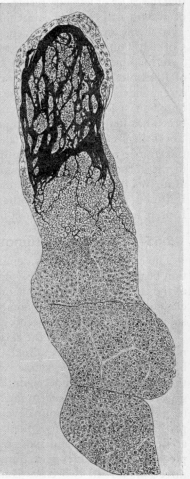

Fig. 229. Hereditary melanoma of *Drosophila* imaginal rudiment.

tumour (most commonly), or as small wing-shaped structures, halteres, or orifices fringed with bristles. Hereditary tumours in butterfly stocks have also been described (Federley). The hereditary tumours of insects offer a most inviting field for future work.

* In an interesting review, Gowen[3] has related the genetically induced tumours of insects to inheritance of resistance to cancer in mammals, and even to inheritance of resistance to infectious diseases. For further information on the benign larval melanomata see Eliz. S. Russell.

† Phenocopies of these melanotic tumours have been produced by Rapoport[2] using As compounds.

It is not generally known that hereditary tumours have been described in fishes (Reed & Gordon; Gordon; Gordon & Smith). Like those of the fruit-fly, these too are melanotic, resembling melano-sarcoma in their final stages; they arise from a hyperplasia of the epidermal macromelanophores, which later invade the underlying muscles. They may appear at any time in life, but if in the late embryonic stages they are invariably lethal. No metastases occur. The tumour-bearing strains which have been most studied occur on crossing Mexican platy-fishes (e.g. *Platypoecilus maculatus, variatus, xiphodium*, etc.). A review by Kosswig on hereditary tumours in fishes is available, and G. M. Smith in

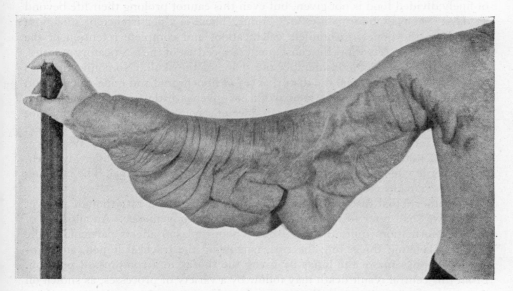

Fig. 230. Neurofibroma simulating elephantiasis (v. Recklinghausen's disease).

describing a tumour carrying a red pigment in *Pseudopleuronectes*, gives a guide to the plentiful literature on fish "cancers", some of which may be suspected to have a hereditary basis.

It will be remembered that in man also hereditary tumours are known, e.g. the neurofibroma simulating elephantiasis in v. Recklinghausen's disease, for which the short note of Anzinger and pathological books such as that of Bland-Sutton[2] may be consulted (see Fig. 230). Into the whole question of the inheritance of resistance to cancer we cannot, of course, attempt to go here, but the review of C. C. Little[3] will be a help in the exploration of the literature.

2·655. Genes producing anomalies of embryonic metabolism

In a sense all the lethal genes which have so far been described might well be included under this heading, for their effects are undoubtedly accompanied by various metabolic changes, and if they affect organiser action directly, as some of them seem to do, this action is likely to be on the chemical properties or relationships of the inductor substances. But there are cases where a single gene has been shown to produce a wide variety of effects, some of which may be shown to be caused by others, but the totality of which is difficult to relate to any common factor. In these "pleiotropic" cases, it is likely that the fundamental anomaly is metabolic, manifesting itself in different ways.

Two such cases have been given intensive study by Grüneberg. He described[1] an autosomal recessive gene[3] in the mouse, *grey-lethal*, which suppresses the formation of all yellow pigment in the coat and kills the homozygotes at the age of 22–30 days. The animals fail to grow normally and die at weaning if a supply of finely divided food is not given, but even this cannot prolong their life beyond 42 days. Skeletal abnormalities are serious. There is a persistence of spongiosa spicules in the bones, incomplete calcification, and complete retention of the teeth[2]. This tooth retention, the consequent reduced use of the masticatory muscles, etc. ends by distorting considerably the shape of the skull. No connection has as yet been found between the disturbances of lipochrome and bone-calcification metabolism respectively, but later work has given a little further information. Thus Grüneberg has shown that the absence of secondary bone absorption is not due to any lack of osteoclasts, and has noted a degeneration of the thymus cortex. Injections of thymus, however, had no good effect. According to Watchorn, *grey-lethal* mice are characterised by low blood phosphate, low liver glycogen, and rather low blood sugar, though the muscle glycogen is normal. The meaning of all this remains enigmatic.

The other lethal discovered by Grüneberg[4] was in the rat; though not yet named, we shall refer to it here as *emphysemic* for convenience. An anomaly of cartilage formation (quite distinct from achondroplasia) affects particularly the thoracic skeleton, thickening the ribs, narrowing the tracheal lumen, and producing emphysema of the lungs by fixing the thorax in a compressed position. From this central result death may follow by a variety of processes, as shown on the accompanying diagram (Fig. 231), taken from Grüneberg. As yet nothing definite is known about the anomaly of the cartilage which sets the whole train in motion, but the abnormal cartilage behaves selfwise when transplanted to a normal body, and *vice versa* (Fell & Grüneberg). In explants the abnormalities do not appear.

A very different sort of pleiotropic gene, the effects of which will be even more difficult to unravel than these mammalian ones, is the *dia* gene of the waxmoth, *Ephestia*, discovered by Strohl & Köhler. This involves depigmentation of the body, shortening of the length of adult life, and a great reduction in fertility of the female moths.

We may now pass to some instances where the effect of the gene seems to have been tracked down with some precision to a definite metabolic anomaly. A disturbance of normal water and protein metabolism seems to be involved in the action of the gene *sticky* of chicks (Byerly & Jull[1]), a lethal which stops hatching. There is no absorption of the amniotic liquid, which contains much protein, and there is no withdrawal of calcium from the shell for the bones, as normally. The bones have a certain chondrodystrophic appearance and calcification is delayed

(T. P. Sun). The gene is recessive, not dominant, like *creeper*. According to Ogorodny & Penionskevitch vitamin-B_2 deficiency is connected with this anomaly.

Disturbances of keratin metabolism seem to be involved in the action of another chick gene, *frizzle*, which has received much study at the hands of Landauer[5] and his school.

Frizzle in the homozygous condition was proved by Landauer & Dunn[1] to be lethal in 1930, though this had previously been suspected. The heterozygote forms the long-known Frizzle breed, of wide geographical distribution. *Frizzle*

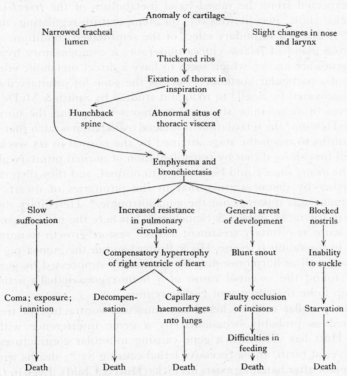

Fig. 231. Multiple effects of the *emphysemic* gene in the rat.

feathers are curled and quite abnormal in appearance; the homozygote has none at all for the first year of its life, and though later it grows a sparse plumage it does not usually reach sexual maturity. Determination of the anomaly is evidently complete by hatching, for skin transplants to normal chicks always behave selfwise (Landauer & de Aberle). When the abnormal physiology of *frizzle* chicks was analysed, it was found that the excessive loss of body heat causes a striking increase of the basal and general metabolism (Benedict, Landauer & Fox). Although *frizzles* show a large increase of food-intake and reduced rate of water-loss, they usually cannot balance their metabolic requirements. Correspondingly, the *frizzle* heart rate is raised much above the normal (Boas & Landauer[1]) and so is the actual weight of the heart (Boas & Landauer[2]). The condition, in fact, resembles Graves' disease (Landauer[20]), though the abnormality is peripheral, and this resemblance extends to the blood picture, which was studied by Landauer & L. T. David[1]. The position of the thyroid gland in the *frizzle* state is complicated (Landauer[13]; Landauer & L. T. David[2]; Landauer[16]; Landauer & de Aberle), it may be secreting too little thyroxin or too much, depending on environmental temperature, and it is generally hypertrophied. A very complete study of the weights of all organs in normal and *frizzle* chicks at different ages has been made by Landauer & Upham; it agreed with what would be expected from the raised basal metabolism of the *frizzles*. Finally, *frizzle* chicks show peculiarities of the temperature-regulating mechanism (Landauer[22]); this is a secondary effect of the semi-naked condition.

The thyroid gland of *frizzle* chicks undergoes a compensatory hypertrophy. But other genes are known which seem to have a direct metabolic effect on the production of a particular hormone. Such is the gene for pituitary dwarfism in mice, first discovered by Snell[1] in 1929 and studied by Smith & McDowell. The growth curves of *dwarf* mice show an enormous lag behind the normals, and according to Dawson[1] the retardation of skeletal ossification is such that the dwarf takes six months to reach the stage attained by the normal in six weeks. Smith & McDowell first showed that by transplantation of normal pituitary glands, the growth of the *dwarf* mice could be restored to normal, and they then carried the analysis further by demonstrating that in the pituitaries of dwarfs only the growth hormone was missing, not the gonadotrophic.* Hereditary dwarfism is known also in the rat (Lambert & Sciuchetti), but here the mechanism seems to be more obscure as pituitary treatment did not restore growth to normal. It is known also in the rabbit (Greene, Hu & Brown) and in the guinea-pig (Sollas).†

Enzymes as well as hormones may be specifically suppressed by genes. Thus Glembosky found the essential cause of a homozygous lethal's action in the caracul sheep to be the absence of gastric rennin.

Lastly, the peculiar cases of hereditary muscular contracture or tremor may be mentioned as probably occasioned by a genic interference with muscle chemistry. Hutt has described a gene causing muscular contractures and the death of calves at birth, and a recessive lethal causing 85 % deaths with violent muscular tremor after hatching exists in chicks (Hutt & Child). Even in *Drosophila* a *shaker* strain has been described (Lüers). Cole & Ibsen's congenital palsy in guinea-pigs presumably belongs here too.

That quite subtle metabolic differences are affected by genes is of course well established. Avoiding eye- and coat-colour examples as somewhat hackneyed, mention may be made of the *Drosophila* strain found by l'Héritier & Teissier

* Prolactin and the thyreotrophic will act instead of, or perhaps because they *are*, the growth hormone here (Bates, Laanes & Riddle). The Californian school nevertheless maintains the absolute separate existence of the growth hormone (Evans; Meamber, Fraenkel-Conrat, Simpson & Evans). Others, e.g. the Montreal school, have suggested that the pituitary hormones may all be prosthetic groups of the same protein.

† Perhaps the most beautifully analysed case of hereditary dwarfism is in plants, not animals. Van Overbeek found that the gene *nana* in maize produces dwarfism by causing excessive peroxidase activity, which is associated with abnormally great auxin destruction, and hence dwarf growth.

which was sensitive to pure CO_2. In general the gas has only a reversible anaesthetic effect on flies, but in this strain the damage done by it was irreversible. Similarly, Lamoreux & Hutt have discovered genetic susceptibility to vitamin-B deficiency in the chick. But to pursue this subject further would be to leave the realm of lethal factors interfering with morphogenesis.

2·656. Genes killing or deforming the embryo by unknown mechanisms

Into this category fall perhaps the majority of lethal genes at present known. Many lethals in *Drosophila* have been reported (e.g. in the papers of N. P. S. Dobzhansky; Kaliss; or J. C. Li[1], or in the monograph of Morgan, Bridges & Sturtevant), but we understand practically nothing about their mode of action. Or we might take as examples the lethal genes in the grouse locust *Apotettix eurycephalus* described by Nabours & Kingsley. Some of these embryos develop normally till four days before hatching, and then die showing no irregularities or deformities whatever. The same sort of mystery surrounds the gene *hooded-lethal* of mice described by Crew & Kon. This recessive gene kills the newly born mice in the second week of lactation by inanition, but no causal mechanism has been found. In birds, too, lethals of unknown action have been described. Dunn[1] found a lethal gene linked with recessive white plumage in Wyandotte fowls which destroys the embryos in the first week of development. Upp & Waters suspect the existence of a sex-linked one in White Leghorns.

It is evident that the field of the embryology of lethal genic action offers a vast scope for work on "reversed" experimental embryology, particularly hopeful owing to the recent advances in experimental embryology itself, and the aid which biochemistry is beginning to be able to give.

2·66. The kinetics of gene action

With the subject of this division we find ourselves right in the middle of the debatable land between embryology and genetics. Those who have no training in the niceties of genetic analysis may be forgiven for abiding in their own regions of biochemistry or morphology as the case may be, but nevertheless some account of the mechanism of action of genes in embryonic development is indispensable in a book of the scope here planned. Readers who wish for more information will of course find it in the standard works on genetics.*

Chemical embryology is involved here for two reasons. In the first place, some chemical reactions at the phenotypic end of the chain of gene causation are fairly well known. But often, as in the case of pigment production, these may not be very interesting from the morphogenetic point of view. Secondly, the effect of external agencies, such as temperature change, on the development of the phenotype, has shown that the intermediate processes in the chain of gene causation are themselves chemical reactions. If the end-result is a morphological change

---

* It may be repeated at this point that in these matters the present book expressly disclaims any pretensions to being an exhaustive examination of the relevant literature.

this fact is therefore very interesting. We shall begin by discussing the second point and finish with the first one. One realises to-day that the development of an organ or part is a complex of processes presided over at every successive stage by the appropriate genes. Perhaps the fullest illustration of this is Waddington's work[23] on wing-differentiation in *Drosophila* in which sixteen separate but not necessarily independent processes can be distinguished, under the control of some forty different genes.

2·661. Phenocopies and rate genes controlling morphological processes

It has already been mentioned that changes in the phenotypic end-result brought about by genetic changes may be imitated by externally applied agencies, or by spontaneous factors normally occurring. Such effects may only be distinguishable from the genetic effects by the fact that they are not inherited. The previous section has afforded a number of examples of these *Phenocopies*, as the following table shows:

Genetic effect	Phenocopy
Hereditary otocephaly (Wright)	Spontaneous teratological otocephaly
Hereditary pseudencephaly (Snell)	Failure of neural tube to close in the absence of sufficient mesoderm (Holtfreter)
Hereditary rumplessness (Landauer)	Rumplessness due to abnormal temperatures (Danforth)
Hereditary failure of cornea induction in Bagg-Little strains (Bonnevie)	Failure of cornea induction in blind cave forms (Schlampp)
Hereditary aristapedia (Balkaschina)	{Aristapedia produced by boron (Rapoport) {Regenerative aristapedia (Brecher)
Hereditary melanoma (Stark; Gordon)	Melano-sarcoma
Hereditary achondroplasia (Landauer)	Experimental achondroplasia (Fell & Landauer)
Hereditary wing pattern (Goldschmidt)	Experimentally produced wing pattern (Goldschmidt)

The last entry in this table leads us from the study of lethality and its deep-seated dislocation of embryonic processes to those lesser changes within the borders of the "normal" which are the study of geneticists.

In the nineties of the last century it was found by a number of workers (e.g. Dixey; Standfuss) that the wing pattern of butterflies can be changed by the action of abnormal temperatures during the pupal period in such a way that the experimentally produced pattern cannot be distinguished from that of other races of the same species. Thus Central European *Papilio machaon* pupae subjected to unusual temperatures produced butterflies almost indistinguishable from the Syrian *Papilio sphyrus* or the *Papilio centralis* of Turkestan. At that time, owing to the influence of vague evolutionary speculations, the real significance of the effect was not noticed,* and it was left for Goldschmidt[2] in 1917 to point out that as temperature might most reasonably be expected to influence rates of chemical reactions, it was probable that the action of genes in producing patterns was due to their influencing such reaction rates. This early work was not confined to insects. Beebe had found in 1907 that changes in the feathering

* Spemann's remark, made in another context, would fit here: "Man kann eine Entdeckung machen ohne es zu wünschen, aber doch nicht ohne es zu wissen."

of doves such as *Scardafella inca* could be produced by the action of unusual humidities during the early post-hatching period, shifting the appearance of the animal towards types of more tropical origin.

Goldschmidt[6] later on made a systematic investigation of the phenocopy phenomenon in *Drosophila*, using high temperatures applied at different times during development. In this way phenocopies of very many known mutations were produced, such as bristle changes, wing changes, and even the production of benign tumours. In every case the exact conditions for producing a large percentage of a given phenocopy are now known. They include the following principal factors: (1) the developmental stage at which the abnormally high temperature is allowed to operate, (2) the time during which it is applied (6–24 hr.), (3) the exact temperature (35–37° C.), (4) the genetic constitution of the "normal" (wild-type) animals used. It has since been shown that exposure to abnormally low temperatures (Gottschewski), to X-rays (Friesen) and to high-frequency electric current (Malov & Friesen), will similarly produce phenocopies, but naturally they are not the same ones, and the exact conditions vary.

That rates of reaction now occupy the centre of the picture is not remarkable in itself, since any interpretation of the facts of embryology must contain the time variable, however disguised. But the essential point is that the normal integration of all the reaction rates in the embryo is subject to interference. Such interference may be brought about by external agencies at the will of the experimentalist, or by internal genic change. Thus in a former section, we had occasion to notice with special interest a case in which the onset of a determination process was shifted with relation to visible morphological differentiation by the action of temperature (Twitty, see p. 337). Later on we shall examine a number of other cases which demonstrate the "dissociability of the fundamental processes in ontogeny" (p. 505). That genes are catalysts or producers of catalysts (enzymes) or producers of inhibitors is a doctrine now generally accepted, largely due to Goldschmidt's advocacy[8]. And if gene changes shift the rates of reactions during embryonic development relatively to one another, so also can external agencies, particularly changes of environmental temperature. In this sense all genes are "rate genes".

A caution might perhaps be interpolated here. Wright[5] and his collaborators, for example, in analysing their remarkable series of gross hereditary morphological anomalies of the head, have tended to use the concept of rate genes to explain the distortions, suppressions, etc. as due to inhibitions of "developmental rate" at various moments of development. But this is a relatively vague notion, much vaguer than that of a definite chemical reaction such as the production of a pigment from a chromogen. And it has in fact been unsatisfactory in so far as it has left out of consideration the primary and second-grade organisers of the head, i.e. certain chemical substances, some of the properties of which we already know. When an evocator acts from an inductor upon a competent reactor there are several types of reaction to be considered; that which synthesises it chemically; that which liberates it from an inactive precursor; that which conveys it to its destination whether by diffusion or otherwise; and finally those which it itself sets in motion or catalyses. In addition to these, there are the actual differentiation rates of the inductor and reactor tissues. A Stockardian or Childian "developmental rate" is unsatisfactory, because it confuses all these things

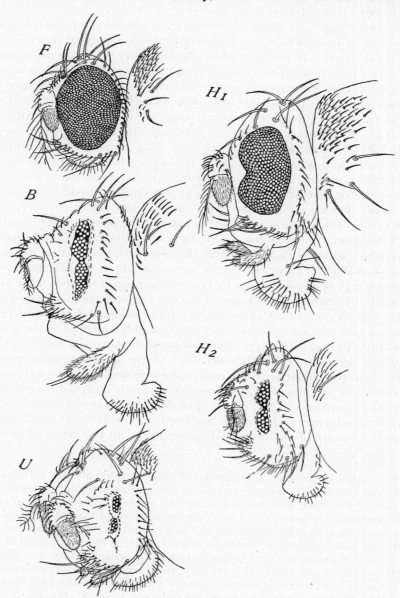

Fig. 232. Effects of *bar* genes in *Drosophila*. *F*, full-eye wild type; *B*, white *bar*; *U*, *ultrabar*; H_1, heterozygous *F/B*; H_2, heterozygous *F/U*.

together. The inhibition of an organiser action is a quite definite concept, and the nature of such inhibition (of which we have already seen several examples, cf. pp. 223 ff., as well as Wright's otocephaly) demands specific study. But rates of reaction will ultimately be involved none the less.

Controlled environmental temperature can not only produce from a given type (the "wild type") of organism, phenocopies presenting the characteristics of other types (e.g. geographical races); it can also produce phenocopies of quantitatively different members of a series of multiple allelomorphs. Of this the *bar* gene in the fruit-fly is the classical example. As is well known, the fruit-fly has a series of genes all producing more or less reduction in the number of facets in the eye (a morphological but quantitative effect). These effects are shown in Fig. 232, taken from Zeleny's account[2] of *ultrabar*, and in Fig. 233 (opp. p. 372) by some photographs recently published by Kikkawa[2].

The *bar* gene was originally discovered by Tice in 1914 and soon afterwards Zeleny[3] and his collaborators began a systematic study of it. Thus, for example, the following reductions of facet number occur at 27° C. in the female:

	Average number of facets present
"Wild type"	810·6
Infrabar	362·0
Bar	61·8
Ultrabar	22·0

Seyster and Krafka then found that the number of facets in the eyes of flies of known genetic composition varied in accordance with the temperature at which they were cultivated, and that the higher the temperature, the fewer facets were formed. Thus it appeared that for every increase in temperature of 1° C. the number of facets decreased by 10 %. From Fig. 234, taken from Hersh's work[1], it can be seen that the relationship is linear on the semi-logarithmic plot, and that a *bar* fly at 30° C. has as few facets as (i.e. is a phenocopy of) an *ultrabar* fly at about 18°. The homozygous state is also, of course, a factor; thus homozygous *ultrabar* at 15° has the same number of facets as the heterozygote at 25°. Such relationships are known as *Thermophenes*.

A considerable literature has arisen in the detailed study of the *bar* allelomorphs; it may be explored through the papers of Zeleny[3]; Hersh; Luce; Driver; Margolis & Robertson, etc. Here we only need to observe that the action both of increased temperature and of the various genes of the series must be to accelerate differentially some chemical reaction inhibitory of the facet-forming mechanism. Again the parallel with the case of determination of ciliary polarity should be drawn, for there (p. 337) increase of temperature delays the onset of the determination relative to the morphological age.

It seems fairly clear that until we know much more than at present about the nature of the facet-forming mechanism itself, speculation about the action of the *bar* gene series or temperature, and even calculations of temperature coefficients and characteristics on various assumptions, will not be really illuminating. The possibilities are numerous. We do not even know how far inductor effects are involved, and as Margolis has pointed out, the nature of the facet-inhibiting reaction might be a removal of the facet material, e.g. by prior determination in other directions. The only relevant study of the necessary embryology is contained in the careful work of T. Y. Chen, who measured the sizes of the imaginal buds of the eye and other structures in the larvae of normal and mutant flies.* The imaginal buds for the eyes occupy a position in the larva in the metathoracic segment just posterior to the antennal buds and anterior to the larval ganglia.

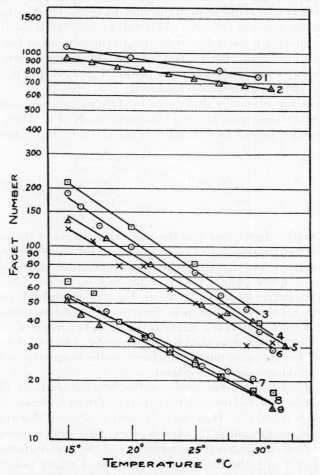

Fig. 234. Thermophene of *bar* eye *Drosophila*. 1, Wild type full; 2, reverted full; 3, unselected *bar*; 4–6, different low selected *bar*; 7–9, various *ultrabar*.

The presumptive facet region is distinguishable in them from a very early time. As the following table shows, the eye buds of the mutant types in which eye size and facet number are reduced, are affected from the earliest time at which their sizes can be measured.

* The elaborate work of F. Bernard on the quantitative embryology (heterauxesis of facets, etc.) of arthropod eyes, should not, however, be overlooked.

	Size of the optic buds in sq. μ	
	48 hr. larva	Mature larva
Normal ("wild-type")	128	535
Lozenge-3 (a gene causing facet-fusion and giving the eyes a glazed appearance)	—	360
Bar	—	232
Eyeless-4 (suppressing eyes in adult)	11	107

In the two last cases other imaginal discs, such as the antennal buds, were demonstrably of normal size, so that the effect is very discriminatory. Chen had some reason for thinking that in the earliest stages the eye buds of *bar* were relatively unaffected, and that the inhibiting influence came on later. This is important in view of the later work of Goldschmidt[6] on *vestigial* wing forms (see below, p. 400).

Chen's work was later amplified by Medvedev[1], who studied the mutant races *lobe-c*, *glass*-2 and *eyeless*-2 which are all characterised by various degrees of reduction of eye size. In addition, the first and third of these genes produce asymmetry in eye suppression, so that one side of the head may have no eye while the other side may have a more or less well-developed one. The results of Medvedev's measurements, summarised in Fig. 235, showed that as early as the 24th hour the differences between the types were visible, but become more pronounced as time goes on. The drop at the 48th hour is due to the fact that before that time the whole cephalic complex (eye bud and antenna bud together) was measured, and after that time the eye buds only. The larval ganglion in the three races and the wild type served as a control; not a trace of difference in size or growth rate was observable in it. It remains to be decided whether inhibition of cell-divisions or some other

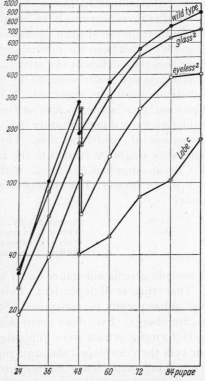

Fig. 235. Logarithmic curves of the growth of the imaginal discs of eye mutants.

mechanism is the basis of this hereditary size limitation of eye buds in mutant *Drosophila* races.

Further light on the production of eye abnormalities by genes in arthropods comes from the work of Wolsky & Huxley[2], who, starting from the fact that in the non-faceted regions surrounding a *bar* or *ultrabar* eye there is a deposition of the usual eye pigment, showed that in these regions ommatidial (sensory) cells exist, but in an atrophied condition. As normal facets only occur in the cuticle

over normal ommatidia, there is a presumptive parallel with lens induction in
vertebrates. The *bar* genes would thus interfere with "eye-cup induction"
(cf. avitaminosis-A, referred to on p. 227, or the otocephalic syndrome, p. 375).
On the strength of this analogy Huxley[14] rectified the interpretation previously
current of Sturtevant's interesting mosaic experiments. Sturtevant[2] found that
if a mosaic spot of genetically wild-type tissue happened to occur in the region
of a *bar* eye, the number of facets was considerably increased as compared with
usual *bar* eyes, as if the wild-type tissue had induced facet formation in adjacent
bar-type areas. Huxley pointed out that as the number of ommatidia was increased
as well, the increase in facet number must have been secondary, and the basic
effect of the wild-type tissue must have been upon the growth rate or determination of the ommatidia in the eye bud at earlier stages.

Parallel investigations have been made on the development of the normal and
mutant eye in the shrimp *Gammarus* (Schatz; Wolsky & Huxley[1]). Some races
(e.g. *red*) differ from the normal only in pigmentation (see below, p. 407); others
have seriously deficient eye structures (e.g. *colourless*).

If facets are really induced by ommatidia, the existence of a facet-forming
substance would be expected. Now transplantation experiments by Beadle &
Ephrussi[2] (see below, p. 408) on *Drosophila* had shown that *bar* eyes, contrary to
wild-type eyes, do not produce the "vermilion substance" required in the pigment
formation of wild-type flies. Thinking therefore that the pigment-forming
substance might be related to the facet-forming substance, Ephrussi, Khouvine
& Chevais supplied *bar* larvae with extracts (prepared from *Calliphora* larvae)
rich in the former. The results were astonishingly clear; as the following table
shows:

	Average facet number
Control *bar* fly	103
With 4 % extract	144
With 8 % extract	244
With 10 % extract	455

A morphogenetic substance under genetic control is therefore probably at work.

That this is independent, however, of the pigment-forming substances
was subsequently demonstrated (Steinberg & Abramovitch; Chevais, Ephrussi
& Steinberg). Eye discs from *bar* larvae which had been fed a *Calliphora*
facet-forming extract were implanted into *vermilion* hosts. Such facet-increased
bar eyes then developed the same pigmentation as eyes from untreated *bar* larvae
(containing many fewer facets) similarly transplanted. Facet number can thus
clearly be affected without affecting the colour eventually attained by the eye,
and the facet-forming or reducing reaction system must be regarded as independent of the pigment-forming system.

The other instructive cases of the interaction of environmental temperature
and genetic constitution take us away from the eye, to the suppression of bristles
and wings. Probably the first mathematical analysis which the phenomenon
received was that devoted to the fruit-fly gene *dichaete* by Plunkett. This gene
removes certain bristles from the fly's thorax, especially the dorso-central ones.
Just as in the case of *bar* eye, increase of temperature was found to intensify the

effect, although the points seem to need not one straight line on arithlog paper but two meeting at an intersection about 25° C. (Fig. 236). Plunkett proposed that there might be an increase, relative to other developmental processes, in the rate of a reaction destroying a bristle-forming substance or forming a bristle-suppressing substance. He was able by calculation to reach the conclusion that the primary effect of the *dichaete* gene was the destruction of an enzyme. But in the absence of any experimental and biochemical evidence these conclusions must remain to some extent hypothetical.* The analysis has since been much refined by G. Child.

When the effects of genes reducing wing size in the fruit-fly were studied, exactly the opposite relation with temperature was found. Roberts[1], and following him, a succession of investigators, showed that increase in temperature during

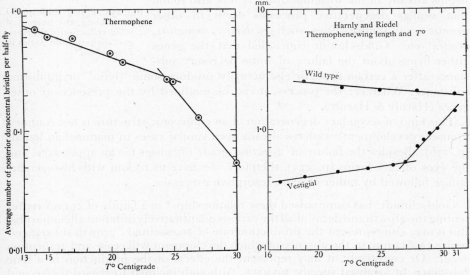

Fig. 236. Thermophene of *dichaete*
Drosophila.

Fig. 237. Thermophene of *vestigial* and
normal *Drosophila*.

the larval stage of *vestigial* flies increased the length of the wing retracing all the steps towards the normal wild-type wing. Fig. 237, drawn from the data of Harnly and Riedel, shows that a sharp increase in length takes place just below 30°. At lower temperatures the breadth is more affected. Wings of wild-type flies, on the contrary, diminish in length with rising temperature, but to a much lesser extent.[†]

The embryology of wing reduction in the allelomorphic series *vestigial* was studied by Goldschmidt[6] with remarkable results. The action of the genes of this series ranges all the way from the normal wing through *notched*, *nicked*, *ragged*, *antlered* and *strap* to *no-wing*. The various stages of the series show greater and greater reductions of wing area starting from the tip and going on until nothing

* Phenocopies of bristle reduction have been produced by Rapoport using Hg compounds.
† See also Hersh & Ward and Li & Tsui.

but a stump is left. These reductions take the form at many stages of a scalloping, i.e. a more intense reduction of areas between the veins. This is illustrated in Fig. 238 by a topographic model based on the data of Goldschmidt (*Physiol. Genetics*, p. 229), where the base represents the normal shape of the wing and the summit the shape of the most highly reduced wing. Goldschmidt now found that the development of the wing proceeds normally at first, but after a certain time degeneration sets in, and a greater or lesser area of the wing is resorbed. The earlier the process begins, the more complete is the abolition of the wing; where *no-wing* is concerned for instance, it begins already in the imaginal disc. It was also found that similar degenerative processes occur in other genetic wing abnormalities, such as *dumpy, miniature, beadex*, etc. Goldschmidt concluded that the genes either bring about the failure of some necessary substance after a certain time, or else actually produce some "lytic" or inhibiting substance. Whatever the process, it can be modified by the presence of other genes (Harnly & Harnly).

Fig. 238. Topographic model illustrating the stages of wing-reduction in the *vestigial* allelomorphic series in *Drosophila*.

This kind of secondary degeneration of an embryonic structure is not confined to insect development; we have already seen similar cases in mammalian lethals (p. 376). Besides the failure of a second-grade organiser for an appendage, and the eyes of cave forms (p. 225), therefore, we have to reckon with histogenetic failure followed by rather obscure resorptive processes.

Goldschmidt[7] has summarised these relationships* in a family of curves representing the growth and decay of all the various quantitatively different allelomorphs. But it may also represent the production rate of an essential "growth substance" such as a vitamin, in the absence of which development will cease and resorption occur. Or conversely, it may represent the effects of the production of a lytic substance, of localised specific toxicity. Although inclined to regard the second alternative as the less likely of the two, Goldschmidt drew attention to a rather striking analogy which would exist between the decay of the wings in the *vestigial* allelomorphs and the destruction of a population of bacteria by bacteriophage. Where bacteria and phage are present together both are increasing ("growing"), but the phage is increasing faster than the bacteria. When, therefore, a certain phage/bacteria ratio is reached, known as the lytic threshold, a rapid lysis of the bacteria begins and continues till all are destroyed. This threshold can be shifted at the will of the experimentalist, e.g. it varies according to the presence or absence of traces of manganese. Hence the bacterial growth curve corresponds to the development of the wings of the insect, and the lytic threshold corresponds to the point of maximum development reached by any given mutant wing before it begins to degenerate. This we might call the *vestigial* threshold. Just as the lytic threshold can be changed by external factors such as metallic ions, so the *vestigial* threshold can be affected by temperature, for, as we have seen, the higher the temperature, the larger the eventual wing, and so the higher the threshold must have been.

* Other workers, it is fair to say, do not interpret them as Goldschmidt does—see the recent paper of Waddington[23].

It may be useful here to state the threshold concept, which is much used in genetical work (cf. Goldschmidt[8], p. 65), in embryological terms. As long as the amount of an inductor (hormone, organiser, evocator) is below a certain level, no effect will be produced (cf. Fig. 45). But conversely, when the competence of the reactor has been completely implemented, no further increases in the amount of the inductor will have any effect. Thus in the present instance, no increase in the growth substance, or decrease in the lytic substance, will give rise to wings larger or better developed than the wild-type wings (except the wings of related geographical races, which may be slightly larger than those of the type chosen as "wild type").

Goldschmidt[8] thinks, probably rightly, that the threshold concept may explain one of the most puzzling of the effects of the *vestigial* allelomorph series, namely the fact that not all these genes show the same statistical uniformity in their results. Some affect only a small percentage of the animals carrying them, others affect all. The following table, taken from Mohr[2], illustrates this; it reads from the least to the greatest effect:

	Arbitrary numerical wing-suppression factor	Percentage flies showing the effect
Wild type	40	—
Nicked-notched	37	0·2
Vestigial-no-wing	36	1·3
Nicked	32	27·1
Notched-notched	30	42·4
Nicked-no-wing	28	70·7
Notched	25	100
Notched-no-wing	21	100
Vestigial	20	100
No-wing	16	100
No-wing-no-wing	12	100

In this particular case, the weakest genes are also the least certain in their action, but this relationship might not always hold. In order to distinguish between these two phenomena, Timofeev-Ressovsky introduced the terms "penetrance" and "expressivity". Penetrance means the extent to which every animal carrying the gene is affected by it (percentage phenotypic effect). Expressivity means the extent of the effect itself. Thus *nicked* has a relatively mild expressivity and low penetrance, while *no-wing* has a severe expressivity and full penetrance.

Goldschmidt's detailed explanation for the wing-reduction series is that in the minor degrees of wing reduction the *vestigial* threshold is very high, and hence rather late in development. Hence in many individuals the dangerous period might safely be passed before the reduction in the amount of growth substance (or the appearance of appreciable amounts of lytic substance) had had time to become serious. Hence the low penetrance. Where the expressivity is high, the differentiation processes must have been interfered with at much earlier stages, and the dangerous periods must have been much longer; hence the statistically higher penetrance.

This point of view is perhaps supported by the circumstance that we have already met with some facts, not connected with genetics at all, which seem to fit in to the framework of penetrance and expressivity. If reference be made to Fig. 94, it will be seen that in general the effect of a given hydrocarbon in neural

induction in amphibia is proportionally upon the number of neural tubes produced in a series *and* the average volume of the neural inductions so produced.
Only one hydrocarbon was found (3:4-benzpyrene) which, under the same
conditions as all the others, consistently gave rise to a rather large number of
relatively small neural tubes. In other words, its expressivity was mild but its
penetrance was high. The other hydrocarbons of course showed regularly
increasing degrees of penetrance and expressivity. We have therefore to recognise
that in the production of a distinctively morphogenetic effect by pure chemical
substances there may be a dissociability between these two forms of biological
reaction. And this would be in general agreement with the view that the action
of the genes is mediated by chemical processes whether of enzymic or hormonal
nature.

Here we cannot go further into the genetics of penetrance, etc., but ample
information is available in the genetical literature.

2·662. Phenocritical periods

Throughout the foregoing discussion, the effect of environmental temperature
in producing phenocopies, etc. has been considered as acting at all stages of
development. But it was early suspected that some stages of development would
be more sensitive to temperature than others, and this proved to be the case.
The stimulus of raised temperature has to be applied during a relatively short
period of larval life preceding pupation. Exact data are now extant for this.
(Driver on *bar* eye reduction and facet suppression; W. F. Stanley on *vestigial* wing
reduction). The sensitive or "phenocritical" period for *bar* begins at the time
when the cephalic complex is separating into the eye bud and the antennal bud,
and ends about the time of formation of the first facets. The sensitive period for
vestigial begins at the third instar (larval moulting period) when the wing buds
are just being formed. Now as rise of environmental temperature can, as we
have seen, counteract the wing-reduction process (either by differentially increasing
the rate of a reaction furnishing growth substance or by differentially increasing
the rate of a reaction destroying lytic substance), it must take effect actually in
the larval wing buds. Phenocritical periods and thermophenes have also been
carefully delimited for *scute* (a bristle-reducing gene in *Drosophila*) by G. Child,
for *short-wing* by Eker, for *mottled-eye* by Surrarrer, for *eyeless* by Baron, etc. etc.

With the phenocritical period we can link up what has been said in earlier
sections regarding lethality (p. 365) and competence (p. 116). The *vestigial*
allelomorphs represent a series of genes the effects of which go farther and farther
back into development in accordance with the severity of their final effects.
No-wing in the homozygous condition is actually lethal. It is lethal because it
gives rise to great reductions in the size of the thorax, and it does this because
it affects the wing rudiments at a time so early that they have not as yet become
separated from the thorax segment material, also contained in the same imaginal
disc. If a gene acts sufficiently early in development, therefore, it will affect
tissue still in a highly competent state, still to a large extent undetermined; and
hence the dislocations which it produces will be all the more severe. In the

vestigial series, only those genes which have effects less severe than *snipped* are affected by temperature changes occurring during the main phenocritical period; the others have already by that time done their work, or at any rate brought it to a point at which it is irreversible. Hence other sensitive periods, much earlier in development, should exist; Goldschmidt[8] suggests that perhaps they occur before each moult.

The phenocritical period has been intensively studied in relation to the formation of pattern, especially in the wings of insects. For a detailed account of this the valuable review of Henke[2,3] should be consulted, and for short summaries the articles of Strohl[3]. Outstanding studies are those of Kühn & Henke[1] on the wing-colour patterns of the flour-moth *Ephestia kühniella* and those of Köhler & Feldotto on butterfly wing patterns (*Vanessa urticae* and *io*). Goldschmidt ([8], pp. 201 ff.) has tried to assimilate these wing patterns to the pattern of morphological structure produced by the primary organiser of vertebrates, but although differences have been found between areas on the wing preceding the deposition of pigment in these areas (Goldschmidt[3]), we hardly know enough as yet of the mechanism of determination in insect metamorphosis to be dogmatic about this. It is at any rate clear that the pattern is not finally determined at the beginning of pupation, for Henke[1] showed that extirpation of pieces of the wing-forming imaginal discs in the butterfly *Philosamia cynthia* led to a regulative rather than a mosaic development of pattern later. Similarly phenocopies of pattern may be produced by treatment of the pupae of *Ephestia* at the pheno-critical period. Thus the action of genes which shift the two symmetrical bands on the wings centrally, and so suppress the central field, can be imitated by subjecting the pupae from 48 to 60 hr. after pupation (at 18°) to a temperature of 45° for 45 min. (Kühn & Henke[1]), see Fig. 239.

Investigators of wing pattern find that pigmentation seems to start from one point at the base of the wing and spread outwards from there. Or the pattern may be polycentric, as in many ladybird beetles. They therefore speak of a "determination stream" of pattern, which recalls the elongation and shortening of the primitive streak in the chick, and the flow anteriorly and posteriorly of neuralisation from the point of first appearance of the neural folds in amphibia. But the deposition of pigment in a pattern must not be confused with the determination of that pattern. Until further work of an experimental kind has been done on insect wings, the expression "determination stream" might well be used with caution or replaced by "differentiation stream". For further details on this interesting subject the reader must be referred to Goldschmidt's book[8] (esp. the elaborate chapter on pattern). It may be that processes such as those involved in the formation of Liesegang rings will in time be found to occur in the origin of pigmentation patterns, as was first suggested by Gebhardt in 1912.

From what has so far been said it will be clear that external agents such as temperature or spontaneous non-genetic internal agents (as in spontaneous terata) can imitate the action of certain genes. They are therefore to be supposed to affect rates of reaction in the embryo, or to produce hormones, catalysts, or inhibitors which do so. But since 1916 Goldschmidt[1] has applied the same kind of explanation to genetic dominance. "A dominant gene controls a reaction of a certain velocity; the recessive allele, a reaction of lower velocity, and the

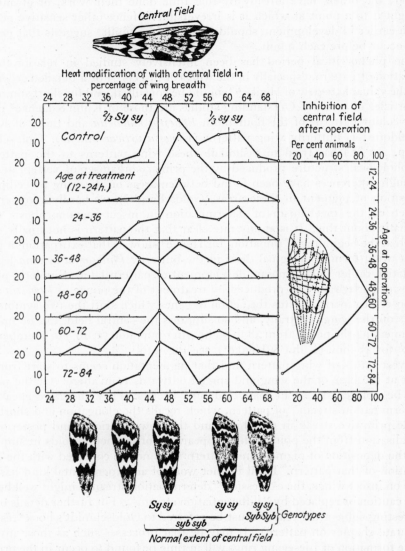

Fig. 239. Suppression of the central pattern-field of *Ephestia* by genic action and by exposure to high temperature during the phenocritical period.

heterozygote an intermediate reaction.... If dominance is controlled by a developmental system, it should be possible to shift the phenotype of the heterozygote by environmental action in the same way as the mutant phenotype may be secured as a phenocopy." And just as external agencies can shift dominance, so there are certain modifying genes which can do so (dominigenes). But these subjects belong to genetics rather than embryology, and for the details, which are convincing enough, the book of Goldschmidt ([8], pp. 99 ff.) must be consulted.

2·663. Rate genes controlling chemical processes

It only remains to add in this section some further examples of genes which have been shown to affect rates of reaction during development. They mostly concern pigmentation* (cf. the monograph of Stern[1] on multiple allelomorphic series).

Wright[1] and Castle & Wright first showed that a series of this kind exists in guinea-pig coat-colour, where all variations between albinism and maximal melanin deposition may be found. Earlier workers such as Cuénot[1] in 1902 had believed that albino animals contained the chromogen but not the melanin-forming enzyme. The direct tests of H. Onslow[1] in 1915 showed, however, that the enzyme itself was missing from recessive unpigmented areas of skin, while in dominant whiteness a specific inhibitor for the enzyme was present. This was confirmed by Gortner[1] and more recently amplified by Koller[2], with the following general conclusion. In recessive whiteness, mammalian skin cannot make melanin from chromogen, nor can it inhibit melanin formation by other skin extracts. In dominant whiteness, mammalian skin cannot make melanin from chromogen, not because the enzyme is absent, but because it is inhibited; and this inhibitor will prevent melanin formation by other skin extracts.† In recessive blackness melanin formation sensitive to the inhibitor occurs, while in dominant blackness the melanin formation is not inhibitable by the dominant-white inhibitor. Little or nothing seems to be known about the chemical nature of the inhibitor, but after death it may disappear so that the blackening of genetically colourless cells follows, as was found in some old but interesting experiments of Riddle & la Mer on the eyes of embryos of different races of ring-doves. This blackening would not take place if death occurred before the time at which eyes of pigmented races begin to acquire their pigment, suggesting that in the earliest stages the enzyme had not appeared. Post-mortem pigment formation required oxygen, was not inhibited by mercuric chloride, but was stopped by boiling the tissue.

All these facts fit in well enough with the changes in coat-colour which can be produced experimentally in adult mammals such as rabbits by shaving, local subjection to low temperatures, etc. (W. Schultz[2]; Iljin; Iljin & Iljina; Engelsmeier; Danneel; Danneel & Lubnow; Danneel & Paul; Danneel & Schaumann). The new formation of enzyme and chromogen is apparently strictly controlled by skin temperature, which forms an intermediate link in the chain of genetic causation. It seems that the enzyme often appears before the chromogen, e.g. in the cow embryo, according to Esskuchen[1].

* The most clear-cut results relating genes to pigmentation have of course been obtained in the study of plant flower colours, but these are outside the plan of the present book (cf. reviews by Scott-Moncrieff and Haldane[3]).

† Compare with this the development of melanin in the connective-tissue cells *in vitro*, also studied by Koller[1]. Embryo extract of black Leghorn embryos would not make white Leghorn cells form pigment. Dorris[2] has studied similar problems in transplantations of chick neural crest cells.

The earlier genetical work was carried on in the absence of very clear ideas about the chemistry of melanin formation, but when this was to some extent cleared up (cf. the reviews of Raper—see the formulae below) the way was open for more satisfying work. Einsele's estimation method[2] for melanin made it possible to investigate the multiple allelomorphic series white-black in the mouse described by Dunn[5]. We know now, therefore, from the work of Dunn & Einsele, that the members of this series differ in a truly quantitative way in terms of absolute weight of melanin produced, the difference being expressed morphologically as a difference of granule size. The absorption spectrum of the melanin of all the members of the series has been shown to be identical by J. Daniel.[*]

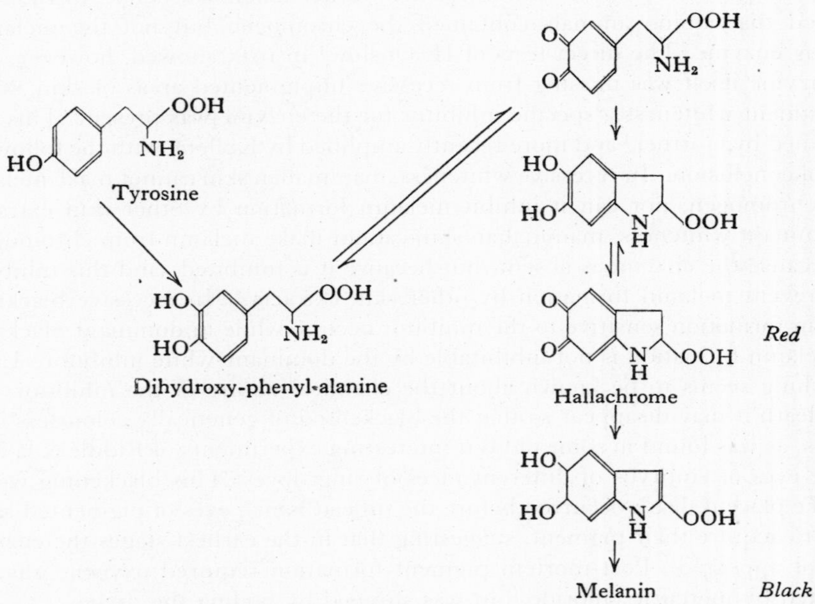

Tyrosine

Dihydroxy-phenyl-alanine

Hallachrome   *Red*

Melanin   *Black*

·N.B. Hallachrome, a pigment found by Mazza & Stolfi in the annelid worm *Halla parthenopea*, is identical with the red intermediate of melanin formation.

The melanin-forming powers of insects have long been recognised to be particularly marked. Insect blood commonly darkens after shedding. Before 1914 Gortner[2] showed that in the potato beetle *Leptinotarsa decemlineata* the unpigmented elytron contains enzyme (tyrosinase) but no chromogen. Extensive studies were subsequently made on the presence of enzyme and substrate in many different kinds of insects (Schmalfuss & Werner; Schmalfuss, Barthmeyer & Hinsch), but the information gained relates rather to comparative biochemistry and the chemistry of tyrosinase action than to the genetico-embryological problem.

[*] Work of a similar kind, though relating rather to the numbers of melanophores produced, has been done by Goodrich & Hansen on fishes and by Twitty[6] on urodele amphibia. On tyrosinase in feather germs of bird embryos, see Charles & Rawles.

This was attacked directly in a careful piece of work by Graubard, who investigated various races of *Drosophila*. Wild type, *yellow* and *black* had a high enzyme concentration at all times and stages; in *ebony* it was lower. Inhibitory factors, however, were often, if not always, present; these could be avoided by treating the flies with chloroform before extraction (liberation of chemical substances from a complex?). Graubard came to the probably rather sound conclusion that genetic factors operated not so much on the actual amounts of substrate or enzyme formed, as on complicated internal conditions, such as *p*H, inhibitors, complex formation, etc., the unravelling of which will require much further work.

Eye-colours, as is well known, have afforded much suitable material for genetic studies. The pigment deposited in the eyes of the shrimp *Gammarus* appears to be melanin, and the black stage is preceded by red stages in which the colour is probably due to the red intermediate indol product (see p. 406). Studies of

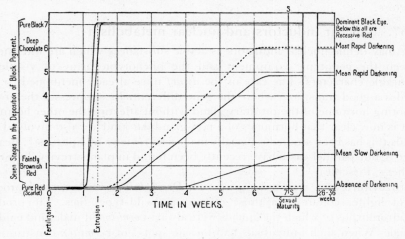

Fig. 240. Changes in eye-colour during the development of *Gammarus*.

various races of the shrimp (Ford & Huxley) showed that they constitute a series of multiple allelomorphs characterised by the rate of darkening, i.e. the rate at which the whole process from substrate to melanin proceeds. An interesting diagram is here reproduced (Fig. 240). The steepest curve represents the dominant black-eye type which is black at hatching. Embryos of this type in earlier stages have pink and then red eyes, though to begin with the eye-rudiment is in all cases colourless. Other genetic types, however, hatch with pink or red eyes and darken much more slowly. Emmart has shown that the reactions producing the melanin are more strongly accelerated by rise of temperature than those producing the red pigment.

By no means all arthropods, however, have melanin deposition as the exclusive pigmentation of their eyes. *Drosophila* has been the subject of important work. Here Johannsen and J. Schultz[2] have found three main pigments, a red, an orange and a yellow, each contained in a special type of granule. In early pupae the eye is still transparent, but about 75 hr. after pupation the whole eye has become tan in colour. At 80 hr. the tan is rapidly changing to red. An aqueous extract of wild-type eyes contains a large amount of a red pigment and smaller amounts of a yellow and a brown. The red pigment can be reversibly oxidised to the yellow, and by treatment of the intact eyes with H_2S the tan colour of the early stages can be immediately changed to red. According to Schultz, all the different eye types (such as *apricot*, *sepia*, *ruby*, *cinnabar*, *vermilion*, etc.) are due to different proportions of the three pigments, two of which seem in effect to be the same pigment in the oxidised and the reduced state.*

By transplantation experiments it has been shown that the colour developed by a *Drosophila* eye may depend not only on its own genetic constitution, but on that of the host into which it is transplanted. In other words hormone-like substances produced by genes and governing eye-colours may be at work. This brings us to the end of one section and the beginning of another—nuclear inductors.

2·67. Nuclear inductors and nuclear metabolism

We may define nuclear inductors as active substances which are intermediate between the genotypic equipment and the phenotypic array of visible and measurable characters. The case of the *dwarf* mice already mentioned (p. 390) provides a good transition, where the gene differentially suppresses the growth-promoting hormone of the pituitary gland without affecting the other hormones. But it is now clear that hormones of a non-glandular kind are also involved. For animals this has been studied especially in insects (reviews by Ephrussi[6]; Beadle[2]; Plagge[6]), but parallel cases have recently been found in plants (review by Stern[3]; Melchers; Pirschle).

In 1920 Sturtevant[1] was studying a gene in *Drosophila*; *vermilion*, which produces eyes of lighter colour than those of normal (wild-type) flies. He produced gynandromorphs in which the gonads were always *vermilion* if male, and wild type if female. When such individuals had mosaic spots[†] of *vermilion* constitution in their eyes, these spots were not *vermilion* if ovaries were present. Hence the simplest explanation was that the wild-type ovary produces a diffusible substance which shifts the eye-colour away from the lightness of *vermilion* to the darkness of the wild type.

Some ten years later, Ephrussi & Beadle[1], who were engaged in successful transplants of ovaries between larvae and pupae of different genetic constitution in *Drosophila*, realising the importance of a study of diffusible substances governing eye pigmentation, found that *vermilion* and *cinnabar* eye discs, transplanted into wild-type hosts, developed perfectly and gave wild-type dark pigmentation. They behaved therefore selfwise from the embryological point of view but neighbour-wise from the point of view of genetics (cf. p. 341), since the determination of the eye was evidently complete, but not of what pigment should be laid down in it. After pupation the transplanted eye disc comes to lie free in the abdomen or occasionally in the thorax. There can therefore be no question but

---

* Phenocopies of the lighter eye types have been produced by Rapoport[2] using Sb compounds.

† See Glossary.

that genetically controlled diffusible substances are involved (Ephrussi & Beadle[3]; Beadle & Ephrussi[1]). In a series of experiments in which a large number of all the possible permutations and combinations of eye-disc transplantations were made between some twenty-five different eye-colour mutant races, it was found that under the conditions of the experiments only *vermilion* and *cinnabar* behaved in this neighbourwise, undetermined, way (Beadle & Ephrussi[1]). Reciprocal transplantations therefore acquired great importance, and here the following results were obtained: a *vermilion* eye disc implanted into a *cinnabar* fly gives a wild-type eye, while a *cinnabar* eye disc implanted into a *vermilion* fly gives a *cinnabar* eye.

Ephrussi & Beadle assumed (and all the later results have fitted in with this view) that two different substances are involved, a v^+ substance which converts the pigment from the *vermilion* to the *cinnabar* stage, and a cn^+ substance which converts the *cinnabar* stage to the fully darkened wild-type stage. The body fluid of the wild-type fly contains both these substances; that of *vermilion* contains neither, and that of *cinnabar* contains only the v^+ substance. Since the cn^+ substance always disappears whenever the v^+ substance disappears in a mutant race, although the latter may be present without the former, it is postulated that the two form part of a common reaction chain, the first link of which is the v^+ substance and the second the cn^+ substance:

$$\longrightarrow v^+ \longrightarrow cn^+ \longrightarrow$$

Thus the mutant genes would interfere at different points in this chain. The reciprocal transplantation results just referred to are now explained as follows: a *vermilion* eye disc implanted into a *cinnabar* host gives a wild-type eye because the *vermilion* disc can make cn^+ substance from v^+ substance. That it never does so if left to itself is because it never has any v^+ substance at its disposal. On the other hand, a *cinnabar* eye disc implanted into a *vermilion* host gives a *cinnabar* eye because the host cannot provide either the v^+ substance or the cn^+ substance. The former of these two cases is particularly remarkable because the eye develops wild-type pigmentation in spite of the fact that normally neither the donor nor the host could have produced the cn^+ substance necessary for wild-type pigmentation (Ephrussi & Beadle[3]; Beadle & Ephrussi[1]).*

If these views are correct it follows that we ought to be able to produce the same results on the eye discs by injecting body fluid from flies of different genetic constitution—the converse of the implant method. This was successfully demonstrated when body fluid from wild-type flies was injected into *vermilion* pupae; the eyes subsequently developed wild-type pigmentation (Ephrussi, Clancy & Beadle; Beadle, Clancy & Ephrussi). It was found that only pupae between 3 and 80 hr. from the beginning of pupation yielded active body fluid; larval and imaginal body fluid was almost inactive, so the substance must be formed and then disappear, at any rate from the circulating medium. The point of eye-colour determination was also ascertained. After 70 hr. pupation the eye-colour could no longer be influenced by the injection of body fluid. Later the cn^+ substance

---

* Note the logical parallelism here with (*a*) *bedeutungsfremde* self-differentiation (p. 157) and (*b*) the relations of enzymes and substrates in the mechanism of mammalian coat-colour production (p. 405).

was shown to behave in the same way (Ephrussi & Harnly; Harnly & Ephrussi) with slight differences in the time limits of existence of the active substance, and in the time limits of competence in the responding eye disc.

First steps towards the chemical elucidation of the substances were taken by Ephrussi & Harnly, who showed that the action of the body fluid was unimpaired after several treatments with liquid air. The living cells in the fluid had therefore nothing to do with the effect (cf. the early work on the primary organiser described on p. 165). Khouvine, Ephrussi & Harnly's extractions of wild-type animals then showed that ethereal extracts had no activity, but that ether-alcohol, alcohol, or aqueous extracts, were active. This was confirmed by Thimann & Beadle, who found that very active aqueous extracts could be obtained if the substances were protected from oxidation by working in nitrogen. The substances were stable to boiling, and after denaturation of the proteins were also stable to oxidation, indicating that the attack had been enzymatic. No differences between the v^+ and the cn^+ substances could be found as regards their stability to temperature (both can be dried at 100° C.) or oxidation by hydrogen peroxide. According to expectation *cinnabar* pupae boiled and extracted proved to contain only the v^+ substance.

Since Ephrussi & Harnly had shown that the substances were present in the blowfly *Calliphora erythrocephala* and the waxmoth *Galleria mellonella* (we shall return to the question of species-specificity; p. 416), Khouvine & Ephrussi worked up large quantities of *Calliphora* pupae. In the ethereal, carbon tetrachloride, or acetone lipin fractions there was nothing active, but the substances could be obtained very active in aqueous medium, and were precipitable with phosphotungstate and basic lead acetate. No pure nitrogenous bases tried, however, were at all active. In the most recent paper (Tatum & Beadle[1]), which confirms all the previous work, we find that though the substances are stable to the temperature of boiling water for an hour, they are completely destroyed by heating to 160° C. for an hour. They are best extracted from dried pupae with 95 % alcohol after a preliminary extraction with boiling chloroform. They are more sensitive to acid than to alkali, readily oxidisable enzymically, and soluble in butyl alcohol at pH 6·0. They can be precipitated by mercuric acetate and regained with H_2S. By a (somewhat inaccurate) diffusion method, their molecular weight is believed to be in the neighbourhood of 500. Lastly, according to Khouvine, Ephrussi & Chevais, the substances may be precipitated with silver along with the purine bases, histidine and arginine, but appear to be destroyed by flavianic, picric and picrolonic acids, which makes further purification difficult.

Working on the basis of these results, Butenandt, Weidel & Becker found that the long-known tryptophane derivative, kynurenin, can carry out the action of the v^+ substance (see formula). Beadle &Tatum find that in its naturally occurring form this is united with two other molecules, of which one is sucrose.

In the meantime more permutations and combinations of transplantations had been going on (Ephrussi & Beadle[3]; Beadle & Ephrussi[2,3]), particularly with regard to the case of *claret*, which is rather hard to interpret. It was shown that the structure of the transplanted eyes develops normally, so that if there is any "inductive" action of the optic ganglion upon the eye, it must occur before the usual time of implantation (Chevais[1]). But the most important line of work was perhaps that of Beadle[1], who identified the fat bodies and the Malpighian tubules as the normal sources or depôts of the substances. Wild-type fat bodies have the v^+ substance but not the cn^+ substance; the Malpighian tubes have both. From larval fat bodies nothing can be extracted, nor from salivary glands, brain, or hind-gut. In the Malpighian tubules the substances are present as soon as 24 hr. after hatching, and larval Malpighian tubules lacking them can take them up if they are injected into the body-cavity. Beadle believes that the Malpighian tubes are actually the site of synthesis of the substances.*

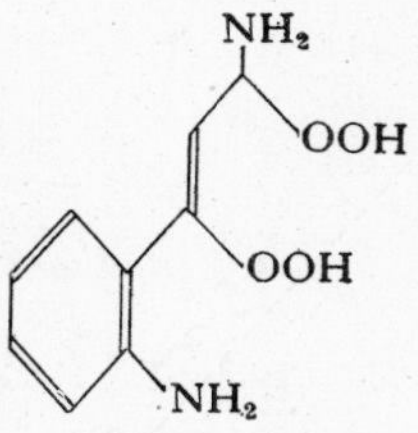

Kynurenin

Next it was found by Beadle & Law that the substances can enter the body of the developing larva by the mouth and still exert their action. *Vermilion* and *cinnabar* larvae fed on cooked wild-type pupae 60–85 hr. after the beginning of embryonic development will later have wild-type eye-colour. This led to an interesting investigation of phenotypic effects exerted by the culture medium (Khouvine, Ephrussi & Chevais). Feeding *Calliphora* extract in agar had the same result, but it was found that starvation also brought about modifications of eye-colour of *vermilion* flies in the wild-type direction. Small doses of dry yeast had this effect, but larger doses suppressed it. Small amounts of peptone gave the same result, and this was abolished by larger amounts. The large doses of yeast extract suppressed the peptone effect but not the effect of feeding *Calliphora* extract. Gelatin peptone was inactive, but became active on the addition of tryptophane. Not one of these effects concerned *cinnabar* flies. Yeast extracts injected into the body-cavity had no action on eye-colour.

These curious phenomena are provisionally explained as follows: normally *vermilion* flies have all the metabolic machinery for making v^+ substance, but at some point or other it is diverted. Starvation restores the mechanism to its normal state. As for the feeding experiments, yeast and peptone must contain some precursor substances which in the protein enzyme systems of the gut are transformed into the active eye-colour inductors. In living yeast (on which *vermilion* flies can be raised) the substance must be either absent or masked.

---

* It is now known that some bacteria can synthesise the v^+ substance or something remarkably like it (Tatum). Cf. the bacterial synthesis of vitamin K (Almquist, Pentler & Mecchi), and the synthesis of vitamin B_1 in refection.

An unmasking mechanism might also account for the fact that if *vermilion* larvae are fed on cooked *vermilion* pupae a shifting of eye-colour towards wild type occurs. The starvation problem was also studied by Beadle, Tatum & Clancy, who found that before 70 hr. from egg-laying starvation would retard larval growth, but not afterwards. The sensitive period for the starvation effect on eye-colour occurs just before this change (Beadle, Tatum & Clancy; Tatum & Beadle[2]; Rudkin).

In what has so far been said, the question of the substrate on which the eye-colour inductors act has been omitted. But eye discs of *vermilion* or *cinnabar*

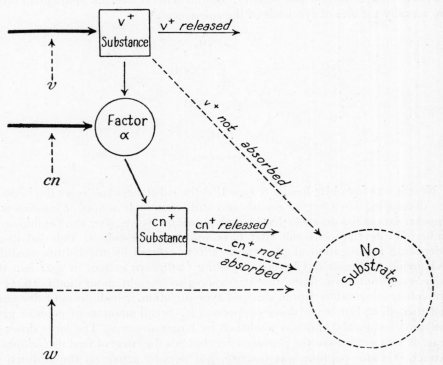

Fig. 241. Scheme of the relations between diffusible substances and eye-pigmentation. (A) case of eyes with no substrate.

pupae transplanted into larvae of the mutant *white* develop into eyes which are wild type in colour. Ephrussi & Chevais concluded therefore that the *white* pupa must possess the two necessary substances, and that the total absence of pigment in its eyes must be due to the disturbance of some mechanism other than that which leads to the formation of the diffusible substances. As Beadle & Ephrussi had previously shown that eyes may themselves give off one or other of the substances, a series of eyes of *white* allelomorphs (*honey*, *cherry*, etc.) were implanted into suitable hosts, and it was found that the release of the v^+ substance was inversely proportional to the intensity of pigmentation of the eyes used. The same proportionality was found for the cn^+ substance. This suggests that the whiteness of the eye may be regarded as a measure of the eye's inability to utilise the substances presented to it.

Ephrussi & Chevais summed up all their results in the two diagrams here reproduced (Figs. 241 and 242). The cn^+ substance is thought to result from the interaction of the v^+ substance with an intermediate product called factor α. Formation of the v^+ substance and that of factor α can be interrupted independently of each other by the genes *vermilion* and *cinnabar*. But the substrate or

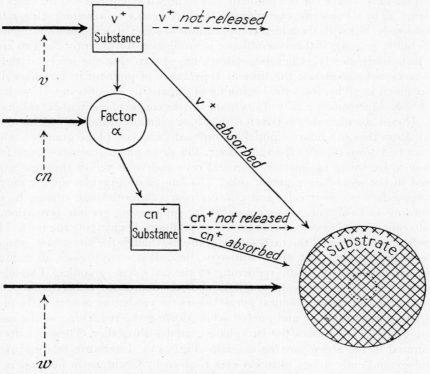

Fig. 242. Scheme of the relations between diffusible substances and eye-pigmentation. (B) case of eyes rich in substrate.

chromogen in the eye is also variable; its appearance is inhibited to a greater or lesser extent by the *white* allelomorph gene series. In the wild-type eye the v^+ and cn^+ substances act upon this substrate. But when the substrate is absent, these substances would hardly be utilised at all by the eye and would therefore thus released be capable of influencing another eye (Fig. 241). When the eye is rich in substrate, the inductors would be utilised, fixed, or consumed, and therefore not released. Intermediate cases of all grades between these two extremes can be imagined.

In sum, then, it appears probable that all the variations of these eye-colours in *Drosophila* are due to the production of varying amounts of a few water-soluble diffusible "nuclear inductors", possibly three in number, certainly not less than two.*

In reviewing the great extension of our knowledge of nuclear inductors just described, the most seriously backward side of the problem seems to be an exact knowledge of the eye pigments themselves and the reactions which lay them down. The reader must often have been reminded of the progressive darkening to black through red described in the eyes of *Gammarus* by Ford & Huxley (see Fig. 240). There the red pigment may well be hallachrome and the black one melanin. The *Drosophila* eye pigment picture seems to resemble this rather suspiciously, but with the added complications of some lipochromes (cf. the work of Schultz, p. 408); if this resemblance is an illusion we urgently need to know the facts more clearly. Chevais[2], indeed, has shown that the effect of the v^+ substance is to accelerate the time of appearance of pigment in *vermilion* flies, by as much as 20 hr. from the beginning of pupation. This fits in well with the work of K. Mori but especially of Cochrane[1] on the course of pigment development.

Cochrane distinguishes between early-phase pigment granules, which are laid down from the 96th hour of pupal life onwards, and late-phase granules, which are not laid down till after the 120th hour. The early-phase granules change from yellow to brown in the primary pigment cells and from yellow through brown to red in the secondary pigment cells. The late-phase granules appear only in the secondary pigment cells and change from yellow through orange to red. According to Cochrane, genes act either by suppressing granule formation, or by altering the rate of colour change, or by qualitatively changing the tints. Thus *vermilion* suppresses all the early-phase granules, though the late-phase granules appear and develop just as in wild-type flies. *White* suppresses all granules completely. *Sepia* stops the reddening of the late-phase granules. The allelomorphic series of *purple* works as follows: *purple*[2] retards the change from yellow to red of the late-phase granules; *purple*[3] slows the reddening of both early-phase and late-phase granules; and *purple*[1] while slowing the reddening of the early-phase granules suppresses the late-phase granules altogether. These results are illustrated in the accompanying diagram (Fig. 243). Literature relating to the histology and embryology of insect eyes is given by Cochrane in her papers.

Attempts have been made to find cases of genetically neighbourwise or undetermined development in other directions by transplantation between genetic races (e.g. body colour by Medvedev[2], and bristle formation by Glancy & Howland) but so far without success.†

---

* The line of distinction here between hormones in the usual sense and precursor-substances is perhaps thin. It has not yet been shown that the nuclear inductors for eye-colours are quantitatively minimal in relation to the pigments produced by their action, as is the case with hormones. They may, therefore, resemble melanin precursors. But all hormones may be precursor substances; the resultant molecules may include them as constituent parts of which very little is needed.

† A case was formerly thought to exist in birds. Willier, Rawles & Hadorn transplanted ectoderm between chick embryos of white and coloured breeds; it was found that while the presumptive coloured feathers never became white, presumptive white ones would develop neighbourwise as far as pigmentation was concerned. Rawles[2] found, however, that this was due to the migration of melanophores. Exactly the same explanation holds good for testis coloration in *Drosophila*

As regards species-specificity, Howland, Glancy & Sonnenblick first found no differences between the various species of *Drosophila* in production of v^+ substance. Then, as has already been mentioned, Ephrussi & Harnly found the v^+ and cn^+ substances present in extracts of blowfly and waxmoth tissues (cf. the

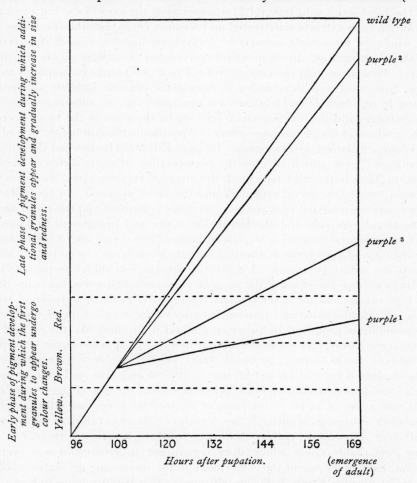

Fig. 243. Action of genes on eye-colour in *Drosophila pseudobscura*.

lack of species-specificity of the primary organiser of vertebrates). Though the question of the identity of various nuclear inductor hormones in different insects is not yet quite settled, it appears that the species-specificity is very low (Beadle, Anderson & Maxwell). Similar diffusible substances have been reported for the

(Stern & Hadorn). But if we include shape as well as pigmentation there is the case of testis form in *Drosophila* reported by Stern[2]. The "genetically spiral" testis of *D. azteca* assumes an ellipsoidal shape if attached to the vas efferens of *D. pseudobscura*, and *vice versa*. This is in striking contrast to the amphibian results described in Section 2·61 where the genetic constitution of the reactor determines the nature of the response to the inductor. It is so far the only case of its kind.

silkworm *Bombyx* (Kikkawa[1]) and for the parasitic wasp *Habrobracon* (Whiting; Whiting & Whiting). Proof of the identity of such substances was given by Becker & Plagge[1]. In the course of work on the red-eyed mutant race of the mealmoth, *Ephestia kühniella*, which will now be described, it had been found that extracts from the black-eyed wild type (AA) would darken the eyes of the red-eyed race (aa). When such extracts were tested on *vermilion Drosophila* the result was just as if wild-type *Drosophila* extracts or transplants had been used. Conversely *aa Ephestia* extracts produced no effect on *cinnabar Drosophila* pupae (Plagge & Becker). This shows that the change from A to a in *Ephestia* corresponds to the change from wild type to *vermilion* in *Drosophila*, because *Ephestia aa* contains nothing from which *Drosophila cinnabar* can make the cn^+ substance. So both the substances and the reaction chain seem to be the same in the two insects.

The analysis of the A-hormone effect in *Ephestia* had begun before that of the eye-colour substances in *Drosophila*. In 1930 Kühn & Henke[2] had described a "pleiotropic" gene which affected the pigmentation of many different parts of the moth. Thus in the wild-type moth the imaginal eyes are black, the testes and the brain brown, the larval eyes dark and the larval skin red. In the red-eyed mutant (aa) the imaginal eyes are red, the testes colourless and the brain almost so, the larval eyes pale and the larval skin white and transparent.* There is, therefore, a general inhibition of pigmentation. Three years later Caspari found that when a wild-type testis is transplanted into an aa larva, the aa testis darkens towards the wild-type colour. A diffusible substance is therefore present in it which can reverse the effect of the aa gene. The same holds true for many other wild-type organs and for other reactive tissues such as the eye (Kühn, Caspari & Plagge), and the substance missing from aa larvae is called the A-hormone.

A-hormone is produced by ovary, testis and brain, and like the v^+ and cn^+ substances of *Drosophila* it can diffuse out of eyes. It appears in the body fluid of wild-type moths during the larval stage. Testis from very young wild-type larvae implanted for a 2-day period into 9-day-old aa pupae and then removed exert a considerable darkening effect upon the eyes of the host (Plagge[1]). Plagge[2] found that a stay of 24 hr. in the body of the aa host is necessary for the effect to take place in some individuals; lesser periods of time do not allow of the diffusion of sufficient A-hormone, and by 72 hr. nearly all the hosts show the effect.

One remarkable effect which does not appear in *Drosophila* was seen in *Ephestia*, namely the use of the A-hormone in the succeeding generation (Kühn, Caspari & Plagge; Kühn & Plagge; Plagge[4]). If a heterozygous *Ephestia* ♀ is back-crossed to an aa ♂, it is found that the homozygous aa offspring have A colours as larvae and aa colours in the adult state. This must be due to the storage of the A-hormone of the heterozygous mother in the eggs, because if testes from wild-type moths are implanted into ♀♀ larvae of the red-eyed race, and if the resulting moths are crossed to red-eyed ♂♂, the offspring again show A larval characters and aa adult ones. The small amounts of A-hormone contained in a heterozygous moth have thus here been replaced by A-hormone from testis transplantations into homozygous aa ♀♀. The hormone may thus exert its effect

* Cochrane[2] finds some association between eye and testis colour in *Drosophila* also.

by "maternal inheritance" (see p. 366), being stored in the eggs and utilised later on.

Periods of competence have been identified in *Ephestia* as in *Drosophila*. On the 9th day after the beginning of pupation the eye-colour is quite undetermined and can be influenced by wild-type testis implantation; but from the 12th day onwards this effect can no longer be obtained (Plagge[1]). This point is earlier relatively than in *Drosophila*. Competence for change in testis pigmentation is already over before the beginning of pupation. Competence is lost from different parts of the eye at slightly different times.

Just as in *Drosophila* two substances are involved, so in *Ephestia* it seems that the *A*-hormone of the gonads is slightly different from that of the brain (Kühn). Implantations of gonads affect eye and gonad pigmentation with equal intensity; implantations of brain affect eye pigmentation without changing gonad pigmentation very much. The two substances must, however, be very closely related. They occur (cf. the lack of species-specificity in the case of *Drosophila*) in many other species, such as *Acidalia virgularia*, *Laspeyresia pomonella*, *Plodia interpunctella* and *Plusia chrysitis* (Plagge[3]). They have been chemically investigated by Becker[1], who found, as in the case of the Beadle-Ephrussi substances, a ready enzymatic oxidation, which could be overcome by making acetone extracts of *A* race abdomina. He concluded that the *A*-hormone cannot be of protein or lipoidal nature, which, so far as it goes, corresponds with what is known of the Beadle-Ephrussi substances.

The drawback about all the preceding researches, valuable though they are, is that the end-results of the nuclear inductors are comparatively trivial, consisting as they do of pigmentation changes. The demonstration of nuclear control over morphogenesis requires end-results of a more clearly morphological character. There exists, it is true, one line of investigation which provides what we are looking for, but unfortunately only for one of the lower plants—the work of Hämmerling on the umbrella alga, *Acetabularia*.

This alga has the form of an umbrella, in the "handle" of which, i.e. at the root or rhizoid end, the single large nucleus is situated. Although the alga is but a single cell it is a giant one, reaching 2 or 3 cm. in height; and the diameter of the umbrella may be as much as 0·6 cm. When the stem is cut at any place, it is found that both the nucleated rhizoid at the "posterior" end and the anucleate umbrella at the "anterior" end are capable of regeneration. The anucleate umbrella end will live for a considerable time before dying, the nucleated rhizoid will of course continue to live indefinitely. Hämmerling[1] found that the regeneration of the umbrella takes place progressively more readily towards the anterior end, and that regeneration of the rhizoid takes place progressively more readily towards the posterior end. The plant is thus polarised along its longitudinal axis. But the amount of regeneration possible for any piece of the plant in isolation from the nucleus is strictly limited, and Hämmerling[4] was able to summarise all his findings in the hypothesis that the nucleus produces two "morphogenetic substances" which diffuse so as to take up positions in a gradient system, there being progressively more of the umbrella-forming substance towards the anterior end and of the rhizoid-forming substance towards the posterior end. During regeneration these substances are used up, and if no nucleus is present, they cannot be replenished. The substances must be rather stable, as a perfect regeneration of an umbrella may occur if the umbrella is cut off from its stem two months after removal of the nucleus from the root. On the other hand, if a nucleated plant is cut at a point fairly near the posterior end, an umbrella will be regenerated at the cut surface although no umbrella would normally have been formed in that region; if now the nucleus be removed, regeneration will continue until the morphogenetic umbrella-forming substance has been used up, after which it will stop, never to be renewed, as the source of the substance has been taken away.

That the nucleus was also the source of genetic characters was proved by Hämmerling[2] using two different species, *Acetabularia mediterranea* and *wettsteinii*. Anucleate fragments could be grafted on to nucleate fragments in both senses. The nucleus then always controlled the differentiation, so that, for example, a *mediterranea* stem upon a *wettsteinii* root forms a *wettsteinii* umbrella.*

The brilliant work of Hämmerling differs from other studies on gradients in plants and animals of stem-like organisation (discussed by him in a separate paper; Hämmerling[3]) because of the decisive advantage possessed by the presence or absence of only one nucleus. But it has also its demerits, more particularly the fact that the morphogenetic substances or nuclear inductors remain hypothetical, and that no chemical work has so far been carried out upon them.†

From all that has been said in the preceding sections, it will have become clear that the genes act in development by producing, inhibiting the production of, or masking and unmasking, hormones, catalysts or inhibitors in more or less diffusible states. Hence their undoubted effects upon rates of reaction in the embryo. *It might even be said that the main difference between genetics and embryology is that in the former study, only those inductors are described which are unable to pass out through the cell-membranes of the cells in which they are formed, while in the latter, only inductors of a more diffusible nature are described.*‡ But this would be to leave out the whole question of competence, which is likely to remain mysterious long after our present knowledge of genes and inductors has quadrupled its present state.

There is, however, one approach to nuclear phenomena in embryology which has not yet been mentioned, namely the direct examination of the egg nucleus by the micro-methods which are now becoming available. This kind of work is only

---

* As Holtfreter points out ([13], p. 417) these results are also in contrast with amphibian xenoplastic transplantations, where the graft, though reacting to the inductor, does so in its own genetically characteristic way (cf. pp. 341 ff.). But there the graft brings its genetic equipment with it.

† In connection with the production of morphogenetic substances by the nucleus, the experiments on the relative power of the amphibian nucleus and cytoplasm in neural induction, mentioned on p. 171, should be recalled. Some cytological evidence also exists pointing to a dispersion of nuclear agents from the "lampbrush chromosomes" of the amphibian oocyte (Waddington[22], p. 101).

‡ Cf. p 140 where intermediate cases are described, in connection with mosaic development, and p. 312 where the inductors of sex-differentiation are described. The same thought recurs on p. 289 in the discussion of Huxley's classification of hormones[13], and on p. 669 where in Whitaker's plant eggs[19] the furthest distant mutual effects are seen. Obviously none of these borderlines can be sharply defined. Medawar's phrase[2] has summarised the matter; "a large part of the physics of developmental mechanics is primarily a physics of diffusion."

just beginning. The early work on oxidation-reduction potential by injection of indicators failed to give any evidence of differences from the cytoplasm (*CE*, p. 846). More important was the work on the metabolic rate of non-nucleate and nucleate fragments of echinoderm eggs obtained by ultra-centrifuging (Shapiro[4]; Navez & Harvey); this will be discussed in detail in the section on embryonic respiration (p. 573), but it clearly showed that there was no particularly strong respiration of the nucleus, and that a non-nucleate fragment could have a much more rapid oxygen uptake than that of the whole egg. The question was examined directly by J. Brachet[15], who measured the CO_2-production (micro-titrimetrically) and the O_2-consumption (Gerard-Hartline method) of the isolated nuclei of *Rana fusca* oocytes nearly at the end of their growth. The metabolic rate of these germinal vesicles turned out to be, in spite of their large size relative to the size of the oocyte, only 1·5 % of the metabolic rate of the latter. Added glucose or cytoplasm gave no increase in the metabolic rate of the nucleus. Later J. Brachet[20] compared the dipeptidase activity (alanylglycine) and the esterase activity (methyl butyrate) of the nucleus and cytoplasm of the amphibian oocyte, using the micro-methods of the Carlsberg school. No difference whatever was detectable between the dipeptidase activity of nucleus and cytoplasm, volume for volume (about $1·2 \times 10^{-3}$ mgm. amino-N/hr.), but the esterase seemed to be more active in the cytoplasm than in the nucleus (in c.mm. N/20 HCl/20 hr./37°; 0·15 for the nucleus and 0·43 for the cytoplasm.

Similar work on echinoderm eggs, carried out by Holter & Linderstrøm-Lang, gave the same result. No perceptible effect of the nucleus could be found in the centrifuged nucleate and non-nucleate portions of eggs of *Dendraster excentricus*. Against the criticism that the germinal vesicle had already dispersed its contents into the cytoplasm, similar experiments on amoebae, in which the nuclei were removed with a micro-manipulator (Holter[2]; Holter & Kopać), gave exactly the same result. It remains of course to be seen whether other enzymes will also be found to have no connection with the nucleus.

In one respect, however, the nucleus shows a striking difference from the cytoplasm. As has already been described (p. 203), J. Brachet[18] has shown how the nucleus of the mature amphibian oocyte gives an intense nitroprusside test after fixation with trichloracetate, while the cytoplasm gives practically none at all. Still more significant, the zone of this reactivity leaves the nucleus during the cleavage stages and ascends to the animal pole of the egg, coming to occupy just before gastrulation precisely that portion of the gastrula which is to invaginate and to possess and exert the inductive power of the primary organiser. As yet we have no idea of the meaning of this distribution of the fixed –SH groups. But it is interesting to note that a large amount of work has given rise to the belief, whether well or ill founded, that the sulphydryl group has something to do with cell-division and even with the control of embryo and adult size. At this point, therefore, it will be worth while to examine this literature, and in so far as the sulphydryl group is believed to be the intermediary between genes and phenotype, it may be regarded as a nuclear inductor elect. What its claims are to assume this office we may see in the following section.

2·68. The control of embryo and adult size

We must first discuss the supposed relation of the sulphydryl group to cell-division. One of the earliest observations made by Hopkins[1] in the discovery of glutathione was that the blastoderm of the chick gives a positive nitroprusside test (in the laid but not incubated egg) while the yolk and white do not. For some time past it has been known that the glutathione content of the embryo falls with age, just as does growth rate, mitotic rate, etc. (evidence in *CE*, Section 12·7).

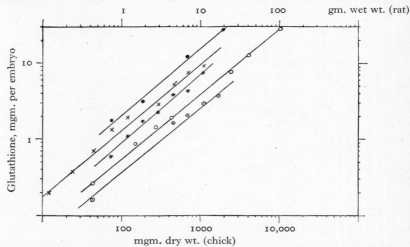

Chick: ⊙ H. A. Murray, * Kamiya, × Yaoi, ● Cahn & Bonot.
Rat: ○ Thompson & Voegtlin

Fig. 244. Glutathione in the embryo of chick and rat (various investigators) plotted on a double logarithmic grid. The decline in glutathione with age (negative heterauxesis) is shown by the fact that the sets of data are all inclined at an angle of less than 45° to the abscissa (see Section 3·22). It may be noted that the heterauxetic constant is very similar in all cases, even that of the rat:

				k
Chick	Murray	...	...	0·84
	Kamiya	...	...	0·90
	Yaoi	...	...	0·87
	Cahn & Bonot		...	0·88
Rat	Thompson & Voegtlin			0·85

The decline with age is also seen in the figures of Bierich & Rosenbohm for rat brain and possibly liver, but they are not suitable for plotting as insufficient weight data are given.

For example, the glutathione of the chick embryo falls from 400 mgm. % dry wt. at the 5th day of incubation to 100 mgm. % dry wt. at the 20th, and that of the rat embryo from 60 mgm. % wet wt. at 0·07 gm. to 36 mgm. % wet wt. at birth. These relationships have been independently confirmed by Kamiya and Tateishi[1]. Fig. 244 illustrates the fall of −SH with age in the chick and rat foetus, in the form of a double logarithmic graph, reduced glutathione being brady-auxetic (see pp. 533 ff.).

Thus the younger the tissue the more glutathione it contains (relatively). If high glutathione content were always associated with cellular proliferation it might be expected that tumour tissue would show high values, but though there has been considerable dispute about this (literature in the reviews of Voegtlin and B. E. Holmes[3]) the upshot seems to be that the glutathione content of mammalian tumours is neither unusually high nor unusually low.

As regards the echinoderm egg, Shearer stated already in 1922 that there was a considerable increase in reducing –SH on fertilisation, as judged by the nitroprusside reaction. This result was accepted with some reserve at the time, but has since been confirmed rather strikingly by the histochemical work of Dulzetto, who found a similar increase in the intensity of the nitroprusside reaction at fertilisation. No perceptible change occurred in its intensity up to the time of the first cleavage. However, histochemical methods could not have the final word in deciding the existence of such a change, and Rapkine[3] measured quantitatively the glutathione (free –SH) content of echinoderm eggs during their early development. He found that the rise at fertilisation is to a level of 40 mgm. –SH/100 gm. wet wt. (from about 30 mgm.). It is followed by a fall to a minimum of 10 mgm. % at 30 min. from fertilisation, and then by a rise to a level of 45 mgm. % just before the first cleavage. This may be regarded as fairly good evidence for his view that segmentation is preceded by increase of sulphydryl groups. Additional evidence is afforded by histochemical nitroprusside reactions tried on ciliates at different stages of division by Chatton, Lvov & Rapkine and by the work of Ephrussi on explanted cells. Ephrussi[1] found that tissue cultures in non-nutritive media gave positive nitroprusside reactions so long as their growth was continuing, but ceased to do so as soon as it stopped. Even if heated to 100° these exhausted cultures gave no nitroprusside reaction, but those still in growth gave an enhanced one owing to the denaturation of the proteins.

Quite independently, certain American workers had been led to envisage a connection of some sort between sulphur metabolism and cell-division. Hammett[1] found that the length of the roots of *Allium cepa* and *Zea mais* growing in very dilute lead nitrate solutions was inversely proportional to the concentration of lead. He also believed he had found a diminution in the number of mitoses, and no effect on cell size. If inhibited roots were transferred to solutions of substances containing the –SH group, there was found an augmentation of length as against the controls. In Hammett's view, these phenomena were to be explained by a combination of the lead with the sulphydryl group and recommencement of mitosis if additional –SH was provided. Normal growth, moreover, was found to be stimulated by the sulphydryl group, and in work with *Paramoecium* cultures, much the same state of affairs was found. In a long series of papers Hammett and his collaborators* extended this theory to wound healing, skin proliferation, crustacean limb regeneration, etc. In later papers of the series, Hammett[3] extended his views to cover the oxidation products of sulphur and claimed that sulphur compounds such as sodium *p*-toluene sulphinate, di-*p*-toluene sulphoxide, di-phenyl-sulphone, etc., exert an inhibiting effect upon growth and mitosis.

Hammett's views were quickly tested by a number of other workers but not with confirmatory results. No effect of sulphydryl compounds on the rate of

* Hammett[2]; Hammett, Anderson & Justice; Chapman; Hammett, Green & Davenport; Hammett & Hammett; Hammett & Justice; Hammett & Reimann; Reimann; Hammett & Smith; Hammett & Wallace; review by Hammett[3,4]; critical review by Hueper.

development of the eggs of the snails *Physa* and *Limnaea* could be detected by Gaunt, who threw doubt on the validity of Hammett's use of amino-acid solutions as controls. T. P. Sun[1], again, was unable to accelerate the rate of cell-division in the sea-urchin egg by treatment with H_2S. Morgulis & Green, who tried the effect of sulphydryl groups on regeneration in the polychaete worm, *Podarke*, found no evidence of any cumulative advance in the regeneration of the experimental animals as compared with the controls. Morgulis & Green brought severe criticisms against Hammett's methods of calculation and expression of results, and particularly against the use of onion-root tips in mitotic counts (cf. the criticisms of Bateman[2]; Moissejeva[1,2]; Richards & Taylor; Taylor & Harvey; Rossmann; von Guttenberg; Gray & Oeillet; Hollander & Claus; etc. against the work on "mitogenetic rays").

Not all the investigations were negative, however. Sulphydryl was found to stimulate growth (cell-division?) in planarian worms (Coldwater[1]), and in tissue cultures (Baker[1]). In connection with the latter, it was found by Sachs, Rapkine & Ephrussi that a large part of the reducing power of normal embryo extract, as prepared for tissue cultures, is due to sulphydryl groups reversibly blockable by iodoacetate. But the most convincing results were those of Voegtlin & Chalkley[1], who found a marked acceleration of nuclear and cell-division in amoebae subjected to low concentrations of glutathione both in the reduced and oxidised form. This work, in which the differences between the control cultures and those receiving glutathione were sometimes very substantial, and which had as a background Chalkley's careful observations[1] of normal maturation and cell-division in amoebae, seems rather reliable. Chalkley & Voegtlin also found that copper in minute concentrations exerts an inhibiting action on cell- and nuclear-division, and that there is an antagonism between copper and glutathione. Later it was shown that mitosis in amoebae is reversibly inhibited by H_2S and HCN but not by CO even in darkness (Voegtlin & Chalkley[2]).

On the whole, however, it cannot yet be said that the causative action of raised –SH concentration on cell-division has been definitely proved. But this does not destroy the basis of Rapkine's belief in a connection between sulphydryl groups and cell-division, which rests on other grounds as well.

The original experiments of Hammett on lead were continued with mercury. Hoadley[4] showed in an interesting paper that very regular inhibitory effects could be obtained by treating sea-urchin eggs with very dilute solutions of mercuric chloride. After fertilisation, he exposed them for short periods to the action of the heavy metal, and then returned them to normal sea-water. As Fig. 245 (drawn from his data) shows, the relation between exposure time and time required for the first egg of the batch to cleave for the first time is practically linear. With higher concentrations of mercury in the protoplasm (longer exposure time) later development is more affected, since the curve relating exposure time to percentage of swimming embryos after 18 hours drops rather suddenly after a certain point is reached. The work was continued by Rapkine[3], who found similarly that with mercuric chloride echinoderm development can be stopped at any given stage in 100 % of the eggs, but also that it is completely reversible if they are transferred to sea-water containing cystein or thioglycollic acid. Controls transferred to sea-water alone or to sea-water plus alanine or cystine manifested a recovery ranging from o to 80 % of that in sea-water plus –SH, with an average of 34 %. So far the experiments of Rapkine were analogous to those of Hammett and Hoadley on other material, but he went further in showing that from echinoderm eggs the thermostable residue of Hopkins & Dixon, i.e. a protein preparation containing fixed –SH, could be made. This lost its reactivity to nitroprusside immediately if treated with mercuric chloride. And mercuric chloride would also inhibit the reduction of methylene blue by thermostable residue and cystine in the Thunberg tube.

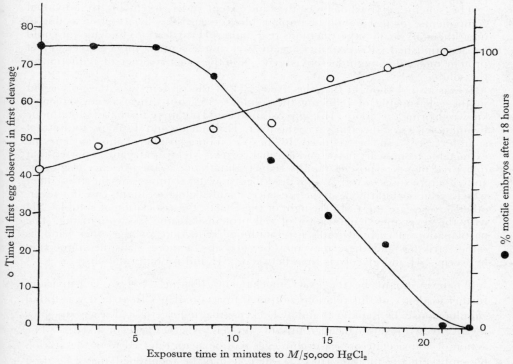

Fig. 245. Effect of Hg on *Arbacia* eggs (Hoadley).

One more link in the chain of reasoning remains. Rapkine supposed that the essential phenomenon causing the increase of –SH before cell-division was the increase of –SH groups attached to protein molecules, i.e. fixed –SH. We know that the denaturation of a protein is often accompanied by the "unmasking" of fixed –SH groups. For some time the phenomenon remained of chemical rather than biological significance, but with the claim of Anson & Mirsky in 1931 that protein denaturation is reversible, it entered a new realm of possibilities. And when it is remembered that among the agents which can cause denaturation of proteins, certain end-products of metabolism, such as urea (Hopkins[2]), are prominent, the significance of the fixed sulphydryl groups in Rapkine's scheme becomes apparent. The concentrations of urea required for protein denaturation are of course very high, but it is by no means unreasonable to suppose the existence of local accumulations in the cell.

The original rather crude form of Rapkine's scheme, then, was as follows: (a) protein catabolism produces urea which accumulates at certain points within the cell, (b) the urea brings about a denaturation of certain proteins, (c) the –SH groups arising therefrom reduce the soluble –SS– and change the oxidation-reduction potential of parts of the cell towards the electro-negative side, (d) the resulting partial anaerobiosis intensifies glycolysis (which may be normally proceeding), and with the energy so produced, cell-division is accomplished.

The scheme as a whole was to some extent independent of its individual components, for if it should be unplausible to regard the denaturation as due to urea, there are many other methods (e.g. light, pH, hydration) by which it could be accomplished. Reversibility, again, was not essential, for a succession of protein molecules might be envisaged. Nor the exact manner of reduction of the soluble –SS–.

It was at first thought that the fixed –SH of the proteins was the only agent in the cell capable of reducing the soluble –SS–, but other systems are now known to bring this about. Hopkins & Elliott could obtain reduction of glutathione by unknown substances in surviving liver. K. Kodama and Tsukano found that soluble –SS– can be reduced by hexose monophosphate plus an enzyme extractable from heart tissue and adrenal cortex. Meldrum, again, reported that intact red blood corpuscles rapidly reduce glutathione in the presence of glucose and other hexoses. And Mann described a water-soluble enzyme from liver which reduces glutathione in the presence of added glucose; in this case, however, other hexoses are inert. In the light of these facts, the association of soluble –SH with the reversible denaturation of cell proteins became less convincing. The participation of carbohydrates in glutathione reduction, on the other hand, is rather striking, and there may now be, perhaps, a more plausible connection between –SH and glycolysis than between –SH and protein catabolism.

Whatever might be the exact mechanisms relating –SH to cell-division, further evidence of this relation continued to accumulate. In 1933 it was found simultaneously by Rapkine[4,5] and by Dickens[1] that iodoacetic acid reacts stoichio-metrically with sulphydryl groups in the cell, whether free or fixed.* The action of iodoacetate on sea-urchin eggs will be again referred to in the section on embryonic respiration (p. 572); here it may be mentioned that Ellis first failed to find any effect at weak concentrations, but later Runnström[17] obtained inhibition of respiration and blockage of cell-division using stronger ones. The best of these studies is that of Rapkine[6] on yeast, who was able to show that after having inhibited all cell-division by carefully controlled doses of iodoacetate, the cells could be made to continue their division by the addition of gluta-thione. The different inhibiting action of iodine could also be reversed by glutathione. Sachs, Rapkine & Ephrussi showed that the –SH groups of chick-embryo extract could be blocked by iodoacetate, thus lowering its reducing power.

* See also p. 204, where the effects of iodoacetate on the primary organiser in amphibia are discussed.

Chalkley[2] later carried out some interesting experiments on the localisation of –SH groups in the amoeba, using the nitroprusside test. Most of the sulphydryl is fixed but nevertheless gives the reaction before protein denaturation; the proteins of the amoeba therefore resemble muscle protein (Mirsky[1]) rather than egg-white protein (L. J. Harris[1]). A strong coloration is always given by the cytoplasm, and a still stronger one by the nucleus, except during anaphase and early telophase. It appears that the sulphydryl is discharged into the cytoplasm at metaphase. In newly formed amoebae the nuclear reaction may not be stronger than the cytoplasmic one. Enucleation had no effect on either cytoplasm or nucleus.

The effects of the sulphydryl group are diverse. As may be seen from Hellerman's review, –SH groups affect the activity of many proteases. Glutathione is now generally regarded as the co-enzyme of glyoxalase (Giršavičius). According to Maver & Voegtlin it inhibits nuclease, a fact which may be related to its effect on nuclear division. The association of glutathione with carbohydrate metabolism and dehydrogenases has recently been strengthened in several different ways; thus Hopkins & Morgan have shown that succinic dehydrogenase depends for its activity on the presence of sulphydryl groups, and Rapkine[7] found that triose-phosphate and triose dehydrogenase is sensitive to sulphydryl groups. When oxidised by iodine, oxidised glutathione or dyes its activity ceases; when reduced by cystein, reduced glutathione or H_2S, it returns. The dehydrogenase activity can also be suppressed by copper and restored by H_2S. Iodoacetate inactivation, however, possible with very minute doses, is irreversible. Lastly, in the non-phosphorylating glucolysis of the chick embryo, there is a stage where sulphydryl groups appear to be necessary though glyoxalase is not involved (Needham & Lehmann[1]).

From the foregoing discussion it may be concluded that some connection probably exists between the sulphydryl group and cell-division. We come now to the work which has led to the suggestion that it is the intermediary between the genotype and hereditary size limitation in the phenotype.*

It began by Painter's demonstration in 1928 that the size differences between Polish and Flemish-giant rabbits are already perceptible in embryonic life (at 12 days' gestation), although the eggs are of exactly the same size. Painter found that the Polish rabbit, which has an adult weight of about 1700 gm., had as 12th day embryo a length of 18·1 mm., while the Flemish-giant rabbit, which has an adult weight of about 5000 gm., had as 12th day embryo a length of 23·1 mm. He considered that the larger form had more cells and had carried out more mitoses. Castle & Gregory confirmed that the eggs of different-sized rabbits have identical dimensions (cf. the monograph of Pincus[1]). Taking observations between 48 and 168 hr. development, they found that at 48 hr. the small race embryo had an average of 14·0 blastomeres, while that of the large race had an average of 21·75. Similarly, the diameter of the blastocysts at 144 hr. was 40·5 μ for the small race and 47·8 μ for the large. Gregory &

* A preliminary review has been given by Lerner[2].

Castle pushed this analysis further back still, to the 30th hour, with the following results:

	Average number of blastomeres	
Hours	Large race	Small race
32¼	4·41	4·06
40	9·94	8·29
41	11·64	8·62

Thus the difference persisted even into the earliest stages. Mitosis must be more rapid in the larger races.

Estimations were now made of the glutathione (first by simple iodine titration and later by technique which allowed for the ascorbic acid present) in new-born rabbits of different size races (Gregory & Goss[1,2]). As Fig. 246 shows, constructed from the data in these papers, the larger the final size, the higher the glutathione concentration in the embryo at birth was found to be. This relation held good even when a Flemish-Polish cross was back-crossed to the large-size parent race, giving a point in the ¾ region[3]. That glutathione might have some connection with the pituitary growth hormone was suggested by the finding that the injection of growth hormone into adult rabbits which had stopped growing caused a rise in the gluta-thione content of their muscles, but no change in the ascorbic acid (Gregory & Goss[4]; Goss & Gregory[2]). This, however,

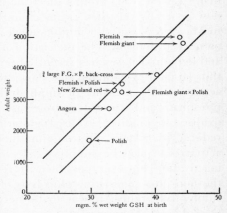

Fig. 246. Relation between size in rabbits and glutathione concentration at birth.

would not apply to stages of development before the appearance of pituitary function. There is also a lactation effect (Goss & Gregory[1]). It appears that the glutathione content of the new-born rat is greatly affected by nursing (e.g. nursed 56·0, not nursed 27·7 mgm. %); the figures on the diagram in Fig. 246 were therefore obtained on non-nursed rabbits. When these figures were subjected to allometric analysis (see p. 533) by Lerner, Gregory & Goss, differences in allometric constant were found as follows:

Race	Adult weight gm.	Allomorphosis constant (see p. 533) glutathione
Flemish giant	4750	1·047
New Zealand Red	3300	1·255
Angora	2750	0·810
Polish	1680	0·617

This means not only that the larger embryos within a race have at birth more glutathione than the smaller ones, but that this is more the case in the races of larger adult size than in those of smaller adult size.

Attention must now be given to the situation in birds. Byerly[3], who had been working on small and large size races of chicks at the same time as the work of

Castle & Gregory on rabbits, had not reached similar conclusions. With White Leghorns (adult wt. 2010 gm.) and Rhode Island Reds (adult wt. 2770 gm.) the small size race was ahead of the large size one in weight before the 10th day of incubation, after which the large size race drew ahead (weight measurements only). Bucciante[2] had stated that within a given race the cell number is the same in chick embryos irrespective of the time taken to reach the stage in question, and Milone found reasons for believing that in chick and sparrow embryos of the same morphological age the same number of cells is present. Measurements of this kind are particularly difficult to appraise. In conformity with Byerly, Henderson got no consistent weight differences between Dark Cornish and White Leghorn embryos at the same age.

Blunn & Gregory, however, considered that weight measurements alone were insufficient to settle the matter. Working with Rhode Island Reds and White Leghorns their results were as follows:

	Rhode Island Red	White Leghorn
Adult weight, gm.	3000	1560
72 hr. incubation: wt. mgm.	20·2	19·8
number of cells in the neural tube at a specified eye-region/embryo	61·6	57·5
mitoses/embryo	19·6	17·9
14 days' incubation: wt. gm.	12·08	11·37
number of cells, etc.	17·2	15·3
mitoses/embryo	2·14	1·68

Though the differences were statistically established, they are evidently nothing like so large as the difference between the adult weights. Again, different conclusions were arrived at by Byerly, Helsel & Quinn, who studied the embryonic growth of Single-Comb Rhode Island Red, Silkie, Single-Comb White Leghorn chicks, and their reciprocal crosses. Between 2 and 20 days of incubation Red embryos and hybrid embryos in Red eggs grew at a practically identical rate. Silkie and hybrid embryos in Silkie eggs also grew at an identical rate, lower than the former. Egg size seems therefore to be a more important factor than genetic constitution in determining embryo size. The weights of the eggs are very different (Red 56·8 gm., Silkie 37·8 gm.). Embryo growth is thus adjusted somehow to the egg in which it is taking place. Cell counts indicated that the four classes of embryos did not differ significantly from one another in cell number.

Correspondingly with this difference between the development of mammals and birds, the results on the glutathione concentration were not as clear-cut in birds as in mammals (Gregory, Asmundson & Goss). No constant differences could be detected between Rhode Island Red embryos and White Leghorn ones in glutathione concentration, though there was perhaps a very slightly greater amount in the former. But it seems that after hatching such differences begin to be perceptible, if we may judge from the comparison between Plymouth Rock chicks (adult wt. 2967 gm.) and White Leghorns (adult wt. 1988 gm.) made by Gregory, Goss & Asmundson, between 2 and 14 days of post-hatching development. Here the glutathione concentration was consistently greater in the large size race, e.g. on the 6th day, Rocks 52 mgm. %, Leghorns 45 mgm. %. Later, Gregory, Asmundson, Goss & Landauer measured the glutathione content of homozygous *Cornish lethal* and heterozygous *creeper* embryos throughout development (see p. 382). The former are retarded in weight, the latter are not; correspondingly the glutathione content was consistently lower than normal in the former, and normal in the latter.

The views of Byerly, Helsel & Quinn were in agreement with the previous work of Kaufman[1,2]. She first found that the logarithmic weight/age curves of hen and pigeon during embryonic life ran closely together, and that the Schmalhausen growth constants (*CE*, Section 2) were for the hen 346, for the pigeon 353. Thus between the 3rd and the 21st day of incubation the instantaneous percentage growth rate of the two birds was identical, and could not account for the differences of body size. Even when the growth rates of the individual organs were compared, the differences between the two birds were but slight. On hatching, however, large differences supervened, no doubt due to genetic differences in nutritional efficiency and body proportions in growth. The size difference was thus thrown back to the period before the 3rd day. Kaufman[3], as the result of *ad hoc* cultivations of blastoderms *in vitro* and measurements of blastoderms of carefully determined age, believed that it lay in a difference in the size of the cells themselves, so that even at the time of the appearance of the primitive streak, the size difference was marked.* The primitive streak appears at the same time in fowls and pigeons, but its length in the two birds is as 1·2 : 1. Hence the volume, and probably the weight, would compare as approximately 2 : 1. In this case, therefore, we have to deal, not with a difference in the integration of growth and differentiation processes, but with a difference in the size attained by a cell before cleavage takes place. Kaufman[4] attributed a discordance between her results and those of earlier authors (e.g. Levi) to the fact that he drew his conclusions solely from linear surface measurements of blastoderms, without taking into account the volumes.

Kaufman's work was all concerned with inter-specific differences, but Byerly's work on large and small breeds of fowls was amplified for large and small races of pigeons by Riddle, Charles & Cauthen, who found that a part of the racial size differences is certainly expressed by the weights at hatching:

	Adult weight gm.	Hatching weight gm.
Runt	565	14·0
Homer	398	14·7
Magpie	296	11·5
Tippler	278	11·3

This would be expected because the eggs are of different sizes. But a large share of the ultimate size differences in these races is attained through differential rate of growth after hatching during the first three weeks.

* On the 3rd day of development, the dimensions of a liver cell are, according to Kaufman, for the chick 974 cubic μ, for the pigeon 596 cubic μ. On the 11th day they are for the chick 1223 cubic μ, for the pigeon 940 cubic μ. And no significant difference exists as regards frequency of mitosis in the embryonic development of the two birds.

In sum, therefore, a definite difference appears to have been established between glutathione concentration of large and small races in mammals, but in birds this is not seen until after hatching. Before hatching, the size of the container is paramount. We cannot but be reminded of the discussions which have taken place about the metabolic peculiarities of the cleidoic egg (e.g. *CE*, p. 1613 and above, p. 48), the difficulties of nitrogen catabolism and the general developmental acceleration of birds over mammals. Even in mammals, however, some kind of special size limitation related to the size of the container comes into play when a large size race embryo is caused to grow in a uterine environment intended only for small size race embryos. In the interesting experiments of Walton & Hammond the large Shire horse (adult wt. 800 kg.) was combined with the small Shetland pony (adult wt. 200 kg.) in reciprocal crosses by artificial insemination. At birth the weight of the foals was found to be proportional to the weight of the mother and equal to the weight of foals of the pure breed of the mother, so that maternal regulation of foetal growth had completely obscured the expected genetic effects:

	Birth weight (kg.)
Pure Shetland	20
Shetland ♀ × Shire ♂	19
Shire ♀ × Shetland ♂	50
Pure Shire	59

After weaning the growth rate established itself as rapid or slow and the characteristic size differences were gradually attained. The mechanism of this regulation remains entirely unknown. But as it is known that the weight of individual rabbits in a litter is inversely proportional to the number produced (Wishart & Hammond; Hammond[4]), although there is room in the uterus for a larger volume of foetal material than is actually there, some regulation of foetal nutrition, possibly endocrine in nature, may be suspected.

The difference between the experiments on rabbits and those on horses lies perhaps in the fact that Gregory and his collaborators were studying a narrower range of birth weights than Walton & Hammond. The difference between the smallest and largest rabbit birth weights seems to be about 1 : 1·57, but that between the small and large size horse birth weights is 1 : 2·75. Were birth-weight differences in rabbits sufficiently extreme they might, in the case of crosses, mask by maternal regulation of size the genetic control of size and what seems to go with it, the glutathione concentration. Though range of birth weight will hardly account for the "maternal" regulation seen in the development of the chick, the conditions of embryonic life are there so much more rigid that the regulation might be expected to function at a lower level of hatching-weight differences (hatching weights: Silkie 24·4, Rhode Island Red 33 gm.).

A glance at a double-logarithmic plot showing birth and hatching weights of many mammals and birds (*CE*, Fig. 72) will show how extremely narrow are the ranges which have so far been studied.* Nevertheless, it may well be said that

---

* What shall we say of size-control in the fishes where the parasitic male is extremely minute, or in ants such as *Carebara* where the queen takes with her, as workers to help in founding the new colony, individuals a thousand times smaller, attached to her legs?

the glutathione effect could only be detected on races likely to be of closely similar chemical constitution.

Before quitting this subject it is worth while to note that among the invertebrates and prochordates there is sometimes a remarkable constancy in cell size, cell number, and body size, as in the rotifers, nematodes and trematodes (H. J. van Cleave) and the ascidians (Berrill[5,6]). Nothing is known regarding the physiological mechanisms underlying these phenomena.

This completes the consideration of the border-line between embryology and genetics. There remains the problem of regeneration and the study of organiser effects in invertebrates.

2·7. Problems of later amphibian life-history

Before ending the discussion of the experimental morphology of vertebrates and dealing with that of the best analysed invertebrate groups, it seems indispensable to say a few words about regeneration and metamorphosis. The phenomena of regeneration, as of limbs or tails, in vertebrates, are important because they raise the following questions. How far may such regenerations be regarded as repetitions of ontogenetic processes which have already occurred? Is determined differentiation preceded by a period of plasticity or pluripotence? What sort of biochemical processes are proceeding in regeneration and do they bear any relation to those proceeding during normal development?

2·71. Regeneration, determination, and the organ-districts

The problem of regeneration has already been met with in this book. The presence or absence of regenerative power seemed to be of significance for the cancer problem (p. 240); regeneration phenomena were referred to in the quotation from Spemann on organ "fields" or districts (p. 286); and the misgivings about the concept of determination voiced by Harrison (p. 110) were partly due to the fresh appearance of pluripotency which has been believed to occur in regeneration.

The aim of the present section is to describe the most important facts and to give a critique of the views which have been based upon them. As in the case of genetics, an exhaustive bibliography will not be required for this purpose, and outlying regions of the very extensive literature can be reached through the interesting reviews of Weiss[4,11,15,16,17]; Guyénot & Ponse[2]; Woerdeman[1]; Poležaiev[3,7,9]; Waddington[8]; Schotté[16] and the book of Korschelt.*

Whether or not regeneration of a limb, for example, occurs in an adult or nearly adult amphibian, has been ascribed (p. 240) to the presence or absence, the persistence or disappearance, of the former Individuation Field of the embryo. As we have seen (p. 288), the main field splits up at the neurula stage into a number of organ districts,† within which the various structures of the adult

---

* I exclude from consideration the regeneration of invertebrates (for which see the reviews of Korschelt; Huxley & de Beer[2]; Stolte[2]) because for us the interest of regeneration lies in its relations to induction, determination, competence, etc. Practically nothing is known about these things in invertebrate development, with the exception of that of the echinoderms. The problem of determination in maturing mosaic eggs will hardly be solved until our knowledge of its more striking features as seen in regulative development is further advanced.

† The same, I shall assume, as Guyénot's "territoires de régéneration".

body arise. This emancipation of the districts is still seen in regeneration; the adult individuation field is such as to ensure that an organised entity is regenerated, but the character of what that is depends on the stump and not on the body as a whole. This was first proved by Guyénot[2] and by Weiss[1], to whom we owe a large part of our knowledge of amphibian regeneration. If a fore-limb of, for instance, a salamander larva, is transplanted homoio-plastically to the region of the hind-limb, or *vice versa*, and the transplant, once well healed into place, is partly re-amputated, what will be formed will depend on the stump and not on the region of the body into which the stump has been put. The stump, not the plan of the body as a whole, controls the character of the regenerate. Fore-limb stumps in hind-limb district give fore-limbs; hind-limb stumps in fore-limb district give hind-limbs; the stump can therefore "carry the field with it" (Weiss[1]; Ruud). Similarly limbs trans-planted to the back or flank will regenerate limbs there (Milojević), or tails will regenerate tails in similar improper and teleologically useless situations (de Giorgi & Guyénot) (Fig. 247, case (1)).

There are also limits to what the body can regenerate at any one time; if too much material be taken away, regeneration fails. The organ districts, in fact, are quite sharply delimited. For example, Schotté[2] found that in *Triton* the removal of the tail anterior to the last sacral vertebra would prevent its regeneration, though it will regenerate when the cut is at any point posterior to this. This will be referred to again (p. 435). Here it need only be said that if the volume of the regeneration blastema is substantially reduced, deficient morphogenesis, such as suppression of digit formation, will follow (Guyénot & Schotté[1]). Yet on the other hand regenerating blastemas can be fused so as to produce single well-shaped limbs until quite a late stage in the regeneration process (Swett[1]; Weiss[1,3]; Milojević & Vlatković). The whole process is, indeed, a remarkable mixture of regulative and determined development.

Its limitations are as instructive as its sometimes noteworthy powers. It can regenerate only the whole structure; not individual components. If the skeletal part is removed from a limb, the bones are never regenerated, but the gap is filled with connective tissue (Schaxel & Böhmel). If now, however, the boneless limb is amputated, regeneration of a perfectly formed limb, *including* the skeletal parts, follows from the boneless stump (Weiss[2]; Bischler; Liosner, Voronzova & Kozmina). Similarly, if the bone which is removed is replaced by bone from another kind of limb, the limb which regenerates distally after amputation has nevertheless the skeleton characteristic of the original one (Weiss[2]; Bischler & Guyénot) (Fig. 247, case (2)). A femur implanted into an arm from which the humerus has been removed will have no deviating effect whatever on the arm-quality of the subsequent distal regenerate. Gaps in a limb cannot be filled in (Schaxel[2]) (Fig. 247, case (3)).

Similar experiments have been made as regards the skin. Weiss[6] removed the epidermis from an amputated stump and encased it in a sleeve of lung tissue, which prevents the appearance of epidermis again upon the stump. Nevertheless, the limb which regenerated was well provided with normal epidermis (Fig. 247,

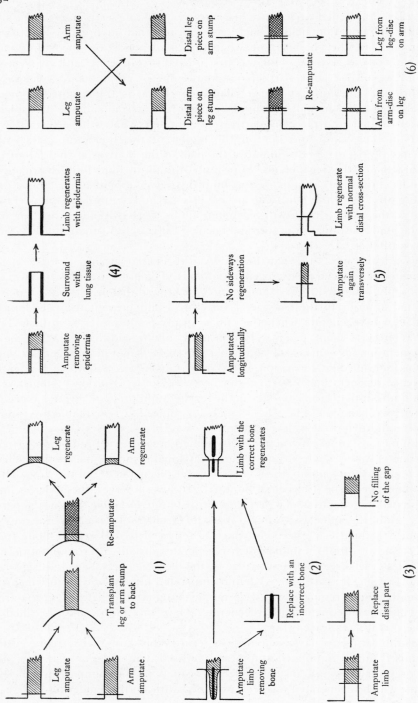

Fig. 247. Diagram to illustrate the laws of regeneration in amphibia.

case (4)). Taube's experiments were even more striking. He replaced the black skin of the normal limb by a sleeve of red skin from the belly, and when this had properly healed, amputated the limb in the middle of the red portion. The regenerate, which appeared normally, was covered with black skin, not red. This has been confirmed by Holzmann. All these facts go to show that the tissues of the stump do not each give rise to the corresponding tissues of the regenerate, but rather that relatively undetermined mesodermal cells come forward and are organised into everything that the new structure needs, and in accordance with the "style" of the old. Exactly how undetermined they are is the most important point of all; we shall return to it before long.

Another remarkable limitation of the regeneration process is that it only works in a distal direction, not sideways. If a limb is amputated not transversely, but longitudinally, a certain amount of growth will subsequently occur on the injured side, but there will be nothing like a completion of the normal cross-section of the limb (Weiss[3]; Gräper[1]). On the other hand, if such a split limb be amputated transversely, a distal regenerate will be formed which will have a normal cross-section (Weiss[5]) (Fig. 247, case (5)).

The quantity of stump material which can carry the limb district quality is remarkably small (Milojević). A thin transverse slice of fore-limb, for example, healed on to the stump of a hind-limb, will give rise to a fore-limb regenerate (Weiss[1]) (Fig. 247, case (6)).

In the foregoing paragraphs it has been said that the complete regeneration process seems to work only distally. If the stump is turned round and healed into place so that regeneration would have to occur from the proximal part of the limb or not at all, only replications of the stump itself occur, and not re-generations of the entire limb. Thus if a knee, with a piece of upper and lower limb, is inverted and implanted into the body so that the upper limb end protrudes, a new knee and lower limb is formed, but never anything in the nature of a foot (Kurz; Gräper[1]; Milojević & Grbić). The same effect has been observed on inverted pieces of tail implanted into the dorsal region (Milojević & Burian). Such phenomena go far to explain many spontaneously occurring limb abnormalities (Przibram[3,6]; Puppe and others).

From the facts so far described it might appear that it is impossible to influence the organ "field" itself; if it works, it works in a unitary way leading to a complete replica of the original structure. Nevertheless, Vogt[4] showed that larvae with defective tails caused by injuries to the embryo in the earlier "mosaic" stage (partial extirpation of the organ district) would subsequently regenerate defective, not whole, tails. Something similar to this occurs in fishes (Nabrit).

It is interesting that the presence of the nervous system is absolutely necessary for the regeneration process; complete severance of the peripheral nerves in the limb region of urodeles leads to failure of regeneration, as was proved, after some divergence of opinion, by Schotté;[*] and complete de-differentiation of the

---

* This was first discovered by Locatelli, but acceptance of the fact was delayed by her erroneous belief in a specific morphogenetic influence of the nerves. Trophic action of the nerves seems to prevent de-differentiation going too far.

stump (Schotté & Butler). Partial denervation slows the regeneration process and late denervation arrests it halfway. It needs the presence of the neural tube and the sympathetic system, but can go on without continuity of the main segmental motor and sensory nerves. Whatever the nature of the nervous influence is,* it cannot override the stump-quality, as the facts already described demonstrate. By deviating the course of nerves to the various organ districts corresponding regenerates or new supplementary growths can be produced (as in the experiments of Guyénot & Schotté[2]; Guyénot & Ponse[2]; Bovet, etc.) but the effect of the nerve is always trophic rather than determinative (Schaxel & Schneider; Samaraiev). Traces of such a trophic action have, indeed, been reported for normal embryonic limb-buds, whether *in situ* (Hamburger[3]) or transplanted (Hamburger & Waugh).

Alone of the organ districts the crest "field" has been successfully stimulated mechanically. On tying a thread round the body to get mechanical duplication of limb blastemas Milojević, Grbić & Vlatković found in a few cases a growth of crest in the neighbourhood of the thread, which was sewn through the skin. This phenomenon is probably not very important.

The question of the origin of the tissues in the regenerated limb, to which reference has already been made, long presented much difficulty, but was eventually solved by the ingenious experiments of G. Hertwig[2]. If a haploid arm is transplanted to a diploid animal, and then after healing into position, amputated, the resulting regenerate is entirely composed of haploid cells. The source of the cells must therefore have been the de-differentiating stump and not the rest of the body.[†] Evidence which supports this conclusion, without being in itself decisive, comes from experiments in which arms of the black axolotl, transplanted on to albino axolotls, regenerate black arms, not white, after amputation (Ruud). Similarly, *Triton* limbs on *Salamandra* hosts, recognisable by differences in nuclear size, regenerate limbs of their own character, not even intermediate between that and the character of those of their hosts (G. Hertwig[2]). Evidence clinching the matter comes from radiation experiments (see below, p. 440). The histology of the regeneration process has been exhaustively described by Naville[1] and Thornton.

Accurate knowledge of the normal rate of regeneration is available. We owe to Grotans[2] an elaborate study of the speed of regeneration of amputated axolotl gills. At first it is rapid, then for several weeks it continues at a steady rate, and thereafter it regularly falls off. With increasing age, the initial rise is slower, the maximum less persistent, and the final fall more gradual. The figures of Murtasi[2], however, show that the speed of regeneration of these axolotl gills temporarily increases again after a certain minimum point is passed. Syngajevskaia's measurements for *Siredon* limbs indicate that their speed of regeneration is maintained constant over a remarkably long range of body weights. And similar quantitative comparisons between the growth of the axolotl tail in normal

---

* Could it be an effect of sympathin? (see p. 211).

† This is in some contrast with conditions in invertebrates, where there appears to be a migration of "reserve" cells from all over the body to the blastema (see the review of Stolte[2]).

development and during regeneration have been made by Godlewski & Latinik-Vetulani. The surprising fact is reported by many observers that regeneration of one extremity (tail or limb) is *accelerated*, not inhibited, by the amputation of another shortly afterwards (Zeleny[1]; Hiller[1]; Blacher, Irichimovitch, Liosner & Voronzova).

With the question of the state of determination of the blastema we enter a debatable land. Is a limb blastema determined from its very first appearance to form a limb, and a tail blastema to form a tail, or can their fates be shifted by implantation into a district foreign to their normal fate? Is their behaviour *ortsgemäss* or *herkunftsgemäss*? It has become customary to say that in their youngest stages they may show *ortsgemäss* (neighbourwise) behaviour; i.e. that they are at first pluripotent or relatively undetermined.

The facts upon which this view rests are the following. From his transplantations of young blastemas between fore-limb and hind-limb districts on the same animal Milojević concluded that if a blastema was placed, while still sufficiently young, in a foreign district, its fate could be changed, and a presumptive hind-limb would become a fore-limb (in contrast with experiments where the old stump is transplanted as well). As there might be difficulties in deciding whether a rather abnormal regenerate was to be regarded as fore- or hind-limb, and in order to see whether the pluripotence was really wide, Weiss[7] proceeded to exchange presumptive tails and limb blastemas. He obtained the *ortsgemäss* effect, a limb from a presumptive tail blastema. The converse experiment was successfully performed by Guyénot & Schotté[3]. Proof of pluripotence appeared to have been given.

But the case is not really as convincing as has been supposed. The number of successful reversals of presumptive fate was small; Weiss described only three, and Guyénot & Schotté only one. True, others were later adduced by Schaxel[4]. Much more serious is the fact that very young blastemas show a marked tendency to resorption. As Weiss himself points out ([16], p. 107), a blastema of less than 2 weeks' development, transplanted to an indifferent situation, such as the flank, is always resorbed. The possibility exists, therefore, that what happens in the cases of apparent reversal of presumptive fate is a resorption of the young blastema, followed by the normal regeneration of the typical local structure. In this case all regeneration would be *herkunftsgemäss*. Now the most obvious way of preventing resorption would be to implant, not one tail blastema, for example, but several, on to the site of an amputated limb; in this way the danger of resorption would be minimised and, if a change of fate occurred, pluripotence would be placed on a much surer basis. This experiment was carried out by Poležaiev[2], with decisive results against pluripotence. With one blastema he was able to confirm the apparent change of fate, but when he implanted from four to eleven blastemas the results were without exception otherwise.* Tail blastema

---

* The fusion of several identical or analogous embryonic parts into one for analytic purposes has been a valuable contribution by the Russian school, cf. the work of Lopaschov[2] already mentioned (p. 283) on putting together several primary organisation centres. It derives perhaps from the observation of Filatov[4] in 1932, that by combining several tail blastemas into one the speed of regeneration of the tail could be greatly accelerated.

material, if present in sufficient amount to guard against resorption, invariably produced tail though in the limb district.* This was later confirmed by Liosner[5].

Poležaiev[5] also showed that the tissues of a limb may be thoroughly minced and will yet, when implanted, regenerate a well-formed limb.[†] In later experiments he found that a limb will regenerate *herkunftsgemäss* if to one blastema, itself incapable of self-differentiation, is added a mass of minced mesoderm from other blastemas, even from young ones themselves incapable of self-differentiation. Resorption must be regarded as due to the lack of a source of mesodermal cells, undetermined within the limits of the limb as a whole, which can give the material required for regeneration. In the earliest stages, this source is, of course, the de-differentiating stump.

A further contribution to the determination problem was made by Lodyženskaia using the method of antero-posterior and dorso-ventral inversions of transplanted limb blastemas. Contrary to the impression gained by other workers, such as Milojević and Weiss, she found that the blastema's spatial dimensions were already determined during the early part of the first week of regeneration. In spite of the much fuller later work of L. David and Schidewski, conclusions here are difficult owing to the resorption process. The general result is nevertheless reached that wide pluripotence probably does not occur in regeneration, and that each organ district in amphibia is capable, if the individuation field is still "intact", of reproducing the organ in question.[‡] Material from such districts cannot be made to form organs characteristic of other districts. That regeneration should not reproduce the condition of wide pluripotence seen in early embryonic development is not after all so very surprising, for, as we have seen, it depends on certain factors quite foreign to the embryo, such as a connection with the nervous system.[§] It is nevertheless still necessary to withhold a final judgment, since Efimov[5] has reported three cases of exchanges between young hind-limb blastemas of the black axolotl to fore-limb site in the white axolotl, in which the

---

* This whole subject has been the cause of a good deal of unfortunate terminology. Weiss[11] and Schaxel[3] have used the word "nullipotent" to describe the status of the very young blastema which on transplantation to an indifferent site is resorbed and performs no morphogenesis. But all that could be said of it was that it was "not capable of differentiation" and hence its potencies could not be tested; hence the word "nullipotent" begged the issue. Since Poležaiev's work[2] we should call it "not capable of self-differentiation" and neither nullipotent nor pluripotent. The only use for the word nullipotent that I can see is to describe such tumours as cannot be made by any known means to perform any morphogenesis but which nevertheless proliferate. Poležaiev himself has also erred, however, for he uses the words "undetermined" and "determined" to describe the status of the blastema before and after the onset of the capacity for self-differentiation. Since nothing whatever, owing to the resorption difficulty, has as yet been found out about the status of the blastema before self-differentiation is possible, this use of the determination concept is unjustified.

† Poležaiev recalled in this connection the reunion of sponges after the destruction of their organisation by filtration (see p. 528).

‡ Regenerative capacity may be remarkably great; according to Schaxel[4], twenty-two regenerations by one animal are possible.

§ Weiss' pluripotence experiments[7] have also been subjected to another criticism by Guyénot[3] and Bovet, who pointed out that in the implantations of the tail blastema on to the limb site the interference with the nerve is a serious source of error, for this alone might be expected to produce a regenerated limb from the host's limb district.

resulting regenerates were black fore-limbs, suggesting that a change of presumptive fate had taken place. It might, however, be said that only the epidermis, or only the melanophores, had been derived from the transplanted blastema, while the remaining tissues were of host origin. Moreover, Savchuk could not repeat Efimov's results, and in new elaborate work Mettetal always obtained *herkunfts-gemäss* differentiation in salamanders.

The question may next be asked: What part do the various tissues of the regenerate play in determining its nature? Important qualities seem to be carried by the epidermis, for Efimov[1] earlier found that if epidermis of the amputated limb stump is replaced by head epidermis, no regeneration ensues (cf. p. 304). It had previously been shown by Tornier[2] and by Godlewski[2] that if an amputation surface is covered *immediately* by any skin, regeneration is inhibited. Efimov[1] then showed that such an inhibition only occurs if the skin is put on before any natural epidermis has had time to form; the effect is therefore by no means mechanical. This was further extended by Poležaiev & Favorina, who found that replacement of limb epidermis by head epidermis at any time during the regeneration process would inhibit it, though naturally as the regeneration proceeded the inhibitory effect became less marked. Poležaiev & Favorina also showed that the removal of the regenerated epidermis alone would itself stop regeneration; a skinned blastema will neither grow nor morphologically differentiate, though its histological complexity may increase (Fig. 248). A close connection between mesoderm and epidermis is therefore a *sine qua non* of regeneration. Correspondingly, I. I. Morosov has shown that if after inhibition of regeneration by skinning, the blastema is put under the skin of the tail, regeneration will proceed. But the longer the period of inhibition has been, the more abnormal are the resulting structures. In the early stages of the regeneration, the action of the epidermis is supposed to be a histolytic one (Poležaiev[1]).*

The mesodermal parts are also very important. Liosner & Voronzova[3], after amputating the hand of an axolotl, dissected the arm and replaced the whole of the arm musculature with a muff of muscle taken from the tail, retaining the nerve and blood-vessel. In many such cases regeneration almost led to the transformation of the structure into a tail, but sometimes structures intermediate between tails and limbs were formed. It is therefore certain that the morphogenetic process of a regenerating limb may be caused to deviate in the direction of tail if tail muscle is present instead of limb muscle.† Liosner & Voronzova[5] described one successful case of the converse experiment. Subsequent work has shown that tail muscle transplanted into a bone-free limb can form fairly typical tail skeletal parts (Liosner[3]). Under such conditions the approximation of the regenerate to a tail is much more complete; the limb bone therefore exerts a limb-quality-retaining influence (Liosner[4]). Just as the muscle mesoderm of a limb blastema can be replaced by tail muscle, so it can be replaced by muscle from various proximal-distal levels of a normal limb. When this is done, the regenerate is found to have a corresponding character, i.e. if muscle from the distal end of the limb is inserted, hands and feet only are regenerated; if muscle

---

* Adova & Feldt have attempted to establish chemical differences between the skin of the back, which will not take part in leg regeneration, and that of the limb, which will do so. The autolytic process was found to differ.

† But though the tail implant will make the surrounding presumptive limb tissues behave neighbourwise as regards their morphogenesis, they behave selfwise as far as the stage in the life-history is concerned. In other words the tail inductor may be from an adult, but the tail produced will have larval characteristics if regenerating on a larva (Voronzova[4]).

from the shank portion of the limb, a full limb is formed (Voronzova[1]). Reciprocal transplantations of fore-limb and hind-limb muscle into young regenerates demonstrated the same kind of specificity; the resulting structure depended to a considerable extent on the origin of the musculature which had been put in it (Voronzova[2]). If musculature from the dorsal part of the body is put into the regenerate instead of its own muscle, the results are variable; regeneration may fail, the fate of the regenerate may be unaltered, or minor modifications of it may occur (Voronzova & Krascheninnikova). Bones of the shoulder girdle can be regenerated from shoulder-girdle musculature, even in abnormal situations, in young larvae (Voronzova[3]). And cartilage of a new limb skeleton can arise from a blastema in which there are no cells of cartilage origin (Thornton).

In tail regeneration Okada and others have shown that the presence of the spinal cord is very important in the determination of normal cross-section. If it is destroyed bone may appear and may differentiate histologically, but the vertebral organisation will not be formed. Without the axial organs a merely finlike structure regenerates.

The part played by bone in the regeneration of a limb has been considered by Liosner[6]; and Nassonov's work has indicated that the action of cartilage cells must not be neglected. First he obtained[1], like Poležaiev[8], duplication of limbs and appearance of digital processes, etc. by implantation of regeneration blastemas destroyed by mincing. He then found[2] he could obtain the same results by implanting regeneration blastemas dried at low temperatures, in which the cells had all been killed, though here the number of positive cases was smaller. This supported the earlier view that an important role in regeneration is played by the decomposition products of the traumatised tissues. Next he found that the addition of an aqueous extract of cartilage would, when added to a tissue culture of connective-tissue fibroblasts, cause the formation of a large amount of hyaline cartilage, sometimes even converting the whole of the culture (Nassonov[3]) (Fig. 249). This could not be confirmed by Rumjantzev & Berezkina (see above, p. 301). Finally, he found that when implanted under the skin in the limb district of the axolotl, a piece of cartilage would give rise to an epidermal prominence resembling a regeneration blastema, as shown in Fig. 250 (Nassonov[4]). Further organisation, however, did not generally occur (cf. the ectodermal proliferations obtained by implantations of chemical substances into the blastocoele cavity, p. 177).

In subsequent papers Nassonov[5] showed that the morphogenetic power of cartilage was destroyed by boiling (confirmed by Kozmina) and that cartilage from anura but not from reptiles or mammals was active. Epidermis, bone and muscle tissue was not, but lung, small intestine and gill showed considerable activity[6]. The products of an alkaline hydrolysis of cartilage can produce new digital appendages with cartilage within them. Among the interesting features of this work (continued by Peredelsky) is that it recalls the induction of limb buds in the embryo by ear-vesicles and nasal placodes (p. 305; Balinsky[6]) though the phenomena in the adult are much less striking morphogenetically.

The cause of the naturally occurring loss in regeneration power (referred to above as disappearance of the individuation field) is interesting but obscure. In the urodeles, as has been known since the 18th century (Barfurth[2], 1895), a strong power of regeneration persists into adult life, but the anura can only regenerate their limbs at the early stages of metamorphosis. Afterwards the power is lost. It seems, indeed, to arise and to disappear in a very definite way. The work of Svetlov[1] and Veizman on anuran tadpole-tail regeneration shows that in the first

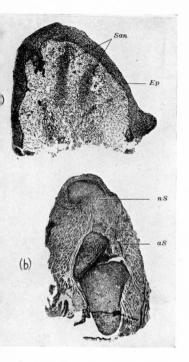

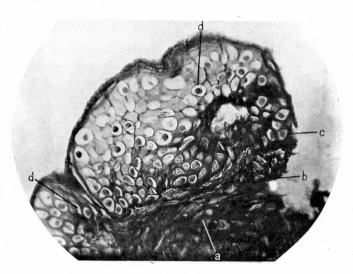

XXXII

Fig. 248. Inhibition of amphibian limb-regeneration by removal of epidermis: (a) normal blastema; (b) skinned blastema. *San*, skeletal rudiments; *Ep*, epidermis; *nS*, newly forming bone; *aS*, remains of old bone.

Fig. 249. Induction of cartilage from fibroblasts by cartilage-extract (Nassonov). *a*, connective tissue fibroblasts; *b*, transitional zone; *c*, rounding fibroblasts; *d*, cartilage.

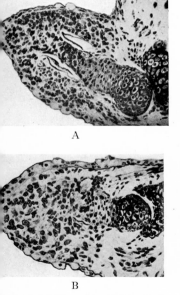

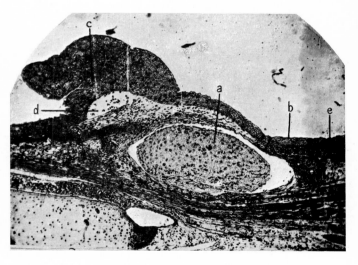

Fig. 251. Inhibition of amphibian limb-regeneration by X-rays. A. Normal blastema. B. Irradiated blastema in which the differentiation of the cartilage is quite inhibited.

Fig. 250. Induction of epidermal prominence by implantation of cartilage (Nassonov). *a*, disintegrating implanted cartilage; *b*, epidermis; *c*, epidermal outgrowth; *d*, cavity containing blood.

tail-bud stage no regeneration is possible, but that the power reaches a maximum about the time of hatching from the egg-membranes. Before metamorphosis it has disappeared. In lizards the tail can regenerate but the limbs not;* Weiss[10] found, however, that regeneration could occur in the limb district if a tail blastema was implanted there. It was strictly *herkunftsgemäss*.

Poležaiev's experiments[4] on limb regeneration are also interesting. By traumatisation of the wound surface (mincing it with a scalpel *in vivo*) he found it possible to obtain limb regeneration at a stage subsequent to that at which the power normally disappears. He explained this by assuming that the traumatisation stimulates the accumulation of the indifferent regenerative cells from which the blastema is formed, and that it is the lack of these which is involved in the normal decline of regenerative power. He showed also that re-regeneration (the regeneration of a limb that has been itself produced by regeneration) may take place quite late in metamorphosis. These facts would indicate that the mechanism of the loss of regeneration power has something to do with the availability of the relatively undetermined mesoderm cells (Poležaiev & Ginzburg). But this does not take us very far.

Much of the discussion about loss of regeneration power has, of course, turned upon the question of whether this loss is brought about by a humoral change in the whole body, or whether it is localised in the limbs. Thus Guyénot[1] first showed that tails and limbs of young frogs which had lost their power of regeneration did not regain it when they were transplanted on to salamander larvae in the proper positions, although salamanders never lose regenerative power. Transplanted and then amputated, they never grew again. Liosner[2] and Naville[2] completed the picture by transplanting backwards and forwards in time (cf. p. 346). From frogs that had lost the power, limbs were transplanted to frogs which still had it; and *vice versa*. In this case also, the limb behaved physiologically *herkunftsgemäss*. If it could regenerate before, it would do so just as well on an older body; but if it had lost the power, transplantation on to a younger body, where the power was still present, did not restore it. All this seems to prove that the change is a local one. But if this is really so, limbs from tadpoles at the regenerative stage, transplanted on to young frogs just about to lose their regenerative power, should lose their regenerative power later than the limbs of the host. On the other hand, if the body as a whole were concerned, the power should be lost in all the limbs approximately simultaneously. This experiment has been tried by Borssuk, with the result that the second of these alternatives was found to hold good. So it would seem that both local and general (hormonal) factors are involved. This agrees with the fact established by Poležaiev[12] that the conditions under which the tadpoles are maintained affect the time at which the regenerative power falls off.

That hormones are involved in amphibian regeneration appears from many experiments, such as those of Schotté[1] on the pituitary and Speidel on the thyroid. Hypophysectomy suppresses regeneration in *Salamandra* larvae and adult *Triton*

---

* There is a very minor degree of regeneration of lizard limbs according to Guyénot & Matthey.

but not in larval *Triton*. The effects of thyroxin are complex.* But if the circulating hormones are in any sense inductors here, the competence of the tissues must vary independently, for we know that the power of tail regeneration in toads, for example, persists for some time after the power of limb regeneration is lost (Guyénot[1]).

In mammals, especially embryos, traces of regeneration power can be observed if we may accept the results of Mitzkevitch and Selye (on limbs) and of Selye & McKeown[1] (on the uterus).

The loss of regeneration power normally occurring in some animals can, as we have seen, be stayed by certain methods. Conversely, it is possible to inhibit regeneration in animals at stages when it is normally vigorous. Inhibition by withdrawal of the trophic action of the nervous system has already been mentioned. Moreover, if the stump is covered with epidermis immediately after amputation, no blastema will form (Tornier[2]; Godlewski[2]). Or if too great a quantity of material is taken away, regeneration will be inhibited. Thus if the whole of the shoulder girdle in adult newts is removed, no regeneration follows (Weiss[2]), though in younger larval stages this regeneration is possible (Swett[3]; Swett & Parsons) and will even go on in unlikely situations such as the hindlimb if the shoulder muscles are transplanted there (Voronzova[3]). There is an age series of graded effects (E. H. Parsons). But the most controllable way of inhibiting regeneration is by the action of radiant energy which can be fairly accurately dosed.

As we learn from the interesting review of Curtis, it was Schäper in 1904 who discovered that exposure to radium emanation would inhibit limb regeneration in *Triton* larvae. After this initial observation nothing was done for thirty years until a series of papers by Butler[1]; Liczko; and Brunst & Scheremetieva cleared up the question and showed that the inhibitory effect of X-rays may be useful in studying the nature of the regeneration process. Ultra-violet irradiation also inhibits it (Tolmatcheva-Melnitchenko). In either case, stump de-differentiation will follow if the irradiation is applied early (Butler & Puckett) or the regeneration will be arrested midway if it is applied later.

In a normally regenerating limb the first component of the blastema to differentiate is the cartilage. In an irradiated larva not only is there no indication of cartilage differentiation, but also the old cartilage within the stump is severely affected by the radiation and finally disappears (de-differentiation).† Butler[2] concluded that the primary effect of the X-radiation was to inhibit the histological differentiation of cartilage. Fig. 251 (opp. p. 438), from Butler, illustrates his results. Liczko's interpretation of what happens was a little different, for he regarded the primary effect as the formation of a mass of mesodermal connective tissue at the distal end of the blastema into which the cartilage and muscle cells were unable to penetrate. Whatever the effect is, it will persist, as Liczko found, in the tissues,

---

* According to B. D. Morosov, mammalian or amphibian embryo extract, injected into the amphibian body-cavity, accelerates tail regeneration.

† Since the effect of irradiation is analogous to the cutting off of the trophic action of the sympathetic nervous system, may there not be a common point of chemical action here?

and will manifest itself if an amputation is made as long as 8 weeks after the irradiation.

The effect is local, not general (Butler[2]; Brunst & Scheremetieva[1]). Irradiation of a single limb will prevent its regeneration, and an unirradiated limb will still retain the capacity for regeneration though a large part of the body has been irradiated. Unirradiated limbs transplanted to irradiated hosts behave *herkunftsgemass* physiologically, for they regenerate though the limbs of the host are unable to do so. Such experiments demonstrate again that the source of the cells of the new limb is the stump and not the other parts of the body. An even greater localisation is possible. By shielded irradiation, regenerative capacity may be preserved in the middle part of a newt's limb while at the same time it is suppressed in the distal and proximal parts of the limb (Scheremetieva & Brunst[2]). Different tissues have different thresholds of sensitivity; thus Scheremetieva & Brunst[1], X-raying the tail of *Pelobates*, found that the regenerative potencies were lost in the following order: notochord > neural tube and musculature > gelatinous connective tissue > blood-vessels and epidermis. By careful choice of dose, it is possible to maintain limbs in normal life for indefinite periods without any capacity for regeneration (Brunst & Scheremetieva[2]).

Umanski[3] saw that the inhibition of regenerative capacity made it possible to follow up in more detail the participation of single tissue components in the regeneration process. He therefore transplanted non-irradiated limb skin to an irradiated limb, and found that after amputation normal regeneration occurred. That this was not due to a lifting of the inhibition on the internal tissues (muscle and skeleton) he proved by using skin from a white axolotl as the graft and an irradiated limb of the black axolotl as the host. The limb which regenerated was entirely free from pigment, which suggests that in this case the epidermis can itself give off cells capable of forming everything that there is in a non-irradiated regenerated limb. When skin from the head, neck or flank was used as transplant no regeneration occurred (cf. the experiment of Voronzova[2] above, p. 438); when tail skin was used, the regenerate was a tail; when skin from the hind-limb region was used but not from the hind-limb itself, regeneration was very imperfect. When non-irradiated femoral skin was transplanted to the tibial region of an irradiated limb, the regenerate after amputation was a fairly normal limb (cf. the principle of exclusively distal regeneration, p. 433).

Umanski[4] also examined the part played by the muscles. Muscle from the dorsal surface of the axolotl was inserted into an irradiated limb in replacement of the proper limb muscles; in agreement with some of Voronzova & Krascheninnikova's experiments no regeneration occurred. Even when non-irradiated dorsal muscle was combined with non-irradiated limb skin on an irradiated limb stump, no regeneration took place.

It is interesting that developing axolotl limb-buds are inhibited by the same doses as are required to inhibit their later regeneration (Brunst), and that lesser doses prevent digit formation and complete differentiation (Puckett[1]).

Summing up what has been said on this subject, we seem to see in the processes of amphibian regeneration something analogous to the induction of limb buds in normal development.* It seems likely that the competence of the reacting "undetermined" mesodermal cells is rather definitely limited, and that they are only undetermined in the sense that they may form any tissue within the scope of the organ district where they originate. Inside these limits, however, processes go on which it is hard to call by any other name than that of induction; for example, the influence of the stump-skeleton on the new cartilage, the influence of the muscles on the new bone and the influence of epidermis on all the internal contents of the limb.† Notable in this connection is the fact found by de Giorgi, namely that young blastemas of limb or tail implanted on to other parts of the body may sometimes avoid resorption but if so they grow without differentiating. Long projections are formed covered with epidermis and filled with nondescript mesodermal cells, and these can even regenerate similar projections if they are amputated.‡ One gains the impression that nothing is missing but the inductor or inductors. Perhaps the whole process of regeneration partakes of the nature of homoiogenetic induction, the mesoderm still retaining wide but definitely limited competences.

If the conclusions reached above are correct, tail blastema in adult life has no limb competence but has re-acquired lens competence (see p. 296; Schotté & Hummel; Schotté[6]) and perhaps ear competence (H. S. Emerson). This would amount to a reversibility of determination—a process which we have already envisaged as taking place in ascidian life histories (p. 110). It may be desirable to have a new word for this reappearance of competence. "*Recuperation*" of competence is suggested, and will be so used elsewhere in this book. According to Schotté[6], ectoderm which has lost lens competence will recuperate it if allowed to regenerate over the eye-cup, and Poležaiev[4] could prolong limb competence in anurans by injury (p. 439). Similarly, competence for specific differentiations stimulated by androgens is much more marked in the regenerating anal fin of the fish *Platypoecilus maculatus* than in the normal anal fin (Grobstein).

2·72. Biochemical aspects of regeneration

We now come to the more chemical researches. Since de-differentiation in the stump plays an essential role in preparing the competent blastema, we have a problem not entirely unlike that of insect metamorphosis (see pp. 470 ff.), and it is not surprising, therefore, that interest has been concentrated on autolysis and the proteases generally.

Okunev[1] studied the hydrogen-ion concentration of the tissues of the regenerating amphibian blastema, using a Biilmann-Lund-Cullen electrode. The pH of the normal tissues was found to be 7·2; during the period of wound-healing, it fell to 6·8, rose again slightly, then fell to 6·6, a minimum at the first appearance of the blastema as a noticeable bump (6 days); after which it

* It is surprising that no one has tried the effect of transplanting ear and nose tissues to regenerating blastemas in view of the work of Balinsky discussed on p. 305.

† So also the induction of the ventral locus by the rachis in avian feather regeneration (see the lit. mentioned on p. 303).

‡ Knowledge of what morphogenesis blastemas are capable of performing in isolation *in vitro* would be valuable, but apparently it is not yet to be had.

continuously rose to regain its original value at 18 days after amputation. This process repeated itself regularly in all Okunev's experiments. A later parallel investigation (Okunev[2]), on a Pacific crab (*Paralithodes kamschatica*) showed that the same acidosis could be observed during the regeneration of the claw. The normal level was here 7·0 and the minimum 6·7. Okunev[3] then studied the buffering power of the amphibian blastema, and found that it was at a maximum during the time at which the pH itself was at a minimum. The change was regular and repeatable. The suggestion was made that substances of the nature of chondroitin sulphuric acid (cf. p. 10), produced by the de-differentiation of the cartilage stump, might be responsible for both the low pH and the high buffering power (cf. the work of Nassonov[4] just referred to, p. 438).

Chondroitin sulphuric acid (Levene) = 2 acetyl chondrosin sulphuric acid.
Chondrosin = 1 hexosamine + 1 glycuronic acid.

Okunev[4] also investigated the oxidation-reduction potential of the blastema tissue. From the normal level of 225 mv. the E_h fell to a minimum of 180 mv. at the 10th day from amputation, after which it rose, regaining its normal value at 30 days. The values were steady and repeatable. Okunev suggested that this result might be connected with variations in the sulphydryl content of the blastema. The higher the concentration of glutathione, the lower the oxidation-reduction potential might be expected to be. This hypothesis was tested by Orechovitch[1], who found 20 mgm. % in the normal axolotl tail but just over 45 mgm. % in the regenerated tail from the 5th to the 10th day after amputation, after which the concentration fell to normal. The concentration was also raised (but variable) in the stump immediately beneath the blastema. These results, in agreement with the nitroprusside test as used by Coldwater[2], were not confirmed by Maluf[1], who finds a slightly lower sulphydryl concentration in the regenerating tail blastema (*Rana clamitans*).

	Glutathione mgm. % wet wt.
Distal tip of tail at amputation	27
After 3 weeks' regeneration:	
Regeneration blastema	18
Adjacent piece of stump	22
Proximal part of tail	30

More work will be required on this subject, which is not without interest in view of the facts described in the section on nuclear inductors (pp. 420 ff.).

The other suggestion made by Okunev[6] in order to explain the low rH of the regenerating blastema was that deficiency of blood-supply brings about deficient oxygenation of the cells. This explanation would not exclude the former one.

Of special interest in the comparison of the blastema with embryonic and neoplasmic tissue is the study of its carbohydrate metabolism. Yet as far as can be ascertained little work has been carried out in this direction. Pentimalli's examination[2] of the respiratory and glycolytic rate was limited to repair tissue in mammalian muscle (he obtained a respiratory rate slightly higher than normal, and a small aerobic glycolysis). For the amphibian blastema all that can be said is that two investigators have found a lactic acid content considerably higher than normal; thus Okunev[5] found 18 mgm. % lactic acid in the tissues of the normal limb, and 39 mgm. % in the as yet undifferentiated limb blastema. Vladimirova[1] made a somewhat fuller study, dropping the blastemas into liquid air to avoid glycolytic activity during the preparation of the specimens. This will account for the fact that her values were lower. In the normal limb she found 9 mgm. %; on the 6th day after amputation, 13 mgm. %; on the 16th, 16 mgm. % (the maximum), after which there was a steady return to normal. This may imply a high anaerobic glycolysis analogous to that of embryonic tissues, or there may exist, as in tumours, an aerobic glycolysis. We are reminded also of the accumulation of lactic acid which occurs in the hen's egg a few days after development has begun (*CE*, p. 1051). The further investigation of the carbohydrate metabolism of the blastema would be well worth while.

Another piece of work of Vladimirova's[2] was the estimation of the free amino-nitrogen in the blastema. As in the case of lactic acid, the figures are reasonable ones, showing also a rise to a maximum followed by a return to normal. A decreased amino-acid oxidation may be involved here as well as an enhanced turnover connected with the formation of the new tissues. Vladimirova's work was confirmed by Orechovitch & Bromley, who obtained the following figures:

Percentage of total	Normal tail	Blastema
Non-protein nitrogen	21·6	36·2
Free amino-nitrogen	16·8	35·1

These workers also adapted the ninhydrin reaction to histochemical use, with results confirmatory of the chemical estimations.

Equally important is the state of the tissue proteases in the blastema, and these also have received attention. Bromley & Orechovitch[1] first found that the intensity of autolysis in the blastema, the normal tail, and the stump, was very unequal. In terms of non-protein nitrogen liberated upon 12 hr. autolysis, the figures were (*a*) for the normal tail, 58 mgm. %, (*b*) for the tail blastema, 100 mgm. %, and (*c*) for the stump tissue, 233 mgm. %. The two latter showed thus a more thorough-going autolysis, indicating a higher activity of the tissue kathepsins. When directly tested (Bromley & Orechovitch[2]), using glycerol extracts of blastema and stump material, acting on normal tail as substrate, this proved to hold good, though the differences were not so large. It appeared also that the proportions of kathepsin bound and free differ in the blastema and the normal tail. Whereas free (lyo-) kathepsin seemed equally present, there seemed to be more bound (desmo-) kathepsin in the blastema. Finally Orechovitch, Bromley & Kozmina made comparative estimations of kathepsin activity (in glycerol extracts) in the blastema at different stages of its regeneration, using gelatine as substrate.* The activity was found to reach a maximum fairly early (3 days) after amputation, while the blastema was as yet extremely small. Later it fell

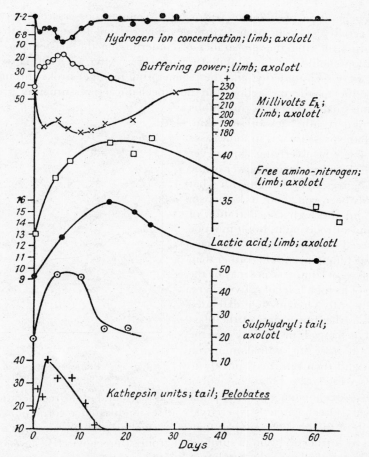

Fig. 252. Biochemical changes in the regenerating amphibian limb and tail blastema.

slowly to normal. The picture was completed by Orechovitch's estimation of sulphydryl compounds in the blastema, already mentioned[1], for as it is known that –SH increases kathepsin activity, the high concentration of it in the blastema would fit in with the increased enzyme activity. This work has been extended by Ryvkina; Ryvkina & Striganova, and Striganova. It agrees very well with the histologically observable fact that de-differentiation occurs in the young blastema.

* A similar piece of work was carried out by Blacher & Tschmutova, but as it involved the mitogenetic ray technique, we shall not consider it here.

Some of the quantitative relationships which have so far been mentioned are placed on the same time scale in Fig. 252, from which it can be seen that there is already a good deal of biochemical regularity observable in the regeneration process. Since both limb and tail regeneration have been studied, the curves are not all exactly comparable, but they show sufficiently well the general course of events.

The regeneration blastema of tail or limb seems to possess considerable histolysing activity. Orechovitch & Bromley covered tail blastemas with flaps of skin, waiting till after the appearance of epidermis on the blastema so that the regeneration should not be inhibited (cf. p. 437). After a certain lapse of time the flaps were seen to decay, and the regenerating member broke through. That this was due to enzyme action and not mechanical pressure was proved by implanting small pieces of normal limbs or tails underneath back or side skin. Although the pressure was in some cases considerable, the overlying skin never gave way. This histolysing power of the blastema agrees well enough with what has already been said about its kathepsin activity and the high concentration of free amino-nitrogen in it, but it must be remembered that the enzyme may have a synthetic function, and this rather than the histolysis of stump tissues may be the meaning of its high level of activity here.

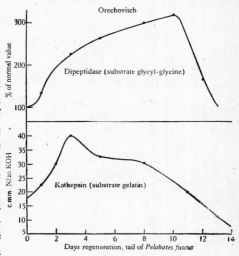

Fig. 253. Protease activities in the regenerating tail blastema.

Subsequently Orechovitch[2] extended his work to the dipeptidase of the blastema. Unchanged in activity on the 1st day it rose to a maximum on the 10th and then returned quickly to normal (see Fig. 253). This behaviour of kathepsin (Orechovitch[5]) and dipeptidase was not found in the regeneration of the avian liver (Orechovitch[3]), where of course no blastema formation occurs.* Kathepsin of the blastema, unlike that of normal amphibian tissue, is not much activated by further addition of sulphydryl groups (Orechovitch[4]), and no amount of extra cystein will raise the autolytic activity (or the kathepsin activity as tested on gelatin) of normal tissues as high as that of the blastema. Mergassova then found that a boiled extract of blastema tissue will activate tissue proteases, but the nature of the thermostable activator is as yet uncertain.

Along this line of thought the question may be raised whether tumour tissue possesses proteases of nature or activity significantly different from normal tissue. There is a considerable literature on this subject, but the best opinion is that no such differences have yet been demonstrated (see Maschmann & Helmert).

* But a certain rise was later found to occur in the early stages of regenerating rat liver (Orechovitch[6]).

In sum, it must be admitted that the chemical work on regeneration is only as yet in the preliminary stages. It will not advance further until some methods are devised (such as the microchemistry of frozen disc sections, of the Copenhagen school) for studying the individual regions of the blastema, rather than the blastema as a whole.

Some of the questions with which this section opened have now been answered. Regeneration is a repetition of ontogenesis in so far as the organ districts involved are the same, but the processes are of necessity somewhat different. There is probably a more restricted set of competences in the reacting material, but within the limits of the organ district in question the material is certainly undetermined. Alterations of protease activity have been shown to accompany regeneration, but we cannot readily compare the biochemical changes with those of normal ontogenetic limb or tail formation since literally nothing is known about these.

2·75. Metamorphosis, competence, and the last inductors

From the standpoint of the present book the interest of amphibian metamorphosis* lies in the fact that the changes occurring during this process are due to the last of the three great inductor types, namely, the developed and functioning glands of internal secretion. The first two of these types have already been described; the nuclear inductors derived from the genes, and the true embryonic inductors, the evocators. The morphological changes of amphibian metamorphosis are the last stages of morphogenesis before the adult form is reached; they correspond, therefore, to the childhood and puberty stages of mammals and man when systematic changes in organ proportions and body form are going on (cf. the reviews of Davenport[2, 3, 4]) under the action of circulating stimulators such as the growth hormone of the pituitary (cf. the reviews of Collip). All such post-natal stages come therefore under the head of "functional differentiation". But in amphibian metamorphosis we may still trace the typical relations of inductor, competence, etc. modified in special ways, which were characteristic of earlier embryonic periods. And the presence of the original organ districts (p. 288) betrays itself in a particularly striking manner, for their reactions to the new circulating stimulator, when it comes, are different, and indeed opposite.

External signs of the metamorphosis of the anuran tadpole are, of course, the growth of the limb buds and the resorption and atrophy of the tail. But there are also internal changes, such as the histolysis and far-reaching reorganisation of the gut, the stomach and the pancreas. The valuable reviews of B. M. Allen[4] and of Aleschin[2] will enable us to omit detailed descriptions of these, and, avoiding any references to secondary problems, we shall only give an account of the essential

---

* I exclude from consideration here the metamorphosis of insects and other invertebrates, which has little in common with that of amphibia except the abstract statement that it *is* the transition between the larval adaptation and the adult form. Something will be found about it, however, above, in connection with genic effects on imaginal discs, etc. (pp. 391 ff.), and below, in connection with determination problems in insects (pp. 463 ff.). Recent work on the biochemistry of the process is discussed on pp. 470 ff.

features of present-day knowledge, regarding metamorphosis in the light of what has been said above on the chain of inductors during development (p. 290).

The pioneer work of Gudernatsch[1] in 1912, in which he demonstrated that thyroid gland tissue fed to tadpoles accelerates metamorphosis (cf. his later review[2]) was followed by the converse demonstration (Allen[1]; Hoskins & Hoskins) that removal of the rudiment of the amphibian thyroid gland inhibits metamorphosis. That iodine was the essential element was soon afterwards shown by Swingle[1], who was able to produce metamorphosis by the action of inorganic iodine not only in normal anuran larvae but also in those from which the thyroid gland had been removed. For urodeles (where the changes concern mainly the loss of the dorsal fin, the disappearance of the external gills and the flattening

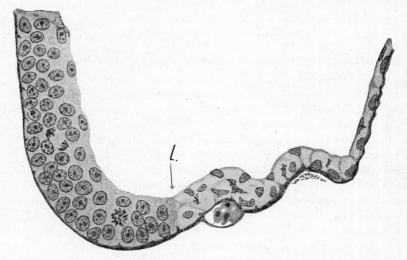

Fig. 254. Differential response of limb-bud and gill tissue to thyroxin; the line separating the two districts, one growing, the other degenerating, is very sharp.

of the body*) the same success was attained, after some difficulties, by Huxley & Hogben in the case of the newt and Blacher & Belkin in the case of the axolotl. But, as will be mentioned later, thyroxin is much more effective than inorganic iodine in stimulating metamorphosis. Again, the converse experiment, that of maintaining tadpoles under conditions of strict iodine starvation, was carried out by Lynn & Brambel; no metamorphosis occurred.†

The most exhaustive study of the changes in the tissues caused by the appearance of thyroxin in threshold concentration in the blood circulation is that of Champy[1]. He made mitotic counts in the various regions, and noted remarkably sharp dividing lines between the different organ districts. For example, Fig. 254,

* Urodele limb buds arise, of course, quite independently of thyroxin (Uhlenluth[4]). Urodele balancers always disappear in accordance with the age of the donor, not the host; this process also appears therefore to be independent of thyroxin (O. Nakamura[5]; Kollross).

† It would be interesting to study the origin and development of the thyroid in tree-frogs, where the tadpole stage is normally suppressed (cf. p. 32); a beginning at this has been made by Glynn on *Eleutherodactylus*.

taken from his paper, shows a section with epidermis of the fore-limb bud on the left and the lining of a branchial cleft on the right. The former has reacted to the thyroid hormone by growth, seen here as numerous mitoses and crowded nuclei; the latter by degeneration, seen here as shrunken nuclei and vacuolation. The limit between the two zones is that between one cell and the next; there is no transitional region whatever.

Many transplantation experiments have shown that the time of metamorphic transformation of an organ depends not on its own age but upon the onset of metamorphosis of the animal into which it has been implanted.* Tissues thus behave invariably *ortsgemäss* with regard to the time of their metamorphic transformations, even if they are considerably younger or older than their host.† This synchrony has been shown for the gills by Kornfeld, for the intestine by Sembrat, for mouth-parts by O. Nakamura[4], for the tongue by Helff[4], and for the skin by Uhlenluth[2] and Voronzova & Liosner. In Vrtelówna's interesting experiments on *Pelobates* it was found that when eyes from younger larvae were transplanted forwards to older ones, or *vice versa*, not only did the morphological changes of the graft eye occur at the same time as that of the host (Uhlenluth[1]), but physiological changes also, such as the reactivity of the iris to light and to drugs, occurred simultaneously. These facts all show that competence to react to the stimulus of thyroxin must arise some time before the threshold concentration in the blood is attained. Otherwise younger, grafted, eyes would not be able to react before their presumptive metamorphic age as they do.

In most of the above cases the grafts act purely passively. If they are taken from stages after the beginning of metamorphosis, however, they may induce metamorphosis in their younger hosts, perhaps because they contain a small amount of thyroxin (Liosner & Voronzova[4]).

On the other hand, many workers (Romeis; Jarisch; Champy[1]) have found that before a certain age is reached the thyroid hormone has no effect on the tadpole's tissues. This fixes the point of onset of competence. It is probable that the orderly succession of steps in metamorphosis (growth of hind-limbs, shortening of intestine, emergence of fore-limbs, loss of chitinous beak, tail atrophy, widening of mouth, etc. etc.) is due to a series of thresholds of response (Blacher; Etkin[1]; Allen[2]), for it is noticeable that the weaker the concentration of thyroxin the more the effects produced are restricted to the earlier members of the series. At the same time, Allen's quantitative experiments[2] showed that time of exposure to thyroxin is also a factor, and that given a sufficiently long time, weaker concentrations will perform as much as stronger ones. This suggests that the metamorphic changes depend on rates of different chemical reactions, catalysed by thyroxin,

* These experiments parallel those of the school of Harrison on "marking-time" and "catching-up" already described (p. 347).

† There is an analogy here with insect metamorphosis. Frew found that isolated imaginal discs of the blowfly would invaginate when explanted into a pupal lymph medium, but would not do so if the medium was made with larval lymph. Hormonic action on pre-existing competence is thus a certainty here also. Analogous work on the silkworm by S. Fukuda[1] shows the same thing; larval ovaries transplanted into pupae undergo a greatly accelerated development.

or for which thyroxin is in some way necessary.* The later phases would thus be those with the slowest reactions (cf. the kinetics of gene action, discussed on pp. 391 ff.). It is interesting in this connection that metamorphosis will not go on at temperatures below 7° (Taniguchi; Belkin) or 5° C. (Huxley[8]) though the larvae live unharmed. At higher temperatures, according to Tchepovetsky, the speed of metamorphosis in the presence of thyroxin is up to a certain point directly proportional to the temperature, though higher temperatures alone have no effect (see also pp. 336 ff., 516).

Some light has been thrown upon the problem of competence by experiments on those forms which have lost the power of metamorphosis, and which reach the stage of sexual maturity while still retaining the "larval" characteristics (neoteny). The evolutionary significance of this fact has been discussed by de Beer[2]. Here we are interested in the mechanism of neoteny. With the classical example before them of the Colorado axolotl,[†] *Amblystoma tigrinum*, which may assume either the neotenic or the metamorphosing state, and in which metamorphosis may be induced by iodine (Hirschler; Blacher & Belkin), a number of workers tried to induce metamorphosis in the perennibranchiate, *Necturus*, but without success.[‡] When investigations of the hormone content of the glands of this animal were made, it was found that the thyroid gland contains active thyroxin[§] (Swingle[2]) and the pituitary contains active pituitrin[||] (Charipper & Corey); cf. the role of the pituitary gland (p. 454 below). The inhibition must therefore lie in the tissues; a loss of competence, not a loss of inductor. This case would therefore clearly join the facts known about the presence of embryonic evocators in adult tissues (p. 172) and the loss of balancer competence in the axolotl with the retention of the balancer inductor (p. 342). But we have also to take into account certain substantial differences between the blood-vessel systems in the axolotl and *Necturus* (Figge[1]; Gilmore & Figge), such as the absence of the sixth branchial arch in the latter. It is thought that the attainment of an efficient pulmonary circulation in perennibranchiates has been rendered difficult by the disappearance of a specific anatomical structure. At the same time, the competence of the external gills for reacting to the thyroid stimulus by atrophy has certainly disappeared. Figge[2] was afterwards able to show that ligating the sixth branchial arch in the axolotl delays, though it may not entirely prevent, metamorphosis induced by thyroxin injections, especially if the temperature is low so that the metamorphic tendency is small. Yet in depulmonated axolotls meta-

---

* It has been found, too, by Voronzova & Liosner and others that, after a certain stage has been reached, metamorphosis will go to completion without further additions of thyroxin. This suggests that the mechanism might be that of a co-enzyme. Skin from an axolotl half-way through metamorphosis, transplanted to an axolotl receiving no thyroid, will complete its metamorphic changes in the new situation.

† The axolotl's metamorphosis, first reported by v. Chauvin in 1885, has a review all to itself, that of L. Marx.

‡ Except for the uncertain effects which Gutman was able to produce by the simultaneous use of thyroxin and adrenalin.

§ Active because when implanted into frog tadpoles metamorphosis was induced.

|| Active because when the anterior lobes were implanted into frog tadpoles metamorphosis was accelerated (thyreotrophic hormone).

morphosis is still possible, cutaneous blood-vessels taking on an enhanced respiratory function (Garber; Koschtojanz & Mitropolitanskaia).

The effect of thyroxin has also been tried on other vertebrates.* It seems that Weiss[8] was able to accelerate tunicate (prochordate) metamorphosis by thyroxin though Bradway could not, and according to Baumann & Pfister dilute thyroxin solutions cause the disappearance of aortic arches in trout embryos. On the other hand, Horton found no effect of thyroxin on the ammocoete → lamprey metamorphosis. The ammocoete thyroid gland contains no iodine and cannot produce metamorphosis in the frog tadpole, though that of the lamprey can.

We come now to the most striking regressive change at metamorphosis in amphibia, namely the resorption and atrophy of the tail. It was originally thought by Barfurth[1] in 1887 that the growing urostyle (the unsegmented posterior portion of the vertebral column) occluded the caudal blood-vessels so that atrophy followed death by asphyxia. This view was accepted as late as 1922 by Bradley[2] and W. Morse, who were the first to regard the tail resorption as an autolysis. But Helff[5] a few years later decisively disproved it by extirpating the urostyle before metamorphosis, after which tail resorption proceeded quite normally. He also showed that ligation of the aorta gave no tail atrophy (Helff[1]), but that injection of small amounts of lactic or butyric acids would do so, autolysis beginning at the site of injection and spreading over the whole tail.

Helff's conclusion was clinched by the demonstration of Helff & Clausen that anuran tails transplanted to the back underwent there the typical atrophy of metamorphosis. The converse of this experiment was made by Lindeman, who transplanted body skin to the tail before metamorphosis, with the result that no atrophy of this particular epidermis took place. Competence to atrophy in response to thyroxin therefore not only appears at a definite time but also only in definite places. Brilliant establishment of this principle came from the work of Schubert, who transplanted limb buds to the tail, and still more from that of Schwind, who transplanted eye-cups there. As in Fig. 255 (opp. p. 464) the eye moved gradually forwards during tail resorption, and finally came to rest in the sacral region at the conclusion of metamorphosis. Clausen, transplanting skin and muscle from different regions of the tail to the back, found that those from the anterior part atrophied more rapidly and completely than those from the posterior part; a finding in agreement with Dunihue's experiments on rate of autolysis *in vitro* at pH 6·6 of parts from anterior and posterior regions. Another interesting experiment was that of O. Nakamura[2], who exchanged, at early tail-bud stages, anuran (*Rana*) with urodele (*Hynobius*) tails. Anuran tails on urodele bodies atrophied completely at metamorphosis; urodele tails on anuran bodies were first covered with anuran epidermis and this then atrophied at metamorphosis, leaving the urodele tissues naked.

* Including the chick embryo (injections into the egg with little result, Guelin-Schedrina[1]). The thyroid of the chick embryo, however, can respond to the thyreotrophic hormone of the pituitary (Studitsky; Woodside[2]). There is a large literature on the development of the thyroid gland and its hormone in birds and mammals, which can be approached through the papers of Daineko[1]; Entin; M. L. Hopkins; Aleschin[3]; Artemov & Valedinskaia; Lelkes; Marza & Blinov. On the correlation of thyroid activity in embryonic life with that of other glands, see M. Aron's review.

If tail resorption is an autolysis, an acidosis, at least local, should be expected. The earlier attempts to demonstrate this were not very convincing, but Aleschin's work[1], which gave a pH of 7·15 for the normal tissues and 6·65 for the tail during resorption, may be accepted. The acidity was perceptible some time before the first sign of morphological change. Parallel work by Helff[9] on the blood pH gave parallel results, as the following figures show:

	pH
Normal pre-metamorphosis	7·50
Intermediate stage I	7·39
Intermediate stage II	7·27
End of metamorphosis	7·20
Three weeks later	7·18
Adult (Kamm; L. H. Hertwig)	7·4–7·6

The acidosis seems therefore to be general, though it may be more extreme locally. We do not yet know whether the tissue reaction of the degenerating gills* is also more acid than normally. It is to be noted that the period of maximal resorption rate of the tail follows immediately after the period of maximal growth rate of the fore- and hind-limbs (Champy[1]; Irichimovitch & Lektorsky; Voitkevitch[2]), see Fig. 256. Any suggestion that the material of the former is necessarily used for the construction of the latter has to reckon, moreover, with the measurements of O. Nakamura[3], who found that limb growth in anuran larvae proceeds just as fast after the tail is removed as normally. During metamorphosis the water-content of the whole body falls slightly (from 92·8 to 89·4 %), as does that of the tail (from 93·8 to 89·6 %), while that of the limb-buds rises (from 75·0 to 86·4 %)— a fact contrary to the association between high water-content and pluripotence, but probably connected with the changing proportions of the skeletal constituents (Irichimovitch[2]). These changes proceed in urodeles, where the tail is retained, just as in anurans, where it is lost. From the interesting weighings of Voitkevitch[2] (see Fig. 256) we see not only that tail resorption follows upon the maximum growth period of the hind-limb, but also that gut reorganisation does so, and that both these processes are immediately preceded by a spurt of growth in the thyroid gland itself. The data from Champy[1] show at the same time that in thyroid-accelerated metamorphosis the same precedence of limb-bud growth over tail resorption is to be seen.

The nitrogen metabolism might well be expected to be affected by tail resorption. Van der Heyde was unsuccessful in finding a corresponding rise in total nitrogen excreted, but this was observed by Blacher & Efimov.† Liosner & Blacher found an 82 % increase of free amino-nitrogen in the tail tissues preceding any morphological changes, and later an increase in polypeptide nitrogen. In the atrophic gills there was a 45 % rise in free amino-nitrogen. Though the total nitrogen of the blood was raised at the beginning of metamorphosis (by protein substances), the free amino-nitrogen was constant, later decreasing. Substantial changes in the activities of enzymes in the gut (Blacher & Liosner; V. Doljanski) and in the stomach (Liosner[1]) during their reorganisation have been observed. The proteases of the blood may also be increased during metamorphosis, according to von Falkenhausen, Fuchs & Schubert, but this requires confirmation.

* Cf. the interesting histological account by Grotans[1].   † See also p. 625.

On the whole the metabolic work so far done tells us little regarding the nature of the characteristic competences for atrophy of tail and gills, but a new attack in the light of modern knowledge could readily be made. It is still uncertain to what extent leucocytic infiltration plays a part in tail resorption. A dissociation

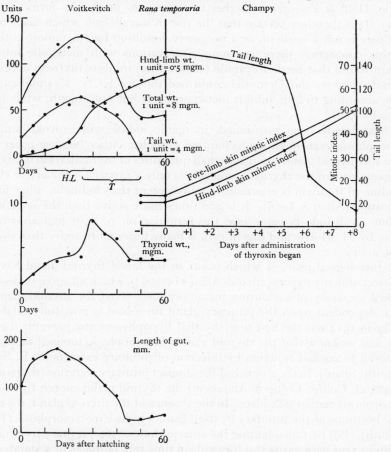

Fig. 256. Temporal relationships of organ changes in amphibian metamorphosis. *HL*, Hind-limb maximal growth-rate period; *T*, tail maximal absorption-rate period.

between regression and development has been worked out by Puckett[2], whose carefully graded X-ray doses favoured resorption of tail and gills but inhibited the differentiation of the limbs.

What effects does the thyroid hormone have on the oxidations of the metamorphosing tadpole? In connection with another problem, that of biological specificity, we have already had occasion to mention this point (p. 264). Groebbels[1] was the first to show that oxygen consumption of the larva rises both absolutely and relatively on metamorphosis, whether natural or provoked by thyroxin, and this has since been fully established (Abelin & Scheinfinkel; Huxley[5]; Helff[2];

Schwartzbach & Uhlenluth; Wills; the sole dissentient voice is Etkin's[2]). But dinitrophenols, which stimulate metabolic rate (cf. p. 570), have no effect on amphibian metamorphosis (Cutting & Tainter). Conversely, acetyl-thyroxin, which has no effect on metabolic rate, still stimulates metamorphosis (Kendall; Swingle, Helff & Zwemer), whether in the mono- or the di- form (Titaiev & Summ). It is therefore certain that the rise in metabolism which takes place in the tadpole is not a cause of, or a necessary condition for, metamorphosis.

In this connection there are some observations with metabolic inhibitors. Hoffman found that methyl-cyanide (acetonitrile) inhibits thyroxin acceleration of metamorphosis, and Demuth[1] confirmed this for KCN. Cyanide and iodo-acetate, according to him, inhibit metamorphosis but not growth, while fluoride inhibits both.

As has already been mentioned, inorganic iodine will provoke amphibian metamorphosis even in thyroidectomised tadpoles, either because other tissues besides the thyroid gland can make small quantities of thyroxin from it, or because the organic part of the thyroxin molecule is only a carrier for the active element. But iodine in thyroxin is enormously more potent than iodine in other forms of combination (Allen & Lein). It is 300 times more active than the same amount of iodine as di-iodo-tyrosine (see the formulae on p. 256) first shown to be effective by M. Morse, and this in turn is many times more active than inorganic salts of iodine.

The histological changes which occur in the larval thyroid gland have been the subject of many papers (cited in Allen's review[4]), which all agree in describing a marked secretory phase during metamorphosis. That the thyroid gland is in its turn dependent upon the pituitary gland for a lead is now fully established. L. Adler in 1914 was the first to prove that hypophysectomy prevents metamorphosis, and we know that the thyroid glands of hypophysectomised tadpoles can be restored to normal function by injection of pituitary extracts (P. E. Smith & I. P. Smith; Spaul). In the absence of the anterior pituitary secretion (thyreotrophic hormone, cf. Collip; Collip & Anderson) the thyroid gland cannot function and metamorphosis cannot take place. In the absence of the thyroid gland, the thyreotrophic hormone of the pituitary by itself cannot induce metamorphosis (Figge & Uhlenluth). But by transplanting the anterior lobe of the pituitary from tadpoles of different ages backwards and forwards in time into tadpoles of a standard pre-metamorphic age, Allen[3] and, later, Etkin[3], were able to show that only those from later stages would induce metamorphosis. Pituitaries from younger stages would not do so.* Hence the initial lead in metamorphosis is given by the pituitary. But upon what "alarm-clock" the pituitary itself depends is as yet quite unknown.†

* Moreover, Etkin & Huth could obtain precocious activation of the thyroid gland by transplanting it to a site near the pituitary. Hence a "field" of thyreotrophic hormone may be envisaged (cf. pp. 312, 418, regarding inductors of varying diffusibility).

† By taking advantage of the unequal distribution of basophile and eosinophile cells in cattle pituitaries, Voitkevitch[3] was able to identify in amphibian experiments the thyreotrophic and gonadotrophic hormones with the former and the growth hormone with the latter. Since the thyreotrophic hormone comes from the anterior lobe, its origin might be referred back to the inductive stimulus from the brain whereby the gland is first formed (see p. 302). In this way the hormone could be linked in with the chain of inductors (cf. Kleinholz; Chen, Oldham & Geiling).

Another interesting experiment was that of Bytinski-Salz[2], who exchanged during early embryonic life the presumptive pituitary regions of the axolotl *Amblystoma mexicanum*, which never normally metamorphoses, and a metamorphosing race of *Amblystoma tigrinum*. No effects on the growth or metamorphosis of the former were found but the latter responded, showing that the pituitary of the non-metamorphosing form contains the inductor, but that its tissues are not competent to react to it even when provided from pituitary tissue known to be effective. But the experiments were not entirely conclusive, for it was observable that in many cases the animals later developed the colour characteristic of hypophysectomised larvae; these on dissection showed that the transplant had disappeared.*

Allen[4] well says that the endocrine glands "take their place alongside the genes, the physiological gradients, and the organiser as an important part of the chain of influences that control development". In the succeeding section we shall see how true this is of insect development, wide though its divergences are from the phenomena in vertebrates which we have been studying in the foregoing pages.

2·8. Determination in insect development and metamorphosis

With this section we leave the domain of vertebrate experimental morphology and come to the invertebrates. As has already been pointed out in the discussion on mosaic and regulation eggs (p. 131), many eggs of invertebrates show mosaic development. Although this may be by no means so absolute as was formerly thought, it is sufficiently striking to have held back considerably our knowledge of the determination processes in invertebrates. Only in certain cases, such as the typically regulation eggs of echinoderms, which will be the subject of the following section, and those of some insects (for among insects all grades of regulation are to be found), has substantial progress been made. Unfortunately up to the present no biochemical work has been done on the problems of determination in insects, but the processes there are so peculiar, and parallel in so interesting a way those which we have seen at work in vertebrate development, that they merit a careful treatment.

Insect embryology has been the subject of many investigations. Those which are relevant but which for reasons of space cannot be referred to here will be found in Seidel's accounts[6, 7] of the work of his school, and in the admirably lucid review of Richards & Miller. The development of an insect is so different from that of an amphibian that a short description of it must first be given.

The yolk of insect eggs is not concentrated at one pole as in the eggs of vertebrates, but occupies a central position (hence the term "centrolecithal eggs") where it is enmeshed in a slight cytoplasmic reticulum and surrounded by a thin layer of peripheral cytoplasm. Cleavage follows the arthropod type; there is a dissociation between nuclear and cytoplasmic division, so that a large number of nuclear divisions occur first, till more than two hundred nuclei have been formed, and then the cell-divisions begin to form radially when the nuclei have migrated into the peripheral cytoplasm. Normally no development will follow if the early cleavage nuclei

---

* It is interesting that the water-regulatory action of the posterior pituitary does not begin till metamorphosis (Howes).

are killed (Reith[1]; Pauli), but in one very exceptional case, reported by Seiler, cell-divisions took place in the absence of any nuclear divisions in hybrids of the moth *Phragmatobia* (cf. cell-division without nuclear division in echinoderms, p. 510, and later, under Dissociability). Insect development is thus fundamentally dependent upon the proper migration of the cleavage nuclei, and as we shall see later, their presence or that of their products in a certain region of the egg is essential for all further development. Many investigations have established the fact that the early nuclear divisions are, as in other organisms, equipotential. Thus neither abnormal nor delayed nuclear distribution, nor even the elimination of one of the first two nuclei, will prevent normal development (Seidel[4]). At first synchronous, the nuclear divisions lose step after some time, usually when they begin to enter the peripheral cytoplasm.

The mechanism of the migration of the nuclei remains very obscure, but Eastham[1], who studied the development of the butterfly *Pieris*, speaks of centrifugal streaming. The nuclei, both during and between mitoses, exist in minute protoplasmic islands, with comet-like tails stretching backwards in the direction of migration. Other observers quoted by Richards & Miller give similar accounts. At present we are far from any understanding of protoplasmic motion. One has only to read the description of the streaming and churning of the protoplasm of the egg of a polychaete worm, *Sabellaria*, filmed by J. E. Harris, for example, to realise the extent of our ignorance of these processes. When to this we add that special nuclei, the vitellophages, no doubt important for yolk absorption, either remain behind in the yolk when most of the other nuclei have entered the peripheral cytoplasm, or else subsequently migrate from the blastoderm into the yolk (Sehl), we appreciate further the complexities of the situation. Fortunately it has been possible, taking the mechanism of these early stages for granted, to find out a good deal about the determination process in those cases where it occurs later.

We have already had occasion (p. 218) to mention the long-known generalisation of Hallez that the polarity of the three embryonic axes is determined already in the ovary before laying. The egg is like a crystal with the same lattice as its mother. This holds good generally for insects but not universally, for in the bug *Pyrrhochoris* Seidel[1] found that the longitudinal embryonic axis may vary from being coincident with the longitudinal egg axis to being transverse to it. In insects such as the grasshopper *Melanoplus*, moreover, the embryo begins to develop in the reverse position, with the head towards the posterior end, where the micropyle is, and attains its normal position later by the active movement known as "blastokinesis" (Slifer[3]). We shall return to this point (p. 463).

2·81. Organiser phenomena in insects

Insects show all variations between mosaic and regulation eggs. The work of Seidel's school[2] has suggested the following series:

Regulative	Incompletely regulative	Mosaic

Odonata—Hemiptera—Orthoptera—Coleoptera—Hymenoptera—Neuroptera—Lepidoptera—Diptera

This must, however, be regarded as somewhat tentative, for practically nothing is known experimentally about the Hemiptera. Moreover, only a single species of Odonata has been studied, two Orthoptera, two higher Hymenoptera, one Neuropteron, five Coleoptera, one Lepidopteron, and three higher Diptera. Possibly considerable variation within the groups may be expected, particularly since the parasitic polyembryonic Hymenoptera (cf. p. 228), which have not so far been studied experimentally, would be expected to be highly regulative. The highly regulative eggs (e.g. the dragon-fly *Platycnemis* (Seidel[3]) and the camel-cricket *Tachycines* (Krause[1])) already possess at fertilisation an irrevocable antero-posterior and dorso-ventral determination. But shortening the longitudinal axis produces perfectly normal dwarf embryos, and rudiments split by injuries easily produce duplications of any organ. This may go so far as to produce one dwarf embryo almost inside another (Seidel[2]) if yolk-fissures are made by bending the egg at the beginning of development. In the incompletely regulative eggs (e.g. the honey-bee *Apis* (Schnetter), the ant *Camponotus* (Reith[2]), the beetle *Bruchus* (Brauer & Taylor), the moth *Ephestia* (Maschlanka) and the alder-fly *Sialis* (du Bois)) power of regulation is lost much sooner than in the previous group, at or shortly after blastoderm formation. In the mosaic eggs (e.g. the beetles *Calligrapha* and *Leptinotarsa* (Hegner); the blowfly *Calliphora* (Pauli); the fruit-fly *Drosophila* (Sonnenblick; Howland & Sonnenblick; Howland & Child); or the beetle *Agelastica alni* (Smreczyński)) injuries always result in losses of the corresponding parts, and all development from fertilisation onwards is self-differentiation.*

To give reality to what has been said, the illustrations of Fig. 257 (taken from Seidel[6]) should be examined. The stage of nuclear divisions is followed by the migration of the nuclei into the peripheral cytoplasm and the formation of two areas of especially intense cleavage which shortly join to form the embryonic blastoderm. The future embryonic axis is then sketched out; it bears in the centre the median plate which later will sink in forming the neural groove and ultimately the chain of ganglia in a way not unlike the formation of the central nervous system in vertebrates, save that this of course is on the ventral, not the dorsal, surface of the embryo. The first sign of segmentation (analogous with the somitic segmentation but of course to be pushed much farther) appears. as a groove between the second maxillary and the first thoracic segments. The embryo is now in two layers. At a later stage the appendages become visible and invaginations at the anterior and posterior ends of the body form the stomodaeum and proctodaeum respectively. The endoderm is formed at this stage, growing backwards from the tip of the stomodaeal and proctodaeal invaginations.† At last the form of the completed insect or larva appears.

The basic plan of primary determination in insect eggs was revealed by Seidel's classical work[3] on the dragon-fly *Platycnemis*. He found that when cauterisations or constrictions of the egg with a knot of child's fine hair were applied, what happened afterwards depended on the exact time at which they were made. This result is shown in Fig. 258. A constriction applied in an early stage

---

* It is rather unfortunate that *Drosophila*, the classical object of genetic research, should happen to have a mosaic type of development. A special monograph, that of Poulson[2], has been devoted to its descriptive embryology.

† The above short description does little justice to the highly complicated nature of the formation of the germ-layers in insects; still an obscure subject, as the reviews of Eastham[2] and of Roonwal[2,3] and the book of Johannsen & Butt show. We have noted already (p. 367) clear cases of genetic interference with normal germ-layer formation in *Drosophila*.

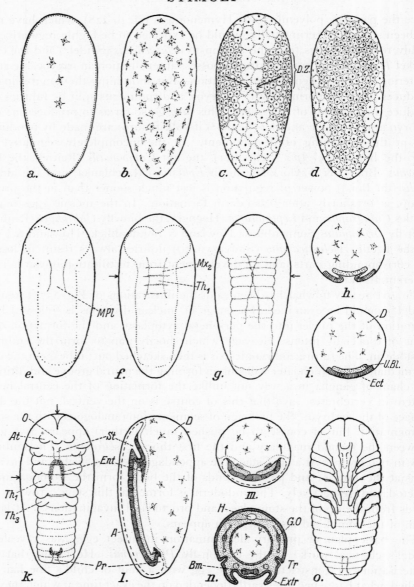

Fig. 257. Development of a typical insect egg. *a*, 4-nucleus stage; *b*, 128-nucleus stage; *c, d*, formation of blastoderm; *e*, formation of embryonic axis with head-folds; *f*, beginning of segmentation; *g*, continuation of segmentation forwards and backwards from differentiation-centre; *h*, cross-section at arrow in *f*; *i*, cross-section at arrow in *g*; *k*, appearance of organ- and limb-rudiments, starting from 2nd maxillary (*Mx₂*) and 1st thoracic segment (*Th₁*); *l*, longitudinal section of *k*; *m*, cross-section at arrow in *k*; *n*, cross-section of *o*; *o*, completed organogenesis.

DZ, Differentiation-centre; *MPl*, middle plate; *D*, yolk; *UBl*, under plate; *Ect*, ectoderm; *St*, stomodaeum; *Ent*, endoderm; *Pr*, proctodaeum; *A*, amnion; *O*, upper lip; *At*, antenna; *H*, heart; *Bm*, ventral neural cord; *Extr*, appendages; *Tr*, trachea; *GO*, gonad.

right at the posterior end (*a*) had no effect and normal development followed (*b*). But if it was applied farther forwards, development was subsequently entirely inhibited (*c*). If, however, it was applied at the same forward point some time after the formation of the blastoderm (*d*), again normal development followed (*e*). Parallel

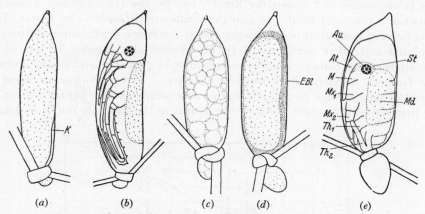

Fig. 258. Effects of constriction on a regulative insect egg (see text).

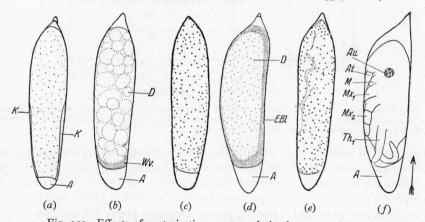

Fig. 259. Effects of cauterisation on a regulative insect egg (see text).
  · (*a*), (*b*) Increasing extent of cauterisation; (*c*) surface view of (*b*); (*d*), (*e*) only the extra-embryonic blastoderm develops; (*f*) normal development when cauterisation is late.

Key for Figs. 258 and 259. *K*, Blastoderm; *At*, antenna; *Au*, eye; *D*, yolk; *EBl*, extra-embryonic blastoderm; *M*, mandible; *Md*, midgut; *Mx*, maxilla; *Th*, thoracic leg; *St*, stomodaeum; *A*, cauterised portion; *Wv*, wound healed.

results were obtained by destroying the posterior regions of the egg with cautery; if this was done early, development was inhibited, if it was done later, development was not inhibited (Fig. 259). The time effect, graphically plotted, gave striking regularities; thus Fig. 260 shows how as time went on the cauterisation had to be pushed farther and farther forwards in order to obtain inhibition of development; cf. the arrow in Fig. 259 (*f*).

These and other facts led to the conceptions of the *Bildungszentrum* and the *Differenzierungszentrum*, well translated respectively by Richards & Miller as "Activation centre" and "Differentiation centre". The existence of these centres has been demonstrated* in a dragon-fly (*Platycnemis*) by Seidel[3], two ants (*Camponotus* and *Lasius*) by Reith[2], two beetles (*Bruchus* and *Sitona*) by Brauer & Taylor and Reith[3], and in the mealworm (*Tenebrio*) by Ewest. They must be very general, probably as general as the single organisation centre is in amphibia and other vertebrates. In *Platycnemis* the anterior border of the activation centre coincides with the presumptive posterior end of the embryo; how far back it extends posteriorly is unknown. Its action depends upon the arrival of the cleavage nuclei, and some consequent reaction between them and the material of the centre. Thus if the arrival of the cleavage nuclei is delayed

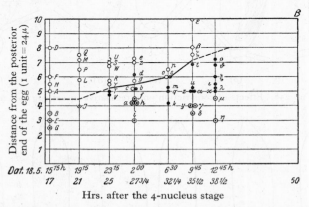

Fig. 260. Progressive change in the position of cautery required to inhibit the development of a regulative insect egg (see text). ○, Inhibition of development; ⊙, normal development; ●, formation of blastoderm; ◉, head defects.

(as the result of killing one of the first two nuclei or constricting the egg and later removing the constriction) the function of the centre is delayed. If, however, the nuclei are absolutely prevented from reaching the centre, no axiation and development ever follow, although an embryonic blastoderm is formed (Fig. 261). It is as if the canvas is ready, but the painter does not arrive (cf. anidian embryos of vertebrates, referred to on p. 223). Partial constrictions prove that the presence of nuclei is essential in *Platycnemis*, but in the ant *Camponotus* Reith[2] has shown that the nuclei are not essential, and that some diffusible substance preceding them is involved. In the two beetles Brauer & Taylor report that nuclei themselves are necessary. It is probably not unsafe to conclude that in some of these cases the active substance, whatever it is, can diffuse out of the nuclei and all over the egg, while in other cases it cannot do so, or not in sufficient quantity, until at the activation centre some releasing change occurs in the permeability of the nuclear membranes. We are strongly reminded of what has already been said (p. 418) regarding nuclear inductors.

* See the review of Krause[2].

Without the activity of the activation centre there will be no axiation and no development. But what precisely does this centre do? From the constriction and cautery experiments already described Seidel[5] concluded that a new substance, essential for development, was produced and diffused forwards towards the point where differentiation first visibly begins. The curve in Fig. 261 traces, indeed, the very progress of this diffusion. The most remarkable thing is that in some eggs this diffusion can actually be followed, for along with it there goes "a structural change in the yolk, which, with the loss of the finer yolk-particles, becomes more solid and more transparent" (Seidel[7], p. 415). Nothing is known about the significance of this colloidal change, but when it reaches the point where visible differentiation will shortly afterwards begin, the differentiation centre, an actual contraction of the yolk occurs. This spread of visible differentiation from the presumptive prothorax is a very general phenomenon in insects

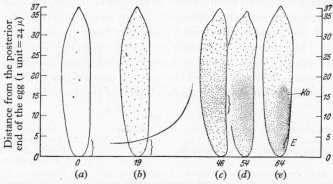

Fig. 261. Diagram showing the part played by the activation centre and the differentiation centre in the development of a regulative insect egg. (a) Activation centre not yet activated; (b) beginning of its function; (c) differentiation centre; (d) blastoderm arising; (e) completion of blastoderm. *Ko*, headfolds.

(Schnetter), so the differentiation centre is probably very general also. In the mosaic types it would presumably be stimulated at or before fertilisation. In the honey-bee, an incompletely regulative type, Schnetter found staining differences during cleavage with dyes such as thionin (cf. above, pp. 133 ff.). Unless the differentiation centre is present uninjured, no differentiation will take place, but before the product of the activation centre has reached it, it is inert. The direct action of the differentiation centre appears at least to involve a wave of contraction in the yolk spreading from it anteriorly and posteriorly, the yolk contracting from the chorion in such a way as to force the sheet of blastoderm cells to fill the vacant space. The work of Krause[1] on the camel-cricket and of Schnetter on the honey-bee confirms in general this picture and suggests that the differentiation centre is concerned in determining the boundaries of the various organ regions which at this time shift out of the centre in a stream backward and forward to their definitive positions. But it must be remembered that our knowledge of the determination process in insects is greatly hampered by the lack of the possibility of transplanting pieces at different stages in *herkunftsgemäss-ortsgemäss* experiments.

Since the establishment of the existence of the two centres in the insect egg there has been much discussion regarding their relation to the amphibian organisation centre. That the insect axis should form on the ventral surface and the amphibian one on the dorsal surface is a similarity, not a difference, since axiation begins on that surface of the embryo which will form the central nervous system. Weiss[16] has made the suggestion that the "activating and organising functions" which in amphibian development are combined in one centre are in the insects separated into two centres. From the context, however, it is probable that he had in mind the distinction made in this book between evocation and individuation, i.e. the chemical stimulus to neural determination and differentiation on the one hand, and the forces, probably of field character, which govern the regional differentiation of the evocated axis. This distinction will not cover the facts of insect development.* On the contrary Richards & Miller's translation of *Bildungszentrum* as activation centre is especially happy, because it is better to regard what happens in the activation centre as analogous to the setting free (activation) of the amphibian primary evocator from its inactive combination (see p. 170). Then instead of having to travel only the very short distance separating its site of origin in the archenteron roof from its site of reaction in the dorsal ectoderm, it has to travel the greater part of the length of the insect egg. These considerations may be made clearer in the form of a table:

		Cause	Effect
Amphibia	Beginning of gastrulation	Change of metabolism	Liberation (activation) of evocator
	Later gastrulation	Evocator	Neural determination and differentiation
Insects	Activation centre	Arrival of nuclei or substance produced by nuclei	Formation of chemical stimulating substance
	Differentiation centre	Arrival of chemical stimulating substance	Contraction of yolk microfibrils, and determination processes

On this scheme the product of the activation centre would logically correspond to the primary evocator. At the same time it invites the thought that in amphibia also the local change in metabolism which occurs at the beginning of gastrulation may be conditioned by chemical products of the nuclei, and as we have already seen (p. 203), the work of Brachet[18] on the fixed –SH distribution is pointing fairly clearly in this direction. It should furthermore be possible to reveal a metabolic change in the activation centre when the migrating nuclei or their product arrive in it. It should also perhaps be possible to activate the differentiation centre of an insect egg by chemical preparations of the activation centre. One would like to know whether denaturing the proteins, for example, of the activation centre, would liberate any substance capable of stimulating the differentiation centre, without any nuclei or their products ever having been involved.

* Spemann's suggestion (made in ch. 18, which was omitted from the English edition of *Embryonic Development and Induction*) was less helpful; he compared Seidel's two centres to the two determinations of double assurance. But the necessity for this concept is now doubtful (see pp. 293, 308).

From what has so far been said it is clear that actual contractile processes of the yolk micro-fibrils (cf. p. 674) play a greater part in insect development than in that of other forms. We are reminded of the curious contractions of the blastoderm in fishes already mentioned (p. 74). And in the later development of some insects, the role of contraction processes is still further notable. As has already been mentioned, in the grasshoppers, embryonic development begins with the cephalic end of the embryo at the posterior end of the egg, but at a relatively later stage, when the length of the embryo is just less than half that of the egg, "blastokinesis" (sometimes called "revolution")* occurs, a process by which the embryo moves round until its head is at the anterior pole of the egg. From the description of Slifer[3], this process, which does not begin to occur until a good deal of liquefaction of the egg-contents has taken place due to imbibition of water (see p. 37), is carried out by waves of contraction originating posteriorly and running along the side of the abdomen. It takes about a day. It ends by a slow movement on the long axis which brings again the ventral surface of the embryo and yolk to lie against the concave surface of the egg. In grasshoppers which go through a diapause, such as *Melanoplus differentialis*, the first contractions begin before the onset of the hibernation period, but die away and are resumed when development begins again five months later. If blastokinesis fails to occur hatching is impossible. Unstriated muscle fibres which might cause it have been described (Slifer[5]).

Although blastokinesis is involved only in a fairly limited group of insect eggs with much yolk (Imms), it shows, taken together with the yolk contractions at the differentiation centre, how important such phenomena may be in some types of development. And it recalls the suggestions made above (p. 146) that even in amphibian development the part played by contractions of fibrils may be at some stages not altogether negligible.

As regards later hints of dependent differentiation (inductions) in insect embryos there is some evidence that the formation of the intestinal tract may be the result of a dependent differentiation (A. G. Richards). Reith[1], too, has found that in such a mosaic egg as that of the house-fly the rudiments of the midgut are the only parts capable of development beyond their prospective significance. Here, if either the stomodaeal or proctodaeal invaginations are absent, the gut will form wholly from one or from the other. It appears, too, that muscles only differentiate properly in the presence of the uninjured ganglion of their segment (Kopeć[2]).

2·82. Post-embryonic inductors in insects

We come now to the facts which are known about development after hatching in insects. It might be logical to rule out such a subject from the field of this book, but for the fact that relatively few insects hatch from their eggs in a fully developed state. The larval forms of the holometabolous insects are but the carriers of those imaginal discs from which, during metamorphosis, the structure of the adult will be formed, and which, in the form of districts, may already have

---

* The insects are not the only group which have a "revolutionary" development—it occurs also in some cephalopods (Portmann[1]).

existed in the egg. Considerable advances have been made in our knowledge of the chemical control of these post-embryonic processes, where, nevertheless, the problems of competence and induction continue to be met with.

Berlese's classification (see Imms) best correlates the manifold phenomena of embryonic and post-embryonic development seen in insects. In the "protopod" stage, the embryo is incompletely differentiated, both externally and internally; the head and thorax are segmented and have rudimentary appendages, but the abdomen lacks them and is poorly segmented or not at all. In the "polypod" stage, the abdomen has become completely segmented with appendages, the tracheal system has been laid out, and other internal organs are well defined. Last is the "oligopod" stage, in which the appendages of head and thorax attain their definitive form, the tracheal system is completed, and the abdominal appendages resorbed except such as are to be retained by the adult.

Hemimetabolous insects pass all these stages within the egg, and emerge in a post-oligopod stage called a "nymph". They then pass through a succession of stages (the "instars") separated by moults, at the last one of which more deep-seated changes take place, leading to the fully adult form. Holometabolous insects hatch at various earlier stages, either in the protopod stage as true embryos (e.g. the primary larvae of the endoparasitic Hymenoptera), or in the polypod state (e.g. the caterpillars of Lepidoptera), or in the oligopod state (e.g. the campodeiform larvae of Coleoptera), or finally in a degenerate oligopod state which is apodous (e.g. the larvae of Diptera). The more precocious the state in which the organism emerges from its egg, the more far-reaching the adaptations for free larval life have to be, and it naturally follows that the reorganisation at the pupal stage (what we commonly think of as insect metamorphosis) is much more severe than the passage from the last nymphal instar to the adult form in hemimetabolous insects. The whole process from the pre-protopod state to that of the adult is, as Richards & Miller rightly say, continuous and indivisible.

The most striking feature of the development of the post-oligopod stages of hemimetabolous insects is the succession of moults which they go through. When the physiology of these had been carefully described (Wigglesworth[2,3]) it proved possible to show that a glandular hormone or circulatory chemical substance is involved and to trace it to its point of origin by decapitations done at varying times during each instar. Wigglesworth[4] found that some organ in the head produces a hormone, not species-specific among the bugs investigated (*Rhodnius*, *Triatoma*, etc.), which induces the moult, by stimulating the competent epidermal cells. The organ is brought into activity by a nervous excitation, derived from proprioceptive fibres which register abdominal distension caused by the food. When the heads of two such blood-sucking bugs have been cut off the bodies may be joined together at the cut surfaces; if the epithelia heal, moulting will be simultaneous, if not, it will nevertheless take place, depending on the concentration of hormone present in the partners, i.e. on the time relations of feeding, decapitation, and joining with the second insect.*

* The two animals may be separated by a glass tube through which the substance diffuses; see Fig. 262.

g. 255. Resistance of the eye to atrophy when transplanted to the anuran tail before metamorphosis.

Fig. 262. Two decapitated 4th-instar nymphs
of *Rhodnius* united by a glass tube.

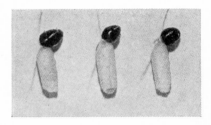

Fig. 263. After ligation of the blowfly larva
(*Calliphora erythrocephala*) pupation proceeds
only in the anterior portion (see p. 467).

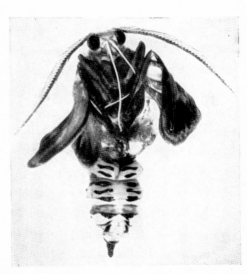

Fig. 264. After ligation of the moth pupa (*Phry-
ganidia californica*), imaginal development
(metamorphosis) proceeds only in the anterior
half (see p. 468).

The change from the last of the nymphal instars (the fifth) to the adult form (metamorphosis) is less straightforward. The simplest hypothesis would be that at the last (metamorphic) moult, the same or some other organ would throw into the circulation a new substance, different from the former. But the facts did not at first lead Wigglesworth[4] to this conclusion. That competence for the moulting hormone exists at the last (metamorphic) moult, could easily be shown by experiments in which fifth-instar nymphs lacking hormones (because of decapitation shortly after feeding) were joined with fourth-instar nymphs containing hormones; the fifth-instar ones moulted instead of metamorphosing. But blood from a fifth-instar nymph transfused back into a fourth-instar or even into a first-instar nymph, produced not nymphal moulting but metamorphosis, so that in the latter case an extremely minute adult was produced. Competence for metamorphosis must therefore be present from the beginning of nymphal life. Something seems, however, to inhibit its expression. And that an inhibition is involved rather than the production of a new substance at the fifth nymphal instar was suggested by the remarkable fact that if fourth-instar nymphs lacking hormones (because of decapitation shortly after feeding) are joined with fourth-instar ones containing hormones, both undergo metamorphosis, though neither would normally have done so. Wigglesworth[4] explained this and similar experiments by saying that at no time does the inhibitory factor reach a sufficient concentration in the blood to prevent the metamorphosis of a second insect. The converse to the production of very minute adults, i.e. the production of giant nymphs, was also successfully obtained.

From the above work it seemed that the head produces not only the moulting substance but also a metamorphosis-inhibitory substance. Simple decapitation at any nymphal instar produces a certain number of nymphs which undergo metamorphosis, developing into miniature adults; this was explained as being due to the secretion of the metamorphosis-inhibitory substance slightly later than the moulting substance. In a certain number of cases, therefore, decapitation will occur after a sufficient amount of the moulting substance has been formed but before there is enough metamorphosis-inhibitory substance.

Wigglesworth's most recent view[6] is that the real nature of the inhibition may be that in nymphal moulting the cuticle is quickly laid down, whereas in metamorphosis this process is delayed so that there is time for the more elaborate remodelling of the epidermal and other organs on which the assumption of the adult state depends. He now speaks of two inductors, the nymphal moult hormone and the metamorphic hormone, and of a competition between them for mastery of the competent tissue, the results of which are experimentally variable.* He believes that the metamorphic hormone is produced in the *corpora allata*, minute glands situated above the oesophagus posterior to the oesophageal ganglion, while the moult hormone is produced in the brain, especially the protocerebrum which contains Hanström's cells. He draws attention to the

---

* This is a relief, since in general one should be reluctant to have recourse to "inhibitory factors". They occupy the same methodological *demi-monde* as "reactive forms" of chemical substances in cell-reactions.

similarity in position of the differentiation centre in development at the pre-protopod stage and the *corpus allatum* at the oligopod stage. This may be but a coincidence, but in holometabolous insects, to which we shall return in a moment, the work of Hachlov and others has shown that visible differentiation again spreads from a centre in the thorax. Of the chemistry of the two substances nothing is as yet known.

Interesting from the standpoint of this book is the light which the hormonic control of moulting and metamorphosis throws upon prothetely and metathetely (phenomena already referred to in another connection; p. 239). Larvae occasionally develop typically pupal structures (prothetely). Pupae sometimes retain typically larval structures (metathetely). Many such cases have been described, as by Goldschmidt[5]; Pruthi[1] in mealworms; Chapman in *Tribolium*; v. Lengerken in other coleoptera; du Bois & Geigy in *Sialis*, etc. Now the production of adult characters in a first-instar nymph by decapitation, just mentioned, is experimental prothetely, and conversely, the production of giant nymphs which may show imperfect adult characters is experimental metathetely. By the interplay of the nymphal moult hormone and the metamorphosis hormone in varying concentrations, all permutations and combinations of morphological effects may, it is thought, be produced.

Similarly, if the development of the pupa is disturbed by centrifuging or by operations on the larva, histogenesis may be retarded and the adult may emerge with its internal tissues still in a stage characteristic of the pupa, according to Guareschi[4] (working on the beetle, *Lina*).

In the holometabolous insects, larval moulting is also almost certainly controlled by a diffusible substance, but less is known about it. As it proceeds in starved animals, moths, for instance, such as *Tineola* (Pruthi[2]; Titschak), the stimulus to the secretion of the substance must be different from the mechanism in *Rhodnius*. The best proof that a substance is involved is contained in the experiments of Bodenstein[1] — analogous to those of workers on amphibian metamorphosis (p. 449) — in which legs from *Vanessa* caterpillars were transplanted on to younger and older caterpillars of the same instar, with the result that moulting occurred in both host and graft simultaneously. The site of production of the hormone has not been ascertained.

A remarkable fact here is that the tissues of the adult conserve their competence to react to the moulting stimulus long after it normally occurs. Furukawa transplanted the extremities of adult *Anisolabis* to larvae and saw that they moulted there; Mauser had the same experience with *Dixippus*. In *Rhodnius* Wigglesworth[5] could get the adult to moult by injecting into it blood from nymphal stages, and Piepho[2] obtained the most striking results by implanting into the fat-body of young *Galleria* larvae bits of epidermis from old larvae ready to pupate. Here, forming balls of skin inside out (cf. human ovarian dermoid cysts, p. 232), they went through all the larval moults over again, the cast skins collecting concentrically in the cyst. Pupal epidermis was willing to do the same thing. What a contrast this is with the loss of competence of adult mammalian tissues (cf. p. 174).

Conversely, the tissues (skin) of the young caterpillar shortly after hatching possess competence to react to the pupation stimulus long before it is normally manifested (Piepho[2]).

After earlier uncertainties we know now that pupation is definitely under hormonic control. Kopeć[1] extirpated the supra-oesophageal ganglion from caterpillars of the gypsy-moth *Lymantria*, and found that no metamorphosis followed. Ligaturing experiments then showed that the ganglion, or a site near it, was the source of the diffusible substance causing pupation.* Similar experiments have been carried out by G. Fraenkel on the blowfly *Calliphora*, Bodenstein on *Drosophila*, Geigy & Ochsé on *Sialis*, etc., so that metamorphosis-inducing hormones are now known for the diptera as well as the lepidoptera and hemiptera (Fig. 263, opp. p. 464). The flies contains no organ which can be homologised with the *corpus allatum*,† but Hadorn & Neel[2] and Burtt showed that Weismann's ring is the source.‡

Progress has been made with the chemical identification of the pupation hormone by Becker & Plagge[2], who were able to separate it completely from the *A*-hormone which governs eye-colour (see p. 416). A quantitative test was worked out using various dilutions injected into tied-off abdomen halves. The hormone is non-species-specific, easily dialysable, soluble in water, alcohol and acetone, but not in chloroform or petrol-ether; stable to heat and acid but not to alkali. Below a critical temperature (6° C. in *Ephestia*) it is not produced at all (Caspari[2]).

It was mentioned above that larval tissues possess competence to react to the pupation stimulus long before it normally occurs. Evidence that the pupal blood possesses hormones which that of the larva does not has been thought to arise from the work of Frew, who claimed that imaginal discs would evaginate *in vitro* if the medium contained pupal lymph but not if it contained only larval lymph. It is unfortunate that this work, which stands alone in the literature, was not adequately described and illustrated. There is no doubt, however, that larval ovaries (S. Fukuda[1]) and eye-discs (Bodenstein[10, 13]), when transplanted prematurely into pupae, differentiate earlier than they would otherwise have done. But this does not show that any hormone other than the pupation hormone is concerned.

Some investigators (e.g. Kühn and Piepho) have postulated a second hormone governing the transition from pupa to imago just as the hormone of the *corpus allatum* or brain governs the transition from larva to pupa. It is indeed true that ligaturing the pupa will prevent the posterior half from metamorphosing further, including any imaginal discs grafted into it, while the anterior half continues to do so (see Fig. 264, opp. p. 464, from Bodenstein's work[8] on the Californian oak-moth).

* In a later paper, Kopeć[3] doubted his previous interpretation, but subsequent work has shown him to have been right. Caspari & Plagge, using sphingid moths, obtained quite parallel results, and by implanting a ganglion into a de-brained caterpillar, restored to it its power to pupate. Injection of the pupal blood will also do (Plagge[5]).

† It appears that the functions of the *corpora allata* are numerous and important, but still obscure (Weed-Pfeiffer; Kühn & Piepho, on *Melanoplus*; Pflugfelder on *Dixippus*; Bounhiol and Plagge[5] on the silkworm and other moths).

‡ Cf. nervous tissue as the source of other diffusible chemical stimulating substances — acetylcholine, sympathin, etc. (cf. p. 211).

But when Bodenstein[8] applied the glass-tube method of Wigglesworth to the problem, no diffusible stimulator passed over, and no head-organ could be found which would act when implanted into the posterior half. Bodenstein[9, 11] then found that the posterior half will complete its metamorphosis if subjected to conditions of high oxygen-tension, a fact which makes it probable that the occlusion of tracheal system and spiracles, rather than any endocrine organ, is responsible for the effects produced by ligaturing pupae. Under conditions of asphyxiation, some organs suffer more than others, e.g. skin more than ovaries (Bodenstein[6]). This is analogous with the work of Wolsky[2] on localised inhibition of development caused by localised CO inhibition (see p. 473).

In this connection we may recall the transplantations made by Bytinski-Salz[1] of non-developing hybrid moth pupal organs into normal pupae, where they developed normally and completely (cf. p. 361). Conversely, normal pupal organs transplanted into non-developing pupae died. The lethal effect may therefore well have been upon the oxygenation conditions in the hybrid pupae.

This brings us to the subject of the determination of the imaginal discs in holometabolous insects. As Fig. 265 shows (reproduced from Tower's description), the imaginal organs are, to begin with, nothing more than ectodermal thickenings on the larval body. From the review of Bodenstein[4] we learn that these ectodermal thickenings are still quite undetermined at the beginning of pupation. Large pieces may be removed from the wing discs, for example, without affecting the form of the wings subsequently produced (Meisenheimer; v. Ubisch[1]), but if the extirpation of the organ district is too large, the wing may be suppressed (D. M. Steinberg). This is analogous to the experiments of Harrison[3] (p. 287), in which discs of varying somite breadths were removed from the limb-bud district in amphibia. The determination of colour patterns, wing-vein patterns, etc. has already been discussed in the section on the kinetics of gene action (p. 403); there it was stated that these forms may be affected only during specific periods in late larval and pupal development by the action of external temperature and similar factors. Bodenstein's review[4] gives details of these specific periods.*

In the moth *Vanessa* the imaginal discs for the legs of the adult are contained in the proximal segment of the anterior caterpillar legs. Bodenstein[2], amputating caterpillar legs at a point half-way along the proximal segment, obtained later adult moth legs smaller than usual but of normal form. Removal of the whole segment suppressed altogether the adult moth legs. The regulation possible, moreover, did not extend to the production of an adult hind-leg from a caterpillar fore-leg grafted on to a caterpillar hind-leg. In this case *herkunftsgemäss* development follows.† These results recall the regeneration of amphibian

---

* Nature seems to perform an interesting defect experiment in the parasitism of the pupae of the ant *Pheidole* by the larvae of the fly *Orasema*. The parasite removes much material from the anterior end of the host, which eventually produces an ant imago with microcephaly.

† Though there is a growth effect, making the transplanted member more resemble a hindlimb in size.

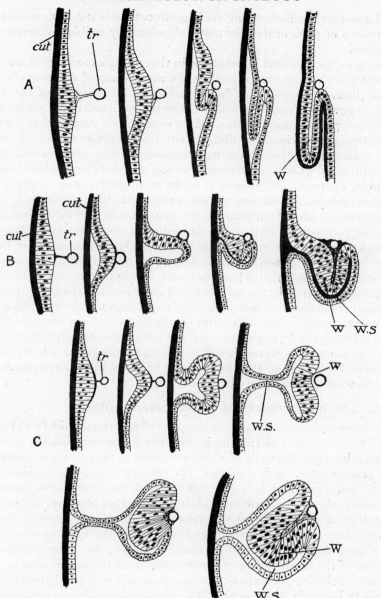

Fig. 265. Development of the imaginal bud of the wing in various insects. A. Orthoptera. B. Lepidoptera. C. Diptera. *cut*, Cuticle; *tr*, trachea; *w*, wing rudiment; *ws*, wing sac.

extremities, where great regulation is possible within the organ district itself, but not the indifferent exchange of organ districts.* Determination is not long

---

* Regeneration of caterpillar legs followed a number of the rules already noted (pp. 430 ff.) in the case of vertebrates (Bodenstein[2]). Lateral defects are not made up; the stump has to be present and controls the axial determination of the regenerate; the new growth may be suppressed by covering with hypodermis, etc. etc.

delayed, however, for Bodenstein[3] could produce double and triple malformations by inversions of parts of the disc material before any visible differentiation had taken place.

Nothing in the genetical literature or in that of the experimental morphology of insects gives us any clue regarding the mechanism of determination in the imaginal discs. It may be that further advances in the study of the diffusible substances governing pupation will elucidate the induction of the adult structures from the competent imaginal discs. Many experiments suggest that the onset of self-differentiation in imaginal discs, at any rate, depends on one or more substances present in the pupal internal medium. Further transplantation experiments will show exactly how undetermined the discs are before this trigger mechanism operates. At present it looks as if their status is rather that of amphibian limb-buds at metamorphosis or amphibian limb blastemas in regeneration than that of amphibian tissue in the gastrula with its vast potentialities.

Looking back on this section, it is interesting to note that a rather large part of what we know about determination in insect development arises from the possibility of interrupting the longitudinal continuum of the elongated insect egg through which stimulating substances have to travel, at varying times (e.g. the tying-off of the activation centre at the pre-protopod stage and the tying-off of the site of production of the pupation hormone at the pupal stage). This method has to some extent compensated for the difficulties of transplantation experiments in insects.

With echinoderm development we reach yet another widely different system. Before discussing it, however, it will be logical to interpolate a brief account of our present knowledge of the biochemistry of insect metamorphosis as a background to what has just been discussed.

2·85. The biochemistry of insect metamorphosis

The metamorphosis of insects has long offered an attractive field for biochemical work. From this point of view its interest lies in the fact that the reorganising pupa is a microcosm the walls of which are closed to the passage of matter except in the gaseous state. It is therefore, like the egg of a bird, a cleidoic system (see *CE*, pp. 1613, and here, pp. 48 ff.), and we find that it possesses the characteristic metabolic qualities of such systems, that is to say, an emphasis on carbohydrate and fat combustion, a relative suppression of protein breakdown, and the storage of the nitrogenous end-product in the form of uric acid. But while a good deal is known about these questions, the other desideratum, a knowledge of the biochemistry of the imaginal discs themselves, has so far been impossible to obtain. New ultra-micro methods will no doubt before long enable an attack to be made on these also. The biochemistry of insect metamorphosis was surveyed in 1931 by D. M. Needham[1] (in *CE*, p. 1685), and about the same time J. Heller[2], in reviewing his own series of publications, gave a useful view of the literature. There is also the interesting, though not easily accessible, review by Tirelli[6]. Here therefore we need only make a summary of the principal biochemical events in the metamorphosis of an insect and then mention briefly the more reliable contributions which have been made since that time.

The main outlines of what happens during the metamorphosis of a holo-metabolous insect may be formulated as follows:

(1) There is a pronounced loss of weight.

(2) The respiratory rate drops to a very low level and later rises again before emergence (the typical U-shaped curve).

(3) There is a marked utilisation of glycogen (stored during larval life) as reserve material.*

(4) There is a marked utilisation of fat (stored during larval life) as reserve material.*

(5) There is little protein breakdown, but accumulation of uric acid† goes on which is left behind with the chrysalis on emergence.

(6) The respiratory quotient is low, indicating a combustion in which fat plays at least a very prominent part.

(7) The histolytic processes may be accompanied by an acidosis (cf. amphibian regeneration, p. 442, and amphibian metamorphosis, p. 452).

These conclusions are illustrated by Table 25, which gives the essential details (in round numbers) collected from the before-mentioned surveys, in abstraction from differences of sex, social specialisation, etc. The names of the many writers

Table 25. *Survey of metabolism in insect metamorphosis*

		Pupation		Percentage loss during pupation
		Beginning	End	
Silkworm moth,	Glycogen	3	0	100
Bombyx mori	Fat	7	2	72
	Protein	28	26·5	5
Hawk-moth, Deilephila	N-free extr.	264	11	96
euphorbiae	Fat	141	77	45
	Protein	378	249	34 too high, cocoon not allowed for
Tent caterpillar moth,	Glycogen	0·8	0·05	94
Malacosoma americana				
Blowfly, Calliphora	Glycogen	0·63	0·17	73
vomitoria	Fat	6·96	3·93	44
	Protein	No change		0 but small amounts of uric acid
Bee, Apis mellifica	Glycogen	9	0·5	94
	Fat	6	0·5	92
	Protein	No change		0 but small amounts of uric acid

Note. The above figures are as follows: for the silkworm, gm./100 individuals; for the hawk-moth, mgm./individual; for the tent caterpillar moth, mgm. % dry wt.; for the blowfly, gm./1336 individuals; for the bee, mgm./individual.

on whose work it rests are not given, since full references to the literature will be found there. In every case (lepidoptera three, diptera one, and hymenoptera one) it will be seen that by far the heaviest losses bear on the glycogen and the fat. But whereas it is probable that all the fat is combusted, some of the glycogen is probably devoted to the formation of the new chitin—a process quantitatively followed in the silkworm pupa by Kuwana[1]. Recently, further surveys of

* See, on this, the beautiful balance-sheet experiments of Białaszewicz[3] on the silkworm and those of Heller[1,7] on the hawk-moth. Fatty substances tend to be laid down before the glycogen.

† See Kuwana[3].

metabolism in metamorphosis have been made (A. C. Evans on the sheep blowfly *Lucilia sericata*; Patton, Hitchcock & Haub on the blowfly *Phormia regina*; Melampy & Olsen on the bee; Yonezawa & Yamafuji on the silkworm; Busnel on the potato beetle); as will be seen from Table 26, their results entirely confirm the previous picture.

Table 27. *Metabolism in insect metamorphosis*

		Pupation		Percentage loss during pupation
		Beginning	End	
Silkworm moth, *Bombyx mori*	Total carbohydrate	23	18	47
	Fat	121	68	44
	Protein	33	28	15 to uric acid
Sheep blowfly, *Lucilia sericata*	Total carbohydrate	80	14	83
	Fat	240	60	75
	Protein	797	743	7 to uric acid
Mealworm beetle, *Tenebrio molitor*	Fat	13	8	38
Potato beetle, *Leptinotarsa decemlineata*	Fat	3·45	1·54	55·5

Note. The above figures are as follows: for the silkworm, gm./1000 individuals; for the sheep blowfly, mgm./100 individuals; for the mealworm and the potato beetle, % wet wt.

It must not be supposed that the whole of protein metabolism is suppressed or unimportant during metamorphosis. On the contrary, the most spectacular part of the process, the spinning of the cocoon (as in the silkworm, for instance) and later on the final moult, are proceedings in which protein metabolism is mainly involved. There is also the fundamental breaking-down of the larval structure and the development of the imaginal discs at the expense of the products. But the important thing is that this transfer of protein building-stones is effected with great efficiency, little waste nitrogen (in the form of uric acid) being produced. Percentage loss of protein in Tables 25 and 26 refers only to such waste nitrogen. That it should be so small an amount, and that it should appear as uric acid, is in agreement with the known properties of cleidoic systems.

The loss of weight during insect metamorphosis is almost entirely due to loss of water by evaporation. Elaborate data for this are available from Kuwana[2] and many other authors, and Heller[5] showed that the rate of water-loss is proportional to the external humidity (cf. that other cleidoic system, the hen's egg, which behaves similarly, *CE*, p. 876).

The U-shaped curve of respiratory rate in metamorphosis has been confirmed so often that it has become a commonplace (see e.g. Ludwig[1] on the Japanese beetle *Popillia japonica*; Janda & Kocián on the mealworm *Tenebrio*; Hitchcock & Haub on the fly *Phormia regina*; Poulson[1] on *Drosophila melanogaster*; Dobzhansky & Poulson on the *pseudobscura* species; Balzam[1] and Bell on *Lymantria* and *Bombyx*; Zolotarev, Lavrova & Tokareva on *Antheraea*, and Crescitelli on the bee moth *Galleria mellonella*).* The respiratory quotient of approximately 0·7 is also always obtained (Ludwig[1]; Hitchcock & Haub; Poulson[1]; Crescitelli; Balzam[1]).

* Contrary to what is generally found in developing systems, the total oxygen consumption of pupae seems definitely smaller at intermediate than at high or low temperatures (Kozhanchikov & Maslova).

Though the older results of Bodine & Orr, who found no difference in respiratory rate between wild-type and *vestigial Drosophila* pupae, are of interest, the persistent belief that sex differences in respiratory rate have been established must be treated with scepticism. Male pupae consumed more oxygen than female ones in Poulson's work, but only in one experiment out of two, and the difference in the same direction found for adult flies by Gowen[1] is the diametrical opposite of the difference reported by Kucera. There is much Japanese work also on sex differences, but it is unconvincing and will not be reviewed here. Melampy & Willis have at any rate established that the queen bee larva and pupa respires more intensely than that of the worker bee.

Of greater interest is the significance of the U-shaped curve. The most simple-minded explanation of it, namely that it represents the metabolic consequences of complete histolysis followed by new tissue differentiation, as from yolk, was criticised by D. M. Needham[1] and cannot be true. Histolysis and imaginal differentiation are simultaneous processes. As Wigglesworth[3] says, the imaginal histoblasts begin to proliferate rapidly; then each tissue that is to suffer histolysis disintegrates and dissolves. It is uncertain whether the cytolytic agents come from the dying tissues themselves or whether they are secreted into the haemolymph by the leucocytes or other cells. Sometimes phagocytes seem to invade apparently healthy tissues, but usually obvious degeneration precedes their appearance; sometimes the whole process may go on without the intervention of phagocytes; sometimes different methods occur in different parts of the same insect. References to the literature on the histology of metamorphosis will be found in the papers of Poyarkov; Tiegs, and Ch'ang; and the books of Wigglesworth[3] and Imms. But it may not be wide of the mark to suggest that the disorganisation, at any rate, of the pupal tissues passes through a maximum and so accounts for the U-shaped curve.*

Since 1931 some new facts have accumulated which we must consider in this connection. That the respiratory quotient is the same at all temperatures in the physiological range (Crescitelli), though interesting, does not immediately help us. The U-shaped curve is compressed at high temperatures, elongated at lower ones. More significant is the observation of Drilhon on a lepidopteran pupa (*Attacus polyphemus*) that the alkali reserve follows a U-shaped curve precisely inverted. At the beginning of metamorphosis 5·9 vol. % CO_2 were found; after 48 hr. in the cocoon there were 28·0 vol. % and after four months the value had sunk again to 8·0 vol. %. The most interesting contribution is that of Wolsky[3], who measured the extent of CO-inhibition during metamorphosis in *Drosophila*. Though it might have been expected (on analogy with embryonic diapause in insects, q.v., p. 578) that the respiration in the trough of the U would be unaffected by the inhibitor; the inhibition was light-sensitive and equally marked at all stages, thus showing that during the initial decline there is no change in the Warburg-Keilin system of respiratory enzymes, nor any change in the saturation of the

---

* There are radiographic studies of insect metamorphosis (Heller & Meisels), but owing to the opacity of the haemolymph and the pupa-case, nothing very illuminating has yet been derived from them.

cytochrome oxidase, but rather a definite destruction of it, followed by its re-formation. Later in the subsequent rise there was a certain change, but it was not large and could be explained by secondary effects. By exposing *Drosophila* pupae to CO and then illuminating them with a narrow beam of light (which would reverse the inhibitory effect locally) Wolsky[2] was able to produce a variety of regional defects. Depression of respiratory metabolism, for instance, hindered the deposition of pigment in the eyes, delayed the development of the wing discs, and reduced the size of bristles, thus producing phenocopies of *stubble* and *stubbloid*. This is the first work in which our knowledge of cell respiration, and all that that implies, has been brought into contact with the development of imaginal discs and the quantitative expression of genes (see p. 398). Its further development will be awaited with interest.

The three other directions in which progress has been made are those of work on the heat-production of pupae, the composition of the pupal haemolymph, and the phosphorus metabolism during metamorphosis.

For pupal heat-production we have studies by Białaszewicz[2] and Balzam[1] on *Lymantria* and *Bombyx* and by Taylor & Crescitelli on *Galleria*. All the results agree in showing that, as was to be expected, the heat-production declines and rises again in the characteristic U-shaped curve. But there are several discrepancies. In *Galleria* calculation of the heat-production expected on the basis of a purely fat oxidation corresponding to the low R.Q. gives a curve keeping at a considerably higher level than the experimental one. More information is needed before the meaning of this can be understood. In both insects the calorific quotients* are far from the theoretical C.Q. of between 3 and 4 assuming normal catabolism of carbohydrate, protein and fat. *Galleria* pupal C.Q. ranged from 1 to 2·8, and while that of *Lymantria* and *Bombyx* in the larval stages is from 4 to 7, in the pupal stage it is again generally just under 3. No explanation is as yet forthcoming for these figures, but Balzam suggests that anaerobic reactions giving out heat may be proceeding during larval development. All the authors are forced to the conclusion that endothermal processes exist on a considerable scale during metamorphosis. It is to be noted, lastly, that during larval life there is at each moult a severe but transient diminution of heat-production (Białaszewicz and Balzam); this gives colour to the idea (which we have already encountered on p. 465 in relation to the course of events in the hemimetabolous bugs) that metamorphosis may be, in a certain sense, a very exaggerated moult, in which the whole larval organisation, and not merely the cuticle, is dispensed with.

The composition of the pupal haemolymph has been studied mostly on the silkworm and the hawk-moth (Florkin; Heller[4]; Heller & Mokłowska; Demjanovsky & Prokofieva; Akao[1]), but also on the wax-moth (Crescitelli & Taylor). The volume of the haemolymph decreases during metamorphosis from 0·35 to 0·1 c.c./pupa, and there is a percentage decrease of dry matter in it, as also of total nitrogen and protein nitrogen. The non-protein nitrogen remains approximately constant (an interesting commentary on the nice balance between histolytic disintegration and imaginal differentiation).† So does the amino-

---

* $\text{C.Q.} = \dfrac{\text{heat-production in gm.cals.}}{\text{oxygen-consumption in mgm.}}$ per unit weight per unit time.

† Note also that the changes in peptone and amino-nitrogen observed by A. C. Evans on the whole pupa of the blowfly were very small.

nitrogen and the uric acid (Florkin), but others find that the uric acid of the haemolymph falls slightly (Akao[1]). Haemolymph glucose, it is agreed, rises at first to a maximum and later falls (Florkin; Demjanovsky & Prokofieva), but the time of the maximum seems uncertain and the values are very divergent. This may be due to race differences (cf. Heller's results[3] on aberrant races of the hawk-moth below). In the wax-moth, according to Crescitelli & Taylor, there is a maximum of free glucose in haemolymph and tissues at the time of spinning, followed by a fall and rise to another maximum (not very well established). A great deal more work will be necessary on many species before we can be said to have a clear picture of the changes in the haemolymph. The establishment or a peak in glucose concentration would imply that glycogen breakdown at first proceeds more rapidly than chitin formation or glucose oxidation. But as Crescitelli & Taylor were unable to detect any glycogen in *Galleria* during larval or pupal life, it may be that, in some insects, the carbohydrate is stored as muco-protein, or that there is conversion of fat to carbohydrate. Some particulars about glutathione are given by Demjanovsky; during pupation the free –SH falls.

Phosphorus metabolism during pupation in *Deilephila euphorbiae* has been studied by Heller[6], who found the significant fact that there is much less adenyl-pyrophosphate P in the pupa than in the moth, though the arginine phosphate P is not greatly different. It is likely that this is a sign of the muscular disintegration. Although Heller analysed a number of organs in the pupa separately for their phosphorus partition the subject would repay further study. For example, there seems to be relatively more protein P in the haemolymph than in any of the solid parts.

As regards inorganic ions, Białaszewicz & Landau have shown that the silk-worm's pupal haemolymph has enormously more Mg″, K′ and Ca″ than human serum and correspondingly less Na′. A large part of the Mg″ is contributed to the cocoon, with a compensatory rise in K′.

The only specific study of fat metabolism is the work of M. Becker on the mealworm *Tenebrio molitor*, incorporated in Table 26. Those interested in the histochemistry of tissues will find a paper on fat in the pupae of this insect by Zakolska.

There remains only for discussion the question of acidosis during histolytic degeneration. The older literature was contradictory, and disagreement still continues. Ludwig[3] on the beetle *Popillio* found a change from pH 7·07 to 6·79 followed by a return to 6·94; Booker on the mud-dauber wasp *Sceliphron* found a fall from 6·7 to 6·4 followed by a return; Taylor, Birnie, Mitchell & Solinger found the tissues of the bee-moth *Galleria* went as low as 5·53, but late rather than early in pupation. On the other hand the silkworm, studied by Demjanovsky, Galzova & Roshdestvenska and by Ongaro[1], showed no acid trend at pupation. It is probable that the degree of acidosis depends on the exact nature of the reorganisation processes, and there is room for a study on much more extensive lines than hitherto comparing the pH of the haemolymph at various stages in many different insects with the presence of phagocytosis and other physiological phenomena.

Pupation may be interrupted by a diapause of varying length (cf. p. 578, where embryonic diapause is discussed in detail). In the hawk-moth *Deilephila euphorbiae* Heller[3] detected eight genetic races differing in their properties in this respect:

	Pupation length days	Glucose in haemolymph %
Two-year	640	—
Latent	300	50
Lethal-latent	Die owing to excessive water-loss	
Protracted	45	—
Subitan	20	50
Rapid	13	—
Lethal soft-wing	Die in pupal stage	220
Lethal black-pupa	Die in pupal stage	380

In pupae showing a diapause the U-shaped respiration curve is correspondingly prolonged, diapause supervening at the bottom of the trough. The weight loss curve is drawn out to an equivalent extent. They are said to differ from *subitan* pupae in that their metabolic rate depends on the temperature at which they pupated, being higher, not lower, the lower the external temperature (Heller[7]). As may be seen from Goldschmidt's book ([9], p. 43) length of diapause is indeed a typically gene-conditioned property.

It will be realised that many interesting questions have perforce been left untouched in this short survey. For example, one would like to know whether the metabolism of the pupa is affected by the vast quantities of bacteria often contained, as in the flies, in the larval gut. According to Balzam[2], they are isolated in certain parts of the pupa during histolysis, and undergo a rather sudden sterilisation at the time of emergence of the imago.

It may be said in conclusion that the typically cleidoic metabolism of the insect pupa stands to-day in clearer relief than ever, but that the comparison of bio-chemical with histological and morphological events during metamorphosis still offers many unsolved problems. Especially valuable would be an approach to the biochemistry of the imaginal discs; their chemical constitution and heter-auxesis (cf. p. 532), their competence and specificity of response to the pupation hormone, and their capacity for utilising the chemical building-stones set free by the disintegration of the larval tissues.

2·9. Determination in echinoderm development and metamorphosis

Echinoderms, like insects, develop first into a larval form, which later gives rise to the adult by a metamorphosis, though nothing could be more different than the plan of organisation on which the two groups of organisms are built. Like the insects, too, the echinoderms in their early development have revealed to experimental analysis the existence of two centres, vaguely analogous to the organisation centre of the amphibia, but not easily to be compared either with that or with the two centres of the insect egg. Research on the determination processes in echinoderms has followed more chemical lines than in the case of insects, but owing to the minute size of the eggs of sea-urchins and star-fishes the chemical work has largely concerned the action of external substances upon the egg, to which it is very susceptible, rather than that of substances which it itself contains. Nevertheless many important facts have been established, mostly by the work of the Swedish school, and no praise can be too high for the technical skill and perseverance with which Hörstadius has extended the method of micro-operations to the echinoderm egg, while Runnström, Lindahl, and their collaborators have sought to unite in a unitary hypothesis their numerous biochemical and morphological results.

2·91. The double gradient system

The essential points may be briefly summarised as follows. Two points of radiating influence (organiser centres, if you like) are distinguishable in the echinoderm egg, one at the animal pole, the other at the vegetal. As first proposed by Runnström[11], their action is best represented in the form of two gradients, one maximal at the animal pole, the other at the vegetal.* Normal development to

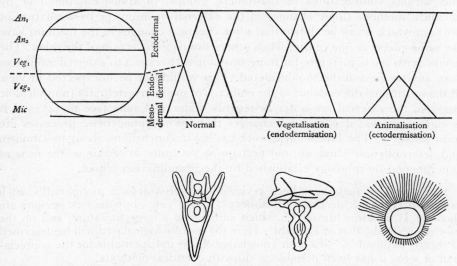

Fig. 266. Diagram to illustrate the double gradient system in the echinoderm egg (see text).

the pluteus depends upon the "co-operation" of these two "gradients", and many, if not all, of the anomalies of development can be traced to changes in their intensity. If one centre or gradient apex is depressed then the development of the embryo veers towards the type "advocated" by the other (see Fig. 266). Since the vegetal pole forms endodermal tissues, it will, if circumstances allow it wholly to control development, bring about a "vegetalisation" or "endodermisation" of the embryo. In this case, presumptive ectoderm is made to form endoderm. Conversely, since the animal pole forms ectodermal tissues, it will, if given a like control of development by circumstances, bring about an "animalisation" or "ectodermisation" of the embryo. In this case, presumptive endoderm is made to form ectoderm. Extreme but typical forms illustrating these effects are shown in Fig. 266. When the gradients are equally

* The concept of gradients really derives from Boveri's study[2] of polarity in the echinoderm egg in 1901. The gradient system of echinoderm determination has been graphically formulated with various refinements by v. Ubisch[5] and Dalcq ([7], p. 150).

powerful, the normal pluteus larva results, with its ectodermal skin, its mesen-chymal spicule skeleton, and its endodermal tripartite gut. When vegetalisation occurs, much material other than presumptive endoderm is deviated to form gut, which in this case is seen hanging out from an exo-gastrula, the ectodermal part of which is reduced to a thin-walled bag containing abnormally arranged spicules. Conversely, when animalisation occurs, no gut or skeleton ever forms at all, and the embryo becomes a thick-walled ectodermal bag provided with a very excessive complement of cilia and a hypertrophied apical tuft. These descriptions will become clearer in what follows.

Now the characteristic thing about echinoderm development is that these changes can be brought about *either* by morphological methods (such as isolations and various combinations of blastomere groups, or transplantations) *or* by chemical methods (ionic changes in the external medium) *or* by a mixture of both, in which it can be shown that some chemical changes in the medium have the same effects as one centre, while others simulate the action of the other. The echinoderm egg is therefore far more unstable with respect to external influences than any we have hitherto considered. This would not be unexpected in view of the nutritional dependence of the embryos of many invertebrates (non-cleidoic eggs) on the external sea-water as regards water and salts (see p. 34), and it is significant that it is to ionic changes that the morphogenetic processes are most sensitive. The biochemical work has led to conclusions about metabolism and determination which are not perhaps as yet quite so certain as the facts of experimental morphology established by the Scandinavian school.

Some of these facts have been reviewed in embryological monographs, such as that of Huxley & de Beer[2];* Caullery[3]; Dalcq[7], etc., but the best reviews are those of Hörstadius himself[6,7], which summarise a large literature, and on the biochemical side, that of Lindahl[7]. Here the morphological part will be described as shortly as possible, but a clear knowledge of it is indispensable for the apprecia-tion of what it has been possible to do with chemical methods.

In the development of the sea-urchin egg (e.g. *Paracentrotus*) the first two furrows are meridional, the second equatorial, so that the egg has eight equal blastomeres (Fig. 267). The next cleavage is unequal, giving one ring of eight animal mesomeres, four large vegetative macromeres, and four very small micro-meres. In the 32-cell stage, the animal half has two rings of eight cells each (an_1 and an_2). In the 64-cell stage, the vegetal half macromeres have also divided into two rings of eight cells each (veg_1 and veg_2). The egg at this important time consists, therefore of an_1, an_2, veg_1, veg_2 and the micromeres, mic.†

Each of these groups of cells has a presumptive significance. An_1 will form the apical ciliary tuft of the blastula and gastrula and the ectoderm of adjoining regions, which is about a third of the ectodermal surface of the pluteus; an_2 will form another third of the ectodermal surface of the pluteus. Veg_1 (contrary to earlier opinion) will not invaginate, but will form the remaining third of the pluteus ectoderm. Veg_2 alone invaginates, forming all the endoderm, and finally

---

* As Hörstadius has cautioned, parts of an important diagram in the book of Huxley & de Beer[2] (their Fig. 150) are transposed.

† The micromeres derive from the most vegetal part of the egg, even in *Arbacia*, where there had been some doubt (Hörstadius[8]).

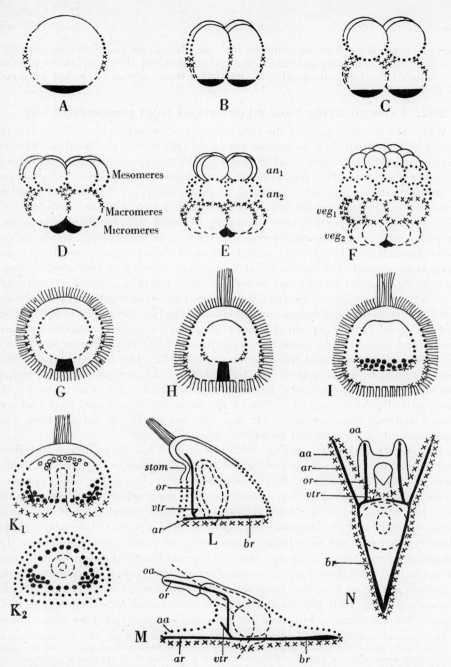

Fig. 267. Normal development of an echinoderm egg (the sea-urchin; *Paracentrotus lividus*). *aa*, Anal arm; *ar*, anal rod of skeleton; *br*, body rod of skeleton; *oa*, oral arm; *or*, oral rod of skeleton; *stom*, stomodaeum; *vtr*, ventral transverse rod of skeleton.

A. Uncleaved egg; B, 4-cell stage; C, 8-cell stage; D, 16-cell stage; E, 32-cell stage; F, 64-cell stage; G, young blastula; H, later blastula, with apical tuft; I, still later blastula, after the formation of the mesoderm; K₁, gastrula; K₂, transverse optical section of K₁; L, prismatic gastrula, stomodaeum invaginating; M, pluteus from left side (broken line indicates position of egg axis); N, pluteus from anal side. Continuous lines, *an₁*; dotted lines, *an₂*; crosses, *veg₁*; broken lines, *veg₂*; black, *micromeres*.

the cells originating from *mic* migrate into the blastocoele cavity before gastrulation has begun, and later give rise to the spicule skeleton. The arms of the pluteus are induced by the skeletal cells, a "formative Reiz" already detected long ago by Herbst[1].*

2·92. Experimentally induced deviations from presumptive fate

When Hörstadius[1] studied the development of isolated animal† and vegetal halves he found that their behaviour was exactly the contrary of what would have been expected had the early views of Driesch (see p. 121) been justified. Regulation would have given small but harmoniously formed larvae. But instead the influences of the two centres seemed to be freed in such conditions. The animal material remained purely ectodermal, producing first ectodermal balls or blastulae with hypertrophied ciliation and later swimming blastulae with bands of cilia, sometimes showing a small stomodaeum as the façade to an absent gut. Animal halves never gastrulate.‡ The vegetal material showed various types, mostly gut and irregularly shaped skeleton enclosed in an oval, mouthless and armless, bag; though the gradient system may re-form so that a proper pluteus is produced. Hörstadius[2] noticed, on cultivating the operated embryos together, that the most poorly developed animal halves corresponded to the best developed vegetal halves and *vice versa*. This he explained by supposing that the animal halves with stomodaeum and ciliated band came from eggs in which the third furrow had struck through nearer than usual towards the vegetal pole. The material which they thus obtained from the vegetal half had inhibited the overgrowth of the apical tuft and had *induced* the stomodaeum and the ciliated band. Correspondingly, the poorly developed vegetal halves of the same embryos would have had less animal material than normally. He was thus led to try the permutations and combinations of the different groups of cells.

The results are shown in Fig. 268. The animal half alone $(an_1 + an_2)$ gives nothing but a ciliated ectodermal ball (A3). But $an_1 + an_2 + veg_1$ gives a ciliated blastula with a stomodaeum, thus proving the correctness of the view expressed above, that the material of veg_1 is necessary for suppression of excessive apical tufts and formation of the stomodaeum (B3). More remarkable is the result obtained by the combination $an_1 + an_2 + veg_2$, namely, a perfect pluteus larva (D3). Here veg_2 behaves not only in accordance with its prospective significance, by invaginating and forming the gut, but it also provides the skeleton by induction or regulation. Then if the four micromeres are implanted in an entire presumptive ectoderm so that we have the combination $an_1 + an_2 + veg_1 + mic$, a perfect pluteus is obtained (E3). Here the micromeres have behaved not only according to their prospective significance, but have also provided the gut, either by induction or

* But the *position* of the arm evaginations is determined earlier, though they will not begin to grow out without the skeleton. For the rather complex developmental mechanics of the skeleton, the reader is referred to the papers of Runnström[8]. Its formation can be selectively suppressed by Ag′ (Bohn & Drzwina[1]) and intensified by Sn″ (Drzwina & Bohn).

† First obtained by Zoja in 1895.

‡ Hence the echinoderm egg is not a harmonious equipotential system in the extreme sense of Driesch. Hörstadius[1] has well said that every part can give rise to every part, but not under its own power.

regulation. Induction is more probable, since the volume of material in the micromeres is so small. Just the same thing happens, but more strikingly, if the

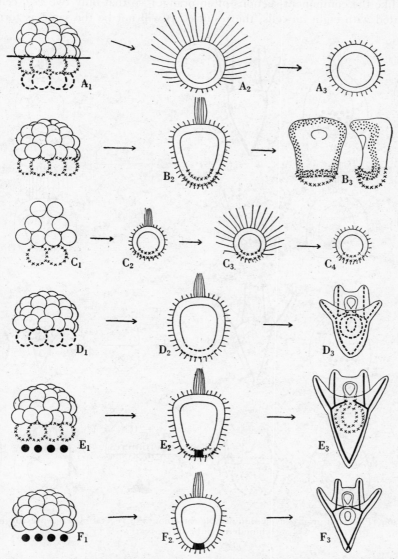

Fig. 268. Effects of permutations and combinations of the various groups of cells in the echinoderm embryo (see text).

micromeres are associated only with an_1 and an_2 $(an_1 + an_2 + mic)$, for again a perfect pluteus is formed (F_3). The four micromeres thus have the power to convert even an_1 material to endoderm. But if they are removed, leaving everything else, the gradient re-forms with veg_2 as the most vegetal part, and a normal pluteus results.

In this series of results there is one strange, and as yet unexplained, phenomenon. If an ectodermal half ($an_1 + an_2 + veg_1$) is divided into four *meridional*
parts (like the component sections of an orange), so that only two veg_1 cells are
associated with eight *an* cells, the result (C4) will not be the same as formerly

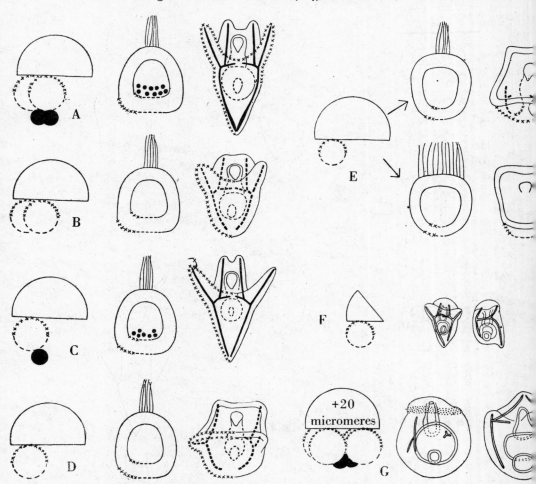

Fig. 269. Effects of progressively diminishing the amount of vegetal material present in the
echinoderm embryo (see text).

(B3). In this former experiment a certain degree of vegetalisation was brought
about by the presence of the veg_1 cells (see above), i.e. the suppression of apical
tuft hypertrophy and the formation of the stomodaeum; here, however, the
animalisation forces have it all their own way. Hörstadius[3] concluded that the
vegetal forces, when quantitatively below a certain level, can no longer resist the
animalising ones.

This experiment leads on to one of the most important principles established by Hörstadius[6], namely that the development of a typical pluteus is dependent, not on the absolute amount of animal or vegetal material present, but on their relative amount. This can be demonstrated (*a*) by gradually diminishing the amount of vegetal material present, and (*b*) by implanting varying numbers of micromeres into different combinations of blastomeres. We will glance first at the former method, then at the latter.

Fig. 269 shows the results of the following combinations: animal halves in contact with two macromeres and two micromeres, or with two macromeres only, or with one macromere and one micromere, or with one macromere only, or finally with only the half of one macromere. There is a definite trend towards animalisation as the amount of vegetal material decreases, so that at the final stage, *E*, animalisation is complete, and the vegetal macromere itself may be animalised. It may also be seen that volume for volume the micromeres are more strongly vegetalising than the macromeres. But now supposing that instead of combining the half-macromere with a full quantity of animal cells, the presumptive ectoderm is reduced by two-thirds, the original balance is regained, and a very minute but well-formed pluteus results (F). Similarly, if the amount of micromere material is greatly increased by adding twenty micromeres to a normal embryo, the whole will differentiate as if it were a vegetal half (G).

Fig. 270 shows the results of transplanting varying numbers of micromeres into isolated parts of the embryo. The differentiation of the isolated layers, if not interfered with, is shown in the left-hand column. An_1 and an_2 form completely animalised blastulae, and veg_1 almost so. Veg_2, the presumptive endoderm, gives oval larvae with a large gut and a few small spicules. This means that it has made some material more animal than itself (ectoderm) by regulation, and also that it has made some material more vegetal than itself (spicules). In certain isolated regions, therefore, there can be a reorganisation of the gradient system leading to new poles.

Micromeres when isolated form at first a small blastula, but this eventually disintegrates as they are too vegetative to be able to form any ectoderm by regulation. Now if one, two, or four micromeres are added to the isolated veg_2 the embryo will be too vegetative to gastrulate. An exo-gastrula will be formed similar to those produced chemically (see below, p. 486). But four micromeres together with veg_1 gives an embryo of the same type as what is produced from veg_2 isolated. In order to get the most pluteus-like larva four micromeres have to be added to an_1; two micromeres have to be added to an_2; one micromere to veg_1; and none to veg_2. This clearly illustrates the balance required between the gradient forces. If the animal material is present in great superiority, the vegetal force of a micromere may be overwhelmed. This happens when one micromere is combined with isolated an_1. First the apical tuft hypertrophy is suppressed, but later animalisation gains the day, and no gut or skeleton ever appears.

Another type of experiment consists in implanting the four micromeres at different levels along the gradient system. Thus in the example shown in Fig. 271 the micromeres were implanted into an embryo between an_1 and an_2. A new

vegetal centre was formed, which attracted mesenchyme cells and gave two
gastrular invaginations instead of one, leading in the end to an induced gut and

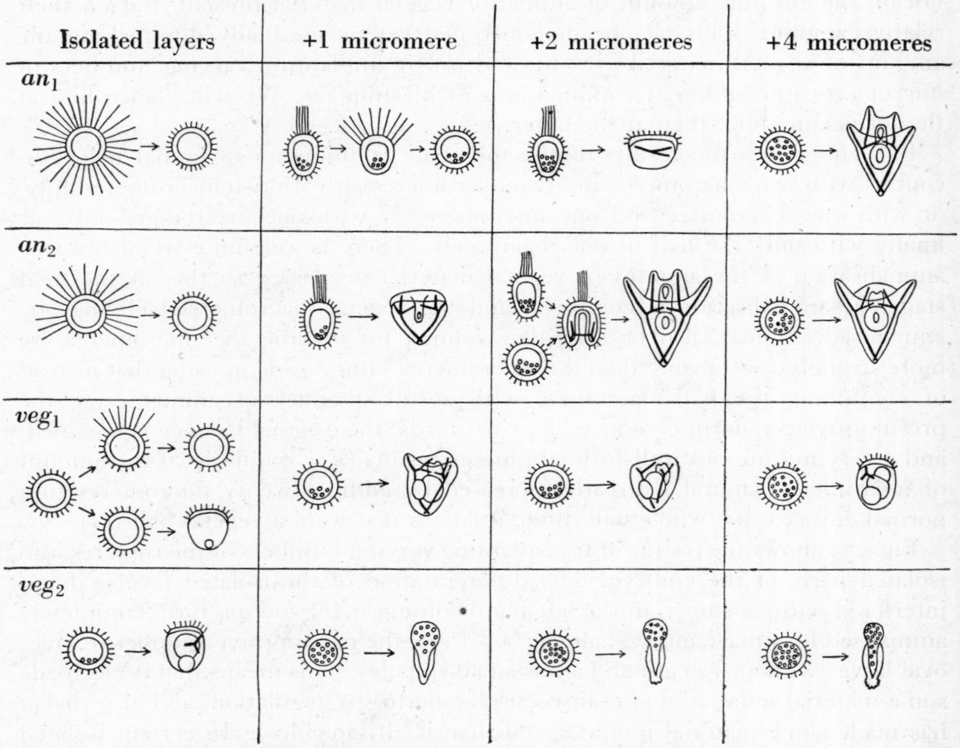

Isolated layers	+1 micromere	+2 micromeres	+4 micromeres
an_1			
an_2			
veg_1			
veg_2			

Fig. 270. Effects of implanting varying numbers of micromeres into various isolated groups of cells in the echinoderm embryo (see text).

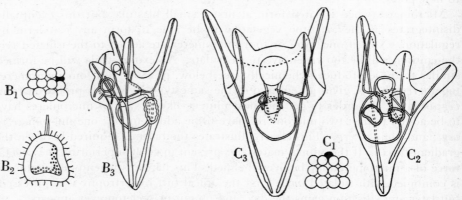

Fig. 271. Implantation of micromeres between an_1 and an_2.

Fig. 272. Implantation of micromeres at the animal pole.

a fairly complete supernumerary skeleton. The nearer to the animal pole the
micromeres are implanted, the smaller the induction will be, while in the other

direction, if they are implanted towards the vegetal pole the induced gut will eventually fuse with the host gut.

One of the extremest forms of this experiment consists in removing the micromeres from the vegetal pole and implanting them at the animal pole (Fig. 272). Here nothing is added to the egg. The result is an enlarged gut and coelom. The micromeres at the animal pole were unable to induce a secondary gut or skeleton, but they weakened the animalising tendency so that some material which normally would have formed ectoderm was deviated to form endodermal derivatives.

The animal-vegetal axis is, as we know, already laid down in the ovary. The animal pole (where the micropyle and the polar bodies are) is opposite to the point of attachment (see Lindahl[2], with literature). This axis must, in a sense, correspond to the yolk gradient of the amphibian egg, but the analogy is difficult and we shall return to it later (p. 500). The animal-vegetal axis is more stable than the dorso-ventral and the dextro-sinistral axes; it is not affected by moderate stretching or by centrifuging (cf. pp. 497, 668), though a strong constriction may divide it into two. A new axis may be formed in an embryo by the implantation of a new vegetal centre (micromeres, in the experiment just described), but the animal centre may also reverse or alter polarities in the embryo. If the vegetal half of an egg be cut off and turned upside down beneath the animal half, the latter may incorporate part of the endodermal material within itself by forcing it to gastrulate in a direction directly opposite to that of its normal polarity. Chemical animalisation, too, may lead to the appearance of a secondary animal centre at what should have been the vegetal pole of the egg, and hence the production of two apical tufts instead of one (Lindahl[7]). Treatment with hypotonic sea-water may, furthermore, as Runnström[3] and Newman[2] found, cause the appearance of several small supplementary invaginations anywhere on the gastrula (Fig. 273). Competence for invagination is evidently widespread. Hörstadius[6] speaks of this as an "anarchy of the axis" (cf. multiple gastrulation in amphibia, p. 257).

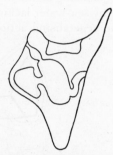

Fig. 273. Supernumerary invaginations in the echinoderm embryo.

2·93. Chemically induced deviations from presumptive fate

We are now in a position to appreciate the effects of various chemical agents on echinoderm determination. The initial step was taken in 1892 by Herbst[1] when he found that exposure of developing sea-urchin embryos to lithium salts resulted in uniform exo-gastrulation. Later he studied the effects of excess and defect in all the ions of sea-water in a series of researches (described in *CE*, p. 1271). Fig. 266 shows one of Herbst's drawings of exo-gastrulae, showing the deviation of most of the material of the egg in an endodermal direction (vegetalisation). If lithium is applied in late stages, after determination is completed, it has, of course, no effect.

From the work of Lindahl[7], Runnström[14] and others, a table can be drawn up showing the chemical agents which have been found to have marked vegetalising and animalising properties.

Table 27. *Morphological effects of external agents on echinoderm embryos*

Vegetalising agents	Animalising agents
Li'	SCN' (best before fertilisation)
Li'+KCN (intensified)	
Li'+CO (intensified)	SO₄"-lack
Mg" (?)	Ca"-lack (slight)*
	I' (best before fertilisation)
	Pyocyanin+SCN'

* Due perhaps to impurities in the reagents (Runnström & Thörnblom).

Among these agents antagonism and reversibility are easily demonstrable. The vegetalising action of lithium (seen, for example, in Fig. 274) is reversed by SCN' or by sea-water lacking the sulphate ion, and *vice versa*. Pyocyanin also reverses the vegetalising action of lithium (Runnström[15]), while cyanide (Lindahl[7]) and

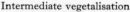

Intermediate vegetalisation   Intermediate animalisation

Fig. 274. Vegetalisation and animalisation. The examples in Fig. 266 are more extreme cases.

carbon monoxide (Runnström[6, 11]) intensify it. Here there is a direct connection with respiratory mechanisms. A typical animalising action, not extreme, brought about by thiocyanate, is seen in Fig. 274; the invagination of the endoderm has been arrested and the apical ciliary tuft exaggerated.

In attempting to analyse the metabolic meaning of these effects Lindahl[5] made manometric measurements of the respiratory rate of embryos in the presence and absence of lithium salts.* He was able to establish that whereas during normal cleavage and gastrulation the respiratory rate increased steadily,† in the presence of lithium this increase of respiratory was inhibited, so that respiration proceeded at a uniform rate, no faster after 8 hr. than during the first hour after fertilisation (see Fig. 275). The inhibition was thus not large (maximally 25 %

* What is known of the respiratory metabolism during normal echinoderm development will be found in Section 3·31.

† This rising curve has been obtained by many reliable observers; cf. *CE*, Fig. 111. Lindahl wishes to regard it as analogous to the rising curves of fermentation after the "induction" period, but I prefer the older explanation that it is due to the formation of new respiring protoplasm from inert yolk. The fact that the respiration is not inhibited at all immediately after fertilisation means, according to him, that the animalising substance is not yet being formed.

at 0·135 N LiCl), but it may be accepted that there is a lithium-sensitive fraction of respiration in the sea-urchin embryo.* Up to a late time it is reversible on replacing the eggs in lithium-free sea-water. Like the morphological action of Li′, it is decreased by the simultaneous presence of excess K′ ions. In brei experiments (cytolysed eggs), the reduction time of methylene blue in the presence of hexose-monophosphate as donator was considerably retarded by lithium. On these grounds, Lindahl[7] put forward the hypothesis, which must still be regarded as tentative, that the metabolism characteristic of the animal pole of the egg, and hence reduced or abolished by vegetalising agents, was of carbohydrate nature.

Animalising agents also inhibit the respiratory rate, as Horowitz[4] has shown for NaSCN. In the same way, Lindahl[7] found that lack of SO_4'' ions inhibits the respiratory rate at the blastula stage, but the fraction so inhibited is not the same as that sensitive to Li′, because the inhibitions sum. Suggesting that the only function which SO_4'' ions could be likely to serve was that of detoxicating and removing the aromatic waste-products of protein breakdown, he framed the hypothesis that the metabolism characteristic of the vegetal pole of the egg (and hence reduced or abolished or rendered toxic by animal-ising agencies) was of protein nature. As is well known, phenol and indoxyl, arising from tyrosine and tryptophane in the intermediary protein catabolism of the higher animals (not only, as was once thought, from bacterial action in the gut; see *CE*, p. 1236), are excreted in combination with sulphur; phenyl-sulphate and indoxyl-sulphate (indi-can), forming the greater part of the so-called "ethereal sulphates" of the

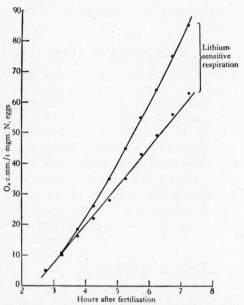

Fig. 275 Respiratory rate of echinoderm embryos in the presence and absence of lithium.

urine. Phenol-sulphatases, of very various origin, plant and animal, especially mollusc, are known (see the review of Fromageot). Lindahl[7] then succeeded in showing that sea-urchin eggs do indeed contain a phenol-sulphatase, capable of hydrolysing 115×10^{-4} mMol. phenyl-sulphate/c.c. egg brei (containing 5 mgm. N) in 15 hr., and hence presumably of synthesising similar compounds in the presence of sufficient inorganic sulphate. It also acts on indoxyl-sulphate (Lindahl & Öhman).

Animalising determination would therefore be associated with carbohydrate breakdown, and in the event of this being in some way inhibited, the vegetalising

* The inhibition is naturally very much larger if calculated only on the respiration-rise above the constant rate.

forces would predominate. Conversely, vegetalising determination would be associated with protein breakdown, and in the event of an inhibition of this, or an intoxication by certain of its products, the animalising forces would predominate. Such an interesting hypothesis deserves a detailed criticism, but

Phenyl sulphate　　　　　　　　　　Indoxyl sulphate

before proceeding to this, we must consider the third of the three main classes of experiment referred to at the beginning, namely the use of a mixture of morphological and chemical methods in order to demonstrate their interchangeability.

2·94. Combinations of the experimental and chemical methods

Only two of the most striking mixed examples will be taken. In the first, Hörstadius[4] studied the course of animal-vegetal determination, using micromeres and lithium in parallel as vegetalising agencies. In the second[5], he was able to show that animal blastomeres, if first vegetalised with lithium, will act as powerfully as the micromeres themselves in vegetalisation.

In the normal whole egg, isolation experiments at varying times after fertilisation show that for the animal half (presumptive ectoderm) determination occurs quite early (8–10 hr. after fertilisation). If an_1 is isolated alone, however, determination is not complete till 14–16 hr. after fertilisation. This shows that an influence spreading from the vegetal pole must be at work. The question now arises, how late in development can animal halves react to a vegetalising stimulus, i.e. the implantation of micromeres or treatment with lithium. Hörstadius answered it by isolating large numbers of animal halves in the 16-cell stage (4 hr. after fertilisation) and then implanting micromeres into them after the lapse of various times (see Fig. 276). In this way many different kinds of larvae are obtained (plutei; plutei with small guts; larvae with guts that do not reach the stomodaeum; blastulae with a ciliated band, stomodaeum, skeleton and arms; similar blastulae without arms; blastulae with skeleton and band only; and finally blastulae with a few spicules or none at all). This is a graded series of vegetalisation. Now if the time elapsing between the isolation of the animal half and the implantation of micromeres into it is long, the effects produced will be poor. If it has been relatively short, they will be better. This shows the onset of animalising determination. The shorter the time, the better the effects will be; hence in Fig. 276 it is possible to draw a line across the table showing the average effects produced in this way (the thick dotted line). For example, implantation of micromeres 6 hr. after fertilisation will vegetalise the animal half to the extent of giving a pluteus, but at 12 hr. a few spicules only will be produced in the resulting blastula. An analogous line can be drawn showing the best effects produced in this way (the thick continuous line). If, however, the animal halves are isolated, not at the 16-cell stage but at various subsequent times, and the micromeres implanted into them immediately after their isolation, the results tend to be much better (the thin dotted line represents the average and the thin continuous line the maximal effect). Under such conditions, an animal half can still respond to a vegetalising influence as late as the stage immediately prior to gastrulation. Hörstadius[4] concluded that the animal half becomes determined more quickly in the animal direction when it is isolated than when it is all the time connected with vegetal material.

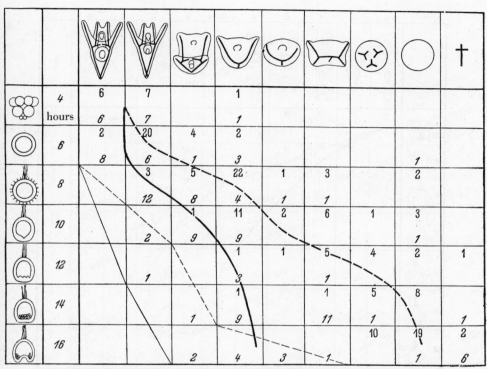

Fig. 276. Plot of the course of animal pole determination in the echinoderm embryo, micromeres being used as the vegetalising influence (see text).

The whole set of experiments was then repeated using Li′ as the vegetalising agency instead of the micromeres. As Fig. 277 shows, the results were very similar. If the animal halves were placed in lithium sea-water immediately after isolation, they could be vegetalised until just before gastrulation; whereas if time was allowed to elapse after the isolation at the 16-cell stage before the lithium treatment was begun, the halves were much more difficult to vegetalise.

Such a far-reaching parallelism between the action of Li′ and that of the micromeres invited a test of the vegetalising power of animal fragments themselves vegetalised with Li′. Such an experiment is shown in Fig. 278. Hörstadius isolated animal halves and vitally stained the vegetative surface of an_2 (D).

Controls developed into typical ectodermal blastulae with ciliary hypertrophy (A, B), showing that he had pure animal material in hand. The cells containing the stained material were then put into lithium sea-water (E) and after 24 hr. the stained part was cut off (F), stained, if necessary, more deeply (H), and attached to the vegetal pole of another animal half in the 32-cell stage (I), which, if left to itself, would have produced nothing but an ectodermal ball and cilia (C, B). On the next day it was clear that the ciliary tuft was under strict vegetal control and not hypertrophying (M), and before long invagination took place (P), leading

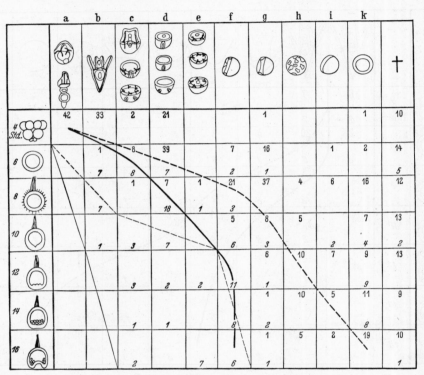

Fig. 277. Plot of the course of animal pole determination in the echinoderm embryo, lithium being used as the vegetalising influence (see text).

eventually to a perfect pluteus (O). It was thus proved that under the influence of lithium, the most vegetal material of the animal half had become as vegetal as any micromeres.* Left to itself, the animal half after the lithium treatment would have given a not very well-formed pluteus (L), but this, nevertheless, is far more vegetal than anything the animal half could form by itself.

* There is a parallel here with the awakening of induction power when a piece of presumptive ectoderm is allowed to remain for a time in the amphibian organisation centre (see p. 156). This was well realised by Hörstadius[5], but he unfortunately referred to such an activated piece of tissue as a "secondary organiser"; from p. 290 it will be seen that this terminology invites confusion and should not be adopted. A closer analogy would be the activation of a piece of presumptive ectoderm by boiling or by treatment with dyes (see p. 189).

That this remarkable effect is not due to the mere transportation of a certain amount of Li′ into the animal half at (I) was proved in a subsidiary experiment

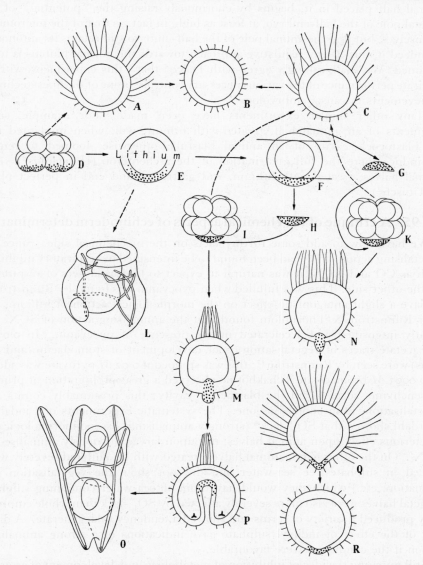

Fig. 278. Demonstration of the vegetalising power of animal fragments themselves vegetalised by lithium (see text).

in which not the most vegetal material of the lithium half-embryo, but the most animal material, was taken, stained vitally (G), and attached to an animal half in the same vegetal position as before (K). Only the slightest vegetalisation resulted from this (N), and later on there was no invagination (Q), and no formation of a pluteus (R). The conclusion must therefore be that Li', in vegetalising the animal half placed in it, begins by enormously raising the "potential" of the vegetal pole of the half-embryo, at least as high, in fact, as that of the micromeres themselves, but that the animal pole of the half-embryo still retains its autonomy, as indeed it must if any regulative approach towards the normal pluteus is to be obtained. We shall willingly agree with Dalcq[7] that "this experiment, with its accurate performance of the various necessary controls, is one of the finest technical achievements in causal embryology".

Many other similar experiments have been made. For example, small fragments of an animal half, treated with lithium, will, when implanted into the blastocoele cavity of an animal blastula, otherwise doomed to remain so, induce there the full invagination of the archenteron (Hörstadius[6]). And animal halves, treated with lithium, will gastrulate and end in perfect plutei (v. Ubisch[2]).

2·95. A critique of the chemical aspects of echinoderm determination

We have now to add some further facts on the biochemical side. Since the vegetalising action of Li' had been found to be intensified by respiratory inhibitors such as CO and KCN, it was natural to expect to find catalysers of respiration on the other side. This was fulfilled when pyocyanin was found by Runnström[15] to have a slight antagonistic effect on the morphological action of lithium, and later Runnström & Thörnblom found that the animalising action of SCN' was greatly intensified and accelerated in the presence of pyocyanin. In 0·03 M iodoacetate traces of vegetalisation (poor development of stomodaeum and oral arms) were seen by Runnström[17]; this was stopped if 0·02 M pyruvate was added. In 0·0003 M bromacetate Tchakhotin[3] observed a great proliferation of primary mesenchyme, so as to fill the blastocoele cavity; this presumably counts as a vegetalisation, if an anomalous one. The systematic experiments of Lindahl & Stordahl showed that SO_4''-lack* (strongly animalising) has no morphological or deleterious effect upon animal halves, nor upon larvae previously animalised by SCN'. On the other hand, animal halves treated with Li' suffered severely when placed in sulphate-free sea-water and did not show the vegetalisation (gut formation, etc.) which they would have done in sea-water containing sulphate. Vegetal halves were also more severely affected by SO_4''-lack than whole embryos; they produced a variety of forms and showed a tendency to degenerate. A direct test of the effect of indoxyl sulphate gave indications of a strong animalising action if the conditions were favourable.

Still pursuing the idea of inhibition of metabolism and development of aromatic substances, Lindahl & Stordahl carried further the well-known observation of Peebles that during the development of echinoderm eggs some substance is given

* Among the minor effects of vegetalisation by SO_4''-lack is the failure of the red pigment to appear in the egg (Lindahl & Örström[1]), and its formation can be intensified by excess of sulphate. A direct connection is difficult to believe, for echinochrome, though at one time thought to contain sulphur, is, according to the latest analyses, sulphur-free (Lederer & Glaser).

off which inhibits the development of other eggs.* They found that in the absence of sulphate more of this inhibitory substance is given off than in its presence (statistical experiments on otherwise normal embryos). Attempts to trace it spectroscopically have not so far succeeded. From the results on animal and vegetal halves just described, Lindahl & Stordahl concluded that the site of formation of the inhibitory substance is the presumptive endoderm and adjacent vegetal regions.

The kinetics of Li′ inhibition were further studied by Lindahl & Öhman. Lithium added at later stages, after a certain amount of the "rising" fraction of respiration has gone on (see Fig. 275), will not inhibit this fraction, but only further increases of it. Hence the effect on the respiration is smaller the later the lithium is applied, and this exactly parallels the morphological effect. On the other hand a remarkable dissociation between the respiratory and morphological effects occurs with rising temperature. At first (between 20 and 24°) both are intensified, and the increase in the "rising" lithium-sensitive fraction of respiration is relatively greater than that in the lithium-insensitive part, but at 26° some new limiting factor comes into play, for the morphological effect is stronger than ever, but the inhibition of respiration disappears. This suggests that the respiratory inhibition is not an indispensable link in the chain of causes leading to vegetalisation. Moreover, the morphological effect begins before the respiratory inhibition is detectable.

Lindahl's earlier experiments[7] on cytolysed egg-breis were carried further by Lindahl & Öhman. Runnström[12] had shown that hexose-monophosphate and hexose-diphosphate would act as hydrogen-donators to methylene blue with egg-brei in anaerobic experiments. He had also demonstrated the presence of co-zymase in sea-urchin eggs. In Lindahl's experiments, the conditions had admittedly to be rather carefully chosen in order to demonstrate any inhibitory effect of Li′. Using only the donators present in the egg-brei itself, Li′ had even an acceleratory effect on methylene blue reduction. On centrifuging, however, these auto-donators could be more or less completely removed, and then an inhibitory effect of Li′ at 0·256 N LiCl could be shown, with hexose-mono-phosphate as donator. But the effect was not very large, and excess Na′ and K′ showed something of the same kind. Both Li′ and K′ activated when dilute and inhibited when concentrated, but the inhibitory effect of the former began before that of the latter. Lindahl & Öhman now put together egg-brei, hexose-diphosphate, co-zymase, and pyocyanin, obtaining a not very large respiration (65 c.mm./1·5 c.c. egg-brei/75 min.). With Li′ this was inhibited, but at the highest concentration used (0·135 M) only 24 %.

Pyocyanin added to respiring eggs accelerated the constant fraction of respiration but not the "rising", lithium-sensitive one. This suggests that the oxygen-activating systems and not the dehydrogenases are the limiting factor in it.

* It must be remembered that Peebles also found stimulatory substances given off, which may explain the slightly faster rate of cleavage of echinoderm eggs in crowded cultures than in sparse ones, persistently claimed by Allee & Evans. Cf. also the work of A. R. Moore[7] on cleavage-stimulating substances within the eggs (p. 364). It is to be feared that both stimulatory and inhibitory substances are given off, and their identification will not be an easy matter.

Urethane, on the other hand, though inhibiting development and respiration, as in the classical experiments (*CE*, p. 625), never upset the gradient system, so that plutei later formed were morphologically normal, even when the urethane concentration was so chosen as to imitate the degree of respiratory inhibition shown by a strong Li′ concentration. Hypotonic and hypertonic sea-water also inhibit respiration (Borei[2]), but they have no effect on the gradient system.

To what has already been said regarding the gradients between the animal and vegetal poles we need only add that if the distance between them is artificially lengthened, the influence takes longer to operate. Lindahl[3] accomplished this by forcing eggs into a glass capillary tube smaller than their normal diameter; in certain cases the longitudinal axis of the "sausage" egg was identical with the animal-vegetal axis. As a result the animal pole of the embryo was more animalised than it should have been (hypertrophy of ciliary tuft, displacement of stomodaeum vegetally, etc.). Hörstadius[10] accomplished it by beautiful constriction experiments, in which he succeeded in ligaturing sea-urchin embryos

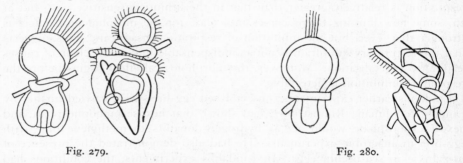

Fig. 279.   Fig. 280.

Fig. 279. Effect of a tight equatorial ligature made at the 16-cell stage (see text).

Fig. 280. Effect of a loose equatorial ligature made at the 16-cell stage and tightened at the late blastula stage (see text).

at early stages with fibres of silk. A tight equatorial constriction in the 16-cell stage led to animalisation of the animal half, and attempted normal development of the vegetal. A loose equatorial constriction in the 16-cell stage, followed by the tightening of the knot in the late blastula stage, gave development of the animal half more in accordance with its prospective significance (see Figs. 279 and 280). All these experiments demonstrate well enough the diffusion of some substance through the embryo, or at least the passage of some change. They are the counterpart of the constriction experiments made on the eggs of amphibia and insects (see pp. 103 and 459), but few would have thought it possible that they could be performed on so minute and fragile a sphere as the developing echinoderm egg.

Viewing critically the mass of work of the Scandinavian school on echinoderm development, we find that it hangs together remarkably well. That there are two centres of morphogenetic influence and a gradient system which re-forms after certain experimental interferences must be regarded as established. It is undeniable also that the actions of these centres may be imitated by a variety of chemical agencies, and it is at least exceedingly probable that the two centres are characterised by two different kinds of metabolism. That they produce each a definite morphogenetic substance or substances is likely. But with the statement that the animal pole is associated with carbohydrate catabolism and the vegetal pole with that of protein, we reach the point at which caution is necessary. The carbohydrate attribution rests mainly on the rather unsatisfactory experiments with hexose-phosphates in breis of cytolysed eggs (Runnström[12]; Lindahl[7]; Lindahl & Öhman); the vegetalisations so far produced by the halogen acetates have been too weak or anomalous to give it strong support. Moreover, Needham & Needham[5] were unable to find any trace of vegetalisation in echinoderm embryos cultured in solutions of dl-glyceraldehyde of strength amply sufficient to inhibit glycolysis. Further experiments will be necessary before convincing proof can be said to have been provided that animal-pole metabolism is carbohydrate breakdown, and it is not too much to hope that with modern methods direct R.Q. measurements on animal and vegetal halves will become possible. Similarly, the appearance of aromatic breakdown-products is really no proof that the catabolism of vegetal regions is predominantly that of protein. Protein breakdown might simply take a different course there. Here again, direct estimations of ammonia excretion should help to clarify the issue (cf. p. 626).

Another important question is the degree of specificity of the vegetalising agents. Exo-gastrulation has been produced by a number of salts (e.g. of K', Cu', Hg", Ca", etc.) by McArthur and by Waterman[1], but exo-gastrulation, as Lindahl[7] points out, need not imply vegetalisation, if the proportions of material hold exactly to their prospective significances.* Thus Lindahl accepts as vegetalisation only the results of Waterman with Mg". On the other hand, he also accepts the vegetalisations obtained by Motomura[1] using auxin fractions, glycogen, and even $KClO_3$. Such unspecificity calls for further investigation.†

There is no space here to go into the question of the nature of the action of Li', but it remains highly obscure. That it actually penetrates into the egg we know from the spectrographic measurements of Ranzi & Falkenheim on the ash of lithium larvae. One clue was provided by Runnström[5], who in connection with the examination of the egg from a colloidal point of view (Runnström[4]; Lindahl)[1], studied it under dark-ground illumination. After fertilisation there arises an orange-yellow ring or fringe, probably of lipoidal nature, and occupying approximately the position of veg_2; during gastrulation this enters at the blastopore when the endoderm is forming by invagination (Fig. 281). In embryos treated with lithium, however, this ring of material extends much farther up the side of the gastrula (Fig. 281), and perhaps is not wholly invaginated. Other clues have been sought for in biochemical experiments on other material. Thus the respiration, fermentation, and hydrolysis of hexose-phosphate by yeast may under certain

* This would presumably rule out the mechanically produced exo-gastrulae of I. Törö, and the thorium and uranium malformations of Okada[1].

† Child[7] even relates that tobacco smoke, kept above the cultures, will cause exo-gastrulation. It would be interesting to know whether the Scandinavian school would accept such exo-gastrulae as showing true vegetalisation. Nicotine in minute concentrations does inhibit certain enzymes. Tobacco ash is said to be rich in lithium.

conditions be inhibited by Li' (Lindahl[6]; Boas). But others have found no influence on yeast proliferation or fermentation (Lasnitzki & Szorenyi) nor on the $Q_L^{N_2}$ of tumour tissue (Lasnitzki[2]) nor on Q_{O_2} of various animal tissues (Kisch[3]). It may indeed be said that up to the present time a biochemical effect of lithium equivalent in magnitude to its morphological effect has been sought for in vain. Turning, therefore, to more biophysical effects, Spek[1] maintained that the primary effect of Li' must be a swelling and precipitation of the outside of the vegetal cells, and it is almost certain that physical action of this kind must play a part in the Li' effect. Runnström[15], examining *Arbacia* and *Echinarachnius* blastulae with micro-dissection needles, found that after treatment with lithium, the physical state of the cells was quite different. Cells of normal blastulae, torn away from their neighbours in the ectoderm, immediately became spherical, but those from lithium blastulae retained for a long time a spindle-shaped appearance. The hyaline plasma layer, however, is softened, not hardened, by lithium.

Equally obscure is the mechanism of the intensification of the Li' effect by CO and KCN. It has long been known that CO to some extent *accelerates* (not inhibits) the respiration of unfertilised echinoderm eggs (Runnström[7]), the respiration of insect embryos during the diapause (Bodine & Boell[1] for *Melanoplus*; Wolsky[1] for *Bombyx*), and the respiration of yeast with formate as substrate (Örström[2]). Similarly, CN' accelerates (not inhibits) the pyocyanin-raised respiration of unfertilised echinoderm eggs (Runnström[14]), the respiration of certain mammalian tissues (Kisch[2]), and the respiration of yeast with dioxyacetone as substrate (Lindahl & Örström[2]). At low concentrations, therefore, the acceleratory effects of CO and KCN may outweigh the inhibitory ones. Dehydrogenases, too, may sometimes be activated, not inhibited, by KCN (Leloir & Dixon). Lindahl & Öhman consider, therefore, that the formation of some kind of complex with heavy metals previously present, themselves the inhibitors of sulphydryl-catalysed reactions, for example, is likely to be the explanation of these effects of KCN and CO.

It must be concluded that echinoderm development offers special attractions for the study of the relations between metabolism and determination.

The gradient hypothesis, as we have seen, has proved not unfruitful in the experimental morphology of echinoderms. But if it had T. Boveri for its father, C. M. Child has been its prophet. Among his earlier investigations were some in which the disintegration of echinoderm embryos in toxic solutions (KCN, ammonia, etc.) was said to begin from the animal pole in the cleavage, blastula, gastrula, or pluteus stages, and to progress from there (Child[2]). Recently he has studied the question again, using the much more acceptable technique of differential anaerobic reduction of vital dyes (Child[6]). With methylene blue or janus green reduction occurs first at the animal pole in the cleavage stages and the blastula. As soon as the micromeres have migrated into the blastocoele cavity, they constitute a second, vegetal, centre of reduction activity. At gastrulation reduction starts simultaneously at the animal pole and the tip of the endodermal invagination, spreading everywhere from these two points. As gastrulation proceeds, the region of rapid decolorisation moves excentrically from the previous

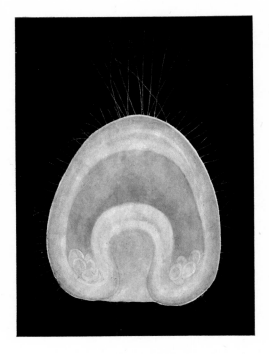

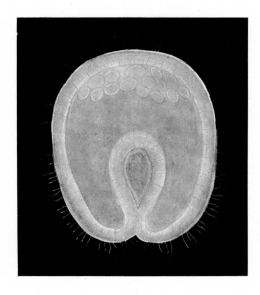

XXXIV

Fig. 281. The orange-yellow fringe at
veg_2 in the echinoderm gastrula.

animal pole to meet the endoderm at the point where the stomodaeum will form. In lithium exo-gastrulae (Child[7]) the reduction seems to begin at the junction between the ectoderm and the endoderm, spreading outwards in both directions, and sometimes the animal pole reduction centre disappears altogether. These interesting results agree more nearly with the views of the Scandinavian school than one would think from Child's discussion of them.*

Work along the same lines was continued on lithium larvae by Ranzi & Falkenheim, using a variety of reducible dyes, with substantially the same results; the animal pole becomes less highly reducing. The indophenol blue reaction, the α-naphthol peroxidase reaction, and the nitroprusside test for fixed –SH, however, varied little as between control and lithium larvae (cf. p. 136). But Pitotti[3] was able to observe definite shifts in the benzidine-peroxidase responding area (characteristic of presumptive endoderm and mesoderm), lithium extending it and thiocyanate restricting it.

2·96. Later determinations

Up to the present we have spoken only of the animal-vegetal gradient system. But this is primarily of larval significance. Ultimately the dorso-ventral axis is more important, since most adult echinoderms possess a radial symmetry. We do not yet know what determines it; as may be seen from the review of Hörstadius[6],

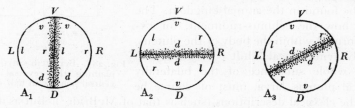

Fig. 282. Determination of the dorso-ventral axis in half-embryos (see text).

its relation to the point of entry of the spermatozoon or the first cleavage plane, for instance, is still an open question. It is certainly much less stable than the animal-vegetal axis, for it may be affected by centrifugal force (Runnström[2]; Lindahl[4]; Pease[1]), and when Lindahl[3] forced eggs through a capillary tube, as already described, producing "sausage" embryos, the front end of the sausage was always afterwards the ventral pole. If heavy vital staining is applied to it, however, this relation can be reversed, and it will become the dorsal pole.

The determination of the dorso-ventral axis in half-embryos presents the remarkable feature that it often involves a complete inversion of the previous polarity (Hörstadius[6]; Hörstadius & Wolsky). If the cut is meridional, separating right and left halves (see Fig. 282), the polarity remains as it was, but if the cut is equatorial or oblique the subsequent dorsal pole always forms at the cut surface, so that the dorsal half has its polarity inverted. A pleasing, if speculative,

---

* Other organisms, invertebrates, protozoa, etc., have been investigated with the dye-reduction method, with results more or less illuminating (Child[5]; Child & Watanabe; Watanabe; Child & Rulon). Cf. also the similar work on the amphibian gastrula (p. 203) and the chick blastoderm (p. 335).

explanation for this has been put forward by Bernstein[2] (see Fig. 283). Supposing there to be a ventral pole substance, and supposing this to be water-soluble, it would be expected to diffuse away from the cut surface. In this case the gradient of it would be reversed in the case of the dorsal half but only steepened in the case of the ventral half. By the time the late blastula stage is reached, inversion of the dorso-ventral axis becomes impossible. That the ventral pole, naturally determined, is characterised by metabolic peculiarities appeared from the work of Förster & Örström, in which the larger of two depressions formed when fertilised but uncleaved eggs are treated with KCN always occurred at a point which vital staining showed later became the ventral pole. This observation was later much extended by Pease[2], who found that on exposure of the egg to a chemical diffusion field, the exposed (i.e. most affected) side always became the dorsal pole. The effective substances were, however, too miscellaneous to allow of any definite biochemical implications.*

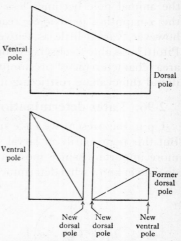

Fig. 283. Bernstein's hypothesis explaining the reversal of dorso-ventral polarity in half-embryos.

In later development, many of the echinoderms undergo a metamorphosis as wonderful as any to be found in the animal kingdom. The details of how the echinus-rudiment, for example, is formed within the body of the pluteus, the radially symmetrical adult converting into itself the whole substance of the bilaterally symmetrical pluteus larva, must of course be found in the classical descriptions (such as that of McBride[2]). But as it is likely that important induction processes go on during this metamorphosis, a brief account of what happens is necessary. The coelom begins as a diverticulum from the archenteron, budded off at its tip, near the stomodaeum, and this bud presently splits into two (cf. Fig. 284 A). The left and right coeloms remain applied closely to the gut and stomach but in time become constricted into anterior and posterior portions (Fig. 284 B). Later, at the echinopluteus stage (see Fig. 284 C, D), the left anterior coelom buds off a vesicle called the hydrocoele, which will later give rise to the adult water-vascular system. The neck by which it remains connected to the coelom will eventually become the stone canal of the adult. Now between two of the larval arms, over the hydrocoele as it approaches the ectoderm, an ectodermal invagination occurs, the amniotic invagination (cf. Fig. 284 C), which before long is cut off from the exterior and becomes a cavity (Fig. 284 F). The hydrocoele soon changes to an annulus (Fig. 284 E), and through the centre of this ring, with the amniotic cavity above it and the coelom below, runs the axis of radial symmetry of the future discoidal sea-urchin. The whole is termed the echinus-rudiment. For the present purpose we need not follow its further development.

* Effective: CN′, Fe(CN)$_6$‴, I′, CH$_2$I.COO′, picric acid, urethane and 2:4-dinitrophenol; ineffective: F′, AsO$_3$′, Cu″, H$_3$P$_2$O$_7$′ and malonate.

The most important point here is that the adult structure arises in normal development only on the left side, not on the right. However, the literature contains a number of descriptions of larvae containing two echinus-rudiments, one on each side, and McBride's observations[1] on such rare naturally occurring larvae in 1911 led him to try to produce them experimentally. This he was able to do (McBride[3]) by the use of hypertonic sea-water, but not in large quantity. It was at any rate clear that competence exists on the right side of the

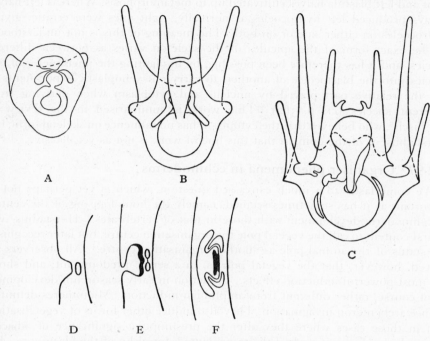

A   B   C

D   E   F

Fig. 284. Sketch of echinoderm metamorphosis (see text).

larva. Noticing, however, that wherever a hydrocoele appeared, there also appeared the amniotic invagination and other later transformations of the ectoderm, such as the formation of the spines, he concluded that the hydrocoele probably gave out a morphogenetic hormone and that the ectodermal changes were a dependent differentiation.* "We can have", he said, "a right hydrocoele without a right amniotic invagination and right dental sacs" (developed from the posterior coelom), "but we cannot have these latter structures without a right hydrocoele." This conclusion was later somewhat weakened by observations

* McBride himself compared the induction of the amniotic invagination in echinoderm metamorphosis with the induction of the lens by the amphibian eye-cup. It is interesting that he wrote "the body of an embryo is not...a mosaic of pieces each destined to form a particular organ, but consists of sheets of indifferent material, without form and void, on which a formative 'something' works and *evokes* the beautiful detail of the adult structure." And he spoke of this something as a hormone. How enlightened his general position in embryological theory was in this earlier period may be seen by a glance at pp. 528, 529 of his *Textbook of Invertebrate Embryology* (1914). Unfortunately, it did not remain so[4].

which show that a slight degree of amniotic invagination may be found without the action of a hydrocoele (Runnström[1]; Runnström & Runnström). But there is general agreement that its development cannot proceed further than the slightest initial stages without the underlying structure (cf. the induction of the oral arms in the pluteus, p. 480 above).*

The Echinopluteus, difficult to culture in the laboratory, has not lent itself to transplantation experiments. Hörstadius[6], however, describes the behaviour of right and left blastula halves cultivated up to metamorphosis. Whereas left halves always produced left hydrocoeles as normally, right ones were erratic, giving hydrocoeles on either side or on both. The meaning of this is not understood.

The exact form of the spicules of the skeleton varies as between different species, and it has therefore been possible, by implanting the micromeres of one species into the blastocoele of another, to carry out xenoplastic experiments to test the relative part played by nucleus and cytoplasm when deriving from different species (cf. p. 362). This work is summarised in the review of v. Ubisch[6], who believes that the cytoplasm has no influence on skeletal form, but Hörstadius[7] is of the opinion that this line of work is not as yet decisive.

2·97. Organiser phenomena in echinoderms

We come lastly to a much canvassed question, which is yet perhaps not so important as it has sometimes seemed, namely the homologising of the centres in echinoderm development with those in that of vertebrates.† Hörstadius[3] was at first content to call the vegetal pole the organisation centre, but later recognised the status of the animal pole as another organisation centre.‡ All observers are agreed, however, that the vegetal pole is, in a sense, predominant, and shows the most powerful induction effects. Induction in early sea-urchin development is, of course, rather different from amphibian induction. Micromeres definitely induce archenteron invagination. They also induce other forms of vegetalisation, and in those cases where they alter the presumptive significance of adjacent material, they act just as does the transplanted dorsal lip of the blastopore. But the formation of many organs (such as the stomodaeum and the band of cilia) will not take place if the embryo is too animal or too vegetal; only if there is a proper balance between the two centres. Hence if, for instance, too many micromeres are put into an animal half, these structures, which would form under balanced conditions, no longer do so because the embryo is now too vegetalised. Again, when the micromeres are transplanted to the animal pole, an archenteron

---

* If either the hydrocoele or the amniotic invagination fails to develop properly, the energy normally used in further metamorphosis will run to the formation of many queer structures never otherwise seen (cf. the description of Runnström[1]).

† It may not be irrelevant, in view of the similarity between the embryos of echinoderms and vertebrates as regards their late determination and hence their immense powers of regulation, to refer to the views of Bateson on the evolutionary connection between these two phyla, so surprisingly confirmed in biochemical experiments by Needham, Needham, Baldwin & Yudkin (see Needham & Needham[4]) and many others.

‡ Spemann lent his authority to this view in the omitted eighteenth chapter of his book[17]— "Man könnte die Sache vielleicht auch so ausdrücken, dass beim Seeigelkeim zwei gegeneinander gerichtete und ineinander greifende Induktionsströme wirksam sind" (p. 273).

may be induced there, but the archenteron at the other pole forms just as well because the most vegetal material takes on the characteristics of the vegetal pole. A new vegetal organiser is in fact formed.

One disadvantage of the echinoderm material as compared with the amphibian is that so far it has proved impossible to test the effects of implantation of material killed in different ways, or of chemical extracts or substances. This is only partially compensated for by the remarkable instability of the echinoderm gradient system towards external ionic factors. Nevertheless, the Scandinavian hypothesis of specific morphogenetic substances, formed at the animal and the vegetal poles respectively, and associated with two different kinds of metabolism, remains the most satisfactory one for future work.

It is not easy to regard the animal-vegetal gradient of the echinoderm egg as analogous to the animal-vegetal yolk gradient of the amphibian egg (A–B in Fig. 285), for no active centre has been found at the vegetal pole of the latter, unless we were to think of the as yet unidentified site of endo-mesodermal determination referred to above (p. 163) in connection with the determination of endo-dermal organs (E). It is not quite clear whether this is in the mind of those (Dalcq & Pasteels; Dalcq[7], p. 147) who identify the influence radiating from the vegetal pole of the echinoderm egg with the yolk gradient of the amphibian egg (cf. p. 220). The same authors regard the influence radiating from the animal pole of the echinoderm egg with the grey-crescent field of the amphibian egg

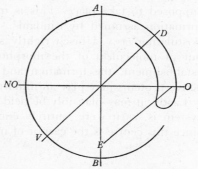

Fig. 285. Diagram of the amphibian gastrula. A–B, animal-vegetal yolk gradient; O–NO, "organiser-non-organiser" gradient; grey-crescent field; E, hypothetical origin of endodermal determination; D–V, dorsal-ventral gradient.

(cf. p. 220). As there is no neural induction in the formation of the pluteus, these identifications leave a feeling of uneasiness in the mind.

Lindahl, on the other hand ([7], p. 355), suggests that the ventral pole, on account of its importance for bilateral symmetry, should be regarded as corresponding to the organiser. He points out that the induction of a secondary embryo by the organiser in amphibia means a doubling of the dorso-ventral axis (D–V), while the induction of an archenteron in echinoderms by vegetal material means a doubling of the egg axis (A–B). Doubling of the dorso-ventral axis can of course occur also in echinoderms. On this view there would be nothing in amphibia corresponding to the animal and vegetal centres.

As against Lindahl, however, Hörstadius ([6], p. 166; [10], p. 236) reminds us that the position of the ventral centre is very unstable, that it has not been shown to have any inducing power, and that there is also the dextro-sinistral axis to be considered. This is not without importance in view of the later asymmetric origin of the echinus-rudiment in metamorphosis. Hörstadius further draws attention to certain analogies between the dorso-ventral axis in the

amphibian gastrula (or the "Organiser-non-organiser" axis of v. Ubisch[5]; O–NO)
and the animal-vegetal axis in echinoderms. Dorsal lip region, isolated, may,
as we have seen (p. 283), give much more differentiation than its prospective
significance would allow. Then a piece of dorsal lip transplanted to the presumptive neural region will turn into neural tube, whereas a piece of presumptive
neural ectoderm transplanted to the dorsal lip region will act as organiser and
become mesoderm after invagination (p. 170). The former phenomenon may be
akin to the suppression of the vegetal centre in an animalisation. The latter may
be akin to the suppression of animal differentiation by vegetalisation. Just as in
the case of the amphibia these experiments are most easily explained by the
liberation or the failure of liberation of an active morphogenetic substance from
inactive combination, so when the gradient re-forms in the echinoderm egg after
removal of animal or vegetal material, a new liberation of substances may be
supposed to take place. This is much the same as the reversible reactions of
formation assumed by Lindahl[7]. During such re-formation, we cannot avoid
postulating, as v. Ubisch rightly says[5], some kind of *Entmischung* (re-sorting
out of molecules) of the morphogenetic substances. In this process of re-
establishment, the formation and destruction of "masking complexes" with
active proteins may play an important part.

Lastly, it may plausibly be held that the echinoderm animal-vegetal gradient
system is, as it were, entirely enclosed in the amphibian organisation centre,
since this centre is the carrier of individuation.

Part Three
THE MORPHOGENETIC MECHANISMS

THE MORPHOGENETIC MECHANISMS

3·05. Introduction

In the development of an animal embryo, proceeding normally under optimum conditions, the fundamental mechanisms constitute a perfectly integrated whole. They fit in with each other in such a way that the final product comes into being by means of a precise co-operation of reactions and events. But it seems to be a very important, if perhaps insufficiently appreciated, fact, that these fundamental mechanisms are not separable only in thought; that, on the contrary, they can be dissociated experimentally or thrown out of gear with one another. This conception of gearing still lacks a satisfactory name, but in the absence of better words, *Dissociability* or disengagement will be used in what follows. It is already clear that embryonic growth can be stopped without abolishing embryonic respiration, and conversely, it is possible that growth or differentiation, under certain conditions, may proceed in the absence of the normal respiratory processes. There are many instances, again, where growth and differentiation are separable. It is as if either of these processes can be thrown out of gear at will, so that although the mechanisms are still intact, one or the other of them is acting as "layshaft" or, in engineering terms, is "idling".

Since one of the principal aims of embryology must be to understand the means whereby in the normal embryo the fundamental mechanisms are integrated, it is probably worth while to devote some attention to the cases in which this integration is artificially broken down. We shall then discuss in succession the *growth-process* itself in its chemical bearings, the various forms of *respiration* and *metabolism*, and the hints which we possess about the way in which *polarity* and *symmetry* properties of embryos arise. All this is involved in the mechanism of morphogenesis.

3·1. Dissociability

Let us first define more clearly the fundamental processes which are under consideration. For the present purpose, the rough classification shown in Table 28 will be adopted.

3·11. Disengagement of growth and differentiation

A certain incompatibility between these two great processes is very widely seen. It is a commonplace that the higher the differentiation of a tissue the fewer cell-divisions take place in it, and conversely. Or a myxomycete plasmodium, for instance (see p. 672), grows chaotically as long as it is fully nourished, but when conditions deteriorate it goes into differentiation and produces beautifully formed fruits. Some authors, such as v. Bertalanffy[3], have urged the existence of a causal relationship between rising differentiation and falling specific growth-rate.

Table 28. *Fundamental processes in morphogenesis*

A. GROWTH: increase in spatial dimensions and in weight	1. *Multiplicative*: increase in number of nuclei) 2. *Multiplicative*: increase in number of cells 3. *Auxetic*: increase of size of cells 4. *Accretionary*: increase in amount of non-living structural matter

B. DIFFERENTIATION: increase in complexity and organisation	1. Increase in number of kinds of cells:
	   (*a*) invisible. *Determination* of fates; gain or loss of *Competences*, etc.
	   (*b*) visible. *Histogenesis*
	2. Increase in morphological heterogeneity; the assumption of form and pattern; *Organogenesis*:
	   (*a*) *individuative*: due to the primary individuation field of the organism before any of its parts have become functional
	   (*b*) *corporative*: due to the mutual interactions of functional parts

C. METABOLISM: the chemical changes in the organism	1. *Respirative*, i.e. oxidation of carbon compounds 2. Fermentation or *Glycolysis*: the non-oxidative catabolism of carbohydrate 3. *Catabolism of protein* 4. *Catabolism of fat* 5. *Characteristic chemical activity*: e.g., pigment formation, or the synthesis of glycogen

Notes to Table 28

(1) The grouping of the processes is difficult and perhaps necessarily arbitrary. Thus, C5 might possibly be placed in B; A4 in C; and A1 and A2 in B.

Although increase in number of nuclei and increase in number of cells is not growth in the sense of size expansion, it is so intimately bound up with growth that it is placed in group A. Nuclear and cell division can occur without size expansion in many organisms, e.g. developing marine invertebrates in the cleavage stages, but here respiring protoplasm grows at the expense of inert yolk.

(2) *Multiplicative* is equivalent to the *Meristic* of the botanists.

(3) B1 needs further comment:

(i) Increase in number of kinds of cells may be either reversible or non-reversible. The specialisation of the histological elements in a tissue such as the kidney (which may be reversed in rapidly growing explants) is reversible (*Modulation*), but the determinations of early development are generally not reversible. Thus the kidney cells can lose their visible differentiation, but they cannot differentiate off into liver cells.

(ii) The notion of *Competence* has been discussed already (pp. 112 ff.).

(iii) Huxley[9] has introduced two further terms—"Histodifferentiation" and "Auxanodifferentiation". He distinguishes two periods in the growth of a part, firstly, a laying down of general form, accompanied by rapid morphological and histological change; secondly, a cessation of histological change, and a reduction of the form change to "quantitative alterations in the proportions of the definitive structural plan". For these two periods he suggests the terms "Histodifferentiation" and "Auxanodifferentiation" respectively. I am unable to see in what way this latter process differs from Growth, since growth, defined as increase in spatial dimensions, need not necessarily take place equally in all of them. The valuable investigations of Huxley on heterauxesis show, indeed, that it hardly ever does. And reference may be made to those directed cell streams, the existence of which is so important for early differentiation, and the nature of which is so obscure (see pp. 145 ff.). Again, if we define differentiation (as in B) as increase of complexity and organisation, "quantitative alterations in proportions" of a fixed plan can hardly be included in it. Such alterations amount to deformations and appertain rather to the sphere of relative Growth. "Histodifferentiation" appears to be covered by B1(*b*) and B2(*a*).

(iv) B1(*b*) is termed by Woodger[1] "Elaboration". "Differentiation" he defines as "a process whereby two (or more) different parts come into being by the separation unequally between them of something previously present in one part (or whole)."

(4) B2(*a*) is given the term *Individuative* after the concept of the individuation field, discussed on pp. 125 ff.

B2(*b*) is given the term *Corporative*, since in the late stages of development (the secondary or functional period) the various parts of the organism react upon one another as members of a corporation.

(5) It may be noted that processes C1–4 are all energy-providing processes, whereas C5 need not be, and may be the reverse.

3·111. Growth without differentiation

Growth without differentiation is likely to occur whenever there is any failure of the formation or liberation of primary or secondary organisers. In a preceding Section (2·371) we have seen many instances of this occurrence, especially the "anidian" embryos of sauropsida in which the primary organiser is suppressed. First discovered by Panum in 1860, such cases are referred to here only to emphasise their logical position. This they have in common with malignant tumours, already discussed in Section 2·38. In connection with this, it is of interest that in birds normal growth of the yolk-sac (Aliberti; Remotti[4]), the allantois (Katzenstein) and the amnion (E. Wolff[1]), will go on after the death of the embryo itself.

In one sense, all developmental arrests belong in this section. There are a great number of well-known malformations due to the failure of some morphogenetic process .(e.g. hare-lip in mammals). Another instance of this, discussed by Gudger and Kyle, is the occasional failure of flat fishes to turn over, thus retaining their bilateral symmetry throughout their lives.

3·112. Differentiation without growth

Another instance of the disengagement of growth and differentiation is afforded by the work of Hoadley[3], who transplanted eye-cups of the chick embryo to the chorio-allantoic membrane and allowed them to develop there until they had reached a degree of differentiation equivalent to that of the controls. Then by making and weighing wax models he ascertained the relative weights, and always found that the controls were much heavier. But there was a direct relation with age, for the younger the transplant at the time of transplantation, the smaller the eventual fully differentiated organ;

At the time of transplantation		After 8 days
Embryo age (hours)	No. of somites	size of control : size of transplant ($1 : x$)
48	28	$1 : 0·219$
35	14	$1 : 0·098$
20	0	$1 : 0·0136$
4	0	$1 : 0·00075$

Thus, in all cases the growth of the graft was very much inferior to that of the control in its normal situation, although differentiation, as judged by the degree of histological development and cellular elaboration, was the same. There seems no obvious mechanical reason why this suppression of growth should occur, for the transplanted fragment becomes surrounded by highly vascular tissue between the epithelial margin of the allantois and the repaired chorionic epithelium. Capillaries of the host membrane ramify freely in the neighbourhood of the transplant, the cells of which should have no difficulty in obtaining from them the nutritive building stones required for growth.

Hoadley explains these results by the assumption of a lag period immediately after transplantation which would be felt more acutely by the youngest and therefore smallest fragments. During this time cell-division would be suppressed, but histogenesis would be proceeding. It remains obscure, however, why the cell-division should be more fully suppressed in the case of the youngest grafts.

"Differentiation", says Hoadley[3], "is not primarily dependent upon any mechanism involved in specific cell-divisions, but to a large extent takes place independently of these. Inasmuch as this is true, and inasmuch as typical development depends not only on the subordinate differentiation of the constituent parts of the embryo, but also on the spatial relations between them and their size, the truth of the following statement is evident—'typical development is the result of the usual balance between morphogenetic (form producing) and histogenetic (cell differentiating) processes'." The derangement of this balance appears again in other instances. Thus Waddington[1], in the course of his extensive studies on the chick organiser *in vitro*, observed that the mere development of the blastoderm outside the egg gave rise to a considerable disengagement of the growth and differentiation processes. Both were slowed, but the former more than the latter. For the formation of one somite under normal conditions *in vivo* between the stages of 2 and 27 somites, 1·04 hr. are required, but *in vitro* the time taken was found to be between 2·44 and 1·15 hr., with an average at 1·65 hr. This reduction in the rapidity of somite formation was approximately constant whatever the age of the embryo at the beginning. On the other hand, the growth (measured as increase in embryo length) was very much slowed. "The stage of differentiation attained by an embryo", writes Waddington, "is very largely independent of the absolute size, and if a blastoderm is explanted at an early age, and thus is affected by the slowing of the growth rate for the whole of its life, an embryo may be formed which, compared with the normal, is very much too small for its stage of differentiation." Thus, an embryo of 19 somites, which *in vivo* is 5·3 mm. long, may be *in vitro* only 2 mm. long.

There are many other reports of experimental dwarfism in the literature. Mention of one recent one only will be made. Bohn & Drzwina[2] subjected the developing eggs of the sea-urchin (*Strongylocentrotus*) to $N/20$ and $N/10$ solutions of phenyldimethylbetaine in sea-water. It was found that if the treatment was begun at the gastrula stage, there was no stop to differentiation, and normal plutei were formed, but less than half the size of the controls. The betaine plutei provided a striking instance of a derangement of gearing as between differentiation and growth.

3·113. Nuclear division without cell-division

The experiments of Hoadley, referred to in the preceding section, were to some extent inspired by the numerous earlier works, dating from about the beginning of this century, in which it was shown that morphological differentiation is possible without either cell-division or nuclear division.

Chabry[2], in 1888, first succeeded in suppressing cytoplasmic division without nuclear division, by subjecting ascidian eggs to pressure, and J. Loeb[1] later, using hypertonic sea-water, obtained multinuclear single cells and blastomeres. The experiments were carried further by Norman, who found that multiple mitosis (11–18 nuclei/cell) regularly took place when the eggs of the sea-urchin (*Arbacia*) or the fish (*Ctenolabrus*) were made to develop in sea-water containing 2 % $MgCl_2$. In the case of *Ctenolabrus* high temperatures (30°) were more effective, and blastodiscs containing many nuclei but without a trace of cell-division could easily be obtained.* Later Morgan[2] criticised some of Norman's conclusions, but for the most part confirmed the possible dislocation of the phenomena of nuclear and cytoplasmic cleavage. Work of his own, moreover (Morgan[3]), on the effect of strychnine on echinoderm eggs, brought to light a similar dislocation, and Driesch[6] found other instances in which abnormal temperatures produced polynuclear undifferentiated masses from fertilised eggs. Barta, too, could encourage the production of giant polynuclear cells in explanted rabbit lymph nodes by subjecting them to conditions of poor oxygenation, and it is said that nuclear division without cell-division often occurs in explanted malignant cells (Warren H. Lewis).

In some forms, especially the arthropods, nuclear division occurs naturally without cell-division in the ordinary course of development. Thus, in the classical case of *Astacus*, the zygote nucleus divides into some twenty descendant nuclei which migrate towards the periphery of the egg. The egg then divides into a series of radially arranged pillars or yolk pyramids each having a nucleus at its peripheral end. The pyramidal blastomeres do not, however, persist, for their intervening planes break down in the central part of the egg, and give a layer of cells surrounding an internal mass of yolk. And a similar state of affairs is found in all insects, arachnids and crustaceans (see p. 455).

The most interesting recent work on this subject is that of Sugiyama. Sugiyama[1] discovered that sterile hen's egg-albumen, dissolved in sea-water and aged for a week at 15°, will completely suppress cytoplasmic cleavage in sea-urchin eggs without interfering with nuclear cleavage. This will occur whether fertilisation is effected naturally or parthenogenetically. The effective substance seems to be a spontaneous breakdown product of the protein; it adsorbs readily on kaolin or charcoal, but will not pass through a Chamberland filter; it is destroyed by moderate heating. The effect may be obtained not only with sea-urchins of well-known species such as *Strongylocentrotus*, but also with such forms as *Diadema*, *Clypeaster* and *Anthocidaris*. Sugiyama then showed that no difference existed in the respiratory rate of normal eggs and eggs with nuclear cleavages only (cf. J. Brachet's experiments[16] below, p. 512). These interesting facts merit further investigation.

Chromatin "growth" and multiplication, without either nuclear or cell-division (*Polyploidy*) may be added to this section. It may be brought about experimentally by the action of many pure chemical substances, especially colchicine (Blakeslee & Avery; Koltzov[5]), acenaphthene (Schmuck; Navashin; Kostov[1]), bromacenaphthene (Schmuck & Kostov), tribromaniline (Favorski) and other compounds (Schmuck & Gusseva); cf. formulae on pp. 248, 256. According

---

* Action of low temperatures, about 0° C., also gave the same results, as was shown by the work of Gerassimov on plant cells and Bucciante[1] on explanted heart fibroblasts. Evidently cytoplasmic division is more susceptible to unfavourable conditions at both ends of the normal range than is nuclear division.

to the last-named authors, nearly all derivatives of α-naphthalene are active but none of β-napthalene. The reader may be referred to an excellent review by Kostov[2], and the genetical importance of the phenomenon need hardly be emphasised. Its physiological mechanism is under study (Schmuck, Gusseva & Iljin). The giant or "polytene" chromosomes of *Drosophila* and other insects furnish another instance of chromatin growth, but without even any attempt at division.

3·114. Cell-division without nuclear division

Until a few years ago it would have been thought impossible to write such a title. Yet, in the remarkable case of parthenogenetic merogony (discussed in Section 2·645), E. B. Harvey has demonstrated the possibility of segmentation and even blastula formation in the complete absence of either paternal or maternal chromatin, in echinoderm eggs. How far this could be generalised to other forms is a question of much interest.

3·115. Differentiation without cleavage

So far we have only described cases in which differentiation was out of play, and cell-division and nuclear division (defined for the present purpose as two components of growth) were disengaged. It would seem very unlikely, *a priori*, that we could have a differentiation proceeding in the absence of either cell-division or nuclear division. Yet such a state of affairs was described by F. R. Lillie[2] in a classical paper in 1902 on the egg of the annelid *Chaetopterus*. After a short exposure to abnormal concentrations of potassium in sea-water, the eggs, without dividing into cells, passed through certain well-defined phases of differentiation, the yolk accumulating in a dense mass in the interior, and the peripheral cytoplasm becoming vacuolated and ciliated. "The ciliated ectoplasm and the yolk-laden endoplasm", said Lillie, "are analogous to the ectoderm and endoderm of the trochophore, and the phases of differentiation resemble some of the normal processes, though the resulting object can by no stretch of the term be properly called a trochophore." The phenomenon could be demonstrated in both fertilised and unfertilised eggs; and the action of the potassium ion was therefore in the first case to suppress certain events which arise normally, in the second case to induce certain events which do not normally occur.

The subject is of such interest that two of Lillie's illustrations are here reproduced. Fig. 286 *a* shows a well-developed ciliated object from his potassium cultures, Fig. 286 *b* a normal trochophore. The ciliated object is radially symmetrical round its long axis, and is divided by a constriction into a smaller and larger hemisphere. The yolk is aggregated in a dense mass in the latter, and a row of large vacuoles completely surrounds the body near the constriction; the proto-plasm of the smaller hemisphere is very dense and granular; the cilia are strong, active, and regularly placed. No cell structure could be detected in the living object. It presents, as Lillie said, "an undeniable resemblance to a trochophore; if the smaller hemisphere be compared to the pre-trochal, and the larger to the post-trochal region, the large vacuoles occupy approximately the position of the prototroch. A similar girdle of vacuoles is found in this position in the trocho-phore. The aggregation of yolk is in a similar position to the gut of the trochophore." And on sectioning it was clear that instead of the hundreds of nuclei visible in the normal trochophore, only one large nucleus was contained in such a ciliated object.

In a later paper Lillie[3] reported that certain other treatments besides abnormal K′ concentrations would produce the same result, i.e. letting the eggs remain abnormally long in ordinary sea-water before fertilisation (cf. p. 255), or exposing them to abnormal temperatures after fertilisation. Moreover, the unsegmented eggs undergoing differentiation might be either mononuclear or polynuclear. Unfertilised eggs treated with KCl would be of the former type, fertilised eggs would be of the latter if they had undergone polyspermy or if they had shown

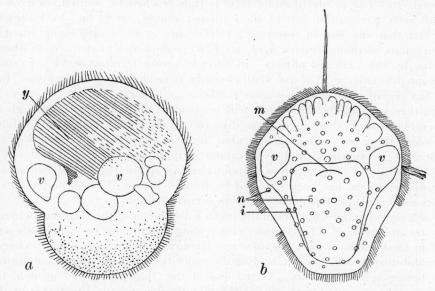

Fig. 286. Differentiation without cleavage in the egg of *Chaetopterus* (rearranged from Lillie) To the left: a "ciliated object" developed from an unfertilised egg (after 24 hr.). Treatment: in 95 parts of sea-water plus 5 parts of 22 N KCl for 1 hr. To the right: a normal trochophore 24 hr. old (same magnification). Reference letters: *i*, intestine: *m*, mouth; *n*, nucleus; *v*, vacuole; *y*, yolk.

an original cleavage followed by fusion of the blastomeres. It was found that the polynuclear unsegmented eggs differentiated into the "ciliated object" condition just as rapidly as normal eggs produced trochophores, whereas the mononuclear eggs took a considerably longer time to develop.

Chaetopterus did not long remain the only animal in which differentiation without cleavage had been observed, for Treadwell described it in another annelid, *Podarke*, and J. W. Scott found a particularly striking case in yet another, *Amphitrite*. Development without cleavage, under Scott's conditions of parthenogenesis, took the form of nuclear divisions, the early differentiation of a layer of ectoplasm, the growth of cilia, the appearance of vacuoles that are found in the fertilised egg of the same age, the development of brownish pigment, amoeboid movements of cytoplasm, and change in the general shape of the egg.

The apical tuft of cilia characteristic of the normal trochophore was always absent, however, from these "ciliated objects" of *Amphitrite*, as also from those of *Chaetopterus*. Certain experiments of Bastian on rotifer eggs could not be explained, according to Lillie[3], except upon the supposition that differentiation without cleavage was a possibility there also.

This curious phenomenon has recently been the subject of physiological research by J. Brachet[16], using *Chaetopterus*. The differentiation without cleavage was inhibited by KCN and iodoacetate. It was possible to show beyond doubt that there is an initial fall of respiratory rate, just as in normal fertilisation (cf. p. 563), but that the gradual rise which follows occurs at only about half the normal rate. The pre-fertilisation level is thus reached by normal embryos at $3\frac{1}{2}$ hr. after fertilisation, but by the "ciliated objects" at $6\frac{1}{2}$ hr. Nevertheless, the fact that any rise in respiratory rate occurs at all is interesting enough. Estimations of thymo-nucleic acid (see p. 633) showed also that the synthesis of this by the "ciliated objects" in 8 hr. following fertilisation or activation amounted to only 30 % of that synthesised by normal developing embryos. But the fact remains that there is such a synthesis.

Further work along these lines will be awaited with interest. Although in general it is true that no nuclear division occurs in differentiation without cleavage, there are numerous peculiar nuclear phenomena (description by J. Brachet), so that nuclear activity cannot be quite excluded.

A mild case of differentiation without cleavage in the egg of the Zebra fish, *Brachydanio rerio*, has been reported by Roosen-Runge. The blastoderm took on its normal shape, but no cell-division occurred within it.

Such considerations lead naturally on to a discussion of the relations between cell size and cleavage. Thus Henrici's interesting experiments on bacteria showed that in fresh media some cause operates favourable to increase of cell material but inhibitory to cell-division, leading the majority of cells to attain an abnormal size before splitting occurs. At the end of the "lag period" of growth, for instance, the average length of *Bacillus megatherium* was six times that of the cells at the time of inoculation. Gates, too, could abolish cell-division in bacteria by ultra-violet light, so that "spaghetti-like" filaments were formed. Similar results have been obtained by G. S. Wilson on other bacteria. But it would take us too far from the point at issue to deal adequately with the relation of cell size and cell number to cleavage, and a reference to the recent reviews of H. J. van Cleave and Berrill[5,6] must suffice.

One of the more interesting recent contributions to this question from the embryological point of view is the work of Bhatia on the muscle cells of the developing trout. Bhatia showed that throughout the free-swimming yolk-sac period after hatching, the growth of the muscle cells is entirely due to growth in size, and no cell-divisions take place. But as soon as the larva begins feeding and the yolk has all been used, cell multiplication begins; the increase in cell size alone cannot now keep pace with body growth. These phenomena call to mind the statement of G. P. Wright[1] and others, that avian yolk inhibits mitosis in explanted cells (cf. p. 631).

On the general question of differentiation without cleavage, the conditions in Protozoa may be remembered. An extraordinarily high degree of differentiation and morphological complication may exist in unicellular organisms (cf. Calkins).

3·116. Differential action of hormones, amino-acids, etc. on differentiation and growth

The endocrine system affords a powerful instrument for the disengagement of growth and differentiation, as was first realised by Gudernatsch[1] in his classical work in 1912 on the thyroid control of amphibian metamorphosis (cf. pp. 447 ff.). "Wir haben es also hier mit Differenzierung ohne Wachstum zu tun", said Gudernatsch[2], reviewing his earlier work in 1929. "Wir wissen, dass in der Normalentwicklung des Individuums zwei Processe parallel laufen, Wachstum und Differenzierung. Im Kaulquappenversuch ist es uns zum ersten Male gelungen, diese Prozesse voneinander zu isolieren." And indeed the premature metamorphosis induced by thyroxin affords the standard example of the disengagement of two fundamental processes (Uhlenluth[3]).*

At the present time attempts are being made to discover a differential action of the amino-acids on growth and organisation. It is possible to rear tadpoles to metamorphosis on a diet in which a single amino-acid is the sole source of nitrogen; carbohydrates, fats and vitamins being given in the form of a basal food. It has thus appeared that the amino-acids differ in their effects on growth and differentiation (Gudernatsch & Hoffman[1]; Hoffman & Gudernatsch). Arginine, cystine, lysine and phenylalanine permit growth, but practically no differentiation. Tyrosine and tryptophane, on the contrary, favour differentiation rather than growth, and produce small well-developed animals. In certain cases both growth and differentiation may be suppressed, the tadpoles simply continuing without much change, e.g. with leucine and alanine. Certain amino-acids, such as histidine and aspartic acid, hardly permit even maintenance. Di-iodotyrosine brought on the usual early metamorphosis, but this could be retarded, it was found, if a typical "growth acid" were given at the same time (O. Hoffman). Gudernatsch & Hoffman[2] are studying the permutations of the amino-acids, and Gudernatsch[3] has summarised the whole of the work. Experiments of a similar nature have been reported by Pincherle & Gelli.

Finally, high oxygen tensions are said to retard the growth of the chick embryo while differentiative processes are accelerated (Demuth[2]).

3·117. Disengagement of determination from differentiation and growth

The determinative factor in differentiation can, it seems, be disengaged from the growth process just as much as the other factors. In the interesting work of Twitty[1] (p. 337), who studied the determination of the ciliary beat on the skin of amphibian embryos, we have seen how a determinative process can be shifted in time experimentally, and thus thrown out of gear with the growth processes.

* But there are many others; the work of Voitkevitch[1] on feather development, for example, or that of Gray & Worthing in which the development of the neural system in the chick embryo is inhibited by tetanus toxin, without the somites being affected.

The well-known observations of Hörstadius[3] on echinoderm eggs also belong here. By various operative procedures (alteration of osmotic pressure, etc.) he was able to shift the relations of cleavage and the determination of regional differences of cleavage type. Cleavage type (i.e. micromere formation) would thus follow the usual course in time in spite of great delays in cleavage, with the result that a micromere could be formed with only one vegetal macromere instead of four and only two animal mesomeres instead of eight. These conclusions have been confirmed in parallel work on *Pomatoceros* by Berrill[5].

3·118. Mutual independence of growth, histogenesis and organogenesis

The technique of tissue culture has afforded many new possibilities of studying the relation between growth and differentiation. It was early found by Strangeways in 1924 that successful cultures of metazoan cells *in vitro* proceeded along one of two directions; either cell-division took place side by side with much "outwandering" from the original explant, or cell-division occurred only in the explant itself. The first of these processes led to no increase in differentiation, and was often called "unorganised growth", the second might involve far-reaching histogenesis and even organogenesis, and was called "organised growth". This "unorganised growth" was formerly thought to involve dedifferentiation, but it is now regarded as mere modulation (see pp. 279, 506). "Organised growth" we have already met with under the name of self-differentiation (p. 103). Self-differentiation may include morphogenesis as well as histogenesis, or it may not. The otocyst is a perfect case of the latter, for all the tissues of the normal finished product are formed, but no spatial arrangement whatever (Fell[1]). Embryonic limb bones show the former type, for a high degree of morphological or even anatomical differentiation is realised in their explants (Fell[2]). In all teratomata we have histogenesis, but organogenesis only in very few.

The independence of histogenetic and organogenetic differentiation has been much emphasised by Ranzi[2, 9, 10]. In the course of his work on the experimental embryology of cephalopods he often observed that the formation of the different sorts of cells characteristic of an organ may proceed quite normally, although the characteristic morphological form of the organ has been suppressed or modified experimentally. Morphological development was profoundly affected if a *Sepia* embryo was removed from its yolk and allowed to go on developing, but histological development proceeded normally.

It is clear, then, that histogenesis can occur without organogenesis; and the converse may probably also be true.

3·12. Disengagement of growth, differentiation and metabolism

3·121. Metabolism without growth

Bacterial phenomena do not come strictly within the field of this book, but it is hard to exclude a reference to them, since they throw some light on the essentials of the present problem. In bacteria differentiation does not come into play, hence the relations between growth and metabolism can be studied without that additional complicating factor. Under natural conditions growth and metabolism are not separated in bacterial life, for if a colony finds sufficient nutrient materials to permit of existence, it will automatically enter a period of cleavage. But there are certain cases in which it has been shown that metabolism can occur in the absence of growth, so that cell-division is quite absent and the cells are simply maintaining themselves as far as possible in the *status quo*.

The first of these is that of "resting bacteria". The term "resting bacteria" was introduced by Quastel & Whetham in 1924 to mean bacteria which were respiring but not growing. The organisms were grown in tryptic broth, separated by centrifuging, washed in saline solution, made up to a thick emulsion with saline, and finally well aerated. Under these conditions no proliferation takes place and the properties of the enzymes present in the cells can be investigated without the complicating factor introduced by the syntheses of growth. The results of such investigations have been summarised by Quastel in a review. How far the resting bacteria were alive, however, was questionable from the beginning, and attention was specifically devoted to this point by Cook & Stephenson. The only available criterion of the life of resting bacteria was viability; would they divide when replaced in nutrient medium? The results of viability counts showed that only from one-third to one-tenth of the resting bacteria were able to divide when given natural conditions. Oxidations and other reactions produced by the resting bacteria, however, were not due to these viable cells alone, for when this viable minority was reduced to 0·02 % of its original value by exposure to ultra-violet light, the rate of reaction was only slightly affected. Nevertheless, the presence of a considerable proportion of viable organisms among the resting bacteria demonstrated a disengagement of the growth and metabolism processes. It appears that under natural conditions most of the bacteria in sewage, though chemically very active, are incapable of growth, i.e. a state similar to the "resting" bacteria (Wooldridge & Standfast).

A second case is that of crowded yeast cells. Adrian Brown showed in 1892 that with yeast growing in malt wort, the medium would only allow the cells to increase up to a certain limit, and that if this concentration were reached or exceeded in the initial sowing, no multiplication took place although fermentation proceeded freely. This observation, an earlier example of the "resting" condition, permitted the study of fermentation apart from growth and enabled Brown to examine the effect of oxygen supply, etc., on the fermentation process alone. For an account of the subsequent work to which this led, reference may be made to the monograph of M. Stephenson.

A third case is that of nitrogen fixation by *Azotobacter*. Iwasaki showed that on a medium free from fixed nitrogen, cell-division may occur without nitrogen fixation and conversely nitrogen fixation without cell-division. The first of these conditions happens when an old culture is diluted with new medium; the respiration rises, and cell-multiplication goes on without nitrogen fixation, while the size of the individual cells decreases. After some time, when the population has greatly increased, cell-division ceases and nitrogen fixation sets in; parallel with this, the cells increase in size. If the culture is very greatly diluted, a multiplication phase occurs as before, but it is followed by a simultaneous multiplication and nitrogen fixation phase, before the colony passes over into the pure nitrogen fixation phase. Owing to the dissociation here between cell-division and increase in size of the cells, the dissociation between growth and characteristic chemical activity is not perfectly clear, but it is obvious that a certain measure of disengagement between metabolism and growth is normal to this organism. The state of affairs has striking resemblances to that referred to below in the discussion of growth and metabolism in explanted cells of higher animals.

The antibody inhibiting cell-division in trypanosomes, but not any of their other normal functions (Taliaferro), might be regarded as a fourth case.

3·122. Metabolism without growth and differentiation

We do not have far to look for phenomena of a similar kind in the embryonic life of the higher animals. Many insect embryos, as described elsewhere in this book (pp. 578 ff.), complete a certain part of their development and then enter upon a state of dormancy or diapause which lasts for a considerable time, and at the end of which a short period of rapid development leads to hatching. The silkworm egg is perhaps the most famous example of this mechanism, and it is very common among the orthoptera. But during the diapause the respiration of the embryo does not completely disappear; on the contrary, there is a small but continuous maintenance metabolism. The growth and differentiation processes have been thrown out of gear with the metabolic processes for the duration of the diapause.

It is not so generally known that something analogous to the diapause occurs in mammalian development. According to the work of many observers, reviewed by Hamlett[3], the embryos of many mammals pass a considerable proportion of their intra-uterine life in a state of suspended development. There can be little doubt but that some metabolism is proceeding also during this "diapause". In reptiles, embryonic hibernation is found in the case of the tuatara lizard, *Sphenodon* (Dendy), and the pond tortoise, *Emys* (Boulenger). In birds it has not so far been reported, but certain English wild birds present suspicious behaviour from this point of view. It is said that the blackbird, for instance, lays one egg each day for a week, but although sitting has begun from the first day, all the eggs hatch out together. There are also some subtle adjustments of incubation times in parasitic birds such as the cuckoos (H. Friedmann), of which the oecological value, but not the mechanism, is obvious.

If natural embryonic hibernation is an instance of the possibility of suppressing differentiation and growth in a developing system, while retaining a maintenance metabolism, so is the viability of metazoan cells at low temperatures. Bucciante[3,4] has examined with care the survival of embryonic cells in these conditions. Taking hen's eggs from the incubator at definite times after the beginning of incubation, he allowed them to remain at standard lower temperatures and measured the time elapsing before it became no longer possible to make successful tissue cultures from the embryonic cells. As the table opposite shows, the various tissues have different times of survival.

But more important for the present purpose is the fact that at these low temperatures metabolic processes were proceeding although growth and differentiation had been totally done away with. Not much is known about the respiration of tissues at 0°C., but the work of E. C. Bate-Smith[1] conclusively showed that it is not negligible. Amphibian muscle at 0° takes up about 0·025 c.c. of oxygen/gm./hr. (wet wt.).

	Time in days before complete loss of viability in tissue culture	
	At 15–20°	at 0°
Skin	24	32
Cornea	18	28
Meninges	21	25
Amnios { epithelium	7	17
{ muscle	5	10
Skeletal muscle	21	22
Spleen { fibrocytes	6	12
{ leucocytes	3	5
Heart	6	10
Liver { endothelial cells	6	10
{ hepatic cells	2	5
Mesonephros	6	12
Mesencephalon	6	10
Aorta	7	10
Iris	6	11

Other workers have carried out experiments similar to those of Bucciante, but their durations of survival have not usually been as great as his, probably owing to the fact that they have placed the embryos or parts of embryos in saline solutions at the low temperatures instead of keeping within the intact egg. Thus, Verne & Odiette used Tyrode solution and found the survival was improved if glucose were added to it. Bucciante[5] himself considers that the saline solution washes out from the tissue certain necessary growth factors. A similar difference of technique explains the long survivals of whole rat embryos reported by Simonin as opposed to the short survivals of S. Kodama. Nicholas[1], again, has successfully kept whole rat embryos at 24° in Ringer solution for 72 hr. with no sign of differentiation or growth, but no doubt considerable respiration.*

One point which deserves mention here is that the development of tissues may be unequally affected by exposure to low temperature. Coghill[2] states that in amphibia ectoderm is much more sensitive than mesoderm.

3·123. Stepwise inhibition of growth, differentiation, fermentation and respiration

So far, abnormal temperatures and naturally adjusted processes alone have entered into the description of the disengagements which may occur between growth and differentiation on the one hand and metabolism on the other. But there are other agents which may be used experimentally to affect the gearing.

Of these the classical example is the work of Warburg[1]. In 1908 he studied the effect of hindering or inhibiting the cleavage of sea-urchin eggs by placing them in hypertonic sea-water. Although profound effects were observed on the segmentation, the respiratory rate was very little changed (0·368 mgm. O_2/hr./ 28 mgm. egg nitrogen in the normal lot, 0·347 in the inhibited lot). In a later paper he described the effect of phenylurethane. Under the influence of this narcotic, segmentation could be wholly abolished, and yet the respiratory rate

* Cf. p. 450.

remain unaltered, if the dose was well chosen. Thus, after 25 min. in ordinary sea-water the astrospheres are visible, but nothing has happened in $N/2000$ phenylurethane. At 40 min. the first cleavage takes place in normal eggs, but in the phenylurethane eggs the astrospheres are only just appearing. At 90 min. the second cleavage takes place in normal eggs, but in the phenylurethane eggs only the equatorial plate stage has been reached. Yet in spite of this great retardation of cell-division, the respiratory rate was in no case affected to a higher degree than 20 % (e.g. 0·450 c.c. O_2/hr./28 mgm. egg nitrogen in the normal lot, 0·438 c.c. in the phenylurethane lot). "The visible changes in the early developing egg", as Warburg said, "are not conditions of the oxygen utilisation following fertilisation. But the oxygenation is a condition of the visible changes, so that those chemical processes, the activity of which we can judge by the amount of oxygen consumed, would seem to underlie the morphological ones."

In this connection it is striking that the cleavage of *Arbacia* eggs is completely suppressed at a CO_2 tension of 120 mm. Hg (Haywood & Root[1]), but their oxygen-consumption is still 30 % of the normal value and remains so up to 180 mm. Hg (Root).

Experiments which show an identical disengagement of the fundamental processes were made in 1930 by Partachnikov. Partachnikov studied the action of quinine on heart fibroblasts of the chick growing *in vitro*, and found that the inhibitory effects on growth and on metabolism (as assessed by the disappearance of glucose from the culture medium) ran by no means parallel.

	Growth (Ebeling index)	Glucose utilisation %
Control	3·96	71
Quinine: 1 in 10,000	0·33	58
1 in 5,000	0·0	34
1 in 2,500	0·0	0

Thus, at a certain concentration of quinine the growth (cell-division, increase in cell size, etc.) was inhibited 100 %, the metabolism only 55 %. Here growth was more vulnerable than carbohydrate breakdown.

A similar stepwise inhibition is seen in experiments with iodoacetate* (Krontovsky[2]; Krontovsky, Jazimirska-Krontovska & Savitzka) and with radium emanation (Mikhailovsky). Mikhailovsky cultured explants of embryo heart in much the same way as Partachnikov, but subjected them to the action of radium emanation for periods ranging from 4 to 20 hours. Again the inhibitions of growth and metabolism were not parallel.

	Area of explant sq. mm.	Glucose utilisation %
Control	14·15	79
4 hr. irradiation	7·5	67
20 hr. irradiation	—	53

So in this case the inhibition of growth was 47 % and that of glucose utilisation 15 %.

* It may be of interest that Soliterman has detected changes in vital staining properties of the cells which coincide with bromacetate inhibition of glycolysis and not with cessation of growth.

Hubert's work on the chick embryo was particularly interesting, in that he measured the inhibitions not only of growth and respiration, but also glycolysis, caused by irradiation with X-rays. He subjected 5th-day chick embryos while still *in ovo* to the action of accurately dosed X-radiation, and then measured glycolysis and respiration by Warburg manometer methods on the same day, and growth by leaving a population of embryos until the 12th day and then observing their fate. Fig. 287 shows the results. It is clear that the effect of X-radiation upon the fundamental processes was not at all the same; growth was the first to be affected, then glycolysis, and finally respiration. The strongest treatment of

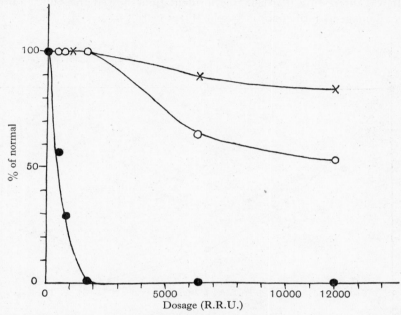

Fig. 287. Effect of X-rays on chick embryo (Hubert). ○ glycolysis, ● growth, × respiration.

all (12,000 Röntgen radiation units) caused only a 15 % decrease of the respiration below normal, but glycolysis was reduced by 45 % and growth entirely stopped.

X-radiation is therefore an agent which at a certain level will abolish growth, but cause no change in the intensity of respiration or fermentation, while at another level it will exercise an inhibiting influence on these metabolic processes as well. The parallel with Warburg's early experiments is striking.

Similar data exist for tumour cells. Crabtree[1] irradiated Jensen rat sarcoma with γ-rays. At 10 hr., respiration, but not aerobic glycolysis, was depressed. At 18 hr. both were depressed. It seems also that doses lethal for tissue cultures of normal mammalian cells do not depress their glycolysis (Krontovsky[1]; Krontovsky & Lebensohn). And this can be seen in *Arbacia* or *Chaetopterus* embryos, where L. Chesley found that developmental anomalies occur at doses of γ-radiation quite insufficient to inhibit normal or dye-stimulated respiration.

So far the examples of this section have all been concerned with developing embryonic cells and tumour cells, but there are examples from unicellular organisms which fit in closely with the argument. Firstly, the effect of X-rays on the fermentation, respiration and division of yeast cells was investigated by Wels & Osann. Here again a marked differential effect was obtained, growth being by far the most susceptible of the fundamental processes, and the other two, indeed, being hardly affected by the doses used. Fig. 288 summarises the data. Just as with the chick embryo in Hubert's experiments, growth (division) is abolished before any effect has become evident upon respiration or fermentation. It was interesting that cell-free fermentation showed the same immunity to X-ray damage as fermentation by the intact cells.

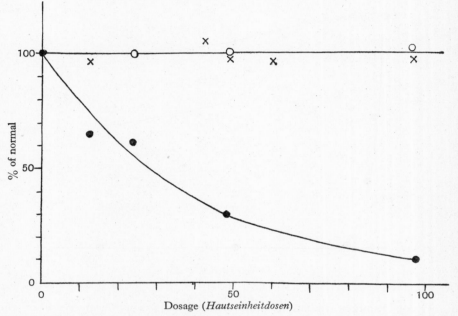

Fig. 288. Effect of X-rays on yeast (Wels & Osann). ○ respiration, × fermentation, ● growth.

Similar experiments to those of Wels & Osann were carried out by Rubner as part of a general survey of the role of water in the living cell. By placing yeast cells in different concentrations of salt, he reduced their water-content and—when the concentration was sufficiently raised—their heat production. But only a mild degree of hypertony was required to stop growth, and as the data in Fig. 289 show, in 4 % salt solution growth is almost completely suppressed while water-content and heat production continue at almost their normal level.

By differential exposure to radiation a veritable "dissection" of metabolism is possible. Using γ-radiation on embryonic rat kidney growing *in vitro*, B. E. Holmes[1] found that protein catabolism (as judged by ammonia production) was much less sensitive than carbohydrate catabolism (as judged by lactic acid production), the former continuing unaffected when the latter was reduced to 50 % of its normal value. She next showed (B. E. Holmes[2]) that in the same tissue, arginase, an enzyme significant in view of the ornithine cycle (p. 52), was unaffected by a dose of γ-rays capable of reducing the activity of lactic dehydrogenase over 50 %. Moreover, glucolysis (with glucose as substrate) is much more severely affected than glycolysis (from hexosediphosphate) in tumour tissue (B. E. Holmes[4]).

Parallel experiments were made by Crabtree[2]. Using mixed β- and γ-radiation on tumour tissue, he discovered a temperature effect. At 37° respiration was much more vulnerable than anaerobic glycolysis; at 0° the reverse was the case. But glyoxalase, like arginase, was unaffected at any temperature. Crabtree[3] extended

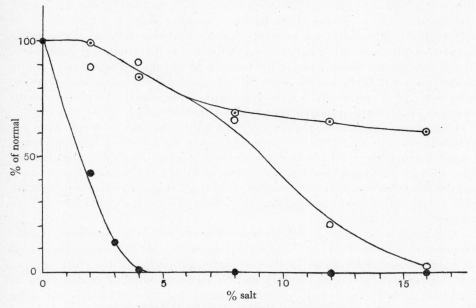

Fig. 289. Effect of hypertonic solutions on yeast (Rubner).
⊙ water content, ○ heat production, ● growth.

the former results on protein catabolism by showing that ammonia production by tumour and normal adult tissue is actually stimulated 100 % by β- and γ-radiation, independent of what has happened to the carbohydrate catabolism, i.e. whether glycolysis has been reduced to nil by irradiation at 0° or whether it is still fully active after irradiation at 37°. Crabtree, however, views the stimulated protein breakdown as compensatory for the lost power of carbohydrate breakdown (cf. work on grasshopper (p. 581) and chick (p. 592) embryos). Stimulation of protein breakdown in this way has also been found in the case of the protozoon *Bodo caudatus* by Lawrie & Robertson. Crabtree & Cramer discovered the interesting fact that the radiosensitivity (diminution of lethal dose) of tumour tissue is increased if respiration is inhibited by anaerobiosis, KCN, or low temperature, but not if glycolysis is inhibited by iodoacetate or fluoride. This fits in with Lipmann's finding[1] that the growth of embryonic heart fibroblasts is inhibited by partial anaerobiosis 50 % when the Q_{O_2} is reduced only 30 %, and suggests again that respiration is more important than glycolysis for growth. Conversely, Kolomjetz found with radium emanation on the same material that growth was reduced only 34 % when glycolysis (as judged by lactic acid production) was decreased 50 %.

3·124. Cell-division and differentiation without oxidative metabolism

Cases of the disengagement of growth or differentiation from metabolism have so far been restricted to the removal of the former while the latter remains intact. But is it possible for growth and differentiation to proceed in the absence of metabolism? In the profoundest and most inclusive sense in which the word metabolism can be used, it is probably not possible, but there are not wanting suggestions that the early stages of development may be facultatively anaerobic. If this were so, the whole series of mechanisms within the cell which carry out its oxidations would be dissociable from the processes of growth and differentiation, which would compulsorily engage only with fermentative reactions or with reactions involving hydrogen acceptors other than oxygen.

Thus, as we shall see (p. 564), many if not all the reactions of fertilisation are possible anaerobically in some marine eggs. On the other hand, experiments suggesting that the echinoderm egg can develop using dyes such as methylene blue instead of oxygen have not been confirmed. The special case of nematode eggs may be relevant here. Although the cleavage stages are in general aerobic (see p. 576), Dyrdovska now claims that the first cleavage can occur in pure nitrogen.

As regards the chick embryo, little is known, in view of the experimental difficulties attending the study of it in the earliest stages. But M. T. Burrows discovered in 1920 a remarkable change in the respiratory properties of the chick embryo with age. Taking explants of chick embryonic heart cells, he placed them in different partial pressures of oxygen and found that their behaviour was quite different according to the stage of development of the chick from which they had been taken. Fibroblasts from 4–5-day embryos would grow and pulsate for 46 hr. or so in pure nitrogen, and for 50 hr. in only 7·5 % of oxygen. Fibroblasts from 10–15-day embryos, however, would not grow at all in nitrogen, and only for 12 hr. in 1·5 % oxygen. This must mean that the younger cells possess some source of energy which the older ones do not have. Burrows' work was later confirmed and extended by Wind, who paid particular attention to strictness of anaerobiosis. In his conditions only the smallest amount of growth proceeded even with 4–5-day embryos, but at an oxygen concentration of 2×10^{-4} vol. % the difference was clear-cut, for cells from 10-day embryos would not grow at all, while those from 5-day embryos grew profusely. Wind also made the interesting observation that if the cells were in each case subcultured for several weeks and then brought under anaerobic conditions, the difference between 5-day and 10-day embryos had disappeared.

There is certainly no truth in the statements of the earlier workers that avian development may proceed to any length anaerobically. But *unorganised* growth in such circumstances undoubtedly occurs (Byerly[1]).

3·125. Selective engagement of growth or differentiation and characteristic chemical activity

In the preceding paragraphs we have been considering the disengagement of growth and metabolism only in the sense of maintenance metabolism. But there are many other chemical reactions performed by the cells of an embryo, such as the storage of glycogen or the elaboration of a characteristic pigment, and it is in order to enquire what relations exist between these special types of metabolic change and the growth processes.

On purely morphological grounds, Peter[1] and Lauche were led to the conclusion that "a cell which is working does not divide, and a cell in mitosis is not working".* The subject entered a new phase when Fischer & Parker described a technique for regulating the intensity of proliferation of tissue cultures, for it was found that osteoblasts, growing under the conditions of "vie ralentie", formed bone, although, as Doljanski[1] had previously shown, they would not do so when cultivated in the normal way. This indicated that growth and characteristic chemical activity might be mutually incompatible in tissue cultures, as if a primary shaft could engage with one of two adjacent gears, but not with both.

These considerations led Doljanski to make a closer study of the relations between growth of explanted iris, and melanin production. Explanted iris, if rapidly growing and repeatedly subcultured at short intervals, had been found by Kapel not to produce pigment. Doljanski[2] now took pure cultures of iris epithelium from the chick embryo and cultivated some in the usual embryo extract, others in adult fowl plasma. The latter grew much more slowly, of course, than the former. But "while the intensely proliferating culture in embryo extract contained at the end of some time few or no pigment granules, the culture whose growth had been inhibited showed compact masses of black pigment. The granules of melanin accumulated in great quantities in the central part of the colony, and were found dispersed throughout the whole cellular membrane." Doljanski deduced an "antagonisme entre la multiplication et la fonction physiologique de la cellule".

Precisely the same state of affairs was found when liver explants were examined (Doljanski[3]). Ordinarily, when proliferation is rapid, the glycogen contained in the original explant soon disappears and no more is formed. But when cultures were transferred to adult fowl plasma, so that growth practically ceased or was very much slowed, glycogen was visible in abundance when the cells were treated with iodine vapour after 12–14 days. Thus hepatic cells in pure culture certainly conserve their glycogenic power, and it only appears to be absent because growth *in vitro* is usually too rapid to permit of the characteristic chemical function of the cells. Support for this view could be found in the work of various

---

* Thus, for example, cells in mitosis are not found in developing kidney tubules if these are functional. And Chlopin found that human embryo epidermis would *either* grow *or* differentiate.

histochemical investigators such as Jordan, who concludes that the amount of glycogen in a tissue is inversely proportional to the intensity of cell-division in it.

Another instance of the disengagement of development and characteristic chemical activity might be found in the work of Pozerski & Krongold, as early as 1914, on intestinal grafts. Krongold first showed that the embryonic intestine of the rat, grafted to a position under the skin of the adult animal, gives rise to a completely differentiated system, all the normally appearing kinds of cells (fusiform cells, muscle fibres, glands of Lieberkühn, mucous cells, Brünner glands, etc.) being arranged around a central lumen, the whole forming a kind of vesicle. But in spite of this display of histological self-differentiation, the characteristic chemical products, saccharase, maltase, lactase, and secretin, were found by Pozerski & Krongold to be entirely missing. Enterokinase alone could be demonstrated, and even this, according to Pozerski & Krongold, was more probably of host origin, and was merely being excreted into the vesicle because of its chance presence in the host's blood stream. Histological differentiation, then, was in this case unaccompanied by the appearance of characteristic chemical activity.

On the general question of this subsection, Fischer[2] has assembled further relevant facts in a review. What was proved by Doljanski for explants of iris and liver finds its parallel, it seems, in explants of nerve, heart, and connective tissue, according to the results of other observers. Berrill[5], too, has amassed evidence from his extensive data on ascidian development that cell-division (growth) and differentiation occur in separate phases.

3·126. Disengagement of growth or differentiation from characteristic chemical composition

The question now arises whether every morphological stage of development is characterised by an absolutely definite chemical composition. In general it may be said that the limits of "play" here are small. As will later be seen in the section on relative chemical growth, or heterauxesis (p. 559), there are as yet relatively few data on the chemical composition of organisms which have been made to grow fast or slow, and which therefore are of the same total weight but differ in age. But they indicate that following a chemical ground-plan of growth, the composition follows size or weight, not age. On the other hand, there is a remarkably clear case, which we owe to Briggs[1], where in amphibian embryos developing at different temperatures the rate of density decrease (due to uptake of water by a non-cleidoic egg, cf. p. 33) follows a different course from the rate of morphogenesis. At both high (29°) and low (10°) temperatures the decrease in density between fertilisation and a given stage of gastrulation is greater than at intermediate temperatures, so that water-content of the system must vary independently of morphogenesis. At the same time it should be remembered that a holoblastic embryo constitutes a mixed system embryo + yolk, and hence it may not be fair to compare it with embryonic tissues proper where the chemical composition is probably more rigidly fixed. Indeed, the data of Galloway, analysed by Briggs, show that in larval life the water-content of the tissues is independent of temperature and would vary only with growth.

3·13. Dedifferentiation and degrowth

The analogy of gearing cannot be followed far without involving the concept of reversibility. And indeed it appears from a survey of the known facts that a considerable degree of reversibility does exist in the cases of growth and differentiation. The question of greatest interest for the present discussion is whether the phenomenon of disengagement is shown when the processes are under reversal, just as it is when they are proceeding normally from homogeneity to heterogeneity and from smallness to largeness.

When an animal organism regresses after reaching a certain stage of development, its regression may, we find, take one of two forms; either it may consist only of a reduction in size, or it may also undergo marked morphological changes. The former of these alternatives could be called pure degrowth, the latter degrowth plus dedifferentiation. Of the first type several examples are known, but no doubt the most famous is that of the starved planarian worm, which was thoroughly worked out by F. R. Lillie[1] in 1900. When *Planaria dorotocephala* is starved, it becomes smaller and smaller until it shrinks to a size less than that at which it originally hatched from its egg. But the only morphological changes which occur are slight alterations of the proportions of the parts, and with regard to these Abeloos has shown that they retrace the alterations of the proportions of parts which accompanied normal growth. Thus measurements of the ratio pre-ocular length/total length fall upon the same curve whether the worm is growing or degrowing, and according to Abeloos & Lecamp the same statement applies to the relative size of the epithelial cells of the gut. This reappearance of the characteristic heterauxesis during degrowth enhances our opinion of the reversibility of the process, but there are senses in which it is not truly reversible. Abeloos has demonstrated that the regressed planarian has not become rejuvenescent as regards growth rate; the high growth rate of normal early life is never regained, and subsequent size increase, brought about by the cessation of starvation conditions, proceeds only slowly.

There is thus no true dedifferentiation in the regression of planarians—unless the term were used to refer to the disappearance of the terminal branches of diverticulated organs, which occurs in the later stages of reduction. On the contrary, such special differentiations as the existence of subsidiary eyes persist during reduction even when the worm has sunk below the size at which it hatched. Lillie was inclined to think that reduction of cell number was alone concerned in regression, but Abeloos & Lecamp show that cell size is reduced too.

More common, perhaps, is the association of dedifferentiation with degrowth. The phenomenon has been known among ascidians for a considerable time; thus, regression in colonial ascidians was discussed by Caullery[1] in 1895 and by Driesch[11] in 1902. For the present line of argument the work of Huxley[6] on regression in the ascidian *Clavellina* is of particular interest. His illustrations of the stages in the regression are reproduced here in Fig. 290. Fig. 290a shows "a perfect little *Clavellina*, highly contractile, with active gill cilia and wide-open syphons, the tissues all as clear as crystal. The first sign of reduction is the permanent contraction, followed by the total closure, of the syphons, accompanied by a general shrinkage (Fig. 290 b). After a time the gill cilia stop working and the gills present somewhat the appearance of buttons. Then follows a marked shrinkage, accompanied by loss of transparency and usually by the appearance of numbers of the characteristic white pigment cells (Fig. 290 c). The opacity obliterates the gill slits in the living animal, and the shrinkage leads to the pharynx and peribranchial cavities pulling away from the ectoderm at the syphons; the one process takes place first in some cases, the other in others. More shrinkage, with consequently

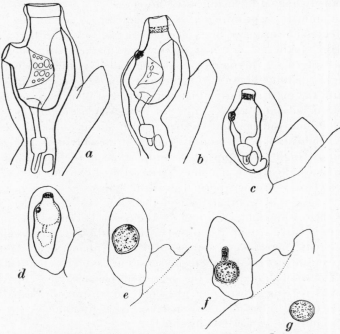

Fig. 290. Regression of the ascidian, *Clavellina* (rearranged from Huxley). (*a*) After 5 days; practically normal and still actively functional. (*b*) After 7 days; exhalant syphon closed, inhalant syphon contracting, gill slits reduced. (*c*) After 9 days; both syphons closed and withdrawn within tunic, gill slits no longer visible. (*d*) After 28 days; heart no longer beating, internal organs no longer clearly visible. (*e*) After 37 days; a pigmented ovoid much reduced in size. (*f*) After 46 days; further shrinkage. (*g*) After 53 days; tunic omitted, slight further shrinkage.

greater opacity, brings us to a somewhat dumpy, sausage-shaped condition (Fig. 290 d). The heart, after beating ever more slowly and more irregularly in the dense mass of loose cells which fill the body-cavity, now ceases its action altogether." The organism will now have a small tail if it began with a fairly long stem stolon, but Fig. 290 e shows a tailless example. "In the next stage, this tail, if present and not too large, is absorbed entirely, and the long oval or slightly dumbbell-shaped body of the creature shrinks into a very dense white ball, spherical, ellipsoidal, or egg-shaped. Reflected light shows it to be of a yellowish cream colour with milk-white pigment cells dotted over it; by transmitted light the pigment cells show black on a greenish ground, and here and there in the dense mass vague clearer spaces are seen, showing that organs and cavities still persist within (Fig. 290 *f* and *g*)." Such is the process of degrowth and dedifferentiation as described by Huxley, and it should be noted that it is not taking place in some organism low in the taxonomic scale, but in no less an animal than one of the Protochordata. Starvation and toxic agents of all kinds will produce regression, but in this particular case mere confinement in an unchanged volume of sea-water was sufficient to produce it.

Now the illustrations of Fig. 290 were all drawn to the same magnification, and it is therefore interesting to note that there is a rather sharp distinction between stages *d* and *e*. At first it would seem that degrowth occurs without dedifferentiation, for stage *d* differs from stage *a* mainly as regards its size, but later on dedifferentiation sets in, and the disappearance of organs and morphological form ensues. The examination of regression from the point of view of a comparison between dedifferentiation rate and degrowth rate would be highly desirable. The data required for any quantitative analysis of this relationship do not seem to exist as yet in the literature.

In Huxley's opinion, the regressive changes observed in *Clavellina* cannot be supposed to represent reversions to stages passed through in embryogenesis. No ascidian tadpole makes its appearance when an ascidian dedifferentiates. The chief cell change which occurs during dedifferentiation is a regression to the cuboidal or spherical state, since such a condition requires the least amount of energy for its maintenance. These conclusions were also reached by Huxley[1] in his work on regression in the colonial ascidian *Perophora*. But regression must be in some sense a true reversibility, since it is a passage from complexity to simplicity, from heterogeneity to homogeneity, from high to low organisation, from largeness to smallness, and therefore the exact opposite of embryonic development.*

Many other studies of dedifferentiation throw light on the points discussed above (e.g. for echinoderm plutei, Runnström; Huxley[2]; for Hydrozoa, Huxley & de Beer[1]; de Beer & Huxley; for Amphibia, Blacher). Some investigators are inclined to regard the regressed conditions as more closely allied to embryonic stages than others will admit. Thus, Hadzi in 1910, subjecting the medusa *Chrysaora* to starvation, obtained "gastraea-like bodies" only 80 μ in diameter, composed of cells quite embryonic in appearance. The most remarkable case, however, is probably that of Davydov, who studied regression in the nemertine worm, *Lineus*. By keeping portions of the worm in darkness in a constant volume of sea-water under starvation conditions for long periods (up to two years) a stage was reached at which the size and shape of all the organs made the animal hardly distinguishable from an embryo in the pilidium. There was no anus, the brain was enormous, and the lateral nerve cords ran immediately under the ectoderm. In the final stage (at which Davydov had only two specimens left) a "blastula" or lump of large cells was obtained.

It is clear, then, that degrowth can exist without dedifferentiation, and that the rates of degrowth and dedifferentiation in a given case of regression by no means

---

* And, moreover, the regressed ascidian will retrace its steps to normality given favourable conditions.

necessarily run parallel. Is dedifferentiation without degrowth possible? It would mean the collapse of cellular organisation and morphological form without any decrease of mass; the biological analogue of what would occur, as Robert Boyle thought, if the divine sustaining power were withdrawn from the universe "reducing it to a sort of Chaos or confused State of shuffled and depraved Things" (1744). It might involve both histological and morphological collapse, or it might involve morphological collapse only. Both cases occur in nature, the former in the metamorphosis of holometabolic insects (discussed in detail on pp. 470 ff.), the latter in the dissociation and reunition of sponges.

Dissociation and reunition in the sponges has been known for a very long time, but the facts are as extraordinary as those of insect metamorphosis. If a sponge is dissociated by being pressed through a sieve of gauze, the individual cells, if sown on to a glass or other surface, will come together again and in time will form a new sponge fully provided with canals, chambers, surface epithelium, skeleton and all the different varieties of cells. Recent papers of Wilson & Penney and Wilson[2] give a clear account of the process. The normal sponge consists of the following histological elements: (a) epithelioid syncytial sheets lining the exterior and canal surfaces, (b) nucleolate mesenchyme cells, (c) "grey" mesenchyme cells, (d) rhabdiferous mesenchyme cells, (e) globiferous mesenchyme cells, (f) fibre cells, (g) reproductive cells, (h) scleroblasts and spongoblasts connected with the skeleton, (i) flagellated collar cells. The pressed-out material is, of course, primarily characterised by the fact that the previous morphological organisation has gone and the cells are isolated, or at any rate only bound together in very small groups. But the number of cell types in it is also limited; it contains only (b), (c) and (i) of the above list, though occasionally (e) or (g) may be present in negligible proportions. In all cases (a), (d), (f) and (h) stay behind with the skeleton on the gauze. Yet in spite of this absence of constituents normally essential to the life of the sponge, the reunition mass succeeds in time in producing new examples of the missing classes, presumably by the specialisation of certain of its elements. Such elements (probably the mesenchyme cells) must possess throughout normal life an undiminished embryonic potency or competence, similar to that which attaches, it is suggested, to those mesodermal cells of the amphibia or other higher animals responsible for regenerations of lost members.

The dedifferentiation of the dissociated sponge, then, consists, first, in the fact that the morphological arrangement is totally destroyed by the passage through the sieve, and must be made anew, and, secondly, in the fact that the population of cells passing through is not representative of the original sponge. We cannot quite be said, therefore, to be dealing with pure morphological dedifferentiation, but the histological dedifferentiation is statistical, not individual, and thus of a very different kind from that of the histolysing cells of the insect pupa. Regeneration after dissociation does not consist primarily in the sorting out of already differentiated cells which take up their former positions as in the normal organism. The collar cells certainly behave in this way, but the mesenchyme cells (though some of them will remain unaltered) metamorphose into the various types of cells whose former representatives have remained behind.

The above account of this fascinating subject should be supplemented by reference to the recent papers of Fauré-Fremiet[2]; Brøndsted[1]; Brien[2]; Brien & Meewis; Meewis.

It was formerly usual to regard the growth of cells in tissue culture as involving dedifferentiation, but many workers (e.g. Fischer[1], Törö[2]) have shown that this is only apparently so. "Cultivated gland cells of the intestine", writes Weiss ([16], p. 468), "and pigment cells of the eye, may become very similar in shape and general conduct, but under proper conditions the former will resume the production of digestive enzymes and the latter that of black pigment." Such reversible changes are called *Modulations* (cf. pp. 279, 506, 514).

3·14. Integration or engagement of the fundamental processes

The use of analogy, dangerous though it undoubtedly proved in mediaeval hands, is in our time perhaps almost underrated. An analogy, says Dingle, severely, is at best worthless and at worst misleading. The analogue, not being identical with the point to be illustrated, has certain relevant and certain irrelevant elements. If the reader does not already understand the point in question, he has no means of distinguishing between them, and if he does understand it, the analogue is superfluous. However, as even Dingle admits, there is something to be said for analogies. Previous experiences spring to the mind when an analogy is put forward, and the art of choosing an analogy consists in making sure that these psychological overtones will as far as possible correspond with the relevant elements of the analogue. And, above all, in certain circumstances an analogy may provide a framework or co-ordinate net, in which a previously inchoate mass of information may advantageously be assembled.

In the course of this section I have attempted to use the analogy of mechanical gears to co-ordinate a considerable number of facts concerning the fundamental processes of ontogenesis. It has been pointed out that:

(1) Growth can occur without differentiation (explanted embryonic cells; anidian sauropsid embryos).

(2) Differentiation can occur without growth (chorio-allantoic grafts; avian development *in vitro*; tissue cultures; various forms of experimental dwarfism).

(3) Nuclear division can occur without cell-division (normally in arthropod eggs; experimentally in echinoderm and fish eggs).

(4) Cell-division can occur without nuclear division (parthenogenetic merogony in echinoderm eggs).

(5) Differentiation can occur without nuclear and cell division (normally in Protozoa; experimentally in annelid and rotifer eggs).

(6) Endocrine factors can exert a differential action on differentiation and growth (amphibian metamorphosis).

(7) Dietary factors can exert a differential action on differentiation and growth (amphibian metamorphosis).

(8) The time of engagement of determinative processes with the processes of growth and differentiation can be altered (ciliary polarity in amphibian development; determination and cleavage in echinoderms).

(9) The relation between growth and differentiation can be affected by genetic and specific factors (large and small races of rabbits and fowls; large and small species of birds; see pp. 420 ff.).

(10) Histogenetic differentiation can occur without organogenetic differentiation (tissue cultures of mammalian and avian cells; yolkless development of cephalopod embryos).

(11) Metabolism can occur without growth (crowded yeast cells; resting bacteria; nitrogen-fixing organisms).

(12) Metabolism can occur without growth and differentiation (insect embryos in diapause; reptile and mammal embryos in hibernation; cells of mammalian and avian embryos surviving at low temperatures).

(13) Cell-division, growth, respiration, fermentation, and heat production show stepwise inhibition under the influence of chemical agents and radiant energy; growth and differentiation disappearing first, then fermentation, then respiration and heat production (yeast cells; echinoderm embryos; bird embryos).

(14) Growth and characteristic chemical activity may be mutually incompatible, one or the other being engageable with the basal metabolism at one time (explants of avian and mammalian tissues; histological analysis of tissues).

(15) Cell-division may occur without oxidative metabolism (nematode embryos; asphyxiated chick embryos).

(16) Growth, differentiation, and metabolism manifest a certain independence of rhythm even in the intact organism (temperature characteristics of growth and respiration, CE, p. 524).

(17) Degrowth can occur without dedifferentiation (planarian worms).

(18) Degrowth and dedifferentiation may not run parallel in regression (colonial ascidians).

(19) Morphological dedifferentiation can occur without degrowth (dissociation and reunion of sponges).

(20) Morphological and histological dedifferentiation can occur without degrowth (metamorphosis of holometabolic insects).

Thus when the evidence is viewed as a whole, no room is left for doubt that a great deal of independence is possible between the fundamental processes of ontogenesis. In considering what we mean by dependence we may distinguish between existential dependence and dependence with regard to properties. In the first case a part or process isolated from the organism would cease to exist altogether; in the second, it would continue to exist but with modified properties. The fundamental processes of ontogenesis are evidently not existentially dependent on the integrity of the whole, but as in the case of the non-segmenting *Chaetopterus* egg, they do not take a wholly normal course when the integration has been interfered with. It is as if each fundamental process represented a layshaft which may or may not be in gear with the primary shaft, and the animal economy is obviously so constituted that more than one secondary gear can be engaged with the primary shaft at one time. But in some cases, as we have seen (p. 523), there is apparently a check on this flexibility, and if growth is in gear, characteristic chemical activity cannot be. And *vice versa*. The engagement must be selective.

Lastly, what is the primary shaft? It is probably not identifiable with one chemical reaction, but may be defined as whatever reaction the cell can carry out which will provide it with the minimum amount of energy necessary to maintain itself as a going concern in the physical world. In autotrophic bacteria it may be the oxidation of ferrous carbonate to ferric hydroxide, or of methane to carbon dioxide and water; in metazoan cells it may be fermentation, the esterification and desmolysis of glucose. Whatever, under the worst environmental conditions, suffices for basal metabolism may be thought of as the primary shaft, or rather, the automotive unit to which the primary gearshaft is attached.

3·2. Heterauxesis

When a living organism grows, its parts, as is well known, do not necessarily grow at exactly the same rate as the organism as a whole. Indeed a part of an organism which correctly reproduces the growth curve of the totality to which it belongs is the exception rather than the rule. "The form of an animal", as d'Arcy Thompson wrote in 1916, "is determined by its specific rate of growth in various directions; accordingly the phenomenon of rate of growth deserves to be studied as a necessary preliminary to the theoretical study of form, and mathematically speaking, organic form itself appears to us as a function of time."* These facts have been well appreciated by the majority of morphologists, and a good deal of information about the growth of parts has become available (cf. the summaries of Scammon[1]; Scammon & Calkins; Fauré-Fremiet[1]; Hofmann; and Needham (*CE*, p. 440)). If it remained somewhat un-coordinated, this was no doubt due to the inadequacy of the terminological and conceptual technique at our disposal. The work of J. S. Huxley[4] (see especially his book, *Problems of Relative Growth*[9]) has, however, led the way towards a new order in this field.

3·21. Embryonic growth in general

Before proceeding further, a word must be said regarding the mathematical approach to embryonic growth in general. This was fairly fully reviewed up to 1931 in *CE*, Section 2, and since that time there have been a number of attempts to make further progress by applying mathematical analysis to growth phenomena. In general, however, there is very little of such progress to report.

The two most important books are those of Kostitzin and Rashevsky[3]. The former gives an excellent account of the use of elementary mathematical technique for the treatment of massed results; but the latter, with more originality, seeks to apply a kind of mathematical physics to the behaviour of individual organic systems in certain specified circumstances and limitations.

A rather ambitious treatment of embryonic growth has been given by Wetzel, who attempts to treat what he calls the "motion of growth" in terms of equations of dynamics, especially by a modification of Lagrange's equations. From the point of view of mechanics the use of Lagrange's equations is an analytical stratagem valid only in certain very clearly defined circumstances. For example, the so-called Lagrangian co-ordinates used must be capable of defining explicitly in terms of themselves and of time the position of every element of the system. But Wetzel, in using a single "generalised co-ordinate" to represent a single organism, is confusing the implicit interrelation between the various elements of an organised system which may be revealed as the *result* of calculation, with the explicitly expressible relations between the actual quantities used to represent these elements. With its supply of constants adequate to fit almost any series of results, Wetzel's treatment fails to give us a conviction of validity. The only

---

* As noted elsewhere (p. 128), embryologists have been very slow to apply d'A. Thompson's co-ordinate distortions to embryonic development, but the first attempt to do this must be welcomed in the work of Richards & Riley. Unfortunately they have so far been almost exclusively occupied with the changes in whole body form during the *later* stages of amphibian development.

precise correlation between experimental observations and the results of calculation is found in the Gayda series of figures for growth and heat-production of the toad, but here inspection of the curves shows that there is not sufficient goodness of fit to justify the precise mathematical system which Wetzel propounds.

Other workers have proposed exponential and logarithmic formulae, but considering the fluidity and plasticity of the equations the fit of the curves leaves much to be desired. The most considerable volume of work of this kind is that of Backman and his school (e.g. Edlén); it cannot be said that their theoretical superstructure is in any way justified by their mathematics. The same remarks apply to some extent to the work of Schmalhausen; Courtis; and v. Bertalanffy[3]. Evidence constituting strong criticism of these workers is presented by Lerner[3], who draws attention to periodic oscillations in the growth rate of the chick embryo and rightly points out the over-simplification inherent in logarithmic plotting. The theory of the "master reaction" is attacked by Burton, whose paper should be read by everyone interested in the subject, and Kavanagh & Richards and Richards[1] have demonstrated that the Crozier master reaction equation is nothing but a function of the hyperbolic tangent. Teissier[7], in a valuable review, draws the following conclusion regarding the Robertson and Crozier formulae: "L'exposé que nous venons de faire suffit amplement à montrer que les théories chimiques de la croissance n'ont donnés de cette dernière que des schémas à peu près vides de tout contenu concret et parfaitement invérifiables."

The empirical formula of v. Hoesslin is in a class by itself, but none the better for that.

The variabilities inherent in biological material have usually been insufficiently appreciated by writers on the mathematics of embryonic growth. From this criticism the writings of Duyff and Janisch are free, however, and the treatment they give to their data (on the chick embryo and insect respectively) is restrained and reasonable.

The upshot of the matter is that the effort devoted to the mathematical analysis of the growth of the whole body has been singularly lacking in fruitfulness or illumination. Perhaps this might have been appreciated long ago, had not the ever-present temptation to prefer paper work to the mastery of difficult technique, led us to forget the vast complexity of the organic systems of which we like to measure the wet weight. When we come, however, to the growth of parts, especially of chemical parts, as compared with the growth of the body as a whole, we recognise more fully this complexity, and in the end, as the succeeding sections show, arrive at certain relations which seem to have a general significance.*

3·22. The theory of heterauxesis

There had been no lack of empirical equations, devised particularly in America, relating the weight of parts to that of the whole by a series of linear equations, in which many special constants were introduced. But in 1924 Huxley[4] first demonstrated a simple and significant relation between the growth of a whole organism and that of a part increasing or decreasing in relative size. It had

* For the critical work embodied in the above paragraphs I am much indebted to two friends, the one a biologist and the other a mathematician, Dr Sydney Smith and the Rev. Christopher Waddams.

already been used, in closely similar form, by Dubois in 1897 and Lapicque in 1898 for the relative growth of the mammalian encephalon, but had remained hidden in the literature. If x be the magnitude of the whole organism (as measured by some standard linear unit, or by its weight), and y the magnitude of the differentially growing organ, the relation between them is

$$y = bx^k \qquad\qquad \text{......(1),}$$

where b and k are constants. The constant b is of little biological significance, since it merely denotes the value of y when x is unity, i.e. the fraction of x which y occupies when x is unity. It is called the "fractional coefficient". On the other hand, k has an important meaning, for it implies that over the range for which the formula holds, the ratio of the growth rate of the part to the growth rate of the whole remains constant.

Such a ratio was termed by Huxley a "constant differential growth ratio", and the whole process by Pézard "heterogony". The process may obviously be either positive, if the relative size of the part increases with time (i.e. "grows more quickly than the whole"), or negative, if the relative size diminishes (i.e. "grows more slowly"). If it attains a very extreme degree, it may be called, in Champy's phrase[2] "dysharmonic growth". Some authors, e.g. Teissier[1], at first preferred to use the word "dysharmony" instead of heterogony, but this was not acceptable since if we liken the organism to a piece of music, the growth of all parts at the same rate would be "unison" and their growth at different rates "harmony", not "dysharmony". Later, after the term "heterogony" had come into general use, Huxley & Teissier proposed its replacement by "allometry" on the ground that it had previously for a long time been employed by sex physiologists to denote a special type of reproductive cycle. But since heterogony after all implied something inherent in the developmental plan of the organism, while allometry seems to refer rather to our metrical methods, Needham & Lerner suggested the word *Heterauxesis*,* with *Isauxesis*, *Bradyauxesis* and *Tachyauxesis* for the three cases formerly called isogony, negative heterogony, and positive heterogony. This terminology will be used here. At the same time, it may be convenient to have a terminology which distinguishes between differences *during* growth and *after* it. Comparisons may be made between organisms of the same group differing in age, size, weight, chemical composition, etc.; this is heterauxesis. But comparisons may also be made between organisms of different groups (races, varieties, species, genera, and the like) differing in size but of the same age, e.g. birth or adult maturity. Huxley, Needham & Lerner suggest that the word *Allomorphosis* should be reserved for this. In many such cases the Huxley equation holds good; for example the relation between egg size and bird weight (in his book), or the work of Hersh[2] on titanotheres, or that of Lerner, Gregory & Goss already mentioned (p. 426) on glutathione, or that of Brody and Kibler on organ sizes. Since such differences arise on account of heterauxesis, a word covering both this and allomorphosis may be useful; *Allometry* will do very well. The word allometry has been used in a similar sense by Osborn, who spoke of the changes in bodily proportions occurring during evolution as allometric changes. And the coordinate distortions of d'Arcy Thompson would thus be said to represent allometric differences.

* Due to my friend Dr A. L. Peck.

The formula given above may also be written

$$\log y = \log b + k \log x \qquad \ldots\ldots(2),$$

from which it follows that any magnitudes obeying the equation will fall along straight lines if plotted on a double logarithmic grid. The appearance of a straight line on double log paper does not establish the sufficiency of the formula for the case in question, since no purely graphical method could do so, but it affords strong evidence of its applicability. If the data were sufficiently good, each point could be verified by calculation. The logarithmic method of plotting, moreover, emphasises an important point which is entirely obscured by the usual method of plotting on absolute scales, namely, the fact that growth is essentially concerned with multiplication. Equal spaces on the logarithmic grid denote equal amounts of multiplication, equal spaces on the ordinary scale denote equal additions. Another great advantage of logarithmic plotting consists in the fact that in embryonic development, where the numerical values on the axes cover such an enormous range, it is impossible in any other way to view the course of development as a whole in one and the same picture. For preliminary purposes, the constant k can be read off from the slope of the straight line, for if a is the angle it makes with the x axis

$$\tan a = k \qquad \ldots\ldots(3).$$

These methods were originally developed to deal with the phenomena of arthropod limb growth, the growth of antlers in some mammals, and other similar morphological magnitudes.* But considerable scope exists for their application to magnitudes of a distinctively chemical nature. The metazoan organism, in its passage from fertilisation to maturity, passes through a long succession of stages which are characterised just as much by changing chemical composition as by changing morphological form. And just as we think of organs or structures as parts of a morphological totality, so we may think of substances or groups of substances as parts of a chemical totality. It is legitimate to enquire into the regularities which the constituents of this chemical totality exhibit during its continuing increase in mass. Now there exist in the literature of chemical embryology and the biochemistry of growth a large number of sets of data regarding such entities as protein nitrogen, glycogen, calcium, etc., in a variety of embryos and animals throughout the phyla. These may be considered from the point of view of heterauxesis. The constant k is here of great value, for it permits us to compare quantitatively the percentage relationships of any substance in any developing organism. A family of curves with meaningless shapes relating a chemical entity to age for several embryos may become a series of straight lines of the same slope when the entity is logarithmically plotted against the totality. This is in fact the case, and arises because the complicating factor of time is not explicit in such graphs. The constant b, on the other hand, though it fixes the percentage composition of the body at unity, is not of great use, since

---

* A specific rate of relative growth, expressible by a heterauxetic constant, may be regarded as the "fate" of an organ just as much as its characteristic histology. Hence we may speak of the determination of a growth-rate, as in the considerable body of work reviewed in Section 2·62.

the time at which unity is reached occurs at very different points in the life cycle owing to the great variation in the sizes of animals, and this is still true no matter what scale is taken.

Teissier[1] in 1931 was the first to apply heterauxetic theory to the chemical development of an organism. Taking as experimental animals the larvae of the mealworm and the waxmoth, he studied the increase in a number of chemical entities, such as water, fat, ash, phosphorus, etc., and found that in all cases Huxley's relation was obeyed as the totality grew. Some of his results will presently be considered in more detail. The behaviour of the same substance in different classes of animals was, however, left over until in the following year, in a paper mainly concerned with the evaluation of k for all the constituents of the chick embryo, I drew attention[9,10] to the rather close correspondence which may be exhibited by the relative "growth" of a given chemical entity in very different animals both in embryonic and later stages.*

Before going on to the experimental, or rather, observational data, reference must be made to the logical status of Huxley's equation. According to the theory of dimensions, the dimensions on the two sides of the relation must be equivalent. In this case,

$$y = bx^k \qquad \qquad \ldots\ldots(1),$$

they are only so when k is 1, which is comparatively rare. Ordinarily, since x is a mass, x^k cannot be. The equation, therefore, has no true physical meaning, that is to say, no new concept can be deduced from it as it stands, in the sense that the concept of acceleration arises from the relation

$$Mf = a \qquad \qquad \ldots\ldots(4),$$

between mass, force, and acceleration. The constant k is enigmatic without further elucidation, and cannot itself provide any theory of growth. The equation essentially gives us a good technical method for the examination of curves.

The point at issue may be made clearer by means of an analogy. The theory of dimensions is not satisfied in the relation

$$\Delta \propto c \qquad \qquad \ldots\ldots(5),$$

where Δ is osmotic pressure and c is the concentration of a solute. Accordingly, until the implications in the relation are made explicit by considerations due to kinetic theory, the relation has no true physical meaning. A relation similar to the heterauxesis relation occurs in magnetism. If a mass of steel be carried through a cycle of flux from a maximum induction B to an equal negative and back again, the energy lost in such a cycle (Joules), per cubic centimetre of material, is

$$J = kf(B) \qquad \qquad \ldots\ldots(6).$$

Now $f(B)$ has the same shape for all steel alloys, which differ solely in the value of k. Further, over a long range of B, $f(B) = c \cdot B^{1\cdot6}$. Hence over that range

$$J = \eta B^{1\cdot6} \qquad \qquad \ldots\ldots(7).$$

* The idea of chemical heterauxesis has also been extended to plant development (Meunier[1]; Petrie & Williams). Refined mathematical methods for detecting genetic differences in k are described by Schmalhausen; Feldstein & Hersh and by Lerner[1].

Here there is no theoretical explanation of the exponent, and η is a constant varying with the alloy. The formula conveniently expresses relative losses, but a "magnetic theory" could not be founded on it, for it is purely an empirically verifiable numerical relation. Ultimately it arises from the fact that work has to be done to orient the particles of the iron. Similarly, in the heterauxesis equation for chemical entities, the ultimate explanation might be in terms of colloidal stability, but nothing can be premised concerning it from the equation, in the absence of further experiment and observation.*

Attempts have already been made to offer a physical interpretation of the heterauxesis equation. Robb suggested that the distribution of "building-stone" molecules between circulating blood and tissue cells proceeds in a similar way to the distribution of solute molecules between two phases. Each organ would thus have a partition coefficient of its own for any given substance, and would grow more rapidly or more slowly than the body as a whole according to the value of this coefficient. Robb regarded the constant k as a measure of the concentration or activity of a substance governing the partition (i.e. capable of altering the partition coefficient), in so far as it was present in the organ in greater or less concentration than in the body as a whole. He also modified the Huxley equation by the insertion of a term, c, to represent the amounts of inert, non-growing substance present in the organ rudiment; thus

$$y = bx^k + c \qquad \ldots\ldots(8).$$

The requirements of the theory of dimensions do not seem to be any better satisfied here than in the original equation. This does not mean that the theory of partition coefficients is inapplicable, only that it does not follow from the empirical regularities so far found to hold. Moreover, it will hardly do where chemical entities are concerned, for something more subtle is required to hold the balance between them. How far the introduction of a correction for non-growing constituents helps is not clear. To this it may be added that Medawar[2] later demonstrated the validity of the Huxley equation for long and short axes when an explant is made to grow in an artificial field of growth-inhibitory substance.

The second attempt to find a physical basis for the heterauxetic relation is very different from that of Robb. O. Glaser & Child noticed that the mathematical series which describes the stacking of hexoctahedra around a central hexoctahedron in successive shells is related to the heterauxetic formula. The orthic hexoctahedron is bounded by six non-contiguous squares and eight hexagons; it is the ideal form taken up by compressed adjacent spheres of equal size so that the material aggregates with no waste spaces, as may be shown experimentally (Matzke).† Polyhedra of this type are actually visible microscopically in yolk, caterpillar tissues, etc. The mathematical relationship arises because the Marvin series, which describes the stacking of the polyhedra, has the form

$$\log S_n = 3 \log (2n+1) - 0{\cdot}301, \qquad \ldots\ldots(9),$$

where S_n is the number of individuals stacked in n layers about a central unit. This is formally identical with the growth formula of Glaser:

$$\log w = k \log (2t+1) + c, \qquad \ldots\ldots(10),$$

* But see the discussion of Lumer. The relations between the heterauxetic equation and the sigmoid growth curve have been discussed by Bernstein[1].

† Cf. the description of d'Arcy Thompson (*Growth and Form*, pp. 336 ff.). The connection of spiral growth with another series, the Fibonacci series, is well known.

where w is the weight of the totality or entity which is growing, t time, and k and c constants (see his full discussion of this). And therefore the stacking series should replace either organic correlative in the heterauxetic equation, and this, according to Glaser & Child, it will do. It remains difficult, however, to understand why chemical substances should obey this law, especially where molecules much smaller than the hypothetical intracellular polyhedra are concerned. As yet it is too early to appraise fully the suggestion of Glaser & Child, which must be left for further investigation.

So far practically all cases of chemical "growth" give linear plots on the double logarithmic grid. O. Richards[1], however, has drawn attention to the fact that this is not always so in the case of morphological entities, where sometimes k seems to be continually changing or oscillating widely about a mean. He sounds a justified caution in accepting any theory of heterauxesis.

Not all the sets of data for the relative growth of chemical constituents are representable, however, by *single* linear plots; frequently two or even three separate straight lines may be found necessary. The breaks between these successive phases obviously indicate a change in one or other of the constants in the equation. If k changes, the slope of the line will alter; if b changes, there will be a period of nil growth of the part, followed by the resumption of growth at the same rate as before. These possibilities are exemplified in Fig. 291, which shows the water-soluble non-protein nitrogen of the chick embryo (CE, p. 1075) plotted against the embryo's dry weight, and the water of the waxmoth larva (Teissier[1]) plotted against the wet weight of the larva. The former entity grows with tachyauxesis ($k = 1·69$) up to the 6th day of development, when the embryo weighs approximately 15 mgm. (dry), after which it suddenly becomes bradyauxetic ($k = 0·83$). The latter entity (water in the waxmoth) grows throughout the whole period with a k of $0·96$, but between the weights of 60 and 100 mgm. (wet) there is a period of adjustment during which b changes from $0·0075$ to $0·0055$.

If the results of Teissier[1] for the insect larvae are compared with those of Needham[10] for the chick embryo, it is at once noticeable that changes of the first type (changes in k) are much more characteristic of the embryonic material than those of the second type, and *vice versa*, although a few examples of the opposite correlation can be found. Any interpretation of this fact might as yet be premature, but it may not unreasonably be connected with the circumstance that the larval insect growth is much less accompanied by profound changes of morphological differentiation than is the growth of the chick embryo. Further examples are needed to test this hypothesis. For the present purpose, however, discussion will be mainly confined to those cases where the data exhibit no breaks, no changes in either b or k. This is not because the breaks may not be perfectly real, but because greater certainty of comparison between different organisms may be attained by the use of data which do not show them. For example, the calcium in the chick embryo has been determined by several investigators (references and plots in Needham[10]), some of whose series give breaks of k. But when the entire mass of data is plotted without distinction of investigator, it is found to lie with equal deviation on both sides of a straight line from the smallest to the very largest embryos. This suggests that with further work, many of the changes in k hitherto recorded may turn out to be illusory. Here, therefore, emphasis will only be laid on sets of data giving the unbroken logarithmic relationship, and on its similarity between different animals.

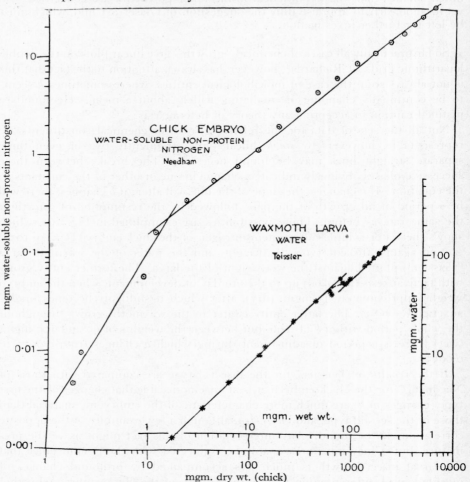

Fig. 291. Examples of change in b, and change in k: for the former; water in the waxmoth larva (Teissier): for the latter; non-protein nitrogen in the chick embryo (Needham).

It may be mentioned, however, that certain investigators have believed in the existence of well-defined critical points at which changes occur in the heterauxetic constants of numerous groups of chemical substances. Thus Teissier[5,6] has located two such critical points in the development of the chick embryo, one at 1·5 gm. and the other at 12 gm. (wet weight) or 9 and 15 days respectively. Variations in k, which may be either in the tachyauxetic or bradyauxetic direction, are said to occur at either or both of these points, in the data for purine nitrogen, non-protein nitrogen, lipid P, and total fat. Meunier[2] added potassium. But it is very doubtful whether all the available data of sufficient reliability support these conclusions, and the matter requires much further investigation.

In the following graphs it will be seen that in all cases the mass of the chemical entity is plotted against the mass of the chemical totality. Although this is quite logical, it implies that in the case of a part growing tachyauxetically, i.e. faster than the whole, a time must eventually come when the mass of the part equals that of the whole, and finally exceeds it. This unfortunate state of affairs is avoided by writing the equation

$$y = b(x-y)^k \qquad\qquad ......(11),$$

and plotting the part against the remainder of the whole rather than the whole itself. This was the method adopted by Huxley[9] in the first instance, who defined

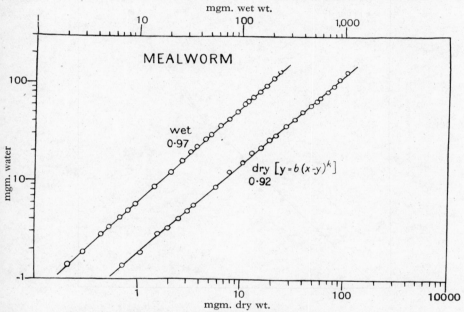

Fig. 292. Comparison of the relation of water in the mealworm larva to wet and dry weight
(Teissier).

x accordingly, although the graphs in his book do not seem to be always consistent with it. Teissier[1], on the other hand, used the formula (1), and its application to chemical heterogony will be continued here. The question is somewhat academic, since growth is not infinite, and a point imagined as travelling along one of the straight lines in the succeeding figures would slow down and finally stop at the adult condition long before the consequences of the convention here adopted became serious. Moreover, the difference in k is not very marked for the two conventions. In the case of a very small part it is quite inappreciable, and even in that of a very large part, such as the water-content of the body, it is comparatively small. Fig. 292 shows the water in the body of the mealworm larva, plotted against the wet weight (x) and the dry weight ($x-y$) respectively. In the former case k is 0·97, in the latter case 0·92.

A word should here be said concerning the reproducibility of graphs of chemical heterauxesis. It seems to be satisfactory. In Fig. 293 the solid line represents the relation of dry weight to wet weight in the chick embryo as determined from the averaged data of thirteen separate investigators (see Fig. 220 in *CE*). Points represent the data subsequently reported by Saccardi & Latini, who appear to have been unaware of the work of their predecessors, and by Penquite. It will be seen that over the greater part of the range the agreement is excellent, although in the earliest stages, when the embryos are very small, the points of the Italian workers show a slight divergence.

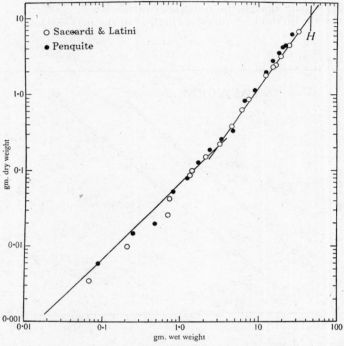

Fig. 293. Relation of dry to wet weight in the chick embryo; confirmation of earlier workers by Saccardi & Latini, and by Penquite.

3·23. Chemical heterauxesis

We may now go on to specific examples. It may be convenient first to consider water-content. Fig. 293 showed that during the first period in the development of the chick embryo, the increase in solids is practically isauxetic, and during the second period strongly tachyauxetic ($k = 0·93–1·06$, then $1·43–1·48$). This is only a more accurate way of stating the well-established fact that the chick embryo becomes drier as it develops (see *CE*, p. 871). For organisms other than birds, numerous sets of data exist covering in some cases the whole life cycle, in others only pre-natal or only post-natal life. A good many of these are plotted in Fig. 294 on one double logarithmic grid, and the references and constants are collected in Table 29. The points of birth or hatching are suitably marked on the chart in each case (Needham[16]).

It is at once obvious that within any one group of animals, there is a marked similarity of slope, that is to say, they get drier at the same absolute rate. The mammals, for instance, all show a k closely approximating to 1·23, and it is remarkable that over the whole growth period of man (from 1 to 66,000 gm. wet wt.) this relationship holds good. Again, the three researches on teleostean fishes give a k of unity, indicating that over the period studied (from some time before hatching until the end of the free-swimming larval period) the tissues are not drying at all. The same statement applies to the woodlouse, which, however, has only been measured (and that some considerable time ago) after hatching. Critical points, where k changes, appear both in the chick, and in some of the selachian fishes, which show the same change from isauxetic growth of the dry weight to marked tachyauxetic growth. On the whole, it may be concluded that dry weight never shows bradyauxesis, that is to say, all organisms dry up with age, and that within one group, the absolute rate at which this process takes place, is very uniform indeed. In Table 29 the ranges of k are given as well as the average values. The birds alone show a wide range, but this may be explained by the existence of two separate phases, and the restricted nature of some of the sets of data available. The range of all k's for all organisms is only 0·27.

In those cases where no drying up of the organism is taking place, we have not as yet sufficient knowledge of the dry weight throughout the life cycle to say whether the process of dehydration has finished or whether it has not yet begun. It is tempting to relate the concurrence of evidence for the isauxesis of the teleostean fishes, with the views of Bidder on the absence of adult size in some aquatic organisms, and the possible absence of death from senility. Although plaice and carp, he suggests, may continue to grow until many times the age of their sexual maturity, and although we have no reason to suppose that they ever die, except by violence, the same could not be true of swiftly moving terrestrial creatures. These must maintain a relation between their weight and the cross-sectional area of their bones and muscles, necessitating a definite adult size, shape, and habit. Old age and death by senility became the necessary fate for animals only when they left the water and attempted swiftness and tallness in a medium 1/800th of their own specific gravity. On the other hand, there is at present no obvious reason why the tissues of selachian fishes should dehydrate with growth, while those of teleosts do not. The subject merits closer attention than it has yet received.

The next example which may be taken is that of fat. Fig. 295 gives the appearance of the data for rat, chick, and mealworm when plotted on the double logarithmic grid (data of Chanutin; H. A. Murray; Cahn & Bonot; Romanov[5]; and Teissier[1]). For the insect and the bird, in spite of the great differences in the absolute weights concerned (e.g. at hatching or birth, mealworm = 3 mgm. dry wt., chick = 7000 mgm. dry wt., rat = 660 mgm.), the correspondence is very good; the rat shows slightly more tachyauxesis. The values of k, together with their range of variation, are shown in Table 30. In all cases, the auxesis is

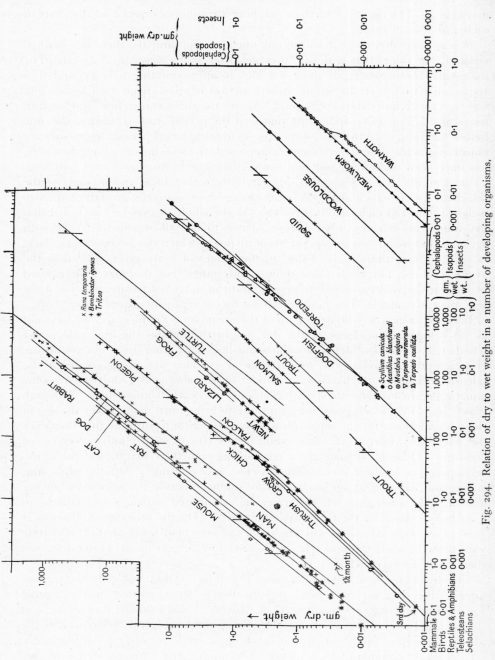

Fig. 294. Relation of dry to wet weight in a number of developing organisms,

Table 29. *Dry weight as function of wet weight*

[Numbers in brackets refer to the notes below the table; superscript numbers are bibliographical.]

Organism	Investigators	Period	k	Range of k	Average k
MAMMALS (1)					
Man (*Homo sapiens*)	Michel; Fehling; Brubacher; Camerer & Söldner; Langstein & Edelstein; Sommerfeld; Moleschott; Bischoff; collected by H. Aron; also Schmitz and von Bezold	Whole life	1·26		
Rabbit (*Lepus cuniculus*)	Fehling; Schkarin; Friedenthal; Steinitz; collected by H. Aron	Whole life	1·28		
Dog (*Canis vulgaris*)	Thomas; Gerhardtz; Orgler; Eckert; collected by H. Aron	Whole life	1·24	0·09	1·23
Cat (*Felis vulgaris*)	K. Thomas	Post-natal life	1·21		
Rat (*Mus norvegicus*)	Lowrey; Donaldson; collected by Donaldson, also Gonzalez	Whole life	1·23		
	Hamilton & Dewar	Whole life (0·072 gm. to 435 gm. wet wt.)	1·19		
Mouse (*Mus vulgaris*)	von Bezold	Whole life	1·21		
Man (*Homo sapiens*)	Iob & Swanson[1]; Hamilton	Embryonic life	1·27	—	—
	Givens & Macy	Embryonic life	1·17	—	—
Rat (*Mus norvegicus*)	Roche & Garcia	Post-natal life (−25 days)	1·33	—	—
Pig (*Sus scrofa*)	Wilkerson & Gortner[2] (2)	Embryonic life	1·07	—	—
BIRDS (3)					
Fowl (*Gallus domesticus*)	Murray; Romanov; Byerly; Mankin; collected in *CE*	Embryonic life	1·42		
Turkey (*Meleagris americana*)	Martin & Insko[1]; Scott & Hughes	Embryonic life	1·20	0·39	1·19
Pigeon (*Columba livia*)	Kaufman[1]	Post-embryonic life	1·19		
Thrush (*Turdus merula*)	Groebbels[2]	Embryonic life	1·03		
Crow (*Corvus corone*)	Groebbels[2]	Embryonic life	1·04		
Hawk (*Buteo buteo*)	Groebbels[2]	Embryonic life	1·28		
REPTILES					
Lizard (*Lacerta viridis*)	von Bezold	Post-embryonic life	1·11	0·09	1·15
Turtle (*Thalassochelys corticata*)	Karashima	Embryonic life	1·20		
AMPHIBIA					
Frog (*Rana temporaria*)	von Bezold	Post-larval life	1·07		
Frog (*Bombinator igneus*)	von Bezold	Post-larval life	1·15	0·08	1·09
Triton (*Triton cristatus*)	von Bezold	Post-larval life	1·07		
TELEOSTEAN FISHES					
Salmon (*Salmo salar*)	Hayes[1]	Embryonic and larval life	1·00		
Trout (*Salmo fario*)	Gray[1]	Embryonic and larval life	1·00		
	Kronfeld & Scheminsky	Embryonic and larval life	1·00	0·01	1·00
Carp (*Cyprinus carpio*)	Nowak	Larval and post-larval life	1·01		
SELACHIAN FISHES					
Dogfish (*Scyllium canicula*)	Ranzi[12]	Embryonic life	1·21		
Torpedo (*Torpedo ocellata*)	Ranzi[12]	Embryonic life	1·00 then 1·33		
Torpedo (*Torpedo marmorata*)	Ranzi[12]	Embryonic life	1·00 then 1·32	0·12	1·27
Dogfish (*Acanthias blanchardi*)	Ranzi[12]	Embryonic life	1·00 then 1·29		
Dogfish (*Mustelus vulgaris*)	Ranzi[12]	Embryonic life	1·00 then 1·23		
INSECTS					
Mealworm (*Tenebrio molitor*)	Teissier[1]	Larval life	1·04	0·01	1·035
Waxmoth (*Galleria mellonella*)	Teissier[1]	Larval life	1·03		
CRUSTACEA (terrestrial)					
Woodlouse (*Oniscus murarius*)	von Bezold	Post-larval life	1·00	—	1·00
MOLLUSCA					
Periwinkle (*Littorina irrorata*)	Newcombe & Gould	Post-larval life	0·95	—	—
Squid (*Sepia officinalis*)	Ranzi[6]	Embryonic life	1·09	—	—

Notes to Table 29

(1) The data of Esskuchen[2] for the cow, though interesting, are excluded as lacking a sufficient number of separate points. Yapp's promised data for this animal have not yet become available, nor have those of Scammon[2] for man as far as I can ascertain, though mentioned in a preliminary note ten years ago. All individual organs show a desiccation, similar to that of the body as a whole (see e.g. Duyff's data on the chick embryo's organs), and so does the blood; thus at 3 months' foetal life in man the blood contains 87·8 % water and at birth only 67·3 % (Takakusu, Kuroda & K. N. Li; Kuroda).

(2) As these authors measured dry weight after acetone extraction their data are not comparable with other sets.

(3) In very early embryonic life birds often show isauxesis or even bradyauxesis of dry weight (e.g. *k* 1·00 for the chick during the first week and 0·67 for the turkey). Cf. the initial rise of the curve in *CE*, Fig. 220. But as not all sets of data show it I am inclined to think it may possibly be due to technical errors in the handling of very small embryos. There must of course be an upper limit of "wetness" probably represented by the gastrula itself, the dry weight of which no one has yet succeeded in measuring.

Table 30. *Chemical heterauxesis in various organisms*

[N.B. Values for entities not listed in this table may be found in Needham [10,16].]

Fig.	Chemical substance or group of substances	Animal	Investigators	Period	k	Range of k	Average k
295	Fat (to dry wt.)	Waxmoth	Teissier[1]	Larval	1·32		
		Mealworm	Teissier[1]	Larval	1·07		
		Chick	H. A. Murray; Cahn & Bonot; Romanov[5]	Embryonic	1·06	0·03	1·07
		Rat	Chanutin	Post-natal	1·09		
		Man	Iob & Swanson[1]	Embryonic	1·07(1)	—	—
305	Fat (to wet wt.)	Chick	Romanov & Faber	Embryonic			
				32° C.	1·84	—	—
				34° C.	1·85	—	—
				36° C.	1·77	—	—
				38° C	1·79	—	—
291	Non-protein nitrogen (to dry wt.)	Mealworm	Teissier[1]	Larval	0·77	—	—
		Chick	Needham[2]	Embryonic	0·83	—	—
—	Purine nitrogen (to total N)	Chick	Lebreton & Schaeffer	Embryonic	0·72		
		Pig	Lebreton & Schaeffer	Embryonic	0·81	0·19	0·71
		Mouse	Lebreton & Schaeffer	Embryonic	0·62		
296	Creatine (to wet wt.)	Chick	J. Mellanby	Embryonic	1·12	0·02	1·11
		Rat	Chanutin	Post-natal	1·10		
244	Glutathione (to wet wt.)	Chick	H. A. Murray	Embryonic	0·84		
			Kamiya	Embryonic	0·90	0·06	0·87
			Yaoi	Embryonic	0·87		
			Cahn & Bonot	Embryonic	0·88		
		Rat	Thompson & Voegtlin	Whole life	0·84	0·01	0·84
		Pig	Wilkerson & Gortner	Embryonic	0·85		
297	Glycogen (to wet wt.)	Chick	Vladimirov & Danilina	Embryonic	1·42		
		Pig	Mendel & Leavenworth[1]	Embryonic	1·35	0·07	1·37
		Rabbit	Lochhead & Cramer	Embryonic	1·35		
298	Liver glycogen (to wet wt. of liver)	Chick	Vladimirov[1]	Embryonic	1·99	0·03	1·97
		Rabbit	Lochhead & Cramer	Embryonic	1·96		
299	Total ash(2) (to dry wt.)	Mealworm	Teissier[1]	Larval	0·82		
		Squid	Ranzi[6]	Embryonic	0·86		
		Dogfishes	Ranzi:[11]				
			Scyllium canicula	Embryonic	0·96		
			Torpedo marmorata	Embryonic	0·96		
			Torpedo ocellata	Embryonic	0·86		
			Mustelus vulgaris	Embryonic	0·85		
		Carp	Nowak	Larval and post-larval	0·96	0·14	0·90
		Chick	H. A. Murray	Embryonic	0·89		
			Romanov[2]	Embryonic	0·89		
			Bishop[2]	Embryonic	0·85		
			Saccardi & Latini	Embryonic	0·94		
			Penquite	Embryonic	0·94		
		Turkey	Martin & Insko[1]	Embryonic	0·93		
		Rat	Chanutin	Post-natal	0·95		
			Buckner & Peter	Post-natal	0·93		
		Man	Givens & Macy	Embryonic	0·96(3)	—	—
			Iob & Swanson[1]	Embryonic	0·99(4)	—	—
—	Sodium (to dry wt.)	Chick	Mankin[3]	Embryonic	0·69(5)	—	—
		Man	Iob & Swanson[1]	Embryonic	0·70	—	—
—	Potassium (to dry wt.)	Trout	Leulier & Paulant	Embryonic and larval	0·90	—	—
		Chick	Leulier & Paulant	Embryonic	0·96(6)	—	—
			Mankin[3]		0·96	—	—
		Man	Iob & Swanson[1]	Embryonic	0·80	—	—
—	Magnesium (to dry wt.)	Rat	Greenberg & Tufts	Post-natal (to 18 wks)	1·20	—	—
		Man	Givens & Macy	Embryonic	1·09	—	—
			Iob & Swanson[1]	Embryonic	0·90	—	—
300	Calcium (to dry wt.)	Trout	McCay, Tunison, Crowell & Paul	Larval	1·16		
		Chick	Mankin[3]; Plimmer & Lowndes; Romanov[1,2]	Embryonic	1·21		
		Turkey	Insko & Lyons[1]; Martin & Insko[1]	Embryonic	1·15	0·08	1·17
		Rat	Sherman & McLeod	Post-natal	1·13		
		Man	Schmitz	Embryonic	1·20		
			Givens & Macy	Embryonic	1·15		
			Iob & Swanson[1]	Embryonic	1·15(7)		
—	Chloride (to dry wt.)	Chick	Mankin[2]	Embryonic	0·65	—	—
			K. Yamada[2]	Embryonic	0·78	—	—
		Man	Iob & Swanson[1]	Embryonic	0·71	—	—

Table 30 (cont.)

Chemical substance or group of substances	Animal	Investigators	Period	k	Range of k	Average k
Total phosphorus (to dry wt.)	Mealworm	Teissier[1]	Larval	0·97		
	Waxmoth	Teissier[1]	Larval	0·95		
	Silkworm	Akao[3]	Larval	0·99		
	Trout	McCay, Tunison, Crowell & Paul	Larval	1·00		
	Turkey	Martin & Insko[1]	Embryonic	0·96		
	Rat	Buckner & Peter; Sherman & Quinn	Post-natal	1·00	0·07(8)	0·97
		Roche & Garcia	Post-natal (to 25 days)	0·86		
	Man	Iob & Swanson[1]	Embryonic	0·93		
Lipin phosphorus (to dry wt.)	Mealworm	Teissier[1]	Larval	0·80	—	—
	Chick	Cahn & Bonot	Embryonic	0·78	—	—
		Kugler	Embryonic	0·74	—	—
Nuclein phosphorus (9) (to dry wt.)	Mealworm	Teissier[1]	Larval	0·80	—	—
Iron (10) (to dry wt.)	Chick	McFarlane & Milne; Szejnman-Rosenberg	Embryonic	0·87	—	—
		Szejnman-Rosenberg	Emb. membranes	1·13	—	—
Copper (to dry wt.)	Chick	McFarlane & Milne	Embryonic	0·85	—	—
(to wet wt.)	Chick	Sümegi	Embryonic	0·96	—	—
Lead (to dry wt.)	Chick	Bishop[1,3,4]	Embryonic		—	—
			High	0·93		
			Low	0·93		

Notes to Table 30

(1) Up to 50 gm. dry wt., $k = 1·82$ afterwards.
(2) Further data could be calculated from Cox & Imboden and Kozhukhar.
(3) In this set of data there are indications of a preliminary early period of tachyauxesis ($k = 1·66$ below 3 gm. dry embryo wt.). The same applies to Nowak's data on the carp (1·28 before 0·04 gm. dry alevin wt.).
(4) Here again there are indications of early tachyauxesis ($k = 1·54$ below 50 gm. dry embryo wt.).
(5) Strictly speaking 0·91 before the 14th day and 0·47 afterwards.
(6) Up to 0·6 gm. dry wt.; above that point the bradyauxesis is more marked ($k = 0·63$).
(7) There are faint indications here of a late period of bradyauxesis ($k = 0·85$ after 300 gm. dry wt.).
(8) Omitting the figure from Roche & Garcia's data, which cover only a short period.
(9) Cf. the figures for purine nitrogen higher in the table; see also pp. 632 ff.
(10) The data of Katsunuma & Nakamura and of Kojima are omitted as remarkably aberrant. Akao[4] has published figures for other elements as well as iron from which heterauxetic constants for the silkworm could be calculated.

positive, that is to say, fat is increasingly present in the body with growth. Other aspects of this phenomenon are often met with, e.g. the development of mutton quality in the sheep (Hammond & Appleton); the chemical development of the mammalian brain (to be referred to later); and the fact that the calorific value of developing tissues (H. A. Murray; Penionskevitch & Schechtman on the chick; Ivlev on the fish *Silurus glanis*) increases tachyauxetically ($k = 1·03$ for the chick, 1·09 for the mealworm).

Next, two tissue extractives, creatine and glutathione, are illustrated in Figs. 296 and 244. The tachyauxesis of creatine in the chick shown by J. Mellanby's (unfortunately rather restricted) data is exactly paralleled by that of the same substance in the rat according to Chanutin. Similarly, the marked bradyauxesis of glutathione is identical for the rat and the pig (data of Thompson & Voegtlin and Wilkerson & Gortner respectively) and also the chick embryo (see Fig. 244 in Section 2·68).* Creatine, then, "grows" more rapidly than the whole body, glutathione more slowly. The range covered by the data in Table 30 is very wide, in the case of the rat from 0·2 to 200 gm. wet weight, in the case of the pig from 1·0 to 1000 gm. wet weight. The well-known decline in nucleoplasmic ratio is seen in the bradyauxetic constants for nuclein nitrogen and phosphorus.

Glycogen is a substance which has a rather special relation to embryonic life. Until the liver is sufficiently developed to store it, it is laid down in

---

* There have been numerous other sets of data for glutathione (e.g. in the chick embryo, Tateishi[1]; Hirano; Castagna & Talenti) but I have found them either inaccessible or useless.

the extra-embryonic structures (mammalian placenta, avian yolk-sac) which perform the functions of a transitory liver (see *CE*, Section 8·5). Nevertheless, there is remarkable regularity between different animals in the rates at which the substance increases in the embryonic body itself. Fig. 297 shows the data for the chick, rabbit and pig, all of which give a k in the near neighbourhood of 1·38. Even when we consider the accumulation of glycogen in the embryonic liver, we find that $k = 1·97$ would hold for both chick and rabbit (Fig. 298) — a remarkable circumstance (when we remember that structures so different as the avian

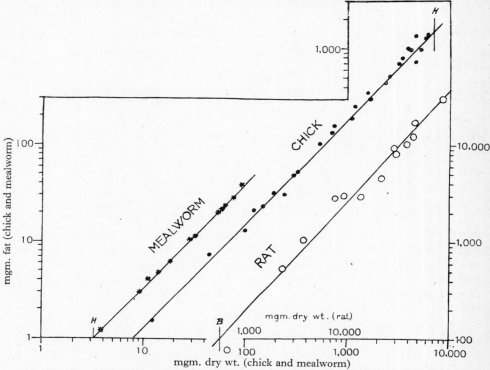

Fig. 295. Fat in the mealworm (larval period; Teissier), the chick (embryonic period; Murray; Cahn & Bonot; and Romanov), and the rat (post-natal life; Chanutin).

yolk-sac and the rodent placenta are handing over glycogen to the foetal liver after serving as temporary depots themselves). It is interesting that after the change-over point the heterauxesis follows a different course, i.e. the process is not strictly reversible (see Needham[10]).

The next graph (Fig. 299) is a particularly interesting one. The total ash of the body is plotted against the dry weight for the chick, rat, squid, mealworm, and several selachian fishes. The chick, the squid and the fishes are all embryonic, the mealworm is larval, and the rat is post-natal. There is a close uniformity in the slopes of the straight lines, in spite of the enormous differences of morphological form and of time taken to accomplish the changes depicted. Thus the chick series is accomplished in three weeks, the development of the squid and

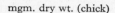

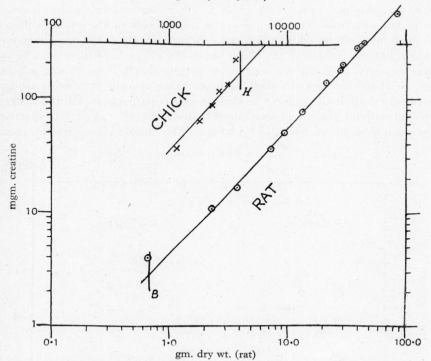

Fig. 296. Creatine in the chick embryo (Mellanby) and in the rat (post-natal life; Chanutin).

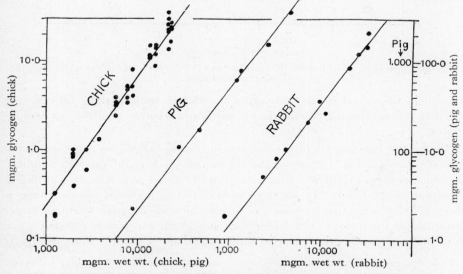

Fig. 297. Glycogen in the chick embryo (Vladimirov & Danilina); in the pig embryo (Mendel & Leavenworth) and in the rabbit embryo (Lochhead & Cramer).

the selachian fishes may take several months, and that of the rat more than a year. Again, the absolute values vary from a few milligrams in the mealworm to kilos in the selachian fishes. Finally, the nutritive factors are quite different. The chick finds all the ash which it requires for its body in the egg interior and the shell, but the embryo of the squid has to absorb 30 % of the inorganic substances required from the sea-water surrounding its egg. As for the selachian fishes, *Scyllium* develops in a shell but nevertheless absorbs 75 % of its ash from the sea-water, while the others, developing ovoviviparously within the maternal body (but without placentoid attachments) absorb 25, 60 and 97 % respectively of their ash from the maternal blood stream. The rat and the mealworm receive

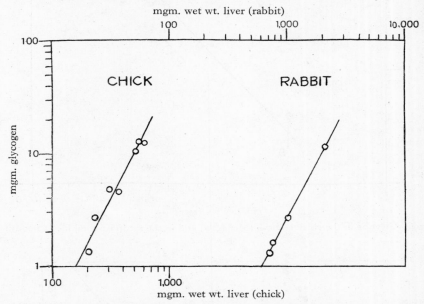

Fig. 298. Glycogen in the liver of the chick embryo (Vladimirov) and of the rabbit embryo (Lochhead & Cramer).

the ash in their food in the ordinary way. But in spite of these far-reaching nutritive differences, the heterauxesis is uniformly negative, as Table 30 demonstrates.

Among the constituents of the ash, calcium and phosphorus may be selected for examination. Fig. 300 shows the calcium in the human embryo (Schmitz), the chick embryo (Mankin[3]; Romanov[2]; and Plimmer & Lowndes) and the rat (Sherman & McLeod). In all three series of data a closely similar slope is observable, showing that with increasing size the body contains relatively larger quantities of the metal. Other ions, such as sodium and potassium, give place to it, just as chloride gives place to bicarbonate (see p. 585).* Thus from 0·01

---

* From this it follows that the K/Ca ratio of the tissues will be changing throughout embryonic life, as was experimentally found by Kaufman & Laskowski. It is interesting in this connection that according to Lasnitzki[1] the respiratory and glycolytic systems of embryonic tissues are much less susceptible to serious deviations of K/Ca balance than adult ones.

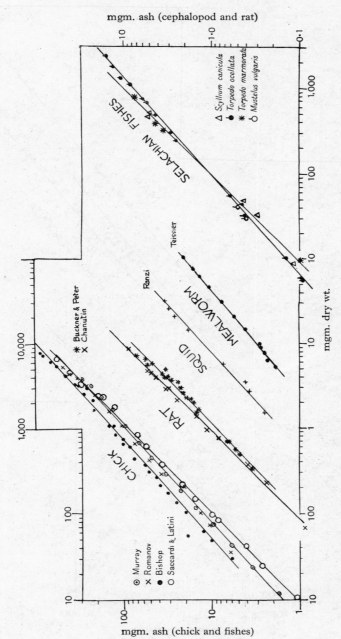

Fig. 299. Total ash in the chick embryo (Murray; Romanov; Bishop; Saccardi & Latini); the rat (post-natal life; Chanutin; Buckner & Peter); the squid embryo (Ranzi); the mealworm (larval life; Teissier) and various selachian fish embryos (Ranzi).

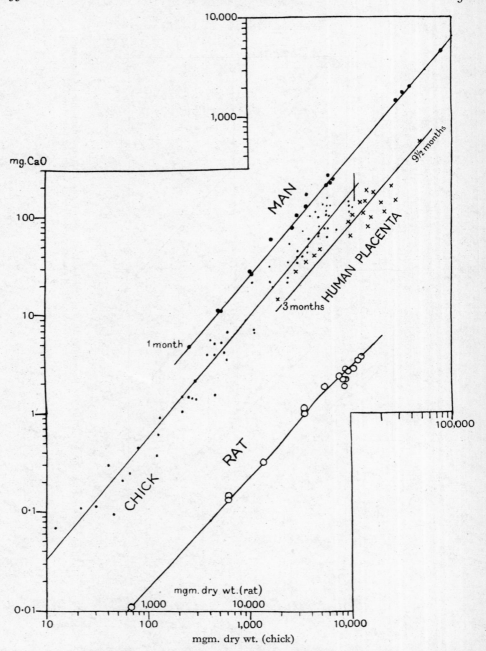

Fig. 300. Calcium in the chick embryo (Mankin; Plimmer & Lowndes; Romanov); in the human embryo (Schmitz) and in the rat (post-natal life; Sherman & McLeod).

to 10·0 gm. dry wt. in the case of the chick embryo,* and from 0·2 to 100·0 gm. dry wt. in the case of the human embryo, the absolute increase rate of calcium is the same. The data of Wehefritz for the calcium in the human placenta have been included on the same graph, but the points are too scattered and too limited in locality to make the drawing of a straight line through them justifiable. It can be seen, however, that a line drawn parallel to the embryo line does not greatly misrepresent them, and it is quite possible that the growth of the placenta follows a similar course to that of other cells of embryonic origin. Here is the requisite conception for a new analysis of the constitution of the placenta.

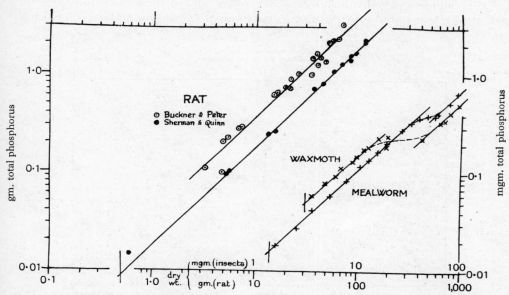

Fig. 301. Total phosphorus in the rat (post-natal life; Buckner & Peter; Sherman & Quinn); in the mealworm and in the waxmoth (larval life; Teissier).

The period of life covered in Fig. 300 is that during which ossification (see p. 555) is proceeding,† and at first sight it may appear strange that the onset of the process is not accompanied by a change in the differential growth ratio. The data clearly indicate, however, that the deposition of calcium in the bones is largely a matter of translocation of the element, and does not involve any relatively greater increase of intake at the same time.

Unlike calcium, phosphorus is an element which shows an isauxetic relation. The data for the post-natal life of the rat (Buckner & Peter; Sherman & Quinn) and for the larval life of the mealworm and waxmoth (Teissier[1]), shown in Fig. 301,

---

* That much of the calcium of the embryonic body is derived from the egg-shell is a well-established fact (*CE*, p. 1265). Glaser & Piehler have actually been able to demonstrate, by staining, the erosion grooves made by the allantoic blood-vessels on the egg-shell as the calcium bicarbonate dissolves in the blood.

† Kamachi[1] tried to alter the rate of deposition of calcium and phosphorus in the chick embryo by injecting various substances, but with little success.

illustrate this clearly. If any heterauxesis is present at all, it is very slightly negative. The insect data here show changes in the constant b, as referred to above. A biochemical interpretation of this behaviour is difficult, especially as total phosphorus includes so many separate fractions (lipin P and nuclein P both of which are bradyauxetic; and inorganic P, which is tachyauxetic, e.g. $k = 1·20$ for the chick (Baldwin & Needham[1]). The interest lies in the close similarity of the mammal and the insects.

So far (except for liver glycogen) we have been discussing the changes occurring in the body of the organism as a whole. It is interesting to examine the development of individual organs and tissues, and of these a good instance for the present purpose is the mammalian brain.* Donaldson & Hatai and Koch &

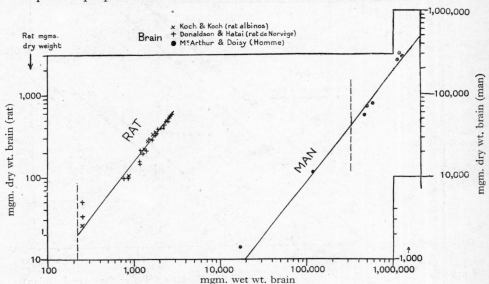

Fig. 302. Dehydration of the brain of the rat (post-natal life; Koch & Koch; Donaldson & Hatai; and man (pre- and post-natal life; McArthur & Doisy).

Koch studied the water content of the developing brain of the rat, while McArthur & Doisy investigated that of man, and Bäcklin that of the rabbit. In Fig. 302 the dry solid in the brain of rat and man is plotted against the wet weight. The correspondence between the slopes of the straight lines resulting is remarkably good ($k = 1·38$ for the rat, $1·33$ for man).

The correspondence between the general pictures of constituents for the mammalian brains is, however, equally noticeable. Fig. 303 gives the data for proteins, phosphatides, cerebrosides, sulphatides, organic and inorganic extractives, and cholesterol. It can be seen at a glance that although the general level of absolute weight is quite different, and the time of occurrence of birth is also different, the principal course of events is closely similar. The slopes agree,

* It is of interest that as long ago as 1898 Lapicque compared the ether extracts of brains of different sizes, taken from a series of mammals. The available technique, however, was too crude to permit of any sure conclusions.

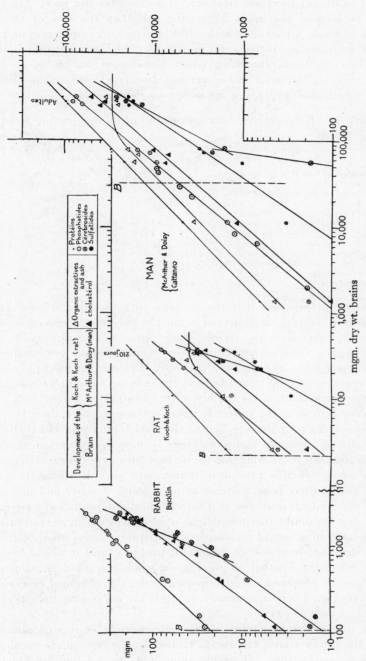

Fig. 303. Growth of the constituents of the brain of the rat (Koch & Koch); the rabbit (Bäcklin) and man (McArthur & Doisy; Cattaneo).

and each constituent increases relatively at a characteristic pace. The increase continues so long as the brain is growing except in the case of the organic extractives and ash, a fraction which seems to come to completion at a certain point, and thereafter to increase no more. Particularly interesting is the very rapid increase of the cerebrosides, where sometimes two phases seem to be indicated; it occurs last of all and is certainly associated with myelinisation. The numerical values and ranges of k are shown in Table 31.

Table 31. *Heterauxesis of the constituents of the mammalian brain*

Fig.	Constituent	k Rat	k Man	k Rabbit	Range of k	Average k
302	Related to wet weight of brain:					
	Dry substance	1·38	1·33	—	0·05	1·35
303	Related to dry weight of brain:					
	Protein	0·94	0·94	—	0·0	0·94
	Phosphatides	1·21	1·17	1·03	0·18	1·14
	Cerebrosides (first phase)	3·11	7·11	1·35	—	—
	(second phase)		1·86	2·35	—	—
	Sulphatides	1·66	1·43	—	0·23	1·54
	Organic extractives and ash (until the point of completion)	0·92	0·95	—	0·03	0·94
	Cholesterol	1·15	1·12	1·28	0·16	1·18

From Table 31 we find that the proteins and the organic and inorganic extractives show definite bradyauxesis, while the phosphatides, cerebrosides and sulphatides, together with the sterols, show pronounced tachyauxesis. This is equivalent to saying that the structure is first laid down on a protein (one might almost say water-soluble) basis, and it is only later on that the lipin and sterol molecules come to take up their places in the nervous system. This is just what we see in the chick embryo, for the histochemical work of Marza[1] indicates that in the nervous system, at the primitive streak and somite stage, there is very little lipoid. It is as if the proteins represented the steel framework, and the lipins the concrete walls, of a modern building. There is, moreover, a certain reversibility in this succession, for R. M. May[1] has shown that during the course of degeneration of nervous tissue (whether produced by traumatic encephalitis or by section of the sciatic nerve) there is an increase in percentage of water and nitrogen and a decrease of lipoid. It is almost as if the chemical differentiation were retracing its course, but no doubt the phenomena of degeneration in nervous tissue are too complicated to admit of so simple a description. Nevertheless, the heterauxesis of degrowth remains an important problem, and it would be of great interest to determine whether, in an organism such as a planarian worm, where degrowth can be observed on an imposing scale, the chemical heterauxesis is strictly reversible. There are already grounds for supposing that it would be found to be so (see p. 525).

Another interesting possibility is that some forms of mental abnormality, especially the idiocy states, are due to failures in the normal chemical growth-plan of the nervous system. The estimations of Ashby & Glynn lend plausibility to this suggestion.

Returning to the question of applying heterauxetic methods within the limits of one organ, the picture for the mammalian brain could be further filled in, if desired, by calculations from the data on lipins and sterols between birth and maturity collected by Lang on the rat, and Schuwirth on man,* and the parallel data on ash collected by Leulier & Bernard; Landa-Glaz; Goldenberg, Landa-Glaz, Voinar & Ostrumova. For the liver there are analyses on the pigeon by Kaufman & Nowotna (though the number of separate points would be rather few), and we have already seen something of liver heterauxesis while discussing glycogen accumulation. For muscle there are a few figures collected by Asmolov. Finally, Meunier[2] has analysed the existing data for the chemical heterauxesis of developing bone, especially those of Hammett[5]. Table 32 gives the results for the femur of the rat, where there is a critical point at 150 mgm. (wet wt.):

Table 32. *Heterauxesis of the constituents of mammalian bone*

	k	
	Before the critical point	After the critical point
Organic substance	1·13	1·13
Total ash	1·12	1·80
Calcium	1·18	1·81
Total P	1·15	1·80
Mg″	—	1·56
Water	0·94	0·46

Thus as ossification proceeds lime and phosphorus displace water, but the milder tachyauxesis of the organic substance persists (see also Burns & Henderson). Carbonate may also tend slightly to take the place of phosphate (Yoshida).

The actual mechanism of the deposition of ash in the bones of the embryo has been the subject of a good deal of work. Robison's phosphatase[1] increases in activity in calcifying bone whether *in vivo* or *in vitro*† (Fell & Robison[1]), though since the ash is continually being deposited, it declines on the dry weight after passing through a peak.‡ Similarly, prospective non-ossifying cartilage produces no phosphatase *in vitro* while prospective membrane bone does (Fell & Robison[2]). Osteoblasts are therefore correlated with an enzyme which makes inorganic phosphate from hexosephosphates, glycerophosphates, etc. and so brings about the supersaturation of the tissue fluid with bone salt. But this is not the only factor; some other mechanism is concerned with promoting deposition from this medium (Robison, McLeod & Rosenheim). The two factors are difficult to separate (Shipley, Kramer & Howland; Robison[2]), but the latter, unlike the former, is sensitive to cyanide, iodoacetate and fluoride (Robison & Rosenheim),

* Here there is also older work by Fraenkel & Dimitz; Mansfeld & Liptak; and Bergamini, which may not be without value.

† There is an interesting connection here between the appearance of phosphatase and self-differentiation. Third day chick embryo femora *in vitro* never develop epiphyses, etc., nor do they form phosphatase; fifth day femora do both. That the formation of bone shape is not primarily conditioned by muscle tractions or other *in vivo* conditions is now clear (Fell & Robison[2] on femora; Fell & Canti on knee-joints; Glasstone on teeth; cf. the monograph of P. D. F. Murray[3]).

‡ It also does this when related to total nitrogen (Roche & Leandri), though the peak comes earlier.

which suggests that the deposition of the ash may require energy derivable from carbohydrate breakdown. Hence the glycogen in growing bone (H. A. Harris[2]) and teeth (Glock) may be of much importance. The calcifying mechanisms appear in development before any visible calcification (Niven & Robison; Fell & Robison[3]). When calcification is complete, excess of calcium or phosphorus in the maternal diet produces no hypercalcification (Booher & Hansmann).

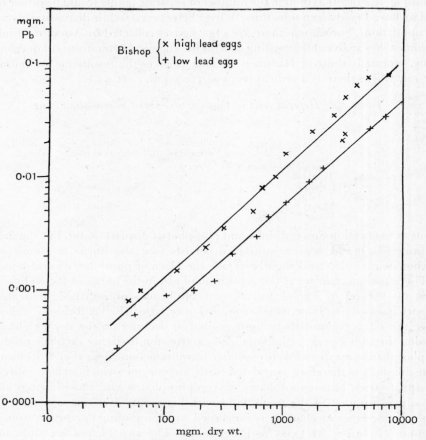

Fig. 304. Lead in the chick embryo (Bishop).

Little is known about the intrinsic nature of calcification. But Cagliotti & Gigante have pointed out that the molecular spacing of hydroxyapatite is the same as that of the peptide chain, and Schour & Hoffman have discovered a basic calcification rhythm at 16μ intervals in teeth, possibly connected with calcospherite formation.

It was remarked in connection with the heterauxesis of total ash (Fig. 299) that the process of chemical differentiation seemed to be independent of the nutritive conditions of the animal. This phenomenon is found again in Fig. 304, which gives Bishop's data[1] for the lead in the chick embryo. The entry of lead into the chick follows the same course no matter what the level of lead in the unincubated egg, fixed by the diet of the laying hen, may be.*

Does temperature affect the heterauxesis of chemical substances? Can the composition of the body at a given moment be varied by changes of external temperature, i.e. by speeding up or slowing down, the rate at which the whole growth process takes place? At present there are insufficient data in the literature to give a satisfactory answer to this question, but a beginning has been made by Romanov & Faber, who have studied the fat content, calcium content, etc., of chick embryos reared at different temperatures from 32° to 40° C. Their results for the former group of substances are shown in Fig. 305, where fat is plotted against *wet* weight for four different temperatures. The experiments, as far as they go, indicate a uniform k (1·79 at 38°, 1·77 at 36°, 1·85 at 34°, 1·84 at 32°).

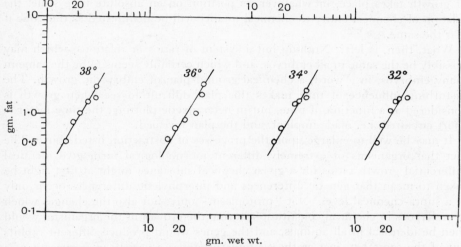

Fig. 305. Fat in the chick embryo incubated at different temperatures (Romanov & Faber).

But they are unsatisfactory; first, because only a very small range of growth was taken (15–35 gm. wet wt.), and secondly, because in the chick embryo large variations of temperature are not possible without the death or damage of the embryo. Work of this kind should be done on amphibian embryos, where the temperature may vary from 8 to 21° and the time of development from 30 to 8 days. A similar experiment to that of Romanov & Faber was carried out by Lerner & Gunns on the leg-bone lengths of the chick embryo incubated at different temperatures; k was not appreciably different. It may be foreseen that temperature, and hence the speed of the whole process, will have little or no effect on chemical heterauxesis, for it is now fairly clear that at least the efficiency (the material stored in relation to that combusted) is, within ranges of temperatures permitting normal development, nearly always the same (see on this the discussion in *CE*, pp. 937 and 973). The question should, however, be experimentally tested.

* The work of Melampy, Willis & McGregor indicates that small differences in chemical heterauxesis may underlie the divisions into "castes" found in the social insects (cf. p. 309).

3·25. The chemical ground-plan of animal growth

Let us now consider what is the general significance of the work reviewed in this account. When we plot the magnitude of a chemical constituent of an organism against the magnitude of the organism as a whole, we have to do with a rather high level of abstraction.

(1) We are abstracting from the morphological form.

(2) We are abstracting from the factors of nutrition.

(3) We are abstracting from the absolute values of the magnitudes. The relation of chemical entity to chemical totality may run an identical course in organisms of widely different size.

(4) We are abstracting from the time factor. At whatever speed the process of growth takes place, or whatever its position on an absolute time scale, the relation of chemical entity to chemical totality will be manifested as the same if it *is* the same.

What, then, is left? Nothing but a system of ratios or relations, which may possibly be the same in all embryos, and which certainly seems to be the same in many embryos, in a word, a chemical ground-plan of embryonic growth. The disturbing influence of time makes this plan difficult to see when growth is considered as a function of time, but in heterauxetic plotting, the time factor is short-circuited, i.e. made implicit, and the plan revealed.

It may be well to enlarge upon the processes of abstraction listed above. The fact that organisms of extremely different morphological form give identical differential growth ratios for a given chemical substance might at first sight be taken to mean that genetic differences and interphyletic differences occur only at a supra-chemical level. Not "protoplasm" only, but also the changes which it undergoes in chemical constitution during the growth of the organism, would then be identical in all animals, and the genes would produce different rabbits out of the same hat, just as the sculptor produces a variety of forms out of a homogeneous substance. But it must be remembered that chemical heterauxesis, as so far discussed, refers to two separate levels of analysis. Out of the seventeen substances which have been discussed as chemical entities in the preceding account, seven are chemically well-defined (e.g. water, glycogen, lead) and ten are more properly speaking groups of substances (e.g. non-protein nitrogen, ash, sulphatides). The problem of how far every well-defined chemical substance in the body will be amenable to heterauxetic treatment must await further work. Meanwhile, it is clear that even in the extreme case where every given group of substances had the same heterauxesis in different animals, there would still be a great deal of scope for genetic differences at the chemical level. "Sulphatide", for example, might show a k identical for supra-oesophageal ganglion and for brain, but its chemical structure (position and nature of side-chains, etc.) might vary considerably as between crustacea and mammals. One has only to think of the specificity of animal pigments or of immunologically active proteins to see that whatever else uniform heterauxetic relationships mean, they cannot possibly mean that "protoplasm" is the same everywhere. The classical work of

Reichert & Brown on haemoglobins, and of Reichert on starches, together with the early studies on the racemisation of proteins (Dakin & Dudley; Dudley & Woodman) established long ago that profound interspecific differences exist in the molecular structure of different proteins. If "protein", therefore, gives us an identical k for different animals, it is much more likely to be due to some question of colloidal stability, as suggested above.

This brings up the question of how far senescence (a process obviously inclusive of that of growth) affects chemical heterauxesis. For the present, although an association between senescence and colloidal stability is guessed at (cf. Dhar), we cannot suppose that age as such has much to do with chemical heterauxesis. After the cessation of growth, there is little change in chemical composition until the end of life, as is well shown in Fig. 306, modified from Moulton's well-known paper. On this subject there is a review by McCay.

In this connection the question may be asked whether if growth were greatly retarded or accelerated heterauxesis would follow suit. When the growth of rats was severely retarded by restricting the calorie intake, heart size was larger and liver size smaller than normal for the weight in question, though no biochemical data were obtained (McCay, Crowell & Maynard). But within limits of growth-rate more near the normal the water-content of tissues (muscle and thymus) varied solely with size and not with age (Moment) as also did the chemical composition of the bones (Outhouse & Mendel) and their length (Lerner & Gunns). The same is true of the water-content of larval amphibia (Galloway); see p. 524. Further investigations on the temporal deformability of heterauxetic relations would be of interest.*

The present treatment constitutes a step towards a theory of biochemical transformations. It cannot be too much emphasised that although the time factor is short-circuited in heterauxetic graphs, it is not completely eliminated from the problem. On the contrary the chemical heterauxesis of different animals should fuse into one common plan if they were reduced to the same time scale and the same absolute size. This has been shown by Waddington[4] in the following way. If we have two animals p and q, and measure in each of them several chemical magnitudes M_p, M_q, N_p, N_q, etc., we find relations of the type

$$\log \frac{M_p}{M_{p_0}} = k \log N_p \qquad \ldots\ldots(12)$$

and
$$\log \frac{M_q}{M_{q_0}} = k \log N_q \qquad \ldots\ldots(13),$$

* The pituitary growth hormone should provide opportunity for testing whether the chemical composition of the body depends on size or on age, but the data so far obtained are not quite clear. In hypophysectomised rats the bradyauxesis of water and the tachyauxesis of fat seen in normal growth cease, and the animal maintains approximately the same composition which it had when growth was arrested by the operation (Lee & Ayres). Thus composition depends on size and not on age. But in rats gigantised by administration of extra growth-hormone the water loss and the fat gain seem to be inhibited and the abnormally large animal has much the same composition as it had at the beginning of the treatment (Lee & Schaffer), i.e. more water and less fat than its normal control of the same age, though if composition depends on size and not on age it should have exactly the opposite, since in terms of weight it is "older". The question needs further analysis.

where M_{p_0} and M_{q_0} are specific constants, and k a general constant relating M and N for all animals. Now M and N are also functions of the time t. If we have

$$\log \frac{M_p}{M_{p_0}} = F(t) \qquad \qquad \dots\dots(14)$$

and

$$\log \frac{M_q}{M_{q_0}} = F(t) \qquad \qquad \dots\dots(15),$$

we can choose another variable ϕ such that

$$F(\phi) = F(t) \qquad \qquad \dots\dots(16).$$

That is to say, by choosing a suitable unit or function for the measurement of time, we can convert the growth curve of M_p into that of M_q; and further, the same system of time measurement will convert all the growth curves of chemical magnitudes of animal p into those of animal q, provided only that in each case

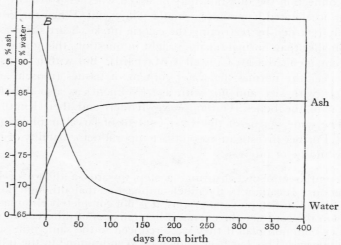

Fig. 306. Percentage of water and ash in the rat throughout the life cycle (modified from Moulton's plot of the data of Hatai; Zuntz; Inaba; Chanutin; Buckner & Peter; Sinclair).

there is a linear relation between the logarithms, with the general constant k. Then we could regard the two systems of time measurement defined by t and ϕ as the relative time scales of chemical development in the two animals. But relative time scales may be derived in other ways, e.g. from morphological development, and from embryonic determination. The morphological stage at which an organ becomes embryologically determined may vary widely in related species. It would be interesting to know how the time scales of morphological, determinative, and chemical development, are related to one another.

If, then, the unitary chemical ground-plan of animal growth exists, we must think of it as deformable in space-time. Just as d'Arcy Thompson was able, by systematic deformations of Cartesian co-ordinates, to transform one morphological shape into another (e.g. the sun fishes, the amphipods, mammalian skulls, etc.), so the chemical ground-plan is deformed in time and space. It can be slowed down or speeded up, and it can vary dimensionally from the extreme of the ichthyosaur to the extreme of the pocket gopher. But for the most part, if the spatial magnitudes are reduced, the temporal magnitudes will be reduced too, and it is here that we touch upon the thought of Lambert & Teissier who have proposed as a fundamental biological law that homologies exist between animals in time as well as in space. There is an equality, they suggest, between the ratio of homologous spatial magnitudes and the ratio of homologous times, and they support their "theory of biological similarity" by the existing empirical data for cardiac frequency, metabolic rate, gestation time, longevity, etc. At one point they closely approach the present discussion, for they suggest that "at homologous instants (in the life cycle) two homologous organs will have the same qualitative and quantitative (chemical) composition". Mouse time must bear the same, or a similar, relation to elephant time as mouse spatial magnitudes to elephant spatial magnitudes. Indeed, unless the time factor is brought into account, we may understand morphological similarity, but we can never hope to understand physiological, still less embryological, similarity. Analogous ideas have found expression in the short note by Carrel[2], the paper of Brody[1] and the books of du Noüy and Lacape, who analyse the notion of "physiological time". Here, with Hoagland's possibly radioactive "physiological clock", we tread upon the borders of psychology, but it may be that the sense of duration will some day be found to be not so far removed from embryonic events as we may at present imagine.

Lambert & Teissier, then, suggest that animals of the same general form, but of different sizes, have the same form in space-time. The investigation of chemical heterauxesis leads to the suggestion that animals of different form and different sizes have the same basic general chemical plan of growth, which is deformable within wide limits in space-time. The process of growth would thus proceed according to a definite plan recognisable in the constitution of the organism at any given stage of its life history. Potentiality offers to Actuality a formula in which substitution may be freely made from a wide, but not infinite, range of values.

As regards the validity of the conclusions which can be drawn from similarities in k, it is clear that the main difficulty lies in knowing how close values of k must be in order to permit a conviction of identity. This is essentially a statistical problem. Unfortunately, from the ranges of k given in Tables 29, 30 and 31, it is impossible to argue far, for in each case k depends, at least partly, on the standard deviation, probable error, etc., of each of the individual sets of data on which it is founded, and the statistical qualities of the data are not often known. Nor could they now be calculated, in view of the peculiar conditions attaching to each individual investigation and the particular chemical method employed. Since the data are so heterogeneous it is surprising that we find as many regularities as we do, and these can hardly be thought the result of chance or coincidence. What is required is a new and extensive investigation, carried out with the same chemical methods and sampling, on a selected number of widely different organisms at all stages of the life cycle and, if possible, under constant, known, and various, environmental conditions. The present account probably goes as far as it is possible to go by simple analysis of the published information.

It seems improbable, however, that the correspondences already noticed are illusory. In Fig. 299, for example, where the range of variation of k is rather wider than usual (0·14), it may be observed that the range for the chick embryo alone is over half the total range, i.e. 0·09. This may be interpreted as favourable to the view that the constants would correspond more closely than they do if the data were better.

The general upshot of this section may be summarised by saying that heterauxetic plotting allows us to recognise similarities between the developmental plans of different organisms which, owing to the great differences in their time-scales, would otherwise remain hidden. We always find the following distribution of relative growth rates:

I.   *Isauxetic constituents*
(to dry weight): protein nitrogen, total phosphorus.

II. *Tachyauxetic constituents*
(to total weight): dry substances.
(to dry weight): fat, creatine, glycogen, calcium, possibly magnesium, inorganic phosphorus, bicarbonate.

III. *Bradyauxetic constituents*
(to total weight): water.
(to dry weight): non-protein nitrogen, purine nitrogen, glutathione, total ash, sodium, potassium, chloride, lipin phosphorus, nuclein phosphorus, iron, copper, lead.

In this way we arrive at a universal ground-plan of chemical development. It is safe to say that when all exceptional cases have been considered, and when all criticisms of the heterauxetic treatment have been taken into account, it nevertheless remains the most convenient way of comparing the chemical development of one organism with that of another.

3·3. Respiration

Preceding pages have often emphasised the fundamental fact that, apart from certain exceptional cases, embryonic growth and morphogenesis cannot proceed in the absence of oxygen. How exactly the oxido-reductive energy-providing machinery engages with the differentiation processes is still very obscure, but must some day be understood; meanwhile a few hints may be remembered (p. 474 and p. 498).

3·31. Respiration of the eggs of Echinoderms, Worms and Molluscs

For over thirty years the alecithic eggs of echinoderms, in view of their uniformity, transparence, and many other convenient qualities, have been a favourite material for studies on the physiology of cell respiration.* Without

---

* The technical difficulties, not so much of work with the eggs of marine invertebrates, as of comparisons between the work of different investigators, should not, however, be minimised. The determination of the volume of the material by centrifugation and the weight by calculations from Kjeldahl analyses, long customary, needed the critique which it has received from Gerard & Rubenstein; Rubenstein & Gerard[1]; Whitaker[8]; and Shapiro[2]. Shapiro[3] has provided a useful nomogram for calculating the force of a given centrifuge in terms of gravities.

going back beyond 1931 we have here to review the most significant recent work, of which there has been an abundance. Although we are still far from a real solution of the essential problem, namely, how the chemical transformations and energy changes proceeding during development are integrated with the morpho-genetic changes, this section should be read in conjunction with that on deter-mination in echinoderm development (2·9).

Perhaps the most classical result of the earlier investigations was the rise in respiratory rate of the sea-urchin egg on fertilisation (*CE*, p. 636). This pheno-menon has been repeatedly confirmed (e.g. by Whitaker[6]; T'ang[1] and L. Chesley on *Arbacia punctulata*; by Borei[1] and Laser & Rothschild on *Psammechinus miliaris*; and by Brock, Druckrey & Herken on *Paracentrotus lividus* and *Sphaerechinus granularis*). It is true that the immediate rise, which may be well above 400 %, is not always maintained, and the respiratory rate settles down to a slower level, though still well above that of the unfertilised egg. But it was known that some species, such as the starfish, showed no rise at all on fertilisation, and this also has been frequently confirmed (e.g. by T'ang[2] on *Asterias glacialis*). When the eggs of a wider range of invertebrates were studied, many different types of behaviour were found. Thus the eggs of the clam, *Cumingia tellinoides*, studied by Whitaker[4],* showed a decrease on fertilisation to about 45 % of the pre-fertilisation rate, and the same was found to hold good for those of the annelid *Chaetopterus* by Whitaker[6]; L. Chesley and J. Brachet[16]. For another annelid, *Nereis*, the position is not quite clear, for while Barron[2] found absolutely no change in respiratory rate at fertilisation, Whitaker[5] found an increase of about 140 %.

Whitaker has essayed a general interpretation by showing that (apart from the observations of T'ang[1] and T'ang & Gerard, which do not fit into his scheme) the respiratory rates of the embryos shortly after fertilisation are roughly of the same order (1·3 to 2·0 c.mm. O_2/hr./10 c.mm. eggs at 21°), although the respiratory rates of the unfertilised eggs may be very different. There would thus be a damping at fertilisation in the mollusc and annelid examples, and a removal of inhibition in the case of the echinoids, while the asteroids would show no change. Perhaps it is too early to decide whether this generalisation meets all the facts. Further investigations are still required.

The respiratory quotient has often been calculated, but as there is much variation in the results obtained, it is likely that the technique of measurement is still deficient. The general tendency of the earlier work (cf. *CE*, pp. 623, 989) placed it at unity or a little below, and this was confirmed by the finding of 0·9 by Borei[1] on *Psammechinus miliaris*, and 0·92 by van Herk on *Sphaerechinus*. But Ephrussi[4], working on *Paracentrotus lividus*, obtained a figure of approximately 0·8 for the period between the two-cell stage and the hatching of the blastula, and this did not vary between 10° and 25° C. On the other hand Laser & Roths-child, using a modern and carefully thought out technique, obtained 0·84 for

---

* Whitaker[3] also determined the change of oxygen consumption in the "eggs" of the alga, *Fucus vesiculosus*, on which important studies of polarity have been made (see p. 668), and found that there is an increase of 190 %. In a later paper, he reported the odd fact that though the Q_{O_2} of clam eggs falls on fertilisation and that of sea-urchin eggs rises, both show increased stainability with certain dyes such as methyl violet and gentian violet after fertilisation (Whitaker[17]).

0–30 min. after fertilisation (*Psammechinus miliaris*) but 0·66 for 5–20 min. afterwards. A still later paper, by Öhman of Runnström's school, gives the R.Q. of *Paracentrotus lividus* eggs as 0·73 in the earliest cleavage stages (1–2 hr. after fert.) and 0·85 after 7–8 hr. (presumably blastula stages). If this is correct, the sea-urchin embryo would join the amphibian embryo (see p. 588) in having a carbohydrate quotient at gastrulation, and the succession of energy-sources (see Section 3·37) would start only from that point onward. There is no doubt that the measurement of the respiratory quotient of sea-urchin eggs is a rather difficult task, partly because of the considerable acid production which occurs shortly after fertilisation (see p. 572).

The effect of temperature upon the respiratory mechanisms of the echinoderm egg was studied by Rubenstein & Gerard[2], who found that the temperature coefficients of respiration before and after fertilisation in the sea-urchin (*Arbacia*) are different (Q_{10}, 4·1 before and 1·8 after). The increase of oxygen consumption on fertilisation therefore varies markedly with temperature, being tenfold at 11° and only double at 30°. By extrapolation, there would at 32° be no rise at all. These results were later confirmed by Korr[1], as will be mentioned later, though the eggs of other invertebrates do not show the effect clearly (Tyler & Humason). Ephrussi[4] sought an answer to the question whether the total amount of oxygen consumed is identical at all viable temperatures, and found (using *Paracentrotus*) that for the period from the two-cell stage to the hatching of the blastula there is between 10° and 23° no variation in this quantity. From fertilisation to the two-cell stage, however, constancy was only found between 17° and 23°; below these temperatures the oxygen consumption fell off to −40 % of the normal at 10° and above them rose to +40 % at 24·3°. The substantial constancy shown throughout development in these results was later confirmed by Tyler[5], as will be mentioned in another connection (p. 574).

Ephrussi[4] also calculated the Q_{10} at different temperatures for respiratory rate, and found that, contrary to the original Loeb-Wasteneys view, it came very close to the Q_{10} for velocity of segmentation. This conclusion, that developmental rate and respiratory rate cannot be dissociated by temperature, was again confirmed by Tyler[5].

The relation between oxygen tension and oxygen consumption by echinoderm eggs was studied on unfertilised *Arbacia* eggs by T'ang[3] and on fertilised *Arbacia* embryos by Amberson and T'ang & Gerard. There is general agreement that respiration continues at 100 % of its normal level from 160 mm. Hg pressure (air) down to about 40 mm. Hg, after which point it rapidly drops, reaching zero when the oxygen pressure becomes negligible. T'ang has reviewed this effect in comparison with data on other respiring systems, and it has received mathematical analysis at the hands of Gerard[2]; Rashevsky[2], and Landahl.

Anaerobiosis of course stops echinoderm development completely (E. B. Harvey[1]), but in some forms, such as the polychaete worm *Nereis* (Barron[2]), fertilisation is still possible under anaerobic conditions. Permeability of the *Arbacia* egg to water is slightly but significantly decreased in the absence of oxygen, according to Kekwick & Harvey, but there is no difference in the equilibrium values attained. Hunter thinks there is no effect at all.

The effect of pH, CO_2, etc. on echinoderm egg respiration has also been the subject of some study. As might be expected, respiration falls off on both sides of the normal external pH (Ashbel[2] on *Paracentrotus* and *Arbacia*), and the same is true of osmotic pressure (Borei[2] on *Asterias glacialis*). Excess CO_2 in the sea-water retards cleavage (Haywood & Root[1] on *Arbacia*); the effect is perceptible at tensions as low as 4 mm. Hg and progressively increases up to 125 mm. Hg, at which point segmentation is suppressed. Respiration of *Arbacia* eggs is diminished by all acids, but by CO_2 more strongly than HCl at the same pH on account of its greater penetrating power (Root). At a given CO_2 tension the presence of bicarbonate in the sea-water lessens the cleavage retardation mentioned above proportionally to its concentration, according to Haywood & Root[2]. In this connection it is interesting that carbonic anhydrase, the enzyme which catalyses the decomposition of bicarbonate, was found by Brinkman to be present in echinoderm eggs.

Acid pH may have curious effects on fertilisation. Thus in the gephyrean worm *Urechis* Tyler & Schultz found that fertilisation can be blocked by sea-water at pH 7·2. If eggs are placed in this medium as soon as 3 min. after insemination, they will show no development, and this block is reversible up to 10 min. but later irreversible. Nevertheless these eggs retain their fertilisability, and upon transfer to ordinary sea-water and re-insemination, polyspermic development will ensue. This was subsequently generalised to the eggs of many invertebrate forms by Tyler & Scheer.

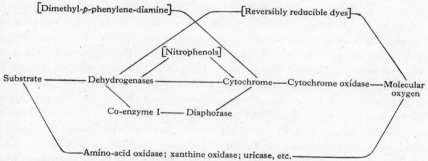

Fig. 307. Scheme of cell respiration. Added agents not occurring naturally are placed in square brackets. Between the dehydrogenases and cytochrome or flavoprotein there intervene the C_4-acid cycles of Szent-Györgyi and Krebs (see Elliott's review, and here, pp. 597, 603, 615).

At this point it is necessary for the understanding of what follows to refer to a scheme of cellular respiration summarising our present knowledge (Fig. 307). It was built partly upon facts derived from the study of echinoderm eggs, and all of it was by no means clear at the time when some of the most important experiments on them were made. As is generally realised, the major part of the respiration of animal cells goes through the dehydrogenase and Warburg-Keilin systems (the central line of the diagram). Hydrogen of the substrate, activated by the dehydrogenases, reacts with the reversibly oxidising-reducing substance cytochrome, and this in turn reacts with molecular oxygen activated by the enzyme cytochrome oxidase (formerly called "indophenol oxidase" or the *Atmungs-*

ferment). Since the Warburg-Keilin system is composed of iron-containing tetrapyrrol substances, respiration is generally inhibitable by KCN and CO. That it is not always so inhibitable, however, shows the existence of alternative tracks, represented in the diagram by coenzyme I (see p. 209) and diaphorase (similar to the yeast flavoprotein formerly called the "yellow enzyme" of Warburg[4]), though these may sometimes go through cytochrome; and mechanisms such as the amino-acid oxidase, xanthine oxidase, etc. which undoubtedly react direct with oxygen.

At the same time a number of substances are known which have the property of greatly increasing the respiratory rate of certain cells. The reversibly oxidisable-reducible dyestuffs are the best known of these; they act by side-tracking the Warburg-Keilin system, as we know from the fact that the respiration catalysed by them is unaffected by KCN and CO, the inhibitors of "ferrous" respiration. Then the substance dimethyl-*p*-phenylene-diamine, known from a much earlier date than the dyes, behaves in a different way. With α-naphthol, as the classical Rohmann-Spitzer reagent, much used in histochemical work (cf. p. 136), an oxidative condensation occurs forming the dye indophenol blue, but even without α-naphthol there is a condensation giving a coloured product (Keilin[2]). It is now clear (Keilin & Hartree) that dimethyl-*p*-phenylene-diamine reacts with cyto-chrome and not directly with the "indophenol oxidase", so that its action may be said to be analogous to introducing a large quantity of fully activated substrate into the system. Thus the dehydrogenases and normal substrates are side-tracked. Last come the compounds of nitrophenol type. We know that these are not autoxidisable; they must act through the Warburg-Keilin system, since the excess respiration catalysed by them is fully inhibitable with KCN and CO (Krahl & Clowes[1]; de Meio & Barron); and they do not affect the nature of the substrates used since they leave the respiratory quotient of the cells unchanged (Krahl & Clowes[1]). What they do is to accelerate the reduction of cytochrome and the oxidation of its substrates (Krahl & Clowes[2]); a stage which must in many cells therefore be normally a limiting factor.

The nitrophenols also accelerate fermentation, but it is likely that their action there is of a different kind (Ronzoni & Ehrenfest). Nevertheless it is interesting that Krahl & Clowes[2] showed that the greatest stimulations of fermentation occur with those yeasts having normally the lowest fermentation rate and *vice versa*.

The foundations of our knowledge of respiratory mechanisms in echinoderm eggs were laid by Runnström[7]. Working on the eggs of the sea-urchin *Paracentrotus lividus*, where the post-fertilisation respiration rate is four or five times that before fertilisation, he established that KCN strongly inhibits the respiration of the fertilised eggs* but has no effect on the unfertilised ones, and that CO strongly inhibits the respiration of the fertilised eggs but has a definite acceleratory effect on the unfertilised ones. It was thus clear that the small respiration of the

---

* KCN of course inhibits cell-division. Brinley[1] has made the interesting claim that a concentration of KCN which will retard cell-division if the eggs are placed in it will not do so if it is micro-injected into them.

unfertilised eggs must proceed along tracks alternative to the Warburg-Keilin system, and this has been confirmed by all subsequent work.* Attention must here be drawn to the parallel cases of the orthopteran embryo in diapause (p. 578), the ventral ectoderm of the amphibian gastrula (p. 198) and the yeasts of high and low fermentation rates (p. 496). Whatever the damping mechanism is before fertilisation, it seems to break down in ageing unfertilised eggs, where it is found that the respiration rises (see p. 572), and in these cases the sensitivity to CO and KCN rises at the same time. Runnström regarded the cytochrome oxidase as unsaturated with its substrate in unfertilised eggs — this concept will not quite integrate with our present knowledge. More probably the substrates, dehydrogenases and cytochrome are held apart.

In the presence of dimethyl-p-phenylene-diamine, however, the respiration of both unfertilised and fertilised eggs was greatly raised (by many hundreds per cent), and now the inhibition by KCN and CO was of the same order. This was what we should expect, since the Warburg-Keilin system would, after the addition of this substance, be working at full pressure. Runnström[10] drew the right conclusion that there could be, at fertilisation, no change in the concentration or affinity of the enzymes of the Warburg-Keilin system, but rather a change in their accessibility.

In the presence of methylene blue, an acceleration of respiratory rate was observed[10] on both unfertilised and fertilised eggs, but much greater in the case of the former than in that of the latter. Hence in the former the Warburg-Keilin system, which is side-tracked by reversible dyes, is much more of a limiting factor than it is in the latter. The anaerobic reduction of methylene blue was as active before fertilisation as afterwards, showing that no change in the dehydrogenases occurs at fertilisation.†

Runnström[7] also studied the effects of abnormal osmotic pressures. Here the results are hard to interpret. Hypotonic media, for example, depress the respiration of fertilised eggs, but accelerate that of unfertilised ones.

The study of dimethyl-p-phenylene-diamine was continued by Runnström[14] and Örström[1]. At 3×10^{-3} M the respiration of unfertilised sea-urchin eggs reaches the level of fertilised ones, but they do not become activated. At 2×10^{-2} M the effect is maximal on both. Hydroquinone can act in the same way as dimethyl-p-phenylene-diamine. The Swedish authors were of the opinion that both these substances could be reduced by the cells as well as oxidised, so that they would

---

* It should be stated, however, that Lindahl[9] believes he has shown that the respiration which proceeds in the presence of CO and KCN is no part of normal respiration. He thinks that the whole of the normal respiration is abolished and that a new quota appears, the same as that which under other circumstances we call the acceleratory effect of CO and KCN. This interpretation is not as yet generally accepted.

† But this is contested by Ballentine[1], who believes that he has shown, with the ferricyanide method, a very large change in dehydrogenase activity on fertilisation; about equivalent, indeed, to the total rise in oxidations. A large increase in the reducing power of the eggs at fertilisation would agree with the observation of Dulzetto ten years ago that fertilised eggs give a much stronger qualitative test for reduced glutathione than unfertilised ones. Ballentine[3] also states that the dehydrogenase activity is associated with granules as well as with cytoplasm as may be seen by studying centrifugally fragmented eggs.

function as a side-track for the dehydrogenases (see diagram), but it is not probable that this view is correct.

Already in 1929 Barron[1], following the pioneer experiments of Harrop & Barron with mammalian red blood cells, had shown that the respiration of starfish and sea-urchin eggs was greatly increased by methylene blue, and that this extra respiration was insensitive to KCN. This was confirmed on unfertilised *Nereis* eggs, fertilised *Arbacia* eggs, and unfertilised *Asterias* eggs by Barron & Hamburger using toluylene blue. Then Creach found that methylene blue would also accelerate the respiration of adult echinoderm tissues, and Friedheim made the interesting discovery that hallachrome (see p. 406) and echinochrome, the pigment of echinoderm eggs themselves (obtained crystalline by Ball; E_h data by Cannan), would increase the respiration of echinoderm eggs some seventeen times.* Barron & Hoffman established that the catalytic power of a dye depends on its position on the E_h scale; those with low potentials (very difficult to oxidise) or those with high ones (very difficult to reduce) will not act as respiration catalysts.† In the proper range comes pyocyanin, the pigment of *B. pyocyaneus* (Friedheim & Michaelis; Michaelis, Hill & Schubert), and this, tried by Runnström[14] on *Arbacia* eggs, behaved in every way similarly to methylene blue. As was expected the increase in the unfertilised eggs was greater than in the fertilised ones (210 against 96 %), and the dye can also catalyse the respiration of cytolysed egg-breis (see p. 493). The further interesting fact was revealed that although the respiration of fertilised eggs can be restored to normal or even above it by pyocyanin after inhibition by KCN, the block on segmentation and development is not removed thereby. This was confirmed by Korr.‡ It may therefore be accepted, and it is quite significant because, as we shall see when the respiration of grasshopper embryos is discussed (p. 578), the progress of active development is associated with ferrous respiration. It almost looks, therefore, as if non-ferrous respiration cannot be geared to morphogenesis.

In the work just mentioned Runnström[14] found an acceleratory effect of KCN on the respiration of the unfertilised egg. The study of this phenomenon and the acceleratory influence of CO was taken up by Örström[2], who was able to make a model of the state of affairs in the egg by using yeast. The auto-respiration of yeast and its respiration in the presence of formate was, he found, accelerated by CO, but its respiration in the presence of all other substrates was inhibited. As suggested elsewhere (p. 496), this action is probably the result of removal of inhibitory heavy metals by KCN or CO. The fact established by Lindahl[8], namely that the acceleratory effect is photosensitive, like the well-known inhibitory effect, may not be in contradiction with this view. Lindahl also finds, by a study of the respiratory quotient, that the acceleration is not due to oxidation of the CO itself.

* This action of echinochrome is contested by Tyler[8].

† In the light of this (confirmed by Ellis using KCN simultaneously) it is hard to know what to make of the stimulus to respiration and lactic acid production in echinoderm eggs given by dyes such as janus green, which should not accelerate respiration, in van Herk's experiments.

‡ Though M. M. Brooks and Ellis have mentioned experiments to the contrary.

Stimulation of respiration by pyocyanin was combined with temperature alterations by Korr[1] in a study of the temperature coefficients of "ferrous" and "non-ferrous" respiration. He first confirmed the finding of Rubenstein & Gerard[2] that the temperature coefficients of pre- and post-fertilisation respiration differ, so that the rise on fertilisation would cease to exist at high temperatures.

Unsymmetrical dimethyl-*p*-phenylene-diamine   α-Naphthol   Indophenol blue

4 : 6-Dinitro-*o*-cresol   4 : 6-Dinitro-carvacrol

2 : 4-Dichlorophenol   2 : 6-Dinitro-4-chlorophenol

2 : 4-Dinitro-*ar*-tetrahydro-α-naphthol   Methylene blue   Pyocyanin

He then showed that after fertilisation the temperature coefficients of normal respiration and of respiration reduced to low levels by KCN and restored to the normal level or above by pyocyanin (i.e. a respiration of non-ferrous type), were the same. But in the unfertilised egg the temperature coefficients are a function of the pyocyanin concentration, i.e. of the respiration rate. All this strengthens further the conclusion that in the unfertilised egg the mechanisms for the transfer of activated hydrogen from the substrates to the Warburg-Keilin system are the limiting factors. It should be noted, however, that the fertilised egg in which the Warburg-Keilin system has been thrown out of gear is not identical with the unfertilised egg, even though their respiratory rates may be raised to the same level by pyocyanin. Korr[3] considered that the damping mechanism in the un-fertilised egg is best pictured as the masking of cytochrome in some inactive form. But it is doubtful, as a result of work now to be discussed, whether the cytochrome of the unfertilised egg can really be regarded as inactive.* In any case, it is clear that the change at fertilisation will probably not be explicable without recourse to colloidal and protein factors.

The action of the nitrophenols on cell-metabolism and cell-division in echinoderm eggs has been the subject of exhaustive investigations (Clowes & Krahl[3]; Krahl & Clowes[1,2]). Clowes and his col-laborators discovered the remarkable fact that at the concentration of dinitro-o-cresol which produces the maximum stimulatory effect on the respiration of the fertilised *Arbacia* egg ($8 \times 10^{-6} M$), cell-division is completely inhibited. At lower concentrations the respiration-stimulating effect is of course less marked, and the block to cell-division less complete. At higher concentrations the respiration-stimulating effect also falls off, but the block to cell-division continues unchanged. Nevertheless, this block is completely reversible at all concentrations except the highest, unlike that produced with maximally respiration-stimulating amounts of the reversible oxidation-reduction dyes. KCN, which reverses the stimulation of respiration, acts additively on the cell-division block (Clowes & Krahl[3]). These effects were shown to be obtainable with a wide variety of compounds of nitrophenol type. Fig. 308 shows data for 4 : 6-dinitro-carvacrol; at the concentration which increases the respiration of the eggs nearly fourfold, the cell-division frequency has fallen to nearly zero. With rise of temperature, the respiration-stimulating effect is decreased but the cell-division

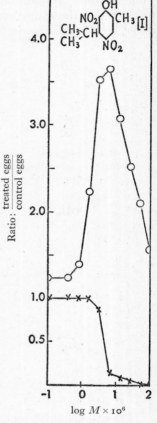

Fig. 308. Stimulation of oxygen consumption and block to cell-division of fertilised *Arbacia* eggs by various concentrations of 4 : 6-dinitro-carvacrol.

---

* Moreover, Korr[4] later revised this opinion, coming to the view that the limiting factor in the unfertilised egg is the provision of the necessary substrates for the dehydrogenases. These them-selves, however, increase in activity on fertilisation (Ballentine[1]). One of the puzzling features of the situation is that those who examine the echinoderm egg directly for cytochrome (Krahl, Keltch & Clowes[2]; Ball & Meyerhof) find only the minutest traces there. There is plenty of cytochrome oxidase, however (Krahl, Keltch, Neubeck & Clowes), and flavin-adenine dinucleo-tide is present (Krahl, Keltch & Clowes[3]).

block is unimpaired. The nitrophenols are active at much lower concentrations than the reversible dyes (about 10^{-6} instead of $10^{-3} M$). In their presence the respiration of the unfertilised egg can never be brought to equal that of the fertilised egg, as is the case with methylene blue and dimethyl-p-phenylene-diamine. But in another paper Clowes & Krahl[1] found that the effect of the nitrophenols is relatively much greater on the unfertilised than on the fertilised eggs.

Krahl & Clowes[3] extended the series to dihalo- and trihalophenols, such as 2:4-dichlorophenol or compounds such as 2:6-dinitro-4-chlorophenol, with the same results. Monohalophenols do not show the effect while the m- and p-mononitrophenols do; but symmetrical trihalophenols do while symmetrical trinitrophenols do not. Hydroxybenzoic acids and hydroxybenzaldehydes are inactive, as also phenols with one aldehydic group (Keltch, Clowes & Krahl); it appears that an active compound must have one phenolic hydroxy group and one or more nitro or halogen group. The activity may be much increased by the insertion of an additional ring, as in the compound 2:4-dinitro-ar-tetrahydro-α-naphthol, which exerts its maximum activity at concentrations far below $10^{-6} M$.

A good deal of argument has centred round the mechanism of action of these compounds. While it is not disputed that the ultimate effect is on the oxidation and reduction of cytochrome, experiments with different CO_2 tensions led to the view that respiration-stimulation is favoured by a high concentration of the anions within the cell, whereas the cell-division block is favoured by a high concentration of the undissociated form (Krahl, Keltch & Clowes[1]; Krahl, Clowes & Taylor). The effect of pH has proved difficult to interpret (Krahl & Clowes[4]; Clowes, Keltch & Krahl; Tyler & Horowitz; Hutchens, Krahl & Clowes), and it has not been possible to dissociate very clearly the two effects of the nitrophenols. Whatever the interpretation may turn out to be, these effects are so clear-cut and well marked that they must be of importance for the future understanding of the integration of respiration and morphogenesis.

To the above, it may be added that reagents poisoning copper catalyses (e.g. potassium dithio-oxalate, diphenylthiocarbazone, or isonitro-acetophenone) give no inhibition of cell-division (Clowes & Krahl[4]). The relations between effect on respiration and effect on cell-division may be summarised as follows:

	Effect on respiration at complete cleavage block Normal ≡ 100 %
Low oxygen tension (see p. 564)	35
CO	35
KCN	35
Phenyl-urethane (see p. 517)	100
γ-(2-methyl-piperidyl)-propyl-benzoate	100
Halo- and nitro-phenols	350

All the extra respiration caused by the substituted phenols is sensitive to cyanide and CO (Krahl & Clowes[5]). Though raising the overall oxygen-consumption, they must inhibit some intermediate step essential for cleavage. Carcinogenic hydrocarbons, tried as their water-soluble choleic acid derivatives (Keltch, Krahl & Clowes), gave very small stimulating effects on cell-division in *Arbacia* eggs.

Among the most important of the other inhibitors of metabolism which have been tried on echinoderm eggs are fluoride and iodoacetate (cf. p. 492). A 0·02 M solution of the latter did not inhibit fertilisation in *Urechis* (Tyler & Schultz), and Ellis found a surprising resistance to the action of both in the fertilised eggs of *Urechis* and *Strongylocentrotus*, as tested on the rate of cell-division. In 0·01 M iodoacetate there was no slowing of the division rate. By direct estimations of the glutathione content of the developing eggs, Ellis was able to prove that the iodoacetate penetrated and combined with the sulphydryl groups. At the same time there seems no reason why development should not proceed at the expense of some substrates other than those involved in glycolysis. Runnström[17], using 0·03 M iodoacetate on *Arbacia* eggs, found no inhibition of fertilisation, although there was a 55 % inhibition of respiration and only the earlier segmentations could be passed through. Pyruvate reversed the respiratory inhibition, but could not restore normal morphogenesis.

From the work of Simon[2] it appears that radium emanation has little action on the respiration of the egg, judging from experiments on the piddock clam, *Barnea candida*. But L. Chesley, using X-radiation on *Arbacia* and *Chaetopterus*, obtained interesting results showing that developmental anomalies appear at doses quite insufficient to affect the respiration either normal or intensified by methylene-blue catalysis.

One of the most obscure phenomena about the early development of the echinoderm egg is the acid production which occurs shortly after fertilisation. At Runnström's suggestion, Ashbel[1] discovered in 1929 a considerable positive pressure produced in manometers by eggs after fertilisation under anaerobic conditions. We know now that this takes place also under aerobic conditions (Borei[1]; Laser & Rothschild) and the fixed acid produced amounts to 8 milliequivalents per c.c. eggs. Unfortunately success has not attended efforts to ascertain what this fixed acid is. The acid production of eggs cytolysed by hypertonic or hypotonic sea-water was found by Runnström[12] to be increased by hexosemonophosphate, suggesting glycolysis, but it was not inhibited by iodoacetate, and when cytolysis was produced by freezing and thawing, hexosemonophosphate had no effect (Runnström[16]). With saponin cytolysis, phloridzin inhibited the acid production (Rothschild). The position thus needs further elucidation.

Before fertilisation, as we have seen, the respiratory rate of the sea-urchin egg is low. On ageing (i.e. if no fertilisation takes place) the respiratory rate gradually rises, and the check to ferrous respiration must slowly be released, as sensitivity to KCN and CO also begins. Ageing is known to be accompanied by permeability changes (Goldforb). The actual length of the fertilisable period is variable according to the species, but always shorter than the time elapsing before the capture of external nourishment by the embryo if fertilisation had occurred. The aged egg dies, therefore, in possession of plenty of raw materials capable of supporting life (cf. pp. 258, 511). According to the older Lillie-Just "fertilisin" hypothesis (F. R. Lillie[5]; E. E. Just[2]), fertilisability ceased because the substance which attracts spermatozoa ("fertilisin") ceases after a time to diffuse out from the eggs. But this does not explain why the eggs should die. Thyroxin was found by Carter to extend considerably the fertilisable life of echinoderm eggs; he believed that it was related chemically to "fertilisin" (see p. 214). The fertilisable life is also known to be extended by anaerobiosis or KCN. To these factors Whitaker[13] has now added 1 % ethyl alcohol, which extends the fertilisable life of *Urechis* eggs 300 %, while 1 % glucose extends it even longer. It seems likely that these substances act nutritionally, for it may well be that the yolk reserves are not accessible to the egg until after the colloidal changes occurring on fertilisation. Tyler, Ricci & Horowitz have shown that some of the rise in respiration can be accounted for by bacterial growth. Sterile conditions prolong the fertilisable life of eggs enormously, but the alcohol at Whitaker's concentration does not act by virtue of its bactericidal properties. Tyler & Dessel say that acid pH prolongs the life of gephyrean eggs. Here again, then, is a subject which needs much further elucidation.

An important advance was that made by Shapiro[4] in his studies on the respiratory rates of fragments of sea-urchin eggs (*Arbacia*) obtained by centrifugation (merogons; see Fig. 211). By strong centrifugal force the eggs were broken into two halves, the upper (light) one white, containing the nucleus, hyaline cytoplasm and oil-cap; the lower (heavy) one red because of the echinochrome granules, and containing most of the yolk. The light half is a little larger than the heavy half (vol. 116,750 cu.μ against 84,700 cu.μ). Manometric measurements of respiratory rate then gave the following figures:

μl. O$_2$/10 c.mm. cells/hr.

	Unfertilised	Fertilised
Light halves	1·23	3·49
Heavy halves	2·81	2·66
Whole eggs	1·49	3·85

Thus the unfertilised light half, containing nucleus but no pigment, respired at approximately the same rate as the whole egg, while the unfertilised heavy half, without nucleus but containing yolk and echinochrome, respired 88 % in excess of it. The combined activity of both halves, allowing for volumes, was thus 29 % greater than that of the intact egg. On fertilisation, only the light protoplasmic halves showed the well-known increase, and the combined activity of both halves after fertilisation was thus 17 % less than that of the intact egg.

These experiments are suggestive in that they associate the "damping" mechanism already described with the protoplasmic part of the egg. The heavy halves produced by centrifuging seem to have escaped from its influence. Examination of the action of inhibitors and stimulators of metabolism on the halves will be awaited with much interest. In the meantime Navez & Harvey have given a short report on the indophenol oxidase in fragmented eggs, from which it seems that there is twice as much of this fundamental enzyme in the heavy halves as in the light ones, as detected by the histochemical method. Other workers, however, find that the increase of respiration due to para-phenylenediamine is more marked in the light than in the heavy halves, (Boell, Chambers, Glancy & Stern).

We come now to a remarkable series of papers by Tyler on the energetics of differentiation, which show a consciousness of the most essential problem, namely the integration of metabolism with morphogenesis. In the first of these Tyler[2] compared the oxygen consumption of "half" and whole embryos of the sea-urchin, basing his argument on the view that when two blastomeres are separated and two identical but smaller embryos develop from an egg which would otherwise have produced only one, there is twice as much morphogenetic work going on for the same amount of energy sources (raw material) as normally. Using *Echinus microtuberculatus*, measurements of respiratory rate (per mgm. nitrogen) showed that it was identical in cultures of normal embryos and dwarf ones. But as has long been known (Driesch[2] for the sea-urchin; Spemann & Falkenberg for the newt; Morgan[6] for *Amphioxus*; Tyler[1] for *Chaetopterus*), dwarf embryos develop with a considerable delay. Though it has been classical to put this down to the exigencies of "regulation", Tyler suggested that it was really due to the necessity for the performance of twice the usual amount of morphogenetic work. In his own experiments the delay was of the order of 38 %. In order to reach a given stage of development, therefore, for example the end of gastrulation, the dwarf embryos actually use considerably more oxygen than the normal ones, i.e. two dwarf embryos require more than one normal one although the amount of living matter is the same.

Unfortunately these interesting experiments teach us relatively little about the metabolic cost of morphogenetic work, for the variables of maintenance and growth are not really ruled out.* In order to keep itself going as a physical system quite without any sort of differentiation, a cell has to expend energy, and since the time taken to reach the given morphological stage was longer, more maintenance must have been required.† And we do not know the indispensable datum, namely, what the respiratory rate would have been during the extra time required to reach the same stage, if the maintenance quota had been subtracted from it. There is also the factor of growth. This is not eliminated in experiments with echinoderm eggs, since no egg is absolutely "alecithic", and as the embryo develops, the yolk, however modest and transparent, will be converted into respiring protoplasm. Hence the rise in respiratory rate during echinoderm development *CE* (Fig. 111).

Nevertheless, Tyler's interpretation of regulation is acceptable enough. In a later paper[4] he showed that giant embryos obtained by fusion of normal eggs

---

* Moreover, the energy needed for the formation of new surfaces is usually grossly overrated. Borsook has calculated, using the equation

$$T\left(\frac{\partial Y}{\partial T}\right)_{\sigma} = Y - \left(\frac{\partial H}{\partial \sigma}\right)_{T},$$

where T, Y and σ refer to absolute temperature, surface tension and surface area, that to make 10^6 sq. cm. of surface needs only 3 gm. cals. even if Y is taken as high as for an air-water interface, i.e. 72 dynes/sq. cm., though for living cells it is more like 2 dynes/sq. cm. If the volume of the chick embryo is taken as 50 c.c. and all its cells assumed to be $\frac{1}{10}$th the volume of erythrocytes, then even if it were a solid cell-mass, the total surface would only be $2\cdot4 \times 10^5$ sq. cm. Yet the total energy turnover of the chick embryo during its development is some 23,000 gm. cals.

† This criticism does not seem to me to be met even in Tyler's recent, otherwise very instructive, monograph[7].

(cf. p. 103) develop even faster than normal-sized ones. This had previously been observed by Spemann & Wessel-Bautzmann on the newt.

In a third paper[4], Tyler studied the effect of temperature on cleavage rate and the rate of gastrulation and pluteus formation, in an attempt to dissociate cleavage and organogenesis, on the echinoderms *Dendraster*, *Lytechnis* and *Strongylocentrotus*, the ascidian *Ciona*, and the gephyrean worm *Urechis*. It proved impossible, however, to do so. With the same species an attempt to dissociate oxygen consumption and developmental rate (Tyler[5]) also failed, but the results are a striking confirmation and extension of the earlier ones of Ephrussi[4], namely that the total oxygen consumption of the eggs of invertebrates throughout their development is independent of the temperature. If membrane elevation is prevented in sea-urchin eggs, gastrulae are formed the walls of which have nearly twice the normal thickness, and there is some delay at gastrula and prism stages. Nevertheless, the rate of respiration was found to be exactly the same (Tyler[6]). But the same arguments apply in this case as in that first discussed above.

As additional proof of the linkage (obscure though its mechanism may be) between respiration and development, Tyler & Horowitz[2] compared the respiration of *Urechis* eggs normally fertilised with others parthenogenetically stimulated in such a way as either to produce cleavages or not. In all cases the respiratory rate rose with time, but more after fertilisation than after parthenogenesis, and more during parthenogenetic cleavage than non-cleavage. Urethane reduced the rate of the fertilised eggs to a rate more approaching the others.

The last question which arises is whether it is possible to distinguish any oscillations of oxygen consumption during the phases of cell-division itself. It has been usual to deny this (*CE*, p. 641), and scepticism on technical grounds is to be recommended, but Runnström[13] has brought forward some experiments which seem to show a periodic variation during the first three cleavages, and P. Reiss supports this by periodic kicks on the E_h readings of a platinum electrode in an egg suspension. Moreover, Trurnit, using delicate thermocouples adjacent to echinoderm, and actually *inside*, amphibian, blastomeres, claims to have found repeatable peaks of heat-production, the rise beginning when nuclear cleavage begins, and attaining its maximum just before the moment of cell cleavage.

The respiratory quotient of *Urechis* eggs during development has been measured by Horowitz[3]. It falls rapidly from unity at 2 hr. after fertilisation to 0·7 at 20 hr. This agrees with the succession of energy-sources seen in so many other organisms, but is remarkable in that the fall in R.Q. begins directly from fertilisation. It could not be arrested by adding glucose to the medium. This may be the place to mention a curious respiratory pigment found in the eggs of the same gephyrean worm by Horowitz[2]. "Urechrome" changes from pink to yellow reversibly on oxygenation, and its *r*H was measured. Since it occurs naturally in both forms its physiological function is particularly obscure (cf. the pigment mentioned on p. 210).

To complete this section we must glance at work done on some non-marine forms. Baldwin[3] has given us a careful study of the respiratory quotient during the development of a pulmonate gastropod, *Limnaea stagnalis*. The measurements were made, after shelling out the eggs from their capsules, by the Dickens-Šimer annular cup manometric method. The R.Q. continued at approximately unity throughout development, but since a synthesis of fat takes place in these forms (as was first shown in the remarkable pioneer experiments of F. G. Burdach in 1853 and confirmed by Baldwin), the catabolic utilisation of protein and fat which probably goes on, at any rate towards the end of development, was masked. The quotient is nevertheless of interest in view of the large stores of polysaccharides in molluscan eggs (see p. 29).

Another piece of work on molluscan eggs is that of Leiner[4], who found a very marked decrease in carbonic anhydrase activity during the development of *Aplysia*, although the respiratory rate was constantly rising. One may have the suspicion that this is connected with the formation of the alkali reserve, referred to below (p. 585).

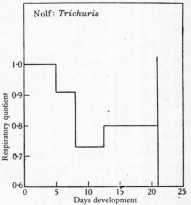

Fig. 309. Respiratory quotient during the development of the nematode embryo (*Trichuris trichiura*).

Additions to our knowledge of the metabolism of cestode eggs have been made by Friedheim & Baer. The metabolism of the adults, as is well known, is very peculiar, volatile fatty acids being given off as end-products (cf. the recent reviews of Krüger[3]). Friedheim & Baer studied the cestode *Diphyllobothrium latum*, which lives in the intestines of man, the dog, cat and fox. The eggs (see p. 30) develop free in water, and the first ciliated larva is transformed into a second larval form in the gut of the fresh-water crayfish. On passing along the food chain into fishes such as the pike, the worm lies inert in the muscles as a "plerocercoid", and at last is eaten by mammals. Investigation showed that the eggs are obligate aerobes though the larvae and adults are facultatively anaerobic. The following data are therefore of interest:

	Oxygen uptake	Q_{O_2}	$Q_L^{O_2}$	$Q_L^{N_2}$
Eggs (dry weight not measured)	23 c.mm./hr.	—	None	None
Plerocercoid	—	− 0·55	+0·28	+1·06
Adult	—	− 2·36	+1·75	+5·5
Adult+glucose	—	−15·01	—	—

The adult data agree with other observations on the tapeworm *Moniezia expansa* by Alt & Tischer. The most peculiar characteristic of the metabolism of the eggs, however, was that it was completely inhibited by KCN but not at all by CO — a unique case. That of the adults was inhibited 60 % by KCN but not by CO. No explanation has as yet been offered for this, but the phenomenon is probably general to cestodes, as the pike's tapeworm *Triaenophorus lucii* also shows it.

A nematode, *Trichuris trichiura*, has afforded data on the R.Q. throughout development in the work of Nolf. As Fig. 309 shows, it declines steadily. This is the first instance of several (p. 603) in which there is apparently a utilisation of carbohydrate preceding the utilisation of protein and fat.

On the eggs of the nematode *Ascaris megalocephala* (see p. 139) Huff & Boell discovered the remarkable fact that disorganisation of the cell by ultra-centrifuging causes an inhibition of the ferrous respiration while leaving the non-ferrous respiration unchanged. This is shown from the following figures:

	c.mm. O_2/hr./10^6 eggs	
	Alone	+KCN $10^{-3} M$
Control eggs	− 102·3	− 11·3
Centrifuged eggs	− 22·3	− 12·8

It will be seen that this is very significant in view of what has already been said about the association of non-ferrous respiration with the low metabolism of the unfertilised echinoderm egg, and what will later be adduced about its association with the low metabolism of orthopteran diapause. The idea is beginning to crystallise that ferrous, cyanide-sensitive, respiration is more connected with normal colloidal conditions and more closely linked with morphogenesis than non-ferrous respiration. The latter may support life, however, in "damped" periods or regions. All the eggs of Huff & Boell's experiments subsequently underwent normal cleavage.

In stages of the life cycle later than the embryo, the metabolic rate of invertebrates, though poikilothermic animals, falls (see *CE*, p. 746, and new confirmation by Davis & Slater on the earthworm, Chapheau on the oyster's hepatopancreas, and Terao on daphnids). The subject will be alluded to again later (p. 600).

3·32. Respiration of the eggs of Arthropods

On the embryos of crustacea there is only one piece of work to be recorded, in which measurements were made of the respiratory quotient during the development of the shore-crab *Carcinus moenas* (Needham[11]). In the cleavage states it is approximately unity, but by the time the yolk has diminished to four-fifths the diameter of the egg, it has fallen to 0·72, after which there is a slow rise to about 0·8 (see Fig. 310). Clearly over the greater part of the developmental period there is a combustion of protein and fat, with a predominant combustion of carbohydrate in the earliest stages. This is the second case of the succession of energy sources, carbohydrate-protein-fat, which we have noticed; another will shortly be adduced. Incidental to this work

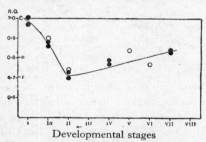

Fig. 310. Respiratory quotient during the development of the shore-crab embryo (*Carcinus moenas*). • at 37°; ○ at 15°. *C*, *P* and *F*, the theoretical quotients for normal combustion of carbohydrate, protein and fat.

was the interesting finding that normal respiration and normal metabolism (so far as the R.Q. shows it) would go on in the embryos of this crab at temperatures above the thermal death-point of the adult.

We come now to the main subject of this section, the advances which have been made in the study of respiration and metabolism in insect embryos. We owe in particular to the work of J. H. Bodine and his collaborators in Iowa one of the most comprehensive and elaborate accounts of the embryonic metabolism of an important group of animals existing in the literature. What follows applies mainly to the grasshopper *Melanoplus differentialis*, but many other species of orthoptera were studied by this school, and the same description, with minor modifications, holds good for all of them. The peculiarity which afforded the means for an unusual insight into embryonic metabolism was the "diapause"

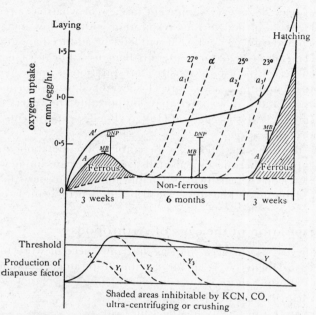

Fig. 311. Diagram illustrating the mechanisms of respiration and diapause in the grasshopper (orthopteran) embryo (see text). *MB*, Methylene blue; *DNP*, dinitrophenol.

After 3 weeks' development the egg of the grasshopper enters a dormant state, which may last for many months, and at the conclusion of this diapause, development proceeds as before, culminating in hatching after a second 3-week period. It is known that during diapause there is a complete cessation of mitosis, growth, yolk utilisation, and movement (Slifer[2]).

Bodine[3] and Burkholder first studied the respiration of the eggs during these periods. As shown diagrammatically in Fig. 311, the respiratory rate first rises to a maximum (curve *A*), then falls to a low and constant level during diapause, finally rising again very rapidly during the last three weeks. Cumulatively, respiration would follow a line such as *A'*. Now it had long been known that diapause could be artificially broken at any time by exposing the eggs to a temperature about freezing-point. On returning to room temperature they would then set about the concluding phases of their development and hatch in three weeks (as represented by curve α in the diagram). On the other hand, if the eggs were maintained at relatively high temperatures directly from the time of laying, then also they would complete their development in a shorter time (according to curves a_1, a_2, a_3); the duration of diapause, but not its existence, being dependent upon external temperature.* Bodine[5] tried to explain these facts by the assumption of a substance which inhibits development on attaining a certain threshold (curve X in Fig. 311). The concentration of this substance at any given time would be the resultant of processes of formation and destruction. Exposure to 0° from the time of laying prevents diapause altogether. We may say, therefore, that the destructive process has been differentially favoured at the expense of the formative process, and so no threshold concentration ever arises (curve Y_1 on the diagram). On the other hand, exposure to 0° at a later date, after diapause has set in (i.e. the threshold concentration has been reached) will break the diapause; here again the destructive process must be less susceptible to cold (have a lesser temperature coefficient) than the formative one and consequently the concentration of the substance will fall below the threshold (curves Y_2 and Y_3 on the diagram) and development will proceed as soon as room temperature is regained. Such an explanation follows the general lines of Goldschmidt's thought (see p. 400). It would be satisfying if some definite evidence could be found, e.g. the control of diapause by injection of extracts of other embryos.†

It should be emphasised at this point that there are no morphological differences between embryos in late pre-diapause, diapause, and early post-diapause; they can only be distinguished by their rate of oxygen-consumption.‡ At all stages the oxygen consumption can be reversibly depressed by exposure of the embryo to hypertonic solutions (Bodine [4,5]; Thompson & Bodine[5]; Bodine & Thompson; Schlipper); on return to isosmotic conditions, the respiration returns to a level low or high depending on the original stage of the embryo. Diapause block can therefore not be affected by this means.§

A discovery of real importance was made when Bodine[7] and Bodine & Boell[1,2] showed that the respiration of the grasshopper egg was sensitive to KCN and CO only when in the actively developing stages, and furthermore that the fraction of respiration insensitive to KCN and CO in these stages was of approximately the same order as the whole of the respiration during diapause. These facts are shown by the hatched areas on Fig. 311. Bodine & Boell[2] drew attention to the parallel between this case and that of the unfertilised and fertilised echinoderm egg (see also p. 199). There is thus a non-ferrous respiration (see diagram on p. 578) running

---

* This is the distinction between diapause and hibernation. Hibernation is caused by low temperature; it is not inherent, as shown by Bodine[1] on the nymphs of another grasshopper, *Chortophaga viridifasciata*. True diapause exists, between laying and the initiation of development, in many other invertebrate groups, such as daphnids (where distinct races exist, one showing it and another not; T. R. Wood) and brine-shrimps.

† Earlier theories postulating intoxication by metabolic waste-products may be disregarded.

‡ E.g. the following figures taken from the later work of Bodine & Boell[9]; for an embryo weighing 0·067 mgm. Q_{O_2} is in pre-diapause $-1·52$, in diapause $-0·59$, and in post-diapause $-1·47$.

§ The differences between the temperature coefficients of diapause and developing embryo respiration (Goodrich & Bodine) are not very marked.

throughout development in the orthoptera, and in addition a respiration of ferrous type exclusively linked with active morphogenesis. Bodine & Boell[2] also found that during diapause there is a certain stimulation of respiration by CO, analogous to the Örström effect on the unfertilised echinoderm egg (p. 496). Inhibitions by KCN cause delays in hatching time (Robbie, Boell & Bodine).*

From inhibitors they turned to stimulators with equally satisfactory results. Before dealing with these, however, it will be convenient to refer to Boell's study of the normal R.Q. during the development of the grasshopper. This is shown in Fig. 312. Beginning at just under unity, it falls to a minimum at 0·6 during the pre-diapause period but is stabilised at 0·7 before diapause has started. This value is maintained, apart from slight oscillations at the beginning of the post-diapause period, till hatching. Thus we have a third case suggesting the carbohydrate-protein-fat succession of energy sources. Some 75 % of the total oxygen consumed corresponds with the loss of fat actually found by Slifer[1,4].

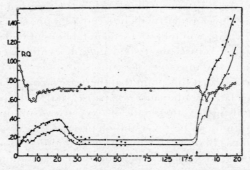

Fig. 312. Respiratory quotient during the development of the grasshopper embryo (*Melanoplus differentialis*). ● O₂ uptake (c.mm./egg/hr.); × CO₂ production (c.mm./egg./hr.); ○ R.Q. The arrow indicates the end of diapause.

The action of methylene blue (Bodine & Boell[8,10]) differs according to the stage at which it is applied. During the actively developing pre-diapause the stimulation of respiration is 2 %, during diapause it is 101 %, and during post-diapause it is 24 %. These results, indicated in Fig. 311, show, as would be expected, that it is possible to side-track the Warburg-Keilin system far more effectively when it is out of action during diapause than at other times. It should also be noticed that just as in other cases (pp. 199, 568) the final level is much the same since the stimulating action is much greater on the phase originally damped.† All these effects hold good for the isolated embryos as well as the intact eggs, generally speaking (Bodine & Boell[9]). Bodine & Boell[11] also studied the effects of 2 : 4-dinitrophenol and 3 : 5-dinitro-*o*-cresol. Here again the stimulation was one of 275 % on diapause embryos and of 150 % on developing embryos (see Fig. 311),

---

* Purely morphological effects of KCN on beetle embryos have been described by Brauer.

† In later papers Bodine & Boell[10] have stated that the respiration stimulated by methylene blue is sensitive to CO though not to KCN. This very odd finding requires elucidation (cf. the work of Friedheim & Baer above, p. 576).

bringing the final levels rather close. As in other cases (p. 571) the excess respiration was fully sensitive to KCN and CO. This demonstrated that the Warburg-Keilin system *can* be made to function during diapause, and suggests that it is inactivated either by colloidal changes or by some inhibition which the nitrophenols can overcome.* Bodine & Boell[11] then noted that the R.Q. of the nitrophenol-stimulated respiration was different from the normal; instead of being 0·7 it was between 0·91 and unity. Investigating this phenomenon they were able to prove that since neither iodoacetate nor fluoride would inhibit it, and since considerably more ammonia was produced by the embryos in presence of nitrophenols than normally, it was due to the catabolism of protein ending in ammonia. This was a case parallel, therefore, to the fluoride-inhibited chick embryo observed by Needham[8] (see p. 592). And it pointed to the interesting conclusion that although the grasshopper embryo during the latter part of its development is combusting fat, it is nevertheless quite capable of combusting protein if circumstances make it necessary that it should do so. In another paper, Bodine & Boell[7] further showed that post-diapause embryos *in vitro* can still utilise added glucose.

Another instructive line of attack employed by Bodine & Boell[5] was ultra-centrifugation. By means of the air ultra-centrifuge the grasshopper egg can be stratified into lipoidal, protoplasmic and proteinaceous layers, and if the rotational force does not exceed 20,000 g. recovery and development are possible. Now the important fact emerged that although in pre-diapause or post-diapause eggs forces of only 1300 g. will decrease the respiratory rate considerably and forces of 20,000 g. will lower it to about 50 % of its normal value (i.e. nearly to the level of the non-ferrous fraction), forces of even 400,000 g. have no effect on the respiration of diapause eggs. The ferrous fraction is therefore sensitive to cell disorganisation while the non-ferrous fraction is not. We have already seen a similar case in the experiments of Huff & Boell on the nematode egg (p. 577). On the other hand, X-rays (Boell, Ray & Bodine) did not differentiate between the various forms of respiration. At the doses used, diapause respiration was unaffected, and developing embryo respiration was only decreased in so far as the development was slowed.

Among other valuable contributions on this subject the following may be noted. The yolk (containing as it does many nuclei and some cells) cannot be overlooked as a seat of respiration. During pre-diapause and diapause the oxygen consumption of the whole egg is double and sometimes more than double that of the embryo alone (Bodine & Boell[7]); this corresponds to the extra-embryonic membrane respiration of amniotes. Moreover, at the onset of post-diapause, the increase in oxygen consumption occurs first in the yolk-cells. Tyrosinase, followed through development (Bodine & Boell[4]), rises to a maximum of activity during pre-diapause and remains there throughout the two succeeding periods. It is mostly contained in the yolk. Yet the naturally occurring substrate appears only in the post-diapause period (Bodine, Allen & Boell). By centrifugation, an acti-

---

* On the other hand pyocyanin, studied by Carlson & Bodine, gives 100 % stimulation of respiration, whether in diapause or not.

vator, the action of which could be imitated by sodium oleate, is removed from the enzyme (Bodine & Allen; Ray & Bodine[2]). During pre-diapause there is a great decrease in the potency of the naturally occurring lipidic protyrosinase activator (Bodine, Ray, Allen & Carlson; Bodine & Carlson[1]). Other properties of the enzyme have been studied by Allen & Bodine[1], and Bodine, Carlson & Ray state that in eggs X-rayed so as to destroy the embryo but not the yolk and serosa cells, protyrosinase and its activator are formed much the same as normally. X-rays cannot substitute for the activator (Ray), but a number of other agents can, such as heat (63–77°), detergents, acetone, urea, urethane, and reduction of salt concentration by dialysis. Denaturation of the enzyme precursor therefore enters the picture (Allen & Bodine[2]).* The precursor has a different ultra-violet spectrophotograph from the enzyme (Bodine & Carlson[2]).

Indophenol oxidase activity was measured by Bodine & Boell[6] by determining the excess respiration of egg brei in the presence of dimethyl-p-phenylene-diamine. Actually this should measure the activity of the whole Warburg-Keilin system (see diagram on p.565). Where, as in egg-brei, the structure of the cells has been destroyed, the respiration was insensitive to KCN and CO, a further proof of the dependence of the Warburg-Keilin system on integrity of cell structure. Egg-brei from diapause eggs, moreover, respires at the same rate as the intact eggs. Nevertheless, when optimal amounts of this activated substrate were added, there was a considerable extra respiration quite cyanide-sensitive. A view of the activity of the Warburg-Keilin system, obtained in this and other ways (T. H. Allen), showed that it resembles the tyrosinase curve in rising somewhat to the end of pre-diapause and then continuing unaltered until post-diapause. But in post-diapause it rises considerably until hatching. These experiments demonstrate that the intrinsic activity of the Warburg-Keilin system is not diminished during diapause, but only in some way inactivated. It is puzzling, however, that no cytochrome c is spectroscopically visible in the embryos.

"Peroxidase" (by the guaiacum test) is absent during pre-diapause, gradually increases during diapause and rapidly during post-diapause (Bodine & Boell[3]). But the enzyme is always in the embryo and not in the yolk. Glutathione (qualitative tests) is equally present throughout development (Evans[1]). Bodine & Wolkin determined the total iron in yolk and embryo; it passes regularly from the former to the latter, and the respiratory rate cannot be dependent on it, for embryos with the same total iron concentration show entirely different oxygen consumptions in accordance with their developmental stage. The relation between oxygen tension and oxygen consumption does not vary with the stage (Bodine[6]); the curve is very similar to that already described for the eggs of echinoderms.† During diapause, respiration and development are less susceptible to ultraviolet light (Ray & Bodine[1]) or X-rays (Bodine & Evans[2]; Evans[2]) than at other times.

---

* One cannot help drawing attention here to the similarity between this case and the liberation of the primary inductor in the dorsal lip of the blastopore.

† Rhythmic pulsations of the body-wall, possibly connected with respiration, and later merged in the heart-beat, occur in orthoptera. Extensive studies have been made on their reaction to temperature (Thompson[1]), electrolytes (Thompson[2]), CO_2 (J. F. Walker), etc.

By way of commentary on the above account it may be said that diapause may occur at any stage of the insect's life cycle* and that exposure to low temperatures is by no means the only stimulus which can break it. There is a wealth of entomological literature (available through the reviews of Uvarov; Roubaud; Cousin and others) which supports this. For example, diapause may occur in nymphal stages, as in the bug *Reduvius* (Readio) and in some grasshoppers, such as *Chortophaga viridifasciata* (Bodine[1]), or in wasps such as the mud-dauber, *Sceliphron caementarium* (Bodine & Evans[1]). What has been said about depression of respiratory rate applies, *mutatis mutandis*, to these cases. In the work of Heller[3] on the hawkmoth, *Deilephila* (p. 476) we have already seen an example of diapause occurring during metamorphosis. Diapause can be broken by many agencies, e.g. cold (see above, also Flemion & Hartzell; Burdick; Bodine & Robbie); friction, electrical or mechanical shocks (Roubaud); various chemicals such as xylol and carbon tetrachloride (Pepper); and parasitism (Salt; Varley & Butler). That it is genetically conditioned in silkworms we know from the work of Uda and Umeya; the characters of the embryo as regards diapause are a case of "maternal inheritance".

The silkworm, *Bombyx mori*, formerly a classical object of research, has received less physiological study than the grasshoppers. Curves for respiratory rate in diapause (univoltine) and non-diapause (bivoltine) races, similar to those in Fig. 311, have, however, been established for the moth *Lymantria monacha* by Tuleschkov and the silkworm by Ashbel[2]. Ashbel[7] finds that in this form also the duration of the diapause depends on the temperature. Thyroxin, added to egg-powders prepared from embryos at different stages, has a much larger effect upon the diapause material than upon that from either pre-diapause or post-diapause (Ashbel[5]). Neither X-rays nor radium emanation affect the length of the diapause (Ashbel[6]). Changes in glutathione (Teodoro; Demjanovsky) are not yet clear and the extremely deviant R.Q.'s of Ongaro[2] may be disregarded. Total oxygen consumption after various forms of parthenogenesis by strong acids[†] in silkworms have been measured by J. Fukuda; the differences were not great (cf. the work of Tyler mentioned on p. 575). Wolsky[4] states that this respiration is insensitive to CO. Kozhanchikov[1] reports that the larvae of the pyralid moth *Loxostege sticticalis* are especially resistant to anaerobiosis during their diapause, and pile up an oxygen debt which is later discharged at the expense of fat oxidation.

For accounts of specialised mechanisms for respiration in insect eggs the reader must be referred to the entomological literature. We may mention two or three, however. The egg of *Oöencyrtus johnsoni*, an encyrtid parasite on the eggs of the cabbage bug *Murgantia histrionica*, has a special "aeroscopic plate" on its

* The insects are not the only group of animals with embryonic diapause. Something remarkably like it occurs in mammals. Hamlett[3] writes: "We know of a dozen mammals whose period of gestation is regularly from 2 to 6 months longer than actual embryonic development", e.g. certain badgers, bears, shrews, etc. and the stonemarten and pinemartin. The diapause occurs at the blastocyst stage preceding implantation (see p. 84). We have already seen (p. 78) that the blastocyst stage may be unusually prolonged in most mammals if lactation is occurring. Hamlett gives a good discussion of the endocrine mechanism (progesterone) and the biological significance of mammalian "diapause".

† Tirelli[7] gives an account of Tichomirov's strange discovery, which preceded the work of J. Loeb, that parthenogenesis in the silkworm may be brought about by treatment with strong acids such as sulphuric or hydrochloric.

case to allow of direct respiration, avoiding diffusion through the shell of the host egg (Maple). Later, the encyrtid larvae become attached to the host's tracheal system (Thorpe[1]). The larvae of *Cryptochaetum grandicorne*, parasitic in the haemocoele of the coccid *Guerinia serratulae*, have remarkable caudal filaments to aid in the absorption of oxygen and nutriment from the host (Thorpe[2]).

3·33. Respiration of the eggs of Fishes

Teleostean fishes alone lend themselves to experimental attack on embryonic respiration and heat production. From the morphological details already given (p. 326) it will be remembered that their eggs are highly lecithic and meroblastic, cleavage being confined to a small germinal area, and the blastoderm then growing round the yolk so as to enclose it in a yolk-sac, while at the same time the embryonic axis is laid down. Observations have been made mostly on freshwater forms such as *Salmo irideus* and *Perca fluviatilis*, besides the American marine minnow or killifish *Fundulus heteroclitus*.

As would be expected from the growth of the embryo at the expense of inert yolk, the oxygen consumption of the whole egg rises regularly throughout development (Schlenk and S. Smith[1] on *Salmo*; Trifonova[3]* on *Perca*; Amberson & Armstrong and Phillips[1] on *Fundulus*; Bežler on cyprinids). In postnatal life metabolic rate falls, as in all animals (Raffy on eels; Keys[1] and N. A. Wells on the killifish). More interesting is the respiratory quotient.[†] Here quite independently two pieces of work (Schlenk and Amberson & Armstrong) have given strong indications of an R.Q. of unity at the beginning of development,

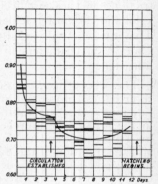

Fig. 313. Respiratory quotient during the development of the minnow embryo (*Fundulus heteroclitus*).

followed by a value of about 0·7 for the major part of the period, although the development of *Salmo* studied by Schlenk takes (till hatching) a little over 40 days, and that of *Fundulus*, studied by Amberson & Armstrong, is complete in 12. The latter set of data is shown in Fig. 313. Teleostean fishes therefore constitute yet another case of an apparent succession of energy sources, carbohydrate preceding protein and protein preceding fat.

The latter half of this process, at any rate, was made clearly visible in S. Smith's work[1], where separate determinations of protein nitrogen and fat were made. The fat (% dry wt.) rises in the yolk (whether by synthesis or as a result of protein utilisation) from 10 at 0 days to 30 at the 80th day, remaining constant in the embryonic tissues themselves from the 30th day on. Total fat drops by 51 % of the initial value (cf. Table 11). Protein drops also, by 49·3 % of the initial value (though 32·5 % is a better figure, as there may have been some starvation after exhaustion of yolk-sac reserves), but in absolute amount, this is a much greater

---

* We may leave out of account the oscillations in the curves of this author, to which undue significance seems to be attributed by her.

† Krüger[2] has designed an apparatus especially for determining R.Q.'s of fish embryos.

loss than that of the fat. The noteworthy point is, however, that fat utilisation does not begin until the 56th day, and goes on till the 75th, while protein utilisation has already become considerable by the 40th.

Careful calorimetric estimations of heat production were also made throughout development. Like oxygen consumption, heat production rises regularly, from about 1 g. cal./100 eggs/hr. to between 30 and 40. The full publication of the data, with a comparison between heat found and heat expected, analogous to the well-known Bohr-Hasselbalch balance-sheet for the chick (*CE*, p. 706), will be awaited with interest.

Trifonova's work[3] on the development of the perch added some interesting points to the above account. The greatest sensitivity of the embryos to anaerobiosis was found to be before segmentation had begun, and at this time also there were maximal amounts of lactic acid in the eggs. Lactic acid declined from an initial high level to a final low one, suggesting an aerobic glycolysis at the earliest stages of development (cf. below, p. 594). Yet intracellular pH, as judged by neutral red, becomes more acid (Trifonova[2]). Attempts to correlate respiratory rates with growth rates at different periods of development cannot, however, be regarded as successful (Trifonova[3]; Privolniev[1]). Privolniev[2] claims to have found rhythms of respiratory rate during early cleavages of the egg of the lamprey (*Lampetra fluviatilis*) — not a meroblastic egg — and these may well be justified, in view of the experiences of other authors, e.g. J. Brachet[4,7] (see below, p. 587), with amphibian eggs.

Fish eggs differ considerably in their resistance to anaerobiosis. It was a classical observation of J. Loeb's that the minnow (*Fundulus*) is very resistant while the cunner (*Ctenolabrus*) is not (confirmed for other eggs, e.g., by Trifonova & Popov using various partial pressures of oxygen). Philips and Vernidub find that this is correlated with their cyanide-sensitivity, though whether development proceeds or not, the respiration falls by 80–90 %. *Fundulus* eggs may therefore possess, like frog eggs (see on, p. 588), the power of piling up an oxygen-debt, while those of other teleosts (cunner, scup, mackerel) do not.

Amberson & Armstrong and Manery, Warbritton & Irving almost simultaneously discovered that an important alkali reserve is formed as teleostean eggs develop. In *Fundulus*, the first form investigated, this increase in bicarbonate amounts to 9 c.c. CO_2/100 gm. eggs in the 12 days of its development. It was thought that there must be a loss of chloride from the eggs to compensate for the new anion and the increase in buffering power, but when the phenomenon was confirmed (by Irving & Manery[1]) on eggs of the trout *Savelinus fontinalis*, it was at the same time found that there is a loss of chloride to the exterior far more than balancing the gain in bicarbonate, indeed seven times its size.* This unexpected unbalance is probably met partly by the new anion of inorganic phosphate arising from phosphatide and phosphoprotein breakdown, and partly by protein on the alkaline side of its isoelectric point. The subject may be further followed in the review of Irving & Manery[2]. Retention of respiratory CO_2 for purposes of

---

* Loss of chloride has also been reported for eggs of invertebrates, such as the gastropod *Hemifusus tuba*, studied by Kumon.

alkali reserve is probably a general phenomenon. It fits in with the marked bradyauxesis of chloride which we have already noted (p. 548, Table 30), and with facts such as those collected by Surber on the optimal CO_2-content of water in which fish eggs are developing.

We come lastly to a special case of much interest, that of the embryos of syngnathid teleosts developing in the paternal brood-pouch, to which Leiner[3] has devoted much study. He first found[1] that the respiration of the embryos of the sea-horse, *Hippocampus brevirostris*, rose gradually to a maximum and then fell off, as follows:

	Q_{O_2}
Embryos from the brood-pouch:	
stage 1	−0·50
stage 2	−1·00
stages 4 and 5	−1·60
at "birth"	−2·84
Young fish some weeks old	−1·54
Adult fish	−0·66
Adult fish with embryos in brood-pouch	−0·85

The rise must represent the disappearance of inert yolk, and the fall the slowing of respiratory rate inherent in all tissues (cf. *CE*, p. 746 and below, p. 601). Leiner also noted a distinct aerobic glycolysis in brood-pouch embryos ($Q_L^{O_2} + 0·53$ for the snake-fish *Nerophis ophidion* and $+ 1·71$ for the sea-horse). But it was now observed[2] that *in vitro* in air the sea-horse embryos respired a good deal less intensely than they did (by calculation of difference between "pregnant" and "non-pregnant" males) in the brood-pouch. Thinking that this might be associated with the fact that the Fe content of the brood-pouch liquid was 30–50 times as high as that of sea-water, Leiner tested the *in vitro* respiration of the embryos in the presence of different concentrations of iron, and found that in this way the respiration of the embryos could be increased 100 %. The effect passes off in the later stages of development. Thus at stage 2*a* the increase may be 450 %, while at stage 5*b* it may be only 20 %. The respiration of the embryos is very sensitive to CO_2 tension and osmotic pressure; the former naturally increases some four times in the brood-pouch liquid during development, and the latter doubles. If these experiments should be confirmed they will establish one of the most remarkable known cases of dependence upon the parent organism, and show that although we may have information in the case of an ovoviviparous animal whether the embryo gains or loses in organic substance during its development within the parental body, much of the interactions between the two generations may still be hidden from us. For in *Hippocampus*, as has already been mentioned (p. 47), the embryo is furnished with all the necessary organic substance by the maternal ovary and takes none from the paternal body.

Another interesting adaptation occurs in the case of the Dipnoi. The eggs of the lung-fish, *Lepidosiren paradoxa*, develop in burrows where the water contains no measurable dissolved oxygen (Carter & Beadle), but the male fish which guards the nest has long vascular filaments on its pelvic fins during the breeding season, and J. T. Cunningham thought that these secreted oxygen into the water around the eggs. In order to test the matter, Cunningham & Reid[1] made a special journey to the Amazon and carried out experiments in the field which showed that the oxygen-content of the water around a fish in such a condition can indeed rise. The emissive function was contested by Foxon, but the advantage of the argument seems on balance to remain with Cunningham & Reid[2].

3·34. Respiration of the eggs of Amphibia

It has been unavoidably necessary to include much of the more interesting work on the respiratory metabolism of amphibian embryos in Part 2 (pp. 193 ff.), as it has so intimate a connection with the physiology of the organiser centre. Here only those studies which concern the metabolism of the embryo and yolk as a whole will be considered.

In the oviduct the amphibian egg discharges its large load of CO_2 (Dalcq, Pasteels & Brachet, confirming older work), and not until a certain level has been reached does the egg become fertilisable. It seems fairly well established that there is no change in respiratory rate at fertilisation (J. Brachet[7]; Stefanelli[1]), and evidence, not unconvincing, has been brought forward showing that there are cyclical variations of oxygen consumption during cleavage (J. Brachet[4,7]; Stefanelli[1]), but to ascertain exactly how they link up with the cytological changes will need further investigations (cf. p. 575). Should polyspermy occur, oxygen consumption after fertilisation may be doubled (J. Brachet[7]).

As would be expected from the growth of the embryonic protoplasm at the expense of inert yolk, the oxygen consumption of the whole egg rises regularly from fertilisation until the end of the tadpole period (J. Brachet[6] on *Rana temporaria*; Stefanelli[2] on the toads *Bufo vulgaris* and *viridis*; Atlas[2] on *Rana pipiens* and *sylvatica*; Wills on the urodeles *Triturus torosus* and *Amblystoma maculatum* and *tigrinum*). Following the respiratory rate until metamorphosis (see p. 453), Wills observed in every case a decline in metabolic rate as soon as the yolk was exhausted. This phenomenon is universal in growing organisms (see below, p. 601) and will be met with again in the discussion of birds and mammals. It is shown with isolated amphibian tissues such as skin (Börnstein & Klee).

The determination of respiratory quotient has been particularly interesting. In a careful study, J. Brachet found that it rose from a low level after fertilisation to a carbohydrate level at gastrulation:

Day	Stage	R.Q.
	Morula	0·66
	Advanced blastula	0·70
1	Gastrula	1·03
2	Neurula	0·98
5	Hatching tadpole	0·97

We have already noted that these values were confirmed subsequently in the work on the R.Q. of the regions of the gastrula by Boell, Koch & Needham. That the quotient in the post-hatching yolk-sac stage must be lower we know from the fact that some fat is combusted during that period. For example, Atlas[2] found that about 250 c.mm. O_2 are required per individual of *R. pipiens* from fertilisation to the end of the yolk-sac period, during which time 0·2 mgm. dry weight are lost.

Were the whole of embryonic and pre-feeding larval life at the expense of carbo-hydrate this figure would have been 149, or for protein 191 and for fat 400. At first sight, therefore, the situation in the amphibia seems to contradict the generalisation that a succession of energy sources occurs during development, carbohydrate preceding protein and protein preceding fat.* But it may be that this succession only starts from gastrulation. In the examples given for fish embryos (p. 584), no R.Q. for pre-gastrulation stages was available. And we have seen some evidence in the more recent work on echinoderm eggs (p. 564) that there also pre-gastrulation R.Q.'s may be low. In view of facts now to be men-tioned regarding the metabolism of amphibian cleavage stages, it may be suspected that low R.Q.'s of pre-gastrulation stages are due rather to special processes of unusual type than to the catabolism of fat or protein.

In the first place it may now be regarded as certain that amphibian embryos in the cleavage stages are exceptionally resistant to anaerobiosis. J. Brachet[6] found that under strictly anaerobic conditions cleavage and the formation of the blastula will proceed in the frog, though gastrulation is incomplete and neurula-tion impossible. The toad *Discoglossus*, however, he found able to gastrulate and neurulate anaerobically. Spirito[5] later observed marked differences between amphibian species as to their power of withstanding anaerobiosis; toads such as *Bufo* being in general able to develop further than frogs in the absence of oxygen. Successful attainment of advanced tail-bud stages has even been reported, by Lallemand. J. Brachet's lactic acid estimations[6] showed the existence of an anaerobic glycolysis, rising with development, and later confirmed by Lenner-strand and on isolated regions of gastrulae by Boell, Needham & Rogers. Under anaerobic conditions glycogen disappears (J. Brachet[6]). Brachet[6] was also able to show experimentally that under these conditions the embryos accumulate an oxygen debt, discharged by an excess oxygen consumption when they are brought out into air or oxygen. As the R.Q. of this excess oxygen consumption is very low (0·3) oxygen must be used to build up an "oxygen reserve", which seems, however, not to be in the form of oxidised cytochrome or glutathione. Anaero-bically CO_2 is certainly produced (J. Brachet[6]; Latinik-Vetulani[2]).

J. Brachet's work[6] on the action of inhibitors revealed that respiration is much more sensitive to KCN than is cleavage (a parallel with anaerobic conditions). At $M/1000$, while segmentation is normal (though gastrulation and neurulation cannot proceed), respiration has been inhibited 90 %. At $M/5000$ even gastrula-tion is possible. The converse effect is seen with phenyl-urethane. This narcotic stops segmentation instantly at $M/1000$ but only reduces respiration 40 % (a parallel with the work of Warburg on echinoderm eggs already referred to, p. 517). Among other facts brought to light by Brachet was the slight aerobic glycolysis shown by amphibian eggs, confirmed by Lennerstrand and (on isolated regions) by Needham, Rogers & Shen. Addition of dimethyl-*p*-phenylene-diamine to the eggs showed a very active Warburg-Keilin system, and tests for the dehydrogenases showed them also to be active. Changes on fertilisation in

* Hutchens has drawn attention to the fact that a similarly declining R.Q. curve is found in the cyclical development of *Chilomonas* cultures growing on acetate and inorganic salts.

this respect have not been found. Strong centrifuging does not affect the respiration of gastrulae, but on crushing gastrulae to a brei there is an enormous, though transient, increase of respiration. This effect does not persist during development; on crushing tadpole tissues, the respiration falls by 50 %. All in all, the most striking speciality of amphibian embryos seems to be their "oxygen store" and resistance to anaerobiosis.

In connection with this, it is interesting that Buchanan has found a special compensatory acceleration possible in amphibian development. Following exposure to low temperatures, or, significantly, to KCN, in moderate concentrations, the treated embryos would develop more rapidly than the controls, eventually attaining equality with them. Thus acceleration of developmental rate may accompany the recovery from KCN inhibition of oxidations. Such experiments should be tried with anaerobiosis also. They are interesting in view of the developmental accelerations and retardations which we have already seen in the transplantation experiments of Harrison and his colleagues (p. 346). With 2:4-dinitrophenol no acceleration could be obtained (cf. p. 454).

Temperature cannot be used to dissociate developmental rate from respiratory rate in amphibian embryos any more than it can in those of echinoderms (Atlas[1], parallel with Tyler[5], p. 575). Amphibian embryo respiration was found by Duryee to be sensitive to external osmotic pressure, being increased 62 % in Ringer and reduced 21 % in distilled water. Slight acceleratory effects of radium emanation on amphibian embryo respiration have been reported by Simon[2].

After the appearance of haemoglobin (see p. 644) the circulatory system of the tadpole begins to play an important part in its respiratory exchanges. As the result of a train of thought which began in the study of the respiration of mammalian embryos (see p. 598), it was discovered that the haemoglobin present in foetal life is by no means the same as that of the adult or of the pregnant organism. The amphibia are the lowest group in which this fact has, up to the present, been verified. Working on the blood of the bullfrog *Rana catesbiana*, McCutcheon[2] was able to show that the dissociation curves of the haemoglobin change markedly as development proceeds. Fig. 314, taken from his paper, illustrates this. The dissociation curve, in the most general terms, moves from left to right and changes from a rectangular hyperbola to a sigmoid curve, so that in the youngest tadpoles the loading capacity is very high, while in the adult it is low. The two forms of haemoglobin (if there are only two) also differ in their reactions to pH. In sum, it appears that the haemoglobin of the youngest stages is more adapted for conditions of anoxaemia (presumably due here to relatively inefficient respiratory mechanisms and to poor environmental oxygenation) than that of adults.

This ontogenetic change seems to be paralleled by certain phylogenetic changes, for McCutcheon & Hall found that species of amphibia differ in the characteristics of their haemoglobin dissociation curves, some approximating to the "embryonic type", others to that of the adult. In this there would be adaptive significance (cf. below, p. 595). Similar adaptive displacements of the dissociation curves of haemoglobin have also been found in fishes and correlated with their oecology by Hall & McCutcheon. Thinking that the forms of haemoglobin in the developing amphibian might derive from different sites of origin (of which amphibia have three; spleen, kidney and bone-marrow), McCutcheon[2] splenectomised bullfrogs, with the result that the dissociation curve of the blood sank. Conversely that of spleen blood was unusually high. But it may be doubted whether a given site of origin is limited to the making of one form of haemoglobin. For the cytology of the blood cell generations, see Cameron[2].

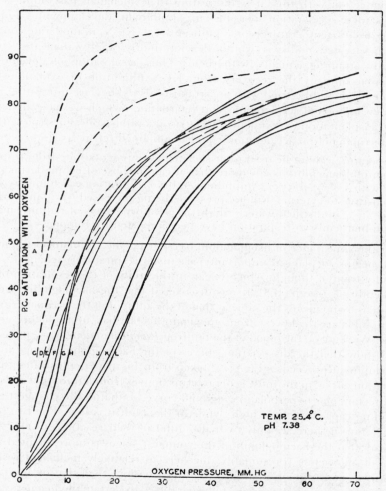

Fig. 314. Changes during the life history of the bullfrog in the dissociation curve of its haemoglobin. *A*, Tadpole 25 mm.; *B*, tadpole 92 mm.; *C*, legs well developed; *D*, respiratory method optional; *E*, tailstub still present; *F*, *G*, metamorphosis complete; *H–L*, adult.

3·35. Respiration of the eggs of Sauropsida

Extremely little attention has been paid to the embryonic respiration of reptiles, the only contribution being that of Zarrow & Pomerat on the eggs of the smooth green snake *Liopeltis vernalis*. Towards the end of development an R.Q. of 0·82 was obtained; after hatching it rose to 0·91 — an effect parallel to that seen with the chick (Fig. 315).

It is interesting that in later stages of the life cycle the metabolic rate (oxygen consumption per unit weight) of poikilothermic vertebrates steadily declines (Kestner; Terroine & Delpech; Terroine, Hée, Roche & Roche; C. M. Liang). Since the surface law can hardly here be important, Kestner suggests that the decline really signifies a decline in the proportion of actively metabolising protoplasm at the expense of "paraplasmatic" reserves and deposits. Senescence in post-developmental stages would thus be, in a sense, the converse of that process whereby inert yolk gives place to respiring protoplasm during embryonic life (see p. 600).

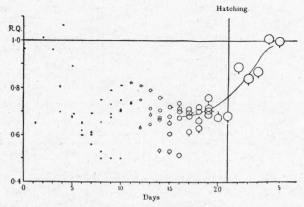

Fig. 315. Respiratory quotient during the development of the chick embryo (*Gallus domesticus*). The more accurate the determination the larger the circle. The sizes of the circles have been obtained by putting $d = w_1/w_2$ where d = diam. of circle, w_1 = weight of living matter in the system, w_2 = weight of inert matter including the shell. ♀ Lussana. o Bohr & Hasselbalch. ○ Hasselbalch. ○- Murray. (All on intact eggs.)

There is no new work on the respiratory exchange of the hen's egg during development, and that described in *CE* (p. 693) must be regarded as classical.* The R.Q. was never easy to obtain owing to the large volume of inert material in relation to the respiring protoplasm, especially in the early stages. Nevertheless, the tendency to obtain high quotients in the neighbourhood of unity during the first week (see Fig. 315) has shown itself again in two new pieces of work along the old lines (Hudiwara and Fukahori).

Investigators have turned more and more, in view of these facts, to manometric methods, for which the chick embryo is suitable up to the 6th day of development. The extra-embryonic membranes are suitable, in view of their thinness favouring diffusion, up to much later periods of incubation. By fixing isolated

---

* Nevertheless Noyons & de Hasselle have in progress a study of the heat production and CO_2 output of the hen's egg throughout its development with new and extremely sensitive apparatus, which is said to distinguish between fertile and infertile eggs. They have already published a preliminary metabolic rate curve.

blastoderms of the first two days to silk discs with a plasma clot their R.Q. also could be obtained. As is shown in Fig. 316 (Needham[7]), the R.Q. of these blastoderms is mainly of a carbohydrate character, but if the *area opaca* alone is used, they fall between 0·7 and 0·8. From the 3rd to the 6th day onwards the R.Q. of the embryo is invariably 1·0, as was also found by Dickens & Šimer[2,4]. In the absence of added glucose, respiratory rate is very little affected, but the R.Q. falls to about 0·85 (cf. the action of fluoride mentioned below). The R.Q. of

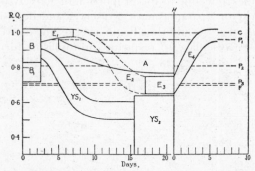

Fig. 316. Respiratory **quotient of embryo and the separate embryonic** membranes during the development of the chick embryo. *H*, Hatching. Theoretical quotients: *C*, carbohydrate; P_1, protein, ending in ammonia; P_2, protein, ending in urea; P_3, protein, ending in uric acid; *F*, fat; *B*, blastoderm, *area pellucida*; B_1, blastoderm, *area opaca*; E_1, embryo, manometric-ally; E_2, embryo, assumed; E_3, embryo, from measurements on intact egg; E_4, chick, newly-hatched; *A*, allantois; YS_1, yolk-sac; YS_2, yolk-sac, late period.

the allantois falls slowly, but its values show that the catabolism of this important membrane must always be mixed; here again the R.Q. is reduced in the absence of glucose. The R.Q. of the yolk-sac falls rapidly to very low levels, the meaning of which is not altogether clear. Working with a number of in-hibitors, Needham[8] found that the respiration of the embryo is inhibited by iodoacetate in moderate doses, but this is reversible by lactate; during such inhibition the R.Q. falls to 0·85. At the same dosage yolk-sac respiration is also inhibited, but irreversibly. Fluoride in doses sufficient to abolish anaerobic glycolysis completely has no effect on the respiration of the embryo, which is only affected by higher concentrations. But no matter how thorough the inhibi-tion, the R.Q. remains in the neighbourhood of unity, and this was quantitatively shown to be due to the occurrence of a protein catabolism ending in ammonia (cf. the similar case of the grasshopper embryo, p. 581). Fluoride also inhibits yolk-sac respiration, but at lower concentrations, and it has no effect on the yolk-sac R.Q. In general, it is clear that the embryo and the membranes during early avian development have quite different metabolic qualities.* In particular, the occurrence of protein breakdown after inhibition of carbohydrate breakdown by fluoride is interesting, because it shows that the preferential catabolism of carbohydrate by the chick embryo during the first week of its development is not

* Kagiyama has studied the development of the Warburg-Keilin system in chick embryos but his results are obscure.

due to any lack of the machinery necessary for catabolising protein. The same thing was demonstrated for ammonia production during respiration without substrate, by Dickens & Greville[3] (NH_3 N γ/mgm. dry wt./hr. with glucose 0·1, with fructose 0·3, without substrate 1·0).

The metabolic rate of the chick embryo had previously been calculated, making no allowance for the respiration of the membranes (*CE*, p. 707). Needham[6] was able to show that when recalculated on the basis of the *in vitro* oxygen consumption of the membranes, it still falls, though only by about 500 c.mm./hr./gm. wet. wt. instead of 1000 as formerly thought. At the beginning of development (6th day) the respiration of the two principal membranes (allantois and yolk-sac) accounts for nearly 35 % of the total respiration of the egg; by the 19th day this has fallen to 5 %. In this paper will be found the Plastic Efficiency Coefficient (*CE*, p. 935) and the Apparent Energetic Efficiency (*CE*, p. 970) of the chick embryo, properly corrected in accordance with the new information on the metabolism of the membranes. In the earliest stages of development the efficiency of the yolk-sac is extraordinarily high, for only some 5 % of its material intake is combusted.

The fall of metabolic rate during embryonic life in the chick can also clearly be seen in the Q_{O_2} of special tissues, e.g. liver (Carroll), lens (Kihara), retina (Kiyohara).* It is continued in the remainder of the life cycle, as indicated by numerous elaborate studies (Mitchell & Card; Riddle, Nussmann & Benedict; Brody, Hall, Ragsdale, Trowbridge, Funk, Kempster, Ashworth, Hogan & Procter; Brody, Procter & Ashworth; see Fig. 321).

Romanov & Romanov[1] have studied the effects of varying concentrations of CO_2 in the air surrounding the eggs during incubation, and Remotti[6] has found that the allantoic blood-vessels can adapt themselves to partial reduction of the shell area over which gas exchange normally occurs. Romijn & Roos, taking up again the question of the air-space (*CE*, p. 719), found that it increases from a few tenths of a c.c. to 11 c.c. during normal development, while its O_2-content falls from 20 vol. % to 12, and its CO_2-content rises from 1 vol. % to 6. There is work by B. Cunningham on the effects of increased oxygen tension on development, but the results are not striking.

As is well known, the anaerobic glycolytic rate of the chick embryo falls sharply during development, and this was found to be true by Needham & Nowiński whether mannose or glucose is used as substrate (Fig. 317). Autoglycolysis also falls slightly. The fall in $Q_L^{N_2}$ occurs early; at $2\frac{1}{4}$ days a value of +35 is found, while at $5\frac{3}{4}$ days it is only +10. The Pasteur effect, i.e. the

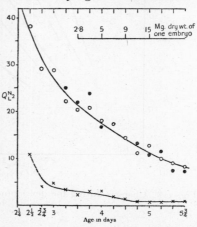

Fig. 317. Fall of $Q_L^{N_2}$ during development in the chick embryo. o Glucose; • mannose; × autoglycolysis.

* The enormous $Q_L^{O_2}$ of the retina is said to arise only after hatching.

inhibition of fermentation by oxygen (discussed more fully in the succeeding section, p. 596), is marked in the embryo, whether glucose or mannose be the substrate. But the rate of oxidative disappearance of lactate is quite insufficient

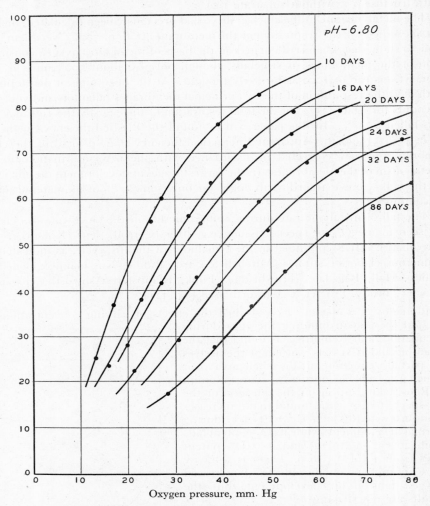

Fig. 318. Changes during the development of the chick embryo in the dissociation curve of its haemoglobin.

to account for the effect of oxygen in reducing glycolysis (Needham, Nowiński, Dixon & Cook), hence the Meyerhof resynthesis theory of the Pasteur effect cannot apply to the chick embryo.

It is usually said that the chick embryo shows no aerobic glycolysis. This statement may need modification in the light of some experiments by Laser[3], from which it appears that the $Q_L^{O_2}$ of the chick embryo may be as high as $+$ 10 at $2\frac{1}{2}$ days' incubation, decreasing to nearly zero at 5 days and then (as tested on slices) rising again somewhat.*

Van Groor has studied the development and localisation of carbonic anhydrase during the chick embryo's development. It appears in the blood at the 11th day and passes adult level on the 15th rising to nearly three times this on the 19th. The organ in which it is earliest detectable is the eye-cup, where there are appreciable quantities as early as the 4th day; it is suggested that this is a protective mechanism against the high aerobic glycolysis of the young retina.

Whether the haemoglobin of the chick embryo is identical with that of adult birds is an interesting question. The dissociation curve of chick-embryo haemoglobin was examined by F. G. Hall[2]. As shown in Fig. 318, there is the same trend of the curves from the left to the right with increasing age, as was seen in the amphibia (p. 590) and as we shall see in mammals (p. 599). Although the oxygen for the embryo is obtained direct from the air, the shell is not easily permeable to gases and the partial pressure of oxygen in the yolk and white is low. Thus when the oxygen tension in the environment of the embryo is low a haemoglobin is present which has a greater affinity for oxygen, but after hatching, when "aquatic" respiration has given place to "aerial", this is replaced by a haemoglobin with less affinity for oxygen. As in the case of the amphibia these ontogenetic changes may have phylogenetic parallels, for Hall, Dill & Barron have shown that the haemoglobins of certain birds native to high altitudes have a significantly greater affinity for oxygen than those of other members of the same group of animals at sea level. As regards the mechanism of the change in the chick embryo, Dawson[2], in an interesting paper, gives tables showing the percentage of various types of cell in the blood during incubation. He suggests that the embryonic type of haemoglobin must be associated rather with the whole output of the yolk-sac's blood islands as opposed to the bone and marrow production of later life, than with the specifically primitive blood-cells of the first half of incubation which are not found at all during the second half.

3·36. Respiration of mammalian embryos

Unfortunately we have little or no information about the respiratory metabolism of mammalian embryos in the earliest stages of their development — we know nothing of the respiration of blastocysts, for example. Only the egg of the cow has been investigated from this point of view, in the work of Dragoiu, Benetato & Opreanu, mentioned elsewhere (p. 201) in another connection. Mammalian eggs may, however, be excellent material for such studies and the full publication of Cartesian diver studies on the rat is awaited with interest (see Boell & Nicholas). Before implantation the respiration rapidly rises, presumably due to utilisation of yolk or externally supplied nourishment. The Q_{O_2} and $Q_L^{N_2}$ of tissue-slices of embryonic organs of rat and rabbit, falls off later with increasing

---

* A curious feature of chick-embryo extract, noted by Frisch & Willheim, is that unlike extracts of tumour and adult tissue, it will not, when added to tissue the anaerobic glycolysis of which has been stopped by reagents such as hydroquinone, restore the $Q_L^{N_2}$ to normal. But Pentimalli[3] thinks it accelerates the $Q_L^{N_2}$ of muscle. These effects are obscure.

age, as would be expected, both before (Hanaoka) and after (Peruzzi; Pearce; P. D. Adams) birth. According to Kisch[1], they are more resistant to certain poisons such as $AlCl_3$ than adult tissues. But such tissue-slices, and also the extra-embryonic membranes of rodents, have given a good deal of information about the Pasteur effect and analogous problems in embryonic tissues, in the work of Dickens and his collaborators.

The Pasteur effect (to which a provocative review has been devoted by Burk) is the name given to the almost universal suppression of fermentation by oxygen. In most tissues, though not of course in tumours or in retina, glycolysis is reduced to low or negligible values on passing from anaerobiosis to aerobiosis, while respiration comes into play. On the Pfeffer-Pflüger theory this was because the intermediate substances of fermentation are directly oxidised. On the Meyerhof theory (which, as we have just seen, does not apply to the chick embryo) the fermentation intermediates are mainly resynthesised to carbohydrate at the expense of energy derived from oxidations. On the Lipmann theory the inter-mediates are not formed at all in the presence of air, the enzymes of fermentation being themselves sensitive to the presence or absence of oxygen. The third of these theories is to-day the most acceptable. It has received substantial support from the finding of Schlayer and Kempner with bird red blood corpuscles that at oxygen concentrations intermediate between aerobic and anaerobic conditions, the inhibition of respiration goes very far before the onset of fermentation begins. The curves are not reciprocal and the rate of glycolysis depends therefore directly upon the oxygen concentration and not upon the rate of respiration. Laser[4] verified this in another way by finding that at low oxygen concentrations, the respiration may be affected very little while the glycolysis is largely increased. This was done on rat chorion and chick-embryo allantois as well as other tissues.

On the other hand, when in the presence of KCN respiration is decreased, glycolysis will correspondingly rise, although the conditions may be perfectly aerobic, as was shown long ago by Warburg on the chick embryo, and the same is true of CO, as Laser[5] has shown. KCN thus not only inhibits the combination of oxygen with the cytochrome oxidase, but also lifts the inhibition of the fermentation enzymes by oxygen. Hence it inhibits the Pasteur effect. Such inhibition has, however, been found to be brought about by many other sub-stances.

Dickens & Greville[2] worked with mammalian embryos (rodents and carnivores) and their yolk-sacs and chorions. Like the chick embryo, such tissue has a large and prolonged autorespiration, indicating the presence of unknown intrinsic substrates. With added glucose its R.Q. is unity, without, it is about 0·85. Dickens & Greville[4] showed that unlike some adult tissues, the respiration of such tissue is not permanently affected by deprivation of substrate. As in the case of the chick embryo, urea formation is negligible, and ammonia is only produced when no carbohydrate substrates are given. Again like the chick embryo (Needham, Nowiński, Dixon & Cook), there is no stimulation of respiration and glycolysis by K' (Dickens & Greville[5]), though otherwise there are similarities with brain cortex (the high $Q_L^{N_2}$ and R.Q. of unity). Phenylhydrazine was now found by

Dickens[2] to inhibit the Pasteur effect in these embryonic tissues completely. So also (as established in a later paper, Dickens[4]) does phenosafranin at only 10^{-5} M and other substances related to pyridine, acridine, and quinoline, such as acriflavin (Dickens).

The mechanism of these effects is still not fully elucidated, but while it seems likely that KCN (Warburg), CO (Laser), 1-amino : 2-naphthol : 6-sulphonic acid (Krah) or glutathione (Bumm & Appel) all act by uniting with iron, it is suggested

Phenylhydrazine

Phenosafranine   Acriflavine

Pyocyanin   Phenazine methochloride

that substances containing heterocyclic nitrogen in six-membered rings and so resembling the co-enzymes (cf. p. 209) act by displacing them on their protein carriers. But some of these, such as pyocyanin and phenazine methochloride, which, as we have seen (p. 568), accelerate respiration, increase the Pasteur effect by abolishing aerobic glycolysis altogether (Dickens[3]).

Turning to the processes of oxidation, we find that there are indications of a special position occupied there also by embryonic tissues. The Szent-Györgyi-Krebs cycles of C_4 compounds are now recognised as playing an important part in oxidations (cf. Fig. 307). It is of interest, therefore, that Blazsó states that embryo rat tissues will not attack either fumarate or oxaloacetate, and that according to Elliott & Greig the succinoxidase activity of the chick embryo is very slight. So is that of the echinoderm embryo (Krahl, Keltch, Neubeck & Clowes). Greig, Munro & Eliot indeed consider that the succinate-fumarate-malate-oxaloacetate cycle cannot occur in the chick embryo, because none of these substances are attacked by it.

The question of anaerobiosis and embryonic life provides a suitable transition here. It has long been suspected that embryonic cells are more resistant to anaerobiosis than those of adult animals (see *CE*, p. 742 and above, p. 522). This has been studied mainly on tissue cultures, but there is some difference of opinion as to what these can do, Laser[2] maintaining that chick embryo heart fibroblasts can grow anaerobically (provided they have been first allowed an aerobic period), though this is contested by Havard & Kendal. At the end of embryonic development in the rat and mouse the foetuses are remarkably resistant to asphyxia (M. Reiss), and a quantitative study was made of this phenomenon by Enzmann & Pincus. In spinal mice the interval between decapitation and the disappearance of all spinal reflexes decreases enormously (from 1200 to 30 sec.) during the first fortnight after birth. If, then, embryonic cells are more resistant to anaerobiosis than they later become, it is of interest not merely because many have thought, as did Pütter in 1905, that anaerobiosis might be regarded as more "primitive" than aerobiosis, but also because there is reason for thinking that towards the end of foetal life, in mammals, conditions of partial physiological anaerobiosis may be encountered.

The clearing up of our ideas on this subject is due mainly to the extensive work of Barcroft and his collaborators, reviewed in several places* such as the monograph of Windle). As we have already seen (p. 79), the placenta does not grow at the same rate as the foetus, but attains its maximum size and blood volume early. But the volume of blood passing through the foetal heart and the oxygen consumption of the foetus continually increase, and this is compensated for by increasing blood pressure during development. In the sheep, for example, the blood pressure of the foetal circulation rises from 20 mm. Hg at 50 days to 76 mm. Hg at birth.[†] But while the foetus grows and the blood flow through the foetal part of the placenta increases, conditions on the maternal side remain comparatively steady. Hence there should be a continually increasing deoxygenation of the maternal blood passing through the placenta, and this has been shown experimentally to be the case (Barcroft, Herkel & Hill; Barcroft, Flexner, Herkel, McCarthy & McClurkin). In the rabbit the percentage saturation of the uterine vein blood falls from 85 to about 15 at the end of pregnancy. The oxygen capacity of the foetal blood rises, but its percentage saturation remains at about 45 (Barcroft, Kramer & Millikan).

It may now be taken as established that diffusion processes alone account for the passage of the blood gases through the placenta (Haselhorst & Stromberger[1,3]; Keys[2]). The more thoroughly, therefore, the maternal haemoglobin gives up its oxygen at the pressure existing on the maternal side, and the more thoroughly the foetal haemoglobin accepts it at the pressure existing on the foetal side, the

---

* Barcroft[3,4,5,6,7]. Here we omit consideration of the genesis of the respiratory movements, with which this subject is connected (see Barcroft & Barron; Barcroft, Barron & Matthews; Barcroft, Barron & Windle; Steele & Windle; Windle, Monnier & Steele; Snyder & Rosenfeld for mammals; Windle & Nelson; Windle, Scharpenberg & Steele; Windle & Barcroft; Z. Y. Kuo; Z. Y. Kuo & T. C. Shen for birds). N.B. also the interesting problem of the closing of the ductus arteriosus (Barclay, Barcroft, Barron & Franklin; Barclay & Franklin).

† Blood-pressure reflexes of mammals do not develop till after birth (G. A. Clark[2]).

more efficient the transfer will be. Apart from the relative amounts of haemo-globin in maternal and foetal bloods, the important factor here will be their respective dissociation curves, and it was when these were studied that the most interesting physiological adaptations were brought to light. Fig. 319, taken from Barcroft[5], shows such curves. The steeper the curve in the middle portion and the further it is to the right, the more readily will the blood yield up oxygen; the less inflected the curve and the further it is to the left, the more readily will

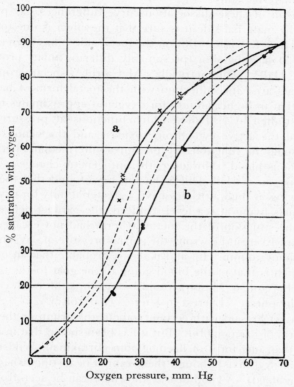

Fig. 319. Typical dissociation curves of (a) foetal, (b) maternal, blood, in a mammal. Dotted lines indicate limits of normality.

the blood take up oxygen. The foetal blood should therefore be to the left, the maternal to the right. This is exactly what was found. Fig. 319 shows the average curves for the goat, and the dotted lines give the normal limits of variation in adults (Barcroft, Elliott, Flexner, Hall, Herkel, McCarthy, McClurkin & Talaat). This divergence of foetal and maternal bloods probably occurs in all mammals; it certainly occurs in man* (Eastman, Geiling & de

* This was for a time uncertain because of the work of Haurowitz, but it has been shown that human haemoglobin is very sensitive to dilution and that when the conditions are adjusted, the difference in the dissociation curves can readily be seen (Hill & Wolvekamp; Likhnitzkaia & Sax).

Lawder; Sax & Likhnitzkaia; Leibson, Likhnitzkaia & Sax), in the cow (Roos & Romijn) and in rodents. It would be of the greatest interest to find whether it occurs also in non-mammalian viviparity and ovoviviparity. These were the findings which stimulated the search for similar displacements in foetal forms of haemoglobin, the results of which have already been mentioned under the headings of amphibia and birds. But in evolution the adaptation to embryonic anoxaemia must have preceded viviparity and hence also the corresponding compensatory maternal shift.

The mechanism of these dissociation-curve differences has provoked much research. On the maternal side it is sure that the effect is due to an increase in blood acidity (Hoag & Kiser). But on the foetal side it has been established that a special form of haemoglobin (presumably differing in its protein moiety) is present (F. G. Hall[1], manometrically; McCarthy[1], spectroscopically). Many interesting points arise in connection with the foetal form of haemoglobin. It may be noted that in higher affinity for oxygen it approximates more to muscle haemoglobin, studied by R. Hill[2,3], which functions at pressures intermediate between that of the active cytochrome oxidase and the venous blood, just as ordinary haemoglobin functions at pressures intermediate between those of arterial and venous blood. Moreover, the ontogenetic change is paralleled by a phylogenetic one, since Hall, Dill & Barron have found that the haemoglobins of mammals living at high altitudes have a greater affinity for oxygen than those of similar animals living at sea level. Indeed, the parallel between life at high altitudes and the situation of the foetus at term had not escaped Barcroft. At the top of Mount Everest, he wrote, the partial pressure of oxygen in the alveolar air would be about 30 mm. Hg, and it seems probable that an oxygen pressure of 24 mm. (which is that of the blood going to the goat foetus towards the end of pregnancy; Barcroft & Mason) would be incompatible with consciousness in man. Hence the phrase "Everest in Utero" (Barcroft[4]).

Nothing can yet be said with certainty about the origin or the disappearance of the foetal form of haemoglobin. But it has been shown that in adult mammals there is more than one form of haemoglobin (Brinkman, Wildschut & Wittermans; Brinkman & Jonxis), though these cannot be distinguished spectrophotometrically (Jongbloed). A good deal is known, as can be seen from the accounts of C. Smith and Enzmann[2], about the blood picture during development in mammals, but so far it has not proved possible to correlate the different forms of haemoglobin with different morphologically distinguishable types of erythrocyte or with different sites of production. There is, however, a change in the rat about the time of birth, whereby the large foetal corpuscles (diam. $9·17\mu$) are replaced by small adult ones (diam. $5·88\mu$), see on, p. 648.

3.365. The rise and fall of metabolic rate

We may find it convenient to regard the metabolic rate of late pre-natal life as one of adaptively depressed metabolic rate. It is sure that in mammals (Barcroft, Kennedy & Mason; Barcroft, Flexner & McClurkin) the total oxygen consumption of the embryo is four or five times less than it is shortly after birth.

We are probably right in viewing this as connected with the respiratory difficulties occurring towards the end of gestation, but the phenomenon is seen in
birds as clearly as in mammals. Fig. 320, taken from the paper of Brody, Hall,
Ragsdale, Trowbridge, Funk, Kempster, Ashworth, Hogan & Procter, shows
the minor peak in metabolic rate after hatching in the chick.* The major peak
of metabolic rate in the life cycle exists at a much earlier stage; in the chick
certainly before the 6th day of incubation (Needham[6]) and probably before the
2nd (Philips). We have seen it in amphibia (Wills), fishes (Leiner) and many
other animals. In embryos with holoblastic cleavage, the peak will appear to be

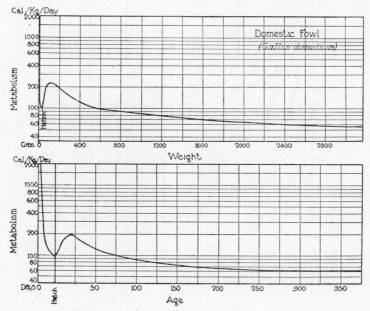

Fig. 320. Metabolic rate during the life-history of the fowl; above, related
to weight; below, to age.

shifted to a relatively later stage in the life cycle than in embryos with
meroblastic cleavage, for the cells of the former are more loaded with yolk
the utilisation of which is a pre-requisite for the manifestation of the peak
and the downward trend. The universality of this downward trend includes
not only the homoiothermic organisms usefully brought together by Brody *et al.*
in Fig. 321, but also poikilothermic organisms, where the loss of heat by radiation
from the steadily decreasing surface area does not apply. It is best to regard the
fall of metabolic rate as due to the same cause as its rise during embryonic
development, namely the relative proportion of respiring protoplasm and inert
material in the cells. At the same time, the surface factor must not be excluded.
It is still required to explain the difference between large and small homoiotherms

* Isolated tissues show this as well as the intact organism, as witness Carroll's figures for liver
Q_{O_2}; chick embryo 6th day -7.53, 20th day -1.53, after hatching -2.35, two years old -1.42.

at equivalent stages in the life cycle (cf. H. Blank). These views do not disagree with the speculations about the rise and fall of metabolic rate which have been made by Jordan-Lloyd and Dhar.

The difficulties of determining the metabolic rate of the chick embryo are of course far greater in the case of mammals, especially man. Apart from the

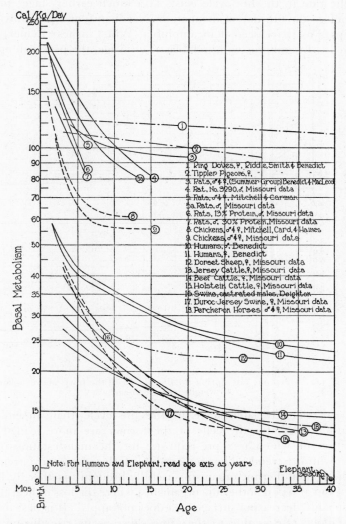

Fig. 321. Metabolic rate during the life-history of a number of birds and mammals.

gasometric determinations of Barcroft and his collaborators, just referred to, there are numerous studies with calorimeters on man (Plass & Yoakam; Schwarz & Drabkin; Rowe & Boyd), but the contributions of foetus and mother are peculiarly difficult to disentangle, and cannot be discussed here. On a review of the data for all mammals, it has been concluded that the production of 1 kg. birth weight of foetus is associated with a gestation heat increment of about 4400 cal. and that the magnitude of this increment varies with approximately the 1·2 power of the offspring's birth weight (Brody[2]; Brody, Riggs, Kaufman & Herring).

Lastly, the subject of the genesis of body temperature regulation in homoiotherms (reviewed since *CE*, p. 772, by Deighton) may be referred to. There is no doubt that in nidicolous mammals (cf. p. 76) there is a remarkably rapid onset of the power of temperature regulation (Pincus, Sterne & Enzmann; Stier; Gulick; Antoschkina). The details must be sought in the original papers.

3·37. The succession of energy-sources

Nearly twenty years ago it was remarked (see *CE*, p. 986) that embryonic development seemed to show a succession of energy-sources, carbohydrate preceding protein and protein preceding fat. If we glance through the foregoing pages we see that this is now established for many organisms, the chick (p. 592); a fish, the minnow (p. 584); an insect, the grasshopper (p. 580); a crab (p. 577); a nematode (p. 576); and a gephyrean worm (p. 575). Where it has been possible, however, to measure the respiratory quotient in the cleavage stages, values more characteristic of protein or fat have been found, e.g. amphibia (p. 587) and sea-urchins (p. 564). If we may assume, though it is by no means proved, that these quotients are to be interpreted in a normal way, we reach the coherent hypothesis that gastrulation is accompanied by carbohydrate catabolism, and that the succession begins, as it were, from that point onwards. The only observations incompatible with this view are those of Horowitz[3] on the gephyrean worm, *Urechis*, in which the fall of R.Q. commences from fertilisation.

3·38. Respiratory enzymes

In the course of the preceding sections of this discussion there have been a good many references to the enzymes concerned in respiration. The purpose of the present section is to add a few points which have escaped mention up to now.

There has never been any systematic description of the dehydrogenases and their quantitative activity during embryonic development,* but it is to be assumed that these arise *pari passu* with the respiring protoplasm. They are certainly weak in some embryonic tissues (Banga and Blazsó for pyruvic dehydrogenase; Breusch for fumaric, succinic and oxaloacetic dehydrogenases; Glezina for succinic dehydrogenase; see also pp. 612, 615). By the 8th day of incubation, succinic, lactic, hexosediphosphate, α-glycerophosphate, and glucose, dehydrogenases are all present in the chick embryo (Booth), but neither xanthine oxidase nor the Schardinger aldehyde oxidase appear till a day or two later (evidence for their identity), though, as E. J. Morgan found, they are both in the yolk-sac already on the 7th. In embryo extract, prepared from 8-day embryos, as for explant media, Sachs, Ephrussi & Rapkine found xanthine oxidase, succinic, lactic and citric dehydrogenases, and glucose dehydrogenase.

* The paper of Simola, Stenberg & Uutela I have found inaccessible.

It is unfortunate that more attention has been paid to the less important auxiliary enzymes, catalase and peroxidase. The function of catalase is now generally allowed to be the elimination of hydrogen peroxide arising as the product of the primary dehydrogenase reactions. Catalase accomplishes this by adding oxygen to the peroxide; peroxidase does the same job, but by adding hydrogen and simultaneously oxidising certain aromatic compounds, either intrinsic or artificially added, such as benzidine or guaiacol.

Contrary to a persistent belief, it is not possible to show any relation between catalase activity and respiratory intensity.* This is illustrated by work on insects, of which there is a good deal (on the silkworm, Yamafuji; Yamafuji & Goto; on the mealworm, Charikova & Mikhailov; on the gipsymoth, Kozhanchikov[2]). The best study is that by M. E. Williams on the grasshopper, *Melanoplus differentialis*. Catalase shows minimal activity till the 8th day of development; the curve then climbs during the pre-diapause respiration peak and reaches a constant level, which it maintains during the earlier part of diapause. At the beginning of post-diapause, however, it is found to be at a lower level, after which it rises steadily until hatching. Or again, in the eggs of amphibia such as the toad *Bufo*, Friggeri has shown that there is no change in the amount of catalase until the external gills appear; it then increases rapidly to the end of the yolk-sac period, when the original activity is about doubled.

Both catalase and benzidine peroxidase (cf. pp. 136 ff.) are widely distributed in the eggs of echinoderms, fishes and amphibia, mostly at the animal pole of the latter, according to Sammartino and Spirito[2,3,4], and evince no change at fertilisation. In the echinoderm egg catalase, like dipeptidase, is associated with the hyaline ground-substance (Holter[2]; Holter & Linderstrom-Lang) but the dehydrogenases are, to a certain extent, tied up with the centrifugable granular elements (Ballentine[3]). The chick embryo shows benzidine peroxidase in the blastoderm, especially the yolk islands, well before the appearance of any haemoglobin.

For the embryos of birds and mammals there are good quantitative studies, showing that the activity of catalase per unit wet weight increases with age (e.g. Kleinzeller & Werner and Kolobkova for the whole chick embryo; Galvialo & Goryukhina for chick embryo blood; Vadova for the rabbit embryo). The former curve rises as if it were the reciprocal of the declining glycolytic rate (*CE*, Fig. 175). On the other hand, in the last half of incubation there seems to be a fall of activity (Schkljar[2,5], who related activity to dry weight). When Blagoveschensky compared the temperature coefficients of catalase from animals of different ages, he found marked and progressive differences. This is interesting, for it would mean that there are changes in the protein component of the enzyme with age, analogous to those we have already seen so well established for the haemoglobins. Vadova subsequently found that the catalase of rabbit tissues changes in this respect just about the time of birth, and Kolobkova reports a continuous fall in its temperature coefficient during the development of the chick embryo.

---

* There is, however, the remarkable fact observed by Yamafuji, that the optimal conditions for breaking the diapause of silkworm eggs (exposure to 20 % HCl for 4 min. at 50° C. on the 12th day) are also those for maximal catalase activity.

3·39. Axial gradients and respiration

It may be advisable to conclude this section by a word concerning the axial gradient work carried out by the school of C. M. Child. This was recently summarised in two reviews by him, and we now also have the valuable bibliography of his papers prepared by Hyman & van Cleave. The standpoint formerly taken (*CE*, pp. 582 ff.) was that although gradients of susceptibility to lethal agents, intensity of vital staining, etc. had been clearly demonstrated in a large number of adult invertebrates and many embryos, no evidence whatever had been brought forward to justify a belief in the existence of "respiratory" or "metabolic"* gradients in embryos, and very little satisfactory evidence for this in the case of adult organisms. I see no reason now to modify these conclusions. Closer analysis strongly indicates an identity of respiratory rate in the organiser region and ventral ectoderm of the amphibian gastrula—a test case for the "respiratory gradient"; cf. p. 199. As regards adult organisms, results have been quite contradictory. Reference may be made to the (somewhat unconvincing) reply to various critics by Watanabe & Child and the subsequent renewal of the attack by Maluf[2].

But when all this has been said, it is sure that embryology owes a considerable debt to Child, who by introducing his gradient concept was one of the first to recognise those invisible dispositions of order with which the organism imposes its organisation on the matter of which it is made. To-day the gradient theory has merged in the general theory of morphogenetic fields, already discussed (pp. 127 ff.). It is not likely we shall be able to do without this in the future progress of embryology.

3·4. Metabolism

In this section we return at last to what was the main subject discussed in *Chemical Embryology*, namely the chemical changes which go on side by side with the process of morphogenesis. But before proceeding to the various special forms of metabolism in their order, there are a few points of a general kind which may receive brief mention. Besides those investigations which have approached the biochemistry of the embryo directly, there are a number of indirect ways of attack which throw more or less light on it.

---

* Some American biologists have been unusually vague in the meanings they have attached to the word "metabolism". Whereas the school of Child regarded the time taken for a piece of a worm to dissolve in KCN solutions, or the intensity of staining with a vital dye, as throwing some light on metabolic differences, certain European writers seem to think as if metabolism were purely oxidative respiration, purely a matter of "fuel" rather than "machine". It is needless to say that in the present book "metabolism" is given its classical definition as the sum total of all the chemical changes going on in the body; it therefore includes the structural alterations of the proteins which are the basis of all morphology. Modern work with isotopes has demonstrated how the proteins constantly participate in metabolism. The term "metabolic rate", as used by the calorimetric physiologists, refers specifically to heat production in calories, always related to the wet weight, dry weight, or surface area of the organism. The school of Child cannot but plead guilty to the charge of having had a concept of metabolism too wide and vague; others, however, have thought of it in an unduly narrow way.

3·41. Indirect approaches

Changes in density of the egg system during development are an example of this. The well-known decrease in density of teleostean eggs during their development (*CE*, p. 822) probably occurs as a result of fat utilisation. Amphibian eggs, as we know from an admirable paper by Briggs[1], also decrease in density during development, from fertilisation to the advanced tadpole stages, though the change is interrupted during the time of closure of the neural folds, when there is a transient density increase. There can be no doubt that the main trend here is due to the absorption of water by a non-cleidoic egg (cf. p. 35) and the transient density increase to the occlusion of the archenteron which amounts to a temporary "solidification" of the system. In molluscan eggs also there are interesting density changes (Comandon & de Fonbrune[1,3]).

Another way in which it has been hoped to gain information about the general course of events in embryonic life is the measurement of the osmotic pressure. According to the description of Backman & Runnström (*CE*, p. 778) which had become classical, the osmotic pressure of the amphibian egg sinks shortly after fertilisation from $\Delta - 0.48°$ to about the level of the fresh-water medium, apparently owing to a binding of nearly 90 % of the osmotically active substances. This account can no longer be accepted, for the work of Krogh, Schmidt-Nielsen & Zeuthen has shown that the fall is only a slight one (20 %); rapid till the first cleavage, slower from then till the beginning of gastrulation. After gastrulation a rise sets in until the original value is approximately regained at the time of maximal development of the external gills. The early fall must be interpreted as again due to the absorption of water by the whole system (see p. 38, Table 6), though there are indications of a slight loss of Cl' ions; the later rise must be due to the production of small organic molecules which cannot escape owing to the very low permeability. During later development there must be many changes in the osmotic pressure of the internal medium, as is shown, for example, in the work of Fontaine & Boucher on eels.

Osmotic pressure measurements, of more or less dependability, are also available for the amniotic and allantoic liquids of birds (Howard[3]), chelonian reptiles (Imamura[1]), etc., but their significance is minor. A careful study of the electrical conductivity, surface tension, and *p*H of the chick's amniotic and allantoic fluids was made by P. A. Walker. In the latter the conductivity falls until the 8th day, is stable till the 15th, and falls thereafter catastrophically, as would be expected from what we know of its ionic composition. The surface tension falls until the 13th day and then rises again to its original value, a fact more difficult to explain. The *p*H is roughly stable at 7·4 till the 13th day after which it falls steadily to 5·6 (see p. 622). In the amniotic liquid the conductivity is high until the entry of albumen through the sero-amniotic connection, when it rapidly falls. Surface tension tends to rise throughout development except after the 16th day, and during this same late period, the *p*H, which earlier has been constant around neutrality or slightly falling, rises to 7·8.

After earlier controversies (described in *CE*, pp. 839) it is now generally accepted that the *p*H of the cell interior, and of egg-cells in particular, is in the close neighbourhood of neutrality (e.g. the value of 6·8 given for the egg of the sea-urchin by Pandit & Chambers and for that of *Fundulus* by Chambers[2]). But where the meroblastic egg has become a large and complex structure, as in birds, the neutrality characteristic of protoplasm is departed from. Thus all observers find the egg-white of fowls' eggs to be as alkaline as *p*H 9·0 and the yolk between 5·0 and 6·0 (Gaggermeier; Moran[4]; Penionskevitch[2]; Berenstein & Penionskevitch).

They also all agree that as development proceeds yolk and white attain neutrality and even overpass it somewhat in the opposite directions. The alkali reserve increases in the albumen and decreases in the yolk (Schkljar[3,4]; Rubinstein) and the titratable acid in the amniotic liquid rises (Remotti[5]). Other terrestrial eggs, such as that of the silkworm moth, also show variations in pH during development (Ongaro[1]; Tirelli[5]).

Since the former discussion (*CE*, pp. 865 ff.) of rH, little that is relevant here has been done on oxidation-reduction potential. The upshot of the body of work done in the decade 1920–1930 (summarised by Chambers[3] and by B. Cohen) was to establish that the E_h of egg-cells under aerobic conditions is of the order of $-0·07$ volt and under anaerobic conditions of the order of $-0·17$ volt. The expectations once entertained as to the value of a knowledge of the overall reduction potential of the embryonic cell have not yet quite materialised, and our knowledge of the properties of heterogeneous systems must, as Korr[2,3] points out, advance a good deal further before this line of approach will again become important.

Analogous work is that on echinoderm eggs dealing with surface tension (K. S. Cole[2]; Cole & Michaelis; E. N. Harvey[1,2,5]), cataphoretic charge (Uspenskaia; Dan[1,2,3]; Dan, Yamagita & Sugiyama; Mazia[1]), and electrical conductivity and impedance (K. S. Cole[1,3,4]; Cole & Cole; Cole & Curtis; Cole & Spencer). It cannot be summarised here. In the case of the hen's egg physical methods have proved of value in studying the changes in water distribution. Orrù[1,3] and Romanov & Romanov[2], who confirm older work (*CE*, p. 878), show that the egg-white is continuously dehydrated, partly by loss through evaporation, and partly because of the uptake of water by the yolk which later loses it to the embryo. The changes in electrical conductivity and refractive index of the egg-white are therefore very considerable (Romanov & Sullivan; Romanov & Grover; Romanov, Smith & Sullivan; Perov[1]; Perov & Dolinov). Electrical conductivity of egg-white drops from a high level to 33 % of its initial value and then slowly rises; that of the yolk starts at a value lower than that of the white, drops to the same relative extent and then rises to nearly its original value; that of the allantoic liquid rises to a plateau, and that of the amniotic liquid has a sharp rise and then a fall to below early values. Refractive index follows a course reciprocal to that of conductivity, first rising greatly, then finally falling off.

3·42. Carbohydrate metabolism

In the course of the argument on the biochemistry of organisers in vertebrates (pp. 189 ff.) we have already had occasion to consider a number of features of the metabolism of carbohydrates in embryonic life. Without repeating these here, therefore, we must give a picture of this department of metabolism as complete as possible. It is convenient to begin with the situation in birds and mammals.

3·421. Carbohydrate metabolism in the development of higher vertebrates

As may be seen in the former summary (*CE*, Section 8·1) the general course of carbohydrate changes in the hen's egg during development was rather confused. There emerged the facts, however, that free glucose decreased to a minimum about the middle of the incubation period and thereafter rose to some extent, while glycogen, present only in traces at the beginning of incubation, increased steadily as time went on, first in the yolk-sac and later in the embryonic body itself. More difficult to trace were the movements of the protein-combined sugar (ovomucoid and the albumens and globulins). These relationships have been clarified by subsequent work.

Donhoffer introduced a new method of fractionation of the carbohydrate groupings, based on alcohol extraction of the whole egg, which he found to contain about 290 mgm. total glucose/50 gm. wet wt. (approximately one egg without shell). The alcohol-soluble fraction, interpreted as all the free glucose, starting at 110 mgm. (in the same units), falls to 10 mgm. at about the 9th day, and rises to 20 at the end of incubation. The water-soluble fraction of the alcohol precipitate, interpreted as mainly ovomucoid glucose, begins at 90 mgm., sinks to 65 at the 9th day and thereafter returns to 90 mgm. by the end of incubation. The water-insoluble fraction of the alcohol precipitate declines more or less steadily throughout incubation from 85 to 40 mgm. On the other hand, glycogen steadily rises from 10 to 60 mgm. Hence the water-insoluble fraction of the alcohol precipitate is regarded as the principal source of the glycogen formed; it is supposed to be sugar in protein combinations other than ovomucoid (see p. 9). In view of what will appear later (p. 616), it is not without interest that the free sugar appeared to contain a certain proportion of fructose as well as glucose[1], and this was confirmed by K. Yamada[1]. The fall of free glucose has also been confirmed by K. Takahashi[1] and Tateishi[1].

One of the most striking events accompanying this great fall is, as has long been known (*CE*, p. 1051), the peak of lactic acid, both in yolk and white, which accumulates till the 5th day reaching a maximum of some 80 mgm. %. K. Takahashi[1] studied the effects on this of injecting carbohydrates and hormones, but even when glucose is injected the peak is quite unaffected (cf. the work on *in vitro* autorespiration, p. 592). Glutathione appears to reduce it slightly, and the injection of 0·2 c.c. 1 % iodoacetate or bromacetate will halve it.

The rise in blood-sugar during the chick's development, mentioned in *CE*, p. 1039, has been further established by Milletti and Zorn & Dalton.

It is now agreed that glycogen appears among organs in the chick embryo first in the heart (Jordan) and only at the 7th–8th day in the liver, up to which time it is mainly in the yolk-sac (Dalton[4]; Muglia & Masuelli; Guelin-Schedrina[2]). Its appearance cannot be accelerated by injections of glucose into the egg (Canevazzi). The earlier conclusion, represented in the review of M. Aron, that it appears first at the 11th–12th day in connection with the beginning of insulin secretion* seems to be wrong, for Guelin-Schedrina[2] injected insulin into the embryonic blood circulation on the 5th day and yet found no liver glycogen present till its usual time, the 8th. She also transplanted 3-day embryos to the chorio-allantoic membranes of older embryos whose blood certainly contained insulin, but although it was proved that functional vascularisation had occurred, liver glycogen never appeared in the grafts before their presumptive time (another case of physiological *herkunftsgemäss* behaviour). By the last week of incubation,

---

* In some mammals the insulin content of the developing pancreas seems to be exceptionally high; in cattle up to 15 times that of the maternal gland, if Fisher & Scott are right.

glycogen is everywhere, though at hatching, liver glycogen rapidly disappears (a well-known fact confirmed by Muglia & Masuelli and Trotti[2]).

The effect of endocrine factors on the glycogen of the chick embryo's liver has been studied in an interesting paper by Gill, who used the Linderstrøm-Lang-Heatley methods. Adrenalin, injected into the embryonic circulation any time after the 11th day, caused a well-marked fall of liver glycogen, but when the liver cells were growing in explants the adrenalin effect could not be obtained, although anaerobiosis caused complete breakdown of the glycogen contained in such explants. The reason for this difference is not yet clear. Adrenalin* does not reduce the muscle glycogen of the embryo until the 18th day. Amylase is said to increase in activity during the development of the chick embryo as a whole (Kolobkova) and its liver (Galvialo & Goryukhina).

The glycogen of the mammalian foetus and placenta is also less available than that of the maternal organism. Insulin has relatively little effect on the foetal glycogen stores in the dog (Schlossmann[4]) and the sheep and goat (Passmore & Schlossmann). In the rat, adrenalectomy, as Corey[4] found, depletes the maternal liver glycogen much more than that of the foetal liver, and neither that nor any other deviation from normal conditions has any effect on the placental glycogen. Like maternal liver glycogen, foetal liver glycogen rises to a maximum after feeding, then falls off again (Stuart & Higgins). But on fasting it never falls so low (remaining e.g. at 5 % instead of 0·3 %), and on breaking the fast, it rises much further than the maternal liver glycogen. It is striking that Passmore & Schlossmann observed in ungulates, where the foetal blood-sugar is normally higher than the maternal, that in hyperglycaemia glucose goes through the barrier and afterwards the levels settle down to their normal inequality. The glycogen in the human placenta cannot, according to Rabe, be increased by hyperglycaemia. Nor can it be decreased even if there is dire need for carbohydrate on the maternal side, as Rabinovitch showed in his study of placental glycogen after Caesarean section on diabetic pregnant women. Moreover, glycogen disappears much more slowly in autolysis of the foetal liver than in that of the maternal (Aloisi[2], working on the guinea-pig), while the glycogen of the human placenta gives only glucose and no lactic acid on autolysis, thus resembling the liver (Davy & Huggett). The conception of the placenta, and also the avian and reptilian yolk-sac, as the "foie transitoire" of Claude Bernard (CE, p. 1018), remains indeed classical (cf. the recent series of glycogen estimations in man by Rabe, in the rat by Corey[3], and in the rabbit by Loveland, Maurer & Snyder). There is, however, the suggestion of Szendi[3,4] that in some mammals the lung, like the placenta, holds more glycogen than the liver at early stages, and acts as a store until the liver can take on its function. Lastly, physico-chemical differences between glycogen preparations from embryo and adult have been described by H. Lund, mainly as regards optical rotation.

---

* Data on the onset of function in the adrenal gland will be found in the papers of Cattaneo[1] and Harman & Derbyshire for the guinea-pig; S. Davies for the cat; and Pankratz for the rat. Lewis & Geiling noted that the adrenal gland cells of the chick embryo continue to form adrenalin in tissue culture.

Precursors of glycogen have been studied by injecting various carbohydrates into the hen's egg by Kataoka. Glucose, fructose, mannose, galactose, maltose and lactose were effective; pentoses and glucosamine were not. Histochemical work on glycogen distributions in tissues, mostly of late embryonic life, always continues; some references are given for those interested (chick, Morello; rodents and carnivores, Aloisi[3]; Komatsu; Manai; human, Aloisi[1]; Mitani; N. Nakamura). Glycogen may have a connection with the calcification of bone (see p. 556).

3·422. The phosphorylation problem

The question next arises whether glycogen is really an immediate source of energy for the embryonic tissues, or a remoter store. It arose originally in a slightly different way. As we have seen (p. 603), there is much evidence for a succession of energy sources during embryonic development, carbohydrate preceding protein and protein preceding fat. In casting about for the meaning of this, it was natural to wonder whether perhaps the breakdown of glucose might not be a simpler mechanism than any of the others and hence would have onto-genetic priority. If deamination and desaturation were out of court, glucose might be broken down in some simple way. Hence much effort has been devoted to determining whether the embryo, like the classical examples of muscle and yeast, phosphorylates its carbohydrate, or whether some simpler mechanism takes the place in it of the complicated cycles now known to exist in phos-phorylating tissues (cf. Parnas; Parnas & Ostern; Meyerhof[2]; D. M. Needham[2]).

The first observation (J. Needham & Nowiński) was that the substrate pre-ference of the 4-day chick embryo under anaerobic conditions is very narrow, for besides glucose, mannose is the only carbohydrate which can be glycolysed to any substantial degree.* All phosphorylated hexoses and glycogen are com-pletely unattacked, though an occasional experiment may show a slight breakdown of glycogen or hexosediphosphate. In such cases there are indications of sum-mation with the glucolysis proceeding at the same time, suggesting a separate pathway. Mannose and glucose compete. Up to the present time no one has succeeded in preparing actively glycolysing extracts of the chick embryo, parallel with muscle extracts, but the glucolytic activity of brei is very considerable. The substrate preference is common to all parts of the embryo, and is shown by brei as clearly as by the intact embryos.

The chick embryo in the first week of its development thus joins that class of tissues, of which tumour and mammalian brain are the most classical examples, which strongly attack glucose and leave glycogen and the phosphorylated hexoses untouched. This division of tissues was first made by Bumm & Fehrenbach, in 1930; we shall shortly meet with it again. In this connection Schönfelder's work is important. Reviewing the relevant physiological literature, he showed that foetal skeletal muscle, the muscles of young animals just after birth, and de-

---

* This may be significant in view of the mannose-containing trisaccharides which, as we have seen (p. 9), are prosthetic groups of egg-proteins.

nervated adult muscle, have several common properties. Thus contractures are given with nicotine, coniine, etc., a special type of tension curve is seen, and recovery is very slow.* No histological peculiarities exist, but the muscles are clearly characterised by the difference in their utilisation of substrates; thus muscle of the foetal type glycolyses glucose, not glycogen, as opposed to ordinary skeletal muscle. Schönfelder found that the change-over occurred during the first fortnight after birth in the rabbit. With Schönfelder's results may be compared those of other workers (Koschtojanz; Koschtojanz & Rjabinovskaia), who show that the phosphagen-content of foetal mammalian muscle is exceedingly low or even (Rjabinovskaia[1]) non-existent. Needham & Nowiński found that chick leg muscle from the 15th day of incubation still showed glucolytic, not glycogenolytic, properties. They did not believe that the difference could be due to difficulties of permeability, as the brei was made by grinding and cytolysis in distilled water. Moreover, as will shortly be mentioned, triosephosphate is formed from hexosediphosphate even when intact embryos are used.

Following the important discovery that *dl*-glyceraldehyde has the property of completely inhibiting anaerobic glucose breakdown even in very small concentrations (Mendel; Mendel, Bauch & Strelitz), while having no effect upon phosphorylated glycolysis, Needham & Nowiński found that the glucolysis of the chick embryo is also powerfully inhibited by *dl*-glyceraldehyde. It was later shown (Needham & Lehmann) that the inhibitory effect is due to the *l*-compound only, which corresponds in its configuration to the *l*-lactic acid deriving from *d*-glucose in the body. Subsequently this was confirmed on other tissues by Mendel, Strelitz & Mundell. Contrary to various statements in the literature, the discriminatory value of glyceraldehyde as between glucolysis and phosphorylating glycolysis holds good if attention is paid to whether the triose is present in the monomeric or dimeric form (Lehmann & Needham[2]).

The subject was carried further in a series of papers by Needham & Lehmann[1]. In spite of the use of the most delicate methods available (the *B. influenzae* glucose dehydrogenase test and the mammalian malic dehydrogenase test), coenzyme I (the cozymase of Harden and v. Euler, see p. 209) could not be demonstrated in the chick embryo. On the other hand, coenzyme II (the hexosemonophosphate codehydrogenase of Warburg) was found to be present in all probability throughout development and even in traces in the yolk.[†] Phosphorus-transporting coenzyme, presumably adenylpyrophosphate, could be demonstrated in the embryo, but in very small amounts. A systematic investigation was then made of the enzymes involved in the phosphorylation cycles (see Fig. 322), and the effect of additions of all kinds of phosphorus transporters, such as adenylic acid, adenylpyrophosphate, and cozymase to the glucolysing but not glycogenolysing brei was tested. It was found that aldolase is present, but that

* See also Rosanova; Arshavsky & Rosanova; Arshavsky & Malkiman.

† Extreme lack of coenzymes in embryonic tissue has also been reported by Bernheim & v. Felsovanyi working on the rat.

the triosephosphate formed by it from hexosediphosphate continually accumulates, even when intact embryos are used.* On the other hand, all the enzymes concerned in the breakdown of phosphoglyceric acid are present in the embryo. All the facts established could be interpreted upon the hypothesis that in the chick embryo there are two separate routes of carbohydrate breakdown: (1) a non-phosphorylating glucolysis mechanism, very active, and closely bound to the cell

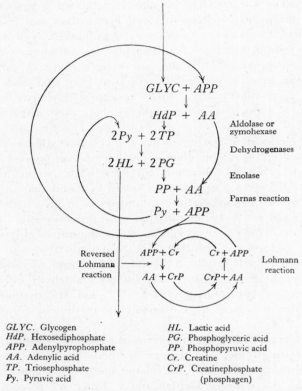

GLYC. Glycogen
HdP. Hexosediphosphate
APP. Adenylpyrophosphate
AA. Adenylic acid
TP. Triosephosphate
Py. Pyruvic acid

HL. Lactic acid
PG. Phosphoglyceric acid
PP. Phosphopyruvic acid
Cr. Creatine
CrP. Creatinephosphate
    (phosphagen)

Fig. 322. Simplified diagram of the cycles of phosphorylating glycolysis. The important role of inorganic phosphate is not shown (see D. M. Needham[2] and Pillai).

structure (desmo-enzyme), and (2) a phosphorylating system closely similar to that in muscle, dealing with glycogen and hexosediphosphate, but of very low activity because deficient in four distinct places, (i) the enzyme esterifying glycogen, (ii) the enzymes forming phosphoglycerate and lactate from triosephosphate and pyruvate, (iii) lack of adenylpyrophosphate, (iv) lack of cozymase.

* This step cannot be involved in normal glucolysis because normal glucolysis is inhibited by *dl*-glyceraldehyde, and this inhibition can be reversed by the simultaneous addition of hexosediphosphate. The dihydroxyacetonephosphate formed from the latter removes the glyceraldehyde, giving hexosemonophosphate. It is therefore difficult to see how normal glucolysis, so fully inhibitable by glyceraldehyde, could involve hexosediphosphate as an intermediate.

It seems, then, that in early embryonic life the phosphorylation machinery has not yet been fully laid down.*

It is of interest that a not dissimilar state of affairs has been described for some tumours. A number of investigators (e.g. Boyland & Mawson; Boyland & Boyland[2]; Scharles, Baker & Salter) have found in tumour tissue the complete phosphorylation system, as in muscle, but most of them admit that the course of carbohydrate breakdown in tumour has certain peculiarities. Thus Caló and Tsuzuki found great inefficiency in the oxido-reduction between pyruvic acid and triosephosphate; precisely the step which is missing in the embryo. Hitchings, Oster & Salter have brought forward further evidence in support of this and have shown that in presence of fluoride, muscle and tumour make quite different intermediate substances. Brain tissue is another case in point. Earlier workers (Ashford & Holmes; E. G. Holmes; Ashford) adduced many experiments in favour of two paths of breakdown there, glucolysis as well as glycolysis of hexosephosphates and glycogen, and later successful extractions of the phosphorylation cycle enzyme system were made (v. Euler, Gunther & Vestin; Mazza & Malaguzzi-Valeri). But Mazza & Lenti[1] went on to study the properties of brain brei from which all or nearly all of the phosphorylation activity had been extracted. The residue showed powerful glucolysis (as active as in intact slices), inhibitable with glyceraldehyde, fluoride and iodoacetate, and unaffected by addition of coenzymes or phloridzin. They were able to exclude methyl glyoxal and pyruvic acid as intermediates. In a later paper[2] they established the fact that no esterification of inorganic phosphorus occurs when brain residue begins a vigorous glucolysis. Nor is any triosephosphate formed. Evidence for non-phosphorylating glucolysis has also accrued for kidney and liver (Wajzer & Lippmann; Lenti[1]), for retina (Califano; Possenti; Lenti[3]) and for placenta (Kutscher, Veith & Sarreither). Both brain and placenta contain strongly acting alkaline phosphatases which would dephosphorylate phosphorylated hexoses (Fleischhacker; Busse).

Further experiments pointing in the direction of a non-phosphorylating glucolysis in the chick embryo were made by Needham & Lehmann[1], who found a differential inhibition by fluoride. At $M/200$ the conversion of phosphoglyceric acid into phosphopyruvic acid by enolase, and hence the phosphorylation cycle, is completely suppressed, while glucolysis in the same material is only 45 % suppressed. Moreover, if by a carefully chosen fluoride concentration, the phosphorylation route is blocked but not the non-phosphorylation route, then the inhibition of glucolysis should be greater in the presence of hexokinase than in its absence, since the hexokinase should shunt the glucose into the phosphorylation route, where its breakdown beyond phosphoglyceric acid should be impossible. Conversely, glyceraldehyde will block the non-phosphorylation

* Meyerhof & Perdigon later succeeded in obtaining breakdown of hexosediphosphate by chick embryo by adding enormous quantities of coenzyme to replace the rapid destruction. Needham, Lehmann & Nowiński then pointed out that embryo brei retains powerful glucolytic activity for as long as 80 hr. at 0° although only a very short time would suffice for the complete destruction of the coenzyme present by the nucleotidases. Such preparations are absolutely incapable of carrying out the dismutation between pyruvic acid and triosephosphate.

route but not the phosphorylation route; hence in the presence of hexokinase the inhibition of glucolysis should be less than in its absence, since all the glucose should be shunted through the phosphorylation route. Yet in all cases fluoride and glyceraldehyde inhibitions were the same whether hexokinase was present or not. This further strengthens the evidence that glucose breakdown in the living embryo proceeds without phosphorylation.

In connection with the earlier results of Geiger with glutathione, it is interesting that Needham & Lehmann[1] obtained some evidence that glutathione can reverse the inactivation of glucolysis brought about when embryo brei is dialysed. That glutathione is the coenzyme of glyoxalase has been known since the work of Giršavičius and Lohmann, but in the chick embryo methyl glyoxal cannot be an intermediate stage in glucolysis, as Needham & Lehmann were able to show that summation occurs when the enzymes are flooded simultaneously with glucose and methylglyoxal.* It seems possible to exclude many other substances also as intermediates, such as gluconic acid, glycerol, glyceric acid, pyruvic acid, and dihydroxyacetone. Additional evidence favouring the non-phosphorylation scheme was obtained by Needham, Nowiński, Dixon & Cook in a study of the partition of the acid-soluble phosphorus at different stages of normal and fluoride-inhibited glucolysis.†

Summing up, it seems clear that in the chick embryo during the first week of its development, we have the most extreme case known of non-phosphorylating glucolysis. This process is most active; phosphorylating glycolysis is weakest. In brain, tumour, etc. the two paths are more equally present. Two objections against the non-phosphorylation hypothesis have been raised; it cannot be accepted until (1) the enzymes responsible have been obtained in extract, (2) the intermediate stages have been discovered. The second objection appears to have more force than the first, for many enzymes (such as cytochrome oxidase, for instance) are closely attached to the cell structure and have long defied attempts to get them off it.‡ There is little that can yet be said about the intermediates of non-phosphorylating glucolysis, but in spite of some serious arguments to the contrary perhaps the most probable suggestion is that the glucose molecule is split into two molecules of glyceraldehyde.§

* The old observation of Neuberg, Kobel & Laser that methylglyoxal is formed from hexose-diphosphate by chick embryo acetone powders was certainly due to the non-enzymic conversion of triosephosphate.

† This may be the place to mention a remarkable peculiarity in the distribution of acid-soluble phosphorus in the chick embryo, namely the high proportion (56·5 %) not precipitable with barium. This fraction, which is very resistant to acid hydrolysis, is not fermentable by yeast, and probably plays no part in the mechanism of carbohydrate breakdown. During later development it declines relatively in amount. It is interesting that a fraction with similar properties and also present in large amount occurs in tumours. The nature of the material is still obscure; it may be the aminoethylalcohol-phosphate of Outhouse; the diphosphoglyceric acid of Greenwald, or the choline phosphate of Beznák & Chain and Plimmer & Burch.

‡ Lenti & Fuortes have made the interesting discovery that the enzymes of the phosphorylation cycles are not destroyed by subjection of tissues to liquid air, but that those of glucolysis are. Presumably this is because the structure is destroyed and with it any enzymes attached to it.

§ There are reports of the isolation of a non-phosphorylated compound intermediate in glucolysis (Mazza & Lenti[2] on brain; Lenti[3] on retina); further progress will be awaited with interest.

During the investigations just described, the behaviour of pyruvic acid during embryonic glucolysis was examined by Needham, Nowiński, Dixon & Cook.* It seems to be more concerned with autoglycolysis than glucolysis. They also measured the rate of oxidative disappearance of lactic acid and found that it is insufficient to account for the effect of oxygen in reducing glycolysis. Meyerhof's resynthesis theory cannot therefore explain the Pasteur effect in the chick embryo, which must more probably be ascribed either to an action of oxygen upon accessibility (K. C. Dixon) or upon the enzymes themselves (Lipmann[3]).

We may now turn to questions relating to phosphagen (see the lower part of the diagram in Fig. 322). Phosphagen (whether phosphocreatine or phospho-arginine) is closely connected with the phosphorylation scheme of carbohydrate breakdown, but it is now known not to be exclusively concerned with muscular motion, as was first thought. One of the most important ways in which this energy reserve is synthesised in the cell is by the Parnas reaction (Ostern, Baranowski & Reis; D. M. Needham & van Heyningen; Meyerhof & Lehmann), in which phosphopyruvic acid gives up its phosphorus to adenylpyrophosphate and the latter hands it on to the guanidine base. Under suitable conditions tissues which contain hardly enough phosphagen to measure satisfactorily may thus be made to demonstrate that they can form it (e.g. Baldwin & D. M. Needham[2] on echinoid jaw muscle; Torres on many mammalian tissues). Baldwin & D. M. Needham[1] first showed that phosphagen is present in the chick embryo as early as the 70th hour of development. Very shortly after its first appearance in measurable quantities, it attains a maximum (as related to total inorganic and labile phosphorus); this coincides exactly with the first beginnings of movement in the embryo, established by the work of Kuo[1] — head vibrations, body vibrations, head lifting and head bending. The heart beat alone has begun much earlier, about the 40th hour. Later the phosphagen "grows" first tachyauxetically ($k = 1·57$) then bradyauxetically ($k = 0·90$). When Lehmann & Needham[2] found the Parnas reaction to be fully present in the chick embryo (see above) it was of interest to trace it back as far as possible, and it was found to be very active already at the 43rd hour when the heart is beating, though only feeble at the 23rd while the neural folds are still open. Further research will be required to ascertain how much significance attaches to the appearance of the Parnas reaction at the same time as the heart beat.

From the point of view of recapitulation it is interesting that not even at the earliest stage investigated could any phosphoarginine be formed by the Parnas reaction. This does not support the claim of Bagdassarianz that phospho-arginine is present in chick and amphibian embryos. Avian yolk, according to K. Takahashi[2], contains a little phosphocreatine, the meaning of which is unknown. In embryonic brain considerable changes in the proportion of adenylpyro-phosphate with age were found by Epelbaum & Khaikina.

* In this connection it should be remembered that Blazsó and Banga could demonstrate no pyruvic dehydrogenase in embryonic muscle.

3·423. Ascorbic acid

Another subject which requires treatment under the heading of carbohydrates is the metabolism of ascorbic acid (vitamin C; formula on p. 266). A vitamin for some mammals, it might better be called a hormone for birds, since they can certainly synthesise it, probably from glucose or other hexoses. It was first found that the hen's egg at the beginning of development is devoid of ascorbic acid (Hauge & Carrick; Machida & Sasaki), and the synthesis during development was then followed by S. N. Ray and later by Martino & Bonsignore and Suomalainen[2]. This proceeds steadily; thus at the 4th day there is 0·012 mgm./embryo, and at the 12th, 1·26 (see p. 335); and it is doubtful whether ascorbic acid shows the bradyauxesis of the other highly reducing substance, glutathione, which is usually estimated at the same time (Bierich & Rosenbohm). Nor is it concerned with hereditary size differences (see p. 426).

It seems, moreover, that the chick is not the only organism capable of synthesising ascorbic acid. The amphibian embryo does so too (Suomalainen[3]). Moreover, it is said that mammals also possess the power in foetal life and afterwards lose it. This was first inferred by Mouriquand & Schoen from the resistance of pregnant guinea-pigs to scorbutic diets, but was later proved more directly by a number of workers (Rohmer, Sanders & Bezssonov; Rohmer, Bezssonov & Stoerr; Giroud, Ratsimamanga, Rabinowicz, Santos-Ruiz & Cesa). From other work (Neuweiler[1,2]; Goldstein & Volkenson; Ammon; Caffier & Ammon) it is clear that the placenta contains appreciable amounts of ascorbic acid, but in relative amount it lags far behind the corpus luteum, which, as Biskind & Glick have found, is second only to the pars intermedia of the pituitary and perhaps the adrenal cortex itself, where Szent-Györgyi first recognised the substance (Bourne[1]; Glick & Biskind). As the ascorbic acid content in the corpus luteum parallels the progesterone content, it is thought that the vitamin may be necessary for the production of the hormone. If so, it would have an important function in maintaining successful pregnancies. Bourne[2] induced corpora lutea artificially by treatment with anterior pituitary hormone; the guinea-pigs were then more resistant to scorbutic diets than their controls.

3·424. The fructose problem

Before leaving the embryos of birds and mammals, something must be said about the curious association of fructose with embryonic life. Since the time of Claude Bernard (1855) it has been known that the sugar of the human amniotic liquid is laevo-rotatory, and as previously stated (*CE*, p. 1054) fructose is commonly found in mammalian foetal blood. It also occurs in traces in normal blood (Okamura[1]), where it is said to be raised somewhat in amount during mammalian pregnancy (Okamura[3]) and during the breeding season of the salmon (Okamura[2]). Itizyo's work indicates that about one-tenth of the human amniotic sugar is fructose, and the ubiquity of fructose was further established when K. Yamada[1] showed that it is definitely present in the amniotic liquid of the chick embryo. As mentioned above, the chick embryo cannot glycolyse fructose

(Needham & Nowiński). In order to probe further into its function Dickens & Greville [1,2] undertook an elaborate study of the fructolytic power of the embryonic scaffolding structures in mammals and birds. It was found that a fructolysis as intense as the glucolysis exists in the amniotic membranes of rat and guinea-pig and the rabbit allantois, and to a lesser extent in placenta of the ferret and the placenta and yolk-sac of the cat. The placentas of rabbit, guinea-pig, rat and mouse, and embryonic tissues in general, do not fructolyse.* Without going into the further complications, the general picture seems to be that fructolysis is characteristic of foetal scaffolding structures. It is regrettable that Dickens & Greville did not test the chick embryo's yolk-sac. And in view of trophoblastic functions it may be significant that tumour tissue also shows fructolysis.

3·425. Carbohydrate metabolism in the development of lower vertebrates and invertebrates

We come now to carbohydrate metabolism in the eggs of lower vertebrates and invertebrates. Taking first the amphibia, reference must again be made to the details already given in connection with organiser phenomena, the glycogen breakdown, its distribution in the gastrula, and the rates of catabolism under different conditions in the different regions. That glycogen undergoes a general diminution all over the embryo throughout development is clearly established (Brachet & Needham [2] on *Rana*; Kataoka & Tsunoo on *Cryptobranchus*; Takamatsu [2] on *Hynobius*). In the first of these cases it falls between fertilisation and hatching from 3·25 mgm./40 eggs to 2·15 mgm. and at the same time the proportion of desmo-glycogen becomes smaller (8·9 to 2·6 %). Total carbohydrate, as Savage [2] has shown, also falls. Expressed as loss/embryo in mgm. this varies from −0·065 in the case of *Rana*, *Bufo* and *Discoglossus* to −0·15 in the case of *Bombina* — somewhat greater than the glycogen losses just referred to.

Nowiński [3] extended the investigation of the course of carbohydrate breakdown to amphibian development. Taking larvae between loss of yolk and beginning of feeding, he established that there is the same substrate preference as that shown by the chick embryo, glycogen and hexosediphosphate being unattacked and glucose strongly broken down. This glucolysis was also inhibitable by *dl*-glyceraldehyde. Tadpole autoglycolysis, like that of the chick, decreases with age. It may be added that an amylase acting on glycogen and a glycerophosphatase have been found in amphibian eggs by I. Takahashi [2] and by Takamatsu [1].

As regards phosphorus compounds, a phosphagen, later shown to be identical with phosphocreatine, was found in frog eggs at fertilisation to the extent of 0·016 mgm./100 eggs by Zieliński [1]. About the 55th hour of development, during the closure of the neural folds, the phosphagen reaches a minimum of about 0·008 mgm./100 eggs, after which it rises to 0·05 mgm. at hatching. A similar rise occurs in fish embryos (Rosenfeld & Bagdassarianz), and the whole course of events in the frog embryo was strikingly confirmed by Verjbinskaia. Total acid-soluble phosphorus rises much further and more continuously, as also does

* There is, however, the observation of Watchorn & Holmes that the ammonia and urea production of explanted embryo rat kidney is reduced in presence of fructose.

the inorganic phosphate, derived, as we shall see (p. 637), from breakdown of lipin substances. The late rise of phosphagen accompanies the onset of muscular motion, but the meaning of the initial fall remains obscure. It might be tempting to relate it to the energy requirements of neural induction. Zieliński[1] also showed that the amphibian embryo contains a good deal more creatine than that which is present as phosphagen. His determinations of the partition of the acid-soluble phosphorus were extended by J. Brachet[22] and Verjbinskaia to the period covering gastrulation, but no marked variations were found to occur during that process. But the rise in total acid-soluble phosphorus, which goes on throughout development, is almost entirely due to the difficultly-hydrolysable fraction which in the neurula accounts for 55 % of the total (cf. the chick embryo, p. 614). Adenyl-pyrophosphate, however, is at the end of the yolk-sac period about double its original amount (Verjbinskaia).

In Dorris' thorough work[1] on the development of the digestive system in tadpoles, it was found that the first enzyme to appear is amylase. This precedes the histological differentiation of the gut, whereas the appearance of pepsin and trypsin follow it.

The only study of carbohydrate changes in a developing fish egg is that of Hayes & Hollett on the teleost, *Salmo salar*. Glycogen is absent at the beginning but increases steadily at the later stages, though not in the liver till after hatching (cf. the "foie transitoire", p. 609). The free glucose concentration also rises. These authors propose a correlation between cleavage type and the nature of the store of carbohydrate initially present. Holoblastic eggs (e.g. those of annelids, insects, echinoderms and amphibia), according to all the data we possess, have glycogen; meroblastic eggs (e.g. those of fishes and sauropsida) have free glucose, which is more readily diffusible. There seems to be nothing against this interesting suggestion, which is indeed supported by the abundant polysaccharide of molluscan eggs (see p. 29), except the glycogen of the meroblastic cephalopod egg, referred to in the following paragraph.

Among invertebrates it is also usual for glycogen to fall markedly during development, but in some cases polysaccharide seems to be a more important egg material than in others. There is agreement that in silkworm development the glycogen almost disappears, falling, e.g., from 1·2 gm. % to 0·05 (Tirelli[3]; Minoya). Histochemical observations on nematode (Giovannola) and sea-urchin (Kurkiewicz) embryos confirm this, and in the grasshopper egg the free glucose falls during pre-diapause (D. Hill) resembling in this way the hen's egg and agreeing with the R.Q. data. For the sea-urchin we have also chemical estimations of glycogen by Zieliński[2] and Örström & Lindberg. The eggs of squids and crabs also have large supplies, according to Chaigne (up to 1·5 %), and we are reminded of the work of F. May and his collaborators, (p. 29), which revealed the quantities of galactogen amassed in the eggs of land-snails. A glance at Table 10 will show, moreover, that the eggs of certain invertebrates do, in fact, consume a relatively large proportion of carbohydrate during their development.

On the other hand, there are a few reports to the contrary. Thus Dyrdowska finds that glycogen does not decrease during early developmental stages of nematodes, and Nouvel states that it actually increases during the embryonic development of orthonectid parasites of ophiuroids, such as *Rhopalura*.

Phosphorus compounds related to carbohydrate metabolism have been studied in the developing orthopteran egg by Thompson & Bodine[2] and Steinhart, in that of the squid by Needham, Needham, Yudkin & Baldwin, and in sea-urchin eggs by Zieliński[2]. The most interesting of these papers is the first. Pyrophosphate phosphorus was found to be high at the beginning and to drop sharply till the beginning of diapause, after which it remained constant till the end. Since the decrease in adenylpyrophosphate-content closely parallels the falling course of the respiratory quotient (see Fig. 312), these observations support the notion of a carbohydrate period in early embryonic life, but in this case under conditions such that the adenylpyrophosphate cannot be resynthesised, or not as quickly as would be necessary to maintain its original level. In the locust egg, according to Steinhart, there is twice as much adenylpyrophosphate at hatching as at the beginning of development. In the squid embryo, *Sepia officinalis*, Needham, Needham, Yudkin & Baldwin found phosphoarginine from the earliest stage examined (36 days). It increased tachyauxetically ($k = 1\cdot54$ on the wet wt.), giving peaks at various stages when related to total water-soluble phosphorus or inorganic plus labile phosphorus. This increase resembles that seen above in the chick embryo (p. 615) and must be analogous to the rise at the later amphibian stages. Zieliński's figures[2] for the sea-urchin cleavage stages again show the relatively large amount of barium-non-precipitable difficultly-hydrolysable material characteristic of chick and frog embryo and tumours. No significant changes were observable in inorganic or adenylpyrophosphate fractions, but it is interesting that after iodoacetate treatment there was no accumulation of hexosediphosphate or decrease of adenylpyrophosphate, only a diminution of inorganic phosphorus and an increase in the non-precipitable fraction. This contrasts with similar experiments on the chick embryo with fluoride (Needham, Nowiński, Dixon & Cook).

In locust eggs Steinhart could find no phosphagen till an advanced stage of development. But the preceding paragraphs have contained a number of indications that in early stages (chick, frog, grasshopper, etc.) phosphocreatine and adenylpyrophosphate seem to be functional. How this affects the possible general occurrence of non-phosphorylating glucolysis at those stages remains to be seen.

3·43. Protein metabolism

Several aspects of the protein metabolism of embryos have already been considered in relation to the cleidoic and non-cleidoic systems constituted by different types of eggs (Section 1·64, pp. 50 ff.). Something will be said later (p. 671) on the structural aspects of proteins. Here, therefore, we shall be mainly concerned with amino-acid and enzyme changes during development.

3·431. Protein metabolism in the development of the higher vertebrates

The first question which arises is that of the relative rates of utilisation of the various proteins of the yolk and white, described on pp. 8, 15. There is work on this by Schenck, whose data, however, are rather erratic, and there is the less ambitious but more convincing paper of T. Onoe[2]. From this it seems that the vitellin and livetin of the yolk are absorbed at about the same rate, while the vitellomucoid (although there is very little of it) is absorbed at a more rapid rate during the second half of incubation. In the white, ovomucoid is absorbed much faster than the albumens (measured as total coagulable protein). Precisely the same relationships hold for the egg of the marine turtle *Thalassochelys* (Kusui[2]; M. Takahashi).

In the former survey of these questions (*CE*, Section 9·2) the rise and fall of certain amino-acids in the whole closed system of the hen's egg was discussed. The problem is a fair one, but unfortunately the changes are relatively very small, and it is unlikely that any of the technical methods which it has been possible so far to apply to them are capable of registering them accurately. In spite of much contradiction, however, the general picture appeared to be a decrease in the monoamino-acids and an increase in the diamino ones, together with specific decreases in tryptophane, tyrosine and cystine, and increases in histidine and lysine. But this is still subject to controversy, as the work of Calvery[3] shows; he believes that he has established a decrease in histidine. A great deal of labour was devoted by Schenck[1] to the analysis of protein samples from the whole embryo, the yolk and the white, at different stages of incubation, in order to prove that the amino-acid composition of the same protein varies during development. This would have fitted in with other studies by the same author in non-embryological fields (Schenck[4]; Schenck & Kunstmann; Schenck & Wollschitt), but his technique was strongly criticised by Abderhalden and he cannot be said to have demonstrated either chemically or statistically that the variations he found were beyond the range of experimental error. The same criticism applies to his claim that the composition of feather keratin varies during embryonic development.

Work of a somewhat more careful kind was that carried out on the developing pig embryo (0·313 to 725 gm. wet wt.) by Wilkerson & Gortner, who used the van Slyke nitrogen distribution method on hydrolysed embryo acetone powders. Amide, humin and cystine nitrogen showed no significant change. As in the whole hen's egg and in the eggs of salamander and teleostean fish (*CE*, pp. 1109 and 1112), there was a slight decrease in the monoamino-nitrogen. Arginine, histidine and tyrosine decreased, lysine increased; though caution should be exercised in accepting these statements, since the powders used were not purified proteins.

Wilkerson & Gortner were interested in the relative decrease in arginine which they observed in the embryo pig because they regarded it as possibly of re-capitulatory significance, since tumour protein (Edlbacher) and proteins of invertebrates are often said to contain much arginine, though this is not in accord with the determinations of Rosedale & Morris. A similar decrease of arginine in the developing chick embryo is reported by Rashba and Palladin & Rashba, but it would hardly be justifiable to regard it as yet established. That the enzyme arginase decreases in activity during embryonic development was first noticed by Edlbacher & Merz on the guinea-pig, and later fully established on the chick (Brachet & Needham[1]; Needham, Brachet & Brown), where it "grows" bradyauxetically with a k of only 0·48. The latter authors also showed that this decrease in arginase activity was not due to a diminution of activating substances, such as glutathione or ascorbic acid.

Injections of arginine into the hen's egg by Kamachi[2] led to increases in histidine, but the experiments were few and unconvincing. So also is the claim of Goldberger that the 7-day chick embryo contains no free amino-acid nitrogen. This subject deserves much more attention than it has yet received, for the problem of whether the proteins of yolk and white are converted into those of the embryo, with or without complete scission to free amino-acids, is a very real one. Perhaps the use of isotopes will help in its solution. Deuterium (Ussing[2]; Krogh & Ussing[2]) and isotopic nitrogen (Schoenheimer & Rittenberg[2]) are built into the protein of the embryonic body.

Finally, proline and oxyproline manifest no change during the development of the chick embryo (R. Ido). Alone among amino-acids, arginine increases the creatinine of the egg (I. Takahashi[1]) when injected, and another Japanese worker, K. Ono, states that injected creatinine is destroyed.

Coming now to the question of nitrogen excretion during the development of birds, there is little to add to the previous discussion (*CE*, pp. 1076 ff.). The successive production of ammonia, urea and uric acid remains a remarkable fact, but it would be very desirable to know to what extent the extra-embryonic membranes contribute to the nitrogen excreted, and in what form. Nothing on these lines has yet been attempted. How exactly the catabolism of protein originates in the chick embryo is unknown; the *in vivo* maximal ammonia production occurs on the 4th day, that of urea on the 9th and that of uric acid on the 11th (see *CE*, Figs. 323 and 324). When the embryo of the 4th–5th day is studied *in vitro*, as by Dickens & Greville[3] and by Needham[8], the ammonia production is considerable in the absence of added substrate (glucose) and the R.Q. is 0·75, but in its presence the R.Q. is unity and the ammonia production very small. Thus the embryo at this age certainly has the ability to catabolise protein, as is shown further by the fact that if fluoride is given as well as glucose, the ammonia production rises to a level just sufficient to account for the whole of the respiration as protein catabolism, though the R.Q. is in this case unchanged (Needham[8]). But it had already proved impossible (Needham[2]) to postpone the normal rise in nitrogen excretion by injecting glucose into the egg at the beginning of development. There is therefore some intrinsic mechanism governing the transition from carbohydrate to protein catabolism, of whose nature we are still ignorant.

In this connection we cannot accept the contention of Glover that no deamination of amino-acids by 5-day chick embryo tissues takes place, all the more so as this is not borne out by his figures, which do show a rise of ammonia above the controls, ranging from 12 % in the case of glutamic acid to 77 % in the case of proline, with an average of 33 % for all amino-acids. No doubt the deaminases are weak. One curious thing about the "protein-sparing" action of glucose in Dickens & Greville's experiments[3] is that it is shown also by fructose, a carbohydrate which is not anaerobically converted to lactic acid by the embryo

(Needham & Nowiński). Rat embryo tissues, according to Dickens & Greville, behave in a similar way to the chick embryo, but the rat yolk-sac produces practically no ammonia whether glucose is present or not, though in absence of glucose its respiratory rate does not fall. This is unexplained. Jensen rat sarcoma behaves like the chick embryo.

If ammonia production may be assumed to be by direct deamination, there was room for much speculation regarding the origin of the chick embryo's urea, till Needham, Brachet & Brown showed that it arises, not from general protein catabolism by way of the ornithine cycle, nor from uric acid by Stransky's complex (uricase, allantoinase and allantoicase, see p. 52), but entirely from the arginine-arginase system. The embryo thus "recapitulates", not the ornithine ureotelism of amphibia and chelonia, nor the uricase complex of amphibia and fishes, but rather the arginase system itself, widely present throughout invertebrates (Baldwin[1]). The production of urea from arginine by arginase is also an adult avian characteristic, and is the sole mode of origin of urea in adult birds. It was furthermore shown by Needham, Brachet & Brown that the urea formed cannot participate in the synthesis of uric acid by the embryo, in agreement with the views of Clementi[2] and his school. By what means the uric acid of the embryo is synthesised remains mysterious, but a discussion of this process in adult sauropsida has already been given (p. 55) in connection with the cleidoic question, to which the reader is referred. During development the blood uric acid, measured by Zorn & Dalton, slowly but steadily rises.

Nitrogen excretion in the chick embryo has, of course, morphological aspects. Boyden[2] was able to correlate the development of the kidney with the development of nitrogen catabolism. Uric acid excretion begins on the 5th day. By the 13th day the allantois contains between 8 and 13 mgm. uric acid dissolved in 6 c.c. water. The mesonephros is therefore clearly functional. Now between the 5th and 14th days the whole Wolffian body increases its volume 12 times. Though there is no increase in glomerular count, the tubule volume increases 16 times, the corpuscle volume 3 times, the collecting tubule volume 10 times but the secretory tubule volume 33 times. The glomerular volume of the 14th day mesonephros is 50 times that of the metanephric glomerulus at the same age. Thus the growth of the chick embryo kidney during the main period of nitrogen excretion (the protein-combustion phase) is located in the secretory portion of the tubules. In 1931 it was suggested by Needham[4] that all the uric acid in the allantois is free. Thus the pH of the allantoic liquid about the 15th day is 6·3, it contains 10 times too little ammonia to account for the acid as ammonium urate, and the inorganic anions balance the inorganic cations (Kamei; Iseki[1]). There might thus be a cycle of base as well as water in the conveyance of uric acid into the allantois, similar to that shown to occur in the Malpighian tubes of insects by Wigglesworth[1]. Uric acid is deposited in the form of crystals during the latter half of incubation, rendering the allantoic liquid milky, and these are rendered harmless by the mucus which Rossi's allantoic vesicles secrete. The existence of the cycle of base was later demonstrated by Białaszewicz & Głogowska, who found that while the absolute amount of total nitrogen in the allantoic contents increased steadily throughout incubation, the amount of sodium increased to a maximum of 25 mgm. on the 12th day and then rapidly fell off almost to zero by the time of hatching. It must be concluded that before the 12th day excretory processes predominate over absorptive ones as far as sodium is concerned, but that after that time the opposite occurs. The potassium does not show this peak but increases steadily until at the end of incubation it reaches an amount of about 7 mgm. The maximal volume of the allantois is reached at the same time as the sodium maximum (10·3 c.c.); at the time of hatching it is only 2·1 c.c.

For mammalian embryos the placental route of excretion is, of course, open (CE, p. 1122), and the absence of urea from foetal bladders cannot therefore mean, as Chouke supposes, that none is excreted by the foetus till birth.

The onset of function in the kidney has often been studied, and the most recent point of view will be found in the paper of Gersh. It is accompanied, according to Flexner & Gersh, by an unusually sudden rise in oxygen-consumption of the tissue, and an increase in histologically detectable cytochrome (Flexner[2]). The one-way excretion of sulphonephthalein dyes through the mesonephric cells discovered by Chambers & Cameron was elegantly studied by Chambers & Kempton, who were able to show the accumulation of dye in explanted tubules which had spontaneously closed themselves. This transfer polarity could be destroyed by cold or respiratory inhibitors (Chambers, Beck & Belkin; Beck & Chambers), and it appeared later in vitro than in vivo. In fishes the approach of blood-vessels to the tubules seems to be the most important limiting factor (Armstrong). The pronephros of amphibia is capable of much more excretory work than it ever has to perform under normal conditions, as is shown by the parabiosis experiments of Hiller[3], in which only one pronephros could take care of two animals.

As we have already seen (p. 72), the proteases of the yolk-sac are many times more active than those of the embryo. The best study of the latter is that of Mystkowski[2], who noted very little change in activity of the kathepsin of the embryonic tissues during the development of the chick embryo, whether activated by cysteine or not (P.Q. 0·27). At first the tail portion has more than the head, and as time goes on, the enzyme becomes mainly localised in liver and kidney. Mystkowski could not find any synthesis of protein by chick embryo kathepsin in vitro. On the other hand, a certain increase of kathepsin activity during the development of embryos has been reported by Kolobkova for the chick and by Goldstein & Millgram for mammals. By the use of micro-methods it has been established that the ectodermal parts of the 3-day chick embryo head have three times as high a dipeptidase activity as the mesodermal parts (Levy & Palmer; Palmer & Levy). Enzymes of the intestinal tract and its derivatives would naturally be expected to increase gradually during development, pari passu with the histological changes, and this seems in fact to happen (Galvialo & Goryukhina on chick embryo stomach proteases; Orechovitch[3] on chick embryo liver kathepsin and dipeptidase; Michejev on rennin in the stomach of cow foetuses; Blum, Jarmoschkevitch & Jakovtschuk and Cardin on the intestinal tract of human embryos). In the last-named case enzymes appeared in the following succession: dipeptidase at the 2nd month; kathepsin at the 3rd; urease at the 4th; and tryptic protease and carboxypolypeptidase at the 5th.

An interesting but isolated observation is that of asparaginase in chick embryo liver at the 8th day by Impallomeni. Since suspicion has fallen upon glutamine as concerned in uric acid synthesis, there is a way open here for some interesting experiments.

3·432. Protein metabolism in the development of lower vertebrates and invertebrates

Passing now to what is known of nitrogen metabolism in amphibia and fishes, we find that the newer work is very fragmentary. That the eggs of all fishes and amphibia excrete a certain amount of ammonia into the surrounding water has long been observed (*CE*, pp. 1132 ff.), but the only up-to-date work on this is that of S. Smith[1], who used the eggs of the trout, and Ivlev, who worked on *Silurus glanis*. As already mentioned (p. 49) the trout embryo resembles other aquatic embryos in combusting a relatively large quantity of protein. Its ammonia nitrogen excretion rises from 20–50γ/100 eggs/hr. to 150γ/100 eggs/hr., accounting for at least 80 % of the excreted nitrogen; the rest seems to include amino-acids, possibly arising from proteolysis of the egg-membrane (cf. hatching enzymes, p. 627). Felix, Baumer & Schörner, also working on the trout, noted the presence of dipeptidase and arginase in the egg at the outset, later falling in activity, and no tryptic protease, kathepsin nor polypeptidase. The amino-acids arginine, tyrosine, histidine and cystine remained constant, but there was a fall in tryptophane and lysine. Little advance has been made with the metabolism of selachian embryos, but the presence of bases in dogfish embryos, such as betaine, choline and trimethylamine-oxide (already discussed in *CE*, p. 345), has been fully confirmed by Spahr and Kutscher, Muller & Spahr.

In Section 2·351 we had occasion to say a good deal about carbohydrate — and even lipin — metabolism in the developing amphibian embryo with reference to organiser phenomena. Protein changes have not been overlooked. According to Graff & Barth, there are no changes in total cystine during cleavage, but between early gastrulation and hatching there is a large increase, perhaps at the expense of methionine. Arginine also increases, but very slightly. Analysis of the protein metabolism of the amphibian gastrula urgently needs continuing. In the experiments of Boell, Needham & Rogers, the anaerobic ammonia production of the dorsal lip region was found to be three times as high as that of the ventral ectoderm:

	Ammonia production μl. $\times 10^{-3}/\gamma$ dry wt/5 hr.
Early gastrula ectoderm	0·97
Middle gastrula dorsal lip region	2·97
Neural tissue from open neural folds	2·70
Neural tissue from just closed neural folds	1·30
Later neural tube and somites	0·98

The ammonia production, therefore, seems to be higher during gastrulation than just before or just afterwards. But this is not detectable when aerobic ammonia production of the intact whole embryo is measured (J. Brachet[22]).

G. Pickford is studying the dipeptidase distribution in the regions of the gastrula; it seems to be inversely proportional to the yolk-content, as would be expected from the results on echinoderm eggs (see pp. 419, 627). Ammonia is not the sole nitrogenous end-product of amphibian eggs (see *CE*, p. 1107). Takamatsu & Kamachi found 4 mgm. % of urea in *Cryptobranchus* embryos at the 4th week, but only traces of uric acid; and Takamatsu[1] reports a similar result for *Hynobius*. A. F. Munro, in an interesting paper, has shown that the ammonia excretion, pre-dominant in the tadpole, gives place during metamorphosis to a urea excretion (dropping from 90 to 19 % of the total N), while at the same time there is a twentyfold increase in the liver arginase, presumably for the ornithine cycle (see p. 52). These changes, followed in *Rana*, *Bufo*, etc., do not occur in the permanently aquatic toad *Xenopus*.

As regards enzymic activities in the amphibian embryo, tryptic protease and arginase, but not urease, were found in *Hynobius* eggs by Takamatsu[1], but none of these could be recognised in those of *Cryptobranchus* by I. Takahashi[2]. The most interesting study in this field was that of Truszkowski & Czuperski, who could find no uricase in frog eggs, but observed its appearance during develop-ment in two distinct increases, the first at the time of disappearance of the external gills, the second at the time of completion of limb-bud growth. The former sudden rise was associated with the onset of liver function, the latter with that of the kidneys. Both organs contain uricase, of course, in the adult frog.

When we come to the invertebrates, we find that interest centres mainly round two forms, the silkworm moth and the sea-urchin, the one a typically terrestrial and the other a typically aquatic egg.

In the developing silkworm egg, by all expectations (see p. 56) uric acid should accumulate. That it does so appears from the figures of a number of workers (Akao[3]; Tanda; Manunta[2]), although the production of uric acid seems to be much larger during larval (Kuwana[3]) and pupal (Brighenti) life. According to Manunta's figures[4], most of the uric acid present at the time of hatching is already contained in the undeveloped egg, suggesting that the maternal organism deposits it there. This curious result demands reinvestigation. So also does the persistent statement, originating from the work of Ashbel[4] and Ongaro[3], that developing silkworm eggs produce as much as 0·5 gm. ammonia/kilo wet wt./hr. This is quite irreconcilable with uricotelic metabolism, especially as Manunta[4] found no trace of uricase in the eggs, and seems likely (if true) to be due to the activities of the abundant bacterial flora with which, according to Masera, silkworm eggs are provided. Until sterile silkworm eggs are available, this question will not emerge from the state of uncertainty surrounding till recently the nitrogen excretion of the blowfly larva (see p. 56). Lastly, an increase of urea in the developing silkworm egg has been reported by Bottero. This also would repay further investigation, as it resembles the appearance of urea during the development of the chick embryo, just discussed (p. 622). Another insect egg, that of *Antheraea pernyi*, analysed by Leifert, showed uric acid as the overwhelmingly preponderant part of the sum of excretory nitrogen. And grasshopper eggs (*Melanoplus*) lose no nitrogen during their development (Bodine & Trowbridge).

The developing echinoderm would be expected to give off a certain amount of ammonia, arising from protein breakdown. This it undoubtedly does. Ashbel[4] first and then Örström[3] found an excretion of ammonia amounting to 1 micromol./c.c. eggs/hr. before fertilisation (*Paracentrotus*). After fertilisation, according to Ashbel[4], the ammonia production continues unchanged, but according to Örström[3] it decreases to zero. Conversely, in anaerobic conditions the unfertilised eggs produce 0·5 micromol, while the fertilised ones produce 1·3 micromols. Other amines may also be excreted (Örström[3]).

But this is by no means the whole story. In 1933 Hayes[2], studying the Floridan species *Echinometra lucunter*, reported a rise in the total nitrogen of the eggs between fertilisation and the pluteus stage (see Fig. 323).* This was generally regarded at the time as incredible, but Örström[4] quite independently discovered in *Paracentrotus* eggs an ability to bind considerable amounts of ammonia from the environment. This binding capacity is in the neighbourhood of 3 micromols ammonia/c.c. eggs/hr. aerobically and 10 anaerobically for the unfertilised eggs, and no less than 50 micromols aerobically and 12 anaerobically for the fertilised eggs. The ammonia taken up goes into organic "unhydrolysable" combination, and the amount can be increased by the simultaneous addition of aspartic and glutamic acids, histidine and ornithine. Örström interprets these facts as meaning that an oxidative amination of keto-

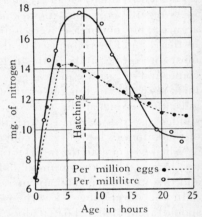

Fig. 323. Nitrogen content of the eggs of the sea-urchin *Echinometra* (see text).

acids occurs after fertilisation. The full publication of his results will be awaited with interest; should they be confirmed, they would add a further remarkable chapter to the natural history of the cleidoic egg (pp. 34 ff.), for we should have to admit that in some non-cleidoic eggs, dependence upon the environment would go as far as a proportion of the potein-building element itself. Örström's work should be extended as soon as possible to eggs of invertebrate species more primitive than the echinoderms.†

---

* No such rise occurs during the development of *Urechis* from egg to trochophore (Horowitz).

† There is another case of this nitrogen-intake recorded in the literature, namely that of the eggs of the gastropod *Hemifusus tuba* reported by Kumon, and in the light of Örström's work it should certainly be reinvestigated. Organic substance was also found to *rise* slightly during development by Kamachi[3] on the cephalopod *Loligo bleekeri* and even by Takamatsu[1] on the amphibian *Hynobius nebulosus*. There is now just a possibility that these workers were not mistaken; cf. the review of Krogh[2] on the dissolved organic substance of sea- and fresh-water.

The nitrogen partition of the developing eggs of the gephyrean worm, *Urechis*, has been followed by Horowitz[1]. One can hardly believe that no appreciable protein catabolism goes on in these embryos, as would appear from the stability of the total nitrogen figures. Ought we perhaps to beware of compensation by a nitrogen intake here?

Another direction in which results of much importance have been derived from experiments on marine invertebrate eggs is that of the location of intra-cellular enzymes. Approaching the subject with the ultra-micro-chemical technique developed by the Carlsberg school (see Linderstrøm-Lang & Holter[2]), Linderstrøm-Lang[1] in 1933 estimated the dipeptidase activity in two marine invertebrates, the gephyrean worm *Urechis* and the sand-dollar *Dendraster*. There appeared to be twice as much alanyl-glycine dipeptidase in the latter as in the former. When *Dendraster* eggs were cut into half, the activity of the two halves was approximately equal to that of the whole egg, showing that the presence of the nucleus was immaterial. On fertilisation in *Urechis* the activity fell, and later rose and fell again. The work was continued by Philipson, who found that the distribution of this enzyme was uniform among all the layers formed in *Psammechinus* eggs by centrifugation. The conclusion, that this dipeptidase is associated with the hyaline ground substance of the egg and not with any of the formed elements, was finally clinched by Holter[1] (using *Arbacia*, *Echinarachnius* and *Chaetopterus*), and this fully agrees with the results of Holter & Kopać on the amoeba.* It seems that some egg dipeptidases need activation by magnesium; this is so for leucyl-glycine dipeptidase in *Tubifex* eggs, but not for alanyl-glycine dipeptidase (Holter, Lehmann & Linderstrøm-Lang).

Particularly interesting was the attempt to follow the dipeptidase through the crucial determination periods. Doyle first found that there is no change in dipeptidase activity during fertilisation (*Psammechinus*), and Holter, Lanz & Linderstrøm-Lang then went on to show that all eight blastomeres at the 8-cell stage have the same amount of the enzyme. This is important, because the third cleavage is equatorial (see p. 479) and separates the future endodermal material and future micromeres from the future ectoderm. Finally Holter & Lindahl found that neither animal and vegetal halves of blastulae and gastrulae, nor ectoderm and endoderm of plutei, differ appreciably in their dipeptidase activity.

3·433. Hatching proteases

Proteases of very specialised function are those which are secreted into the perivitelline cavity in nearly all phyla by embryos towards the time of hatching (*CE*, p. 1595). These digest the protein of the egg-case or membrane. Proteolytic hatching is of very wide occurrence; apart from examples previously given, it is reported, with much detail, for amphibia (Cooper; Nedler), selachian fishes (Wintrebert & T. Y. Wang; T. Y. Wang), teleostean fishes (Bourdin), ascidians (Berrill[1]), orthoptera (Slifer[6]), cephalopods (K. C. Yung; Jecklin; Hibbard[2]),

---

* This is in contrast to the amylase of the amoeba which was shown by Holter & Doyle to be clearly associated with the mitochondria.

echinoderms (Ishida), and trematodes (Onorato & Stunkard). In many cases, the enzyme is secreted by special organs or glands which disappear after hatching, such as the organ of Hoyle in cephalopods or Stunkard's glands in trematodes, or the pleuropodia (organs whose function had been a mystery for a hundred years) in grasshoppers. The pleuropodia were proved by Slifer's ligature experiments to be the origin of the hatching protease. Proteolytic hatching seems to be a later development in evolution than the earliest method, osmotic hatching (cf. *CE*, p. 1600), and it lasted on until the hard-shelled sauropsid egg necessitated more violent measures. This problem solved itself, however, since the cleidoic egg permitted the embryo to remain within it until it could hatch, as it does in the gallinaceous fowls, in a state almost of adult perfection, and hence of considerable muscular power. Moreover, as the review of Sergeiev reminds us, there are special adaptations to the hard-shelled egg such as the transient ectodermal proliferations known as "egg-teeth" or "egg-tubercles".

3·434. The growth-promoting factor

In a paragraph at the beginning of this section, reference was made to the problem of whether the proteins of yolk and white are broken down to simple amino-acids before being built into the embryonic body, or whether this transformation takes place rather by way of intermediate substances of proteose type. This question would be important enough if its embryological interest alone were considered, but it has also remoter repercussions.* The possibility arises that such a proteose is identical with that which has been thought to be the cause of the growth-promoting power of embryo extract. In order to explain this, a very brief summary of the development of tissue-culture technique is necessary.†

R. G. Harrison's first tissue cultures[2] in 1907 were made with adult lymph, and it was Carrel[1] who introduced in 1913 the practice of using saline extracts or press juices of chick embryos, which were found to be far more effective in promoting the growth of the explanted cells.‡ Since that time, the effect of age has been studied in two ways, by varying the age of the tissue explanted, and conversely by varying the age of the donor of the extract. It may now be said that the effectiveness both of growth-promoting power, and of capacity to grow, decrease with advancing age. Taking the latter point first, Cohn, Murray & Rosenthal showed that the growth capacity of cells from the chick embryo falls off very regularly during development, while the latent period increases (see *CE*, Fig. 62§); this has frequently been confirmed (Hoffman, Goldschmidt & Dol-

---

* It could almost certainly be solved by the introduction of amino-acids labelled with the heavy nitrogen isotope into the hen's egg at the beginning of development, along the lines to be mentioned later in connection with lipin metabolism (p. 638).

† For further details, consult the reviews of A. Fischer[4,5] and E. Mayer[2].

‡ Of course even in the presence of embryo extract cultures eventually cease to grow. Baker[2] has shown that this is probably due to an exhaustion of the nutrient materials in the serum, for growth will continue indefinitely if fresh serum is provided from time to time.

§ Significantly, it is said that cancer cells, when explanted, show practically no latent period (Doljanski & Hoffman).

janski; Medawar*), but if the cells are grown for some time *in vitro* and sub-cultured, it can no longer be demonstrated (Parker[2]). But the former point is of more interest here. That the growth-promoting power of hen plasma falls off with age was proved by Carrel & Ebeling, and it does so more or less *pari passu* with the cicatrisation power of wounds, the declining curve for which was worked out by du Noüy. Confirmatory evidence for this was provided by Parker[1] and others, and it became generally accepted that the period between 7 and 10 days of incubation was the optimum time for the preparation of the embryo extract. However, this has since been modified by the elaborate work of Gaillard[1]; Andai; and Fowler, who agree that the maximal growth-promoting power is reached not at that time but somewhat later, between the 11th and 17th day of incubation, though explanted tissues differ a certain amount as to the optimal age of the embryo providing the extract. Gaillard & Varossieau have also shown that the age of the donor embryo affects the direction in which *in vitro* differentiation proceeds. It makes no difference from what region of the chick embryo the extract is prepared (Fowler; Hueper, Allen, Russell, Woodward & Platt).

Much work has been done on the distribution of the growth-promoting factor and its effects. It may be prepared from chick embryo blood (Zifferblatt & Seelaus) and in smaller amount from the maternal tissues of pregnant mammals (Centanni; Pybus & Fawns); it accelerates the growth of planarians (Pettibone & Wulzen) but not that of mammals in the post-natal period (Kaufman[5]). Certain adult tissues possess more of the factor than others, e.g. brain, spleen, thymus and thyroid (Trowell & Willmer; R. S. Hoffman). Pomerat & Willmer have sought to explain this on the basis of a common possession of glucolysis inhibitable by glyceraldehyde (p. 611), but as the glyceraldehyde inhibition of tissue cultures is unspecific (shown by all aldehydes) it is not certain that this is a useful line of thought. The growth-promoting factor in embryo extract can be completely inhibited by lipin preparations, such as the petrol-ether extract of mammalian brain used by E. Mayer[1], but this inhibition is not reversed by heparin. Indications of specificity are shown by findings such as that of Horning & Byrne, namely that tissue juice of tumour-bearing animals is more favourable for *in vitro* growth of tumour tissue than tissue juice from normal animals. The most interesting recent results here are those of Willmer and his collaborators, who have established that embryo extract will start growth again after tissue in culture has stopped growing (Willmer & Jacoby), and that the continued presence of embryo extract is not necessary for the mitoses and migration which it produces (Jacoby, Trowell & Willmer). Embryo extract speeds migration, increases the number of mitoses, and shortens the time taken by each mitosis. The presence of 15 % embryo extract in Tyrode solution for 1 hr. is sufficient to produce a large crop of mitoses many hours later, and within limits the higher the concentration of the extract the shorter the time needed for its effect. If it is present for less than 10 hr. only one crop of mitoses occurs. This reminds us of the problem of the time taken by an inductor to produce its effect (pp. 156, 291).

* Medawar[2] used an interesting method; he found that the younger the cells, the higher the dose of mesoderm-inhibitory factor of malt extract (see p. 359) was required to stop their growth.

What is the nature of the growth-promoting factor? It is partly, though not entirely, destroyed at 70° C. (Carrel[1]; I. Lasnitzki). It was early realised that proteins alone would not do (Smyth; Swezy) and Baker & Carrel[1] went on to show that mixtures of amino-acids were equally ineffective. Indeed, Mayer & Fischer found later that after culturing on amino-acids alone, tissues even lose their power to respond to embryo extract. Polysaccharides were ruled out by Hueper, Allen, Russell, Woodward & Platt. But Carrel & Baker then discovered the important fact that peptic digests of proteins (e.g. egg-albumen, edestin, casein, fibrin) showed considerable growth-promoting power, though by prolonged digestion this was destroyed. The activity of proteoses was later confirmed by Fischer & Demuth and Willmer & Kendal, though as they are ineffective in the absence of plasma or embryo extract,* it is clear that they require some further factor. This was conjectured to be the protease of the plasma, and following this line of thought Borger & Zenker[1] obtained some experimental evidence that the growth-promoting power of embryo extract was proportional to its protease activity.† On the other hand, when Borger & Peters examined the enzymes in 10-day embryo extract, though dipeptidase and amino-polypeptidase were present, their activity was very weak, and no protease could be demonstrated. Later, Hueper and his colleagues did find a weakly acting kathepsin. But it is not necessary to postulate protease activity; some carrier substance of nucleotide type might meet the case as well or better.

The next important step was that taken by Baker & Carrel[2], who showed that the growth-promoting power of proteoses (protein digests) plus thymo-nucleic acid was almost the equal of that of embryo extract. The level of embryo extract was at last reached when Baker[1] made a mixture of casein digest, glycine, thymo-nucleic acid, glutathione and small amounts of haemoglobin, but even here, the activity was not maintained for so long as that of embryo extract. Baker & Carrel[3] further showed that α-proteoses and β-proteoses were equally effective, but after this (1928) the work of the New York school diverged in another direction and this line of work lapsed. Although the success of the organ-perfusion method (Carrel & Lindbergh) is striking and can lead to hyperplasia *in vitro* (Foot, Baker & Carrel[4]), the media used for it are very complex mixtures, including many vitamins and hormones (Baker & Carrel[4]; Baker & Ebeling), so that we obtain no further knowledge of the growth-promoting factor.

The subject was, however, revived in what may be important work by A. Fischer[7], of which only an abstract has yet appeared. From cow embryos he isolated by the Hammarsten method for nucleoproteins a substance which shows great effectivity. The composition of this nuclein fraction seems to be unusual, but it contains nucleic acids of both thymo- and phyto-nucleoprotein type, and it is with the latter of these that the growth-promoting power seems to be associated. If this is so, some light will be thrown on the occurrence of phyto-nucleic acids in animal embryo tissue (Jorpes; Levene & Jorpes; Caspersson).

* Embryo extract need only be added to the proteoses at concentrations of 1 % or less to maintain full activity, suggesting the presence of carrier substances.

† Coagulation factors in embryo extract have been studied by J. T. King.

The active substance is destroyed by short tryptic digestion and by boiling, but a thermostable substance (possibly the same as that described by Willmer & Kendal) is also involved. In the dried state the substance may maintain its activity for many months (Borger & Zenker[2]; Peacock & Shukov; Hetherington & Craig).

We are thus led to a hypothesis which would link together many apparently unrelated facts. Substances of nuclein type are now found to intervene, apart from their role in the cell nucleus, at many previously unsuspected points. Certain nucleotides in the cytoplasm are essential coenzymes in glycolysis, phosphorus transfer, hydrogen transfer, and amino-acid oxidation* (see p. 208). Many plant viruses, and the Rous sarcoma virus, and even a closely similar substance which can be prepared from the chick embryo itself (see p. 269) contain nucleotides. Proteoses plus nucleic acid are most effective, as we have seen, in promoting the growth of explanted cells. Is it not possible that a substance of nucleotide structure might act either as a carrier for the requisite peptone "bundle" entering the embryo at the absorptive surface of the yolk-sac and so protect it from the further action of proteases, or else possibly as a "bricklaying" mechanism at the site of protein synthesis, and that thus in combination with the peptones it might be in fact the growth-promoting proteose? Consideration of such a question requires some knowledge of the behaviour of nucleins during embryonic development, and this will be the subject of the succeeding section. Two facts may, however, be noted at this point. First, the growth-promoting factor is to some extent dialysable (G. P. Wright; Tazima; Jacoby[2]) and although 8th day egg-yolk itself is inhibitory on account of the lipins in it, G. P. Wright[1] was able to prepare a very active dialysate from it. Secondly, the peptone bundle pictured as the lowest building-stone of proteins must be quite small, since the growth-promoting factor of embryo extract is quite non-species-specific (again like inductors), as Kiaer[1] and others have shown.

What may be an important addition to the argument here is the work of Loofbourow and his collaborators on growth-promoting substances produced by yeast cells when damaged by ultra-violet irradiation. The wave-length most lethal for the cells is the same as that of the maximum absorption of nuclein derivatives (Loofbourow & Heyroth). The absorption spectrum of the cell-free filtrate of yeast killed by ultra-violet light or mechanically (Loofbourow, Cook, Dwyer & Hart) differs entirely from that of normal cells, and shows both active growth-promoting power for yeast and an absorption characteristic of cyclic nitro-genous bases (Sperti, Loofbourow & Dwyer). The same is true of breis of the chick embryo, irradiated and non-irradiated, tested on chick embryo heart fibroblasts in explant (Sperti, Loofbourow & Lane; Loofbourow, Cueto & Lane) and of mammalian embryos (Loofbourow, Dwyer & Lane). The active factor can be dried at 65° without injury (Loofbourow, Dwyer & Morgan), and contains guanine, adenine (not uracil), pentose (not desoxy-pentose) and phosphorus (Loofbourow, Cook & Stimson). Since the substance is most amply

---

* Even also in decarboxylation, for, as Baumann & Stare point out, thiamin pyrophosphate is quite possibly linked with ribose and adenine in the cell, thus forming a nucleotide like the other coenzymes.

given off by cells cultured in the most favourable medium, it is viewed rather as a physiological response to injury than as a disintegration product of dead cells (Loofbourow & Dwyer). There can be little room for doubt that nucleotides are here again involved with growth-promoting power. Another indication pointing in the same direction is the curious phenomenon of "chromidial extrusion" noted by many cytologists (Ludford; Horning & Richardson; Horning & Miller) in tumour tissue. Chromatin appears in the cytoplasm and even in the intercellular material; and the more malignant the neoplasm (i.e. the faster its growth) the more pronounced the process is. Tumour tissue contains growth-promoting factor (Horning & Richardson). We shall shortly return to the significance of cytoplasmic nucleotides (p. 636). Allied with this is the observation of not a few cytologists that rapidly growing embryonic tissues frequently show pycnotic nuclei and degenerating cells. In spite of the criticisms of Peter[2]; Glücksmann[1], for example, seems to make good his case for a "*morphogenetische Degeneration.*" * This might mean that the construction of nuclei sometimes normally overruns the amount of cytoplasmic nucleotide, and that the balance is redressed in this way.

A connection between irradiation products and animal tumours has now been made by Loofbourow, Cueto, Whelan & Lane. If minced rat or mouse embryos are irradiated with lethal ultra-violet, and the cell suspension then freed from débris by a Berkefeld filtration before subcutaneous injection into an adult rat, tumour-like masses arise. These are stated to consist of connective tissue and striated muscle. The detailed publication of this work will be awaited with interest.

Some reference must be made here to the acceleratory action of embryo-extract on the healing of wounds, in favour of which there is a surprisingly unanimous opinion (Carnot; Carnot & Terriss; Roulet; Wallich; Bergami; Kiaer[2]; Morosov, Striganova & Skomorovskaia; and Schloss). Its action is presumably due to the growth-promoting substance usually tested in other ways. The same acceleration has also been reported with A. Fischer's purified preparation (Nielsen; Waugh).

3·44. Nuclein metabolism

During the past ten years our knowledge of the nuclein and nucleoprotein metabolism in embryonic development has made considerable advances, mainly centering round the existence of phyto-nucleic acid as well as thymo-nucleic acid in young growing tissues.

According to the classical descriptions phyto- or zymo-nucleic acid carried a pentose as the central part of its nucleotides, while thymo- or animal-nucleic acid had a hexose instead. But the work of Levene, Mikeska & Mori established the fact that in both cases the sugar was a pentose, only in thymo-nucleic acid it is a pentose with two OH groups (2-desoxy-*d*-ribose), while in zymo- or phyto-nucleic acid it is a pentose with three (*d*-ribose)—see diagram.† In the

* So also T. K. Chang[2].

† The nomenclature here is now very unsatisfactory, for plant as well as animal tissues contain both kinds of nucleic acid. But perhaps phyto- and thymo- may be retained on account of brevity as well as for historical reasons.

nucleic acids four nucleotides are joined together, either in chain form or cyclically, more probably the former, by their phosphoric acid groups. Further details will be found in the monograph of Levene & Bass.

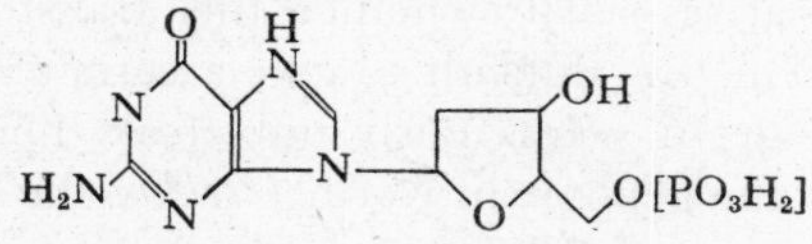

Guanine-2-desoxy-d-ribose nucleotide,
typical of thymo-nucleic acid

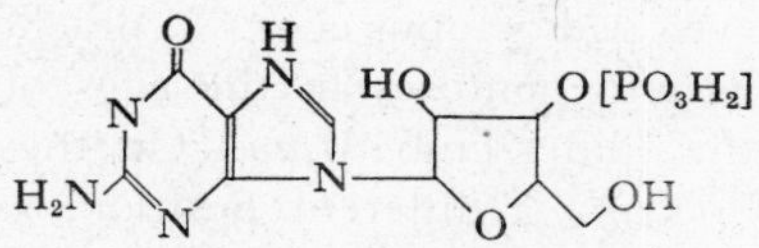

Guanine-d-ribose nucleotide,
typical of phyto-nucleic acid

The role of nucleins in embryonic development may best be approached by way of the sea-urchin egg. Long ago Godlewski[1] and J. Loeb[2] observed an unmistakable increase in total nuclear material during its development from fertilised egg to free-swimming pluteus. Up to the blastula stage alone the increase is of the order of 50 times. But as time went on a succession of workers (Masing[1,2]; Shackell; Needham & Needham[2]; J. Brachet[5]) found that neither in nucleoprotein phosphorus nor in purine nitrogen is there any change during the development of the echinoderm egg. Nevertheless J. Brachet[1], who applied the Feulgen reaction, which is specific for thymo-nucleic acid, to the different stages of development, found a consistent rise. He confirmed this further[2] by the Dische colorimetric method for thymo-nucleic acid, obtaining the following figures:

	Mgm./gm. dry wt.
Fertilised eggs	0·1
Gastrulae	6·2
Plutei	13·6

There was thus a plain contradiction between the results of the various methods. It was resolved by J. Brachet himself[3] when he estimated the pentose-content* of the embryos, which was found to decline from 6·35 mgm./gm. dry wt. in the fertilised eggs to 2·6 mgm./gm. dry wt. in the plutei. The conception was thus developed of an initial store of phyto-nucleic acid nucleoprotein which during development is gradually transformed into thymo-nucleic acid.† The former

* Reference here and later in this section to estimations of "pentose" does not include the desoxy-pentose of thymo-nucleic acid, which gives no furfurol under the same conditions.

† And during maturation a converse process would have been going on (Brachet[23]). The conclusions of J. Brachet were questioned by two investigators. G. Schmidt[1] believed he could isolate thymo-nucleic acid from unfertilised echinoderm eggs, but he relied only on the Feulgen test, and after criticism from Brachet[9], later withdrew his claim[12]. The Feulgen test, in the absence of proper precautions, is given also by aldehydic phosphatides ("plasmal"), see p. 190; if possible it should never be used in vitro. Hen egg-yolk, as was found by Marza & Marza[2], gives the plasmal Feulgen reaction, though it certainly contains no nuclein. Blanchard also isolated small amounts of thymo-nucleic acid from unfertilised echinoderm eggs, but in this case his material was almost certainly contaminated by ovarian tissue.

must be in the cytoplasm;* the latter appears in the increasing total nuclear volume (Brachet[5]). Moreover, the undeveloped egg contains no free purines (Brachet[11]).

The question next arises as to whether this transformation[†] is a general phenomenon. At present it is difficult to give a certain answer, but for marine invertebrates, at any rate, it seems fairly universal. The work of Needham & Needham[2] extended to the gephyrean worm *Urechis*, the sand-crab *Emerita*, and the brine-shrimp *Artemia*, as well as several species of echinoderms, and in all cases between 35 and 100 % of the eventual complement of nuclein phosphorus was to be found at the beginning of development. Henze significantly isolated 1 % of pentose from undeveloped octopus eggs. J. Brachet[11] found abundance of nucleoprotein phosphorus and pentoses in the eggs of the opisthobranchiate gastropod *Aplysia* and the spider-crab *Maia*. On the other hand, terrestrial invertebrates seemed to occupy a different position, as the classical work of Tichomirov[1] on the silkworm had shown, by estimation of purine bases, the synthesis of 90 % of the nuclein present at hatching. But the newer figures of Landi seem to show only a 31 % increase. Amboni finds about the same for nuclein phosphorus, and Caspersson & Schultz[1], applying ultra-violet absorption technique to the undeveloped *Drosophila* egg, find evidence of considerable amounts of pyrimidine-containing nucleins. They also report strongly positive qualitative tests for pentoses there.

The lower vertebrates occupy the same doubtful position. Large amounts of purine bases were isolated by earlier workers (Levene & Mandel; Tchernorutsky; Steudel & Takahashi[1]) from teleostean eggs, e.g. cod, herring, carp, pike, etc., which would suggest that transformation rather than synthesis occurs during their development. But v. d. Ghinst[2], working on the trout egg in J. Brachet's laboratory, found not only the expected appearance and rise of the material yielding a positive Feulgen reaction, but also a clear rise of total pentose. Moreover, Rosenheim, C· šavičius, Ashford & Stickland, in preliminary experiments, had noted a rise in nucleoprotein P in trout eggs during development (0 to 7 % of the total P); this was confirmed by v. d. Ghinst. A rise of pentose and nucleoprotein P was also found by J. Brachet[11] in the case of the dogfish, *Scyllium*. It is possible, therefore, that in fish eggs we have to deal with an initial store of purine and pyrimidine bases, followed by a synthesis both of phyto- and thymo-nucleic acid. As we shall see, the entire synthesis of all these compounds is carried out by the chick embryo. Before passing to this, however, the amphibian embryo must be mentioned. Again it seems to occupy an intermediate position, for Plimmer & Kaya found only 28 % of the final nuclein phosphorus present at the beginning, while Graff & Barth recently reported an increase in purine nitrogen in the frog egg beginning at early gastrulation and continuing with increasing speed till at the tadpole stage it had increased 100 % and they ceased to follow it.

---

* This explains the early results of van Herwerden, who in 1912 described granules in echinoderm egg cytoplasm which nuclease preparations digested like chromidia.

† Such a transformation is not very compatible with Just's otherwise ingenious attempt[3] to link the appearance of nuclein compounds with the loss of morphogenetic potencies.

Close confirmation of this is to be found in the figures of Takamatsu[1] on *Hynobius*. Then v. d. Ghinst[1] and J. Brachet[13] found an increase in thymo-nucleic acid of 4·5 mgm./1000 eggs (by the Dische method) and a decrease in pentose during the development of the frog, followed later by an increase.

Events in the chick embryo, at least, are fairly clear, for it represents the other extreme from the sea-urchin. Kossel's failure[1] in 1884 to isolate any purine base from the unincubated hen's egg was repeated by numerous later investigators (references in *CE*, Section 10·1), who all demonstrated its synthesis by the embryo. Plimmer & Scott, moreover, found that 93 % of the nuclein phosphorus present at the end of development was newly formed during the process. But Mendel & Leavenworth[2] also observed an increase of total pentose during development, and later this was explained by Calvery's isolation[1] of phyto-nucleoprotein from the chick embryo and his demonstration that this is synthesised as well as the thymo-nucleoprotein itself. The phyto-nucleoprotein of the chick embryo contains guanine and adenine as purine bases, with cytosine and uracil as pyrimidines, instead of the thymine which the thymo-nucleic acid contains. Zinkernagel has lately established the existence of a thorough-going synthesis of purines in the egg of the snake, *Tropidonotus natrix* (about 1 mgm. purine N per egg).

We may summarise therefore the ways in which nuclein metabolism may proceed, in four types of embryonic behaviour:

(1) Simple transformation from stored phyto-nucleic acid to thymo-nucleic acid — Echinoderms, worms, molluscs, arthropods

(2) Transformation, but synthesis of some thymo-nucleic acid — Amphibia.

(3) Store of purine and pyrimidine bases, followed by a synthesis of both forms of nucleic acid — Teleostean and selachian fishes.

(4) Complete synthesis of both phyto- and thymo-nucleic acid from non-cyclic precursors — Birds and reptiles.

The provisional nature of this classification, drawn from data which are, after all, still fragmentary, should not be forgotten. One should always remember that the cleavage and pre-larval stages of the eggs of invertebrates are the most convenient for study, while in the higher organisms often only the later stages are easily had. There is nothing in the above table, therefore, inconsistent with the view that in *all* animals there is first a transformation from phyto- to thymo-nucleotides, followed by a synthesis of both types of molecule. In any case, the suggestion earlier made by Needham & Needham[3], that only those eggs which possess uricotelic metabolism can synthesise nuclein (cf. *CE*, p. 1158), must evidently now be abandoned.

It may next be asked whether the nucleotides can be identified as intermediate products between phyto- and thymo-nucleic acid. J. Brachet[13] examined this question in the eggs of the sea-urchin *Paracentrotus*, but there was no change at any stage in the relations between the mono-nucleotides and the protein-combined nucleins. The former certainly do not disappear as thymo-nucleic acid is formed. In this paper Brachet also showed by a balance-sheet that the phyto-nucleic acid disappears at every stage in quantitative correspondence with the thymo-nucleic acid formed, both in the sea-urchin and in the frog. Because of Edlbacher's views concerning the synthesis of the arginine-containing protein of nucleoprotein, he examined echinoderm eggs both for arginase and nucleotidase (nucleophosphatase), but the results were negative. On the other hand nucleophosphatase was found in amphibian eggs (I. Takahashi[2] on *Cryptobranchus*; Takamatsu[1] on *Hynobius*).

As to the well-established fact that the chemical nucleoplasmatic ratio (*CE*, p. 1150) declines steadily during development, the above data throw no new light on it. We have already seen it in Table 30, where nuclein P and purine N show bradyauxesis. It is stated by v. Euler & Schmidt that tumour tissue shows the high nucleoprotein content characteristic of young embryonic tissue, but this requires further confirmation.

That phyto-nucleins occur in animal tissues is thus now far beyond doubt. Pancreas nucleoprotein, suspected as far back as 1894 to include nucleins of "pentose" type, was proved to do so by Jorpes and Levene & Jorpes thirty-five years later. The phyto-nucleins and nucleotides are becoming associated with the cytoplasm rather than the nucleus.* The work of Caspersson & Schultz[†] on the cytoplasm of *Drosophila* eggs has already been referred to, and these workers suggest that the cytoplasm of all growing tissues is characterised by ultra-violet absorption spectra which reveal the presence of the cyclic nitrogenous bases of the nucleins. Apart from plant examples, they find[2] that the imaginal discs of the *Drosophila* larva show a strongly absorbing cytoplasm, while the mature larval gut cells show no trace of the peak. They further suggest that the cytoplasmic phyto-nucleins are responsible for the well-known histological basophily of embryonic tissues.[‡] In this connection it is interesting that the first nucleotide of phyto-nuclein type to be isolated from animal tissues, inosinic acid, or hypoxanthine-*d*-ribose nucleotide (Liebig in 1847), originates, as is now known, directly from the adenylpyrophosphate which is one of the coenzymes of glycolysis (see p. 209). Moreover, in many if not in all developing eggs adenylpyrophosphate is synthesised (cf. pp. 617 ff.). Enzymes which may be responsible have been studied by v. Euler & Skarżyński.

All this goes to prove the essential nature of the phyto-nucleins for embryonic development, and to it we must add Claude's isolation[1] of phyto-nuclein material similar to that of the Rous sarcoma factor from chick embryos (see p. 269), and Fischer's identification[7] of phyto-nucleins in the growth-promoting factor in chick embryo extract (p. 630). If nothing has been said in this section about the role of nucleins as the carriers of genetic influences in the nucleus, it is only because as yet this, perhaps their most important function, has not been brought into direct chemical relation with embryonic metabolism. But it is only necessary to glance through Sections 2·64 and 2·67 to see how direct this relation will probably become and how important for morphogenesis itself. In the fact that

---

* Though the nucleolus is said to be ribo-nuclein (Caspersson & Schultz[3]).

† Description of methods in a paper by Caspersson.

‡ Cf. the remarks above (p. 632) on chromidial extrusion in rapidly growing neoplasms.

both viruses and genes are self-perpetuating and that both are nucleoproteins we feel that we have found an essential key to the understanding of the reproduction of living organisms in all its aspects. And for this reason it is significant that thymo-nucleic acid is a highly anisometric micelle or molecule (300 times as long as it is thick) with great optical anisotropy — a fibrous macro-molecule (Signer, Caspersson & Hammarsten; Greenstein & Jenrette). It is believed that poly-peptide chains run lengthwise along the chromosome, with their basic arginine side-chains united with the phosphoric acid groups of the nucleotides (Astbury[2]) which have exactly the same spacing. But this is as much as there is room to say here on this subject, which may be followed in reviews such as those of Astbury[1] and Astbury & Bell. We need only add the curious observation of these workers, interesting in view of what has been said above, that phyto-nucleic acid does not give optically anisotropic solutions nor a fibrous product on drying. If we could have a full knowledge of the physico-chemical properties of the two nucleins, the whole contents of this section would probably fall into a rational order.

3·45. Lipin and sterol metabolism

The earlier work on the lipin and sterol metabolism of the hen's egg during development, summarised in *CE* (Sections 12·1 and 12·4), led, apart from differences in detail, to three rather simple conclusions. In the first place the total fat of the egg decreases, on account of the considerable fat combustion, mainly during the last week of incubation, with a total loss of some 55 % of the original store by the time of hatching (*CE*, Fig. 359). Secondly, the phosphatides present in the yolk at the beginning also suffer a regular fall, similar in its time incidence and of about the same relative magnitude (*CE*, Figs. 373 and 374). At the same time there occurs a correspondingly large increase of inorganic phosphorus (deposited in the new bones) and a lesser increase (as already men-tioned) of nuclein phosphorus. To these phosphorus is also contributed from the breakdown of the phosphoprotein (cf. p. 15) in the yolk at the beginning of development. Alone of all the phosphorus fractions the water-soluble organic shows little change. Thirdly, in contrast to the great utilisation of neutral fat and phosphatides, the total sterol, mostly cholesterol, shows practically no change throughout incubation (*CE*, Fig. 389), although there is a clearly marked increase of cholesterol esters and decrease of free cholesterol.

With these well-established facts as background, we may see how the newer methods of approach have further illuminated the processes going on. The general disappearance of neutral fat has been confirmed by a number of investi-gators, e.g. Romanov[5]; Kusui[1]; Jost & Sorg, though as they did not always follow incubation through till quite the end, their figures for total fat combusted are somewhat lower than that given in Table 10. Jost & Sorg paid special attention to the parallelism between loss of fat and loss of phosphatide. They found that on the 11th day 6·3 % of fat and 5·04 % of phosphatides had been lost, while on the 19th day the corresponding figures were 42·1 and 42·5 %. This they inter-preted as favouring the view, for which much other evidence has been brought forward, that fatty acid glycerides must be combined with phosphoric acid before they can be oxidised. Measurements of iodine values of the fat and phosphatides of the whole egg and the embryo and yolk separately made by Kusui as well as Jost & Sorg are in agreement with such a view. Thus throughout development the iodine value of the neutral fat remains steady at 75–76, while that of the phosphatides sinks from about 71 to 55. Moreover, this latter process occurs to a greater extent in the yolk than in the embryo, and all workers agree that the fat of the embryo is more unsaturated than that of the yolk. It may be allowable, therefore, to picture a preferential absorption of unsaturated fatty acids into the embryo, and a preferential combustion of those phosphatide molecules containing unsaturated fatty acids. This reinforces the impression gained from the earlier work (*CE*, p. 1171). Kusui adds the interesting facts that free fatty acids in yolk reach a peak on the 3rd day of incubation (cf. the lactic acid peak about the same time, already referred to, p. 608) at 3 % of the total, while in the embryo itself they are maximally present on the 14th day at 9·7 % of the total. Volatile fatty acids are never present in any part of the egg in more than traces.

An observation of Romanov's[9] the significance of which is not entirely clear is that while the refractive index of the neutral fat from the yolk remains constant at 1·4690 throughout incubation, that of the embryo fat declines very regularly from 1·4890 at the 9th day to 1·4720 at hatching. This is quite in agreement with the more unsaturated nature of the embryo fat, but it may also mean that the fatty acids of the embryo have relatively longer carbon chains than those of the yolk.

Just as Jost & Sorg investigated the relations between the quantities of neutral fat and phosphatide, so Kugler examined the relation between the two most important substances constituting the group of phosphatides, lecithin and kephalin. As noted already from Table 30 phosphatides are built into the body of the embryo bradyauxetically with a k of about 0·75. But during the process, according to Kugler, a rather strict ratio of 3 of lecithin to 1 of kephalin is observed, both in the waning yolk and in the waxing embryo. Whether this ratio obtains in all birds' eggs throughout development seems doubtful, if the figures of Masuda & Hori are trustworthy.

The question now arises as to whether the neutral fats, phosphatides and sterols of the yolk are transferred as such into the embryo or are first broken down to simpler molecules and then rebuilt. In recent years it has been possible by the use of molecules labelled with isotope elements to gain answers — at any rate provisionally convincing — to all three of these questions (see the review of Hevesy). Taking first the phosphatides, Hevesy, Levi & Rebbe injected radio-active sodium phosphate into hen's eggs before incubation for varying periods of time. It was found that the phosphatide phosphorus extracted from the embryo showed a high activity while the phosphatide phosphorus extracted from the yolk showed none. Phosphatide molecules present in the embryo must therefore have been synthesised and not taken over directly from the store in the yolk; this constitutes strong support for the "carrier" theory represented by Jost & Sorg, namely that combination of fatty acids with phosphorus assists their oxidation, since it shows that a complete breakdown does intervene between yolk and embryo phosphatide. It may also be mentioned here that the acid-soluble phosphorus and the nucleoprotein phosphorus also showed much radioactivity, thus demonstrating that they also had arisen by a synthesis for which inorganic phosphate had been employed. Similar results were obtained when labelled hexosemonophosphate was introduced into the embryo at the beginning of incubation, showing that it had been broken down to inorganic phosphate and that this had then been used for the various syntheses.

In sharp contrast with these findings were those of workers who introduced heavy hydrogen isotope into the egg at the beginning of incubation (see the review of Schoenheimer[3]). After the injection of a sufficient amount of heavy water into the egg the amount of deuterium in the fatty acids of the chick embryo towards the time of hatching was measured (Schoenheimer & Rittenberg[1]). It was found that the fatty acids had taken up no deuterium whatever, in contrast to mice synthesising fatty acids on a carbohydrate-rich diet, where the fatty acids rapidly take up deuterium. The conclusion was therefore reached that fatty acids are simply transferred as such from yolk to embryo and that no intermediate breakdown occurs.* Exactly the same result was obtained for the cholesterol of the embryo by Rittenberg & Schoenheimer, again in contrast to the entry of deuterium into the cholesterol of adult mice.

We must therefore picture the fat and cholesterol stores of the embryo being composed of molecules taken directly from the yolk, while the embryo's phosphatide molecules are new. Neutral fat and phosphatide in the whole egg both greatly decrease, the latter probably acting as a carrier for fatty acids destined for combustion, and itself supplying phosphorus for many synthetic purposes, including the provision of phosphatides for the embryonic tissues.

The general conclusion that no synthesis or destruction of cholesterol occurs during the development of the chick embryo is supported by the work of Skarżynski[2] and Serono, Montezemolo & Balboni, though like so many other investigators (see *CE*, pp. 1222 ff.) they got fairly consistently the slight rise, in Skarżynski's experiments an average of 9·5 %. That this is significant has never been demonstrated. A new observation was the definite decrease in ergosterol content (from 0·23 to 0·12 % total sterol) and the absence of any change in the dihydrocholesterol content. It is thought by these authors that the sex-hormones are formed from cholesterol, but this still awaits a definite proof. According to Riboulleau the fertilised hen's egg has 2γ/gm. oestrone, and it is certainly synthesised fairly steadily by the embryo (Serono, Montezemolo & Balboni); see p. 316. Dalton[3] has again studied the appearance of the anisotropic droplets of cholesterol esters (*CE*, p. 1219) in the chick embryo liver; according to him they appear normally on the 11th day of incubation, but their appearance may be hastened considerably by grafting the liver on the chorio-allantois of older embryos. The blood cholesterol of the chick embryo, measured by Zorn &

---

* This is substantially, not absolutely, true, because in later experiments of Schoenheimer's, where the embryo was carefully separated from the spare yolk at the end of development, a minute amount of deuterium was found in the fatty acids, indicating very slight synthetic processes.

Dalton, rises to a maximum of double the adult value at the 18th day of incubation, but returns nearly to normal at hatching. Infiltration of the liver with neutral fat begins somewhat before, about the 15th day, in Trotti's description[1]. At this time liver lipase is increasing in activity (Galvialo & Goryukhina), while the lipase and tributyrinase activity of the whole embryo increase steadily throughout (Ammon & Schütte; Kolobkova — see Fig. 72).

A view of the chick's fat metabolism at hatching has been given by Entenmann, Lorenz & Chaikov. The bird hatches with marked lipaemia and a "fatty liver" containing enormous amounts of cholesterol esters. After hatching, the lipids and sterols of liver and blood return fairly quickly to normal levels.

To the above account need only be added a mention of numerous histochemical papers on the appearance of fat in embryonic organs and in explanted pieces of them, to which reference must be made by those interested (Konopacka[1,2]; Zweibaum; Szantroch; Rix; Haszler; Hadjioloff; Hadjioloff & Usunoff).

Nothing comparable with the above-described advances has been made in the case of the mammalian embryo, which almost certainly derives the greater part of its fat from the maternal blood stream, and whose relative fat combustion has so far proved impossible to measure. Interest has mainly centred round the supply of neutral fat; and the plan of altering the degree of unsaturation of the fat in the maternal diet, and then examining whether the foetal fat has changed in the same direction and to the same extent as the maternal fat, has been tried again and again. The previous conclusion, that the foetal fat *can* be changed in this way, but not so readily as the maternal fat, remains unaltered. Different workers have had different degrees of success. Sinclair[2]* found that after cod-liver oil feeding, the iodine values of the neutral fat and the phosphatides of rat foetuses were significantly raised, the former nearly 100 % and the latter 25 %. Similar results were reported for rabbits by Bickenbach & Rupp[2]. Continuing Sinclair's work on rats, Chaikov & Robinson found the same variation in the foetus, but noted that it was distinctly less than that in the mother, which reached 400 %. Substantially the same conclusions were reached by Miura[2], who also found that elaidic acid can be incorporated into the phosphatides of the foetus. With this McConnell & Sinclair agree, but point out that the transfer of elaidic acid appears to take place much more effectively by way of the milk than through the placenta.

The characteristics of the fat and phosphatides normally present in the mammalian foetus at different stages have been studied by many workers (Torrisi on the dog; Chaves on the hedgehog; Martinoli and Roussel & Deflandre[2] on the cow and sheep; J. Suzuki on the rabbit; Lovern[1] on the porpoise; and Cattaneo on man). The details, as yet fragmentary, hardly lend themselves to any generalisation, and must be found in the original papers.

According to Schoenheimer & v. Behring the meconium of the human foetus contains surprisingly large quantities of cholesterol; and cetyl alcohol, a substance (see p. 249) which has been suggested by Schoenheimer & Hilgetag to be a natural endogenous purgative, has been found in meconium by Schoenheimer[2]. Another isolated point of interest is that curious tissue, Wharton's jelly, of the umbilical

---

* Sinclair attributes to me the view that the mammalian placenta is impermeable to fat and hence that the foetus must synthesise all it contains from other substances. Reference to *CE*, pp. 1190 and 1524, will show that I never put forward any such suggestion.

cord, which has been found to be almost devoid of fat, phosphatide, or sterols (Boyd[2]; Mangili).

Chievitz & Hevesy have found that radioactive phosphorus is taken up by the rat foetus and the human placenta, probably in the synthesis of phosphatides such as we have seen to occur in the chick embryo.

The question of possible changes in the protein carriers of enzymes concerned with fat metabolism was examined by Bamann, Mahdihassan & Laeverenz, with negative results, for they found no changes between the 6-month human foetus and the adult in the optical specificity of liver esterase, tested on racemic mandelic acid ethyl ester. This is in contrast with catalase (p. 604) and haemoglobin (p. 600), which do change.

The lipin metabolism of the eggs of lower vertebrates has unfortunately received little study. Working on those of the trout, S. Smith[1] found that the percentage of fat in the yolk-sac rises from 10 to 30 %, either because of a synthesis of fat or more probably because of preferential protein utilisation, up to the 80th day of development. Over the whole period there is a definite loss of fat from the system, as shown in Table 11, analogous to the hen's egg. Protein combustion predominates from the 40th day and goes on till the 70th; fat combustion begins at the 56th day and lasts till the end of the yolk-sac period. When the last of the stored reserves are used, the alevin lives on its muscle protein. A similar study was made by Hayes & Ross on the eggs of the salmon. Here the fat loss in percentage of the fat originally present was found to be high, 61·9 %. These authors believe a synthesis of fat occurs in the earlier half of development, but their figures do not clearly show this. They also believe that the maximal period of fat absorption corresponds with the maximal period of fat combustion, contrary to what is the case in the chick embryo (see CE, Fig. 251); this may well be so. Whether it is correct to identify the peak in fat-absorption rate shown by the salmon 30 days after hatching with the peak on the 9th day of incubation in the chick seems doubtful, though Hayes & Ross bring forward interesting morphological comparisons suggesting that 10 days of salmon embryo time approximates closely to 1 day chick embryo time. Masuda & Hori state that the lecithin-kephalin ratio differs a good deal in fish eggs from its avian value.

That a synthesis, and hence a total overall increase, of total fat can occur in embryonic development is an old observation in chemical embryology (see CE, p. 1188) and has been noted in a urodele amphibian, teleostean fishes, and especially pulmonate gastropods (cf. p. 576, where the effect of this on the R.Q. was mentioned). The urodele which had shown it was the American salamander *Cryptobranchus alleghaniensis*, but this has not been confirmed on the Japanese species, *japonicus*, where Tomita & Fujiwara and Kataoka & Takahashi find a distinct fall of total fat, as noted in Table 11. Moreover, the two pieces of work above mentioned did not show striking evidence of a synthesis in teleostean eggs. The subject therefore remains obscure.

In all amphibian eggs there is certainly a combustion of fat (see CE, p. 1173) and Takamatsu's figures[2] on *Hynobius* illustrate it again. The work of Kataoka & Takahashi on the salamander during development showed that no change occurs in the iodine value of the fat, and that of Iseki & Kumon indicates that half the phosphatide is broken down, as in the hen's egg. There seems to be a slight fall in the total cholesterol, and it is interesting that there is an increase in cholesterol esters and a decrease of free cholesterol, just as in the chick embryo. Differences in the cholesterol content of tadpoles raised in iodine-containing and iodine-free waters have been reported by Brambel & Lynn; these require confirmation. Tributyrinase was found by Takamatsu[1] in *Hynobius*, but not by I. Takahashi[2] in *Cryptobranchus*, embryos.

Invertebrates also, as would be expected, for the most part show considerable fat combustions. Pelluet counted the fat globules in the blastomeres of echinoderm eggs and found that their number steadily decreases from unfertilised egg to blastula (from 14 to 10·75 granules/cu.μ). The loss of fat was confirmed and substantiated chemically by Hayes[3], who found the total loss up to 40 hr. (the pluteus stage) in the case of *Arbacia* to be 42·5 % of the initial fat store, a figure higher than that previously current (see Table 11). Hayes represents his data by a curve which descends to a minimum, then rises again and finally falls, but it seems more likely that a general downward trend is all that one would be justified in asserting, since much depends on a single high point at o hr. and a single low one at 43 hr. In consequence of this, his rather elaborate argument, directed to showing that the hitherto unidentified acid produced after fertilisation (see p. 572) is really a fatty acid, has little force. Interesting, in view of facts already mentioned with regard to the chick and amphibia, is Hayes' finding that during the development of *Arbacia* no change occurs in the total cholesterol.

In insect eggs it has several times been shown that events take place in a way very similar to the hen's egg as far as fat (e.g. Busnel on the potato beetle *Leptinotarsa*) and phosphatide phosphorus is concerned. Inorganic and nucleoprotein phosphorus increase at the expense of phosphatide and phosphoprotein phosphorus (Akao[2] and better Amboni on the silkworm moth *Bombyx*; Thompson & Bodine[2] on the grasshopper, *Melanoplus*). But while Akao (whose results in general are, it must be admitted, erratic and puzzling) found only a slight increase of total cholesterol, Murao, in a more convincing paper, found regular and persistent increases of some 30 % in total sterol during the silkworm's development. Should this be confirmed, it will constitute an interesting exception to the rule that in general sterols are not synthesised by embryos. It may be significant in this connection that a special form of sterol, bombicesterol, occurs in the eggs of the silkworm moth.

3·455. Choline esterase and the development of nervous function.

Since choline is the nitrogenous base in lecithin, a few words must be said in this section on the subject of choline-esterase. As we have already seen (p. 211), acetylcholine is given off from nerve-endings of the "cholinergic" type, probably by sudden liberation from a bound form. It may perhaps have special significance also in the placenta (p. 86). But those who have studied choline-esterase, the enzyme which restores the *status quo* by destroying the stimulating compound, have not neglected to examine the appearance of this enzyme during embryonic development.

In the brain, of the chick for example, it rises steadily during the incubation period (Nachmansohn[1]) from Q_{ChE} 1·3 to 20·8, reaching the adult level a few days later. On the other hand, in striated muscle the Q_{ChE} rises to a level much higher (12·0) than that characteristic of adult muscle, after which it falls to the adult level. The peak is reached by the chick embryo about the 18th day of development (Nachmansohn[2]) and by the rabbit about the 7th day of post-natal life (Leibson). It is explicable by the relative proportions of motor nerve-endings and muscle fibres in the developing muscle. Choline-esterase is localised round the former (Marnay & Nachmansohn; Couteaux & Nachmansohn; T. P. Fêng & Y. C. Ting). Hence if the nerve-endings multiply at first more than the muscle fibres, a peak in choline-esterase would be expected. Cardiac muscle, however, behaves more like brain, in that no peak is seen, the choline-esterase content steadily rising to the adult level (reached at the 10th day of post-natal life in the rabbit; Shamarina). In tissue cultures of embryonic heart muscle, nerve fibres and endings soon disappear; choline-esterase disappears too (Thomas & Nachmansohn). Other aneural but enzymatically very active tissues, such as the yolk-endoderm and *area vasculosa* ectoderm of the chick, examined by these authors, gave, as would be expected, extremely low values for Q_{ChE}.

Nachmansohn[3] summarises all these results in the statement that choline-esterase increases during development *pari passu* with synapses and motor nerve-endings. This is borne out by independent investigations of Rjabinovskaia[2] and of Sessunin, who found that during development the properties of embryo brain extracts change from adrenalin-like to acetylcholine-like. One fact of real interest is that, according to Youngstrom, choline-esterase activity in amphibia reaches a considerable value before the differentiation of the central nervous system has gone far and before movements have been initiated.* Estimations of acetylcholine are reported for the whole chick embryo by Z. Y. Kuo[7]. At 2·5 days' incubation the embryo contains about 0·33 γ/gm. and by 5 days this has risen to 0·95 γ/gm., after which the amount remains constant till at least the 12th day.

3·46. Pigment metabolism

Among the synthetic powers of developing embryos none is more striking than the elaboration of pigments which they perform. The amounts of pigment in the membranes and jellies of eggs are relatively small; and the egg-cell itself, or in higher forms, the yolk, contains nothing but carotenoids which are, so far as we

---

* It is felt that the literature on the initiation of movements, reflexes, and the like, is too physiological to be reviewed in the present book. Recourse must be had to the monograph of Coghill and the numerous papers of Windle; Z. Y. Kuo; L. Carmichael, and their colleagues. Here it must suffice to mention the important fact that muscular movements and nervous co-ordination arise by self- rather than functional differentiation. For if amphibian embryos are allowed to develop in anaesthetic solutions which maintain them immobile, they will manifest, upon being returned to fresh-water, all the performances which normal animals show at the same stage of differentiation (Matthews & Detwiler; L. Carmichael). Recent work has not given much support to the old idea that myelinisation of the central nervous system is a limiting factor in the development of movements.

know, unessential to the life and morphogenesis of the new individual. The synthesis of haemoglobin is, it has been said, the most spectacular of all the syntheses. And although we certainly have more light than we had ten years ago, its mechanism still remains very obscure.

This section will divide itself naturally into (a) the pyrrol pigments (comprising both porphyrin metabolism and iron and copper metabolism), (b) the carotenoid pigments, (c) the flavines, (d) the pterines, and (e) melanin.

3·461. Pyrrol derivatives

If we glance at a scheme of the formulae of the porphyrins we can see that owing to the great complexity of these tetrapyrrol compounds, there are a large number of possible variations, differing in the symmetry of the positions at which the side-chains are attached.* Yet out of the various theoretical possibilities, only those porphyrins of types I and III occur in nature. It is not believed that they can be converted one into the other without breakdown. The aetioporphyrins have as side-chains methyl and ethyl groups, the coproporphyrins have methyl and propionic acid side-chains, the uroporphyrins have acetic acid and propionic acid side-chains, and protoporphyrin has methyl, vinyl and propionic acid side-chains. As shown in the diagram chlorophyll is derived from aetioporphyrin III and the haem of haemoglobin from protoporphyrin III. In chlorophyll the pyrrol nitrogen atoms are linked by a magnesium atom; in haemoglobin they are linked by an iron atom, and in turacin, a red pigment from the feathers of a family of tropical birds,[†] the pyrrol nitrogen atoms of uroporphyrin I are linked by a copper atom. Uroporphyrin I also occurs in pathological conditions, such as ochronosis, accumulating in the bones.

The physiology of the porphyrins of the type I series is still, indeed, to some extent a mystery. Coproporphyrin I seems to have no functional role in the adult,[‡] but the impression is growing that just as the bile pigments are stages in the breakdown of haemoglobin, so the porphyrins of series I are by-products in the synthesis of those of series III. Thus the former are always found to appear whenever an intensive haemopoiesis is going on,[§] and as we shall see, this evidence is reinforced by what happens during the development of the chick embryo.

A word must first be said, however, about the porphyrin which is deposited in the shell of the hen's egg. A number of workers had observed that egg-shells give a red fluorescence in the Wood's (ultra-violet) light (Derrien; Tapernoux; Bierry & Gouzon; Gouzon; Dhéré), and this was shown to be due to a porphyrin,

---

* For porphyrin nomenclature see Fischer & Stangler, and for porphyrin metabolism in general, the monograph of Carrié and the reviews of Lemberg[3,4].

† This pigment is water-soluble. It is worthy of being put on record that Sir F. G. Hopkins had some of these feathers, which he used to shake about in water before his class. Anyone who was not fascinated by the beautiful red pigment which dissolved out would, he said, never make a biochemist.

‡ But it is the principal porphyrin synthesised by yeast (Carrié & Mallinkrodt-Haupt).

§ And even in normal adult life, very small amounts of coproporphyrin I are constantly excreted (Dobriner).

at first called ooporphyrin, but later identified by Fischer & Kögl and Fischer
& Lindner as protoporphyrin III. On very long keeping, the fluorescence goes

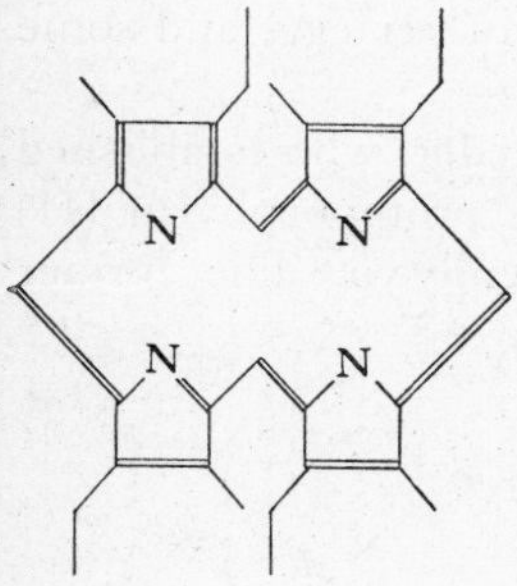

Aetioporphyrin I

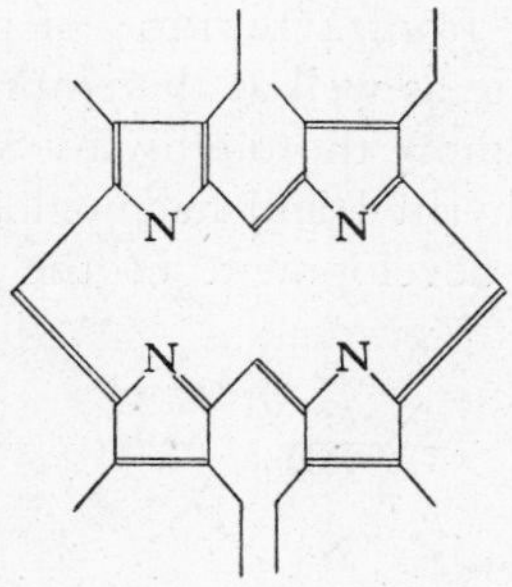

Aetioporphyrin III

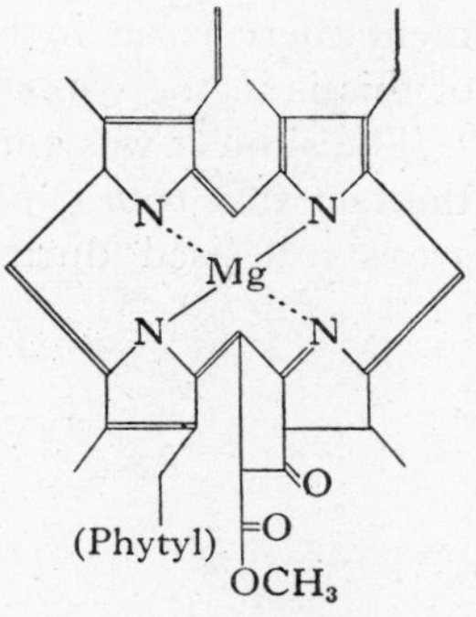

(Phytyl)

Chlorophyll α

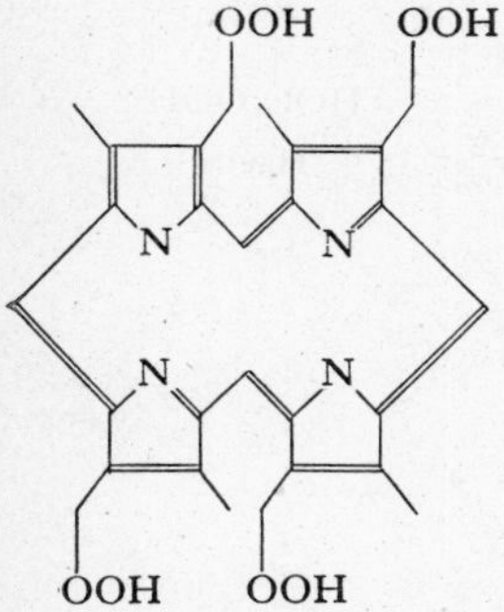

Coproporphyrin I

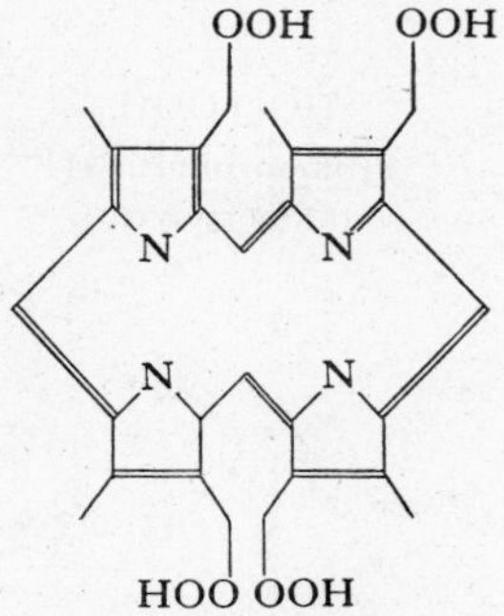

Coproporphyrin III

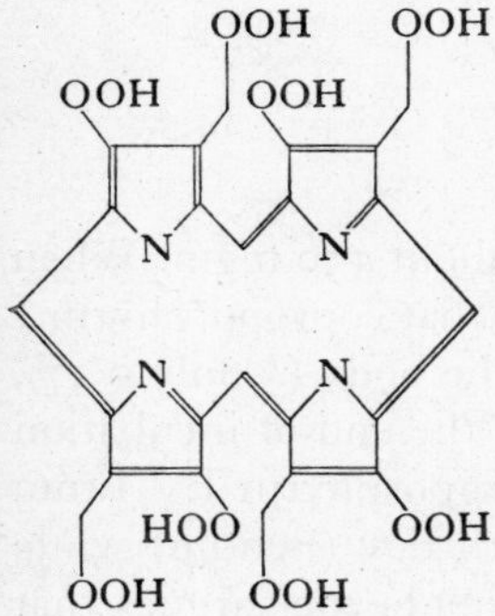

Uroporphyrin I

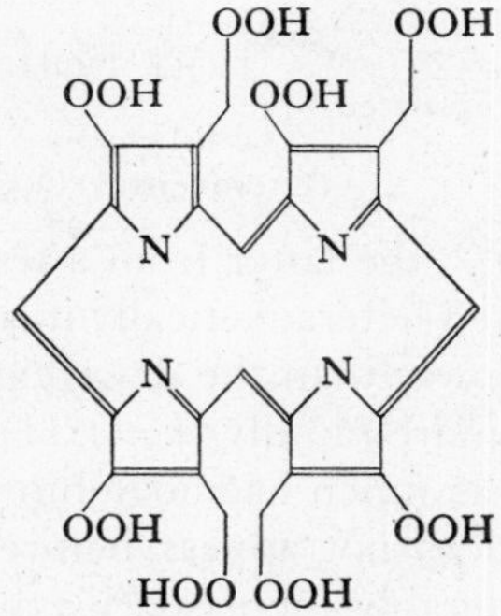

Uroporphyrin III

over partly into the egg-white (Baetsle) and then disappears altogether (Straub,
van Stijgeren & Kabos). There is said to be a trace of porphyrin in the shell-

membrane (Klose & Almquist). On repeating Buckner's experiment* with opened egg-shells, van den Bergh & Grotepass found that no porphyrin leaves the shell within a few weeks, but nevertheless within the first week of development there is an increase of from 4 to 140 γ of porphyrin per egg, and some of this is in the yolk and white as well as the embryo.

The subject was gone into more thoroughly by Schønheyder, who established the fact that *both* coproporphyrin I and haemoglobin (i.e. protoporphyrin III) are synthesised during the development of the chick embryo. The former

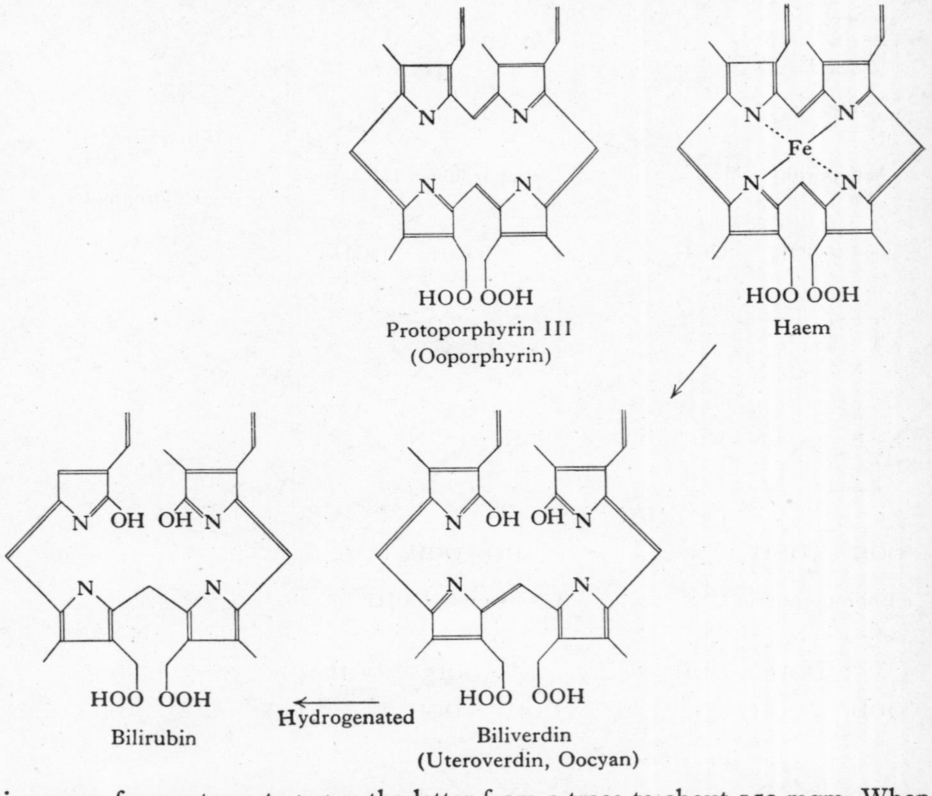

HOO OOH
Protoporphyrin III
(Ooporphyrin)

HOO OOH
Haem

HOO OOH
Bilirubin

←——— Hydrogenated

HOO OOH
Biliverdin
(Uteroverdin, Oocyan)

increases from a trace to 7·5 γ; the latter from a trace to about 150 mgm. When the rates of increase are plotted heterauxetically it is found that coproporphyrin I is synthesised much more slowly than the dry weight of the body (*k* only 0·37), while haemoglobin is made more rapidly (*k* = 1·11), till at the end of incubation there is about 20,000 times as much haemoglobin as coproporphyrin I.[†] From the fact that bile pigment does not appear before the 7th day (Sendju) while coproporphyrin I has considerably increased by then, Schønheyder argues that

* This consists in opening an egg, removing the contents, and bubbling CO_2 through a quantity of water placed in it (*CE*, p. 1268).

† Were it not for the great disproportion between the amounts of porphyrins of the two series formed, it would be tempting to analogise this dual synthesis with the formation of *both* kinds of nucleotide (p. 635).

it is not likely to have been formed by the breakdown of red blood corpuscles. No coproporphyrin III was found in any of the fractions, and there was no coproporphyrin I in the shell.

In mammalian embryos, especially those of rodents, it has often been observed that the bones are rich in a red pigment, and this is known to be either uro- or copro-porphyrin (Derrien; Fikentscher[1]; Borst & Konigsdorfer). After the end of foetal life the colour in the bones fades away. This fits in with a number of facts. Thus there is porphyrin (either coproporphyrin I or III, not protoporphyrin) in human foetal blood, according to Fikentscher[2], but never in maternal blood nor normal adult blood. The placenta is reported to be impermeable to porphyrins (E. Fraenkel). Porphyrin is found in small amounts, about 5γ %, in human amniotic liquid (Fikentscher[1]), and (an old observation) in the meconium (H. Gunther; Stockvis), where it has been identified as coproporphyrin I by Waldenström. Herold finds that it is excreted by the new-born human infant reaching a maximum of 8γ per day on the 4th day of extra-uterine life. Everything points to a physiological porphyria in foetal life, probably not unconnected with the vigorous haemoglobin synthesis then going on.

Hence the observations of W. J. Turner acquire much interest. The Pennsylvanian fox-squirrel, *Sciurus niger*,* has long been known to have bright red bones, and Turner was able to identify the pigment as uroporphyrin I. He associated the formation of the pigment with the megaloblasts of the bone-marrow, and suggests that in this animal we have an exceptional persistence into adult life of what is the general normal way of haemoglobin synthesis in foetal life.

Since there are none of the cytochromes nor much catalase in the undeveloped hen's egg, it is obvious that these haem compounds must also be synthesised by the developing embryo. Unfortunately, no exact studies have been made of cytochrome synthesis (that of Ito is purely qualitative). Catalase synthesis is referred to elsewhere (p. 604).

Turning now to the bile pigments, it will be seen from the formulae that they consist of four pyrrol nuclei no longer united in a ring. Bilirubin is the hydrogenated form of biliverdin, which may arise from haem by way of the intermediate compound verdo-haemochromogen, where the tetrapyrrol chain is opened but the iron atom still retained (see the discussion of Lemberg[3]).

Biliverdin is generally found wherever intensive breakdown of haemoglobin is going down. It was described by Lemberg & Barcroft (under the name of uteroverdin; Lemberg, Barcroft & Keilin) in the characteristic green regions of the dog's placenta, where it occurs in such quantities as almost to equal the total amount of haemoglobin in the body of the foetal dog. Uteroverdin was later identified with biliverdin by Lemberg[2]. Less easy to account for is its presence in the egg-shells of birds, from which it was isolated by Lemberg[1] (under the name of oocyan). The blue colour of the eggs of many species of birds, such as the gulls, is due to this pigment, and Punnett has found that its deposition in the

---

* The specific name of this animal is perhaps no coincidence. If it were not heavily pigmented it would probably suffer from photosensitivity, as human porphyrinurics do.

egg-shell occurs also in certain races of domestic fowls, among which it seems to have arisen as a dominant mutation in South America. In some egg-shells, such as that of the cassowary, oocyan, or some pigment extremely like it, is present in combination with protein, according to Dinelli.

In mammalian embryos bile-pigment production may play a more important part, as the well-known jaundice of the new-born human infant (*icterus neonatorum*) shows. As we have already seen (p. 598) the mammalian foetus, towards the end of its intra-uterine life, enters into a condition of anoxaemia similar to that experienced at high altitudes. To this it reacts, as Anselmino & Hoffmann[2] have pointed out, in several ways; the pulse rate increases,* the heart hypertrophies, the haemoglobin content of the blood, the red blood corpuscle count and the blood volume all rise, and there is a shift in the dissociation curve of the haemoglobin. All these processes occur to more or less the same extent in acclimatisation to high altitudes. That the haemoglobin and the erythrocytes increase appears from many studies. In the rat, for example, the haemoglobin content rises from 5·5 to 10·78 gm./100 c.c., and the red blood corpuscles from 694,000 to 2,600,000/c.mm. In man, the haemoglobin content reaches 22·5 gm./ 100 c.c. blood, higher than at any subsequent time. Now after birth this load of haemoglobin becomes excessive and is rapidly got rid of. What happens to the iron is uncertain; some believe that it is stored in the infant's liver (Gladstone; Ramage, Sheldon & Sheldon), others think that it is lost from the body (Stearns; Stearns & Singer; Stearns & McKinley). But the porphyrin undoubtedly breaks down, for Yllpö discovered a physiological post-natal bilirubinaemia, and Anselmino & Hoffmann[2†] put forward the plausible view that if the destruction of porphyrin goes on faster than the bile pigments can be excreted, *icterus neonatorum* will follow. At the same time it must not be supposed that no breakdown of haemoglobin is going on before birth; on the contrary bile pigments are always accumulating to a certain extent, and the breakdown process can even be carried out, according to Sümegi & Csaba, by embryonic cells cultured *in vitro*.

All workers agree, as we have seen, that during embryonic life, the haemoglobin content of the blood and the erythrocyte count increase steadily up to the normal adult level, and then greatly overpass it, doubtless as an adaptation to the relatively anoxaemic conditions of late foetal life (cf. p. 600), falling gradually to normal after birth. This occurs not only in mammals (as witness Kindred & Corey and Nicholas & Bosworth on the rat; Zeidberg and Kunde, Green, Changnon & Clark on the rabbit; Jones, Shipp & Gonder on the pig; and

---

* Usually foetal heart rate exceeds maternal heart rate; cf. the following figures from Hartman, Squier & Tinklepaugh (beats/min.):

	Foetal	Maternal
Man	135	70
Cow	161	50
Dog	150	100

And in the chick, as Bogue has shown, the rate increases from 130 to 235, the high level being attained by the 10th day, i.e. before the respiratory difficulties become acute.

† Some of their views were contested by Haselhorst & Stromberger[2] but the main idea holds good (Anselmino & Hoffmann[3]).

Wintrobe & Shumacker[2] on many mammals especially thoroughly), but also in the chick, where, as Sümegi and Zorn & Dalton show,* the adult level, both as regards haemoglobin content and erythrocyte count, is passed about the 12th day of incubation, and by the time of hatching almost doubled. It has also been demonstrated, by Holmes, Pigott & Campbell, that the chick shows the great decrease in blood haemoglobin following hatching, just as mammals do following birth.† In all these forms, as Sabin, Miller, Smithburn, Thomas & Hummel have shown, the approach to adult leucocyte content is very much slower, adult level not being reached in the rabbit, for instance, until the end of the 6th month of extra-uterine life. This enhances the plausibility of the view that the high haemoglobin content is a direct adaptation to the embryo's respiratory difficulties.‡

But now Wintrobe & Shumacker, in their elaborate studies just referred to, observed that as the number of red blood cells in the circulation increases, their individual diameter and volume greatly decrease. Thus in the pig the number of red blood cells climbs from 0·2 million/c.mm. in early foetal life to 3·6 million/c.mm., but during this time the average diameter of an erythrocyte has fallen from 16μ to 7μ, and its average volume from 360 cu.μ to 70 cu.μ. In other words, the foetus, as they expressed it[1], exhibits in the earlier stages a hyperchromic macrocytic anaemia, accompanied by the presence of many immature erythrocytes. And they pointed out that this condition resembles so exactly the blood picture in patients suffering from pernicious anaemia that it is difficult to distinguish the slides under the microscope. Curves showing the return to normal after the beginning of liver treatment may also be superimposed upon those drawn against foetal age. They suggest, therefore, that perhaps the haematopoietic tissues of the foetus are being influenced by the anti-anaemia principle of the liver, as part of the physiology of normal development.

As is generally known, the work of Castle, Townsend & Heath showed that the anti-anaemia principle of adult mammalian liver is formed from two sources, an extrinsic factor in the diet, and an intrinsic factor provided by the wall of the stomach.§ For the foetus the principle can hardly come through the placenta (Last & Hays), though it has been found there by Minot. Should the drain of the foetus be too great, pernicious anaemia, either of the maternal organism (Strauss) or of the new-born (L. G. Parsons), may develop. The principle has been found in foetal liver by Wintrobe, Clark, Trager & Danziger, towards the end of pregnancy.

Extremely little is yet known about the chemical nature, either of the principle or the two factors. It seems very likely, however, that the pterines (complex purines; see below, p. 655) are involved at some stage or other, for W. Jacobson[3]

* Confirmed by Ogorodny[3] and Fetischeva on ducks.

† In fowls there is a marked peak of mortality at the 6th day after hatching; could it be connected with these adjustments?

‡ It should also be remembered that, as we have already seen (p. 594), there are shifts in the dissociation curve of the chick embryo's haemoglobin analogous to those found in mammalian foetuses.

§ Practically nothing is known about the stomach in embryonic life, except that in mammals gastric acidity is slight until some time after birth (Manville & Lloyd).

has shown that the argentaffin cells of the cardiac and pyloric portions of human and pig stomach, which agree topographically with the distribution of the intrinsic factor, and which give a brilliant yellow fluorescence in ultra-violet light, contain a pterine, probably xanthopterine or uropterine. That xanthopterine possesses some anti-anaemia activity in the rat has been stated (Tschesche & Wolf) and denied (Mazza & Penati[2]). Active preparations prepared from liver (B. M. Jacobson & Subbarow; Subbarow & Jacobson; Mazza & Penati[1]) all contained pterine fractions among others. The preparations of Dakin's group (Dakin & West; Dakin, Ungley & West), however, were of purely polypeptide nature, and there were apparently purines but not pterine in the active extracts of the Swiss school (Karrer, Frei & Fritzsche; Karrer, Frei & Ringier). It does not follow, however, that the active principle itself must contain pterine or purine. In view of what has been said above (p. 631) regarding the function of the purine-containing nucleotides, it would be reasonable to think that the work of the polymeric purines, the pterines, might lie only in the formation of the principle. That the principle is complex is suggested from the report of Eisler, Hammarsten & Theorell that two active fractions with different effects can be separated by electrodialysis. As for the extrinsic factor, whatever it is, it is very stable to heat and to treatment with acids (W. B. Castle), and it is present in the undeveloped hen's egg (Miller & Rhoads).

In the light of all these facts, it is clear that the pterine metabolism of developing embryos in full haematopoiesis urgently invites study.

If, as indicated above, the blood picture in pernicious anaemia resembles that in early foetal life, it might be expected that the dissociation curve of pernicious anaemia haemoglobin would resemble the foetal type in having greater affinity for oxygen (cf. p. 599). This is actually the case. Korjuev & Belousov showed that the dissociation curve of the whole blood resembles that of foetal blood, and later Belousov established that this is a property of the haemoglobins concerned.

Many works have, of course, been devoted to the histology and morphology of the embryonic origin of haemoglobin and the blood circulation. Unfortunately, they have up to the present time contributed little towards our understanding of the mechanism of the process. As examples may be mentioned the papers of John on fishes, Słonimski[1] on amphibia and Hughes[1,2] and Grodziński[4] on the chick, to which the reader is referred for further information along such lines. Słonimski's work[2] was interesting in another connection, namely, as establishing that the presumptive blood-forming area on the ventral surface is determined at an early stage in gastrulation (see pp. 301, 378 and Table 22).

This may be the place to mention those invertebrates which possess porphyrin pigments for respiratory purposes or otherwise (cf. the monograph of Baldwin and the review of Barcroft[3]). One of these, the gephyrean worm, *Urechis*, has haemoglobin in its corpuscles (not in the plasma) and this turns to haematin giving them a brownish black colour. When the eggs are formed, granules of haematin are deposited in them, according to Baumberger & Michaelis, but whether or not as a store of porphyrin to relieve the strain of synthesis on the embryo is as yet unknown. The embryological aspect of all these porphyrin-containing organisms deserves much further study.

Another way of approaching the problem of haemoglobin synthesis is to investigate the metabolism of iron and copper in embryonic life, and this has been done on a considerable scale. The latter metal is important because it is now an established fact that copper is in some way connected with the mechanism of haemoglobin synthesis (Waddell, Elvehjem, Steenbock & Hart).

Estimations by many workers (e.g. Lesné, Zizine & Briskas; Imaizumi, etc.) show that the yolk of a hen's egg contains about 170 mgm. iron/kilo dry wt. and of copper about 7 mgm./kilo dry wt. There is no iron in the white, but about 5 mgm. copper/kilo dry wt. All of this is absorbed into the embryo by the end of incubation and most of it is stored in the liver. Thus Loeschke found 1·3 mgm./ kilo wet wt. of copper in the whole egg at the beginning of incubation; at the end the liver had 13·5 mgm./kilo wet wt. and the remainder of the body only 1·0 mgm./kilo wet wt. Now we possess certain careful studies, such as those of McFarlane & Milne and Szeynman-Rosenberg, from which the rate of accumulation of iron and copper in the growing embryonic body may be calculated. Both the metals show a bradyauxesis. On the dry weight, for the whole body, iron gives $k = 0·87$ and copper $k = 0·85$ (cf. Table 30); for the liver alone the bradyauxesis is more marked; iron gives $k = 0·62$ and copper $k = 0·67$. This, of course, implies nothing about the physiological importance of the metals at different stages, for in the case of entities such as these which cannot be formed synthetically by the embryo, bradyauxesis may only mean that there is not sufficient originally present to allow of tachyauxesis, unless a very rapid absorption were to take place, followed by complete cessation of absorption.* And this appears not to happen. On the other hand a bradyauxesis of copper in the embryonic liver of the pig has been established by Wilkerson (or rather, it may be inferred from his data, as McFarlane & Milne point out), although in this case, where the maternal organism might have preferential access to foods rich in copper, the limitation has no such obvious origin.

That both iron and copper accumulate in embryonic livers, however, is the agreed conclusion of many investigators (Loeschke; Lenti[2]; McFarlane & Milne; Szeynman-Rosenberg, and Imaizumi on the chick; Mayeda & Yamanuchi on the rabbit; Gruzewska & Roussel[1,2]; Roussel & Deflandre[1,3,5] on the cow; Hahn & Fairman; Sheldon & Ramage[1]; Ramage, Sheldon & Sheldon; Adler & Adler; Nitzescu; Iob & Swanson[2] on man; and Lintzel & Radeff on a variety of mammals). In some cases the liver of the foetus may contain 20 times the amount of iron or copper ever present in the adult liver. Unfortunately as the data given are often insufficiently full to allow of the necessary comparisons, and as all the estimation methods used are not equally worthy of complete confidence, we are in this, as in so many other similar problems, not as far advanced as the volume of the literature would at first sight suggest. Some points are, however, clear enough. Thus McFarlane & Milne and Lenti[2] showed that the non-haematin iron in the chick embryo liver rises from 10 % of the total iron present in the liver at the 11th day to 66 % at the time of hatching. This gives a picture of the destruction of blood pigment and the formation of bile pigments. Another side-light on the bradyauxesis of the metals is obtained by the work of Kamegai, who estimated the iron and copper content of the blood of the chick

* All inorganic raw materials of the chick embryo are not, of course, in this position. It by no means exhausts its lime supply.

embryo from the 9th day of incubation till hatching. Both declined, but copper more markedly than iron before hatching and both equally afterwards.

After hatching or birth, the total amount of iron and copper in the body temporarily declines.

3·462. Lipochromes

Leaving now the red porphyrin pigments we come to the yellow carotenoids.* Instead of the flat box-like tetrapyrrol molecule, we have to think of the long partly unsaturated chains ending either free or in ionone rings (see formulae on p. 249). Connected with this topic, too, is the position of vitamin A.

The yellow pigments of the yolk of the hen's egg, which are very readily affected by the feed of the laying hen (Albright & Thompson; Henderson & Wilcke; etc.), have been known for many years to be carotenoids, but Kühn, Winterstein & Lederer and Kühn & Smakula definitively identified the mixture (previously called egg-yolk lutein) as two-thirds xanthophyll, identical with that in leaves, and one-third zeaxanthin. Normally egg-yolks contain practically no carotin. Brockmann & Völker and Titus, Fritz & Kauffmann were able to show that in order to get into the yolk a carotenoid pigment must apparently have two or more –OH groups, since neither violaxanthin, carotin, nor lycopin will enter the yolk when fed, except in so far as the hen can, according to Virgin & Klussmann, convert some of the carotin into xanthophyll. Brockmann & Völker state that certain eggs, such as those of gulls and storks, may contain a little astacin (see p. 249 and below). When given physaliene (the palmitic derivative of zeaxanthin, which occurs in certain plants) the hen cannot break it down, but will readily put zeaxanthin itself into the yolk (Kühn & Brockmann). Quantitative estimations by Terenyi; Mattikov and v. Euler & Klussmann [1] indicate that one yolk contains about 1·5 mgm. of the mixed carotenoids, with as little as 0·7 γ of carotin. Besides these Gillam & Heilbron identified small quantities of cryptoxanthin, especially if the hen's diet had contained maize. Lizard eggs follow this sauropsid type, if we may generalise from the observations of Manunta [5] on the chameleon.

That the embryo can develop quite normally and hatch after the almost complete suppression of carotenoid pigments from the egg is now a classical observation (cf. *CE*, p. 1379). Particularly interesting in this connection is the work of Wald & Zussman on the origin of the carotenoid pigments of the chick's retina. By the time of hatching the retina contains three kinds of droplets; red ones containing astacin, golden-yellow ones containing xanthophyll, and greenish-yellow ones containing a hitherto unidentified hydrocarbon. All these appear during the last week of incubation, and the first and third types must certainly involve synthesis on the part of the embryo.

Astacin has already been mentioned in some detail in other connections (see pp. 210, 250). Green when combined with protein, it is pink when free.† Dis-

---

* See the monograph of Zechmeister.

† The name "ovoverdin" is proposed for the green combination. The pigment described by Yamamoto (p. 210) in the eggs of the worm *Ceratocephale* is probably a similar lipochrome-protein; its colour changes reversibly from lemon-yellow to green according to the conditions of illumination.

covered in lobster eggs by Kühn & Lederer, it was found by Fabre & Lederer and Stern & Salomon in the eggs of a wide variety of crustacea, and by Kühn, Lederer & Deutsch in those of the spider-crab, *Maia squinado*, where it is accompanied, not by xanthophyll, but by β-carotin. The same combination occurs in teleostean fish eggs such as that of the cod, according to Emmerie, van Eekelen & Wolff, and v. Euler, Gard & Hellström. That this cannot hold good for all teleosts, however, would seem to follow from the work of Manunta[7], who studied the eggs of two races of goldfish, *Carassius*, a red and a yellow. Both kinds of egg had mainly xanthophyll esters, but the red one had astacin as well. The eggs of daphnid crustacea would bear further investigation in this connection; all we know from a note of Teissier[3] is that some races have red eggs (due to the presence of haemoglobin or erythrocruorin?) and others have green ones, due to a carotenoid pigment the nature of which is unknown, but which is only formed when carotin is present in the diet. Finally, the eggs of the silkworm moth have been the subject of a good deal of Italian work. First described by T. Teodoro; Duce, and Tirelli[4], the work of Manunta[1] has shown that all races have xanthophyll and some have carotin as well. But the most interesting fact, and one which presumably accounts for the wide variations in egg-colour which silkworm races exhibit, is the presence of the plant pigments, flavones, in the eggs of certain races discovered by Jucci; Jucci & Manunta, and Manunta[1,3,6]. The bright green pigment, formerly called bombichlorine, thus turned out to be a flavone. Such an explanation may also account for Gerould's bright blue caterpillars, a mutant race of the normally green pierid *Colias philodice*.

Very little is known of the carotenoid metabolism of the mammalian embryo.* Xanthophyll has been identified by v. Euler & Klussmann[2] in the human placenta, and Clausen & McCoord have noted that the carotin content of human foetal blood seems to be exceptionally low. Kühn & Brockmann found that at the beginning of gestation there are equal amounts of xanthophyll and carotin in the placenta; towards the end there is much more of the former than the latter. All this remains at present quite enigmatic.

The metabolism of vitamin A (formula on p. 249) in eggs seems to be rather independent of the other polyene compounds. Brockmann & Völker found that the amount of it in the yolk was not much varied by the revolutions which they effected in the pigments of the diet. On the other hand, there is no doubt that the nature of the diet of the laying hen can profoundly affect the vitamin A content of the egg (Russell & Taylor; König, Kramer & Payne; Bearse & Miller; Cruickshank & Moore). At the beginning of development an average yolk will contain some ninety colour units of the vitamin, and as development proceeds this is all absorbed by the embryo, though very slowly, so that the greater part of the embryo's supply is still to be found in the spare yolk at the time of hatching (Holmes, Tripp & Campbell). As we have already seen (p. 23 and pp. 227 ff.), a proper supply of vitamin A is essential for normal morphogenesis, development and hatchability. This explains the great loss (45 % of the initial store) of the vitamin during the development of the chick, noted by Suomalainen[1] as well as

* The monograph of Lütgerath has not been accessible to me.

by the American workers. Aykroyd & Sankaran, testing the growth-promoting power of plasma from vitamin-A-deficient rabbits on chick embryo explants, found it to be noticeably low, but it is extremely doubtful whether the whole effect of vitamin A on development is explainable in terms of growth.

As regards mammals, something has already been said (p. 77) about transmission of vitamin A from maternal organism to foetus or new-born. Placentas contain it (Gaehtgens[2]; Dann[2]). The reserves of the vitamin in the liver are liable to be low at birth (Busson & Simonnet on dogs; L. K. Wolff[2] on man). Mason[2] has made an elaborate study of the interferences with gestation in the rat which subnormality in vitamin A entails; they differ from those characteristic of vitamin E deficiency (Evans, Burr & Althausen; Urner) in various ways, partly because the uterine wall in A deficiency generally becomes infected during the resorptive changes.

One very interesting problem concerning vitamin A is its synthesis by teleostean fishes. Just as for the chick vitamin C is a synthesisable hormone rather than a vitamin, so it appears that for the cod embryo vitamin A is a synthesisable hormone. McWalter & Drummond found, in conformity with what has been said above, that the eggs of the trout contain a carotenoid very similar if not identical with carotin, and that this decreases continuously until the end of the yolk-sac stage, while the vitamin A content continuously increases. Such a power of synthesis is needed to explain the appearance of vitamin A (as in cod liver oil) at the fish stage in the marine food-chain (cf. Elton).

3·463. Flavines

The next group of pigments to be discussed is that of the greenish yellow fluorescent water-soluble flavines. As has already been noticed (p. 208) they are found free as well as in various combinations with phosphoric acid and protein which have co-enzyme and vitamin properties (R. Kühn[1]). Ovoflavin was isolated from egg-white of the hen's egg by Kühn, György & Wagner-Jauregg almost simultaneously with the discovery of lactoflavin in milk by Ellinger & Koschara. They are now considered to be the same substance: 6 : 7-dimethyl-9-(d-ribityl)-benz-iso-alloxazine (Kühn & Wagner-Jauregg; Karrer & Schöpp; see formula on p. 209). Without the ribitol, the alloxazine is a reversibly reducible substance (K. G. Stern[1]); plus ribitol, the compound (lacto- or ovo-flavin) is identical with vitamin B_2 (hence the significance of the presence of this vitamin in the egg-white); vitamin B_2 plus phosphate is the co-enzyme, and hence (on the Warburg view) the prosthetic group, of yeast flavoprotein, i.e. an important dehydrogenase (R. Kühn[2]). Like other flavines, such as urochrome (Stern & Greville), ovoflavin catalyses cell respiration (Adler & v. Euler). Nothing further is known about its function or metabolism in the chick embryo, but Neuweiler[3] finds 5γ/gm. in the human foetal liver (of which 70 % is bound) as against 2γ/gm. in the placenta (of which only 40 % is bound). Much remains to be done on this subject. Apart from this we have only a few fragmentary observations on the presence of flavines in the embryos of selachian fishes (Fontaine & Gurevitch) and lepidoptera (Drilhon & Busnel) to which as yet it is impossible to attach meaning.

3·464. Pterines

So far as we know at present, the pterine pigments are confined to insects, although small amounts probably occur in all animals, where quite possibly they have some important function not yet clear.* In 1889 F. G. Hopkins[3] made the classical discovery that pigments which are derivatives of purines occur in the wings of butterflies. These pigments account for much of the cream and yellow coloration of insects, such as that of the wings of the cabbage-white butterfly, *Pieris*, or the yellow bands of the wasp, *Vespa*; and they must have some relation with the well-known uricotelic and guanotelic metabolism of insects and arachnids (see p. 50 above). The cream pigment was at first thought to be uric acid itself, but later the school of Wieland (cf. Wieland, Metzger, Schöpf & Bülow) considered the pterines to be tripurine polymers or condensation products. The most recent view, however, is that xanthopterine and leucopterine differ from purines only in that the pyrimidine ring is condensed with a pyrazine instead of an iminazol ring (Wieland & Purrmann; Purrmann). It will be noted that they form thus an intermediate group between the purines and the alloxazine ring of the flavines (see p. 209).

All that we know about the ontogenetic aspect of the pterines is derived from the work of Becker[3], who studied the formation of xanthopterine in the developing wasp, and found that it arises much later than the melanin. But there is an interesting parallel in this field to the action of inductors across genetic boundaries

Uric acid   Xanthopterine   Leucopterine

(see p. 341). In 1895 Hopkins[4] showed that when an insect mimics the colour of another insect, it does so by using the pigments characteristic of its own group, e.g. pterines instead of carotenoids. Here the stimulus is that of natural selection, and if the organism reacts, it can only do so in terms of its own enzyme proteins. Fifty years later, Sir F. G. Hopkins[5] continues still these studies.

3·465. Melanin

A word may finally be said about melanin, though the reader must be referred to the genetical section (pp. 405 ff.), where melanin formation was discussed in another connection. The subject, after much work, is still in a rather unsatisfactory state, since apart from histological and morphological studies which, alone, are incapable of advancing it much (e.g. Thumann on fishes; Dorris[2] on the chick; du Shane on amphibia; Peck on the rabbit; Makarov on the chick embryo retina), we know practically nothing of the nature or origin of the substrate of tyrosinase in embryos, nor what mechanisms control the formation and deposition of melanin. According to Dawes, the amount of melanin in the frog egg increases considerably between blastula and neurula stages.†

* For example, a connection with the anti-anaemia factor in the stomach of primates is suspected, as we have just seen (p. 650).

† In this connection the experiments of M. R. Lewis[2] and Figge[3] are interesting. They found that certain indophenol dyes, such as phenol-indophenol or *o*-cresol-indophenol, suppressed the normal development of melanin in pigmented amphibia such as the axolotl. Figge proved that the effect is not an indirect one by way of the pituitary, and later that the dyes exert an inhibitory effect on tyrosinase itself (Figge[4]). In the course of these experiments it appeared that melanin itself is capable of being reversibly reduced to a light brown or tan colour (Figge[5,6]).

The darkening and hardening of the insect cuticle at various developmental stages has been the subject of much recent work. The protein has been studied both chemically (Trim) and physically (Fraenkel & Rudall). Pryor believes that a veritable tanning process takes place, a dihydroxyphenol being secreted together with the protein, and then denaturing it by forming aromatic cross-linkages. The tanning process would be closely connected with the melanin formation since the parent substance for both would be the same; dihydroxyphenylalanine brought in the blood stream.

3·5. Polarity

In concluding this book, it is essential to say something about the biochemical approach to the fundamental and difficult question of polarity, perhaps the central puzzle of embryonic development. How do the axes of symmetry originate, and how does morphogenesis conform to them? Although I shall avoid going over the ground already covered in a former book (*Order and Life*, ch. 3), a certain amount of repetition will hardly be avoidable. The argument there put forward was as follows.

A logical analysis of the concept of organism leads us to look for organising relations at all the levels, higher and lower, coarse and fine, of the living structure.* Biochemistry and morphology should, then, blend into each other instead of existing, as they tend to do, on each side of an enigmatic barrier. The chemical structure of molecules, the colloidal conditions in the cell, and the morphological patterns so arising, are inextricably connected. It is easy to find instances of the way in which organisation may appear already at the chemical level. We are driven to the view that the living cell possesses as complex a set of interfaces, oriented catalysts, molecular chains, reaction vessels, etc. as the organs, tissues and other anatomical structures of the whole organism.

3·51. The organisation of cell protoplasm

Indications of the organised nature of the cell protoplasm are so numerous and well known that only a few may be mentioned. Thus it is significant that Vlès & Gex, examining the transparent sea-urchin egg in the ultra-violet spectrometer, obtained an absorption curve which was not typical of protein, although such substances must undoubtedly form by far the largest proportion of the solid matter present in the eggs. Only upon cytolysis and the death of the cell did the characteristic protein curve appear — an indication that some curious structural state is present in the egg. Then much of value has come from the vigorous

---

* An amusing essay, in the style of Veblen or Mencken, might be written on the influence of sectional interests on the development of scientific terminology. Morphologists have long tended to speak of the "higher" (morphological) as opposed to the "lower" (chemical) levels of organisation, but the balance may be redressed by speaking of "coarser" and "finer" levels. So also physical chemists, wishing to avoid the undertones of the word "superficial", speak of "interfacial" reactions, as opposed to those going on in the "bulk phase", thus ingeniously giving the impression that the classical methods of chemistry are really only fit for peasants. "Bulk analysis" is a term also used by morphologists who wish to suggest that most of biochemistry consists in analysing dead materials on the "raw fibre" and "humin nitrogen" level. It is regrettable that this temptation seduces even the elect (e.g. Weiss[16], *Principles of Development*, p. 187). And it is to be hoped that the present book will have helped to demonstrate the absurdity of such an outlook.

crop of researches which followed the perfection of the micro-manipulator, largely due to the work of Robert Chambers, as an efficient tool for handling isolated single cells. When Hiller[2] made experiments on amoebae with the narcotics, ethyl alcohol, chloretone, chloroform and ether, he found that, although their action was marked on immersion, they had much less action when micro-injected into the cytoplasm. In all concentrations chloretone increased the ·fluidity and streaming movements of the interior. Alcohol might cause a reversible coagulation. Still more remarkable was the result of Pollack, who micro-injected solutions of picric acid, and found that concentrations which coagulated proteins in the test-tube had no such action in the interior of the living *Amoeba*. Only if there was a local injury at the point of injection did the picric acid coagulate the protoplasm. In contrast to the non-toxicity of the reagent within was its extreme toxicity when applied to the external surface. An analogous observation is that if an oil-drop be micro-injected into a cell, it can be swollen and shrunk by changing the pressure at will so long as the cell lives, but immediately upon cytolysis a film of adsorbed protein forms on the droplet causing a wrinkling if oil is withdrawn into the pipette, (Chambers & Kopać[2]).

Moreover, the importance of intra-cellular organisation in explaining certain metabolic and respiratory phenomena cannot be overlooked. It has been well emphasised in a review by Korr[3], who mentions most of the classical cases. Disorganisation of the cell structure in cytolysis destroys some former enzyme-substrate contiguities necessary for the life of the organism, and sets up some new ones, usually not geared to the maintenance of life, nor capable of being so. Thus in sea-urchin eggs cytolysis almost entirely destroys respiration. In lysed bacteria the haematin catalysts remain active but the dehydrogenases are destroyed, hence the respiration also falls nearly to zero. Conversely, where respiration is normally restrained, interference with cell structure may much increase it, as in the bruised portion of a mushroom, or the amphibian gastrula in cytolysis (p. 589). In this connection we shall recall the "damped" metabolism of the unfertilised echinoderm egg, the grasshopper embryo in diapause, and the ventral ectoderm of the amphibian gastrula (p. 198). To these examples the following may be added. The echinochrome granules of the sea-urchin egg show no fading when the egg is subjected to anaerobiosis, although the dye is known to be reversibly reducible, and the intra-cellular reduction potential is known to change. The hermidin of dog-mercury (*Mercurialis perennis*) exists in the plant in a state of 95 % reduction, although bubbles of oxygen are actually being formed photosynthetically in the same cells. Luminous animals at death give out a continual steady luminescence, the "death glow", indicating that the luciferin-luciferase contiguity can no longer be controlled. Many plants darken to blackness on death, e.g. the false indigo (*Baptisia tinctoria*), but attempts to produce this on the living leaf by forcing oxygen into it under high pressure (1500 lb./sq. in.) fail because, until the cell structure breaks down, chromogen and oxidase are held firmly apart. Lastly may be mentioned the numerous cases in which important enzymes, such as cytochrome, desmokathepsin, or the glucolytic enzyme system, are so closely associated with cell structure. This suggests a micro-morphological arrangement of them within the cell.

Against the conception of high organisation in protoplasm might arise, however, the fear of a revival of completely discredited theories of the "biogen molecule" type, were it not for the fact that we now possess much more powerful means of rendering living wholes "transparent". Among these the ultra-violet spectrometer, which has achieved such success in the location of nucleic acids within the cell (p. 636), may be mentioned, and especially the X-ray analysis of polysaccharides (Sponsler) and of proteins (Astbury; Astbury & Bell; Astbury & Street; Astbury & Woods). The study of highly anisometric molecules by other methods will also be of great importance for embryology; in this respect the reviews of Staudinger should be consulted.

3·52. Fibre-molecules and "dynamic structure"

The importance of the work on the crystal structure of animal fibres can hardly be overestimated. Is not biology as a whole very largely the exploration of fibre properties? It is now certain, for instance, that in the case of muscle the contractile mechanism is essentially a molecular contraction, and with this there ends a state of ignorance not much less intense than that which existed in the time of Descartes and Borelli. But we are especially interested in the egg-cell. It may be a far cry from the macroscopic fibres hitherto mostly studied to the intracellular lattice of protein molecules forming the cell architecture, but there is no reason to suppose that the one is less open to investigation than the other. And the subject has real importance in connection with the polarity and symmetry of the fertilised egg.

Embryologists have long been impressed with the capacity of eggs to develop normally after their contents have been thoroughly stratified by centrifugal force. The centrifuged amphibian egg, for instance, will show a mass of yolk-granules at the centrifugal pole, a cap of fat at the opposite one, and an intermediate transparent protoplasmic layer. Conklin[2], in particular, has drawn attention to the possible existence of a "spongioplasmic framework" not destroyed by centrifugation, in spite of the egg's "ballast" being movable through its meshes.[*] There is much evidence, indeed, that the egg contains a "framework of viscid protoplasm which is so elastic or contractile that it recovers its normal form after distortion". On the other hand, egg cytoplasm can be remarkably fluid, showing churning movements as cinematographed by J. E. Harris; and Chambers' descriptions[4] of the blastomeres during cleavage show convincingly that the cytoplasm may for a time at any rate flow in currents like a liquid.[†] The protein chains of the cell's web or lattice must therefore be pictured rather as connected at many points by residual valencies and relatively loose attachments, *so that they can, as it were, snap back after disarrangement*, whether the disarrangement is normally periodic or experimentally induced. We may call this "dynamic structure". It is probably connected with those isothermal reversible sol-gel transformations which go under the name of thixotropy (described in Freundlich's monograph),

---

[*] There is of course also the special structure of the egg's surface, reviewed by Harvey & Danielli.
[†] Or perhaps like a liquid crystal?

though doubtless much more complex than any of the "inorganic" examples containing few components. A thixotropic gel liquefies if shaken or stirred, later returning to its previous consistency. Angerer has shown quantitatively that amoeboid protoplasm does just this.

Like Pascal's being suspended between two infinites, protoplasm with its polarities hovers between the perfectly liquid and the perfectly solid. At the other extreme from the liquid cytoplasm just mentioned are the eggs of the brine-shrimp, *Artemia*, which can be held in high vacuum over phosphorus pentoxide for a year (Needham & Pirie) or cooled to the temperature of liquid air or held at 10^{-6} mm. Hg for six months (Whitaker[20]), and yet subsequently give normal development. We are reminded of the crystallisation of some of the plant virus nucleoproteins. In certain cases, therefore, the "dynamic structure" can pack into highly dehydrated form without permanent injury.

The thixotropy of protoplasm is strikingly illustrated by the remarkable and well-known observation of Kühne, who in 1863 saw a small nematode worm moving actually within an amphibian muscle fibre. "The movements of the worm", writes Freundlich, "caused characteristic changes in the aspect of the striations; if the animal moved parallel to the fibre axis, the striations moved like the hairs of a brush, touched slightly by a solid body; when the nematode had pierced a striation, the latter closed up again, so to say, behind the worm's tail. The ease of movement seems to indicate that the mass of the interior is fluid; the well-defined, reversibly rearranged orientation of the striations looks more like the behaviour of a solid body. Both facts are explained by assuming a thixotropic state." Other later observers have reported similar facts; thus Takagi found an actively moving bacterial parasite within the protoplasm of the protozoon *Spirostomum*, which seemed to have no deleterious influence on its host. Protoplasmic thixotropy has also been studied by Fauré-Fremiet[4] on the amoebocytes of marine annelids, which have well-marked active and quiescent states. He added the significant finding that only in the "sol" state does the cytoplasm show the nitroprusside reaction for free sulphydryl groups, not in the "gel" state. Reversible denaturation of the proteins may therefore be involved (cf. p. 423).

Elsewhere Conklin[5] describes how the nuclei, centrospheres, yolk inclusions, etc. of invertebrates' eggs usually come back slowly to their normal places after the action of pressure or centrifugal force has ceased. Wintrebert[8], too, speaks of the need for the hypothesis of a "cyto-squelette" or "trame spongioplasmique". Fig. 324 shows the familiar failure of centrifuging to affect the micromere pole in echinoderm eggs. Schleip, again, at the conclusion of his encyclopaedic work on embryonic determination, says: "In every attempt at the explanation of polarity and symmetry in the egg, some as yet unknown property of the protoplasm has to be introduced. To avoid giving it any new name, which could only be arbitrary and tentative, I will call it 'intimate structure' (*Intimstruktur*)." Or again: "An intimate structure is present, i.e. a morphologically invisible, specific property of the cytoplasm, the manner of working of which we do not yet understand. To this must the difference between spiral and radial cleavage be referred, this determines too the direction of bilateral-symmetrical cleavage, and governs the planes of subsequent mitoses." Przibram[5], in his work on the crystal analogy, is still bolder, and does not hesitate to hypostatise the Thompsonian co-ordinates (p. 531). "Such systematic deformations", he says, "may most easily be explained on the assumption of an organic space-lattice. Just as by the substitution of radicals in eutopic series of crystals, deformations may be obtained

because the effect is the same on all parts of the lattice, so the substitution of one protein combination for another would bring about a parallel change in the organic space-lattice."

This is not nearly so extravagant as it sounds, if we remember the elastic properties of protoplasm. An entanglement or "brush heap" of interlacing crystalline fibres or amino-acid chains would be elastic; an emulsion of spherical particles would not (Seifriz[3]).

The "cell skeleton" is, however, by no means easy to study. Among interesting recent attempts to do so are those of Howard[1] and of Vlès. Howard studied the apparent viscosity of sea-urchin egg protoplasm at various rates of shear by observing the rate of movement of granules under centrifugal force. No unequivocal plasticity was shown by the egg at rates of shear so low that the granules were moving at velocities comparable to their own migration velocities (i.e. their velocities of return to place after centrifuging). No continuous structure, it was concluded, was present which could significantly affect diffusion, nor, a fortiori, be the basis of polarity. But it is uncertain, as Pfeiffer[1]

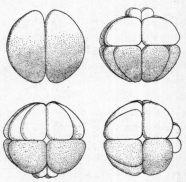

Fig. 324. Persistence of the primary axis in the echinoderm egg (*Arbacia*) in spite of the rearrangement of visible substances in the cytoplasm. After centrifuging, the egg becomes stratified with fat at the centripetal pole, then clear cytoplasm, then yolk-granules with increasing amounts of pigment. The first cleavage (top left) is always at right angles to the stratification, but the micromeres are always formed at the vegetal end of the original axis, whether this coincides with the centripetal pole of the centrifuged egg (top right), its centrifugal pole (bottom left), or its side (bottom right). (From Morgan, redrawn by Huxley & de Beer.) Cf. Fig. 267.

has pointed out, whether what we are looking for could be studied by any centrifuge method. Vlès, again, investigated the moduli of elasticity and rigidity in the sea-urchin egg as it becomes a sphere after deformation to a sausage-like shape. His results do not exclude the possibility of the structure we are envisaging. One thing at least is clear, namely that proteins within the protoplasmic structure do not behave like isolated proteins in Svedberg sedimentations. The classical observations of Beams & King[1] showed that *Ascaris* eggs may be subjected to centrifugal forces of 400,000 × *g* for a short time, and 150,000 × *g* for four days continuously, and yet later respire and develop normally. Gortner[3] has acutely remarked that if we had no microscopes with which such eggs had been examined, they would have been described as uniform-sized protein micelles of enormous particle-weight, and classed among the globular proteins. Baitsell[7] replied that this is no criticism of either protein chemistry or embryology, and that we are indeed forced to envisage the egg-cell as a single extremely elaborate protein complex, holding within itself many other types of molecule, as even the crystalline virus "molecules" or micelles almost certainly do.*

The extraordinary tendency of protozoa to swim in spirals, some right and some left, leads Schaeffer to postulate a stereochemical difference in the protein framework, and the sinistrality and dextrality of molluscan eggs in their cleavage, which corresponds with their shell rotation, may be referable to stereochemical differences in their proteins, though this has not been proved (Boycott, Garstang & Diver). Rand & Hsu have a passage of particular interest. Describing the motion of amoeboid cells, they say: "In watching the nucleus one is perplexed by the fact that it is so freely movable within limits which are in no way visibly defined. Apparently freely immersed in a labile and actively flowing protoplasm, why is its position not completely at the mercy of the currents? It must possess some highly elastic anchorage. It is as if a slightly buoyant sphere, immersed in a strongly flowing stream, were anchored by relatively slender and extremely elastic cables."

Protoplasm undoubtedly shows anomalous viscosity. Pfeiffer[3] and others have found that the velocity of its forced shearing flow, as in a capillary tube, increases with increasing force at first slowly, later more quickly. For Newtonian liquids the relation, however, is linear. Also protoplasm has a "yield value", i.e. requires a specially large initial force to start the flow (cf. thixotropy). The viscosity of true liquids is unaffected by pressure, but that of protoplasm greatly decreases as the pressure increases. The anomalous flow of protoplasm is paralleled by many non-living colloids, as Lawrence's paper[1] shows, if they consist of anisometric particles which mechanically interfere with each other's motion.

3·53. The paracrystalline state

The aspect of molecular pattern which seems to have been most underestimated in the consideration of biological phenomena is that found in liquid crystals (cf. the books of Rinne and W. J. Schmidt[1,5]). It is likely that progress in the discovery of the "leptonic fibres"† of which we have been speaking will be delayed until more knowledge is available of the peculiar form of order which liquid crystals exhibit. Liquid crystals, it is to be noted, are not important for biology and embryology because they manifest certain properties which can be regarded as analogous to those which living systems manifest (models), but because living systems actually *are* liquid crystals, or, it would be more correct to say, the paracrystalline state undoubtedly exists in living cells. The doubly refracting portions of the striated muscle fibre are, of course, the classical instance of this arrangement, but there are many other equally striking instances, such as cephalopod spermatozoa (W. J. Schmidt[2]) or the axons of nerve cells (Schmitt; Schmitt & Bear) or cilia (McKinnon & Vlès) or birefringent phases in molluscan eggs (Pfeiffer[2]) or in nucleus and cytoplasm of echinoderm eggs

---

* Cf. Pirie's thoughtful discussion of this point.

† This expression is taken from the suggestion of Rinne (Faraday Society Symposium, *Liquid Crystals*, p. 1029) that for international purposes parallel words are needed for the German "*Feinbaulehre*", "*Feinbau*". He suggests "leptology", "leptonic", from λεπτος, delicate, fine.

(W. J. Schmidt[4]). Even in the amphibian egg the yolk-platelets are anisotropic (W. J. Schmidt[2]). And it is very probable that the paracrystalline state exists in many phases in the cell whose scarcity or position has so far rendered them

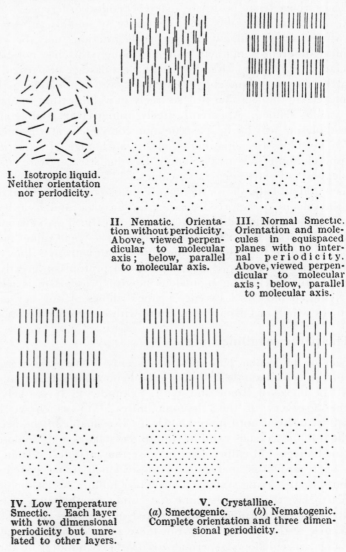

I. Isotropic liquid. Neither orientation nor periodicity.

II. Nematic. Orientation without periodicity. Above, viewed perpendicular to molecular axis; below, parallel to molecular axis.

III. Normal Smectic. Orientation and molecules in equispaced planes with no internal periodicity. Above, viewed perpendicular to molecular axis; below, parallel to molecular axis.

IV. Low Temperature Smectic. Each layer with two dimensional periodicity but unrelated to other layers.

V. Crystalline. (a) Smectogenic. (b) Nematogenic. Complete orientation and three dimensional periodicity.

Fig. 325. Types of liquid crystals.

immune from investigation and invisible when the living cell is placed in polarised light. As Rinne points out, the Langmuirian oriented film of fatty acid on water may be regarded as paracrystalline, and this is certainly a close relation of the oriented films within the cell or the egg.

According to our ordinary conceptions, when a solid is heated it melts on reaching a certain temperature, passing immediately into the state of an isotropic liquid, where the molecules are arranged at random according to kinetic theory. This, however, is much too simple for many substances, which on the contrary pass through a succession of "mesoforms", conditions where a certain amount of liquid-like flow is possible, and yet where there is within the liquid a definite and regular arrangement of molecules. These mesoforms are now classified as follows. Remembering that all paracrystalline substances have elongated molecules several different cases may arise (see Fig. 325). In I the molecules are completely at random; this is the true isotropic liquid state. In II all the molecular axes are parallel, but the centres of the molecules are as irregularly arranged and as free to move as in the first case. This is called the *nematic* state. In III also the molecular axes are parallel, but the centres of the molecules have lost one degree of freedom and are now restricted to a set of regularly spaced parallel surfaces. This is the *smectic* state. In IV the smectic state is still further ordered by having the molecules regularly arranged within each layer. V gives the conditions in the crystalline solid state. The molecules are parallel and their centres form a regular three-dimensional network. A smectogenic crystal has its molecules arranged in planes (V a); a nematogenic crystal has them arranged so as to interleave each other (V b). In addition to these states there may be others, and it is interesting that one has long been recognised, the most complicated of all, called *cholesteric*, since it is typically shown by the fatty acid esters of cholesterol. The liquid crystals of plant viruses seem to be intermediate between type II and type III (Bernal & Fankuchen), since the molecules are packed with hexagonal two-dimensional regularity at right angles to their length, but there is still no evidence of regularity of molecular arrangement in the direction of their length. With pure substances, such as ethyl-*p*-azoxybenzoate, paracrystalline phenomena are most easily to be observed in melting, but with two or more components in a system, the temperature factor may be replaced by changes in the concentration of the solvent or dispersing medium. Most of the protein, fat and myelinic substance of the cell probably exists in these states, but this is only directly visible when all the molecules are oriented in enormous swarms in one direction, as in muscle fibrils. The paracrystalline state seems the most suited to biological functions, as it combines the fluidity and diffusibility of liquids while preserving the possibilities of internal structure characteristic of crystalline solids (Lawrence[2]).

Ethyl-*p*-azoxybenzoate

This has been very well put by Bernal. "The biologically important liquid crystals", he said, "are plainly systems of two or more components. At least one must be a substance tending to paracrystallinity and another will in general be water. This variable permeability of liquid crystals enables them to be as effective for chemical reactions as true liquids or gels, as against the relative impenetrability of solid crystals. On the other hand, liquid crystals possess *internal structure* lacking in liquids, and *directional properties* not found in gels. These two properties have far-reaching consequences. In the first place, a liquid crystal in a cell, through its own structure, becomes a *proto-organ* for mechanical or electrical activity, and when associated in specialised cells (with others) in higher animals gives rise to true organs, such as muscle and nerve. Secondly, and perhaps more fundamentally, the oriented molecules in liquid crystals furnish an ideal medium for catalytic action, particularly of the complex type needed to account for growth and reproduction. Lastly, a liquid crystal has the possibility of its own structure, singular lines, rods and cones, etc. Such structures belong to the liquid crystal as a unit and not to its molecules, which may be replaced by others without destroying them, and they persist in spite of the complete fluidity of the substance. They are just the properties to be required for a degree of organisation between that of the continuous substance, liquid or crystalline solid, and even the simplest living cell." *

It has been known for a long time that many constituents of the living cell manifest remarkable interfacial properties when in contact with water (lipins and sterols); this knowledge long antedates our information on the crystalline character of protein chains. The "myelin figures" produced by wetted lecithin are classical. They are undoubtedly connected with the solubility of the phosphoric acid and glycerol groups in the molecule and the insolubility of the hydrocarbon chains of the fatty acids. First, spheres with optically positive radii are formed, and from these there grow out excrescences and long sluggishly writhing masses, often with remarkable turns, twists, spiral wrappings and swellings. As the water-content is increased the double refraction falls to a low value. Leathes, who pleads for the view that protein structure alone will not be adequate for the understanding of intra-cellular structure, has cinematographed these myelin forms under many different conditions.

3·54. Rigidity in paracrystalline mesoforms and in embryonic structures undergoing determination of spatial axes

It has been pointed out by Lawrence & Rawlins that the mesoforms can be defined in terms of dimensions. A solid is rigid in three dimensions, a smectic mesoform in two, a nematic mesoform in one, an isotropic liquid in none. This can easily be appreciated by reference to Fig. 325. On heating, melting occurs in the three dimensions in turn; the three different melting-points being the temperatures at which the three vectorial forces of adhesion are overcome by thermal agitation. This formulation, though not quite accurate, draws attention to the fact that the basic observation on mesoforms is that several melting-points succeed one another in the transition from solid to liquid. Now there is a similarity here between these successive stages of dimensional rigidity and the very curious phenomena seen in the determination of limb-buds in amphibia.

* Cf. Staudinger's "Darüber hinaus aber weisen makromolekulare Stoffe eine Fülle von *nur bei ihnen* möglichen Reaktionen auf. Darauf fasst ihre Bedeutung für den Biologen."

In order to explain this we shall have to take up again the story of limb-bud differentiation at the point at which it was left on p. 308. From the fundamental work of R. G. Harrison[5] in 1921 we know that when discs of the outer wall of the body, representing the buds of the future limbs, are transplanted at a certain stage, it is found that the original anterior edge of the bud always produces the pre-axial part of the limb. So a limb-bud of the left side, planted the right way up on the right side of the embryo, will develop into a limb with the elbow pointing forwards instead of backwards, if it is a forelimb. On the other hand, if the disc is rotated so that the original dorsal edge of the bud is ventral, the original anterior edge will be anterior again, and a normal limb develops. Therefore the dorso-ventral axis can be inverted without producing rearrangement, but not so the antero-posterior axis. The medio-lateral axis can also be inverted at this stage with impunity, i.e. it does not matter whether the limb-bud is attached to its new situation proximally or distally. This means that the dorso-ventral and medio-lateral axes are still plastic, while the antero-posterior axis is determined. Later on they also become determined. Here, then, we have successive stages of dimensional determination. Just as the liquid crystal passes through stages of rigidity in one, two and three dimensions, so the limb-bud passes through stages of determination in one, two and three dimensions. The analogy here may, of course, be superficial, but it is sufficiently striking to warrant attention.

To the above account some further facts may be added. Harrison's work was continued by one of the investigators of his school. In Swett's[2] first paper* the actual times at which the successive determinations take place are given us:

Dimension of space	Time of determination Harrison stage
Antero-posterior axis	Before 20 (probably from the first appearance of the limb district)
Dorso-ventral axis	Newt 29; axolotl 33–34
Medio-lateral axis	36–37

The stages vary somewhat according to the species. The determination of the dorso-ventral axis cannot be delayed by transplanting the limb-bud to a younger host (Swett[4]), nor can it be accelerated by transplantation to an older host (Swett[6]). But arising out of an earlier experiment of Nicholas, it was found that the determination of dimensional polarity does to some extent depend upon the site to which the limb-bud is transplanted. If placed in non-harmonic orientation on the flank at stages prior to the determination shown above, the limb-bud will behave *ortsgemäss* as it should, and hence will plastically assume harmonic relations; but if placed more dorsally over the somites, it will behave *herkunfts-gemäss* at an unexpectedly early stage (Swett[7]). This seems to be due to a failure of the re-orienting influences, which otherwise would operate, to reach the back from the flank. It afforded Swett[8] an opportunity of finding out how long it takes for the prospective polarity of a limb-bud to be changed in the direction of harmonic relations after transplantation on to the flank. Limb-buds were transplanted to the flank and then removed after varying intervals of time to the back. The result showed that the determination requires very little time, being completed almost as soon as the limb-bud has healed into its first new position.

* Swett[5] has also contributed a valuable review of this subject.

The successive determination of the spatial axes seen in the limb buds also holds good of the ear-vesicle; see Fig. 326 (Harrison[8,15]; E. K. Hall[1]). At its first appearance, the axolotl ear-vesicle is "isotropic", a state which continues up to stage 19, when determination of the antero-posterior axis sets in. At stage 25 the dorso-ventral axis becomes determined, and finally, probably at stage 29 or later, the medio-lateral axis is determined.

Several points of some importance arise out of this. In the first place, it is rather remarkable that the axes, or at any rate the first two, are determined before the ear-vesicle ceases to be an equipotential system. As late as stage 24 it is possible to fuse two ear-vesicles into a single complete organ. Secondly, it is

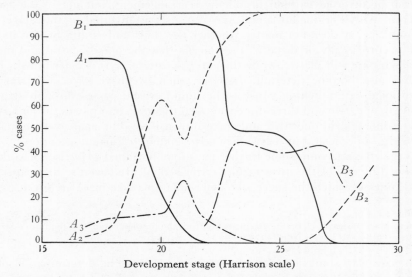

Fig. 326. Determination of spatial axes in the ear-vesicle of *Amblystoma*. (i) Antero-posterior axis only reversed. A_1, Ears with reversed asymmetry (neighbourwise; undetermined); A_2, ears with original asymmetry (selfwise; determined); A_3, ears showing reduplication anomalies. (ii) Dorso-ventral axis only reversed. B_1, Ears with reversed asymmetry (neighbourwise; undetermined); B_2, ears with original asymmetry (selfwise; determined); B_3, ears showing vesicular anomalies.

remarkable that the limb-bud and the ear should both conform to the same succession of axis determination, in view of the radical difference between them as embryonic systems. "The limb", writes Harrison[15], "is a mesenchymatic rudiment, out of which muscles and skeleton are segregated; the ear is an epithelial system, consisting at first of a circular plate of cells, or placode, from which the mature organ develops by invagination and folding. While these changes can be accomplished only by a complicated shifting of cells, roughly speaking, a circular plate of cells becomes transformed into a sphere." Since the tail also conforms to the same succession of axis determinations (Morikawa), we are forced to think there is some fundamental meaning in it. Even in the neural plate itself there may be a succession of axial determinations. There is a stage at which hardly any transverse determination is present, although the antero-posterior axis is quite fixed (Cummings). Similar considerations apply to the pronephros rudiment (T. C. Tung[3]).

A third point of great interest which arises out of Harrison's work on the ear is that reduplicated or enantiomorphic twin ear-vesicles sometimes appear. They occur statistically most frequently (up to 27 % of the total number of cases) very near the change-over point of antero-posterior determination (cf. Fig. 326). This is as if the short transitional period was one in which the transplanted tissue genuinely "did not know what to do". And correspondingly around the change-over point of dorso-ventral determination, curious vesicular anomalies occur most frequently.

"One is led from these facts", writes Harrison, "to the conclusion that the transformation from the first to the second phase, and probably, in some measure at least, from the second to the third phase, involve changes in the orientation of ultra-microscopic elements.... With the newer knowledge of the crystalline structure of organic fibres in mind, it may not be too rash to expect that X-ray diffraction photographs may some day reveal the changes in finer structure which underlie differentiations like those described as taking place in the development of the ear." An attempt has been made (by Harrison, Astbury & Rudall) to demonstrate molecular orientation as the basis of axial determination, by direct X-ray photography of pieces of amphibian tissue, but the presence of water, yolk-material, etc. has unfortunately so far prevented any correlation. When it becomes possible to cut sections sufficiently fine for the electron microscope, that tool may have an important part to play in revealing the fine structure for which we are looking. In any case, it remains a remarkable fact that just as simple rigidity in one, two or three dimensions occurs in liquid crystal states, so determination (which is itself surely a very complex form of rigidity) in one, two or three dimensions occurs in the organs of the developing embryo. It suggests that we have to deal with an intra-cellular, but at the same time super-cellular, pattern in three dimensions, irreversibly polarisable along fairly independent axes. The dynamic structure in the cell, or "cell skeleton", would be involved in this.

Somewhat analogous to the foregoing facts of organ development in amphibia are the conclusions reached by Hörstadius[2] regarding determination of axes in the eggs of invertebrates, especially the worm *Cerebratulus*. Here the animal-vegetal axis is determined between fertilisation and an early cleavage stage, while the dorso-ventral axis is not determined till after the end of the cleavage stages. Conversely, in the sea-urchin, the unfertilised egg already has a dorso-ventral axis, but the full determination of ectoderm and endoderm is not attained till the beginning of gastrulation; hence animal-vegetal determination happens later than in *Cerebratulus*. It will be remembered that dorso-ventral polarity can be shifted experimentally while the animal-vegetal cannot (in echinoderms, Runnström[2]; Lindahl[3]; in other invertebrates, Pease; but not in amphibia, Motomura[2]).

3·55. Modifications of polarity by external influences

One important property possessed by animal eggs is that their polarity is very fixed and almost impossible to alter. The only case in which the primary polarity of an animal egg has been determined or altered by centrifuging is that of C. V. Taylor's work[2] on the gephyrean worm *Urechis*. With long-continued centrifuging (up to 18 hr.) at a force which would be regarded to-day as relatively low ($4800 \times g$) he was able to shift the primary egg axis, and the longer the centrifuging was continued the more the polar bodies tended to occur at the centrifugal pole. But these experiments have been criticised (Morgan & Tyler; Costello[2]) and appear to be of doubtful significance. Since the introduction of the air-turbine centrifuge, it has been possible to apply much greater centrifugal forces to cells (up to $800,000 \times g$), after which normal development has been shown to proceed (Costello[2] on nudibranch eggs; Beams & King[1] on *Ascaris* eggs; McDougald, Beams & King on chick embryo heart fibroblasts). That cleavage polarity can be altered experimentally by pressure without producing any alteration in the primary embryonic axis is of course an old observation, known since Morgan's work[2] on *Arbacia* in 1893. Moreover, the dorso-ventral axis of echinoderm eggs is sensitive to external treatments, such as centrifuging, compression or local action of chemical substances (see pp. 497, 498).

It is obvious that when cell materials have been redistributed in the egg, the amount of actual ground substance in that part of the egg occupied by yolk-platelets or oil will be greatly reduced. Costello[1] has calculated that probably less than 25 % of the material in such a packed region is hyaline cytoplasm. The "cyto-skeleton", therefore, must be thought of, not as any series of protoplasmic strands, but as something molecular which can reform in time to permit of normal development. Hence the remark of F. R. Lillie[4] in 1909 is still relevant: "The existence of polarity and bilaterality in an optically homogeneous medium, and the persistence of both as to orientation under experimental conditions that seriously modify the quantitative relations of the oriented medium in different regions...seem to me to argue for a molecular basis of the fundamental principle of vital organisation."

It is interesting, however, that in plant "eggs" or spores the polarity is much more determinable by external influences. As long ago as 1885 it was noted by Stahl that differential exposure to light determined polarity in the spores of *Equisetum*. An interesting series of papers has been devoted by Whitaker to the eggs of the seaweed *Fucus*. The first division of these eggs gives rise to two cells of different shape, making the whole look like a pear. One, which includes the rhizoidal protuberance, is the parent cell for the formation of the rhizoid, the other gives rise by divisions to the thallus. The polarity of the eggs is directed by the presence of near-by neighbours, if more than a dozen are present, so that at the outskirts of a batch of eggs in a dish the pear-shaped embryos are all pointing inwards towards the main mass (Whitaker[2]). It was then found that the rhizoid protuberance always protrudes in the acid direction of a pH gradient (Whitaker[9]), unless the acid side is too acid, in which case complete reversal will occur and the protuberance will always point the other way (Whitaker[16]). This suggests that the mutual induction effect is due either to CO_2 or some other acid produced by the mass of cells, especially as the effect shows no species-specificity. If placed at one end of quartz capillary tubes, so that the gradient in one direction falls off more sharply than the other, the eggs will, when pear-shaped, point in the more acid direction, i.e. down the capillary tube (Whitaker & Lowrance[2]). One interesting effect is that when the sea-water is acidified to pH 6·0 all sizes of aggregations will carry out mutual inductions, and even two eggs in proximity will send out the protuberance in each other's direction.* Temperature and electric currents are also effective; the rhizoid protuberance appearing always on the warmed side (Lowrance) and on the side toward the positive pole (E. J. Lund[1]). So is light. Diffuse white light speeds up the early development (Whitaker[10]) and directed white light (during a sensitive period) causes the rhizoid protuberance to appear on the dark side (Whitaker & Lowrance[1]). If the egg is deformed into elongated shape by being sucked into a capillary tube, the rhizoid always forms at one of the ends (Whitaker[19]; cf. the determination of the ventral pole in echinoderm eggs, p. 497). Finally, the polarity in this case can be affected by centrifuging, for Whitaker[12] found that the protuberance always arises at the centrifugal pole, away from the packed chloroplasts, though after redistribution of the materials this tendency disappears. This is true whether the centrifuging has been mild ($3000 \times g$) or severe ($150,000 \times g$)—(Whitaker[15]). But if the centrifuging is done in acid sea-water the polarity is 100 % reversed and the rhizoid protrudes at the centripetal pole[18].

Of course the significance of these interesting observations for the problem of polarity in the eggs of animals is not quite clear.† The protuberance of the rhizoid seems to be due to a softening of the cellulose wall in that region, and the rhizoid itself is sheathed with cellulose. After earlier failures, du Buy & Olson succeeded in finding auxin in *Fucus* eggs, and it is therefore possible that all the above facts may be explained by variations of the position of active auxin within the cell. This is supported by the occurrence of rhizoid protuberances at the side of the eggs to which hetero-auxin is applied in sufficient concentration (Olson & du Buy); but it does not help us in understanding polarity in animal eggs.‡ In fact it reinforces the idea that in these we have to deal, not usually with any movable polarity determiner, but with a fixed molecular three-dimensional co-ordinate system, easily capable of being broken, but persistently re-forming itself.§

---

* This mutual effect has been analogised by Whitaker with the mutual effect of cohering echinoderm blastomeres in suppressing pluripotence. We thus have action only within the single cell (nuclear inductors, pp. 408 ff.), then action across cell-boundaries (evocators, pp. 148 ff.) and lastly action at a distance which may be ¼ mm.

† There is certainly a similarity between the directed appearance of the rhizoid protuberance from the *Fucus* egg, and the directed appearance of the axon from the neuroblast in vertebrates, as Weiss points out. Moreover, the determination of polarity by electric currents is not absolutely unknown in animals. It is said that *Obelia* hydranths appear mainly at the anodic pole and stolons at the kathodic (E. J. Lund[2], confirmed by Barth[4]).

‡ Might it be that the reason why the polarity of the plant egg is so readily altered, while that of the animal egg is so fixed, is because the relevant orienting micelles are in the former case wholly or partly cellulose, and in the latter, protein?

§ Cf. the interesting paper of Pollister on the orientation of mitochondria in cells, seemingly according to an invisible lattice of fibre-molecules.

3·56. Molecular orientation and contractility as the basis of response to morphogenetic inductors

Not only gastrulation, but neurulation also, must be thought of as involving changes in the orientation of protein micro-fibres. It is a demonstration of our ignorance that although there has been so much to say in Section 2 about the stimulus of neural induction, so little has been said about the process itself, the intimate nature of that lengthening of cell and nucleus which constitutes the principal feature in the change from dorsal cuboidal ectoderm cell to neural tube cell. The only attempt to make some headway with this problem is contained in the important but too much neglected papers of O. Glaser[2,3] (1914 and 1917), and the abstract recognition of the problem of induction as one of molecular orientation by Rashevsky[1] (1932). During the folding for the neural plate to form the neural tube, the number of nuclei does not increase (confirmed later by Boerema), but there is a remarkable concentration of them in the half of the wall furthest from the lumen of the developing tube.* The main factor is the change in cell form, possibly due to extension of protein micro-fibres. There is also a great increase in the volume of the neural area, as compared with the non-neural ectoderm. In the second paper, Glaser sought to correlate the vesiculation of the cephalic end of the neural tube with longitudinal compression, improving the old speculations of His on this matter.

It is possible that some light might be obtained on the way in which evocator substances affect the protein structure of the neural cells which they induce by recourse to surface film technique. When a naked blastula or gastrula is brought to an air-water surface, it immediately disintegrates, the formation of the powerfully spreading lipo-protein film being accompanied by a series of explosions, easily visible under the dissecting microscope, as one cell after another cytolyses. This bane of embryologists might be turned to good account if the effects of various substances, including evocators, were to be tested on the physicochemical properties of such monolayers, especially if films derived from ventral ectoderm and from dorsal blastopore lip or presumptive neural plate were tested separately. The substances could be injected under the monolayers in the manner of Rideal & Schulman. Following this suggestion, a beginning has been made by X-raying built-up newt embryo multilayers (Gatty & Schulman).

The probability is that molecular contractility plays a far greater part in development than embryologists have generally been willing to admit. Apart from the examples already given, we ought to remember the very extreme contractions of the chromosomes during mitosis, the blastodermal contractions of teleosts (p. 74), and in insect eggs the curious phenomena of blastokinesis, or even more important, the wave of yolk contraction at the differentiation-centre (p. 461).

---

* Later, as Sauer[1] has convincingly shown, the nuclei approach the lumen before dividing, and afterwards come away again. No explanation is forthcoming for this behaviour.

3·57. Anisometric protein molecules in eggs and embryos

Another way of getting some light on the problem would be to ask whether the molecular shape of the protein molecules of the egg hyaloplasm is globular (as respiratory pigment molecules, for instance, seem to be) or fibrillar. If the latter were true, it would be easier to imagine the dynamic lattice of protein chains.

Besides the protein of muscle, many other groups of highly anisometric molecules in the body are now known, e.g. the stroma protein of erythrocytes (Böhm), where the molecular length is 1000 times the breadth. The most direct method would be to prepare the protein of "yolkless" eggs (such as those of sea-urchins) and examine it, either in the ultra-centrifuge or with co-axial rotating cylinders. v. Muralt & Edsall made such a study (now classical) of the properties of muscle globulin (myosin). Having separated it with the utmost care from the other constituents of muscle, they enclosed it in a space between two concentric cylinders, one of which revolves so that the shear force orients the particles in a uniform direction. Then, observing the flow birefringence through crossed Nicol prisms in a direction longitudinal to the cylinders, they found the "cross of isocline". This optical phenomenon means that the particles are all *rods* of nearly the same size, identical, in fact, with the crystals previously postulated to account for the double refraction of the muscle fibre. Similar work upon egg-proteins has already begun.

Before describing it, we may recall that all eggs appear to contain two main classes of protein (p. 15). On the one hand there are the vitellins or ichthulins, phosphoproteins containing much serine phosphoric acid; on the other hand there are the livetins or thuicthins, which are pseudo-globulins not unrelated to the myosin of muscle. It might be that the former are primarily stores of nitrogen and phosphorus for the subsequent elaboration of the embryo's architecture, and that the latter (much less usually in amount) are the molecules constituting the "cyto-skeleton".

In the first place direct observation of echinoderm eggs has succeeded in revealing phases of birefringence within the cell (W. J. Schmidt[4]; Moore & Miller). Pfeiffer[3] submitted mollusc eggs to deformation within capillary tubes, and described interference striations from which the existence of fibrillar molecules was deduced. Longitudinal striations have also been reported by Seifriz[2] in living amoeboid protoplasm by the aid of the Spierer oil-immersion dark-ground ultra-microscope. Mirsky's examination[2] of echinoderm egg-proteins before and after fertilisation is also rather significant. He brought about the disintegration of the cell structure by freezing the eggs in solid CO_2 before placing them in a high-vacuum desiccator; by this method (believed to distort the protein molecules less than any other), he found distinctly more of the soluble protein fraction before fertilisation than afterwards. The protein that coagulates during fertilisation he showed to consist of very fibrillar molecules, for it had a high viscosity and showed flow birefringence.

Some work has also been done on the hen's egg. Böhm & Signer found a very marked flow birefringence in ovoglobulin preparations (probably mixtures of ovomucin and ovomucoid) but none in ovoalbumen. Wöhlisch & Belonoschkin also found that ovoglobulin gave a value for the Gans effect (depolarisation of the Tyndall beam) higher even than myosin itself, again indicating the presence of very long molecules. As the yolk might be regarded as of greater embryological significance than the white, Needham & Robinson examined livetin preparations in an improved co-axial cylinder apparatus also capable of acting as a Couette viscosimeter* (Robinson[1]; Lawrence & Robinson). They found flow birefringence though weaker than that of myosin, and an accompanying thixotropy and rheopexy. Vitellin and vitellomucoid did not show these properties. However, when this work was continued by Lawrence, Needham & Shen with an improved Couette apparatus on amphibian material, the total globulin fraction (containing vitellin) showed regular anomalous viscosity at low rates of shear, closely resembling that of tobacco-mosaic virus. The pseudo-globulin, on the other hand, though showing at first normal flow, formed polyfilms at the air-water interface after some rotation from which particles giving anomalous flow appeared to be split off. Much further work is needed along these lines. For example, the ultracentrifuge may be expected to permit the separation of the protein fractions containing anisometric molecules.

It must also be significant that the particles of many plant viruses are highly fibrillar (X-ray evidence of Bawden, Pirie, Bernal & Fankuchen; viscosity and flow birefringence evidence of Robinson[2]). W. M. Stanley emphasises the importance of this as linking chemical with morphological forms, but it has not yet been possible to form a picture of exactly how the fibrillar molecules are fitted together in the cell. Since the viruses are nucleoproteins, the suggestion that chromosomes are paracrystalline protein macro-molecules, is of great interest (see p. 637). It was first made by Koltzov[1] in 1912 with nothing more to go upon than the polypeptide chains of Emil Fischer, and since that time the proposal has been developed with new evidence in a remarkable way (Wrinch[2]; Koltzov[2,3,4]).

Less important embryologically, but still not without interest, are the observations of W. J. Schmidt[3] on the complex crystal structure of *Ascaris* egg-shells, in which chitin fibrils are involved.

One cannot leave the subject of fibres and micro-fibres without referring to a type of living matter at first sight somewhat far removed from the animal embryo — the plasmodium or slime-mould. According to a remarkable observation of Arthur Lister in 1888 such a mass of protoplasm will flow through cotton-wool: "I placed some wet cotton-wool in front of the still dingy plasmodium; this it readily penetrated, and afterwards emerged possessing its normal yellow colour, leaving the wool charged with spores and other debris." Recently Moore[4] has pushed further the analysis of protoplasmic structure along these lines. *Physarum polycephalum*, a slime-mould, was not killed by being ground in a mortar with quartz sand, but was killed if finely crushed glass was used instead. Both soft and hard filter papers permitted the passage of the plasmodium, and it readily passed through parchment, where the pores were of an average diameter

* The inner cylinder is supported on a torsion wire, and the deflection at different speeds read off with a mirror.

of 5×10^{-5} mm. Yet if pressure was applied, as by forcing it through textiles or filter paper, the protoplasm was destroyed at far larger pore sizes. Thus to be squeezed through fine silk (pore size, 5×10^{-2} mm.) was lethal for it, and pores as large as 0·25 mm. had to be used before it could be safely pressed through. "These results suggest", says Moore, "that the plasmodium contains long threads of living material essential to its existence, and if these be broken too short, life is impossible." From the rough data just given such micro-fibrils would be 5×10^{-5} mm. in diameter and 2000 times as long, i.e. about 100 times the diameter of the protein chains envisaged by Peters. However, the general suggestiveness of these experiments cannot be questioned.

In subsequent papers Moore[5,6] tells us that filtration treatments which are lethal or almost lethal for the plasmodium reduce its respiratory rate 50 %. Severe injury may also be caused by centrifuging at 75,000 × g for a few minutes.

3·58. The mechanism of gastrulation

That embryology may become largely the study of protein fibres and their planned distortion and orientation may be illustrated by further examples. Formerly there was much discussion about what was called the inadequacy of the cell theory of development (Sedgwick; Whitman). But whereas this mainly turned on the meaning of the histological evidence, i.e. the existence and nature of cell boundaries, in our day the importance of the cell has largely evaporated, because we tend to think more in terms of embryonic *regions*. Determination operates, as far as individual cells are concerned, in a statistical way. Whether a given cell is just inside or just outside the boundary of a field or district influence is not a matter of much importance. But recently the question has again been brought to the fore by the experimental demonstration of the existence of cell bridges in the early development of echinoderm eggs (A. R. Moore[1]; M. M. Moore; Moore & Moore; Huang) and the suggestion that these may have a function of some importance in gastrulation. Here is a fundamental field for the application of X-ray analysis. It might be well worth while to know something of the crystal structure of the protein chains of these cell bridges. Is it not significant, in view of what we now know of molecular contractility, that they possess considerable elasticity?

A long and interesting chapter in embryology has been made by efforts to understand the process of gastrulation. Forty years ago Rhumbler[2] regarded invagination as being due to a change of shape in the cells of the vegetal pole of the blastula so that their internal bases were broadened, but in most cases no such shape has ever been observed. Then Bütschli thought that differential absorption of water might be responsible, but Moore's analysis of lithium exo-gastrulation[2] has fairly clearly disproved this possibility. Moore considers Assheton's suggestion[2] more hopeful, namely, that if the cells of the blastula attract each other and if the centres of attraction are eccentric, any increase in pressure will result in an invagination if the centres are peripheral or an evagina-tion if the centres are located in the basal parts of the cells. The nuclei themselves cannot be, as Assheton thought, these centres of attraction, since in most echino-derms, for example, they are disposed towards the periphery at both poles of the blastula, i.e. at the presumptive non-invaginating part as well as at the pre-sumptive site of invagination. Moore's study[1] of cell bridges by micro-dissection, however, has shown that they might perform this function. They certainly seem to account for the tendency to curl at the edges, which cell plates (e.g. in *Dendraster* developing without the fertilisation membrane) exhibit.

Nothing quite similar to this has been found in amphibian embryos, but Waddington's examination[24] of the forces involved in gastrulation, though only preliminary, shows the way for further work. The invaginating mesoderm was made to act upon a steel ball subject to the opposing force of a magnet; in this way it was found that the invaginating material could successfully oppose a force of 0·34 mgm./sq. mm. Another point to be remembered is the suspicion already referred to (p. 146) that the initiation of gastrulation, at any rate, is connected with contractility on the part of the piriform cells of the endoderm. Holtfreter[21], too, has studied the natural curling propensities of isolated pieces of the amphibian gastrula and neurula. The neural plate curls in concave fashion, the epidermis convex. Speed and intensity of curling are sensitive to osmotic pressure, and the curling tendency is present before it occurs in the normal course of development.

The subject has been pursued by Moore & Burt, who report that the invagina-tion tendency of the echinoderm gastrula is inherent in the gastral plate. If the gastral plates are excised from sea-urchin embryos, the edges curl up and may eventually meet and fuse, but nevertheless invagination continues. If there is not enough ectodermal material for this, the gastral plate will eventually form a naked endodermal sphere. These facts show that the invagination tendencies are intrinsic to the invaginating area. Other experiments of Moore & Burt indicate that the composition of the blastocoele fluid can have little to do with the process because it is almost identical with sea-water. The only changes in it concern the pH, but this is not till after gastrulation. At first more acid than the environmental sea-water, it rises when the mesenchyme cells first appear to a level above that of sea-water (Hirabayashi; in confirmation of Rapkine & Prenant, not of Chambers & Pollack; see *CE*, p. 847). If the pressure of the blastocoele fluid is raised to a sufficient degree (as by the use of sucrose-containing sea-water), gastrulation may be delicately inhibited (Moore[8]).

3·59. The fibres of the mesoderm

We come lastly to the fibrils of connective tissue ground substance. It has long been known that the organs first formed in the embryo (neural tube, notochord, somites, etc.) are held apart by a "ground substance", the precursor of connective tissue, which at first contains no cells. This has long been known and often observed and figured, as by Merkel; von Szily; and in the beautiful series of papers by Baitsell. It has been suggested (*CE*, p. 566) that this material, which histochemically contains a high proportion of mucoprotein (Bianchi), and which occurs in relatively largest amount in the earlier stages of development, accounts for the high level of protein-combined carbohydrate found during amphibian

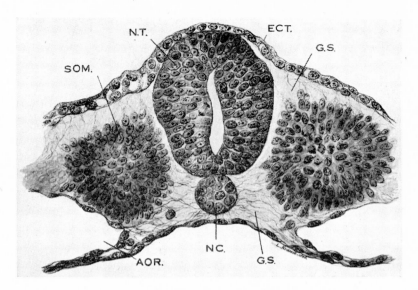

XXXV

Fig. 327. Connective tissue ground substance fibrils in the early chick embryo. *AOR*, Aorta; *SOM*, somites; *NC*, notochord; *NT*, neural tube; *ECT*, ectoderm; *GS*, ground substance.

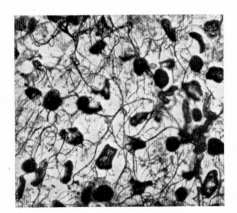

Fig. 328. Primitive connective-tissue fibrils of the exocoelom of the chick embryo.

development. The important point is that wavy fibrils appear, as if by a sort of crystallisation, in this ground substance some considerable time before any cells are present in it (Fig. 327). Cells migrate into it later (Baitsell[4,5]; Bauer). Analogous processes have been shown to be at work in the development of clot material during wound healing (Baitsell[2]), and in plasma clots subjected to traction where there seems to be a continuity between fibrin filaments and the later fibres (Baitsell[3]),* and also in the origin of fibrous tissue arising at sites of tuberculous infection (Baitsell[6]; Baitsell & Mason). The "crystallisation" of anisometric protein molecules must certainly be involved in these processes. Fig. 328, taken from Snessarev's review, shows a micro-photograph of the outer layer of the exocoelom of a chick embryo of 72 hr. incubation; the primitive connective-tissue fibrils can be clearly seen. The appearance of the wavy fibrils in tissue cultures has been also studied a good deal (Baitsell[1]; Rumjantzev & Suntzova; Berezkina; Litvać). The observation of Berezkina that the old fibres serve as a basis for the new ones suggests a crystallisation process originating from "nuclei" provided. If the ground substance is a secretion of the embryonic cells, as it must be, it should be possible to detect some cytological evidence of this; hence the interest of Cameron's contention that the Eberth bodies of amphibian embryonic epidermis are concerned in the secretion of ground substance. Furthermore, Litvać has shown that the fibrils formed in the ground substance of tissue cultures are birefringent, and hence probably paracrystalline in nature.

The subject has been much amplified from the experimental angle by Weiss[9], who first showed that if a membrane of blood plasma is subjected to mechanical tension, e.g. by placing it in frames of different geometrical forms, then fibroblast cultures growing upon it will show an oriented migration directed along the main axes of tension.† This is interpreted to mean (Weiss[12]) that the micelles or fibrillar protein molecules of the substratum are oriented, and that the cells divide and migrate accordingly. Similar results were obtained by Huzella. Later Weiss[13] extended these ideas to the growth of nerve fibres, which probably occurs more as the result of following the fibrillar ultra-structure than on account of any other factors.

But while we cannot fail to admit the importance of ultra-structure in guiding morphogenetic processes, we are forced back on the mystery of how the ultra-structure itself is formed. Probably the only way in which we can think of this is as a kind of extension of the pattern ("dynamic structure") possessed by the cells of the primitive organs themselves. We may recall W. B. Hardy's theory[1,2] of trophic action, which still remains, indeed, the only attempt to explain that curious phenomenon. Tonus and heat production of muscles depend on continuity of nerve-supply, but it is not believed that the nerve fibre here acts by conveying impulses to the muscle. Hardy[2] suggested that its effect might be due

---

* It is generally agreed, however, that fibrin is by no means *necessary* as a precursor of fibril formation.

† The exact physical nature and origin of this micellar orientation is a matter of discussion (see the note of Pfeiffer[2]).

to orientation of molecules. A layer of molecules oriented by an interface might, he thought, be of the order of at least a thousand molecules in thickness. This, he wrote, would be a humble example to place beside the colossal examples furnished by living matter, "but when one has for years contemplated a scientific problem towards the explanation of which nothing could be advanced, even the slightest clue is welcome".

This suggestion derived from the work of Hardy & Nottage in which it was shown that orientation effects may be transmitted from a surface such as that of a metal capable of influencing a surrounding film for a distance of 7 to 8 μ. "Here we have at once", wrote Peters, commenting upon this later, "a means by which a mosaic may radiate its effects throughout the cell. It is possible to appreciate how a co-ordinate structure may be maintained even in a medium which is apparently entirely liquid. We have in effect chains of liquid crystals which need not be more than a few molecules thick. Once we have seen this, it becomes possible to develop some dim conception as to how the miracle occurs that colloid molecules, which have apparently no particular location in the cell, can yet control the synthetic and other reactions which are proceeding. This theory also is all that is needed to enable us to understand how substances can reach a special site in the cell. Between the chains of molecules, fixed by the radiating webs, there will exist paths from the external to the internal surface."

Related to the connective-tissue fibrils mentioned above are the collagen or elastoidin fibres of tendons, fins, etc., which show birefringence and are certainly built up of anisometric molecules. These have been much studied by Fauré-Fremiet and his school (Fauré-Fremiet[5,6]; Fauré-Fremiet & Garrault[2]; Fauré-Fremiet & Baudouy[1]; Fauré-Fremiet & Woelfflin; Garrault[2]) and X-ray photographs have been made (Champetier & Fauré-Fremiet). Ingenious methods of micro-dissection using protozoa which do not attack the fibres have been used (Fauré-Fremiet, Kruszynski & Mazoué).

We do not yet know, of course, the normal equilibrium configuration of most of these fibres. We could guess, however, that the Weissian connective-tissue ground substance fibres are usually contracted, like the chains of wool, while the cell bridges of echinoderms and the yolk fibres of insects are usually expanded, like the chains of silk. On such a view, the elongation of the cells during neural plate formation might be connected with the expansion of molecules previously contracted, and the clamping of them in their new condition. But that is as far as speculation can well go in the present state of our knowledge.

3·595. Dynamic structure at the molecular level

Throughout this book in general, and the foregoing section in particular, it has been necessary to emphasise our reliance upon the proteins as the responsible structures in competence (pp. 114, 119, 203), cell-architecture (p. 670), dynamic structure (p. 658), and so on. But it would be a great mistake to think of them as static solid material like that which is used by the human architect or statuary. The protein molecule in the living body, though not when isolated and purified, is in a condition of constant flux and flow, and if it participates in living architecture, it participates in metabolism too. This had indeed long been known, but the extraordinary speed at which the replacement of atoms in molecules within the living cell occurs was not revealed until the employment of radio-active and other isotopes permitted the use of marked or tagged atoms (reviews by Schoenheimer & Rittenberg[2]; Hevesy; van Heyningen[2] and others). Hence the protein molecule itself, no less than the cell-structure of which it is a part, is a pattern the components of which are in perpetual motion. In visualising this, it is difficult not to think of it as a prefiguration of the mutual collaboration of social units in maintaining patterns at far higher levels of organisation.

CONCLUSION

The foregoing sections, which conclude the plan of the present work, have emphasised successfully, one may hope, that the polarity and symmetry properties of embryos will never be understood except in the light of a knowledge of the dynamic structure of the egg-cell. The problem of these properties has led us directly to the fundamental problem of biology, the nature of cytological and morphological organisation. The whole object of the book was, after all, to show that the fields of chemistry and morphology are not so sundered as is often supposed. Organising relations are found at the molecular level and at the colloidal and paracrystalline level, the level of protein macro-molecules and highly polymerised substances, just as clearly as at the anatomical level itself. Although we are still in the earliest stages of any real theory of living organisation, we can yet see that biological order, like crystal order, but on a much more complicated plane, is a natural consequence of the properties of matter, and one characteristic mode of their manifestation. But when this is said and accepted, the task of embryologists is far from being on the point of ending. On the contrary, only now, freed from the vestiges of primitive philosophy, can they look forward to see their real task clearly in the future, the revealing of the causal sequence whereby the organisation of the fully formed organism arises from the lesser organisation of its zygote. Human science is yet young and the task of embryologists is really only beginning. Human society itself, in that slow but persistent and inevitable rise in level of organisation which biological and social evolution represents, mirrors, as the great Victorian biologists recognised very well, the ontogenetic rise of the individual organism in organisational level. Let us hope that human society will before long attain a level of organisation such as to encourage and facilitate the great enterprise of embryology. The study of the beginnings of life has something in it antithetical to death, the harmonious relations of morphogenesis something inimical to human disunity and hatred, the co-operation of embryonic processes something utterly alien to human oppression and war. But we are not as men without hope, for just as the greater number of ontogenies safely achieve their triumphant ends, so biological and social evolution, in spite of many grave setbacks, give us the right to look forward with confidence to the *Regnum Dei* of man's full stature.

GLOSSARY OF SOME OF THE SPECIAL TERMS
USED IN THIS WORK

(Cf. the "Terminologie" of Roux, Correns, Fischer & Küster)

Activation. The liberation of naturally occurring evocators from inactive combination. *See also* Fertilisation.

Activation centre (*Bildungszentrum*). One of the two organisation centres in the insect egg (see p. 460).

Affinity. The extent to which tissues from early embryos tend to cling together when in isolation from their normal surroundings; ≡ Roux's Cytarme.

Alecithic. Having little yolk; said of e.g. echinoderm eggs.

Allantois. The sac developing from the hindgut in amniotes, which receives the foetal excretions, and, in birds and some mammals, has important respiratory and nutritional functions.

Allometry. Allomorphosis and Heterauxesis.

Allomorphosis. The relation of parts of organisms at some definite age to wholes or parts also at some definite age, but of different groups (races, varieties, species, genera); e.g. egg size or hatching weight to adult size or weight.

Amnion. The sac filled with liquid immediately surrounding the embryo.

Analogous. Said of organs or structures which have the same function but are of different evolutionary origin. *See* Homologous.

Androgen. A chemical substance with activity similar to that of the male sex hormones.

Androgenesis. The fertilisation of an egg, the nucleus of which has been so treated that nuclear fusion cannot ensue, by normal sperm.

Andro-merogony. *See* Merogony.

Anidian. A blastoderm on which no trace of embryonic axis appears.

Animal pole. That pole of the egg which contains most cytoplasm and least yolk.

Animalisation (*Ectodermisation*). The shifting of the presumptive fate of tissue which should have been endodermal (in echinoderm development).

Anlage (*Rudiment, Ébauche*). A group of cells, not necessarily distinguishable from their surroundings, which will later give rise to a given organ.

Area. A morphogenetic area is one of the constituent regions on a Fate-Map.

Area opaca. The opaque yolky extra-embryonic blastoderm of the sauropsids, which later forms the yolk-sac.

Area pellucida. The transparent central part of the sauropsidan blastoderm, from which the axiated embryo arises.

Auflagerung. The laying of competent ectoderm upon a dead inductor in order to test inductive power.

Autonomous induction. An induction in which the inductor takes no part itself in what is formed (opp. of *Complementary induction*). All chemical inductions are autonomous inductions.

Autoplastic transplantation. Transplantation between different sites on the same individual.

Autotomy. The spontaneous shedding of an appendage (arthropods).

Auxesis. Growth in size by cell expansion without cell-division.

Bauchstück. The ventral half of an amphibian gastrula, from which all the organiser region has been removed; hence it never forms any neural axis.

Bedeutungsfremde Selbstdifferenzierung. Self-differentiation in accordance neither with the tissue's original presumptive fate nor with the presumptive fate implied by the new surroundings, i.e. neither selfwise nor neighbourwise.

Bildungszentrum. *See* Activation centre.

Blastema. The small bud composed of competent cells from which begins the regeneration of an organ or appendage.

Blastocoele. The cavity in the blastula.

Blastocyst. The mammalian embryo at the time of implantation into the uterine wall.

Blastoderm. The flat sheet formed by the embryo's cell-layers when cleavage is discoidal.

Blastokinesis (*Revolution*). A process of cephalo-caudal reversal in the egg, accomplished in some animals by movement during development (e.g. insects).

Blastomere. One of the cells into which the egg divides during cleavage; the term is not usually applied after the blastula stage. When blastomeres or sets of blastomeres differ in size, the terms *macromere* and *micromere* are used.

Blastula. The stage in development at which a central cavity is formed in the mass of dividing cells (the *Blastocoele* cavity).

Bradyauxesis. *See* Heterauxesis.

Carcinogen. A chemical substance which has the property of giving rise to cancer when administered to animals by external or internal application.

Cecidogen. A chemical substance, usually secreted by an insect, which has the property of giving rise to a gall on a plant.

Centrolecithal. Having yolk accumulated at the centre of the egg (e.g. arthropods).

Chimaera. A compound organism or tissue obtained by grafting or other means so that it is composed of tissues of different kinds (e.g. an andro-merogonic head on a normal diploid tail) or of different species or genera.

Chorio-Allantoic Grafting. The placing of small pieces of avian or mammalian embryos upon the chick embryo's allantois under the inner shell-membrane (the bird's egg has no chorion), in which position they become vascularised from the allantoic circulation and develop, though not in a strictly neutral medium.

Cleavage. The processes of cell-division by which the original egg-cell gives rise to all the cells of the organism; if the cells are from the beginning of approximately equivalent size, the cleavage is *holoblastic*; if some are much larger than others owing to yolk accumulation or other reasons, the cleavage is *meroblastic*; and if cell-division occurs only at the surface of an enormous yolk-mass, the cleavage is *discoidal*. Cleavage in some species follows a definite pattern, where it is said to be *determinate* (this has permitted the tracing of "*cell lineages*"); in other species the pattern is quite lost after the first few cell-divisions.

Cleidoic. Highly isolated from the environment, e.g. the eggs of birds, with walls permeable only to matter in the gaseous state.

Colostrum. The first secretion from the mammalian mammary glands after the birth of young; it has quite different properties from the milk which it precedes.

Competence. The state of reactivity of a part of an embryo, enabling it to react to a given morphogenetic stimulus by determination and differentiation in a given direction. An embryonic cell develops as many competences as it has prospective potencies or possible morphogenetic fates. Generally speaking, determination to proceed to one morphogenetic destination cancels the other competences previously possessed by the cell. Competences appear in a succession at definite stages of development and later disappear, whether or not any of them have been implemented by the appropriate morphogenetic stimulus. Some competences exist in the adult tissues, if we may so term the states of reactivity to hormone action, though these effects are generally reversible (cf. *Modulation*).

Complementary induction. An induction in which the inductor itself is built into the morphological complex which it has induced (opp. of *Autonomous induction*).

Compression. The acceleration of development or extension of the period before hatching, so that the larval stages are accomplished or omitted before the organism hatches from the egg.

Corporative. Differentiation due to the physiological functioning of parts, as opposed to *Individuative*, due to the action of morphogenetic fields.

Cyclopia. The development of one median eye instead of two lateral ones.

Cytotaxis. The spontaneous attractive or repulsive movements of cell-groups. *See also* Affinity *and* Incompatibility.

Dedifferentiation (*Einschmelzung, Catachony*). Loss of differentiation, as of organelles by protozoa before cleavage, or of starving ascidians, or of the vertebrate limb stump in blastema formation.

Degrowth. Decrease in spatial dimensions and weight of an organism.

Dependent differentiation (*Correlative or abhängige Differenzierung*; *Différentiation provoquée*). The differentiation of a tissue partly as a response to factors existing outside itself, i.e. differentiation which requires the morphogenetic stimulus of an inductor from a neighbouring part of the embryo.

Determination. The fixation of the fate of a part of an embryo, so that (if it develops further) it will perform one kind of morpho- or histo-genesis, and one kind, only, irrespective of the situation in which it may find itself. In using this concept one must bear in mind that all possible situations have usually not been tested; hence the word must be handled with due regard to the limitations of our knowledge. Ideally, the situations should be specified in every case, but definition will continue to be the victim of convenience until some rapid notation is devised. The term should not be used for the status of the tissue following the process, and it should have reference to an immediately subsequent developmental change. It must be remembered that determination is embryological, not genetical. This is seen with special force when ontogenetic stimulus-response systems are considered in which the stimulator and the reactor are of genetically different tissue. The sense in which the word "determination" is here used must be distinguished from other uses, e.g. (1) the "determination" of the original axes of the

Determination (*cont.*)

egg by external factors (Roux), (2) the " determinate " cleavage of eggs in which the cleavage pattern is very definite and permits of accurate descriptions and predictions of cell lineage (Conklin); (3) "determinants", hypothetical units in the germ-plasm (Weismann), a use now obsolete. *See also* Dynamic determination, Material determination.

Diapause. A state of dormancy in the normal development of numerous animals, especially insects, and distinguished from hibernation in that it must be passed through, no matter what the temperature or other external conditions may be.

Differentiation. Increase in complexity and organisation. Increase in number of kinds of cells; invisibly, as by determination and losses of competence; visibly, as by histogenesis. Increase in morphological heterogeneity by the appearance of form and pattern. *See also* Self-differentiation, Dependent differentiation (Correlative differentiation), Functional differentiation (Corporative differentiation), Regional differentiation.

Differentiation centre (*Differenzierungszentrum*). One of the two organisation centres in the insect egg (see p. 460).

Diploid. Containing the normal number of chromosomes.

Direct induction. The induction of a new neural axis in competent ventral ectoderm by a chemical compound acting in the same way as the naturally occurring evocator.

District (*Territoire*). A minor morphogenetic field, possessing a general determination for production of a certain structure or organ, and undergoing progressive regional specification of detail, e.g. limb district or gill district.

Double assurance (*doppelte Sicherung*). The ability of an embryo to perform a morphogenetic process by a method not normally employed, if the usual method is inhibited by experimental means (Spemann). Reasons are given on pp. 157, 294, 308 for believing that this is not a true category of morphogenetic behaviour. The normal concurrence of many factors towards the production of a given morphogenetic result may, however, be admitted (cf. the *synergetical principle* of development, the *kombinative Einheitsleistung* of Lehmann).

Dynamic determination. Determination of a part of an embryo to carry out a certain set of morphogenetic movements.

Ectoderm. The outermost of the three cell- or germ-layers of classical embryology, see Fig. 55. Strictly speaking, the term ectoderm should not be used until the definitive ectoderm and mesoderm have fully formed, but the loose usage avoids multiplicity of terms.

Eidogen. A chemical substance which has the power of modifying the form of an embryonic organ which has been induced by other means. The term is intended to cover such substances as may be involved, beside the primary evocator, in bringing about regional differentiation of the neural axis; but it also would be applicable to such substances as are involved in *Nachbarschaft* or *Contiguity* effects, including those which diffuse from diploid into haploid tissue (cf. p. 360), and it grades, without a very sharp dividing line, into the category of the second-grade inductor.

Einsteckung. The implantation of a tissue, living or dead, or a piece of implant medium containing one or more chemical substances, into the blastocoele cavity of the gastrula, in order to test induction power.

Emancipation. The process whereby the various organ districts acquire their boundaries in arising out of the primitive individuation field.

Embryoma. *See* Teratoma.

Endoderm. The innermost of the three cell- or germ-layers of classical embryology.

Epiboly. The expansion of the ectoderm to surround the endoderm and mesoderm, whether by their invagination or its own active envelopment of them.

Epigenesis. The development of the organism by the new appearance of structures and functions, as against the unfolding or growth of entities already fully present in the egg at the beginning of development (*Preformation*).

Evocation. That part of the morphogenetic effect of an organiser which can be referred back to the action of a single chemical substance, the *Evocator*.

Evocator. The chemical substance, acting as the whole or part of a morphogenetic stimulus, emitted by an organiser.

Exogastrulation. Formation of a Teras by gastrular evagination instead of the normal invagination.

Explantation. The development and growth of cells outside the organism from which they were derived, and in isolation from other tissue, i.e. *in vitro*.

Expressivity. The extent to which an animal carrying a gene and showing it is affected by it (the relative severity of the effect). *See also* Penetrance.

Fate-map. A map of an embryo in an early stage of development, e.g. the late blastula stage, indicating the various regions whose prospective significance has been established by marking methods. The regions are known as *Areas*, e.g. the area of presumptive notochord material.

Fertilisation. The essential stimulus to development given by the spermatozoon to

Fertilisation (*cont.*)

the egg. In artificial *Parthenogenesis* this stimulus may be replaced by a wide variety of agents, some of which do not effect more than a few of the preliminary cytological changes preceding cleavage; this is known as *Activation*; others stimulate more profoundly so as to put in motion the cleavage processes (*Regulation*).

Field. A morphogenetic field is a system of order such that the positions taken up by unstable entities in one part of the system bear a definite relation to the positions taken up by other unstable entities in other parts of the system. The field effect is constituted by their several equilibrium positions. A field is bound to a material substratum from which a dynamic pattern arises. It is heteroaxial and heteropolar, has recognisably separate districts, and can, like a magnetic field, maintain its pattern when its mass is either reduced or increased. It can fuse with a similar pattern entering with new material if the axial orientation is favourable. The morphogenetic *gradient* is a special limited case of the morphogenetic field.

Foetus. Medical synonym for embryo.

Freemartin. A mammalian intersex due to the masculinisation of a female twin by its male partner when the foetal circulations are continuous.

Functional differentiation. The differentiation of tissues as the result of forces arising out of the functions which they are performing.

Gastrulation. The processes whereby the endoderm, mesoderm, and ectoderm come to occupy their eventual position in the embryo.

Genotype. The genetic equipment of an organism.

Gestalt. A patterned whole distinguished from its surroundings.

Grade. *See* Organiser. Organisers are said to be of first-, second-, or third-grade according to the time, whether early or late, at which they act during development.

Gradient. A system of order involving progressively increasing or decreasing entities or intensities from one pole or point of a morphogenetic field to another. It is a two-dimensional special case of field pattern. *See* Field.

Growth. Increase in spatial dimensions and weight of the organism. Increase in number of nuclei and cells (*Multiplicative* or *Meristic growth*). Increase of cell-size (*Auxetic growth*). Increase of non-living structural matter (*Accretionary growth*). The increase of living respiring protoplasm at the expense of inert yolk during embryonic development in holoblastic eggs should probably be regarded as growth although the total weight of the system may be decreasing owing to loss of water and carbon dioxide in combustion of energy sources.

Gynandromorph. A sex-mosaic. *See* Mosaic (2). Contrast with *Intersex*, all of whose cells have a genetic constitution intermediate between male and female.

Gynogenesis. The fertilisation of an egg by sperm so treated that nuclear fusion cannot ensue.

Gyno-merogony. *See* Merogony.

Haemotrophe. The sum total of nutritive substances supplied to the embryo from the maternal blood stream in viviparous animals.

Haploid. Containing half the normal number of chromosomes, i.e. one of each of the necessary kinds.

Herkunftsgemäss. *See* Selfwise.

Heterauxesis. The relation of the growth rate of a part of the developing organism (whether morphological or chemical) to the growth rate of the whole, or of another part. The part may grow more rapidly than the whole (*Tachyauxesis*) or it may grow more slowly than the whole (*Bradyauxesis*), or at the same rate (*Isauxesis*). Heterauxesis is a comparison between organisms of the same group but of different ages and hence sizes. *See also* Auxesis, Merisis, Allometry, Allomorphosis.

Heterogametic sex. That sex the gametes of which are divided into two populations, one of which possesses the *X*-chromosome, governing sex differentiation, and the other does not.

Heteromorphosis. The appearance, whether in embryonic development or in regeneration, of an organ or appendage inappropriate to its site, e.g. antenna instead of leg or eye. Embryonic heteromorphosis was termed by Bateson *Homoeosis*; Goethe knew it as *Metamorphy*.

Heteroplastic transplantation. Transplantation between individuals of different species within the same genus.

Heterotopic transplantation. Transplantation between different sites.

Hippomanes. Packets of uterine secretion surrounded by trophoblast and invaginated into the allantoic cavity.

Histogenesis (*Elaboration*). The development of histological differentiation.

Histolysis. Tissue destruction, as in insect metamorphosis.

Histotrophe. The sum total of nutritive substances supplied to the embryo in viviparous animals from sources other than the maternal blood stream, e.g. uterine glands and the like.

Holoblastic. *See* Cleavage.

Homoiogenetic induction. The power acquired by a part of an embryo, at the

Homoiogenetic induction (*cont.*)
same time as it is determined for a certain differentiation, to induce another of the same kind. Thus at the same time as the neural plate is induced by the underlying mesoderm, it acquires the power of inducing another neural plate if it is removed and implanted into the blastocoele cavity of another embryo.

Homoioplastic transplantation. Transplantation between individuals of the same species.

Homoiotopic transplantation. Transplantation between corresponding (identical) sites.

Homologous. Said of organs or structures which have the same evolutionary origin though their functions may be very different. *See* Analogous.

Hybridisation. The fertilisation of an egg of one species by the sperm of another.

Implantation. The addition of tissue, grafts, chemical fractions in an implant medium, etc. to an embryo, without the removal of anything from it.

Incompatibility. Opposite of Affinity (q.v.), ≡ Roux's Cytochorism.

Indirect induction. The induction of a new neural axis in competent ventral ectoderm by a chemical compound, not acting in the same way as the naturally occurring evocator, but by liberating the masked evocator in the reacting tissue.

Individuation. That part of the morphogenetic effect of the primary organiser in neural induction, especially the regional differentiation, which cannot with certainty be referred back to the action of a single chemical substance, the *Evocator*. Since it seems probable that forces of field type are involved in the regional differentiation of the evoked axis, the *Individual field* is spoken of. The history of the word has already been discussed (p. 125). It has, however, another embryological use which was not appreciated when it was introduced into causal morphology. Neurologists have spoken of the differentiation of local reflex systems, behaviour patterns, and the like, out of the total undifferentiated diffuse response of the early embryo to stimuli, as "individuation" (Coghill[3]). This is not perhaps a very good use, since it applies to *parts* of the organism, hence not to true *individuals*. The word "emancipation" (q.v.) would be preferable.

Induction. The morphogenetic effect brought about by an organiser, inductor, evocator, etc. acting on competent tissue. *See also* Homoiogenetic, Direct, Indirect, Autonomous, Complementary, Palisade.

Inductor. A piece of living tissue, dead tissue, a chemical fraction, or a chemical substance of known constitution, which carries out an *induction* similar to that performed by an *organiser*. The word can thus be used indiscriminately for a number of types of inducing agent. To the primary and secondary organisers correspond primary and secondary inductors. (N.B. the term "secondary" here must be sharply distinguished from its use for the induced member of a twin, e.g. the "secondary" embryo as opposed to the "host" embryo.) *See also* Nuclear inductor.

Infection (*Weckung*). The appearance of inductive power in a piece of tissue not originally possessing it, when that tissue has been placed for some time in the normal organiser centre.

Ingression. An inward movement of cells in the amphibian blastula, especially the yolk endoderm. (In coelenterate embryology the word has another meaning.)

Intermediate. *See* Mosaic.

Intersex. *See* Gynandromorph.

Localisation (*Sonderung*). The separation of phases ("plasms") in mosaic eggs, distinguishable cytologically, and each associated with a specific subsequent differentiation. Such phases or regions are often characterised by pigments or granules which led to the old term "organ-forming substances". Since the cytoplasm of these regions, rather than the visible elements within them, must be regarded as responsible for the morphogenetic specificity, and since evocators now deserve rather to be called "morphogenetic substances", the old term should be abandoned.

Macromere. *See* Blastomere.

Material determination. Determination of a part of an embryo to perform a certain type of histogenesis or morphogenesis; as opposed to Dynamic determination (q.v.).

Maturation. The sum total of processes whereby the egg is prepared for fertilisation, including oogenesis, the growth of the ovum, and the final nuclear changes.

Merisis. Growth in size due to cell-division.

Meroblastic. *See* Cleavage.

Merogony. The development of fragments of eggs. A fragment of an egg, containing the nucleus, may be fertilised by sperm (*Diploid merogony*). A fragment of an egg, not containing the nucleus, may be fertilised by sperm (*Andro-merogony*). A fragment of an egg, containing the nucleus, may be removed from the remainder, after fertilisation by sperm but before nuclear fusion (*Gyno-merogony*), or the same experiment may be made using artificial fertilisation (*Parthenogenetic gyno-merogony*). Lastly, a fragment of an egg, not containing the egg-nucleus, can be fertilised by artificial means (*Parthenogenetic merogony*).

Mesoderm. The middle of the three cell- or germ-layers of classical embryology.

Metabolism. The sum total of chemical changes in the organism, including all catabolic and all synthetic processes, respiration, fermentation, etc.

Metathetely. The appearance of structures normally associated with an earlier stage of development, at a later stage of development, e.g. the retention of larval organs by insect pupae.

Metamorphosis. The sum total of changes, often very profound, by which the larval organism is transformed into the adult form (as tadpole into frog, caterpillar into butterfly, or pluteus into sea-urchin).

Metastasis. Spontaneous dispersion of a tumour from its original site in the body to other sites.

Micromere. *See* Blastomere.

Modulation. The reversible histological differentiations which occur as the final stages of the development of a tissue.

Morphogenesis. The coming-into-being of characteristic and specific form in living organisms.

Morphogenetic stimulus. A stimulus exerted by one part of the developing embryo upon another, leading to morphogenesis in the part reacting.

Morphallaxis (*Reparation*). Morphological modification or Regeneration without the formation of any new cells (opp. of true *Regeneration* or *Epimorphosis*).

Morphogenetic movements (*Gestaltungsbewegungen*). The modelling movements in differentiation which change the shape of tissues, e.g. the cell streams, whereby, as in gastrulation, the eventual positions of embryonic cells are attained. They include invagination, expansion, longitudinal extension, and dorsal convergence.

Mosaic. (1) as opposed to *Regulative*. In mosaic development the fates of all the parts of the embryo are already fixed at or before fertilisation, so that any localised ablation or injury will later be manifested as a complete lack of the part in question. In regulative development, however, injuries can be repaired, deficiencies made up, and indifferent added tissue organised into the pattern of the body. Mosaic eggs cannot form giant embryos by egg fusion nor dwarf ones by blastomere separation, and, generally speaking, twinning is much more restricted in them.

(2) in Genetics. Genetic mosaics are individuals the tissues of which contain cells of more than one genetic constitution. They may be due to somatic mutation, somatic crossing-over, or chromosome elimination.

(3) as opposed to *Intermediate*. Organisms vary not only as to the presence or absence of organs, but also to the extent to which any one organ is formed. An organism might have all its organs complete save one, but that one entirely absent; conversely, it might have them all, but imperfectly developed. The former are, in this sense, mosaics; the latter intermediates. This sense of the word is used mainly by workers on Aphids.

Nachbarschaft (*Contiguity effects*). The morphogenetic effects produced by neighbouring tissues or structures on a developing embryonic organ.

Neighbourwise (*Ortsgemäss*). Behaving after transplantation in accordance with the new situation in which the graft finds itself, i.e. plastic, pluripotent, undetermined. "Falling into step" with the new neighbour cells and going along with them to their morphogenetic destination.

Neoplasm, Neoplastic. A tumour, tumorous.

Neoteny. Sexual maturity occurring in the larval period.

Neurogen. A term which may be used for the primary evocator and all other chemical substances which bring about neural induction in vertebrates (cf. *Carcinogen*, *Oestrogen*, etc.). Not to be confused with adjectives for muscular contractions, said to be neurogenic or aneurogenic (myogenic) according as to whether they require nervous stimulus or not.

Neurula. The stage in development at which the neural axis is fully formed and histogenesis rapidly proceeding.

Neutral medium. A situation for the culture of young undetermined parts of embryos such that they are not exposed to the action of unknown inductors (e.g. Holtfreter salt solution, as against chorio-allantoic grafting, yolk-sac implantations, or culture in eye- or body-cavity).

Nidicolous. Birds which hatch in a relatively undeveloped stage and hence have to remain in the nest highly dependent on their parents for some considerable time.

Nidifugous. Birds which hatch in a relatively highly developed stage, and can therefore leave the nest at once or almost so.

Nuclear inductor. A morphogenetic stimulating substance given off by the nuclear material bearing the factors of heredity (genes and chromosomes), usually operating only within the cell in question, but in some cases more diffusible.

Oestrogen. A chemical substance with activity analogous to that of the female sex hormones in the oestrous cycle.

Organiser. A living part of an embryo which exerts a morphogenetic stimulus upon another part or parts, bringing about their determination and the following histological and morphological differentiation. The organiser which acts first in development is known as the *Primary* or *First-Grade* Organiser; those acting at successive stages are known as *Secondary* (*Second-grade*) and *Tertiary* (*Third-grade*) Organisers. The appearance of the latter is generally de-

Organiser (*cont.*)
pendent upon the prior function of the primary organiser, which therefore *seems* to be more manifold in its inductive effects. The chemical substance emitted by an organiser is called an *Evocator*. *See also* Inductor.

Organogenesis. The development of morphological differentiation.

Orthotopic transplantation. Transplantation between corresponding (identical) sites in such a way that the graft maintains its normal orientation.

Ortsgemäss. *See* Neighbourwise.

Oviparity. The bringing forth of young from eggs developing outside the maternal body.

Ovoviviparity. The bringing forth of young from eggs developing within the maternal body, but having little or no connection with the maternal tissues.

Palisade induction. A neural induction in which the stimulus has not been sufficiently strong, or the mechanical conditions not sufficiently favourable, to allow of the formation of a well-shaped neural tube. Instead of this, the histologically differentiated neural cells are arranged in palisade fashion around the implant.

Parabiosis. The grafting together of two or more embryos, usually side by side, in order to study their later mutual morphogenetic and endocrine effects.

Parthenogenesis. The stimulation to development, more or less complete, of an egg by artificial means.

Penetrance. The percentage of animals carrying a gene which manifest its effects. *See also* Expressivity.

Phenocopy. The imitation of a morphological or physiological character due to one genotype in an organism possessing a different genotype, by the action of external physical and chemical factors, not leading to any inheritable change.

Phenocritical period. A period in the development of an organism during which an effect produced by a gene can most readily be influenced or initiated by externally applied factors.

Phenotype. The sum total of morphological and physiological characters produced in an organism by the influence of its genotype.

Placenta. The organ formed at the junction between foetal and maternal tissues and blood streams. Its various types have been classified morphologically and physiologically (see p. 80).

Plasm. A region of cytoplasmic localisation in mosaic eggs which gives rise later to a definite organ or organ system.

Plasmatic localisation. *See* Localisation.

Plasticity. Equivalent to *Pluripotence*; the readiness of a part of the embryo for a large number of possible fates.

Pleiotropism. The causation of multiple effects by one genetic factor, probably not directly, but because of an influence on metabolism which manifests itself in many different ways.

Pluripotent. In a condition of having a large number of possible fates; undetermined.

Polyploid. Containing integral multiples of the haploid number of chromosomes.

Polyploidogen. A chemical substance which, by inhibiting certain phases of nuclear division, gives rise to polyploid cells and tissues.

Potency. The use of this word should carry the undertone of "possibility", not "power" (in correct accord with its Aristotelian origin). Potency differs from competence in that it characterises the whole period of development, including the time before any competences have arisen. It makes no reference to any causal process. The expressions "self-differentiating potency" and "inductive potency" should therefore be avoided unless it is made clear that these qualities have not yet appeared, but may do so later. *See* Prospective Potency.

Presumptive. Having regard to the normal fate of the part in question, e.g. presumptive skin or presumptive ear-vesicle.

Prospective potency (*prospective Potenz*; *Potentialité totale*). The sum total of possible fates of any part of the embryo at the beginning of development. In regulative eggs prospective potency is much wider than prospective significance; in mosaic eggs the two are identical.

Prospective significance (*prospektive Bedeutung*; *Potentialité réelle*). The actual normal fate of any part of the embryo at the beginning of development.

Prothetely. The appearance of structures normally associated with a later stage of development, at an earlier stage of development, e.g. pupal organs in larval insects.

Recapitulation. The appearance, during the development of a higher organism, of structures or functions characteristic of developmental stages of lower organisms.

Recuperation. The reappearance of competence at late stages of development, as in e.g. the regenerative power of the adult ascidian after its mosaic embryonic development, or the lens competence believed to exist in the regenerating amphibian limb blastema.

Regeneration. The formation of part or whole of a new organ in place of one that has been removed. *See also* Wolffian regeneration.

Regional differentiation. The appearance of regional differences within the individuation field, e.g. the differentiation of an embryonic axis into cephalic and caudal ends, with all that that implies; or within organ districts such as that of the limb.

Regulation. The power of continuing normal or approximately normal development or regeneration in spite of experimental interference by ablation, addition (implantation),

Regulation (*cont.*)

exchange (transplantation), fusion, etc. The processes thus revealed operate in all embryos to a greater or lesser extent and may be regarded as part of the sum total of processes whereby the organism is rendered more or less independent of its environment. *See also* Fertilisation.

Regulative. *See* Mosaic.

Reintegration. The restoration to the organism of control by its individuation field, through the action of hormonal and neural factors during the period of functional differentiation, after the period of self-differentiation.

Reunition. The reconstitution of an organism after it has been separated into its component cells, not all types of which may be available, as in sponges.

Second-grade. *See* Organiser.

Secondary. This adjective has two senses; sometimes it is used as the equivalent of "second-grade", of organisers, but also for the embryo produced by an implanted organiser, as opposed to the host, or primary, embryo.

Self-differentiation (*Différentiation spontanée*). The differentiation of a tissue solely as the result of factors existing within itself, i.e. differentiation after determination has been accomplished (*see* Dependent differentiation). Any differentiation in explanted tissue is self-differentiation (*see* Bedeutungsfremde Selbstdifferenzierung).

Selfwise (*Herkunftsgemäss*). Behaving after transplantation in accordance with the old situation in which the graft had previously been, i.e. in accordance with its prospective significance, i.e. determined, self-differentiating, and no longer pluripotent. Refusing to "fall into step" with the new neighbour cells and go along with them to their morphogenetic destination.

Spaltung. A twin embryo in which the two neural axes are united posteriorly; it is due to a splitting of the morphogenetic movements into two streams, and may simulate an induction.

Stepwise inhibition. The successive inhibition of different organic processes by successively stronger applications of an externally acting agent.

Substrate. The substance acted on by an enzyme.

Tachyauxesis. *See* Heterauxesis.

Telolecithal. Having yolk accumulated at one point of the egg.

Teras. A monster caused by developmental deviation.

Teratology. The study of terata and their origin.

Teratoma. A more or less malignant assembly of tissues, well differentiated histologically, but showing varying degrees of morphological differentiation, and embedded in the body of an otherwise normal fully developed organism. Some pathologists distinguish between the histologically differentiated type (*Embryoma*) and that showing both histological and morphological differentiation (*Teratoma* proper).

Thermophene. A curve relating quantitatively the effects of a gene to the temperature at which development has taken place.

Transplantation. The transfer of a piece of tissue from one embryo to another; if the transfer is reciprocal between two embryos, we speak of *Exchange* (*Austausch*). *See also* Autoplastic, Homoioplastic, Heteroplastic, Xenoplastic, Homoiotopic, Heterotopic, Orthotopic.

Trophallaxis. The exchange of secretions etc. between individual members of a community of social insects, whereby social cohesion and caste differentiation are thought to be effected.

Trophic. An action of the nervous system in the absence of which muscle tonus fails and regeneration is impossible.

Trophoblast. The non-embryonic part of the mammalian blastocyst, which effects its implantation into the uterine wall, and part of which later develops into the foetal portion of the placenta.

Umhüllung. The wrapping of an inductor in sheets of competent ectoderm, in order to test inductive power.

Vegetal pole. That pole of the egg which contains most yolk and least cytoplasm.

Vegetalisation (*Endodermisation*). The shifting of the presumptive fate of tissue which should have been ectodermal, in echinoderm development.

Verlagerung. Shuffling of the blastomeres of an egg.

Verschmelzung. Fusion of two eggs together.

Vitelline membrane. The non-cellular membrane surrounding the yolk before development in the hen's egg, eventually superseded by the yolk-sac.

Viviparity. The maintenance of embryos during their development within the maternal body in relatively close contact with the maternal tissues or blood stream.

Wolffian regeneration. The regeneration of a new lens in the eye after the removal of the former one, from the upper margin of the iris.

Xenoplastic transplantation. Transplantation between individuals of different genera or widely distant species.

Yolk-sac. The extension of the blastoderm into a membrane surrounding the yolk in amniotes, specialised for absorption of nutritive materials.

TERMS AND CONCEPTS THE USE OF WHICH
IS NOT RECOMMENDED

	See discussion on p.
Auxano-differentiation	506
Blastogen or **Blastomogen** (for Carcinogen)	—
Closed term	112
Determinant	101
Differential dichotomy	112
Double assurance (*doppelte Sicherung*)	157, 294, 308
Entelechy	119
Harmonious equipotential system	121, 140, 477
Heterogonic growth (for Heterauxetic)	533
Histo-differentiation	506
Isotropic (for Pluripotent)	112
Labile determination	112, 157
Mitogenetic rays	422
Nullipotent	436
Open term	112
Organ-forming substances	131
Segregation	112
Self-origination	112
Specific potency	112
Substrate (in a morphological sense, for Substratum)	153
Totipotent	121

Bibliography

"Parthians and Medes and Elamites, and the dwellers in Mesopotamia, in Judaea and Cappadocia, in Pontus and Asia, in Phrygia and Pamphylia, in Egypt and the parts of Libya about Cyrene, and sojourners from Rome, both Jews and proselytes, Cretans and Arabians, we do hear them speaking in our tongues the mighty works of God." Acts of the Apostles ii. 9–11.

NOTES

(1) In order to save space, the names of the journals are printed in abbreviated form. A list of these arbitrary abbreviations is given below.

(2) The difficulties caused by haphazard spelling of Slavonic names are overcome here by the adoption of standard forms, as follows:

Russian endings: -vitch, -vsky, -ev, -ov.
Polish endings: -wicz, -wski.
Czech and Yugo-slav endings: -vić, -vać.
Bulgarian endings: -eff, -off.

(3) Chinese names are romanised as far as possible according to Wade's system, and the hsing or family name precedes the hyphenated ming-dze.

KEY

TO ABBREVIATIONS USED IN THE BIBLIOGRAPHY

A	Analyst
AA	Anat. Anzeiger
AAB	Annals of Applied Biol.
AAFS	Atti d. R. Accad. dei Fisiocrit. d. Siena
AAHE	Archives d'Anat., d'Histol. & d'Embryol.
AAIUS	Arbeiten a. d. Anat. Instit. d. k. Univ. Sendai
AAL	Atti d. R. Accad. (Naz.) dei Lincei
AAM	Archives d'Anat. Mic.
AAP	Archiv f. Anat. u. Physiol. (Meckel's, later Müller's, later du Bois Reymond's)
AAPA	Annales d'Anat. pathol.
AASF	Annales Acad. Sci. Fenn. A
AAST	Atti d. R. Accad. d. Sci. Torino
AAUSSR	Archiv Anat., Histol. & Embryol. (U.S.S.R.) (formerly Russkiy Archiv...)
AB	Acta Biotheoretica
ABE	Acta Biol. Exp. (Warszawa)
ABL	Acta Biol. Latvica (formerly Bull. Soc. Biol. Latvia)
ABN	Acta Brevia Nederland. de Physiol. Pharmacol. & Microbiol.
ABSRSMNB	Annales & Bull. Soc. Roy. Sci. Med. & Nat. Bruxelles
ABSUSSR	Archiv Biol. Nauk. (U.S.S.R.) (or Archives des Sci. Biol.)
AC	Acta Cancrologica (Budapest)
ACZAA	Annals Czechoslovak Acad. Agric. (Sbornik Ceskoslovenske Akad. Zemedelske)
ADB	Archives de Biol.
ADC	Annalen d. Chem. (Liebig's)
ADS	Archives of Dermatol. & Syphilol.
AEPUP	Archiv f. exp. Pathol. u. Pharmacol. (Schmiedeberg's)
AESA	Ann. Entomol. Soc. Amer.
AEZF	Archiv f. exp. Zellforsch. (Erdmann's)
AF	Archivio di Fisiol.
AFBQ	Ann. Farmacol. & Bioquim. (Buenos Aires)
AFDS	Archiv f. Dermatol. u. Syphilol.
AFEWM	Archiv f. Entwicklungsmechanik (Roux's)
AFGF	Archiv f. Geflügelkunde
AFO	Archiv f. Ophthalmol.
AFS	Archivio di Farmacol. sper.
AFZ	Arkiv för Zool. (Upsala)
AG	Archiv f. Gynaekol.
AGC	Angewandte Chem.
AGMNT	Archiv f. d. Geschichte d. Math., Naturwiss. u. Technik
AGUSSR	Acta Gynaecol. U.S.S.R.
AGWG	Abhandl. d. kgl. Gesell. d. Wiss. zu Göttingen
AH	Anat. Hefte
AHB	Archiv f. Hydrobiol. u. Planktonkunde
AIAE	Archivio Ital. di Anat. & Embriol.

AIB	Archives Ital. de Biol.
AIM	Annals of Internal Medicine
AIOP	Ann. de l'Instit. Océanogr. Paris
AIP	Archives internat. de Physiol. (Liège)
AIPS	Ann. de l'Institut Pasteur
AIPT	Archives internat. de Pharmacodynamie & Thérapie
AIUC	Acta of the Internat. Union against Cancer
AJA	Amer. Journ. Anat.
AJB	Amer. Journ. Bot.
AJBMS	Australian Journ. Exp. Biol. & Med. Sci.
AJC	Amer. Journ. Cancer
AJDC	Amer. Journ. Dis. Children
AJMS	Amer. Journ. Med. Sci.
AJOG	Amer. Journ. Obstet. & Gynaecol.
AJO	Amer. Journ. Ophthalmol.
AJOH	Amer. Journ. Hyg.
AJOP	Amer. Journ. Physiol.
AJOPA	Amer. Journ. Pathol.
AJOPS	Amer. Journ. Psychol.
AJPA	Amer. Journ. Physical Anthropol.
AK	Allattani Közlemenyek (Budapest)
AKC	Archiv. f. klin. Chirurg.
AKS	Archiv d. J. Klaus Stiftung f. Vererbungsforsch., Sozialanthropol. u. Rassenhyg. (Zürich)
ALRFS	Annali Lab. Ric. Fermenti Spallanzani
AMA	Archiv f. mik. Anat.
AMAD	Annales des maladies des Appareils Digestifs
AMBUSSR	Advances in Modern Biology (U.S.S.R.)
AMGE	Archives de Morphol. gén. & exp.
AMN	Amer. Midland Naturalist
AMNH	Annals & Mag. Nat. Hist.
AMS	Acta Med. Scand.
AMUO	Arbeiten a. d. med. Univ. Okayama
AN	Amer. Naturalist
ANM	Acta Nederland. Morphol.
ANP	Archives Néerland. de Physiol.
ANVH	Abhandl. (a. d. Gebiete d. Naturwiss. hrsg. vom) Naturwiss. Verein in Hamburg
ANYAS	Ann. New York Acad. Sci.
ANZJS	Austral. & New Zealand Journ. Surgery
AO	Annali di Ostet. & Ginecol.
AOB	Annals of Bot.
AOG	Archivio di Ostet. & Ginecol.
AOGS	Acta Obstet. & Gynaecol. Scand.
AOP	Archives of Pathol.
AP	Archiv f. d. g. Physiol. (Pflüger's)
APANL	Atti d. Pontif. Accad. Sci. Nuovi Lincei
APB	Archives de Physique Biol.
APBPC	Annales de Physiol. & Physico-chimie Biol.

APD	Acta Pediatrica
APPA	Archiv f. path. Anat. u. Physiol. (Virchow's)
AR	Anat. Record
ARB	Annual Rev. Biochem.
ARBA	Annual Reps. Brit. Assoc. Advancement of Sci.
ARCS	Annual Reps. Chem. Soc.
ARFA	Rapp. de l'Assoc. Français pour l'Avancement des Sci.
ARP	Annual Rev. Physiol.
ART	Amer. Rev. Tuberculosis
ASB	Archivio di Sci. Biol.
ASBA	Annales de la Station Biol. d'Arcachon
ASBSP	Annuario d. Staz. Bacol. Sper. Padova
ASFFF	Acta Soc. pro Fauna & Flora Fennica
ASNZ	Annales des Sci. Nat. (Zool.)
ASOJ	Acta Soc. Ophthalmol. Jap.
ASRZB	Annales de la Soc. Roy. Zool. & Malacol. Belg.
AT	Arbeiten. d. Ungar. Biol. Forschungsinstit. Tihany
ATT	Archiv f. Tierernährung u. Tierzucht
AZ	Acta Zoologica (Stockholm)
AZEG	Archives de Zool. exp. & gén.
AZF	Archiv f. Zellforsch.
AZI	Archivio Zool. Ital.
AZJ	Annotationes Zool. Jap.
AZSZBFV	Ann. Zool. Soc. Zool. Bot. Fenn. Vanamo
B	Biomorphosis
BAA	Bull. de l'Assoc. des Anat.
BAEC	Bull. de l'Assoc. pour l'Étude de Cancer
BAM	Bull. de l'Acad. de Méd. Paris
BAMNH	Bull. Amer. Mus. Nat. Hist.
BAMR	Boll. d. R. Accad. Med. Roma
BAPS	Bull. internat. Acad. Pol. Sci. & Lett. (Kraków)
BARBS	Bull. de l'Acad. Roy. Belg.
BARMB	Bull. de l'Acad. Roy. de Méd. Belg.
BASUSSR	Bull. Acad. Sci. U.S.S.R.
BB	Biol. Bull.
BBFB	Bull. Biol. Fr. & Belg. (formerly Bull. Sci. Fr. & Belg.)
BBMEUSSR	Bull. Biol. & Med. Exp. U.S.S.R.
BC	Biol. Centralbl.
BCC	Biochem. Centralbl.
BDBG	Ber. d. deutsch. Bot. Gesell.
BDCG	Ber. d. deutsch. Chem. Gesell.
BDJ	Brit. Dental Journal
BDZ	Boll. di Zool.
BER	Bull. Entomol. Res.
BFASUK	Bull. Fac. Sci. Agric. Univ. Kyushu
BG	Biol. Generalis (Wien)
BGA	Bibliographia Genetica ('s Graavenhage)
BHA	Bull. de l'Histol. appl.
BIZUR	Boll. Instit. Zool. Univ. Roma
BJ	Biochem. Journ.
BJEP	Brit. Journ. Exp. Pathol.
BJHH	Bull. Johns Hopkins Hospital (Baltimore)
BJR	Brit. Journ. Radiol.

BJS	Brit. Journ. Surgery
BL	Biol. Listy (Prag)
BLL	Bratislavske Lekarsky Listy
BLZAM	Boll. Lab. Zool. Agrar. Milano
BM	Biometrika
BMA	Bergens Mus. Aarbog
BMDIMBL	Bull. Mt. Desert Island Marine Biol. Lab.
BME	Bruxelles Méd.
BMHN	Bull. Mus. d'Hist. Nat. Paris
BMJ	Brit. Med. Journ.
BMKDVS	Biol. Medd. Kgl. Dansk Videnskab Selskab
BMLZACUG	Boll. Mus. & Lab. Zool. & Anat. Comp. Univ. Genova
BN	Brain
BNYAM	Bull. New York Acad. Med.
BOG	Bot. Gazette
BOGG	Ber. d. Oberhess. Gesell. f. Natur. u. Heilk. zu Giessen.
BOR	Bot. Rev.
BOZ	Bot. Zeitung
BOZJ	Bot. & Zool. Jap.
BP	Bull. Math. Biophysics
BPAAP	Beitr. z. pathol. Anat. u. allg. Pathol. (Ziegler's)
BR	Biol. Rev.
BSAP	Bull. Soc. Anthropol. Paris
BSBG	Bull. Soc. Belg. Gynaecol.
BSCB	Bull. Soc. Chimie Biol.
BSCF	Bull. Soc. Chimie de Fr.
BSHA	Bull. Soc. Hyg. alimentaire
BSIBS	Boll. Soc. Ital. Biol. Sper.
BSNM	Bull. Soc. Naturalistes Moscow
BSNN	Boll. Soc. Naturalisti Napoli
BSPP	Bull. Soc. Philomath. Paris
BSRNPPE	Bull. Soc. Roumaine de Neurol., Psychiatr., Psychol. & Endocrinol.
BSSAR	Bull. Sect. Sci. Acad. Roumaine
BSZF	Bull. Soc. Zool. de Fr.
BTBC	Bull. Torrey Bot. Club
BTS	Biochim. & Terap. Sper.
BUSSR	Biochemia (U.S.S.R.)
BZ	Biochem. Zeitschr.
BZUSSR	Biol. Zhurn. U.S.S.R.
C	Cytologia (Tokyo)
CA	Chem. Abstr.
CAC	Centralbl. f. agric. Chem. u. rat. Landw. (Biedermann's)
CAP	Centralbl. f. allg. Pathol.
CAR	Cancer Rev.
CB	Centralbl. f. Bakteriol. u. Parasitenkunde
CBSSF	Commentat. Biol. Soc. Sci. Fenn. (Helsinki)
CBTI	Contrib. Boyce-Thompson Instit.
CCE	Carnegie Contrib. to Embryol.
CCP	Colorado College Pub.
CDE	Chemie. d. Erde
CG	Centralbl. f. Gynaekol.
CI	Chem. & Industry (Journ. Soc. Chem. Industry)
CIWP	Carnegie Instit. of Washington Publications

CIWY	Carnegie Instit. of Washington Year-book	*FZ*	Fortschritte d. Zool.
CJEB	Chinese Journ. Exp. Biol.	*FZP*	Frankfurter Zeitschr. f. Pathol.
CJOP	Chinese Journ. Physiol.	*G*	Growth
CJR	Canadian Journ. Res.	*GA*	Genetica ('s Graavenhage)
CMJ	China Med. Journ.	*GC*	Ginecologia (Firenze)
CMW	Centralbl. f. med. Wiss.	*GHR*	Guy's Hospital Reps. (London)
CN	Collecting Net (Woods Hole)	*GO*	Gynécologie & Obstétrique
CO	Clinica Ostet.	*GR*	Genesis (Roma)
CP	Chem. Products	*GS*	Genetics (Princeton)
CR	Chem. Rev.		
CRAA	Comptes Rend. Assoc. Anat.	*H*	Hilgardia (Journ. Agric. Sci. Calif. Agric. Exp. Sta.)
CRAS	Comptes Rend. hebdomadaires Acad. Sci. Paris	*HB*	Human Biol.
CRASUSSR	Comptes Rend. Acad. Sci. U.S.S.R.	*HCA*	Helvetica Chem. Acta
CRSB	Comptes Rend. Soc. Biol.	*HJ*	Hibbert Journ.
CRSSV	Comptes Rend. Soc. Sci. Varsovie (Warszawa)	*IEC*	Industrial & Eng. Chem. (formerly Journ. Industr. & Eng. Chem.)
CRTLC	Comptes Rend. Trav. Lab. Carlsberg		
CS	Current Sci. (Bangalore)	*IJMR*	Indian Journ. Med. Res.
CSHSQB	Cold Spring Harbor Symposia on quantitative Biology	*IRGHH*	Internat. Rev. d. ges. Hydrobiol. u. Hydrogr.
CSM	Colloid Symposium Monogr.	*IRJMS*	Irish Journ. Med. Sci.
CT	Centralbl. f. Tierernährung	*ISCJS*	Iowa State Coll. Journ. Sci.
CV	Cornell Veterinarian	*IZPCB*	Internat. Zeitschr. f. physik.-chem. Biol. (dead)
CW	Chem. Weekblad		
D	Discovery	*JA*	Journ. Anat. (continuation of Journ. Anat. & Physiol.)
DAKM	Deutsches Archiv f. klin. Med.		
DB	Der Biologe	*JACS*	Journ. Amer. Chem. Soc.
DMW	Deutsche Med. Wochschr.	*JACSJ*	Journ. Agric. Chem. Soc. Jap.
		JAMA	Journ. Amer. Med. Assoc.
E	Ecology	*JAOAC*	Journ. Assoc. Official Agric. Chemists
EAEG	Ergebn. d. Anat. u. Entwicklungs-gesch.	*JAP*	Journ. Anat. & Physiol.
EAPPA	Ergebn. d. allg. Pathol. u. pathol. Anat.	*JAPA*	Journ. Amer. Pharmacol. Assoc.
		JAPP	Journ. Applied Physics (formerly Physics)
EB	Ergebn. d. Biol.	*JAR*	Journ. Agric. Res.
EEF	Ergebn. d. Enzymforsch.	*JAS*	Journ. Agric. Sci.
EIMK	Ergebn. d. inn. Med. u. Kinderheil-kunde	*JAVA*	Journ. Amer. Vet. Assoc.
		JBC	Journ. Biol. Chem.
EK	Erkenntnis	*JCCP*	Journ. Cellular & Comp. Physiol.
EM	Ecological Monographs	*JCE*	Journ. Chem. Educ.
EMJ	Edinburgh Med. Journ.	*JCI*	Journ. Clin. Investigation
EN	Encéphale (Journ. mensuel de Neurol. & Psychiatr.)	*JCMA*	Journ. Chosen Med. Assoc.
		JCN	Journ. Comp. Neurol.
ENE	Endokrinologie (Leipzig)	*JCP*	Journ. de Chim. Physique
ENY	Endocrinology (Glendale, Calif.)	*JCPS*	Journ. Comp. Psychol.
EP	Ergebn. d. Physiol.	*JCR*	Journ. Cancer Res.
EVHF	Ergebn. d. Vitamin- u. Hormon-Forsch.	*JCS*	Journ. Chem. Soc.
		JCSJ	Journ. Chem. Soc. Jap.
EXSR	Exp. Sta. Record	*JCST*	Journ. Coll. Sci. Imp. Univ. Tokyo
EZ	Enzymologia	*JDAP*	Journ. de l'Anat. & Physiol.
		JDR	Journ. Dental Res.
FAJ	Folia Anat. Jap.	*JE*	(Brit.) Journ. Endocrinol.
FF	Forschungen u. Fortschritte	*JEB*	Journ. Exp. Biol. (formerly Brit. Journ. Exp. Biol.)
FFF	Fermentforsch.		
FG	Folia Gynaecol.	*JECB*	Journ. Econ. Biol.
FGR	Forschungen a. d. Geb. d. Röntgen-strahl.	*JEE*	Journ. Econ. Entomol.
		JEM	Journ. Exp. Med.
FH	Folia Haematol.	*JEZ*	Journ. Exp. Zool.
FLSH	Finska Läkaresällskapets Handlingar	*JFI*	Journ. Franklin Instit. Philadelphia
FMH	Folia Morphol. Hispanica (Valencia)	*JFSHIU*	Journ. Fac. Sci. Hokkaido Imp. Univ.
FMP	Folia Morphol. Polonica (Warszawa)	*JFSUT*	Journ. Fac. Sci. Imp. Univ. Tokyo
FPJ	Folia Pharmacol. Jap.	*JG*	Journ. Genetics

JGP	Journ. Gen. Physiol.
JGPS	Journ. Gen. Psychol.
JH	Journ. Heredity
JI	Journ. Immunol.
JID	Journ. Infectious Dis.
JJB	Journ. Biochem. (Tokyo)
JJCR	Gann; Jap. Journ. Cancer Res.
JJEM	Jap. Journ. Exp. Med.
JJG	Jap. Journ. Genetics
JJMSA	Jap. Journ. Med. Sci. (I, Anat.)
JJMSB	Jap. Journ. Med. Sci. (II, Biochem.)
JJMSP	Jap. Journ. Med. Sci. (III, Biophysics)
JJOG	Jap. Journ. Obstet. & Gynaecol.
JJZ	Jap. Journ. Zool.
JK	Jahrb. f. Kinderheilkunde
JLSZ	Journ. Linn. Soc. (Zool.)
JM	Journ. Morphol. (for a period was Journ. Morphol. & Physiol.)
JMBA	Journ. Marine Biol. Assoc. (Great Britain)
JMFUS	Jahrb. d. med. Fak. d. Univ. Sofia
JMM	Journ. Mammalol.
JMR	Journ. Med. Res.
JMS	Journ. Mental Sci.
JMSHNY	Journ. Mt Sinai Hosp. New York
JN	Journ. Nutrition
JNCI	Journ. Nat. Cancer Instit. (U.S.)
JNP	Journ. Neurophysiol.
JNPP	Journ. Neurol. & Psychopathol.
JNYES	Journ. New York Entomol. Soc.
JO	Journ. f. Ornithol.
JOGBE	Journ. Obstet. & Gynaecol. of the Brit. Empire
JOH	Journ. Hyg.
JOP	Journ. Physiol.
JP	Journ. Pediatr.
JPA	Journ. Parasitol.
JPB	Journ. Pathol. & Bacteriol.
JPC	Journ. Physical Chem.
JPE	Journ. de Psychol.
JPET	Journ. Pharmacol. & exp. Therapeutics
JPPG	Journ. de Physiol. et de Pathol. Gén.
JPS	Journ. Philos. Studies
JRMS	Journ. Roy. Mic. Soc.
JSHU	Journ. Sci. Hiroshima Univ.
JSI	Journ. Sci. Instruments
JSM	Journ. State Med.
JUGP	Jahresber. ü. d. ges. Physiol. u. exp. Pharmacol.
JWBO	Jahrb. f. wiss. Bot.
JZN	Jenaische Zeitschr. f. Naturwiss.
K	Kosmos (Journ. Soc. Pol. Nat. Kopernik)
KB	Kolloid Beihefte (formerly Kolloid Chem. Beihefte)
KIZ	Kyoto Igakkwai Zasshi (Kyoto Med. Journ.)
KJM	Keijo Journ. Med.
KMA	Klin. Monatsbl. f. Augenheilkunde
KNVSF	Kgl. Norske Videnskab Selskabs Forhandlinger
KOKK	Közlemények az összehesonlitó élet-és Kórtan Köreböl

KW	Klin. Wochschr. (formerly Berlin. Klin. Wochschr.)
KZ	Kolloid Zeitschr.
L	Lancet
LC	La Cellule (Louvain)
LJB	Bull. Scripps Oceanograph. Instit. La Jolla, Calif.
LT	Landbouwkundig Tidschr.
LUA	Lunds Univ. Årsskrift
MAPS	Mem. Acad. Pol. Sci.
MARBS	Mém. Acad. Roy. Belg. (Sci.)
MASIB	Mem. d. R. Accad. Sci. dell' Istit. di Bologna
MASU	Mem. Acad. Sci. Ukraine (U.S.S.R.)
MC	Monatshefte f. Chemie
MCSK	Mem. Coll. Sci. Kyoto Imp. Univ. (B)
MDS	Medycyna Doswiadczalna i Społeczna (Warszawa)
ME	Micro-entomology (Contrib. to Entomol. from Stanford Univ.)
MGG	Monatschr. f. Geburtshilfe u. Gynaekol.
MINPERP	Mem. Instit. Nat. Pol. d'Écon. Rurale de Puławy
MJ	Morphol. Jahrb. (Gegenbaur's)
MJA	Med. Journ. Australia
MJBG	Mitt. d. Jap. Biochem. Gesell.
MJGG	Mitt. d. Jap. Gesell. f. Gynaekol.
MK	Monatschr. f. Kinderheilkunde
MMAK	Mitt. a. d. med. Akad. Kyoto
MMFUT	Mitt. a. d. med. Fak. Univ. Tokyo
MMRHNB	Mémoires du Mus. Roy. d'Hist. Nat. Belg.
MOG	Monit. Ostet. & Ginecol.
MRAI	Mem. d. R. Accad. Ital. (Roma)
MSPEUSSR	Materials for the Study of the Pathol. of Emb. Devel. in Birds (U.S.S.R.)
MZI	Monit. Zool. Ital.
N	Nature
NAV	North American Veterinarian
NEJM	New England Journ. Med.
NGWG	Nachr. v. d. Gesell. d. Wiss. zu Göttingen
NIZ	Nagasaki Igakkwai Zasshi (Nagasaki Med. Journ.)
NJMS	Nagoya Journ. Med. Sci.
NTG	Nederlandsch Tijdschrift v. Geneeskunde
NTVKI	Nordisk Tidskr. för Vetenskap, Konst. o Industrie
NW	Naturwissenschaften
O	Osiris
OIZ	Okayama Igakkwai Zasshi (Okayama Med. Journ.)
OJDI	Oriental Journ. Dis. Infants
OJS	Ohio Journ. Sci.
OJVSAI	Onderstepoort Journ. Vet. Sci. & Anim. Industry
OMaly	Ber. ü. d. ges. Physiol. u. Pharmacol. (Maly's)

OO	Orvoskepzes (Budapest)
OPAAA	Occasional Pub. Amer. Assoc. Adv. Sci.
P	Protoplasma
PA	Parasitology
PAAAS	Proc. Amer. Acad. of Arts & Sciences
PAG	Philippine Agric.
PAHUSSR	Probl. of Animal Husbandry (U.S.S.R.)
PAPS	Proc. Amer. Philos. Soc.
PARNMED	Proc. Assoc. for Res. in Nervous & Mental Disease
PAS	Proc. Aristot. Soc.
PASAP	Proc. Amer. Soc. Animal Production
PCAS	Proc. Cambridge Antiquarian Soc.
PCPS	Proc. Cambridge Philos. Soc. (Biol.)
PESP	Proc. Entomol. Soc. Philadelphia
PEUSSR	Probl. d. Ernährung (U.S.S.R.)
PGL	Polska Gazeta Lekarska
PIAS	Proc. Indian Acad. Sci.
PIAT	Proc. Imp. Acad. Tokyo
PIUSSR	Probl. d. Inkubation (U.S.S.R.)
PKAWA	Proc. Kon. Akad. Wetenskap Amsterdam
PLSNSW	Proc. Linnean Soc. New South Wales
PM	La Presse Médicale
PMA	Prensa Med. Argentina
PNASW	Proc. Nat. Acad. Sci. (Washington)
PNHB	Peiping Nat. Hist. Bull.
PNSIS	Proc. Nova Scotia Instit. Sci.
PP	Phytopathology
PPP	Plant Physiol.
PPS	Proc. Physical Soc. (London)
PR	Physiol. Rev.
PRPSE	Proc. Roy. Physical Soc. Edinburgh
PRS	Proc. Roy. Soc. (London)—before division into two parts
PRSA	Proc. Roy. Soc. A (London)
PRSB	Proc. Roy. Soc. B (London)
PRSE	Proc. Roy. Soc. Edinburgh
PRSM	Proc. Roy. Soc. Med. (London)
PS	Poultry Sci.
PSEBM	Proc. Soc. Exp. Biol. & Med.
PSR	Psychol. Rev.
PSSW	Proc. Sci. Soc. Warsaw (Warszawa)
PSZN	Pubbl. d. Staz. Zool. Napoli (formerly Mitteilungen a. d.)
PTRSA	Philos. Trans. Roy. Soc. A. (London)
PTRSB	Philos. Trans. Roy. Soc. B. (London)
PZ	Physiol. Zool.
PZEUSSR	Probl. d. Zootechn. & Exp. Endocrinol. (U.S.S.R.)
PZS	Proc. Zool. Soc. (London)
PZUSSR	Physiol. Zhurn. (U.S.S.R.)
QJEP	Quart. Journ. Exp. Physiol.
QJMS	Quart. Journ. Mic. Sci. (formerly Journ. Mic. Sci.)
QJPP	Quart. Journ. Pharmacy & Pharmacol.
QRB	Quart. Rev. Biol.
R	Redia (Giornale di Entomol.; Firenze)
RA	Radiology
RAL	Rendiconti d. R. Accad. (Naz.) dei Lincei
RCT	Rassegna di Clin. Terap. e. Sci. affini
RDB	Riv. di Biol.
RG	Radiobiol. Gen.
RGS	Rev. gén. des Sci.
RIG	Riv. Ital. di Ginecol.
RLAUR	Ricerche d. Lab. Anat. Univ. Roma
RM	Ricerche d. Morfol. (Roma)
RPS	Riv. di Patol. sper.
RS	La Ricercha Sci.
RSAB	Rev. Soc. Argentina de Biol.
RSEIQ	Riv. Sudamer. Endocrinol., Immunol. & Quimoterap.
RSZ	Rev. Suisse de Zool.
RTBN	Recueil des Trav. Bot. Nederland
RTCPB	Recueil des Trav. Chim. des Pays-Bas
RUB	Rev. de l'Univ. de Bruxelles
RZR	Rev. Zool. Russe
S	Science (New York)
SA	Scientific Agriculture
SAIMRP	South African Instit. Med. Res. Pub.
SAJS	South African Journ. Sci.
SAWB	Sitzungsber. d. preuss. Akad. Wiss. Berlin
SAWH	Sitzungsber. d. Akad. Wiss. Heidelberg
SAWW	Sitzungsber. d. Akad. Wiss. in Wien
SC	Scientia
SGMPM	Sitzungsber. d. Gesell. f. Morphol. u. Physiol. in München
SGNFB	Sitzungsber. d. Gesell. Naturforsch. Freunde zu Berlin
SGO	Surgery, Gynaecol. & Obstet.
SGY	Surgery
SIDT	Stud. Instit. Div. Thom. (Cincinnati)
SIKMJ	Sei-i-Kwai Med. Journ.
SK	Suomen Kemistilehti (Acta Chem. Fennica)
SKAP	Skand. Archiv f. Physiol.
SKT	Svensk Kemisk Tidskrift
SM	Scientific Monthly
SMA	Semana Med. (Buenos Aires)
SMW	Schweiz. Med. Wochschr.
SNGB	Sitzungsber. d. naturforsch. Gesell. Bern
SNVAO	Skr. udg. av det Norske Videnskaps Akad. i Oslo
SP	Science Progress
SPHNGCR	Comptes Rend. Soc. Phys. Hist. Nat. Genève
SPHNGM	Mem. Soc. Phys. Hist. Nat. Geneva
SPMGW	Sitzungsber. d. phys.-med. Gesell. zu Würzburg
SRICRS	Sci. Reps. Imp. Cancer Res. Fund (London)
SRMRC	Spec. Reps. Med. Res. Council (Great Britain)
SRTIU	Sci. Reps. Tohoku Imp. Univ.
SSR	School Sci. Rev.
TAFS	Trans. Amer. Fisheries Soc.
TAMS	Trans. Amer. Mic. Soc.
TB	Tabulae Biol. (Period.)
TDDUSSR	Trans. on the Dynamics of Devel. (U.S.S.R.)

TES   Trans. Entomol. Soc. (London)
TFS   Trans. Faraday Soc.
TIEMUM   Trans. Instit. Exp. Morphogen. Univ.
            Moscow (U.S.S.R.)
TH   Tidskr. för Hermetikind.
TIN   Trav. de l'Instit. Nencki (Warszawa)
TIPNUSSR Trudy Instit. Physiol. Narkompros
            (U.S.S.R.)
TIZBASU   Trav. de l'Instit. Zool. & Biol. Acad.
            Sci. Ukraine (U.S.S.R.)
TJEM   Tohoku Journ. Exp. Med.
TJPS   Trans. Jap. Pathol. Soc.
TLEBZM   Trans. Lab. Exp. Biol. Zoopark
            Moscow (U.S.S.R.)
TLZEMAM   Trav. du Lab. Zool. & Exp.
            Morphogén. Anim. Moscow
            (U.S.S.R.)
TNDV   Tijdschr. d. Nederlandsche Dierkund.
            Vereeniging (continued as Archives
            Néerland. de Zool.)
TNISI   Trans. Nat. Instit. Sci. India
TOSL   Trans. Obstet. Soc. (London)
TPRSA   Trans. & Proc. Roy. Soc. Australia
TRSC   Trans. Roy. Soc. Canada (Sect. V)
TSBE   Trans. Soc. Brit. Entomologists
TSBR   Trav. de la Stat. Biol. Roscoff
TSZW   Trav. de la Stat. Zool. Wimereux
TZS   Trans. Zool. Soc. (London)

U   Umschau (Frankfurt)
UBZ   Ukrainian Biochem. Zhurn. (later
            Biochem. Zhurn. Acad. Sci. Ukraine,
            U.S.S.R.)
UCM   Mem. Univ. Calif.
UCPP   Univ. Calif. Pub. in Physiol.
UCPZ   Univ. Calif. Pub. in Zool.
UL   Ugeskrift for Laeger (Danish Med.
            Journ.)
ULF   Upsala Läkareförenings Forhand-
            lingar
UMN   Unterrichtsblätter f. Math. u. Natur-
            wiss.
UMZ   Med. Zhurn. Acad. Sci. Ukraine,
            U.S.S.R.
UOB   Univ. Oklahoma Bull.
UPB   Univ. Pittsburgh Bull.
USBFIR   United States Fisheries Bureau In-
            vestigations Reports (formerly Bull.
            U.S. Fish Commission)
USNMP   United States National Mus. Proc.
USPHR   United States Public Health Reports
UW   Unsere Welt (Godesberg)

VDAG   Verhandl. d. deutsch. Anat. Gesell.
            (suppl. vols of Anat. Anzeiger)
VDPG   Verhandl. d. deutsch. Pathol. Gesell.
            (suppl. vols of Centralb. f. allg.
            Pathol.)
VDZG   Verhandl. d. deutsch. Zool. Gesell.
            (suppl. vols. of Zool. Anzeiger)
VIVTAL   Verhandl. d. Internat. Vereinigung
            f. theor. u. angew. Limnologie

VNGB   Verhandl. d. naturforsch. Gesell. in
            Basel
VNGZ   Vierteljahrschr. d. naturforsch. Gesell.
            in Zürich
VOUSSR   Vopros. Onkol. (U.S.S.R.)
VPMGW   Verhandl. d. phys.-med. Gesell. zu
            Würzburg
VRUSSR   Vestnik Röntgenol. & Radiol.
            (U.S.S.R.)
VSNFG   Verhandl. d. Schweiz. Naturforsch.
            Gesell.

W   Wszechswiat (Warszawa)

YJBM   Yale Journ. Biol. & Med.
YMaly   Ber. ü. d. wiss. Biol. (Maly's)

ZA   Zoologica (Sci. Contribs. New York
            Zool. Soc.)
ZAE   Zeitschr. f. Anat. u. Entwicklungs-
            geschichte
ZAP   Zeitschr. f. allg. Physiol. (dead)
ZAZ   Zool. Anzeiger
ZB   Zeitschr. f. Biol.
ZE   Zeitschr. f. angewandte Entomol.
ZEBUSSR   Zhurn. Exp. Biol. (U.S.S.R.)
ZF   Zeitschr. f. Fischerei u. d. Hilfswiss.
ZGEM   Zeitschr. f. d. ges. exp. Med.
ZGG   Zeitschr. f. Geburtshilfe u. Gynäkol.
ZH   Zeitschr. f. Hyg.
ZIAV   Zeitschr. f. induktive Abstamm. u.
            Vererbungslehre
ZJAO   Zool. Jahrb. (Abt. f. Anat. u.
            Ontog.)
ZJS   Zool. Jahrb. (Abt. f. Systematik)
ZJZP   Zool. Jahrb. (Abt. f. allg. Zool. u.
            Physiol.)
ZKF   Zeitschr. f. Krebsforsch.
ZKH   Zeitschr. f. Kinderheilkunde
ZKL   Zeitschr. f. Konstitutionslehre
ZKM   Zeitschr. f. klin. Med.
ZMA   Zeitschr. f. Morphol. u. Anthropol.
ZMAF   Zeitschr. f. mik. Anat. Forsch. (Jahrb.
            f. Morphol. u. mik. Anat. Abt.
            II)
ZMJ   Zool. Mag. Jap.
ZMOT   Zeitschr. f. Morphol. u. Ökol. d. Tiere
ZP   Zoologica Poloniae
ZPC   Zeitschr. f. physiol. Chem. (Hoppe-
            Seyler's)
ZPCUSSR   Zhurn. Physik. Chem. U.S.S.R.
ZTZ   Zeitschr. f. Tierzüchtung u. Züch-
            tungsbiol. (now Zeitschr. f. Züch-
            tung, Abt. B; Tierzücht. u. Züch-
            tungsbiol. einschl. Tierernährung)
ZUNG   Zeitschr. f. Untersuch. Nahr. u.
            Genuss-Mittel (now Zeitschr. f.
            Untersuch. d. Lebensmittel)
ZVP   Zeitschr. f. vergl. Physiol.
ZWZ   Zeitschr. f. wiss. Zool.
ZZFMA   Zeitschr. f. Zellforsch. u. mik. Anat.
ZZUSSR   Zool. Zhurn. U.S.S.R.

BIBLIOGRAPHY

Abderhalden, E. *ZPC*, 1936, **240**, 237.

Abe, M. (1). *JJOG*, 1931, **14**, 2 and 28; *OMaly*, 1932, **63**, 810 and 811; *YMaly*, 1931, **19**, 484.

Abe, M. (2). *JJOG*, 1932, **15**, 44 and 284.

Abe, M. (3). *JJOG*, 1932, **15**, 289.

Abe, M. (4). *JJOG*, 1933, **16**, 225.

Abelin, I. & Scheinfinkel, N. *AP*, 1923, **198**, 151.

Abeloos, M. *CRSB*, 1928, **98**, 917; *BBFB*, 1930, **64**, 1.

Abeloos, M. & Lecamp, M. *CRSB*, 1929, **101**, 899.

Abeloos, M. & Toumanov, K. *BSZF*, 1926, **51**, 281.

Abels, H. *KW*, 1922, 1785.

Abercrombie, M. (1). *JEB*, 1937, **14**, 302.

Abercrombie, M. (2). *N*, 1939, **144**, 1091.

Abercrombie, M. (3). Private communication to the author, 1941.

Abercrombie, M. & Waddington, C.H. *JEB*, 1937, **14**, 319.

de Aberle, S. *AN*, 1925, **59**, 327; *AR*, 1925, **29**, 343, 344; *AJA*, 1927, **40**, 219.

de Aberle, S., Morse, A.H., Thompson, W.R. & Pitney, E.H. *AJOG*, 1930, **20**, 397.

Abraham & Robinson. *N*, 1937, **140**, 24.

Achard, C., Levy, J. & Georgiakakis, N. *AMAD*, 1934, **24**, 785; *OMaly*, 1935, **84**, 22.

Ackermann, D. *ZPC*, 1932, **209**, 12.

Adams, A. E. *JEZ*, 1924, **40**.

Adams, P.D. *ADS*, 1937, **86**, 606.

Adamstone, F.B. (1). *JM*, 1931, **52**, 47.

Adamstone, F.B. (2). *AJC*, 1936, **28**, 540.

Adelmann, H.B. (1). *AJA*, 1922, **31**, 55; *JM*, 1926, **42**, 371; 1932, **54**, 1.

Adelmann, H.B. (2). *AFEWM*, 1928, **113**, 704.

Adelmann, H.B. (3). *JEZ*, 1929, **54**, 249, 291; 1930, **57**, 223; 1934, **67**, 217; 1937, **75**, 199; *QRB*, 1936, **11**, 161 and 284; *AJO*, 1934, **17**, 890.

Adelmann, H.B. & McLean, B.L. *AR*, 1934, **58**, 249, 273; *JEZ*, 1935, **72**, 101; *AR*, 1939, **75**, 155.

Adler, E. & von Euler, H. *ZPC*, 1934, **225**, 41.

Adler, K. & Adler, M. *ZGG*, 1931, **101**, 128.

Adler, L. *AFEWM*, 1914, **39**, 21.

Adler, P. *P*, 1932, **15**, 15.

Adova, A.N. & Feldt, A.M. *CRASUSSR*, 1939, **25**, 43.

Agduhr, E. *ULF*, 1932, **38**, 1; *Proc. 2nd Internat. Congr. Sex Research*, 1931, p. 20; *YMaly*, 1932, **22**, 210; 1933, **23**, 196.

Aiyar, R.G. & Nalini, K.P. *PIAS*, 1938, **7**, 252.

Akao, A. (1). *JCMA*, 1931, **21**, 769.

Akao, A. (2). *KJM*, 1932, **3**, 250.

Akao, A. (3). *KJM*, 1932, **3**, 360; *YMaly*, 1933, **25**, 174.

Akao, A. (4). *JJB*, 1939, **30**, 303.

Albano, G. *GO*, 1933, **27**, 416.

Albaum, H.G. & Nestler, H.A. *JEZ*, 1937, **75**, 1.

Albright, W.P. & Thompson, R.B. *PS*, 1935, **14**, 373.

Alcock, R.S. *PR*, 1936, **16**, 1.

Alderman, A.L. *JEZ*, 1935, **70**, 205.

Aleschin, B. (1). *BZ*, 1926, **171**, 79.

Aleschin, B. (2). *BZUSSR*, 1935, **4**, 461.

Aleschin, B. (3). *AZ*, 1936, **17**, 1.

Alexander, L.E. *JEZ*, 1937, **75**, 41.

Aliberti, G. *BMLZACUG*, 1933, **13**.

Allan, H. & Wiles, P. *JOP*, 1932, **75**, 23.

Allee, W.C. & Evans, G. *AR*, 1934, **60** (Suppl.), 33; *E*, 1937, **18**, 337; *JCCP*, 1937, **10**, 15; *BB*, 1937, **72**, 217; *S*, 1937, **85**, 59.

Allen, B.M. (1). *S*, 1916, **44**, 755.

Allen, B.M. (2). *AR*, 1932, **54**, 45.

Allen, B.M. (3). *AR*, 1932, **54**, 65.

Allen, B.M. (4). *BR*, 1938, **13**, 1.

Allen, B.M. & Lein, A. Cit. in B.M. Allen, *BR*, q.v. p. 3.

Allen, C.R.K. *PNSIS*, 1932, **18**, 34.

Allen, E. *PSEBM*, 1927, **24**, 608.

Allen, E., Pratt, J.P. & Doisy, E.A. *JAMA*, 1925, **85**, 399.

Allen, E., Whitsett, J.W., Hardy, J.W. & Kneibert, F.L. *PSEBM*, 1924, **21**, 500.

Allen, Russell. Private communication to the author.

Allen, T.H. *JCCP*, 1940, **16**, 149.

Allen, T.H. & Bodine, J.H. (1). *JCCP*, 1939, **14**, 183.

Allen, T.H. & Bodine, J.H. (2). *JGP*, 1940, **24**, 99.

Allen, W.M. & Wintersteiner, O. *S*, 1934, **80**, 190.

Allsopp, C.B. *N*, 1940, **145**, 303.

Almquist, H.J. (1). *Univ. Calif. Coll. Agric. Bull.* No. 561, 1933.

Almquist, H.J. (2). *PS*, 1934, **13**, 375.

Almquist, H.J. & Burmester, B.R. *PS*, 1934, **13**, 116.

Almquist, H.J. & Givens, J.W. *PS*, 1935, **14**, 182.

Almquist, H.J., Givens, J.W. & Klose, A. *IEC*, 1934, **26**, 847.

Almquist, H.J. & Holst, W.F. *H*, 1931, **6**, 61.

Almquist, H.J., Lorenz, F.W. & Burmester, B.R. *JBC*, 1934, **106**, 365.

Almquist, H.J., Pentler, C.F. & Mecchi, E. *PSEBM*, 1938, **38**, 336.

Aloisi, M. (1). *MZI*, 1932, **43**, 173; *YMaly*, 1933, **24**, 44.

Aloisi, M. (2). *MZI*, 1933, **44** (Suppl.), 118; *YMaly*, 1934, **29**, 162.
Aloisi, M. (3). *AIAE*, 1933, **32**, 25; *YMaly*, 1934, **29**, 161.
Alpatov, W.W. (1). *JEZ*, 1930, **56**, 63.
Alpatov, W.W. (2). *JEZ*, 1932, **63**, 85.
Alphonse, P. & Baumann, G. *CRSB*, 1934, **116**, 1277.
Alt, H.L. & Tischer, O.A. *PSEBM*, 1931, **29**, 222.
Altekrüger, W. *AFEWM*, 1932, **126**, 1.
Altmann, M. & Hutt, F.B. *ENY*, 1938, **23**, 793.
Amberson, W.R. *BB*, 1928, **55**, 79.
Amberson, W.R. & Armstrong, P.B. *JCCP*, 1933, **2**, 381.
Amboni, B. *ASBSP*, 1937, **49**, 45.
Amemiya, I. & Murayama, S. *PIAT*, 1931, **7**, 176; *YMaly*, 1931, **18**, 703.
Ammon, R. *BZ*, 1936, **288**, 93.
Ammon, R. & Schütte, E. *BZ*, 1935, **275**, 216.
d'Amora, T. *ASB*, 1937, **23**, 513.
Ancel, P. & Vintemberger, P. (1). *CRSB*, 1925, **92**, 172 and 1401.
Ancel, P. & Vintemberger, P. (2). *CRSB*, 1932, **111**, 43.
Ancel, P. & Vintemberger, P. (3). *CRSB*, 1934, **117**, 697.
Ancel, P. & Vintemberger, P. (4). *CRSB*, 1938, **128**, 95, 98, 101, 412, 414, 417, 1212; 1938, **129**, 241; *CRAS*, 1938, **206**, 1196, 1325.
Andai, G. *AEZF*, 1932, **12**, 301 and 307.
André, E. *CRAS*, 1927, **184**, 901.
André, E. & Canal, H. *BSCF*, 1929 (sér. 4), **41**, 921; **45**, 498 and 511.
Andrewes, C.H. *L*, 1934 (ii), 63, 117; *PRSM*, 1939, **33**, 75.
Angerer, C.A. *JCCP*, 1936, **8**, 329.
Anissimova, V. *BBMEUSSR*, 1938, **6**, 390.
Ankel, W.E. *VDZG*, 1937, **10**, 77.
Annetts, M. *BJ*, 1936, **30**, 1815.
Anonymous. *CAR*, 1930, **5**, 510.
Anselmino, K.J. & Hoffmann, F. (1). *AG*, 1931, **143**, 505.
Anselmino, K.J. & Hoffmann, F. (2). *KW*, 1931, **10**, 97; *AG*, 1931, **143**, 477.
Anselmino, K.J. & Hoffmann, F. (3). *AG*, 1931, **147**, 69.
Anson, M.L. & Mirsky, A.E. *JPC*, 1931, **35**, 185.
Antoniani, C. & Clerici, A.S. *BTS*, 1936, **23**, 129.
Antoschkina, E.D. *PZUSSR*, 1939, **26**, 1 and 16.
Anzinger, F.P. *JH*, 1933, **24**, 309.
Aoki, K. *JFSHIU*, 1939, **7**, 27 and 87.
Arey, L.B. *AR*, 1922, **23**, 245 and 253.
Armstrong, P.B. *AJA*, 1932, **51**, 157.
Arnold, E. *P*, 1938, **29**, 321.
Arnold, N.K. *BMDIMBL*, 1941, **43**, 31.
Aron, H. "Biochemie des Wachstums des Menschen und der höheren Tiere." Fischer, Jena, 1913.

Aron, M. *BBFB*, 1931, **65**, 438.
Arshavsky, I.A. & Malkiman, I.V. *BBME USSR*, 1938, **5**, 135.
Arshavsky, I.A. & Rosanova, V.D. *BBME USSR*, 1938, **5**, 129.
Artemov, N.M. & Valedinskaia, L.K. *BSNM*, 1938, **47**, 194.
Arthus, A. & Provoost, M. *CRSB*, 1937, **124**, 345.
Aschheim, S. & Hohlweg, W. *DMW*, 1933, **59**, 12.
Aschner, B. *AG*, 1913, **99**, 534.
Ashbel, R. (1). *BSIBS*, 1929, **4**, 492.
Ashbel, R. (2). *P*, 1930, **11**, 97; 1932, **15**, 177; *BSIBS*, 1931, **6**, 659; *BDZ*, 1932, **3**, 3.
Ashbel, R. (3). *PSZN*, 1931, **11**, 194; *BSIBS*, 1931, **6**, 670.
Ashbel, R. (4). *PSZN*, 1931, **11**, 204; *BSIBS*, 1928, **3**, 1310.
Ashbel, R. (5). *ASB*, 1935, **21**, 351 and 356; *N*, 1935, **135**, 343.
Ashbel, R. (6). *Verhdl.* 1. *Internat. Kongr. Elektro-Radio-Biol.* 1935, **1**, 635; *YMaly*, 1935, **35**, 425.
Ashbel, R. (7). *ASB*, 1935, **21**, 183.
Ashby, W.R. & Glynn, A. *JNPP*, 1935, **15**, 193.
Ashford, C.A. *BJ*, 1933, **27**, 903; 1934, **28**, 2229.
Ashford, C.A. & Holmes, E.G. *BJ*, 1929, **23**, 748; 1931, **25**, 2028.
Asmolov, E. *BBMEUSSR*, 1936, **1**, 119.
Asmundson, V.S. (1). *SA*, 1931, **11**, 590, 662 and 775; *YMaly*, 1931, **19**, 44.
Asmundson, V.S. (2). *JH*, 1936, **27**, 401.
Asmundson, V.S. & Burmester, B.R. *JEZ*, 1935, **72**, 225.
Asmundson, V.S. & Jervis, J.G. *JEZ*, 1933, **65**, 395.
Assheton, R. (1). *JAP*, 1898, **32**, 362.
Assheton, R. (2). "The geometrical relation of the Nuclei in an invaginating Gastrula" in "Growth in Length". Cambridge University Press, 1916, p. 65.
Astaurov, B.L. *GA*, 1935, **17**, 409.
Astbury, W.T. (1). "Fundamentals of Fibre structure." Oxford University Press, 1933.
Astbury, W.T. (2). *TFS*, 1933, **29**, 193; *SP*, 1933, **28**, 210; 1939, **34**, 1; *ARB*, 1939, **8**, 113; *ARCS*, 1938, **35**, 198; *TB*, 1939, **17**, 90.
Astbury, W.T. & Bell, F.O. *CSHSQB*, 1938, **6**, 109.
Astbury, W.T. & Street, A. *PTRSA*, 1931, **230**, 75.
Astbury, W.T. & Woods, H.J. *PTRSA*, 1933, **232**, 333.
Atkeson, F.W. & Warren, T.R. *JH*, 1935, **26**, 331.
Atlas, M. (1). *PZ*, 1935, **8**, 290.
Atlas, M. (2). *PZ*, 1938, **11**, 278.
Atwell, W.J. & Taft, J.W. *PSEBM*, 1940, **44**, 53.
Austoni, G. *AAL*, 1935, VI, **22**, 544.

Avery, G.S. *OJS*, 1937, **37**, 317.

Avrech, V.V. & Heronimus, E.S. *BBME USSR*, 1937, **4**, 493.

Axelsson, J. *LUA*, 1932, **28**, 1.

Aykroyd, W.R. & Roscoe, M.H. *BJ*, 1929, **23**, 483.

Aykroyd, W.R. & Sankaran, G. *IJMR*, 1936, **23**, 929.

Baas-Becking, L.G.M., Karstens, W.K.H. & Kanner, M. *P*, 1936, **25**, 32.

Bachmann, W.E., Cook, J.W., Dansi, A., de Worms, C.G.M., Haslewood, G.A.D., Hewett, C.L. & Robinson, A.M. *PRSB*, 1937, **123**, 343.

Bäcklin, E. *ULF*, 1930, **35**, 105.

Backman, E.L. & Runnström, J. *CRSB*, 1909, **67**, 414; *BZ*, 1909, **22**, 290; *AP*, 1912, **144**, 287.

Backman, G. *EP*, 1931, **33**, 883; *SKAP*, 1932, **64**, 127; *AFEWM*, 1938, **138**, 37 & 59; *LUA*, 1939, **34**, no. 5; 1939, **35**, nos. 7, 8, 11 & 12.

Badger, G.M., Cook, J.W., Hewett, C.L., Kennaway, E.L., Kennaway, N.M., Martin, R.H. & Robinson, A.M. *PRSB*, 1940, **129**, 439.

von Baer, K.E. (1). "Ueber Entwickelungsgeschichte der Thiere", Pt. I, Königsberg, 1828, Pt. II, Königsberg, 1837; "Nachrichten ü. Leben u. Schriften d. Dr K.E.v. Baer, mitgetheilt von ihm Selbst". Braunschweig, 1886.

von Baer, K.E. (2). *AAP*, 1834, **1**, 481; *Bull. Sci. pub. par l'Acad. Imp. des Sci. St Pétersbourg*, 1835, **1**, 4. 9.

von Baer, K.E. (3). *Nova Acta Phys.-Med. Acad. Caes. Leopoldino-Carol. nat. cur.* 1828, **14**, 827.

von Baer, K.E. (4). *AAP*, 1827, **10**, 576; *Mem. de l'Acad. Imp. des Sci. St Pétersbourg* (6th ser.), 1838, **3**, Bull. no. 2; 1845, **6**, 79 and 179.

von Baer, K.E. (5). *Bull. Cl. Phys. Math. de l'Acad. Imp. Sci. St Pétersbourg*, 1847, **5**, 231.

von Baer, K.E. (6). *Bull. de l'Acad. Imp. des Sci. de St Pétersbourg*, 1862, **4**, 215.

Baernstein, H.D. *JBC*, 1936, **115**, 25 and 33.

Baetsle, M. *AFBQ*, 1937, **8**, 5; *CA*, 1937, **31**, 5468.

Bagdassarianz, G.T. *UBZ*, 1938, **10**, 803.

Bagg, H.J. (1). *AJA*, 1926, **36**, 275.

Bagg, H.J. (2). *AJA*, 1929, **43**, 167.

Bagg, H.J. (3). *AJC*, 1936, **26**, 69.

Bagg, H.J. (4). *Occasional Pub. Amer. Assoc. Adv. Sci.* 1937, **4**, 92.

Bagg, H.J. (5). *AJC*, 1936, **27**, 542.

Bagg, H.J. & Little, C.C. *AJA*, 1924, **33**, 119.

Bailey, S.W. *JEZ*, 1937, **76**, 187.

Baird, I.C. & Prentice, J.H. *A*, 1930, **55**, 20.

Baitsell, G.A. (1). *JEM*, 1915, **21**, 455.

Baitsell, G.A. (2). *JEM*, 1916, **23**, 739.

Baitsell, G.A. (3). *AJOP*, 1917, **44**, 109; *PSEBM*, 1916, **13**, 194.

Baitsell, G.A. (4). *PSEBM*, 1920, **17**, 207; *PNASW*, 1920, **6**, 77; *AJA*, 1921, **28**, 447.

Baitsell, G.A. (5). *QJMS*, 1925, **69**, 571.

Baitsell, G.A. (6). *PNASW*, 1927, **13**, 481.

Baitsell, G.A. (7). *AN*, 1940, **74**, 5.

Baitsell, G.A. & Mason, K.E. *ART*, 1930, **21**, 593; 1934, **29**, 587.

Baker, A.Z. & Wright, M.D. *BJ*, 1935, **29**, 1802.

Baker, L.E. (1). *JEM*, 1929, **49**, 163.

Baker, L.E. (2). *JEM*, 1939, **69**, 625.

Baker, L.E. & Carrel, A. (1). *JEM*, 1926, **44**, 387 and 397.

Baker, L.E. & Carrel, A. (2). *JEM*, 1928, **47**, 353.

Baker, L.E. & Carrel, A. (3). *JEM*, 1928, **48**, 533.

Baker, L.E. & Carrel, A. (4). *JEM*, 1939, **70**, 29.

Baker, L.E. & Ebeling, E.H. *JEM*, 1939, **69**, 365.

Baldes, E.J. *PRSB*, 1934, **114**, 436.

Baldwin, E.H.F. (1). *BJ*, 1935, **29**, 252.

Baldwin, E.H.F. (2). *BJ*, 1935, **29**, 1538.

Baldwin, E.H.F. (3). *JEB*, 1935, **12**, 27.

Baldwin, E.H.F. (4). "An Introduction to Comparative Biochemistry." Cambridge, 1937.

Baldwin, E.H.F. & Bell, D.J. *JCS*, 1938, 1461.

Baldwin, E.H.F. & Needham, D.M. (1). *JEB*, 1932, **10**, 105.

Baldwin, E.H.F. & Needham, D.M. (2). *PRSB*, 1937, **122**, 197; *N*, 1936, **138**, 506.

Baldwin, E.H.F. & Needham, J. *BJ*, 1934, **28**, 1372.

Baldwin, W.M. *AR*, 1915, **9**, 365; *BB*, 1919, **37**, 294.

Balfour, F.M. & Sedgwick, A. *PRS*, 1878, 443; *QJMS*, 1879, **19**, 1.

Balinsky, B.I. (1). *AFEWM*, 1925, **105**, 718; 1926, **107**, 679.

Balinsky, B.I. (2). *AFEWM*, 1927, **110**, 63 and 71.

Balinsky, B.I. (3). *AFEWM*, 1931, **123**, 566.

Balinsky, B.I. (4). *MASU*, 1931, **8**, 273; "Induction of Limbs in Amphibia", Kiev, 1936.

Balinsky, B.I. (5). *AFEWM*, 1933, **130**, 704; *TIZBASU*, 1934, **1**, 91; *AA*, 1935, **80**, 136.

Balinsky, B.I. (6). *AFEWM*, 1937, **136**, 221 & 250.

Balinsky, B.I. (7). *CRASUSSR*, 1937, **17**, 503; *TIZBASU*, 1938, **4**, 73.

Balinsky, B.I. (8). *CRASUSSR*, 1938, **20**, 215; *TIZBASU*, 1938, **4**, 3.

Balinsky, B.I. (9). *TIEMUM*, 1938, **6**, 33.

Balinsky, B.I. (10). *CRASUSSR*, 1939, **23**, 196 and 199.

Balkaschina, E.I. *AFEWM*, 1929, **115**, 448.

Ball, E.G. *BB*, 1934, **67**, 327.

Ball, E.G. & Meyerhof, B. *BB*, 1939, **77**, 321.
Ballentine, R. (1). *BB*, 1938, **75**, 368; *BB*, 1939, **77**, 328; *JCCP*, 1940, **15**, 217.
Ballentine, R. (2). *JCCP*, 1940, **15**, 121.
Ballentine, R. (3). *JCCP*, 1940, **16**, 39.
Balls, A.K. & Swenson, T.L. *JBC*, 1934, **106**, 409; *IEC*, 1934, **26**, 570.
Baltzer, F. (1). *VSNFG* (Neuenburg), 1920, 1; (Bern), 1922, 248.
Baltzer, F. (2). *RSZ*, 1928, **35**, 225; 1932, **39**, 281; 1933, **40**, 243; *PSZN*, 1937, **16**, 89.
Baltzer, F. (3). *RSZ*, 1930, **37**; *VDZG*, 1933.
Baltzer, F. (4). *RSZ*, 1934, **41**, 405; 1938, **45**, 391.
Baltzer, F. (5). *CN*, 1935, **10**, 3.
Baltzer, F. (6). *NW*, 1940, **28**, 177 & 196; *BSIBS*, 1939, **15**, 39.
Baltzer, F. & de Roche, V. *RSZ*, 1936, **43**, 495.
Baltzer, F., Schönmann, W., Lüthi, H.R. & Boehringer, F. *AEZF*, 1939, **22**, 276.
Balzam, N. (1). *AIP*, 1933, **37**, 317.
Balzam, N. (2). *AIPS*, 1937, **58**, 181.
Balzam, N. (3). *APBPC*, 1937, **13**, 370.
Bamann, E., Mahdihassan, S. & Laeverenz, P. *ZPC*, 1933, **215**, 142.
Bamann, E. & Salzer, W. *EEF*, 1938, **7**, 28.
Banga, I. *ZPC*, 1937, **249**, 209.
Bank, O. (1). *CRSB*, 1932, **110**, 389.
Bank, O. (2). *P*, 1932, **14**, 556; 1932, **17**, 476.
Bánki, Ö. *VDAG* [*Anat. Anz.* (Ergänzungsh.)], 1927, **63**; *NTG*, 1928, **72**.
Barancheev, L.M. *PAHUSSR*, 1936, **7**, 147.
Barbas, W.C. *LT*, 1936, **48**, 669; *CA*, 1937, **31**, 3531.
Barclay, A.E., Barcroft, J., Barron, D.H. & Franklin, K.J. *BJR*, 1938, **11**, 570.
Barclay, A.E. & Franklin, K.J. *JOP*, 1938, **94**, 256.
Barcroft, J. (1). *PR*, 1925, **5**, 596.
Barcroft, J. (2). *JOP*, 1932, **76**, 443.
Barcroft, J. (3). *L*, 1933, II, 1021; "Features in the Architecture of Physiological Function", *CUP*. 1934; *N*, 1934, **134**, 705.
Barcroft, J. (4). *RSAB*, 1935, **10**, 164.
Barcroft, J. (5). *PRSB*, 1935, **118**, 242 (Croonian Lecture).
Barcroft, J. (6). *PR*, 1936, **16**, 103.
Barcroft, J. (7). *SC*, 1938, **63**, 76; *IRJMS* (Purser Lecture), 1935, 3.
Barcroft, J. & Barron, D.H. *JOP*, 1936, **88**, 56; *Proc. XVth Int. Congr. Physiol.* (Moscow), 1935, 21.
Barcroft, J., Barron, D.H. & Matthews, B.H.C. *JOP*, 1936, **86**, 29*P*.
Barcroft, J., Barron, D.H. & Windle, W.F. *JOP*, 1936, **87**, 73.
Barcroft, J., Elliott, R.H.E., Flexner, L.B., Hall, F.G., Herkel, W., McCarthy, E.F., McClurkin, T. & Talaat, M. *JOP*, 1934, **83**, 192.
Barcroft, J., Flexner, L.B., Herkel, W., McCarthy, E.F. & McClurkin, T. *JOP*, 1934, **83**, 215.

Barcroft, J., Flexner, L.B. & McClurkin, T. *JOP*, 1934, **82**, 498.
Barcroft, J. & Gotseff, T. *JOP*, 1937, **90**, 27*P*.
Barcroft, J., Herkel, W. & Hill, R.M. *JOP*, 1933, **77**, 194.
Barcroft, J. & Kennedy, J.A. *JOP*, 1939, **95**, 173.
Barcroft, J., Kennedy, J.A. & Mason, M.F. *JOP*, 1938, **93**, 21*P*; 1939, **95**, 159, 173 and 269; 1940, **97**, 347.
Barcroft, J., Kramer, K. & Millikan, G.A. *JOP*, 1937, **90**, 28*P*.
Barcroft, J. & Mason, M.F. *JOP*, 1938, **93**, 22*P*.
Barcroft, J. & Rothschild, P. *JOP*, 1932, **76**, 447.
Barcroft, J. & Stevens, J.G. *JOP*. 1928, **66**, 32
Barfurth, D. (1). *AMA*, 1887, **29**, 35.
Barfurth, D. (2). *AFEWM*, 1895, **1**, 91, 117.
Barnard, W.G. *JPB*, 1931, **34**, 389.
Barnett, S.A. & Bourne, G. *JA*, 1941, **75**, 251.
Barnum, G.L. *JN*, 1935, **9**, 621.
Baron, A.L. *JEZ*, 1935, **70**, 461.
Baronovsky, I.D. & Schechtman, A.M. *PSEBM*, 1938, **39**, 209.
Barott, H.G., Byerly, T.C. & Pringle, E.M. *JN*, 1936, **11**, 191.
Barron, E.S.G. (1). *JBC*, 1929, **81**, 445.
Barron, E.S.G. (2). *BB*, 1932, **62**, 42 and 46.
Barron, E.S.G. & Hamburger, M. *JBC*, 1932, **96**, 299.
Barron, E.S.G. & Harrop, G.A. *JBC*, 1928, **79**, 65.
Barron, E.S.G. & Hoffman, L.A. *JGP*, 1930, **13**, 483.
Barry, G. & Cook, J.W. *AJC*, 1934, **20**, 58.
Barry, G., Cook, J.W., Haslewood, G.A.D., Hewett, C.L., Hieger, I. & Kennaway, E.L. *PRSB*, 1935, **117**, 318.
Barta, E. *AEZF*, 1926, **2**, 6.
Barth, L.G. (1). *BB*, 1934, **67**, 244.
Barth, L.G. (2). *AR*, 1935, **61** (Suppl. no. 1), 4 (abstr.); *BB*, 1937, **73**, 346 (abstr.); *PZ*, 1939, **12**, 22.
Barth, L.G. (3). *CN*, 1939, **14**, 53; *JEZ*, 1941, **87**, 371
Barth, L.G. (4). *PZ*, 1934, **7**, 340; *BR*, 1940, **15**, 405.
Barth, L.G. & Graff, S. *CSHSQB*, 1938, **6**, 385.
Barthélemy, H. *CRAS*, 1922, **175**, 1102 and 1248.
Bartuszek, S. *BZ*, 1932, **253**, 279.
Bastian, H.C. Cit. in F.R. Lillie (3) without ref.
Bataillon, E. (1). *CRAS*, 1910, **150**, 996; *AZEG*, 1910 (5e sér.), **6**, 101.
Bataillon, E. (2). *CRAS*, 1911, **152**, 920 and 1271; 1913, **156**, 812; *RGS*, 1911, **22**, 786; *ASNZ*, 1912 (9th ser.), **16**, 249; 1919 (10th ser.), **3**, 1.
Bataillon, E. (3). *CRAS*, 1911, **152**, 1120.
Bataillon, E. (4). *AFEWM*, 1929, **115**, 707.

Bataillon, E. & Ch'ou Su (Tchou-Su). *AFEWM*, 1929, **115**, 779.

Bate-Smith. *See* Smith, E.C.

Bateman, J.B. (1). *JEB*, 1932, **9**, 322.

Bateman, J.B. (2). *BR*, 1935, **10**, 42.

Bates, R.W., Laanes, T. & Riddle, O. *PSEBM*, 1935, **33**, 446.

Bates, R.W., Lahr, E.L. & Riddle, O. *AJOP*, 1935, **111**, 361.

Bates, R.W. & Riddle, O. *JPET*, 1935, **55**, 365.

Bates, R.W., Riddle, O. & Lahr, E.L. (1). *PSEBM*, 1934, **31**, 1223.

Bates, R.W., Riddle, O. & Lahr, E.L. (2). *AJOP*, 1935, **113**, 259.

Bateson, William. "Materials for the Study of Variation." Macmillan, London, 1894.

Baudouy, C. *JCP*, 1938, **35**, 268.

Bauer, C. *ZMAF*, 1934, **35**, 362.

Baumann, A. (1). *AG*, 1933, **153**, 584.

Baumann, A. (2). *CRSB*, 1935, **119**, 456.

Baumann, A. & Witebsky, E. *CRSB*, 1934, **116**, 10.

Baumann, C.A. & Stare, F.J. *PR*, 1939, **19**, 353.

Baumann, G. & Pfister, C. *CRSB*, 1936, **122**, 1156.

Baumberger, J.P. & Michaelis, L. *BB*, 1931, **61**, 417.

Baurmann, M. *KMA*, 1922, **68**, 73.

Bautzmann, H. (1). *AFEWM*, 1926, **108**, 283.

Bautzmann, H. (2). *AFEWM*, 1927, **110**, 631.

Bautzmann, H. (3). *AFEWM*, 1928, **114**, 177.

Bautzmann, H. (4). *AFEWM*, 1929, **119**, 1.

Bautzmann, H. (5). *NW*, 1929, **17**, 818; *SGMPM*, 1929, **39**, 1.

Bautzmann, H. (6). *CAP*, 1935, **63**, 82; *VDPG*, 17 and 142.

Bautzmann, H., Holtfreter, J., Spemann, H. & Mangold, O. *NW*, 1932, **20**, 971. Illustrations for Bautzmann's contribution were published afterwards in *VDAG* (*Anat. Anz.*; Erg.-Band), 1935, **81**, 118 and 283; *VDPG* (*Centrlbl. f. allg. Pathol.*; Erg.-Band), 1935, **63**, 17 and 142.

Bawden, F.C. & Pirie, N.W. "Liquid crystalline preparations of plant viruses", *Trav. Congr. Palais de la Découverte, Paris*, 1938, p. 377.

Bawden, F.C., Pirie, N.W., Bernal, J.D. & Fankuchen, I. *N*, 1936, **138**, 1051.

Bayer, G. & Wense, T. *AFEWM*, 1937, **137**, 372.

Bayliss, W.M. & Starling, E.H. *JOP*, 1902, **28**, 325.

Bayon, H. *L*, 1912 (ii), 1579.

Beadle, B.W., Conrad, R.M. & Scott, H.M. *PS*, 1938, **17**, 372.

Beadle, G.W. (1). *PNASW*, 1937, **23**, 146; *GS*, 1937, **22**, 587.

Beadle, G.W. (2). *AN*, 1937, **71**, 120.

Beadle, G.W., Anderson, R.L. & Maxwell, J. *PNASW*, 1938, **24**, 80.

Beadle, G.W., Clancy, C.W. & Ephrussi, B. *PRSB*, 1937, **122**, 98.

Beadle, G.W. & Ephrussi, B. (1). *CRAS*, 1935, **201**, 620; *PNASW*, 1935, **21**, 642.

Beadle, G.W. & Ephrussi, B. (2). *GS*, 1936, **21**, 225.

Beadle, G.W. & Ephrussi, B. (3). *GS*, 1937, **22**, 76; *AR*, 1937, **71**, 91.

Beadle, G.W. & Law, L.W. *PSEBM*, 1938, **37**, 621.

Beadle, G.W. & Tatum, E.L. *AN*, 1941, **75**, 107.

Beadle, G.W., Tatum, E.L. & Clancy, C.W. *BB*, 1938, **75**, 447; 1939, **77**, 407.

Beams, H.W. & King, R.L. (1). *BB*, 1937, **73**, 99; *S*, 1936, **84**, 138; *JRMS*, 1940, **60**, 240.

Beams, H.W. & King, R.L. (2). *JM*, 1938, **63**, 477.

Beams, H.W., King, R.L. & Risley, P.L. *PSEBM*, 1934, **32**, 181.

Beams, H.W. & Meyer, R.K. *PZ*, 1931, **4**, 486.

Beard, J.W. & Wyckoff, R.W.G. *S*, 1937, **85**, 201; **86**, 92; *JBC*, 1938, **123**, 461.

Bearse, G.E. & Miller, M.W. *PS*, 1937, **16**, 39.

Beatty, R.A., de Jong, S. & Zieliński, M.A. *JEB*, 1939, **16**, 150.

Beck, H. & Truszkowski, R. *MDS*, 1930, **11**, 36 (in Polish).

Beck, L.V. & Chambers, R. *JCCP*, 1935, **6**, 441.

Becker, E. (1). *NW*, 1937, **25**, 507.

Becker, E. (2). *ZPC*, 1937, **246**, 177.

Becker, E. (3). *ZMOT*, 1937, **32**, 672.

Becker, E. & Plagge, E. (1). *NW*, 1937, **25**, 809.

Becker, E. & Plagge, E. (2). *BC*, 1939, **59**, 326.

Becker, M. *BZ*, 1934, **272**, 227.

Beckwith, C. *JEZ*, 1927, **49**, 217.

Bednjakova, T.A. *CRASUSSR*, 1937, **14**, 471.

Beebe, C.W. *ZA*, 1907, **1**, 1.

de Beer, G.R. (1). *BR*, 1926, **2**, 137.

de Beer, G.R. (2). "Embryology and Evolution." Oxford University Press, 1930.

de Beer, G.R. (3). "Embryology and Evolution" essay in "Evolution". Goodrich Presentation Volume. Oxford University Press, 1938.

de Beer, G.R. & Huxley, J.S. *QJMS*, 1924, **68**, 471.

Behmel, G. *ZPC*, 1932, **208**, 62.

Beijerinck, M.W. (1). *BOZ*, 1888, **46**, 1.

Beijerinck, M.W. (2). "Verzammelte Geschriften", 1897, vol. 3, p. 199, esp. p. 203.

Belkin, R. *CRASUSSR*, 1934, **4**.

Bell, G.H. & Robson, J.M. *QJEP*, 1937, **27**, 205.

Bell, J. *PZ*, 1940, **13**, 73.

Bellamy, A.W. *BB*, 1919, **37**, 312; 1921, **41**, 351; *AJA*, 1922, **30**, 473.

Bellamy, A.W. & Child, C.M. *PRSB*, 1924, **96**, 132.

Belogolowy, G. *AFEWM*, 1918, **43**, 556.

Belousov, A.P. *BBMEUSSR*, 1938, **6**, 225.

Benazzi, M. "Attualità Zoologica", **1**, 1 [Suppl. to *AZI*, 1933, **19**]; *YMaly*, 1934, **28**, 724.

Benda, C. *EAPPA*, 1895 (ii), 541.

Benedict, F.G. "The Physiology of Large Reptiles." Washington, 1932, pp. 86 ff., 130 ff. Nitrogen; *D*, 1932, **13**, 147; *Carnegie Instit. Wash. News Service Bull.* 1932, **2**, 193.

Benedict, F.G. & Fox, E.L. *AP*, 1933, **231**, 455.

Benedict, F.G., Landauer, W. & Fox, E.L. *Bull. Storrs Agric. Exp. Sta.* No. 177, 1932.

Benzinger, T. & Krebs, H.A. *KW*, 1933, **12**, 1206.

Berenblum, I. *JPB*, 1929, **32**, 425; 1931, **34**, 731; 1935, **40**, 549.

Berenblum, I. & Kendal, L.P. *BJ*, 1936, **30**, 429.

Berenblum, I., Kendal, L.P. & Orr, J.W. *BJ*, 1936, **30**, 709.

Berenblum, I. & Wormall, A. *BJ*, 1939, **33**, 75.

Berenstein, F.J. & Penionskevitch, E.E. *PZUSSR*, 1935, **18**, 655; *YMaly*, 1936, **36**, 347.

Berezkina, L.F. *CRASUSSR*, 1938, **21**, 209.

Bergami, G. *RAL*, 1925, **2**, 140.

Bergamini, M. *BTS*, 1924, **11**, 410; *OMaly*, 1925, **31**, 281.

van den Bergh, A.A.H. & Grotepass, W. *CRSB*, 1936, **121**, 1253.

Bergmann, W. (1) *JBC*, 1934, **104**, 317 and 553.

Bergmann, W. (2). *ZKF*, 1939, **48**, 546.

Berkesy, L. & Gönczi, K. *AEPUP*, 1933, **171**, 260.

Berkowitz, P. *AR*, 1938, **71**, 161; *JEZ*, 1941, **87**, 233.

Berlese, A. *R*, 1913, **9**, 121.

Bernal, J.D. Remarks on F. Rinne's paper, *Faraday Soc. Symp.* 1933, p. 1082.

Bernal, J.D. & Fankuchen, I. *N*, 1937, **139**, 923.

Bernard, F. *BBFB*, 1937. Supplement No. 23.

Bernheim, F. & Bernheim, M.L.C. *CSHSQB*, 1939, **7**, 174.

Bernheim, F. & v. Felsovanyi, A. *S*, 1940, **91**, 76.

Bernstein, F. (1). *CSHSQB*, 1934, **2**, 209.

Bernstein, F. (2). *CN*, 1936, **11**, 265.

Berrier, H. *CRSB*, 1937, **124**, 1319; 1937, **126**, 453; *CRAS*, 1937, **205**, 1009; *BBFB*, 1940, Supplement no. 27.

Berrill, N.J. (1). *PTRSB*, 1929, **218**, 37.

Berrill, N.J. (2). *PTRSB*, 1931, **219**, 281.

Berrill, N.J. (3). *PTRSB*, 1935, **225**, 255.

Berrill, N.J. (4). *PTRSB*, 1935, **225**, 327.

Berrill, N.J. (5). *JM*, 1935, **57**, 353.

Berrill, N.J. (6). *G*, 1937, **1**, 211.

Bersin, T., Willfang, G. & Nafziger, H. XVIth International Congress of Physiology, Zürich, 1938, *Kongressbericht*, II, p. 106.

von Bertalanffy, L. (1). "Kritische Theorie d. Formbildung." Bornträger, Berlin, 1928, p. 223.

von Bertalanffy, L. (2). *EK*, 1930, **1**, 361.

von Bertalanffy, L. (3). *AFEWM*, 1934, **131**, 613; *BC*, 1933, **53**, 639; *HB*, 1938, **10**, 181.

von Bertalanffy, L. (4). *UW*, 1936, **28**, 161; *YMaly*, 1936, **39**, 249.

von Bertalanffy, L. & Woodger, J.H. "Modern Theories of Development." Oxford University Press, 1933.

Berutti, E. *GC*, 1936, **2**, 407; *OMaly*, 1936, **95**, 655.

Bethe, A. *ZPC*, 1895, **20**, 472.

Bethke, R.M., Record, P.R. & Wilder, F.W. *JN*, 1936, **12**, 309.

Beynon, J.H., Heilbron, I.M. & Spring, F.S. *N*, 1936, **138**, 1017.

Bežler, F.I. *CRASUSSR*, 1939, **23**, 101.

Beznák, A.B.L. & Chain, E. *QJEP*, 1937, **26**, 201.

von Bezold, A. *ZWZ*, 1857, **8**, 487; 1858, **9**, 246.

Bhatia, D. *ZZFMA*, 1931, **12**, 430.

Białaszewicz, K. (1). *P*, 1933, **19**, 350.

Białaszewicz, K. (2). *AIP*, 1933, **37**, 1.

Białaszewicz, K. (3). *ABE*, 1936, **10**, 352; 1937, **11**, 20.

Białaszewicz, K. & Błędowski, R. *PSSW*, 1915, **8**, 429.

Białaszewicz, K. & Głogowska, H. *ABE*, 1938, **12**, 50.

Białaszewicz, K. & Landau, C. *ABE*, 1938, **12**, 307.

Białaszewicz, K. & Lewin, M. *ABE*, 1938, **12**, 265.

Bianchi, G. *BSIBS*, 1939, **15**, 194.

Bickenbach, W. & Rupp, H. (1). *ZGG*, 1931, **100**, 1.

Bickenbach, W. & Rupp, H. (2). *ZGG*, 1932, **101**, 632; *KW*, 1931, **10**, 63.

Bickenbach, W. & Rupp, H. (3). *ZGG*, 1932, **103**, 170.

Bickenbach, W. & Rupp, H. (4). *ZGG*, 1932, **103**, 183.

Bidder, G.P. *N*, 1925, **115**, 155 and 495; *BMJ*, 1932 (ii), 583.

Bidone, M. *AO*, 1930, **52**, 1136; *OMaly*, 1931, **60**, 475.

Biereigel, R.O. *ZZFMA*, 1938, **28**, 341.

Bierich, R. & Rosenbohm, A. *ZPC*, 1935, **231**, 47.

Bierry, H. & Gouzon, B. *CRAS*, 1932, **194**, 653.

Bijtel, J.H. *AFEWM*, 1931, **125**; 1936, **134**.

Billroth, T. "Über die Einwirkungen lebender Pflanzen- und Tier-Zellen auf einander." Wien, 1890.

Bills, C.E. *PR*, 1935, **15**, 1.

Birguer, E. & Afanasiev, M. *BBMEUSSR* 1936, **2**, 273.

Bisbey, B., Appleby, V., Weis, A. & Cover, S. *Res. Bull. Missouri Agric. Exp. Sta.* 1934, No. 205.

Bischler, V. *RSZ*, 1926, **33**, 431.

Bischler, V. & Guyénot, E. *CRSB*, 1925, **92**, 678 and 774; 1926, **94**, 968.

Bischoff, C., Grab, W. & Kapfhammer, J. *ZPC*, 1932, **207**, 57.

Bishop, H.S. & Morgan, A.F. *PSEBM*, 1928, **25**, 438.

Bishop, W.B.S. (1). *MJA*, 1928, **15** (i), 480.

Bishop, W.B.S. (2). *MJA*, 1929, **16** (i), 96.

Bishop, W.B.S. (3). *MJA*, 1929, **16** (i), 792.

Bishop, W.B.S. (4). *MJA*, 1931, **18** (ii), 130.

Bishop, W.B.S. (5). *MJA*, 1931, **18** (ii) 197.

Bishop, W.B.S. & Cooksey, T. *MJA*, 1929, **16** (ii), 660.

Biskind, G.R. & Glick, D. *JBC*, 1936, **113**, 27.

Blacher, L.J. *TLZEMAM*, 1928, **3**, 172.

Blacher, L.J. & Belkin, R.O. *TLZEMAM*, 1927, **3**, 97.

Blacher, L.J. & Efimov, M.I. *BC*, 1930, **50**, 271.

Blacher, L.J., Irichimovitch, A.I., Liosner, L.D. & Voronzova, M.A. *AFEWM*, 1932, **127**, 370.

Blacher, L.J. & Liosner, L.D. *BC*, 1930, **50**, 285.

Blacher, L.J., Liosner, L.D. & Voronzova, M.A. *BAPS*, 1934, B (ii), 325.

Blacher, L.J. & Tschmutova, A.P. *ABS USSR*, 1934, **35**, 43; *YMaly*, 1935, **34**, 518.

Blackwood, J.H. & Wishart, G.McF. *BJ*, 1934, **28**, 550.

Blagoveschensky, A.V. *EZ*, 1938, **2**, 203.

Blakeslee, A. & Avery, A. *S*, 1937, **86**, 36.

Blanchard, K.C. *JBC*, 1935, **108**, 251.

Bland-Sutton, J. (1). *JAP*, 1885, **19**, 464.

Bland-Sutton, J. (2). "Tumours, Innocent and Malignant." London, 1922, p. 490.

Bland-Sutton, J. (3). *BMJ*, 1923 (ii), 847.

Blandau, R.J. & Young, W.C. *AJA*, 1939, **64**, 303.

Blank, H. *AP*, 1934, **234**, 310.

Blazsó, A. *ZPC*, 1936, **244**, 138.

Bloch, B. & Dreifuss, W. *SMW*, 1921 (ii), 1033.

Bloch, S. *RSZ*, 1938, **45**, 157.

Block, R.J. *JBC*, 1934, **104**, 343; 1934, **105**, 455, 663; *YJBM*, 1935, **7**, 235; 1937, **9**, 445.

Block, R.J., Darrow, D.C. & Cary, K. *JBC*, 1934, **104**, 347.

Bloom, W. *PR*, 1937, **17**, 589.

Blum, E., Jarmoschkevitch, A.I. & Jakovtschuk, A.I. *BBMEUSSR*, 1936, **1**, 113.

Blunck, H. *ZWZ*, 1914, **111**.

Blunn, C.T. & Gregory, P.W. *JEZ*, 1935, **70**, 397.

Boas, B.Z. *BZ*, 1921, **117**.

Boas, E.P. & Landauer, W. (1) *AJMS*, 1933, **185**, 654.

Boas, E.P. & Landauer, W. (2). *AJMS*, 1934, **188**, 359.

Bodenstein, D. (1). *AFEWM*, 1933, **128**, 564.

Bodenstein, D. (2). *AFEWM*, 1933, **130**, 746; *ZAZ*, 1933, **103**, 209; *JEZ*, 1941, **87**, 31.

Bodenstein, D. (3). *AFEWM*, 1935, **133**, 156; *BC*, 1934, **54**, 181.

Bodenstein, D. (4). *EB*, 1936, **13**, 174.

Bodenstein, D. (5). *AFEWM*, 1937, **136**, 745.

Bodenstein, D. (6). *BC*, 1938, **58**, 329.

Bodenstein, D. (7). *AFEWM*, 1938, **137**, 474.

Bodenstein, D. (8). *AFEWM*, 1938, **137**, 636.

Bodenstein, D. (9). *PNASW*, 1939, **25**, 14.

Bodenstein, D. (10). *JEZ*, 1939, **82**, 1.

Bodenstein, D. (11). *JEZ*, 1939, **82**, 329.

Bodenstein, D. (12). *GS*, 1939, **24**, 494.

Bodenstein, D. (13). *JEZ*, 1940, **84**, 23.

Bodine, J.H. (1). *JEZ*, 1923, **37**, 457.

Bodine, J.H. (2). *PZ*, 1929, **2**, 459.

Bodine, J.H. (3). *PZ*, 1932, **5**, 538 and 549.

Bodine, J.H. (4). *PZ*, 1933, **6**, 150.

Bodine, J.H. (5). *PNASW*, 1934, **20**, 640.

Bodine, J.H. (6). *PZ*, 1934, **7**, 459.

Bodine, J.H. (7). *JCCP*, 1934, **4**, 397.

Bodine, J.H. & Allen, T.H. *JCCP*, 1938, **11** 409.

Bodine, J.H., Allen, T.H. & Boell, E.J. *PSEBM*, 1937, **37**, 450.

Bodine, J.H. & Boell, E.J. (1). *JCCP*, 1934, **4**, 475.

Bodine, J.H. & Boell, E.J. (2). *JCCP*, 1934, **5**, 97.

Bodine, J.H. & Boell, E.J. (3). *PSEBM*, 1935, **32**, 783.

Bodine, J.H. & Boell, E.J. (4). *JCCP*, 1935, **6**, 263.

Bodine, J.H. & Boell, E.J. (5). *JCCP*, 1936, **7**, 455.

Bodine, J.H. & Boell, E.J. (6). *JCCP*, 1936, **8**, 213.

Bodine, J.H. & Boell, E.J. (7). *JCCP*, 1936, **8**, 357.

Bodine, J.H. & Boell, E.J. (8). *PSEBM*, 1936, **34**, 629.

Bodine, J.H. & Boell, E.J. (9). *PZ*, 1937, **10**, 245.

Bodine, J.H. & Boell, E.J. (10). *PSEBM*, 1937, **36**, 21.

Bodine, J.H. & Boell, E.J. (11). *JCCP*, 1938, **11**, 41; *PSEBM*, 1936, **35**, 504.

Bodine, J.H. & Carlson, L.D. (1). *JCCP*, 1940, **16**, 71.

Bodine, J.H. & Carlson, L.D. (2). *JGP*, 1941, **24**, 423.

Bodine, J.H., Carlson, L.D. & Ray, O. M. *BB*, 1940, **78**, 437.

Bodine, J.H. & Evans, T.C. (1). *BB*, 1932, **63**, 235.

Bodine, J.H. & Evans, T.C. (2). *PZ*, 1934, **7**, 550.

Bodine, J.H. & Orr, P.R. *BB*, 1925, **48**, 1.

Bodine, J.H., Ray, O.M., Allen, T.H. & Carlson, L.D. *JCCP*, 1939, **14**, 173.
Bodine, J.H. & Robbie, W.A. *PZ*, 1940, **13**, 391.
Bodine, J.H. & Thompson, V. *JCCP*, 1935, **6**, 255.
Bodine, J.H. & Trowbridge, C. *BB*, 1940, **79**, 452.
Bodine, J.H. & Wolkin, J.E. *PZ*, 1934, **7**, 464.
Boehringer, F. *AFEWM*, 1939, **138**, 376.
Boell, E.J. *JCCP*, 1935, **6**, 369.
Boell, E.J., Chambers, R., Glancy, E.A. & Stern, Kurt G. *BB*, 1940, in press.
Boell, E.J., Koch, H. & Needham, J. *PRSB*, 1939, **127**, 374.
Boell, E.J. & Needham, J. (1). *PRSB*, 1939, **127**, 356.
Boell, E.J. & Needham, J. (2). *PRSB*, 1939, **127**, 363.
Boell, E.J., Needham, J. & Rogers, V. *PRSB*, 1939, **127**, 322.
Boell, E.J. & Nicholas, J.S. *AR*, 1939, **73** (2nd Suppl.), 9.
Boell, E.J., Nicholas, J.S. & Sawyer, C.H. Private communication to the author.
Boell, E.J. & Poulson, D.F. *AR*, 1940, in press.
Boell, E.J., Ray, M. & Bodine, J.H. *RA*, 1937, **29**, 533.
Boerema, I. *AFEWM*, 1929, **115**, 601.
Bogaert, R. *JOP*, 1937, **90**, 67P.
Bogucki, M. (1). *TIN*, 1921, **1**, no. 2, no. 6 and no. 16; 1923, **2**, no. 32; *CRSB*, 1923, **89**, 1357.
Bogucki, M. (2). *TIN*, 1926, **3**, no. 50; *P*, 1930, **9**, 432.
Bogucki, M. (3). *ABE*, 1928, **2**, no. 2.
Bogue, J.Y. *JEB*, 1932, **9**, 351; 1933, **10**, 286.
Böhm, G. *BZ*, 1935, **282**, 32.
Böhm, G. & Signer, R. *HCA*, 1931, **14**, 1370.
Bohn, G. & Drzwina, A. (1). *CRAS*, 1931, **193**, 491.
Bohn, G. & Drzwina, A. (2). *CRAS*, 1931, **193**, 1478.
du Bois, A.M. *RSZ*, 1938, **45**, 1.
du Bois, A.M. & Geigy, R. *RSZ*, 1935, **42**, 169.
de Boissezon, P. *CRSB*, 1933, **114**, 487.
Bonnevie, K. (1). *JEZ*, 1934, **67**, 443; *ZIAV*, 1932, **62**, 73.
Bonnevie, K. (2). *SNVAO*, 1936 (Kl. I, Math.-Naturvid), no. 9.
Bonnevie, K. (3). *GA*, 1936, **18**, 105; *OMaly*, 1937, **98**, 45.
Bonney, W.F.V. *TOSL*, 1903, **44**, 354.
Booher, L.E. & Hansmann, G.H. *JBC*, 1932, **94**, 195.
Booker, W.M. *AR*, 1932, **54** (Suppl.), 47.
Boone, E. & Baas-Becking, L.G.M. *JGP*, 1931, **14**, 753.
Booth, V.H. *BJ*, 1935, **29**, 1743.
Borchardt, E. *AFEWM*, 1927, **110**, 366.
Borei, H. (1). *ZVP*, 1933, **20**, 258.
Borei, H. (2). *ZMOT*, 1935, **30**, 97.

Borger, G. & Peters, T. *ZPC*, 1933, **214**, 91.
Borger, G. & Zenker, R. (1). *VDPG*, **26**, 1931, 124 and 135.
Borger, G. & Zenker, R. (2). *AEZF*, 1932, **12**, 347.
Born, G. *AFEWM*, 1897, **4**, 349.
Börner, R. *AP*, 1928, **220**, 716.
Bornmann, J.H. *JAOAC*, 1931, **14**, 416; *CA*, 1932, **26**, 779.
Börnstein, K. & Klee, E.E. *AEZF*, 1927, **3**, 395.
Borovanský, L. *BL*, 1938, **23**, 23.
Borsook, H. *EEF*, 1935, **4**, 1.
Borssuk, R.A. *AFEWM*, 1935, **133**, 349.
Borst, M. & Königsdorfer, "Untersuchungen über Porphyrie." Leipzig, 1929.
Botella-Llusiá, J. (1). *AG*, 1935, **159**, 27.
Botella-Llusiá, J. (2). *AG*, 1936, **160**, 467.
Bottero, G.D. *B.SIBS*, 1937, **12**, 635.
Boucek, C.M. & Renton, A.D. *SGO*, 1932, **54**, 906.
Boulenger, G.A. *N*, 1898, **58**, 619.
Bounhiol, J.J. *CRAS*, 1936, **203**, 388, 504 and 1182; 1937, **205**, 175; 1938, **206**, 773; *CRSB*, 1937, **126**, 1189; *BBFB*, 1938, Suppl. **24**, 1.
Bourdin, J. *CRSB*, 1926, **95**, 1149, 1183, 1239 and 1242.
Bourne, G. (1). *N*, 1932, **132**, 859.
Bourne, G. (2). *N*, 1935, **135**, 148.
Boursnell, J.C. & Wormall, A. *BJ*, 1939, **33**, 1191.
Boveri, T. (1). *SGMPM*, 1889, **5**.
Boveri, T. (2). *VPMGW*, 1901, **34**, 145.
Boveri, T. (3). "Zur Frage der Entstehung maligner Tumoren." Fischer, Jena, 1914. Eng.tr. by Marcella Boveri, Baltimore, 1929.
Bovet, D. *RSZ*, 1930, **37**, 83.
Boycott, A.E., Diver, C., Garstang, S.L. & Turner, F.M. *PTRSB*, 1931, **219**, 51.
Boycott, A.E., Garstang, S.L. & Diver, C. *JG*, 1925, **15**, 113.
Boyd, E.M. (1). *BJ*, 1935, **29**, 985.
Boyd, E.M. (2). *JBC*, 1935, **111**, 667.
Boyd, E.M. (3). *JOP*, 1937, **91**, 394.
Boyd, E.M. & Wilson, K.M. *JCI*, 1935, **14**, 7; *OMaly*, 1935, **88**, 477.
Boyden, E.A. (1). *JEZ*, 1924, **40**, 437.
Boyden, E.A. (2). *PSEBM*, 1931, **28**, 625.
Boyland, E. (1). *L*, 1932, 1108.
Boyland, E. (2). *BJ*, 1933, **27**, 791.
Boyland, E. & Boyland, M.E. (1). *BJ*, 1934, **28**, 244.
Boyland, E. & Boyland, M.E. (2). *BJ*, 1935, **29**, 1097 and 1910.
Boyland, E. & Boyland, M.E. (3). *BJ*, 1937, **31**, 454.
Boyland, E. & Mawson, C.A. *BJ*, 1934, **28**, 1409.
Boyle, Robert. "Collected Works." London, 1744, vol. 5, p. 519.
Boysen-Jensen, P. (1). "Growth Hormones in Plants." New York, 1936.
Boysen-Jensen, P. (2). *ARB*, 1938, **7**, 513.

Brachet, A. (1). *ADB*, 1911 (**24**?), **26**, 337.
Brachet, A. (2). "Traité d'Embryologie." Ed. A. Dalcq & P. Gérard. Paris, 1935.
Brachet, J. (1). *ADB*, 1929, **39**, 677.
Brachet, J. (2). *CRSB*, 1931, **108**, 813.
Brachet, J. (3). *CRSB*, 1931, **108**, 1167.
Brachet J. (4). *CRSB*, 1932, **110**, 562.
Brachet, J. (5). *ADB*, 1933, **44**, 519.
Brachet, J. (6). *ADB*, 1934, **45**, 611.
Brachet, J. (7). *ADB*, 1934, **46**, 1; *CRSB*, 1932, **110**, 562.
Brachet, J. (8). *ADB*, 1934, **46**, 25; *CRSB*, 1934, **116**, 658.
Brachet, J. (9). *ABSRSMNB*, 1935, 73.
Brachet, J. (10). *CRSB*, 1936, **122**, 108.
Brachet, J. (11). *BSCB*, 1936, **18**, 305.
Brachet, J. (12). *MMRHNB*, 1936 (2nd ser.), **3**, 481.
Brachet, J. (13). *ADB*, 1937, **48**, 529.
Brachet, J. (14). *RUB*, 1937, **3**, 261.
Brachet, J. (15). *S*, 1937, **86**, 225.
Brachet, J. (16). *BB*, 1938, **74**, 93; *ADB*, 1937, **48**, 561.
Brachet, J. (17). "Le Métabolisme de l'œuf en voie de développement" in "Trav. du Congrès du Palais de la Découverte". Hermann, Paris, 1938.
Brachet, J. (18). *BARBS*, 1938, 499; *ADB*, 1940, **51**, 168.
Brachet, J. (19). "Le Rôle Physiologique et Morphogénetique du Noyau." Hermann, Paris, 1938. Also *ABSRSMNHB*, 1938, 91.
Brachet, J. (20). *CRSB*, 1938, **127**, 1455; *BB*, 1937, **73**, 343; *AEZF*, 1939, **22**, 541.
Brachet, J. (21). *CRSB*, 1938, **129**, 18.
Brachet, J. (22). *ADB*, 1939, **50**, 233; *BSCB*, 1939, **21**, 115.
Brachet, J. (23). *ADB*, 1940, **51**, 151.
Brachet, J. (24). *CRSB*, 1940, **133**, 88 & 90.
Brachet, J. & Needham, J. (1). *CRSB*, 1935, **118**, 840.
Brachet, J. & Needham, J. (2). *ADB*, 1935, **46**, 821.
Brachet, J. & Rapkine, L. *CRSB*, 1939, **131**, 789.
Brachet, J. & Shapiro, H. *JCCP*, 1937, **10**, 133.
Braconnier-Fayemendy, M. *ASBA*, 1933.
Bradley, H.C. (1). *JBC*, 1907, **3**, 151.
Bradley, H.C. (2). *PR*, 1922, **2**, 415.
Bradway, W. *JEZ*, 1935, **72**, 213.
Bragg, A.N. *ZZFMA*, 1938, **28**, 154.
Brambel, C.E. *AJA*, 1933, **52**, 397.
Brambel, C.E. & Lynn, W.G. *AR*, 1935, **64** (1st Suppl.), 46.
Brambell, F.W.R. *PTRSB*, 1925, **214**, 113.
Brandes, J. *BARBS*, 1938 (Sér. 5ᵉ), **24**, 92; *ASRZB*, 1939, **70**, 153; *ADB*, 1940, **51**, 219.
Brandstrup, E. (1). *AOGS*, 1930, **10**, 251.
Brandstrup, E. (2). *AOGS*, 1931, **11**, 85.
Brandt, W. *AFEWM*, 1928, **114**, 54.
Branion, H.D. & Smith, J.B. *PS*, 1932, **11**, 261.
Brauer, A. *PZ*, 1938, **11**, 249.

Brauer, A. & Taylor, A.C. *JEZ*, 1936, **73**, 127; *AR*, 1934, **60** (Suppl.), 61.
Braun, W. *GS*, 1940, **25**, 143.
Braus, H. (1). *MJ*, 1906, **35**, 509.
Braus, H. (2). *AFEWM*, 1906, **22**, 564.
Brecher, L. *AFEWM*, 1924, **102**, 549.
Brehme, K.S. *PSEBM*, 1937, **37**, 578.
Breusch, F.L. *BZ*, 1937, **295**, 101; 1938, **295**, 125.
Bridges, C.B. (1). *GS*, 1916, **1**, 107.
Bridges, C.B. (2). "Genetics of Sex in Drosophila" in "Sex and Internal Secretion." 1932, p. 207.
Bridges, C. B. & Dobzhansky, T. *AFEWM*, 1933, **127**, 575.
Brien, P. (1). *BBFB*, 1933, **67**, 100.
Brien, P. (2). *ADB*, 1937, **48**, 185.
Brien, P. & Meewis, H. *ADB*, 1938, **49**, 177.
Briggs, R.W. (1). *JCCP*, 1939, **13**, 77.
Briggs, R. W. (2). *N*, 1940, **146**, 29.
Brighenti, A. *BSIBS*, 1939, **15**, 198.
Brinkman, R. *JOP*, 1933, **80**, 171.
Brinkman, R. & Jonxis, J.H.P. *JOP*, 1935, **85**, 115; 1936, **88**, 162.
Brinkman, R., Wildschut, A. & Wittermans, A. *JOP*, 1934, **80**, 377.
Brinley, F.J. (1). *PZ*, 1930, **3**, 366.
Brinley, F.J. (2). *TB*, 1938, **16**, 51.
Brinley, F.J. & Jenkins, G.B. *PZ*, 1939, **12**, 31.
Broad, C.D. *PAS*, 1919, **19**, 86, p. 123.
Brock, N., Druckrey, H. & Herken, H. *AEPUP*, 1938, **188**, 436 and 451.
Brockmann, H. *EVHF*, 1938, **2**, 55.
Brockmann, H. & Völker, O. *ZPC*, 1934, **224**, 193.
Brody, S. (1). *G*, 1937, **1**, 60.
Brody, S. (2). *Bull. Univ. Missouri Agric. Exp. Sta.* 1938, No. 283.
Brody, S., Funk, E.M. & Kempster, H.L. *PS*, 1932, **11**, 133.
Brody, S., Hall, W.C., Ragsdale, E.C., Trowbridge, E.A., Funk, E.M., Kempster, H.L., Ashworth, U.S., Hogan, A.G. & Procter, R.C. *Bull. Univ. Missouri Agric. Exp. Sta.* No. 166 and No. 176, 1932.
Brody, S. & Kibler, H.H. *Bull. Univ. Missouri Agric. Exp. Sta.* 1941, No. 328.
Brody, S., Procter, R.C. & Ashworth, U.S. *Bull. Univ. Missouri Agric. Exp. Sta.* 1934, No. 220.
Brody, S., Riggs, J., Kaufman, K. & Herring, V. *Bull. Univ. Missouri Agric. Exp. Sta.* 1938, No. 281.
Bromley, N.W. & Orechovitch, W.N. (1). *BZ*, 1934, **272**, 324.
Bromley, N.W. & Orechovitch, W.N. (2). *CRASUSSR*, 1934, **2**, 44; *BG*, 1935, **11**, 317.
Brøndsted, H.V. (1). *AZ*, 1936, **17**, 75.
Brøndsted, H.V. (2). *P*, 1937, **28**, 123.
Bronkhorst, J.J. & Hall, G.O. (1). *PS*, 1935, **14**, 42.
Bronkhorst, J.J. & Hall, G.O. (2). *PS*, 1935, **14**, 112.

Brooks, J. & Pace, J. *PRSB*, 1938, **126**, 196.

Brooks, M.M. *CN*, 1933, **8**, 64.

Brooksby, J.B. & Newton, W.H. *JOP*, 1938, **92**, 136.

Broom, R. *PTRSB*, 1914, **206**, 1 (Croonian).

Brouha, L. & Simonnet, H. *CRSB*, 1927, **97**, 459.

Brown, Adrian. *JCS*, 1892, **61**, 369.

Brown, A.W.A. (1). *JEB*, 1936, **13**, 131.

Brown, A.W.A. (2). *JEB*, 1937, **14**, 87.

Brown, A.W.A. (3). *BJ*, 1938, **32**, 895 and 903.

Brown, A.W.A. & Farber, L. *BJ*, 1936, **30**, 1107.

Brown, G.L. *PR*, 1937, **17**, 485.

Brown, N.A. & Gardner, F.E. *PP*, 1936, **26**, 708.

Browning, C.H., Cohen, J.B., Cooper, K.E., Ellingworth, S. & Gulbranson, R. *PRSB*, 1933, **113**, 300.

Browning, C.H., Gulbranson, R. & Niven, J.S.F. *JPB*, 1936, **42**, 155.

Brues, A.M. & Cohen, A. *BJ*, 1936, **30**, 1363.

Brues, A.M., Jackson, E.B. & Aub, J. *PSEBM*, 1936, **34**, 270.

Bruger, M. *JBC*, 1935, **108**, 463.

Brunel, A. *BSCB*, 1937, **19**, 805, 1027 and 1683.

Bruns, E. *AFEWM*, 1931, **123**, 682.

Brunst, V.V. *BBMEUSSR*, 1938, **6**, 527 and 530.

Brunst, V.V. & Scheremetieva, E.A. (1). *AFEWM*, 1933, **128**, 181; *AZEG*, 1936, **78**, 57.

Brunst, V.V. & Scheremetieva, E.A. (2). *BBMEUSSR*, 1937, **3**, 397.

Bruynoghe, G. *CRSB*, 1934, **115**, 441.

Bucciante, L. (1). *AEZF*, 1928, **5**, 1.

Bucciante, L. (2). *AFEWM*, 1929, **115**, 396.

Bucciante, L. (3). *AEZF*, 1931, **11**, 397; 1933, **14**, 56.

Bucciante, L. (4). *MZI*, 1931, **42**, 243; *YMaly*, 1932, **20**, 660.

Bucciante, L. (5). *AAL*, 1931 (ser. 6ᵉ), **14**, 356; *YMaly*, 1932, **21**, 275.

Buchanan, J.W. *JEZ*, 1938, **79**, 109.

Buckner, G.D., Martin, J.H. & Peter, A.M. *AJOP*, 1924, **71**, 543.

Buckner, G.D. & Peter, A.M. *JBC*, 1922, **54**, 5.

Budde, M. *BPAAP*, 1926, **75**, 357.

Bumm, E. & Appel, H. *ZPC*, 1932, **210**, 79.

Bumm, E. & Fehrenbach, K. *ZPC*, 1930, **193**, 238; 1931, **195**, 101.

Burch, A.B. *PSEBM*, 1938, **38**, 608; 1939, **40**, 341.

Burdach, F.G. "Experimenta quaedam de commutatione substantiarum proteinacearum in adipem." Inaug. Diss. Königsberg, 1853.

Burdick, H.C. *PZ*, 1937, **10**, 156.

Burdon, E.R. *JECB*, 1908 **2**, 119.

Burk, D. *OPAAA*, 1937, **4**, 121.

Burkholder, J.R. *PZ*, 1934, **7**, 247; *PSEBM*, 1933, **30**, 1113.

Burmester, B.R. *JEZ*, 1940, **84**, 445.

Burmester, B.R. & Card, L.E. *PS*, 1939, **18**, 138.

Burnet, F.M. "The Use of the developing Egg in Virus Research." *SRMRC*, 1936, No. 220.

Burns, C.M. & Henderson, N. *BJ*, 1936, **30**, 1202 and 1207.

Burns, R.K. (1). *AR*, 1935, **61** (2nd suppl.), 8.

Burns, R.K. (2). *AR*, 1938, **71**, 447.

Burns, R.K. (3). *AR*, 1939, **73**, 73.

Burns, R.K. (4). *AN*, 1938, **72**, 207.

Burrows, H. (1). *BJS*, 1935, **23**, 191.

Burrows, H. (2). *JPB*, 1935, **41**, 43, 218, 423; 1936, **42**, 161; *BJS*, 1934, **21**, 507; 1936, **23**, 191, 658; *JOP*, 1935, **85**, 159; *PRSB*, 1935, **118**, 485; *N*, 1934, **134**, 570; *AJC*, 1935, **23**, 490.

Burrows, H. & Kennaway, N.M. *AJC*, 1934, **20**, 48.

Burrows, H. & Mayneord, W.V. *AJC*, 1937, **31**, 484.

Burrows, M.T. *PSEBM*, 1920, **18**, 133.

Burton, A.C. *JCCP*, 1936, **9**, 1.

Burtt, E.T. *PRSB*, 1937, **124**, 13; 1938, **126**, 210.

Busnel, R.G. *CRAS*, 1937, **205**, 1177.

Buşniţa, T. & Gavrilescu, N. *BSSAR*, 1932, **15**, 208; *CA*, 1934, **28**, 1109.

Busse, O. *ZPC*, 1936, **242**, 271.

Busson, A. & Simonnet, H. *CRSB*, 1931, **109**, 1253.

Butenandt, A., Weidel, W. & Becker, E. *NW*, 1940, **28**, 63.

Butler, E.G. (1). *JEZ*, 1933, **65**, 271; *AR*, 1935, **62**, 295; *SM*, 1934, **39**, 511.

Butler, E.G. (2). "Effects of Radium and X-rays on embryonic Development" in "Biological Effects of Radiation." New York, 1936.

Butler, E.G. & Puckett, W.O. *JEZ*, 1940, **84**, 223.

Bütschli, O. *SAWH*, 1915, **4**.

Buxton, P.A. (1). "Animal Life in Deserts." London, 1923, pp. 84 and 97.

Buxton, P.A. (2). *BR*, 1932, **7**, 275.

du Buy, H.G. & Olson, R.A. *AJB*, 1937, **24**, 609.

Buzzi, B. *RIG*, 1932, **13**, 549; *OMaly*, 1933, **69**, 384.

Byerly, T.C. (1). *AR*, 1926, **32**, 249; 1926, **33**, 319.

Byerly, T.C. (2). *Proc. IVth World Poultry Congress (London)*, 1930, p. 174.

Byerly, T.C. (3). *JM*, 1930, **50**, 341.

Byerly, T.C. (4). *JEB*, 1932, **9**, 15.

Byerly, T.C. (5). *Proc. Vth World Poultry Congress (Rome)*, 1934, **2**, 373.

Byerly, T.C. (6). *PS*, 1938, **17**, 200.

Byerly, T.C., Helsel, W.G. & Quinn, J.P. *JEZ*, 1938, **78**, 185.

Byerly, T.C. & Jull, M.A. (1). *JEZ*, 1932, **62**, 489.

Byerly, T.C. & Jull, M.A. (2). *PS*, 1935, **14**, 217.

Byerly, T.C. & Olsen, M.W. (1). *PS*, 1931, **10**, 281.

Byerly, T.C. & Olsen, M.W. (2). *S*, 1934, **80**, 247.

Byerly, T.C. & Olsen, M.W. (3). *PS*, 1936, **15**, 158 and 163.

Byerly, T.C., Titus, H.W. & Ellis, N.R. (1). *JN*, 1933, **6**, 225.

Byerly, T.C., Titus, H.W. & Ellis, N.R. (2). *JAR*, 1933, **46**, 1.

Byerly, T.C., Titus, H.W., Ellis, N.R. & Landauer, W. *PSEBM*, 1935, **32**, 1542.

Bytinski-Salz, H. (1). *AFEWM*, 1933, **129**, 356.

Bytinski-Salz, H. (2). *JEZ*, 1935, **72**, 51.

Bytinski-Salz, H. (3). *Comptes Rend. XII Congrès Int. de Zool. Lisbon*, 1936, p. 595.

Bytinski-Salz, H. (4). *AIAE*, 1937, **39**, 178; *RS*, 1937, **2**, 200; *YMaly*, 1938, **45**, 468.

Bytinski-Salz, H. & Gunther, A. *ZIAV*, 1930, **53**, 153.

Caffier, P. & Ammon, R. *CG*, 1936, 7; *OMaly*, 1936, **92**, 480.

Cagliotti, V. & Gigante, D. *AAL*, 1936 (6e ser.), **23**, 878; *CA*, 1937, **31**, 7501.

Cahn, T. & Bonot, A. *APBPC*, 1928, **4**, 399 and 435.

Califano, L. *RAL*, 1937 (ser. 6e), **25**, 93.

Calkins, G.N. "Biology of the Protozoa." London, 1926.

Caló, A. *AC*, 1935, **1**, 1; *Communications de la 2e Congr. Internat. de la lutte contre le Cancer*, 1936, p. 116.

Calvery, H.O. (1). *JBC*, 1928, **77**, 489.

Calvery, H.O. (2). *JBC*, 1932, **94**, 613.

Calvery, H.O. (3). *JBC*, 1932, **95**, 297.

Calvery, H.O. (4). *JBC*, 1933, **100**, 183.

Calvery, H.O. (5). *JBC*, 1933, **102**, 73.

Calvery, H.O. (6). *JBC*, 1935, **112**, 171.

Calvery, H.O., Block, W.D. & Schock, E.D. *JBC*, 1936, **113**, 21.

Calvery, H.O. & Schock, E.D. *JBC*, 1936, **113**, 15.

Calvery, H.O. & Titus, H.W. *JBC*, 1934, **105**, 683; *Proc. XIVth Internat. Congr. Physiol. Rome*, 1932, p. 45.

Calvery, H.O. & White, A. *JBC*, 1932, **94**, 635.

Calzoni, M. (1). *RDB*, 1930, **12**, 14; *OMaly*, 1930, **60**, 389.

Calzoni, M. (2). *PSZN*, 1935, **15**, 169.

Cameron, J.A. (1). *JM*, 1936, **60**, 279.

Cameron, J.A. (2). *JM*, 1941, **68**, 231.

Campbell, J.A. *JOP*, 1936, **87**, 68P.

Campbell, N.R. *PRSB*, 1940, **129**, 527.

Campbell, N.R., Dodds, E.C. & Lawson, W. *N*, 1938, **141**, 78; 1938, **142**, 1121; *PRSB*, 1940, **128**, 253.

Canevazzi, O. *AIAE*, 1940, **43**, 187.

Cannan, R.K. *BJ*, 1927, **21**, 184.

Cannon, H.G. *PRSB*, 1923, **94**, 232.

Cantarov, A., Stuckert, H. & Davis, R.C. *SGO*, 1933, **57**, 63.

Capraro, V. & Fornaroli, P. *ASB*, 1939, **25**, 117.

Card, L.E. (1). *PS*, 1921, **1**, 1.

Card, L.E. (2). *PS*, 1929, **8**, 328.

Card, L.E. (3). *PS*, 1932, **11**, 339.

Card, L.E., Mitchell, H.H. & Hamilton, T.S. *PS*, 1930, **8**, 328; *Proc. Internat. Poultry Congress*, 1930, **4**, 301.

Cardin, A. *ASB*, 1933, **19**, 69 and 76.

Cardot, H. *CRSB*, 1924, **91**, 1019.

Carl, W. *CAP*, 1913, **24**, 436.

Carlson, L.D. & Bodine, J.H. *JCCP*, 1939, **14**, 159.

Carmichael, E. B. & Rivers, T. D. *E*, 1932, **13**, 375.

Carmichael, J. *N*, 1933, **131**, 878.

Carmichael, L. *PSR*, 1926, **33**, 51.

Carnap, R. (1). *EK*, 1931, **2**, 219.

Carnap, R. (2). "The Unity of Science", tr. M. Black. Psyche Miniature Series, Kegan Paul, London, 1934, pp. 68 ff.

Carnot, P. *CRSB*, 1925, **94**, 637.

Carnot, P. & Terriss, E. *CRSB*, 1926, **95**, 655.

Carpenter, E. *JEZ*, 1937, **75**, 103.

Carr, R.H. & James, C.M. *AJOP*, 1931, **97**, 227.

Carrel, A. (1). *JEM*, 1913, **17**, 14.

Carrel, A. (2). *S*, 1931, **74**, 618.

Carrel, A. & Baker, L.E. *JEM*, 1926, **44**, 503.

Carrel, A. & Ebeling, A.H. *JEM*, 1921, **34**, 599.

Carrel, A. & Lindbergh, C.A. "The Culture of Organs." New York, 1938.

Carrié, C. "Die Porphyrine." Thieme, Leipzig, 1936.

Carrié, C. & v. Mallinkrodt-Haupt, A. *AFDS*, 1934, **170**, 521.

Carroll, M.J. *AEZF*, 1939, **22**, 592.

Carter, G.S. *JEB*, 1930, **7**, 41; 1931, **8**, 176 and 194; 1932, **9**, 238, 249, 253, 264 and 378.

Carter, G.S. & Beadle, L.C. *JLSZ*, 1930, **37**, 197.

Carver, J.S., Robertson, E.I., Brazie, D., Johnson, R.H. & St John, J.L. *Bull. Washington Agric. Exp. Sta.* 1934, No. 299.

Caspari, E. (1). *AFEWM*, 1933, **130**, 353.

Caspari, E. (2). *JEZ*, 1941, **86**, 321.

Caspari, E. & Plagge, E. *NW*, 1935, **23**, 751.

Caspersson, T. *JRMS*, 1940, **60**, 8.

Caspersson, T. & Schultz, Jack (1). *N*, 1938, **142**, 294.

Caspersson, T. & Schultz, Jack (2). *N*, 1939, **143**, 602.

Caspersson, T. & Schultz, Jack (3). *PNASW*, 1940, **26**, 507.

Castagna, S. & Talenti, M. *AFS*, 1933, **55**, 28; *YMaly*, 1933, **26**, 82.

Castelnuovo, G. *BDZ*, 1932, **3**, 291; *YMaly*, 1933, **25**, 192.

Castigli, G. *PSZN*, 1939, **17**, 183.

Castle, G.B. *S*, 1934, **80**, 314.

Castle, W.B. *Harvey Lectures*, 1935, **30**, 37.

Castle, W.B., Townsend, W.C. & Heath, C.W. *AJMS*, 1929, **178**, 748 and 764; 1930, **180**, 305.

Castle, W.E. *JH*, 1933, **24**, 81.

Castle, W.E. & Gregory, P.W. *JM*, 1929, **48**, 81.

Castle, W.E. & Little, C.C. *S*, 1910, **32**, 868.

Castle, W.E. & Wright, S. *CIWP*, 1916, No. **241**.

Cattaneo, L. (1). *AO*, 1931, **53**, 407; *YMaly*, 1932, **20**, 481.

Cattaneo, L. (2). *AO*, 1931, **53**, 1755; *YMaly*, 1932, **23**, 133.

Cattaneo, L. (3). *AIB*, 1932, **86**, 1 and 33; *AO*, 1931, **53**, 253; *OMaly*, 1931, **62**, 628.

Cattaneo, L. (4). *AIB*, 1932, **87**, 199; 1932, **88**, 39; *Proc. XIVth Internat. Congr. Physiol. (Rome)*, 1932, 49.

Cattaneo, L. (5). *ASB*, 1933, **18**, 372; *YMaly*, 1934, **27**, 669.

Caullery, M. (1). *BBFB*, 1895, **27**, 1.

Caullery, M. (2). *RSZ*, 1936, **43**, 467.

Caullery, M. (3). "Les Progrès récents de l'Embryologie expérimentale." Flammarion, Paris, 1939.

Cavers, J.R. & Hutt, F.B. *JAR*, 1934, **48**, 517.

Cazzaniga, A. *BDZ*, 1935, **6**, 83; *YMaly*, 1935, **34**, 138; *OMaly*, 1935, **88**, 376.

Centanni, G. *RIG*, 1931, **12**, 479; *YMaly*, 1932, **20**, 275.

Chabry, L. (1). *JDAP*, 1887, **23**.

Chabry, L. (2). *CRSB*, 1888, **40**, 589.

Chaigne, M. Inaug. Diss. Bordeaux, 1934; *CRSB*, 1933, **114**, 1103; 1933, **115**, 174.

Chaikov, I.L. & Robinson, A. *JBC*, 1933, **100**, 13.

Chaletzkaia, P.M. (1). *ABSUSSR*, 1935, **38**, 631; 1936, **40**, 101 and 107.

Chaletzkaia, P.M. (2). *BBMEUSSR*, 1938, **6**, 387.

Chalkley, H.W. (1). *USPHR*, 1931, **46**, 1736.

Chalkley, H.W. (2). *P*, 1937, **28**, 489.

Chalkley, H.W. & Voegtlin, C. *USPHR*, 1932, **47**, 535; *JNCI*, 1940, **1**, 63.

Chalmers, J.G. *BJ*, 1934, **28**, 1214.

Chalmers, J.G. & Peacock, P.R. *BJ*, 1936, **30**, 1242.

Chambers, R. (1). *JGP*, 1923, **5**, 821; *BB*, 1930, **58**, 344.

Chambers, R. (2). *JCCP*, 1931, **1**, 65.

Chambers, R. (3). *CSHSQB*, 1933, **1**, 205.

Chambers, R, (4). *AN*, 1938, **72**, 141; *JCCP*, 1938, **12**, 149.

Chambers, R., Beck, L.V. & Belkin, M. *JCCP*, 1935, **6**, 425.

Chambers, R. & Cameron, G. *JCCP*, 1932, **2**, 99.

Chambers, R. & Kempton, R.T. *JCCP*, 1933, **3**, 131.

Chambers, R. & Kopać, M.J. (1) *JCCP*, 1937, **9**, 331 and 345.

Chambers, R. & Kopać, M. J. (2). Private communication to the author. Sept. 1940.

Champetier, G. & Fauré-Fremiet, E. *JCP*, 1937, **34**, 197; 1938, **35**, 223; *CRAS*, 1937, **204**, 1901.

Champy, C. (1). *AMGE*, 1922, **4**, 1.

Champy, C. (2). "Sexualité et Hormones." Paris, 1924.

Champy, C. & Champy, Ch. *CRSB*, 1935, **118**, 861; *BAEC*, 1935, **24**, 206.

Champy, C., Lavedan, J.P. & Marquez, J. *AZEG*, 1939, **80**, 389.

Ch'ang, Yung-Tai (Tchang). *BBFB*, 1929, **12** (Suppl.), 1.

Chang, Hsi-Chun (1). *CJOP*, 1933, **7**, 171.

Chang, Hsi-Chun (2). *PSEBM*, 1935, **32**, 1001; 1936, **34**, 665; *CJOP* (*Proc. Chinese Physiol. Soc.*), 1934, 1935, 1936.

Chang, Hsi-Chun (3). *CJOP*, 1938, **13**, 145.

Chang, Hsi-Chun & Gaddum, J.H. *JOP*, 1933, **79**, 255.

Chang, Hsi-Chun, Wen, I-Chuan & Wong, Amos. *Proc. XVth Int. Congr. Physiol.* (*Moscow*), 1935, 54.

Chang, Hsi-Chun & Wong, Amos. *CJOP*, 1933, **7**, 151.

Chang, Tso-Kan (1). *PNHB*, 1940, **14**, 119.

Chang, Tso-Kan (2). *PNHB*, 1940, **14**, 159.

Channon, H.J. *BJ*, 1926, **20**, 400.

Channon, H.J. & Saby. *BJ*, 1932, **26**, 2021.

Chanutin, A. *JBC*, 1931, **93**, 31.

Chapheau, M. *Inaug. Diss. Bordeaux.*

Chapman, R.N. *JEZ*, 1926, **45**, 292.

Chapman, S.S. *G*, 1937, **1**, 299.

Charikova, A.F. & Mikhailov, W.A. *TIPN USSR*, 1936, **2**, 464; *YMaly*, 1937, **42**, 373.

Charipper, H.A. & Corey, E.L. *AR*, 1930, **45**, 258.

Charles, D.R. & Rawles, M.E. *PSEBM*, 1940, **43**, 55.

Charleton, Walter. "A Ternary of Paradoxes; the Magnetick Cure of Wounds, Nativity of tartar in wine, Image of God in Man; written originally by Joh. Bapt. van Helmont; translated, illustrated, and ampliated by W.C." London, 1650.

Chase, H.B. & Chase, E.B. *JM*, 1941, **68**, 279.

Chatton, E., Lvov, A. & Rapkine, L. *CRSB*, 1931, **106**, 626.

von Chauvin, M. *ZWZ*, 1885, **41**, 365.

Chaves, P.R. *CRSB*, 1920, **83**, 879.

Chen, G., Oldham, F.K. & Geiling, E.M.K. *PSEBM*, 1940, **45**, 810.

Chen, K'o-K'uei & Chen, Amy-Ling. *JPET*, 1933, **47**, 295.

Chen, K'o-K'uei, Chen, Amy-Ling & Anderson, R.C. *JAPA*, 1936, **25**, 579.

Chen, Tse-Yin. *JM*, 1929, **47**, 136.

Ch'êng, T.H. *PNHB*, 1933, **7**, 1.

Chesley, L. *BB*, 1934, **67**, 259.

Chesley, P. *PSEBM*, 1932, **29**, 437; *JEZ*, 1935, **70**, 429.

Chesley, P. & Dunn, L.C. *GS*, 1936, **21**, 525.

Chevais, S. (1). *AAM*, 1937, **33**, 107.

Chevais, S. (2). *CRSB*, 1938, **127**, 1005.

Chevais, S., Ephrussi, B. & Steinberg, A.G. *PNASW*, 1938, **24**, 365.

Chiarugi. *MZI*, 1929, **40**, 146.

Chick, H., Copping, A.M. & Roscoe, M.H. *BJ*, 1931, **24**, 1748.

Chievitz, O. & Hevesy, G. *BMKDVS*, 1937, **13**, no. 9; *YMaly*, 1937, **43**, 268; *N*, 1935, **136**, 754.

Child, C.M. (1). "Individuality in Organisms." Chicago University Press, 1915.

Child, C.M. (2). *AJOP*, 1915, **37**, 302; *JM*, 1916, **28**, 65; *BB*, 1916, **30**, 391.

Child, C.M. (3). *JM*, 1927, **44**, 467.

Child, C.M. (4). *P*, 1928, **5**, 447; *AFEWM*, 1929, **117**, 21.

Child, C.M. (5). *PSEBM*, 1934, **32**, 34; *P*, 1934, **22**, 377.

Child, C.M. (6). *AFEWM*, 1936, **135**, 426.

Child, C.M. (7). *AFEWM*, 1936, **135**, 457; *PZ*, 1940, **13**, 4.

Child, C.M. & Rulon, O. *JEZ*, 1936, **74**, 427.

Child, C.M. & Watanabe, Y. *PZ*, 1935, **8**, 395.

Child, G. *GS*, 1935, **20**, 109 and 127; 1936, **21**, 808; 1940, **25**, 354.

Chlopin, N.G. *AFEWM*, 1932, **126**, 69.

Choi, M.H. *FAJ*, 1932, **10**, 29.

Chomković, G. *ACZAA*, 1928, **3A**, 1.

Ch'ou Su (Tchou-Su) (1). *AAM*, 1931, **27**, 1; *CRAS*, 1936, **202**, 242.

Ch'ou Su (Tchou-Su) (2). *CRAS*, 1938, **207**, 599.

Chouke, K.S. *PSEBM*, 1930, **28**, 43.

Christoffersen, A.K. *CRSB*, 1934, **117**, 641.

Chuang, Hsiao-Hui (1). *BC*, 1938, **58**, 472; *AFEWM*, 1939, **139**, 556.

Chuang, Hsiao-Hui (2). *AFEWM*, 1940, **140**, 25.

Chubb, G.C. *PTRSB*, 1906, **198**, 447.

Chun, C. Leuckart-Festschrift, 1892.

Ciaccio, G. *MZI*, 1935, **45** (Suppl.), 295.

Clark, F.H. *PNASW*, 1932, **18**, 654; 1935, **21**, 150; *AR*, 1934, **58**, 225.

Clark, G.A. (1). *JOP*, 1932, **74**, 391.

Clark, G.A. (2). *JOP*, 1934, **83**, 229.

Clark, G.A. & Holling, H.E. *JOP*, 1932, **73**, 305.

Clarke, M.F. & Smith, A.H. *JN*, 1938, **15**, 245.

Clauberg, C. "Abhdl. a. d. Geb. d. Inn. Sekretion." Ed. W. Berblinger. Barth, Leipzig, 1937. *OMaly*, 1938, **104**, 455.

Claude, A. (1). *JEM*, 1935, **61**, 27 and 33; 1937, **66**, 59; *AJC*, 1937, **30**, 742; 1939, **37**, 59; *S*, 1937, **85**, 294; 1938, **87**, 467; 1939, **90**, 213.

Claude, A. (2). *PSEBM*, 1938, **39**, 398.

Claude, A. (3). *S*, 1940, **91**, 77.

Claude, A. (4). *JEM*, 1939, **69**, 641.

Claude. A. & Rothen, A. *JEM*, 1940, **71**, 619.

Clausen, H.J. *AR*, 1929, **44**, 218; *BB*, 1930, **59**, 199.

Clausen, S.W. & McCoord, A.B. *JBC*, 1937, **119**, xviii.

van Cleave, C.D. *PZ*, 1938, **11**, 168.

van Cleave, H.J. *QRB*, 1932, **7**, 59.

Clement, A.C. *JEZ*, 1938, **79**, 435.

Clementi, A. (1). *AIB*, 1932, **86**, 70.

Clementi, A. (2). *EZ*, 1937, **4**, 205.

Cloete, J.H.L. *OJVSAI*, 1939, **13**, 417.

Clowes, G.H.A. & Bachman, E. *PSEBM*, 1921, **18**, 120; *JBC*, 1921, **46**, xxxi.

Clowes, G.H.A., Davis, W.W. & Krahl, M.E. *AJC*, 1939, **36**, 98; 1939, **37**, 453.

Clowes, G.H.A., Keltch, A.K. & Krahl, M.E. (abstrs.) *BB*, 1937, **73**, 359, 375, 376.

Clowes, G.H.A. & Krahl, M.E. (1). "Eli Lilly Research Laboratories Dedication Volume." Indianapolis, 1934, p. 73.

Clowes, G.H.A. & Krahl, M.E. (2). *BB*, 1934, **67**, 332.

Clowes, G.H.A. & Krahl, M.E. (3). *S*, 1934, **80**, 384; *JGP*, 1936, **20**, 145.

Clowes, G.H.A. & Krahl, M.E. (4). *JGP*, 1940, **23**, 401.

Cochrane, F. (1). *PRSE*, 1937, **57**, 385; *JG*, 1936, **32**, 183; 1938, **36**, 1.

Cochrane, F. (2). *JG*, 1938, **36**, 11.

Coghill, G.E. (1). "Anatomy and the Problem of Behaviour." Cambridge University Press, 1929.

Coghill, G.E. (2). *PSEBM*, 1936, **35**, 71.

Coghill, G.E. (3). "Early embryonic somatic movements in birds and mammals other than man", Monogr. for the Study of Child Devel. 1940, **5**, no. 2; *S*, 1933, **78**, 131; 1938, **88**, 351; *AJOPS*, 1938, **51**, 759; *JGPS*, 1930, **3**, 431.

Cohen, A. (1). *JEZ*, 1938, **79**, 461.

Cohen, A. (2). *BB*, 1938, **75**, 340; *CN*, 1938, **13**, 87.

Cohen, A. & Berrill, N.J. *BB*, 1936, **70**, 78.

Cohen, B. *CSHSQB*, 1933, **1**, 195 and 214.

Cohn, A.E., Murray, H.A. & Rosenthal, A. *JEM*, 1925, **42**, 275.

Cohn, L. *AA*, 1904, **24**.

Coldwater, K.B. (1). *PSEBM*, 1930, **27**, 1031.

Coldwater, K.B. (2). *JEZ*, 1933, **65**, 43.

Cole, F.J. "Early Theories of Sexual Generation." Oxford, 1930.

Cole, H.H., Hart, G.H., Lyons, W.R. & Catchpole, H.R. *AR*, 1933, **56**, 275.

Cole, K.S. (1). *JGP*, 1928, **12**, 29 & 37; 1935, **18**, 877.

Cole, K.S. (2). *JCCP*, 1932, **1**, 1.

Cole, K.S. (3). *TFS*, 1937, **33**, 966.

Cole, K.S. (4). *N*, 1938, **141**, 79.

Cole, K.S. & Cole, R.H. *JGP*, 1936, **19**, 609 and 625.

Cole, K.S. & Curtis, H.J. *JGP*, 1938, **21**, 591; (abstr.) *BB*, 1937, **73**, 362.

Cole, K.S. & Jahn, T. *JCCP*, 1937, **10**, 265.

Cole, K.S. & Michaelis, E.M. *JCCP*, 1932, **2**, 105.

Cole, K.S. & Spencer, J.M. *JGP*, 1938, **21**, 583; (abstr.) *BB*, 1937, **73**, 362.

Cole, L.J. & Ibsen, H.L. *AN*, 1920, **54**, 130.

Cole, L.J. & Rodolfo. *PASAP*, 1924, 116.

Cole, R.K. *PS*, 1937, **16**, 356; *AR*, 1938, **71**, 349.

Cole, W.H. *JEZ*, 1922, **35**, 383; *PNASW*, 1922, **8**, 29.

Coleridge, S.T. "The Theory of Life" (1848). Ed. T. Ashe. Bell, London, 1885.

Collip, J.B. *JMSHNY*, 1934, **1**, 28; *CSHSQB*, 1937, **5**, 210; *EMJ*, 1938, **45**, 782.

Collip, J.B. & Anderson, E.M. *JOP*, 1934, **82**, 11; *JAMA*, 1935, **104**, 965.

Collip, J.B., Thomson, D.L., Browne, J.S.L., McPhail, M.K. & Williamson, J.E. *ENY*, 1931, **15**, 315.

Colucci, V.L. *MASIB*, 1891.

Comandon, J. & de Fonbrune, P. (1). *CRSB*, 1931, **106**, 181; *YMaly*, 1931, **18**, 548.

Comandon, J. & de Fonbrune, P. (2). *CRSB*, 1931, **106**, 248.

Comandon, J. & de Fonbrune, P. (3). *AAM*, 1935, **31**, 79.

Comes, O.C. (1). *BSIBS*, 1934, **9**, 1256.

Comes, O.C. (2). *PSZN*, 1936, **15**, 339.

Common, R.H. *JAS*, 1932, **22**, 576.

Commoner, B. *JCCP*, 1938, **12**, 171; *BB*, 1937, **73**, 383.

Conklin, E.G. (1). *JEZ*, 1905, **2**, 185; *AFEWM*, 1906, **21**, 727.

Conklin, E.G. (2). *JEZ*, 1917, **22**, 311.

Conklin, E.G. (3). *JEZ*, 1931, **60**, 2.

Conklin, E.G. (4). *AN*, 1933, **67**, 289.

Conklin, E.G. (5). "Cellular Differentiation" in Cowdry's "General Cytology", 1933, p. 562.

Conrad, R.M. & Phillips, R.E. *PS*, 1938, **17**, 143.

Conrad, R.M. & Scott, H.M. *PR*, 1938, **18**, 481.

Contardo, G.B. *BSIBS*, 1931, **6**, 766 and 770; *OMaly*, 1932, **66**, 120; **67**, 161.

Cook, E.S., Hart, M.J. & Joly, R.A. *AJC*, 1939, **35**, 543.

Cook, J.W. (1). *PRSB*, 1933, **113**, 277.

Cook, J.W. (2). *N*, 1934, **134**, 758; *BDCG*, 1936, **69**, 38; *BSCF*, 1937, 792.

Cook, J.W. (3). *EVHF*, 1938, **2**, 213.

Cook, J.W. & Dodds, E.C. *N*, 1935, **135**, 793 and 959.

Cook, J.W., Dodds, E.C. & Greenwood, A.W. *PRSB*, 1934, **114**, 286.

Cook, J.W., Dodds, E.C., Hewett, C.L. & Lawson, W. *PRSB*, 1934, **114**, 272.

Cook, J.W., Dodds, E.C. & Lawson, W. *PRSB*, 1936, **121**, 133.

Cook, J.W., Dodds, E.C. & Warren, F.L. *N*, 1935, **136**, 912.

Cook, J.W. & Haslewood, G.A.D. *CI*, 1933, **38**, 758; *JCS*, 1934, 428; 1935, 767 and 770.

Cook, J.W., Haslewood, G.A.D., Hewett, C.L., Hieger, I., Kennaway, E.L. & Mayneord, W.V. *Rep. IInd Internat. Cancer Congr.* 1936, **1**, 1; also Supplement in *AJC*, 1937, **29**, 219.

Cook, J.W., Hewett, C.L. & Hieger, I. *JCS*, 1933 (i), 395.

Cook, J.W. & Kennaway, E.L. *AJC*, 1938, **33**, 50; 1940, **39**, 381 and 521.

Cook, J.W., Kennaway, E.L. & Kennaway, N.M. *N*, 1940, **145**, 627.

Cook, R.P. & Stephenson, M. *BJ*, 1928, **22**, 1368.

Cooper, K.W. *PNASW*, 1936, **22**, 433.

Copenhaver, W.M. *PNASW*, 1927, **13**, 484; *JEZ*, 1930, **55**, 293; 1933, **65**, 131.

Coppini, R. *MZI*, 1937, **48**, 106; *YMaly*, 1937, **44**, 12.

Corda, G.M. *AO*, 1932, **54**, 29; *OMaly*, 1932, **69**, 167.

Corey, E.L. (1). *PZ*, 1932, **5**, 36.

Corey, E.L. (2). *AR*, 1933, **56**, 195.

Corey, E.L. (3). *AJOP*, 1935, **112**, 263.

Corey, E.L. (4). *AJOP*, 1935, **113**, 450.

Cori, C.F. *JEM*, 1927, **45**, 983.

Cori, M. *PSZN*, 1936, **15**, 368.

Corner, G.W. (1). *CCE*, 1921, **13**, 63.

Corner, G.W. (2). *AJOP*, 1928, **86**, 74.

Corner, G.W. (3). *PR*, 1938, **18**, 154.

Corradi, A., "Dell' antica autoplastica Italiana" in Atti d. r. Istit. Lombardo, 1874, p. 226.

Correll, J.T. & Hughes, J.S. *JBC*, 1933, **103**, 511.

Costello, D.P. (1). *PZ*, 1939, **12**, 13.

Costello, D.P. (2). *JEZ*, 1939, **80**, 473.

Cotronei, G. & Guareschi, C. *RAL*, 1930, **12**, 180; 1931, **13**, 44 & 368.

Cotronei, G. & Perri, T. *AAL*, 1934 (6ᵉ ser.), **20**, 346.

Cotronei, G. & Spirito, A. *AAL*, 1929 (6ᵉ ser.), **10**, 212.

Courtis, S.A. *G*, 1937, **1**, 155.

Cousin, G. *BBFB*, 1933 (Suppl.), **15**, 1.

Couteaux, R. & Nachmansohn, D. *N*, 1938, **142**, 481.

Cox, W.M. & Imboden, M. *JN*, 1936, **11**, 177.

Cozzi, L. *AOG*, 1932, **39**, 61.

Crabtree, H.G. (1). *SRICS*, 1932, **10**, 33.

Crabtree, H.G. (2). *BJ*, 1936, **30**, 1622.

Crabtree, H.G. (3). *BJ*, 1937, **30**, 2140.

Crabtree, H.G. (4). *JPB*, 1940, **51**, 303.

Crabtree, H.G. & Cramer, W. *PRSB*, 1933, **113**, 226 and 238.

Cramer, W. (1). *CAR*, 1932, 241.

Cramer, W. (2). *BMJ*, 1938 (i), 829.

Crampton, H.E. *AFEWM*, 1896, **3**, 1.

Creach, P. *CRSB*, 1934, **117**, 367.

Creech, H.J. & Franks, W.R. *AJC*, 1937, **30**, 555; 1939, **35**, 203.

Crescitelli, F. *JCCP*, 1935, **6**, 351.

Crescitelli, F. & Taylor, I.R. *JBC*, 1935, **108**, 349.

Crew, F.A.E. (1). *PRSB*, 1923, **95**, 228.

Crew, F.A.E. (2). *BMJ*, 1936 (i), 993.

Crew, F.A.E. & Auerbach, C. (1). *PRSE*, 1937, **57**, 255.

Crew, F.A.E. & Auerbach, C. (2). *JG*, 1941, **41**, 267.

Crew, F.A.E. & Kon, S.K. *JG*, 1933, **28**, 25.

Crew, F.A.E. & Mirskaia, L. *JG*, 1931, **25**, 17.

le Cron, W.L. *AJA*, 1906, **6**.

Croone, W. *PTRS*, 1672, **7**, 5080.

Crosman, A.M. *PSEBM*, 1936, **34**, 360.

Crowther, C. & Raistrick, H. *BJ*, 1916, **10**, 434.

Crozier, W.J. & Enzmann, E.V. *JGP*, 1935, **19**, 249.

Cruickshank, E.M. *BJ*, 1934, **28**, 965.

Cruickshank, E.M. & Moore, T. *BJ*, 1937, **31**, 179.

Cuénot, L. (1). *AZEG*, 1902 (3ᵉ sér.), **10**, Notes and Rev. xxvii; 1903 (4ᵉ sér.), **1**, Notes and Rev. xxxiii; 1907 (4ᵉ sér.), **6**, Notes and Rev. i; 1911 (5ᵉ sér.), **8**, Notes and Rev. xl.

Cuénot, L. (2). *AZEG*, 1905 (4ᵉ sér.), **3**, cxxiii; 1908 (4ᵉ sér.), **9**, vii.

Cuénot, L. (3). *CRAS*, 1921, **172**.

Cummings, F. Inaug. Diss. Yale, 1938.

da Cunha, D.P. *CRSB*, 1937, **124**, 1023.

da Cunha, D.P. & Jacobsohn, K.P. *CRSB*, 1936, **123**, 609.

Cunningham, B. *S*, 1934, **80**, 99.

Cunningham, B. & Huene, E. *AN*, 1938, **72**, 380.

Cunningham, B. & Hurwitz, A.P. *AN*, 1936, **70**, 590.

Cunningham, B., Woodward, M.W. & Pridgeon, J. *AN*, 1939, **73**, 285.

Cunningham, J.T. *PRSB*, 1929, **105**, 484.

Cunningham, J.T. & Reid, D.M. (1). *PRSB*, 1932, **110**, 234.

Cunningham, J.T. & Reid, D.M. (2). *N*, 1933, **131**, 913.

Curry, H.A. *RSZ*, 1931, **38**, 401; *AFEWM*, 1936, **134**, 694.

Curtis, W.C. "Effects of X-Rays and Radium on Regeneration" in "The Biological Effects of Radiation." McGraw-Hill, New York, 1936.

Curtis, W.C., Cameron, J.A. & Mills, K.O. *S*, 1936, **83**, 354.

Cutler, I.E. *JH*, 1925, **16**, 352.

Cutting, C.C. & Tainter, M.L. *PSEBM*, 1934, **31**, 97.

Dąbrowska, W. *CRSB*, 1932, **110**, 1091; *MINPERP*, 1932, **13**, 276.

Daineko, L.N. (1). *BBMEUSSR*, 1938, **6**, 97.

Daineko, L.N. (2). *BBMEUSSR*, 1939, **8**, 301.

Dakin, H.D. & Dudley, H.W. *JBC*, 1913, **15**, 2.

Dakin, H.D., Ungley, C.C. & West, R. *JBC*, 1936, **115**, 771.

Dakin, H.D. & West, R. *JBC*, 1935, **109**, 489; *JCI*, 1935, **14**, 708.

Dalcq, A. (1). *AAM*, 1929, **25**, 336; *CRSB*, 1930, **104**, 1055.

Dalcq, A. (2). *AAM*, 1933, **29**, 389.

Dalcq, A. (3). *CRSB*, 1935 (Réun. Plen.), **119**, 1420.

Dalcq, A. (4). "L'Organisation de l'Œuf chez les Chordés." Gauthier-Villars, Paris, 1935.

Dalcq, A. (5). *BARMB*, 1936, 495.

Dalcq, A. (6). *ASRZB*, 1937, **68**, 69.

Dalcq, A. (7). "Form and Causality in early Development." Cambridge University Press, 1938.

Dalcq, A. (8). *CRAA* (Basel meeting), 1938, p. 1.

Dalcq, A. (9). *ADB*, 1938, **49**, 397; *TSZW*, 1938, **13**, 101.

Dalcq, A. (10). *ASRZB*, 1938, **69**, 97.

Dalcq, A. (11). *ABSRSMNB*, 1938, 1.

Dalcq, A. (12). *BME*, 1939, no. 28, May 14.

Dalcq, A. & Pasteels, J. *ADB*, 1937, **48**, 669; *BARMB*, 1938 (6ᵉ sér.), **3**, 261.

Dalcq, A., Pasteels, J. & Brachet, J. *MMR HNB*, 1936 (2nd ser.), **3**, 882.

Dalcq, A. & Simon, S. *CRAA* (Amsterdam meeting), 1930, p. 1; *P*, 1932, **14**, 497; *ADB*, 1931, **42**, 107.

Dalcq, A. & Tung, Ti-Chao. Unpublished experiments, cit. in Dalcq, "L'Org. de l'œuf etc.", p. 152.

Dalton, A.J. (1). *AR*, 1934, **58**, 321.

Dalton, A.J. (2). *JEZ*, 1935, **71**, 17.

Dalton, A.J. (3). *AR*, 1937, **67**, 431.

Dalton, A.J. (4). *AR*, 1937, **68**, 393.

Dan, K. (1). *JCCP*, 1933, **3**, 477.

Dan, K. (2). *BB*, 1934, **66**, 247.

Dan, K. (3). *PZ*, 1936, **9**, 43 and 58.

Dan, K., Yanagita, T. & Sugiyama, M. *P*, 1937, **28**, 66.

Danchakova, V. (1). *ZAE*, 1924, **74**, 401.

Danchakova, V. (2). "La Cellule Germinale dans le dynamisme de l'Ontogenèse." Hermann, Paris, 1934.

Danchakova, V. (3). *BBFB*, 1938, **72**, 187; *CRSB*, 1937, **126**, 177 and 275.

Danchakova, V. (4). *BC*, 1938, **58**, 302; *CRAS*, 1937, **204**, 1897; 1938, **206**, 1411; *CRSB*, 1937, **126**, 851.

Danchakova, V. (5). *BBFB*, 1937, **71**, 269; *CRSB*, 1937, **127**, 1255.

Danchakova, V. (6). *CRSB*, 1938, **127**, 1259.

Danchakova, V. (7). *CRAS*, 1937, **205**, 424; *CRSB*, 1938, **128**, 895; *AFEWM*, 1938, **138**, 465.

Danchakova, V. (8). *CRSB*, 1938, **128**, 1116.

Danchakova, V. & Kinderis, A. (1). *CRSB*, 1937, **124**, 308.

Danchakova, V. & Kinderis, A. (2). *CRSB*, 1938, **127**, 602.

Danchakova, V. & Massavičius, J. *BBFB*, 1936, **70**, 241.

Danchakova, V. & Vaškovičuté, A. *CRSB*, 1936, **121**, 755.

Danchakova, V., Vaškovičuté, A., Jankovsky, M., Massavicius, J. & Kinderis, A. "Histoire d'un Coq, sa cinétique sexuelle." Hermann, Paris, 1936.

Danforth, C.H. (1). *AJA*, 1930, **45**, 275.

Danforth, C.H. (2). *PSEBM*, 1932, **30**, 143.

Daniel, Janet. *JG*, 1938, **36**, 139.

Daniel, F.J. (1). *PSEBM*, 1936, **34**, 608.

Daniel, F.J. (2). *QRB*, 1937, **12**, 391.

Daniel, F.J. (3). Private communication to the author, Sept. 1940.

Daniel, F.J. & Schechtman, A.M. *Comptes Rend. XII Congr. Internat. Zool.* (*Lisbon*), 1936, p. 591.

Daniel, F.J. & Yarwood, E.A. *UCPZ*, 1939, **43**, 321.

Danielli, J.F. & Harvey, E.N. *JCCP*, 1935, **5**, 483.

Danilova, A.K. & Nefedjova, W.A. *CAC*; *YMaly*, 1936, **37** 490.

Dankmeijer, J. *AR*, 1935, **62**, 179.

Dann, W.J. (1). *BJ*, 1932, **26**, 1072.

Dann, W.J. (2). *BJ*, 1934, **27**, 1998; 1934, **28**, 634 and 2141.

Dann, W.J. (3). *BJ*, 1936, **30**, 1644.

Danneel, R. (1). *BC*, 1934, **54**, 287.

Danneel, R. (2). *NW*, 1938, **31**, 55.

Danneel, R. & Lubnow, E. *BC*, 1936, **56**, 572.

Danneel, R. & Paul, H. *BC*, 1940, **60**, 79.

Danneel, R. & Schaumann, K. *BC*, 1938, **58**, 242.

Dareste, C. "Recherches sur la production artificielle des Monstruosités." Paris, 1877.

Darlington, C.D. "Recent Advances in Cytology." London, 1937.

Darwin, Charles. "Origin of Species." London, 1891, p. 8; "Variation of Animals and Plants under Domestication." London, 1893, pp. 311 ff.

Davenport, C.B. (1). *PSEBM*, 1920, **17**, 75.

Davenport, C.B. (2). *PNASW*, 1933, **19**, 783; *HB*, 1934, **6**, 1; *AJPA*, 1933, **17**, 333.

Davenport, C.B. (3). *PNASW*, 1934, **20**, 359; *ZMA*, 1934, **34**, 76.

Davenport, C.B. (4). *SM*, 1934, **39**, 97; *PAPS*, 1931, **70**, 381; 1935, **75**, 537; 1937, **78**, 61; *CIWP* (Contrib. to Embryol. No. 169), 1938, **496**, 271.

David, Lore. *AFEWM*, 1932, **126**, 457.

David, L.T. *ZZFMA*, 1932, **14**, 616.

David, P.R. *AFEWM*, 1936, **135**, 521.

Davidson, F.A. & Shostrom, D.E. *USBFIR*, 1936, **33**, 1; *CA*, 1936, **30**, 4222.

Davidson, J. (1). *AJBMS*, 1931, **8**, 143.

Davidson, J. (2). *AJBMS*, 1932, **10**, 65.

Davidson, J. (3). *AJBMS*, 1933, **11**, 9.

Davies, S. *QJMS*, 1937, **80**, 81.

Davies, W.L. *BJ*, 1939, **33**, 898.

Davis, J.G. & Slater, W.K. *BJ*, 1928, **22**, 338.

Davy, A. & Huggett, A. St G. *JOP*, 1934, **81**, 183.

Davydov, C. *CRAS*, 1924, **179**, 1222 and 1361.

Dawes, B. *JEB*, 1941, **18**, 26.

Dawson, A.B. (1). *AR*, 1935, **61**, 485.

Dawson, A.B. (2). *ZZFMA*, 1936, **24**, 256.

Dawson, A.B. (3). *JEZ*, 1938, **78**, 101.

Deakin, A. *N*, 1936, **137**, 619.

Dean, I.L., Shaw, M.E. & Tazelaar, M.A. *JEB*, 1928, **5**, 309.

Deanesly, R. & Parkes, A.S. *QJEP*, 1937, **26**, 393.

Deelman, H.T. *Proc. Conference Leeuwenhoek-Vereeniging*, 1935, **4**, 37.

Deighton, T. *PR*, 1933, **13**, 427, p. 437.

Dejust, L.H. & Vignes, H. *CRSB*, 1925, **93**, 314.

Delage, B. *BSCB*, 1935, **17**, 927, 938.

Delage, Y. *Verhdl. d.* VIII *Internat. Kongr. d. Zoologie* (Graz), 1910, p. 100.

Delaunay, H. *BR*, 1931, **6**, 265.

Demereć, M. *CSHSQB*, 1934, **2**, 110.

Demjanovsky, S. *Summaries of Communications.* XV *Internat. Congr. Physiol.* (Moscow), 1935, p. 73.

Demjanovsky, S., Galzova, R. & Roshdestvenska, W. *BZ*, 1932, **247**, 386.

Demjanovsky, S. & Prokofieva, E. *BZ*, 1935, **275**, 455.

Dempster, W.T. (1). *BB*, 1930, **58**, 182.

Dempster, W.T. (2). *JEZ*, 1933, **64**, 495.

Demuth, F. (1). *AFEWM*, 1933, **130**, 340.

Demuth, F. (2). *PKAWA*, 1939, **42**, 506.

Dendy, A. *N*, 1898, **58**, 609.

Dennis, E.A. *PZ*, 1936, **9**, 204.

Deobald, H.J., Lease, E.J. Hart, E.B. & Halpin, J.G. *PS*, 1936, **15**, 179.

Derrick, G.E. (1). *UOB*, 1928, **8**, 100.

Derrick, G.E. (2). *JM*, 1937, **61**, 257.

Derrien, E. *CRSB*, 1924, **91**, 634.

van Deth, J.H.M.G. *ANM*, 1940, **3**, 151.

Detlefsen, J.A. *AR*, 1923, **24**, 417.

Detwiler, S.R. (1). *QRB*, 1926, **1**, 61.

Detwiler, S.R. (2). *JEZ*, 1931, **60**, 141; 1932, **61**, 245; *JCN*, 1932, **56**, 465.

Detwiler, S.R. (3). *BR*, 1933, **8**, 269.

Detwiler, S.R. (4). "Neuroembryology." Macmillan, New York, 1936.

Detwiler, S.R. (5). *JEZ*, 1938, **79**, 361.

Detwiler, S.R. & van Dyke, K.H. *JEZ*, 1934, **69**.

Dhar, N.H. *QRB*, 1932, **7**, 70.

Dhéré, C. *CRSB*, 1933, **112**, 1595.

Dickens, F. (1). *N*, 1933, **131**, 130; *BJ*, 1933, **27**, 1141.

Dickens, F. (2). *BJ*, 1934, **28**, 537.

Dickens, F. (3). *BJ*, 1936, **30**, 1064.

Dickens, F. (4). *BJ*, 1936, **30**, 1233; *N*, 1935, **135**, 762.

Dickens, F. & Greville, G.D. (1). *BJ*, 1932, **26**, 1251 and 1546.

Dickens, F. & Greville, G.D. (2). *BJ*, 1933, **27**, 832.

Dickens, F. & Greville, G.D. (3). *BJ*, 1933, **27**, 1123.

Dickens, F. & Greville, G.D. (4). *BJ*, 1933, **27**, 1134.

Dickens, F. & Greville, G.D. (5), *BJ*, 1935, **29**, 1468.

Dickens, F. & Šimer, F. (1). *BJ*, 1930, **24**, 905.

Dickens, F. & Šimer, F. (2). *BJ*, 1930, **24**, 1301.

Dickens, F. & Šimer, F. (3). *BJ*, 1931, **25**, 973.

Dickens, F. & Šimer, F. (4). *BJ*, 1931, **25**, 985.

Dieckmann, W.J. & Davis, M.E. *AJOG*, 1933, **25**, 623; *OMaly*, 1934, **76**, 529.

Dierks, K. & Becker, M. *AG*, 1933, **152**, 679.

Dijkstra, C. *ZMAF*, 1933, **34**, 75.

Dimter, A. *ZPC*, 1932, **208**, 55.

Dinelli, D. *AAL*, 1935 (6th ser.), **22**, 464.

Dingemanse, E. & Mühlbock, O. *ABN*, 1939, **9**, 95.

Dingle, H. *N*, 1932, **130**, 183.

Diver, C. & Andersson-Kottö, I. *JG*, 1938, **35**, 447.

Dixey, F.A. *TES*, 1893, 69.

Dixon, K.C. *BR*, 1937, **12**, 431.

Dixon, M. "Manometric Methods." Cambridge University Press, 1934.

Dobó, J. In "Données Numériques de Biol." Ed. Terroine & Janot. Paris, 1930, p. 1648.

Dobriner, K. *PSEBM*, 1937, **36**, 757.

Dobzhansky, N.P.S. *AFEWM*, 1927, **109**, 535.

Dobzhansky, T. & Duncan, F.N. *AFEWM*, 1933, **130**, 109.

Dobzhansky, T. & Poulson, D.F. *ZVP*, 1935, **22**, 473.

Dodds, E.C. "Harvey Lectures", 1935, p. 119; *EP*, 1935, **37**, 264; *HCA*, 1936, **19**, E 49; "Goulstonian Lectures" in *L*, 1934 (i), pp. 931, 987 & 1048; 1937, **233**, 1.

Dodds, E.C., Golberg, L., Lawson, W. & Robinson, R. *N*, 1938, **141**, 247; **142**, 34 and 211; *PRSB*, 1939, **127**, 140.

Dodds, E.C. & Lawson, W. (1). *N*, 1936, **137**, 996.

Dodds, E.C. & Lawson, W. (2). *N*, 1937, **139**, 627 and 1068.

Dodds, E.C. & Lawson, W. (3). *PRSB*, 1938, **125**, 222.

Döderlein, G. *AG*, 1934, **155**, 22.

Doisy, E.A., Ralls, J.O., Allen, E. & Johnston, C.G. *JBC*, 1924, **61**, 711.

Dolinov, K. *Trudy Lab. Belka & Belkovkogo Obmena v Organisme*, 1934, p. 66; *Arb. Lab. Proteinforsch. (Referate)*, *Moscow*, 1936, pp. 15, 43.

Doljanski, L. (1). *ZZFMA*, 1929, **8**.

Doljanski, L. (2). *CRSB*, 1930, **105**, 343.

Doljanski, L. (3). *CRSB*, 1930, **105**, 504.

Doljanski, L. & Hoffman, R.S. *N*, 1940, **145**, 857.

Doljanski, V. *APPA*, 1933, **291**, 418.

Dollo, Louis, *BARBS*, 1903, no. 8, **51**.

Dols, M.J.L. *ANP*, 1937, **22**, 372.

Domm, L.V. & Dennis, E.A. *PSEBM*, 1937, **36**, 766.

Domm, L.V. & van Dyke, K.H. *PSEBM*, 1932, **30**, 349 and 351.

Donahue, J.K. & Jennings, E.D. *PSEBM*, 1939, **42**, 220; *ENY*, 1937, **21**, 690.

Donaldson, H.H. "The Rat." Wistar Institute, Philadelphia, 1915 and 1924.

Donaldson, H.H. & Hatai, S. *JCN*, 1911, **21**, 417.

Doneddu, F.P. *AFS*, 1934, **57**, 105; *OMaly*, 1934, **80**, 332.

Donegan, J.F. Private communication to the author, May 1936.

Donhoffer, C. *BJ*, 1933, **27**, 806.

Dorfman, W.A. (1). *P*, 1932, **16**, 56.

Dorfman, W.A. (2). *P*, 1934, **21**, 235.

Dorfman, W.A. (3). *P*, 1934, **21**, 245.

Dorfman, W.A. (4). *P*, 1936, **25**, 427 and 465; *BBMEUSSR*, 1936, **1**, 135.

Dorfman, W.A. & Grodsensky. *BBMEUSSR*, 1937, **4**, 265.

Dorris, F. (1). *JEZ*, 1935, **70**, 491; *AA*, 1934, **78**, 435.

Dorris, F. (2). *AFEWM*, 1938, **138**, 323; *JEZ*, 1939, **80**, 315.

Dove, W.F. *JAR*, 1935, **50**, 923.

Doyle, W.L. *JCCP*, 1938, **11**, 291; *CRTLC*, 1938, **21**, no. 22.

Doyle, W.L. & Harding, J.P. *JEB*, 1937, **14**, 462.

Drage, J.S. & Sandström, W.H. *PSEBM*, 1936, **34**, 376.

Dragoiu, J., Benetato, G. & Opreanu, R. *CRSB*, 1937, **126**, 1044.

Dragomirov, N. (1). *AFEWM*, 1929, **116**, 633.

Dragomirov, N. (2). *ZEBUSSR*, 1931, **7**, 284; *YMaly*, 1932, **21**, 652.

Dragomirov, N. (3). *AFEWM*, 1932, **126**, 636; 1933, **129**, 522.

Dragomirov, N. (4). *AFEWM*, 1936, **134**, 716.

Dragomirov, N. (5). *CRASUSSR*, 1937, **15**, 57; *BASUSSR*, 1938, **1**, 48.

Dragomirov, N. (6). *CRASUSSR*, 1939, **23**, 399.

Dragomirov, N. (7). *CRASUSSR*, 1940, **27**, 189.

Drea, W.F. *JN*, 1935, **10**, 351.

Driesch, H. (1). *ZWZ*, 1891, **53**, 160.

Driesch, H. (2). *ZWZ*, 1893, **55**, 1; *AFEWM*, 1900, **10**, 361.

Driesch, H. (3). *AFEWM*, 1900, **10**, 411.

Driesch, H. (4). *AFEWM*, 1895, **2**, 169; 1903, **17**, 41; 1905, **20**, 1.

Driesch, H. (5). *PSZN*, 1895, **11**; *AFEWM*, 1897, **4**, 75, esp. p. 113; 1902, **14**, 500, esp. p. 517.

Driesch, H. (6). *AFEWM*, 1899, **8**, 35

Driesch H. (7). "The Science and Philosophy of the Organism." Gifford Lectures, 1st edn. 1908, 2nd edn. 1929, Black, London. (References in the text are to the later issue.)

Driesch, H. (8). "Der Begriff der Organischen Form." Bornträger, Berlin, 1919.

Driesch, H. (9). *BC*, 1927, **47**, 641, esp. 651.

Driesch, H. (10). *AB*, 1935, **1**, 185 (190); 1937, **3**, 51.
Driesch, H. (11). *AFEWM*, 1902, **14**, 256.
Drilhon, A. *CRSB*, 1935, **118**, 131.
Drilhon, A. & Busnel, R.G. *CRAS*, 1938, **207**, 92.
Driver, E.C. *JEZ*, 1931, **59**, 1.
Droogleever-Fortuyn, A.B. *GA*, 1939, **21**, 97.
Druckrey, F. *ZKF*, 1937, **47**, 13.
Drzwina, A. & Bohn, G. *AAM*, 1931, **193**, 491.
Dubois, E. *BSAP*, 1897 (ser. 4), **8**, 337.
Duce, W. *BSIBS*, 1931, **6**, 511; *OMaly*, 1932, **64**, 683.
Dudley, H.W. & Rosenheim, O. *BJ*, 1925, **19**, 1034.
Dudley, H.W. & Woodman, H.E. *BJ*, 1915, **9**, 97.
Dulzetto, F. *ADB*, 1931, **41**, 221; *AZI*, 1931, **16**, 461.
Dulzetto, F. & li Volsi, N. *BDZ*, 1932, **3**, 45 and 53; *YMaly*, 1932, **22**, 278.
Dumansky, A.V. & Strukova, E.P. *ZPC USSR*, 1929, **61**, 381.
Dunihue, F.W. *AR*, 1935, **64** (Suppl. no. 1), **87**.
Dunn, L.C. (1). *AN*, 1923, **57**, 345.
Dunn, L.C. (2). *JH*, 1925, **16**, 127.
Dunn, L.C. (3). *AFEWM*, 1927, **110**, 341.
Dunn, L.C. (4). *PNASW*, 1934, **20**.
Dunn, L.C. (5). *JG*, 1936, **33**, 443.
Dunn, L.C. & Coyne, J. *BC*, 1935, **55**, 385.
Dunn, L.C. & Einsele, W. *JG*, 1938, **36**, 145.
Dunn, L.C. & Landauer, W. (1). *AN*, 1926, **60**, 574.
Dunn, L.C. & Landauer, W. (2). *JG*, 1934, **29**, 217; 1936, **33**, 401.
Duran-Reynals, F. *YJBM*, 1939, **11**, 613.
Dürken, B. (1). *AFEWM*, 1926, **107**, 727; *ZAZ* (Erg. Bd.), 1925, p. 84.
Dürken, B. (2). "Experimental Analysis of Development." Tr. H.G. & A.M. Newth. London, 1932.
Dürken, B. (3). *ZWZ*, 1933, **144**, 123.
Dürken, B. (4). *ZWZ*, 1935, **147**, 295.
Duryee, W.R. *S*, 1932, **75**, 520; *ZVP*, 1936, **23**, 208.
Duyff, J.W. *ANM*, 1939, **2**, 153 and 165.
Dwinnell, L.A. *PSEBM*, 1939, **42**, 264.
van Dyke, H.B. *EP*, 1936, **38**, 836.
Dyrdowska, M. *CRSB*, 1931, **108**, 593; *BAA*, 1931, **25**, 172, 175; *YMaly*, 1932, **22**, 100 and 101.

Eakin, R.M. (1). *UCPZ*, 1933, **39**, 191; 1939, **43**, 185.
Eakin, R.M. (2). *PSEBM*, 1939, **41**, 308; *G*, 1940, **3**, 373.
Eakin, R.M. (3). *AFEWM*, 1939, **139**, 274.
Eastcott, E.V. *JPC*, 1928, **32**, 1094; *JACS*, 1929, **51**, 2773.
Eastham, L.E.S. (1). *QJMS*, 1927, **71**, 353.
Eastham, L.E.S. (2). *BR*, 1930, **5**, 1.

Eastman, N.J. & Dippel, A.L. *BJHH*, 1933, **53**, 288.
Eastman, N.J., Geiling, E.M.K. & de Lawder, A.M. *BJHH*, 1933, **53**, 246.
Eastman, N.J. & McLane, C.M. *BJHH*, 1931, **48**, 261.
Eaton, O.N. *JH*, 1937, **28**, 353.
Eberth, C.J. *APPA*, 1868, **44**, 12.
Eckles, C.H. (1). *Res. Bull. Univ. Missouri Agric. Exp. Sta.* No. 26, 1916.
Eckles, C.H. (2). *Res. Bull. Univ. Missouri Agric. Exp. Sta.* No. 35, 1919.
Edlbacher, S. "Die Chemie der Wachstumvorgänge." University Press, Basel, 1934.
Edlbacher, S. & Kutscher, F. *ZPC*, 1931, **199**, 200 and 211.
Edlbacher, S. & Merz, K.W. *ZPC*, 1927, **171**, 252.
Edlén, Å. *LUA*, 1937, **34**, no. 1; *AFEWM*, 1938, **137**, 804.
Edson, N.L., Krebs, H.A. & Model, A. *BJ*, 1936, **30**, 1380.
Edwards, C.L. *AJOP*, 1902, **6**, 351.
Effkeman, G. *AG*, 1936, **162**, 148.
Efimov, M.I. (1). *ZEBUSSR*, 1931, **7**, 1; *BZUSSR*, 1933, **2**.
Efimov, M.I. (2). *BBMEUSSR*, 1938, **6**, 75.
Efroimson, W.P. & Rilova, K.N. *BZUSSR*, 1936, **5**, 625.
Eigenmann, C.H. *AFEWM*, 1899, **8**, 545; *TAMS*, 1900, **21**, 49; *PNASW*, 1902, **4**, 533. Mark Anniversary Volume, 1903, p. 167. *Proc. 7th Int. Zool. Congr.* 1907.
Eigenmann, C.H. & Denny, W.E. *BB*, 1900, **2**, 33.
Einsele, W. (1). *AFEWM*, 1930, **123**, 279.
Einsele, W. (2). *JG*, 1937, **34**, 1.
Eisler, B., Hammarsten, E. & Theorell, H. *NW*, 1936, **24**, 142.
van Ekenstein, W.A. & Blanksma, J.J. *CW*, 1907, **4**, 407; *BCC*, 1907, **6**, 708.
Eker, R. *JG*, 1935, **30**, 357.
Ekman, G. (1). *AFEWM*, 1925, **106**, 320; 1929, **116**, 327.
Ekman, G. (2). *AASF*, 1936, **45**, no. 4.
Elderfield, R.C. *CR*, 1935, **17**, 187.
Elford, W.J. & Ferry, J.D. *BJ*, 1936, **30**, 84.
Ellinger, P. & Koschara, W. *BDCG*, 1933, **66**, 315 and 808.
Elliott, K.A.C. *PR*, 1941, **21**, 267.
Elliott, K.A.C. & Greig, M.E. *BJ*, 1938, **32**, 1407.
Elliott, R.H., Hall, F.G. & Huggett, A. St G. *JOP*, 1934, **82**, 160.
Ellis, E.L. *JCCP*, 1933, **4**, 127.
Ellis, N.R., Miller, D., Titus, H.W. & Byerly, T.C. *JN*, 1933, **6**, 243.
Elton, C. "Animal Ecology." London, 1927, p. 55.
Elvehjem, C.A., Hart, E.B. & Sherman, W.C. *JBC*, 1933, **103**, 61.
Ely, F., Hull, F.E. & Morrison, H.B. *JH*, 1939, **30**, 105.

Emerson, A.E. (1). *EM*, 1938, **8**, 247.
Emerson, A.E. (2). *AMN*, 1939, **21**, 182; *EM*, 1939, **9**, 287.
Emerson, G.A., Lu, Gwei-Djen, & Emerson, O. Private communication to the author.
Emerson, H.S. *JEZ*, 1940, **83**, 191.
Emery, C. *BC*, 1893, **13**, 397.
Emery, F.E. *AJOP*, 1935, **111**, 392.
Emmart, E.W. *JEZ*, 1936, **74**, 353.
Emmens, C.W. & Parkes, A.S. *N*, 1939, **143**, 1064.
Emmerie, A., van Eekelen, M. & Wolff, L.K. *ABN*, 1934, **4**, 5; *CA*, 1934, **28**, 4492.
Enderlein, G. "Bakterien-Cyclogenie." Berlin 1925.
Engel, R.W., Phillips, P.H. & Halpin, J.G. *PS*, 1940, **19**, 135.
Engelsmeier, W. *ZIAV*, 1935, **68**, 361.
Entenmann, C., Lorenz, F.W. & Chaikov, I.L. *JBC*, 1940, **133**, 231.
Entin, T.I. *AAUSSR*, 1936, **15**, 104; *YMaly*, 1937, **41**, 312.
Enzmann, E.V. (1). *AR*, 1933, **56**, 345.
Enzmann, E.V. (2). *AJOP*, 1934, **108**, 373.
Enzmann, E.V. (3). *AR*, 1935, **62**, 31.
Enzmann, E.V. & Crozier, W.J. *JGP*, 1935, **18**, 791.
Enzmann, E.V. & Haskins, C.P. (1). *AFEWM*, 1938, **138**, 159.
Enzmann, E.V. & Haskins, C.P. (2). *AFEWM*, 1938, **138**, 161.
Enzmann, E.V. & Haskins, C.P. (3). *AN*, 1939, **73**, 470.
Enzmann, E.V. & Pincus, G. *JGP*, 1934, **18**, 163.
Enzmann, E.V., Saphir, N.R. & Pincus, G. *AR*, 1932, **54**, 325.
Epelbaum, S. & Khaikina, B. *UBZ*, 1938, **11**, 277 and 292; 1939, **13**, 261.
Ephrussi, B. (1). *CRAS*, 1931, **192**, 1763.
Ephrussi, B. (2). "La Culture des Tissus." Gauthier-Villars, Paris, 1932.
Ephrussi, B. (3). *ABSRSMNB*, 1931, 15.
Ephrussi, B. (4). *ADB*, 1933, **44**, 1.
Ephrussi, B. (5). *JEZ*, 1935, **70**, 197; *CRAS*, 1933, **197**, 96.
Ephrussi, B. (6). *AN*, 1938, **72**, 5.
Ephrussi, B. & Beadle, G.W. (1). *BBFB*, 1935, **69**, 492.
Ephrussi, B. & Beadle, G.W. (2). *CRAS*, 1935, **201**, 98 and 1148.
Ephrussi, B. & Beadle, G.W. (3). *BBFB*, 1937, **71**, 54, 75; *GS*, 1937, **22**, 65, 479; *JG*, 1936, **33**, 407.
Ephrussi, B. & Chevais, S. *BBFB*, 1938, **72**, 48; *PNASW*, 1937, **23**, 428.
Ephrussi, B., Clancy, C.W. & Beadle, G.W. *CRAS*, 1936, **203**, 545.
Ephrussi, B. & Harnly, M.H. *CRAS*, 1936, **203**, 1028.
Ephrussi, B., Khouvine, Y. & Chevais, S. *N*, 1938, **141**, 204.

Ephrussi, B. & Lacassagne, A. *CRSB*, 1933, **113**, 976.
Erdmann, R. *AFEWM*, 1927, **112**, 739.
Erdmann, R. & Schmerl, E. *AEZF*, 1926, **2**, 280.
Ermakov, M.W. (1). *UMZ*, 1934, **4**, 383; *YMaly*, 1935, **32**, 675.
Ermakov, M.W. (2). *UMZ*, 1935, **4**, 659; *OMaly*, 1936, **92**, 555; *YMaly*, 1936, **36**, 667.
Erxleben, H. *EP*, 1935, **37**, 186.
Esskuchen, E. (1). *ZTZ*, 1930, **19**, 268; *OMaly*, 1931, **60**, 386.
Esskuchen, E. (2). *ATT*, 1931, **5**, 589; *YMaly*, 1931, **19**, 328.
Etkin, W. (1). *PZ*, 1932, **5**, 275; *JEZ*, 1935, **71**, 317.
Etkin, W. (2). *PZ*, 1934, **7**, 129.
Etkin, W. (3). *AR*, 1935, **64** (Suppl.), 34.
Etkin, W. & Huth, T. *JEZ*, 1939, **82**, 463.
Etkin, W. & Lotkin, R. *PSEBM*, 1940, **44**, 471.
Eufinger, H. *MGG*, 1932, **92**, 273.
von Euler, H., Gard, U. & Hellström, H. *SKT*, 1932, **44**, 191; *CA*, 1933, **27**, 1921.
von Euler, H., Gunther, G. & Vestin, R. *ZPC*, 1936, **240**, 265.
von Euler, H. & Klussmann, E. (1). *ZPC*, 1932, **208**, 50.
von Euler, H. & Klussmann, E. (2). *BZ*, 1932, **250**, 1.
von Euler, H. & Schmidt, G. *ZPC*, 1934, **223**, 215.
von Euler, H. & Skarżyński, B. *ZPC*, 1940, **263**, 259.
von Euler, H. & Zondek, B. *BZ*, 1934, **271**, 64.
Evans, A.C. *JEB*, 1932, **9**, 314.
Evans, H.M. *PARNMED*, 1936, **17**, 175.
Evans, H.M., Burr, G.O. & Althausen, T.L. *UCM*, 1927, **8**, 1.
Evans, H.M., Emerson, G.A. & Eckert, J.E. *JEE*, 1937, **30**, 642.
Evans, T.C. (1). *PSEBM*, 1933, **30**, 1214.
Evans, T.C. (2). *PZ*, 1934, **7**, 556; 1935, **8**, 521; 1936, **9**, 443; 1937, **10**, 258.
Everett, J.W. (1). *PSEBM*, 1933, **31**, 77.
Everett, J.W. (2). *JEZ*, 1935, **70**, 243.
Ewest, A. *AFEWM*, 1937, **135**, 689.

Fabre, R. *CRSB*, 1933, **113**, 1380.
Fabre, R. & Lederer, E. *BSCB*, 1934, **16**, 105; *CRSB*, 1933, **113**, 344.
Falaschino, A. *BTS*, 1936, **23**, 289; *OMaly*, 1937, **99**, 481.
Falin, L.I. *AJC*, 1940, **38**, 199.
Falin, L.I. & Gromzeva, K.E. *BBMEUSSR*, 1938, **6**, 394.
von Falkenhausen, M., Fuchs, H.J. & Schubert, M. *BZ*, 1928, **193**, 269.
Falkenheim, M. *PSZN*, 1937, **16**, 212.
Fancello, O. *PSZN*, 1937, **16**, 80.
Fankhauser, G. (1). *AFEWM*, 1925, **105**, 501.
Fankhauser, G. (2). *AFEWM*, 1930, **122**, 116.

Fankhauser, G. (3). *JEZ*, 1934, **67**, 159 and 349; 1934, **68**, 1.

Fankhauser, G. (4). *JEZ*, 1937, **75**, 413.

Fankhauser, G. (5). *JH*, 1937, **28**, 3.

Farès, A. Inaug. Diss. Paris, 1935.

Fasten, N. *JH*, 1932, **23**, 421.

Fauré-Fremiet, E. (1). "La Cinétique du développement." Presse Universitaire, Paris, 1925.

Fauré-Fremiet, E. (2). *AAM*, 1931, **27**, 421; 193 **28**, 1 and 121.

Fauré-Fremiet, E. (3). *P*, 1933, **19**, 63.

Fauré-Fremiet, E. (4). *AEZF*, 1934, **15**, 79 and 373.

Fauré-Fremiet, E. (5). *CRSB*, 1933, **113**, 715; *CRAA*, 1933, 1935; *AAM*, 1936, **32**, 249.

Fauré-Fremiet, E. (6). *JCP*, 1936, **33**, 681; 1937, **34**, 125.

Fauré-Fremiet, E. (7). *AAM*, 1938, **34**, 23.

Fauré-Fremiet, E. & Baudouy, C. (1). *BSCB*, 1937, **19**, 1134.

Fauré-Fremiet, E. & Baudouy, C. (2). *BSCB*, 1938, **20**, 14.

Fauré-Fremiet, E. & Garrault, H. (1). *APBPC*, 1928, **2**, 218.

Fauré-Fremiet, E. & Garrault, H. (2). *AAM*, 1937, **33**, 81.

Fauré-Fremiet, E. & Garrault, H. (3). *BSCB*, 1938, **20**, 24.

Fauré-Fremiet, E. & Kaufman, L. *APBPC*, 1928, **1**, 64.

Fauré-Fremiet, E., Kruszynski, J. & Mazoué, H. *CRAA*, 1935.

Fauré-Fremiet, E. & Woelfflin, R. *JCP*, 1936, **33**, 666, 695 and 801.

Favorski, M.V. *CRASUSSR*, 1939, **25**, 71.

Fawcett, D.W. *BB*, 1939, **77**, 174.

Federley, H. *FLSH*, 1935, **78**, 241.

Fee, A.R., Marrian, G.F. & Parkes, A.S. *JOP*, 1929, **67**, 377.

Fekete, E. & Green, C.V. *AJC*, 1936, **27**, 513.

Feldstein, M.J. & Hersh, A.H. *AN*, 1935, **69**, 344.

Felix, K., Baumer, L. & Schörner, E. *ZPC*, 1936, **243**, 43.

Fell, H.B. (1). *AEZF*, 1929, **7**, 69.

Fell, H.B. (2). *AEZF*, 1931, **11**, 245.

Fell, H.B. (3). *JA*, 1932, **66**, 157.

Fell, H.B. (4). *JRMS*, 1940, **60**, 95.

Fell, H.B. & Canti, R.G. *PRSB*, 1934, **116**, 316.

Fell, H.B. & Grüneberg, H. *PRSB*, 1939, **127**, 257.

Fell, H.B. & Landauer, W. *PRSB*, 1935, **118**, 133.

Fell, H.B. & Robison, R. (1). *BJ*, 1929, **23**, 767.

Fell, H.B. & Robison, R. (2). *BJ*, 1930, **24**, 1905.

Fell, H.B. & Robison, R. (3). *BJ*, 1934, **28**, 2243.

Fellner, O.O. *KW*, 1925, **4**, 1651.

Fêng, Teh-P'ei & Ting, Yen-Chieh. *CJOP*, 1938, **13**, 141.

Fenn, W.O. *AJOP*, 1927, **80**, 327.

Fernandez, M. *AA*, 1915, **48**, 305.

Ferris, G.F. & Usinger, R.L. *ME* (Contrib. Entomol. Stanford Univ.), 1939, **4**, 1.

Ferry, R.M. & Levy, A.H. *JBC*, 1934, **105**, xxvii.

Ferwerda, F.P. *GA*, 1928, **11**, 1.

Fetischeva, M.I. *MSPEUSSR*, 1939, **2**, 69.

Feulgen, R. In "Abderhalden Handbuch d. biol. Arbeitsmethoden", 1932, v, 2, ii, 1055.

Feulgen, R. & Bersin, T. *ZPC*, 1939, **260**, 217.

Feulgen, R., Bersin, T. & Behrens, M. xvith *International Congress of Physiology* (Zürich), 1938, Kongressbericht ii, p. 105.

Fieser, L.F. (1). "The Chemistry of Natural Products related to Phenanthrene." New York, 1936; Supplement, New York, 1937.

Fieser, L.F. (2). *OPAAA*, 1937, **4**, 51.

Fieser, L.F. (3). *AJC*, 1938, **34**, 37.

Fieser, L.F., Fieser, M., Hershberg, E.B., Newman, M.S., Seligman, A.M. & Shear, M.J. *AJC*, 1937, **29**, 260.

Figge, F.H.J. (1). *JEZ*, 1930, **56**, 241.

Figge, F.H.J. (2). *PZ*, 1934, **7**, 149.

Figge, F.H.J. (3). *JEZ*, 1938, **78**, 471; *CN*, 1937, **12**.

Figge, F.H.J. (4). *PSEBM*, 1938, **39**, 569

Figge, F.H.J. (5). *PSEBM*, 1939, **41**, 127; 1940, **44**, 293.

Figge, F.H.J. (6). *JCCP*, 1940, **15**, 233.

Figge, F.H.J. & Uhlenluth, E. *PZ*, 1933, **6**, 450.

Fikentscher, R. (1). *AG*, 1933, **154**, 129.

Fikentscher, R. (2). *KW*, 1935, **14**, 569.

Filatov, D. (1). *RZR*, 1916, **1**, 48.

Filatov, D. (2). *AFEWM*, 1925, **105**, 475.

Filatov, D. (3). *AFEWM*, 1927, **110**, 1; 1930, **121**, 272.

Filatov, D. (4). *ZJZP*, 1932, **51**, 589.

Filatov, D. (5). *AFEWM*, 1933, **127**, 776.

Filatov, D. (6). *BZUSSR*, 1934, **3**, 261.

Filatov, D. (7). *TDDUSSR*, 1935, **10**, 333.

Filatov, D. (8). *ZJZP*, 1937, **58**, 1; *BZUSSR*, 1937, **6**, 385.

Filatov, D. (9). *BASUSSR*, 1937, 955.

Filhol, J. & Garrault, H. *AAM*, 1938, **34**, 104.

Finkelstein, E.A. & Schapiro, E.M. (1). *UMZ*, 1937, **3**, 5 and 10; *OMaly*, 1937, **102**, 213.

Finkelstein, E.A. & Schapiro, E.M. (2). *BBMEUSSR*, 1938, **6**, 149.

Fischel, A. "Grundriss d. Entwicklung des Menschen." Springer, Berlin, 1937.

Fischer, A. (1). *AP*, 1929, **223**, 163.

Fischer, A. (2). *P*, 1931, **14**, 307 and 461.

Fischer, A. (3). *AFEWM*, 1931, **125**, 203.

Fischer, A. (4). *EP*, 1933, **35**, 82.

Fischer, A. (5). *EP*, 1933, **35**, 98.

Fischer, A. (6). *AJC*, 1937, **31**, 1.

Fischer, A. (7). *N*, 1939, **144**, 113. *CP*, 1940, **3**, 79.

Fischer, A. & Demuth, F. *AEZF*, 1927, **5**, 131.

Fischer, A. & Parker, R.C. *AEZF*, 1929, **8**, 297 and 325.

Fischer, F.G. *VDZG*, 1935, 171.

Fischer, F.G. & Hartwig, H. (1). *ZVP*, 1936, **24**, 1.

Fischer, F.G. & Hartwig, H. (2). *BC*, 1938, **58**, 568.

Fischer, F.G. & Wehmeier, E. *NW*, 1933, **21**, 518.

Fischer, F.G., Wehmeier, E. & Jühling, L. *NGWG*, 1934 (VI), **9**, 394.

Fischer, F.G., Wehmeier, E., Lehmann, H., Jühling, L. & Hultzsch, K. *BDCG*, 1935, **68**, 1196.

Fischer, H. & Kögl, F. *ZPC*, 1923, **131**, 241; 1924, **138**, 262.

Fischer, H. & Lindner, F. *ZPC*, 1925, **142**, 141.

Fischer, H. & Stangler, G. *ADC*, 1927, **459**, 62.

Fischer-Wasels, B. In "Handbuch d. norm. u. path. Physiol.", 1927, **14**, 1211.

Fisher, A.M. & Scott, D.A. *JBC*, 1934, **106**, 305.

Fisher, I. *AJOP*, 1906, **15**, 417.

Fittig, R. & Ostermayer, E. *BDCG*, 1872, **5**, 933. *ADC*, 1873, **166**, 361.

Fleischhacker, H.H. *JMS*, 1938, **84**, 947.

Fleming, A. *PRSM*, 1932, **26**, 71.

Flemion, F. & Hartzell, A. *CBTI*, 1936, **8**, 167; *OMaly*, 1937, **98**, 43.

Flexner, L.B. (1). *AJOP*, 1938, **124**, 131.

Flexner, L.B. (2). *JBC*, 1939, **131**, 703.

Flexner, L.B. & Gersh, I. *CIWP* (Contrib. to Embryol. No. 157), 1937, **479**, 121.

Flexner, L.B. & Pohl, H.A. *PSEBM*, 1940, **44**, 345; *JBC*, 1941, **139**, 163; *JCCP*, 1941, **18**, 49.

Flexner, L.B. & Roberts, R.B. *AJOP*, 1939, **128**, 154.

Flexner, L.B. & Stiehler, R.D. *JBC*, 1938, **126**, 619.

Florkin, M. *MARBS*, 1937 (sér. 2ᵉ), **16**, 3; *AIP*, 1937, **45**, 17.

Floyd, H. Private communication to the author.

Fol, H. *CRAS*, 1877, **84**, 357; *SPHNGM*, 1879, **26**, 89.

Folley, F.J. & White, P. *N*, 1937, **140**, 505.

Fomina, P.I. *AG*, 1935, **160**, 333.

Fontaine, M. & Boucher, M. *Proc.* xvth *Int. Congr. Physiol.* (Moscow), 1935, p. 99.

Fontaine, M. & Gurevitch, A. *CRSB*, 1936, **123**, 443.

Foot, N.C., Baker, L.E. & Carrel, A. *JEM*, 1939, **70**, 39.

Foote, C.L. & Witschi, E. *AR*, 1938, **72** (Suppl.), 120; 1939, **75**, 75.

Ford, E.B. & Huxley, J.S. *JEB*, 1927, **5**, 112; *N*, 1925, **116**, 861.

Förster, M. & Örström, Å. *TSBR*, 1933, **11**, 63.

Foulds, L. (1). *SRICRS*, 1934, **11** (Suppl.), 1.

Foulds, L. (2). *AJC*, 1937, **29**, 761.

Foulds, L. (3). *AJC*, 1937, **31**, 404.

Fowler, O.M. *JEZ*, 1937, **76**, 235.

Fox, H.M. (1). *N*, 1936, **138**, 839; *PZS*, 1938, **108**, 501.

Fox, H.M. (2). Private communication to the author, 1937.

Foxon, G.E.H. *N*, 1933, **131**, 732.

Fraenkel, E. *APPA*, 1923, **248**, 125.

Fraenkel, G. *N*, 1934, **133**, 834; *PRSB*, 1935, **118**, 1.

Fraenkel, G. & Rudall, K.M. *PRSB*, 1940, **129**, 1.

Fraenkel, S. & Dimitz, L. *BZ*, 1910, **28**, 295.

de Francesco, C. *PSZN*, 1933, **13**, 279.

Francisco, A.A., Chan, G.S. & Fronda, F.M. *PAG*, 1934, **22**, 685.

Franke, K.W., Moxon, A.L., Poley, W.E. & Tully, W.C. *AR*, 1936, **65**, 15.

Franke, K.W. & Tully, W.C. *PS*, 1935, **14**, 273; 1936, **15**, 316.

Fraps, R.M. & Juhn, M. *PZ*, 1936, **9**, 293, 319 & 378.

Fraser, D.T., Jukes, T.H., Branion, H.D. & Halpern, K.C. *JI*, 1934, **26**, 437.

Frederici, E. *ADB*, 1926, **36**, 465.

Freed, S.C. & Coppock, A. *PSEBM*, 1935, **32**, 1589.

French, C.S. Proc. XVI Internat. Congr. Physiol. 1938, p. 257; *JGP*, 1940, **23**, 469 & 483.

Freund, J. *JI*, 1930, **18**, 315.

Freundlich, H. "Thixotropy." Hermann, Paris, 1935.

Frew, J.G.H. *JEB*, 1928, **6**, 1.

Friedheim, E.A.H. *CRSB*, 1932, **111**, 505.

Friedheim, E.A.H. & Baer, J.G. *BZ*, 1933, **265**, 329.

Friedheim, E.A.H. & Michaelis, L. *JBC*, 1931, **91**, 355.

Friedmann, E. (1). *N*, 1935, **135**, 622; **136**, 108.

Friedmann, E. (2). "Sterols and related Compounds." Heffer, Cambridge, 1937.

Friedmann, H. *BB*, 1927, **53**, 343; *USNMP*, 1932, **80**, no. 18.

Friesen, H. *AFEWM*, 1936, **134**, 147; *BZUSSR*, 1935, **4**.

Friggeri, A. *BSIBS*, 1938, **13**, 478; *RDB*, 1938, **26**, 434.

Frisch, C. & Willheim, R. *BZ*, 1935, **277**, 148; 1936, **287**, 198.

Fröhlich, K. *AFEWM*, 1936, **134**, 348.

Fromageot, C. *EEF*, 1938, **7**, 50.

Fugo, N.W. *AR*, 1938, **72** (Suppl.), 117.

Fugo, N.W. & Witschi, E. *ABL*, 1938, **8**, 73.

Fujita, A. & Ebihara, T. *BZ*, 1939, **301**, 229.

Fukahori, Y. *NIZ*, 1933, **11**, 288; *YMaly*, 1933, **25**, 696.

Fukuda, J. *PIAT*, 1936, **12**, 269; *YMaly*, 1937, **44**, 45.

Fukuda, S. (1). *PIAT*, 1939, **15**, 19.

Fukuda, S. (2). *JJB*, 1939, **30**, 125.
Fukuda, S. (3). *JJB*, 1939, **30**, 135.
Funk, E.M. *Bull. Univ. Missouri Agric. Exp. Sta.* 1934, no. 341.
Furreg, E. *BC*, 1931, **51**, 162.
von Fürth, O., Herrmann, H. & Scholl, R. *BZ*, 1934, **271**, 395.
Furuhashi, Y. *JJMSB*, 1936, **3**, 227.
Furukawa, H. *PIAT*, 1935, **11**, 158.

Gaarenstroom, J.H. Inaug. Diss. Amsterdam, 1938; *JEZ*, 1939, **82**, 31. *JE*, 1940, **2**, 47.
Gabritschevsky, E. & Bridges, C.E. *ZIAV*, 1928, **46**, 248.
Gaehtgens, G. (1). *AG*, 1937, **164**, 398.
Gaehtgens, G. (2). *AG*, 1937, **164**, 588.
Gaggermeier, G. *AFGF*, 1930, **4**, 453.
Gaillard, P.J. (1). Inaug. Diss. Leiden, 1931; *ABN*, 1932, **2**, 1; *YMaly*, 1932, **20**, 659.
Gaillard, P.J. (2). *P*, 1935, **23**, 145.
Gaillard, P.J. (3). *ANM*, 1937, **1**, 3.
Gaillard, P.J. & Varossieau, W.W. *ANM*, 1938, **1**, 313.
Galen of Pergamos. "On the Natural Faculties", *ca.* A.D. 160. Ed. and tr. with introduction by A.J. Brock. Loeb Classics, Heinemann, London, 1916.
Gallien, L. *CRAS*, 1937, **205**, 375; *BBFB*, 1938, **72**, 269.
Galloway, T.W. *AN*, 1900, **34**, 949.
Galtsov, P.S. *BB*, 1938, **74**, 461; 1939, **75**, 286.
Galvialo, M.J. & Goryukhina, T.A. *PZUSSR*, 1937, **22**, 215; *YMaly*, 1937, **45**, 49.
Ganfini, G. *BSIBS*, 1930, **5**, 949; *CA*, 1931, **25**, 2488.
Garber, S.T. *PZ*, 1930, **3**, 373.
Gardner, J.A. & Gainsborough, H. *BJ*, 1927, **21**, 130 and 141.
Garofalo, A. *CO*, 1931, **33**, 65; *OMaly*, 1931, **61**, 769.
Garrasi, G. *AO*, 1933, **55**, 729; *OMaly*, 1934, **75**, 710.
Garrault, H. (1). *CRSB*, 1933, **113**, 384.
Garrault, H. (2). *CRAA*, 1935; *CRAS*, 1935, **200**, 1248; *APBPC*, 1936, **12**, 291; *AAM*, 1936, **32**, 105; 1937, **33**, 167.
Garrault, H. & Filhol, J. *CRSB*, 1937, **126**, 773.
Garufi, G. *ASB*, 1935, **21**, 233; *OMaly*, 1936, **91**, 404.
Gates, F. *S*, 1933, **77**, 350.
Gatty, O. & Schulman, J.H. *TFS*, 1939, **35**, 1510.
Gaujoux, E. & Krijanowski, A. *CRSB*, 1932, **110**, 1083.
Gaunt, R. *PSEBM*, 1931, **28**, 660.
Gause, G.F. (1). *BC*, 1931, **51**, 209.
Gause, G.F. (2). *EB*, 1936, **13**, 54.
Gause, G.F. & Alpatov, W.W. *AFEWM*, 1934, **131**, 188.
Gebauer-Fuelnegg, E. *PSEBM*, 1932, **29**, 529.

Gebhardt, F.A. *VDZG*, 1912, **22**.
Gedroyć, W., Cichocka, J. & Mystkowski, E.M. *BZ*, 1935, **281**, 422.
Geiger, A. *BJ*, 1935, **29**, 811.
Geigy, R. *AFEWM*, 1931, **125**, 406; *RSZ*, 1931, **38**, 187.
Geigy, R. & Ochsé, W. *RSZ*, 1940, **47**, 193.
Geinitz, B. *AFEWM*, 1925, **106**, 357.
Gelfan, S. *PSEBM*, 1931, **29**, 58.
Genell, S. *BZ*, 1931, **232**, 335.
Gerard, R.W. (1). *AJOP*, 1927, **82**, 381.
Gerard, R.W. (2). *BB*, 1931, **60**, 245.
Gerard, R.W. & Hartline, H.K. *JCCP*, 1934, **4**, 141.
Gerard, R.W. & Rubenstein, B.B. *JGP*, 1934, **17**, 375.
Gerassimov, I.I. *ZAP*, 1902, **1**, 220.
Gerber, L. & Carr, R.H. *JN*, 1931, **3**, 245.
Gerould, J.H. *JEZ*, 1921, **34**, 385.
Gersch, M. (1). *AFEWM*, 1937, **136**, 210.
Gersch, M. (2). *AEZF*, 1939, **22**, 548.
Gersch, M. & Ries, E. *AFEWM*, 1937, **136**, 169.
Gerschenson, A.O. *ZKH*, 1931, **51**, 20; *OMaly*, 1931, **63**, 299.
Gersh, I. *CIWP* (Contrib. to Embryol. No. 153), 1937, **479**, 33.
Gey, G.O., Seegar, G.E. & Hellman, L.M. *S*, 1938, 306.
Gheorghiu, I. *AIPS*, 1938, **60**, 549.
van der Ghinst, M. (1). *ADB*, 1933, **44**, 545.
van der Ghinst, M. (2). *ADB*, 1934, **45**, 729.
van der Ghinst, M. (3). *BHA*, 1935, **12**, 257.
Ghiron, V. Cit. in Cook, Kennaway & Kennaway.
Gibbons, R.A. *JOGBE*, 1932, **39**, 539.
Giersberg, H. *VDZG*, 1935, **37**, 160.
Gilchrist, F.G. (1). *PZ*, 1928, **1**, 231.
Gilchrist, F.G. (2). *QRB*, 1929, **4**, 544.
Gilchrist, F.G. (3). *JEZ*, 1933, **66**, 15.
Gilchrist, F.G. (4). *QRB*, 1937, **12**, 251.
Gill, P.M. *BJ*, 1938, **32**, 1792.
Gillam, A.E. & Heilbron, I.M. *BJ*, 1935, **29**, 1064.
Gilmore, R.J. & Figge, F.H.J. *CCP*, 1929, No. 161.
Giolitti, G. *BSIBS*, 1939, **12**, 679.
de Giorgi, P. *RSZ*, 1924, **31**, 1.
de Giorgi, P. & Guyénot, E. *CRSB*, 1923, **89**, 488.
Giovannola, A. *JPA*, 1936, **22**, 207.
Giroud, A., Ratsimamanga, R., Rabinowicz, M., Santos-Ruiz, A. & Cesa, I. *CRSB*, 1936, **123**, 1038.
Giršavičius, O. *BZ*, 1933, **260**, 278.
Givens, J.W., Almquist, H.J. & Stokstad, E.L.R. *IEC*, 1935, **27**, 972.
Givens, M.H. & Macy, I.G. *JBC*, 1933, **102**, 7.
Gladstone, S.A. *AJDC*, 1932, **44**, 81.
Glancy, E.A. & Howland, R.B. *BB*, 1938, **75**, 99.
Glaser, O. (1). *BB*, 1914, **26**, 387.
Glaser, O. (2). *AR*, 1914, **8**, 525.

Glaser, O. (3). *AR*, 1917, **12**, 195.
Glaser, O. (4). *AN*, 1921, **55**, 368; *BB*, 1921, **41**, 63.
Glaser, O. (5). *AR*, 1922, **24**, 382.
Glaser, O. (6). *BR*, 1938, **13**, 20. *G*, 1939, **3** (Suppl.), 53.
Glaser, O. & Child, G. *BB*, 1937, **73**, 205.
Glaser, O. & Piehler, E. *BB*, 1934, **66**, 351.
Glasstone, S. *PRSB*, 1938, **126**, 315.
Glaus, H. *PSZN*, 1933, **13**, 39.
Glebova, M.S. *BBMEUSSR*, 1939, **7**, 16.
Glembosky, I. *BZUSSR*, 1935, **4**, 355.
Glezina, O. *UBZ*, 1939, **13**, 116.
Glick, B. *AR*, 1931, **48**.
Glick, D. & Biskind, G.R. *JBC*, 1935, **110**, 1; 1936, **115**, 551.
Glock, G.E. *JOP*, 1940, **98**, 1.
Glover, E.C. *CRSB*, 1931, **107**, 1603.
Glücksmann, A. (1). *ZAE*, 1930, **93**, 35.
Glücksmann, A. (2). Private communication to the author, 1938.
Glücksohn-Schoenheimer, S.S. (1). *GS*, 1938, **23**, 573.
Glücksohn-Schoenheimer, S.S. (2). *GS*, 1940, **25**, 391.
Glynn, W. *BB* (abstr.), 1937, **73**, 373.
Godfrey, A.B. *PS*, 1936, **15**, 294.
Godlewski, E. (1). *AFEWM*, 1908, **26**, 279.
Godlewski, E. (2). *AFEWM*, 1928, **114**, 108.
Godlewski, E. & Latinik-Vetulani, I. *BAPS*, 1930, B 11, 79.
Goecke, H. *AG*, 1936, **161**, 295.
Goecke, H., Wirz, P. & Daners, H. *AG*, 1933, **153**, 233.
Goeldi, E.A. *ZJS*, 1898, **10**, 640.
Goerttler, K. (1). *AFEWM*, 1927, **112**, 517.
Goerttler, K. (2). *VDAG*, 1939, **46**, 22 (Anat. Anz. **87**).
Goldberger, J., Wheeler, G.A., Lillie, R.D. & Rogers, L.M. *USPHR*, 1928, **43**, 1385.
Goldberger, S. *BSIBS*, 1931, **6**, 70; *OMaly*, 1931, **62**, 280.
Goldby, F. *JA*, 1928, **72**, 135.
Goldenberg, E.E., Landa-Glaz, R.I., Voinar, E.O. & Ostrumova, W.A. *Proc. xvth Internat. Congr. Physiol.* (Moscow), 1935, p. 117.
Goldforb, A.J. *JGP*, 1935, **19**, 149.
Goldhammer, H. & Kuen, F.M. *BZ*, 1933, **267**, 417.
Goldschmidt, R. (1). *PNASW*, 1916, **2**, 53.
Goldschmidt, R. (2). *AN*, 1917, **52**, 28.
Goldschmidt, R. (3). *AFEWM*, 1920, **47**, 1.
Goldschmidt, R. (4). *ZIAV*, 1920, **23**, 1; 1922, **29**, 145; 1923, **31**, 100; 1929, **49**, 169; 1930, **56**, 275; *QRB*, 1931, **6**, 125.
Goldschmidt, R. (5). *AMA*, 1923, **98**, 292.
Goldschmidt, R. (6). *ZIAV*, 1935, **69**, 38 and 70; *BC*, 1935, **55**, 535; *UCPZ*, 1937, **41**.
Goldschmidt, R. (7). *PNASW*, 1937, **23**, 219.
Goldschmidt, R. (8). "Physiological Genetics." McGraw, New York, 1938.

Goldschmidt, R. (9). "The Material Basis of Evolution." Yale Univ. Press, New Haven, 1940.
Goldstein, B. (1). *UBZ*, 1935, **8**, 87.
Goldstein, B. (2). *EZ*, 1938, **2**, 193.
Goldstein, B. & Ginzburg, M. *EZ*, 1937, **1**, 369; *UBZ*, 1936, **9**, 341; 1938, **11**, 65.
Goldstein, B. & Millgram, E. *UBZ*, 1935, **8**, 105 and 139.
Goldstein, B. & Volkenson, D. *UBZ*, 1939, **13**, 311.
Gomez, E.T. & Turner, C.W. *PSEBM*, 1936, **35**, 59.
y Gonzalez, A. *AR*, 1932, **52**, 117.
Goodale, H.D. *AJA*, 1911, **12**, 173.
Goodrich, H.B. & Hansen, I.B. *JEZ*, 1931, **59**, 337.
Goodrich, J.P. & Bodine, J.H. *PZ*, 1939, **12**, 312.
Gorbunova, G.P. *CRASUSSR*, 1939, **23**, 298.
Gordon, M. *AJC*, 1931, **15**, 732 and 1495; 1937, **30**, 362.
Gordon, M. & Smith, G.M. *AJC*, 1938, **34**, 255 and 543.
Gortner, R.A. (1). *AN*, 1910, **44**, 497; *JBC*, 1911, **10**, 113.
Gortner, R.A. (2). *AN*, 1912, **45**, 743.
Gortner, R.A. (3). "Outlines of Biochemistry." 2nd edition, New York, 1938, p. 458.
Goss, H. & Gregory, P.W. (1). *JEZ*, 1935, **71**, 311.
Goss, H. & Gregory, P.W. (2). *PSEBM*, 1935, **32**, 681.
Gosteieva, M. *BZUSSR*, 1935, **4**, 447.
Gottschewski, G. *ZIAV*, 1934, **67**, 477.
Gouzon, B. *CRSB*, 1934, **116**, 925.
Gowen, J.W. (1). *JGP*, 1931, **14**, 463.
Gowen, J.W. (2). *AOP*, 1934, **17**, 638.
Gowen, J.W. (3). *CSHSQB*, 1934, **2**, 128.
Gowen, J.W. & Gay, E.H. *AN*, 1932, **66**, 289.
Graff, S. & Barth, L.G. *CSHSQB*, 1938, **6**, 103.
Graff, S. & Graff, A.M. *JBC*, 1937, **121**, 79.
Grafflin, A.L., Gould, R.G. & Spence, G. *BB*, 1936, **70**, 16.
Grah, H. *ZJZP*, 1937, **57**, 356.
Graham, W.R. *SA*, 1932, **12**, 427; *YMaly*, 1932, **22**, 650.
Graham, W.R., Smith, J.B. & McFarlane, W.D. *Bull. Ontario Agric. Coll.* 1931, no. 362.
Graham-Kerr, J. "Textbook of Embryology, vol. 11, Vertebrata." London, 1919. For vol. 1, *see* McBride.
Grandori, R. (1). *BLZAM*, 1931, **2**, 22; *YMaly*, 1932, **22**, 676.
Grandori, R. (2). *BLZAM*, 1932, **3**, 156; *YMaly*, 1933, **23**, 730.
Gräper, L. (1). *AFEWM*, 1922, **51**, 587.
Gräper, L. (2). *AFEWM*, 1926, **107**, 154 and 162.

Grasset, E. *SAIMRP*, 1929, **4**, 171.

Grasset, E. & Zoutendyk, A. *SAIMRP*, 1929, 4, 377.

Graubard, M.A. *JG*, 1933, 27, 199.

Gravier, C. *ASNZ*, 1931 (10ᵉ sér.), 14, 303.

Gray, J. (1). *JEB*, 1926, 4, 215.

Gray, J. (2). "A Textbook of Experimental Cytology." Cambridge University Press, 1931.

Gray, J. (3). *JEB*, 1932, 9, 277.

Gray, J. & Oeillet, C. *PRSB*, 1933, 114, 1.

Gray, P. *AFEWM*, 1939, 139, 732.

Gray, P. & Worthing, H. *JEZ*, 1941, 86, 423.

Green, C.V. *JH*, 1936, 27, 181.

Greenberg, D.M. & Tufts, E.V. *JBC*, 1936, 114, 135.

Greene, H.S.N., Hu, Ch'uan-K'uei & Brown, W.H. *S*, 1934, 79, 487.

Greene, H.S.N. & Saxton, J.A. *JEM*, 1939, 69, 301.

Greene, R.S., Burrill, M.W. & Ivy, A.C. *AJOG*, 1938, 36, 1038.

Greenstein, J.P. & Jenrette, W.V. *JNCI*, 1940, 1, 77 & 91.

Greenwald, I. *JBC*, 1925, 63, 339.

Greenwood, A.W. *JEB*, 1925, 2, 165.

Greenwood, A.W. & Blyth, J.S.S. *PSEBM*, 1931, 29, 38.

Gregory, J.G. *JPS*, 1927, 2, 301.

Gregory, P.W., Asmundson, V.S. & Goss, H. *JEZ*, 1936, 73, 263.

Gregory, P.W., Asmundson, V.S., Goss, H. & Landauer, W. *G*, 1939, 3, 75.

Gregory, P.W. & Castle, W.E. *JEZ*, 1931, 59, 199.

Gregory, P.W. & Goss, H. (1). *AN*, 1933, 67, 180.

Gregory, P.W. & Goss, H. (2). *JEZ*, 1933, 66, 155.

Gregory, P.W. & Goss, H. (3). *JEZ*, 1933, 66, 335.

Gregory, P.W. & Goss, H. (4). *JEZ*, 1934, 69, 13; *G*, 1939, 3, 159.

Gregory, P.W., Goss, H. & Asmundson, V.S. *PSEBM*, 1935, 32, 966; *G*, 1937, 1, 89.

Greig, M.E., Munro, M.P. & Elliott, K.A.C. *BJ*, 1939, 33, 443.

Gresser, E.B. & Breder, C.M. *ZA*, 1940, 25, 113.

Greulich, W.W. *AJPA*, 1934, 19, 392.

Grevenstuk, A. *EP*, 1929, 28, 171.

Grigaut, M.A. *BSCB*, 1935, 17, 1031.

Grobstein, C. Inaug. Diss. Los Angeles, 1940; *UCPZ*, 1940, 47, no. 1; *PSEBM*, 1940, 45, 484.

Grodziński, Z. (1). *BAA*, 1931, 25, 210.

Grodziński, Z. (2). *AEZF*, 1932, 12.

Grodziński, Z. (3). *AFEWM*, 1933, 129, 502; 1934, 131, 653.

Grodziński, Z. (4). *BAPS*, 1934, 415; 1935, 305.

Grodziński, Z. (5). *BAPS*, 1939, 317.

Groebbels, F. (1). *ZB*, 1922, 75, 155.

Groebbels, F. (2). *JO*, 1927, 75, 225 and 376.

Groebbels, F. (3). *ZVP*, 1933, 19, 574.

Grollman, A. (1). *JBC*, 1929, 81, 267.

Grollman, A. (2). *BZ*, 1931, 238, 408.

van Groor, H. *ABN*, 1940, 10, 37.

Gross, F. *QJMS*, 1936, 79, 57.

Grosser, O. (1). "Frühentwicklung, Eihautbildung und Placentation." Bergmann, München, 1927.

Grosser, O. (2). *L*, 1933 (i), 224, 999 and 1053.

Grosser, O. (3). *VDAG*, 1936, 43, 15.

Grossfeld, J. *ZUNG*, 1935, 70, 82; *YMaly*, 1936, 36, 294.

Grotans, A. (1). *ABL*, 1934, 4.

Grotans, A. (2). *ABL*, 1934, 4; *AZ*, 1934, 15, 215.

Grüneberg, H. (1). *PRSB*, 1935, 118, 321; *JH*, 1936, 27, 105.

Grüneberg, H. (2). *JA*, 1937, 71, 236.

Grüneberg, H. (3). *JG*, 1938, 36, 153.

Grüneberg, H. (4). *PRSB*, 1938, 125, 123.

Grünwald, P. *AFEWM*, 1937, 136, 786.

Gruzewska, Z. & Roussel, G. (1). *APBPC*, 1935, 11, 176; *CRSB*, 1934, 117, 863; *Proc. xvth Internat. Congr. Physiol.* (Moscow), 1935, p. 133.

Gruzewska, Z. & Roussel, G. (2). *CRSB*, 1934, 115, 951; 1936, 122, 613; 1936, 123, 377.

Grynberg, M.Z. & Kisiel, S. *BZ*, 1932, 253, 146.

Guareschi, C. (1). *BIZUR*, 1928, 6, 1.

Guareschi, C. (2). *MRAI*, 1932, 3, 1.

Guareschi, C. (3). *AZI*, 1933, 18, 265.

Guareschi, C. (4). *AAL*, 1934 (6ᵉ ser.), 19, 648; 1934 (6ᵉ ser.), 20, 56.

Guareschi, C. (5). *MZI*, 1935, Suppl. 45, 315.

Guariglia, G. *BSIBS*, 1937, 12, 690.

Gudernatsch, F. (1). *AFEWM*, 1912, 35, 457; *AJA*, 1914, 15, 431.

Gudernatsch, F. (2). *ZAE*, 1926, 80, 764; "Handbuch d. inn. Sekretion", 1930, vol. ii, 1491.

Gudernatsch, F. (3). *CSHSQB*, 1934, 2, 94; Barell Jubilee Volume, Basel, 1936, p. 453.

Gudernatsch, F. & Hoffman, O. (1). *AJOP*, 1931, 97, 527; 1932, 101, 47; 1933, 105; 1934, 109; 1935, 113; *AR*, 1933 (Suppl.), 55, 4 and 57; 1934 (Suppl.), 58, 65.

Gudernatsch, F. & Hoffman, O. (2). *AFEWM*, 1936, 135, 136; *KW*, 1931, 10, 1802.

Gudger, E.W. *JM*, 1935, 58, 1.

Guelin-Schedrina, A. (1). *APBPC*, 1934, 10, 453; *CRSB*, 1933, 113, 717.

Guelin-Schedrina, A. (2). *CRSB*, 1936, 121, 144.

Guercia, T. *AOG*, 1935, 42, 1; *OMaly*, 1935, 87, 164.

Guérithault, B. *BSHA*, 1927, 15, 386; *CA*, 1928, 22, 809.

Guerrant, N.B., Kohler, E., Hunter, J.E. & Murphy, R.R. *JN*, 1935, 10, 167.

Guha, B.C. & Pal, J.C. *N*, 1936, 137, 946; 1937, 138, 843.

Guilbert, H.R. & Goss, H. *JN*, 1932, **5**, 251.

Gulick, A. *Proc. xvth Internat. Congr. Physiol.* (Moscow), 1935, p. 139.

Gunther, H. *EAPPA*, 1922, **20**, 608.

Gurwitsch, A. *AFEWM*, 1921, **52**, 383; 1927, **112**, 433.

Guthmann, H. & May, W. *MGG*, 1932, **91**, 306.

Gutman, A.B. *AR*, 1926, **34**, 133.

Gutman, C. & Levy-Solal, E. *BSCB*, 1934, **16**, 720.

von Guttenberg, H. *BC*, 1928, **48**, 31.

Guyénot, E. (1). *RSZ*, 1927, **34**, 1.

Guyénot, E. (2). *RSZ*, 1927, **34**, 127.

Guyénot, E. (3). *SPHNGCR*, 1927, **44**.

Guyénot, E. & Matthey, R. *AFEWM*, 1928, **113**, 520.

Guyénot, E. & Ponse, K. (1). *CRSB*, 1923, **89**, 4.

Guyénot, E. & Ponse, K. (2). *BBFB*, 1930, **64**, 251.

Guyénot, E. & Schotté, O. (1). *CRSB*, 1923, **89**, 491.

Guyénot, E. & Schotté, O. (2). *CRSB*, 1926, **94**, 1050.

Guyénot, E. & Schotté, O. (3). *SPHNGCR*, 1927, **44**, 21.

Guyer, M.F. *S*, 1907, **25**, 910.

Guyer, M.F. & Smith, E.A. *JEZ*, 1924, **38**, 449.

Gye, W.E. *BMJ*, 1938 (i), 551.

Gye, W.E. & Purdy, W.J. "The Cause of Cancer." London, 1931.

György, P. *JBC*, 1939, **131**, 733.

György, P., Rose, C.S., Hofmann, K., Melville, D.B. & du Vigneaud, V. *S*, 1940, **92**, 62 and 609.

Haagen-Smit, A.J. *EVHF*, 1938, **2**, 347.

Haberling, W. *AGMNT*, 1927, **10**, 166.

Hachlov, V. *AFEWM*, 1931, **125**, 26.

Haddow, A. (1). *AIUC*, 1937, **2**, 376; 1938, **3**, 342.

Haddow, A. (2). *JPB*, 1938, **47**, 553 and 567.

Haddow, A. (3). *JPB*, 1938, **47**, 581.

Haddow, A. & Robinson, A.M. (1). *PRSB*, 1937, **122**, 442.

Haddow, A. & Robinson, A.M. (2). *PRSB*, 1939, **127**, 277.

Haddow, A., Scott, C.M. & Scott, J.D. *PRSB*, 1937, **122**, 477.

Hadjioloff, A. *JMFUS*, 1931, **11**, 1.

Hadjioloff, A. & Usunoff, G. *CRSB*, 1933, **114**, 578; *JMFUS*, 1932, **12**, 1; *YMaly*, 1936, **36**, 18.

Hadley, C.E. *JEZ*, 1929, **54**, 127.

Hadley, F.B. & Warwick, B.L. *JAVA*, 1927, **70**, 492.

Hadley, Philip. *JID*, 1937, **60**, 129.

Hadorn, E. (1). *AFEWM*, 1932, **125**, 495.

Hadorn, E. (2). *AFEWM*, 1934, **131**, 238; *RSZ*, 1934, **41**, 411.

Hadorn, E. (3). *AFEWM*, 1937, **136**, 400; *RSZ*, 1935, **42**, 417.

Hadorn, E. (4). *PSEBM*, 1937, **36**, 632; *RSZ*, 1938, **45**, 425.

Hadorn, E. (5). *PNASW*, 1937, **23**, 478.

Hadorn, E. & Neel, J. (1). *GS*, 1938, **23**, 151.

Hadorn, E. & Neel, J. (2). *AFEWM*, 1939, **138**, 281.

Hadzi, P. *Verhdl. d. Internat. Kongr. Zool.* (Graz), 1910, **8**, 578.

Hagan, H.R. *JM*, 1931, **51**, 3.

Hahn, L. & Hevesy, G. *N*, 1937, **140**, 1059.

Hahn, P.F. & Fairman, E. *JBC*, 1936, **113**, 161.

Haigh, L.D., Moulton, C.R. & Trowbridge, P.F. *Res. Bull. Univ. Missouri Agric. Exp. Sta.* no. 38, 1920.

Hain, A.M. *QJEP*, 1935, **25**, 131 and 303; 1936, **26**, 29.

Haldane, J.B.S. (1). *AN*, 1932, **66**, 5.

Haldane, J.B.S. (2). *JPB*, 1934, **38**, 507.

Haldane, J.B.S. (3). *CRSB*, 1935 (Réunion Plénière), **119**, 1481; "The Biochemistry of the Individual" in "Perspectives in Biochemistry." Hopkins Presentation Volume. Cambridge, 1937, p. 1.

Hale, F. *JH*, 1933, **24**, 105; *Press Bull. Texas Agric. Exp. Sta.* 1934, Jan. 15th.

Hale, H.P. & Hardy, W.B. *PRSB*, 1933, **112**, 473.

Hall, E.K. (1). *JEZ*, 1937, **75**, 11.

Hall, E.K. (2). *AFEWM* (abstr.), 1932, **127**, 573; 1937, **135**, 671.

Hall, F.G. (1). *JOP*, 1934, **80**, 502; 1934, **82**, 33.

Hall, F.G. (2). *JOP*, 1934, **83**, 222.

Hall, F.G., Dill, D.B. & Barron, E.S.G. *JCCP*, 1936, **8**, 301.

Hall, F.G. & McCutcheon, F.H. *JCCP*, 1938, **11**, 205.

Hall, G.O. & van Wagenen, A. *PS*, 1936, **15**, 501.

Hall, J.A. *CR*, 1937, **20**, 305.

Hallez, P. *CRAS*, 1886, **103**, 606.

Hamburger, V. (1). *JEZ*, 1935, **70**, 43.

Hamburger, V. (2). *PSEBM*, 1939, **41**, 13; *PZ*, 1941, **14**, 355.

Hamburger, V. (3). *AFEWM*, 1928, **114**, 272.

Hamburger, V. & Waugh, M. *PZ*, 1940, **13**, 367.

Hamilton, B. *APD*, 1936, **18**, 272; *OMaly*, 1936, **94**, 30.

Hamilton, B. & Dewar, M.M. *G*, 1938, **2**, 13.

Hamlett, G.W.D. (1). *QRB*, 1933, **8**, 348.

Hamlett, G.W.D. (2). *AR*, 1935, **62**, 279.

Hamlett, G.W.D. (3). *QRB*, 1935, **10**, 432.

Hämmerling, J. (1). *BC*, 1931, **51**, 633; 1932, **52**, 42; *AFEWM*, 1934, **131**, 1; *ZIAV*, 1933, **62**, 92.

Hämmerling, J. (2). *NW*, 1934, **22**, 829; *AFEWM*, 1934, **132**, 424; *BC*, 1934, **54**, 650.

Hämmerling, J. (3). *ZJZP*, 1936, **56**, 440.

Hämmerling, J. (4). *BC*, 1939, **59**, 158.

Hammett, D.W. & Hammett, F.S. *P*, 1932, **15**, 59.

Hammett, F.S. (1). *P*, 1928, **4**, 183; 1928, **5**, 535 and 547; 1929, **7**, 297 and 535; 1930, **11**, 382; 1931, **13**, 331; 1933, **19**, 510.

Hammett, F.S. (2). *AOP*, 1929, **8**, 575.

Hammett, F.S. (3). *PAPS*, 1930, **68**, 151; 1930, **69**, 217.

Hammett, F.S. (4). Art. in Roffo Presentation Volume. Buenos Aires, 1935, p. 501.

Hammett, F.S. (5). *JBC*, 1925, **64**, 410 & 685.

Hammett, F.S., Anderson, J. & Justice, E. *P*, 1931, **12**, 190.

Hammett, F.S. & Hammett, D.W. *P*, 1932, **16**, 253.

Hammett, F.S. & Justice, E. *P*, 1928, **5**, 543.

Hammett, F.S. & Reimann, S.P. *JEM*, 1929, **50**, 445; *PSEBM*, 1929, **27**, 20.

Hammett, F.S. & Smith, D.W. *P*, 1931, **13**, 261.

Hammett, F.S. & Wallace, V.L. *JEM*, 1928, **48**, 659.

Hammond, J. (1). "Reproduction in the Cow." Cambridge University Press, 1927.

Hammond, J. (2). *JEB*, 1934, **11**, 140.

Hammond, J. (3). *TDDUSSR*, 1935, **10**, 93.

Hammond, J. (4). *SSR*, 1937, 548.

Hammond, J. & Appleton, A.B. "Growth and Development of Mutton Qualities in Sheep." Edinburgh, 1932.

Hanaoka, M. *JJCR*, 1936, **30**, 458; *OMaly*, 1937, **98**, 41.

Hardy, W.B. (1). Guthrie Lecture; *PPS*, 1916, **28** (ii), 99; *JCS*, 1925, **127**, 1207; *CSM*, 1928, **6**, 7.

Hardy, W.B. (2). *JGP*, 1927, **8**, 641.

Hardy, W.B. & Nottage, M.E. *PRSA*, 1928, **118**, 209.

Harman, M.T. *AR*, 1922, **23**, 363.

Harman, M.T. & Derbyshire, R.C. *AJA*, 1931, **49**, 335.

Harnly, M.H. (1). *JEZ*, 1930, **56**, 363.

Harnly, M.H. (2). *GS*, 1936, **21**, 84.

Harnly, M.H. & Ephrussi, B. *GS*, 1937, **22**, 393.

Harnly, M.H. & Harnly, M.L. (1). *JEZ*, 1935, **72**, 75.

Harnly, M.H. & Harnly, M.L. (2). *JEZ*, 1936, **74**, 41.

Harris, H.A. (1). *JA*, 1930, **64**, 303.

Harris, H.A. (2). *N*, 1932, **130**, 996.

Harris, J.E. *JEB*, 1935, **12**, 65.

Harris, L.J. (1). *PRSB*, 1923, **94**, 426.

Harris, L.J. (2). "Vitamins and Vitamin Deficiencies." 8 vols. Churchill, London, vol. I, 1938.

Harris, L.J. (3). "Vitamins." In *Brit. Encyclo. of Med. Pract.* 1939, **12**, 570.

Harrison, R.G. (1). *AFEWM*, 1898, **7**, 430; *AMA*, 1903, **63**, 35.

Harrison, R.G. (2). *PSEBM*, 1907, **4**, 140.

Harrison, R.G. (3). *JEZ*, 1918, **25**, 413.

Harrison, R.G. (4). *PSEBM*, 1920, **17**, 199.

Harrison, R.G. (5). *JEZ*, 1921, **32**, 1.

Harrison, R.G. (6). *PNASW*, 1924, **10**, 69.

Harrison, R.G. (7). *JCN*, 1924, **37**, 123.

Harrison, R.G. (8). *S*, 1924, **59**, 448; also "Linacre Lecture," Cambridge University, 6 May 1939

Harrison, R.G. (9). *AFEWM*, 1929, **120**, 1.

Harrison, R.G. (10). *S*, 1931, **74**, 575.

Harrison, R.G. (11). "Harvey Lectures", 1933–4, p. 116; *Proc. xth Internat. Congr. Zool.* (Budapest), 1929, p. 642.

Harrison, R.G. (12). *AN*, 1933, **67**, 306.

Harrison, R.G. (13). *AR*, 1935, **64** (Suppl. no. 1), 38.

Harrison, R.G. (14). *PRSB*, 1935, **118**, 155 (Croonian lect.).

Harrison, R.G. (15). *PNASW*, 1936, **22**, 238.

Harrison, R.G. (16). *CN*, 1936, **11**, 217.

Harrison, R.G. (17). *S*, 1937, **85**, 369.

Harrison, R.G. (18). *VDAG*, 1938, **85**, 4.

Harrison, R.G., Astbury, W.T. & Rudall, K.M. *JEZ*, 1940, **85**, 339.

Harrop, G.A. & Barron, E.S.G. *JEM* 1928, **48**, 207.

Hartman, C.G. (1). *JMM*, 1927, **8**, 96.

Hartman, C.G. (2). *AJOG*, 1930, **19**, 511.

Hartman, C.G. (3). *JAMA*, 1931, **97**, 1863.

Hartman, C.G., Squier, R.R. & Tinklepaugh, O.L. *PSEBM*, 1930, **28**, 285.

Hartmann, M. "Die methodologischen Grundlagen der Biologie." Meiner, Leipzig, 1933; also "Wesen und Wege d. biol. Erkenntnis", *NW*, 1936, **24**, 705; also "Die Kausalität in Physik u. Biologie", *Verh. Akad. Wiss.* Berlin, 1937.

Hartmann, M., Schartau, O., Kühn, R. & Wallenfels, K. *NW*, 1939, **27**, 433; 1940, **28**, 144; *BC*, 1940, **60**, 398.

Harvey, E.B. (1). *BB*, 1927, **52**, 147; 1930, **58**, 288.

Harvey, E.B. (2). *BB*, 1932, **62**, 155; 1933, **64**, 125; 1933, **65**, 389; 1934, **66**, 228.

Harvey, E.B. (3). *BSIBS*, 1934, **9**, 484.

Harvey, E.B. (4). *AR*, 1935, **64** (Suppl.), 57.

Harvey, E.B. (5). *BB*, 1935, **69**, 287.

Harvey, E.B. (6). *BB*, 1936, **71**, 101; *S*, 1935 (II), 277.

Harvey, E.B. (7). *BB*, 1938, **75**, 170; 1940, **78**, 412; 1940, **79**, 166.

Harvey, E.B. (8). *BB*, 1939, **76**, 384.

Harvey, E.B. & Hollaender, A. *BB*, 1938, **75**, 258.

Harvey, E.N. (1). *BB*, 1931, **60**, 67.

Harvey, E.N. (2). *BB*, 1931, **61**, 273.

Harvey, E.N. (3). *JEB*, 1931, **8**, 267; *JFI*, 1932, **214**, 1; *S*, 1930, **72**, 42.

Harvey, E.N. (4). *BB*, 1932, **62**, 141.

Harvey, E.N. (5). *JCCP*, 1933, **4**, 35.

Harvey, E.N. (6). *AEZF*, 1939, **22**, 463.

Harvey, E.N. & Danielli, J.F. *BR*, 1938, **13**, 319.

Harvey, E.N. & Fankhauser, G. *JCCP*, 1933, **3**, 463.

Harvey, E.N. & Schoepfle, G. *JCCP*, 1939, **13**, 383.

Harvey, E.N. & Shapiro, H. *JCCP*, 1934, **5**, 255.

Haselhorst, G. *ZKL* [Zeitschr. f. d. ges. Anat. Abt. II], 1931, **15**, 177.

Haselhorst, G. & Stromberger, K. (1). *ZGG*, 1930, **95**, 400; 1930, **98**, 49.

Haselhorst, G. & Stromberger, K. (2). *AG*, 1931, **147**, 65.

Haselhorst, G. & Stromberger, K. (3). *ZGG*, 1932, **100**, 48; 1932, **102**, 16.

Haszler, K. *BPAAP*, 1933, **92**, 101; *YMaly*, 1934, **28**, 309.

Hatt, P. (1). *CRSB*, 1933, **113**, 246; *AAM*, 1934, **30**, 131.

Hatt, P. (2). "Les Mouvements Morphogénétiques dans le développement des Vertébrés." Hermann, Paris, 1935.

Hauge, S.M. & Carrick, C.W. *PS*, 1926, **5**, 166.

Hauptstein, P. *AG*, 1932, **151**, 262.

Haurowitz, F. *ZPC*, 1935, **232**, 125; *Proc. xvth Int. Congr. Physiol.* (Moscow), 1935, p. 148.

Havard, R.E. & Kendal, L.P. *BJ*, 1934, **28**, 1121.

Hayes, F.R. (1). *BJ*, 1930, **24**, 723 and 735.

Hayes, F.R. (2). *CIWY*, 1931, **31**, 284; *CIWP*, no. 435, 1933, 181.

Hayes, F.R. (3). *BB*, 1938, **74**, 267.

Hayes, F.R. & Hollett, A. *CJR*, 1940, **18**, 53.

Hayes, F.R. & Ross, D.M. *PRSB*, 1936, **121**, 358.

Hays, F.A. & Nicolaides, C. *PS*, 1934, **13**, 74.

Haywood, C. & Root, W.S. (1). *BB*, 1930, **59**, 63.

Haywood, C. & Root, W.S. (2). *JCCP*, 1932, **2**, 177.

Hearne-Creech, E.M. *N*, 1936, **138**, 291; *AJC*, 1939, **35**, 191; 1940, **39**, 149.

Heatley, N.G. (1). *BJ*, 1935, **29**, 2568.

Heatley, N.G. (2). *JSI*, 1940, **17**, 197.

Heatley, N.G., Berenblum, I. & Chain, E. *BJ*, 1939, **33**, 53 and 68.

Heatley, N.G. & Lindahl, P.E. *PRSB*, 1937, **122**, 395.

Heatley, N.G., Waddington, C.H. & Needham, J. *PRSB*, 1937, **122**, 403.

Heaton, T.B. *JPB*, 1926, **29**, 293; 1929, **32**, 565.

Hegner, R.G. *BB*, 1908, **16**, 19; 1911, **20**, 237.

Heidermanns, C. (1). *TB* (data), 1937, **14**, 209; *NW* (review), 1938, **26**, 263 and 279.

Heidermanns, C. (2). *ZJZP*, 1937, **58**, 57.

Heidermanns, C. (3). *VDZG*, 1938, **11**, 55; *FF*, 1938, **14**, 370.

Heilbrunn, L.V. *BB*, 1936, **71**, 299.

Heilbrunn, L.V. & Daugherty, K. *PZ*, 1938, **11**, 383.

Heilbrunn, L.V. & Mazia, D. Contrib. to "Biological Effects of Radiation." McGraw Hill, N.Y. 1936, chap. xviii, p. 625.

Heilbrunn, L.V. & Wilbur, K.M. *BB*, 1937, **73**, 557.

Hektoen, L. & Cole, A.G. (1). *JID*, 1928, **42**, 1.

Hektoen, L. & Cole, A.G. (2). *PSEBM*, 1937, **36**, 97.

Helff, O.M. (1). *AR*, 1926, **34**, 129.

Helff, O.M. (2). *JEZ*, 1926, **45**, 69.

Helff, O.M. (3). *JEZ*, 1926, **45**, 400; *AR*, 1924, **29**, 102.

Helff, O.M. (4). *PZ*, 1929, **2**, 334.

Helff, O.M. (5). *AR*, 1928, **41**, 40; 1930, **47**, 177.

Helff, O.M. (6). *PZ*, 1928, **1**, 463; *BB*, 1934, **66**, 38.

Helff, O.M. (7). *JEZ*, 1931, **59**, 179.

Helff, O.M. (8). *BB*, 1931, **60**, 11; 1933, **65**, 304.

Helff, O.M. (9). *BB*, 1932, **63**, 405.

Helff, O.M. (10). *AR*, 1934, **59**, 201.

Helff, O.M. (11). *JEZ*, 1934, **68**, 305.

Helff, O.M. (12). *BB*, 1934, **66**, 38.

Helff, O.M. (13). *JEB*, 1937, **14**, 1.

Helff, O.M. (14). *JEB*, 1939, **16**, 96.

Helff, O.M. (15). *JEB*, 1940, **17**, 45.

Helff, O.M. & Clausen, H.J. *PZ*, 1929, **2**, 575.

Helff, O.M. & Stark, W. *JM*, 1941, **68**, 303.

Heller, H. & Holtz, P. *JOP*, 1932, **74**, 134.

Heller, J. (1). *AP*, 1925, **210**, 736; *BZ*, 1926, **169**, 208; 1926, **172**, 59; *ZVP*, 1928, **8**, 99; 1930, **11**, 448.

Heller, J. (2). *ABE*, 1928, **1**, 225.

Heller, J. (3). *BC*, 1931, **51**, 259.

Heller, J. (4). *BZ*, 1932, **255**, 205.

Heller, J. (5). *ZVP*, 1933, **18**, 796.

Heller, J. (6). *CRSB*, 1936, **121**, 414.

Heller, J. (7). *ABE*, 1938, **12**, 99.

Heller, J. & Aremówna, H. *ZVP*, 1932, **16**, 362.

Heller, J. & Meisels, E. *BC*, 1927, **47**, 257; *FGR*, 1927, **36**, 104.

Heller, J. & Mokłowska. *BZ*, 1930, **219**, 473.

Hellerman, L. *PR*, 1937, **17**, 454.

Henderson, E.W. *Bull. Missouri Agric. Exp. Sta.* 1930, no. 149.

Henderson, E.W. & Penquite, R. *Atti d. v Congr. Mond. di Pollicoltura*, 1933, no. 11.

Henderson, E.W. & Wilcke, H.L. *PS*, 1933, **12**, 266.

Henke, K. (1). *AFEWM*, 1933, **128**, 15.

Henke, K. (2). *VDZG*, 1935, 176; *NW*, 1933, **21**, 633.

Henke, K. (3). "Allgemeine Genetik, einschl. Genphysiol." *FZ*, 1937, **1**, 501; 1938, **3**, 418.

Henrici, A.T. *PSEBM*, 1921, **19**, 132; 1923, **21**, 215, 343, 345.

Henze, M. *ZPC*, 1908, **55**, 433.

Hepburn, J.S. & Miraglia, P.R. *JBC*, 1934, **105**, xxxviii.

Herbst, C. (1). *ZWZ*, 1892, **55**, 446.

Herbst, C. (2). *BC*, 1894, **14**, 657, 689, 727, 753 and 800; 1895, **15**, 721, 753, 792, 817 and 849; also "Formative Reize in d. Ontogenese." Leipzig, 1901.

Herbst, C. (3). *AFEWM*, 1896, **2**, 544.

Herbst, C. (4). *SAWH*, 1928, no. 2; 1929, no. 16; *NW*, 1932, **20**, 375; *AFEWM*, 1935, **132**, 576; 1936, **134**, 313; 1937, **135**, 178; 1939, **139**, 282.

Herd, J.D. *BJ*, 1936, **30**, 1743; 1937, **31**, 1478 and 1484.

Heringa, G.C. & Valk, S.H. van K. *PKAWA*, 1930, **33**, 530; *OMaly*, 1931, **57**, 396.

l'Héritier, P. & Teissier, G. *CRAS*, 1937, **205**, 1099.

van Herk, A.W.H. *ANP*, 1933, **18**, 578.

Herlitzka, A. *AFEWM*, 1896, **4**, 624.

Herold, L. *AG*, 1934, **158**, 213.

Hersh, A.H. (1) *JEZ*, 1930, **57**, 283.

Hersh, A.H. (2). *AN*, 1934, **68**, 537.

Hersh, A.H. & Ward, E. *JEZ*, 1932, **61**, 223.

Hertwig, G. (1). *AMA*, 1911, **77**, 165; 1913, **81**, 87.

Hertwig, G. (2). *AFEWM*, 1925, **105**, 294; 1927, **III**, 292.

Hertwig, G. & Hertwig, P. "Regulation von Wachstum, Entwicklung und Regeneration durch Umweltsfaktoren." In *Handb. d. norm. u. pathol. Physiol.* (Bethe), 1930, **16**, Hälfte 1.

Hertwig, L.H. *AP*, 1927, **216**, 796.

Hertwig, O. (1). *AMA*, 1895, **44**, 285; "Gegenbaur Festschrift", 1896, **2**, 89.

Hertwig, O. (2). *AMA*, 1899, **53**, 514.

Hertwig, O. (3). *AMA*, 1911, **77**, 1; 1913, **82**, 1.

Hertwig, P. (1). *AMA*, 1916, **87**, 63; *AFEWM*, 1924, **100**, 41.

Hertwig, P. (2). "Artbastarde bei Tieren." In *Handbuch f. Vererbungslehre*, 1936, **21**.

van Herwerden, M.A. *AZF*, 1912, **10**, 431.

Hetherington, D.C. & Craig, J.S. *PSEBM*, 1940, **44**, 282.

Hevesy, G. *JCS*, 1939, 1213; *EZ*, 1938, **5**, 138.

Hevesy, G. & Hahn, L. *BMKDVS*, 1938, **14**, no. 2.

Hevesy, G., Levi, H.B. & Rebbe, O.H. *BJ*, 1938, **32**, 2147.

van der Heyde, *BC*, 1922, **42**, 419.

Heydenreich, F. *AFEWM*, 1935, **132**, 600.

Heyl, H.H. *S*, 1939, **89**, 540.

van Heyningen, W.E. (1). Unpublished observations communicated to the author, 1938.

van Heyningen, W.E. (2). *BR*, 1939, **14**, 420.

Heys, F. *QRB*, 1931, **6**, 1.

Hibbard, H. (1). *ADB*, 1927, **38**, 3.

Hibbard, H. (2). *BB* (abstr.), 1937, **73**, 385.

Hildebrand, S.F. *JH*, 1938, **29**, 243.

Hilgenberg, F.C. *ZGG*, 1931, **98**, 291.

Hill, A.V. (1). *TFS*, 1931, **26**, 667.

Hill, A.V. (2). "Adventures in Biophysics." Oxford University Press, 1931, pp. 55 ff.

Hill, C.J. & Hill, J.P. *TZS*, 1933, 413.

Hill, D. Inaug. Diss. Iowa State, 1940.

Hill, L. & Burdett, E.F. *N*, 1932, **130**, 540.

Hill, Robin (1). *PRSB*, 1930, **107**, 205.

Hill, Robin (2). *PRSB*, 1936, **120**, 472.

Hill, Robin (3). "Haemoglobin." In Hopkins Presentation Volume "Perspectives in Biochemistry." Cambridge, 1937, p. 127.

Hill, Robin & Wolvekamp, H.P. *PRSB*, 1936, **120**, 484.

Hill, R.T. *ENY*, 1937, **21**, 495 and 633.

Hill, R.T. & Strong, M.T. *ENY*, 1938, **22**, 663.

Hiller, S. (1). *BAPS*, 1926.

Hiller, S. (2). *PSEBM*, 1927, **24**, 427.

Hiller, S. (3). *CRAA*, 1931, 1.

Hinrichs, M.A. *PZ*, 1938, **11**, 155.

Hinrichs, M.A. & Genther, I.T. *PZ*, 1931, **4**, 461.

Hirabayashi, K. *AZJ*, 1937, **16**, 205.

Hiraiwa, Y.K. *JEZ*, 1927, **49**, 441; *FAJ*, 1930, **8**, 157.

Hirano, Y. *OJDI*, 1935, **17**, 7; *CA*, 1937, **31**, 7974.

Hirschler, J. *AFEWM*, 1922, **51**, 482.

His, W. "Unsere Körperform." Vogel, Leipzig, 1874, p. 28.

Hitchcock, F.A. & Haub, J.G. *AR*, 1935, **64** (Suppl.), 61; *AESA*, 1941, **34**, 17.

Hitchings, G.H., Oster, R.H. & Salter, W.T. *BJ*, 1938, **32**, 1389.

Hoadley, L. (1). *BB*, 1924, **46**, 281; *JEZ*, 1926, **43**, 151; *ADB*, 1926, **36**, 225; *JEZ*, 1927, **48**, 459.

Hoadley, L. (2). *JEZ*, 1928, **52**, 7.

Hoadley, L. (3). *AFEWM*, 1929, **116**, 278.

Hoadley, L. (4). *BB*, 1930, **58**, 123.

Hoadley, L. (5). *ADB*, 1931, **42**, 325.

Hoadley, L. (6). *BB*, 1934, **67**, 484.

Hoadley, L. (7). *G*, 1938, **2**, 25.

Hoadley, L. & Brill, E.R. *G*, 1937, **1**, 234.

Hoag, L.A. & Kiser, W.H. *AJDC*, 1931, **41**, 1054; *OMaly*, 1931, **63**, 101.

Hoagland, H. *JGPS*, 1933, **9**, 267.

Hobson, R.P. *JEB*, 1932, **9**, 128.

von Hoesslin, H. *ZB*, 1930, **90**, 615.

Hoffman, O. *CSHSQB*, 1934, **2**, 106; *JPET*, 1935, **54**, 146.

Hoffman, O. & Gudernatsch, F. *ENY*, 1933, **17**, 239; *ENE*, 1936, **18**, 96; *PSEBM*, 1931, **28**, 731; *AJOP*, 1933, **105**, 54; 1934, **109**, 1935, **113**.

Hoffman, R.S. *G*, 1940, **4**, 361.

Hoffman, R.S., Goldschmidt, J. & Doljanski, L. *G*, 1937, **1**, 228; *CRSB*, 1937, **126**, 389 & 744.

Hofmann, P.B. *BBMEUSSR*, 1936, **2**, 396.

Hoge, M.A. *JEZ*, 1915, **18**, 241.

Hollander, A. & Claus, W.D. *National Research Council Bull.* no. 100, 1937.

Holmes, A.D., Doolittle, A.W. & Moore, W.B. *PS*, 1926, **5**; *JAPA*, 1927, **16**, 518; *AFGF*, 1928, **2**, 30.

Holmes, A.D., Pigott, M.G. & Campbell, P.A. *JBC*, 1934, **105**, xli; *PS*, 1935, **14**, 183.
Holmes, A.D., Tripp, F. & Campbell, P.A. *JN*, 1936, **11**, 119.
Holmes, B.E. (1). *BJ*, 1933, **27**, 391.
Holmes, B.E. (2). *BJ*, 1935, **29**, 2285.
Holmes, B.E. (3). *ARB*, 1935, **4**.
Holmes, B.E. (4). *BJ*, 1937, **31**, 1730; *PRSB*, 1939, **127**, 223.
Holmes, E.G. *BJ*, 1933, **27**, 523.
Holmes, S.J. *AFEWM*, 1904, **17**, 265.
Holmgren, E. *AFEWM*, 1933, **129**, 199.
Holst, W.F. & Almquist, H.J. (1). *H*, 1931, **6**, 45.
Holst, W.F. & Almquist, H.J. (2). *H*, 1931, **6**, 49.
Holst, W.F. & Almquist, H.J. (3). *PS*, **11**, 81.
Holst, W.F., Almquist, H.J. & Lorenz, F.W. *PS*, 1932, **11**, 144.
Holt, E. *AR*, 1921, **22**, 207.
Holter, H. (1). *JCCP*, 1936, **8**, 179; *AR*, 64 (1st Suppl.), 57.
Holter, H. (2). *AEZF*, 1937, **19**, 232.
Holter, H. & Doyle, W.L. *CRTLC*, 1938, **22**, 219.
Holter, H. & Kopać, M.J. *JCCP*, 1937, **10**, 423.
Holter, H., Lanz, H. & Linderstrøm-Lang, K. *CRTLC*, 1938, **23**, no. 1; *JCCP*, 1938, **12**, 119.
Holter, H., Lehmann, F.E. & Linderstrøm-Lang, K. *ZPC*, 1937, **250**, 237; *CRTLC*, 1938, **21**, no. 18, 259.
Holter, H. & Lindahl, P.E. Oral communication to the xvith Internat. Physiol. Congr. Zürich, 1938 (in the press). Cit. in K. Linderstrøm-Lang, "Harvey Lecture", 1939, p. 214.
Holter, H. & Linderstrøm-Lang, K. *SAWW*, 1936 iib, **145**, 898.
Holtfreter, J. (1). *AFEWM*, 1929, **117**, 421; *VDZG*, 1929, 174.
Holtfreter, J. (2). *AFEWM*, 1931, **124**, 404; *VDZG*, 1931, 158.
Holtfreter, J. (3). *SGMPM*, 1933, **42**, 2.
Holtfreter, J. (4). *AFEWM*, 1933, **127**, 591.
Holtfreter, J. (5). *AFEWM*, 1933, **127**, 619.
Holtfreter, J. (6). *AFEWM*, 1933, **128**, 584.
Holtfreter, J. (7). *NW*, 1933, **21**, 766.
Holtfreter, J. (8). *AFEWM*, 1933, **129**, 669.
Holtfreter, J. (9). *BC*, 1933, **53**, 404.
Holtfreter, J. (10). *AEZF*, 1934, **15**, 281.
Holtfreter, J. (11). *AFEWM*, 1934, **132**, 225.
Holtfreter, J. (12). *AFEWM*, 1934, **132**, 307.
Holtfreter, J. (13). *AFEWM*, 1935, **133**, 367.
Holtfreter, J. (14). *AFEWM*, 1935, **133**, 427; *SGMPM*, 1935, **44**, 1.
Holtfreter, J. (15). *AR*, 1935, **64** (Suppl.), 41.
Holtfreter, J. (16). *AFEWM*, 1936, **134**, 467.
Holtfreter, J. (17). "Grundphänomene d. Embryogenese." In *Trav. du Congrès du Palais de la Découverte*, Hermann, Paris, 1938, p. 447; *CS*, 1939, Suppl. no. 4, 23.
Holtfreter, J. (18). *AFEWM*, 1938, **138**, 163.

Holtfreter, J. (19). *AFEWM*, 1938, **138**, 522.
Holtfreter, J. (20). *AFEWM*, 1938, **138**, 657.
Holtfreter, J. (21). *AFEWM*, 1939, **139**, 110 & 227.
Holtfreter, J. (22). *AEZF*, 1939, **23**, 169.
Holzmann, O.G. *BBMEUSSR*, 1939, **8**, 128 & 132.
Hooker, S.B. & Boyd, W.C. *JI*, 1936, **30**, 41.
Hopkins, F.G. (1). *BJ*, 1921, **15**, 286.
Hopkins, F.G. (2). *N*, 1931, **126**, 328 and 383.
Hopkins, F.G. (3). *JCS*, 1889, **5**, 117.
Hopkins, F.G. (4). *PTRSB*, 1895, **186**, 661.
Hopkins, F.G. (5). *PRSB*, 1942, **130**, 359.
Hopkins, F.G. & Dixon, M. *JBC*, 1922, **54**, 529.
Hopkins, F.G. & Elliott, K.A.C. *PRSB*, 1931, **109**, 58.
Hopkins, F.G. & Morgan, E.J. *BJ*, 1938, **32**, 611.
Hopkins, H.S. *JEZ*, 1930, **56**, 209.
Hopkins, M.L. *JM*, 1935, **58**, 585.
Hoppe-Seyler. "Medizinische-Chemische Untersuchungen." 1866.
Höpping, R. *MJ*, 1937, **79**, 123; *OMaly*, 1937, **102**, 306; *YMaly*, 1937, **43**, 330.
Horning, E.S. & Byrne, J.M. *AJBMS*, 1929, **6**, 1.
Horning, E.S. & Miller, I.D. *AJBMS*, 1930, **7**, 151.
Horning, E.S. & Richardson, K.C. *MJA*, 1930, Feb. 22.
Horning, E.S. & Scott, G.H. *AR*, 1932, **52**, 351.
Horowitz, N.H. (1). *JCCP*, 1939, **14**, 189.
Horowitz, N.H. (2). *PNASW*, 1940, **26**, 161.
Horowitz, N.H. (3). *JCCP*, 1940, **15**, 299.
Horowitz. N.H. (4). *JCCP*, 1940, **15**, 309.
Hörstadius, S. (1). *AFEWM*, 1927, **112**, 239.
Hörstadius, S. (2). *AZ*, 1928, **9**, 1.
Hörstadius, S. (3). *PSZN*, 1935, **14**, 251; *AFZ*, 1931, **23** B, 1.
Hörstadius, S. (4). *AFEWM*, 1936, **135**, 1.
Hörstadius, S. (5). *AFEWM*, 1936, **135**, 40.
Hörstadius, S. (6). *CN*, 1936, **11**, 236; *BR*, 1939, **14**, 132; *CS*, 1939, Suppl. no. 4, 44.
Hörstadius, S. (7). *MMRHNB*, 1936 (sér. 2ᵉ), **3**, 803.
Hörstadius, S. (8). *BB*, 1937, **73**, 295.
Hörstadius, S. (9). *BB*, 1937, **73**, 317.
Hörstadius, S. (10). *AFEWM*, 1938, **138**, 197.
Hörstadius, S. & Wolsky, A. *AFEWM*, 1936, **135**, 69.
Horton, F.M. *JEB*, 1934, **11**, 257.
Hoskins, E.R. & Hoskins, M.M. *AR*, 1917, **11**, 363; *JEZ*, 1919, **29**, 1.
Hoskins, F.M. & Snyder, F.F. *AJOP*, 1933, **104**, 530.
Hosoi, K. *JJMSB*, 1937, **3**, 249.
Houston, J. & Kon, S.K. *N*, 1939, **143**, 558.
Howard, E. (1). *JCCP*, 1932, **1**, 355.
Howard, E. (2). *JGP*, 1932, **16**, 107.
Howard, E. (3). *JCCP*, 1933, **3**, 291; *AJOP*, 1933, **105**, 56.
Howes, N.H. *JEB*, 1940, **17**, 128.
Howland, R.B. *JEZ*, 1921, **32**, 355.

Howland, R.B. & Bernstein, A. *JGP*, 1931, 14, 339.

Howland, R.B. & Child, G.P. *JEZ*, 1935, 70, 415.

Howland, R.B., Glancy, E.A. & Sonnenblick, B. *GS*, 1937, 22, 196.

Howland, R.B. & Sonnenblick, B. *JEZ*, 1936, 73, 109.

Hoyle, W.E. *PRPSE*, 1888, 10, 58.

Hsiao, Chih-De. Private communication to the author, Oct. 1940.

Huang Hsi-Tsan (Whong, S.H.). *P*, 1931, 12, 123.

Hubbard, M.J. & Rothschild, V. *PRSB*, 1939, 127, 510.

Hubbard, R.S. & Wright, F.R. *JBC*, 1922, 50, 361.

Hubert, R. *AP*, 1929, 223, 333.

Huddleston, O.L. & Whitehead, R.W. *JPET*, 1931, 42, 274.

Hudiwara, K. *OIZ*, 1935, 47, 68; *OMaly*, 1935, 88, 367; *YMaly*, 1936, 36, 72.

Hueper, W.C. *AOP*, 1934, 17, 218.

Hueper, W.C., Allen, A., Russell, M., Woodward, G. & Platt, M. *AJC*, 1933, 17, 74.

Hueper, W.C. & Russell, M. *AEZF*, 1932, 12, 407.

Huff, G.C. *PR*, 1940, 20, 68.

Huff, G.C. & Boell, E.J. *PSEBM*, 1936, 34, 626.

Hughes, A.F.W. (1). *JA*, 1935, 70, 76.

Hughes, A.F.W. (2). *JA*, 1937, 72, 1.

Hughes, J.S. & Scott, H.M. *PS*, 1936, 15, 349.

Hughes, J.S., Titus, R.W. & Smits, B.L. *S*, 1927, 65, 264.

Hughes, W. *BB*, 1927, 52, 121; *AR*, 1929, 41, 213.

Hummel, F.C., Hunscher, H.A., Bates, M.F., Bonner, P. & Macy, I.G. *JN*, 1937, 13, 263.

Humphrey, R.R. *JEZ*, 1933, 65, 243.

Humphrey, R.R. & Burns, R.K. *JEZ*, 1940, 81, 1.

Hunscher, H.A., Donelson, E., Nims, B., Kenyon, F. & Macy, I.G. *JBC*, 1933, 99, 507.

Hunt, H.R., Mixter, R. & Permar, D. *GS*, 1933, 18, 335.

Hunt, H.R. & Permar, D. *AR*, 1928, 41, 117.

Hunt, T.E. (1). *PSEBM*, 1929, 27, 84; 1931, 28, 626; *JEZ*, 1931, 59, 395; *AR*, 1932, 52, 19 (Suppl.); 1932, 55, 41.

Hunt, T.E. (2). *AR*, 1937, 68, 349.

Hunter, F.R. *JCCP*, 1936, 9, 15.

Huskins, C.L. *Proc.* vith *Internat. Bot. Congress*, Amsterdam, 1935.

Hutchens, J.O. *CN*, 1939, 14 (unpaged offprint); *JCCP*, 1941, 17, 321.

Hutchens, J.O., Krahl, M.E. & Clowes, G.H.A. *JCCP*, 1939, 14, 313.

Hutchinson, A.H. & Ashton, M.R. *CJR*, 1933, 9, 49.

Hutt, F.B. (1). *JH*, 1934, 25, 41.

Hutt, F.B. (2). *CV*, 1934, 24, 1.

Hutt, F.B. (3). *GS*, 1938, 23, 152.

Hutt, F.B. & Boyd, W.L. *ENY*, 1935, 19, 398.

Hutt, F.B. & Cavers, J.R. *PS*, 1931, 10, 403.

Hutt, F.B. & Child, G.P. *JH*, 1934, 25, 341.

Hutt, F.B. & Greenwood, A.W. (1). *PRSE*, 1928, 49, 118.

Hutt, F.B. & Greenwood, A.W. (2). *PRSE*, 1929, 49, 145.

Hutt, F.B. & Pilkey, A.M. *PS*, 1934, 13, 3.

Hutt, F.B. & Sturkie, P.D. *JH*, 1938, 29, 371.

Huxley, J.S. (1). *QJMS*, 1921, 65, 643.

Huxley, J.S. (2). *BB*, 1922, 43, 210.

Huxley, J.S. (3). *S*, 1932, 58, 291.

Huxley, J.S. (4). *N*, 1924, 114, 895.

Huxley, J.S. (5). *PRSB*, 1925, 98, 113.

Huxley, J.S. (6). *PSZN*, 1926, 7, 1.

Huxley, J.S. (7). *AFEWM*, 1927, 112, 480.

Huxley, J.S. (8). *N*, 1929, 123, 712.

Huxley, J.S. (9). "Problems of Relative Growth." London, 1932.

Huxley, J.S. (10). *N*, 1932, 129, 166.

Huxley, J.S. (11). "Problems in Experimental Embryology." Robert Boyle Lecture, Oxford, 1935.

Huxley, J.S. (12). *TDDUSSR*, 1935, 10, 269.

Huxley, J.S. (13). *BR*, 1935, 10, 427.

Huxley, J.S. (14). *BZUSSR*, 1935, 4, 421.

Huxley, J.S. & de Beer, G.R. (1). *QJMS*, 1923, 67, 473.

Huxley, J.S. & de Beer, G.R. (2). "Elements of Experimental Embryology." Cambridge University Press, 1934.

Huxley, J.S. & Hogben, L.T. *PRSB*, 1922, 93, 36.

Huxley, J.S. & Teissier, G. *CRSB*, 1936, 121, 934; *N*, 1936, 137, 780.

Huzella, T. *AA*, 1929, 67, 36.

Hyman, L.H. & van Cleave, C.D. *PZ*, 1938, 11, 105.

Hyre, H.M. & Hall, G.O. *PS*, 1932, 11, 166.

Ibsen, H.L. & Steigleder, E. *AN*, 1917, 51, 740.

Ichikawa, M. (1). *MCSK*, 1933, 9, 47.

Ichikawa, M. (2). *PIAT*, 1934, 10, 683; *ZMJ*, 1936, 48, 954; *PIAT*, 1938, 14, 21.

Ido, R. *OIZ*, 1931, 43, 1097; *OMaly*, 1931, 63, 426; *JJMSB*, 1935, 3 (18).

von Ihering, R. *ZAZ*, 1937, 120, 45.

Ikeda, Y. (1). *AAIUS*, 1934, 16, 1, 47, 69; 1935, 17, 11; 1936, 18, 1, 17; 1937, 20, 17; 1938, 21, 1; 1939, 22, 27.

Ikeda, Y. (2). *JFSUT*, 1934, IV, 3, 499.

Ikeda, Y. (3). *JFSUT*, 1937, 4, 307 and 313.

Ilberg, G. *ZGG*, 1937, 114, 174.

Iljin, N.A. *TLEBZM*, 1926, 1, 1; 1927, 3, 183.

Iljin, N.A. & Iljina, W.N. *TLEBZM*, 1929, 5, 217.

Imaizumi, M. *JJB*, 1937, 26, 433.

Imamura, G.I. (1). *JJB*, 1939, 29, 391 and 403.

Imamura, G.I. (2). *JJB*, 1940, 31, 303.

Imms, A.D. "A General Textbook of Entomology." London, 1934.

Impallomeni, S. *GR*, 1931, **11**, 43; *YMaly*, 1932, **21**, 562.

Ing, H.R. *SP*, 1935, **30**, 252.

Ingle, D.J. & Fisher, G.T. *PSEBM*, 1938, **39**, 149.

Insko, W.M. & Lyons, M. (1). *JN*, 1933, **6**, 507.

Insko, W.M. & Lyons, M. (2). *Bull. Kentucky Agric. Exp. Sta.* 1936, no. 363.

Insko, W.M. & Martin, J.H. *PS*, 1935, **14**, 361.

Iob, V. & Swanson, W.W. (1). *AJDC*, 1934, **47**, 302.

Iob, V. & Swanson, W.W. (2). *JBC*, 1938, **124**, 263.

Irichimovitch, A.I. (1). *BC*, 1935, **56**, 639.

Irichimovitch, A.I. (2). *ZAZ*, 1936, **115**, 288; *CRASUSSR*, 1940, **27**, 94.

Irichimovitch, A.I. & Lektorsky, J.N. *BC*, 1935, **55**, 98.

Irving, L. & Manery, J.F. (1). *JCCP*, 1934, **4**, 483; *AJOP*, 1933, **105**, 57.

Irving, L. & Manery, J.F. (2). *BR*, 1936, **11**, 287.

Irwin, M.R. *PSEBM*, 1932, **29**, 850.

Irwin, M.R. & Cole, L.J. *JEZ*, 1936, **73**, 85 and 309.

Irwin, M.R., Cole, L.J. & Gordon, C.D. *JEZ*, 1936, **73**, 285.

Iseki, T. (1). *ZPC*, 1930, **188**, 189.

Iseki, T. (2). *JJB*, 1934, **19**, 1.

Iseki, T. & Kumon, T. *JJB*, 1933, **17**, 409.

Iseki, T., Kumon, T., Takahashi, I. & Yamasaki, F. *JJB*, 1933, **17**, 413.

Ishida, J. *AZJ*, 1936, **15**, 453.

Itizyo, M. *JJMSB*, 1934, **2**, 359.

Ito, S. *TJPS*, 1930, **20**, 360; *YMaly*, 1931, **18**, 60.

Ivlev, V.S. *CRASUSSR*, 1939, **25**, 87.

Iwasaki, K. *BZ*, 1930, **226**, 32.

Jacobi, E.F. & Baas-Becking, L.G.M. *TNDV*, 1933 (3rd), **3**, 145.

Jacobs, W.A. & Elderfield, R.C. (1). *JBC*, 1935, **108**, 497.

Jacobs, W.A. & Elderfield, R.C. (2). *ARB*, 1938, **7**, 449.

Jacobson, B.M. & Subbarow, Y. *JCI*, 1937, **16**, 573.

Jacobson, W. (1). *JM*, 1938, **62**, 415.

Jacobson, W. (2). *JM*, 1938, **62**, 445.

Jacobson, W. (3). *JPB*, 1939, **49**, 1.

Jacoby, F. *AEZF*, 1937, **19**, 241.

Jacoby, F., Trowell, O.A. & Willmer, E.N. *JEB*, 1937, **14**, 255; *AEZF*, 1937, **19**, 240.

Jaffé, R.H. *PR*, 1931, **11**, 277 and 279.

Jahn, T.L. (1). *JCCP*, 1935, **7**, 23; *PSEBM*, 1935, **33**, 159.

Jahn, T.L. (2). "Respiratory Metabolism." in *Protozoa in Biological Research.* Ed. G.N. Calkins, 1940.

Janda, V. & Kocián, V. *ZJZP*, 1933, **52**, 561.

Janicki, J. *BZ*, 1937, **289**, 348.

Janisch, E. *AB*, 1935, **1**, 47.

Janke, A. & Jirak, L. *BZ*, 1934, **271**, 309.

Jarisch, A. *AP*, 1920, **179**, 159.

Jaschik, A. & Kieselbach, J. *ZUNG*, 1931, **62**, 572; *A*, 1932, **57**, 105; *CA*, 1932, **26**, 3853.

Jecklin, L. *RSZ*, 1934, **41**, 593; *YMaly*, 1935, **32**, 542; and Inaug. Diss. Geneva, 1934.

Jenkinson, J.W. "Experimental Embryology." Oxford University Press, 1909, p. 277.

Job, T.T., Leibold, G.J. & Fitzmaurice, H.A. *AJA*, 1935, **56**, 97.

Jobling, J.W., Sproul, E.E. & Stevens, S. *AJC*, 1937, **30**, 667, 685.

Johannsen, O.A. *JM*, 1924, **39**, 337.

Johanssen, O.A. & Butt, F.H. "Embryology of Insects and Myriapods." New York and London, 1941.

Johlin, J.M. (1). *JGP*, 1933, **16**, 605.

Johlin, J.M. (2). *JGP*, 1935, **18**, 481.

John, C.C. *PRSB*, 1932, **110**, 112.

Johnson, C.G. *JEB*, 1937, **14**, 413; *TSBE*, 1934, **1**, 1.

Johnson, E.A., Pilkey, A.M. & Edson, A.W. *PS*, 1935, **14**, 16.

Johnston, J.A., Hunscher, H.A., Hummel, F.C., Bates, M.F., Bonner, P. & Macy, I.G. *JN*, 1938, **15**, 513.

Jollos, V. & Péterfi, T. *BC*, 1932, **43**.

Jolly, J. *B*, 1939, **1**, 385.

Jonen, P. *ZGG*, 1932, **103**, 192.

Jones, D.F. *S*, 1935, **81**, 75; *PNASW*, 1935, **21**, 90.

Jones, J.M., Shipp, M.E. & Gonder, T.A. *PSEBM*, 1936, **34**, 873.

Jongbloed, J. *JOP*, 1938, **92**, 229.

Jordan, P. *ZZFMA*, 1928, **6**, 558.

Jordan-Lloyd, D. *BR*, 1932, **7**, 254.

Jorpes, E. *AMS*, 1928, **68**, 503.

Jost, H. & Sorg, K. *AP*, 1932, **231**, 143.

Jowett, M. *BJ*, 1931, **25**, 1991.

Jucci, C. *AAL*, 1930, **11**, 86; *CA*, 1930, **24**, 3053; *BSIBS*, 1930, **5**, 160; 1932, **7**, 163 and 573; *CA*, 1930, **24**, 4337; 1932, **26**, 4885 and 5666.

Jucci, C. & Manunta, C. *BSIBS*, 1932, **7**, 162.

Jucci, C. & Ponseveroni, N. *BSIBS*, 1930, **5**, 1056; *CA*, 1931, **25**, 2770.

Juhn, M., Faulkner, G.H. & Gustavson, R.G. *JEZ*, 1931, **58**, 69.

Juhn, M. & Gustavson, R.G. *JEZ*, 1930, **56**, 31.

Jukes, T.H. (1). *JBC*, 1933, **103**, 425.

Jukes, T.H. (2). *JBC*, 1939, **129**, 225.

Jukes, T.H., Fraser, D.T. & Orr, M.D. *JI*, 1934, **26**, 353.

Jukes, T.H. & Kay, H.D. (1). *JBC*, 1932, **98**, 783.

Jukes, T.H. & Kay, H.D. (2). *JEM*, 1932, **56**, 469.

Jukes, T.H. & Kay, H.D. (3). *JN*, 1932, **5**, 81.
Just, E.E. (1). *BB*, 1919, **36**, 11.
Just, E.E. (2). *P*, 1930, **10**, 300.
Just, E.E. (3). *AN*, 1936, **70**, 267.

Kaan, H.W. *JEZ*, 1930, **55**, 263; 1938, **78**, 159.
Kaboth, T. *AG*, 1929, **137**, 727 and 752.
Kaestner, S. *AAP*, 1907, 250.
Kagiyama, S. *JJB*, 1933, **17**, 135.
Kahane, E. & Levy, J. *BSCB*, 1937, **19**, 777.
Kaliss, N. *GS*, 1939, **24**, 244.
Kamachi, T. (1). *JJB*, 1935, **22**, 189.
Kamachi, T. (2). *JJB*, 1935, **22**, 199.
Kamachi, T. (3). *ZPC*, 1936, **238**, 91.
Kamegai, S. *JJB*, 1939, **30**, 33.
Kamei, T. *ZPC*, 1928, **171**, 101.
Kamenov, R.J. *JM*, 1935, **58**, 117.
Kamiya, T. *NJMS*, 1930, **5**, 1.
Kamm, B. *AP*, 1930, **223**, 214.
Kapel, O. *AEZF*, 1929, **8**, 35.
Karashima, J. *JJB*, 1929, **10**, 375.
Karrer, P. (1). *HCA*, 1936, **19**, E 33.
Karrer, P. (2). *EVHF*, 1938, **2**, 381.
Karrer, P., Frei, P. & Fritzsche, H. *HCA*, 1937, **20**, 622.
Karrer, P., Frei, P. & Ringier, B.H. *HCA*, 1938, **21**, 314.
Karrer, P. & Helfenstein. *HCA*, 1931, **14**, 78.
Karrer, P. & Ringier, B.H. *HCA*, 1939, **22**, 334 and 610.
Karrer, P. & Schöpp, K. *HCA*, 1934, **17**, 735.
Kasansky, W.J. *ZAZ*, 1928, **75**, 235; 1936, **115**, 89.
Kataoka, E. *ZPC*, 1931, **203**, 272.
Kataoka, E. & Takahashi, I. *JJB*, 1933, **17**, 419.
Kataoka, E. & Tsunoo, S. *JJB*, 1933, **17**, 417.
Katsu, Y. *JJOG*, 1931, **14**; 1933, **16**, 2, 10, 21.
Katsunuma, S. & Nakamura, H. *NJMS*, 1932, **6**, 107.
Katzenstein, W.F. *AEZF*, 1925, **1**, 173.
Kaufman, L. (1). *MINPERP*, 1926, **7**, 92.
Kaufman, L. (2). *AFEWM*, 1930, **122**, 395; *AAM*, 1929, **25**, 325.
Kaufman, L. (3). *ABE*, 1930, **5**, 33.
Kaufman, L. (4). *CRAA*, 1930, **25**, 1.
Kaufman, L. (5). *CRSB*, 1932, **110**, 1094.
Kaufman, L. (6). *CRSB*, 1932, **111**, 881.
Kaufman, L. (7). *AFEWM*, 1934, **131**, 193.
Kaufman, L. (8). *MINPERP*, 1938, **17**, 1.
Kaufman, L. & Dąbrowska, W. *CRAA*, 1931; *YMaly*, 1932, **22**, 70.
Kaufman, L. & Laskowski, M. *BZ*, 1931, **242**, 424; *CRAA*, 1931.
Kaufman, L. & Nowotna, A. *AP*, 1934, **235**, 247.
Kavanagh, A.J. & Richards, O.W. *AN*, 1934, **68**, 54.
Kawakami, I. *BOZJ*, 1936, **6**, 1842 and 2006.
Kawamura, T. *JSHU*, 1939, **6**, 115.
Kay, H.D. *JBC*, 1931, **93**, 727.
Kay, H.D. & Marshall, P.G. *BJ*, 1928, **22**, 1264.

Kaylor, C.T. *JEZ*, 1937, **76**, 375. *BB*, 1939, **77**, 334.
Kearns, P.J. *AJOG*, 1934, **27**, 870.
Kedrovsky, B. *BHA*, 1934, **11**, 288; *ZZFMA*, 1937, **25**, 694 and 708; 1937, **26**, 21; *BZUSSR*, 1937, **6**, 1137.
Keeler, C.E. & Castle, W.E. *PNASW*, 1934, **20**, 273.
Keilin, D. (1). *AZEG*, 1916, **55**, 393.
Keilin, D. (2). *PRSB*, 1929, **104**, 206.
Keilin, D. (3). *Trav. Congr. Palais de la Découverte*, Paris, 1938, p. 357.
Keilin, D. & Hartree, E. *PRSB*, 1938, **125**, 171.
Kekwick, R.A. & Cannan, R.K. *BJ*, 1936, **30**, 227 and 235.
Kekwick, R.A. & Harvey, E.N. *JCCP*, 1934, **5**, 43.
Keller, K. & Niedoba, T. *ZTZ*, 1937, **37**, 245.
Keltch, A.K., Clowes, G.H.A. & Krahl, M.E. *BB*, 1936, **71**, 399.
Keltch, A.K., Krahl, M.E. & Clowes, G.H.A. *BB*, 1937, **73**, 377.
Kemp, T. *CRSB*, 1925, **92**, 1318.
Kempner, W. *JCCP*, 1937, **10**, 339; *PSEBM*, 1936, **35**, 148.
Kendall, E.C. *AJOP*, 1919, **49**, 136.
Kennaway, E.L. *BMJ*, 1925 (ii), 1.
Keppel, D.M. & Dawson, A.B. *BB*, 1938, **76**, 153.
Kesselyak, A. *AFEWM*, 1936, **134**, 331.
Kestner, O. *AP*, 1934, **234**, 290.
Keys, A.B. (1). *LJB*, 1931, **2**, 417 and 457.
Keys, A.B. (2). *JOP*, 1934, **80**, 491.
Khouvine, Y. & Ephrussi, B. *CRAS*, 1937, **124**, 885.
Khouvine, Y., Ephrussi, B. & Chevais, S. *BB*, 1938, **75**, 425.
Khouvine, Y., Ephrussi, B. & Harnly, M.H. *CRAS*, 1936, **203**, 1542.
Kiaer, S. (1). *AEZF*, 1925, **1**, 115.
Kiaer, S. (2). *AKC*, 1927, **149**, 146.
Kido, I. (1). *CG*, 1936, **60**, 1162; *YMaly*, 1936, **39**, 642.
Kido, I. (2). *CG*, 1937, **61**, 1551; *YMaly*, 1937, **44**, 493.
Kihara, Y. *ASOJ*, 1933, **37**, 835; *YMaly*, 1934, **28**, 147.
Kikkawa, H. (1). *ZMJ*, 1937, **49**, 348.
Kikkawa, H. (2). *JH*, 1938, **29**, 395.
Kindred, J.E. & Corey, E.L. *PZ*, 1931, **4**, 294.
King, E.J., Stantial, H. & Dolan, M. *BJ*, 1933, **27**, 1002.
King, H.D. *AFEWM*, 1905, **19**, 85.
King, J.T. *PSEBM*, 1932, **29**, 1112, 1114; 1933, **30**, 1384.
King, R.L. & Beams, H.W. *JEZ*, 1938, **77**, 425; *N*, 1937, **139**, 369.
King, R.L. & Slifer, E. *JM*, 1934, **56**, 603.
Kingsbury, B.F. & Adelmann, H.B. *QJMS*, 1924, **68**, 239.
Kinosita, R. *TJPS*, 1937, **27**, 665.
Kirkham, W.B. (1). *AR*, 1917, **11**, 480.

Kirkham, W.B. (2). *JEZ*, 1919, **28**, 125.
Kirkham, W.B. & Haggard, H.W. *AR*, 1916, **10**, 537.
Kirsanov. *Soviet Bird-Breeding* (*U.S.S.R.*), 1935, **3**.
Kirschbaum, J.D. & Jacobs, M.B. *SGO*, 1940, **71**, 297.
Kisch, B. (1). *BZ*, 1931, **238**, 370.
Kisch, B. (2). *BZ*, 1933, **263**.
Kisch, B. (3). *BZ*, 1934, **271**.
Kisch, B. & Remertz, O. *IZPCB*, 1914, **1**, 354.
Kitamura, K. *MMAK*, 1929, **3**, 183; *CA*, 1931, **25**, 5713.
Kiyohara, K. *CRSB*, 1931, **106**, 920.
Kiyono, K. & Sueyoshi, Y. *KIZ*, 1917, **14**, 598.
Klaus, K. *CG*, 1933, **57**, 558.
Kleinenberg, H.E., Neufach, S.A. & Shabad, L.M. *BBMEUSSR*, 1939, **8**, 25 & 125.
Kleinenberg, N. (1). *QJMS*, 1879, **19**, 206.
Kleinenberg, N. (2). *ZWZ*, 1886, **44**, 212.
Kleinholz, L.H. *BB*, 1940, **79**, 432.
Kleinzeller, A. & Werner, H. *BJ*, 1939, **33**, 291.
Kliger, I.J. & Olitzki, L. *ZH*, 1929, **110**, 459.
Klose, A.A. & Almquist, H.J. *PS*, 1937, **16**, 173.
Kluyver, A.J. "The Chemical Activities of Micro-Organisms." London University Press, 1931, p. 102.
Knight, F.C.E. *AFEWM*, 1938, **137**, 461.
Knox, C.W. & Godfrey, A.B. *PS*, 1934, **13**, 18; 1938, **17**, 159.
Kobayashi, S. *KIZ*, 1926, **23** (two papers); Inaug. Diss. Kyoto, 1926.
Koch, C., Schreiber, B. & Schreiber, G. *BAEC*, 1939, **28**, 852.
Koch, F.C. *BNYAM*, 1938, **14**, 655; "Harvey Lectures", 1938, p. 205 and "The Biochemistry of Androgens" in "Sex and Internal Secretions". 1939, p. 807.
Koch, F.J. Inaug. Diss. Erlangen, 1937.
Koch, M. *VDPG*, 1904, **7**, 136.
Koch, Waldemar & Koch, M.L. *JBC*, 1913, **15**, 423.
Kodama, K. *JJB*, 1932, **15**, 300.
Kodama, S. *C*, 1931, **2**, 77.
Kogan, R.E. *CRASUSSR*, 1939, **23**, 307.
Kögl, F. *BDCG*, 1935, **68**, 16; *ARBA*, 1933, 600; *AGC*, 1933, **46**, 469; *NW*, 1933, **21**, 17.
Kögl, F., Erxleben, H. & Haagen-Smit, A.J. *ZPC*, 1934, **225**, 215.
Kögl, F., Haagen-Smit, A.J. & Erxleben, H. *ZPC*, 1933, **220**, 137 (p. 152).
Kögl, F. & von Hasselt, W. (1). *ZPC*, 1936, **242**, 74.
Kögl, F. & von Hasselt, W. (2). *ZPC*, 1936, **243**, 189.
Kögl, F. & Tönnis, B. *ZPC*, 1936, **242**, 43.
Köhler, Wilhelm & Feldotto, W. *AFEWM*, 1937, **136**, 313; *AKS*, 1935, **10**, 314.

Köhler, Wolfgang. "Gestalt Psychology", London, 1930; Germ. edn. "Psychologische Probleme", Berlin, 1933; "Die physische Gestalten im Ruhe u. stationäres Zustand", 1920; *JUGP*, 1922, **3**, 512; *AFEWM*, 1927, **112**, 315.
Kojima, K. *NJMS*, 1930, **5**, 49.
Koller, P. (1). *AEZF*, 1929, **8**, 490.
Koller, P. (2). *JG*, 1930, **22**, 103.
Kollross, J.J. *JEZ*, 1940, **85**, 33.
Kolobkova, E.V. *BUSSR*, 1939, **4**, 295 and **302**.
Kolomjetz, M.J. *AEZF*, 1934, **16**, 260.
Kolster, R. *AH*, 1907, **34** (1), 403.
Koltzov, N.K. (1). "Priroda" (in Russian), 1912.
Koltzov, N.K. (2). *BC*, 1928, **48**, 345.
Koltzov, N.K. (3). "Physiologie du Développement et Génétique." Hermann, Paris, 1935. See also "Organisatsia Kletki" (in Russian), Moscow, 1936.
Koltzov, N.K. (4). "Les Molécules Hérédi-taires." Hermann, Paris, 1939.
Koltzov, N.K. (5). *CRASUSSR*, 1939, **23**, 482.
Komai, P. *GS*, 1926, **11**, 280.
Komatsu, H. *NIZ*, 1935, **13**, 477; *YMaly*, 1935, **35**, 151.
Kondo, K. *CRTLC*, 1938, **22**, 275.
Kondo, K., Yamada, T. & Nagashima, M. *JCSJ*, 1937, **58**, 108; *CA*, 1937, **31**, 3161.
König, M.C., Kramer, M.M. & Payne, L.F. *PS*, 1935, **14**, 178.
Konopacka, B. (1). *BAA*, 1931, **25**, 306; *CRAA*, 1931, Varsovie; *YMaly*, 1932, **21**, 652.
Konopacka, B. (2). *ADB*, 1933, **44**, 251; *BAPS*, 1932, 643.
Konopacka, B. (3). *PGL*, 1936, **15**, 1.
Konopacka, B. (4). *PSZN*, 1937, **16**, 327.
Konopacka, B. (5). *BAPS*, B II, 1935, 163.
Konopacki, M. (1). *CRSSV*, 1929, **22**, 1 (in Polish).
Konopacki, M. (2). *BAPS*, B II, 1931, 351.
Konopacki, M. (3). *BAPS*, B, 1933, 51; *K*, 1933, **58**, 133.
Konopacki, M. (4). *CRSB*, 1936, **122**, 139.
Konopacki, M. (5). *PGL*, 1936, **15**, 1.
Konopacki, M. & Ereciński, K. *BAPS*, B, 1932, 141.
Konopacki, M. & Konopacka, B. *BAPS*, B, 1926, 229; *CRSB*, 1924, **91**.
Kopeć, S. (1). *BB*, 1922, **42**, 322; 1924, **46**, 1.
Kopeć, S. (2). *JEZ*, 1923, **37**, 15.
Kopeć, S. (3). *BG*, 1927, **3**, 375.
Kopeć, S. (4). *MINPERP*, 1922, **2**, 214; 1923, **4**, 173 and 218; 1925, **6**, 288; 1926, **7**, 261; 1929, **10**, 224 and 475; 1930, **11**, 335.
Kopeć, S. (5). *AFEWM*, 1932, **126**, 575.
Kopeć, S. & Latyszewski, M. *MINPERP*, 1929, **10**, 509; 1930, **11**, 299.
Kopsch, F. "Untersuchungen über Gastrulation und Embryobildung bei Chordaten." Leipzig, 1904.

Korenchevsky, V. *EVHF*, 1938, **2**, 418.
Korenchevsky, V. & Hall, K. *JPB*, 1937, **45**, 681.
Korjuev, N. & Belousov, A.P. *BBMEUSSR*, 1936, **1**, 139.
Kornfeld, W. *AFEWM*, 1914, **40**, 369.
Korr, I.M. (1). *JCCP*, 1937, **10**, 461; *BB*, 1937, **73**, 357.
Korr, I.M. (2). *JCCP*, 1938, **11**, 233.
Korr, I.M. (3). *CSHSQB*, 1939, **7**, 74.
Korr, I.M. (4). Private communication to the author, Sept. 1939.
Korschelt, E. "Regeneration und Transplantation." Bornträger, Berlin, 1927–31.
Korschelt, E. & Heider, K. "Lehrbuch d. vergl. Entwicklungsgeschichte." Fischer, Jena, 1902–10; 2nd ed. 1936.
Kosaka, T. *JJOG*, 1932, **15**, 97; *YMaly*, 1933, **24**, 209.
Koschtojanz, C.S. *AMBUSSR*, 1935, **4**, 187; *PZUSSR*, 1935, **19**, 187.
Koschtojanz, C.S. & Mitropolitanskaia, R. *BBMEUSSR*, 1936, **1**, 207.
Koschtojanz, C.S. & Rjabinovskaia, A. *AP*, 1935, **235**, 416; *BZUSSR*, 1935, **4**, 242.
Kossel, A. (1). *ZPC*, 1884, **8**, 407.
Kossel, A. (2). *SAWH*, 1921, **1** (Math. Nat. Kl.), 1.
Kosswig, C. *ZIAV*, 1931, **59**, 61.
Kostitzin, V.A. "Biologie Mathematique," Colin, Paris, 1937. Eng. tr. T. Savory, Harrap, London, 1939.
Kostov, D. (1). *CRASUSSR*, 1938, **19**, 197.
Kostov, D. (2). *CS*, 1938, **6**, 549.
Kostov, D. & Kendall, J. *JG*, 1929, **21**, 113.
Kozelka, A.W. & Gallagher, T.F. *PSEBM*, 1934, **31**, 1143.
Kozhanchikov, I. (1). *CRASUSSR*, 1935, **2**, 326.
Kozhanchikov, I. (2). *CRASUSSR*, 1940, **27**, 80.
Kozhanchikov, I. & Maslova, E. *ZJZP*, 1935, **55**, 219.
Kozhukhar, E.M. *UBZ*, 1937, **10**, 663; 1939, **13**, 607.
Kozmina, N.A. *CRASUSSR*, 1940, **26**, 504.
Krafka, J. *JGP*, 1920, **2**, 409.
Krah, E. *BZ*, 1930, **219**, 432.
Krahl, M.E. & Clowes, G.H.A. (1). *PSEBM*, 1934, **32**, 226.
Krahl, M.E. & Clowes, G.H.A. (2). *JBC*, 1935, **111**, 355.
Krahl, M.E. & Clowes, G.H.A. (3). *JGP*, 1936, **20**, 173; *PSEBM*, 1935, **33**, 477.
Krahl, M.E. & Clowes, G.H.A. (4). *JCCP*, 1938, **11**, 1 and 21.
Krahl, M.E. & Clowes, G.H.A. (5). *JGP*, 1940, **23**, 413.
Krahl, M.E., Clowes, G.H.A. & Taylor. *BB*, 1936, **71**, 400.
Krahl, M.E., Keltch, A.K. & Clowes, G.H.A. (1). *PSEBM*, 1937, **36**, 700.

Krahl, M.E., Keltch, A.K. & Clowes, G.H.A. (2). *BB*, 1939, **77**, 318.
Krahl, M.E., Keltch, A.K. & Clowes, G.H.A. (3). *PSEBM*, 1940, **45**, 719.
Krahl, M.E., Keltch, A.K., Neubeck, C.E. & Clowes, G.H.A. *JGP*, 1941, **24**, 597.
Krämer, W. *AFEWM*, 1934, **131**, 220.
Krause, G. (1). *AFEWM*, 1934, **132**, 115; *ZMOT*, 1938, **34**, 1.
Krause, G. (2). *BC*, 1939, **59**, 495.
Krebs, H.A. (1). *BZ*, 1931, **238**, 174 and 191.
Krebs, H.A. & Henseleit, K. *ZPC*, 1932, **210**, 33.
Křižinecký, J. *AFEWM*, 1914, **37**, 629.
Krogh, A. (1). *BR*, 1931, **6**, 412.
Krogh, A. (2). "Osmotic Regulation in Aquatic Animals." Cambridge, 1939.
Krogh, A., Krogh, Agnes & Wernstedt, C. *SKAP*, 1938, **80**, 214.
Krogh, A., Schmidt-Nielsen, K. & Zeuthen, E. *ZVP*, 1938, **26**, 230.
Krogh, A. & Ussing, H.H. (1). *JEB*, 1937, **14**, 35.
Krogh, A. & Ussing, H.H. (2). *CRTLC*, 1938, **22**, 282.
Kronfeld, P. & Scheminsky, F. *AFEWM*, 1926, **107**, 129.
Krongold, S. Inaug. Diss. Paris, 1914.
Krontovsky, A.A. (1). *AEZF*, 1931, **11**, 93.
Krontovsky, A.A. (2). *CRSB*, 1932, **109**, 188.
Krontovsky, A.A., Jazimirska-Krontovska, M.C. & Savitzka, H.P. *CRSB*, 1932, **109**, 190.
Krontovsky, A.A. & Lebensohn, E.G. *AEZF*, 1932, **13**, 407.
Krüger, F. (1). *AFEWM*, 1930, **122**.
Krüger, F. (2). *ZVP*, 1934, **21**, 249.
Krüger, F. (3). *PKAWA*, 1935, **38**, 101; *ZJZP*, 1936, **57**, 1.
Kucera, W.G. *PZ*, 1934, **7**, 449.
Kugler, O.E. *AJOP*, 1936, **115**, 287.
Kuhl, H. *ZVP*, 1935, **22**, 32.
Kuhl, W. *VDZG*, 1938, 137.
Kühn, A. *NGWG*, 1936, **2**, 239.
Kühn, A., Caspari, E. & Plagge, E. *NGWG*, 1935, **2**, 1.
Kühn, A. & Henke, K. (1). *AGWG*, 1929, **15**, 3; 1932, **15**, 127; 1936, **15**, 225.
Kühn, A. & Henke, K. (2). *AFEWM*, 1930, **122**, 204.
Kühn, A. & Piepho, A. *BC*, 1938, **58**, 12; *NGWG*, 1936, 2.
Kühn, A. & Plagge, E. *BC*, 1937, **57**, 113.
Kühn, Philalethes & Sternberg, K. "Über Bakterien und Pettenkoferien." Fischer, Jena, 1931.
Kühn, R. (1). *BSCB*, 1935, **17**, 905.
Kühn, R. (2). *NW*, 1937, **25**, 225.
Kühn, R. & Brockmann, H. *ZPC*, 1932, **206**, 41.
Kühn, R., György, P. & Wagner-Jauregg, T. *BDCG*, 1933, **66**, 317 and 576.
Kühn, R. & Lederer, E. *BDCG*, 1933, **66**, 488.

Kühn, R., Lederer, E. & Deutsch, A. *ZPC*, 1933, **220**, 229.
Kühn, R., Moewus, F. & Jerchel, D. *BDCG*, 1938, **71**, 1541.
Kühn, R. & Smakula, A. *ZPC*, 1931, **197**, 161.
Kühn, R. & Sørensen, N.A. *BDCG*, 1938, **71**, 1879.
Kühn, R. & Wagner-Jauregg, T. *BDCG*, 1930, **66**, 1577.
Kühn, R. & Wendt, G. *BDCG*, 1938, **81**, 780.
Kühn, R., Winterstein, A. & Lederer, E. *ZPC*, 1931, **197**, 141.
Kühne, W. *APPA*, 1863, **26**, 222.
Kumanomido, S. *BZ*, 1928, **195**, 79.
Kumon, T. *JJB*, 1933, **18**, 145.
Kunde, M.M., Green, F., Changnon, E. & Clark, E. *AJOP*, 1931, **99**, 463.
Kuo, Tsing-Yang [Z.Y.] (1). *JEZ*, 1932, **61**, 395.
Kuo, Tsing-Yang [Z.Y.] (2). *JEZ*, 1932, **62**, 453.
Kuo, Tsing-Yang [Z.Y.] (3). *JCPS*, 1932, **13**, 245; 1932, **14**, 109.
Kuo, Tsing-Yang [Z.Y.] (4). *PSR*, 1932, **39**, 499.
Kuo, Tsing-Yang [Z.Y.] (5). *JCPS*, 1937, **24**, 49.
Kuo, Tsing-Yang [Z.Y.] (6). *AJOPS*, 1938, **51**, 361.
Kuo, Tsing-Yang [Z.Y.] (7). *JNP*, 1939, **2**, 488.
Kuo, Tsing-Yang [Z.Y.] & Shen, Tsi-Chen R. *JCPS*, 1937, **24**, 49.
Kurkiewicz, T. *CRAA*, 1928, **23**, 265; *BAA*, 1931, **25**, 343; *YMaly*, 1932, **22**, 96.
Kuroda, K. *KJM*, 1935, **6**, 41; 1935, **6**, 23; 1934, **5**, 140 and 111; *CA*, 1935, **29**, 6312; 1935, **29**, 6292; 1934, **28**, 7334.
Kurz O. *AFEWM*, 1922, **50**, 186.
Kusche, W. *AFEWM*, 1929, **120**, 192.
Küster, E. "Die Gallen d. Pflanzen." Leipzig, 1911; and in *Handbuch d. norm. u. pathol. Physiol.* 1927, **14**, 1195 ff.
Kusui, K. (1). *JJB*, 1932, **15**, 319.
Kusui, K. (2). *JJB*, 1932, **15**, 325.
Kutscher, F., Müller, E. & Spahr, W. *ZB*, 1933, **93**, 239.
Kutscher, W., Veith, G. & Sarreither, W. *ZPC*, 1938, **251**, 124.
Kuwana, Z. (1). *PIAT*, 1933, **9**, 280.
Kuwana, Z. (2). *JJZ*, 1937, **7**, 311.
Kuwana, Z. (3). *JJZ*, 1937, **7**, 305; *YMaly*, 1937, **43**, 270.
Kyle, H.M. *PTRSB*, 1921, **211**, 75.

van der Laan, P.A. *RTBN*, 1934, **31**, 691.
Laboulbène, A. *CRAS*, 1892, **114**, 720.
Lacape, R.S. "À la Recherche du Temps Vécu." Hermann, Paris, 1935.
Lacassagne, A. (1). *CRAS*, 1932, **195**, 630; *AJC*, 1936, **27**, 217; **28**, 735; *CRSB*, 1933, **114**, 427; 1934, **115**, 937; **116**, 95; 1935, **120**, 685, 833, 1156; 1936, **121**, 607; **122**, 183.

Lacassagne, A. (2). *EVHF*, 1938, **2**, 259.
Lahr, E.L. & Riddle, O. *AJOP*, 1938, **123**, 614.
Lallemand, S. *CRSB*, 1932, **110**, 722.
Lambert, R. & Teissier, G. *APBPC*, 1927, **2**, 212.
Lambert, W.V. & Sciuchetti, A. *S*, 1935, **81**, 278; *JH*, 1935, **26**, 91.
Lamoreux, W.F. & Hutt, F.B. *GS*, 1937, **22**, 198.
Landa-Glaz, R.J. "Bases Physico-Chimiques de l'activité Nerveuse." *Trav. Sect. Physiol. Syst. Nerv. Institut Bechterev*, Leningrad, 1935, p. 213.
Landahl, H.D. *G*, 1937, **1**, 263.
Landauer, W. (1). *AFEWM*, 1927, **110**, 195; *KW*, 1928, **7**, 2047.
Landauer, W. (2). *JH*, 1928, **19**, 453.
Landauer, W. (3). *APPA*, 1928, **271**, 534.
Landauer, W. (4). *PS*, 1929, **8**, 301.
Landauer, W. (5). *Proc. IVth World Poultry Congress*, 1930, p. 128.
Landauer, W. (6). *ZMAF*, 1931, **25**, 115.
Landauer, W. (7). *JG*, 1932, **25**, 367.
Landauer, W. (8). *AFO*, 1932, **129**, 268.
Landauer, W. (9). *Bull. Storrs Agric. Exp. Sta.* 1932, no. 179; *EXSR*, 1933, **68**, 803.
Landauer, W. (10). *JG*, 1932, **26**, 285; *N*, 1933, **132**, 606.
Landauer, W. (11). *ZMAF*, 1933, **32**, 359.
Landauer, W. (12). *JH*, 1933, **24**, 293.
Landauer, W. (13). *ENE*, 1933, **12**, 260.
Landauer, W. (14). *Bull. Storrs Agric. Exp. Sta.* 1934, no. 193.
Landauer, W. (15). *AJA*, 1934, **55**, 229.
Landauer, W. (16). *AIPT*, 1934, **49**, 130.
Landauer, W. (17). *JG*, 1935, **30**, 303.
Landauer, W. (18). *JG*, 1935, **31**, 237.
Landauer, W. (19). *AR*, 1936, **64**, 267.
Landauer, W. (20). *AJMS*, 1937, **194**, 667.
Landauer, W. (21). *AR*, 1937, **69**, 247.
Landauer, W. (22). *AIPT*, 1937, **56**, 121.
Landauer, W. (23). *Bull. Storrs Agric. Exp. Sta.* 1937, no. 216.
Landauer, W. (24). *GS*, 1938, **23**, 155.
Landauer, W. (25). *Bull. Storrs Agric. Exp. Sta.* 1939, nos. 232 and 233.
Landauer, W. (26). *HB*, 1939, **11**, 447.
Landauer, W. & de Aberle, S. *AJA*, 1935, **57**, 99.
Landauer, W. & David, L.T. (1). *FH*, 1933, **50**, 1.
Landauer, W. & David, L.T. (2). *AIPT*, 1934, **49**, 125.
Landauer, W. & Dunn, L.C. (1). *JH*, 1930, **21**, 291.
Landauer, W. & Dunn, L.C. (2). *JG*, 1930, **23**, 397.
Landauer, W. & Landauer, A.B. *AN*, 1931, **65**, 492.
Landauer, W. & Thigpen, L.W. *FH*, 1929, **38**, 1.
Landauer, W. & Upham, E. *Bull. Storrs Agric. Exp. Sta.* 1936, no. 210.

Landauer, W., Upham, E., Rubin, F. & Robison, R. *JBC*, 1934, **108**, 121.

Landi, E. *ASBSP*, 1934, **47**, 23; *YMaly*, 1935, **35**, 425.

Lang, A. *ZPC*, 1937, **246**, 219.

Langdon-Brown, W. "The Integration of the Endocrine System." Horsley Lecture, Cambridge, 1935.

de Lange, D. *B*, 1938, **1**, 163.

Lankester, E. Ray. *QJMS*, 1877, **17**, 399.

Lapicque, L. *CRSB*, 1898, **50**, 62 and 856.

Laser, H. (1). *BZ*, 1932, **251**, 2.

Laser, H. (2). *BZ*, 1933, **264**, 72; 1934, **268**, 451; *KW*, 1933, **12**, 754.

Laser, H. (3). Inaug. Diss. Cambridge, 1936.

Laser, H. (4). *BJ*, 1937, **31**, 1671.

Laser, H. (5). *BJ*, 1937, **31**, 1676.

Laser, H. & Rothschild, V. *PRSB*, 1939, **126**, 539.

Laskowski, M. *BZ*, 1935, **275**, 293; **278**, 345; 1936, **284**, 318; *BJ*, 1938, **32**, 1176.

Lasnitzki, A. (1). *KW*, 1933, **12**, 1224; 1934, **13**, 565; *P*, 1934, **22**, 274.

Lasnitzki, A. (2). *BZ*, 1933, **264**, 292.

Lasnitzki, A. & Rosenthal, O. *BZ*, 1929, **207**, 120; 1933, **262**, 203; 1933, **264**, 285; 1935, **281**, 395; 1936, **285**, 101.

Lasnitzki, A. & Szorenyi, E. *BJ*, 1934, **28**, 1678; 1935, **29**, 580.

Lasnitzki, I. (Lasnitzki-Glücksmann). *SKAP*, 1937, **76**, 303.

Last, J.H. & Hays, E.E. *PSEBM*, 1941, **46**, 194.

Lathrop, A.E.C. & Loeb, L. *JCR*, 1916, **1**, 1.

Latinik-Vetulani, I. (1). *BAPS*, 1928, p. 623.

Latinik-Vetulani, I. (2). *BAPS*, 1935, p. 273; *CA*, 1936, **30**, 3101.

Lauche, A. *CAP*, 1925, **36**.

Laufberger, V. *BLL*, 1935, **15**, 325; *OMaly*, 1935, **87**, 640.

Laulanié, M. *CRSB*, 1886, **38**, 87.

de Laurentis, G. *MOG*, 1929, **1**, 830; *YMaly*, 1931, **17**, 105.

Lawrence, A.S.C. (1). *PRSA*, 1935, **148**, 59; "Anomalous Viscosity", in "The Science of Petroleum." Oxford University Press, 1937.

Lawrence, A.S.C. (2). *JRMS*, 1938, **58**, 30.

Lawrence, A.S.C., Needham, J. & Shen Shih-Chang. *N*, 1940, **146**, 104.

Lawrence, A.S.C. & Rawlins, F.I.G. *SP*, 1933, **28**, 339.

Lawrence, A.S.C. & Robinson, J.R. *PRSA* in the press.

Lawrie, N.R. *BJ*, 1935, **29**, 588.

Lawrie, N.R. & Robertson, M. *BJ*, 1935, **29**, 1017.

Lazarev, N.I. *BBMEUSSR*, 1939, **8**, 513.

Leathes, J.B. "Croonian Lectures", *L*, 1925, 803, 853, 957 and 1019.

Lebedeva, M.N. *CG*, 1934, **58**, 1449.

Lebour, M.V. *JMBA*, 1935, **20**, 373.

Lebreton, E. & Schaeffer, G. *Travaux de l'Instit. Physiol. du Fac. Méd. Univ. Strasbourg*, 1923.

Lecamp, M. *BBFB*, 1935, Supplementary vol. **19**.

Lederer, E. & Glaser, R. *CRAS*, 1938, **207**, 454.

Lee, M.O. & Ayres, G.B. *ENY*, 1936, **20**, 489.

Lee, M.O. & Schaffer, N.K. *JN*, 1934, **7**, 337.

Lees, J.C. *QJEP*, 1937, **27**, 161, 171 and 181.

Leeson, H.S. & Mellanby, K. *N*, 1933, **131**, 363.

Legrand, R. *BSBG*, 1936, **12**, 17; *OMaly*, 1936, **95**, 655; *BME*, 1936, **16**, 1131.

Lehmann, F.E. (1). *AFEWM*, 1926, **108**, 243.

Lehmann, F.E. (2). *AFEWM*, 1929, **117**, 312.

Lehmann, F.E. (3). *AFEWM*, 1932, **125**, 566.

Lehmann, F.E. (4). *BC*, 1933, **53**, 471.

Lehmann, F.E. (5). *AFEWM*, 1934, **131**, 333; *RSZ*, 1933, **40**, 251; *SNGB*, 1933; *YMaly*, 1934, **29**, 373.

Lehmann, F.E. (6). *NW*, 1933, **21**, 737; 1937, **25**, 124; *RSZ*, 1935, **42**, 405; *VSNFG*, 1934, 360.

Lehmann, F.E. (7). *RSZ*, 1936, **43**, 535.

Lehmann, F.E. (8). *NW*, 1936, **24**, 401.

Lehmann, F.E. (9). *AFEWM*, 1936, **134**, 166.

Lehmann, F.E. (10). *ABSRSMNB*, 1936, p. 1.

Lehmann, F.E. (11). *RSZ*, 1937, **44**, 1.

Lehmann, F.E. (12). *YMaly*, 1937, **42**, 184.

Lehmann, F.E. (13). *AFEWM*, 1937, **136**, 112.

Lehmann, F.E. (14). *AFEWM*, 1938, **138**, 106.

Lehmann, F.E. (15). *VNGZ*, 1938, **73**, 187.

Lehmann, F.E. (16). *NW*, 1940, **28**, 231.

Lehmann, F.E. & Ris, H. *RSZ*, 1938, **45**, 419.

Lehmann, H. & Needham, J. (1). *JEB*, 1937, **14**, 483.

Lehmann, H. & Needham, J. (2). *EZ*, 1938, **5**, 95.

Lehmann, O. *EP*, 1918, **16**, 256.

Lehrman, L. & Kabat, E.A. *JACS*, 1937, **59**, 1050.

Leibson, R.G. *BBMEUSSR*, 1938, **7**, 514.

Leibson, R.G., Likhnitzkaia, I.I. & Sax, M.G. *JOP*, 1936, **87**, 97.

Leiby, R.W. *JM*, 1922, **37**, 195.

Leifert, H. *ZJZP*, 1935, **55**, 131 and 171.

Leiner, M. (1). *ZVP*, 1936, **23**, 147; *ZAZ*, 1934, **108**, 273.

Leiner, M. (2). *ZVP*, 1937, **24**, 143.

Leiner, M. (3). "Die Physiologie d. Fischatmung." Akad. Verlag, Leipzig, 1938.

Leiner, M. (4). *NW*, 1940, **28**, 165.

Lelkes, Z. *ENE*, 1933, **13**, 35; *CA*, 1933, **27**, 5397.

Lell, W.A. *AR*, 1931, **51**, 119.

Lell, W.A., Liber, K.E. & Snyder, F.F. *AJOP*, 1932, **100**, 21.

Leloir, L.F. & Dixon, M. *EZ*, 1937, **2**, 81.

Lemberg, R. (1). *ADC*, 1931, **488**, 74; *OMaly*, 1931, **63**, 431.

Lemberg, R. (2). *BJ*, 1934, **28**, 978.

Lemberg, R. (3). "The Disintegration of Haemoglobin in the Animal Body." In Hopkins Presentation Volume "Perspectives in Biochemistry", Cambridge, 1938.

Lemberg, R. (4). *ARB*, 1938, **7**, 421.

Lemberg, R. & Barcroft, J. *PRSB*, 1932, **110**, 362.

Lemberg, R., Barcroft, J. & Keilin, D. *N*, 1931, **128**, 967.

Lemeland, H.J. & Delétang, R. *CRSB*, 1934, **116**, 953.

von Lengerken, H. *ZAZ*, 1924, **58**, 179; **59**, 323.

Lenner, A. *AOGS*, 1934, **14** (Suppl.), 1.

Lennerstrand, Å. *ZVP*, 1933, **20**, 287.

Lennox, F.G. *N*, 1940, **146**, 268.

Lenti, C. (1). *ASB*, 1938, **24**, 182.

Lenti, C. (2). *ASB*, 1939, **25**, 1.

Lenti, C. (3). *ASB*, 1939, **25**, 455.

Lenti, C. & Fuortes, M. *AAST*, 1939, **74**, 3.

Leontiev, H. & Alexandrovsky, V. *ZB*, 1935, **96**, 146.

Lepeschinskaia, O.P. In "Probl. d. Theor. Biol." Timiriazev Institute, Moscow, 1935, p. 291; *C*, 1935, **6**, 294.

Lepkovsky, S., Taylor, L.W., Jukes, T.H. & Almquist, H.J. *H*, 1938, **11**, 559.

Lereboullet, A. *CRAS*, 1855. **40**, 916.

Lerner, I.M. (1). *AN*, 1936, **70**, 595.

Lerner, I.M. (2). *H*, 1937, **10**, 511.

Lerner, I.M. (3). *S*, 1939, **89**, 16.

Lerner, I.M., Gregory, P.W. & Goss, H. *PSEBM*, 1936, **35**, 283.

Lerner, I.M. & Gunns, C.A. *G*, 1938, **2**, 261.

Lesné, E., Zizine, P. & Briskas, S. *CRSB*, 1938, **128**, 935.

Lettré, H. & Imhoffen H.H. "Über Sterine, Gallensäuren und Verwandte Naturstoffe." Enke, Stuttgart, 1936.

Leulier, A. & Bernard. *BSCB*, 1937, **19**, 664.

Leulier, A. & Paulant, F. *CRSB*, 1935, **118**, 254; *BSCB*, 1935, **17**, 1124.

Levene, P.A. & Bass, L.W. "Nucleic Acids." American Chem. Soc. Monograph, New York, 1931.

Levene, P.A. & Jorpes, E. *JBC*, 1930, **86**, 389.

Levene, P.A. & Lopez-Suarez, J. *JBC*, 1916, **26**, 373.

Levene, P.A. & Mandel, J.A. *ZPC*, 1907, **49**, 262; *JBC*, 1905, **1**, 425.

Levene, P.A., Mikeska, L.A. & Mori, T. *JBC*, 1930, **85**, 785.

Levene, P.A. & Schormuller, A. *JBC*, 1933, **103**, 537.

Levine, I. *See* le Vine.

Levine, M. *BOR*, 1936, **2**, 439.

Levy, M. & Palmer, A.H. *JBC*, 1938, **123**, lxxiv; 1940, **136**, 415.

Levy-Solal, E., Dalsace, J. & Gutman, C. *CRSB*, 1934, **115**, 269.

Levy-Solal, E., Walther, P. & Dalsace, J. *CRSB*, 1934, **115**, 272.

Lewis, M.R. (1). *AJOH*, 1930, **12**, 288.

Lewis, M.R. (2). *JEZ*, 1932, **64**, 57.

Lewis, M.R. & Geiling, E.M.K. *AJOP*, 1935, **113**, 529.

Lewis, M.R. & Lewis, Warren H. *AJC*, 1932, **16**, 333.

Lewis, Warren H. (1). *AJA*, 1903, **3**, 505; *JEZ*, 1905, **2**, 431.

Lewis, Warren H. (2). *AJA*, 1907, **6**, 473; 1907, **7**, 259.

Lewis, Warren H. (3). *AJA*, 1907, **7**, 137.

Lewis, Warren H. (4). *S*, 1935, **81**, 545; *AEZF*, 1939, **23**, 8.

Lewis, Warren H. (5). *G*, 1939, **3** (Suppl.), 1.

Li, Ju-Chi (1). *GS*, 1927, **12**, 1.

Li, Ju-Chi (2) *PNHB*, 1935, **9**, 57.

Li, Ju-Chi & Tsui, Yu-Lin. *GS*, 1936, **21**, 248.

Li, Tai. *CRSB*, 1935, **120**, 219.

Liang, Chung-Mou. *AP*, 1934, **234**, 302.

Liczko, E.J. *CRASUSSR*, 1932, **7**, 65; *TLZE MAM*, 1934, **3**, 101; 1935, **4**, 169.

von Liebig, J. *ADC*, 1847, **62**, 317.

Light, S.F., Hartman, O. & Emerson, O.H. Unpublished work.

Likhnitzkaia, I.I. & Sax, M.G. *BBMEUSSR*, 1938, **5**, 320.

Lillie, F.R. (1). *AN*, 1900, **34**, 173.

Lillie, F.R. (2). *AFEWM*, 1902, **14**, 477.

Lillie, F.R. (3). *JEZ*, 1906, **3**, 153.

Lillie, F.R. (4). *BB*, 1909, **16**, 54.

Lillie, F.R. (5). *JEZ*, 1914, **16**, 523.

Lillie, F.R. (6). *S*, 1916, **43**, 611; *JEZ*, 1917, **23**, 371.

Lillie, F.R. (7). "Problems of Fertilisation." Chicago University Press, 1919, ch. VII.

Lillie, F.R. (8). "The Development of the Chick." 2nd ed. New York and London, 1919.

Lillie, F.R. (9). *AFEWM*, 1929, **118**, 499.

Lillie, F.R. (10). *PZ*, 1940, **13**, 143.

Lillie, F.R. & Juhn, M. *PZ*, 1932, **5**, 124; 1938, **11**, 434.

Lillie, F.R. & Just, E.E. "Fertilisation." In Cowdry's "General Cytology", 1924, p. 449.

Lillie, F.R. & Wang, Hsi. *PNASW*, 1940, **26**, 67.

Lillie, R.S. (1). *JEZ*, 1908, **5**, 375; *JGP*, 1925, **8**, 339; 1927, **10**, 703; *BB*, 1931, **60**, 288; 1934, **66**, 361; *PZ*, 1941, **14**, 239.

Lillie, R.S. (2). *AN*, 1938, **72**, 389.

Lindahl, P.E. (1). *P*, 1932, **16**, 378.

Lindahl, P.E. (2). *AFEWM*, 1932, **126**, 373.

Lindahl, P.E. (3). *AFEWM*, 1932, **127**, 300.

Lindahl, P.E. (4). *AFEWM*, 1932, **127**, 323.

Lindahl, P.E. (5). *AFEWM*, 1933, **128**, 661; *AFZ*, 1935, **28** B, 4.

Lindahl, P.E. (6). *NW*, 1934, **22**, 105.

Lindahl, P.E. (7). *AZ*, 1936, **17**, 179; *YMaly*, 1937, **42**, 290; *AFZ*, 1935, **28** B, 4.

Lindahl, P.E. (8). *NW*, 1938, **26**, 709.

Lindahl, P.E. (9). *ZVP*, 1939, **27**, 136 and 233.

Lindahl, P.E. & Öhman, L.O. *BC*, 1938, **58**, 179; *NW*, 1936, **24**, 157.

Lindahl, P.E. & Örström, Å. (1). *P*, 1932, **17**, 25.

Lindahl, P.E. & Örström, Å. (2). *NW*, 1936, **24**, 142.

Lindahl, P.E. & Stordahl, Å. *AFEWM*, 1937, **136**, 44.

Lindeman, V.F. *AR*, 1928, **41**, 39; *PZ*, 1929, **2**, 255.

Linderstrøm-Lang, K. (1). *ZPC*, 1933, **215**, 167; *CRTLC*, 1933, **19**, no. 13.

Linderstrøm-Lang, K. (2). *N*, 1937, **140**, 108.

Linderstrøm-Lang, K. & Glick, D. *CRTLC*, 1938, **22**, 300.

Linderstrøm-Lang, K. & Holter, H. (1). *CRTLC*, 1933, **19**, no. 14.

Linderstrøm-Lang, K. & Holter, H. (2). *EEF*, 1934, **3**, 309.

Lintzel, W. *ATT*, 1931, **7**, 42; *OMaly*, 1932, **65**, 530.

Lintzel, W. & Radeff, T. *ATT*, 1931, **6**, 313.

Liosner, L.D. (1). *BC*, 1930, **50**, 308.

Liosner, L.D. (2). *AFEWM*, 1931, **124**.

Liosner, L.D. (3). *BBMEUSSR*, 1937, **4**, 150.

Liosner, L.D. (4). *BBMEUSSR*, 1937, **4**, 153.

Liosner, L.D. (5). *BBMEUSSR*, 1938, **6**, 262.

Liosner, L.D. (6). *BBMEUSSR*, 1939, **8**, 14.

Liosner, L.D. & Blacher, L.J. *BC*, 1932, **52**, 697.

Liosner, L.D. & Voronzova, M.A. (1). *BAPS*, 1935, B (ii), 231.

Liosner, L.D. & Voronzova, M.A. (2). *BBMEUSSR*, 1937, **4**, 303.

Liosner, L.D. & Voronzova, M.A. (3). *ZAZ*, 1935, **110**, 286; *AAM*, 1937, **33**, 313.

Liosner, L.D. & Voronzova, M.A. (4). *BBMEUSSR*, 1938, **5**, 439; 1939, **7**, 224.

Liosner, L.D. & Voronzova, M.A. (5) *CRASUSSR*, 1940, **26**, 819.

Liosner, L.D., Voronzova, M.A. & Kozmina, N.A. *AFEWM*, 1936, **134**, 738.

Lipmann, F. (1). *BZ*, 1933, **261**, 157.

Lipmann, F. (2). *BZ*, 1933, **262**, 3 and 9.

Lipmann, F. (3). *BZ*, 1933, **265**, 131; **268**, 205.

Lipmann, F. & Levene, P.A. *JBC*, 1932, **98**, 109.

Lippincott, W.A. & du Puy, P.L. *PS*, 1923, **3**, 25.

Lipschütz, A. & Vargas, L. *L*, 1939, (i), 1313; 1940, (i), 541.

Lipsett, H. *JEZ*, 1941, **86**, 441.

Lison, L. "Histochimie Animale." Gauthier-Villars, Paris, 1936.

Lissmann, H.W. & Wolsky, A. *ZVP*, 1933, **19**, 555; *ZAZ*, 1935, **110**, 92.

Lister, Arthur. *AOB*, 1888, **2**, 2.

Little, C.C. (1). *AN*, 1915, **49**, 727.

Little, C.C. (2). *AN*, 1931, **65**, 370.

Little, C.C. (3). *JAMA*, 1936, **106**, 2234.

Little, C.C. & Bagg, H.J. *JEZ*, 1924, **41**, 45.

Little, C.C. & McPheters, B.W. *GS*, 1932, **17**, 674.

Litvać, M.A. *AAM*, 1937, **33**, 151.

Litwer, G. *ZMAF*, 1932, **30**, 599; *YMaly*, 1932, **23**, 297.

Ljubitzky, A.I. *ZJZP*, 1935, **54**, 405.

Ljubitzky, A.I. & Svetlov, P. *BC*, 1934, **54**, 195.

Locatelli, P. *ASB*, 1924, **5**, 362.

Lochhead, J. & Cramer, W. *PRSB*, 1908, **80**, 263; *JOP*, 1907, **35**, xi.

Lockhart-Mummery, J.P. "The Origin of Cancer." London, 1934.

Lodyženskaia, V. *CRASUSSR*, 1928, 99; *TLZEMAM*, 1932, **1**, 61.

Loeb, J. (1). *JM*, 1892, **2**; *AFEWM*, 1895, **2**.

Loeb, J. (2). *UCPP*, 1907, **3**, 61.

Loeb, J. (3). "Die kunstliche Parthenogenese." *Handbuch d. Biochemie* (Oppenheimer's), 1909, II, i, 79.

Loeb, L. (1). *CAP*, 1907, **18**, 563; *JAMA*, 1908, **50**, 1897.

Loeb, L. (2). *JMR*, 1919, **40**, 477; *JCR*, 1924, **8**, 274.

Loeb, L. (3). *PR*, 1930, **10**, 547.

Loeb, L. (4). *AIM*, 1931, **4**, 669.

Loeb, L. (5). *JAMA*, 1935, **104**, 1597.

Loeb, L., Burns, E.L., Suntzev, V. & Moskop, M. *AJC*, 1937, **30**, 47.

Loeschke, A. *ZPC*, 1931, **199**, 125.

Loeser, A. *AG*, 1932, **148**, 118; *CG*, 1932, **56**, 206.

Loewe, S. (1). *ZGG*, 1926, **90**, 380.

Loewe, S. (2). *CG*, 1926, 551.

Loewe, S., Lange, F. & Käer, E. *ENE*, 1929, **5**, 177.

Loewe, S., Raudenbusch, W., Voss, H.E. & van Heurn, W.C. *BZ*, 1932, **244**, 347.

Loewi, O. *AP*, 1937, **239**, 430.

Lohmann, K. *BZ*, 1932, **254**, 332.

Löhner. *AP*, 1924, **203**, 524.

Longsworth, L.G., Lannau, R.K. & McInnes, D.A., *JACS*, 1940, **62**, 2580.

Loofbourow, J.R., Cook, E.S., Dwyer, C.M. & Hart, M.J. *N*, 1939, **144**, 553.

Loofbourow, J.R., Cook, E.S. & Stimson, M.M. *N*, 1938, **142**, 573.

Loofbourow, J.R., Cueto, A.A. & Lane, M.M. *AEZF*, 1939, **22**, 607.

Loofbourow, J.R., Cueto, A.A., Whelan, D. & Lane, M.M. *N*, 1939, **144**, 939.

Loofbourow, J.R. & Dwyer, C.M. *N*, 1939, **143**, 725; 1940, **145**, 185.

Loofbourow, J.R., Dwyer, C.M. & Lane, M.M. *BJ*, 1940, **34**, 432.

Loofbourow, J.R., Dwyer, C.M. & Morgan, M.N. *SIDT*, 1939, **2**, 137.

Loofbourow, J.R. & Heyroth, F.F. *N*, 1934, **133**, 909.

Loosli, M. *PSZN*, 1934, **15**, 16.

Lopaschov, G.V. (1). *CRASUSSR*, 1935, **4**, 55.

Lopaschov, G.V. (2). *BC*, 1935, **55**, 606; *YMaly*, 1936, **37**, 177.

Lopaschov, G.V. (3). *ZJZP*, 1935, **54**, 299.

Lopaschov, G.V. (4). *BZUSSR*, 1935, **4**, 429; *YMaly*, 1936, **37**, 550.

Lopaschov, G.V. (5). *N*, 1935, **136**, 835; *BZUSSR*, 1936, **5**, 463.

Lopaschov, G.V. (6). *CRASUSSR*, 1937, **15**, 283.

Lopaschov, G.V. (7). *CRASUSSR*, 1937, **15**, 286.

Lopaschov, G.V. (8). *CRASUSSR*, 1939, **24**, 205.

Lorenz, F.W., Almquist, H.J. & Hendry, G.W. *S*, 1933, **77**, 606.

Lorenz, F.W., Taylor, L.W. & Almquist, H.J. *PS*, 1934, **13**, 14.

Lotka, A.J. "Elements of Physical Biology." Williams & Wilkins, Baltimore, 1925, pp. 77, 146, 294.

Loughlin, W.J. *BJ*, 1933, **27**, 99.

Lounsbury, C.P. *SAJS*, 1915, **12**, 33.

Louvier, R. (1). *CRSB*, 1930, **103**, 1168.

Louvier, R. (2). *CRSB*, 1932, **109**, 1116.

Love, W.H. *MJA*, 1929 (Sept. 21st).

Loveland, G., Maurer, E.E. & Snyder, F.F. *AR*, 1931, **49**, 265.

Lovern, J.A. (1). *BJ*, 1934, **28**, 394.

Lovern, J.A. (2). *BJ*, 1936, **30**, 20.

Lowrance, E.W. *PSEBM*, 1937, **36**, 590; *JCCP*, 1937, **10**, 321.

Lu, Gwei-Djen. *See* Emerson, Lu & Emerson.

Lucas, W.P. & Dearing, B.F. *AJDC*, 1921, **21**, 96.

Luce, W.M. *JEZ*, 1935, **71**, 125.

Luciani, F., Filomeni, M. & Severi, L. *RDB*, 1930, **12**, 136; *YMaly*, 1932, **21**, 217.

Luck, J.M. & Ritter, R.C. *PSEBM*, 1931, **28**, 829.

Lucke, B. (1). *JCCP*, 1932, **2**, 193.

Lucke, B. (2). *Proc. xvth Internat. Physiol. Congr.* (Moscow), 1935, p. 247; *AJC*, 1934, **20**, 352; 1934, **22**, 326; 1938, **34**, 15.

Lucke, B. (3). *JEM*, 1939, **70**, 269.

Lucke, B. & McCutcheon, M. *PR*, 1932, **12**, 68.

Lucke, B. & Schlumberger, H. *JEM*, 1939, **70**, 257; 1940, **72**, 311 and 321.

Ludford, R.J. *PRSB*, 1925, **98**, 457 and 557; *JRMS*, 1925, 249.

Ludwig, D. (1). *JEZ*, 1931, **60**, 309.

Ludwig, D. (2). *PZ*, 1932, **5**, 431; *AJOP*, 1928, **85**, 389.

Ludwig, D. (3). *AESA*, 1934, **27**, 429.

Lüers, H. *ZIAV*, 1937, **72**, 119.

Lumer, H. *AN*, 1939, **73**, 339.

Lund, E.J. (1). *BOG*, 1923, **76**, 288.

Lund, E.J. (2). *JEZ*, 1921, **34**, 471; 1924, **39**, 357.

Lund, H. *CRSB*, 1932, **110**, 1121.

Lund, W.A., Heiman, V. & Wilhelm, L.A. *PS*, 1938, **17**, 372.

Lunde, G., Kringstad, H. & Olsen, A. *TH*, 1938, **1**, 81 and 184; *CA*, 1938, **32**, 6301.

Luntz, A. In "Probl. d. Theor. Biol.", Timiriazev Institute, Moscow, 1935, p. 152.

Lütgerath, F. "Carotenoid Metabolism in the human foetus." Inaug. Diss. München, 1937.

Luther, A. *CBSSF*, 1925, **2**, 1.

Luther, W. (1). *AFEWM*, 1934, **131**, 532.

Luther, W. (2). *BC*, 1935, **55**, 114.

Luther, W. (3). *AFEWM*, 1936, **135**, 359; *VDZG*, 1936, 73.

Luther, W. (4). *AFEWM*, 1936, **135**, 384.

Luther, W. (5). *AFEWM*, 1937, **137**, 404.

Luther, W. (6). *AFEWM*, 1937, **137**, 425.

Luther, W. (7). *CS*, 1939, Suppl. no. 4, 33.

Lüthi, H.R. *AFEWM*, 1939, **138**, 423.

Lvov (Lwoff), A. & Roukhelman, N. *CRAS*, 1926, **183**, 156.

Lynn, W.G. *AR*, 1938, **70**, 597.

Lynn, W.G. & Brambel, C.E. *AR*, 1935, **64**, 46.

Lyons, M. & Insko, W.M. *Bull. Kentucky Agric. Exp. Sta.* 1937, no. 371; *S*, 1937, **86**, 328.

Lyons, W.R. (1). *PSEBM*, 1937, **37**, 207.

Lyons, W.R. (2). *CSHSQB*, 1937, **5**, 198.

Lyons, W.R. & Page, E. *PSEBM*, 1935, **32**, 1049.

McArthur, C.G. & Doisy, E.A. *JCN*, 1919, **30**, 445.

McArthur, J.W. *BB*, 1924, **46**.

McBride, E.W. (1). *QJMS*, 1911, **57**, 235.

McBride, E.W. (2). "Textbook of Embryology, I. Invertebrata." Macmillan, London, 1914.

McBride, E.W. (3). *PRSB*, 1918, **90**, 323.

McBride, E.W. (4). *D*, 1934, **15**, 218.

McCammon, R.B., Pittman, M.S. & Wilhelm, L.A. *PS*, 1934, **13**, 95.

McCarthy, E.F. (1). *JOP*, 1933, **80**, 206.

McCarthy, E.F. (2). *JOP*, 1938, **93**, 81.

McCay, C.M. "Chemical Aspects of Ageing" in "Problems of Ageing," N.Y. 1939, p. 572.

McCay, C.M., Crowell, M.F. & Maynard, L.A. *JN*, 1935, **10**, 63; *Proc. xvth Internat. Physiol. Congr.* p. 249.

McCay, C.M., Tunison, A.V., Crowell, M. & Paul, H. *JBC*, 1936, **114**, 259.

McClendon, J.F. *BB*, 1907, **12**, 142; *AFEWM*, 1908, **26**, 662.

McClung, C.E. *BB*, 1902, **9**, 75.

McConnell, K.P. & Sinclair, R.G. *JBC*, 1937, **118**, 123.

McCorquodale, D.W., Steenbock, H. & Adkins, H. *JACS*, 1930, **52**, 2512.

McCutcheon, F.H. (1). *JCCP*, 1936, **8**, 63.

McCutcheon, F.H. (2). *JEB*, 1938, **15**, 431.

McCutcheon, F.H. & Hall, F.G. *JCCP*, 1937, **9**, 191.

McDougald, T.J., Beams, H.W. & King, R.L. *PSEBM*, 1937, **37**, 234.

McFarlane, W.D. *BJ*, 1932, **26**, 1038 and 1061.

McFarlane, W.D., Fulmer, H.L. & Jukes, T.H. *BJ*, 1930, **24**, 1611.

McFarlane, W.D. & Milne, H.I. *JBC*, 1934, **107**, 309.

McGowan, J.P. *BZ*, 1934, **272**, 9.

McHargue, J.S. *JAR*, 1924, **27**, 417.

McIntosh, J. *BJEP*, 1933, **14**, 422.

McIntosh, J. & Selbie, F.R. *BJEP*, 1939, **20**, 49.

McKinnon, D.L. & Vlès, F. *CRAS*, 1908, **147**.

McLeod, J. *HJ*, 1918, **16**, 210.

McNally, E. (1). *PSEBM*, 1933, **30**, 1254.

McNally, E. (2). *PSEBM*, 1934, **31**, 946.

McWalter, R.J. & Drummond, J. *BJ*, 1933, **27**, 1415.

Macchiarulo, O. *AG*, 1935, **159**, 349 and 355.

Macciotta, M. *FG*, 1932, **29**, 481; *OMaly*, 1934, **75**, 538.

Macé, S. *TSBR*, 1936, **14**, 91.

Machabeli, A.I. *CRASUSSR*, 1939, **23**, 967.

Machebœuf, M. "État des Lipides dans la matière Vivante; les Cénapses et leur importance biologique." Hermann, Paris, 1936.

Machebœuf, M. & Sandor, G. *BSCB*, 1932, **14**, 1168.

Machemer, H. *AFEWM*, 1929, **118**, 200.

Machida, S. & Sasaki, T. *JACSJ*, 1937, **13**, 305.

Macy, I.G. & Hunscher, H.A. *AJOG*, 1934, **27**, 878.

Macy, I.G., Hunscher, H.A., Nims, B. & McCosh, S.S. *JBC*, 1930, **86**, 17.

Magiotti, R. "Renitenza certissima dell' Acqua alla Compressione, dichiarata con varii scherzi in occasione d' altri problemi curiosi." Moneta, Rome, 1648.

Makarov, P. *AAUSSR*, 1929, **8**, 255; *OMaly*, 1931, **62**, 52.

Makepeace, A.W., Fremont-Smith, F., Dailey, M.E. & Carroll, M.P. *SGO*, 1931, **53**, 635.

Malan, A.I., Malan, A.P. & Curson, H.H. *OJVSAI*, 1937, **9**, 205.

Malan, A.P. & Curson, H.H. *OJVSAI*, 1935, **4**, 481; 1936, **7**, 239, 251, 261; 1937, **8**, 417.

Mall, F.P. *JM*, 1908, **19**, 1.

Malov, M.N. & Friesen, H. *BZUSSR*, 1936, **5**, 561.

Maluf, N.S.Rustum (1). *N*, 1936, **138**, 75.

Maluf, N.S.Rustum (2). *BC*, 1936, **56**, 429.

Maluf, N.S.Rustum (3). *PR*, 1938, **18**, 28.

Maluf, N.S.Rustum (4). *JGP*, 1940, **24**, 151.

Manahan, C.P. & Eastman, N.J. *BJHH*, 1938, **52**, 478.

Manai, A. *Studi Sassar.* 1935, **13**, 195; *YMaly*, 1935, **35**, 151.

Manderscheid, H. *BZ*, 1933, **263**, 245.

van Manen, E. & Rimington, C. *OJVSAI*, 1935, **5**, 329.

Manery, J.F. & Irving, L. *JCCP*, 1935, **5**, 457.

Manery, J.F., Warbritton, V. & Irving, L. *JCCP*, 1933, **3**, 277.

Mangili, C. *AO*, 1929, **51**, 1474.

Mangold, O. (1). *VDZG*, 1922, **27**, 51; *AFEWM*, 1923, **100**, 198.

Mangold, O. (2). *VDZG*, 1925, 50.

Mangold, O. (3). *NW*, 1926, **14**, 1169.

Mangold, O. (4). *NW*, 1928, **16**, 387 and 661.

Mangold, O. (5). *EB*, 1928, **3**, 152.

Mangold, O. (6). *EB*, 1929, **5**, 290.

Mangold, O. (7). *AFEWM*, 1929, **117**, 586.

Mangold, O. (8). *EB*, 1931, **7**, 193.

Mangold, O. (9). *NW*, 1931, **19**, 475.

Mangold, O. (10). *NW*, 1931, **19**, 905.

Mangold, O. (11). *NW*, 1932, **20**, 371.

Mangold, O. (12). *NW*, 1933, **21**, 761.

Mangold, O. (13). *NW*, 1936, **24**, 753.

Mangold, O. (14). *DB*, 1936, 82. *CS*, 1939, Suppl. no. 4, 1.

Mangold, O. (15). *Comptes Rend.* xiie *Congrès International de Zoologie.* Lisbon, 1936, p. 65.

Mangold, O. (16). *ASFFF*, 1937, **60**, 3.

Mangold, O. & Seidel, F. *AFEWM*, 1927, **111**, 593.

Mangold, O. & Spemann, H. *AFEWM*, 1927, **111**, 341.

Mankin, W.R. (1). *MJA*, 1928, **15** (ii), 87.

Mankin, W.R. (2). *MJA*, 1929, **16** (ii), 358.

Mankin, W.R. (3). *MJA*, 1930, **17** (ii), 41.

Mann, I. "Developmental Abnormalities of the Eye." Cambridge University Press, 1937.

Mann, P.J.G. *BJ*, 1932, **26**, 785.

Mann, P.J.G., Tennenbaum, M. & Quastel, J.H. *BJ*, 1938, **32**, 243; 1939, **33**, 1506.

Mansfeld, G. & Liptak, P. *AP*, 1913, **152**, 68.

Manton, S.M. & Heatley, N.G. *PTRSB*, 1937, **227**, 411.

Manuilova, N.A. (1). *AAUSSR*, 1935, **14**, 504. *CRASUSSR*, 1938, **18**, 689.

Manuilova, N.A. (2). *CRASUSSR*, 1939, **23**, 311 & 972.

Manuilova, N.A. & Kislov, M.N. *ZJZP*, 1934, **53**.

Manuilova, N.A., Machabeli, A.I. & Sikharulidze, T.A. *CRASUSSR*, 1938, **18**, 693.

Manunta, C. (1). *BSIBS*, 1933, **8**, 1278.

Manunta, C. (2). *BSIBS*, 1934, **9**, 766; *RAL*, 1934, **20**, 283; *CA*, 1935, **29**, 3043.

Manunta, C. (3). *BSIBS*, 1936, **11**, 50.

Manunta, C. (4). *AZI*, 1937, **24**, 369; *YMaly*, 1938, **45**, 168.

Manunta, C. (5). *BSIBS*, 1937, **12**, 33.

Manunta, C. (6). *BSIBS*, 1937, **12**, 626.

Manunta, C. (7). *BSIBS*, 1937, **12**, 628 and 629.

Manville, I.A. & Lloyd, R.W. *AJOP*, 1932, **100**, 394.

Maple, J.D. *AESA*, 1937, **30**, 123 and 144; *SP*, 1937, **32**, 127.

Marchlewski, L. & Wierzuchowska, J. *BAPS*, A, 1928, 471.

Margen, S. & Schechtman, A.M. *PSEBM*, 1939, **41**, 47.

Margolis, O.S. *GS*, 1935, **20**, 156 and 207; 1936, **21**.

Margolis, O.S. & Robertson, C.W. *GS*, 1937, **22**, 318.

Marlow, H.W. & King, H.H. *PS*, 1936, **15**, 377.

Marnay, A. & Nachmansohn, D. *CRSB*, 1937, **125**, 41; *JOP*, 1938, **92**, 37.

Marrack, J. "Specificity of Antigens and Antibodies." Medical Research Council. London, 1934, pp. 74 ff.

Marshall, W. & Cruickshank, D.B. *JAS*, 1938, **28**, 24.

Marston, H.R. *BJ*, 1923, **17**, 851.

Martella, N.A. *RDB*, 1935, **18**, 197.

Martin, J.H., Erikson, S.E. & Insko, W.M. *Bull. Kentucky Agric. Exp. Sta.* 1930, no. 304.

Martin, J.H. & Insko, W.M. (1). *Bull. Kentucky Agric. Exp. Sta.* 1935, no. 359.

Martin, J.H. & Insko, W.M. (2). *PS*, 1935, **14**, 152.

Martino, E. & Bonsignore, A. *BTS*, 1934, **21**, 169; *YMaly*, 1935, **32**, 437.

Martinoli, M. *BSIBS*, 1932, **7**, 354.

Marx, A. (1). *AFEWM*, 1925, **105**, 20.

Marx, A. (2). *AFEWM*, 1931, **123**, 333.

Marx, L. *EB*, 1935, **11**, 244.

Marza, V.D. (1). *CRSB*, 1929, **102**, 195.

Marza, V.D. (2). *Proc. 2nd Internat. Congress for Sex Research*, London, 1930, p. 100.

Marza, V.D. (3). *QJMS*, 1935, **78**, 192; *CRSB*, 1934, **117**, 1278.

Marza, V.D. (4). "Histophysiologie de l'Ovogenèse." Hermann, Paris, 1938.

Marza, V.D. & Blinov, A.V. *BSRNPPE*, 1934, **6**, 3.

Marza, V.D. & Chiosa, L.T. *CRSB*, 1934, **117**, 524; 1935, **120**, 345; *BHA*, 1935, **12**, 58; 1936, **13**, 153.

Marza, V.D. & Golaescu, M. *CRSB*, 1935, **118**, 1470.

Marza, V.D. & Marza, E.V. (1). *BHA*, 1932, **9**, 313.

Marza, V.D. & Marza, E.V. (2). *BHA*, 1934, **11**, 65.

Marza, V.D. & Marza, E.V. (3). *QJMS*, 1935, **78**, 134, 160, 172 and 181.

Marza, V.D., Marza, E.V. & Chiosa, L.T. *BHA*, 1932, **9**, 213.

Marza, V.D., Marza, E.V. & Guthrie, M.J. *BB*, 1937, **73**, 67.

Masamune, H. & Hoshino, S. *JJB*, 1936, **24**, 219.

Maschkovzev, A. *MJ*, 1934, **73**, 551.

Maschlanka, H. *AFEWM*, 1938, **137**, 714.

Maschmann, E. & Helmert, E. *ZPC*, 1933, **216**, 161.

Masera, E. *RDB*, 1935, **18**, 98; 1939, **27**, 245.

Masing, E. (1). *ZPC*, 1910, **67**, 161.

Masing, E. (2). *AFEWM*, 1914, **40**, 666.

Mason, K.E. (1). *AJA*, 1935, **57**, 303.

Mason, K.E. (2). "Relation of the Vitamins to the Sex Glands." In "Sex and Internal Secretions", 2nd edn. 1939, p. 1149.

Mason, K.E. & Melampy, R.M. *PSEBM*, 1936, **35**, 459.

Masuda, Y. & Hori, T. *JACSJ*, 1937, **13**, 200; *CA*, 1937, **31**, 7548.

Matsuhita, T. *MMAK*, 1935, **15**, 341, 354 and 582.

Matthews, S.A. & Detwiler, S.R. *JEZ*, 1926, **45**, 279.

Matthias, P. *BSZF*, 1931, **55**, 421; *YMaly*, 1931, **18**, 217.

Mattikov, M. *PS*, 1932, **11**, 83.

Matzke, E.B. *AJB*, 1939, **26**, 288.

Mauser, F. *BG*, 1938, **14**, 179.

Maver, M.E. & Voegtlin, C. *AJC*, 1935, **25**, 780.

Maxwell, J.P., Hu, C.-H. & Turnbull, H.M. *JPB*, 1932, **35**, 419.

May, F. (1). *ZB*, 1932, **92**, 325.

May, F. (2). *ZB*, 1934, **95**, 277, 401, 606 and 614.

May, F. & Kordovitch, F. *ZB*, 1933, **93**, 233.

May, F. & Stadelmann, L. *ZB*, 1939, **99**, 462.

May, F. & Weinbrenner, H. *ZB*, 1938, **99**, 199.

May, R.M. (1). *BSCB*, 1929, **11**, 312; 1930, **12**, 934; *EN*, 1930, **25**, 447.

May, R.M. (2). "La Transplantation Animale." Gauthier-Villars, Paris, 1932.

May, R.M. (3). *APBPC*, 1930, **6**, 206; *PM*, 1935, p. 3; *AAHE*, 1936, **21**, 31.

May, R.M. (4). *BSPP*, 1936, **119**, 15.

Mayeda, S. *NIZ*, 1932, **10**, 564 and 582; *OMaly*, 1933, **69**, 384.

Mayeda, S. & Yamanuchi, S. *NIZ*, 1934, **12**, 1159; *YMaly*, 1935, **32**, 119.

Mayer, B. (1). *AFEWM*, 1935, **133**, 518.

Mayer, B. (2). *NW*, 1939, **27**, 277.

Mayer, E. (1). *SKAP*, 1936, **75**, 1.

Mayer, E. (2). *TB*, 1939, **19**, 65.

Mayer, E. & Fischer, A. *SKAP*, 1937, **75**, 268.

Mayer, F.K. *CDE*, 1931, **6**, 239; *CA*, 1932, **26**, 5151.

Mazia, D. (1). *S*, 1933 (II), **78**, 107.

Mazia, D. (2). *JCCP*, 1937, **10**, 291.

Mazza, F.P. *ASB*, 1931, **15**, 12; *BSIBS*, 1929, **4**, 1151; *OMaly*, 1931, **59**, 680.

Mazia, D. & Davis, J.O. Private communication to the author, Oct. 1940.

Mazza, F.P. & Lenti, C. (1). *ASB*, 1938, **24**, 203; *YMaly*, 1939, **50**, 630; *AAST*, 1938, **73**, 228.

Mazza, F.P. & Lenti, C. (2). *ASB*, 1939, **25**, 447.

Mazza, F.P. & Malaguzzi-Valeri, C. *BSIBS*, 1935, **10**, 722; *ASB*, 1935, **21**, 443.

Mazza, F.P. & Penati, F. (1). *ASB*, 1937, **23**, 443.

Mazza, F.P. & Penati, F. (2). *ASB*, 1938, **24**, 83.

Mazza, F.P. & Stolfi. *ASB*, 1931, **16**, 182.

Meamber, D.L., Fraenkel-Conrat, H.L., Simpson. M.E. and Evans, H.M. *S*, 1939, **90**, 18.

Medawar, P.B. (1). *QJEP*, 1937, **27**, 147.

Medawar, P.B. (2). Inaug. Diss. Oxford, 1940; *PRSB*, 1940, **129**, 332.

Medvedev, N.N. (1). *ZIAV*, 1935, **70**, 55.

Medvedev, N.N. (2). *CRASUSSR*, 1937, **14**, 45.

Medvedev, N.N. (3). *CRASUSSR*, 1938, **20**, 319.

Medvedeva, N.B. *ZVP*, 1927, **5**, 547.

Meewis, H. *ADB*, 1938, **50**, 3.

Meier, R. *BZ*, 1931, **231**, 247 and 253.

de Meio, R.H. & Barron, E.S.G. *PSEBM*, 1934, **32**, 36.

Meisenheimer, J. "Experimentelle Studien zur Soma u. Geschlechts-differenzierung." Jena, 1909.

Melampy, R.M. & Jones, D.B. *PSEBM*, 1939, **41**, 382.

Melampy, R.M. & Olsen, R.D. *PSEBM*, 1940, **45**, 754.

Melampy, R.M. & Stanley, A.J. *S*, 1940, **91**, 457.

Melampy, R.M. & Willis, E.R. *PZ*, 1939, **12**, 302.

Melampy, R.M., Willis, E.R. & McGregor, S.E. *PZ*, 1940, **13**, 283.

Melandri, V. *AIAE*, 1934, **33**, 859; *OMaly*, 1935, **86**, 305; *YMaly*, 1935, **33**, 481.

Melchers, G. *BC*, 1937, **57**, 568; *BDBG*, 1939, **57**, 29.

Meldrum, N.U. *BJ*, 1932, **26**, 817.

Mellanby, E. (1). *JPB*, 1938, **46**, 447.

Mellanby, E. (2). *JPB*, 1938, **47**, 47.

Mellanby, J. *JOP*, 1907, **36**, 474.

Mellanby, K. (1). *PRSB*, 1932, **111**, 376.

Mellanby, K. (2). *N*, 1933, **132**, 66.

Mellander, O. *BZ*, 1935, **277**, 305.

Melvin, R. *AESA*, 1931, **24**, 485; *CA*, 1932, **26**, 529.

Mencl, E. *AFEWM*, 1903, **16**, 328.

Mendel, B. *KW*, 1929, **8**, 169.

Mendel, B., Bauch M. & Strelitz, F. *KW*, 1931, **10**, 118.

Mendel, B., Strelitz, F. & Mundell, D. *N*, 1938, **141**, 288.

Mendel, L.B. & Leavenworth, C.S. (1). *AJOP*, 1907, **20**, 117.

Mendel, L.B. & Leavenworth, C.S. (2). *AJOP*, 1908, **21**, 77.

Menke, J.F. *S*, 1940, **92**, 290.

Menschick, W. & Page, I.H. *ZPC*, 1932, **211**, 246.

Menville, L.J. & Ané, J.N. *PSEBM*, 1932, **29**, 1045.

Mergassova, H.P. *BZUSSR* 1937, **5**, 927; *YMaly*, 1937, **42**, 183.

Merkel, F. *AH*, 1909, **38**, 321.

Merriman, D. *JEB*, 1935, **12**, 297.

Mestscherskaia, K.A. *AAUSSR*, 1935, **14**, 656 and 720; *YMaly*, 1937, **41**, 9; *OMaly*, 1937, **99**, 41.

Mettetal, C. *AAHE*, 1939, **28**, 3.

Meunier, P. (1). *BSCB*, 1936, **18**, 636.

Meunier, P. (2). *BSCB*, 1937, **19**, 244.

Meyer, K. *CSHSQB*, 1938, **6**, 91.

Meyer, K., Palmer, J.W. & Smith, E.M. (1). *JBC*, 1936, **114**, 689.

Meyer, K., Palmer, J.W. & Smith, E.M. (2). *JBC*, 1937, **119**, 501.

Meyerhof, B. see Ball & Meyerhof.

Meyerhof, H.L. & Zironi, A. *JOP*, 1940, **97**, 495.

Meyerhof, O. (1). *BZ*, 1931, **242**, 244.

Meyerhof, O. (2). *CS*, 1936, **4**, 669.

Meyerhof, O. & Lehmann, H. *NW*, 1935, **21**, 337.

Meyerhof, O. & Perdigon, E. *CRSB*, 1939, **132**, 186; *EZ*, 1940, **8**, 289.

Michaelis, L., Hill, E.S. & Schubert, M.P. *BZ*, 1932, **255**, 66.

Michalovsky, I. *CAP*, 1926, **38**; *APPA*, 1928, **267**, 27; 1929, **274**, 319.

Michejev, M. *Arb. Lab. Proteinforsch. (Referate); Lenin Akad. für Landwirtsch. Wiss. Moscow*, 1936, **1**, 23.

Michelbacher, A.E., Hoskins, W.M. & Herms, W.B. *JEZ*, 1932, **64**, 109.

Miescher, K., Wettstein, A. & Tschopp, E. *SMW*, 1936, **66**, 310; *CI*, 1936, **55**, 238; *BJ*, 1937, **30**, 1970 and 1977; *YMaly*, 1937, **41**, 202.

Migliavacci, A. *RDB*, 1930, **12**, 18.

Mikami, Y. (1). *BOZJ*, 1937, **5**, 45.

Mikami, Y. (2). *ZMJ*, 1938, **50**, 223.

Mikami, Y. (3). *PIAT*, 1938, **14**, 195.

Mikami, Y. (4). *MCSK*, 1939, **15**, 135.

Mikami, Y. (5). *ZMJ*, 1939, **51**, 253.

Mikhailovsky, B. *VOUSSR*, 1929, **2**, 148; *OMaly*, 1930, **55**, 14.

Miller, D.K. & Rhoads, C.P. *NEJM*, 1934, **211**, 921.

Miller, H.M. *PSEBM*, 1932, **29**, 1124.

Miller, R.A. & Smith, H.B. *CIWP*, 1931, **413**, 47.

Miller, W. Lash. *JCE*, 1930, **7**, 257.

Miller, W. Lash, Eastcott, E.V. & Maconachie, J.E. *JACS*, 1933, **55**, 1502.

Milletti, M. *BSIBS*, 1937, **12**, 673.

Milojević, B.D. *AFEWM*, 1924, **103**, 80.

Milojević, B.D. & Burian, H. *CRSB*, 1926, **95**, 989.

Milojević, B.D. & Grbić, N. *CRSB*, 1925, **93**, 649.

Milojević, B.D., Grbić, N. & Vlatković, B. *CRSB*, 1926, **95**, 984.

Milojević, B.D. & Vlatković, B. *CRSB*, 1926, **94**, 685.

Milone, S. *AIAE*, 1923, **20**, 417.

Minot, G.R. *L*, 1935, **228**, 361.

Minoura, T. *JEZ*, 1921, **33**, 1.

Minoya, K. *JJMSB*, 1937, **3**, (75).

Mirakel, R.C. (a pseudonym?). *BC*, 1933, **53**, 614.

Mirsky, A.E. (1). *JGP*, 1936, **19**, 559.

Mirsky, A.E. (2). *S*, 1936, **84**, 333.

Mitani, S. *MJGG*, 1935, **30**, 33; *OMaly*, 1936, **95**, 548.

Mitchell, H.H. & Card, L.E. (Urbana, Ill.). "The Basal Heat Production of Cockerels, Pullets and Capons of different Ages." Unpaged reprint with no indication of origin.

Mitchell, H.H., Carroll, W.E., Hamilton, T.S. & Hunt, G.E. *Illinois Agric. Exp. Sta. Bull.* 1931, no. 375, p. 467.

Mitchell, M.L. *AJBMS*, 1931, **8**, 237.

Mitrophanov, P. *AFEWM*, 1894, **1**, 347.

Mitzkevitch, M. *BZUSSR*, 1936, **5**, 1055; *CRASUSSR*, 1936, **10**, 181.

Miura, K. (1). *JJMSA*, 1931, **2**, 105.

Miura, K. (2). *JJB*, 1937, **25**.

Mixter, R. & Hunt, H.R. *GS*, 1933, **18**, 367.

Miyachi, S. *JJMSB*, 1937, **3**, 267.

Miyajima, S. *SIKMJ*, 1935, **54**, 2089 and 8 (English summary); *CA*, 1936, **30**, 3101.

Miyamori, S. *NJMS*, 1933, **8**, 176; *CA*, 1937, **31**, 7098.

Mizutani, K. *JJZ*, 1931, **3** (126).

Moewus, F. *JWBO*, 1938, **86**, 753; *YMaly*, 1939, **49**, 56 and 57.

Mohr, O.L. (1). *ZIAV*, 1926, **41**, 59.

Mohr, O.L. (2). *Proc. 6th Internat. Congr. Genetics* (Ithaca), 1932, **1**, 190.

Mohr, O.L. & Wriedt, C. *JG*, 1928, **19**, 315.

Mohs, H. *AG*, 1931, **147**, 532.

Moissejeva, M. (1). *BZ*, 1931, **241**, 1; **243**, 67; 1932, **251**, 132.

Moissejeva, M. (2). *BZUSSR*, 1937, **6**, 437; *YMaly*, 1937, **44**, 136.

Molliard, M. *CRAS*, 1912, **155**, 1531.

Mollitor, A. *ZJZP*, 1937, **57**, 324.

Moment, G.B. *PSEBM*, 1933, **30**, 686; *JEZ*, 1933, **65**, 359.

Monroy, A. (1). *AFEWM*, 1937, **136**, 580.

Monroy, A. (2). *AFEWM*, 1937, **137**, 25.

Montalenti, G. *PZ*, 1933, **6**, 329.

Moore, A.R. (1). *P*, 1930, **9**, 9 and 18.

Moore, A.R. (2). *P*, 1930, **9**, 25.

Moore, A.R. (3). *JEB*, 1933, **10**, 230.

Moore, A.R. (4). *SRTIU*, 1933, **8**, 189.

Moore, A.R. (5). *JCCP*, 1935, **7**, 113; *PSEBM*, 1934, **32**, 174.

Moore, A.R. (6). *SC*, 1935, 92.

Moore, A.R. (7). *PSEBM*, 1938, **38**, 162.

Moore, A.R. (8). *JEZ*, 1940, **84**, 73; 1941, **87**, 101.

Moore, A.R. & Burt, A.S. *JEZ*, 1939, **82**, 159.

Moore, A.R. & Miller, W.A. *PSEBM*, 1937, **36**, 835.

Moore, A.R. & Moore, M.M. *ADB*, 1931, **42**, 375.

Moore, C.R. *BB*, 1916, **31**, 137.

Moore, J.A. (1). *E*, 1939, **20**, 459.

Moore, J.A. (2). *JEZ*, 1941, **86**, 405.

Moore, M.M. *AFEWM*, 1932, **125**, 487.

Moran, T. (1). *PRSB*, 1925, **98**, 436.

Moran, T. (2). *PRSB*, 1935, **118**, 548.

Moran, T. (3). *JEB*, 1936, **13**, 41.

Moran, T. (4). *CI*, 1937, **56**, 96T.

Moran, T. & Hale, H.P. *JEB*, 1936, **13**, 35.

Morelli, E. & Dansi, A. *N*, 1939, **143**, 1021.

Morello, G. *MZI*, 1937, **48**, 35; *YMaly*, 1937, **43**, 450.

Morgan, C.L. & Mitchell, J.H. *PS*, 1938, **17**, 99.

Morgan, E.J. *BJ*, 1930, **24**, 410.

Morgan, T.H. (1). *AA*, 1893, **9**, 141.

Morgan, T.H. (2). *AFEWM*, 1899, **8**, 448.

Morgan, T.H. (3). *S*, 1900, **11**, 178.

Morgan, T.H. (4). "Experimental Embryology." Columbia University Press, New York, 1927.

Morgan, T.H. (5). *BB*, 1935, **68**, 268, 280 and 296.

Morgan, T.H. (6). *AFEWM*, 1901, **13**, 416.

Morgan, T.H., Bridges, C.B. & Sturtevant, A.H. *BGA*, 1925, **2**, 1, esp. p. 54.

Morgan, T.H. & Tyler, A. *JEZ*, 1935, **70**, 301.

Morgulis, S. & Green, D.E. *PSEBM*, 1931, **28**, 797; *P*, 1931, **14**, 161.

Mori, J. *OIZ*, 1933, **45**, 1871 and 2113; *OMaly*, 1934, **76**, 437; *YMaly*, 1934, **28**, 512 and 644.

Mori, K. *JJG*, 1937, **13**, 81.

Morikawa, H. *ZMJ*, 1940, **52**, 287.

Morita, S. *CRSB*, 1935, **120**, 1027; *AA*, 1936, **82**, 81; 1937, **84**, 81.

Mörner, C.T. *ZPC*, 1937, **250**, 25.

Morogami, S. *JJZ*, 1940, **8**, (52).

Morosov, B.D. *CRASUSSR*, 1935, 333; *CA*, 1935, **29**, 4460; *AFEWM*, 1935, **133**, 310.

Morosov, B.D., Striganova, A. & Skomorovskaia, R.L. *JPPG*, 1934, **32**, 1148.

Morosov, I.I. *CRASUSSR*, 1938, **20**, 207.

Morrell, J.A., McHenry, E.W. & Powers, H.H. *ENY*, 1930, **14**, 28.

Morrell, J.A., Powers, H.H. & Varley, J.R. *ENY*, 1930, **14**, 28.

Morse, M. *JBC*, 1914, **19**, 95.

Morse, W. *BB*, 1918, **34**, 149.

Morton, A.A., Clapp, D.B. & Branch, C.F. *S*, 1935, **88**, 134; *AJC*, 1936, **26**, 754.

Mossman, H.W. *CIWP*, 1937, **479**, 129.

Moszkowski, M. *AMA*, 1903, **61**, 348.

Motomura, I. (1). *SRTIU*, 1934 (4th ser.), **9**, 123.

Motomura, I. (2). *SRTIU*, 1935, **10**, 211.

Mottram, J.C. *BJEP*, 1934, **15**, 71.

Moulton, C.R. *JBC*, 1923, **57**, 79.

Mouriquand, G. & Schoen, C. *CRAS*, 1933, **197**, 203.

Moxon, A.L. & Poley, W.L. *PS*, 1938, **17**, 77.

Moyle-Needham, see Needham, D.M.

Mozołowski, W. *W*, 1937, **4**, 99 (in Polish).

Mřsić, J. *AFEWM*, 1923, **98**, 129.

Muglia, G. & Masuelli, L. *BSIBS*, 1934, **8**, 1772.

Mühlbock, O. *ZPC*, 1937, **250**, 139; *L*, 1939, 634; *ABN*, 1940, **10**, 1.

Mull, J.W. *JCI*, 1936, **15**, 513.

Muller, H.J. *BR*, 1939, **14**, 261 (see esp. p. 272).

Müller, J. (1). *AAP*, 1829, **12**, 65.
Müller, Johannes (2). "Über den glatten Hai des Aristoteles, u. über die Verschiedenheiten unter den Haifischen und Rochen in der Entwickelung des Eies." Berlin, 1842.
Munro, A.F. Inaug. Diss. Aberdeen, 1941; *BJ*, 1939, **33**, 1957.
Munro, S.S. (1). *SA*, 1932, **13**, 97.
Munro, S.S. (2). *PS*, 1938, **17**, 17.
Münzel, P. *ZJZP*, 1938, **59**, 113.
von Muralt, A.L. & Edsall, J.T. *JBC*, 1930, **89**, 289, 315 and 351; *TFS*, 1930, **26**, 837.
Murao, S. *JJB*, 1938, **28**, 251.
Murphy, R.R., Hunter, J.E. & Knandel, H.C. *Bull. Pennsylvania Agric. Exp. Sta.* 1934, no. 303; 1936, no. 334.
Murray, H.A. *JGP*, 1926, **9**, 405 and 603.
Murray, J.A. (1). *SRICRS*, 1908, **3**, 41.
Murray, J.A. (2). *PRSB*, 1933, **113**, 268.
Murray, M.M. *JOP*, 1936, **87**, 388.
Murray, P.D.F. (1). *PRSB*, 1934, **115**, 380.
Murray, P.D.F. (2). *PRSB*, 1935, **116**, 434, 452.
Murray, P.D.F. (3). "Bones." Cambridge University Press, 1936.
Murray, P.D.F. & Selby, D. *JEB*, 1930, **7**, 404.
Murtasi, F. (1). *BBMEUSSR*, 1938, **6**, 79.
Murtasi, F. (2). *BBMEUSSR*, 1938, **6**, 269.
Mutscheller, F. *BC*, 1935, **55**, 615.
Mystkowski, E.M. (1). *BZ*, 1935, **278**, 240.
Mystkowski, E.M. (2). *BJ*, 1936, **30**, 765.
Mystkowski, E.M., Stiller, A. & Zysman, A. *BZ*, 1935, **281**, 231.

Nabours, R.K. & Kingsley, L.J. *BGA*, 1929, **5**, 91.
Nabrit, S.M. *BB*, 1939, **77**, 336.
Nachmansohn, D. (1). *CRSB*, 1938, **127**, 670; *JOP*, 1938, **93**, 2P.
Nachmansohn, D. (2). *CRSB*, 1938, **128**, 599.
Nachmansohn, D. (3). *YJBM*, 1940, **12**, 565; *JNP*, 1940, **3**, 396.
Nadson, G. & Stern, E. *VRUSSR*, 1934, **13**, 35.
Naeslund, J. (1). *AOGS*, 1931, **11**, 293 and 474; *OMaly*, 1932, **65**, 629.
Naeslund, J. (2). *AOGS*, 1934, **14**, 143.
Nagayama, A. *NIZ*, 1937, **15**, 2708; *OMaly*, 1938, **104**, 643.
Nakagawa, T. *CB*, 1936, **136**, 147.
Nakamura, N. *APPA*, 1931, **253**, 286.
Nakamura, O. (1). *PIAT*, 1935, **11**, 121; *BOZJ*, 1935, **5**, 432.
Nakamura, O. (2). *BOZJ*, 1935, **5**, 973.
Nakamura, O. (3). *BOZJ*, 1936, **6**, 711.
Nakamura, O. (4). *BOZJ*, 1936, **6**, 887.
Nakamura, O. (5). *BOZJ*, 1936, **6**, 1052.
Nakano, T. *FPJ*, 1934, **19**, 50 and 131; *YMaly*, 1935, **33**, 403.
Nassonov, N.V. (1). *CRASUSSR*, 1934, **9**, 259.
Nassonov, N.V. (2). *CRASUSSR*, 1934, **9**, 325.

Nassonov, N.V. (3). *CRASUSSR*, 1934, **9**, 202.
Nassonov, N.V. (4). *CRASUSSR*, 1935, **10**, 413.
Nassonov, N.V. (5). *CRASUSSR*, 1936, **11**, 207; 1936, **13**, 97 and 101; 1937, **15**, 111 and 381.
Nassonov, N.V. (6). *CRASUSSR*, 1938, **19**, 127, 133 and 137.
Natan-Larrier, L., Eliava, G. & Richard, L. *CRSB*, 1931, **106**, 794.
Natan-Larrier, L. & Grimard-Richard, L. (1). *CRSB*, 1932, **110**, 1242.
Natan-Larrier, L. & Grimard-Richard, L. (2). *CRSB*, 1933, **113**, 257.
Natan-Larrier, L., Noyer, B. & Richard, L. *CRSB*, 1931, **107**, 14.
Natan-Larrier, L. & Richard, L. (1). *CRSB*, 1931, **106**, 897.
Natan-Larrier, L. & Richard, L. (2). *CRSB*, 1931, **107**, 945.
Navashin, M. *CRASUSSR*, 1938, **19**, 193.
Navez, A.E. *BB*, 1939, **77**, 323.
Navez, A.E. & Harvey, E.B. *BB*, 1935, **69**, 342.
Naville, A. (1). *ADB*, 1922, **32**, 37; 1924, **34**, 235.
Naville, A. (2). *SPHNGCR*, 1924, **41**, 17.
Navratil, E. *ZGG*, 1937, **114**, 146.
Nedler, D. *RGS*, 1933, **44**, 3.
Needham, D.M. (Moyle-Needham) (1). In "Chemical Embryology", 1931, p. 1685; *BR*, 1929, **4**, 307.
Needham, D.M. (2). *ARB*, 1937, **6**, 395; *EZ*, 1938, **5**, 158.
Needham, D.M. & van Heyningen, W.K. *BJ*, 1935, **29**, 2040.
Needham, D.M., Needham, J., Baldwin, E. & Yudkin, J. *PRSB*, 1932, **110**, 260.
Needham, J. (1). *EP*, 1926, **25**, 1.
Needham, J. (2). *JEB*, 1927, **4**, 114, 145 and 258; **5**, 6.
Needham, J. (3). "Chemical Embryology." 3 vols. Cambridge University Press, 1931.
Needham, J. (4). *N*, 1931, **128**, 152.
Needham, J. (5). *JEB*, 1931, **8**, 330.
Needham, J. (6). *PRSB*, 1932, **110**, 46; *CRSB*, 1932, **109**, 611.
Needham, J. (7). *PRSB*, 1932, **112**, 98.
Needham, J. (8). *PRSB*, 1932, **112**, 114.
Needham, J. (9). *N*, 1932, **130**, 845.
Needham, J. (10). *BSPP*, 1932, **115**, 11.
Needham, J. (11). *JEB*, 1933, **10**, 79.
Needham, J. (12). *BR*, 1933, **8**, 180.
Needham, J. (13). *JOP*, 1933, **77**, 41 P.
Needham, J. (14). *PRSB*, 1933, **113**, 429.
Needham, J. (15). "A History of Embryology." Cambridge University Press, 1934.
Needham, J. (16). *BR*, 1934, **9**, 79.
Needham, J. (17). *AMBUSSR*, 1935, **4**, 239.
Needham, J. (18). *BJ*, 1935, **29**, 238.
Needham, J. (19). *BMJ*, 1936, ii, 701.
Needham, J. (20). *BMJ*, 1936, ii, 892.

Needham, J. (21). *AMBUSSR*, 1936, **5**, 27.

Needham, J. (22). *SP*, 1936, **31**, 41.

Needham, J. (23). *PRSM*, 1936, **29**, 31.

Needham, J. (24). "Order and Life." Yale University Press, 1936; Cambridge University Press, 1936.

Needham, J. (25). "Chemical Aspects of Morphogenetic Fields." In "Perspectives in Biochemistry." Hopkins Presentation Volume, Cambridge University Press, 1937, p. 66.

Needham, J. (26). *BUSSR*, 1937, **2**, 479.

Needham, J. (27). *AMBUSSR*, 1937, **6**, 499.

Needham, J. (28). "Integrative Levels; a Revaluation of the Idea of Progress." Herbert Spencer Lecture, Oxford, 1937.

Needham, J. (29). *FMP*, 1938, **8**, 1.

Needham, J. (30). "Morphogénèse et Métabolisme des hydrates de carbone." In *Trav. du Congrès du Palais de la Découverte*, Hermann, Paris, 1938.

Needham, J. (31). *BR*, 1938, **13**, 225.

Needham, J. (32). *G*, 1939, **3** (Suppl.), 45.

Needham, J. & Boell, E.J. *BJ*, 1939, **33**, 149.

Needham, J., Boell, E.J. & Rogers, V. *N*, 1938, **141**, 973; *PSEBM*, 1938, **39**, 287.

Needham, J., Brachet, J. & Brown, R.K. *JEB*, 1935, **12**, 321.

Needham, J. & Lehmann, H. (1). *BJ*, 1937, **31**, 1210, 1227 and 1238; *N*, 1937, **139**, 368.

Needham, J. & Lehmann, H. (2). *BJ*, 1937, **31**, 1913; *N*, 1937, **140**, 198.

Needham, J., Lehmann, H. & Nowiński, W.W. *CRSB*, 1940, **133**, 6.

Needham, J. & Lerner, I.M. *N*, 1940, **146**, 618.

Needham, J. & Needham, D.M. (1). *JEB*, 1930, **7**, 7.

Needham, J. & Needham, D.M. (2). *JEB*, 1930, **7**, 317.

Needham, J. & Needham, D.M. (3). *CRSB*, 1930, **104**, 671.

Needham, J. & Needham, D.M. (4). *SP*, 1932, **26**, 626.

Needham, J. & Needham, D.M. (5). *JEB*, 1940, **17**, 147; *BMDIMBL*, 1941, **43**, 1.

Needham, J., Needham, D.M., Yudkin, J. & Baldwin, E. *JEB*, 1932, **9**, 212.

Needham, J. & Nowiński, W.W. *BJ*, 1937, **31**, 1165.

Needham, J., Nowiński, W.W., Dixon, K.C. & Cook, R.P. *BJ*, 1937, **31**, 1185, 1196 and 1199; *N*, 1936, **138**, 462.

Needham, J. & Pirie, N.W. Unpublished observations, 1931, cit. in Needham, J. *SC*, 1932, **54**, 43.

Needham, J. & Robinson, J.R. *CRSB*, 1937, **126**, 163.

Needham, J., Rogers, V. & Shen, Shih-Chang. *PRSB*, 1939, **127**, 576.

Needham, J., Shen Shih-Chang, Needham, D.M. & Lawrence, A.S.C. *N*, 1941, **147**, 766.

Needham, J. & Smith, M. *JEB*, 1931, **8**, 286.

Needham, J., Smith, M., Shepherd, J., Stephenson, M. & Needham, D.M. *CRSB*, 1932, **109**, 688.

Needham, J., Stephenson, M. & Needham, D.M. *JEB*, 1931, **8**, 319.

Needham, J., Waddington, C.H. & Needham, D.M. *PRSB*, 1934, **114**, 393; *AEZF*, 1934, **15**, 307; *N*, 1933, **132**, 239; *NW*, 1933, **21**, 771.

Nestler, R.B., Byerly, T.C., Ellis, N.R. & Titus, H.W. *PS*, 1936, **15**, 67.

Neuberg, C., Kobel, M. & Laser, H. *ZKF*, 1930, **32**, 92.

Neuberger, A. *BJ*, 1938, **32**, 1435.

Neufach, S.A. & Shabad, L.M. *BBMEUSSR*, 1938, **6**, 259.

Neuweiler, W. (1). *SMW*, 1935 (i), 539; *OMaly*, 1935, **88**, 477.

Neuweiler, W. (2). *AG*, 1936, **162**, 384.

Neuweiler, W. (3). *ZPC*, 1937, **249**, 225.

Newcombe, C.L. & Gould, S.A. *BG*, 1937, **13**, 474.

Newman, H.H. (1). "The Biology of Twins." Chicago, 1917; "The Physiology of Twinning." Chicago, 1923.

Newman, H.H. (2). *JEZ*, 1921, **33**.

Newman, H.H. & Patterson, J.T. *JM*, 1910, **21**, 359.

Newton, W.H. (1). *JOP*, 1935, **84**, 196.

Newton, W.H. (2). *PR*, 1938, **18**, 419. "Problems of Endocrine Function in Pregnancy" in "Sex and Internal Secretions," 1939, p. 720.

Nicholas, J.S. (1). *PSEBM*, 1931, **29**, 188.

Nicholas, J.S. (2). *AR*, 1938, **70**, 199.

Nicholas, J.S. & Bosworth, E.B. *AJOP*, 1928, **83**, 499.

Nicholas, J.S. & Rudnick, D. (1). *PSEBM*, 1931, **29**, 325.

Nicholas, J.S. & Rudnick, D. (2). *JEZ*, 1933, **66**, 193.

Nicholas, J.S. & Rudnick, D. (3). *PNASW*, 1934, **20**, 656; *JEZ*, 1938, **78**, 205.

Nicholson, G.W. (1). *JPB*, 1929, **32**, 365; 1931, **34**, 711; 1936, **43**, 209.

Nicholson, G.W. (2). *GHR*, 1930, **80**, 384; 1934, **84**, 140 and 389; 1935, **85**, 8 and 379; 1937, **87**, 46 and 391; 1938, **88**, 263.

Nielsen, E. *UL*, 1939, 1071.

Nikitenko, M.F. (1). *CRASUSSR*, 1937, **16**, 477.

Nikitenko, M.F. (2). *BBMEUSSR*, 1939, **8**, 18 and 136.

Nishikawa, H. *Pub. Sericultural Exp. Sta. Suigen, Chosen*, 1931.

Nishimoto, U. *PIAT*, 1934, **10**, 578; *YMaly*, 1935, **34**, 139.

Nitzescu, I.I. *CRSB*, 1931, **106**, 1176.

Niven, J.S.F. & Robison, R. *BJ*, 1934, **28**, 2237.

Noble, G.K. *ANYAS*, 1927, **30**, 31.

Noble, G.K. & Brady, M.K. *N*, 1933, **132**, 971; *ZA*, 1933, **11**, 8.

Nolf, L.O. *AJOH*, 1932, **16**, 288.

Nordby, J.E. *JH*, 1929, **20**, 229; 1930, **21**, 499.

Norman, W.W. *AFEWM*, 1896, **3**, 106.

Norris, L.C., Wilgus, H.S., Ringrose, A.T., Heiman, V. & Heuser, G.F. *Bull. Cornell Univ. Agric. Exp. Sta.* 1936, no. 660.

Northrop, J.H. *BR*, 1935, **10**, 263; *Trav. Congrès du Palais de la Découverte*, Paris, 1938, p. 365.

Nouvel, H. *CRAS*, 1935, **200**, 972.

du Noüy, L. *Proc.* XIV*th Internat. Physiol. Congrèss*, 1932, p. 153; also "Biological Time." Methuen, London, 1936.

Novikov, A.B. (1). *BB*, 1938, **74**, 198 and 211.

Novikov, A.B. (2). *CN*, 1939, **14**, 182; *JEZ*, 1940, **85**, 127.

Nowak, W. *ZF*, 1935, **33**, 319, 331 and 503; *CA*, 1934, **28**, 6856; 1935, **29**, 5930; *ACZAA*, 1934, **10**, 38 and 177.

Nowiński, W.W. (1). *Odbita z Pamiętnika* XIV *Zjazdu Lekarzy i Przyrodników Polskich w Poznaniu*, 1933, p. 337; *PSZN*, 1934, **14**, 110.

Nowiński, W.W. (2). *D*, 1939, **2**, 392.

Nowiński, W.W. (3). *BJ*, 1939, **33**, 978.

Noyons, A.K.M. & de Hasselle, P.M.P. *ABN*, 1939, **9**, 170; *BSIBS*, 1939, **15**, 119.

Nürnberger, L. *AG*, 1930, **142**, 93.

Nuttall, G.H.F. "Blood Immunity and Blood Relationship." Cambridge, 1904.

O'Connor, R.J. (1). *JA*, 1938, **73**, 145; 1939, **74**, 34; 1940, **75**, 95.

O'Connor, R.J. (2). *JA*, 1940, **74**, 301.

O'Donoghue, C.H. *AA*, 1910, **37**, 530.

Oberst, F.W. & Woods, E.B. *AJOG*, 1935, **30**, 232.

Ogorodny, J.M. (1). *PAHUSSR*, 1936, **2**, 108; *PZUSSR*, 1936, **20**, 741.

Ogorodny, J.M. (2). *MSPEUSSR*, 1939, **2**, 30.

Ogorodny, J.M. (3). *MSPEUSSR*, 1939, **2**, 56.

Ogorodny, J.M. & Penionskevitch, E.E. *MSPEUSSR*, 1939, **2**, 5.

Öhman, L.O. *AFZ*, 1940, **32**A, no. 15.

Okada, K. *SRTIU*, 1935, **9**; 1935, **10**; 1936, **11**, 49.

Okada, Y.K. (1). *ZAZ*, 1928, **75**, 157.

Okada, Y.K. (2). *G*, 1938, **2**, 49; *MCSK*, B, 1938, **14**, 301; *JJ*, 1940, **8**, (37).

Okada, Y.K. (3). *AZJ*, 1938, **17**, 339.

Okada, Y.K. & Mikami, Y. *PIAT*, 1937, **13**, 283.

Okamura, H. (1). *JJMSB*, 1933, **2**, 313.

Okamura, H. (2). *JJMSB*, 1935 **3**, 85.

Okamura, H. (3). *JJMSB*, 1938, **4**, 15.

Oku, M. *JJOG*, 1930, **13**, 472; *OMaly*, 1932, **63**, 808.

Okunev, N. (1). *BZ*, 1928, **195**, 421.

Okunev, N. (2). *BZ*, 1929, **208**, 328.

Okunev, N. (3). *BZ*, 1929, **212**, 1.

Okunev, N. (4). *BZ*, 1932, **255**, 387.

Okunev, N. (5). *BZ*, 1933, **257**, 242.

Okunev, N. (6). *TLZEMAM*, 1932, **1**, 121; 1934, **3**, 55.

Oliver, G. & Schäfer, E.A. *JOP*, 1895, **18**, 230.

Olsen, M.W. & Byerly, T.C. *PS*, 1935, **14**, 46.

Olson, R.A. & du Buy, H.G. *AJB*, 1937, **24**, 611.

Ongaro, D. (1). *ASBSP*, 1931, **46**, 331, 341 and 347.

Ongaro, D. (2). *ASBSP*, 1934, **47**, 3; *AZI*, 1932, **16**, 1285.

Ongaro, D. (3). *ASBSP*, 1934, **47**, 27; *YMaly*, 1935, **35**, 7.

Ongaro, D. (4). *ASBSP*, 1934, **47**, 30.

Ono, K. *MJBG*, 1932, **7**, 335.

Ono, T. (1). *JACSJ*, 1932, **8**, 788; *CA*, 1932, **26**, 5222.

Ono, T. (2). *JJMSB*, 1937, **3** (74).

Onoe, T. (1). *JJB*, 1936, **24**, 1.

Onoe, T. (2). *JJB*, 1936, **24**, 9.

Onorato, A.R. & Stunkard, H.W. *BB*, 1931, **61**, 120.

Onslow, H. (1). *PRSB*, 1915, **89**, 36.

Onslow, H. (2). *BJ*, 1923, **17**, 334 and 564.

van Oordt, G.J. & Rinkel, G.L. *AFEWM*, 1940, **140**, 59.

Oppenheimer, J.M. (1). *PSEBM*, 1934, **31**, 1123; *PNASW*, 1934, **20**, 536.

Oppenheimer, J.M. (2). *JEZ*, 1936, **72**, 247.

Oppenheimer, J.M. (3). *JEZ*, 1936, **72**, 409.

Oppenheimer, J.M. (4). *PSEBM*, 1936, **34**, 461.

Oppenheimer, J.M. (5). *JEZ*, 1936, **73**, 405; *S*, 1935, **82**, 598; *PNASW*, 1935, **21**, 551.

Oppenheimer, J.M. (6). *O*, 1936, **2**, 124.

Oppenheimer, J.M. (7). *JEZ*, 1938, **79**, 185.

Oppenheimer, J.M. (8). *JEZ*, 1939, **80**, 391.

Oppenheimer, J.M. (9). *QRB*, 1940, **15**, 1.

Orechovitch, W.N. (1). *CRASUSSR*, 1933, **1**, 27; *ZPC*, 1934, **224**, 61.

Orechovitch, W.N. (2). *BZ*, 1936, **286**, 91.

Orechovitch, W.N. (3). *BZ*, 1936, **286**, 248.

Orechovitch, W.N. (4). *BZ*, 1936, **286**, 285.

Orechovitch, W.N. (5). *BBMEUSSR*, 1937, **3**, 177.

Orechovitch, W.N. (6). *BBMEUSSR*, 1938, **6**, 233.

Orechovitch, W.N. & Bromley, N.W. *BC*, 1934, **54**, 523; *CRASUSSR*, 1934, **2**, 249.

Orechovitch, W.N., Bromley, N.W. & Kozmina, N. *BZ*, 1935, **277**, 186.

Orrù, A. (1). *AAL*, 1931 (ser. 6ᵉ), **14**, 523; *YMaly*, 1932, **23**, 329.

Orrù, A. (2). *BSIBS*, 1932, **7**, 1491.

Orrù, A. (3). *BSIBS*, 1933, **8**, 284.

Orrù, A. (4). *BSIBS*, 1933, **8**, 286.

Orrù, A. (5). *BSIBS*, 1933, **8**, 668.

Orrù, A. (6). *BSIBS*, 1933, **8**, 1386.

Orrù, A. (7). *ASB*, 1933, **18**, 361; *YMaly*, 1933, **26**, 123.

Orrù, A. (8). *AAL*, 1935 (6ᵉ ser.), **22**, 458; *YMaly*, 1936, **37**, 484; *AAL*, 1936 (6ᵉ ser.), **23**, 959; *OMaly*, 1937, **98**, 359.

Orrù, A. (9). *AAL*, 1936 (6ᵉ ser.), **23**, 954. *ASB*, 1940, **26**, 32.

Örström, Å. (1). *P*, 1932, **15**, 566; *AFZ*, 1932, **24**, 1.

Örström, Å. (2). *P*, 1935, **24**, 177.

Örström, Å. (3). *AFZ*, 1935, **28** B, no. 6; *YMaly*, 1936, **37**, 177.

Örström, Å. (4). *NW*, 1937, **25**, 300; *CA*, 1937, **31**, 8032.

Örström, Å. & Lindberg, O. *EZ*, 1940, **8**, 367.

Örström, Å., Örström, M. & Krebs, H.A. *BJ*, 1939, **33**, 990.

Örström, Å., Örström, M., Krebs, H.A. & Eggleston, L.V. *BJ*, 1939, **33**, 995.

Orts, F. *FMH*, 1938, **1**, 6.

Osborn, H.F. *AN*, 1932, **66**, 52.

Osborne, W.A. *AJBMS*, 1931, **8**, 239.

Ostern, P., Baranowski & Reis. *BZ*, 1935, **279**, 85.

Otte, H.G. *SMA*, 1934, **2**, 1013; *CA*, 1935, **29**, 207.

Ouang, see Wang.

Outhouse, E.L. *TRSC*, 1935, **29**, 77; *BJ*, 1936, **30**, 197; 1937, **31**, 1459.

Outhouse, J. & Mendel, L.B. *JEZ*, 1933, **64**, 257.

van Overbeek, J. *PNASW*, 1935, **21**, 292.

Packard, C. *QRB*, 1931, **6**, 253.

Padoa, E. (1). *BSIBS*, 1937, **12**, 397; *AIAE*, 1938, **40**, 122.

Padoa, E. (2). *B*, 1939, **1**, 337.

Paechtner, J. *Hdbch. d. Biochem. d. Mensch. u. Tier.* (*Abderhalden's*), 1934, Ergänzungswerk Bd. 2, p. 366.

Painlevé, J., Wintrebert, P. & Yung, Ko-Ching. *CRAS*, 1929, **189**, 208.

Painter, T.S. *JEZ*, 1938, **50**, 441.

Palladin, A.V. & Rashba, E.Y. *UBZ*, 1937, **10**, 193; *CA*, 1937, **31**, 7499; *YMaly*, 1938, **45**, 293.

Palmer, A.H. & Levy, M. *JBC*, 1938, **123**, xc; 1940, **136**, 407 and 629.

Palmer, L.S. & Eckles, C.H. *JBC*, 1914, **17**, 223.

du Pan, R.M. & Ramseyer, M. *CRSB*, 1935, **119**, 1236.

Panagiotou, P.P. *Neas Iatrikes*, 1939; also as a monograph "The neo-mechanistic interpretation of embryonic development." Athens, 1939.

Pandit, C.G. & Chambers, R. *JCCP*, 1932, **2**, 243.

Pankratz, D.S. *AR*, 1931, **49**, 31; *YMaly*, 1931, **18**, 512.

Panum, C.L. "Untersuchungen ü. d. Entstehung d. Missbildungen." Kiel, 1860.

Parat, M. *CRSB*, 1933, **112**, 1134.

Parker, F. & Tenney, B. *ENY*, 1938, **23**, 492; *CA*, 1938, **32**, 9228.

Parker, G.H. (1). *PTRSB*, 1931, **219**, 381.

Parker, G.H. (2). *PAAAS*, 1940, **73**, 165; *QRB*, 1935, **10**, 251.

Parker, R.C. (1). *S*, 1931, **74**, 181.

Parker, R.C. (2). *JEM*, 1933, **58**, 401.

Parker, R.C. (3). *S*, 1933, **77**, 544; *JEM*, 1934, **60**, 351.

Parkes, A.S. & Bellerby, C.W. *JOP*, 1927, **62**, 385.

Parmenter, C.L. *JEZ*, 1933, **66**, 409.

Parnas, J.K. *BSCB*, 1936, **18**, 53.

Parnas, J.K. & Krasinska, Z. *BZ*, 1921, **116**, 108.

Parnas, J.K. & Ostern, P. *BSCB*, 1936, **18**, 1471.

Parr, T. *Bull. Yale Univ. Sch. Forestry*, 1940, no. 46.

Parsons, E.H. *JEZ*, 1929, **54**, 23.

Parsons, L.D. *JPB*, 1936, **43**, 1.

Parsons, L.G. *JAMA*, 1931, **97**, 973.

Partachnikov, M. *CRSB*, 1930, **104**, 1163.

Pasquini, P. (1). *AAL*, 1927 (6ᶜ), **5**, 453.

Pasquini, P. (2). *RAL*, 1931, **14**, 56.

Pasquini, P. & Reverberi, G. *BIZUR*, 1929, **7**, 1.

Passmore, R. & Schlossmann, H. *JOP*, 1938, **92**, 459.

Pasteels, J. (1). *BARBS*, 1929 (5ᵉ sér.), **15**, 421; *ADB*, 1930, **40**, 247; 1931, **42**, 389; 1935, **46**, 229; *BARBS*, 1938 (5ᵉ sér.), **24**, 721; *TSZW*, 1938, **13**, 515.

Pasteels, J. (2). *ADB*, 1932, **43**, 521.

Pasteels, J. (3). *CRSB*, 1934, **117**, 1231.

Pasteels, J. (4). *AAM*, 1936, **32**, 303.

Pasteels, J. (5). *BARBS*, 1936 (5ᵉ sér.), **22**, 737.

Pasteels, J. (6). *ADB*, 1936, **47**, 205; *CRSB*, 1933, **113**, 425; *CRAA*, 1934, 1 (Brussels meeting).

Pasteels, J. (7). *ADB*, 1936, **47**, 631; *CRSB*, 1936, **121**, 1390, 1394 and 1584.

Pasteels, J. (8). *ADB*, 1937, **48**, 105; *BARBS*, 1935 (5ᵉ sér.), **21**, 88.

Pasteels, J. (9). *ADB*, 1937, **48**, 381 and 463; *BARBS*, 1936 (5ᵉ sér.), **22**, 736.

Pasteels, J. (10). *AAM*, 1937, **33**, 279; *CRSB*, 1938, **129**, 59 and 62.

Pasteels, J. (11). *CRAA* (Basel meeting), 1938, 1; *ADB*, 1940, **51**, 103 and 335.

Pasteels, J. (12). *ADB*, 1938, **49**, 629.

Pasteels, J. (13). *ADB*, 1939, **50**, 291.

Pasteels, J. (14). *BARBS*, 1939 (5ᵉ sér.), **25**.

Pasteels, J. (15). *BR*, 1940, **15**, 59.

Pasteels, J. & Léonard, G. *BHA*, 1935, **12**, 293.

Patel, M.D. *PZ*, 1936, **9**, 129.

Patterson, J.T. *QRB*, 1927, **2**, 399.

Patton, A.R. *JN*, 1937, **13**, 123.

Patton, A.R. & Palmer, L.S. *JN*, 1936, **11**, 129.

Patton, M.B., Hitchcock, F.A. & Haub, J.G. *AESA*, 1941, **34**, 26 and 32.

Pauli, M.E. *ZWZ*, 1927, **129**, 483.

Payne, L.F. & Hughes, J.S. *Bull. Kansas Agric. Exp. Sta.* 1933, no. 34.

Payne, N.M. *E*, 1930, **11**, 500.

Peacock, A.D. *N*, 1939, **144**, 1036.
Peacock, P.R. (1). *Rep.* iiird *Conf. Brit. Emp. Cancer Campaign*, London, 1934.
Peacock, P.R. (2). *BJEP*, 1936, **17**, 164.
Peacock, P.R. & Shukov, R.I. *N*, 1940, **146**, 30.
Pearce, J.M. *AJOP*, 1936, **114**, 255.
Pearl, R. & Curtis, M.R. *JEZ*, 1912, **12**, 99; 1914, **17**, 395.
Pease, D.C. (1). *BB*, 1938, **75**, 409; *JEZ*, 1939, **80**, 225.
Pease, D.C. (2). *JEZ*, 1941, **86**, 381.
Pease, D.C. & Marsland, D.A. *JCCP*, 1939, **14**, 407.
Peck, E.S. *PCAS*, 1934, **34**, 34.
Peck, S.M. *ADS*, 1931, **23**, 705.
Pedersen, K.O. & Waldenström, J. *ZPC*, 1937, **245**, 152.
Peebles, F. *BB*, 1929, **57**, 176.
Pelluet, D. *QJMS*, 1938, **80**, 285.
Penionskevitch, E.E. (1). *AFGF*, 1934, **8**, 182.
Penionskevitch, E.E. (2). *AFGF*, 1934, **8**, 273.
Penionskevitch, E.E. (3). *PIUSSR*, 1934, **1**, 6; *AFGF*, 1935, **9**, 93.
Penionskevitch, E.E. (4). *PIUSSR*, 1935, **2**, 1; *AFGF*, 1936, **10**, 437.
Penionskevitch, E.E. (5). *AFGF*, 1937, **11**, 1.
Penionskevitch, E.E. & Retanov, A.N. *AFGF*, 1934, **8**, 369.
Penionskevitch, E.E. & Schechtman, L.I. *MSPEUSSR*, 1939, **2**, 76.
Penners, A. (1). *VDZG*, 1922, **27**, 46; *ZGZP*, 1924, **41**, 91.
Penners, A. (2). *VDZG*, 1931, **34**, 67.
Penners, A. (3). *AFEWM*, 1929, **116**, 53; *ZWZ*, 1936, **148**, 189.
Penners, A. & Schleip, W. *ZWZ*, 1928, **130**, 305; 1928, **131**, 1.
Penquite, R. *Res. Bull. Iowa State Agric. Exp. Sta.* 1938, no. 232.
Penquite, R. & Thompson, R.B. *PS*, 1934, **13**, 303.
Pentimalli, F. (1). *ZKF*, 1914, **14**, 627.
Pentimalli, F. (2). *ZKF*, 1927, **25**, 347.
Pentimalli, F. (3). *BZ*, 1931, **242**, 233.
Pentimalli, F. (4). *OO*, 1935, **5**, 729.
Pepper, J.H. *JEE*, 1937, **30**, 380.
Peredelsky, A.A. *CRASUSSR*, 1940, **26**, 495 and 499.
Perov, S.S. (1). *Trudy Lab. Belka & Belkovkogo Obmena v Organisme*, 1931, p. 14; *Arb. Lab. Proteinforsch. (Referate)*, Moscow, 1936, p. 7.
Perov, S.S. (2). *Trudy Lab. Belka & Belkovkogo Obmena v Organisme*, 1935, p. 55; *Arb. Lab. Proteinforsch. (Referate)*, Moscow, 1936, p. 61.
Perov, S. & Dolinov, K. *Trudy Lab. Belka & Belkovkogo Obmena v Organisme*, 1932, p. 3; *Arb. Lab. Proteinforsch (Referate)*, Moscow, 1936, p. 19.
Perri, T. (1). *AAL* (6ᵉ ser.), 1930, **12**, 66; 1931, **14**, 229; *AFEWM*, 1932, **126**, 512.

Perri, T. (2). *AFEWM*, 1934, **131**, 113, esp. p. 127.
Peruzzi, P. *BSIBS*, 1935, **10**, 489; *OMaly*, 1935, **89**, 287.
Peschen, K.E. *ZJZP*, 1939, **59**, 430.
Peter, K. (1). *ZZFMA*, 1929, **9**, 129.
Peter, K. (2). *ZAE*, 1936, **105**, 409.
Peter, K. (3). *AA*, 1938, **86**, 94; *ZMAF*, 1938, **43**, 362 and 416.
Peters, R.A. *JSM*, 1929, **37**, 1; *TFS*, 1930, **26**, 797.
Peters, R.A. & Wakelin, R.W. *BJ*, 1938, **32**, 2290.
Peterson, W.H. & Skinner, J.T. *JN*, 1931, **4**, 419.
Petrie, A.H.K. & Williams, R.F. *AJBMS*, 1938, **16**, 347.
Pettibone, M. & Wulzen, R. *PZ*, 1934, **7**, 192.
Pettit, A. & Vaillant, L. *BMHN*, 1902, **8**, 301.
Peyron, A. *CRSB*, 1940, **133**, 203.
Peyron, A. & Limousin, H. *CRAS*, 1938, **207**, 87 and 646; *CRSB*, 1936, **123**, 409; 1938, **129**, 822.
Pézard, A. *BBFB*, 1918, **52**, 1.
Pfaundler, M. *ZKH*, 1935, **57**, 185; *YMaly*, 1935, **35**, 331.
Pfeffer, W. "Pflanzenphysiologie." Leipzig, 1881, vol. ii, p. 163.
Pfeiffer, H.H. (1). *NW*, 1935, **23**, 800.
Pfeiffer, H.H. (2). *N*, 1936, **138**, 1054; *VDZG*, 1937, 106; *P*, 1937, **27**, 442.
Pfeiffer, H.H. (3). *C*, 1937, Fujii Jubilee Vol. p. 701.
Pflüger. *AP*, 1877, **15**, 61.
Pflugfelder, O. *ZWZ*, 1937, **149**, 477; *VDZG*, 1937, **10**, 121; *ZWZ*, 1938, **150**, 451; *VDZG*, 1938, **11**, 127; *ZWZ*, 1938, **151**, 149; 1939, **152**, 159 and 384.
Philipp, E. *CG*, 1929, **53**, 2386.
Philipp, E. & Huber, H. *CG*, 1936, **60**, 2706.
Philips, F.S. (1). *BB*, 1940, **78**, 256.
Philips, F.S. (2). 1940, In the press.
Philipson, T. *ZPC*, 1934, **223**, 119; *CRTLC*, 1934, 20, no. 4.
Phillips, P.H., Halpin, J.G. & Hart, E.B. *JN*, 1935, **10**, 93.
Piccardo, A. *AOG*, 1932, **39**, 117; *OMaly*, 1933, **69**, 380.
Pick, L. & Poll, H. *KW*, 1903, **40**, 518, 546 and 572.
Pickford, Grace. Private communication to the author, Oct. 1940.
Pictet, A. & Ferrero, A. *RSZ*, 1940, **47**, 193.
Piepho, H. (1). *BC*, 1938, **58**, 90.
Piepho, H. (2). *NW*, 1938, **26**, 841; 1939, **27**, 302.
Pierantoni, U. *BDZ*, 1930, **1**, 277; *YMaly*, 1931, **18**, 16.
Pierce, M.E. *JEZ*, 1933, **65**, 443.
Piettre, M. *CRAS*, 1924, **178**, 91; 1936, **202**, 699.
Pillai, R.K. *BJ*, 1938, **32**, 1087 and 1961.

Pincherle, M. & Gelli, G. *BSIBS*, 1931, **6**, 733; *CA*, 1932, **26**, 1985.

Pincus, G. (1). "The Eggs of Mammals." Macmillan, New York, 1936.

Pincus, G. (2). *CSHSQB*, 1937, **5**, 44.

Pincus, G. & Enzmann, E.V. *JM*, 1937, **61**, 351.

Pincus, G. & Shapiro, H. *PAPS*, 1940, **83**, 631.

Pincus, G., Sterne, G. de R. & Enzmann, E.V. *PNASW*, 1933, **19**, 729.

Pincus, G. & Werthessen, N.T. (1). *AJOP*, 1937, **120**, 100; 1938, **124**, 484.

Pincus, G. & Werthessen, N.T. (2). *JEZ*, 1938, **78**, 1.

Pincus, G. & Werthessen, N.T. (3). *PRSB*, 1938, **126**, 330.

Pine, L. *JAOAC*, 1924, **8**, 57.

Pirie, N.W. (1). "The Meaninglessness of the terms Life and Living." In Hopkins Presentation Volume "Perspectives in Biochemistry", Cambridge, 1937, p. 11.

Pirie, N.W. (2). "Criteria of Purity used in the Study of large Molecules of biological origin." *BR*, 1940, **15**, 377.

Pirschle, K. *ZIAV*, 1939, **76**, 512.

Pitotti, M. (1). *PSZN*, 1936, **15**, 217.

Pitotti, M. (2). *RAL*, 1937 (6th ser.), **24**, 526; *PSZN*, 1938, **17**, 20.

Pitotti, M. (3). *PSZN*, 1939, **17**, 193.

Plagens, G.M. *JM*, 1933, **55**, 151.

Plagge, E. (1). *AFEWM*, 1935, **132**; *BC*, 1936, **56**, 406.

Plagge, E. (2). *ZIAV*, 1936, **72**, 127.

Plagge, E. (3). *NGWG*, 1936, **2**, 251.

Plagge, E. (4). *NW*, 1938, **26**, 4; *U*, 1938, **16**, 1.

Plagge, E. (5). *BC*, 1938, **58**, 1.

Plagge, E. (6). *EB*, 1939, **17**, 105.

Plagge, E. & Becker, E. *BC*, 1938, **58**, 231.

Plass, E.D. & Yoakam, A. *AJOG*, 1929, **18**, 556; *OMaly*, 1930, **55**, 670.

Plehn, M. *ZKF*, 1906, **4**, 525.

Plimmer, J. *PZS*, 1912, 235; 1913, 142.

Plimmer, R.H.A. & Burch, W.J.N. *BJ*, 1937, **31**, 398.

Plimmer, R.H.A. & Kaya, R. *JOP*, 1909, **39**, 45.

Plimmer, R.H.A. & Lowndes, J. *BJ*, 1924, **18**, 1163.

Plimmer, R.H.A., Raymond, W.H. & Lowndes, J. *BJ*, 1933, **27**, 58.

Plimmer, R.H.A. & Scott, F.H. *JOP*, 1909, **38**, 247.

Plough, H.H. *BB*, 1927, **52**.

Plum, K. *ZVP*, 1935, **22**, 155.

Plunkett, C.R. *JEZ*, 1926, **46**, 181.

Pohlmann, A.G. *AR*, 1909, **3**, 75.

Poležaiev, L.V. (1). *BZUSSR*, 1933, **2**; *ZZUSSR*, 1936, **15**, 277.

Poležaiev, L.V. (2). *CRASUSSR*, 1934, **13**, 465; *BBFB*, 1936, **70**, 54; *YMaly*, 1935, **33**, 575.

Poležaiev, L.V. (3). *BZUSSR*, 1935, **4**, 1117.

Poležaiev, L.V. (4). *CRASUSSR*, 1935, **14**, 673; 1939, **22**, 648; *AAUSSR*, 1935, **14**, 384 and 510; *AAM*, 1936, **32**, 437.

Poležaiev, L.V. (5). *CRASUSSR*, 1936, **10**, 273; *BZUSSR*, 1936, **5**, 489.

Poležaiev, L.V. (6). *CRASUSSR*, 1936, **13**, 387; 1939, **22**, 142.

Poležaiev, L.V. (7). *AMBUSSR*, 1936, **5**, 101.

Poležaiev, L. (8). *CRASUSSR*, 1937, **15**, 387.

Poležaiev, L.V. (9). *AMBUSSR*, 1938, **8**, 467.

Poležaiev, L.V. (10). *CRASUSSR*, 1938, **21**, 357.

Poležaiev, L.V. (11). *CRASUSSR*, 1938, **21**, 361.

Poležaiev, L.V. (12). *CRASUSSR*, 1939, **22**, 644; 1939, **25**, 538 and 543.

Poležaiev, L.V. & Favorina, W.N. *AFEWM*, 1935, **133**, 701.

Poležaiev, L.V. & Ginzburg, G.I. *CRASUSSR*, 1939, **23**, 733.

Pollack, H. *PSEBM*, 1927, **25**, 145.

Pollard, C.B. & Carr, R.H. *AJOP*, 1924, **67**, 589.

Polletini, B. *BSIBS*, 1936, **11**, 951.

Pollister, A.W. *PZ*, 1941, **14**, 268.

Pomerat, C.M. & Willmer, E.N. *JEB*, 1939, **16**, 232.

Pommerenke, W.T. *JCI*, 1936, **15**, 485.

Ponomareva, W.N. (1). *BBMEUSSR*, 1936, **2**, 325.

Ponomareva, W.N. (2). *AAUSSR*, 1938, **18**, 345 and 478.

Popov, V.V. (1). *CRASUSSR*, 1936, **11**, 347.

Popov, V.V. (2). *BBMEUSSR*, 1936, **2**, 245.

Popov, V.V. (3). *ZJZP*, 1937, **58**, 23.

Popov, V.V. (4). *BBMEUSSR*, 1938, **6**, 399.

Popov, V.V. (5). *BBMEUSSR*, 1938, **6**, 515.

Popov, V.V. (6). *BZUSSR*, 1938, **7**, 488.

Popov, V.V. (7). *CRASUSSR*, 1939, **24**, 720.

Popov, V.V., Eudokimova, S.P. & Krymova, A.G. *CRASUSSR*, 1937, **16**, 241.

Popov, V.V., Kislov, M.N., Nikitenko, M.F. & Canturisvili, P.S. *CRASUSSR*, 1937, **16**, 245.

Popov, W.W. & Nikitenko, M.F. *BBMEUSSR*, 1937, **3**, 395.

Porter, K.R. *BB*, 1939, **77**, 233.

Portmann, A. (1). *AZEG*, 1933, **76** (Notes et Revue), 24.

Portmann, A. (2). *RSZ*, 1935, **42**, 395.

Portmann, A. (3). *AB*, 1935, **1**, 59; *VSNFG*, 1936, 224.

Portmann, A. (4). *B*, 1938, **1**, 49 and 109.

Portmann, A. (5). *RSZ*, 1938, **45**, 273; 1939, **46**, 385.

Portmann, A. & Jecklin, L. *RSZ*, 1933, **40**, 179; *YMaly*, 1933, **25**, 558.

Possenti, G. *RPS*, 1935, **15**, 183.

Posternak, S. & Posternak, T. *CRAS*, 1927, **184**, 306 and 909; **185**, 615; 1928, **187**, 313; 1933, **197**, 429.

Posternak, T. *HCA*, 1935, **18**, 1351.

Poulson, D.F. (1). *ZVP*, 1935, **22**, 466.

Poulson, D.F. (2). "The Embryonic Development of *Drosophila melanogaster*." Herrmann, Paris, 1937.
Poulson, D.F. (3). *PNASW*, 1937, **23**, 133; *JEZ*, 1940, **83**, 271.
Pourbaix, Y. *AIUC*, 1938, **3**, 31; 1939, **4**, 719.
Poyarkov, P. *AAM*, 1910, **13**, 1.
Pozerski, E. & Krongold, S. *CRSB*, 1914, **77**, 278 and 330.
Pratt, E.F. & Williams, R.J. *JGP*, 1939, **22**, 637.
Prell, H. *AZI*, 1932, **16**, 950.
Preto, V. *AIAE*, 1938, **41**, 165.
Price, J.W. (1). *OJS*, 1934, **34**, 287 and 399; 1935, **35**, 40.
Price, J.W. (2). *JGP*, 1940, **23**, 449.
Pritzker, I.J. *CRASUSSR*, 1939, **24**, 823.
Privolniev, T.I. (1). *CRASUSSR*, 1935, **8**, 419; *YMaly*, 1936, **37**, 183.
Privolniev, T.I. (2). *CRASUSSR*, 1936, **13**, 433; *YMaly*, 1937, **42**, 542; *AAUSSR*, 1938, **18**, 165 and 255.
Pruthi, H.S. (1). *PCPS*, 1924, **1**, 139.
Pruthi, H.S. (2). *JEB*, 1925, **3**, 1.
Pryor, M.G.M. *PRSB*, 1940, **128**, 378 and 393.
Przibram, H. (1). *JEZ*, 1907, **5**, 259.
Przibram, H. (2). *AFEWM*, 1910, **29**, 587.
Przibram, H. (3). *AFEWM*, 1921, **48**, 205.
Przibram, H. (4). *JEB*, 1926, **3**, 313.
Przibram, H. (5). "Die anorganische Grenzgebiete d. Biol." Bornträger, Berlin, 1926; "Aufbau mathematische Biol." Bornträger, Berlin, 1923.
Przibram, H. (6). "Connecting Laws in Animal Morphology." London University Press, 1931.
Przyłęcki, S.J. (1). *AIP*, 1926, **26**, 33; 1926, **27**, 159.
Przyłęcki, S.J. (2). *EEF*, 1935, **4**, 111.
Przyłęcki, S.J. (3). *SAWW*, 1936, **145**, 849; *MC*, 1936, **69**, 243.
Przyłęcki, S.J., Andrzejewski, H. & Mystkowski, E.M. *KZ*, 1935, **71**, 325.
Przyłęcki, S.J. & Białek, W. *BZ*, 1932, **253**, 288.
Przyłęcki, S.J., Cichocka, J. & Rafałowska, H. *BZ*, 1936, **284**, 169.
Przyłęcki, S.J. & Dobrowolska, S. *BZ*, 1932, **245**, 388.
Przyłęcki, S.J. & Frajberger, S. *BAPS*, 1935, 407.
Przyłęcki, S.J., Gedroyć, W. & Rafałowska, H. *BZ*, 1935, **280**, 286.
Przyłęcki, S.J. & Grynberg, M.Z. (1). *BZ*, 1932, **248**, 16.
Przyłęcki, S.J. & Grynberg, M.Z. (2). *BZ*, 1932, **251**, 248; 1933, **258**, 389; 1933, **260**, 395.
Przyłęcki, S.J., Grynberg, M.Z. & Szrajber, D. *BZ*, 1932, **244**, 190.
Przyłęcki, S.J. & Gurfinkel, I. *BJ*, 1930, **24**, 179.

Przyłęcki, S.J. & Hofer, E. *BZ*, 1936, **288**, 303.
Przyłęcki, S.J., Hofer, E. & Frajberger-Grynberg, S. *BZ*, 1935, **282**, 362.
Przyłęcki, S.J., Kasprzyk, K. & Rafałowska, H. *BZ*, 1936, **286**, 360.
Przyłęcki, S.J. & Kisiel, S. *BZ*, 1932, **247**, 1.
Przyłęcki, S.J. & Majmin, R. (1) *BZ*, 1931, **240**, 98; 1934, **271**, 168; 1934, **273**, 262; 1935, **277**, 1 and 420.
Przyłęcki, S.J. & Majmin, R. (2). *BZ*, 1934, **271**, 174; 1935, **280**, 413.
Przyłęcki, S.J. & Rafałowska, H. *BZ*, 1935, **277**, 416; 1935, **280**, 92.
Przyłęcki, S.J., Rafałowska, H. & Cichocka, J. *BZ*, 1935, **281**, 420.
Przyłęcki, S.J. & Rogalski, L. *AIP*, 1928, **29**, 423.
Przyłęcki, S.J. & Targońska, J. *BZ*, 1932, **255**, 406.
Puccioni, L., de Niederhausen, A. & Roncallo, P. *BSIBS*, 1933, **7**, 1312 and 1314; *CA*, 1934, **28**, 1753.
Puckett, W.O. (1). *AR*, 1934, **58** (Suppl.), 32; *JM*, 1936, **59**, 173.
Puckett, W.O. (2). *JEZ*, 1937, **76**, 303.
Punnett, R.C. *JG*, 1933, **27**, 465.
Puppe, A. *AFEWM*, 1925, **104**, 125.
Purjesz, B., Berkesy, L. & Gönczi, K. *AEPUP*, 1933, **173**, 553; *CA*, 1934, **28**, 1392.
Purjesz, B., Berkesy, L., Gönczi, K. & Kovács-Oskolás, M. *AEPUP*, 1934, **176**, 578; *CA*, 1935, **29**, 1492.
Purrmann, R. *ADC*, 1940, **544**, 182; **546**, 98.
Pütter, A. *ZAP*, 1905, **5**, 566.
Pybus, F.C. & Fawns, H.T. *JPB*, 1931, **34**, 39.

Quastel, J.H. *JH*, 1928, **28**, 139.
Quastel, J.H. & Whetham, M.D. *BJ*, 1924, **18**, 519.

Rabe, E. Inaug. Diss. Kiel, 1930; *OMaly*, 1932, **66**, 243.
Rabinovitch, I.M. *JOGBE*, 1931, **38**, 601.
Rabl, C. *ZWZ*, 1898, **63**, 496.
Rae, J.J. *BJ*, 1934, **28**, 152.
Rafałowska, H., Krasnodebski, J. & Mystkowski, E.M. *BZ*, 1935, **280**, 96.
Raffy, A. *CRAS*, 1933, **196**, 374.
Ragozina, M.N. *AFEWM*, 1937, **137**, 317; *BBMEUSSR*, 1936, **2**, 329; *BZUSSR*, 1936, **5**, 1073; *N*, 1937, **140**, 199.
Rainey, R.C. *AAB*, 1938, **25**, 822.
Raja, M. *AZI*, 1935, **21**, 447.
Ramage, H. *BJ*, 1934, **28**, 1500.
Ramage, H., Sheldon, J.H. & Sheldon, W. *PRSB*, 1933, **113**, 308.
Rammer, W. *AHB*, 1933, **25**, 692; *YMaly*, 1933, **27**, 468.
Ramon, G. *CRSB*, 1928, **99**, 1473.
Ramseyer, M. & du Pan, R.M. *AEZF*, 1937, **20**, 117.
Rand, H.W. & Hsu, S. *S*, 1927, **65**, 261.

Ransom, W.H. *PRS*, 1854, **7**, 168; *AMNH*, 1866 (3rd ser.), **18**, 249; *PTRS*, 1867, **157**, 431; *JAP*, 1867, **1**, 237.

Ranzi, S. (1). *BSNN*, 1926, **38**, 99.

Ranzi, S. (2). *AAL*, 1928 (6ᵉ ser.), **8**, 425.

Ranzi, S. (3). *PSZN*, 1928, **9**, 82; *BSIBS*, 1926, **1**, 343; *APANL*, 1929, **82**, 74; *AAL*, 1926 (6ᵉ ser.), **6**, 239.

Ranzi, S. (4). *AFEWM*, 1928, **114**, 364; *BDZ*, 1930, **1**, 131.

Ranzi, S. (5). *AAL*, 1929 (6ᵉ ser.), **10**, 111.

Ranzi, S. (6). *RAL*, 1929, **9** (ser. 6), 1171; *AFEWM*, 1930, **121**, 345; *BDZ*, 1930, **1**, 35.

Ranzi, S. (7). *AAL*, 1930 (ser. 6ᵉ), **12**, 468.

Ranzi, S. (8). *PSZN*, 1931, **11**, 86; *AZI*, 1931, **16**, 403; *APANL*, 1931, **84**, 40.

Ranzi, S. (9). *PSZN*, 1931, **11**, 104.

Ranzi, S. (10). *APANL*, 1931, **85**, 27; *AZI*, 1929, **13**, 21.

Ranzi, S. (11). *APANL*, 1932, **86**, 36.

Ranzi, S. (12). *PSZN*, 1932, **12**, 209; *BSIBS*, 1931, **6**, 357; *AZI*, 1931, **16**, 401; *BDZ*, 1931, **3**, 39; *Proc.* xivth *Internat. Physiol. Congr. Rome*, 1932, p. 214.

Ranzi, S. (13). *PSZN*, 1934, **13**, 332 and 357; *BSIBS*, 1933, **8**, 1137; *MZI*, 1933, **43** (Suppl.), 232.

Ranzi, S. (14). *AZI*, 1934, **20**, 569.

Ranzi, S. (15). *APANL*, 1934, **87**, 100.

Ranzi, S. (16). *APANL*, 1934, **87**, 347.

Ranzi, S. (17). *RAL*, 1935, **22** (6th ser.), 605.

Ranzi, S. (18). *SC*, 1935, **14**, 280 and 333.

Ranzi, S. (19). *NW*, 1936, **24**, 642; *BDZ*, 1935, **6**, 153; *Comptes Rend.* xiiᵉ *Congrès Internat. Zool. Lisbon*, 1935, p. 242.

Ranzi, S. (20). *ASB*, 1936, **22**, 80.

Ranzi, S. (21). *RAL*, 1936, **23** (6th ser.), 365.

Ranzi, S. (22). *RAL*, 1937, **24** (6th ser.), 528; *AZI*, 1937, **24**, 169.

Ranzi, S. (23). *APANL*, 1937, **1**, 43; *AEZF*, 1939, **22**, 557.

Ranzi, S. (24). *RS*, 1938, **9**, 11.

Ranzi, S. & Falkenheim, M. *PSZN*, 1937, **16**, 436; *NW*, 1938, **26**, 44; *BSIBS*, 1938, **13**, 120.

Ranzi, S. & Tamini, E. *NW*, 1939, **27**, 566.

Ranzi, S. & Zezza, P. *PSZN*, 1936, **15**, 355.

Raper, H.S. *FFF*, 1927, **9**, 206; *EEF*, 1932, **1**, 270; *JCS*, 1938, 125.

Rapkine, L. (1). *CRSB*, 1925, **93**, 1429.

Rapkine, L. (2). *APBPC*, 1931, **7**, Proc. Réunion des Physiologistes.

Rapkine, L. (3). *APBPC*, 1931, **7**, 382; *CRAS*, 1929, **188**, 650; 1930, **191**, 871.

Rapkine, L. (4). *CRSB*, 1933, **112**, 790 and 1294.

Rapkine, L. (5). *JCP*, 1936, **33**, 493.

Rapkine, L. (6). *JCP*, 1937, **34**, 416.

Rapkine, L. (7). *BJ*, 1938, **32**, 1729; *CRAS*, 1938, **207**, 301.

Rapkine, L. (8). *Trav. Congrès du Palais de la Découverte*, Paris, 1938, p. 471.

Rapkine, L. (9). "Oxydations Cellulaires", *Rev. Ann. Physiol.* (ed. Terroine), Paris, 1939.

Rapoport, J.A. (1). *BBMEUSSR*, 1938, **6**, 725.

Rapoport, J.A. (2). *BBMEUSSR*, 1939, **7**, 415.

Rappoport, S. *BZ*, 1937, **289**, 420.

Rashba, H. *UBZ*, 1938, **11**, 31, 395 and 402; 1939, **13**, 575 and 591.

Rashevsky, N. (1). *P*, 1931, **14**, 99; *JGP*, 1932, **15**, 289; *BP*, 1940, **2**, 109.

Rashevsky, N. (2). *P*, 1933, **20**, 125.

Rashevsky, N. (3). "Mathematical Biophysics." Univ. Press. Chicago, 1938.

Rass, T.R. *CRASUSSR*, 1935, **2**, 597.

Raven, C.P. (1). *PKAWA*, 1933, **36**, 566.

Raven, C.P. (2). *AFEWM*, 1933, **130**, 517; 1935, **132**, 509.

Raven, C.P. (3). *PKAWA*, 1935, **38**, 1107.

Raven, C.P. (4). *PKAWA*, 1935, **38**, 1109; *AFEWM*, 1938, **137**, 661.

Raven, C.P. (5). *AB*, 1938, **4**, 51.

Raven, C.P. (6). *ANM*, 1938, **1**, 337.

Rawles, M.E. (1). *JEZ*, 1936, **72**, 271.

Rawles, M.E. (2). *JG*, 1939, **38**, 517; *PNASW*, 1940, **26**, 86.

Ray, O.M. *RA*, 1938, **31**, 428.

Ray, O.M. & Bodine, J.H. (1). *PZ*, 1938, **11**, 267.

Ray, O.M. & Bodine, J.H. (2). *JCCP*, 1939, **14**, 43.

Ray, S.N. *BJ*, 1933, **28**, 189.

Raynaud, A. *BBFB*, 1938, **72**, 297.

da Re, O. *AF*, 1931, **30**, 147; *YMaly*, 1932, **20**, 480.

Readio, P.A. *AESA*, 1931, **24**, 19.

de Réaumur, A.R.F. "Mémoires pour servir à l'histoire des Insectes." 1740, pp. 127 ff.

Redenz, E. *ZGG*, 1937, **114**, 185.

Redfield, H. *GS*, 1926, **11**, 482.

Reed, H.D. & Gordon, M. *AJC*, 1931, **15**, 1524.

Reed, L.L., Mendel, L.B., Vickery, H.B. & Carlisle, P. *AJOP*, 1932, **102**, 285.

Reedman, E.J. & McHenry, E.W. *BJ*, 1938, **32**, 85.

Regnier, M.T. *CRSB*, 1935, **120**, 1089.

Reichert, E.T. *CIWP*, no. 173, 1913.

Reichert, E.T. & Brown, A.P. *CIWP*, no. 116, 1909.

Reimann, S.P. *P*, 1930, **10**, 82.

Reiss, M. *ZGEM*, 1931, **79**, 345; *CA*, 1932, **26**, 4879.

Reiss, P. *CRSB*, 1934, **116**, 1407.

Reith, F. (1). *ZWZ*, 1925, **126**, 181.

Reith, F. (2). *ZWZ*, 1931, **139**, 664; *AFEWM*, 1932, **127**, 283.

Reith, F. (3). *ZWZ*, 1935, **147**, 77.

Reith, F. (4). *ZWZ*, 1937, **150**, 179.

Remotti, E. (1). *RM*, 1927, **7**, 199; *YMaly*, 1928, **7**, 384.

Remotti, E. (2). *Sunti Comm.* xxi *Riun. Soc. It. Progr. Sci.* 1930, p. 163; *BMLZACUG*, 1935, **15**, 1.

Remotti, E. (3). *RM*, 1931, **11**, 1; *MZI*, 1932, **42** (Suppl.), 219; *YMaly*, 1933, **27**, 341.
Remotti, E. (4). *RM*, 1933, **12**, 1; *BMLZA CUG*, 1933, **13**, 3.
Remotti, E. (5). *BMLZACUG*, 1932, **12**; 1932, **13**.
Remotti, E. (6). *MZI*, 1935, **45** (Suppl.), 258; *YMaly*, 1935, **35**, 396.
Reverberi, G. *BIZUR*, 1929, **7**, 1.
Revoutskaia, P. *BBMEUSSR*, 1939, **8**, 413.
Reynolds, S.R.M. & Foster, F.I. *AJOP*, 1939, **127**, 343.
Rheinberger, M.B. *JBC*, 1936, **115**, 343.
Rhumbler, L. (1). *AFEWM*, 1897, **4**, 659.
Rhumbler, L. (2). *AFEWM*, 1902, **14**, 401.
Riboulleau, J. *CRSB*, 1938, **129**, 914.
Richards, A. (1). *AJA*, 1935, **56**, 355.
Richards, A. (2). "Outline of Comparative Embryology." Wiley, New York, 1931.
Richards, A. & Porter, R.P. *AJA*, 1935, **56**, 365.
Richards, A. & Schumacher, B.L. *AJA*, 1935, **56**, 395.
Richards, A.G. *JM*, 1932, **53**, 433.
Richards, A.G. & Miller, A. *JNYES*, 1937, **45**, 1 and 149.
Richards, M.H. & Furrow, E.Y. *BB*, 1925, **48**, 243.
Richards, O.W. (1). *CIWP*, 1935, no. 452, p. 171.
Richards, O.W. (2). *JEZ*, 1940, **83**, 401.
Richards, O.W. & Riley, G.A. *JEZ*, 1937, **77**, 159.
Richards, O.W. & Taylor, G.W. *BB*, 1932, **63**, 113.
Richards, V. & King, D. *SGY*, 1940, **8**, 409.
Richardson, K.C. *PTRSB*, 1935, **225**, 149.
Riddle, O. (1). *JEZ*, 1910, **8**, 163.
Riddle, O. (2). *PAPS*, 1927, **66**, 497.
Riddle, O. (3). *AJOP*, 1930, **94**, 535.
Riddle, O. (4). *CSHSQB*, 1937, **5**, 218; *SM*, 1938, **47**, 97; *PAPS*, 1935, **75**, 521.
Riddle, O. & Bates, R.W. "Preparation, Assay and Actions of the Lactogenic Hormone." In "Sex and Internal Secretions", 1939, p. 1088.
Riddle, O., Bates, R.W. & Dykshorn, S.W. *PSEBM*, 1932, **29**, 1211; *AJOP*, 1933, **105**, 191.
Riddle, O., Bates, R.W. & Lahr, E.L. *AJOP*, 1935, **111**, 352.
Riddle, O. & Braucher, P.F. *AJOP*, 1931, **97**, 617.
Riddle, O., Charles, D.R. & Cauthen, G.E. *PSEBM*, 1932, **29**, 1216.
Riddle, O. & Dotti, L.B. *S*, 1936, **84**, 557.
Riddle, O. & Dykshorn, S.W. *PSEBM*, 1932, **29**, 1213.
Riddle, O. & Kříženecký, J. *AJOP*, 1931, **97**, 343.
Riddle, O. & la Mer, V.K. *AJOP*, 1918, **47**, 103.

Riddle, O., Nussmann, T.C. & Benedict, F.G. *AJOP*, 1932, **101**, 251.
Rideal, E.K. & Schulman, J.H. *N*, 1939, **144**, 100.
Riedel, H. *AFEWM*, 1935, **132**, 463.
Riemenschneider, R.W., Ellis, N.R. & Titus, H.W. *JBC*, 1938, **126**, 255.
Ries, E. (1). *PSZN*, 1937, **16**, 363.
Ries, E. (2). "Grundriss d. Histophysiologie." Leipzig, 1938; *VDZG*, 1936, 245.
Ries, E. (3). *AEZF*, 1939, **22**, 569.
Ries, E. (4). *AEZF*, 1939, **23**, 95.
Ries, E. & Gersch, M. *PSZN*, 1936, **15**, 223.
Riesinger, E. *AFEWM*, 1933, **129**, 445.
Rimington, C. *BJ*, 1927, **21**, 273.
Rinne, F. "Investigations and considerations concerning Paracrystallinity", *Faraday Soc. Symp. on Liquid Crystals*, 1933; "Grenzfragen des Lebens." Quelle & Meyer, Leipzig, 1931.
Rischbieth, H. & Barrington, A. "Dwarfism." In "Treasury of Human Inheritance", Pts. VII and VIII, London, 1912.
Risley, P.L. *JEZ*, 1939, **80**, 113.
Rittenberg, D. & Schoenheimer, R. *JBC*, 1937, **121**, 235.
Rix, E. *AEZF*, 1933, **13**, 517.
Rizzo, A. *RLAUR*, 1899, **7**, 171.
Rjabinovskaia, A. (1). *BBMEUSSR*, 1936, **1**, 305; *Proc. xvth Internat. Congr. Physiol.* (Moscow), 1935, p. 344; *UBZ*, 1936, **9**, 761.
Rjabinovskaia, A. (2). *CRASUSSR*, 1940, **26**, 826.
Robb, R.C. *JEB*, 1929, **6**, 311.
Robbie, W.A., Boell, E.J. & Bodine, J.H. *PZ*, 1938, **11**, 54.
de Robertis, E. *PMA*, 1937, no vol. no., repaged reprint.
Roberts, E.E. (1). *JEZ*, 1918, **27**, 157.
Roberts, E.E. (2). *AR*, 1924, **29**, 141; 1926, **34**, 172.
Roberts, E.E. & Carroll, W.E. *JH*, 1931, **22**, 125.
Robinson, J.R. (1). Inaug. Diss. Cambridge, 1937; *PRSA*, 1939, **170**, 519.
Robinson, J.R. (2). *N*, 1939, **143**, 923.
Robinson, T.W. & Woodside, G.L. *JCCP*, 1937, **9**, 241; *AR*, 1935, **64** (Suppl.), 33.
Robinson, W. *JPA*, 1935, **21**, 354.
Robison, R. (1). *BJ*, 1923, **17**, 286.
Robison, R. (2). *BJ*, 1926, **20**, 388.
Robison, R., McLeod, M. & Rosenheim, A.H. *BJ*, 1930, **24**, 1927.
Robison, R. & Rosenheim, A.H. *BJ*, 1934, **28**, 684.
Robson, J.M. *JOP*, 1933, **78**, 309; **79**, 83; 1936, **86**, 171; 1937, **88**, 100; **90**, 145.
Robson, J.M. & Bell, G.H. *JOP*, 1937, **88**, 312; 1938, **92**, 131.
Robson, J.M. & Bonser, G.M. *N*, 1938, **142**, 836.
Robson, J.M. & Schild, H.O. *JOP*, 1938, **92**, 1.

Rocco, M.L. *CRAS*, 1938, **207**, 1006.

Roche, A. & Garcia, I. *CRSB*, 1933, **112**, 1686; 1934, **116**, 1029.

Roche, J. & Leandri, A. *CRSB*, 1935, **119**, 1141.

de Roche, V. *AFEWM*, 1937, **135**, 620.

Rodolfo, A. *JEZ*, 1934, **68**, 215.

Roepke, R.R. & Bushnell, L.D. *JI*, 1936, **30**, 109.

Roepke, R.R. & Hughes, J.S. *JBC*, 1935, **108**, 79.

Röhlich, K. *AFEWM*, 1929, **118**, 164.

Rohmer, P., Bezssonov, N. & Stoerr, E. *BAM*, 1934, 871.

Rohmer, P., Sanders, U. & Bezssonov, N. *N*, 1934, **134**, 142.

Romankevitch, N.A. *ZZFMA*, 1934, **21**, 110.

Romanov, A.L. [Romanoff]. (1). *JEZ*, 1929, **54**, 343.

Romanov, A.L. (2). *Memoirs Cornell Univ. Agric. Exp. Sta.* no. 132, 1930.

Romanov, A.L. (3). *AR*, 1931, **48**, 185.

Romanov, A.L. (4). *BJ*, 1931, **25**, 994.

Romanov, A.L. (5). *BB*, 1932, **62**, 54.

Romanov, A.L. (6). *S*, 1933, **77**, 393.

Romanov, A.L. (7). *PS*, 1934, **13**, 283.

Romanov, A.L. (8). *JAS*, 1935, **25**, 318.

Romanov, A.L. (9). *PS*, 1936, **15**, 311.

Romanov, A.L. (10). *Bull. Cornell Univ. Agric. Exp. Sta.* 1934, no. 616; 1938, no. 687.

Romanov, A.L. & Faber, H.A. *JCCP*, 1933, **2**, 457.

Romanov, A.L. & Grover, H.J. *JCCP*, 1936, **7**, 425.

Romanov, A.L. & Romanov, A.J. (1). *Memoirs Cornell Univ. Agric. Exp. Sta.* 1933, no. 150.

Romanov, A.L. & Romanov, A.J. (2). *AR*, 1933, **55**, 271.

Romanov, A.L., Smith, L.L. & Sullivan, R.A. *Memoirs Cornell Univ. Agric. Exp. Sta.* 1938, no. 216.

Romanov, A.L. & Sullivan, R.A. *IEC*, 1937, **29**, 117.

Romeis, B. *AFEWM*, 1924, **101**, 382.

Romijn, C. & Roos, J. *JOP*, 1938, **94**, 365; *ABN*, 1939, **9**, 82.

Rominger, E. *EVHF*, 1938, **2**, 104.

Ronzoni, E. & Ehrenfest, E. *JBC*, 1936, **115**, 749.

Roonwal, M.L. (1). *BER*, 1936, **27**.

Roonwal, M.L. (2). *PTRSB*, 1936, **226**, 391; 1937, **227**, 175.

Roonwal, M.L. (3). *TNISI*, 1939, **2**, 1.

Roos, J. & Romijn, C. *JOP*, 1938, **92**, 249; *PKAWA*, 1937, **40**, 803; *ABN*, 1938, **8**, 140.

Roosen-Runge, E.C. *AR*, 1939, **74**, 349.

Root, W.S. *BB*, 1930, **59**, 48.

Rosahn, P.D. & Greene, H.S.N. *JEM*, 1936, **63**, 901.

Rosahn, P.D., Greene, H.S.N. & Hu, Ch'uan Ku'ei. *JEZ*, 1935, **72**, 195.

Rosanova, V.D. *BBMEUSSR*, 1938, **5**, 126 and 132.

Rose, M. *BBFB*, 1939, **73**, 336.

Rose, M. & Berrier, H. *CRAS*, 1935, **201**, 357; 1936, **203**, 496.

Rose, M.S., Vahlteich, E.McC. & McLeod, G. *JBC*, 1934, **104**, 217.

Rose, S.M. *AR*, 1937, **70** (Suppl. no. 1), 102; *BB*, 1939, **77**, 216.

Rosedale, J.L. & Morris, J.P. *BJ*, 1930, **24**, 1294.

Rosenbloom, D. *PSEBM*, 1935, **32**, 908.

Rosenblueth, A. *PR*, 1937, **17**, 514.

Rosenfeld, L.E. & Bagdassarianz, A. *Proc. xvth Internat. Congr. Physiol.* 1935, p. 348.

Rosenheim, A.H., Giršavičius, J.O., Ashford, C.A. and Stickland, L.H. Unpublished work, with the author, 1928.

Rosenheim, O. & King, H. *ARB*, 1934, **3**, 87.

Ross, H. & Hedicke, H. "Die Pflanzengallen." Jena, 1927.

Ross, R. "The Prevention of Malaria." 2nd ed. 1911, p. 679.

Ross, W.D. Introduction and Commentary in "Aristotle's Metaphysics". Oxford University Press, 1924.

Ross, W.D. "Aristotle." Methuen, London, 1930.

Rossi, F. *MZI*, 1933, **43** (Suppl.), 316; *YMaly*, 1933, **26**, 395 and 735; *ZAE*, 1933, **100**, 735.

Rössig, H. *ZJS*, 1904, **20**, 19.

Rossmann, B. *AFEWM*, 1928, **113**, 346; **114**, 583.

Rostand, J. "La Parthogenèse des Vertébrés." Hermann, Paris, 1938.

Rothbard, S. & Herman, J.R. *AOP*, 1939, **28**, 212.

Rothschild, V. *JEB*, 1939, **16**, 49.

Rotini, O.T. *ALRFS*, 1931, **2**, 249; *OMaly*, 1932, **70**, 768.

Rotmann, E. (1). *AFEWM*, 1931, **124**, 747; 1933, **129**, 85; 1935, **133**, 225.

Rotmann, E. (2). *AFEWM*, 1934, **131**, 702 (abstr.); 1935, **133**, 193.

Rotmann, E. (3). *VDZG*, 1935, 76.

Rotmann, E. (4). *AFEWM*, 1939, **139**, 1.

Rotmann, E. (5). *CS*, 1939, Suppl. no. 4, 16.

Rotmann, E. & MacDougald, T.J. *VDZG*, 1936, 88.

Roubaud, E. *BBFB*, 1922, **56**, 455; *ASNZ*, 1935 (sér. 10), **18**, 39.

Roulet, F. *CRSB*, 1926, **95**, 390 and 1340.

Rous, P. (1). *JEM*, 1910, **12**, 696.

Rous, P. (2). *AJC*, 1936, **28**, 233.

Rous, P. & Kidd, J.G. *S*, 1936, **83**, 468.

Rous, P. & Murphy, J.B. *JAMA*, 1911, **56**, 741.

Roussel, G. & Deflandre, D. (1). *AAPA*, 1931, **8**, 139.

Roussel, G. & Deflandre, D. (2). *AAPA*, 1931, **8**, 1241; *YMaly*, 1932, **23**, 331.

Roussel, G. & Deflandre, D. (3). *CRSB*, 1931, **106**, 260.

Roussel, G. & Deflandre, D. (4). *CRSB*, 1931, **106**, 529.

Roussel, G. & Deflandre, D. (5). *AAPA*, 1932, **9**, 757; *YMaly*, 1933, **25**, 475.

Roux, W. (1). "Über die Bedeutung d. Kerntheilungsfiguren." Engelmann, Leipzig, 1883. In "Gesammelte Abhandlungen über Entwickelungsmechanik d. Organismen." Engelmann, Leipzig, 1895, vol. II, p. 125.

Roux, W. (2). *APPA*, 1888, **114**, 113; and in "Gesammelte Abhandlungen...", vol. II, p. 419.

Roux, W. (3). *AA*, 1888, **3**; and in "Gesammelte Abhandlungen...", vol. II, p. 522.

Roux, W. (4). *AMA*, 1887, **29**; *AA*, 1903, **23**; and in "Gesammelte Abhandlungen...", vol. II.

Roux, W. (5). *BC*, 1893, **13**, 612; and in "Gesammelte Abhandlungen...", vol. II, pp. 873, 882 and 907 ff.

Roux, W. (6). "Gesammelte Abhandlungen." Vol. II, p. 987.

Roux, W., Correns, C., Fischel, A. & Küster, E. "Terminologie der Entwicklungsmechanik der Tiere u. Pflanzen." Leipzig, 1912.

Rowe, A.W. & Boyd, W.C. *JN*, 1932, **5**, 551.

Rubaschev, S.J. *AZ*, 1935, **16**, 387; 1937, **18**, 345.

Rubenstein, B.B. & Gerard, R.W. (1). *PSEBM*, 1933, **31**, 282.

Rubenstein, B.B. & Gerard, R.W. (2). *JGP*, 1934, **17**, 677.

Rubinstein, M. *CRSB*, 1932, **111**, 58, 60 and 63; *APB*, 1933, **11**, 40.

Rubner, M. *BZ*, 1924, **148**, 187.

Rudkin, G.T. *PNASW*, 1939, **25**, 594.

Rudnick, D. & Rawles, M.E. *PZ*, 1937, **10**, 381.

Rudy, H. "Die biologische Feldtheorie." Bornträger, Berlin, 1931.

la Rue, C.D. (1). *AJB*, 1933, **20**, 1 and 159; 1935, **22**, 908.

la Rue, C.D. (2). *BTBC*, 1937, **64**, 97.

Ruffini, A. *AAFS*, 1906 (ser. 4), **18**, 8; *AIAE*, 1907, **6**, 129; *AA*, 1907, **31**, 448; also "Fisiogenia; la Biodinamica dello Sviluppo ed i fondamentali problemi morfologici dell' embriologia generale." Milan, 1925.

Rugh, R. *PAPS*, 1939, **81**, 447.

Rugh, R. & Exner, F. *PAPS*, 1940, **83**, 607.

Rulon, O. *P*, 1935, **24**, 346.

Rumjantzev, A.W. & Berezkina, L.F. *CRAS USSR*, 1938, **20**, 753; *AAUSSR*, 1937, **17**, 179.

Rumjantzev, A.W. & Suntzova, V.V. *AEZF*, 1935, **17**, 360.

Runge, H. *VDAG*, 1936, 80.

Runnström, J. (1). *AFEWM*, 1914, **40**, 526; 1915, **41**, 1; 1918, **43**, 223, 409 and 532.

Runnström, J. (2). *AFZ*, 1925, **18** A, 4.

Runnström, J. (3). *AZ*, 1926, **7**, 117.

Runnström, J. (4). *P*, 1928, **4**, 388.

Runnström, J. (5). *AFEWM*, 1928, **113**, 556.

Runnström, J. (6). *AZ*, 1928, **9**, 365.

Runnström, J. (7). *P*, 1930, **10**, 106.

Runnström, J. (8). *AFEWM*, 1929, **117**, 123; 1931, **124**, 273.

Runnström, J. (9). *P*, 1932, **15**, 448.

Runnström, J. (10). *P*, 1932, **15**, 532.

Runnström, J. (11). *AFEWM*, 1933, **129**, 442; *NTVKI*, 1933, **9**, 279 (in Swedish).

Runnström, J. (12). *BZ*, 1933, **258**, 257.

Runnström, J. (13). *P*, 1933, **20**, 1.

Runnström, J. (14). *BB*, 1935, **68**, 327.

Runnström, J. (15). *BB*, 1935, **68**, 378.

Runnström, J. (16). *BB*, 1935, **69**, 345.

Runnström, J. (17). *BB*, 1935, **69**, 351.

Runnström, J. & Runnström, S. *BMA*, 1918, no. 5.

Runnström, J. & Thörnblom, D. *NW*, 1936, **24**, 447; *ABL*, 1938, **8**, 97.

Rusoff, L.L. & Gaddum, L.W. *JN*, 1938, **15**, 169.

Russell, E.S. "Form and Function." Murray, London, 1916.

Russell, Eliz.S., *JEZ*, 1940, **84**, 363.

Russell, W.C. & Taylor, M.W. *JN*, 1935, **10**, 613.

Ruud, G. *AFEWM*, 1929, **118**, 308.

Ruud, G. & Spemann, H. *AFEWM*, 1922, **52**, 95.

Ruzicka, L. (1). *CR*, 1937, **20**, 69; *JCE*, 1936, **13**, 3; *N*, 1936, **137**, 260; *NW*, 1935, **23**, 44.

Ruzicka, L. (2). "Sur l'Architecture des Polyterpènes." In *Trav. Congr. Palais de la Découverte*, Paris, 1938, p. 311.

Ryvkina, D.E. *CRASUSSR*, 1940, **27**, 380.

Ryvkina, D.E. & Striganova, A.(1). *BASUSSR*, 1939, 445.

Ryvkina, D.E. & Striganova, A.(2). *BASUSSR*, 1939, 799.

St John, J.L. (1). *JACS*, 1931, **53**, 4014.

St John, J.L. (2). *PS*, 1936, **15**, 79.

Sabin, F.R., Miller, F.R., Smithburn, K.C., Thomas, R.M. & Hummel, L.E. *JEM*, 1936, **64**, 97.

Saccardi, P. & Latini, P. *RDB*, 1931, **13**, 9; *OMaly*, 1931, **68**, 656.

Sacharov, N.L. *E*, 1930, **11**, 505.

Sachs, D. *AEZF*, 1934, **15**, 179.

Sachs, D., Ephrussi, B. & Rapkine, L. *CRSB*, 1933, **113**, 708.

Sachs, D., Rapkine, L. & Ephrussi, B. *CRSB*, 1933, **113**, 829.

Sagara, J.I. *JJB*, 1930, **12**, 473.

Sala, S.L. *RSEIQ*, 1934, **17**, 634; *CA*, 1935, **29**, 207.

Salle, A.J. & Shechmeister, I. *PSEBM*, 1936, **34**, 603.

Salt, G. *JEZ*, 1927, **48**, 223; 1931, **59**, 133.

Salvatori, A. *RDB*, 1936, **21**, 16 and 76.

Samaraiev, V.N. *BBMEUSSR*, 1939, **8**, 505 and 509.

Sammartino, U. *AFS*, 1935, **59**, 49; **60**, 342 and 372; *AIB*, 1935, **93**, 131; **94**, 21 and 121; *CA*, 1935, **29**, 3043; *YMaly*, 1935, **33**, 479; 1936, **36**, 347; **38**, 311.

Sandstrom, C.J. (1). *PZ*, 1932, **5**, 354; *AR*, 1932, **52**, 69; 1933, **54**, 59.

Sandstrom, C.J. (2). *AR*, 1935, **64** (Suppl. no. 1), 49.

Sandstrom, C.J. & Kauer, G.L. *AR*, 1933, **57**, 119.

Sandstrom, C.T. & Kauer, J.T. *AR*, 1933, **57**, 105.

Sandstrom, R.H. *PZ*, 1934, **7**, 226.

Sannié, C. *BAEC*, 1935; *BJ*, 1936, **30**, 704.

Sánta, G. *KOKK*, 1931, **24**, 1; *YMaly*, 1931, **17**, 655.

Sasaki, K. *JI*, 1932, **23**, 1.

Sato, I. *PIAT*, 1933, **10**, 378.

Sato, T. (1). *AFEWM*, 1930, **122**, 451.

Sato, T. (2). *AFEWM*, 1933, **128**, 342.

Sato, T. (3). *AFEWM*, 1933, **130**, 19.

Sato, T. (4). *AFEWM*, 1935, **133**, 322.

Sauer, F.C. (1). *JM*, 1936, **60**, 1; *JCN*, 1935, **62**, 377.

Sauer, F.C. (2). *G*, 1939, **3**, 381.

Saunders, J.T. *PCPS*, 1923, **1**, 30.

Savage, R.M. (1). *PZS*, 1937, **107**, 249.

Savage, R.M. (2). *PZS*, 1938, **108**, 465.

Savchuk, M. *CRASUSSR*, 1938, **20**, 701.

Saviano, M. *BSIBS*, 1936, **11**, 146.

Sax, M.G. & Leibson, R.G. *BBMEUSSR*, 1937, **4**, 496.

Sax, M.G. & Likhnitzkaia, I.I. *BBMEUSSR*, 1938, **5**, 523.

Scammon, R.E. (1). "The Measurement of Man." Minneapolis, 1930.

Scammon, R.E. (2). *APD*, 1930, **11**, 354.

Scammon, R.E. & Calkins, L.A. "Growth of the Human Body in the Foetal Period." Minneapolis, 1929.

Schaeffer, A.A. *S*, 1931, **74**, 47.

Schaible, P.J., Davidson, J.A. & Moore, J.M. *PS*, 1936, **15**, 298.

Schäper, A. *AA*, 1904, **25**, 298 and 326.

Scharles, F.H., Baker, M.D. & Salter, W.T. *BJ*, 1935, **29**, 1927; *AJC*, 1935, **25**, 122.

Scharpenack, A.E. & Jerjomin, G.P. *PE USSR*, 1935, **4**, 11.

Scharrer, B. & Hadorn, E. *PNASW*, 1938, **24**, 236.

Scharrer, K. & Schropp, W. *CT*, 1932, **4**, 249.

Schatz, E. *ZWZ*, 1929, **135**, 539.

Schaxel, J. (1). "Die Leistungen der Zellen bei d. Entwicklung d. Metazoen." Jena, 1915.

Schaxel, J. (2). "Untersuchungen ü. d. Formbildung d. Tiere", Arb. a. d. Geb. d. exp. Biol. 1921, vol. I.

Schaxel, J. (3). *RDB*, 1922, **4**, 302; *AFEWM*, 1922, **50**, 498.

Schaxel, J. (4). *CRASUSSR*, 1934, **4**, 174 and 246.

Schaxel, J. & Böhmel, W. *ZAZ*, 1928, **78**, 157.

Schaxel, J. & Schneider, G. *CRASUSSR*, 1939, **23**, 962.

Schechtman, A.M. (1). *UCPZ*, 1932, **36**, 325.

Schechtman, A.M. (2). *UCPZ*, 1934, **39**, 277.

Schechtman, A.M. (3). *UCPZ*, 1934, **39**, 303; *PSEBM*, 1935, **32**, 1072.

Schechtman, A.M. (4). *UCPZ*, 1934, **39**, 393.

Schechtman, A.M. (5). *PSEBM*, 1937, **37**, 153.

Schechtman, A.M. (6). *PSEBM*, 1938, **38**, 430.

Schechtman, A.M. (7). *PSEBM*, 1938, **39**, 236.

Schechtman, A.M. (8). *PSEBM*, 1939, **41**, 48.

Schechtman, A.M. (9). Private communication to the author, Aug. 1940; *JEZ*, 1941, **87**, 1.

Scheminsky Fe. & Scheminsky, Fr. (1). *AP*, 1933, **232**, 808.

Scheminsky, Fe. & Scheminsky, Fr. (2). *BZ*, 1937, **293**, 256.

Schenck, E.G. (1). *ZPC*, 1932, **211**, 111.

Schenck, E.G. (2). *ZPC*, 1932, **211**, 153.

Schenck, E.G. (3). *ZPC*, 1932, **211**, 160.

Schenck, E.G. (4). *EIMK*, 1934, **46**, 269; *NW*, 1930, **18**, 824.

Schenck, E.G. & Kunstmann, H.K. *ZPC*, 1933, **215**, 87.

Schenck, E.G. & Wollschitt, H. *AEPUP*, 1933, **170**, 151.

Scheremetieva, E.A. & Brunst, V.V. (1). *AFEWM*, 1933, **130**, 771; *RG*, 1935, **4**, 57; *YMaly*, 1936, **37**, 181.

Scheremetieva, E.A. & Brunst, V.V. (2). *BBMEUSSR*, 1938, **6**, 723.

Schidewski, G. *AFEWM*, 1935, **132**, 57.

Schkljar, N.M. (1). *AFGF*, 1935, **9**, 213.

Schkljar, N.M. (2). *PZUSSR*, 1935, **18**, 648; *YMaly*, 1936, **36**, 347.

Schkljar, N.M. (3). *UBZ*, 1937, **10**, 379 and 406; *CA*, 1937, **31**, 7500.

Schkljar, N.M. (4). *UBZ*, 1938, **12**, 161.

Schkljar, N.M. (5). *UBZ*, 1938, **12**, 424.

Schlack, H. & Scharfnagel, W. *MK*, 1931, **51**, 273; *CA*, 1932, **26**, 2756.

Schlampp, K.W. *ZWZ*, 1892, **53**, 537.

Schlayer, C. *BZ*, 1937, **293**, 94.

Schleip, W. "Die Determination der Primitiv Entwicklung." Akad. Verlag, Leipzig, 1929.

Schlenk, W. *BZ*, 1933, **267**, 424.

Schlick, M. *Proc. VIIth Internat. Congr. Philosophy*, Oxford, 1930.

Schlipper, A.L. *PZ*, 1938, **11**, 40.

Schloss, W. *AKC*, 1928, **151**, 701.

Schlossmann, H. (1). *AEPUP*, 1932, **166**, 74.

Schlossmann, H. (2). *AEPUP*, 1932, **166**, 81.

Schlossmann, H. (3). *EP*, 1932, **34**, 741. Later as "Der Stoffaustausch zwischen Mutter u. Frucht durch die Placenta." Bergmann, München, 1933.

Schlossmann, H. (4). *JOP*, 1938, **92**, 219.

Schmalfuss, H., Barthmeyer, H. & Hinsch, W. *ZIAV*, 1931, **58**, 332.

Schmalfuss, H. & Werner, H. *ZIAV*, 1926, **41**, 258.

Schmalhausen, I. *AFEWM*, 1930, **123**, 153; 1931, **124**, 82; *BC*, 1930, **50**, 292; 1931, **51**, 379.

Schmalhausen, O.I. *CRASUSSR*, 1940, **28**, 277.

Schmidt, G. (1). *ZPC*, 1934, **223**, 81.

Schmidt, G. (2). *EZ*, 1937, **4**, 40.

Schmidt, G.A. (1). *AFEWM*, 1930, **122**, 663; 1933, **129**, 1.

Schmidt, G.A. (2). *ABSUSSR*, 1935, **39**, 309.

Schmidt, G.A. (3). *BZUSSR*, 1936, **5**, 135.

Schmidt, G.A. (4). *BZUSSR*, 1936, **5**, 145; *YMaly*, 1936, **39**, 199.

Schmidt, G.A. (5). *ZZUSSR*, 1936, **15**, 259.

Schmidt, G.A. (6). *ZAZ*, 1936, **116**, 323; *CRASUSSR*, 1936, **13**, 440; *BZUSSR*, 1936, **5**, 145; *YMaly*, 1936, **39**, 199.

Schmidt, G.A. (7). *BBMEUSSR*, 1936, **2**, 322.

Schmidt, G.A. (8). *ZAZ*, 1937, **117**, 26; *BBMEUSSR*, 1936, **2**, 319.

Schmidt, G.A. (9). *BZUSSR*, 1936, **5**, 633; *ZMOT*, 1937, **32**, 650.

Schmidt, G.A. (10). *ZAZ*, 1937, **117**, 59.

Schmidt, G.A. (11). *BZUSSR*, 1937, **6**, 513.

Schmidt, G.A. (12). *N*, 1937, **140**, 199; *ZAZ*, 1937, **120**, 146.

Schmidt, G.A. (13). *ZAZ*, 1937, **120**, 155; *ADB*, 1937, **48**, 361.

Schmidt, G.A. (14). *AAM*, 1938, **34**, 5.

Schmidt, G.A. & Jankovskaia, L.A. *AZEG*, 1938, **79**, 487.

Schmidt, G.A. & Ragozina, M.N. *BBME USSR*, 1936, **2**, 204; *ZAZ*, 1937, **120**, 143.

Schmidt, W.J. (1). "Die Bausteine des Tierkörpers in polarisiertem Lichte." Cohen, Bonn, 1924.

Schmidt, W.J. (2). *ZJ*, 1928, **45**, 177.

Schmidt, W.J. (3). *ZZFMA*, 1936, **25**, 181.

Schmidt, W.J. (4). *BOGG*, 1936, **17**, 140; *NW*, 1936, **24**, 463; *AEZF*, 1937, **19**, 352.

Schmidt, W.J. (5). "Die Doppelbrechung von Karyoplasma, Zytoplasma und Metaplasma." Bornträger, Berlin, 1937. (Protoplasma Monographs, no. 11.)

Schmidt-Nielsen, S., Aas, G., Astad, A. & Leonardsen, R. *KNVSF*, 1933, **6**, 150; *OMaly*, 1934, **81**, 51.

Schmidt-Nielsen, S. & Stene, J. *KNVSF*, 1931, **4**, 100; *CA*, 1932, **26**, 2248.

Schmidt-Thomé, J. *EP*, 1937, **39**, 192.

Schmitt, F.O. *PR*, 1939, **19**, 270; *JAPP*, 1938, **9**, 109.

Schmitt, F.O. & Bear, R.S. *BR*, 1939, **14**, 27.

Schmitz, E. *AG*, 1923, **121**, 1.

Schmuck, A. *CRASUSSR*, 1938, **19**, 189.

Schmuck, A. & Gusseva, A. *CRASUSSR*, 1939, **24**, 441; 1940, **26**, 460 and 674; *BUSSR*, 1940, **5**, 129.

Schmuck, A., Gusseva, A. & Iljin, G. *BUSSR*, 1939, **4**, 482.

Schmuck, A. & Kostov, D. *CRASUSSR*, 1939, **23**, 263.

Schnetter, M. *AFEWM*, 1934, **131**, 285; *ZMOT*, 1934, **29**, 114; *VDZG*, 1936, 82.

Schoenheimer, R. (1). *S*, 1931, **74**, 579.

Schoenheimer, R. (2). *JBC*, 1934, **105**, lxxvi.

Schoenheimer, R. (3). "Harvey Lectures", 1937, p. 122.

Schoenheimer, R. (4). Private communication to the author, 1940.

Schoenheimer, R. & von Behring, H. *ZPC*, 1930, **192**, 110.

Schoenheimer, R. & Breusch, F. *JBC*, 1933, **103**, 439.

Schoenheimer, R. & Dam, H. *ZPC*, 1932, **211**, 241.

Schoenheimer, R. & Evans, E.A. *ARB*, 1937, **6**, 139.

Schoenheimer, R. & Hilgetag, G. *JBC*, 1934, **105**, 73.

Schoenheimer, R. & Rittenberg, D. (1). *JBC*, 1936, **114**, 381; *S*, 1938, **87**, 221.

Schoenheimer, R. & Rittenberg, D. (2). *PR*, 1940, **20**, 218.

Schönfelder, H. *AEPUP*, 1935, **180**, 24.

Schønheyder, F. *JBC*, 1938, **123**, 491.

Schönmann, W. *AFEWM*, 1939, **138**, 345.

Schooley, J.P., Riddle, O. & Bates, R.W. *PSEBM*, 1937, **36**, 408.

Schotté, O. (1). *SPHNGCR*, 1926, **43**, 67.

Schotté, O. (2). *SPHNGCR*, 1926, **43**, 126.

Schotté, O. (3). *RSZ*, 1926, **33**, 1; *SPHNGCR*, 1922, **39**, 67 and 134; 1923, **40**, 160; 1924, **41**, 45; *CRSB*, 1926, **94**, 1128.

Schotté, O. (4). *AFEWM*, 1930, **123**, 179.

Schotté, O. (5). *CN*, 1938, **13**, 1.

Schotté, O. (6). *G*, 1939, **3** (Suppl.), 59.

Schotté, O., J.B. McK. Arthur & J.S. Kobler. *AR*, 1940, **76** (Suppl.), 49.

Schotté, O. & Butler, E.G. Private communication to the author, 1940; *JEZ*, 1941, **87**, 279.

Schotté, O. & Edds, M.V. *JEZ*, 1940, **84**, 199.

Schotté, O. & Hummel, K.P. *JEZ*, 1939, **80**, 131; *S*, 1937, **85**, 438.

Schour, I. & Hoffman, M.M. *JDR*, 1939, **18**, 91.

Schrader, F. & Schrader, S.H. *QRB*, 1931, **6**, 411.

Schrek, R. *AJOPA*, 1936, **12**, 525.

Schubert, M. *ZMAF*, 1926, **6**, 162; 1927, **8**, 640.

Schuler, W. & Reindel, W. *ZPC*, 1933, **221**, 209 and 231; 1935, **234**, 63.

Schulman, J.H. *TFS*, 1937, **33**, 1116.

Schulman, J.H. & Hughes, A.H. *BJ*, 1935, **29**, 1236.

Schulman, J.H., Stenhagen, E. & Rideal, E.K. *N*, 1938, **141**, 785.

Schultz, Jack (1). *GS*, 1929, **14**, 366.

Schultz, Jack (2). *AN*, 1935, **69**, 30.

Schultz, Julius. "Die Maschinentheorie d. Lebens." Bornträger, Berlin, 1929.

Schultz, W. (1). *AFEWM*, 1918, **43**, 361.

Schultz, W. (2). *AFEWM*, 1915, **41**, 535; 1917, **42**, 139, 222; 1920, **47**, 43; 1922, **51**, 337.

Schultze, O. *AFEWM*, 1894, **1**, 269.

Schulz, F.N. & Becker, M. *BZ*, 1935, **280**, 217.

Schulz, F.N. & Ditthorn, F. *ZPC*, 1900, **29**, 373; 1901, **32**, 428.

Schürch, O. & Winterstein, A. *ZPC*, 1935, **236**, 79.

Schuwirth, K. *ZPC*, 1940, **263**, 25.

Schwalbe, E. "Morphologie u. Missbildungen d. Menschen u. d. Tiere." Jena, 1907.

Schwartzbach, S.S. & Uhlenluth, E. *ENE*, 1936, **16**, 412.

Schwarz, E. *ZKF*, 1923, **20**, 353.

Schwarz, O.H. & Drabkin, C. *AJOG*, 1931, **22**, 571; *OMaly*, 1932, **65**, 293.

Schwind, J.L. *JEZ*, 1933, **66**, 1.

Scott, H.H. *JPB*, 1927, **30**, 61; *CAR*, 1927, **2**, 154.

Scott, H.M. & Hughes, J.S. *Biennial Rep. Kansas Agric. Exp. Sta.* 1932, p. 80.

Scott, H.M., Hughes, J.S. & Warren, D.C. *PS*, 1937, **16**, 53.

Scott, J.P. *JEZ*, 1937, **77**, 123; *JM*, 1938, **62**, 299.

Scott, J.W. *JEZ*, 1906, **3**, 49.

Scott-Moncrieff, R. "Biochemistry of Flower-Colour Variation." In "Perspectives in Biochemistry", Hopkins Presentation Volume, Cambridge, 1937.

Scrimshaw, N.S. Private communication to the author, Oct. 1940.

Secher, K. *ZKF*, 1919, **16**, 297.

Seckel, H. *JK*, 1929, **126**, 83.

Sedgwick, Adam. *QJMS*, 1895, **37**, 87.

Sehl, A. *ZMOT*, 1931, **20**, 533.

Seidel, F. (1). *ZMOT*, 1924, **1**, 429.

Seidel, F. (2). *BC*, 1926, **46**, 321; 1928, **48**, 230; 1929, **49**, 577.

Seidel, F. (3). *AFEWM*, 1929, **119**, 322.

Seidel, F. (4). *AFEWM*, 1932, **126**, 213.

Seidel, F. (5). *AFEWM*, 1934, **131**, 135.

Seidel, F. (6). *VDZG*, 1936, 291.

Seidel, F. (7). "Entwicklungsphysiologie, 1935." *FZ*, 1937, p. 406.

Seifriz, W. (1). *AN*, 1926, **60**, 124; *S*, 1931, **73**, 648; *BOR*, 1935, **1**, 18.

Seifriz, W. (2). *IEC*, 1936, **28**, 136.

Seifriz, W. (3). "The Physical Properties of Protoplasm." Contrib. to "Colloid Chemistry, Theor. and Appl." Ed. J. Alexander, New York, 1928, vol. II, p. 403; "Protoplasm." McGraw-Hill, New York, 1936.

Seiler, J. *BC*, 1024, **44**, 68.

Self, J.T. *ZZFMA*, 1937, **26**, 673.

Selye, H. *JA*, 1934, **68**, 289.

Selye, H., Collip, J.B. & Thomson, D.L. *ENY*, 1935, **19**, 151.

Selye, H. & Friedman, S. *AJC*, 1940, **38**, 558.

Selye, H., Harlow, C. & McKeown, T. *PSEBM*, 1935, **32**, 1253.

Selye, H. & McKeown, T. (1). *JA*, 1934, **69**, 79.

Selye, H. & McKeown, T. (2). *PRSB*, 1935, **119**, 1.

Sembrat, K. *CRSB*, 1924, **90**, 894.

Sendju, Y. *JJB*, 1927, **7**, 191.

Sengupta, B. *AEZF*, 1935, **17**, 281.

Sereni, E., Ashbel, R. & Rabinowitz, D. *BSIBS*, 1929, **4**, 746; *OMaly*, 1930, **53**, 585.

Sergeiev, A.M. *BASUSSR*, 1940, p. 28.

Sergueev, G. *BBMEUSSR*, 1940, **8**, 449.

Serono, C. & Montezemolo, R. *BAMR*, 1936, **62**, 153; *OMaly*, 1937, **99**, 481.

Serono, C., Montezemolo, R. & Balboni, G. *RCT*, 1936, **35**, 341; *CA*, 1937, **31**, 1478.

Šesler, S. *AGUSSR*, 1930, **1**, 102; *OMaly*, 1931, **62**, 391.

Sessunin, P.A. *BBMEUSSR*, 1938, **6**, 175.

Severinghaus. *JEZ*, 1930, **56**.

Seyster, E.W. *BB*, 1919, **37**, 168.

Shabad, L.M. *BBMEUSSR*, 1938, **5**, 3.

Shackell, L.F. *S*, 1911, **34**, 573.

Shamarina, N.M. *BBMEUSSR*, 1939, **8**, 67.

du Shane, G.P. *AR*, 1934, **60**, 62, 63; *JEZ*, 1935, **72**, 1.

Shapiro, H. (1). *BB*, 1932, **63**, 456.

Shapiro, H. (2). *BB*, 1935, **68**, 363.

Shapiro, H. (3). *IEC*, 1935, **7**, 25.

Shapiro, H. (4). *JCCP*, 1935, **6**, 101; *AR*, 1935, **64** (Suppl.), 59.

Sharp, D.G., Taylor, A.R., Finkelstein, H. & Beard, J.W. *PSEBM*, 1939, **42**, 459 and 462.

Sharp, P.F. & Powell, C.K. *IEC*, 1931, **23**, 196.

Shaw, A.O. *JH*, 1938, **29**, 319.

Shear, M.J. (1). *AJC*, 1936, **26**, 322.

Shear, M.J. (2). *JBC*, 1936, **114**, xc; *AJC*, 1936, **28**, 334; 1937, **29**, 269.

Shearer, C. *PRSB*, 1922, **93**, 213 and 410.

Sheldon, J.H. & Ramage, H. (1). *BJ*, 1931, **25**, 1608.

Sheldon, J.H. & Ramage, H. (2). *BJ*, 1933, **27**, 674.

Shen, Shih-Chang (1). *JEB*, 1939, **16**, 143.

Shen, Shih-Chang (2). Inaug. Diss. Cambridge, 1940.

Shen, Shih-Chang & Needham, J. Unpublished work.

Shen, Tun-Hui. *AFEWM*, 1934, **131**, 205.

Sherman, C.C. & Sherman, H.C. *ARB*, 1937, **6**, 335.

Sherman, H.C. & McLeod, F.L. *JBC*, 1925, **64**, 429.

Sherman, H.C. & Quinn, E.J. *JBC*, 1926, **67**, 667.

Sherman, W.C., Elvejhem, C.A. & Hart, E.B. *JBC*, 1934, **107**, 289 and 383.

Sherrington, C.S. "The Integrative Action of the Nervous System." Yale Univ. Press, New Haven, 1926.

Sherwood, R.M. & Fraps, G.S. *Bull. Texas Agric. Exp. Sta.* 1932, no. 468; 1934, no. 493.

Shibata, S. & Murata, A. *Bull. Imp. Zootechn. Exp. Sta. Chiba-Shi* (Japan), 1936, no. 13.

Shimasaki, Y. *JJMSA*, 1931, **2**, 291.

Shimkin, M.B. & Grady, H.G. *JNCI*, 1940, **1**, 119.

Shipley, P.G., Kramer, B. & Howland, J. *BJ*, 1926, **20**, 379.

Shrewsbury, J.F.D. *L*, 1933 (i), 415.

Shull, A.F. *BB*, 1937, **72**, 259.

Shumway, W. *QRB*, 1932, **7**, 93.

Sidorov, O.A. *AAUSSR*, 1937, **16**, 25 and 145; *YMaly*, 1938, **48**, 421.

Siedlecki, M. *BC*, 1909, **29**.

Siegert, F. & Neumann, S. *CG*, 1930, **54**, 1630.

Signer, R., Caspersson, T. & Hammarsten, E. *N*, 1938, **141**, 122.

Sikharulidze, T.A. *CRASUSSR*, 1939, **23**, 975.

Silvestri, U. *MOG*, 1936, **8**, 358; *OMaly*, 1937, **98**, 143.

Simola, P.E., Stenberg, M. & Uutela, E. *SK*, 1937, **10** (B), 34.

Simon, S. (1). *CRSB*, 1930, **104**, 1052.

Simon, S. (2). *ADB*, 1939, **50**, 95.

Simonin, C. *CRSB*, 1931, **107**, 1029.

Sinclair, H.M. *CI*, 1938, **57**, 471.

Sinclair, R.G. (1). *JBC*, 1930, **88**, 575.

Sinclair, R.G. (2). *AJOP*, 1933, **103**, 73.

Sinnott, E.W. (1). *AN*, 1936, **70**, 245; 1937, **71**, 113; *GS*, 1935, **20**, 12; *PNASW*, 1937, **23**, 224.

Sinnott, E.W. (2). *S*, 1937, **85**, 61.

Sinnott, E.W. & Dunn, L.C. (1). "Principles of Genetics." McGraw Hill Book Co., New York and London, 1932.

Sinnott, E.W. & Dunn, L.C. (2). *BR*, 1935, **10**, 123.

Skarżyński, B. (1). *N*, 1933, **131**, 766; *BAPS*, 1933, 347.

Skarżyński, B. (2). *BAPS*, 1936, 437.

Skinner, J.T. & Peterson, W.H. *JBC*, 1930, **88**, 347.

Skowron, S. & Skarżyński, B. *CRSB*, 1933, **112**, 1604.

Skubiszewski, L. *MDS*, 1926, **5**, 305.

Sladden, D.E. *PRSB*, 1930, **106**, 318; 1932, **112**, 1.

Slifer, E.H. (1). *PZ*, 1930, **3**, 503.

Slifer, E.H. (2). *JM*, 1931, **51**, 613.

Slifer, E.H. (3). *BC*, 1932, **52**, 223; *JM*, 1932, **53**, 1.

Slifer, E.H. (4). *PZ*, 1932, **5**, 448.

Slifer, E.H. (5). *JEZ*, 1934, **67**, 137.

Slifer, E.H. (6). *QJMS*, 1937, **79**, 493; *JM*, 1938, **63**, 181.

Slifer, E.H. (7). *QJMS*, 1938, **80**, 437.

Slizyński, B.M. *N*, 1936, **137**, 536.

Słonimski, P. (1). *FMP*, 1930, **2**, 162 (in Polish); *YMaly*, 1931, **18**, 770.

Słonimski, P. (2). *ADB*, 1931, **42**, 415.

Smallwood, W.M. *AA*, 1905, **26**, 652.

Smith, B.G. *JM*, 1922, **36**.

Smith, Christianna. *JPB*, 1932, **35**, 717.

Smith, D.E. *CIWP*, 1931, **413**, 41.

Smith, E.C. Bate- (1). *PRSB*, 1931, **108**, 553.

Smith, E.C. Bate- (2). Lecture delivered at Low Temperature Research Station, Cambridge, 27 November 1936.

Smith, E.L. *S*, 1940, **91**, 199.

Smith, G.M. *AJC*, 1934, **21**, 596.

Smith, G.M. & Coates, C.W. *ZA*, 1938, **23**, 93.

Smith, Homer W. (1). *QRB*, 1932, **7**, 1.

Smith, Homer W. (2). *BR*, 1936, **11**, 49.

Smith, Homer W. (3). "Studies in the Physiology of the Kidney." University of Kansas Porter Lectures. Lawrence, Kansas, 1939.

Smith, K:M. *AAB*, 1920, **7**, 40.

Smith, L.I. *CR*, 1940, **27**, 287.

Smith, Michael (1). *JEB*, 1931, **8**, 312.

Smith, Michael (2). *JEB*, 1934, **11**, 228.

Smith, Michael & Shepherd, J. *JEB*, 1931, **8**, 293.

Smith, P.E. & McDowell, E.C. *AR*, 1930, **46**, 249; 1931, **50**, 85.

Smith, P.E. & Smith, I.P. *JMR*, 1922, **43**, 267; *ENY*, 1923, **7**, 579.

Smith, R.M. *Bull. Arkansas Agric. Exp. Sta.* 1933, no. 293.

Smith, Sydney (1). Inaug. Diss. Cambridge, 1936.

Smith, Sydney (2). Private communication to the author.

Smith, S.E. & Bogart, R. *GS*, 1939, **24**, 474.

Smith, W.W. & Smith, H.W. *JBC*, 1938, **124**, 107.

Smorodintzev, I.A., Pavlova, P.I. & Ivanova, A.F. *CRASUSSR*, 1936, **3**, 29.

Smreczyński, S. *ZIZP*, 1938, **59**, 1.

Smyth, H.F. *JMR*, 1914, **31**, 255.

Snell, G.D. (1). *PNASW*, 1929, **15**, 133.

Snell, G.D. (2). *GS*, 1935, **20**, 545.

Snell, G.D., Bodemann, E. & Hollander, W. *JEZ*, 1934, **67**, 93.

Snell, G.D. & Picken, D. *JG*, 1935, **31**, 213.

Snessarev, P. *EAEG*, 1932, **29**, 614.

Snyder, F.F. (1). *BJHH*, 1934, **54**, 1.

Snyder, F.F. (2). *PR*, 1938, **18**, 578.

Snyder, F.F. & Rosenfeld, M. (1). *PSEBM*, 1937, **36**, 45.

Snyder, F.F. & Rosenfeld, M. (2). *AJOP*, 1937, **119**, 153; 1938, **121**, 242; *JAMA*, 1937, **108**, 1946; *AJOG*, 1938, **36**, 363.

Snyder, F.F. & Speert, H. *AJOG*, 1938, **36**, 579.

Sobotka, H. (1). *CR*, 1934, **15**, 311.

Sobotka, H. (2). "The Chemistry of the Sterids." New York and London, 1937.

Solacolu, T. & Constantinesco, D. *CRAS*, 1936, **203**, 437; 1937, **204**, 290.

Solberg, A.N. (1). *Progressive Fish Culturist*, 1938, no. 40. Offprints obtainable from Bureau of Fisheries, Department of Commerce, Washington, D.C.

Solberg, A.N. (2). *JEZ*, 1938, **78**, 417 and 441.

Soliterman, P.L. *AEZF*, 1935, **17**, 106.

Sollas, I.B.J. *Rep. Evolution Committee Roy. Soc.* 1909, **5**, 51.

Sonnenblick, B.P. *AR*, 1934, **60** (Suppl.), 60.

Sontag, L.W., Pyle, S.I. & Cape, J. *AJDC*, 1935, **50**, 337.

Sørensen, M. *BZ*, 1934, **269**, 271; *CRTLC*, 1934, **20**, no. 3.

Sørensen, M. & Haugaard, G. *CRTLC*, 1933, **19**, no. 12.

Sørensen, N.A. *ZPC*, 1935, **235**, 8; *CA*, 1937, **31**, 6741.

Soule, S.D. *AJOG*, 1938, **35**, 309.

Spadola, J.M. & Riemenschneider, R.W. *JBC*, 1937, **121**, 787.

Spahr, W. *ZB*, 1937, **98**, 43.

Spanner, R. *ZAE*, 1935, **105**, 163.

Späth, B. *MGG*, 1936, **102**, 167; *OMaly*, 1936, **95**, 655.

Spaul, E.A. *JEB*, 1923, **1**, 313; 1924, **2**, 33 and 427; 1927, **5**, 166 and 212; 1930, **7**, 49.

Spear, F.G. *AEZF*, 1929, **7**, 484.

Speicher, B.R. *UPB*, 1933, **30**.

Speidel, C.C. *AJA*, 1929, **43**, 107

Spek, J. (1). *KB*, 1918, **9**, 259.

Spek, J. (2). *AFEWM*, 1926, **107**, 54.

Spek, J. (3). *P*, 1930, **9**, 370.

Spek, J. (4). *P*, 1933, **18**, 497.

Spek, J. (5). *P*, 1934, **21**, 394.

Spek, J. (6). *P*, 1934, **21**, 561.

Spek, J. (7). *AFEWM*, 1934, **131**, 362.

Spek, J. (8). *P*, 1938, **30**, 352.

Spemann, H. (1). *AFEWM*, 1901, **12**, 224; 1902, **15**, 448; 1903, **16**, 551; *VDZG*, 1914, 216; *NW*, 1924, **12**, 1092; *ZWZ*, 1928, **132**, 105.

Spemann, H. (2). *SPMGW*, 1901; *VDAG*, 1901, 61; *AA*, 1903, **23**, 457.

Spemann, H. (3). *ZAZ*, 1905, **28**, 419.

Spemann, H. (4). *ZAZ*, 1907, **31**, 379.

Spemann, H. (5). *VDZG* (Stuttgart meeting), 1908, **18**.

Spemann, H. (6). *ZJZP*, 1912, **32**, 1.

Spemann, H. (7). *VDZG*, 1914, 216.

Spemann, H. (8). *AFEWM*, 1918, **43**, 448.

Spemann, H. (9). *NW*, 1919, **7**.

Spemann, H. (10). *AFEWM*, 1921, **48**, 533.

Spemann, H. (11). Rectoral Address. *Jahreshefte d. Univ. Freiburg i./B.* 1923, p. 1.

Spemann, H. (12). "Organisers in Animal Development", Croonian Lecture, *PRSB*, 1927, **102**, 177; German version with references, in *NW*, 1927, **15**, 946; *JEB*, 1926, **2**, 493.

Spemann, H. (13). *AFEWM*, 1931, **123**, 389.

Spemann, H. (14). *VDZG* (Zool. Anz. Suppl. 5), 1931, **34**, 129.

Spemann, H. (15). "Neueste Ergebnisse d. Entwicklungsphysiol. Forschungen." Freiburger Wissenschaftliche Gesellschaft, Heft 23. Speyer & Kärner, Freiburg i./B. 1934.

Spemann, H. (16). *SMW*, 1937, II, 849.

Spemann, H. (17). "Embryonic Development and Induction." Yale University Press, New Haven, 1938. English edition (with ch. XVIII omitted) of "Experimentelle Beiträge zu einer Theorie der Entwicklung", Springer, Berlin, 1936.

Spemann, H. & Falkenberg, H. *AFEWM*, 1919, **45**, 371.

Spemann, H., Fischer, F.G. & Wehmeier, E. *NW*, 1933, **21**, 505.

Spemann, H. & Geinitz, B. *AFEWM*, 1927, **109**, 129.

Spemann, H. & Mangold, H. *AFEWM*, 1924, **100**, 599.

Spemann, H. & Schotté, O. *NW*, 1932, **20**, 463.

Spemann, H. & Wessel-Bautzmann, E. *AFEWM*, 1927, **110**, 557.

Spencer, Herbert. "Principles of Biology", Sect. 327.

Sperti, G.S., Loofbourow, J.R. & Dwyer, C.M. *N*, 1937, **140**, 643; *SIDT*, 1937, **1**, 163.

Sperti, G.S., Loofbourow, J.R. & Lane, M.M. *S*, 1937, **86**, 611.

Spirito, A. (1). *AFEWM*, 1932, **127**, 61; *AAL* (6e ser.), 1930, **12**, 183; 1931, **14**, 154 and 361.

Spirito, A. (2). *MZI*, 1935, **45** (Suppl.), 322; *YMaly*, 1935, **35**, 306; *ASB*, 1934, **20**, 442; *YMaly*, 1935, **34**, 87.

Spirito, A. (3). *RDB*, 1935, **19**, 437; *YMaly*, 1936, **38**, 311.

Spirito, A. (4). *AZI*, 1936, **22**, 223; *YMaly*, 1936, **38**, 93.

Spirito, A. (5). *AAL*, 1936 (6e ser.), **23**, 907; 1937 (6e ser.), **26**, 37; *OMaly*, 1937, **97**, 50; *CA*, 1937, **31**, 8029; *ASB*, 1937, **23**, 185 and 517; *YMaly*, 1937, **44**, 665; 1938, **47**, 558; *BSIBS*, 1939, **13**, 920.

Spitzer, J.M. *ZJZP*, 1937, **57**, 458.

Sponsler, O.L. *JGP*, 1925, **9**, 221 and 677; "Living Matter: a molecular approach." U. of Cal. lecture, Los Angeles, 1934.

Spratt, N.T. & Willier, B.H. *TB*, 1939, **17**, 1.

Sprawson, E. *BDJ*, 1937, 1 (Feb. 15th).

Stahl, E. *BDBG*, 1885, **3**, 334.

Standfuss, M. "Handbuch d. palaearktischen Grossschmetterlinge." Jena, 1896.

Stanley, W.F. *JEZ*, 1935, **69**, 459.

Stanley, W.M. *PR*, 1939, **19**, 524.

Stark, M.B. (1). *JCR*, 1918, **3**, 279; *JEZ*, 1919, **27**, 509.

Stark, M.B. (2). *PNASW*, 1919, **5**, 573; *PSEBM*, 1919, **17**, 51.

Stark, M.B. (3). *AJC*, 1937, **31**, 253.

Stark, M.B. & Bridges, C.B. *GS*, 1926, **11**, 249.

Starkenstein, E. *PZUSSR*, 1937, **22**, 593.

Staudinger, H. *Zangger Festschrift*, 1934, **2**, 939; *YMaly*, 1935, **35**, 472; *TFS*, 1935; *UMN*, 1936, **43**, 33.

Stcherbatov, I.I. *BBMEUSSR*, 1938, **6**, 511.

Stearns, G. *PR*, 1939, **19**, 415

Stearns, G. & McKinley, J.B. *JN*, 1937, **13**, 143.

Stearns, G. & Stinger, D. *JN*, 1937, **13**, 127.

Steele, A.G. & Windle, W.F. *JOP*, 1939, **94**, 531.

Stefanelli, A. (1). *JEB*, 1937, **14**, 171; *BSIBS*, 1937, **12**, 284; 1938, **13**, 475; *YMaly*, 1938, **45**, 207; *RDB*, 1937, **23**, 33.

Stefanelli, A. (2). *ASB*, 1938, **24**, 411.

Stefanelli, A. (3). *BSIBS*, 1939, **13**, 918; *AF*, 1939, **39**, 176.

Steinberg, A.G. & Abramovitch, M. *PNASW*, 1938, **24**, 107.

Steinberg, D.M. *BZUSSR*, 1938, **7**, 295.

Steinhart, K.M. Cit. in Kreps, E.M. *AMBUSSR*, 1935, **19**, 216; also *PZUSSR*, 1935.

Steinmüller, O. *AFEWM*, 1937, **137**, 13.

Stenhagen, E. & Teorell, T. *N*, 1938, **141**, 415.

Stephenson, Marjory. "Bacterial Metabolism." 2nd ed. London, 1938.

Stepp, W., Feulgen, R. & Voit, K. *BZ*, 1927, **181**, 284.

Sterling, E.B. *USPHR*, 1927, **42**, 717.

Stern, Curt (1). "Multiple Allelie." In *Handbuch d. Vererbungswiss.* 1930, **1**, 1.

Stern, Curt (2). *N*, 1938, **142**, 158.

Stern, Curt (3). *G*, 1939, **3** (Suppl.), 19.

Stern, Curt & Hadorn, E. *GS*, 1939, **24**, 162.

Stern, Kurt G. (1). *BJ*, 1934, **28**, 949.

Stern, Kurt G. (2). *CSHSQB*, 1938, **6**, 286.

Stern, Kurt G. & Greville, G.D. *NW*, 1933, **21**, 720.

Stern, Kurt G. & Salomon, K. *BB* (abstract), 1937, **73**, 378; *S*, 1937, **86**, 310; *JBC*, 1938, **122**, 461.

Steudel, H. & Takahashi, E. (1). *ZPC*, 1923, **127**, 210.

Steudel, H. & Takahashi, E. (2). *ZPC*, 1923, **131**, 99.

Stewart, G.F. (1). *PS*, 1935, **14**, 24.

Stewart, G.F. (2). *PS*, 1936, **15**, 119.

Stier, T.J.B. *PNASW*, 1933, **19**, 725.

Stiles, K.A. *BB*, 1938, **74**, 430.

Stiven, D. *BJ*, 1930, **24**, 172.

Stockard, C.R. *AJA*, 1921, **28**, 115.

Stockvis, B.Y. *CMW*, 1873, 211; *ZKM*, 1895, 1.

Stolfi, G. (1). *RAL*, 1933 (ser. 6), **18**, 516; *YMaly*, 1934, **29**, 372.

Stolfi, G. (2). *BSIBS*, 1934, **9**, 1315; *OMaly*, 1935, **87**, 60.

Stolte, H.A. (1). "Formgestaltung im Tierreiche." Kohlhammer, Berlin, 1934.

Stolte, H.A. (2). *BR*, 1936, **11**, 1.

Stone, L.S. (1). *JEZ*, 1922, **35**, 421; *AR*, 1921, **21**, 85; 1923, **25**, 114; 1927, **35**, 24. *JCN*, 1924, **38**, 73; 1928, **47**, 91 and 117; 1933, **57**, 507; 1937, **68**, 83; *S*, 1931, **74**, 577; *PSEBM*, 1933, **30**, 1258.

Stone, L.S. (2). *AR*, 1922, **23**, 39; 1925, **29**, 375; 1932, **51**, 267; *JEZ*, 1926, **44**, 95; 1932, **62**, 109; *PSEBM*, 1927, **24**, 945; *AFEWM*, 1929, **118**, 40.

Stone, L.S. (3). *JEZ*, 1930, **55**, 193.

Stone, L.S. & Sapir, P. *AR*, 1938, **70** (Suppl.), 75; *JEZ*, 1940, **85**, 71.

Stough, H.B. *JM*, 1931, **52**, 535.

Strack, E. & Geissendörfer, H. *AG*, 1936, **160**, 544.

Strack, E., Geissendörfer, H. & Neubaur, E. *ZPC*, 1934, **229**, 25.

Strack, E. & Loeschke, A. *ZPC*, 1931, **194**, 269.

Strangeways, T.S.P. "Tissue Culture in Relation to Growth and Differentiation." Heffer, Cambridge, 1924.

Strangeways, T.S.P. & Fell, H.B. *PRSB*, 1926, **100**, 273.

Stransky, E. *BZ*, 1933, **266**, 287.

Strasburger, E.H. *ZIAV*, 1936, **71**, 538.

Straub, J. *KZ*, 1933, **62**, 13; 1933, **64**, 72; *OMaly*, 1934, **77**, 573.

Straub, J. & Donck, C.M. *CW*, 1934, **31**, 461.

Straub, J. & Hoogerdyun, M.J.J. *RTCPB*, 1929, **48**, 49.

Straub, J., van Stijgeren, G.A. & Kabos, W.J. *CW*, 1937, **34**, 1.

Strauss, M.B. *JAMA*, 1934, **102**, 281.

Streeter, G.L. *CIWP*, 1930, no. 414.

Striganova, A. *CRASUSSR*, 1940, **27**, 385 and 388; *BASUSSR*, 1939, 831.

Ströer, W.F.H. *AFEWM*, 1933, **130**, 131.

Stroganov, N.S. *BZUSSR*, 1938, **7**, 525.

Strohl, J. (1). In Winterstein's *Handbuch d. vergl. Physiol.* 1914, **2**, 443.

Strohl, J. (2). *ASNZ*, 1925 (10ᵉ sér.), **8**, 105.

Strohl, J. (3). "L'Embryogénie physiologique et l'Organisation des Insectes, suivis de propos sur la Métamorphose." Bouvier Jubilee Volume. Paris, 1936, p. 329; *BSZF*, 1938, **63**, 95.

Strohl, J. & Köhler, W. *NGWG*, 1935, **2**, 31.

Stuart, H.A. & Higgins, G.M. *AJOP*, 1935, **111**, 590.

Studitsky, A.N. *CRASUSSR*, 1938, **20**, 493 and 497.

Stuhlman, F.L. *ANVH*, 1887, **10**.

Stunkard, H.W. *BAMNH*, 1923, **48**, 165.

Sturm, E., Gates, F.L. & Murphy, J.B. *JEM*, 1932, **55**, 441.

Sturtevant, A.H. (1). *PSEBM*, 1920, **17**; *Proc.* vith *Internat. Congr. Genetics*, 1932, **1**, 304.

Sturtevant, A.H. (2). *JEZ*, 1927, **46**, 493.

Subbarow, Y. & Jacobson, B.M. *JBC*, 1936, **114**, 102.

Sueyoshi, Y. *JJB*, 1931, **13**, 145.

Sueyoshi, Y. & Furukobo, T. *JJB*, 1931, **13**, 155 and 177.

Sugiura, K. & Benedict, S.R. *JCR*, 1920, **5**, 373.

Sugiyama, M. (1). *JFSUT*, 1938 (4), **4**, 471, 489 and 495.

Sugiyama, M. (2). *JFSUT*, 1938, **5**, 127.

Sümegi, S. *FZP*, 1932, **43**, 565.

Sümegi, S. & Csaba, M. *AEZF*, 1931, **11**, 339; *YMaly*, 1931, **19**, 19.

Sumner, F.B. *AFEWM*, 1904, **17**, 92.

Sun, T.-P. (1). *AR*, 1930, **47**, 309.

Sun, T.-P. (2). *PZ*, 1932, **5**, 375 and 384.

Suomalainen, P. (1). *SK*, 1939, **12** (unpaged offprint).

Suomalainen, P. (2). *AASF*, 1939, **53**, no. 8.

Suomalainen, P. (3). *AASF*, 1939. **53**, no. 9.

Suomalainen, P. & Toivonen, S. *AASF*, 1939, **53**, no. 6.

Surber, E.W. *TAFS*, 1935, **65**, 194; *CA*, 1936, **30**, 4223.

Surrarrer, T.C. *GS*, 1935, **20**, 357.

de Suto-Nagy, G. *PSEBM*, 1940, **43**, 674.

Suzuki, J. In "Données Numériques de Biol." ed. Terroine & Janot, Paris, 1930, pp. 1647 ff.

Suzuki, M. *JJB*, 1939, **30**, 19 and 23.

Suzuki, S. *AFEWM*, 1928, **114**, 372.

Svedberg, T. *N*, 1931, **128**, 999; *CR*, 1937, **20**, 81.

Svetlov, P. (1). *CRASUSSR*, 1932, 125; *YMaly*, 1932, **23**, 222; *AFEWM*, 1934, **131**, 672; *TLZEMAM*, 1934, **3**, 165; 1935, **4**, 29.

Svetlov, P. (2). *BBMEUSSR*, 1937, **4**, 449.

Swanson, E.E. & Chen K'o-K'uei. *QJPP*, 1939, **12**, 657.

Swanson, P.P. & Smith, A.H. *JBC*, 1932, **97**, 745.

Swanton, E.W. "British Plant Galls." London, 1912, p. 24.

Swenson, T.L. & Mottern, H.H. *S*, 1930, **72**, 98.

Swett, F.H. (1). *AR*, 1924, **27**, 273; 1928, **40**, 297.

Swett, F.H. (2). *JEZ*, 1927, **47**, 385.

Swett, F.H. (3). *JEZ*, 1929, **53**, 35.

Swett, F.H. (4). *JEZ*, 1930, **55**, 87; 1937, **75**, 143.

Swett, F.H. (5). *QRB*, 1937, **12**, 322.

Swett, F.H. (6). *JEZ*, 1938, **78**, 47.

Swett, F.H. (7). *JEZ*, 1938, **78**, 81.

Swett, F.H. (8). *JEZ*, 1939, **81**, 127; 1939, **82**, 305.

Swett, F.H. & Parsons, E.H. *JEZ*, 1929, **53**, 13.

Swezy, O. *BB*, 1915, **28**, 47.

Swingle, W.W. (1). *JGP*, 1919, **2**, 161.

Swingle, W.W. (2). *AR*, 1922, **23**, 100; 1924, **27**, 220; *JEZ*, 1922, **36**, 397.

Swingle, W.W., Helff, O.M. & Zwemer, R.L. *AJOP*, 1924, **70**, 208.

Syngajevskaia, K. *ZJZP*, 1936, **56**, 487.

Sypniewski, J. *BHA*, 1931, **8**, 229; *OMaly*, 1932, **67**, 431.

Szantroch, Z. *AA*, 1932, **75**, 46; *AEZF*, 1933, **13**, 600; *AAL*, 1932 (6th), **15**, 904.

Szejnman-Rozenberg, A. *ABE* (in Polish), 1933, **8**, 32.

Szendi, B. (1). *ZAE*, 1933, **101**, 791.

Szendi, B. (2). *AG*, 1934, **157**, 389.

Szendi, B. (3). *AG*, 1934, **158**, 409; 1936, **162**, 27.

Szendi, B. (4). *MK*, 1936, **66**, 128.

Szendi, B. & Papp, G. *AG*, 1935, **159**, 432.

Szepsenwol, J. (1). *CRSB*, 1931, **106**, 1199.

Szepsenwol, J. (2). *CRSB*, 1933, **112**, 116; *AAM*, 1933, **29**, 5.

Szepsenwol, J. (3). *CRSB*, 1935, **119**, 782.

von Szily, A. *AH*, 1908, **35**, 649.

Szuman, J.G. *CRAS*, 1925, **181**, 257.

Tafuri, G.B. *BSIBS*, 1928, **3**, 641.

Tafuri, G.B. & Testa, M. *BSIBS*, 1927, **2**, 893; 1928, **3**, 310.

Takagi, S. *AZJ*, 1938, **17**, 170.

Takahashi, I. (1). *ZPC*, 1933, **219**, 31.

Takahashi, I. (2). *JJB*, 1935, **22**, 45.

Takahashi, K. (1). *MMAK*, 1935, **14**, 159, 173, 199 and 205.

Takahashi, K. (2). *MMAK*, 1935, **14**, 211.

Takahashi, M. *JJB*, 1929, **10**, 443.

Takakusu, S., Kuroda, K. & Li, Ki-Nei. *KJM*, 1937, **8**, 58.

Takamatsu, M. (1). *JJB*, 1935, **22**, 203.

Takamatsu, M. (2). *ZPC*, 1936, **238**, 96.

Takamatsu, M. & Kamachi, T. *JJB*, 1935, **22**, 185.

Takamine, J. *JOP*, 1901, **27**, xxix.

Takano, S. & Iijima, K. *Rep. Govt. Sugar Exp. Sta. Tainan, Formosa*, 1937, **4**, 195; *CA*, 1937, **31**, 8038.

Takaya, H. *MCSK*, 1938, **14**, 321.

Taliaferro, W.H. *QRB*, 1926, **1**, 246; *JI*, 1938, **35**, 303.

Tanaka, S.I. *PIAT*, 1934, **10**, 689; *YMaly*, 1935, **34**, 88.

Tanda, M.L. *BSIBS*, 1932, **7**, 659 and 665; *CA*, 1932, **26**, 6022.

T'ang, P'ei-Sung (1). *BB*, 1931, **60**, 242.

T'ang, P'ei-Sung (2). *BB*, 1931, **61**, 468.

T'ang, P'ei-Sung (3). *QRB*, 1933, **8**, 260; 1941, **16**, 173.

T'ang, P'ei-Sung & Gerard, R.W. *JCCP*, 1932, **1**, 503.

Taniguchi, T. *FAJ*, 1931, **9**, 81.

Tannreuther, G.W. *AR*, 1919, **16**, 355.

Tapernoux, A. *CRSB*, 1930, **105**, 405.

Tapfer, S. *AG*, 1937. **164**, 435.

Tapfer, S. & Haslhofer, L. *AG*, 1935, **159**, 313.

Tateishi, S. (1). *MMAK*, 1931, **5**, 212 and 2040; 1932, **6**, 233 and 382; *OMaly*, 1932, **66**, 207 and **68**, 73; *YMaly*, 1932, **21**, 485.

Tateishi, S. (2). *MMAK*, 1932, **6**, 534 and 899.

Tatum, E.L. *PNASW*, 1939, **25**, 486.

Tatum, E.L. & Beadle, G.W. (1). *JGP*, 1938, **22**, 239.

Tatum, E.L. & Beadle, G.W. (2). *BB*, 1939, **77**, 415; *S*, 1940, **91**, 458.

Taube, E. *AFEWM*, 1921, **49**, 269; 1923, **98**, 98; 1925, **105**, 581.

Taylor, C.V. (1). *PZ*, 1928, **1**, 1.

Taylor, C.V. (2). *PZ*, 1931, **4**, 423.

Taylor, G.W. & Harvey, E.N. *BB*, 1931, **61**, 280.

Taylor, H.B. *MJA*, 1930, **17** (i), 675 (May 24th).

Taylor, H.F. U.S. Dept. of Commerce, Bureau of Fisheries Document No. 989, 1925.

Taylor, I.R., Birnie, J.H. & Mitchell, P.H. & Solinger, J.L. *PZ*, 1934, **7**, 593; *AR*, 1932, **54** (Suppl.), 39.

Taylor, I.R. & Crescitelli, F. *JCCP*, 1937, **10**, 93.

Taylor, Jeremy. "Works." 1835, vol. 11, p. 62.

Taylor, L.W. *PS*, 1934, **13**, 378.

Taylor, L.W., Gunns, C.A. & Moses, B.D. *Bull. Univ. Calif. Coll. Agric.* 1933, no. 550.

Taylor, T.C. & Nelson, J.M. *JACS*, 1920, **42**, 1726.

Taylor, T.C. & Sherman, R.T. *JACS*, 1933, **55**, 258.

Tazima, M. *TJEM*, 1940, **38**, 8.

Tchakhotin, S. (1). *BSIBS*, 1933, **8**, 623.

Tchakhotin, S. (2). *CRSB*, 1935, **119**, 830.

Tchakhotin, S. (3). *CRSB*, 1938, **127**, 1195

Tchepovetsky, G. *PZEUSSR*, 1934, **1**, 396

Tchernorutsky, H. *ZPC*, 1912, **80**, 194.

Teissier, G. (1). *TSBR*, 1931, **9**, 27.

Teissier, G. (2). *ASNZ*, 1931 (10ᵉ sér.), **14**, 5.

Teissier, G. (3). *CRSB*, 1932, **109**, 813.

Teissier, G. (4). *BSZF*, 1932, **57**, 160.

Teissier, G. (5). "Dysharmonies et Discontinuités dans la Croissance." Hermann, Paris, 1934.

Teissier, G. (6). *APBPC*, 1934, **10**, 359; 1936, **12**, 529.

Teissier, G. (7). "Les Lois Quantitatives de la Croissance." Hermann, Paris, 1937.

Tello, F.J. *AFEWM*, 1923, **33**, 1.

Tencatehoedemaker, N.J. *ZZFMA*, 1933, **18**, 299; *YMaly*, 1934, **27**, 168; *TNDV*, 1938 (Suppl.), **3**, 89.

Tennant, R., Liebow, A.A. & Stern, K.G. *PSEBM*, 1941, **46**, 18.

Tennent, D.H., Gardiner, M.S. & Smith, D.E. *CIWP*, 1931, **413**, 1.

Tenney, B. *AJOG*, 1935, **29**, 819.

Teodoro, G. *BDZ*, 1932, **3**, 129; *YMaly*, 1932, **20**, 402.

Teodoro, T. *BDZ*, 1931, **2**, 33; *YMaly*, 1931, **18**, 507.

Terao, A. *PIAT*, 1931, **7**, 23; *YMaly*, 1932, **20**, 501.

Terenyi, A. *ZUNG*, 1931, **62**, 566; *A*, 1932, **57**, 106.

Terroine, E.F. & Delpech, G. *APBPC*, 1931, **7**, 341.

Terroine, E.F., Giaja, A. & Bayle, L. *CRAS*, 1931, **193**, 956.

Terroine, E.F., Hatterer, C. & Roehrig, P. *BSCB*, 1930, **12**, 682.

Terroine, E.F., Hée, A., Roche, J. & Roche, A. *AIP*, 1931, **34**, 282.

Terroine, E.F. & Mourot, G. *CRAS*, 1932, **195**, 1424.

Tesauro, G. *BSIBS*, 1935, **10**, 325; *OMaly*, 1935, **88**, 276.

Teutschlander, O. *ZKF*, 1920, **17**, 285.

Thadani, K.I. *JH*, 1934. **25**, 483.

Theorell, H. *BZ*, 1930, **223**, 1.

Thérèsa, S. *BBMEUSSR*, 1939, **7**, 544.

Thilenius, G. *SAWB*, 1899, 247.

Thimann, K.V. (1). *ARB*, 1935, **4**, 545; *PPP*, 1938, **13**, 437.

Thimann, K.V. (2). *PNASW*, 1936, **22**, 511; *CS*, 1936, **4**, 716.

Thimann, K.V. & Beadle, G.W. *PNASW*, 1937, **23**, 143.

Thimann, K.V. & Bonner, J. *PR*, 1938, **18**, 524.

Thomas, J.A. (1). *AAM*, 1930, **26**, 252; *BBFB*, 1930, **64**, 332.

Thomas, J.A. (2). *BSPP*, 1932, **115**, 70.

Thomas, J.A. (3). *CRAS*, 1933, **197**, 425; 1934, **199**, 886; 1935, **200**, 1140 and 1360; **201**, 988 and 1431; *CRSB*, 1933, **112**, 1206; 1934, **117**, 758; 1935, **118**, 1312, 1315 and 1430; *ASNZ*, 1938 (11ᵉ sér.), **1**, 209.

Thomas, J.A. (4). *CRAA* (Lisbon), 1933, **28**, 1; (Montpellier), 1935, **30**, 1; *AEZF*, 1934, **15**. 131; 1937, **19**, 299; *APBPC*, 1935, **11**, 1000; 1936, **12**, 13.

Thomas, J.A. & Nachmansohn, D. *CRSB*, 1938, **128**, 577.

Thomas, K. *AAP*, 1911, 9.

Thompson, d'Arcy W. "On Growth and Form." Cambridge University Press, 1917.

Thompson, H.E. & Pommerenke, W.T. *JCI*, 1938, **17**, 609.

Thompson, J.W. & Voegtlin, C.G. *JBC*, 1926, **70**, 793.

Thompson, V. (1). *PZ*, 1937, **10**, 21.

Thompson, V. (2). *JEZ*, 1938, **78**, 19.

Thompson, V. & Bodine, J.H. (1). *PZ*, 1936, **9**, 455.

Thompson, V. & Bodine, J.H. (2). *JCCP*, 1938, **12**, 247.

Thomson, D.L. *N*, 1932, **130**, 543.

Thornton, C.S. *JM*, 1938, **62**, 17 and 219.

Thorpe, W.H. (1). *QJMS*, 1934, **77**, 273.

Thorpe, W.H. (2). *PA*, 1936, **28**, 517.

Thumann, M.E. *ZMAF*, 1931, **25**, 50.

Thunberg, T. *BUSSR*, 1937, **2**, 413.

Tice, S.C. *BB*, 1914, **26**, 221.

Tichomirov, A. (1). *ZPC*, 1882, **9**, 518.

Tichomirov, A. (2). *AAP*, 1886 (Suppl.), 35.

Tiegs, O.W. *TPRSA*, 1922, **46**, 319.

von Tiesenhausen, M. *APPA*, 1909, **195**, 154.

Tillmans, J. & Alt., A. *BZ*, 1925, **164**, 160.

Timofeev-Ressovsky, N.W. *NW*, 1931, **19**, 493.

Timpe, O. *AG*, 1931, **146**, 232.

Tirelli, M. (1). *ZVP*, 1931, **14**, 737.
Tirelli, M. (2). *ZVP*, 1931, **14**, 742.
Tirelli, M. (3). *ZVP*, 1931, **15**, 148.
Tirelli, M. (4). *ASBSP*, 1936, **48**, 3.
Tirelli, M. (5). *ASBSP*, 1936, **48**, 13; *APANL*, 1929, **82**, 123.
Tirelli, M. (6). *ASBSP*, 1936, **48**, 103.
Tirelli, M. (7). "Un Precursore di J. Loeb; Alessandro Tichomirov", *ASBSP*, 1936, **48**, 827.
Tiselius, A. & Eriksson-Quensel, I.B. *BJ*, 1939, **33**, 1752.
Titaiev, A.A. & Summ, B.R. *BZ*, 1930, **220**, 62.
Titlebaum, A. *PNASW*, 1928, **14**, 285.
Titschak, E. *ZWZ*, 1926, **128**, 509.
Titus, H.W., Byerly, T.C. & Ellis, N.R. *JN*, 1933, **6**, 127.
Titus, H.W., Fritz, J.C. & Kauffman, W.R. *PS*, 1938, **17**, 38.
Toennies, G. *G*, 1937, **1**, 337.
du Toit, P.J. *JZN*, 1913, **49** (N.F. **42**), 149.
Toivonen, S. (1). *AZSZBFV*, 1938, **5**, no. 8.
Toivonen, S. (2). *AZSZBFV*, 1938, **6**, no. 5; *AASF*, 1940, **55**, no. 6.
Tokin, B.P. Unpublished experiments.
Tolmatcheva-Melnitchenko, E.P. *BBME USSR*, 1939, **8**, 21, 120 and 220.
Tomita, M. (1). *JJB*, 1929, **10**, 351.
Tomita, M. (ed.) (2). "Untersuchungen über Embryochemie, w-amino-Säuren, vergl. Biochemie, u. verschiedenes" (1921–1934). Nagasaki, 1934.
Tomita, M. & Fujiwara, H. *JJB*, 1933, **17**, 401.
Tomita, M. & Kumon, T. *ZPC*, 1936, **238**, 101.
Tompsett, S.L. *BJ*, 1934, **28**, 1537.
Töndury, G. (1). *AFEWM*, 1936, **134**, 1.
Töndury, G. (2). *AFEWM*, 1938, **137**, 510.
Töndury, G. (3). *RSZ*, 1940, **47**, 161.
de Toni, G. "Chem. Bestandteile d. fet. Blutes." *TB*, 1935, **11**, 119.
Tonutti, E. & Plate, E. *AG*, 1937, **164**, 385.
Toomey, J.A. *AJDC*, 1934, **47**, 521.
Tornier, G. (1). *AFEWM*, 1906, **20**, 76.
Tornier, G. (2). *AFEWM*, 1906, **22**.
Törö, E. (1). *VDAG*, 1931.
Törö, E. (2). *AEZF*, 1934, **15**, 312.
Törö, E. (3). *JEZ*, 1938, **79**, 213.
Törö, I. *AFEWM*, 1939, **139**, 303.
Torres, I. *BZ*, 1935, **283**, 128.
Torrisi, D. *BSIBS*, 1932, **7**, 40; *ASB*, 1934, **19**, 398; *CA*, 1932, **26**, 3560; 1935, **29**, 207.
Toverud, K.U. *AJDC*, 1933, **46**, 954.
Tower, W.L. *ZJAO*, 1909, **17**, 517.
Townsend, Grace. *BB*, 1938, **75**, 363 and 364; *CN*, 1939, **14**, 176.
Townsend, G.F. & Lucas, C.C. *S*, 1940, **92**, 43; *BJ*, 1940, **34**, 1155.
Toyama, K. *JG*, 1913, **2**, 351.
Treadwell, A.L. *BB*, 1902, **3**, 5.
Trethewie, E.R. *JOP*, 1938, **94**, 11P.

Trifonova, A.N. (1). *AZ*, 1934, **15**, 183.
Trifonova, A.N. (2). *ABSUSSR*, 1935, **37**, 757.
Trifonova, A.N. (3). *AZ*, 1937, **18**, 375; *BZUSSR*, 1937, **6**, 243; *YMaly*, 1937, **44**, 280; *AAUSSR*, 1939, **22**, 94.
Trifonova, A.N. & Popov, N.A. *Uchonyie Zapiski L.G.U.*, 1937, **15**, 335.
Triggerson, C.J. *AESA*, 1914, **7**, 1.
Trim, A.R.H. Inaug. Diss. Cambridge, 1941; *N*, 1941, **147**, 115.
Trimble, H.C. & Keeler, C.E. *JH*, 1938, **29**, 281.
Trotti, L. (1). *BMLZACUG*, 1934, **14**, 1; *YMaly*, 1935, **35**, 396.
Trotti, L. (2). *BMLZACUG*, 1935, **15**, 3; *YMaly*, 1936, **37**, 182.
Trowell, O.A. & Willmer, E.N. *JEB*, 1939, **16**, 60.
Trurnit, H.J. *NW*, 1939, **27**, 805.
Truszkowski, R. *BJ*, 1928, **22**, 1299.
Truszkowski, R. & Chajkinówna, S. *BJ*, 1935, **29**, 2361.
Truszkowski, R. & Czuperski, H. *BJ*, 1933, **27**, 66.
Truszkowski, R. & Goldmanówna, C. *BJ*, 1933, **27**.
Tsang, Yü-Ch'üan. *JCN*, 1939, **70**, 1.
Tschesche, R. (1). *ZPC*, 1931, **203**, 263.
Tschesche, R. (2). *EP*, 1936, 8.
Tschesche, R. & Wolf, H.J. *ZPC*, 1936, **244**, i; 1937, **248**, 34.
Tsukano, L. *JJB*, 1932, **15**.
Tsuzuki, K. *JJB*, 1936, **23**, 421.
Tuleschkov, K. *ZE*, 1935, **22**, 97; *YMaly*, 1935, **35**, 201.
Tung, Ti-Chao (1). *ADB*, 1933, **44**.
Tung, Ti-Chao (2). *CRSB*, 1934, **115**, 1375; *AAM*, 1934, **30**, 381.
Tung, Ti-Chao (3). *PNHB*, 1936, **10**, 115.
Tung, Ti-Chao & Tung, Fêng-Yeh. (1). *ADB*, 1940, **51**, 203.
Tung, Ti-Chao & Tung, Fêng-Yeh. (2). *CJEB*, 1936, **1**, 97.
Tur, J. (1). *BAPS*, 1933, B II, 43; *YMaly*, 1934, **28**, 226.
Tur, J. (2). *MAPS*, (B), 1935, **1**; *ZP*, 1935, **1**, 1; *FMP*, 1929, 1.
Tur, J. (3). *BAPS*, 1935, B II, 183.
Turner, C.L. (1). *JM*, 1933, **55**, 207.
Turner, C.L. (2). *JM*, 1936, **59**, 313.
Turner, C.W. *NAV*, 1927, **8**, 27.
Turner, W.J. *JBC*, 1937, **118**, 519.
Twiesselmann, F. *ADB*, 1938, **49**, 285; *CRSB*, 1935, **119**, 1169.
Twitty, V.C. (1). *JEZ*, 1928, **50**, 319.
Twitty, V.C. (2). *JEZ*, 1932, **61**, 333.
Twitty, V.C. (3). *CSHSQB*, 1934, **2**, 148.
Twitty, V.C. (4). *PSEBM*, 1935, **32**, 1283.
Twitty, V.C. (5). *AR*, 1935, **64** (Suppl.), 37.
Twitty, V.C. (6). *JEZ*, 1936, **74**, 239 and 264; *AN*, 1937, **71**, 127.
Twitty, V.C. (7). Private communication to the author, Aug. 1940.

Twitty, V.C. & Bodenstein, D. *JEZ*, 1939, 81, 357; 1941, 86, 343 and 362.

Twitty, V.C. & Elliott, H.A. *JEZ*, 1934, 68, 247.

Twitty, V.C. & Johnson, H.H. *S*, 1934, 80, 78.

Twitty, V.C. & Schwind, J.L. *JEZ*, 1931, 59, 61.

Twitty, V.C. & van Wagtendonk, W.J. *G*, 1940, 4, 349.

Tyler, A. (1). *JEZ*, 1930, 57, 347.

Tyler, A. (2). *PSZN*, 1933, 13, 155.

Tyler, A. (3). *BB*, 1935, 68, 451.

Tyler, A. (4). *BB*, 1936, 71, 59.

Tyler, A. (5). *BB*, 1936, 71, 82.

Tyler, A. (6). *JEZ*, 1937, 76, 395.

Tyler, A. (7). "The Energetics of Embryonic Differentiation." Hermann, Paris, 1939.

Tyler, A. (8). *PNASW*, 1939, 25, 523.

Tyler, A. (9). *PNASW*, 1940, 26, 249; *BB*, 1940, 78, 159.

Tyler, A. & Dessel, F.W. *JEZ*, 1939, 81, 459.

Tyler, A. & Fox, S.W. *BB*, 1940, 79, 153.

Tyler, A. & Horowitz, N.H. (1). *PNASW*, 1937, 23, 369; *BB*, 1937, 73, 377; 1938, 75, 209.

Tyler, A. & Horowitz, N.H. (2). *BB*, 1938, 74, 99.

Tyler, A. & Humason, W.D. *BB*, 1937, 73, 261.

Tyler, A., Ricci, N. & Horowitz, N.H. *JEZ*, 1938, 79, 129; *BB*, 1937, 73, 395.

Tyler, A. & Scheer, B.T. *JEZ*, 1937, 75, 179.

Tyler, A. & Schultz, Jack. *JEZ*, 1932, 63, 509.

von Ubisch, L. (1). *AFEWM*, 1911, 31, 637.

von Ubisch, L. (2). *ZWZ*, 1925, 124, 361, 457 and 469; *AFEWM*, 1929, 117, 80.

von Ubisch, L. (3). *ZWZ*, 1927, 129, 214.

von Ubisch, L. (4). *AFEWM*, 1938, 138, 18.

von Ubisch, L. (5). *BC*, 1938, 58, 370.

von Ubisch, L. (6). *BR*, 1939, 14, 88.

Uchida, T. *PIAT*, 1935, 11, 69.

Uda, H. *GS*, 1923, 8, 322.

Ueda, K. *JJOG*, 1931, 14, 225; *OMaly*, 1932, 65, 291.

Uhlenluth, E. (1). *AFEWM*, 1913, 36, 211.

Uhlenluth, E. (2). *JEZ*, 1918, 24, 237.

Uhlenluth, E. (3). *JGP*, 1919, 1, 525.

Uhlenluth, E. (4). *BB*, 1921, 41, 307.

Umanski, E. (1). *ZAZ*, 1932, 97, 286; 1933, 104, 119.

Umanski, E. (2). *ZAZ*, 1935, 112, 205.

Umanski, E. (3). *BBMEUSSR*, 1938, 6, 141.

Umanski, E. (4). *BBMEUSSR*, 1938, 6, 383; 1939, 8, 115.

Umeya, Y. *Bull. Sericult. Exp. Sta. Chosen*, 1926, p. 1.

Upp, C.W. & Waters, N.F. *PS*, 1935, 14, 372.

Urner, J.A. *AR*, 1931, 50, 175.

Uspenskaia, V.D. *P*, 1935, 23, 613.

Ussing, H.H. (1). *SKAP*, 1935, 72, 192.

Ussing, H.H. (2). *SKAP*, 1937, 77, 107.

Utevsky, A.M. & Levantzeva, N.S. *BBME USSR*, 1938, 5, 75.

Utevsky, A.M. & Osinskaia, V.O. *BBME USSR*, 1938, 5, 78.

Uvarov, B.P. *TES*, 1931, 79, 1.

Vadova, A.V. *BBMEUSSR*, 1939, 7, 554.

Vajropala, K. *N*, 1935, 136, 145.

Vandebroek, G. (1). *ADB*, 1936, 47, 499.

Vandebroek, G. (2). Unpublished results, cit. in Dalcq's monograph "Form and Causality", pp. 37 and 104; also *AEZF*, 1937, 19, 411.

Varley, G.C. & Butler, C.G. *PA*, 1933, 25, 263.

Vassin, B.N. *CRASUSSR*, 1939, 24, 577.

Veizman, V. *BBMEUSSR*, 1936, 2, 207.

Verjbinskaia, M. Cit. in Kreps, E.M., *AMBUSSR*, 1935, 19, 216; *PZUSSR*, 1935.

Verne, J. & Odiette, D. *AAPA*, 1931, 8, 681; *YMaly*, 1932, 20, 660.

Vernidub, M.P. *AAUSSR*, 1939, 22, 105 and 119.

Vickery, H.B. & Shore, A. *BJ*, 1932, 26, 1101.

le Vine, I. & Wolf, I.J. *AJOG*, 1940, 40, 327.

Vintemberger, P. (1). *CRSB*, 1932, 111, 48.

Vintemberger, P. (2). *CRSB*, 1935, 118, 52.

Vintemberger, P. (3). *CRSB*, 1936, 122, 927.

Virchow, R. "Reizung und Reizbarkeit" in *APPA*, 1858, 14, 1, pp. 44 and 63.

Virgin, E. & Klussmann, E. *ZPC*, 1932, 213, 16.

Vladimirov, G.E. (1). *BZ*, 1930, 224, 79.

Vladimirov, G.E. (2). *JOP*, 1931, 72, 411.

Vladimirov, G.E. & Danilina, M.J. *BZ*, 1930, 224, 69.

Vladimirova, E.A. (1). *TLZEMAM*, 1935, 4, 119; *CRASUSSR*, 1934, 3, 478.

Vladimirova, E.A. (2). *TLZEMAM*, 1935, 4, 163.

Vlès, F. *AZEG*, 1933, 75, 421.

Vlès, F. & Gex, M. *APB*, 1928, 6, 255.

Vöchting, H. "Über Organbildung im Pflanzenreich." Bonn, vol. I, 1878; vol. II, 1884.

Voegtlin, C. *PR*, 1937, 17, 92.

Voegtlin, C. & Chalkley, H.W. (1). *USPHR*, 1930, 45, 3041.

Voegtlin, C. & Chalkley, H.W. (2). *P*, 1935, 24, 365.

Vogt, W. (1). *SGMPM*, 1926, 37.

Vogt, W. (2). *AFEWM*, 1925, 106, 542; 1929, 120, 385.

Vogt, W. (3). *VDAG*, 1927, 63, 126; 1928, 66, 139; *VDZG*, 1928, 26; *RSZ*, 1932, 39, 309.

Vogt, W. (4). *VDAG*, 1931, 39, 141.

Vogt, W. (5). *VDAG*, 1938, 85, 216; *YMaly*, 1937, 47, 218.

Voitkevitch, A.A. (1). *BC*, 1935, 55, 449.

Voitkevitch, A.A. (2). *BC*, 1937, 57, 196.

Voitkevitch, A.A. (3). *BBMEUSSR*, 1938, 6, 85.

Voorhoeve, H.C. Inaug. Diss. Amsterdam, 1926.

Voronzova, M.A. (1). *BBMEUSSR*, 1937, **4**, 156.

Voronzova, M.A. (2). *BBMEUSSR*, 1937, **4**, 187.

Voronzova, M.A. (3). *BBMEUSSR*, 1938, **6**, 82.

Voronzova, M.A. (4). *BBMEUSSR*, 1939, **8**, 227.

Voronzova, M.A. & Krascheninnikova, W.P. *BBMEUSSR*, 1937, **4**, 189.

Voronzova, M.A. & Liosner, L.D. *ZJZP*, 1936, **56**, 107.

Voss, H. *ZAE*, 1931, **94**, 712; *YMaly*, 1931, **17**, 790.

Votquenne, M. *ADB*, 1934, **45**, 79.

Vrtelówna, S. *AFEWM*, 1925, **105**, 45.

Wachs, H. (1). *AFEWM*, 1914, **39**, 384.

Wachs, H. (2). *SGNFB*, 1920, 133.

Waddell, J., Elvehjem, C.A., Steenbock, H. & Hart, E.B. *JBC*, 1928, **77**, 794.

Waddington, C.H. (1). *PTRSB*, 1932, **221**, 179; *N*, 1930, **125**, 924; *AEZF*, 1934, **15**, 302.

Waddington, C.H. (2). *AFEWM*, 1933, **128**, 502.

Waddington, C.H. (3). *JEB*, 1933, **10**, 38.

Waddington, C.H. (4). *N*, 1933, **131**, 134.

Waddington, C.H. (5). *JEB*, 1934, **11**, 211.

Waddington, C.H. (6). *N*, 1933, **131**, 275; *JEB*, 1934, **11**, 218.

Waddington, C.H. (7). *JEB*, 1934, **11**, 224.

Waddington, C.H. (8). *SP*, 1934, **29**, 336.

Waddington, C.H. (9). *N*, 1935, **135**, 606.

Waddington, C.H. (10). *JEZ*, 1935, **71**, 273.

Waddington, C.H. (11). *N*, 1936, **138**, 125.

Waddington, C.H. (12). *JEB*, 1936, **13**, 75.

Waddington, C.H. (13). *JEB*, 1936, **13**, 86.

Waddington, C.H. (14). *JEB*, 1937, **14**, 229.

Waddington, C.H. (15). *JEB*, 1937, **14**, 232.

Waddington, C.H. (16). *ADB*, 1937, **48**, 273.

Waddington, C.H. (17). *JEB*, 1938, **15**, 371.

Waddington, C.H. (18). *JEB*, 1938, **15**, 377.

Waddington, C.H. (19). *JEB*, 1938, **15**, 382.

Waddington, C.H. (20). "Morphogenetic Substances in Early Development." In *Trav. du Congrès du Palais de la Découverte*, Hermann, Paris, 1938, p. 437.

Waddington, C.H. (21). *PRSB*, 1938, **125**, 365.

Waddington, C.H. (22). "An Introduction to Modern Genetics." Allen & Unwin, London, 1939.

Waddington, C.H. (23). *PNASW*, 1939, **25**, 299; *JG*, 1940, **41**, 75.

Waddington, C.H. (24). *N*, 1939, **144**, 637.

Waddington, C.H. (25). *G*, 1939, **3** (Suppl.), 37; *CS*, 1939, Suppl. no. 4, 39.

Waddington, C.H. (26). "Organisers & Genes." Cambridge Univ. Press, 1940.

Waddington, C.H. (27). *PZS*, in the press.

Waddington, C.H. & Cohen, A. *JEB*, 1936, **13**, 219.

Waddington, C.H. & Needham, D.M. *PRSB*, 1935, **117**, 310.

Waddington, C.H. & Needham, J. *PKAWA*, 1936, **39**, 887.

Waddington, C.H., Needham, J. & Brachet, J. *PRSB*, 1936, **120**, 173.

Waddington, C.H., Needham, J., Nowiński, W.W. & Lemberg, R. *PRSB*, 1935, **117**, 289; *N* (with D.M. Needham), 1934, **134**, 103.

Waddington, C.H., Needham, J., Nowiński, W.W., Lemberg, R. & Cohen, A. *PRSB*, 1936, **120**, 198.

Waddington, C.H. & Schmidt, G.A. *AFEWM*, 1933, **128**, 522.

Waddington, C.H. & Shukov, R.I. In Waddington's "Organisers & Genes," p.118.

Waddington, C.H. & Taylor, J. *JEB*, 1937, **14**, 335.

Waddington, C.H. & Waterman, A.J. *JA*, 1933, **67**, 355.

Waddington, C.H. & Wolsky, A. *JEB*, 1936, **13**, 92.

van Wagenen, A. & Hall, G.O. *PS*, 1936, **15**, 405.

van Wagenen, A., Hall, G.O. & Wilgus, H.S. *JAR*, 1937, **54**, 767.

van Wagenen, Gertrude. *S*, 1935, **81**, 366; *AR*, 1935, **63**, 387.

van Wagenen, Gertrude & Folley, S.J. *JE*, 1939, **1**, 367.

van Wagenen, Gertrude & Newton, W.H. *AJOP*, 1940, **129**, 485.

Wahren, H. & Rundquist, O. *KW*, 1937, **16**, 1498.

Wajzer, J. & Lippmann, R. *BSCB*, 1938, **20**, 312.

Wald, G. *N*, 1935, **136**, 832 and 913; 1937, **139**, 587 and 1017; 1937, **140**, 197 and 545; *JGP*, 1935, **18**, 905; **19**, 351; 1936, **19**, 781; **20**, 45; 1938, **21**, 795.

Wald, G. & Clark, A.B. *JGP*, 1937, **21**, 93.

Wald, G. & Zussman, H. *JBC*, 1938, **122**, 449; *N*, 1937, **140**, 197.

Waldenström, J. *ZPC*, 1936, **239**, 4.

Walker, J.F. *JCCP*, 1935, **6**, 317.

Walker, P.A. Inaug. Diss. Harvard, 1936.

Wallich, R. *CRSB*, 1926, **95**, 1480 and 1482.

Walsh, J. *PESP*, 1863, **3**, 543; 1866, **6**, 223.

Walton, A. & Hammond, J. *PRSB*, 1938, **125**, 311.

Wan, Shing & Wu, Hsien. *CJOP*, 1931, **5**, 53.

Wang, Shih-Chêng. *AFEWM*, 1933, **130**, 243.

Wang, Teh-Yo (=Ouang Te-Yio). *CRAS*, 1931, **193**, 545; *CRSB*, 1930, **103**, 116; *CRAA*, 1929, **24**, 1.

Warburg, O. (1). *ZPC*, 1908, **57**, 1.

Warburg, O. (2). "Stoffwechsel d. Tumoren." Springer, Berlin, 1926. Engl. edn. "The Metabolism of Tumours", tr. F. Dickens, London.

Warburg, O. (3). *BZ*, 1926, **172**, 432.

Warburg, O. (4). "Chemische Constitution d. Fermenten." *Trav. Congrès du Palais de la Découverte*, Paris, 1938, p. 345.

Warren, D.C. *Bull. Kansas Agric. Exp. Sta.* 1934, no. 37.

Warren, D.C. & Scott, H.M. *PS*, 1935, **14**, 195; *JAR*, 1935, **51**, 565.

Warren, H., Kuenen, D. & Baas-Becking, L.G.M. *PKAWA*, 1938, **41**, 873.

Watanabe, Y. *PZ*, 1935, **8**, 417.

Watanabe, Y. & Child, C.M. *PZ*, 1933, **6**, 542.

Watchorn, E. *JG*, 1938, **36**, 171.

Watchorn, E. & Holmes, B.E. *BJ*, 1931, **25**, 843.

Waterman, A.J. (1). *ADB*, 1932, **43**, 471.

Waterman, A.J. (2). *AJA*, 1933, **53**, 317.

Waterman, A.J. (3). *AR*, 1935, **64** (Suppl. no. 1), 44; *AJA*, 1932, **50**, 451; 1934, **54**, 347; 1936, **58**, 27.

Waterman, A.J. (4). *PNASW*, 1935, **21**, 635.

Waterman, A.J. (5). *BB*, 1937, **73**, 401.

Waterman, A.J. (6). *BB*, 1939, **76**, 162.

Waters, N.F. *PS*, 1935, **14**, 208; *S*, 1935, **82**, 66.

Watt, L.J. *JMM*, 1934, **15**, 185.

Waugh, W.G. *BMJ*, 1940, (i), 249 and 263.

Weber, A. *AAM*, 1931, **27**, 230.

Weed-Pfeiffer, I.G. *PSEBM*, 1936, **34**, 885. *JEZ*, 1939, **82**, 439.

Weekes, H.C. *PLSNSW*, 1927, **52**, 499; 1929, **54**, 34; 1930, **55**, 550; *YMaly*, 1929, **12**, 554; 1931, **18**, 185.

Wehefritz, E. *AG*, 1925, **127**, 106.

Wehefritz, E. & Gierhake, E. *AG*, 1930, **142**, 602.

Wehmeier, E. *AFEWM*, 1934, **132**, 384.

Weidel, W. *NW*, 1940, **28**, 137.

Weigl, R. *AFEWM*, 1913, **36**.

Weil, L. *JBC*, 1941, **138**, 375.

Weinberg, M. & Guelin, A. *CRSB*, 1936, **122**, 1229.

Weinstein, P. *ZUNG*, 1933, **66**, 48.

Weisman, A.I., Coates, C.W. & Moses, R.L. *ENY*, 1936, **20**, 561.

Weismann, A. "Die Kontinuität des Keimplasmas als Grundlage d. Vererbung." Jena, 1885.

Weiss, P. (1). *AFEWM*, 1923, **99**, 150; 1924, **102**, 673.

Weiss, P. (2). *AFEWM*, 1924, **104**, 359.

Weiss, P. (3). *AFEWM*, 1924, **104**, 395.

Weiss, P. (4). "Morphodynamik." Bornträger, Berlin, 1926; also *ZIAV*, 1928, 1567; *JUGP*, 1922, 65; 1924, 77; 1926, 107; 1928, 70.

Weiss, P. (5). *AFEWM*, 1927, **107**, 1.

Weiss, P. (6). *AFEWM*, 1927, **109**, 584.

Weiss, P. (7). *AFEWM*, 1927, **111**, 316.

Weiss, P. (8). *BC*, 1928, **48**, 69.

Weiss, P. (9). *BC*, 1928, **48**, 551; *AFEWM*, 1929, **116**, 438.

Weiss, P. (10). *AFEWM*, 1930, **122**, 379.

Weiss, P. (11). "Entwicklungsphysiologie der Tiere." Steinkopf, Dresden, 1930.

Weiss, P. (12). *AN*, 1933, **67**, 322.

Weiss, P. (13). *JEZ*, 1934, **68**, 393.

Weiss, P. (14). *PR*, 1935, **15**, 639.

Weiss, P. (15). *BR*, 1936, **11**, 494.

Weiss, P. (16). "Principles of Development." Holt, New York, 1939.

Weiss, P. (17). *CS*, 1939, Suppl. no. 4, 28.

Weissenberg, R. *AA*, 1936, **82**, 20.

Weldon, W.F.R. *BM*, 1902, **1**, 365.

Wells, H. Gideon. *AOP*, 1940, **30**, 535.

Wells, N.A. *PZ*, 1935, **8**.

Wels, P. & Osann, M. *AP*, 1924, **207**, 156.

Welti, E. *RSZ*, 1928, **35**, 75.

Wen, I-Chuan, Chang, Hsi-Chun & Wong, Amos. *CJOP*, 1936, **10**, 559.

Went, F.W. *BOR*, 1935, **1**, 162.

Went, F.W. & Thimann, K.V. "Phytohormones." New York, 1937.

Werber, E.I. *AR*, 1915, **9**, 529; 1916, **10**, 258; *JEZ*, 1916, **21**, 347 and 485; *BB*, 1918, **34**, 219.

Wermel, J. *BC*, 1934, **54**, 313.

Wessel, E. *AFEWM*, 1926, **107**, 481.

Wetzel, N.C. *PSEBM*, 1932, **30**, 224, 227; 233, 354, 359, 361 and 1044; 1934, **32**, 127; *JP*, 1933, **3**, 252; 1934, **4**, 465; *PNASW*, 1934, **20**, 183; *G*, 1937, **1**, 6.

Wheeler, W. Morton (1). "Ants." New York, 1910.

Wheeler, W. Morton (2). "The Social Insects." London, 1928.

Whitaker, D.M. (1). *BB*, 1929, **57**, 159.

Whitaker, D.M. (2). *BB*, 1931, **61**, 294.

Whitaker, D.M. (3). *JGP*, 1931, **15**, 167.

Whitaker, D.M. (4). *JGP*, 1931, **15**, 183.

Whitaker, D.M. (5). *JGP*, 1931, **15**, 191.

Whitaker, D.M. (6). *JGP*, 1933, **16**, 475.

Whitaker, D.M. (7). *JGP*, 1933, **16**, 497.

Whitaker, D.M. (8). *S*, 1935, **82**, 68.

Whitaker, D.M. (9). *PSEBM*, 1935, **33**, 472; *JGP*, 1937, **20**, 491.

Whitaker, D.M. (10). *BB*, 1936, **70**, 100.

Whitaker, D.M. (11). *PSEBM*, 1936, **34**, 708.

Whitaker, D.M. (12). *BB*, 1937, **73**, 249.

Whitaker, D.M. (13). *JEZ*, 1937, **75**, 155.

Whitaker, D.M. (14). *ARB*, 1937, **6**, 469.

Whitaker, D.M. (15). *PNASW*, 1938, **24**, 85.

Whitaker, D.M. (16). *JGP*, 1938, **21**, 833.

Whitaker, D.M. (17). *G*, 1939, **3**, 153.

Whitaker, D.M. (18). *JCCP*, 1940, **15**, 173.

Whitaker, D.M. (19). *BB*, 1940, **78**, 111.

Whitaker, D.M. (20). *JEZ*, 1940, **83**, 391.

Whitaker, D.M. & Lowrance, E.W. (1). *JCCP*, 1936, **7**, 417.

Whitaker, D.M. & Lowrance, E.W. (2). *JGP*, 1937, **21**, 57. *BB*, 1940, **78**, 407.

White, J. & White, A. *PSEBM*, 1938, **39**, 527; *JBC*, 1939, **131**, 149; *YJBM*, 1940, **12**, 427.

Whiting, P.W. *BB*, 1932, **63**, 296.

Whiting, P.W. & Whiting, A.R. *JG*, 1934, **29**, 311.

Whitlock, J.H. *ISCJS*, 1935, **9**, 667.

Whitman, C.C. *JM*, 1893, **8**.

Whong, *see* Huang.

Whyte, L.L. Private communication to the author.

Widdows, S.T., Lowenfeld, M.F., Bond, M., Shiskin, C. & Taylor, E.I. *BJ*, 1935, **29**, 1145.

Wieland, H. *BDCG*, 1926, **59**, 2067.

Wieland, H. & Dane, E. *ZPC*, 1933, **219**, 240.

Wieland, H., Metzger, H., Schöpf, C. & Bülow, M. *ADC*, 1933, **507**, 226.

Wieland, H. & Purrmann, R. *ADC*, 1939, **539**, 179; 1940, **544**, 63.

Wiesner, B.P. (1). *JOGBE*, 1935, **41**, 867; 1935, **42**, 8.

Wiesner, B.P. (2). "Sex." Butterworth, London, 1936.

Wigglesworth, V.B. (1). *JEB*, 1931, **8**, 411, 428 and 443.

Wigglesworth, V.B. (2). *QJMS*, 1933, **76**, 269.

Wigglesworth, V.B. (3). "Insect Physiology." Methuen, London, 1934; "Principles of Insect Physiology." Methuen, London, 1939.

Wigglesworth, V.B. (4). *QJMS*, 1934, **77**, 191; 1937, **79**, 91; *N*, 1935, **136**, 338.

Wigglesworth, V.B. (5). *NW*, 1939, **27**, 301.

Wigglesworth, V.B. (6). *N*, 1939, **144**; *JEB*, 1940, **17**, 180 and 201.

Wilander, O. *SKAP*, 1938, **81** (Suppl. no. 15), 3.

Wilburg, J. *BAPS*, (B) 1937, 131.

Wilder, H.H. *AJA*, 1908, **8**, 355.

Wilder, O.H.M., Bethke, R.M. & Record, P.R. *JN*, 1933, **6**, 407.

Wildiers, E. *LC*, 1901, **18**, 313.

Wilkerson, V.A. *JBC*, 1934, **104**, 541.

Wilkerson, V.A. & Gortner, R.A. *AJOP*, 1932, **102**, 153; *JBC*, 1932, **97**, lxi.

Williams, M.E. *PZ*, 1936, **9**, 231.

Williams, R.J. *BR*, 1941, **16**, 49.

Williams, R.J., Finkelstein, J., Folkers, K., Harris, S.A., Keresztesy, J.C., Mitchell, H.K., Snell, E.E., Stanbery, S.R., Stiller, E.T. & Weinstock, H.H. *JACS*, 1940, **62**, 1776, 1779, 1784, 1785 and 1791.

Williams, R.J., Lyman, C.M., Goodyear, G.H., Truesdail, J.H. & Holaday, D. *JACS*, 1933, **35**, 2912.

Williams, R.J. & Major, R.T. *S*, 1940, **91**, 246.

Williams, R.J., Truesdail, J.H., Weinstock, H.H., Rohrmann, E., Lyman, C.M. & McBurney, C.H. *JACS*, 1938, **60**, 2719.

Williams, W.R. "Natural History of Cancer." London, 1908.

Willier, B.H. (1). "The Embryological Foundations of Sex in Vertebrates" in "Sex and Internal Secretions", 1932, p. 94; 1939, p. 64; *AN*, 1933, **67**, 1.

Willier, B.H. (2). *AFEWM*, 1933, **130**, 616.

Willier, B.H. (3). *S*, 1937, **86**, 409.

Willier, B.H. (4). *AR*, 1937, **70**, 89.

Willier, B.H., Gallagher, T.F. & Koch, F.C. *PZ*, 1937, **10**, 101; *PNASW*, 1935, **21**, 625; *JBC*, 1935, **109**, xcix, c.

Willier, B.H. & Rawles, M.E. (1). *JEZ*, 1931, **59**, 429.

Willier, B.H. & Rawles, M.E. (2). *AR*, 1931, **48**, 277.

Willier, B.H. & Rawles, M.E. (3). *PSEBM*, 1935, **32**, 1293.

Willier, B.H., Rawles, M.E. & Hadorn, E. *PNASW*, 1937, **23**, 542.

Willier, B.H., Rawles, M.E. & Koch, F.C. *PNASW*, 1938, **24**, 176.

Willier, B.H. & Yuh, E.C. *JEZ*, 1928, **52**, 65.

Willis, R.A. (1). "The Spread of Tumours in the Human Body." London, 1934, p. 109.

Willis, R.A. (2). *JPB*, 1935, **40**, 1.

Willis, R.A. (3). *PRSB*, 1936, **120**, 496.

Willis, R.A. (4). *JPB*, 1937, **45**, 49.

Willis, R.A. (5). *ANZJS*, 1939, **9**, 119.

Willis, R.A. (6). *JPB*, 1939, **49**, 571.

Willmer, E.N. & Jacoby, F. *JEB*, 1936, **13**, 237.

Willmer, E.N. & Kendal, L.P. *JEB*, 1932, **9**, 149.

Wills, I.A. *JEZ*, 1936, **73**, 481.

Willstätter, R. & Rohdewald, M. (1). *ZPC*, 1932, **208**, 258.

Willstätter, R. & Rohdewald, H. (2). *ZPC*, 1934, **225**, 103.

Wilms, J. *DAKM*, 1895, **55**, 289.

Wilson, C.B. *AFEWM*, 1897, **5**, 615.

Wilson, E.B. (1). *Stud. Biol. Lab. Johns Hopkins Univ.* 1882.

Wilson, E.B. (2). *JEZ*, 1904, **1**, 1 and 197.

Wilson, E.B. (3). *AFEWM*, 1929, **117**, 179.

Wilson, G.S. *JOH*, 1926, **25**, 150.

Wilson, H.V. (1). *USBFIR*, 1889, **9**, 1.

Wilson, H.V. (2). *AN*, 1932, **66**, 159.

Wilson, H.V. & Penney, J.T. *JEZ*, 1930, **56**, 73.

Wind, F. *BZ*, 1926, **179**, 384.

Windaus, A. (1). Cit. in Lettré & Imhoffen.

Windaus, A. (2). "Sterine als Ausgangstoffe für Hormone, Vitamine und andere physiologisch wichtige Verbindungen." Offprint with no mark of origin, pp. 21–34; also *NGWG*, 1935, **1**, 59.

Windaus, A. & Stange, O. *ZPC*, 1936, **244**, 218.

Windle, W.F. "Physiology of the Foetus." Philadelphia, 1940.

Windle, W.F. & Austin, M.F. *JCN*, 1936, **63**, 431.

Windle, W.F. & Barcroft, J. *AJOP*, 1938, **121**, 684.

Windle, W.F. & Baxter, R.E. *JCN*, 1936, **63**, 173.

Windle, W.F., Becker, R.F., Barth, E.E. & Schulz, M.D. *SGO*, 1939, **69**, 705; 1940, **70**, 603.

Windle, W.F., Monnier, M. & Steele, A.G. *PZ*, 1938, **11**, 425.

Windle, W.F. & Nelson, D. *AJOP*, 1938, **121**, 700; *JCCP*, 1938, **11**, 325.

Windle, W.F., Scharpenberg, L.G. & Steele, A.G. *AJOP*, 1938, **121**, 692.

Winterstein, H. "Kausalität u. Vitalismus vom Standpunkt d. Denkökonomie." Springer, Berlin, 1928.

Winterstein, H. & Hirschberg, E. *AP*, 1927, **216**, 271.

Wintersteiner, O. & Smith, P.E. *ARB*, 1938, **7**, 253.

Wintrebert, P. (1). *JPE*, 1921, **18**, 353.

Wintrebert, P. (2). *ARFA*, 1921.

Wintrebert, P. (3). *CRAS*, 1931, **193**, 493.

Wintrebert, P. (4). *CRAA* (Warsaw meeting), 1931, 1.

Wintrebert, P. (5). *CRAS*, 1932, **195**, 908; 1933, **196**, 571;, **197**, 602, 655; 1934, **198**, 1181; *CRSB*, 1934, **116**, 694; *TSZW*, 1938, **13**, 762.

Wintrebert, P. (6). *CRSB*, 1929, **102**, 997; 1930, **105**, 273, 520, 701 and 764; 1931, **106**, 724, 784 and 908; 1933, **114**, 1301; 1934, **115**, 1580; *CRAS*, 1929, **189**, 1198; 1931, **192**, 891; 1933, **196**, 1833; *AZEG*, 1933, **75**, 501.

Wintrebert, P. (7). *CRSB*, 1930, **104**, 1229 and 1234; 1931, **107**, 1214 and 1443; 1932, **109**, 833; 1934, **115**, 1575; 1935, **119**, 1420; *CRAS*, 1931, **193**, 447; 1932, **194**, 1013 and 2104; *CRAA* (Bordeaux meeting), 1929, 1; (Amsterdam meeting), 1930, 1; (Nancy meeting), 1932, 1; *BAA*, 1931, 1.

Wintrebert, P. (8). *CRSB*, 1931, **106**, 439.

Wintrebert, P. & Wang, Teh-Yo. *CRAS*, 1931, **193**, 350; *CRSB*, 1931, **107**, 1447.

Wintrebert, P. & Yung, Ko-Ching. *CRAS*, 1926, **183**, 455; 1929, **189**, 208.

Wintrobe, M.M., Clark, D.A., Trager, W. & Danziger, L. *JCI*, 1937, **16**, 667.

Wintrobe, M.M. & Shumacker, H.B. (1). *JCI*, 1935, **14**, 837.

Wintrobe, M.M. & Shumacker, H.B. (2). *AJA*, 1936, **58**, 313.

Wipprecht, C. & Horlacher, W.R. *JH*, 1935, **26**, 363.

Wishart, J. & Hammond, J. *JAS*, 1933, **23**, 463.

Wislocki, G.B. *AR*, 1935, **63**, 183.

Witebsky, E. & Szepsenwol, J. *CRSB*, 1934, **115**, 921 and 1109.

Witherspoon, J.T. (1). *PSEBM*, 1933, **30**, 1367.

Witherspoon, J.T. (2). *JH*, 1935, **26**, 15.

Witschi, E. (1). *AFEWM*, 1924, **102**, 168; *VNGB*, 1922, **34**, 33; *PSEBM*, 1930, **27**, 475; 1934, **31**, 419.

Witschi, E. (2). *JEZ*, 1929, **52**, 235 and 267.

Witschi, E. (3). *AN*, 1932, **66**, 108.

Witschi, E. (4). *BR*, 1934, **9**, 460.

Witschi, E. (5). *BC*, 1935, **55**, 168.

Witschi, E. (6). *AR*, 1936, **66**, 483; *ABL*, 1936, **5**, 79; *JEZ*, 1937, **75**, 313.

Witschi, E. (7). *PNASW*, 1937, **23**, 35.

Witschi, E. (8). "Sex Differentiation." In "Sex and Internal Secretions", 1939, p. 145; *SC*, 1940, in press.

Witschi, E. & Crown, E.N. *AR*, 1937, **70** (Suppl.), 121.

Woerdeman, M.W. (1). *ABSRSMNB*, 1932, 33.

Woerdeman, M.W. (2). *PKAWA*, 1933, **36**, 189; *NTG*, 1933, **77**, 3621.

Woerdeman, M.W. (3). *PKAWA*, 1933, **36**, 423.

Woerdeman, M.W. (4). *PKAWA*, 1933, **36**, 842.

Woerdeman, M.W. (5). *PKAWA*, 1935, **38**, 364.

Woerdeman, M.W. (6). *PKAWA*, 1936, **39**, 306.

Woerdeman, M.W. (7). *NTG*, 1937, **81**, 1614.

Woerdeman, M.W. (8). *PKAWA*, 1938, **41**, 336; *NTG*, 1938, **82**, 3801.

Woerdeman, M.W. (9). *B*, 1938, **1**, 323.

Woerdeman, M.W. (10). *PKAWA*, 1939, **42**, 290.

Woerdeman, M.W. & Hampe, J.F. *PKAWA*, 1933, **36**, 477.

Wöhlisch, E. & Belonoschkin, B. *BZ*, 1936, **284**, 353.

Wolbach, S.B. *S*, 1937, **86**, 569.

Wolf, G. *ZVP*, 1933, **19**, 1.

Wolff, Étienne (1). *CRSB*, 1932, **111**, 740.

Wolff, Étienne (2). *CRSB*, 1935, **120**, 1312 and 1314; 1936, **121**, 1474; 1936, **123**, 237; 1938, **128**, 420.

Wolff, Étienne (3). *CRSB*, 1937, **126**, 1217.

Wolff, Étienne (4). *CRSB*, 1938, **128**, 420.

Wolff, Étienne & Ginglinger, A. *CRSB*, 1935, **120**, 901 and 903; *AAHE*, 1935, **20**, 219.

Wolff, Étienne & Kantor, S. *CRSB*, 1937, **126**, 707.

Wolff, Étienne & Stoll, R. *CRSB*, 1937, **126**, 1215.

Wolff, Étienne & Wolff, Émilienne (1). *CRSB*, 1936, **123**, 1191.

Wolff, Étienne & Wolff, Émilienne (2). *CRSB*, 1937, **124**, 367.

Wolff, G. *AFEWM*, 1894, **1**, 380.

Wolff, L.K. (1). Cit. in Straub & Hoogerduyn, p. 59.

Wolff, L.K. (2). *L*, 1932, **223**, 617.

Wolsky, A. (1). *AK*, 1937, **34**, 65.

Wolsky, A. (2). *N*, 1937, **139**, 1069.

Wolsky, A. (3). *JEB*, 1938, **15**, 225.

Wolsky, A. (4). *AT*, 1939, **11**, 375.

Wolsky, A. (5). *VIVTAL*, 1935, **7**, 449.

Wolsky, A. & Huxley, J.S. (1). *PRSB*, 1934, **114**, 364.

Wolsky, A. & Huxley, J.S. (2). *PZS*, 1936, 485.

Wolsky, A., Tazelaar, M.A. & Huxley, J.S. *PZ*, 1936, **9**, 265.

Womack, E.B. & Koch, F.C. *ENY*, 1932, **16**, 273.

Wong, Amos & Chang, Hsi-Chun. *CMJ*, 1933, **47**, 987.

Wood, A.H. *JEB*, 1932, **9**, 271.

Wood, T.R. *AN*, 1932, **66**, 277.

Wood, T.R. & Banta, A.M. *IRGHH*, 1937, **35**, 229.

Woodger, J.H. (1). *QRB*, 1930, **5**, 1 and 438; 1931, **6**, 178, p. 202.

Woodger, J.H. (2). *PAS*, 1932, **32**, 95, p. 107.

Woodruff, A.M. & Goodpasture, E.W. *AJOPA*, 1931, **7**, 209.

Woodside, G.L. (1). *JEZ*, 1937, **75**, 259.

Woodside, G.L. (2). *AR*, 1937, **67**, 423.

Woodward, A.E. *JEZ*, 1918, **26**, 459.

Wooldridge, W.R. & Standfast, A.H.P. *N*, 1932, **130**, 664; *BJ*, 1933, **27**, 183.

Worley, L.G. *JGP* 1933, **16**, 841.

Worster-Drought, C., Carnegie-Dickson, W.E. & McMenemey, W.H. *BN*, 1937, **60**, 85.

Wottge, K. *P*, 1937, **29**, 31.

Wriedt, C. & Mohr, O.L. *JG*, 1928, **20**, 187.

Wright, G.P. (1). *PSEBM*, 1926, **23**, 603.

Wright, G.P. (2). *JEM*, 1926, **43**, 591.

Wright, S. (1). *JH*, 1917, **8**, 224, 373, 426, 473, 476, 521 and 561; 1918, **9**, 33, 87, 139 and 227; *GS*, 1925, **10**, 223.

Wright, S. (2). *AN*, 1926, **60**, 552; *AR*, 1931, **51**, 115; *JH*, 1934, **25**, 359; *GS*, 1935, **20**, 84.

Wright, S. (3). *GS*, 1931, **16**, 97; *Proc. 6th Internat. Congr. Genetics* (Ithaca), 1932, **1**, 356; *JG*, 1935, **30**, 257.

Wright, S. (4). *GS*, 1934, **19**, 471.

Wright, S. (5). *CSHSQB*, 1934, **2**, 137.

Wright, S. & Eaton, O.N. *JAR*, 1923, **26**, 161.

Wright, S. & Wagner, K. *AJA*, 1934, **54**, 383.

Wrinch, D. (1). *PAS*, 1922, **22**, 208; 1929, **29**, 112.

Wrinch, D. (2). *N*, 1934, **134**, 978; **135**, 788; 1935, **136**, 68; 1936, **137**, 411; **138**, 241; *P*, 1936, **25**, 550.

Wu, Hsien, Liu, Szu-Chih & Chen, J. *Proc. xvth Internat. Congr. Physiol.*, *Leningrad and Moscow*, 1935, p. 440.

Wu, Hsien, Liu, Szu-Chih & Chou, Chi-Yuan. *CJOP*, 1931, **5**, 309.

Wunder, W. (1). *EB*, 1931, **7**, 118.

Wunder, W. (2). *EB*, 1932, **8**, 180.

Wunder, W. (3). *EB*, 1934, **10**, 1.

Wunder, W. (4). *EB*, 1937, **14**, 280.

Wyckoff, R.W.G. & Corey, R.B. *PSEBM*, 1936, **34**, 285.

Yamada, K. (1). *JJMSB*, 1933, **2**, 47, 93 and 107.

Yamada, K. (2). *JJMSB*, 1933, **2**, 71 and 81; *YMaly*, 1934, **27**, 757.

Yamada, T. (1). *AFEWM*, 1937, **137**, 151; *YMaly*, 1937, **46**, 241.

Yamada, T. (2). *FAJ*, 1938, **17**, 369.

Yamada, T. (3). *JFSUT*, 1938, **5**, 133.

Yamada, T. (4). *FAJ*, 1939, **18**, 565 and 569; *JJZ*, 1939, **8**, 265.

Yamada, T. (5). *FAJ*, 1940, **19**, 131.

Yamafuji, K. *EZ*, 1936, **1**, 268; 1937, **2**, 147; *JACSJ*, 1934, **10**, 112 and 116.

Yamafuji, K. & Goto, S. *JACSJ*, 1936, **12**, 1.

Yamagiwa, K. & Ichikawa, K. *MMAK*, 1915, **15**, 295; *APPA*, 1921, **233**, 235; *JCR*, 1924, **8**, 119.

Yamaguchi, M. *AG*, 1932, **148**, 475; *NIZ*, 1929, **18**, 865, 1044 and 1675; *JJZ*, 1931, **3** (123).

Yamamoto, T. (1). *JFSUT* (IV. Zool.), 1931, **2**, 147 and 153; 1933, **3**, 105 and 111; 1934, **3**, 287; 1936, **4**, 221 and 233; 1938, **5**, 37; *PIAT*, 1938, **14**, 393; *AZJ*, 1940, **19**, 69.

Yamamoto, T. (2). *JFSUT* (IV. Zool.), 1934, **3**, 275; *PIAT*, 1938, **14**, 149.

Yamamoto, T. (3). *JFSUT*, 1935, **4**, 99; 1938, **5**, 51.

Yamamoto, T. (4). *JFSUT*, 1936, **4**, 249.

Yaoi, H. *JJEM*, 1928, **7**, 135.

Yapp, W.W. *PASAP*, 1931, 133.

Yasumaru, A. *MMAK*, 1931, **5**, 603; *OMaly*, 1931, **62**, 275.

Yasumaru, A. & Sugiyama, K. *JJMSB* (III. Biophysics), 1931, **2**, 30*; *YMaly*, 1932, **22**, 581.

Yatsu, N. *AZJ*, 1912, **8**; *JCST*, 1912, **32** (3).

Yllpö. *ZKH*, 1913, **9**, 208.

Yokoyama, Y. *PIAT*, 1934, **10**, 582; *YMaly*, 1935, **34**, 140.

Yonezawa, Y. & Yamafuji, K. *Bull. Sci. Fak. Terkultura* (*Agric.*) *Kyushu Imp. Univ.* 1935, **6**, 126; *CA*, 1935, **29**, 4838.

Yoshida, I. *AMUO*, 1931, **2**, 326; *CA*, 1932, **26**, 193.

Yosida, S. *JJMSB*, 1937, **3**, 241.

Young, E.G. (1). *JBC*, 1937, **120**, 1.

Young, E.G. (2). *N*, 1940, **145**, 1021.

Young, E.G., Conway, C.F. & Crandall, W.A. *BJ*, 1938, **32**, 1138.

Young, E.G. & Dreyer, N.B. *JPET*, 1933, **49**, 162.

Young, E.G. & Inman, W.R. *JBC*, 1938, **124**, 189.

Young, E.G. & Musgrave, F.F. *BJ*, 1932, **26**, 941.

Young, E.G., Musgrave, F.F. & Graham, H.C. *CJR*, 1933, **9**, 373.

Youngstrom, K.A. *JNP*, 1938, **1**, 357.

Yung, Ko-Ching. *CRAA* (Bordeaux), 1929, **24**, 1; *AIOP*, 1930, **7**, 300.

Zaitschek, A. *CT*, 1934, **6**, 102; *CA*, 1934, **28**, 5102.

Zakolska, Z. *K, A*, 1929, **53**, 779.

Zanoni, G. *BMLZACUG*, 1933, **13**, 3.

Zarrow, M.X. & Pomerat, C.M. *G*, 1937, **1**, 103.

Zavadskaia, N.D., Koboziev, N. & Vereten-nikov, S. *AZEG*, 1934, **76**, 249.

Zawadzki, B. *P*, 1933, **19**, 485.

Zechmeister, L. "Carotinoide." Springer, Berlin, 1934.

Zeidberg, L.D. *AJOP*, 1929, **90**, 172.

Zeleny, C. (1). *JEZ*, 1909, **7**, 477 and 513.
Zeleny, C. (2). *JGP*, 1919, **2**, 69; *JEZ*, 1920, **30**, 293.
Zeleny, C. (3). *BB*, 1923, **44**, 105; *AN*, 1928, **62**, 88; *S*, 1933, **77**, 177.
Zezza, P. *BSIBS*, 1937, **12**, 74.
Zhinkin, L.N. (1). *CRASUSSR*, 1938, **18**, 213.
Zhinkin, L.N. (2). *CRASUSSR*, 1939, **24**, 620.
Zhinkin, L.N. (3). *CRASUSSR*, 1939, **24**, 623.
Zieliński, M.A. (1). *Bull. Internat. Acad. Pol. Sci. and Lett.* 1935, p. 293; *ABE*, 1935, **9**, 131; *JEB*, 1937, **14**, 48.
Zieliński, M.A. (2). *ABE*, 1939, **13**, 35; *CRSSV*, 1938, **31**, 160.
Zifferblatt, A.H. & Seelaus, H.K. *AR*, 1931, **48**, 367; *YMaly*, 1931, **18**, 326.
da Zilva, S.S., Golding, J., Drummond, J.C. & Coward, K.H. *BJ*, 1921, **15**, 427.
Zimmerman, L. & Rugh, R. *JM*, 1941, **68**, 329.
Zimmermann, K. *ZIAV*, 1933, **64**, 176.
Zinkernagel, R. *BZ*, 1939, **301**, 321.
Zoja, R. *AFEWM*, 1895, **2**, 1.
Zolotarev, E.C., Lavrova, N.P. & Tokareva, L.V. *ZZUSSR*, 1940, **19**, 46.
Zondek, B. *SKAP*, 1934, **70**, 133.
Zorn, C.M. & Dalton, A.J. *AJOP*, 1937, **119**, 627; *PSEBM*, 1936, **35**, 451.

Zuckerkandl, F. & Messiner-Klebermass, L. *BZ*, 1931, **236**, 19.
Zuckerman, S. (1). *L*, 1936 (i), 135; *BMJ*, 1936 (ii), 864.
Zuckerman, S. (2). Lecture to the Cambridge University Medical Society, 2 November 1938; *BR*, 1940, **15**, 231.
Zuckerman, S. & Groome, J.R. *JA*, 1940, **74**, 171.
Zuckerman, S. & Krohn, P.L. *PTRSB*, 1937, **228**, 147.
Zuckerman, S. & van Wagenen, Gertrude. *JA*, 1935, **69**, 497.
Zuntz, L. (1). "Weibliche Geschlechtsorgane u. Placenta." In Oppenheimer's *Handbuch d. Biochem. d. Menschen u. Tiere* (Erg.werk), 1934, **2**, 355.
Zuntz, L. (2). "Fruchtwasser." In Oppenheimer's *Handbuch d. Biochem. d. Menschen u. Tiere* (Erg.werk), 1934, **2**, 676.
Zuntz, L. (3). "Stoffaustausch zwischen Mutter u. Frucht." In Oppenheimer's *Handbuch d. Biochem. d. Menschen u. Tiere* (Erg.werk), 1936, **3**, 101.
Zuntz, L. (4). "Stoffwechsel d. Weibes." In Oppenheimer's *Handbuch d. Biochem. d. Menschen u. Tiere* (Erg.werk), 1936, **3**, 319.
Zweibaum, J. *AEZF*, 1933, **14**, 391.
Zwilling, E. (1). *PSEBM*, 1934, **31**, 933; *JEZ*, 1940, **84**, 291.
Zwilling, E. (2). *JEZ* 1941, **86**, 333.

INDEXES

By Margaret Miall

GENERAL INDEX

N.B. Figures in heavy type indicate the principal point of reference.

INDEX OF ANIMALS

INDEX OF PLANTS

INDEX OF GENES